COLD SPRING HARBOR SYMPOSIA ON QUANTITATIVE BIOLOGY

VOLUME LXIII

COLD SPRING HARBOR SYMPOSIA ON QUANTITATIVE BIOLOGY

VOLUME LXIII

Mechanisms of Transcription

COLD SPRING HARBOR LABORATORY PRESS
1998

COLD SPRING HARBOR SYMPOSIA ON QUANTITATIVE BIOLOGY VOLUME LXIII

©1998 by Cold Spring Harbor Laboratory Press
Cold Spring Harbor, New York
International Standard Book Number 0-87969-550-1 (cloth)
International Standard Book Number 0-87969-552-8 (paper)
International Standard Serial Number 0091-7451
Library of Congress Catalog Card Number 34-8174

Printed in the United States of America

COLD SPRING HARBOR SYMPOSIA ON QUANTITATIVE BIOLOGY
Founded in 1933 by
REGINALD G. HARRIS
Director of the Biological Laboratory 1924 to 1936

Previous Symposia Volumes

Front Cover (*Paperback*): Model for the structure of the complex TBP, TFIID, and the adenovirus major late promoter. A sequence element (TFIIB recognition element, BRE; *yellow*) located immediately upstream of the TATA element (*orange*) affects the ability of TFIIB to enter transcription complexes and support transcription initiation. The sequence element is recognized by a canonical helix-turn-helix DNA-binding motif (*green*) within TFIIB. The figure was prepared starting from the crystallographic structure of the TFIIB-TBP-TATA complex (Nikolov et al., *Nature 377:* 119 [1995]) and extending the DNA upstream and downstream from the TATA element as B-DNA. TFIIB is *green* and *blue*; TBP is *gray;* DNA is *yellow, orange,* and *red;* the transcription start is *white* (for details, see page 86 and Lagrange et al., *Genes Dev. 12:* 34 [1998]). (Courtesy of D. Reinberg and R. Ebright.)

Back Cover (*Paperback*): Structure of GalR dimer modeled after PurR and LacI (Adhya et al., p. 6, this volume). The two subunits are shown in *red* and *pink*, respectively. The amino-terminal helix-turn-helix containing DNA-binding domain (*top*) is connected to the carboxy-terminal core (*bottom*) by a hinge. *Blue circles* in the core show the amino acid residues, which when mutated, give rise to GalR variants that bind to DNA but fail to form the repressosome involved in transcription repression. (Courtesy of George Vasmatzis.)

All Cold Spring Harbor Laboratory Press publications may be ordered directly from Cold Spring Harbor Laboratory Press, 10 Skyline Drive, Plainview, NY 11803-2500 (Phone: 1-800-843-4388 in Continental U.S. and Canada). All other locations: (516) 349-1930. FAX: (516) 349-1946. E-mail: cshpress@cshl.org. For a complete catalog of all Cold Spring Harbor Laboratory Press publications, visit our World Wide Web Site http://www.cshl.org/

Symposium Participants

ABATE-SHEN, CORY, Center for Advanced Biotechnology and Medicine, UMDNJ-Robert Wood Johnson Medical School, Piscataway, New Jersey

ADHYA, SANKAR, Lab. of Molecular Genetics, National Cancer Institute, National Institutes of Health, Bethesda, Maryland

AGARWAL, SUNITA, Metabolic Diseases Branch, NIDDK, National Institutes of Health, Bethesda, Maryland

AIBA, HIROJI, Dept. of Molecular Biology, Nagoya University Graduate School of Science, Chikusa, Nagoya, Japan

AKSAN, ISIL, Eukaryotic Transcription Laboratory, Marie Curie Research Institute, Oxted, United Kingdom

ALBER, TOM, Dept. of Molecular and Cell Biology, University of California, Berkeley

ALDABE, RAFAEL, Dept. of Molecular Biology, Mount Sinai School of Medicine, New York, New York

ALLIS, C. DAVID, Dept. of Biology, University of Rochester, Rochester, New York.

ALWINE, JAMES, Dept. of Microbiology, University of Pennsylvania School of Medicine, Philadelphia

ANDERSSON, MONIKA, Lab. of Developmental Biology, Dept. of Cell and Molecular Biology, Karolinska Institute, Stockholm, Sweden

ANSARI, ASEEM, Prog. in Molecular Biology, Memorial Sloan-Kettering Cancer Center, New York, New York

AOYAGI, NORIKAZU, Cell Biology and Metabolism Branch, NICHD, National Institutes of Health, Bethesda, Maryland

ARANGO, NELSON, Dept. of Molecular Genetics, M.D. Anderson Cancer Center, University of Texas, Houston

ARTHUR, TERRY, Dept. of Bacteriology, McArdle Laboratory for Cancer Research, University of Wisconsin, Madison

ASO, TEIJIRO, Dept. of Molecular Genetics, Institute of Physical and Chemical Research, Tsukuba, Japan

AUBLE, DAVID, Dept. of Biochemistry and Molecular Genetics, Health Sciences Center, University of Virginia, Charlottesville

AUGUSTIN, MARTIN, Center for Medical Biotechnology, Fraunhofer ITA, Hannover, Germany

AXELROD, AMY, *Cell Press*, Cambridge, Massachusetts

BAGGA, RAJESH, Dept. of Regulatory Biology, The Salk Institute for Biological Studies, La Jolla, California

BAI, YONGLI, Dept. of Biochemistry, University of New Hampshire, Durham

BANDYOPADHYAY, RAM, Cancer Research Center, Boston University School of Medicine, Boston, Massachusetts

BANIK, UTPAL, Dept. of Molecular Physiology and Biophysics, Vanderbilt University, Nashville, Tennessee

BAO, YONGPING, Dept. of Biochemistry, Institute of Food Research, Norwich, Norfolk, United Kingdom

BARBARIC, SLOBODAN, Lab. of Biochemistry, Faculty of Food Technology and Biotechnology, Zagreb, Croatia

BARDWELL, VIVIAN, Dept of Biochemistry, Cancer Center and Institute of Human Genetics, University of Minnesota, Minneapolis

BARKER, MELANIE, Dept. of Bacteriology, University of Wisconsin, Madison

BARLOW, CARROLEE, Lab. of Genetic Disease Research, National Institutes of Health, Bethesda, Maryland

BARRERA-SALDAÑA, HUGO, Dept. of Biochemistry, ULIEG, School of Medicine, UANL, Monterrey, Mexico

BARTHOLOMEW, BLAINE, Dept. of Medical Biochemistry, Southern Illinois University, Carbondale, Illinois

BELL, ADAM, Lab. of Molecular Biology, NIDDK, National Institutes of Health, Bethesda, Maryland

BELOTSERKOVSKAYA, RIMMA, The Wistar Institute, Philadelphia, Pennsylvania

BENECKE, ARNDT, IGBMC, Illkirch, Strasbourg, France

BENNANI-BAITI, IDRISS, Prog. in Molecular Biology, Memorial Sloan-Kettering Cancer Center, New York, New York

BENTLEY, DAVID, Dept. of Medical Biophysics, Amgen Institute, University of Toronto, Toronto, Ontario, Canada

BERK, ARNOLD, Molecular Biology Institute, University of California, Los Angeles

BEVAN, MICHAEL, Dept. of Molecular Genetics, John Innes Centre, Norwich, United Kingdom

v

BHAGWAT, ASHOK, Dept. of Chemistry, Wayne State University, Detroit, Michigan

BINA, MINOU, Dept. of Chemistry, Purdue University, West Lafayette, Indiana

BIRKENBERGER, LORI, DG Respiratory, Hoechst Marion Roussel AG, Bridgewater, New Jersey

BJORKLUND, STEFAN, Dept. of Medical Biochemistry and Biophysics, Umea University, Umea, Sweden

BLACK, BRUCE, Insecticide Basic Research, American Cyanamid, Inc., Princeton, New Jersey

BLOBEL, GERD, Dept. of Pediatrics and Hematology, Children's Hospital of Philadelphia, Philadelphia, Pennsylvania

BLOMQUIST, PATRIK, Dept. of Molecular Biology and Genomics, Pharmacia and Upjohn, Stockholm, Sweden

BOHMANN, DIRK, Lab. of Differentiation, European Molecular Biology Laboratory, Heidelberg, Germany

BONDS, WESLEY, Dept. of Genetics, Yale University School of Medicine, New Haven, Connecticut

BOWEN, BEN, Div. of Metabolic and Cardiovascular Disease, Novartis Pharmaceuticals Corp., Summit, New Jersey

BOYER, THOMAS, Molecular Biology Institute, University of California, Los Angeles

BRANDRISS, MARJORIE, Dept. of Microbiology and Molecular Genetics, UMDNJ-New Jersey Medical School, Newark, New Jersey

BREEN, GAIL, Dept. of Molecular and Cell Biology, University of Texas at Dallas, Richardson, Texas

BREGMAN, DAVID, Dept. of Pathology, Albert Einstein College of Medicine, Bronx, New York

BREMNER, ROD, Molecular Division, Eye Research Institute of Canada, Toronto, Canada

BRESNICK, EMERY, Dept. of Pharmacology, University of Wisconsin Medical School, Madison

BROWN, THOMAS, Dept. of Molecular Sciences, Pfizer Central Research, Groton, Connecticut

BRUNAK, SOREN, Center for Biological Sequence Analysis, Technical University of Denmark, Lyngby

BRYANT, GENE, Prog. in Molecular Biology, Memorial Sloan-Kettering Cancer Center, New York, New York

BUCKLEY, NOEL, Dept. of Pharmacology, University College of London, London, United Kingdom

BURATOWSKI, STEPHEN, Dept. of Biological Chemistry and Molecular Pharmacology, Harvard Medical School, Boston, Massachusetts

BURGESS, RICHARD, McArdle Laboratory for Cancer Research, University of Wisconsin, Madison

BURLEY, STEPHEN, Dept. of Molecular Biophysics, Howard Hughes Medical Institute, Rockefeller University, New York, New York

BURTON, ZACHARY, Dept. of Biochemistry, Michigan State University, East Lansing, Michigan

CABLE, MICHAEL, Dept. of Structural Chemistry, Schering-Plough Research Institute, Kenilworth, New Jersey

CAI, RICHARD, Dept. of Molecular and Cell Biology, Novartis Pharmaceuticals, East Hanover, New Jersey

CAIRNS, WILLIAM, Lab. of Gene Expression Sciences, SmithKline Beecham, Harlow, Essex, United Kingdom

CALLIGARIS, RAFFAELLA, Dept. of Molecular Cell Biology, Biocenter, University of Frankfurt, Germany

CAMACHO, ANA, Centro de Biologia Molecular Severo Ochoa, Universidad Autonoma, Madrid, Spain

CAMATO, RINO, Research and Development, Geneka Biotechnology, Inc., Montreal, Canada

CAO, XU, Dept. of Pathology, University of Alabama, Birmingham

CAO, YIXUE, Dept. of Microbiology and Immunology, State University of New York, Brooklyn

CAREY, MICHAEL, Dept. of Biological Chemistry, University of California, Los Angeles

CARLES, CHRISTOPHE, Service de Biochimie et Génétique Moléculaire, CEA/Saclay, Gif-sur-Yvette, France

CARR, KIMBERLY, *Nature Magazine*, London, United Kingdom

CASKEY, C. THOMAS, Merck Genome Research Institute, Merck Research Laboratories, West Point, Pennsylvania

CHAMBERS, ROSS, Dept. of Molecular Cardiology, Southwestern Medical Center, Dallas, Texas

CHAN, DOUG, Dept. of Biological Sciences, University of Calgary, Calgary, Alberta, Canada

CHANG, JU-FANG, Dept. of Biochemistry, University of Cambridge, Cambridge, United Kingdom

CHANG, SIMON, Dept. of Biological Sciences, Louisiana State University, Baton Rouge, Louisiana

CHEN, GUOQING, Dept. of Chemistry and Biochemistry, University of California, Los Angeles

CHIN, WILLIAM, Dept. of Genetics, Brigham and Women's Hospital, Boston, Massachusetts

COHEN, DALIA, Dept. of Molecular and Cellular Biology, Novartis Pharmaceuticals, East Hanover, New Jersey

COIN, FREDERIC, IGBMC, CNRS/INSERM, Illkirch, Strasbourg, France

CONAWAY, JOAN, Prog. in Molecular and Cell Biology, Howard Hughes Medical Institute, Oklahoma Medical Research Foundation, Oklahoma City, Oklahoma

CONAWAY, RONALD, Prog. in Molecular and Cell Biology, Oklahoma Medical Research Foundation, Oklahoma City, Oklahoma

COOKE, ROBERT, *Newsday*, Melville, New York

CORCORAN, ANNE, Lab. of Molecular Biology, Protein and Nucleic Acid Chemistry, Cambridge, United Kingdom

CÔTÉ, JACQUES, Laval University Cancer Research Center, Hotel-Dieu de Quebec, Quebec, Canada

COULOMBE, BENOIT, Dept. of Biology, University of Sherbrooke, Sherbrooke, Quebec, Canada

CRAMER, PATRICK, Grenoble Outstation, European Molecular Biology Laboratory, Grenoble, France

DAHMUS, MICHAEL, Dept. of Molecular and Cellular Biology, University of California, Davis

DANIELS, MARC, Dept. of Molecular Physiology and Biophysics, Vanderbilt University Medical Center, Nashville, Tennessee

DARIMONT, BEATRICE, Dept. of Cellular and Molecular Pharmacology, University of California, San Francisco

DARST, SETH, Rockefeller University, New York, New York

DAVIDSON, IRWIN, IGBMC, CNRS/INSERM, Illkirch, Strasbourg, France

DEGRAFFENRIED, LINDA, Dept. of Medical Oncology, University of Texas Health Sciences Center, San Antonio

DEJONG, JEFF, Dept. of Biology, University of Texas at Dallas, Richardson, Texas

DE RUWE, MARJOLEIN, Lab. for Physiological Chemistry, Utrecht University, Utrecht, The Netherlands

DESTERRO, JOANA, School of Biomedical Sciences, University of St. Andrews, St. Andrews, Scotland, United Kingdom

DILLON, NIALL, Clinical Sciences Centre, Hammersmith Hospital, Medical Research Council, London, United Kingdom

DING, HAO, Samuel Lunenfeld Institute, Mount Sinai Hospital, Toronto, Ontario, Canada

DOETZLHOFER, ANGELIKA, Institute of Molecular Biology, University of Vienna, Austria

DOLAN, COLLETTE, Dept. Haematology and Oncology, St. James Hospital, University of Dublin, Dublin, Ireland

DOMBROSKI, ALICIA, Dept. of Microbiology and Molecular Genetics, University of Texas Medical School, Houston

DONZE, DAVID, NICHD, National Institutes of Health, Bethesda, Maryland

DORSETT, DALE, Prog. in Molecular Biology, Memorial Sloan-Kettering Cancer Center, New York, New York

DORVAL, KIM, Dept. of Laboratory Medicine and Pathobiology, Eye Research Institute of Canada, University of Toronto, Toronto, Ontario, Canada

DOVE, SIMON, Dept. of Microbiology and Molecular Genetics, Harvard Medical School, Boston, Massachusetts

DU, JIAN, Dept. of Microbiology and Immunology, Health Science Center, State University of New York, Brooklyn

DUTNALL, ROBERT, Dept. of Biochemistry, University of Utah School of Medicine, Salt Lake City

DYNAN, WILLIAM, Dept. of Molecular Medicine and Genetics, Medical College of Georgia, Augusta, Georgia

EBRIGHT, RICHARD, Dept. of Chemistry, Howard Hughes Medical Institute, Waksman Institute, Rutgers University, Piscataway, New Jersey

EGLY, JEAN-MARC, IGBMC, CNRS/INSERM, Illkirch, Strasbourg, France

EICK, DIRK, Institute for Clinical Molecular Biology and Tumor Genetics, GSF-Research Center for Health and Environment, Munich, Germany

EISENMAN, ROBERT, Div. of Basic Sciences, Fred Hutchinson Cancer Research Center, Seattle, Washington

ELLWOOD, KATE, Dept. of Biological Chemistry, University of California, Los Angeles

EMERSON, BEVERLY, Dept. of Regulatory Biology, The Salk Institute for Biological Studies, La Jolla, California

EVANS, RONALD, Howard Hughes Medical Institute, The Salk Institute for Biological Studies, San Diego, California

FAUSTMAN, DENISE, Dept. of Immunobiology, Massachusetts General Hospital-East, Harvard Medical School, Charlestown, Massachusetts

FELSENFELD, GARY, Lab. of Molecular Biology, NIDDK, National Institutes of Health, Bethesda, Maryland

FERNANDEZ-PEREZ, FRANCISCO, Dept. of Protein X-ray Crystallography, EMBL Outstation, Hamburg, Germany

FESSING, MICHAEL, Dept. of Pharmaceutical Sciences, St. Jude Children's Research Hospital, Memphis, Tennessee

FISCHER, DAVID, Dept. of Molecular Genetics, Leiden University, Leiden, The Netherlands

FLEISSNER, ERWIN, Div. of Science and Mathematics, Hunter College, City University of New York, New York

FRANKLIN, NAOMI, Dept. of Biology, University of Utah, Salt Lake City

FRANTZ, MARTHA, Dept. of Microbiology, Dartmouth Medical School, Lebanon, New Hampshire

FRASS, BEATE, Institute for Cell Biology, University Hospital of Essen, Essen, Germany

FREE, ANDREW, Institute of Cell and Molecular Biology, University of Edinburgh, Edinburgh, Scotland, United Kingdom

FREEDMAN, LEONARD, Prog. in Cell Biology, Memorial Sloan-Kettering Cancer Center, New York, New York

FRETER, ROLF, Dept. of Medicine and Medical Oncology, Columbia University, New York, New York

FRITZ, STEFAN, Dept. of Genetics, University of Erlangen-Nuremberg, Erlangen, Germany

FRY, CHRISTOPHER, Dept. of Oncology, University of Wisconsin Medical School, Madison

GANN, ALEXANDER, Dept. of Biological Sciences, Lancaster University, Lancaster, United Kingdom

GARBER, MITCH, Dept. of Regulatory Biology, The Salk Institute for Biological Studies, La Jolla, California

GASSER, SUSAN, Dept. of Cancer Research, ISREC, Epalinges, Switzerland

GAUDREAU, LUC, Dept. of Biology, University of Sherbrooke, Sherbrooke, Quebec, Canada

GAVVA, NARENDER, Sect. of Molecular and Cellular Biology, University of California, Davis

GEIDUSCHEK, E. PETER, Dept. of Biology, Center for Molecular Genetics, University of California at San Diego, La Jolla

GEORGIEVA, SOFIA, Institute of Gene Biology, Russian Academy of Sciences, Moscow

GILEADI, OPHER, Dept. of Molecular Genetics, Weizmann Institute of Science, Rehovot, Israel

GILMOUR, DAVID, Dept. of Biochemistry and Molecular Biology, Pennsylvania State University, University Park, Pennsylvania

GLOVER, MARK, Dept. of Biochemistry, University of Alberta, Edmonton, Alberta, Canada

GOLDFARB, ALEX, Public Health Research Institute, New York, New York

GONG, DA-WEI, Diabetes Branch, NIDDK, National Institutes of Health, Bethesda, Maryland

GOODRICH, JAMES, Dept. of Chemistry and Biochemistry, University of Colorado, Boulder

GOTTESFELD, JOEL, Dept. of Molecular Biology, The Scripps Research Institute, La Jolla, California

GOTTESMAN, SUSAN, Dept. of Molecular Biology, National Cancer Institute, National Institutes of Health, Bethesda, Maryland

GOURSE, RICHARD, Dept. of Bacteriology, University of Wisconsin, Madison

GRANT, PATRICK, Dept. of Biochemistry and Molecular Biology, Pennsylvania State University, University Park, Pennsylvania

GRAVES, BARBARA, Dept. of Oncological Sciences, Huntsman Cancer Institute, University of Utah School of Medicine, Salt Lake City

GREEN, MAURICE, Dept. of Molecular Virology, St. Louis University Health Sciences Center, St. Louis, Missouri

GREEN, MICHAEL, Prog. in Molecular Medicine, Howard Hughes Medical Institute, University of Massachusetts Medical Center, Worcester

GREENBLATT, JACK, Banting and Best Dept. of Medical Research, University of Toronto, Toronto, Ontario, Canada

GREENLEAF, ARNO, Dept. of Biochemistry, Duke University Medical Center, Durham, North Carolina

GROSS, CAROL, Dept. of Stomatology, University of California, San Francisco

GROSSCHEDL, RUDOLF, Dept. of Microbiology and Immunology, Howard Hughes Medical Institute, University of California, San Francisco

GRUNDSTRÖM, THOMAS, Dept. of Cell and Molecular Biology, University of Umeå, Umeå, Sweden

GRUNSTEIN, MICHAEL, Dept. of Biological Chemistry, Molecular Biology Institute, University of California, Los Angeles

GU, WEIGANG, Dept. of Biology, Emory University, Atlanta, Georgia

GUO, YULI, Dept. of Chemistry and Biochemistry, University of California, Los Angeles

HACHE, ROBERT, Ottawa Civic Hospital, Loeb Research Institute, University of Ottawa, Ottawa, Canada

HAHN, STEVEN, Howard Hughes Medical Institute, Fred Hutchinson Cancer Research Center, Seattle, Washington

HALL, BENJAMIN, Dept. of Genetics, University of Washington, Seattle

HAMAMORI, YASUO, Dept. of Biochemistry and Molecular Biology, University of Southern California, Los Angeles, California

HAMICHE, ALI, LBME/IBCG, National Center of Scientific Research, Toulouse, France

HAMPSEY, MICHAEL, Dept. of Biochemistry, UMDNJ-Robert Wood Johnson Medical School, Piscataway, New Jersey

HAN, SANG-JUN, Lab. of Molecular Genetics, Samsung Biomedical Research Institute, Seoul, Korea

HANDA, HIROSHI, Dept. of Bioscience and Biotechnology, Tokyo Institute of Technology, Yokohama, Japan

HARDINGHAM, GILES, Div. of Neurobiology, MRC Laboratory of Molecular Biology, Cambridge, United Kingdom

HARTZOG, GRANT, Dept. of Genetics, Harvard Medical School, Boston, Massachusetts

HASDAY, JEFFREY, Dept. of Medicine, University of Maryland School of Medicine, Baltimore

HASWELL, ELIZABETH, Dept. of Biochemistry and Biophysics, University of California, San Francisco

HAUSNER, WINFRIED, Institute for General Microbiology, University of Kiel, Kiel, Germany

HAYES, PATRICK, Dept. of Medicine and Developmental and Molecular Biology, Mount Sinai School of Medicine, New York, New York

HEARD, DAVID, Dept. of Molecular Genetics, Novo Nordisk A/S, Bagsvaerd, Denmark

HEINRICH, JULIA, Molecular Genetics Screen Design, American Cyanamid, Inc., Princeton, New Jersey

HENRY, WILLIAM, James Laboratory, Cold Spring Harbor Laboratory, Cold Spring Harbor, New York

HERNANDEZ, NOURIA, James Laboratory, Cold Spring Harbor Laboratory, Cold Spring Harbor, New York

HERR, WINSHIP, James Laboratory, Cold Spring Harbor Laboratory, Cold Spring Harbor, New York

HERRMANN, MATTHIAS, Zentrale Biotechnologie, Hoechst Marion Roussel GmbH, Martinsried, Germany

HIROSE, SUSUMU, Dept. of Developmental Genetics, National Institute of Genetics, Mishima, Shizuoka-ken, Japan

HOCHSCHILD, ANN, Dept. of Microbiology and Molecular Genetics, Harvard Medical School, Boston, Massachusetts

HODIN, RICHARD, Dept. of Surgery, Beth Israel Deaconess Medical Center, Boston, Massachusetts

HOLSTEGE, FRANK, Dept. of Biology, Whitehead Institute for Biomedical Research, Massachusetts Institute of Technology, Cambridge, Massachusetts

HORIKOSHI, MASAMI, Institute of Molecular and Cellular Biosciences, University of Tokyo, Japan

HOWE, MARTHA, Dept. of Microbiology and Immunology, University of Tennessee, Memphis

HU, WEIMING, Dept. of Plant Biotechnology, American Cyanamid Inc., Princeton, New Jersey

HUI, ARIELA, Dept. of Molecular Biology, Amgen, Thousand Oaks, California

HUYNH, KHANH, Dept. of Biochemistry, University of Minnesota, Minneapolis

IKEDA, KEIKO, Dept. of Biology, Jichi Medical School, Tochigi, Japan

IMIOLEK, ADAM, Biochemicals Business Development, Boehringer Mannheim, Indianapolis, Indiana

INAMOTO, SUSUMU, Dept. of Microbiology, Keio University School of Medicine, Tokyo, Japan

IZUMO, MARIKO, Dept. of Biology, Vanderbilt University, Nashville, Tennessee

JACKSON, STEPHEN, Wellcome Trust, CRC Institute of Cancer and Developmental Biology, University of Cambridge, Cambridge, United Kingdom

JACOBSON, SANDRA, Dept. of Molecular, Cellular and Developmental Biology, University of Colorado, Boulder

JAEHNING, JUDITH, Dept. of Biochemistry and Molecular Genetics, University of Colorado Health Science Center, Denver

JAIN, ANJALI, Dept. of Microbiology, Immunology and Molecular Genetics, Howard Hughes Medical Institute, University of California, Los Angeles

JAYNES, JAMES, Kimmel Cancer Institute, Thomas Jefferson University, Philadelphia, Pennsylvania

JIN, DING, Dept. of Molecular Biology, National Cancer Institute, National Institutes of Health, Bethesda, Maryland

JOHNSON, ALEXANDER, Dept. of Microbiology and Immunology, University of California, San Francisco

JONES, KATHERINE, Dept. of Regulatory Biology, The Salk Institute for Biological Studies, La Jolla, California

JONES, MICHAEL, Gene Search Group, Chugai Institute for Molecular Medicine, Ibaraki, Japan

JUAN, ELVIRA, Dept. of Genetics, University of Barcelona, Barcelona, Spain

KADONAGA, JAMES, Dept. of Biology, University of California at San Diego, La Jolla

KANNO, MASAMOTO, Dept. of Immunology and Parasitology, Hiroshima University School of Medicine, Hiroshima, Japan

KAPANIDIS, ACHILLES, Dept. of Chemistry, Howard Hughes Medical Institute, Waksman Institute, Rutgers University, Piscataway, New Jersey

KAPPEL, ANDREAS, Dept. of Molecular Cell Biology, Max-Planck-Institute, Bad Nauheim, Germany

KASHLEV, MIKHAIL, ABL-Basic Research Program, NCI-Frederick Cancer Research and Development Center, Frederick, Maryland

KASSAVETIS, GEORGE, Dept. of Biology, Center for Molecular Genetics, University of California at San Diego, La Jolla

KATSAFANAS, GEORGE, Lab. of Viral Diseases, NIAID, National Institutes of Health, Bethesda, Maryland

KERKMANN, KATJA, Max-Delbrück Laboratory, Max-Planck-Institute, Cologne, Germany

KIM, JAE, Center for Cancer Research, Massachusetts Institute of Technology, Cambridge, Massachusetts

KIM, YOUNG-JOON, Lab. of Molecular Genetics, Samsung Biomedical Research Institute, Seoul, Korea

KIMURA, AKATSUKI, Lab. of Developmental Biology, IMCB, The University of Tokyo, Tokyo, Japan

KINGSTON, ROBERT, Dept. of Molecular Biology, Massachusetts General Hospital, Boston, Massachusetts

KIREEVA, MARIA, ABL-Basic Research Program, NCI-Frederick Cancer Research and Development Center, Frederick, Maryland

KNUTSON, ANDERS, Dept. of Surgery, University Hospital, Uppsala University, Uppsala, Sweden

KODADEK, THOMAS, Dept. of Internal Medicine and Biochemistry, University of Texas Southwestern Medical Center, Dallas

KOKUBO, TETSURO, Div. of Gene Function in Animals, Nara Institute of Science and Technology, Ikoma, Nara, Japan

KOMISSAROVA, NATALIA, ABL-Basic Research Program, NCI-Frederick Cancer Research and Development Center, Frederick, Maryland

KOOP, RONALD, Institute for Molecular Biology and Tumor Research, Philipps University, Marburg, Germany

KORNBERG, ROGER, Dept. of Structural Biology, Stanford University School of Medicine, Stanford, California

KRAUS, W. LEE, Dept. of Biology, University of California at San Diego, La Jolla

KRUMM, ANTON, Dept. of Basic Sciences, Fred Hutchinson Cancer Research Center, Seattle, Washington

KUNDU, TAPAS, Dept. of Biochemistry and Molecular Biology, Rockefeller University, New York, New York

KUO, MIN-HAO, Dept. of Biology, University of Rochester, Rochester, New York

KURTZ, DAVID, Dept. of Pharmacology, Medical University of South Carolina, Charleston

KUSTU, SYDNEY, Dept. of Plant and Microbial Biology, University of California, Berkeley

KUTTER, ELIZABETH, Dept. of Science, Technology, and Health, Evergreen State College, Olympia, Washington

LADIAS, JOHN, Dept. of Medicine, Harvard Medical School, Boston, Massachusetts

LADURNER, ANDREAS, Medical Research Council Centre, Cambridge Centre for Protein Engineering, Cambridge, United Kingdom

LANDICK, ROBERT, Dept. of Bacteriology, University of Wisconsin, Madison

LANIEL, MARC-ANDRE, Dept. of Molecular Endocrinology, CHUL Research Center, Ste.-Foy, Quebec, Canada

LASPIA, MICHAEL, Dept. of Microbiology, Dartmouth Medical School, Lebanon, New Hampshire

LAURENT, BREHON, Dept. of Microbiology and Immunology, Health Science Center, State University of New York, Brooklyn

LEBLANC, BENOIT, Lab. of Cell and Developmental Biology, NIDDK, National Institutes of Health, Bethesda, Maryland

LEE, DONG-KI, Dept. of Biochemistry, Cornell University, Ithaca, New York

LEHMAN, ALAN, Dept. of Molecular and Cellular Biology, University of California, Davis

LEI, LEI, Dept. of Biochemistry, Michigan State University, East Lansing, Michigan

LEMAIRE, MARC, Biochimie Medicale, CMU, Geneva, Switzerland

LEROY, GARY, Dept. of Biochemistry, UMDNJ-Robert Wood Johnson Medical School, Piscataway, New Jersey

LEVENS, DAVID, Lab. of Pathology, DCS, National Cancer Institute, Bethesda, Maryland

LEWIS, BRIAN, Div. of Hematology and Oncology, Howard Hughes Medical Institute, Children's Hospital, Boston, Massachusetts

LIEBERMAN, PAUL, Dept. of Molecular Genetics, The Wistar Institute, Philadelphia, Pennsylvania

LINN, THOMAS, Dept. of Microbiology and Immunology, University of Western Ontario, London, Ontario, Canada

LIS, JOHN, Dept. of Biochemistry, Molecular and Cell Biology, Cornell University, Ithaca, New York

LIU, FANG, Prog. in Cell Biology, Memorial Sloan-Kettering Cancer Center, New York, New York

LIU, YI, Dept. of Microbiology and Molecular Cancer Biology, Duke University Medical Center, Durham, North Carolina

LOCHOWSKA, ANNA, Dept. of Microbial Biochemistry, Institute of Biochemistry and Physics, Polish Academy of Science, Warsaw, Poland

LONGACRE, ANGELIKA, Div. of Biological Sciences, University of Montana, Missoula

LOSICK, RICHARD, Dept. of Biology, Harvard University, Cambridge, Massachusetts

LUSE, DONAL, Dept. of Molecular Biology, Lerner Research Institute, Cleveland Clinic Foundation, Cleveland, Ohio

LUTTER, LEONARD, Dept. of Molecular Biology Research, Henry Ford Hospital, Detroit, Michigan

LUTZ, PIERRE, INSERM, Paris, France

MADEYSKI, KATJA, Dept. of Molecular Biology, Göteborg University, Gothenburg, Sweden

MAH, THIEN-FAH, Dept. of Medical Genetics and Microbiology, University of Toronto, Toronto, Ontario, Canada

MAITY, SANKAR, Dept. of Molecular Genetics, M.D. Anderson Cancer Center, University of Texas, Houston

MANFREDI, JAMES, Ruttenberg Cancer Center, Mount Sinai School of Medicine, New York, New York

MANIATIS, THOMAS, Dept. of Molecular and Cellular Biology, Harvard University, Cambridge, Massachusetts

MARKOV, DMITRIY, Dept. of Genetics, Waksman Institute, Rutgers University, Piscataway, New Jersey

MARR, MICHAEL, Dept. of Biochemistry, Cell and Molecular Biology, Cornell University, Ithaca, New York

MARSH, KATHERINE, National Cancer Institute, National Institutes of Health, Bethesda, Maryland

MARSHALL, NICK, Dept. of Molecular and Cellular Biology, University of California, Davis

MARTIN, CRAIG, Dept. of Chemistry, University of Massachusetts, Amherst

MARTIN, ROBERT, NIDDK, National Institutes of Health, Bethesda, Maryland

MARX, STEPHEN, Dept. of Genetics and Endocrinology, National Institutes of Health, Bethesda, Maryland

MAYR, BERNHARD, Dept. of Clinical Endocrinology, Medizinische Hochschule Hannover, Hannover, Germany

MCGHEE, JAMES, Dept. of Medical Biochemistry, University of Calgary, Calgary, Canada

MEININGHAUS, MARK, Institute for Clinical Molecular Biology and Tumor Genetics, GSF-Research Center for Health and Environment, Munich, Germany

MEISTERERNST, MICHAEL, Lab. for Molecular Biology-Genzentrum, University of Munich, Munich, Germany

MELNIKOVA, IRENA, Dept. of Molecular Medicine, University of Texas Health Science Center, San Antonio

MENCIA, MARIO, Dept. of Biological Chemistry and Molecular Pharmacology, Harvard Medical School, Boston, Massachusetts

MICHELS, CORINNE, Dept. of Biology, Queens College, City University of New York, Flushing, New York

MITSUI, AKIRA, Center for Cancer Research, Massachusetts Institute of Technology, Cambridge, Massachusetts

MIYAJI, MARY, Dept. of Biomolecular Engineering, Tokyo Institute of Technology, Yokohama, Japan

MIYAO, TAKENORI, Dept. of Molecular Genetics and Microbiology, UMDNJ-Robert Wood Johnson Medical School, Piscataway, New Jersey

MIZUGUCHI, GAKU, Lab. of Molecular Cell Biology, National Cancer Institute, National Institutes of Health, Bethesda, Maryland

MIZZEN, CRAIG, Dept. of Biology, University of Rochester, Rochester, New York

MOON, NAM SUNG, Dept. of Biochemistry, McGill University, Montreal, Quebec, Canada

MOQTADERI, ZARMIK, Dept. of Biological Chemistry and Molecular Pharmacology, Harvard Medical School, Boston, Massachusetts

MORENO-ROCHA, J. CLAUDIO, Dept. of Biochemistry, ULIEG, Medicine School, UANL, Monterrey, Mexico

MORSE, RANDALL, Dept. of Molecular Genetics, Wadsworth Center, Albany, New York

MUCHARDT, CHRISTIAN, Dept. of Biotechnologie, CNRS, Institut Pasteur, Paris, France

MUNSHI, NIKHIL, Dept. of Biochemistry, Columbia University, New York, New York

MURAKAMI, SEISHI, Dept. of Molecular Oncology, Cancer Research Institute, Kanazawa University, Kanazawa, Ishikawa, Japan

MYER, VIC, Whitehead Institute for Biomedical Research, Massachusetts Institute of Technology, Cambridge, Massachusetts

NÄÄR, ANDERS, Dept. of Molecular Biology, Howard Hughes Medical Institute, University of California, Berkeley

NAKATANI, YOSHIHIRO, Lab. of Molecular Growth Regulation, NICHD, National Institutes of Health, Bethesda, Maryland

NARYSHKIN, NIKOLAI, Howard Hughes Medical Institute, Waksman Institute, Rutgers University, Piscataway, New Jersey

NATARAJAN, KRISHNAMURTHY, Lab. of Eukaryotic Gene Regulation, NICHD, National Institutes of Health, Bethesda, Maryland

NEKHAI, SERGEI, Dept. of Biochemistry and Molecular Biology, George Washington University Medical Center, Washington, D.C.

NEPVEU, ALAIN, Molecular Oncology Group, McGill University, Montreal, Canada

NEVADO, JULIAN, Prog. in Molecular Biology, Memorial Sloan-Kettering Cancer Center, New York, New York

NIGHTINGALE, KARL, Prog. in Gene Expression, European Molecular Biology Laboratory, Heidelberg, Germany

NIKOLAJCZYK, BARBARA, Dept. of Biology, Brandeis University, Waltham, Massachusetts

NILSSON, JEANETTE, Dept. of Molecular Biology, Göteborg University, Gothenburg, Sweden

NIU, WEI, Dept. of Chemistry, Howard Hughes Medical Institute, Waksman Institute, Rutgers University, Piscataway, New Jersey

NORBY, PEDER, Dept. of Molecular Genetics, Novo Nordisk, Bagsvaerd, Denmark

OHKUMA, YOSHIAKI, Div. of Cellular Biology, Institute for Molecular and Cell Biology, Osaka University, Osaka, Japan

OLIVIERO, SALVATORE, Dept. of Molecular Biology, University of Siena, Siena, Italy

OMER, CHARLES, Dept. of Cancer Research, Merck Research Laboratories, West Point, Pennsylvania

OPITZ, OLIVER, Gastrointestinal Unit, Massachusetts General Hospital, Harvard Medical School, Boston, Massachusetts

O'SHEA, ERIN, Dept. of Biochemistry and Biophysics, University of California, San Francisco

OSLEY, MARY ANN, Prog. in Molecular Biology, Memorial Sloan-Kettering Cancer Center, New York, New York

OSTLUND FARRANTS, ANN-KRISTIN, Dept. of Cell Biology, Stockholm University, Stockholm, Sweden

PAPE, LOUISE, Dept. of Chemistry, New York University, New York, New York

PAREKH, BHAVIN, Dept. of Molecular and Cellular Biology, Harvard University, Cambridge, Massachusetts

PARVIN, JEFFREY, Dept. of Pathology, Brigham and Women's Hospital, Harvard Medical School, Boston, Massachusetts

PÄTZOLD, ANDREAS, Max-Delbrück Laboratory, Max-Planck-Institute, Cologne, Germany

PAUL, DIETER, Dept. of Cell Biology, Fraunhofer Institute of Toxicology, Hannover, Germany

PEARSON, JAMES, Dept. of Molecular Microbiology and Immunology, University of Missouri, Columbia

PEDERSEN, ANDERS, Center for Biological Sequence Analysis, Technical University of Denmark, Lyngby

PENDERGRAST, P. SHANNON, James Laboratory, Cold Spring Harbor Laboratory, Cold Spring Harbor, New York

PERSSON, CHRISTINE, Dept. of Cell and Molecular Biology, Immunology Unit, University of Lund, Lund, Sweden

PETERSON, CRAIG, Prog. in Molecular Medicine, Dept. of Biochemistry and Molecular Biology, University of Massachusetts Medical Center, Worcester

PFITZNER, EDITH, Tumor Biology Center, Institute for Experimental Cancer Research, Freiburg, Germany

PIETZ, BRADLEY, Dept. of Bacteriology, McArdle Laboratory for Cancer Research, University of Wisconsin, Madison

PINTEL, DAVID, Dept. of Microbiology, University of Missouri School of Medicine, Columbia

POELLINGER, LORENZ, Dept. of Cell and Molecular Biology, Karolinska Institute, Stockholm, Sweden

POLITIS, PANAGIOTIS, Dept. of Biochemistry, University of Oxford, Oxford, United Kingdom

PRELICH, GREGORY, Dept. of Molecular Genetics, Albert Einstein College of Medicine, Bronx, New York

PRICE, DAVID, Dept. of Biochemistry, University of Iowa, Iowa City

PRIOLEAU, MARIE-NOELLE, Lab. of Molecular Biology, NIDDK, National Institutes of Health, Bethesda, Maryland

PTASHNE, MARK, Prog. in Molecular Biology, Memorial Sloan-Kettering Cancer Center, New York, New York

QURESHI, SOHAIL, Dept. of Neurology, University of Wisconsin, Madison

RAMACHANDRAN, ARUNA, Lab. of Eukaryotic Gene Expression, National Institute of Immunology, New Delhi, India

RECILLAS, FELIX, Lab. of Molecular Biology, NIDDK, National Institutes of Health, Bethesda, Maryland

REECE, RICHARD, School of Biological Sciences, University of Manchester, Manchester, United Kingdom

REESE, JOSEPH, Dept. of Biochemistry and Molecular Biology, Pennsylvania State University, University Park, Pennsylvania

REINBERG, DANNY, Dept. of Biochemistry, UMDNJ-Robert Wood Johnson Medical School, Piscataway, New Jersey

REINES, DANNY, Dept. of Biochemistry, Emory University School of Medicine, Atlanta, Georgia

REMENYI, ATTILA, Lab. of Gene Expression, European Molecular Biology Laboratory, Heidelberg, Germany

ROBERT, FRANCOIS, Dept. of Biology, University of Sherbrooke, Sherbrooke, Quebec, Canada

ROBERTS, JEFFREY, Dept. of Biochemistry, Molecular and Cell Biology, Cornell University, Ithaca, New York

ROBERTSON, MIRANDA, Science Division, *Garland Publishing*, London, United Kingdom

ROBYR, DANIEL, Institute of Animal Biology, University of Lausanne, Lausanne, Switzerland

RODRIGUEZ, MANUEL, School of Biomedical Sciences, University of St. Andrews, St. Andrews, Scotland, United Kingdom

ROEDER, ROBERT, Lab. of Biochemistry and Molecular Biology, Rockefeller University, New York, New York

ROOPRA, AVTAR, Wellcome Laboratory for Molecular Pharmacology, University College of London, London, United Kingdom

ROSS, WILMA, Dept. of Bacteriology, University of Wisconsin, Madison

ROTH, SHARON, Dept. of Biochemistry and Molecular Biology, M.D. Anderson Cancer Center, University of Texas, Houston

ROTHBLUM, LAWRENCE, Dept. of Cellular and Molecular Physics, Henry Hood Research Program, Pennsylvania State College of Medicine, Danville, Pennsylvania

ROTHMAN-DENES, LUCIA, Dept. of Molecular Genetics and Cell Biology, University of Chicago, Chicago, Illinois

ROY, ANANDA, Dept. of Pathology, Div. of Immunology, Tufts University School of Medicine, Boston, Massachusetts

ROY, NIVEDITA, Dept. of Molecular Reproduction, Development and Genetics, Indian Institute of Science, Bangalore, India

ROY, SIDDHARTHA, Lab. of Molecular Biology, National Cancer Institute, National Institutes of Health, Bethesda, Maryland

RYAN, MICHAEL, Dept. of Genetic Disorders, Wadsworth Center, Albany, New York

RYAN, ROBERT, The Wistar Institute, University of Pennsylvania, Philadelphia

SAMBADE, MARIA, Dept. of Biochemistry, University of Texas Southwestern Medical Center, Dallas

SANTANGELO, THOMAS, Dept. of Biochemistry, Cellular and Molecular Biology, Cornell University, Ithaca, New York

SARTORELLI, VITTORIO, Dept. of Biochemistry, University of Southern California, Los Angeles, California

SASSONE-CORSI, PAOLO, IGBMC, CNRS/INSERM, Université Louis Pasteur, Illkirch, Strasbourg, France

SASTRY, SRIN, Dept. of Molecular Genetics, Rockefeller University, New York, New York

SATOH, MASAHIKO, Chul Research Center, Laval University, Ste.-Foy, Quebec, Canada

SCHMIDT, EDWARD, Dept. of Human Genetics, Eccles Institute, University of Utah, Salt Lake City

SCHOENHERR, CHRISTOPHER, Dept. of Molecular Biology, Princeton University, Princeton, New Jersey

SCHÖLER, HANS, Lab. of Gene Expression, European Molecular Biology Laboratory, Heidelberg, Germany

SCHROEDER, STEPHANIE, Dept. of Molecular Physiology and Biophysics, Vanderbilt University, Nashville, Tennessee

SCHULTZ, DAVID, Dept. of Molecular Biology, The Wistar Institute, Philadelphia, Pennsylvania

SCHWARTZBAUER, GARY, Dept. of Endocrinology, Children's Hospital of Pittsburgh, Pittsburgh, Pennsylvania

SEGAL, DAVID, Dept. of Molecular Biology, The Scripps Research Institute, La Jolla, California

SEISER, CHRISTIAN, Dept. of Molecular Biology, University of Vienna, Vienna, Austria

SENTENAC, ANDRÉ, Service de Biochimie et de Génétique Moléculaire, Centre d'Etude Nucleaires de Saclay, Gif-sur-Yvette, France

SEPEHRI, SETAREH, James Laboratory, Cold Spring Harbor Laboratory, Cold Spring Harbor, New York

SERFLING, EDGAR, Dept. of Molecular Pathology, Institute of Pathology, University of Wurzburg, Wurzburg, Germany

SERIZAWA, HIROAKI, Dept. of Biochemistry and Molecular Biology, University of Kansas Medical Center, Kansas City

SHARP, PHILLIP, Center for Cancer Research, Massachusetts Institute of Technology, Cambridge, Massachusetts

SHIEKHATTAR, RAMIN, Dept. of Molecular Genetics, The Wistar Institute, Philadelphia, Pennsylvania

SHILATI, A., Dept. of Biochemistry, St. Louis University School of Medicine, St. Louis, Missouri

SIGLER, PAUL, Dept. of Molecular Biophysics and Biochemistry, Yale University, New Haven, Connecticut

SIMMONS, DAVID, Dept. of Molecular Neurobiology Research, SmithKline Beecham, Harlow, Essex, United Kingdom

SINGH, HARINDER, Howard Hughes Medical Institute, University of Chicago, Chicago, Illinois

SINGH, JAGMOHAN, Dept. of Yeast Developmental Genetics, Institute of Microbial Technology, Chandigarh, India

SLOMIANY, BEATRIX, Dept. of Pharmacology, Medical University of South Carolina, Charleston

SMALE, STEPHEN, Dept. of Microbiology, Immunology, and Molecular Genetics, Howard Hughes Medical Institute, Medical Biology Institute, University of California, Los Angeles

SPYCHALA, JOZEF, Dept. of Pharmacology, University of North Carolina, Chapel Hill

STAMMINGER, THOMAS, Institute for Clinical and Molecular Virology, University of Erlangen-Nuremberg, Erlangen, Germany

STARGELL, LAURIE, Dept. of Biochemistry and Molecular Biology, Colorado State University, Fort Collins, Colorado

STEITZ, THOMAS, Dept. of Molecular Biophysics and Biochemistry, Howard Hughes Medical Institute, Yale University, New Haven, Connecticut

STERNER, DAVID, The Wistar Institute, Philadelphia, Pennsylvania

STEUERNAGEL, ARND, Dept. of Biology, Emory University, Atlanta, Georgia

STILLMAN, BRUCE, James Laboratory, Cold Spring Harbor Laboratory, Cold Spring Harbor, New York

STRUHL, KEVIN, Dept. of Biological Chemistry and Molecular Pharmacology, Harvard Medical School, Boston, Massachusetts

STUMPH, WILLIAM, Dept. of Chemistry, San Diego State University, San Diego, California

SUDBECK, PETER, Institute of Human Genetics, University of Freiburg, Freiburg, Germany

SUN, ZU-WEN, Dept. of Biochemistry, UMDNJ-Robert Wood Johnson Medical School, Piscataway, New Jersey

SUNE, CARLOS, Dept. of Pharmacology and Cancer Biology, Duke University Medical Center, Durham, North Carolina

SUSKE, GUNTRAM, IMT, Philipps-University, Marburg, Germany

SVETLOV, VLADIMIR, McArdle Laboratory for Cancer Research, University of Wisconsin, Madison

SZENTIRMAY, MARILYN, Dept. of Molecular Genetics, M.D. Anderson Cancer Center, University of Texas, Houston

TADAYYON, MOHAMMAD, Dept. of Vascular Biology, SmithKline Beecham, Harlow, Essex, United Kingdom

TAGAMI, HIDEAKI, Dept. of Biological Science, Nagoya University, Chikusa, Nagoya, Japan

TAYLOR, DAVID, Dept. of Cell Biology, Parke-Davis Research, Ann Arbor, Michigan

THANOS, DIMITRIS, Dept. of Biochemistry and Molecular Biophysics, Columbia University, New York, New York

THOMAS, ROSS, Institute of Reproduction and Development, Monash University, Melbourne, Australia

TIMMERS, MARC, Dept. of Physiological Chemistry, Utrecht University, Utrecht, The Netherlands

TJIAN, ROBERT, Dept. of Molecular Biology, Howard Hughes Medical Institute, University of California, Berkeley

TORA, LASZLO, IGBMC, CNRS/INSERM, Illkirch, Strasbourg, France

TREISMAN, RICHARD, Transcription Laboratory, Imperial Cancer Research Fund, London, United Kingdom

TRIEZENBERG, STEVEN, Dept. of Biochemistry, Michigan State University, Albany, Michigan

TSAI, SHIH-CHANG, Dept. of Molecular Oncology, Moffitt Cancer Center, University of South Florida, Tampa, Florida

TZAMARIAS, DIMITRIS, Institute of Molecular Biology and Biotechnology, Foundation of Research and Technology, Heraklion-Crete, Greece

UDALOVA, IRINA, Dept. of Pediatrics, University of Oxford, Oxford, United Kingdom

URLINGER, STEFANIE, Institute for Microbiology, University of Erlangen-Nuremberg, Erlangen, Germany

VAN DUREN, CATHELYNE, Dept. of Biosciences, University of Hertfordshire, Hatfield, Hertfordshire, United Kingdom

VAN HAASTEREN, GOEDELE, Foundation for Medical Research, Geneva, Switzerland

VANDEN BERGHE, WIM, Department of Molecular Biology, VIB, Ghent, Belgium

VERRIJDT, GUY, Dept. of Biochemistry, Faculty of Medicine, Catholic University of Leuven, Leuven, Belgium

VERSHON, ANDREW, Dept. of Molecular Biology and Biochemistry, Waksman Institute, Rutgers University, Piscataway, New Jersey

VERTEGAAL, ALFRED, Dept. of Molecular Cell Biology, Leiden University Medical Center, Leiden, The Netherlands

VOIGT, KIMBERLY, Dept. of Pharmacology and Toxicology, Forest Laboratories, New York, New York

WADA, TADASHI, Faculty of Bioscience and Biotechnology, Tokyo Institute of Technology, Yokohama, Japan

WALKER, AMY, Center for Blood Research, Harvard University Medical School, Boston, Massachusetts

WANG, EDITH, Dept. of Pharmacology, University of Washington, Seattle

WATHELET, MARC, Dept. of Molecular and Cellular Biology, Harvard University, Cambridge, Massachusetts

WEI, PING, Dept. of Regulatory Biology, The Salk Institute for Biological Studies, La Jolla, California

WEISBERG, ROBERT, Dept. of Molecular Genetics, NICHD, National Institutes of Health, Bethesda, Maryland

WERNER, MICHEL, Service de Biochimie et Génétique Moléculaire, CEA/Saclay, Gif-sur-Yvette, France

WEST, ADAM, Lab. of Molecular Biology, NIDDK, National Institutes of Health, Bethesda, Maryland

WESTIN, GUNNAR, Dept. of Surgery, University Hospital, Uppsala University, Uppsala, Sweden

WHITE, KEVIN, Dept. of Pharmacology, University of Washington, Seattle

WHITELAW, BRUCE, Dept. of Molecular Biology, Roslin Institute, Roslin, Midlothian, United Kingdom

WILLIS, IAN, Dept. of Biochemistry, Albert Einstein College of Medicine, Bronx, New York

WILLY, PATRICIA, Dept. of Biology, University of California at San Diego, La Jolla

WINROW, CHRISTOPHER, Dept. of Cell Biology and Anatomy, University of Alberta, Edmonton, Alberta, Canada

WINSTON, FRED, Dept. of Genetics, Harvard Medical School, Boston, Massachusetts

WOLFFE, ALAN, Lab. of Molecular Embryology, NICHD, National Institutes of Health, Bethesda, Maryland

WOODARD, ROBIN, Dept. of Gene Regulation, Medical College of Georgia, Augusta

WORKMAN, JERRY, Dept. of Biochemistry and Molecular Biology, Howard Hughes Medical Institute, Pennsylvania State University, University Park, Pennsylvania

WOYCHIK, NANCY, Dept. of Molecular Genetics and Microbiology, UMDNJ-Robert Wood Johnson Medical School, Piscataway, New Jersey

WRIGHT, BARBARA, Div. of Biological Sciences, University of Montana, Missoula

WU, CARL, Lab. of Molecular Cell Biology, National Cancer Institute, National Institutes of Health, Bethesda, Maryland

WU, HAI-YOUNG, Dept. of Pharmacology, Wayne State University, Detroit, Michigan

WU, SHWU-YUAN, Dept. of Biochemistry, University of Illinois, Urbana

WU, WEI-HUA, Dept. of Biochemistry, UMDNJ-Robert Wood Johnson Medical School, Piscataway, New Jersey

XIAO, HUA, Lab. of Molecular Microbiology, NIAID, National Institutes of Health, Bethesda, Maryland

YAMAMOTO, KAZUO, Dept. of Oncology, Nagasaki University School of Medicine, Nagasaki, Japan

YAMAMOTO, KEITH, Dept. of Biochemistry and Biophysics, University of California, San Francisco

YAMAMOTO, YU-ICHI, Dept. of Oncology, Nagasaki University School of Medicine, Nagasaki, Japan

YANG, SHEN-HSI, Dept. of Biochemistry and Genetics, University of Newcastle upon Tyne, Newcastle upon Tyne, United Kingdom

YANG, XIANGLI, Dept. of Pathology, University of Alabama, Birmingham

YEUNG, KAM, Dept. of Molecular Cell Biology and Biochemistry, Brown University, Providence, Rhode Island

YOUNG, RICHARD, Dept. of Biology, Whitehead Institute for Biomedical Research, Massachusetts Institute of Technology, Cambridge, Massachusetts

YU, LIUNING, Div. of Genetic Disorders, Wadsworth Center, Albany, New York

YUDKOVSKY, NATALYA, Dept. of Molecular and Cellular Biology, Fred Hutchinson Cancer Research Center, Seattle, Washington

ZHANG, J. JILLIAN, Dept. of Molecular Cell Biology, Rockefeller University, New York, New York

ZHANG, JIE LIN, Dept. of Infectious Diseases, Beth Israel Deaconess Medical Center, Harvard Medical School, Boston, Massachusetts

ZHANG, LI, Dept. of Biochemistry, New York University Medical Center, New York, New York

ZHOU, JUMIN, Dept. of Molecular and Cell Biology, Div. of Genetics, University of California, Berkeley

First row: A. Krainer, T. Maniatis; J. Roberts, T. Steitz; N. Hernandez
Second row: P. Sigler, M. Robertson; K. Yamamoto
Third row: R. Treisman; R. Evans, R. Roeder; M. Ptashne, I. Moresano
Fourth row: R. Tjian, R. Losick; W. Herr; M. Grunstein, D. Reinberg

First row: P. Sassone-Corsi, R. Kingston; D. Reinberg, M. Green, M. Carey
Second row: A Näär, E. O'Shea; T. Grodzicker, R. Losick, S. Gottesman
Third row: D. Bohmann, S. Jackson; J. Witkowski, A. Wolffe; K. Struhl
Fourth row: J. Workman; A. Krainer, J. Watson, L. Miller, T. Maniatis, B. Stillman

First row: B. Stillman, R. Roeder; T. Steitz, R. Burgess
Second row: P. Sassone-Corsi; R. Treisman, R. Kingston, T. Grodzicker, M. Green
Third row: B. Emerson, R. Martin; S. Hahn, R. Tjian, R. Eisenman
Fourth row: J. Watson, R. Evans; L. Miller

First row: A. Greenleaf; H. Aiba, A. Hochschild; J. Kadonaga, P. Sharp
Second row: W. Dynan; R. Tjian, R. Morse, T. Maniatis; S. Smale
Third row: M. Carey, E. O'Shea; K. Struhl, T. Steitz, G. Felsenfeld
Fourth row: R. Young, T. Steitz; K. Yamamoto

Seminar at CSHL—Hiroshi Handa

Mark Ptashne

Anders M. Näär, W. Lee Kraus, Stephan Fritz

Bob Roeder

View of Cold Spring Harbor

Drawings by Lewis Miller
at the 1998 Symposium

Rich Losick

Drawing rendered in the dark

Bob Tjian

Lunchtime at Blackford

Foreword

The field of gene transcription is intertwined with many other areas of modern biological research because of its fundamental significance. It has been some time since a Cold Spring Harbor Laboratory Symposium focused solely on this important topic, in part because transcription is usually included in all Symposia in one form or another. But because of the significant advances that have occurred over the last few years in our understanding of how genes are transcribed, it seemed appropriate to hold a meeting on this topic again. The dramatic advances made in this field during past 20 years have led to the identification of a vast set of basal transcription factors in eukaryotes. Moreover, this field has encompassed the exciting progress in related areas such as structural biology and chromatin structure and function.

Nearly 40 years have passed since the discovery of the first RNA polymerase by Sam Weiss, an event noted in a later page of this volume. Today we are in the midst of a very stimulating era that is focused on understanding the mechanisms of gene transcription, and, at this Symposium, we listened to 67 talks that reflected this excitement.

Many previous Cold Spring Harbor Laboratory Symposia have incorporated aspects of gene transcription and gene regulation into the program, as this topic in many ways represents the core of biology. Most noteworthy was the famous 1977 Symposium on Chromatin, where the fundamental understanding of gene structure in eukaryotes was overturned.

It was also most fitting that a Symposium on gene transcription be held this year as we celebrated the 30th year that Jim Watson has worked at the Laboratory, first as Director for 25 years and now as President. At the beginning of this meeting, I presented to Jim, on behalf of the Laboratory, a gold model of the DNA double helix. Five years ago, Jim presented the twin of this golden helix to Francis Crick at a Cold Spring Harbor meeting on the occasion of the 40th anniversary of their discovery of the structure of DNA. What better time to make this presentation to Jim than when many of his former students and colleagues were present? Prior to his coming to Cold Spring Harbor, Jim's laboratory at Harvard had made some of the key contributions to the early work on the mechanisms of gene transcription, including the discovery of σ factor.

One of the many responsibilities of the Director is to organize the annual Symposium, and this Jim did for many years. The first Symposium he organized, in 1970, was appropriately enough on "Transcription of Genetic Material." It celebrated, among other things, the discovery 2 years before of the σ factor for RNA polymerase in bacteria. At this meeting, David Baltimore also presented the exciting news of reverse transcriptase.

Jim noted in his foreword to the Symposium volume of that year: "The final result was a compromise between a desire to hear everyone with relevant data and the need to restrict the talks to a number ingestible within a week's time." In this respect, nothing has changed in the last 28 years. It was necessary to make the usual hard choices in selecting speakers. For the organization of this meeting, my colleagues Winship Herr, Nouria Hernandez, Bob Tjian, Carol Gross, Rich Losick, and a number of others provided valuable advice.

The Symposium started with a fascinating first night of introductory talks from Carol Gross, Robert Tjian, Tom Maniatis, and David Allis. Bob Roeder, who has made many seminal contributions to the biochemistry of transcription, presented this year's Reginald Harris Lecture. The formal scientific program consisted of 67 oral presentations and a record 155 poster presentations, and the meeting attracted 437 participants. I thank Rich Losick for agreeing to summarize the meeting and writing such a marvelous and thoughtful paper, matching the great summaries of previous Symposia.

We were particularly fortunate this year to have Lewis Miller as an artist in residence during the meeting. Lewis, who hails form Melbourne, Australia, won the prestigious Archibald Portrait Prize in 1998. While here, he sketched many of the Symposium participants, and Blackford Hall is now graced with some of these marvelous portraits.

Essential funds to run this meeting were obtained from the National Cancer Institute, a branch of the National Institutes of Health. In addition, financial help from the Corporate Sponsors of our meetings program is essential for these Symposia to remain a success and we are most grateful for their continued support. The following are the 1998 Symposium sponsors:

Corporate Sponsors: Amgen Inc.; BASF Bioresearch Corporation; Bayer Corporation; Bristol-Myers Squibb Company; Chiron Corporation; Chugai Research Institute for Molecular Medicine, Inc.; Diagnostic Products Corporation; Forest Laboratories, Inc.; Genentech, Inc.; Genetics Institute; Hoechst Marion Roussel; Hoffmann-La Roche Inc.; Johnson & Johnson; Kyowa Hakko Kogyo Co., Ltd.; Eli Lilly and Company; Merck Genome Research Institute; Novartis Pharma Research; Novo Nordisk Biotech, Inc./ZymoGenetics, Inc.; OSI Pharmaceuticals, Inc.; Pall Corporation; Parke-Davis Pharmaceutical Research; The Perkin-Elmer Corporation, Applied Biosystems Division; Pfizer Inc.; Pharmacia & Upjohn, Inc.; Research Genetics, Inc.; Schering-Plough Corporation; SmithKline Beecham Pharmaceuticals; Wyeth-Ayerst Research; Zeneca Group PLC. *Plant Corporate Associates:* American Cyanamid Company; Monsanto Company; Pioneer Hi-Bred International, Inc.; Westvaco Corporation. *Foundation Associates:* Albert B. Sabin Vaccine Institute at Georgetown University. *Corporate Contributors:* Genome Systems; Lexicon Genetics; Qiagen, Inc.

I thank Diane Tighe, Mary Smith, and Nancy Weeks in our meetings and courses office, and Herb Parsons and his audiovisual staff, under the talented direction of David Stewart, for their efficient organization of this meeting. Wendy Crowley efficiently handled the various grant applications. The organization of this meeting would not have occurred without great help from my assistant Delia King. It was again a pleasure to work with the Laboratory Press, under the direction of John Inglis, particularly Joan Ebert and Dorothy Brown.

Bruce Stillman
December 1999

Dedication and Historical Note

We dedicate this volume to the memory of Samuel D. Weiss, who died on December 1, 1997 near Ancona, in Italy. Sam Weiss' special place in the history of transcription will be generally remembered, particularly among the participants at this Symposium. His discovery of the RNA-synthetic activity of RNA polymerase was communicated tersely nearly 40 years ago in a letter to the *Journal of the American Chemical Society*. The letter is reproduced below. (Many participants at the 63rd Cold Spring Harbor Symposium may recall with pleasure that eukaryotic polymerase rather than a bacterial enzyme holds chronological pride of place.)

A MAMMALIAN SYSTEM FOR THE INCORPORATION OF CYTIDINE TRIPHOSPHATE INTO RIBONUCLEIC ACID[1]

Sir:

A mammalian preparation that incorporates cytidylate from cytidine-P^{32}-P-P into RNA[2] and is markedly stimulated by ATP, UTP, and GTP, is described in this report.

CMP-5′-P^{32} was prepared by a modified procedure for the phosphorylation of 2′,3′-benzylidene-O-cytidine.[3] Labeled CTP was prepared from P^{32}-CMP by a cytidylate kinase isolated from brewers' yeast. Twice washed nuclei were prepared from a 20% rat liver homogenate in 0.25 molar sucrose and centrifuged for six minutes at 600 × g.

When P^{32}-CTP was incubated with the 600 × g preparation, in the presence of all the ribonucleoside triphosphates, a significant amount of label was incorporated into the RNA fraction. Omission of any one of the triphosphates resulted in a reduction of 85% or more of P^{32}-CTP incorporation (Table I). Desoxyribonuclease depresses the incorporation slightly, whereas ribonuclease causes a marked reduction. In other experiments, a requirement for Mg^{++} was demonstrated and a fivefold excess of deoxy-CTP did not reduce the incorporation of P^{32}-CTP.

When P^{32}-labeled RNA (70,000 total counts), formed by this system, was isolated and hydrolyzed with alkali, the mononucleotides separated on Dowex-1-Cl contained these counts: 2′(3′)-CMP, 22,890; 2′(3′)-AMP, 8,680; 2′(3′)-GMP, 8,800; 2′(3′)-UMP, 21,940.

TABLE I
REQUIREMENTS FOR P^{32}-CYTIDINE TRIPHOSPHATE INCORPORATION INTO RNA

Reaction mixture	Radioactivity RNA (total counts)
Complete	3872
Omit ATP	636
Omit UTP	280
Omit GTP	104
Omit ATP, UTP and GTP	60
Complete + 20γ ribonuclease	828
Complete + 20γ desoxyribonuclease	2940
Complete + 10 μmole inorganic pyrophosphate	100
Complete in 100 μmole Pi buffer, pH 7.5 (no TRIS)	4030
Complete: ADP, UDP, GDP in place of ATP, UTP, GTP	1620
Complete: AMP, UMP, GMP in place of ATP, UTP, GTP	128

The complete system contained in 10 μmole $MgCl_2$, 100 μmole TRIS·HCl, *p*H 8.0, 0.1 μmole P^{32}-CTP (16.8 × 10^6 counts/μmole), 0.1 μmole ATP, 0.1 μmole UTP, 0.1 μmole GTP, 100 μmole KCl, 40 μmole NaF, 10 μmole cysteine, and 10–12 mg. of twice washed nuclei (dry weight), in a total volume of 2.0 ml. After incubation at 37° for 12 minutes 5 ml. of cold 5% TCA was added. The acid insoluble material was washed 3 times with 5% TCA, 2 times with ethanol–ether (3:1), and extracted 3 times with 2 ml. of 10% NaCl at 100°, *p*H 8.0, with 2 mg. of yeast RNA added. The combined extracts were precipitated twice with 2 volumes of ethanol. The residue was dissolved in 4 ml. of water and 1.0 ml. was dried and assayed in a windowless flow counter.

The appearance of label in all the mononucleotides, after hydrolysis, suggests strongly that P^{32}-CTP is incorporated into the interpolynucleotide linkages of RNA rather than terminally.[4] The requirement for the four ribonucleoside triphosphates, which are more than twice as effective as the corresponding diphosphates, as well as the inhibition by pyrophosphate but not by inorganic phosphate, suggests that the reaction mechanism resembles that described for DNA formation[5] rather than RNA synthesis by polynucleotide phosphorylase.[6,7]

(1) Supported by funds from the Argonne Cancer Research Hospital, operated by the University of Chicago, for the United States Atomic Energy Commission.
(2) Abbreviations: DNA, deoxyribonucleic acid; RNA, ribonucleic acid; CTP, ATP, UTP and GTP for the tri- and ADP, UDP and GDP for the di- and CMP, AMP, UMP and GMP for the monophosphates of cytidine, adenosine, uridine and guanosine; TRIS, tris-(hydroxymethyl)-aminomethane; TCA, trichloroacetic acid; Pi, inorganic phosphate.
(3) J. Baddiley, J. G. Buchanan and A. R. Sanderson, *J. Chem. Soc.*, 3107 (1958).
(4) (a) E. S. Canellakis, *Biochim. Biophys. Acta*, **25**, 271 (1957); (b) M. Edmonds and R. Abrams, *ibid.*, **26**, 226 (1957); (c) L. I. Hecht, P. C. Zamecnik, M. L. Stephenson and J. F. Scott, *J. Biol. Chem.*, **233**, 954 (1958).
(5) M. Bessman, I. R. Lehman, E. S. Simms and A. Kornberg, *Fed. Proc.*, **16**, 153 (1957).
(6) M. Grunberg-Manago, P. J. Oritz and S. Ochoa, *Science*, **122**, 907 (1955).
(7) R. J. Hilmoe and L. A. Heppel, This Journal, **79**, 4810 (1957).

ARGONNE CANCER RESEARCH HOSPITAL AND
DEPARTMENT OF BIOCHEMISTRY   SAMUEL B. WEISS
UNIVERSITY OF CHICAGO   LEONARD GLADSTONE
CHICAGO, ILLINOIS
                        RECEIVED JUNE 11, 1959

(Reprinted, with permission, from *J. Am. Chem. Soc.* **81**: 4118–4119 [copyright, 1959, American Chemical Society].)

Contents

RNA Polymerase

Elongation and Termination

Repression Mechanisms

COLD SPRING HARBOR SYMPOSIA
ON QUANTITATIVE BIOLOGY

VOLUME LXIII

Transcription Regulation by Repressosome and by RNA Polymerase Contact

S. Adhya, M. Geanacopoulos, D.E.A. Lewis, S. Roy, and T. Aki
*Laboratory of Molecular Biology, National Cancer Institute, National Institutes of Health,
Bethesda, Maryland 20892-4255*

The nucleoid structure in the bacterium *Escherichia coli* contains a circular DNA molecule of 4.7 million base pairs present in highly condensed form (see Fig. 1) (Ryter and Chang 1975; Pettijohn 1976). The condensation is mediated by DNA supercoiling and the binding of several small histone-like proteins, e.g., HU, IHF, FIS, and HNS (Rouviere-Yaniv 1978; Murphy and Zimmerman 1997). These proteins are known to bind DNA either nonspecifically or with low specificity (Broyles and Pettijohn 1986; Gualerzi et al. 1986; Johnson et al. 1986; Mendelson et al. 1991; Skorupski et al. 1994; Froelich et al. 1996; Nash 1996; Atlung and Ingmer 1997), bend DNA, or bind to bent DNA (Rouvière-Yaniv and Yaniv 1979; Thompson and Landy 1988; Koturko et al. 1989; Betermier et al. 1994). It was suggested that these proteins were mainly responsible for the compaction of DNA in a way that distinguishes the bacterial nucleoid from eukaryotic chromatin (Trun and Marko 1998). The bacterial nucleoid is also associated with the machinery of macromolecular biosynthesis, including RNA polymerase, and specific gene regulatory DNA-binding proteins, e.g., repressors and activators (Murphy and Zimmerman 1997). Although the structure of the bacterial nucleoid is not known, its density of 30 mg/ml is three times higher than the density that would exist if the chromosomal DNA were evenly distributed throughout the cell (10 mg/ml) (Kellenberger 1990). This high density would very likely prevent access of RNA polymerase to genes deep within the nucleoid interior. Thus, the mechanism whereby RNA polymerase accesses a promoter is an interesting question. Several theories can be put forward to address

the issue: (1) It has been suggested that nearly all transcription in the cell occurs on the nucleoid surface, which is in contact with the cytoplasm (Ryter and Chang 1975; Kellenberger 1990), and there is a continuous rearrangement of the nucleosome so that parts of the genomic DNA periodically emerge at the surface of the nucleoid, perhaps as loops. (2) Some genes are only transcribed following replication of the chromosome segment, where they are located (Guptasarma 1995); after passage of the replication fork, there is a period of time when the daughter DNA segments are free from the condensing proteins. This provides a window for RNA polymerase to transcribe the encoded genes before the DNA segments are repackaged as nucleoid. (3) The organization of the nucleoid structure is fixed. Highly expressed genes are present on the surface and less active genes are buried. (4) The problem of access of RNA polymerase to a promoter is simplified because it exists largely in the DNA-bound form, and promoters are accessed by a process of linear diffusion and interstrand transfer.

Our research on the mechanisms by which Gal repressor (GalR) represses transcription in the *gal* operon in *E. coli* has suggested different models of transcription repression. These mechanisms provide two new scenarios of how RNA polymerase accesses a promoter in the nucleoid in response to an inducing signal to initiate transcription:

1. In the first model, a specialized multiprotein-DNA structure called a *repressosome* is responsible for inhibition of transcription initiation at a promoter. Repressosomes, which are part of the condensed nucleoid, consist of supercoiled DNA, the specific repressors, and the histone-like proteins, e.g., HU and IHF (Fig. 2a). In the presence of an inducing signal, the repressosome collapses, thus making the promoter available for RNA polymerase.
2. In the second model, RNA polymerase is already bound to a repressed promoter within the nucleoid structure (Fig. 2b). The promoter-bound RNA polymerase is kept inhibited by a direct contact with a specific repressor (*contact inhibition*). In this scenario, the inducer interacts with the repressor and releases the inhibitory contact nucleoid remodeling.

In this paper, we first describe experimental results that support these two molecular models of repression. We also discuss the issue of RNA polymerase access to promoter when the repression is lifted in the two cases.

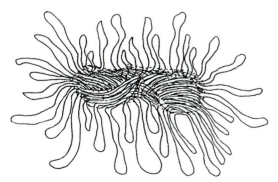

Figure 1. Cartoon of the bacterial nucleoid structure. (Adapted from Ryter and Chang 1975.)

a. Repressosome formation

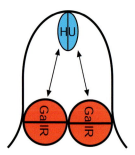

b. Contact inhibition

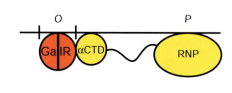

Figure 2. Repression mechanisms. (*a*) Repressosome structure consisting of gene-specific regulator (*red*) and histone-like protein (*blue*) interacting to cognate DNA sites and to each other. (*b*) Contact inhibition. Operator-bound repressor (*red*) making contact with a component, e.g., αCTD, of promoter-bound RNA polymerase (*yellow*). See text for details.

THE *gal* OPERON

Transcription of the *gal* operon is driven by two partially overlapping promoters, *P*1 and *P*2 (Musso et al. 1977; Adhya and Miller 1979). As shown in Figure 3, *P*1 (start site of transcription at position +1) and *P*2 (start site of transcription at −5) are regulated by the Gal repressor (GalR). GalR dimers bind to two operators, O_E (position −60.5) and O_I (position +53.5), that flank the two promoters (Irani et al. 1983). O_I is within the first structural gene of the operon. When GalR binds to both operators, the two GalR dimers associate to form a DNA loop of the intervening 113-bp promoter DNA segment. This structure brings about repression of both *P*1 and *P*2 (Aki et al. 1996; Lyubchenko et al. 1997). On the other hand, when a GalR dimer binds to O_E without forming a DNA loop, the regulator becomes bifunctional. It enhances transcription from *P*2 and represses transcription from *P*1 (Choy and Adhya 1992; Choy et al. 1997). As discussed below, O_E- and O_I-mediated simultaneous repression of *P*1 and *P*2 involves repressosome formation, whereas O_E-mediated repression of *P*1 involves contact inhibition.

REPRESSOSOME

The following experimental results gave rise to the repressosome model. Because of the requirement of dual *gal* operators, it was proposed that O_E- and O_I-bound GalR dimers interact and form a DNA loop from which repression ensues (Irani et al. 1983; Majumdar and Adhya 1984). Although genetic analysis strongly supported DNA-looping-mediated repression (Haber and Adhya

1988; Mandal et al. 1990), GalR binding to O_E and O_I was not sufficient for DNA looping and repression of both *P*1 and *P*2 in vitro, suggesting that an additional factor may be required (Choy and Adhya 1992). From crude extracts of *E. coli*, we purified a protein that when present along with GalR, caused concurrent repression of the two *gal* promoters. The purification of the DNA-looping factor has been reported elsewhere (Aki et al. 1996). The native molecular mass of the factor was 18 kD and was composed of two subunits each with a mass of 9 kD. We determined the sequence of the amino-terminal 20 amino acids by Edman degradation. As shown in Figure 4, the determined sequence was identical to that of the *E. coli* histone-like protein, HU. HU has two heterotypic subunits, α and β, of 9 kD each (Mende et al. 1978; Rouvière-Yaniv and Kjeldgaard 1979). Note that two amino acids were recovered in sequencing cycles of the looping factor, each time the residues differed in the respective position of Hu-α and Hu-β. The identity of the looping factor was confirmed by our demonstration that an authentic HU protein substituted the looping factor in transcription assays in vitro with identical results. As shown in Figure 5, HU acted as a cofactor of GalR in simultaneous repression of *P*1 and *P*2 in transcription assays. In the presence of HU, GalR repressed both promoters; whereas, in its absence, GalR repressed only *P*1 transcription and stimulated transcription from *P*2 (Fig. 5, compare panels A and B). As expected, the repression that depends on both HU and GalR being present was sensitive to inducer D-galactose (Fig. 5C). HU by itself did not alter the levels of *gal* transcription. The involvement of HU in the DNA-loop-mediated repression of the *gal* promoters has been confirmed in vivo (Lewis et al. 1998). GalR-mediated *P*1 repression in the absence of HU is discussed later. The following experiments showed the essentials of GalR-HU mediated repression.

1. *GalR-operator interactions.* GalR-HU mediated repression requires both O_E and O_I to be functional. Mutation of either operator resulted in the loss of simultaneous repression (Fig. 5D,E), suggesting that GalR must bind to both operators for the effect. The requirement of both operators in *gal* repression in vivo has been published previously (Irani et al. 1983).

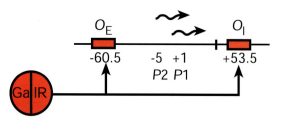

Figure 3. *gal* regulatory region showing the two promoters, *P*1 and *P*2, and two operators, O_E and O_I, the binding sites of GalR.

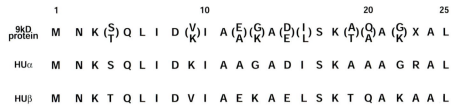

Figure 4. Amino-terminal amino acid sequence of the looping factor determined by Edman degradation. The amino-terminal amino acid sequence of α and β subunits of HU (Mende et al. 1978; Rouvière-Yaniv and Kjeldgaard 1997). At each position, α and β differed in their sequence, the looping factor gave two signals at the corresponding position.

2. *DNA supercoiling.* HU-dependent *gal* repression required that the *gal* DNA template be supercoiled. HU effect was abolished by linearization of DNA template in vitro (Aki et al. 1996). Similarly, addition of coumermycin, a DNA gyrase inhibitor, to growing cells derepressed *P2*, but not *P1*, as predicted from the *in vitro* results (Gellert et al. 1976a,b; Lewis et al. 1998).

3. *DNA looping.* Supercoiled DNA circles with the wild-type genotype ($O_E^+O_I^+$) when mixed with GalR and HU showed formation of DNA loops by interaction of O_E- and O_I-bound GalR as observed by atomic force microscopy (Lyubchenko et al. 1997).

Mechanisms of HU action. We proposed four mechanisms to explain how HU acts as a cofactor for GalR-mediated repression of both promoters (Fig. 6).

a. Multiple HU molecules cover, perhaps solenoidally (Pettijohn 1976), the *gal* promoters, creating a steric problem for RNA polymerase binding to the *gal* promoters.

b. HU helps DNA binding by acting as an adapter between two operator-bound GalR dimers that are otherwise incapable of interaction.

c. HU promotes DNA looping by bending the DNA and thereby stabilizing a weak interaction between two DNA-bound GalR dimers. A similar architectural role for histone-like proteins, including HU, has been previously proposed in site-specific recombination complexes of bacteriophage Mu transpososomes (Lavoie and Chaconas 1993, 1994; Lavoie et al. 1996) and of bacteriophage λ intasome (Goodman et al. 1992; Segall et al. 1994; Nash 1996).

d. HU aids DNA looping by bending DNA as well as contacting GalR.

We distinguished between these models by studying the binding of HU to *gal* DNA.

Site-specific binding of HU. We tested the potential *gal* DNA-HU interactions by chemically converting HU into a nuclease (Ebright et al. 1992; Ermácora et al. 1992; Lavoie and Chaconas 1993). HU modified with 2-iminothiolane and EPD-Fe^{++}, i.e. (EDTA-2-aminoethyl)-2-pyridyl disulfide-iron complex, generates hydroxy radicals in the presence of sodium ascorbate and hydrogen peroxide and thus cleaves DNA within 10 Å distance (Tullius et al. 1987). When the reactions were carried out in the presence of GalR and HU, a specific segment of *gal* DNA between the two operators was cleaved, indicating binding of HU to DNA (Fig. 7, lane 4). The cleavage pattern shown here and in other experiments demonstrated that a single HU dimer binds at a specific region, *hbs*, between O_E and O_I, with the center of binding approximately at position +6.5. HU binding depended on the presence of a functional GalR (cooperativity); cleavages did not occur in the absence of GalR or occurred in much reduced form in the presence of GalR and D-galactose (Fig. 7, lanes 3 and 5). Table 1 summarizes the conditions of cooperative binding of HU to *hbs*. The results showed that HU binding to *hbs* in *gal* DNA required GalR binding to both operators and was dependent on DNA being supercoiled, consistent with the requirements of HU-mediated repression of transcription.

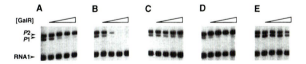

Figure 5. Effect of GalR and HU on *gal* transcription in vitro was carried out as described previously (Choy and Adhya 1992). Supercoiled DNA (2 nM) was preincubated at 37°C for 1 minute after each addition of proteins in 20 mM Tris buffer (pH 7.8), 10 mM Mg-acetate, 200 mM K-glutamate, 1 mM DTT, 5% glycerol, 1 mM ATP, 0.1 mM of GTP and CTP, and 0.01 mM UTP, and 10 MCi of [α-^{32}P]UTP. When present, HU was 80 nM, D-galactose 10 mM. GalR concentration varied from 0 to 40 nM. Transcription was initiated by adding RNA polymerase at 20 nM. After 5 min at 37°C, RNA synthesis was analyzed by gel electrophoresis. (A–C) $O_E^+O_I^+$ DNA; (D) $O_E^+O_I^-$; (E) $O_E^-O_I^+$. (B–E) HU and (C) D-galactose. *gal* RNA made from *P2* and *P1* are indicated. The control RNA1, as indicated, was made from the same templates.

Table 1. Summary of Conditions for the Cooperative Binding of HU to *hbs*

Repressor	D-galactose	Operators[a]	DNA[b]	HU binding[c]
GalR	–	$O_E^+O_I^+$	sc	++
–	–	$O_E^+O_I^+$	sc	–
GalR	+	$O_E^+O_I^+$	sc	+/–
GalR	–	$O_E^+\Delta O$	sc	–
GalR	–	$O_E^+O_I^-$	sc	+
GalR	–	$O_E^-O_I^+$	sc	+
GalR	–	$O_E^-O_I^-$	sc	–
GalR	–	$O_E^+O_I^+$	rel	–
GalS	–	$O_E^+O_I^+$	sc	–
LacI	–	$O_E^LO_I^L$	sc	–

[a]ΔO stands for deletion and O^- for base-substitution mutations of the marked operator; O^L refers to LacI-binding site engineered at the two *gal* operator sites.

[b](sc) supercoiled DNA; (rel) relaxed DNA.

[c]HU-binding results were obtained as shown in Fig. 7.

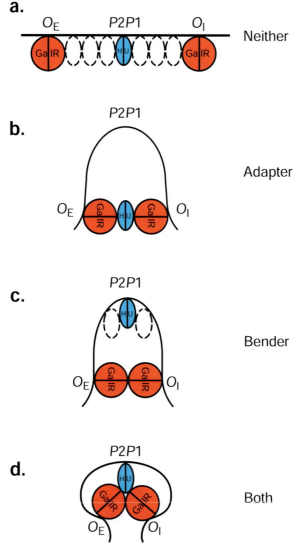

Figure 6. Proposed roles of HU as a cofactor of repression of *gal* promoters by GalR. See text for details.

Tripartite cooperativity. Whereas deletion of an operator eliminated HU binding, base-substitution mutations in either operator, with no intrinsic affinity for GalR, restored measurable HU binding (Table 1). Such findings showed that in the presence of HU, GalR binding to a wild-type operator can rescue repressor binding to a mutant operator. In summary, an interaction of two DNA-bound GalR dimers helped site-specific binding of HU and vice versa, reflecting a tripartite cooperativity. Binding of GalR to O_E and O_I without HU is noncooperative (Brenowitz et al. 1990).

Strength and specificity of HU binding. HU has been described as a nonspecific DNA-binding protein with a low affinity (micromolar) (Pettijohn 1976). The effect of HU in *gal* repression occurred even at 20-fold lower concentrations (Aki et al. 1996). HU binding to *hbs* was similarly stronger. Once bound, HU could not be competed out by excess unbound HU. HU is also specific for repression of *gal* promoters binding to the unique *hbs* site

on *gal* DNA. HU cannot be replaced by other histone-like proteins, IHF, HNS, or FIS. Conversely, repressors homologous to GalR, like GalS and LacI, could not replace GalR in HU binding to the promoter as observed by DNA cleavage reactions (Table 1). In the LacI experiment, the two *gal* operators were replaced by *lac* operator DNA sequences.

Repressosome structure. The binding of a single HU dimer to *hbs* in the *gal* promoter with significantly higher affinity ruled out the solenoidal role (model a) and a simple adapter role (model b) of HU discussed above. Clearly, a bender role of HU (model c) is strongly supported for the repressosome complex. The tripartite cooperativity between GalR and HU, and the specificity in the participation of these two proteins in the formation of the DNA-multiprotein complex suggested that direct protein-protein interactions between GalR and HU (model d) may also exist in the repressosome structure as indicated in Figure 2a.

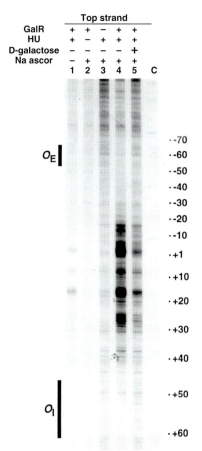

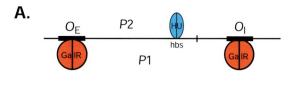

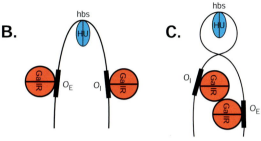

Figure 7. HU-nuclease-mediated cleavage pattern of *gal* DNA. HU-nuclease reactions were carried out as described by Lavoie and Chaconas (1993). Lanes *1* and *3–5,* modified HU; lane *2,* unmodified HU. Conditions were same as in Fig. 5, except that after hydroxy radical reactions, nicked DNA was purified and used as a template to synthesize complementary DNA strands with α-^{32}P-end labeled oligonucleotide primers. Dideoxy chain terminating reactions were performed with the same *gal* DNA template and primer using dideoxy CTP shown on the rightmost lane (*C*).

Figure 8. Role of DNA supercoiling in repressosome formation. (*A*) Face of DNA double helix to which GalR (*red*) and HU (*blue*) bind. (*B,C*) DNA binding by HU at the *hbs* site without and with DNA crossovers between O_E-*hbs* and *hbs*-O_I segments, respectively. The structure in *C* will permit two DNA-bound GalR dimers to interact with each other.

The pattern of HU-mediated DNA cleavage sites showed that HU binds to one face of DNA centered at position +6.5. Assuming a helical pitch typical of B DNA, this face is opposite to that face of O_E and O_I to which GalR binds (Fig. 8A) (Majumdar and Adhya 1989). As shown in Figure 8B, this creates a topological constraint for DNA looping by GalR-GalR interaction assisted by HU acting as a bender. DNA supercoiling may be needed simply to rephase the two operators relative to *hbs*. A DNA crossover between the O_E-*hbs* and *hbs*-O_I DNA segments will overcome the geometric barrier for DNA looping (Fig. 8C).

Mutational analysis of the repressosome structure. Confirmation of the proposed repressosome structure clearly depends on the demonstration of protein-protein interaction between two DNA-bound GalR dimers and/or between DNA-bound GalR and *hbs*-bound HU by biochemical and physicochemical means. In addition, a

search for GalR and HU mutants specifically defective in *gal* transcription repression, but not in DNA binding, will identify the potential interacting partners and the corresponding amino acid residues involved. We isolated three such mutants of GalR. These mutant GalR repressors bound to the *gal* operators normally but failed to show DNA-looping-mediated repression both in vivo and in vitro (M. Geanacopoulos et al., in prep.). We modeled the structures of GalR on the basis of the known structures of PurR and LacI (Schumacher et al. 1994; Lewis et al. 1996). The mutations are in surface-exposed amino acids in the carboxy-terminal subdomain of GalR and far away from the amino-terminal DNA-binding domain (Fig. 9). We expect that these and other similar amino acid substitution mutations in GalR and their potential suppressor mutations in GalR or HU will define interactions between the two GalR dimers and/or between GalR and HU.

Mechanism of repression by repressosome. We have shown that repressosome structure prevents DNA strand separation by RNA polymerase at *P*1 and *P*2 as detected by Cu(II)-phenanthroline footprinting experiments (Aki et al. 1996). DNase protection analyses showed that HU-GalR interactions prevent RNA polymerase from forming heparin-resistant open complexes (Aki and Adhya 1997). Whether the repressosome structure hinders closed complex formation remains to be studied. Whatever the biochemical level at which repressosome blocks RNA polymerase activity, it is clear that the structure makes the promoter inadequate for transcription.

CONTACT INHIBITION

As shown in Figure 5, GalR binding to the upstream operator, O_E, is sufficient to repress transcription from the *P*1 promoter in the absence of loop formation. Under

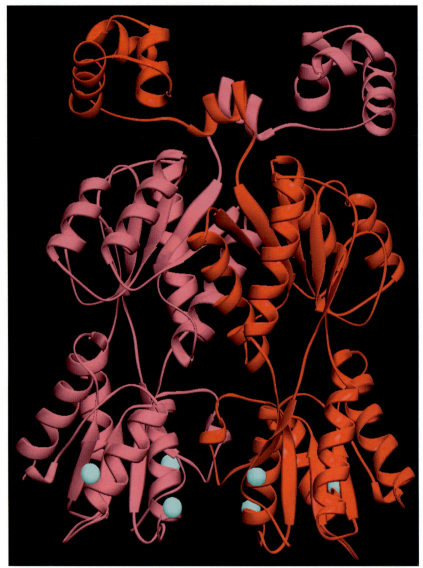

Figure 9. Structure of GalR core dimer modeled after PurR and LacI (M. Geanacopoulos et al., in prep.). The two subunits are shown in *red* and *pink*, respectively. Top is the amino-terminal DNA-binding domain connected to the core. (*Blue circles*) Amino acid residues which when mutated give rise to GalR variants that bind to DNA but fail to form DNA loop.

these conditions, GalR was unable to repress $P1$ if transcription was carried out with RNA polymerase containing specific mutations in the carboxy-terminal domain of the α subunits (αCTD) (Choy et al. 1997). These results suggested that a protein-protein contact between O_E-bound GalR and αCTD mediates repression of $P1$ (see Fig. 2b). GalR also repressed a heterologous promoter, *lacUV5*, by making contact with αCTD (Ryu and Adhya 1998). When the GalR-binding site, O_E, was positioned at –61 from the transcription start site, GalR repressed transcription from the *lacUV5* promoter. The repression was not observed if the RNA polymerase was missing the last 73 amino acids in αCTD.

The GalR contact with αCTD can change the RNA polymerase activity allosterically by inducing a conformational change. The contact inhibition of RNA polymerase by GalR can also be explained by comparing the

action of GalR with that of an enzyme, i.e., GalR acts by changing the energetics of transcription initiation steps. A regulator, like an enzyme, stabilizes one or more of the intermediate and transition states of the initiation pathway (Roy et al. 1998). Depending on the nature of the protein-protein contacts and their duration during the reaction (differential contact), a regulator can either enhance (activation) or hinder (repression) one or more of the initiation steps. For example, the regulator-RNA polymerase contact may be specific for a step, selectively stabilizing the preceding intermediate and lowering its free energy with respect to unbound RNA polymerase or may persist throughout the subsequent steps and lower the free energy level of all connected states. In this way, the regulator can enhance or inhibit initiation by decreasing or increasing the energetic barriers of one or more steps. Decreasing an energetic barrier will result in activation, whereas increasing the barrier will cause repression. Depending on the geometry and DNA sequence of the promoters, changes in RNA polymerase conformation during the steps may facilitate differential contacts favoring activation or repression. An energetic explanation of contact inhibition of RNA polymerase at the $P1$ promoter of GalR is shown in Figure 10. In the free energy diagram, transcription from the $P1$ promoter proceeds without substantial energetic barrier. In our model, GalR by contacting RNA polymerase lowers the free energy of the closed complex $(R \cdot P)_c$ only and not of any other intermediate, including the transition state. This will decrease the ΔG_c and increase the $\Delta G^{\ddagger}$. Although a decrease in ΔG_c will increase K_a of $(R \cdot P)_c$ formation resulting in the accumulation of the latter, large increase in $\Delta G^{\ddagger}$ will create an energy barrier for the isomerization step, i.e., decrease in k_f. If the RNA polymerase concentration is higher than K_a^{-1}, a decreased k_f will bring about repression at a post RNA-polymerase-binding step. We have demonstrated that at $P1$, GalR forms a characteristic GalR·$P1$·RNA polymerase-stable ternary complex (Choy et al. 1997). This model of repression can be tested further by characterizing this ternary complex, and by isolating mutations in GalR which will help define the reaction intermediates that contact GalR. RNA polymerase (αCTD) mutants that are defective in repression have already been described (Choy et al. 1997).

IMPLICATIONS OF TRANSCRIPTION REPRESSION BY REPRESSOSOME FORMATION OR CONTACT INHIBITION

Repressosome

It has been demonstrated that RNA polymerase initiates transcription at a higher rate from condensed nu-cleoid than from decondensed chromosomal DNA (Giorno et al. 1975). How does free RNA polymerase find a repressed promoter more efficiently following an inducing signal despite the nucleoid structure? Although how the DNA becomes condensed by histone-like proteins is unknown, the repressosome structure in the gal operon described above provides a model of how histone-like proteins interact with a promoter to condense it locally and how such a promoter becomes free (remodeled) for transcription initiation. Since the binding of HU in forming the repressosome structure at the gal operon is entirely dependent on the binding of GalR, HU binding is sensitive to inducer D-galactose, whose presence makes the promoter free of any structure. We propose that the regulatory DNA segments in the nucleoid are composed of organized structures ("condensed"). Bacterial histones bind in a site-specific manner to such DNA to form specialized structures, such as the repressosome, acting as architectural proteins in concert with gene-specific regulatory proteins, e.g., repressors. By this mechanism, such structures become responsive to a specific signal. The structure and composition of a repressosome complex will vary from promoter to promoter. HU has been shown to be involved in the GlpR-mediated repressor of the $glpD$ operon and IHF in the GlpR-mediated repression of the $glpTQ$ operon in $E. coli$ (Yang and Larson 1996; Yang et al. 1997). In forming a repressosome structure, the role of an architectural protein could also be performed by a specific gene regulatory protein. For example, in many of the promoters regulated by CytR repressor, CytR itself plays the part of an architectural protein. CytR binds to a site between two DNA-bound CRP molecules in creating a repression structure as shown in model d (Fig. 6) (Søgaard-Andersen et al. 1991). Cooperative binding of HU and LacI repressors to the lac promoter of $E. coli$ has also been reported, although its physiological significance remains unknown (Flashner and Gralla 1988).

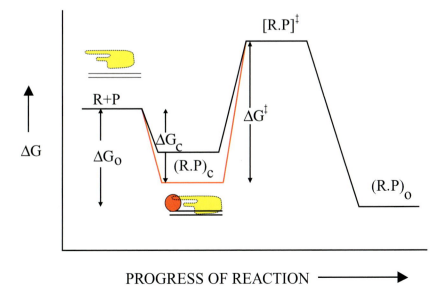

Figure 10. Free energy changes associated with open complex formation during transcription initiation at the $P1$ promoter of gal in the absence and presence of GalR. (*Yellow icon*) RNA polymerase; (*red icon*) GalR.

Contact Inhibition

A simple way to solve the problem of how RNA polymerase accesses a repressed promoter after induction is to have an RNA polymerase molecule sequestered at a promoter and prevented from initiating by a DNA-bound repressor. The repressor may inhibit transcription at any of the steps of initiation either by altering the promoter DNA or by contacting the bound RNA polymerase. Conformational change of the repressor by an inducing signal will remove the inhibitory effect and permit transcription initiation. Inhibition of RNA polymerase bound to the $P1$ promoter of gal by GalR described here provides an example of repression by regulator–RNA polymerase contact. Inhibition of a promoter-bound RNA polymerase activity by a repressor has also been observed in other systems. For example, MerR protein inhibits isomerization of $(R \cdot P)_c$ to an open complex $(R \cdot P)_o$ by binding to its cognate operator located at the -18 position. In this case, it has been suggested that the sequesterization of RNA polymerase at the mer promoter by MerR is effected by changing DNA structure and not by MerR-RNA polymerase contact (O'Halloran et al. 1989; Frantz and O'Halloran 1990; Heltzel et al. 1990; Anasari et al. 1995). The binding of the inducer Hg^{++} to MerR converts the regulator into an activator which releases the inhibitory effect on RNA polymerase.

SUMMARY

The original model of repression of transcription initiation is steric interference of RNA polymerase binding to a promoter by its repressor protein bound to a DNA site that overlaps the promoter. From the results described here, we propose two other mechanisms of repressor action, both of which involve formation of higher-order DNA-multiprotein complexes. These models also explain the problem of RNA polymerase gaining access to a promoter in the condensed nucleoid in response to an inducing signal to initiate transcription.

ACKNOWLEDGMENTS

We thank George Vasmatzis for creating Figure 9 and Angela Fox for critical reading of the manuscript.

REFERENCES

Adhya S. and Miller W. 1979. Modulation of the two promoters of the galactose operon of *Escherichia coli*. *Nature* **279**: 492.

Aki T. and Adhya S. 1997. Repressor induced site-specific binding of HU for transcriptional regulation. *EMBO J.* **16**: 3666.

Aki T., Choy H.E., and Adhya S. 1996. Histone-like protein HU as a specific transcriptional regulator: Co-factor role in repression of *gal* transcription by GAL repressor. *Genes Cells* **1**: 179.

Ansari A.Z., Bradner J.E., and O'Halloran T.V. 1995. DNA-bend modulation in a repressor to activator switching. *Nature* **374**: 371.

Atlung T. and Ingmer H. 1997. H-NS: A modulator of environmentally regulated gene expression. *Mol. Microbiol.* **24**: 7.

Betermier M., Galas D.J., and Chandler M. 1994. Interaction of Fis protein with DNA: Bending and specificity of binding. *Biochimie* **76**: 958.

Brenowitz M., Jamison E., Majumdar A., and Adhya S. 1990. Interaction of the *Escherichia coli gal* repressor protein with its DNA operators in vitro. *Biochemistry* **29**: 3374.

Broyles S.S. and Pettijohn D.E. 1986. Interaction of the *Escherichia coli* HU protein with DNA. Evidence for formation of nucleosome-like structures with altered DNA helical pitch. *J. Mol. Biol.* **187**: 47.

Choy H.E. and Adhya S. 1992. Control of *gal* transcription through DNA looping: Inhibition of the initial transcribing complex. *Proc. Natl. Acad. Sci.* **89**: 11264.

Choy H.E., Hanger R.R., Aki T., Mahoney M., Murakami K., Ishihama A., and Adhya S. 1997. Repression and activation of promoter-bound RNA polymerase activity by Gal repressor. *J. Mol. Biol.* **272**: 293.

Ebright Y.W., Chen Y., Pendergrast P.S., and Ebright R.H. 1992. Incorporation of an EDTA-metal complex at a rationally selected site within a protein: Application to EDTA-iron DNA affinity cleaving with catabolite gene activator protein (CAP) and Cro. *Biochemistry* **31**: 10664.

Ermácora M.R., Delfino J.M., Cuenoud B., Schepartz A., and Fox R.O. 1992. Conformation-dependent cleavage of staphylococcal nuclease with a disulfide-linked iron chelate. *Proc. Natl. Acad. Sci.* **89**: 6383.

Flashner Y. and Gralla J.D. 1988. DNA dynamic flexibility and protein recognition: Differential stimulation by bacterial histone-like protein HU. *Cell* **54**: 713.

Frantz B. and O'Halloran T.V. 1990. DNA distortion accompanies transcription activation by the metal responsive gene regulatory protein MerR. *Biochemistry* **29**: 4747.

Froelich J.M., Phuong T.K., and Zyskind J.W. 1996. Fis binding in the dnaA operon promoter region. *J. Bacteriol.* **178**: 6006.

Gellert M., O'Dea M.H., Itoh T., and Tomizawa J.-I. 1976a. Novobiocin and coumermycin inhibit DNA supercoiling catalyzed by DNA gyrase. *Proc. Natl. Acad. Sci.* **73**: 4474.

Gellert M., O'Dea M.H., Mizuuchi I., and Nash H. 1976b. DNA gyrase, an enzyme that introduces superhelical turns into DNA. *Proc. Natl. Acad. Sci.* **78**: 3872.

Giorno R., Stamato T., Lyderson B., and Pettijohn D. 1975. Transcription in vitro of DNA in isolated bacterial nucleoid. *J. Mol. Biol.* **96**: 217.

Goodman S.D., Nicholson S.C., and Nash H.A. 1992. Deformation of DNA during site-specific recombination of bacteriophage lambda: Replacement of IHF protein by HU protein or sequence-directed bends. *Proc. Natl. Acad. Sci.* **89**: 11910.

Gualerzi C.O., Losso M.A., Lammi M., Friedrich K., Pawlik R.T., Canonaco, M.A., Gianfranceschi, G., Pingoud A., and Pon C.L. 1986. Proteins from the prokaryotic nucleoid. Structural and functional characterization of *Escherichia coli* DNA-binding proteins NS(HU) and H-NS. In *Bacterial chromatin*: (ed. C.O. Gualerzi and C.L. Pon), p. 101. Springer-Verlag, Heidelberg, Germany.

Guptasarma P. 1995. Does replication-induced transcription regulate synthesis of the myriad low copy number proteins of *Escherichia coli*. *BioEssays* **17**: 987.

Haber R. and Adhya S. 1988. Interaction of spatially separated protein-DNA complexes for control of gene expression: Operator conversions. *Proc. Natl. Acad. Sci.* **85**: 9683.

Heltzel A., Lee I.W., Totis P.A., and Summers A.O. 1990. Activation dependent preinduction binding of σ-70 RNA polymerase at the metal regulated *mer* promoter. *Biochemistry* **29**: 9572.

Irani M., Orosz L., and Adhya S. 1983. A control element within a structural gene: The *gal* operon of *Escherichia coli*. *Cell* **32**: 783.

Johnson R.C., Bruist M.F., and Simon M.I. 1986. Host-protein requirements for *in vitro* site specific DNA inversion. *Cell* **46**: 531.

Kellenberger E. 1990. Intracellular organization of the bacterial genome. In *The bacterial chromosome* (ed. K. Drlica and M.Riley), p. 173. ASM Press, Washington, D.C.

Koturko L.D., Daub E., and Murialdo H. 1989. The interaction of E. *coli* integration host factor and λcos DNA: Multiple complex formation and protein-induced bending. *Nucleic Acids Res.* **17**: 317.

Lavoie B.D. and Chaconas G. 1993. Site-specific HU binding in the Mu transpososome: Conversion of a sequence-independent DNA-binding protein into a chemical nuclease. *Genes Dev.* **7:** 2510.

———. 1994. A second high affinity HU binding site in the phage Mu transposome. *J. Biol. Chem.* **269:** 15571.

Lavoie B.D, Shaw G.S., Millner A., and Chaconas G. 1996. Anatomy of a flexer-DNA complex inside a higher order transposition intermediate. *Cell* **85:** 761.

Lewis D.E.A., Geanacopoulos M., and Adhya S. 1998. Role of HU and DNA supercoiling in transcription repression: Specialized nucleoprotein complex at *gal* promoters in *E. coli*. *Mol. Microbiol.* (in press).

Lewis M., Chang G., Horton N.C., Kercher M.A., Pace H.C., Schumacher M.A., Brennan R.G., and Lu P. 1996. Crystal structure of lactose operon repressor and its complexes with DNA and inducer. *Science* **271:** 1247.

Lyubchenko Y.L., Shlyakhenko L.S., Aki T., and Adhya S. 1997. Atomic force microscopic demonstration of DNA looping by GalR and HU. *Nucleic Acids Res.* **25:** 873.

Majumdar A. and Adhya S. 1984. Demonstration of two operator elements in *gal: In vitro* repressing binding studies. *Proc. Natl. Acad. Sci.* **81:** 6100.

———. 1989. Effect of ethylation of operator phosphates on Gal repressor binding: DNA contortionary repressor. *J. Mol. Biol.* **208:** 217.

Mandal N., Su W., Haber R., Adhya S., and Echols H. 1990. DNA looping in cellular repression of transcription of the galactose operon. *Genes Dev.* **4:** 410.

Mende L., Timm N., and Subramanian A.R. 1978. Primary structure of two homologous ribosome-associated DNA-binding proteins of *Escherichia coli*. *Eur. J. Biochem.* **96:** 395.

Mendelson I., Gottesman M., and Oppenheim A.B. 1991. HU and integration host factor and function as auxiliary proteins in cleavage of phage lambda cohesive ends by terminase. *J. Bacteriol.* **173:** 1670.

Murphy D.L. and Zimmerman S.B. 1997. Isolation and characterization of spermine nucleoids from *Escherichia coli*. *J. Struct. Biol.* **119:** 321.

Musso R., deLauro R., Adhya S., and deCrombrugghe B. 1997. Dual control for transcription of the galactose operon by cyclic AMP and its receptor protein at two interspersed promoters. *Cell* **12:** 847.

Nash H.A. 1996. The HU and IHF proteins: Accessory factors for complex protein-DNA assemblies. In *Regulation of gene expression in* Escherichia coli (ed. E.C.C. Lin and A. Lynch), p. 149. R.G. Landes, Austin, Texas.

O'Halloran, T.V., Frantz B., Shin M.K., Ralston D.M., and Wright J.G. 1989. The MerR heavy metal receptor mediates positive activation in a topologically novel transcription complex. *Cell* **56:** 119.

Pettijohn D. 1976. Prokaryotic DNA in nucleoid structure. *CRC*

Crit. Rev. Biochem. **4:** 175.

Rouvière-Yaniv J. 1978. Localization of HU protein on the *Escherichia coli* nucleoid. *Cold Spring Harbor Symp. Quant. Biol.* **42:** 439.

Rouvière-Yaniv J. and Kjeldgaard N.O. 1979. Native *Escherichia coli* HU protein is a heterotypic dimer. *FEBS Letters* **106:** 297.

Rouvière-Yaniv J. and Yaniv M. 1979. E. coli DNA binding protein HU forms nucleosome-like structure with circular double-stranded DNA. *Cell* **17:** 265.

Roy S., Garges S., and Adhya S. 1998. Activation and repression of transcription by differential contact: Two sides of a coin. *J. Biol. Chem.* **273:** 14059.

Ryter A. and Chang A. 1975. Localization of transcribing genes in the bacterial cell by means of high resolution autoradiography. *J. Mol. Biol.* **98:** 797.

Ryu S. and Adhya S. 1998. GalR-mediated repression and activation of hybrid *lacUV* promoter: Differential contact with RNA polymerase. Gene (in press).

Schumacher M.A., Choi K.Y., Zalkin H., and Brennan R.G. 1994. Crystal structure of LacI member, PurR, bound to DNA. *Science* **266:** 763.

Segall A.M., Goodman S.D., and North H.A. 1994. Architectural elements in nucleoprotein complexes: Interchangeability of specific and nonspecific DNA binding proteins. *EMBO J.* **13:** 4536.

Skorupski K., Sauer B., and Sternberg N. 1994. Faithful cleavage of the P1 packaging site (*pac*) requires two phage proteins, PacA and PacB, and two *Escherichia coli* proteins, IHF and HU. *J. Mol. Biol.* **243:** 268.

Søgaard-Andersen L., Pedersen H., Holst B., and Valentin-Hansen P. 1991. A novel function of the cAMP-CRP complex in *Escherichia coli:* cAMP-CRP functions as an adapter for the CytR repressor in *deo* operon. *Mol. Microbiol.* **5:** 969.

Thompson J.F. and Landy A. 1988. Empirical estimation of protein-induced DNA bending angles: Applications to lambda site-specific recombination complexes. *Nucleic Acids Res.* **16:** 9687.

Trun N.J. and Marko J.F. 1998. Architecture of a bacterial chromosome. *ASM News* **64:** 276.

Tullius T.D., Dombroski B.A., Churchill M.E.A., and Kam L. 1987. Hydroxyl radical footprinting: A high-resolution method for mapping protein-DNA contacts. *Methods Enzymol.* **155:** 537.

Yang B. and Larson T. 1996. Action at a distance for negative control of transcription of the *glpD* gene encoding sn-glyceral-3-phosphate of *Escherichia coli* K-12. *J. Bacteriol.* **178:** 7090.

Yang B., Gerhardt S.G., and Larson T. 1997. Action at a distance for *glp* repressor control of *glpTQ* transcription in *Escherichia coli* K-12. *Mol. Microbiol.* **24:** 511.

RNA Polymerase-DNA Interaction: Structures of Intermediate, Open, and Elongation Complexes

R.H. EBRIGHT

Howard Hughes Medical Institute, Waksman Institute, and Department of Chemistry,
Rutgers University, Piscataway, New Jersey 08854

RNA polymerase (RNAP) is the enzyme responsible for the first step in gene expression and is the target, directly or indirectly, of most regulation of gene expression. Understanding RNAP structure and function is essential for understanding gene expression and regulation of gene expression and, as such, is a fundamental objective of structural biology.

Bacterial RNAP, archaeal RNAP, eukaryotic RNAPI, eukaryotic RNAPII, eukaryotic RNAPIII, and chloroplast RNAP constitute a protein family termed the "multisubunit RNAP family" (for review, see Young 1991; Sentenac et al. 1992; Archambault and Friesen 1993). Members of the multisubunit RNAP family exhibit unambiguous sequence, structural, and functional similarities. Each contains a conserved largest subunit with a molecular mass of approximately 160 kD (β' in bacterial RNAP), a conserved second-largest subunit with a molecular mass of approximately 150 kD (β in bacterial RNAP), and a third conserved subunit with a molecular mass of approximately 40 kD (α in bacterial RNAP). Bacterial RNAP and chloroplast RNAP contain only the conserved subunits. Archaeal RNAP and eukaryotic RNAPI, RNAPII, and RNAPIII contain both the conserved subunits and additional subunits. However, the conserved subunits account for fully 50–80% of overall molecular mass, and, at least for *Saccharomyces cerevisiae* RNAPII, the conserved subunits appear to be sufficient for core catalytic functions (i.e., transcription initiation, transcript elongation; E. Blatter, P. Kolodziej, and R.H. Ebright, unpubl.).

Members of the multisubunit RNAP family are molecular machines that carry out complex series of reactions. Transcription initiation involves the following reactions (Fig. 1) (for review, see Record et al. 1996; deHaseth et al. 1998).

1. RNAP, together with initiation factors, binds to promoter DNA, interacting solely with DNA upstream of the transcription start, to yield an RNAP-promoter closed complex (RP$_c$, also referred to as RP$_{c1}$).
2. RNAP then wraps promoter DNA around its circumference, capturing and interacting with DNA downstream from the transcription start, and RNAP undergoes a protein conformational change, clamping tightly onto DNA, to yield an RNAP-promoter intermediate complex (RP$_i$, also referred to as RP$_{c2}$ and I$_2$).
3. RNAP then melts approximately 14 bp of promoter DNA surrounding the transcription start, rendering accessible the genetic information in the template strand of

DNA, to yield an RNAP-promoter open complex (RP$_o$).
4. RNAP then begins RNA synthesis as an RNAP-promoter initial transcribing complex (RP$_{ITC}$). Typically, RNAP fails to achieve productive RNA synthesis on its first attempt and enters into abortive cycles of synthesis and release of short RNA products—RNA products 2, 3, 4, 5, 6, 7, or 8 nucleotides in length. (RNA products less than 9 nucleotides in length are not stably retained within the RNAP active center and dissociate at rates comparable to those of nucleotide addition.)
5. When, by chance, RNAP succeeds in synthesizing an RNA product 9 nucleotides in length, RNAP breaks its interactions with initiation factors, breaks its interactions with promoter DNA, and begins to translocate along DNA, precessively synthesizing RNA as an RNAP-DNA elongation complex with stably bound nascent RNA (RD$_e$).

Each of these reactions—as well as subsequent reactions in elongation, pausing, arrest, and termination—is a potential target for regulation by repressors or activators (Record et al. 1996; deHaseth et al. 1998; Mooney et al. 1998). To understand regulation, it will be necessary to define the structures of *each* of the relevant complexes, to define the structural transitions in RNAP and nucleic acids at each step, to define the kinetics at each step, and to define mechanisms of regulation at each step.

This is a daunting task. The number of complexes to be analyzed is large. The complexes themselves are large, with molecular masses of approximately 0.4 MD to more than 2 MD. The complexes are kinetic intermediates, with fleeting lifetimes under standard transcription conditions. In addition, all steps subsequent to formation of

Figure 1. Reactions in transcription initiation (for review, see Record et al. 1996; deHaseth et al. 1998). (R) RNAP plus initiation factor(s); (P) promoter; (RP$_c$) closed complex; (RP$_i$) intermediate complex; (RP$_o$) open complex; (RP$_{ITC}$) initial transcribing complex; (RD$_e$) elongation complex; (NN, NNN,... NNNNNNNN) 2- to 8-nucleotide RNA abortive products.

the RNAP-promoter open complex are asynchronous, with different molecules in the population occupying different conformational and reaction states (Uptain et al. 1997; Gelles and Landick 1998).

Given these technical challenges, it is apparent that multiple experimental methods will be needed to attack the problem; i.e., high-resolution structure determination using X-ray crystallography and nuclear magnetic resonance (NMR) spectroscopy; low-resolution structure determination using electron crystallography, electron microscopy, and force microscopy; measurement of distances using fluorescence resonance energy transfer; identification of interactions using crosslinking, affinity-cleaving, footprinting, and residue scanning; and real-time, single-molecule observations using optical microsopy and force microscopy. Furthermore, it is apparent that molecular modeling will play a critical part in integrating results from these experimental methods.

Recently, a first detailed molecular model for the structure of an RNAPII intermediate complex has been proposed (Kim et al. 1997). The remainder of this chapter will summarize the structural, biochemical, and imaging data that serve as the basis for the model, will present the model, and will present extensions of the model to the RNAPII open complex, the RNAPII elongation complex, and the corresponding bacterial RNAP complexes.

DATA

Structure of RNAPII

Using electron crystallography, Darst et al. (1991) have determined a 16 Å resolution structure of *S. cerevisiae* RNAPII Δ4/7—a derivative of *S. cerevisiae* RNAPII lacking the fourth- and seventh-largest subunits,

which are not essential for transcription initiation and elongation (Fig. 2) (Edwards et al. 1991). The structure indicates that RNAPII Δ4/7 has dimensions of 140 × 136 × 100 Å and has two prominent, finger-like projections that define a channel with a diameter of 25 Å—i.e., a diameter that could accommodate double-stranded nucleic acid. Darst et al. (1991) have proposed that the active center of RNAPII is located within this channel.

Using difference electron crystallography, specific sites within the structure of RNAPII Δ4/7 have been mapped (Fig. 2). Meredith et al. (1996) have mapped the location of the carboxy-terminal domain (CTD), a tandem heptapeptide repeat at the carboxyl terminus of the largest subunit, implicated in transcription regulation and pre-mRNA capping, splicing, cleavage, and polyadenylation (for review, see Dahmus 1995; Carlson 1997; Corden and Patturajan 1997; Shuman 1997; Steinmetz 1997). Jensen et al. (1998) have mapped the locations of the binding sites for the fourth- and seventh-largest subunits (and shown that RNAP Δ4/7 and intact RNAPII have substantially similar conformations). Leuther et al. (1996) have mapped the locations in x/y projection of the binding sites of transcription factor IIB in the RNAP Δ4/7-IIB complex, and transcription factor IIE in the RNAP Δ4/7-IIE complex.

Transcription factor IIB interacts with promoter DNA in the −30 region in an initiation complex (Lee and Hahn 1995; Nikolov et al. 1995; Lagrange et al. 1996).[1] It is likely that the RNAPII active center interacts with promoter DNA in the +1 region in an initiation complex (cf. Zaychikov et al. 1997). On the basis of these considerations, Leuther et al. (1996) and Kornberg (1996) have presented a schematic model for the structure of an initiation complex; in the model, the promoter DNA segment between positions −30 and +1 interacts with RNAPII along the line connecting the location of the binding site of IIB and the proposed location of the active center (Fig. 2, dashed line). Consistent with the model, the length of this line in x/y projection is approximately 100 Å, which is the expected length of an approximately 30-bp B-form DNA segment (Leuther et al. 1996). Since the binding site of IIB has been mapped only in x/y projection—not in three dimensions—two versions of the model are possible: one in which the promoter DNA segment between positions −30 and +1 interacts with the "front" face of RNAPII, and one in which this DNA segment interacts with the "rear" face of RNAPII (Leuther et al. 1996).

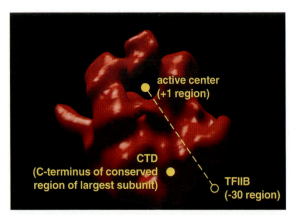

Figure 2. Structure of yeast RNAPII Δ4/7 at 16 Å resolution, showing the location of the amino terminus of CTD, the location in x/y projection of the binding site for transcription factor IIB (and thus the location in x/y projection of the promoter −30 region in an initiation complex), and the proposed location of the active center (and thus the proposed location of the promoter +1 region in an initiation complex) (Darst et al. 1991; Leuther et al. 1996; Meredith et al. 1996).[1] The distance in x/y projection between the location of the binding site for IIB and the proposed location of the active center is about 100 Å, which corresponds to approximately 30 bp of B-form DNA (Leuther et al. 1996).

[1]In mammalian RNAPII-dependent promoters, the distance between the first position of the TATA element and melted region in an open complex is about 20 bp, and the distance between the first position of the TATA element and the transcription start site is about 31 bp (Corden et al. 1980; Wang et al. 1992). In *S. cerevisiae* RNAPII-dependent promoters, the distance between the first position of the TATA element and the melted region in an open complex is about 20 bp, but the distance between the first position of the TATA element and the transcription start is variable, reflecting an RNAPII translocation process subsequent to open-complex formation ("scanning"; Giardina and Lis 1993 and references therein). In this paper, numbered positions of RNAPII-dependent promoters refer to a promoter having 31 bp between the first position of the TATA element and the transcription start. This corresponds to the normal TATA-start distance in mammalian promoters and the minimal TATA-start distance in *S. cerevisiae* promoters.

Asturias et al. (1997) have characterized a conformational isomer of RNAPII Δ4/7 in which the two finger-like projections are separated by an additional 20–30 Å, allowing unhindered access to the proposed DNA-binding channel. These authors propose that this "unclamped" conformational isomer of RNAPII mediates the initial association of RNAPII with promoter DNA (corresponding to an RNAPII-promoter closed complex) and subsequently isomerizes to yield the "clamped" conformational isomer (corresponding to RNAPII-promoter intermediate, open, and elongation complexes). The unclamped conformational isomer presumably also mediates the dissociation of RNAPII from DNA in transcription termination.

Recently, Fu et al. (1998) have reported that X-ray diffraction data at 4 Å resolution suggest the existence of an approximate twofold symmetry axis in RNAPII Δ4/7. Fu et al. conclude that a large fraction of the mass of RNAPII Δ4/7 consists of an approximately repeated, approximately twofold-symmetrically arrayed, substructure and propose that the two largest subunits of RNAPII have similar three-dimensional structures (despite the absence of significant sequence similarity).

Crosslinking Studies of the RNAPII-TBP-IIB-IIF-Promoter Complex

Kim et al. (1997) and Forget et al. (1997) have carried out protein-DNA photocrosslinking within complexes containing human RNAPII, TATA-element-binding protein (TBP), IIB, IIF, and promoter DNA. Kim et al. (1997) analyzed a set of 80 DNA fragments having photoactivatible crosslinking agents incorporated on DNA phosphates (i.e., each DNA phosphate, on each strand, from position –55 to position +25 of the adenovirus major late promoter; Fig. 3). Forget et al. (1997) analyzed a set of seven DNA fragments having photoactivatible crosslinking agents incorporated on DNA bases.

The results establish that the largest subunit of RNAPII, the second-largest subunit of RNAPII, the fifth-largest subunit of RNAPII, TBP, IIB, the RAP30 subunit of IIF, and the RAP74 subunit of IIF contact, or are close to, promoter DNA in the RNAPII-TBP-IIB-IIF-promoter complex and define the positions of these polypeptides relative to promoter DNA (Fig. 3) (Forget et al. 1997; Kim et al. 1997). The results permit five conclusions regarding RNAPII-DNA interaction in the RNAPII-TBP-IIB-IIF-promoter complex (Kim et al. 1997):

1. RNAPII interacts with approximately 70 bp, or about 240 Å, of DNA (positions –53 to +19).
2. RNAPII wraps DNA at least one half to two thirds of the way around its circumference. (The length of the RNAPII-promoter interaction [~240 Å] is approximately 1.7 times the longest dimension of RNAPII [140 Å; Darst et al. 1991].)
3. RNAPII interacts with positions –53 to –6 through a shallow, relatively open, channel (with RNAPII-DNA crosslinking occurring at half of the tested phosphates in this DNA segment). RNAPII interacts with positions –5 to +19 through a deep, nearly completely en-

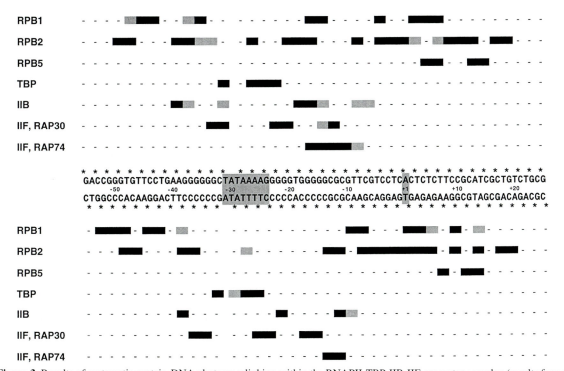

Figure 3. Results of systematic protein-DNA photocrosslinking within the RNAPII-TBP-IIB-IIF-promoter complex (results for non-template strand above sequence; results for template strand beneath sequence) (Kim et al. 1997). Phosphates analyzed are indicated by asterisks. Crosslinks observed are indicated by shaded bars. RPB1, RPB2, and RPB5 are the largest, second-largest, and fifth-largest subunits of RNAPII. RAP30 and RAP74 are the two subunits of transcription factor IIF.

closed, channel (with RNAPII-DNA crosslinking occurring at nearly all tested phosphates in this DNA segment).

4. The interface between the largest and second-largest subunits of RNAPII forms the DNA-binding channel.

5. The largest subunit of RNAPII interacts predominantly with one face of the DNA helix (the face with major grooves in the −11, −1, and +10 regions). The second-largest subunit interacts predominantly with the opposite face of the DNA helix, except between positions −4 and +3, where the second-largest subunit nearly surrounds the DNA helix.

Electron Microscopy of the RNAPII-TBP-IIB-IIF-Promoter Complex

Kim et al. (1997) and Forget et al. (1997) have used electron microscopy to image complexes containing human RNAPII, TBP, IIB, IIF, and promoter DNA. The results permit three conclusions regarding the path of DNA within the complex (Kim et al. 1997):

1. The protein component of the complex compacts DNA by the equivalent of approximately 50 bp (or ~170 Å), indicating that the protein component of the complex wraps DNA at least two thirds of the way around its circumference.

2. The protein component of the complex bends DNA. The mean observed DNA bend angle is 70°, but the distribution of observed DNA bend angles is broad and multimodal (0–160°; presumably due to different orientations of complexes on sample grids, and/or to the influence of hydrodynamic forces upon adsorption to sample grids).

3. The observed handedness of overall DNA writhe is clockwise (proceeding from upstream at left to downstream at right, with the apex of the DNA bend at top).

MODELS

Intermediate Complex

On the basis of the structural data for RNAPII (Darst et al. 1991; Leuther et al. 1996; Meredith et al. 1996) and the photocrosslinking and electron microscopy data for the RNAPII-TBP-IIB-IIF-promoter complex (Forget et al. 1997; Kim et al. 1997), Kim et al. (1997) have proposed a specific model for the path of DNA in the RNAPII-TBP-IIB-IIF-promoter complex (Figs. 4A and 5). The model invokes two DNA bends: (1) an approximately 80° DNA bend centered in the −30 region and phased to maximize interactions between RNAPII and the DNA segment upstream of −30, and (2) a second, approximately 70° DNA bend, centered in the +1 region and phased to maximize interactions between RNAPII and the DNA segment at and downstream from +1.[1] The model places the DNA segment between the −30 region and the +1 region on the "front" face of RNAPII and places positions −53 to −6 within a shallow, relatively

open, channel (Fig. 4A) and positions −5 to +19 within a deep, nearly completely enclosed, channel (Fig. 5).

In the model, TBP interacts with DNA in the −30 region, making sequence-specific interactions with the TATA element and, as a DNA-bending protein, reinforcing/stabilizing the DNA bend in the −30 region (J. Kim et al. 1993; Y. Kim et al. 1993; Nikolov et al. 1995). IIB interacts with DNA in the −35 and −25 regions (Lee and Hahn 1995; Nikolov et al. 1995; Lagrange et al. 1996), making sequence-specific interactions with the −35 region, and inserting a helix-turn-helix DNA-binding motif into the −35 region major groove (Lagrange et al. 1998; Qureshi and Jackson 1998). IIF RAP30 and IIF RAP74 interact with DNA in the −20 and −10 regions, respectively (Coulombe et al. 1994; Robert et al. 1996; Kim et al. 1997). The modeled locations of TBP, IIB, and IIF are on the same face of RNAPII as CTD and are close to CTD (Fig. 4A), consistent with possible interactions between these proteins and CTD and/or CTD-associated factors (Usheva et al. 1992; Chambers et al. 1995; Kang and Dahmus 1995). The model is able to accommodate the crystallographic structures of the TBP-IIB-DNA complex (Fig. 4A) (Nikolov et al. 1995) and the TBP-IIA-DNA complex (not shown; Geiger et al. 1996; Tan et al. 1996) and appears to have ample space to accommodate TBP-associated factors, IIF RAP30, and IIF RAP74 (Fig. 4A).

Kim et al. (1997) note that under the reaction conditions of the photocrosslinking and electron microscopy studies, the RNAPII-TBP-IIB-IIF-promoter complex has the hallmarks of an RNAP-promoter intermediate complex (RP$_i$, also referred to as RP$_{c2}$ and I$_2$; see Record et al. 1996; deHaseth et al. 1998): i.e., (1) stability to nondenaturing gel electrophoresis (Kim et al. 1997 and references therein), (2) resistance to challenge with nonspecific competitor DNA (Kim et al. 1997 and references therein), (3) protein-DNA interactions extending about 50 bp upstream and about 20 bp downstream from the transcription start (Kim et al. 1997 and references therein), and (4) the absence of a stable melted region (Holstege et al. 1996). Therefore, Kim et al. (1997) offer the model in Figures 4A and 5 as a working model for the structure of an RNAPII-promoter intermediate complex.

Kim et al. (1997) further note that other members of the multisubunit RNAP family exhibit unambiguous similarities to RNAPII in three-dimensional structure (Schultz et al. 1993; Polyakov et al. 1995) and in DNA crosslinking, DNA bending, and DNA wrapping (Chenchick et al. 1981; Heumann et al. 1988; Bartholomew et al. 1993, 1994; Meyer-Almes et al. 1994; Rippe et al. 1997). Therefore, Kim et al. (1997) suggest that the model in Figures 4A and 5 may apply generally to members of the multisubunit RNAP family.

Figure 6A presents an extension of the model in Figures 4A and 5 to the structure of a bacterial RNAP-promoter intermediate complex. In the model, the carboxyl terminus of the β′-subunit of bacterial RNAP corresponds in location to the carboxyl terminus of the conserved region of the largest subunit of RNAPII and the start of CTD, and the bacterial initiation factor σ corresponds in location, and possibly in function, to IIB and

IIF (cf. Figs. 4A and 6A). Thus, σ region 4 makes sequence-specific interaction with the –35 region and inserts a helix-turn-helix DNA-binding motif into the –35 region major groove (Gross et al. 1992), analogously to IIB, and σ region 2 interacts with the –10 region (Gross et al. 1992), analogously to IIF RAP74.

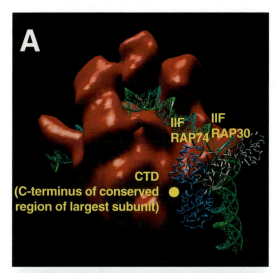

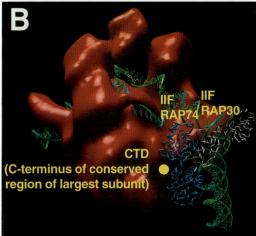

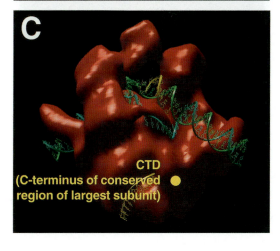

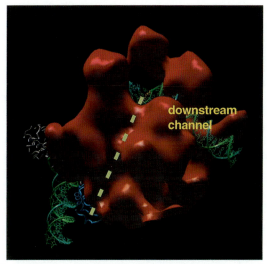

Figure 5. Rear view of the model for the structure of the RNAPII intermediate complex (Fig. 4A; Kim et al. 1997), showing the rear face of RNAPII, the proposed path of downstream DNA ("downstream channel"; a deep, nearly completely enclosed, channel), and an alternative possible path of downstream DNA (*dashed line*; a second deep channel).

Open Complex

In the RNAP-promoter open complex (RP_o), approximately 14 bp of promoter DNA, between positions –11 and +3, are melted, yielding the "transcription bubble" (Melnikova et al. 1978; Siebenlist et al. 1980; Sasse-Dwight and Gralla 1989; Wang et al. 1992).[1] The template strand of the transcription bubble remains in the active center; the nontemplate strand of the transcription bubble is removed from the active center (for review, see deHaseth and Helmann 1995).

The model in Figure 4A accounts for these observations in terms of a 20 × 15 × 15-Å knob-like projection that rises from the floor of the proposed DNA-binding channel between positions –11 and +3 and that ap-

Figure 4. Models for the structures of an RNAPII intermediate complex (the RNAPII-TBP-IIB-IIF-promoter complex; Kim et al. 1997) (*A*), an RNAPII open complex (*B*), and an RNAPII elongation complex (*C*). The models are oriented with upstream DNA at right and downstream DNA at left, such that transcription would proceed right to left. RNAPII is red (CTD indicated in yellow), TBP is white, IIB is blue, IIF RAP30 and IIF RAP74 are indicated in yellow, the DNA nontemplate strand is turquoise (transcription start in yellow), the DNA template strand is green, and the nascent RNA in *C* is yellow. Atomic coordinates for TBP-IIB-TATA at 2.7 Å resolution (Nikolov et al. 1995) were obtained from S. Burley (Rockefeller University). Modeled atomic coordinates for the DNA segments upstream and downstream from the TATA element were generated in INSIGHTII (MSI). Electron density data for yeast RNAPII Δ4/7 at 16 Å resolution (Darst et al. 1991) were obtained from S. Darst (Rockefeller University), edited to eliminate lattice-neighbor density, converted to INSIGHTII contour-file format, and rendered as an INSIGHTII contour object (1.6 σ electron-density isocontour). Atomic structures and contour objects were viewed and manipulated in INSIGHTII.

proaches—and nearly touches—one of the two prominent finger-like projections that form the walls of the proposed DNA-binding channel, forming a "gapped arch" (Kim et al. 1997). Molecular modeling suggests that, by rotation of phosphodiester bonds at positions −12/−11 and +3/+4, the nontemplate strand can be separated from the template strand at position −11, pass around the base of the 20 × 15 × 15-Å knob-like projection, pass under the gapped arch, and rejoin the template strand at position +3 (Fig. 4B). I offer the model in Figure 4B as a working model for the structure of an RNAPII-promoter open complex and the corresponding model in Figure 6B as a working model for the structure of a bacterial RNAP-promoter open complex.

The model in Figure 6B places σ region 2 close to positions −11 to −7 of the nontemplate strand, consistent with experimental data (Huang et al. 1997; Marr and Roberts 1997). It is possible to build into the model the crystallographic structure of σ region 2 (Malhotra et al. 1996), making all identified contacts between amino acids of σ and nucleotides of DNA (Gross et al. 1992; Malhotra et al. 1996), satisfying constraints from affinity cleaving analysis of σ-DNA interaction (Owens et al. 1998), and avoiding steric clash (not shown).

Elongation Complex

The RNAP-DNA elongation complex (RD$_e$) contains a transcription bubble with the same dimensions as those of the transcription bubble in the RNAP-promoter open complex (Gamper and Hearst 1982a,b; Lee and Landick 1992; Komissarova and Kashlev 1997a,b; Nudler et al. 1997) and contains a tightly bound nascent RNA. The eight to ten most recently synthesized nucleotides of the nascent RNA are engaged in Watson-Crick hydrogen bonding with the DNA template strand as an RNA-DNA hybrid (Meliknova et al. 1978; Hanna and Meares 1983; Lee and Landick 1992; Komissarova and Kashlev 1997a,b; Nudler et al. 1997; Sidorenkov et al. 1998), and the five to eight next most recently synthesized nucleotides of the nascent RNA are engaged in interactions with RNAP and are protected by these interactions from enzymatic digestion and oligonucleotide hybridization (Lee and Landick 1992; Reeder and Hawley 1996; Komissarova and Kashlev 1997a,b). The RNA segment engaged in interactions with RNAP and the DNA segment 3–9 bp downstream from the active center can be crosslinked to the same 30-amino-acid region of the largest subunit of RNAP, indicating that these two nucleic acid segments are separated by no more than approximately 20 Å (if the region is folded) to approximately 90 Å (if the region is extended) (Nudler et al. 1998).

The model in Figure 4B accounts for these observations in terms of a remarkable, 40 Å long, 8–10-Å wide tunnel that has one opening on the floor of the proposed DNA-binding channel nine nucleotides upstream of the active center, that passes through the center of mass of RNAP, and that has a second opening on the surface of RNAP near the carboxyl terminus of the conserved region of the largest subunit (Kim et al. 1997). Molecular

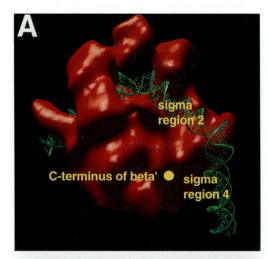

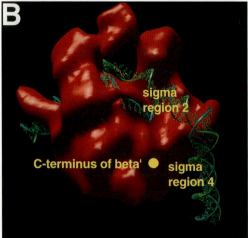

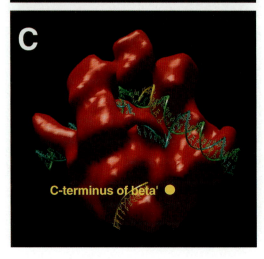

Figure 6. Models for the structures of a bacterial RNAP intermediate complex (*A*), a bacterial RNAP open complex (*B*), and a bacterial RNAP elongation complex (*C*). RNAP is red (carboxyl terminus of β′ indicated in yellow); σ region 4 and σ region 2, which recognize the −35 element and −10 element, respectively (Gross et al. 1992), are indicated in yellow; the DNA nontemplate strand is turquoise (transcription start in yellow); DNA template strand is green; and the nascent RNA in *C* is yellow.

modeling suggests that the nine most recently synthesized nucleotides of the nascent RNA can be accommodated as an RNA-DNA hybrid with the DNA template strand and that the five (if extended) to eight (if stacked) next most recently synthesized nucleotides of the nascent RNA can enter the tunnel, pass through the center of mass of RNAP, and be poised to emerge on the surface of RNAP near the carboxyl terminus of the conserved region of the largest subunit (Fig. 4C). Because the tunnel diverges by nearly 90° from the DNA upstream of the active center, and because the DNA is bent by nearly 90° at the active center, the proposed path of the nascent-RNA segment in the tunnel is approximately parallel to, and close to (within ~30 Å), the proposed path of the DNA segment 3–9 bp downstream from the active center, consistent with crosslinking data (Nudler et al. 1998). I offer the model in Figure 4C as a working model for the structure of an RNAPII-DNA elongation complex and the corresponding model in Figure 6C as a working model for the structure of a bacterial RNAP-DNA elongation complex.

An important implication of the model in Figure 4C is that the nascent RNA emerges from the RNAPII-DNA elongation complex adjacent to CTD, the tandem heptapeptide repeat at the carboxyl terminus of the largest subunit of RNAPII. CTD contains sites for factors involved in regulation of elongation and for factors involved in pre-mRNA capping, splicing, cleavage, and polyadenylation (for review, see Dahmus 1995; Carlson 1997; Corden and Patturajan 1997; Shuman 1997; Steinmetz 1997). Therefore, the model in Figure 4C provides a simple structural basis for spatial and temporal coordination of elongation, regulation of elongation, and pre-mRNA processing: i.e., a tunnel connects, and thereby couples, the active center and RNA-DNA hybrid to CTD.

A further important implication of the model is that the proposed RNA-exit tunnel is only 8–10 Å wide and therefore cannot accommodate double-stranded nucleic acid. Factor-independent transcription termination by bacterial RNAP involves formation of an RNA hairpin structure comprising the nascent-RNA segment 10–19 nucleotides from the 3′ end and a complementary nascent-RNA segment further from the 3′ end (for review, see von Hippel et al. 1984; d'Aubenton Carafa et al. 1990; Roberts 1995). The model in Figure 6C provides a simple structural basis for factor-independent transcription termination by bacterial RNAP: i.e., the termination hairpin nucleates just beyond the end of the RNA-exit tunnel and, unable to penetrate the RNA-exit tunnel, extends by extracting nascent RNA out through the tunnel, concomitantly extracting nascent RNA away from the DNA template strand, leading to transcription-bubble collapse and DNA release.

Subunit Boundaries

Results of protein-DNA photocrosslinking experiments with the RNAPII-promoter intermediate complex indicate that the interface between the largest and second-largest subunits of RNAPII forms the DNA-binding channel, that the largest subunit interacts predominantly with the face of the DNA helix which presents the DNA major groove in the –11, –1, and +10 regions, and that the second-largest subunit interacts predominantly with the opposite face of the DNA helix—except between positions –4 and +3, where the second-largest subunit nearly surrounds the DNA helix (Kim et al. 1997; see DATA, Crosslinking Studies of the RNAPII-TBP-IIB-IIF-Promoter Complex). Results of protein-RNA photocrosslinking experiments with RNAP-DNA elongation complexes indicate that the interface between the largest and second largest subunits also forms the channel for the nascent RNA segment 10–18 nucleotides from the 3′ end (Hanna and Meares 1983; Liu and Hanna 1995; Nudler et al. 1998). In the context of the models in Figures 4 and 6, these results imply that (1) the right finger-like projection and the right approximately one third of the front face of RNAP correspond to the largest subunit, and (2) the left finger-like projection (including the "fingertip" proposed to reach over and enclose the active center) and the left approximately two thirds of the front face of RNAP (including the knob-like projection proposed to separate the strands of the transcription bubble and the horn-like projection at extreme left) correspond to the second-largest subunit.

Results of protein-RNA photocrosslinking further indicate that the third largest conserved subunit (α-subunit in bacterial RNAP) interacts with the nascent RNA 24 nucleotides from the 3′ end (Liu and Hanna 1995). In the context of the models in Figures 4C and 6C, this result implies that the third-largest conserved subunit is close to the end of the proposed RNA-exit tunnel (within ~6–9 nucleotides, or ~20–50 Å). Placement of the third-largest conserved subunit close to the end of the proposed RNA-exit tunnel is attractive in that it places this subunit at the trailing edge of RNAP, consistent with interactions of bacterial RNAP α-subunit with σ (McMahan and Burgess 1994) and with upstream DNA and upstream-binding transcription activators (for review, see Busby and Ebright 1994). Placement of the third-largest conserved subunit close to the end of the proposed RNA-exit tunnel, and thus close to the carboxyl terminus of the conserved region of the largest subunit, also is attractive in that it suggests that the flexible, DNA-binding, regulator-binding carboxy-terminal extensions of the third-largest subunit in bacterial RNAP (αCTD; Busby and Ebright 1994) and of the largest subunit in RNAPII (CTD; Dahmus 1995; Carlson 1997) may emanate from essentially the same point on RNAP and may have fundamentally similar roles.

The above assignments of subunit boundaries are generally consistent with results of immunoelectron microscopy defining locations of subunits of bacterial RNAP (Tichelaar et al. 1983) and are completely consistent with results of immunoelectron microscopy defining locations of segments of the largest, second-largest, and third-largest conserved subunits in the low-resolution structure of *S. cerevisiae* RNAPI (respectively, on the rear face, on the left finger-like projection, and on the bottom face; Klinger et al. 1996).

Uncertainties

The models in Figures 4–6 are speculative. One specific area of uncertainty is whether the DNA segment between positions +1 and +20 should be modeled in the deep, nearly completely enclosed, channel on the rear face of RNAP ("downstream channel" in Fig. 5; Figs. 4–6; Kim et al. 1997) or, instead, in a second deep channel on the rear face of RNAP (Fig. 5, dashed line). Four lines of evidence favor modeling the downstream DNA segment in the deep, nearly completely enclosed, channel: (1) Although present in the structure of RNAPII Δ4/7 (Darst et al. 1991), the second deep channel does not appear to be present in the structure of intact RNAPII (Jensen et al. 1998); (2) the second deep channel does not appear to be present in the structure of RNAPI (Schultz et al. 1993); (3) the second deep channel does not appear to be present in the structure of bacterial RNAP (Polyakov et al. 1995); and (4) binding within the second deep channel would result in a substantially higher overall DNA bend angle in RNAP-promoter intermediate and open complexes than is observed by electron and scanning-probe microscopy (~180° vs. 70–110° [Kim et al. 1997; Rippe et al. 1997]). Nevertheless, because of the low resolutions of the structures of intact RNAPII, RNAPI, and bacterial RNAP, and because of the broad, multimodal distributions of overall DNA bend angles observed by microscopy, these lines of evidence are not definitive. It will be important to determine directly whether the downstream DNA segment should be modeled in the deep, nearly completely enclosed, channel or in the second deep channel—e.g., by use of fluorescence resonance energy transfer (Selvin 1995) to measure distances between fluorescent probes site-specifically incorporated into promoter DNA at positions –50 and +30 (which would be separated by >100 Å in the first case, but by only ~50 Å in the second case).

A second specific area of uncertainty is the relationship to the data of Fu et al. (1998) (see DATA, Structure of RNAPII). These authors have presented evidence that the largest and second-largest subunits of RNAP may have similar three-dimensional structures and may be related by an approximate twofold symmetry axis within RNAP. Fu et al. have hypothesized that, in RNAP-DNA complexes, the putative twofold axis within RNAP is perpendicular to the DNA-helix axis of the DNA segment between positions –30 and –1 and perpendicular to the DNA-helix axis of the DNA segment between positions +1 and +30. The models in Figures 4–6 appear to be incompatible with the hypothesis that a putative twofold axis within RNAP is perpendicular to the DNA-helix axes of the DNA segments between positions –30 and –1 and positions +1 and +30 (since, in the models in Figs. 4–6, these DNA segments are not parallel to each other, but, rather, are related by an angle of ~80°). However, the models in Figures 4–6, and the assignments of subunit boundaries in the preceding section, would be compatible with the following alternative orientations of a putative twofold axis within RNAP:

1. On the *y* axis in Figures 4–6, passing through the RNAP center of mass and the RNAP active center. (In this orientation, the putative twofold axis would relate the structures of the two prominent finger-like projections of RNAP, the paths of the DNA segments between positions –30 and –1 and positions +1 and +30, and the two strands of DNA, or of DNA-RNA, at the active center.)

2. Approximately on the *x* axis in Figures 4–6, passing through the RNAP center of mass.

3. Approximately on the *z* axis in Figures 4–6, passing through the RNAP center of mass.

It will be important to establish whether RNAP in fact contains a twofold axis and, if so, to determine its orientation and functional significance.

PROSPECT

The models in Figures 4–6 provide a framework for designing experiments and interpreting results on transcription initiation, elongation, and regulation. In addition, the models provide a framework for incorporating high-resolution structures of individual components of transcription complexes as they become available.

The models in Figures 4–6 make numerous testable predictions regarding distances between pairs of specific sites in RNAP, in initiation factors, in DNA, and in RNA. My laboratory has developed methods to site-specifically introduce fluorescent probes at polypeptide carboxyl termini, at polypeptide hexahistidine tags, in DNA, and in RNA and is using fluorescence resonance energy transfer (for review, see Selvin 1995) to determine distances between pairs of fluorescent probes site-specifically introduced into transcription complexes (A. Kapanidis, V. Mekler, W. Niu, Y. Ebright, and R.H. Ebright, unpubl.). One priority is measurement of distances between the upstream and downstream segments of wrapped promoter DNA in intermediate and open complexes. Another priority is measurement of distances between the nontemplate and template strands of the transcription bubble in open complexes. Still another priority is measurement of distances between the carboxyl terminus of the conserved region of the largest subunit of RNAP—which has been mapped within the structure of RNAP and therefore can serve as a defined reference point (Figs. 2 and 4–6) (Leuther et al. 1996)—and specific sites within RNAP, initiation factors, DNA, and RNA. Preliminary results provide basis for optimism that these experiments will permit decisive, near-term tests of the models and, importantly, will be adaptable for single-molecule, real-time analyses of transcription initiation, elongation, and regulation (I. Kanevsky V. Mekler, Y. Ebright, Y. Jia, R. Hochstrasser, and R.H. Ebright, unpubl.).

ACKNOWLEDGMENTS

I thank S. Burley, S. Darst, P. Sigler, and T. Richmond for structural data; G. Parkinson for assistance in reformatting electron-density files; and D. Reinberg for discussion and for access to unpublished data. This work was supported by an HHMI investigatorship and by National Institutes of Health grants GM-41376 and GM-53665.

REFERENCES

Archambault J. and Friesen J. 1993. Genetics of eukaryotic RNA polymerases I, II, and III. *Microbiol. Rev.* **57:** 703.

Asturias F., Meredith G., Poglitsch C., and Kornberg R. 1997. Two conformations of RNA polymerase II revealed by electron crystallography. *J. Mol. Biol.* **272:** 536.

Bartholomew B., Braun B., Kassavetis G., and Geiduschek E.P. 1994. Probing close contacts of RNA polymerase III transcription complexes with the photoactive nucleoside 4-thiodeoxythymidine. *J. Biol. Chem.* **269:** 18090.

Bartholomew B., Durkovich D., Kassavetis G., and Geiduschek E.P. 1993. Orientation and topography of RNA polymerase III in transcription complexes. *Mol. Cell. Biol.* **13:** 942.

Busby S. and Ebright R. 1994. Promoter structure, promoter recognition, and transcription activation in prokaryotes. *Cell* **79:** 743.

Carlson M. 1997. Genetics of transcriptional regulation in yeast: Connections to the RNA polymerase II CTD. *Annu. Rev. Cell Dev. Biol.* **13:** 1.

Chambers R., Wang B., Burton Z., and Dahmus M. 1995. The activity of COOH-terminal domain phosphatase is regulated by a docking site on RNA polymerase II and by the general transcription factors IIF and IIB. *J. Biol. Chem.* **270:** 14962.

Chenchick A., Beabealashvilli R., and Mirzabekov A. 1981. Topography of interaction of *Escherichia coli* RNA polymerase subunits with *lac*UV5 promoter. *FEBS Lett.* **128:** 46.

Corden J. and Patturajan M. 1997. A CTD function linking transcription to splicing. *Trends Biochem. Sci.* **22:** 413.

Corden J., Wasylysk B., Buchwalder A., Sassone-Corsi P., Kedinger C., and Chambon P. 1980. Promoter sequences of eukaryotic protein-coding genes. *Science* **209:** 1406.

Coulombe B., Li J., and Greenblatt J. 1994. Topological localization of the human transcription factors IIA, IIB, TATA box-binding protein, and RNA polymerase II-associated protein 30 on a class II promoter. *J. Biol. Chem.* **269:** 19962.

Dahmus M. 1995. Phosphorylation of the C-terminal domain of RNA polymerase II. *Biochim. Biophys. Acta* **1261:** 171.

Darst S., Edwards A., Kubalek E., and Kornberg R. 1991. Three-dimensional structure of yeast RNA polymerase II at 16 Å resolution. *Cell* **66:** 121.

d'Aubenton Carafa Y., Brody E., and Thermes C. 1990. Prediction of rho-independent *Escherichia coli* transcription terminators. *J. Mol. Biol.* **216:** 835.

deHaseth P. and Helmann J. 1995. Open complex formation by *Escherichia coli* RNA polymerase: The mechanism of polymerase-induced strand separation of double helical DNA. *Mol. Microbiol.* **16:** 817.

deHaseth P., Zupancic M., and Record M. 1998. RNA polymerase-promoter interactions. *J. Bacteriol.* **180:** 3019.

Edwards A., Kane C., Young R., and Kornberg R. 1991. Two dissociable subunits of yeast RNA polymerase II stimulate the initiation of transcription at a promoter *in vitro*. *J. Biol. Chem.* **266:** 71.

Forget D., Robert F., Grondin G., Burton Z., Greenblatt J., and Coulombe B. 1997. RAP74 induces promoter contacts by RNA polymerase II upstream and downstream of a DNA bend centered on the TATA box. *Proc. Natl. Acad. Sci.* **94:** 7150.

Fu J., Gerstein M., David P., Gnatt A., Bushnell D., Edwards A., and Kornberg R. 1998. Repeated tertiary fold of RNA polymerase II and implications for DNA binding. *J. Mol. Biol.* **280:** 317.

Gamper H. and Hearst J. 1982a. A topological model for transcription based on unwinding angle analysis of *E. coli* RNA polymerase binary, initiation and ternary complexes. *Cell* **29:** 81.

———. 1982b. Size of the unwound region of DNA in *Escherichia coli* RNA polymerase and calf thymus RNA polymerase II ternary complexes. *Cold Spring Harbor Symp. Quant. Biol.* **47:** 455.

Geiger J., Hahn S., Lee S., and Sigler P. 1996. Crystal structure of the yeast TFIIA/TBP/DNA complex. *Science* **272:** 830.

Gelles J. and Landick R. 1998. RNA polymerase as a molecular motor. *Cell* **93:** 13.

Giardina C. and Lis J. 1993. DNA melting on yeast RNA polymerase II promoters. *Science* **261:** 759.

Gross C., Lonetto M., and Losick R. 1992. Bacterial sigma factors. In *Transcriptional regulation* (ed. S. McKnight and K. Yamamoto), p. 129. Cold Spring Harbor Laboratory Press, Cold Spring Harbor, New York.

Hanna M. and Meares C. 1983. Topography of transcription: Path of the leading end of nascent RNA through the *Escherichia coli* transcription complex. *Proc. Natl. Acad. Sci.* **80:** 4238.

Heumann H., Ricchetti M., and Werel W. 1988. DNA-dependent RNA polymerase of *Escherichia coli* induces bending or an increased flexibility of DNA by specific complex formation. *EMBO J.* **7:** 4379.

Holstege F., van der Vliet P.C., and Timmers H.T. 1996. Opening of an RNA polymerase II promoter occurs in two distinct steps and requires the basal transcription factors IIE and IIH. *EMBO J.* **15:** 1666.

Huang X., Lopez de Saro F., and Helmann J. 1997. σ Factor mutations affecting the sequence-selective interaction of RNA polymerase with –10 region single-stranded DNA. *Nucleic Acids Res.* **25:** 2603.

Jensen G., Meredith G., Bushnell D., and Kornberg R. 1998. Structure of wild-type yeast RNA polymerase II and location of Rpb4 and Rpb7. *EMBO J.* **17:** 2353.

Kang M. and Dahmus M. 1995. The photoactivated cross-linking of recombinant C-terminal domain to proteins in a HeLa cell transcription extract that comigrate with transcription factors IIE and IIF. *J. Biol. Chem.* **270:** 23390.

Kim J., Nikolov D., and Burley S. 1993. Co-crystal structure of TBP recognizing the minor groove of a TATA element. *Nature* **365:** 520.

Kim T.-K., Lagrange T., Wang Y.-H., Griffith J., Reinberg D., and Ebright R. 1997. Trajectory of DNA in the RNA polymerase II transcription preinitiation complex. *Proc. Natl. Acad. Sci.* **94:** 12268.

Kim Y., Geiger J., Hahn S., and Sigler P. 1993. Crystal structure of a yeast TBP/TATA-box complex. *Nature* **365:** 512.

Klinger C., Huet J., Song D., Petersen G., Riva M., Bautz E., Sentenac A., Oudet P., and Schultz P. 1996. Localization of yeast RNA polymerase I core subunits by immunoelectron microscopy. *EMBO J.* **15:** 4643.

Komissarova N. and Kashlev M. 1997a. RNA polymerase switches between inactivated and activated states by translocating back and forth along the DNA and the RNA. *J. Biol. Chem.* **272:** 15329.

———. 1997b. Transcriptional arrest: *Escherichia coli* RNA polymerase translocates backward, leaving the 3′ end of the RNA intact and extruded. *Proc. Natl. Acad. Sci.* **94:** 1755.

Kornberg. R. 1996. RNA polymerase II transcription control. *Trends Biochem. Sci.* **21:** 325.

Lagrange T., Kapanidis A., Tang H., Reinberg D., and Ebright R. 1998. New core promoter element in RNA polymerase II-dependent transcription: Sequence-specific DNA binding by transcription factor IIB. *Genes Dev.* **12:** 34.

Lagrange T., Kim T.K., Orphanides G., Ebright Y., Ebright R., and Reinberg D. 1996. High-resolution mapping of nucleoprotein complexes by site-specific protein-DNA photocrosslinking: Organization of the human TBP-TFIIA-TFIIB-DNA quaternary complex. *Proc. Natl. Acad. Sci.* **93:** 10620.

Lee D. and Landick R. 1992. Structure of RNA and DNA chains in paused transcription complexes containing *Escherichia coli* RNA polymerase. *J. Mol. Biol.* **228:** 759.

Lee S. and Hahn S. 1995. A model for TFIIB binding to the TBP-DNA complex. *Nature* **376:** 609.

Leuther K., Bushnell D., and Kornberg R. 1996. Two-dimensional crystallography of TFIIB- and IIE-RNA polymerase II complexes: Implications for start site selection and initiation complex formation. *Cell* **85:** 773.

Liu K. and Hanna M. 1995. NusA interferes with interactions between the nascent RNA and the C-terminal domain of the alpha subunit of RNA polymerase in *Escherichia coli* transcription complexes. *Proc. Natl. Acad. Sci.* **92:** 5012.

Malhotra A., Severinova E., and Darst S. 1996. Crystal structure of a σ^{70} subunit fragment from *E. coli* RNA polymerase. *Cell* **87:** 127.

Marr M. and Roberts J. 1997. Promoter recognition as measured by binding of polymerase to nontemplate strand oligonucleotide. *Science* **276:** 1258.

McMahan S. and Burgess R. 1994. Use of aryl azide cross-linkers to investigate protein-protein interactions: An optimization of important conditions as applied to *Escherichia coli* RNA polymerase and localization of a σ^{70}-α cross-link to the C-terminal region of α. *Biochemistry* **33:** 12092.

Melnikova A., Beabealashvilli R., and Mirzabekov A. 1978. A study of unwinding of DNA and shielding of the DNA grooves by RNA polymerase by using methylation with dimethylsulphate. *Eur. J. Biochem.* **84:** 301.

Meredith G., Chang W.-H., Li Y., Bushnell D., Darst S., and Kornberg R. 1996. The C-terminal domain revealed in the structure of RNA polymerase II. *J. Mol. Biol.* **258:** 413.

Meyer-Almes F.-J., Heumann H., and Porschke D. 1994. The structure of the RNA polymerase-promoter complex: DNA-bending angle by quantitative electrooptics. *J. Mol. Biol.* **236:** 1.

Mooney R., Artsimovitch I., and Landick R. 1998. Information processing by RNA polymerase: Recognition of regulatory signals during RNA chain elongation. *J. Bacteriol.* **180:** 3265.

Nikolov D., Chen H., Halay E., Usheva A., Hisatake K., Lee D.K., Roeder R., and Burley S. 1995. Crystal structure of a TFIIB-TBP-TATA-element ternary complex. *Nature* **377:** 119.

Nudler E., Mustaev A., Lukhtanov E., and Goldfarb A. 1997. The RNA-DNA hybrid maintains the register of transcription by preventing backtracking of RNA polymerase. *Cell* **89:** 33.

Nudler E., Gusarov I., Avetissova E., Kozlov M., and Goldfarb A. 1998. Spatial organization of transcription elongation complex in *Escherichia coli. Science* **281:** 424.

Owens J.T., Chmura A.J., Murakami K., Fujita N., Ishihama A., and Meares C.F. 1998. Mapping the promoter DNA sites proximal to conserved regions of σ^{70} in an *Escherichia coli* RNA polymerase-*lac*UV5 open promoter complex. *Biochemistry* **37:** 7670.

Polyakov A., Severinova E., and Darst S. 1995. Three-dimensional structure of *E. coli* core RNA polymerase: Promoter binding and elongation conformations of the enzyme. *Cell* **83:** 365.

Qureshi S. and Jackson S. 1998. Sequence-specific DNA binding by the *S. shibatae* TFIIB homolog, TFB, and its effect on promoter strength. *Mol. Cell* **1:** 389.

Record M.T., Reznikoff W., Craig M., McQuade K., and Schlax P. 1996. *Escherichia coli* RNA polymerase ($E\sigma^{70}$), promoters, and the kinetics of the steps of transcription initiation. In *Escherichia coli and* Salmonella: *Cellular and molecular biology* (ed. F. Neidhart), p. 792. ASM Press, Washington, D.C.

Reeder T. and Hawley D. 1996. Promoter proximal sequences modulate RNA polymerase II elongation by a novel mechanism. *Cell* **87:** 767.

Rippe K., Guthold M., von Hippel P., and Bustamante C. 1997. Transcriptional activation via DNA-looping: Visualization of intermediates in the activation pathway of *E. coli* RNA polymerase-σ^{54} holoenzyme by scanning force microscopy. *J. Mol. Biol.* **270:** 125.

Robert F., Forget D., Li J., Greenblatt J., and Coulombe B. 1996. Localization of subunits of transcription factors IIE and IIF immediately upstream of the transcriptional initiation site of the adenovirus major late promoter. *J. Biol. Chem.* **271:** 8517.

Roberts J. 1995. Transcription termination and its control. In *Regulation of gene expression in* Escherichia coli (ed. E.C.C. Lin and A. Simon). R.G. Landes, Austin, Texas.

Sasse-Dwight S. and Gralla J. 1989. KMnO$_4$ as a probe for *lac* promoter DNA melting and mechanism *in vivo. J. Biol. Chem.* **264:** 8074.

Schultz P., Celia H., Riva M., Sentenac A., and Oudet P. 1993. Three-dimensional model of yeast RNA polymerase I determined by electron microscopy of two-dimensional crystals. *EMBO J.* **12:** 2601.

Selvin P. 1995. Fluorescence resonance energy transfer. *Methods Enzymol.* **246:** 301.

Sentenac A., Riva M., Thuriaux P., Buhler J.-M., Treich I., Carles C., Werner M., Ruet A., Huet J., Mann C., Chiannilkulchai N., Stettler S., and Mariotte S. 1992. Yeast RNA polymerase subunits and genes. In *Transcriptional regulation* (ed. S. McKnight and K. Yamamoto), p. 27. Cold Spring Harbor Laboratory Press, Cold Spring Harbor, New York.

Shuman S. 1997. Origins of mRNA identity: Capping enzymes bind to the phosphorylated C-terminal domain of RNA polymerase II. *Proc. Natl. Acad. Sci.* **94:** 12758.

Sidorenkov I., Komissarova N., and Kashlev M. 1998. Crucial role of the RNA:DNA hybrid in the processivity of transcription. *Mol. Cell* **2:** 55.

Siebenlist U., Simpson R., and Gilbert W. 1980. *E. coli* RNA polymerase interacts homologously with two different promoters. *Cell* **20:** 269.

Steinmetz E. 1997. Pre-mRNA processing and the CTD of RNA polymerase II. *Cell* **89:** 491.

Tan S., Hunziker Y., Sargent D., and Richmond T. 1996. Crystal structure of a yeast TFIIA/TBP/DNA complex. *Nature* **381:** 127.

Tichelaar W., Schutter W., Arnberg A., van Bruggen E., and Stender W. 1983. The quaternary structure of *Escherichia coli* RNA polymerase studies with (scanning) transmission (immuno)electron microscopy. *Eur. J. Biochem.* **135:** 363.

Uptain S.M., Kane C.M., and Chamberlin M.J. 1997. Basic mechanisms of transcription elongation and its regulation. *Annu. Rev. Biochem.* **66:** 117.

Usheva A., Maldonado E., Goldring A., Lu H., Houbavi C., Reinberg D., and Aloni Y. 1992. Specific interaction between the nonphosphorylated form of RNA polymerase II and the TATA-binding protein. *Cell* **69:** 871.

von Hippel P., Bear D., Morgan W., and McSwiggen J. 1984. Protein-nucleic acid interactions in transcription: A molecular analysis. *Annu. Rev. Biochem.* **53:** 389.

Wang W., Carey M., and Gralla J. 1992. Polymerase II promoter activation: Closed complex formation and ATP-driven start site opening. *Science* **255:** 450.

Young R. 1991. RNA polymerase II. *Annu. Rev. Biochem.* **60:** 689.

Zaychikov E., Denissova L., Meier T., Gotte M., and Heumann H. 1997. Influence of Mg^{2+} and temperature on formation of the transcription bubble. *J. Biol. Chem.* **272:** 2259.

The Initiator Element: A Paradigm for Core Promoter Heterogeneity within Metazoan Protein-coding Genes

S.T. SMALE, A. JAIN, J. KAUFMANN, K.H. EMAMI, K. LO, AND I.P. GARRAWAY
Howard Hughes Medical Institute, Molecular Biology Institute, and Department of Microbiology, Immunology, and Molecular Genetics, University of California, Los Angeles, California 90095-1662

In metazoans, an enormous number of protein-coding genes must be differentially expressed in specific cell types, during developmental processes, and in response to an array of extracellular signals. Combinatorial gene regulation strategies must be employed to generate these diverse expression patterns because only a limited number of transcription factors can be encoded by a genome of limited size. Combinatorial regulation involves multiple layers of events: For a gene to be activated, higher-order chromatin structures must first be unpacked, individual nucleosomes must be remodeled, transcription factors must bind to promoters and enhancers, and the transcription factors must then communicate with the general transcription machinery in concert with appropriate coactivator proteins. Another fundamental feature of combinatorial regulation is the requirement for several distinct transcription factors to activate a gene. By combining each transcription factor encoded by the genome with various combinations of other factors, the number of gene expression patterns that can be achieved in an animal is greatly enhanced.

Although considerable effort has been applied toward the elucidation of combinatorial gene regulation mechanisms, one potential contributor has received relatively little attention, namely, the core promoter, which is directly responsible for the formation of an RNA polymerase-II-containing preinitiation complex.

Several years ago, we were provided with an opportunity to study core promoter architecture when we found that the lymphocyte-specific terminal transferase (TdT) gene lacks a TATA box, but instead contains an initiator (Inr) element overlapping the transcription start site (Smale and Baltimore 1989). Our early hypothesis was that the TdT promoter contained an Inr, not simply as a replacement for the TATA box, but as an important contributor to the combinatorial regulation of TdT expression, even though basal Inr activity is not regulated in a cell-specific or developmental stage-specific manner.

In this paper, we first discuss hypothetical models by which core promoter architecture and the general transcription machinery might contribute to combinatorial regulation. We then describe our progress toward understanding the mechanism of Inr activity and the relevance of core promoter structure for regulated expression.

MODELS THAT LINK CORE PROMOTER ARCHITECTURE TO COMBINATORIAL GENE REGULATION

Most early discussions of combinatorial gene regulation mechanisms excluded the core promoter and general transcription machinery because almost all metazoan genes analyzed in the late 1970s and early 1980s contained consensus TATA boxes, regardless of their mode of regulation (Breathnach and Chambon 1981). The apparent homogeneity suggested that the core promoter and general transcription machinery had no significant role in regulation. It now is known, however, that protein-coding genes possess a wide variety of core promoter architectures: Some appear to contain only a TATA box, but many contain an Inr in addition to a TATA box, an Inr instead of a TATA box, a TATA or Inr combined with an important element downstream from the start site, or neither a TATA box nor an Inr, with the important core promoter elements undefined (Smale 1994, 1997). Furthermore, some genes exhibit a single transcription start site, whereas others exhibit multiple start sites that range from a few tightly clustered sites to dozens of sites spanning hundreds of nucleotides. The reason that, 15–20 years ago, it was thought that almost all genes contained consensus TATA boxes was that the first cellular genes isolated were expressed at very high levels (e.g., the globin, histone, and ovalbumin genes). The most abundantly expressed genes contain TATA boxes, whereas typical genes contain core promoters with a variety of distinct architectures.

The considerable core promoter heterogeneity that is now apparent suggests that the core promoter and general transcription machinery might in fact have an active role in gene regulation and might contribute to combinatorial regulation strategies. A simple scenario by which core promoter architecture might contribute to gene regulation is as follows: A hypothetical activator protein, A, might be expressed in a particular cell type, for example, mature B lymphocytes. The activator, a member of a multiprotein family such as the Ets family, might bind with high affinity to the upstream promoter regions for two different genes, *X* and *Y*, only the first of which is expressed in B cells. One strategy that could be used to confer specificity is to provide the activator with an activation domain that functions only in cooperation with a specific core promoter structure. If gene *X*, but not gene *Y*, possesses that structure, only gene *X* would be activated. Gene *Y* presumably would be expressed in another cell type, following activation by a different Ets protein that can cooperate with its unique core promoter structure. This basic strategy would of course need to be superimposed on the many other events that contribute to combinatorial regulation.

The hypothetical strategy described above could play an important part in combinatorial gene regulation even if identical components of the transcription machinery were involved in transcription of all genes. This could be accomplished if different core promoter architectures conferred different rate-limiting steps to the transcription initiation reaction. A simple TATA-Inr promoter, for example, might be regulated initially by the recruitment of TFIID. If TFIID remains stably bound to the template following the first initiation event, the rate-limiting step during reinitiation might be the recruitment of another factor, such as the RNA polymerase II holoenzyme. In contrast, a promoter containing an Inr element instead of a TATA box might bind less stably to TFIID. The recruitment of TFIID might therefore remain the rate-limiting step even during reinitiation if the TFIID dissociates from the template after each initiation event. Thus, an activator that selectively recruits TFIID would be expected to stimulate reinitiation from Inr-containing promoters, but not TATA-containing promoters.

Although variable rate-limiting steps would allow core promoters to contribute to combinatorial regulation, even greater flexibility would exist if different core promoters required different general transcription factors. Indeed, as discussed below, Inr-containing promoters appear to require proteins that are not required for all promoters. Furthermore, new TFIID-like complexes, TFIID subcomplexes, and cell-specific TAFs have been identified that may allow different core promoter structures to contribute to combinatorial gene regulation (Parvin and Sharp 1991; Jacq et al. 1994; Dikstein et al. 1996; Hansen et al. 1997; Wieczorek et al. 1998).

Despite the potential importance of mechanisms like those hypothesized above, the hypotheses have not been adequately tested, largely because there is very little fundamental knowledge of transcription initiation mechanisms mediated by core promoter elements other than the TATA box. The Inr element provides a starting point for acquiring this knowledge.

BACKGROUND

The term "initiator" was first used in 1980 by Grosschedl and Birnstiel to describe a DNA region spanning the start site of the sea urchin histone H2A gene, which contains a TATA box between nucleotides –25 and –30 (relative to the start site) (Grosschedl and Birnstiel 1980). A deletion of this region had little effect on the location of the transcription start site, which remained 25 bp downstream from the TATA box, but promoter strength was reduced. Early mutant analyses of other TATA-con-

taining promoters revealed variable phenotypes (see, e.g., Corden et al. 1980; Wasylyk et al. 1980; Breathnach and Chambon 1981; Hu and Manley 1981). In some cases, start site mutations had little effect on the efficiency or accuracy of transcription; in other instances, promoter strength was reduced significantly or heterogeneous start sites emerged. Although the initiator region was critical for accurate transcription of some genes (see, e.g., Hen et al. 1982; Concino et al. 1984), almost all subsequent studies of transcription initiation mechanisms focused on the TATA box.

Experiments performed with TATA-containing promoters from *Saccharomyces cerevisiae* by Chen and Struhl (1985) and others (Hahn et al. 1985; McNeil and Smith 1985; Nagawa and Fink 1985) provided evidence that initiator regions can help determine the location of a transcription initiation site. The properties of TATA boxes are somewhat different in *S. cerevisiae* and metazoans, in that *S. cerevisiae* RNA polymerase II initiates transcription 40–120 bp downstream from the TATA. Mutant analyses revealed that the DNA sequence surrounding the start sites is responsible for the precise start site location; when the start site regions were altered, transcription began at other locations 40–120 bp downstream from the TATA box. Thus, in *S. cerevisiae*, the initiator region assists the TATA box in dictating the location of the start site.

Our studies of the promoter for the murine TdT gene, initially performed in the laboratory of Dr. David Baltimore, provided two additional advances that inspired us to study core promoter architecture in greater detail. First, we found that a metazoan Inr was capable of functioning on its own as a core promoter element and did not merely assist an upstream TATA box (Smale and Baltimore 1989). This property was apparent from the observation that the TdT promoter lacks a TATA box and that the Inr was sufficient for core promoter function in vivo and in vitro. The second important property of the Inr from the initial analysis was its delineation as a discrete 17-bp element that could function in a heterologous context and whose properties could therefore be studied in depth.

General Properties of Inr Elements

Additional studies of Inr elements, analyzed largely in the context of synthetic promoters, revealed several relevant properties. First, although the Inr could direct transcription at a very low level by itself, strong Inr-mediated transcription could be detected when activators, such as Sp1 or GAL4-VP16, were bound to multiple sites upstream of the core promoter (Smale and Baltimore 1989; Emami et al. 1995). Furthermore, TATA and Inr elements strongly synergized with each other when both were present in the same promoter and were separated by 25 bp (Smale et al. 1990). Interestingly, if the two elements were separated from each other by 30, 35, or 40 bp, no synergy was observed (O'Shea-Greenfield and Smale 1992). However, if the elements were separated by only 15 or 20 bp, the strong synergy was retained, but almost all of the transcription initiated from a location 25 bp downstream from the TATA box and not from within the Inr. The inability of the two elements to be separated from each other by more than 25 bp suggests that the proteins that interact with them are stereospecifically constrained. The retention of synergy when the elements are moved closer together, with a TATA dominance in determining the start site, suggests some degree of flexibility in the recognition, although the mechanistic basis of this phenomenon remains unresolved.

The relevance of the Inr for determining the direction of RNA polymerase II transcription has been addressed (O'Shea-Greenfield and Smale 1992). These studies revealed that the Inr is unidirectional, but that it has little relevance for determining the direction of transcription. Instead, directionality appears to be dictated by the relative orientations of activator elements and core promoter elements. When activator elements were placed approximately 20 bp from a TATA box, for example, transcription consistently began 25 bp from the the opposite side of the TATA box, regardless of the orientation of the TATA box. Transcription initiation was not detected on the same side of the TATA box as the activator binding sites. When activator binding sites were placed at a reasonable distance from an Inr, the Inr contributed to promoter strength and start site localization, but only when the Inr was oriented correctly. When the Inr faced toward the activator binding sites, it remained nonfunctional.

One other property of the Inr that has been examined is its role in directing transcription from an adenosine, which almost always is found at a central location of an Inr. Although transcription usually initiates at the adenosine within a consensus Inr, experimental manipulations that retain Inr function can alter the location of the transcription start site. One method was mentioned above, in which the TATA box was moved closer to the Inr (O'Shea-Greenfield et al. 1992). In this case, the Inr still stimulated transcription, but the start site was 5 or 10 bp downstream from the adenosine. Another manipulation was to reduce the concentration of ATP within the in vitro transcription reactions (Zenzie-Gregory et al. 1992). With low concentrations of ATP, promoters containing Sp1 sites and an Inr in the absence of a TATA box directed transcription from thymidine residues at nucleotides +3 and +4, although the Inr contributes to promoter strength. Finally, dinucleotide-directed initiation led to initiation from other nucleotides within the Inr (Zenzie-Gregory et al. 1992). These results suggest that initiation at the central adenosine is not an intrinsic property of an Inr.

A delineation of the sequence requirements for Inr activity was perhaps the most important advance for the subsequent mechanistic studies (Fig. 1). This information was essential for identifying potential Inr elements and potential transcription start sites within newly discovered genes. More importantly, this knowledge was necessary for evaluating the functional relevance of candidate Inr-binding proteins that have been identified.

Visual inspections of the start site sequences from several TATA-containing promoters provided the first predictions regarding the sequences that might be required for Inr activity (Grosschedl and Birnstiel 1980; Breathnach and Chambon 1981). The start site sequences appeared to be poorly conserved, but a central CA was usually present and was often surrounded by a few pyrimidines, with RNA synthesis often beginning at the adenosine. More recently, attempts have been made to define a start site consensus sequence by weight-matrix analysis of hundreds of promoter sequences from vertebrates and invertebrates (Bucher 1990). The computed consensus sequence is similar to that derived from the visual inspections, but a preference for a T at +3 was noted (Fig. 1).

The first functional consensus sequence for Inr activity was derived from an analysis of several dozen randomly generated Inrs and specific Inr mutants (Javahery et al. 1994). The mutants were tested in the context of a synthetic promoter lacking a TATA box, but containing multiple upstream Sp1 sites; the mutant promoters were analyzed in an in vitro transcription assay using HeLa nuclear extracts. This analysis revealed a consistent consensus sequence, Py Py A+1 N T/A Py Py (Py = pyrimidine) (Fig. 1). The strong similarity between this and the computationally derived consensus supports the validity of the functional assay used to perform the mutant analysis. More recently, a representative subset of the mutants were compared using in vivo and in vitro assays, and in two different promoter contexts (in the context described above and in the context of an upstream TATA box in the absence of an activator) (Lo and Smale 1996). The mutants were also analyzed in *Drosophila* extracts (Lo and Smale 1996). In all of these assays, the sequence requirements for Inr activity were virtually identical, strongly suggesting that consensus Inr elements are recognized by the same protein (or family of proteins) in different types of promoters, in different types of assays, and in both vertebrates and invertebrates.

A detailed examination of the results of the mutant analyses reveals that, within the consensus sequence, an A at +1, T or A at +3, and pyrimidine at –1 are most important for determining Inr strength (Javahery et al. 1994; Lo and Smale 1996). The sequence CANT is not sufficient for Inr activity, however. Pyrimidines must surround this core sequence in at least a few positions, although pyrimidines are not needed in all four of the positions indicated in the consensus.

The existence of a very loose consensus sequence for Inr activity raises the question of how a sequence matching this consensus could lead to accurate initiation, given the frequency with which sequences matching the consensus will be present in the genome. A similar question

Functional In Vivo and In Vitro Consensus Py Py A N T/A Py Py

Consensus from Sequence Comparisons Py C A N T Py Py

Figure 1. Strong similarities between the functional Inr consensus sequence and the consensus derived from sequence comparisons. The functional consensus was defined by analysis of a large series of Inr mutants in various assays and promoter contexts (Javahery et al. 1994; Lo and Smale 1996). The consensus from sequence comparisons was derived by Bucher (1990) by weight-matrix analysis.

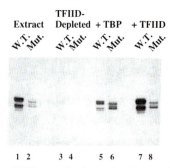

Figure 2. The TFIID complex is required for basal Inr activity. In vitro transcription reactions were performed with TATA-Inr and TATA-Inr mutant promoters in crude nuclear extracts (lanes *1* and *2*), TFIID-depleted extracts (lanes *3* and *4*), or depleted extracts supplemented with TBP (lanes *5* and *6*) or purified TFIID complex (lanes *7* and *8*). (Reprinted, with permission from Kaufmann and Smale 1994.)

can be raised with regard to the TATA box, since most A/T-rich sequences of at least 6 bp can impart TATA activity. The likely explanation is that genomic sequences capable of functioning as TATA or Inr elements will do so only if they are localized appropriately within a promoter. A consensus TATA or consensus Inr at a random location will not be functional because it will not be recognized efficiently by the relevant proteins (i.e., the TFIID complex). Most of the genome is assembled into nucleosomes, which may be inaccessible to TFIID. Before TFIID can bind, it seems likely that specific activator proteins will need to bind to other promoter elements and then recruit the TFIID to the core promoter.

Recognition of the Consensus Inr by TFIID

Several years ago, two experiments yielded results which suggested that the TFIID complex might recognize consensus Inr elements. First, DNase I footprinting experiments using TATA-Inr promoters (e.g., the adenovirus major later promoter [AdMLP]) revealed protection of the TATA box, the Inr, and sequences extending downstream from the start site (Sawadogo and Roeder 1985; Nakajima et al. 1988; Zhou et al. 1992; Purnell and Gilmour 1993). Second, as shown in Figure 2, Inr activity could not be detected in the presence of TATA-box-binding polypeptide (TBP), but the intact TFIID complex restored efficient Inr activity (Smale et al. 1990; Kaufmann and Smale 1994; Martinez et al. 1994; Verrijzer et al. 1995). Despite these findings, early experiments to determine if TFIID binds directly to consensus Inr elements, such as gel-shift and DNase I footprinting experiments with probes containing only a consensus Inr, yielded negative results (J. Kaufmann and S.T. Smale, unpubl.). Furthermore, although TFIID protected the Inr element of the AdMLP in DNase I footprinting experiments, Inr point mutations had no noticeable effect on protection (Chiang et al. 1993; Emami et al. 1997). This result suggested that TFIID interactions at the Inr might not be needed for the Inr protection observed.

Despite the previous uncertainty, our laboratory and four others have now found that TFIID binds directly and specifically to consensus Inr elements. First, in DNase I footprinting with TATA-Inr promoters and purified TFIID, a direct TFIID-Inr interaction has been observed (Fig. 3, TATA-Inr promoter) (Kaufmann and Smale 1994; Emami et al. 1997). Importantly, this interaction was strongly diminished by mutations that diminish Inr function (e.g., Fig. 3, TATA promoter). In fact, a careful mutant analysis revealed that the nucleotides required for the TFIID-Inr interaction correspond precisely with the nucleotides required for Inr function (Kaufmann and Smale 1994; Emami et al. 1997). Highly purified TFIID also bound to Inr elements in the absence of a TATA box, but the interaction was very weak and only detected if the TFIID was recruited to the promoter by an activator bound upstream (Fig. 3, Inr promoter). Similar results were obtained by other laboratories using partially purified TFIID (Wang and Van Dyke 1993; Bellorini et al. 1996). Apparently, the Inr dependence of TFIID binding was not observed in the earlier studies of the AdMLP because strong TFIID interactions downstream from the start site obscured the Inr interaction (see Emami et al. 1997).

A second approach to demonstrate a direct TFIID-Inr interaction was used by Purnell et al. (1994). In footprinting experiments, *Drosophila* TFIID protects the TATA box, Inr, and downstream region of the hsp70 promoter. To determine if a specific start site sequence was required for the interaction, a binding site selection approach was used. The optimal TFIID-binding site that was obtained matched the Inr consensus sequence that we previously defined in our functional assays.

A third set of experiments that has confirmed the direct TFIID-Inr interaction was reported by Burke and Kadonaga (1996). Using purified *Drosophila* TFIID with a series of TATA-less promoters, efficient core promoter binding by TFIID was detected. The stable interaction required the Inr at the start site, plus a sequence element, the downstream promoter element (DPE), located downstream from the start site. Like the TBP interaction at the TATA box, a TBP-associated factor (TAF) interaction at the DPE appears to stabilize the interaction between a component of the TFIID complex and the Inr.

Two other features of TFIID binding to the Inr are noteworthy. First, for synergistic binding of the TFIID complex to TATA and Inr elements, or to Inr and DPE elements, the spacing between the elements cannot be altered (Burke and Kadonaga 1997; Emami et al. 1997). These stringent spacing requirements suggest that the interactions occur in a stereospecific manner, with little flexibility. Second, for efficient recognition of the Inr within a TATA-Inr promoter, TFIIA must be included in the binding reactions (Emami et al. 1997). This result, using an Mg/agarose gel-shift assay, is reproduced in Figure 4. When TFIIA is omitted, TFIID binds to the TATA-Inr and TATA promoters with comparable affinities. In contrast, when TFIIA is added, the affinity of TFIID for the TATA-Inr promoter appears to be substantially higher than for the TATA promoter. Since TFIIA does not itself appear to bind the Inr, the simplest interpretation of this result is that TFIIA induces a conformational

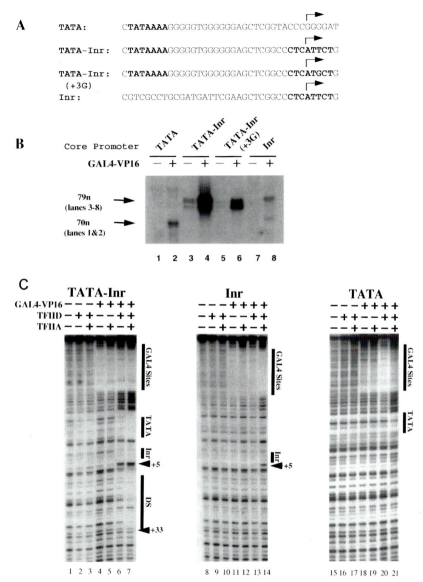

Figure 3. Inr-dependent interactions of TFIID. (*A*) Sequences of the TATA, TATA-Inr, TATA-Inr(+3G), and Inr core promoters. (*B*) In vitro transcription reactions were performed in HeLa nuclear extracts with promoters containing multiple GAL4-binding sites upstream of the four core promoters. Reactions were performed in the absence or presence of GAL4-VP16 as indicated. (*C*) DNase I footprinting experiments were performed with probes containing five GAL4-binding sites upstream of the TATA-Inr (lanes *1–7*), Inr (lanes *8–14*), and TATA (lanes *15–21*) core promoters. Binding reactions contained no protein or various combinations of purified TFIID, purified TFIIA, and GAL4-VP16, as indicated. (Reprinted, with permission, from Emami et al. 1997.)

change in the TFIID complex that allows the TAF-Inr interaction to occur with an affinity that is sufficient to influence the overall affinity of TFIID for the core promoter (Emami et al. 1997). In the absence of TFIIA, TFIID appears to interact weakly with the Inr, but the interaction does not add to the overall affinity of TFIID for the promoter, which is largely dictated by the TBP-TATA interaction (Fig. 4).

The specific subunit of the TFIID complex that binds the Inr remains unknown. At this time, the prime candidate for Inr recognition is TAF250, based on three lines of evidence. First, the TFIID complex used in our laboratory for the TATA-Inr footprinting experiments appears to contain only small quantities of TAF150, suggesting

that this TAF is not responsible for Inr recognition, despite the fact that it has been found to bind DNA (Kaufmann and Smale 1994; Emami et al. 1997; Kaufmann et al. 1998). Second, studies from the Tjian laboratory demonstrated that a trimeric TBP-TAF150-TAF250 complex is sufficient for Inr function (Verrijzer et al. 1995). Because TBP and TAF150 do not appear to bind the Inr, TAF250 is the likely candidate within this complex. Third, UV cross-linking experiments performed in the Roeder laboratory with purified TFIID and the AdMLP revealed strong cross-linking of TAF250 to the Inr element (Oelgeschläger et al. 1996). Although this result merely demonstrates that TAF250 is near the Inr at the time of cross-linking, it is highly suggestive when

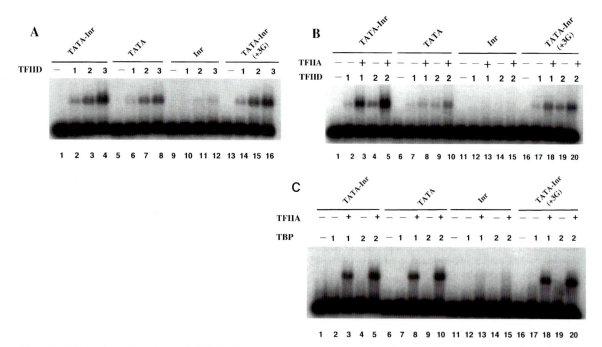

Figure 4. TFIIA selectively enhances TFIID binding to core promoters containing a functional Inr element as well as a TATA box. (*A*) Mg^{++} agarose gel-shift assays were performed with 0 (lanes *1,5,9,13*), 1 (lanes *2,6,10,14*), 2 (lanes *3,7,11,15*), and 3 (lanes *4,8,12,16*) μl of purified TFIID. Radioactively labeled probes were prepared by PCR from plasmids containing the TATA-Inr (lanes *1–4*), TATA (lanes *5–8*), Inr (lanes *9–12*), and TATA-Inr(+3G) (lanes *13–16*) core promoters. (*B*) Mg^{++} agarose gel shifts were performed with the TATA-Inr (lanes *1–5*), TATA (lanes *6–10*), Inr (lanes *11–15*), and TATA-Inr(+3G) (lanes *16–20*) probes. Reactions contained 0 (lanes *1,6,11,16*), 1 (lanes *2,3,7,8,12,13,17,18*), or 2 (lanes *4,5,9,10,14,15,19,20*) μl of purified TFIID and either 0 (lanes *1,2,4,6,7,9,11,12,14,16,17,19*) or 200 (lanes *3,5,8,10,13,15,18,20*) ng of TFIIA. (*C*) Mg^{++} agarose EMSAs were performed as in *B*, but with 1 (lanes *2,3,7,8,12,13,17,18*) or 2 (lanes *4,5,9,10,14,15,19,20*) μl of a 2 ng/μl solution of TBP, and with 10 ng of TFIIA (lanes *3,5,8,10,13,15,18,20*). (Reprinted, with permission, from Emami et al. 1997.)

combined with the previous two results. Furthermore, addition of TFIIA to the cross-linking experiments greatly enhanced TAF250 cross-linking to the Inr, consistent with the enhancement of the overall affinity of the TFIID-Inr interaction observed in our experiments (Emami et al. 1997).

Although the above studies strongly suggest that TFIID recognizes consensus Inr elements, the involvement of other proteins in Inr recognition cannot yet be ruled out. This possibility is highlighted by the observation that the TFIID-Inr interaction is weaker than the TFIID-TATA interaction. TFIID binding to a TATA box can be detected in a gel-shift assay, but binding to a consensus Inr cannot, unless the probe also includes an upstream TATA box (Kaufmann and Smale 1994; Emami et al. 1997). This result suggests that the affinity of TFIID for an Inr may be considerably lower than for a TATA box. Despite this apparent affinity difference, Inr activity and TATA activity are similar both in vitro and in vivo (Smale et al. 1990; Emami et al. 1995). This discrepancy between activity and binding affinity suggests that the TFIID-Inr interaction may be insufficient for Inr activity and that another protein may be needed to stabilize TFIID binding.

A second issue is whether the TFIID complex that binds Inr elements is the same as the complex that binds promoters containing TATA boxes. There is little doubt that the TFIID that binds TATA boxes contains a component that recognizes Inr elements, as the affinity of TFIID for a TATA box is greatly enhanced when an Inr is present at the start site, strongly suggesting that the same complex binds both elements (Kaufmann and Smale 1994; Purnell et al. 1994; Emami et al. 1997). However, a different TFIID complex may interact with Inr-containing promoters that lack a TATA box. Different TFIID complexes have indeed been reported (Parvin and Sharp 1991; Jacq et al. 1994; Dikstein et al. 1996; Hansen et al. 1997; Wieczorek et al. 1998). Moreover, our own studies using purified TFIID do not rule out the possibility that different TFIID complexes are involved, since the purified TFIID preparation may contain multiple distinct complexes.

A final noteworthy issue regarding Inr recognition is how the essential RNA polymerase II–Inr interaction event fits into the mechanism of Inr function. Clearly, at some step during transcription initiation from an Inr-containing promoter, RNA polymerase II must interact with the Inr and then initiate RNA synthesis. Studies by Carcamo et al. (1991) and Weis and Reinberg (1997) have suggested an extreme model regarding this essential event, namely, the RNA polymerase II itself may be responsible for the initial recognition of Inr elements. A sequence-specific interaction between RNA polymerase and Inr elements was first suggested by in vitro transcription experiments containing only purified RNA polymerase II (Carcamo et al. 1991). The polymerase supported a low level of accurate transcription from the Inr. In addition, a gel-shift complex was detected on a probe containing an Inr, but lacking a TATA box, using pure RNA polymerase II, TBP, TFIIB, and TFIIF (Carcamo et al. 1991; Weis and Reinberg 1997).

Although the above data demonstrate that RNA polymerase II can recognize transcription start sites, the step during initiation at which the interaction becomes important remains unknown. RNA polymerase might contribute to the initial recognition of Inr elements in conjunction with TFIID. Alternatively, TFIID alone might be responsible for the initial recognition of Inr elements, with the polymerase-Inr interaction important at a later step of the initiation reaction, after polymerase is recruited to the promoter. Polymerase might in fact displace TFIID from the Inr when it enters the complex. Therefore, the ability of RNA polymerase II to bind to Inr elements in vitro may reflect a recognition event that normally occurs as a late step during preinitiation complex formation.

Mechanism of Initiator Activity

The above studies suggest that subunits of TFIID and RNA polymerase II may be important for Inr recognition. Although the specific subunits that bind the Inr remain to be established, insight into the mechanism of Inr activity has been provided by other biochemical studies. Some experiments performed in our laboratory have involved a comparison of the biochemical parameters of TATA- and Inr-mediated transcription (O'Shea-Greenfield and Smale 1992; Zenzie-Gregory et al. 1992). For example, two defined concentrations of Sarkosyl are known to block specific events during preinitiation complex formation on a TATA-containing promoter (Hawley and Roeder 1985). With an Inr-containing promoter, the Sarkosyl sensitivities were identical to those observed with a TATA-containing promoter, revealing some similarities between the initiation mechanisms with the two types of promoters (Zenzie-Gregory et al. 1992). Analysis of other biochemical parameters revealed more extensive similarities.

One hypothesis which can explain the biochemical similarities is that both TATA and Inr elements recruit TFIID to the core promoter and that the events occurring after recruitment are identical. To test this model, an important issue is whether the TBP subunit of TFIID contacts the –30 region of a TATA-less promoter during transcription initiation. One study performed in our laboratory to address this issue suggested that TBP must indeed contact the –30 region (Zenzie-Gregory et al. 1993). In this study, the activities of several Inr-containing promoters with variable sequences between –25 and –30 were compared, revealing that promoter activity is extremely sensitive to the –25/–30 nucleotide sequence. A comparison of the affinities of TBP for the –25/–30 sequence revealed a direct correlation between TBP affinity and promoter strength, even though most of the sequences bound TBP with an extremely low affinity: Given the low affinity, one would have predicted that the

TBP interaction would have been irrelevant, unless an interaction at the –25/–30 sequence was essential for promoter function. These results suggested that TBP must contact the –25/–30 region during transcription initiation from Inr-containing promoters that lack TATA boxes.

A more recent analysis of this issue by the Berk and Roeder laboratories supports the hypothesis that TBP binding to the –25/–30 region is required for transcription from some TATA-less, Inr-containing promoters, whereas others are independent of TBP binding (Martinez et al. 1995). HeLa cells were stably transfected with an epitope-tagged TBP protein containing a mutation in the DNA-binding domain. As expected, the epitope-tagged TFIID complex purified from this cell line is incapable of binding a TATA box with high affinity. As predicted by our earlier study, this complex was unable to support transcription from some TATA-less, Inr-containing promoters, supporting the hypothesis that TBP must contact the –25/–30 region in the absence of a TATA sequence. Interestingly, the mutant TFIID supported efficient transcription from the TdT promoter, suggesting that TBP binding to the –30 region is not necessary in this instance. These results suggest that some promoters may not require a stable TBP interaction. It is not yet known why the TBP interaction is dispensable on the TdT promoter, but one attractive hypothesis is that TFIID binding is stabilized by an interaction at a critical element that exists downstream from the TdT Inr (Smale and Baltimore 1989).

The studies described above provide insight into the mechanisms of Inr activity, but to study the mechanism in detail, Inr activity must be reconstituted in an in vitro transcription reaction with purified proteins. This goal has not yet been realized, largely because of the difficulty of working with the intact TFIID complex, which is not needed for the reconstitution of TATA activity. Nevertheless, one initial attempt to reconstitute Inr activity has been published by our laboratory (Kaufmann et al. 1996), and another attempt by the Roeder laboratory has been completed, but not yet published (R. Roeder, pers. comm.).

In our study (Kaufmann et al. 1996), TFIID, RNA polymerase II, TFIIB, RAP30, and RAP74 were purified from mammalian cells or Escherichia coli. These proteins, as expected, supported TATA-mediated transcription from supercoiled templates. However, they were insufficient for Inr activity from either a TATA-Inr promoter or an Inr promoter (i.e., lacking a TATA box). The Inr promoter was completely inactive, and the TATA-Inr was no more active than a TATA promoter. These results strongly suggested that additional proteins are needed for Inr activity. A complementation assay was used to identify a factor called CIF (cofactor of Inr function), which restored Inr activity to the TATA-Inr promoter (Fig. 5). In other words, when CIF was added, the TATA-Inr promoter was much stronger than the TATA promoter. Although our results suggested that the CIF fraction contains multiple proteins, purification of one component yielded a unique 150-kD protein (Kaufmann

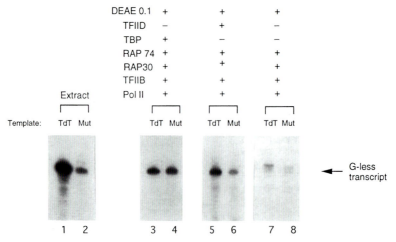

Figure 5. A CIF-containing fraction imparts Inr activity in the presence of the TFIID complex. In vitro transcription assays were performed with RNA polymerase II, TFIIB, RAP30, and RAP74 in combination with a CIF-containing fraction (DEAE 0.1 fraction) and either TBP (lanes *3* and *4*) or immunopurified TFIID (lanes *5* and *6*). Each combination was tested with plasmids containing the TATA box and either the wild-type or mutant (+3C-1A) Inr elements as indicated. Control reactions were performed with HeLa nuclear extract (lanes *1* and *2*) or general factors and the DEAE 0.1 fraction in the absence of TBP and TFIID (lanes *7* and *8*). (Reprinted, with permission, from Kaufmann et al. 1996.)

et al. 1996). By antibody cross-reactivity, peptide sequencing, and cloning, this protein was found to be the human homolog of *Drosophila* TAF150 (Kaufmann et al. 1996, 1998).

TAF150 was first reported by Verrijzer et al. (1994) as an integral component of *Drosophila* TFIID, but a human homolog had not been found in the original studies of human TFIID. The experiments performed by Verrijzer et al. (1994, 1995) suggested that *Drosophila* TAF150 is essential for Inr activity. Furthermore, they showed that recombinant TAF150 binds to core promoters, although the nucleotides needed for TAF150 binding do not appear to match the Inr consensus (Verrijzer et al. 1995; Kaufmann et al. 1996). Our purification of the human TAF150 protein using a complementation assay for Inr function confirms its critical role in mediating Inr activity, although its precise role remains unknown. The DNA-binding activity of TAF150 reported by Verrijzer et al. (1994, 1995) may be important for Inr activity. In addition, Kaufmann et al. (1998) reported that TAF150 stabilizes the binding of TFIID to core promoters. The TFIID stabilization activity is not Inr-dependent, but this activity may nevertheless be relevant for Inr function.

A final issue regarding human TAF150 is its apparent absence from the purified TFIID complex. Although this TAF was undetectable following gel electrophoresis and silver staining of purified TFIID preparations, immunoblot analyses have revealed that it is present in the purified TFIID preparations in small amounts (A. Jain and S.T. Smale, unpubl.). However, greater than 95% of TAF150 in a nuclear extract is not tightly associated with TFIID. The precise stoichiometry of TAF150 in the TFIID preparation remains unknown, but the fact that only a small percentage is present in the purified complex suggests that it is substoichiometric. Nevertheless, given the fact that the purification procedure requires stringent wash conditions, it remains possible, if not likely, that human TAF150 is efficiently associated with the human TFIID complex within a cell.

A working model consistent with all of the data described above is shown in Figure 6. The model depicts a proposed mechanism of Inr activity on a TATA-Inr promoter, rather than an Inr promoter, because most of our major mechanistic advances to date have involved a comparison of a TATA-Inr promoter to a TATA promoter. Our expectation is that the mechanism of Inr activity in the absence of a TATA box will be similar, but perhaps more complicated, than the mechanism of Inr activity in the presence of a TATA box.

In the model shown, a human TFIID complex recognizes both the TATA box and the Inr in a sequence-specific manner. The TBP subunit of TFIID binds the TATA box, and an unknown TAF, most likely TAF250, binds the Inr. An efficient interaction between the TAF and the Inr requires TFIIA, which presumably induces a conformational change in the TFIID complex, exposing the Inr-binding domain of the relevant TAF. TAF150 does not need to be present in the TFIID complex for TATA-Inr binding, but it may be present in the in vivo TFIID complex. Although TAF150 association is not essential for sequence-specific recognition of the Inr, it is needed for Inr function. The DNA-binding or TFIID stabilization activities of TAF150 may contribute to its essential function. In addition to the TAF that binds the Inr and TAF150, other proteins appear to be essential for Inr activity that are not needed for TATA activity. One protein, TIC1, has been purified by the Roeder lab (R. Roeder, pers. comm.) and is essential for Inr activity in the presence of a TATA box, as indicated in the figure. Two other proteins purified by the Roeder laboratory, TIC2 and TIC3, are needed in addition to TIC1 for Inr activity in the absence of a TATA box (R. Roeder, pers. comm.). After TFIID, TAF150, and perhaps TIC1 associate with the promoter, it is reasonable to expect that most or all of the

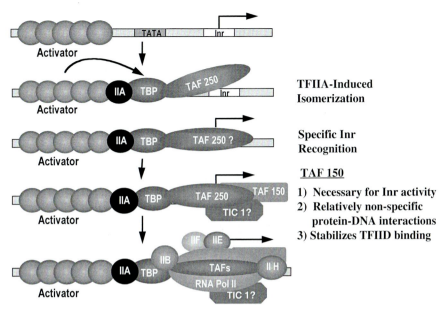

Figure 6. Working model for the mechanism of Inr activity from a synthetic promoter containing multiple activator binding sites, a TATA box, and an Inr. The steps depicted in the schematic model are described in the text.

other general factors required for TATA-mediated transcription will assemble onto the promoter, forming a preinitiation complex that is competent for transcription initiation.

Relevance of Core Promoter Heterogeneity within Eukaryotic Protein-coding Genes

The studies described above have provided valuable insight into the mechanism of Inr activity, but they do not yet answer the questions raised at the beginning of this article: Why do core promoters within eukaryotic protein-coding genes exhibit heterogeneous architectures? Does the heterogeneity contribute to combinatorial gene regulation? These questions are particularly relevant because the functional and mechanistic similarities between TATA- and Inr-mediated transcription suggest, at first glance, that they are analogous to one other. For example, the TFIID complex binds to both elements (see above), and simple activators, such as Sp1 and GAL4-VP16, activate transcription to a similar extent through core promoters containing either element (Smale et al. 1990; Emami et al. 1995). This latter result suggests that the elements may be interchangeable within a core promoter; i.e., the promoter for a cellular gene may be of equivalent strength with either an Inr or a TATA box at the core promoter region.

Despite the extensive similarities between TATA- and Inr-mediated transcription, recent studies have revealed that TATA-less promoters are preferentially responsive to specific activation domains. The first activation domains found to stimulate transcription preferentially through an Inr were the glutamine-rich activation domains of Sp1 (Emami et al. 1995). When these domains were tested as a GAL4 fusion protein for their ability to activate transcription in vivo, they were found to be potent activators of transcription when bound upstream of an Inr, but activation was not detected when the fusion protein bound upstream of a TATA box. In contrast, GAL4-VP16 or a GAL4 fusion protein containing a full-length Sp1 protein stimulated transcription to similar extents from either core promoter (Emami et al. 1995).

The above experiments revealed that a specific activation domain can prefer an Inr element, consistent with the hypotheses proposed at the beginning of this paper regarding possible mechanisms by which core promoter architecture might influence combinatorial gene regulation. However, the relevance of the Sp1 results remains unclear because the full-length Sp1 protein exhibited no Inr preference. Since the artificial, truncated form of Sp1 used in those experiments does not exist in normal cells, the Inr preference observed may have little biological relevance.

A second example of an Inr preference, which is likely to be of greater biological relevance, was provided by an analysis of the natural TdT promoter. To determine why the TdT promoter contains an Inr instead of a TATA box, a consensus TATA box was introduced between –25 and –30, resulting, not surprisingly, in a moderate enhancement of promoter activity (Garraway et al. 1996). However, when the TATA box was inserted and the Inr mutated, TdT promoter activity was undetectable in vitro and in vivo (Garraway et al. 1996). When the TdT promoter region upstream of the core promoter was replaced with multiple Sp1-binding sites, the core promoter containing the TATA insertion and Inr mutation exhibited activity equal to that of the natural core promoter containing the Inr element. These results demonstrate that the TdT promoter contains an Inr because one or more transcription factors that activate the gene only function in the presence of the Inr.

Recent studies suggest that an Ets protein which interacts with an important control element 60 bp upstream of the transcription start site preferentially activates Inr-mediated transcription (P. Ernst and S.T. Smale, unpubl.). Moreover, an element located downstream from the start site exhibits an Inr preference (I.P. Garraway and S.T. Smale, unpubl.), similar to the Inr preference of the DPE described by Burke and Kadonaga (1996).

A detailed analysis will be needed to determine the mechanistic bases of the Inr preferences observed with Sp1 and the TdT promoter. Furthermore, it remains to be determined if most or all natural promoters rely on a specific promoter structure and if transcriptional activators with core promoter preferences possess distinctive features that will allow them to be categorized.

FUTURE PROSPECTS

This paper began with the hypothesis that core promoter heterogeneity might contribute to combinatorial gene regulation. The experiments described above, particularly the final experiments with the natural TdT promoter, strongly suggest that the function and proper regulation of at least some promoters depend on a specific core promoter structure. These findings demonstrate a key prediction of the hypothesis, namely, that the transcription factors which regulate a gene may function only in the context of a particular type of core promoter. To further understand the relevance of core promoter structure to combinatorial regulation, the next step must be an elucidation of the mechanistic basis of the Inr requirement for TdT transcription. After the transcription factors that contribute to the Inr requirement are clearly established, experiments can be designed to determine if the core promoter preference plays an important part in combinatorial regulation. These studies must of course be performed in conjunction with a more detailed analysis of the biochemical mechanisms of Inr-mediated transcription.

The studies described focused almost exclusively on an analysis of Inr elements and a comparison of Inr-mediated transcription to TATA-mediated transcription. Despite this focus, core promoters for RNA polymerase II are undoubtedly much more heterogeneous, with contributions from elements downstream from the start site, between the TATA box and start site, and perhaps other elements overlapping the start site and TATA region. The proteins that contribute to transcription from the heterogeneous core promoters are also likely to be quite diverse and include multiple TFIID-like complexes, subcomplexes, and other factors that are unique to particular types of core promoters. Given these facts, the above description of our studies of Inr elements should not be viewed as a testament to our advanced knowledge of core promoter heterogeneity. Instead, these studies should provide a clear indication of our paucity of knowledge of this important issue.

REFERENCES

Bellorini M., Dantonel J.C., Yoon J.B., Roeder R.G., Tora L., and Mantovani R. 1996. The major histocompatibility complex class II Ea promoter requires TFIID binding to an initiator sequence. *Mol. Cell. Biol.* **16:** 503.

Breathnach R. and Chambon P. 1981. Organization and expression of eukaryotic split genes coding for proteins. *Annu. Rev. Biochem.* **50:** 349.

Bucher P. 1990. Weight matrix descriptions of four eukaryotic RNA polymerase II promoter elements derived from 502 unrelated promoter sequences. *J. Mol. Biol.* **212:** 563.

Burke T.W. and Kadonaga J.T. 1996. *Drosophila* TFIID binds to a conserved downstream basal promoter element that is present in many TATA-box-deficient promoters. *Genes Dev.* **10:** 711.

———. 1997. The downstream core promoter element, DPE, is conserved from *Drosophila* to humans and is recognized by TAFII60 of *Drosophila*. *Genes Dev.* **11:** 3020.

Carcamo J., Buckbinder L., and Reinberg D. 1991. The initiator directs the assembly of a transcription factor IID-dependent transcription complex. *Proc. Natl. Acad. Sci.* **88:** 8052.

Chen W. and Struhl K. 1985. Yeast mRNA initiation sites are determined primarily by specific sequences, not by the distance from the TATA element. *EMBO J.* **4:** 3273.

Chiang C.-M., Ge H., Wang Z., Hoffmann A., and Roeder R.G. 1993. Unique TATA-binding protein containing complexes and cofactors involved in transcription by RNA polymerases II and III. *EMBO J.* **12:** 2749.

Concino M.F., Lee R.F., Merryweather J.P., and Weinmann R. 1984. The adenovirus major late promoter TATA box and initiation site are both necessary for transcription in vitro. *Nucleic Acids Res.* **12:** 7423.

Corden J., Wasylyk B., Buchwalder A., Sassone-Corsi P., Kedinger C., and Chambon P. 1980. Promoter sequences of eukaryotic protein-coding genes. *Science* **209:** 1405.

Dikstein R., Zhou S., and Tjian R. 1996. Human TAFII 105 is a cell type-specific TFIID subunit related to hTAFII130. *Cell* **87:** 137.

Emami K.H., Jain A., and Smale S.T. 1997. Mechanism of synergy between TATA and initiator: synergistic binding of TFIID following a putative TFIIA-induced isomerization. *Genes Dev.* **11:** 3007.

Emami K.H., Navarre W.W., and Smale S.T. 1995. Core promoter specificities of the Sp1 and VP16 transcriptional activation domains. *Mol. Cell. Biol.* **15:** 5906.

Garraway I.P., Semple K., and Smale S.T. 1996. Transcription of the lymphocyte-specific terminal deoxynucleotidyltransferase gene requires a specific core promoter structure. *Proc. Natl. Acad. Sci.* **93:** 4336.

Grosschedl R. and Birnstiel M.L. 1980. Identification of regulatory sequences in the prelude sequences of an H2A histone gene by the study of specific deletion mutants in vitro. *Proc. Natl. Acad. Sci.* **77:** 1432.

Hahn S., Hoar E.T., and Guarente L. 1985. Each of three "TATA elements" specifies a subset of the transcription initiation sites at the CYC-1 promoter of *Saccharomyces cerevisiae*. *Proc. Natl. Acad. Sci.* **82:** 8562.

Hansen S.K., Takada S., Jacobson R.H., Lis J.T., and Tjian R. 1997. Transcription properties of a cell type-specific TATA-binding protein, TRF. *Cell* **91:** 71.

Hawley D.K. and Roeder R.G. 1985. Separation and partial characterization of three functional steps in transcription initiation by human RNA polymerase II. *J. Biol. Chem.* **260:** 8163.

Hen R., Sassone-Corsi P., Corden J., Gaub M.P., and Chambon P. 1982. Sequences upstream of the T-A-T-A box are required in vivo and in vitro for efficient transcription from the adenovirus serotype 2 major late promoter. *Proc. Natl. Acad. Sci.* **79:** 7132.

Hu S.L. and Manley J.L. 1981. DNA sequence required for initiation of transcription in vitro from the major late promoter of adenovirus 2. *Proc. Natl. Acad. Sci.* **78:** 820.

Jacq X., Brou C., Lutz Y., Davidson I., Chambon P., and Tora L. 1994. Human $TAF_{II}30$ is present in a distinct TFIID complex and is required for transcriptional activation by the estrogen receptor. *Cell* **79:** 107.

Javahery R., Khachi A., Lo K., Zenzie-Gregory B., and Smale S.T. 1994. DNA sequence requirements for transcriptional initiator activity in mammalian cells. *Mol. Cell. Biol.* **14:**116.

Kaufmann J. and Smale S.T. 1994. Direct recognition of initiator elements by a component of the transcription factor IID complex. *Genes Dev.* **8:** 821.

Kaufmann J., Verrijzer C.P., Shao J., and Smale S.T. 1996. CIF, an essential cofactor for TFIID-dependent initiator function. *Genes Dev.* **10:** 873.

Kaufmann J., Ahrens K., Koop R., Smale S.T., and Müller R. 1998. CIF150, a human cofactor for transcription factor IID-dependent initiator function. *Mol. Cell. Biol.* **18:** 233.

Lo K. and Smale S.T. 1996. Generality of a functional initiator consensus sequence. *Gene* **182:** 13.

Martinez E., Chiang C.M., Ge H., and Roeder R.G. 1994. TAFs in TFIID function through the initiator to direct basal transcription from a TATA-less class II promoter. *EMBO J.* **13:** 3115.

Martinez E., Zhou Q., L'Etoile N.D., Oelgeschlager T., Berk A.J., and Roeder R.G. 1995. Core promoter-specific function of a mutant transcription factor TFIID defective in TATA-box binding. *Proc. Natl. Acad. Sci.* **92:** 11864.

McNeil J.B. and Smith M. 1985. *Saccharomyces cerevisiae* CYC1 mRNA 5′-end positioning: Analysis by in vitro mutagenesis, using synthetic duplexes with random mismatch base pairs. *Mol. Cell. Biol.* **5:** 3545.

Nagawa F. and Fink G.R. 1985. The relationship between the TATA sequence and transcription initiation sites at the HIS4 gene of *Saccharomyces cerevisiae*. *Proc. Natl. Acad. Sci.* **82:** 8557.

Nakajima N., Horikoshi M., and Roeder R.G. 1988. Factors involved in specific transcription by mammalian RNA polymerase II: Purification, genetic specificity, and TATA box-promoter interactions of TFIID. *Mol. Cell. Biol.* **8:** 4028.

Oelgeschläger T., Chiang C.-M., and Roeder R.G. 1996. Topology and reorganization of a human TFIID-promoter complex. *Nature* **382:** 735.

O'Shea-Greenfield A. and Smale S.T. 1992. Roles of TATA and initiator elements in determining the start site location and direction of RNA polymerase II transcription. *J. Biol. Chem.* **267:** 1391.

Parvin J.D. and Sharp P.A. 1991. Identification of novel factors which bind specifically to the core promoter of the immunoglobulin heavy chain gene. *J. Biol. Chem.* **266:** 22878.

Purnell B.A., and Gilmour D.S. 1993. Contributions of sequences downstream of the TATA element to a protein-DNA complex containing the TATA-binding protein. *Mol. Cell. Biol.* **13:** 2593.

Purnell B.A., Emanuel P.A., and Gilmour D.S. 1994. TFIID sequence recognition of the initiator and sequence farther downstream in *Drosophila* class II genes. *Genes Dev.* **8:** 830.

Sawadogo M. and Roeder R.G. 1985. Interaction of a gene-specific transcription factor with the adenovirus major late promoter of the TATA box region. *Cell* **43:** 165.

Smale S.T. 1994. Core promoter architecture for eukaryotic protein-coding genes. In *Transcription: Mechanisms and regulation* (ed. C. Conaway and J.W. Conaway), p. 63. Raven Press, New York.

———. 1997. Transcription initiation from TATA-less promoters within eukaryotic protein-coding genes. *Biochim. Biophys. Acta* **1351:** 73.

Smale S.T. and Baltimore D. 1989. The "initiator" as a transcription control element. *Cell* **57:** 103.

Smale S.T., Schmidt M.C., Berk A.J., and Baltimore D. 1990. Transcriptional activation by Sp1 as directed through TATA and initiator: Specific requirement for mammalian transcription factors. *Proc. Natl. Acad. Sci.* **87:** 4509.

Verrijzer C. P., Chen J.-L., Yokomori K., and Tjian R. 1995. Binding of TAFs to core elements directs promoter selectivity by RNA polymerase II. *Cell* **81:** 1115.

Verrijzer C.P., Yokomori K., Chen J.-L., and Tjian R. 1994. *Drosophila* $TAF_{II}150$: Similarity to yeast gene TSM-1 and specific binding to core promoter DNA. *Science* **264:** 933.

Wang J.C. and Van Dyke M.W. 1993. Initiator sequences direct downstream promoter binding by human transcription factor IID. *Biochim. Biophys. Acta* **1216:** 73.

Wasylyk B., Derbyshire R., Guy A., Molko D., Roget A., Teoule R., and Chambon P. 1980. Specific in vitro transcription of conalbumin gene is drastically decreased by single-point mutation in TATA box homology sequence. *Proc. Natl. Acad. Sci.* **77:** 7024.

Weis L. and Reinberg D. 1997. Accurate positioning of RNA polymerase II on a natural TATA-less promoter is independent of TATA-binding protein-associated factors and initiator-binding proteins. *Mol. Cell. Biol.* **17:** 2973.

Wieczorek E., Brand M., Jacq X., and Tora L. 1998. Function of TAF(II)-containing complex without TBP in transcription by RNA polymerase II. *Nature* **393:** 187.

Zenzie-Gregory B., O'Shea-Greenfield A., and Smale S.T. 1992. Similar mechanisms for transcription initiation mediated through a TATA box or an initiator element. *J. Biol. Chem.* **267:** 2823.

Zenzie-Gregory B., Khachi A., Garraway I.P., and Smale S.T. 1993. Mechanism of initiator-mediated transcription: Evidence for a functional interaction between the TATA-binding protein and DNA in the absence of a specific recognition sequence. *Mol. Cell. Biol.* **13:** 3841.

Zhou Q., Liebermann P.M., Boyer T.G., and Berk A.J. 1992. Holo-TFIID supports transcriptional stimulation by diverse activators and from TATA-less promoters. *Genes Dev.* **6:** 1964.

X-ray Crystallographic Studies of Eukaryotic Transcription Factors

S.K. BURLEY

Laboratories of Molecular Biophysics and Howard Hughes Medical Institute, The Rockefeller University, New York, New York 10021

Eukaryotes have three distinct RNA polymerases (forms I, II, and III) that catalyze transcription of nuclear genes (Sentenac 1985). Despite their structural complexity, these multisubunit enzymes require sets of auxiliary proteins known as general transcription initiation factors to initiate transcription from corresponding class I, II, and III nuclear gene promoters (Gabrielson and Sentenac 1991; Roeder 1991; Reeder 1992; Maldonado and Reinberg 1995). TATA-box-binding protein (TBP), first identified as a component of the class II initiation factor TFIID, participates in transcription by all three nuclear RNA polymerases (for review, see Nikolov and Burley 1994). Thus,

TBP is the first universal transcription initiation factor component (a situation formally analogous to that of essential subunits common to the three RNA polymerases).

The role of TBP in transcription initiation and its regulation are best understood for genes transcribed by RNA polymerase II (pol II) (for review, see Roeder 1991; Maldonado and Reinberg 1995; shown schematically in Fig. 1). In this setting, TBP is tightly associated with other polypeptides known as TBP-associated factors or TAFs (for review, see Burley and Roeder 1996). This multiprotein complex (TFIID) is a general initiation factor (Matsui et al. 1980) that binds to the TATA element, coordinating

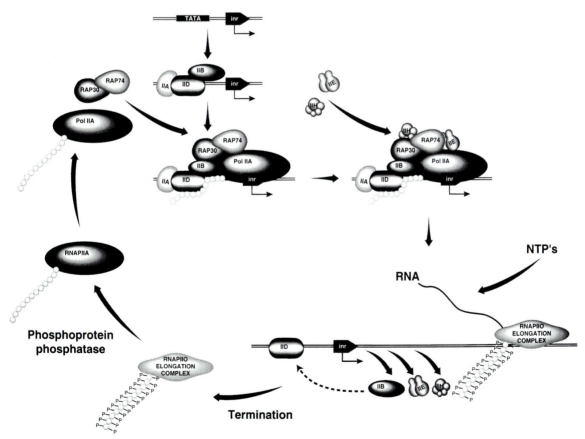

Figure 1. Preinitiation complex assembly begins with TFIID recognizing the TATA element, followed by coordinated accretion of TFIIA, TFIIB, the nonphosphorylated form of pol II and TFIIF (RAP30/RAP74), TFIIE, and TFIIH. Prior to elongation, pol II is phosphorylated by TFIIH. Following termination, a phosphatase recycles pol II to its nonphosphorylated form, allowing the enzyme to reinitiate transcription. TBP (and TFIID) binding to the TATA box is an intrinsically slow step, yielding a long-lived protein-DNA complex. Efficient reinitiation of transcription can be achieved if recycled pol II reenters the preinitiation complex before TFIID dissociates from the core promoter. (Adapted from Zawel et al. 1995.)

accretion of class II initiation factors (TFIIB, -D, -E, -F, -H) and pol II into a functional preinitiation complex (PIC) (for review, see Roeder 1991; Zawel and Reinberg 1993). Although incapable of mimicking TFIID in vivo (at least in higher eukaryotes; for review, see Roeder 1991), recombinant TBP alone is competent for PIC assembly and basal or core promoter-dependent transcription in the presence of the other general class II factors (Buratowski et al. 1989). TBP engages in physical and functional interactions with the general initiation factors TFIIA and TFIIB, the carboxyl terminus of the large subunit of pol II, some negative cofactors (NC1, NC2, DR1) that inhibit PIC formation, some transcriptional activators, and an initiator-binding factor (TFII-I) that may be important for transcription initiation from TATA-less promoters. TFIIB is the second general transcription factor to enter the PIC, creating a TFIIB-TFIID(TBP)-DNA platform that is in turn recognized by a complex of pol II and TFIIF (pol/F). In vitro studies with negatively supercoiled templates demonstrated that transcription initiation can be reconstituted with TBP, TFIIB, and pol II, suggesting that together TBP and TFIIB position pol II (Parvin and Sharp 1993). Mutants of TFIIB alter pol II start sites in yeast, providing compelling evidence for its function as a precise spacer/bridge between TBP and pol II on the core promoter that determines the transcription start site.

In vivo and under different conditions in vitro, pol II transcription initiation depends on TFIIE, TFIIF, and TFIIH, and possibly TFIIA. Once PIC assembly is complete and in the presence of nucleoside triphosphates, strand separation at the transcription start site occurs to give an open complex, the carboxy-terminal domain of the large subunit of pol II is phosphorylated, and pol II initiates transcription and is released from the promoter. During elongation in vitro, TFIID can remain bound to the core promoter supporting rapid reinitiation of transcription by pol II and the other general factors (Fig. 1) (for review, see Zawel et al. 1995). Core promoter binding by the TBP subunit of TFIID is an intrinsically slow step because of the dramatic DNA deformation induced in the TATA element (for review, see Kim and Burley 1994). An abbreviated PIC assembly mechanism has also been proposed, following recent discoveries of various pol II holoenzymes containing many if not all of the general initiation factors except for TFIID (for review, see Koleske and Young 1995).

Reconstitution of the pol II preinitiation complex in vitro has proved remarkably successful for mechanistic studies of basal transcription initiation. However, pol-II-mediated transcription is considerably more complex in vivo. A large number of other transcription factors, both cellular and viral in origin, regulate the precise level of mRNA production from class II nuclear gene promoters (for review, see Hori and Carey 1994). These proteins are often referred to as transcriptional activators. They modulate transcription by recognizing promoter proximal and/or distal enhancer DNA targets and participating in highly specific protein-protein interactions with components of the PIC and with each other. Efficiency of RNA production from pol II promoters depends, at least in part,

on the half-life of the promoter-specific transcription complex, and much effort is now being devoted to establishing good in vitro models of activator-dependent transcription initiation.

Studies of the mechanisms of action of TBP in nuclear gene transcription by RNA polymerases I (pol I) and III (pol III) are also well-advanced (for review, see Reeder 1992; Hernandez 1993). A defined TBP-TAF complex, known as SL1 (selectivity factor 1), has been implicated in pol I transcription, and its three TAFs are believed to be distinct from the pol II TAFs found in TFIID (Comai et al. 1992). TFIIIB is the pol-III-specific TBP-TAF complex consisting of at least two TAFs, one of which is similar to TFIIB (for review, see Wang and Roeder 1995).

My laboratory has been studying some of the mechanistic aspects of eukaryotic transcription initiation using X-ray crystallography and other biophysical methods. Our work has yielded structures of TBP, complexes of two TBPs with the TATA element of the adenovirus major late promoter, and a ternary complex of TFIIB recognizing a preformed TBP-DNA complex. Most recently, we described the structure of a complex of two pol-II-specific TAFs that resemble the histone H3/H4 heterotetramer.

TATA-BOX-BINDING PROTEIN

Apo-TBP: A Quasisymmetric Molecular Saddle

In 1992, we reported the structure of TBP isoform 2 (TBP2) from *Arabidopsis thaliana* at 2.6 Å resolution (Nikolov et al. 1992). Further progress on crystallographic studies of uncomplexed TBPs includes additional refinement of TBP2 at 2.1 Å resolution (Nikolov and Burley 1994), and a molecular replacement structure of the carboxy-terminal 180 residues of yeast TBP (Chasman et al. 1993). The structure of TBP2 determined at 2.1 Å resolution is illustrated in Figure 2. Both apo-TBP structures are very similar, with two α/β structural domains of 89–90 amino acids related by approximate intramolecular twofold symmetry. TBP2 has a relatively flexible 18-amino-acid amino-terminal segment. The carboxy-terminal or core region of TBP binds to the TATA consensus sequence (TATAa/tAa/t) with high affinity and slow off rate, recognizing minor groove determinants and promoting DNA bending. The amino-terminal portion of TBP varies in length, shows little or no conservation among different organisms, and is largely unnecessary for transcription in certain yeast strains.

TBP resembles a molecular saddle with approximate maximal dimensions 32 Å x 45 Å x 60 Å. DNA binding is supported by the concave underside of the saddle, which is lined by the central eight strands of the ten-stranded antiparallel β-sheet. The convex upper surface of TBP2 is composed of the four α helices, the basic peptide linking the two domains, parts of strands S1 and S1´, and the nonconserved 18 amino-terminal residues. This extensive upper surface binds various components of the transcription machinery (for review, see Nikolov and Burley 1994). Each domain or structural repeat comprises approximately half of the phylogenetically conserved

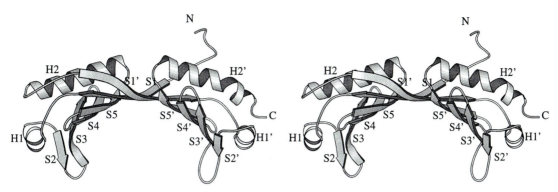

Figure 2. MOLSCRIPT (Kraulis 1991) stereodrawing of the structure of TBP2 viewed perpendicular to the internal pseudodyad axis. The amino and carboxyl termini of the protein are indicated. The α helices are shown as ribbon spirals and labeled H, the β strands are shown as ribbon arrows (S), and loops and turns are drawn as double lines. The symbol ´ refers to the second structural domain or repeat. (Adapted from Nikolov et al. 1992.)

carboxyl terminus of TBP, consisting of a five-stranded, curved antiparallel β-sheet and two α helices. The two helices, lying approximately perpendicular to each other, abut the convex side of the sheet forming the hydrophobic core of each domain. The two structural domains of TBP2 are topologically identical with root mean square (rms) deviation between equivalent α-carbon atomic positions = 1.1 Å, corresponding to the two imperfect repeats in amino acid sequence (30% identical at the amino acid level and 50% identical at the nucleotide level). The ancestor of TBP may therefore have functioned as a dimer, with gene duplication and fusion giving rise to a monomeric, quasisymmetric TBP (for review, see Nikolov and Burley 1994).

The two crystal forms of apo-TBP each have two copies of TBP in the asymmetric unit. For TBP2, this appears to result from weak molecular self-association (buried surface area = 1,700 Å^2 and measured K_d = 1 μM; D.B. Nikolov and S.K. Burley, unpubl.), which can be disrupted by dilution or addition of duplex oligonucleotides bearing a TATA element (Nikolov et al. 1992). There is also a report of human TBP and TFIID forming dimers at physiologic intranuclear concentration (Coleman et al. 1995).

TBP-DNA: Minor Groove Recognition and DNA Bending

Structures of TBP2 complexed with the adenovirus major late promoter (AdMLP) TATA element (TATAAAAG) (J.L. Kim et al. 1993; Kim and Burley 1994), the carboxyl terminus of yeast TBP complexed with the yeast CYC1–52 TATA element (TATATAAA) (Y. Kim et al. 1993), and the carboxyl terminus of human TBP complexed with the AdMLP TATA element (Nikolov et al. 1996) have been reported (Fig. 3). Although the three cocrystal structures differ slightly in detail, they all demonstrate an induced-fit mechanism of protein-DNA recognition. DNA binding is mediated by the protein's curved, eight-stranded, antiparallel β-sheet, which provides a large concave surface for minor groove and phosphate-ribose contacts with the 8-bp TATA element. The 5´ end of standard B-form DNA enters the underside of the molecular saddle, where the carboxy-

terminal portion of TBP produces an abrupt transition to an unprecedented, partially unwound form of the right-handed double helix that is induced by insertion of two phenylalanine residues into the first T:A base step. Thereafter, the widened minor groove face of the unwound, smoothly bent DNA is approximated to the underside of the molecular saddle, burying a total surface area of about 3100 Å^2, permitting direct interactions between protein side chains and the minor groove edges of the central 6 bp. A second large kink is induced by insertion of two phenylalanine residues in the base step between the last 2 bp of the TATA element,

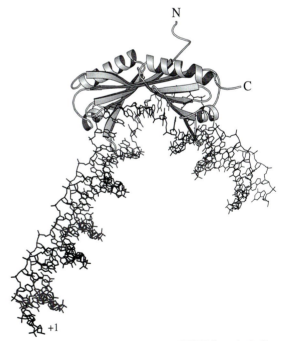

Figure 3. Three-dimensional structure of TBP2 from *A. thaliana* complexed with the AdMLP TATA element. The molecular saddle (amino- and carboxy-terminally labeled) is depicted with a ribbon drawing, and the DNA is shown as a stick figure with the transcription start site labeled +1. When TBP recognizes the minor groove of the TATA element, the DNA is kinked and unwound to present the minor groove edges of the bases to the underside of the molecular saddle. The coding strand is denoted with solid bonds. (Adapted from J.L. Kim et al. 1993.)

and there is a corresponding abrupt return to B-form DNA. Despite this massive distortion, Watson-Crick base pairing is preserved throughout, and there appears to be no strain induced in the DNA, because partial unwinding has been compensated for by right-handed supercoiling of the double helix. Side chain–base contacts are restricted to the minor groove, including the four phenylalanines described above plus five hydrogen bonds and a large number of van der Waals contacts. There are no water molecules mediating side chain–base interactions, and the majority of the hydrogen bond donors and acceptors on the minor groove edges of the bases remain unsatisfied (13/17 in the AdMLP TATA box). Detailed analysis of the TBP2-DNA cocrystal structure at 1.9 Å resolution demonstrates that the protein also undergoes a modest conformational change on DNA binding, involving a twisting motion of one domain with respect to the other (Kim and Burley 1994).

Other biophysical methods have been used to study interactions between TBP and DNA. Site selection experiments with *Acanthamoeba* TBP showed a marked preference for a site very similar to those studied crystallographically (TATATAAG) (Wong and Bateman 1994). DNA bending by TBP in solution was confirmed using circular permutation assays (Starr et al. 1995). TBP binding was also shown to be enhanced by the prebending of DNA toward the major groove (Parvin et al. 1995). TBP-DNA association kinetics have been studied by various techniques (Hoopes et al. 1992; Coleman and Pugh 1995; Perez-Howard et al. 1995; Parkhurst et al. 1996), and three of the four studies gave results consistent with simultaneous binding and bending with a single second-order rate constant of about 10^5 $M^{-1}s^{-1}$. Coleman and Pugh (1995) opted for a dramatically different model, involving dissociation of a tight human TBP dimer, tight nonspecific DNA binding by TBP, and sliding of TBP on DNA. In addition, a novel chemical probe was used to demonstrate that core promoter distortion transiently extends beyond the confines of the TATA box during TBP binding (Sun and Hurley 1995).

TFIIB-TBP-DNA: Recognition of a TBP-DNA Complex

The crystal structure of a TFIIB-TBP-TATA element ternary complex has been determined at 2.7 Å resolution (Nikolov et al. 1995). Core TFIIB (cTFIIB) is a two-domain α-helical protein that resembles cyclin A (Jeffrey et al. 1995). The ternary complex is formed by cTFIIB clamping the acidic carboxy-terminal stirrup of TBP2 (S2´-S3´) in its cleft, interacting with H1´, the carboxyl terminus, and the phosphoribose backbone upstream and downstream from the center of the TATA element (Fig. 4). Although the two domains of cTFIIB have the same fold, they do not have chemically identical surfaces and cannot make equivalent interactions with TBP2. Contacts between cTFIIB and the carboxy-terminal stirrup of TBP2 are made by BH3, BH4, and BH5. The interdomain peptide interacts with H1´ and the carboxy-terminal stirrup of TBP2. cTFIIB's BH2´-BH3´ loop interacts with the same

stirrup and the carboxyl terminus of TBP2. Despite the very extensive intermolecular contacts visualized in the ternary complex structure (total buried surface area ~5600 Å²), the structure of the TBP2-TATA element complex itself is essentially unchanged. cTFIIB recognizes the preformed TBP-DNA complex, including the path of the phosphoribose backbone created by the unprecedented DNA deformation induced by binding of TBP. In addition to stabilizing the TBP-DNA complex, TFIIB binding contributes to the polarity of TATA element recognition. If TBP were to bind to the quasisymmetric TATA box in the wrong orientation (i.e., the amino-terminal half of the molecular saddle interacts with the 5´ end of the TATA element), the basic/hydrophobic surface of the amino-terminal stirrup (S2-S3) would make unfavorable electrostatic interactions with the basic cleft of TFIIB. It is remarkable that the nuclear magnetic resonance (NMR) structure of cTFIIB displays a slightly different arrangement of the two domains (Bagby et al. 1995), suggesting that cTFIIB undergoes a modest conformational change on recognizing the TBP-DNA complex. The first domain of cTFIIB forms the downstream surface of the cTFIIB-TBP-DNA ternary complex where, together with the putative Zn++-binding, amino-terminal domain of full-length TFIIB, it could readily act as a bridge between TBP and pol II, fixing the transcription start site. The remaining solvent-accessible surfaces of TBP (~7900 Å²) and the TFIIB (~8300 Å²) are very extensive, providing for an ample number of recognition sites for binding of TAFs, other class II initiation factors, and transcriptional activators and coactivators.

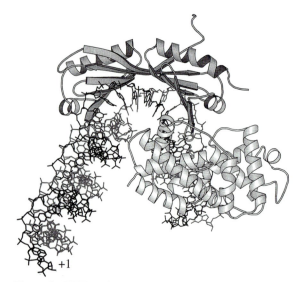

Figure 4. cTFIIB and TBP interacting with the AdMLP. Three-dimensional structure of the ternary complex of human cTFIIB recognizing TBP2 from *A. thaliana* complexed with the AdMLP TATA element. CTFIIB (*light gray*) and TBP (*dark gray*) are depicted as shaded ribbons, and the DNA is shown as a stick figure with the transcription start site labeled +1. The coding strand is denoted with solid bonds. The view is identical to that shown in Fig. 3. When cTFIIB recognizes the TBP-DNA complex, there is essentially no change in trajectory of the negatively charged phosphoribose backbone. (Adapted from Nikolov et al. 1995.)

HISTONE-LIKE TBP-ASSOCIATED FACTORS

Primary structure analyses of some of the pol-II-specific TAFs (TAF$_{II}$s) have revealed considerable amino acid sequence identity with nonlinker histone proteins (Kokubo et al. 1994; Baxevanis et al. 1995; Hisatake et al. 1995; Mengus et al. 1995; Hoffmann et al. 1996). In

Drosophila, dTAF$_{II}$42 and dTAF$_{II}$62 appear to be H3 and H4 homologs, respectively, corresponding to hTAF$_{II}$31 and hTAF$_{II}$80 in humans. Both *Drosophila* and human TFIID also contain putative histone H2B homologs (dTAF$_{II}$30α/22 and hTAF$_{II}$20/15), but they appear to lack histone H2A homologs. A direct connection between components of the eukaryotic transcription apparatus and

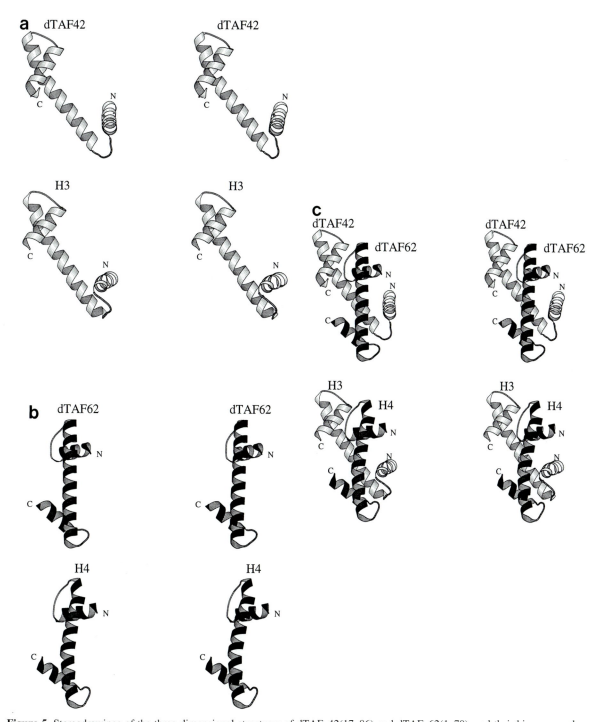

Figure 5. Stereodrawings of the three-dimensional structures of dTAF$_{II}$42(17–86) and dTAF$_{II}$62(1–70), and their binary complex dTAF$_{II}$42(17–86)/dTAF$_{II}$62(1–70). The corresponding views of histones H3 and H4 have been included for comparison (Arents et al. 1991). (*a*) dTAF$_{II}$42(17–86) (*left*) and H3 (*right*). The additional amino-terminal helix of H3 visualized in the structure of the histone octamer core has been omitted for clarity. (*b*) dTAF$_{II}$62(1–70) (*left*) and H4 (*right*). (*c*) dTAF$_{II}$42(17–86)/dTAF$_{II}$62(1–70) (*left*) and H3/H4 (*right*). (Adapted from Xie et al. 1996.)

of the machinery of DNA packaging has already been demonstrated for the linker histones. The cocrystal structure of the DNA-binding domain of the liver-specific transcription factor HNF3-γ (Clark et al. 1993) is virtually identical to the structure of the chicken erythrocyte linker histone H5 obtained without DNA (Ramakrishnan et al. 1993). Moreover, HNF-3α, a related factor, stabilizes a precisely positioned nucleosomal array in the liver-specific enhancer of the mouse albumin gene, where it may function as a sequence-specific linker histone (McPherson et al. 1993).

dTAF$_{II}$42/dTAF$_{II}$62 Heterodimer

The cocrystal structure of a complex of two *Drosophila melanogaster* TAFIIs (dTAF$_{II}$42/dTAF$_{II}$62) has been determined at 2.0 Å resolution (Xie et al. 1996). dTAF$_{II}$42(17–86) and dTAF$_{II}$62(1–70) are illustrated in Figure 5 with their respective histone homologs. Both dTAF42$_{II}$(17–86) and dTAF62$_{II}$(1–70) are folded into a classical histone core protein motif, consisting of a long central α-helix flanked on each side by a random coil segment and a short α-helix (Fig. 5). Truncation of dTAF$_{II}$42 for crystallization removed H3's additional amino-terminal α-helix, which is present in the histone octamer core structure (Arents et al. 1991). The rms deviations between α-carbon atomic positions for dTAF$_{II}$42(22–83) and H3(68–130) and for dTAF$_{II}$62(9–70) and H4(31–93) are 1.6 Å and 1.6 Å, respectively. These values compare favorably with those obtained by comparing individual

histone proteins with one another (Arents and Moudrianakis 1995), reflecting differences in the trajectory of the long α-helix. In H3, this helix is nearly straight, whereas it is somewhat kinked in dTAF$_{II}$42(17–86) near its carboxyl terminus. The converse is true for H4 and dTAF$_{II}$62(1–70). H3 and H4 demonstrate a single cooperative unfolding transition (Karantza et al. 1996), and the ternary structures of dTAF$_{II}$42(17–86) and dTAF$_{II}$62(1–70) are almost certainly not folded in the absence of one another (there are only a small number of intramolecular polar and nonpolar contacts between segments of each polypeptide chain).

Figure 5 illustrates the structure of the dTAF$_{II}$42 (17–86)/ dTAF$_{II}$62(1–70) heterodimer. As in the H3/H4 heterodimer, also depicted in Figure 5, the two polypeptide chains adopt the histone-fold and interact with one another in a head-to-tail fashion (Arents and Moudrianakis 1995). Stabilizing contacts between dTAF$_{II}$42 (17–86) and dTAF$_{II}$62(1–70) are largely hydrophobic, span the entire length of both molecules, and are conserved with H3 and H4. Binary complex formation buries about 3390 Å^2 of solvent accessible surface area (56% of the buried surface is hydrophobic, with the remainder either polar or charged).

dTAF$_{II}$42/dTAF$_{II}$62 Heterotetramer

The structure of the dTAF$_{II}$42(17–86)/dTAF$_{II}$62(1–70) heterotetramer is depicted in Figure 6. Like the histone core octamer structure (Arents et al. 1991; Luger et al.

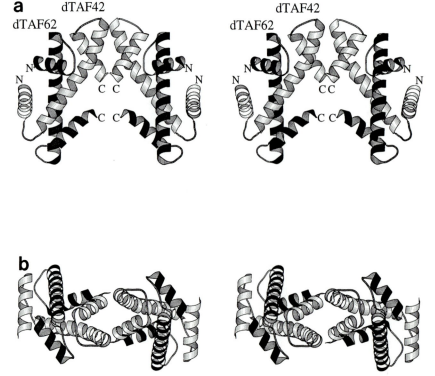

Figure 6. Stereodrawing ribbon representations of the dTAF$_{II}$42(17–86)/dTAF$_{II}$62(1–70) heterotetramer, generated by twofold crystallographic symmetry. (*a*) View perpendicular to the twofold symmetry axis; (*b*) view along the twofold symmetry axis. (Adapted from Xie et al. 1996.)

1997), the symmetry axis within the TAF$_{II}$ tetramer coincides with a crystallographic twofold. Interactions between α helices of the H3 homolog, dTAF$_{II}$42(17–86), stabilize the tetramer, burying about 670 Å^2 of solvent accessible surface area (48% of the buried surface is hydrophobic, with the remainder either polar or charged). These values are typical for biologically-productive protein-protein molecular recognition events (for review, see Janin 1995) and are entirely consistent with the measured equilibrium dissociation constant of 10^{-6} M (Xie et al. 1996). Analysis of the dTAF$_{II}$42(17–86)/dTAF$_{II}$62(1–70) heterotetramer reveals a configuration of surface accessible residues similar to that found in the H3/H4 heterotetramer (for review, see Klug et al. 1980; Arents and Moudrianakis 1993; Pruss et al. 1995; Luger et al. 1997), suggesting that it may be capable of interacting with DNA.

Histone-like Octamer in TFIID

Our crystallographic study of the dTAF$_{II}$42(17–86)/dTAF$_{II}$62(1–70) complex suggests that TFIID contains a (dTAF$_{II}$42/dTAF$_{II}$62)$_2$ heterotetramer. Compelling, albeit indirect, support for this assertion comes from the results of recent studies of the human TAF$_{II}$ homolog of histone H2B. The measured hTAF$_{II}$20:TBP ratio in TFIID is 4:1, and a histone-like pattern of protein-protein interactions has been demonstrated for hTAF$_{II}$31, hTAF$_{II}$80, and hTAF$_{II}$20 (Hoffmann et al. 1996). Thus, TFIID may contain a TAFII substructure that resembles the histone octamer and mediates some of TFIID's nonspecific interactions with DNA.

CONCLUSIONS AND PERSPECTIVES

X-ray crystallographic studies of apo-TBP and its complexes with two TATA elements have revealed a new quasisymmetric protein fold, an unprecedented protein-induced DNA deformation, and a novel induced-fit mechanism of DNA recognition via contacts with the minor groove. This work set the stage for the first structure determination of a protein recognizing a preformed protein-DNA complex the TFIIB-TBP-TATA element ternary complex. These cocrystal structures provided direct views of two critical steps early in the assembly of the preinitiation complex required for correct initiation of transcription by pol II. They have contributed significantly to our understanding of the precise biochemical mechanisms responsible for controlling mRNA production in eukaryotes and continue to serve as a structural foundation from which to plan and interpret studies of class II nuclear gene expression. Finally, with the results of X-ray and biochemical studies of the histone-like TAF$_{II}$s, we have documented a structural connection between DNA packaging and transcription that may be functionally relevant.

ACKNOWLEDGMENTS

I am grateful to J.L. Kim, D.B. Nikolov, and X. Xie for providing illustrations for this review.

REFERENCES

Arents G. and Moudrianakis E. 1993. Topography of the histone octamer surface: Repeating structural motifs utilized in the docking of nucleosomal DNA. *Proc. Natl. Acad. Sci.* **90:** 10489.

———. 1995. The histone fold: A ubiquitous architectural motif utilized in DNA compaction and protein dimerization. *Proc. Natl. Acad. Sci.* **92:** 11170.

Arents G., Burlingame R.W., Wang B.-C., Love W.E., and Moudrianakis E.N. 1991. The nucleosomal core histone octamer at 3.1Å resolution: A tripartite protein assembly and a left-handed superhelix. *Proc. Natl. Acad. Sci.* **88:** 10148.

Bagby S., Kim S., Maldonado E., Tong K., Reinberg D., and Ikura M. 1995. Solution structure of the C-terminal core domain of human TFIIB: Similarity to cyclin A and interaction with TATA-binding protein. *Cell* **82:** 857.

Baxevanis A., Arents G., Moudrianakis E., and Landsman D. 1995. A variety of DNA-binding and multimeric proteins contain the histone fold motif. *Nucleic Acids Res.* **23:** 2685.

Buratowski S., Hahn S., Guarente L., and Sharp P.A. 1989. Five intermediae complexes in transcription initiation by RNA polymerase II. *Cell* **56:** 549.

Burley S.K. and Roeder R.G. 1996. Biochemistry and structural biology of transcription factor IID. *Annu. Rev. Biochem.* **65:** 769.

Chasman D., Flaherty K., Sharp P.A., and Kornberg R. 1993. Crystal structure of yeast TATA-binding protein and a model for interaction with DNA. *Proc. Natl. Acad. Sci.* **90:** 8174.

Clark K.L., Halay E.D., Lai E., and Burley S.K. 1993. Co-crystal structure of the HNF-3/fork head DNA-recognition motif resembles histone H5. *Nature* **364:** 412.

Coleman R. and Pugh B.F. 1995. Evidence for functional binding and stable sliding of the TATA binding protein on nonspecific DNA. *J. Biol. Chem.* **270:** 13850.

Coleman R., Taggart A., Benjamin L., and Pugh B. 1995. Dimerization of TATA binding protein. *J. Biol. Chem.* **270:** 13842.

Comai L., Tanese N., and Tjian R. 1992. The TATA-binding protein and associated factors are integral components of RNA polymerase I transcription factor, SL1. *Cell* **68:** 965.

Gabrielson O. and Sentenac A. 1991. RNA polymerase III (C) and its transcription factors. *Trends Biochem. Sci.* **16:** 412.

Hernandez N. 1993. TBP, a universal transcription factor? *Genes Dev.* **7:** 1291.

Hisatake K., Ohta T., Takada R., Guermah M., Horikoshi M., Nakatani Y., and Roeder R.G. 1995. Evolutionary conservation of human TBP-associated factors TAF31 and TAF80 and interactions of TAF80 with other TAFs and with general transcription factors. *Proc. Natl. Acad. Sci.* **92:** 8195.

Hoffmann A., Chiang C.-M., Oelgeschlager T., Xie X., Burley S.K., Nakatani Y., and Roeder R.G. 1996. A histone octamer-like structure within TFIID. *Nature* **380:** 356.

Hoopes B., LeBlanc J., and Hawley D. 1992. Kinetic analysis of yeast TFIID-TATA box complex formation suggests a multi-step pathway. *J. Biol. Chem.* **267:** 11539.

Hori R. and Carey M. 1994. The role of activators in assembly of RNA polymerase II transcription complexes. *Curr. Opin. Genet. Dev.* **4:** 236.

Janin J. 1995. Elusive affinities. *Proteins* **21:** 30.

Jeffrey P., Russo A., Polyak K., Gibbs E., Hurwitz J., Massague J., and Pavletich N. 1995. Mechanism of CDK activation revealed by the structure of a cyclin A-CDK2 complex. *Nature* **376:** 313.

Karantza V., Friere E., and Moudrianakis E. 1996. Thermodynamic studies of the core histones: pH and ionic strength effects on the stability of the (H3-H4)/(H3-H4)2 system. *Biochemistry* **35:** 2037.

Kim J.L. and Burley S.K. 1994. 1.9 Å resolution refined structure of TBP recognizing the minor groove of TATAAAAG. *Nat. Struct. Biol.* **1:** 638.

Kim J.L., Nikolov D.B., and Burley S.K. 1993. Co-crystal structure of TBP recognizing the minor groove of a TATA element. *Nature* **365:** 520.

Kim Y., Geiger J.H., Hahn S., and Sigler P.B. 1993. Crystal structure of a yeast TBP/TATA-box complex. *Nature* **365:** 512.

Klug A., Rhodes D., Smith J., Finch J.T., and Thomas J.O. 1980. A low resolution structure for the histone core of the nucleosome. *Nature* **287:** 509.

Kokubo T., Gong D.-W., Wootton J., Horikoshi M., Roeder R.G., and Nakatani Y. 1994. Molecular cloning of *Drosophila* TFIID subunits. *Nature* **367:** 484.

Koleske A. and Young R. 1995. The RNA polymerase II holoenzyme and its implications for gene regulation. *Trends Biochem. Sci.* **20:** 113.

Kraulis P.J. 1991. MOLSCRIPT: A program to produce both detailed and schematic plots of protein structures. *J. Appl. Crystallogr.* **24:** 946.

Luger K., Mader A.W., Richmond R.K., Sargent D.F., and Richmond T.J. 1997. Crystal structure of the nucleosome core particle at 2.8Å resolution. *Nature* **389:** 251.

Maldonado E. and Reinberg D. 1995. News on initiation and elongation of transcription by RNA polymerase II. *Curr. Opin. Cell Biol.* **7:** 352.

Matsui T., Segall J., Weil P., and Roeder R.G. 1980. Multiple factors required for accurate initiation of transcription by purified RNA polymerase II. *J. Biol. Chem.* **255:** 11992.

McPherson C., Shin E.-Y., Friedman D., and Zaret K. 1993. An acitve tissue-specific enhancer and bound transcription factors existing in a precisely position nucleosomal array. *Cell* **75:** 387.

Mengus G., May M., Jacq X., Staub A., Tora L., Chambon P., and Davidson I. 1995. Cloning and characterization of hTAFII18, hTAFII20 and hTAFII28: Three subunits of the human transcription factor TFIID. *EMBO J.* **14:** 1520.

Nikolov D.B. and Burley S.K. 1994. 2.1Å resolution refined structure of a TATA box-binding protein (TBP). *Nat. Struct. Biol.* **1:** 621.

Nikolov D.B., Chen H., Halay E.D., Hoffmann A., Roeder R.G., and Burley S.K. 1996. Crystal structure of a human TATA box-binding protein/TATA element complex. *Proc. Natl. Acad. Sci.* **93:** 4956.

Nikolov D.B., Chen H., Halay E., Usheva A., Hisatake K., Lee D., Roeder R.G., and Burley S.K. 1995. Crystal structure of a TFIIB-TBP-TATA element ternary complex. *Nature* **377:** 119.

Nikolov D.B., Hu S.-H., Lin J., Gasch A., Hoffmann A., Horikoshi M., Chua N.-H., Roeder R.G., and Burley S.K. 1992. Crystal structure of TFIID TATA-box binding protein. *Nature* **360:** 40.

Parkhurst K., Brenowitz M., and Parkhurst L. 1996. Simulta-neous binding and bending of promoter DNA by TBP: Real time kinetic measurements. *Biochemistry* **35:** 7459.

Parvin J. and Sharp P. 1993. DNA topology and a minimal set of basal factors for transcription by RNA polymerase II. *Cell* **73:** 533.

Parvin J., McCormick R., Sharp P., and Fisher D. 1995. Prebending of a promoter sequence enhances affinity for the TATA-binding factor. *Nature* **273:** 724.

Perez-Howard G., Weil P., and Beechem J. 1995. Yeast TATA binding protein interaction with DNA: Fluorescence determination of oligomeric state, equilibrium binding, on-rate, and dissociation kinetics. *Biochemistry* **34:** 8005.

Pruss D., Hayes J., and Wolffe A. 1995. Nucleosomal anatomy-where are the histones? *BioEssays* **17:** 161.

Ramakrishnan V., Finch J., Graziano V., and Sweet R.M. 1993. Crystal structure of the globular domain of histone H5 and its implications for nucleosome binding. *Nature* **362:** 219.

Reeder R. 1992. Regulation of transcription by RNA polymerase I. In *Transcriptional regulation* (ed. S. McKnight and K.R. Yamamoto), p. 315. Cold Spring Harbor Laboratory Press, Cold Spring Harbor, New York.

Roeder R.G. 1991. The complexities of eukaryotic transcription initiation: Regulation of preinitiation complex assembly. *Trends Biochem. Sci.* **16:** 402.

Sentenac A. 1985. Eukaryotic RNA polymerases. *CRC Crit. Rev. Biochem.* **18:** 31.

Starr D., Hoopes B., and Hawley D. 1995. DNA bending is an important component of site-specific recognition by the TATA binding protein. *J. Mol. Biol.* **250:** 434.

Sun D. and Hurley L. 1995. TBP unwinding of the TATA box induces a specific downstream unwinding site that is targeted by pluramycin. *Chem. Biol.* **2:** 457.

Wang Z. and Roeder R.G. 1995. Structure and function of a human transcription factor TFIIIB subunit that is evolutionarily conserved and contains both TFIIB- and high-mobility-group protein 2 domains. *Proc. Natl. Acad. Sci.* **92:** 7026.

Wong J. and Bateman E. 1994. TBP-DNA interactions in the minor groove discriminate between A:T and T:A base pairs. *Nucleic Acids Res.* **22:** 1890.

Xie X., Kokubo T., Cohen S.L., Hoffmann A., Chait B.T., Roeder R.G., Nakatani Y., and Burley S.K. 1996. Structural similarity between TAFs and the heterotetrameric core of the histone octamer. *Nature* **380:** 316.

Zawel L. and Reinberg D. 1993. Initiation of transcription by RNA polymerase II: A multi-step process. *Prog. Nucleic Acid Res. Mol. Biol.* **44:** 67.

Zawel L., Kumar K., and Reinberg D. 1995. Recycling of the general transcription factors during RNA polymerase II transcription. *Genes Dev.* **9:** 1479.

Transcription in Archaea

S.D. BELL AND S.P. JACKSON

*Wellcome Trust/Cancer Research Campaign Institute, of Cancer and Developmental Biology, and
Department of Zoology, Cambridge University, Cambridge CB2 1QR, England, United Kingdom*

Before the late 1970s, life was viewed as being divided into two distinct domains, the prokaryotes and the eukaryotes. The eukaryotes possessed nuclei, extensive subcellular structures, and compartmentalization and contained complex RNA polymerases. In contrast, the prokaryotes lacked these structures and had simpler RNA polymerases. However, in 1977, this dogma was challenged by the discovery, based on phylogenetic analysis of 16S and 18S rRNA sequences, that the prokaryotes were composed of two distinct groups: the Bacteria and the Archaea (Woese and Fox 1977). Subsequent studies have led to a confirmation of the conclusion that the Archaea are a coherent domain of life as distinct from bacteria as they are from Eucarya. Divergence of the three lineages appears to have occurred very early in the evolution of life. Moreover, it has been proposed recently that bacteria diverged before the Eucarya and Archaea split (Fig. 1) (Doolittle and Brown 1994). Thus, Archaea and Eucarya shared a period of common evolutionary history distinct from Bacteria, leading to the establishment of biochemical pathways in the progenitors of Archaea and Eucarya distinct from analogous processes in Bacteria. Perhaps one of the clearest examples of this lies in the striking similarities between the archaeal and eucaryal basal transcription machineries and their differences from those in the bacterial systems.

Archaea constitute a diverse range of organisms that exist in a wide variety of environments and constitute an important component of the biosphere. For example, a recent survey indicated that archaeal species constitute approximately one third of the picoplankton in Antarctic waters (DeLong et al. 1994). Perhaps the best known archaeal species, however, are those which are capable of existing under various extreme conditions, such as high temperature, pressure, and extremes of pH. The hyperthermophilic Archaea have been of particular interest since the inherent thermal stability of their proteins has important implications for studies of protein folding and structure. This inherent protein stability also serves as a useful tool for the biochemist, greatly facilitating the production and purification of recombinant proteins. A second advantage that archaeal species possess is their relatively small genomes, a number of which have now been sequenced in their entirety. For example, the recently sequenced genomes of *Methanococcus jannaschii* and *Archaeoglobus fulgidus* are 1.66 Mb and 2.18 Mb, respectively (Bult et al. 1996; Klenk et al. 1997). These small genomes each have the coding potential for 1500–2500 proteins. This low protein complexity, when compared with eukaryotes, greatly simplifies purification of native archaeal proteins. In addition to being amenable to biochemical approaches, an increasing range of systems allowing the genetic manipulation of archaeal organisms have been developed (Cline et al. 1989; Aagaard et al. 1996)

ARCHAEAL TRANSCRIPTION SYSTEMS

In vitro transcription systems have been developed using extracts prepared from a variety of archaeal species. In particular, studies of transcription in the *Methanococcus* and *Sulfolobus* systems have shed considerable light on the sequence and factor requirements for archaeal transcription (Frey et al. 1990; Hudepohl et al. 1990; Reiter et al. 1990; Hausner et al. 1991). Initial work with these in vitro systems using unfractionated cell extracts led to extensive analysis of the structure of archaeal promoters (Fig. 2). The first promoters characterized were those for the *Sulfolobus shibatae* 16S rRNA and *Methanococcus vanielli* tRNA^{Val} (Frey et al. 1990; Hudepohl et al. 1990; Reiter et al. 1990). Intriguingly, the most important motif in these promoters was a TATA-like element, termed box A or TATA box, approximately 30 nucleotides upstream of the transcription initiation site. Point mutations within the TATA box reduce transcription 5–30-fold. Additionally, mutagenesis of purine-rich sequences immediately upstream of the TATA element resulted in strong reduction of promoter activity, with di- and mononucleotide substitutions in this region of the *S. shibatae* 16S rRNA promoter reducing promoter activity up to 25-fold. Mutation of sequences in the vicinity of the

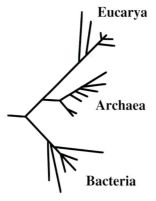

Figure 1. Phylogentic tree of life indicating the divergence of the three principal domains of life: Bacteria, Archaea, and Eucarya. (Modified from Olsen and Woese 1997.)

Figure 2. Diagram of the general structure of an archaeal promoter. A purine-rich motif is found upstream of a TATA box; 25–30 bp downstream from the TATA element, transcription initiates at a weakly conserved initiator element (Inr), usually at a purine preceded by a pyrimidine.

start site also abrogated promoter activity, suggesting a defined sequence requirement for start site recognition. Finally, insertion or deletion of bases between the *S. shibatae* 16S rRNA TATA-like element and the transcription start site produced transcripts initiating at multiple sites, almost invariably at purine nucleotides. However, the major initiation site in these studies was at a position that retained the wild-type distance from the TATA element. Similar results were obtained with the *M. vanelli* tRNA^{Val} promoter, indicating that transcription starts preferentially at purine nucleotides 22–27 bp downstream from the TATA box (Hausner et al. 1991).

The above in vitro experiments characterizing promoter structure have been supported by in vivo studies performed in the halophilic archaeon, *Haloferax volcanii* (Palmer and Daniels 1994, 1995). Plasmid-based transfection protocols have been developed for this organism, allowing the introduction of a reporter construct, a modified yeast tRNA^{Pro} gene fused to the promoter for the *H. volcanii* tRNA^{Lys} gene promoter. Extensive mutagenesis of the TATA-box motif of this promoter confirmed the key role of this element in transcription initiation. Additionally, block substitution of a purine-rich element immediately upstream of the TATA box abolished promoter activity.

With the development of in vitro transcription assays, it became apparent that, although high ratios of purified *Sulfolobus* RNA polymerase to template can effect transcription initiation at multiple sites, including the wild-type start site, accurate initiation at lower ratios of polymerase to template requires a factor that can be separated from polymerase during sucrose gradient centrifugation (Hudepohl et al. 1990). Fractionation of the *Methanococcus* extract revealed the requirement for two distinct transcription factors (Frey et al. 1990; Hausner and Thomm 1993). More recent work in these two systems has led to the identification of these transcription factors (see below).

ARCHAEAL TATA-BOX-BINDING PROTEIN

The key element in the majority of archaeal promoters is a TATA box located 25–30 nucleotides upstream of the transcription start site. As discussed above, mutagenesis of the TATA element greatly reduces or abolishes transcription initiation in all promoters studied thus far. The central role of the TATA element suggested that this motif is functionally analogous to the TATA box of eucaryal RNA polymerase II promoters and might be bound by a homolog of the eucaryal TBP (eTBP). A significant step forward in understanding the identity of archaeal transcription factors came with the identification of archaeal

homologs of eTBP, confirming this hypothesis (Marsh et al. 1994; Rowlands et al. 1994). Homologs of TBP have now been identified in a wide range of Archaea and are generally 35–40% identical to the conserved carboxy-terminal domain of eTBPs (Fig. 3) (Qureshi et al. 1995b; Rashid et al. 1995; Bult et al. 1996; Klenk et al. 1997; Smith et al. 1997; Soppa and Link 1997). A number of archaeal TBPs (aTBPs) possess a characteristic short highly acidic carboxy-terminal tail. The role, if any, of this tail is unknown. Whereas eucaryal TBPs have an amino-terminal extension of variable length, in archaeal TBPs, this region is greatly reduced, with at most only eight amino acid residues preceding the carboxy-terminal domain.

The X-ray crystal structure of *P. woesei* TBP has been determined, demonstrating that *Pyrococcus* TBP has a structure similar to that of eTBPs, possessing a saddle-like shape (DeDecker et al. 1996). The outer, convex face of the saddle comprises four α helices, and the inner DNA-binding face comprises an antiparallel β-sheet. Gel mobility shift assays and DNase I footprinting have demonstrated that archaeal TBPs bind to the TATA box (Rowlands et al. 1994; Hausner et al. 1996). Additionally, titration calorimetry experiments have revealed that the binding of *P. woesei* TBP to DNA is enhanced by high temperature and elevated salt concentrations, indicating that the TBP-DNA interaction is essentially hydrophobic in nature (DeDecker et al. 1996). This is in full agreement with the observation that *P. woesei* grows at 105°C and has an intracellular salt concentration equivalent to 800 mM potassium phosphate. Notably, however, a striking difference between aTBP and eTBPs is observed in the electrostatic charge potential distribution on the proteins. Yeast TBP has a highly asymmetric electrostatic charge potential distribution, with negative potential at the carboxy-terminal stirrup loop and positive potential at the amino-terminal stirrup loop. In contrast, the electrostatic charge potential is symmetrically distributed in aTBP, with both stirrup loops having negative potential. This higher degree of charge symmetry in aTBP is reflected in a greater structural symmetry between the two halves of the protein than is found in eTBP. Indeed, *Pyrococcus* TBP has approximately 40% sequence identity between repeats, compared to 28–30% identity between the sequence repeats in eTBP. As discussed below, these observations have important relevance for the assembly of the preinitiation complex. Evidence for a direct role for TBP in archaeal transcription came from the observation that both yeast and human TBPs were able to function in in vitro transcription using partially purified *Methanococcus* fractions. Furthermore, the eucaryal TBPs were able to substitute for one of the partially purified archaeal factors (Wettach et al. 1995). Additionally, *Thermococcus celer* TBP was able to function in the *Methanococcus* system (Hausner and Thomm 1995). Finally, it was shown that immunodepletion of aTBP from *S. shibatae* extracts abrogates the ability of the extract to mediate transcription and that transcriptional ability is restored to the extract by addition of recombinant *S. shibatae* TBP (Qureshi et al. 1995b).

TFB, AN ARCHAEAL HOMOLOG OF TFIIB

The identity of the second archaeal transcription factor was suggested with the identification of an archaeal homolog of TFIIB (Ouzounis and Sander 1992). As Archaea possess only one polymerase, this factor is termed TFB. Genes encoding TFB have now been identified in a variety of Archaea, and TFB is approximately 30% identical to TFIIB (Creti et al. 1993; Qureshi et al. 1995a; Bult et al. 1996; Klenk et al. 1997; Smith et al. 1997). In vitro transcription experiments indicate that TFB is required for transcription from a range of promoters (Hethke et al.

1996; Qureshi et al. 1997). Like TFIIB, TFB is composed of two distinct domains, an amino-terminal metal-binding domain and a larger carboxy-terminal domain containing an imperfect direct repeat. The carboxy-terminal domain (TFBc) can bind to a TBP-DNA complex to generate a stable ternary complex (Qureshi et al. 1995a). The structure of the *Pyrococcus* ternary complex has been solved using X-ray crystallography (Kosa et al. 1997). This structure indicates that TFBc contacts aTBP and DNA in a manner highly similar to the way in which TFIIB contacts the eTBP-DNA complex (Nikolov et al. 1995). TFB interacts with the carboxy-terminal stirrup loop of TBP and also

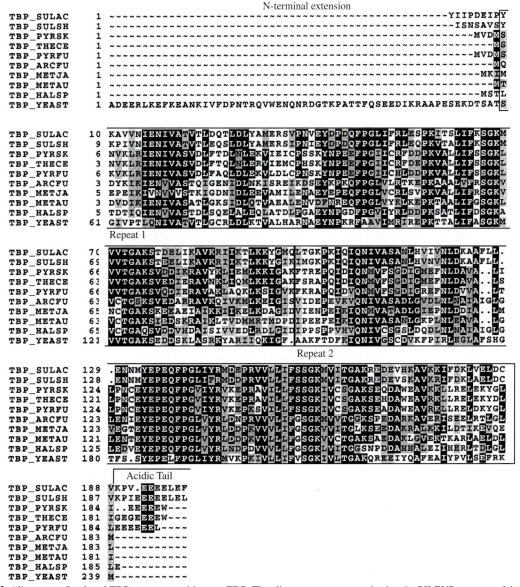

Figure 3. Alignment of archaeal TBP sequences with yeast TBP. The aligment was generated using the PILEUP program of the GCG package and shaded using the program BOXSHADE. Reverse shading indicates sequence identity, and gray shading indicates a conservative substitution. Accession numbers for the sequences are TBP_SULAC (*S. acidocaldarius*), X94935; TBP_SULSH (*Sulfolobus shibatae*), Q55031; TBP_PYRSK (*Pyrococcus* sp. KOD1), Q52366; TBP_THECE (*Thermococcus celer*), Q56253; TBP_PYRFU (*P. furiosus*), Q57050; TBP_ARCFU (*Archaeoglobus fulgidus*), B69394; TBP_METJA (*Methanococcus jannaschii*), Q57930; TBP_METAU (*Methanobacterium thermoautotrophicum*), D69084; TBP_HALSP (*Halobacterium* sp NRC-1), AF016485; TBP_YEAST (*Saccharomyces cerevisiae*), P13393. The amino-terminal extension is indicated as is the highly acidic carboxy-terminal tail present in some archaeal TBPs. The two imperfect direct repeats in the TBP sequence are boxed.

with the DNA backbone on either side of the TATA box. In agreement with the observation that yeast and human TBPs can substitute functionally for *Methanococcus* TBP (Wettach et al. 1995), the residues in TBP that contact TFB and TFIIB are conserved between archaeal and eucaryal TBPs. Footprints of the TFB-TBP-DNA complex show that TFB extends the region protected from DNase I cleavage on both sides of the TATA box (Hausner et al. 1996; Qureshi and Jackson 1998). This, combined with the observation that regions immediately upstream of the TATA box are important for transcription in a range of archaeal promoters, led to the hypothesis that TFB may recognize DNA sequences in this region. Indeed, mutagenesis of the upstream region of the T6 gene promoter of the *S. shibatae* virus, SSV1, demonstrated that in this promoter as well, the region within the first 6 bp upstream of the TATA box are of key importance in defining promoter activity (Qureshi and Jackson 1998). Furthermore, chemical modification interference studies were performed on TBP-TFB-DNA ternary complexes; these experiments indicated that modification of nucleotides in this region interfered with complex formation. Finally, in vitro selection experiments using randomized oligonucleotides in TBP-TFBc complex formation indicated a bias in selection for sequences upstream of the TATA box (Qureshi and Jackson 1998). Together, these data suggest that TFB recognizes DNA upstream of the TATA box in a sequence-dependent manner. Accordingly, this region has been termed BRE (TF*B* *r*esponsive *e*lement; Qureshi and Jackson 1998). It is interesting to note that similar conclusions have been reached by the Ebright and Reinberg laboratories on the binding of TFIIB to DNA in the eucaryal RNA polymerase II system (Lagrange et al. 1998).

As discussed above, the purine-rich BRE is important for promoter activity in a range of promoters of varying strengths. What then might be the function of this element? One possible explanation is that the BRE may have a role in determining the orientation of the preinitiation complex. As noted above, aTBP is a highly symmetrical molecule, and in the crystal structure of TBP-TFBc-TATA box, the protein DNA contacts made between TBP and the TATA box are essentially rotationally symmetric (Kosa et al. 1997). This poses the question of how the orientation of the preinitiation complex on DNA is determined. In studies of eTBP binding to DNA, using eTBP specifically derivatized with a DNA scission reagent, it has been observed that eTBP binds with a relatively modest (60:40) orientational preference (Cox et al. 1997). It is not understood fully how this directionality is attained, but differences in the sequences of the two halves of eTBP have been proposed to result in a preferential choice orientation (Juo et al. 1996); it has also been proposed that the asymmetric charge potential distribution observed in eTBP may be of importance (Kosa et al. 1997). However, as discussed above, archaeal TBP has an essentially symmetrical charge potential distribution. Furthermore, the minor differences between the two halves of eTBP which have been proposed to discriminate in the orientation of eTBP binding are not present in aTBP. It therefore seems possible that aTBP may be in-

capable of defining the orientation with which it binds the TATA box. In light of this, we have tested the hypothesis that the role of the BRE is to define the directionality of the ternary complex, and therefore the orientation of the RNA polymerase. Recently, we have shown that the principal determinants of the orientation of archaeal transcription lie within the 6 bp immediately upstream of the TATA box. Specifically, performing a reciprocal swap of the 6 bp upstream of the TATA box with the 6 bp downstream results in a qualitative shift in the polarity of transcription. In contrast, inverting the TATA box orientation has no detectable effect on the polarity of transcription. Thus, it appears that the BRE is of key importance in defining the absolute polarity of transcription on the promoters studied (S. Bell et al., in prep.).

Although the carboxy-terminal core domain of TFB can interact with the TBP-DNA complex, the recruitment of RNA polymerase requires the full-length TFB protein. Therefore, the amino-terminal domain presumably contacts the RNA polymerase. The structure of the amino-terminal domain of *Pyrococcus* TFB has been solved by nuclear magnetic resonance (Zhu et al. 1996) and was found to be a zinc ribbon structure, similar to that found in the eucaryal elongation factor, TFIIS (Qian et al. 1993). In the eucaryal TFIIB, this amino-terminal zinc-binding motif is thought to recruit the RNA polymerase. The principal contact between TFIIB and the polymerase does not appear to be direct; rather, it is mediated by a further factor TFIIF. However, TFIIF is not absolutely required and it has been possible to reconstitute transcription in the absence of this factor (Parvin and Sharp 1993). As Archaea do not possess any obvious homologs of TFIIF, it is currently assumed that the contact between TFB and RNA polymerase is direct.

The role of aTBP and TFB is to mediate recruitment of the RNA polymerase to the start site (Fig. 4). A fully defined archaeal transcription system has been developed using recombinant TBP and TFB and highly purified RNA polymerase for both the *Pyrococcus* and *Sulfolobus* systems (Hethke et al. 1996; Qureshi et al. 1997). Four promoters were examined in the *Sulfolobus* study. The promoters for the T5, T6, and 16S rRNA genes demonstrated an absolute requirement for TBP, TFB, and RNA polymerase. An intriguing exception, however, was noted with the promoter for the TFB gene of *S. shibatae*. When the transcripts produced for this gene were mapped from cellular RNA, two start sites were observed, approximately 25 nucleotides apart. Furthermore, both start sites are utilized in the reconstituted in vitro system. However, although initiation at the upstream start site required TBP, TFB, and the RNA polymerase, the downstream start site could be recognized and utilized by the RNA polymerase alone, in the absence of TBP and TFB. It therefore appears that the archaeal RNA polymerase may possess an intrinsic ability to recognize certain DNA sequences.

ARCHAEAL RNA POLYMERASE

The first archaeal RNA polymerase to be characterized was that of *Sulfolobus acidocaldarius*. It was immedi-

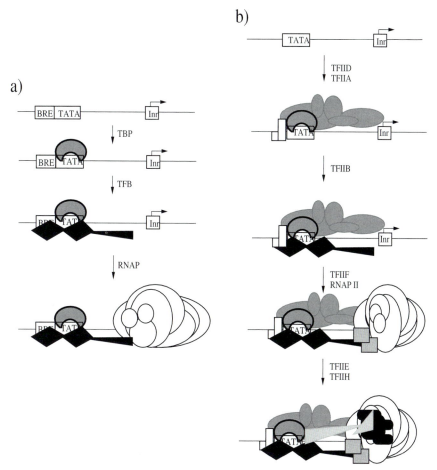

Figure 4. Model of preinitiation complex assembly on archaeal (*a*) and eucaryal RNA polymerase II (*b*) promoters.

ately apparent from the subunit composition of the purified enzyme that the polymerase is more reminiscent of eucaryal RNA polymerases, rather than bacterial RNA polymerases (Zillig et al. 1979). The similarity to eucaryal enzymes was demonstrated further by immunoblotting assays, which suggested that eucaryal and archaeal enzymes shared components not detected in bacterial enzymes (Huet et al. 1983). Subsequent work prin-

cipally by the laboratory of Wolfram Zillig has led to the identification of genes encoding the majority of the RNA polymerase subunits, a summary of which is shown in Table 1. This work has demonstrated that in addition to having a subunit composition analogous to that of eucaryal RNA polymerases, many of the smaller archaeal subunits are homologous to subunits found in eucaryal, but not bacterial, enzymes (Langer et al. 1995).

Table 1. Comparison of RNA Polymerase Subunits between Archaeal (*Sulfolobus acidocaldarius*), Eucaryal (*S. cerevisiae*), and Bacterial (*E. coli*) RNA Polymerases

S. acidocaldarius RNA pol subunit	Size (kD)	Yeast RNA pol I homolog	Yeast RNA pol II homolog	Yeast RNA pol III homolog	*E. coli* RNA pol homolog
B	122	A135	B150	C128	β
A′	101	A190 amino-terminal 2/3	B220 amino-terminal 2/3	C160 amino-terminal 2/3	β′
A″	44	A190 carboxy-terminal 1/3	B220 carboxy-terminal 1/3	C160 carboxy-terminal 1/3	β′
D	30	AC40	B44	AC40	α
E	27		B16	C25	
G	14				
H	12	ABC27	ABC27	ABC27	
I	10				
K	10	ABC23	ABC23	ABC23	
L	10	AC19	B13.6	AC19	α
N	7.5	ABC10β	ABC10β	ABC10β	

The largest subunit of the *Sulfolobus* RNA polymerase, termed subunit B, is homologous to the second largest subunit of the eucaryal enzymes and, more distantly, to the β-subunit of bacterial RNA polymerases. In methanogenic and halophilic Archaea, this subunit is split into two separate polypeptides, B′ and B″. The next two largest *Sulfolobus* RNA polymerase subunits, A′ and A″, together correspond to the largest subunit of eucaryal nuclear RNA polymerases and the β′-chain of bacterial RNA polymerase. The archaeal RNA polymerase subunits A′ and A″ correspond to the amino-terminal two thirds and the carboxy-terminal one third of the largest eucaryal subunits, respectively. The archaeal subunits appear most homologous to the equivalent subunits of eucaryal RNA polymerases II and III. Archaeal subunits D and L both contain copies of the "alpha motif," a sequence which is highly conserved among the α subunits of bacterial RNA polymerases. However, without this short motif, these subunits are related more closely to subunits of eucaryal RNA polymerases: subunit D to AC40 and B44, and subunit L to AC19 and B13.6. The alpha motif has been implicated in the homodimerization of the bacterial α subunits. Furthermore, AC40 and AC19 have been shown to interact via the alpha motif which they both possess (Lalo et al. 1993). It then might be expected that D and L should also interact. This is indeed the case. Specifically, dissociation of the archaeal enzyme in 6 M urea and 25% formamide leads to individual subunits being released and with the exception of subassemblies of D-L and of E-I (Lanzendorfer et al. 1994). Far Western blotting approaches using recombinant D and L also indicate a strong interaction between these proteins (S. Bell et al., unpubl.). It was observed that the D-L complex has a characteristic yellow color, with an absorbance maximum at 400 nm (Lanzendorfer et al. 1994). We note that the sequence of the D subunit has a region of high homology with ferredoxin, an iron-sulfur protein which also has a characteristic absorbance at 400 nm. However, the role in transcription, if any, of this motif is currently unknown.

Archaeal RNA polymerase subunit E contains several regions of homology with different proteins. The amino-terminal 100-amino-acid residues contain homology with the amino-terminal halves of the C25 and B19 subunits of RNA polymerases II and III, and a region closely related to the RNA-binding motif of the ribosomal protein S1 is located from position 100 to 170. Finally, the carboxy-terminal 60-amino-acid residues of E contain a putative zinc ribbon that has weak homology with the BRF component of TFIIIB. In methanogenic Archaea, subunit E is split into two polypeptides encoded by separate genes E′ and E″. E′ contains the C25/B19 homology and S1 motif, whereas E″ constitutes the zinc ribbon motif. In the *S. acidocaldarius* polymerase, subunit E interacts tightly with subunit I, although the identity of I is not yet known. The gene for subunit G has been sequenced but has no significant homologs in the available databases. Intriguingly, there are no apparent homologs of G in any of the completed archaeal genomes. If this is a true subunit of the *Sulfolobus* enzyme, then it would appear to be unique to this genus. Subunit H is homologous over its entire length to the carboxy-terminal third of the ubiquitous eucaryal subunit ABC27. Subunit K is homologous to the carboxy-terminal half of ABC23, a subunit shared by all three eucaryal RNA polymerases. Finally, subunit N is highly similar to ABC10β.

OTHER POTENTIAL ARCHAEAL BASAL TRANSCRIPTION FACTORS

When the gene for the *S. acidocaldarius* RNA polymerase subunit L gene was sequenced, a downstream open reading frame (ORF) was detected that contained homology with the eucaryal elongation factor, TFIIS (Langer and Zillig 1993). Related genes have been identified subsequently in a range of archaeal genomes. It has become apparent that although this short ORF (encoding a polypeptide of 11 kD) is homologous to TFIIS, it is more closely related to the A12.2 and B12.6 components of RNA polymerases I and II (Kaine et al. 1994). Whether this protein is a component of the archaeal polymerase, for example, corresponding to the approximately 10-kD subunit I, or a distinct transcription factor is currently under investigation. We note that the sequence of subunit I (~10 kD) is still unknown.

With the completion of several archaeal genomes, potential homologs of the α-subunit of TFIIE have been identified (Klenk et al. 1997). These putative homologs each has a mass of approximately 20 kD, and as can be seen in Figure 5, they are homologous to the leucine-rich and zinc-binding amino-terminal 185-amino-acid residues of yeast and human TFIIE-α. Extensive deletion analyses of human TFIIE-α have been performed and truncated proteins assayed for their ability to support transcription. Remarkably, it is possible to delete all sequences carboxy-terminal of position 174 and retain activity in basal transcription assays (Ohkuma et al. 1995). These observations were supported by in vivo assays using truncation mutants of yeast TFIIE-α (Kuldell and Buratowski 1997). This work revealed that removal of all sequences carboxy-terminal to position 212 yields yeast that are viable, although susceptible to cold shock. Thus, the archaeal protein appears to correspond to the minimal functional domain of TFIIE-α. There is no apparent archaeal homolog of the second subunit of TFIIE, TFIIE-β. Work is currently under way to determine whether the putative archaeal TFIIE-α subunit has a partner protein and what role, if any, this factor has in transcription. Although it is clear that the protein is not required in the reconstituted transcription systems on the promoters studied (Qureshi et al. 1997), it is possible that this factor is only required under certain physiological conditions or, as may be the case in yeast, for a limited subset of promoters.

ORFs have been detected in a variety of archaeal species which encode putative homologs of the bacterial termination/antitermination factors NusA and NusG. Recent work has demonstrated that these archaeal NusG homologs are also related to the Spt5 component of the eucaryal transcription elongation factor, DSIF (Wada et al. 1998). It therefore seems likely that these archaeal proteins will have a role in transcription elongation.

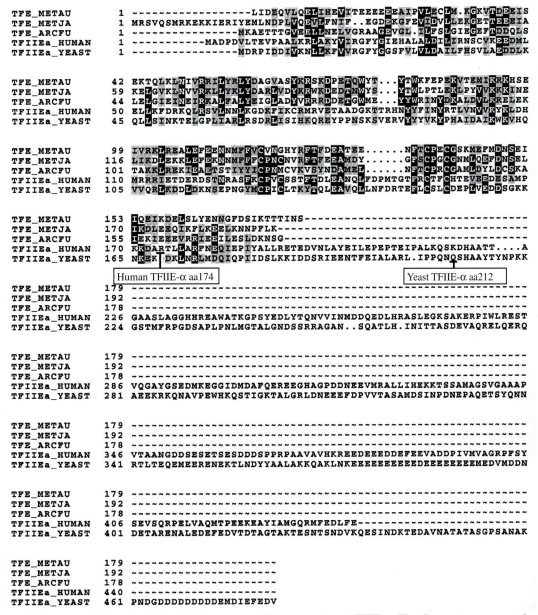

Figure 5. Alignment of archaeal putative TFIIE sequences with human and yeast TFIIE-α. The alignment was generated using the PILEUP program of the GCG package and shaded using the program BOXSHADE. Reverse shading indicates sequence identity, and gray shading indicates a conservative substitution. The positions of yeast TFIIE-α residue 212 and human TFIIE-α residue are indicated with a **T**-shape. Accession numbers for the sequences are TFE_METAU (*M. thermoautotrophicum*), A69090; TFE_METJA (*M. jannaschii*), Q58187; TFE_ARCFU (*A. fulgidus*), E69344; TFIIEa_HUMAN (*Homo sapiens*), P29083; TFIIEa_YEAST (*S. cerevisiae*), P36100.

REGULATION OF ARCHAEAL TRANSCRIPTION

The requirements for basal transcription in the archaeal system have been fairly well characterized, but the manner in which differential gene regulation is attained is still poorly understood. Examination of the completed sequence of various archaeal genomes reveals the presence of several homologs of bacterial transcription regulatory proteins (Bult et al. 1996; Klenk et al. 1997; Smith et al. 1997). For example, the *Archaeoglobus fulgidus* genome contains four members of the ArsR family and nine mem-

bers of the AsnC family of bacterial transcription factors. In addition, homologs of the bacterial global regulator LRP have been identified in many archaeal species (Kyrpides and Ouzounis 1995; Bult et al. 1996; Charlier et al. 1997; Klenk et al. 1997; Smith et al. 1997). However, it is not known which genes these potential transcription factors might control.

A few regulated genes have been identified in Archaea, and these are discussed briefly below. A lysogenic virus has been identified in *S. shibatae*, SSV1 (Martin et al. 1984). Replication of the virus is induced following UV irradiation and, concomitant with this, the pattern of transcription clearly changes. Thus, 2–3 hours after UV irradiation, a short transcript is produced from the T_{ind} promoter (Reiter et al. 1987). This promoter is highly unusual in that it does not have a discernible TATA box. How the activation of this gene is attained is not yet known, although it is tempting to speculate that the RNA polymerase is recruited by a mechanism independent of TBP.

Discovery of another archaeal gene regulatory system came from the study of nitrogen metabolism in methanogenic Archaea. These organisms are capable of fixing atmospheric nitrogen or utilizing ammonium in the medium. It has been shown that growth on ammonium represses transcription of the nitrogen fixation (*nif*) genes. Intriguingly, studies in *Methanococcus maripaludis* have revealed that a palindromic element immediately downstream from the transcription initiation site of the *nifH* gene is required for repression of *nifH* expression (Cohen-Kupiec et al. 1997). This element binds an unidentified factor that is present in extracts prepared from cells grown in the presence of ammonium but absent from extracts made from nitrogen-grown cells. These data fit with a model in which a repressor molecule binds to the gene in the presence of ammonium. Preliminary evidence has been obtained for a repressor of *nifH* expression in the distantly related *Methanosarcina barkeri*. However, it is currently unclear where this repressor binds, and furthermore, there is no apparent palindrome such as that found in the *M. maripaludis nifH* gene (Yueh-tyng et al. 1998).

Certain halophilic Archaea synthesize gas vesicles when they reach stationary phase. A cluster of 14 genes is responsible for the synthesis of these structures. In *Haloferax mediterranei*, when cells reach stationary phase in medium containing 25% salt, induction of transcription of the genes in this cluster is observed. Transfection experiments have indicated a key role for the *gvpE* gene product of the cluster as being a positive regulator for the transcription of these genes (Roder and Pfeifer 1996).

The study of the heat-shock response has been of great value in elucidating the mechanisms of transcription activation in Eucarya and in Bacteria. Recent studies have begun to dissect the analogous process in Archaea. The heat-shock genes of *Haloferax volcanii* show strong transcriptional induction following heat shock (Kuo et al. 1997). Mapping of the promoter elements that mediate the heat shock response has revealed that regions flanking the TATA box are of key importance in mediating this response (Thompson and Daniels 1998). As these sequences essentially overlap the BRE, the TFB recognition element, it is tempting to speculate that a heat-shock-specific TFB may be involved in the heat shock response, perhaps in a manner analogous to the differential use of σ factors in the bacterial heat shock response.

ARCHAEAL CHROMATIN AND DNA TOPOLOGY

A major step in the regulation of eucaryal transcription lies in the modification of chromatin structure to facilitate preinitiation complex assembly (Grunstein 1997). It will

thus clearly be of great interest to examine archaeal chromatin structure and to determine its effect on transcription. Studies to date suggest that there are two distinct systems of DNA compaction used in Archaea. Archaea can be divided into two highly diverged phylogenetic groups, the Euryarchaeota and the Crenarchaeota. The Euryarchaeota, which includes the *Methanogens* and *Pyrococcus* species, possess homologs of eucaryal histones termed Hmf (for review, see Reeve et al. 1997; Zlatanova 1997). These proteins have been demonstrated to form "nucleosome-like structures," perhaps corresponding to eucaryal $(H3-H4)_2$ tetramers (Pereira et al. 1997). The structure of the Hmf monomer has been solved, and it is highly similar to the histone-fold (Starich et al. 1996). The archaeal histones, however, seem to correspond to purely this histone-fold, with the important lysine containing amino-terminal extensions in eucaryal histones being absent in Archaea. It will be of great interest to examine the interplay of archaeal histones and the transcription machinery. In contrast, the Crenarchaeota characterized to date do not appear to have histone homologs. Instead, these organisms, which include many of the sulfur-utilizing Archaea, such as *Sulfolobus*, have small proteins typified by Sac7d. The X-ray crystal structure of Sac7d on DNA reveals that this protein binds to the minor groove of DNA (Robinson et al. 1998) and induces a sharp kink by intercalation of certain amino acids, in a manner reminiscent of the binding of TBP to DNA (J.L. Kim et al. 1993; Y.C. Kim et al. 1993).

A striking feature of the hyperthermophilic Archaea is that although the DNA of Eucarya and mesophilic Archaea is negatively supercoiled, the DNA of hyperthermophiles is topologically relaxed or even positively supercoiled (Forterre et al. 1996). It is thought that this adaptation to prevent localized denaturation of DNA at the high temperatures at which these organisms grow. Extensive work in the eucaryal RNA polymerase II has revealed that the topology of the DNA template can have important effects on the template requirements for in vitro transcription reaction. Specifically, some promoters, when present on negatively supercoiled templates, can be transcribed by eTBP, TFIIB, and RNA polymerase alone (Parvin and Sharp 1993; Goodrich and Tjian 1994; Timmers 1994). However, the same promoters, when on relaxed templates, show a requirement for TFIIE and TFIIH. Since the work determining the factor requirements in the defined *Sulfolobus* transcription system used promoter constructs on negatively supercoiled templates, it was possible that this work overlooked factors that might be required for transcription on more physiological templates. To test this idea, in vitro transcription assays were performed on templates that ranged from highly negatively supercoiled to highly positively supercoiled (S. Bell et al., in prep.). Remarkably, even the highly positively supercoiled templates were efficiently transcribed by TBP, TFB, and RNA polymerase at the physiological temperature of 75°C. This suggests strongly that no additional factors are required by the RNA polymerase for efficient initiation. Furthermore, transcription was insensitive to AMP-PNP, suggesting that there is no requirement for a helicase-like activity analogous to TFIIH. However, when transcription was performed at a lower temperature, 48°C, only negatively supercoiled templates were used efficiently. The block to initiation on the nonpermissive templates has been shown to be at the level of promoter opening (S. Bell et al., in prep.). This observation correlates with the observation that on cold shock, hyperthermophilic Archaea rapidly induce negative supercoiling of their genomes (Marguet et al. 1996; Lopez-Garcia and Forterre 1997). It is possible that this response serves as a global mechanism to maintain the genome in a state whereby gene expression can be maintained efficiently at the lower temperature.

SUMMARY AND PERSPECTIVES

During the period of evolutionary history following the divergence of Bacteria but before the division of Archaea and Eucarya, the basic components of the transcription machinery appear to have evolved in a manner distinct from that of the bacterial lineage. As a consequence, Archaea and Eucarya share many common features in their transcription apparatus that are not found in Bacteria. These features are exemplified by complex multisubunit polymerases, and the presence of TBP and TFIIB homologs. The archaeal transcription machinery is essentially composed of the most central players in the eucaryal RNA polymerase II complex. This is in keeping with the relative simplicity of the Archaea, which are single-celled organisms having small genomes that presumably require less regulation than those of Eucarya. This apparent simplicity, combined with the ease of biochemistry and the growing powers of genetic manipulation, makes Archaea a highly attractive model system to study the basic processes of transcription initition. But the Archaea are more than just simplified Eucarya. It has become apparent with the sequencing of archaeal genomes that a considerable proportion of archaeal proteins have no homologs in either bacterial or eucaryal genomes. Thus, Archaea appear to have evolved a diversity of important and unique functions, the elucidation of which will greatly contribute to our understanding of these organisms and their important role in the biosphere.

ACKNOWLEDGMENTS

This work was funded by the Cancer Research Campaign and the Wellcome Trust. We also thank Marc Nadal, Christine Jaxel, Peter Kosa, and Paul Sigler for many useful discussions.

REFERENCES

Aagaard C., Leviev I., Aravalli R.N., Forterre P., Prieur D., and Garrett R.A. 1996. General vectors for archaeal hyperthermophiles—Strategies based on a mobile intron and a plasmid. *FEMS Microbiol. Rev.* **18**: 93.

Bult C.J., White O., Olsen G.J., Zhou L., Fleischmann R.D., Sutton G.G., Blake J.A., FitzGerald L.M., Clayton R.A., Gocayne J.D., Kerlavage A.R., Dougherty B.A., Tomb J.F., Adams M.D., Reich C.I., Overbeek R., Kirkness E.F., Weinstock K.G., Merrick J.M. Glodek A., Scott J.L., Geoghagen

N.S.M., and Venter J.C. 1996. Complete genome sequence of the methanogenic archaeon, *Methanococcus jannaschii. Science* **273**: 1058.

Charlier D., Roovers M., ThiaToong T.L., Durbecq V., and Glansdorff N. 1997. Cloning and identification of the *Sulfolobus solfataricus lrp* gene encoding an archaeal homologue of the eubacterial leucine responsive global transcriptional regulator Lrp. *Gene* **201**: 63.

Cline S.W., Lam W.L., Charlebois R.L., Schalkwyk L.C., and Doolittle W.F. 1989. Transformation methods for halophilic archaebacteria. *Can. J. Microbiol.* **35**: 148.

Cohen-Kupiec R., Blank C., and Leigh J.A. 1997. Transcriptional regulation in Archaea: *In vivo* demonstration of a repressor binding site in a methanogen. *Proc. Natl. Acad. Sci.* **94**: 1316.

Cox J.M., Hayward M.M., Sanchez J.F., Gegnas L.D., vanderZee S., Dennis J.H., Sigler P.B., and Schepartz A. 1997. Bidirectional binding of the TATA box binding protein to the TATA box. *Proc. Natl. Acad. Sci.* **94**: 13475.

Creti R., Londei P., and Cammarano P. 1993. Complete nucleotide sequence of an archaeal (*Pyrococcus woesei*) gene encoding a homologue of eukaryotic transcription factor IIB (TFIB). *Nucleic Acids Res.* **21**: 2942.

DeDecker B.S., O'Brien R., Fleming P.J., Geiger J.H., Jackson S.P., and Sigler P.B. 1996. The crystal structure of a hyperthermophilic archaeal TATA box binding protein. *J. Mol. Biol.* **264**: 1072.

DeLong E.F., Wu K. Y., Prezelin B.B., and Jovine R.V.M. 1994. High abundance of Archaea in antarctic marine picoplankton. *Nature* **371**: 695.

Doolittle W.F. and Brown J.R. 1994. Tempo, mode, the progenote, and the universal root. *Proc. Natl. Acad. Sci.* **91**: 6721.

Forterre P., Bergerat A., and Lopezgarcia P. 1996. The unique DNA topology and DNA topoisomerases of hyperthermophilic archaea. *FEMS Microbiol. Rev.* **18**: 237.

Frey G., Thomm M., Brudigam B., Gohl H.P., and Hausner W. 1990. an archaebacterial cell free transcription system the expression of transfer RNA genes from *Methanococcus vannielii* is mediated by a transcription factor. *Nucleic Acids Res.* **18**: 1361.

Goodrich J.A. and Tjian R. 1994. Transcription factors IIE and IIH and ATP hydrolysis direct promoter clearance by RNA polymerase II. *Cell* **77**: 145.

Grunstein M. 1997. Histone acetylation in chromatin structure and transcription. *Nature* **389**: 349.

Hausner W. and Thomm M. 1993. Purification and characterization of a general transcription factor, aTFB, from the archaeon *Methanococcus thermolithotrophicus. J. Biol. Chem.* **268**: 24047.

———. 1995. The translation product of the presumptive *Thermococcus celer* TATA binding protein sequence is a transcription factor related in structure and function to *Methanococcus* transcription factor B. *J. Biol. Chem.* **270**: 17649.

Hausner W., Frey G., and Thomm M. 1991. Control regions of an archaeal gene a TATA box and an initiator element promote cell free transcription of the tRNA(Val) gene of *Methanococcus vannielii. J. Mol. Biol.* **222**: 495.

Hausner W., Wettach J., Hethke C., and Thomm M. 1996. Two transcription factors related with the eucaryal transcription factors TATA binding protein and transcription factor IIB direct promoter recognition by an archaeal RNA polymerase. *J. Biol. Chem.* **271**: 30144.

Hethke C., Geerling A.C.M., Hausner W., De Vos W.M., and Thomm M. 1996. A cell free transcription system for the hyperthermophilic archaeon *Pyrococcus furiosus. Nucleic Acids Res.* **24**: 2369.

Hudepohl U., Reiter W.D., and Zillig W. 1990. *In vitro* transcription of two rRNA genes of the archaebacterium *Sulfolobus sp. B12* indicates a factor requirement for specific initiation. *Proc. Natl. Acad. Sci.* **87**: 5851.

Huet J., Schnabel R., Sentenac A., and Zillig W. 1983. Archaebacteria and eukaryotes possess DNA dependent RNA polymerases of a common type. *EMBO J.* **2:** 1291.

Juo Z.S., Chiu T.K., Leiberman P.M., Baikalov I., Berk A.J., and Dickerson R.E. 1996. How proteins recognize the TATA box. *J. Mol. Biol.* **261:** 239.

Kaine B.P., Mehr I.J., and Woese C.R. 1994. The sequence, and its evolutionary implications, of a *Thermococcus celer* protein associated with transcription. *Proc. Natl. Acad. Sci.* **91:** 3854.

Kim J.L., Nikolov D.B., and Burley S.K. 1993. Co-crystal structure of TBP recognising the minor-groove of a TATA element. *Nature* **365:** 520.

Kim Y.C., Geiger J.H., Hahn S., and Sigler P.B. 1993. Crystal structure of a yeast TBP TATA-box complex. *Nature* **365:** 512.

Klenk H.P. Clayton R.A., Tomb J.F., White O., Nelson K.E., Ketchum K.A. Dodson R.J., Gwinn M., Hickey E.K., Peterson J.D., Richardson D.L., Kerlavage A.R., Graham D.E., Kyrpides N.C., Fleischmann R.D., Quackenbush J., Lee N.H., Sutton G.G., Gill S., Kirkness E.F., Dougherty B.A., McKenney K., Adams M.D., Loftus B., Venter J.C., et al. 1997. The complete genome sequence of the hyperthermophilic, sulphate reducing archaeon *Archaeoglobus fulgidus. Nature* **390:** 364.

Kosa P.F., Ghosh G., DeDecker B.S., and Sigler P.B. 1997. The 2.1 Å crystal structure of an archaeal preinitiation complex: TATA box binding protein/transcription factor (II)B core/TATA-box. *Proc. Natl. Acad. Sci.* **94:** 6042.

Kuldell N.H. and Buratowski S. 1997. Genetic analysis of the large subunit of yeast transcription factor IIE reveals two regions with distinct functions. *Mol. Cell. Biol.* **17:** 5288.

Kuo Y.P., Thompson D.K., St. Jean A., Charlebois R.L., and Daniels C.J. 1997. Characterization of two heat shock genes from *Haloferax volcanii:* A model system for transcription regulation in the Archaea. *J. Bacteriol.* **179:** 6318.

Kyrpides N.C. and Ouzounis C.A. 1995. The eubacterial transcriptional activator Lrp is present in the archaeon *Pyrococcus furiosus. Trends Biochem. Sci.* **20:** 140.

Lagrange T., Kapanidis A.N., Tang H., Reinberg D., and Ebright R.H. 1998. New core promoter element in RNA polymerase II dependent transcription: Sequence specific DNA binding by transcription factor IIB. *Genes Dev.* **12:** 34.

Lalo D., Carles C., Sentenac A., and Thuriaux P. 1993. Interactions between three common subunits of yeast RNA polymerase I and polymerase III. *Proc. Natl. Acad. Sci.* **90:** 5524.

Langer D. and Zillig W. 1993. Putative TFIIS gene of *Sulfolobus acidocaldarius* encoding an archaeal transcription elongation factor is situated directly downstream of the gene for a small subunit of DNA dependent RNA polymerase. *Nucleic Acids Res.* **21:** 2251.

Langer D., Hain J., Thuriaux P., and Zillig W. 1995. Transcription in archaea similarity to that in eucarya. *Proc. Natl. Acad. Sci.* **92:** 5768.

Lanzendorfer M., Langer D., Hain J., Klenk H.P., Holz I., Arnoldammer I., and Zillig W. 1994. Structure and function of the DNA dependent RNA polymerase of *Sulfolobus. Syst. Appl. Microbiol.* **16:** 656.

Lopez-Garcia- P. and Forterre P. 1997. DNA topology in hyperthermophilic archaea: Reference states and their variation with growth phase, growth temperature, and temperature stresses. *Mol. Microbiol.* **23:** 1267.

Marguet E., Zivanovic Y., and Forterre P. 1996. DNA topological change in the hyperthermophilic archaeon *Pyrococcus abyssi* exposed to low temperature. *FEMS Microbiol. Lett.* **142:** 31.

Marsh T.L., Reich C.I., Whitelock R.B., and Olsen G.J. 1994. Transcription factor-IID in the archaea—Sequences in the *Thermococcus celer* genome would encode a product closely-related to the TATA-binding protein of eukaryotes. *Proc. Natl. Acad. Sci.* **91:** 4180.

Martin A., Yeats S., Janekovic D., Reiter W.D., Aicher W., and Zillig W. 1984. SAV 1, a temperate UV inducible DNA virus like particle from the archaebacterium isolate B 12. *EMBO J.* **3:** 2165.

Nikolov D.B., Chen H., Hala E.D., Usheva A.A., Hisatake K., Lee D.K., Roeder R.G., and Burley S.K. 1995. Crystal structure of a TFIIB-TBP-TATA element ternary complex. *Nature* **377:** 119.

Ohkuma Y., Hashimoto S., Wang C.K., Horikoshi M., and Roeder R.G. 1995. Analysis of the role of TFIIE in basal transcription and TFIIH mediated carboxy terminal domain phosphorylation through structure function studies of TFIIE-α. *Mol. Cell. Biol.* **15:** 4856.

Olsen G.J. and Woese C.R. 1997. Archaeal genomics: An overview. *Cell* **89:** 991.

Ouzounis C. and Sander C. 1992. TFIIB, an evolutionary link between the transcription machineries of archaebacteria and eukaryotes. *Cell* **71:** 189.

Palmer J.R. and Daniels C.J. 1994. A transcriptional reporter for *in vivo* promoter analysis in the archaeon *Haloferax volcanii. Appl. Environ. Microbiol.* **60:** 3867.

———. 1995. *In vivo* definition of an archaeal promoter. *J. Bacteriol.* **177:** 1844.

Parvin J.D. and Sharp P.A. 1993. DNA topology and a minimal set of basal factors for transcription by RNA polymerase II. *Cell* **73:** 533.

Pereira S.L., Grayling R.A., Lurz R., and Reeve J.N. 1997. Archaeal nucleosomes. *Proc. Natl. Acad. Sci.* **94:** 12633.

Qian X.Q., Jeon C.J., Yoon H.S., Agarwal K., and Weiss M.A. 1993. Structure of a new nucleic acid binding motif in eukaryotic transcriptional elongation factor TFIIS. *Nature* **365:** 277.

Qureshi S.A. and Jackson S.P. 1998. Sequence specific DNA binding by the *S. shibatae* TFIIB homolog, TFB, and its effect on promoter strength. *Mol. Cell* **1:** 389.

Qureshi S.A., Bell S.D., and Jackson S.P. 1997. Factor requirements for transcription in the archaeon *Sulfolobus shibatae. EMBO J.* **16:** 2927.

Qureshi S.A., Khoo B., Baumann P., and Jackson S.P. 1995a. Molecular cloning of the transcription factor TFIIB homolog from *Sulfolobus shibatae. Proc. Natl. Acad. Sci.* **92:** 6077.

Qureshi S.A., Baumann P., Rowlands T., Khoo B., and Jackson S.P. 1995b. Cloning and functional analysis of the TATA binding protein from *Sulfolobus shibatae. Nucleic Acids Res.* **23:** 1775.

Rashid N., Morikawa M., and Imanaka T. 1995. An abnormally acidic TATA binding protein from a hyperthermophilic archaeon. *Gene* **166:** 139.

Reeve J.N., Sandman K., and Daniels C.J. 1997. Archaeal histones, nucleosomes, and transcription initiation. *Cell* **89:** 999.

Reiter W.D., Hudepohl U., and Zillig W. 1990. Mutational analysis of an archaebacterial promoter: Essential role of a TATA box for transcription efficiency and start site selection *in vitro. Proc. Natl. Acad. Sci.* **87:** 9509.

Reiter W.D., Palm P., Yeats S., and Zillig W. 1987. Gene expression in archaebacteria physical mapping of constitutive and UV inducible transcripts from the *Sulfolobus* virus like particle SSV1. *Mol. Gen. Genet.* **209:** 270.

Robinson H., Gao Y.-G., McCrary B.S., Edmondson S.P., Shriver J.W., and Wang A.H.-J. 1998. The hyperthermophile chromosomal protein Sac7d sharply kinks DNA *Nature* **392:** 202.

Roder R. and Pfeifer F. 1996. Influence of salt on the transcription of the gas vesicle genes of *Haloferax mediterranei* and identification of the endogenous transcriptional activator gene. *Microbiology* **142:** 1715.

Rowlands T., Baumann P., and Jackson S.P. 1994. The TATA binding protein a general transcription factor in eukaryotes and archaebacteria. *Science* **264:** 1326.

Smith D.R., Doucette-Stamm L.A., Deloughery C., Lee H., Dubois J., Aldredge T., Bashirzadeh R., Blakely D., Cook R., Gilbert K., Harrison D., Hoang L., Keagle P., Lumm W., Pothier B., Qiu D., Spadafora R., Vicaire R., Wang Y., Wierzbowski J., Gibson R., Jiwani N., Caruso A., Bush D., Reeve J.N. et al. 1997. Complete genome sequence of *Methanobacterium thermoautotrophicum* deltaH: Functional analysis and comparative genomics. *J. Bacteriol.* **179:** 71355.

Soppa J., and Link T.A. 1997. The TATA-box-binding protein

(TBP) of Halobacterium salinarum: Cloning of the *tbp* gene, heterologous production of TBP and folding of TBP into a native conformation. *Eur. J. Biochem.* **249:** 318.

Starich M.R., Sandman K., Reeve J.N., and Summers M.F. 1996. Nmr structure of Hmfb from the hyperthermophile, *Methanothermus fervidus*, confirms that this archaeal protein is a histone. *J. Mol. Biol.* **255:** 187.

Thompson D.K. and Daniels C.J. 1998. Heat shock inducibility of an archaeal TATA like promoter is controlled by adjacent sequence elements. *Mol. Microbiol.* **27:** 541.

Timmers H.T.M. 1994. Transcription initiation by RNA polymerase II does not require hydrolysis of the β γ phosphoanhydride bond of ATP. *EMBO J.* **13:** 391.

Wada T., Takagi T., Yamaguchi Y., Ferdous A., Imai T., Hirose S., Sugimoto S., Yano K., Hartzog G.A., Winston F., Buratowski S., and Handa H. 1998. DSIF, a novel transcription elongation factor that regulates RNA polymerase II processivity, is composed of human Spt4 and Spt5 homologs. *Genes Dev.* **12:** 343.

Wettach J., Gohl H.P., Tschochner H., and Thomm M. 1995. Functional interaction of yeast and human TATA-binding proteins with an archaeal RNA polymerase and promoter. *Proc. Natl. Acad. Sci.* **92:** 472.

Woese C.R. and Fox G.E. 1977. Phylogenetic structure of the prokaryotic domain: The primary kingdoms. *Proc. Natl. Acad. Sci.* **74:** 5088.

Yueh-tyng C., Helmann J.D., and Zinder S.H. 1998. Interactions between the promoter regions of nitrogenase structural genes (*nifHDK2*) and DNA binding proteins from N2 and ammonium grown cells of the archaeon *Methanosarcina barkeri* 227. *J. Bacteriol.* **180:** 2723.

Zhu W.L., Zeng Q.D., Colangelo C.M., Lewis L.M., Summers M.F., and Scott R.A. 1996. The N terminal domain of TFIIB from *Pyrococcus furiosus* forms a zinc ribbon. *Nat. Struct. Biol.* **3:** 122.

Zillig W., Stetter K.O., and Janekovic D. 1979. DNA-dependent RNA polymerase from the Archaebacterium *Sulfolobus acidocaldarius. Eur. J. Biochem.* **96:** 597.

Zlatanova J. 1997. Archaeal chromatin: Virtual or real? *Proc. Natl. Acad. Sci.* **94:** 12251.

Polarity of Transcription on Pol II and Archaeal Promoters: Where Is the "One-way Sign" and How Is It Read?

F.T.F. Tsai,* O. Littlefield,*† P.F. Kosa,*‡ J.M. Cox,§ A. Schepartz,§ and P.B. Sigler*†

*Departments of *Molecular Biophysics and Biochemistry and of §Chemistry and the †Howard Hughes Medical Institute, Yale University, New Haven, Connecticut 06511*

The current view of the preinitiation assembly of RNA polymerase II (pol II) promoters appears to be growing at about 30-kD per month. The ever-increasing number of protein participants is matched by the dazzling array of new functions, many of which were addressed at the Symposium this year. However, the current understanding of how these functions are biochemically coordinated to produce regulated transcription is lagging behind and, to some extent, being buried by the bewildering size and complexity of the system. Our approach is to define the stereochemistry that underlies some of the general properties of regulated gene expression from pol II promoters. One of these properties is the polarity of transcription, which is the focus of this paper.

Clearly, pre-mRNA synthesis is unidirectional, which implies the generally accepted view that the preinitiation complex (PIC) is highly asymmetrical. It is also generally believed that the binding of TFIID through its core component, the TATA-box-binding protein (TBP), to the key identity element of the promoter, the TATA box, is the nucleating event in PIC assembly. However, as detailed below, both the TBP and DNA surfaces that form this interface are nearly symmetrical. Thus, we address three questions: (1) Is transcription from a specific promoter unidirectional and, if so, to what extent does the transcription apparatus devote itself to the production of pre-mRNA rather than noncoding transcripts of opposite polarity? (2) If transcription is distinctly unidirectional, then what promoter element or elements, respectively, provide the "one-way sign" for the polar assembly of the PIC? (3) If such DNA elements exist, what protein or proteins "read" the one-way sign and what is the stereochemistry of this recognition process? Some of the studies reviewed here address transcription in Archaea, which appears to be a robust, simplified "stripped down," mimic of eukaryal pol II transcription and, as such, may be a convenient system with which to study structural principles common to both.

We invoke evidence from four types of ongoing experimental programs in our own laboratories and elsewhere that bear on these questions: (1) *chemical experiments* addressing the orientation with which a protein binds to DNA is assessed by the cleavage pattern generated by a radical species attached to one or the other side of the protein (Cox et al. 1997, 1998); (2) *binding and transcription experiments* in which TATA-flanking sequences are implicated in binding TFIIB (Lee and Hahn 1995; Lagrange et al. 1996, 1998) or its archaeal homolog, TFB (Qureshi and Jackson 1998); (3) *transcription experiments* in which the direction of transcription is assessed in response to inverting segments of the promoter; and (4) *X-ray crystallographic studies* of the carboxy-terminal cores of TBP and TFIIB (and the archaeal homolog, TFB) bound to promoter fragments that provide TATA-flanking segments which are long enough to include nucleotide sequences likely to specify polarity of assembly.

IMPRECISION IN THE TBP/TATA BOX INTERACTION

Structural Observations

Most eukaryal pol II promoters contain a TATA box centered about 25–30 bp upstream of the transcription start site. This element is generally accepted as the focus for the assembly of the PIC on TATA box containing promoters. Mutagenesis and structural studies indicate that the TATA box is the target of the highly conserved, 180-amino-acid carboxy-terminal domain of the TBP (TBPc). When the structure of *Arabidopsis thaliana* TBPc was solved (Nikolov et al. 1992; Nikolov and Burley 1994), the sequence repeat (30% identity) was reflected in a nearly twofold symmetrical saddle-like structure. The initial crystal structures of the TBP promoter complex showed a highly distorted promoter in which TBPc was bound to the widened minor groove of an 8-bp element which unwound a third of a helical turn and bent 80° toward the major groove (J.L. Kim et al. 1993; Y. Kim et al. 1993). Surprisingly, the polar groups of all but the central 2 bp formed van der Waals contacts with the hydrophobic, concave under-surface of the saddle. Thus, 12 of the 16 hydrogen-bond acceptors on the TATA-box's minor groove interface were buried and were not compensated for their loss of hydration by hydrogen-bonded interactions with the protein.

The most intriguing feature of the TBPc/TATA box interface is its almost perfect symmetry. Figure 1 shows that the residues which comprise the interface with the TATA box's minor groove form a nearly perfect twofold symmetrical pattern. The notable exception is the proline residue in the carboxy-terminal stirrup of eukaryal TBPc.

---

‡*Present address:* Department of Biological Chemistry and Molecular Pharmacology, Harvard Medical School, Boston, Massachusetts 02115.

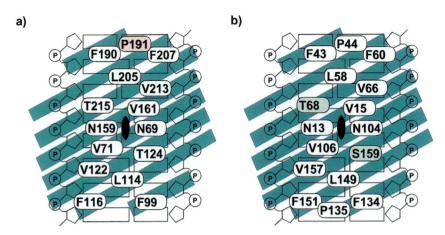

Figure 1. Schematic view of the TBP side-chain–DNA minor groove base interactions. (*a*) Yeast TBP interactions with the *CYC-1* TATA box (Y. Kim et al. 1993). (*b*) *Pyrococcus furiosus* TBP interactions with the *ef1α* TATA box (Kosa et al. 1997). (*Broad blue bars*) β-strands of TBP. The DNA of the TATA box is represented by open black rectangles for bases, circles for phosphates, and pentagons for sugars. (*Ovals*) Residues that make contact to the bases in the minor groove; (*white*) residues that are identical between the two repeats; (*green*) residues that are similar; (*pink*) residues that are unique to one repeat. With the exception of a single proline residue, the interactions of the yeast TBP are completely symmetric between the two repeats. The pyrococcal TBP/DNA interface is even more symmetrical with the only major difference between the repeats being a single methyl group on residue T68. In addition, the binding surfaces of the eukaryal and the archaeal TBPs are nearly identical.

But, even if the binding surface of TBP had a distinctive asymmetry, the fact that the minor groove van der Waals' surface and hydrogen-bonding patterns are almost the same for both A·T and T·A base pairs deprives the TATA element's minor groove of orientation-specific features to be recognized by TBP. We have pointed out that there may be an asymmetry of the TATA box's deformability due to the tendency of the more flexible TATA sequence to occur in the upstream half of the element and the more rigid AAAN sequence to occur in the downstream half (Y. Kim et al. 1993). This feature would require the more positively charged amino-terminal half of TBPc to be aligned so that it could bind with presumably higher affinity and deform the more resistant downstream end of the TATA box. But, many TATA boxes have symmetrical or nearly symmetrical sequences. In short, the protein-DNA interface of the TBP/TATA complex would seem to contribute little to specifying the orientation of the nucleating event in PIC assembly. Notwithstanding the apparent lack of structural features to impose directionality on TBP/DNA interaction, all crystal structures containing the TBP/DNA complex (except for one notable exception to be discussed below) show TBP to be oriented in one direction relative to the start site of transcription, namely, with the carboxy-terminal stirrup in contact with the upstream end of the TATA box and the amino-terminal stirrup in contact with the downstream end. Later, we revisit this apparent contradiction between expectation and observation.

In addition to the potential for orientational ambiguity, the similarity in minor groove stereochemistry of A·T and T·A base pairs permits imprecision in "axial" positioning of TBPc on the TATA box. Typically, most transcription factors bind to targets of more variable base composition. An axial displacement of the protein's DNA-binding surface by 1 bp upstream or downstream results in juxtapo-

sition of the protein's binding surface with a radically different array of major groove functional groups. In effect, misregistration of 1 bp produces a "nonspecific" interface that is as deleterious to specific complementary as a displacement of 1000 bp. The mismatches arise from bases within the target as well as from intrusion upon "nonspecific" base pairs at the identity element's boundary. However, in the minor groove of the TATA box, a displacement of the protein's binding surface imposes no mismatches within the A·T/T·A-containing segment and imposes mismatches only by intrusion on the flanking DNA. Moreover, sequence-dependent deformation and deformability, now established as important determinants of specific affinity, tend to be local (often a single base pair step) in DNA targets of nonuniform base sequences. Deformation and deformability factors are more uniform across the TATA box (Guzikevich-Guerstein and Shakked 1996; Lebrun et al. 1997; El Hassan and Calladine 1998).

Axial imprecision of TBP/TATA was demonstrated dramatically in the differences between the yeast TBPc/TATA box interfaces observed in the TBPc/*CYC-1* promoter complex (Y. Kim et al. 1993) and that observed in the ternary complex, yeast TFIIA/TBPc/*CYC-1* promoter (Geiger et al. 1996; Tan et al. 1996). In the binary complex, it is the first 8 bp of the sequence TATATAAAac that are used as a TATA box[1], whereas in the latter complex, the binding site was shifted 2 bp downstream, i.e., taTATAAAAC. Nevertheless, the structures of the two TBPc/TATA box complexes were superimposable (rmsd = 0.63 Å for all backbone atoms).

---

[1]Presumably caused by the steric interference of a loop in the stem-loop construct used only in the crystallization of the binary TBPc/TATA-box complex.

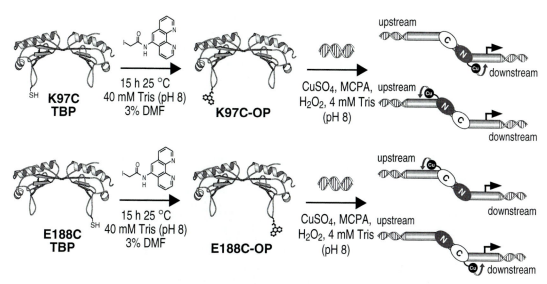

Figure 2. Scheme illustrating modification of wild-type and K97C and E188C variants of yeast TBPc with the affinity cleavage reagent 5-iodoacetamido-1,10-phenanthroline, and the regions of the promoter that are expected to be cleaved when K97C-OP and E188C-OP are bound in the orientations shown. N and C refer to the pseudosymmetrical halves of TBPc formed by the amino- and carboxy-terminal repeats, respectively. The lengths of the flanking DNA are not drawn to scale in this schematic. (Adapted from Cox et al. 1997.)

Affinity Cleavage Experiments

The orientational and axial precision exhibited by yeast TBPc when it binds to the TATA box was measured on two promoters: the yeast *CYC-1* promoter and the adenovirus major late promoter (AdMLP) (Cox et al. 1997, 1998). The strategy is outlined in Figure 2. Two variants of yTBPc were designed in which a cysteine residue replaced a polar residue in either the carboxy-terminal (amino acid residue E188) or amino-terminal (amino acid residue K97) stirrup. Phenanthroline moieties were appended by alkylation to these sulfhydryl groups with negligible changes in the modified TBPs' affinity for the promoter. The orientation of TBPc binding was assessed by examining the pattern of cleavage produced by the nondiffusible oxidants generated upon addition of cupric sulfates, H_2O_2, and mercaptopropionic acid. The results shown in Figure 3 were essentially the same for both TBPc constructs on either promoter. TBPc bound to the TATA box with only an approximate 60:40 preference for the orientation seen in the eukaryal crystal structures. Moreover, the "breadth" of the cutting pattern produced by this nondiffusible reagent suggested a substantial axial imprecision in TBP/promoter complex formation. The latter result was substantiated by a sharpening of the cutting pattern when a hairpin loop was used (as in the yTBPc/TATA crystal structure) to sterically restrict the axial positioning of the bound TBP (Cox et al. 1997, 1998).

Further affinity cleavage experiments were carried out which showed that TFIIA and TFIIBc influenced the ori-

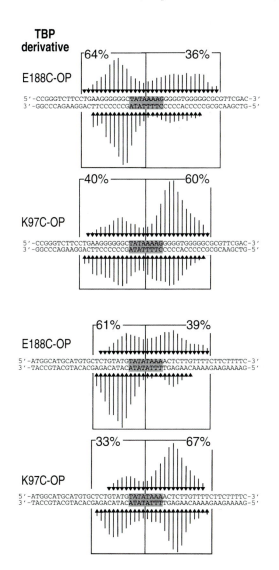

Figure 3. Histograms illustrating DNA cleavage by E188C-OP and K97C-OP TBPc variants at each base pair in the AdMLP and *CYC-1* promoters. The extent of cleavage at each position above a cupric ion, hydrogen peroxide, mercaptopropionic acid control is proportional to the length of the arrow. (Adapted from Cox et al. 1997.)

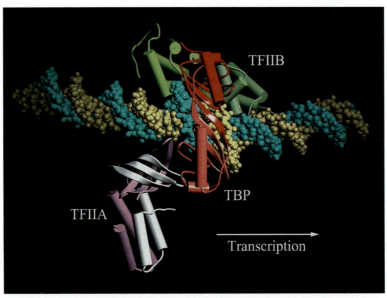

Figure 4. Presumed model of the TFIIBc/TFIIA/TBP/promoter complex assembled from the crystal structures of the hTFIIBc/hTBPc/TATA box (F. Tsai et al., in prep.) and the yTFIIA/yTBPc/TATA-box complex (Geiger et al. 1996; Tan et al. 1996), respectively. TBP is shown in *red*, TFIIB in *green*, and TFIIA in *magenta* (LSU) and *gray* (SSU). The two strands of the model B-form DNA are colored in *yellow* and *cyan*, respectively, and are depicted as a space-filling model. The quaternary complex was modeled by superimposition of yTBPc onto hTBPc, allowing the structure of yTFIIA to be overlaid onto the structure of the hTFIIBc/hTBPc/TATA box complex.

entation and axial positioning of TBP bound to the promoter. The TBPc/promoter complex nucleates assembly of the PIC by forming a unique nucleoprotein unit in which both the protein and DNA contribute to a characteristic distribution of functional groups and electrostatic charge potential. This nucleoprotein unit is the target of TFIIA, TFIIB, and, possibly, the entire holoenzyme complex. Crystal structures of the TFIIBc(TFBc)/TBPc/DNA complex show that the carboxy-terminal core of human TFIIB (or TFB) clamps the carboxy-terminal stirrup of TBPs (Nikolov et al. 1995; Kosa et al. 1997). TFIIBc (and TFBc) can interact with the DNA segments that flank both the upstream and downstream ends of the TATA box because they are brought close together by the severe bend imposed by TBP upon the promoter. Similarly, the crystal structures of the TFIIA-TBPc-DNA complex show that TFIIA binds to the amino-terminal stirrup on the downstream end of the TATA box and reaches back across the arched promoter to make extensive contact with the flanking DNA upstream of the TATA box (Geiger et al. 1996; Tan et al. 1996). As shown in Figure 4, these basal factors do not interfere with one another when bound to the promoter, nor do they change the structure of the TBP/TATA box complex.

To minimize steric interference between TFIIA or TFIIBc and the Cu^{++}-phenanthroline moieties, affinity cleavage was carried out only with constructs alkylated on the stirrup not bound by the general factor. The results showed that both factors enhance the asymmetry of binding on both promoters by 0.6–0.7 kcal·mol^{-1}, shifting the ratio of the two orientations from 60:40 to approximately 80:20 in favor of that seen in the crystal structures (Cox et al. 1997,

1998). This change in specificity suggests that the orientation of TBPc binding as seen in the crystal structures of eukaryal complexes is the one that nucleates a productive PIC but that the discrimination between the bound axial and rotational isomers, at least by each of these yeast factors, is still incomplete.

If a more precise orientation is to be achieved in the transcription process, interactions involving other components of the eukaryal PIC (TBP-associated factors, polymerase, etc.) are likely to provide the discriminatory free energy of binding that fixes the complex in the correct orientation. It is also important to note that only yeast factors were used in these affinity cleavage experiments. In view of the recent work on the IIB/TFB recognition element reviewed below (Qureshi and Jackson 1998; Lagrange et al. 1998), it may be important to revisit these experiments with human factors.

ARCHAEAL TRANSCRIPTION

About a decade ago, evidence began to mount rapidly that transcription in Archaea was strikingly similar to pol II transcription in Eukarya (for an excellent review, see Bell and Jackson, this volume). First, archaeal promoters had the same architecture containing a TATA-like element (the "A-box") centered about 27 bp upstream of the transcription start site. Second, Archaea employ nearly exact replicas of eukaryal TBP and TFIIB to execute specific, highly efficient, regulatable transcription by an RNA polymerase whose subunit composition and sequences are almost identical to that of pol II. As pointed out by Langer et al. (1995), human pol II has greater sequence similarity to archaeal RNA polymerase than to human pol I or pol III.

Biochemical experiments (Frey et al. 1990; Hausner and Thomm 1993; Hethke et al. 1996; Qureshi et al. 1997) and examination of complete archaeal genome sequences indicate archaeal transcription uses *only* TBP and the archaeal TFIIB homolog, TFB, to form a specific, regulatable PIC. Thus, to the extent that archaeal transcription shares basic stereochemical features with pol II, its simplicity provides an attractive alternative to the elaborate eukaryal complex for the structure-function analysis of underlying mechanisms. In addition, Archaea survive a wide range of environmental stresses, such as temperature, salinity, and pH, and thereby enlarge the parameter range with which to study the thermodynamics and kinetics of their behavior. Their resistance to environmental extremes also suggests a robustness suitable for structural studies. Finally, the archaeal life kingdom is a significant if not the dominant life form on the planet.

ARCHAEAL BASAL FACTORS: EXACTLY THE SAME STRUCTURES WITH A BIG DIFFERENCE

The crystal structure of TBP from the extreme thermophile, *P. furiosus*[2], revealed a fold that was essentially identical to that of eukaryal TBPc (Fig. 5), except for a curious disulfide bridge in its amino-terminal stirrup and a difference in the segment that contacts TFIIA in yeast TBPc (DeDecker et al. 1996). The latter difference is not surprising since archaeal genomes contain no TFIIA homolog. The crystal structure of the ternary complex of pyrococcal TBP-TFBc bound to the TATA box of an archaeal promoter was similar structurally to its eukaryal counterpart, including the severe deformation of the 8-bp TATA element and a seemingly specific interface in which the carboxy-terminal stirrup was clamped between the two cyclin domains of TFBc (Kosa et al. 1997). The big difference was the orientation of the archaeal proteins, which in the pyrococcal case was rotated 180° from the orientation seen in all of the eukaryal structures; i.e., the carboxy-terminal stirrup of TBP with its bound TFBc was now facing the *downstream* end of the TATA box.

Although it was tempting to speculate from differences in in vitro footprinting (Hausner et al. 1996) and the fact that archaeal TATA boxes often have sequences that are rotated 180° relative to the canonical eukaryal sequence (Zillig et al. 1993) that, indeed, physiologically, Archaea use a PIC in which the initial binding events are inverted to that seen in Eukarya, there was reason to interpret the inversion in the crystal structure with caution. First, the direct sequence repeats that give rise to the pseudo-symmetry of TBP are even more exact in Archaea than in Eukarya, which is consistent with the idea that the more slowly evolving Archaea approximate the gene duplica-

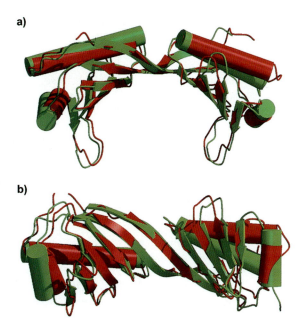

Figure 5. Comparison of the protein folds of TBPc from *P. furiosus* and eukaryal TBPc. (*Green*) Pyrococcal TBP; (*red*) yeast TBP. Both are displayed as ribbon diagrams. (*a*) Side view of the two TBPs superimposed; (*b*) view of the superimposed TBPs from the underside.

tion event that gave rise to the primitive TBP progenitor. In fact, the surface of TBP that contacts the minor groove is essentially twofold symmetrical; moreover, unlike the asymmetry of charge potential surrounding eukaryal TBPc, the charge potential around archaeal TBP is rotationally symmetric (DeDecker et al. 1996). Thus, the TBP/TATA interface in Archaea is virtually devoid of orientation clues and likely to impose even less directionality than its eukaryal counterpart. More importantly, in light of subsequent studies reviewed below, TFIIB and TFB exhibit a highly asymmetrical preference for sequence-specific interactions up to 7 bp upstream of the TATA box. The DNA fragments used in the crystal structure determinations of the TBP/TFIIBc/promoter complex (Nikolov et al. 1995) and its archaeal homolog (Kosa et al. 1997) present only three and two covalently contiguous base pairs, respectively, upstream of the TATA box, thereby obviating the influence exerted by TFIIBc or TFBc on the orientation of the bound proteins.

TFIIB AND TFB: POTENTIAL DETERMINANTS OF POLARITY

Qureshi and Jackson (1998) and Lagrange et al. (1998) have presented convincing evidence in Archaea and vertebrates, respectively, that a conserved sequence segment immediately upstream of the TATA box is a site for TFB/TFIIB binding and is important in transcription. They implicate TFBc and its homolog, TFIIBc, respectively, as the proteins that recognize this segment, which they refer to as the TFB/IIB recognition element (BRE). In the case of Archaea, 6 bp immediately upstream of the

---

[2]*P. furiosus* and *P. woesei*, the name used in previous publications, are different names for the same organism.

TATA box were implicated by TFB-binding assays, as well as mutational sensitivity in transcription and chemical modification interference studies. Using human factors and a TFIIBc affinity selection protocol, Lagrange et al. (1998) identified a 7-bp consensus sequence adjacent to the 5′ end of the TATA box. This region shows remarkably high affinity for TFIIBc in either the presence or absence of TBPc ($K_D = 45$ nM) and sensitivity to mutation in transcription assays. The authors also note that in the crystal structure of the TFIIBc/TBPc/AdMLP complex, helices H4′ and H5′ in the second cyclin lobe of TFIIBc have the structure and sequence signature of the helix-turn-helix motif seen in the DNA-binding interface of several bacterial transcription factors. In fact, the mode of interaction between the helix-turn-helix motif and the BRE is remarkably similar to that seen in the *trp* repressor/operator complex, although an analysis of the details of the interactions is precluded by the fact that the DNA fragment in the crystal structure of TBPc/TFIIBc/AdMLP complex contains only 3 bp upstream of the TATA box. Their analysis of eukaryal promoter sequences shows a highly significant correlation with the selected sequence in vertebrates but not necessarily in plants and fungi.

BRE AS A DETERMINANT OF DIRECTIONALITY IN TRANSCRIPTION

Within the elaborate assembly of the PIC on pol II promoters, there are many specific protein-DNA interactions that might contribute to unidirectional binding. Specific protein interactions with DNA elements such as the "initiator" sequence may combine with the recognition of the BRE to augment unidirectional assembly and transcription. On the other hand, Qureshi et al. (1997) have shown that efficient unidirectional in vitro transcription in Archaea required only recombinant TBP, recombinant TFB, and highly purified polymerase. Thus, the simplicity of this system recommends its use in defining the stereochemical basis of unidirectional transcription.

In collaboration with Stephen Jackson's Laboratory (CRC-Wellcome Laboratory, Cambridge, United Kingdom), in vitro transcription experiments were initiated with the two recombinant factors, pure polymerase from *Sulfolobus* and three different archaeal promoters in which various sequence segments were inverted. Preliminary evidence suggests that consistent with the affinity cleavage experiments in Eukarya (Cox et al. 1997, 1998), the orientation of the TATA box is irrelevant; rather, the direction of transcription was determined by the 6 bp adjacent to the upstream end of the TATA box that Qureshi and Jackson (1998) had identified as the target for TFB. Since Lagrange et al. (1998) had identified a similar target (although a difference consensus) for TFIIB in higher eukaryotes, and in view of the homology between the factors and the polymerase in the two systems, it is very likely that the specificity and affinity of the TFIIB-BRE interface contributes significantly to the polarity of transcription in Eukarya as well. The degree to which this presumption is true will require further experimentation.

STRUCTURAL STUDIES

To establish the stereochemical basis for the apparently specific interaction between the BRE and eukaryal TFIIBc, as well as its archaeal counterpart TFBc, we have initiated crystallographic studies of the appropriate ternary complexes.

Pyrococcal TFBc/TBPc/Promoter Complex

TFBc and TBPc from *P. furiosus* were prepared as described in Kosa et al. (1997), and the complex was cocrystallized with a blunt-ended base pair fragment of the SSV1 T6 promoter containing the TATA box flanked by 9 bp upstream and 6 bp downstream. Large, well-ordered cubic crystals (Fig. 6) were grown that diffracted to 4.0 Å resolution at home and 2.5 Å resolution at a synchrotron radiation light source. A molecular replacement solution has been found using the ternary complex of Kosa et al. (1997) that indicates the presence of one complex in the asymmetric unit. The preliminary model shows that helix 5′ of TFBc is lying in the major groove of the 6-bp BRE in a manner that would enable the side chains to reach the functional groups of the bases. So far, the result is consistent with findings of Qureshi and Jackson (1998) and the preliminary report of Bell (pers. comm.) that the BRE in the flanking segment upstream of the TATA element participates in a specific interaction that determines the direction of transcription.

Human Factors on a Modified MLP; hTFIIBc/hTBPc/AdMLP* Complex

Human TFIIBc (hTFIIBc) and hTBPc were prepared as described by F. Tsai et al. (in prep.), and the proteins

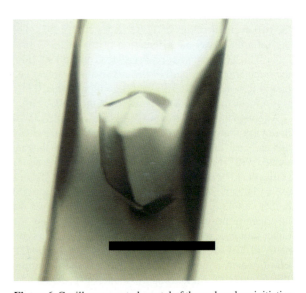

Figure 6. Capillary mounted crystal of the archaeal preinitiation complex. The complex includes TBP and TFBc both from *P. furiosus*, and a DNA oligonuleotide containing the TATA box and the TFB recognition element (BRE). Crystals of this complex reached a typical size of 0.4 × 0.4 × 0.4 mm³ in 7–10 days. For comparison, the scale of the bar shown is 0.5 mm.

Figure 7. Crystal of the hTFIIBc/hTBPc/promoter complex suspended in a cryoloop. The crystals were grown by the hanging-drop vapor diffusion method and reached a typical size of 0.6 × 0.1 × 0.05 mm³ after 4–6 weeks.

were cocrystallized with an 18-bp fragment of the AdMLP containing one version of the selection-optimized BRE sequence reported by Lagrange et al. (1998). The promoter segment contained 6 bp upstream and 4 bp downstream from the TATA box. The ends of the oligonucleotide have a 5′ overhang that was found to base pair at the abutting ends of adjacent promoter fragments in the crystal structure and thereby produced the seventh base pair of the BRE in good helical register. Well-ordered monoclinic crystals were grown (Fig. 7) that diffracted to 2.65 Å at a synchrotron radiation light source. The structure was determined by molecular replacement using the structure of the *A. thaliana* TBPc/hTFIIBc/DNA complex (Nikolov et al. 1995) as a search

model. Five ternary complexes were found in the asymmetric unit, which stacked end to end and were held in helical register by the G·C base pairs of the 5′ overhang. Figure 8 shows the current partially refined model of the structure in which helix 5′ of hTFIIBc is placed in the major groove of the BRE. It is noteworthy that the 78-amino-acid acidic activating domain of the herpes simplex virion protein VP16 was included in the crystallization. However, at the current stage of the refinement, no density could be assigned to the polypeptide.

DISCUSSION

The affinity cleavage experiments confirm the inference drawn from the symmetrical nature of the TBP/TATA interface—that the binding of the TBPc itself confers only a marginal degree of asymmetry on the assembly of the PIC on eukaryal pol II promoters. Thus, it is likely, but not yet shown, that the archaeal TBP/TATA-box complex, which has a more symmetrical interface than its eukaryal counterpart, will form with even less orientational preference. This supposition is supported by the finding that in the crystal structure of the *P. furiosus* complex containing TFBc, TBP, and a promoter fragment essentially devoid of the BRE, the proteins were bound in an orientation inverted to that seen in the eukaryotic complexes (Fig. 9), as well as that observed in the ongoing crystal structure analysis of an archaeal ternary complex in which a longer promoter fragment containing the BRE segment was used for cocrystallization (Fig. 10). We now believe that in the original structure of Kosa et al. (1997) favorable crystal packing forces defined the orientation rather than physiological interactions.

It is quite clear from the initial results using in vitro transcription experiments with archaeal factors, RNA

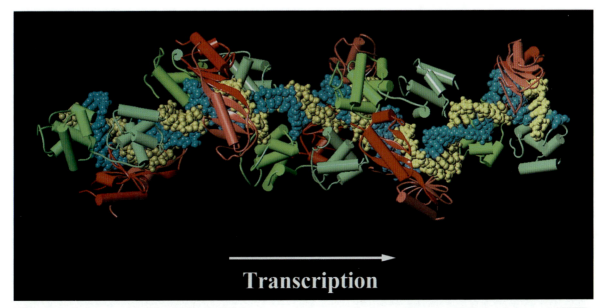

Figure 8. Preliminary model of the hTFIIBc/hTBPc/promoter complex as seen in the crystal structure (F. Tsai et al., in prep.). hTFIIBc is *light green* and *green*, and hTBPc is in *pink* and *red*, indicating the amino- and carboxy-terminal repeats, respectively. The two strands of the DNA are shown in *yellow* and *cyan*, and are depicted as a space-filling model. The figure depicts the orientation of all five ternary complexes present in the asymmetric unit of the crystal structure. The complexes are stacked end to end and were held in helical register by the G·C base pairs of the 5′ overhang.

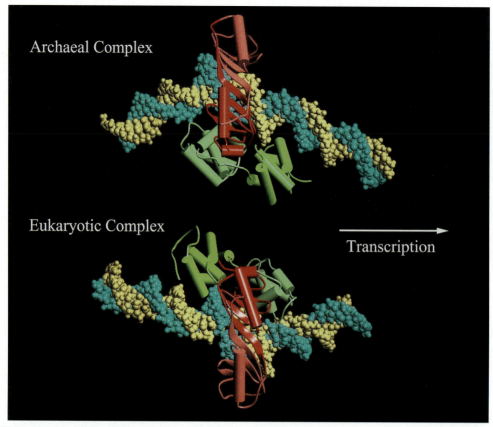

Figure 9. Ribbon drawings of the archaeal and eukaryal TFIIB(TFB)/TBP/DNA ternary complex as seen in the crystal structures (Kosa et al. 1997; F. Tsai et al., in prep.). TFIIB is colored in *light green* (amino-terminal repeat) and *green* (carboxy-terminal repeat), and TBP is shown in *pink* (amino-terminal repeat) and *red* (carboxy-terminal repeat). The DNA is *yellow* and *cyan* and is extended in both directions outside the oligonucleotide used in crystallization by modeling B-form DNA. The figure depicts the misoriented archaeal complex (*top*) bound to a promoter element that lacks the BRE, and the correctly orientated eukaryal complex (*bottom*) bound to a DNA segment in which the BRE was present. In both structures, the carboxy-terminal repeat of TBP interacts with TFIIB, although in the archaeal case, the TFBc/TBP complex arranged itself on the promoter in an inverted orientation relative to that seen for its eukaryal homolog.

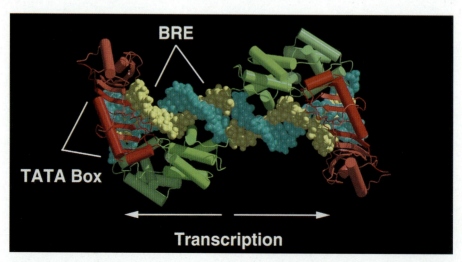

Figure 10. Current model of the *P. furiosus* TFBc/TBP/DNA complex (O. Littlefield et al., in prep.). The figure shows two copies of the model related around a crystallographic twofold. The arrows indicate the direction of transcription in the two models. The upstream ends of the oligonuleotides align in a pseudocontinuous helix. The two strands of the DNA are shown in *yellow* and *cyan*, respectively. The TBP molecule is *pink* and *red* for the amino- and carboxy-terminal repeats, and the TFBc molecule is *light green* and *green* for the amino- and carboxy-terminal cyclin repeats. The carboxy-terminal repeat of TFBc makes a close approach to the upstream BRE.

polymerase, and three different archaeal promoters that in vitro transcription in Archaea is unidirectional. Moreover, the direction in which transcription initiates is indifferent to the orientation of the TATA box, but very dependent on the promoter segment immediately upstream of the TATA box. Earlier studies in both Archaea and Eukarya had implicated the same 6- and 7-bp fragments, namely, the BRE, as an essential site for transcriptional activity. It has therefore become clear that this segment is the likely one-way sign in Archaea and an important determinant of transcriptional polarity in the pol II promoters of higher eukaryotes. Lagrange et al. (1998) have suggested that the BRE is recognized in much the same way as a helix-turn-helix domain interacts with several bacterial regulatory elements. Preliminary results of X-ray diffraction studies now indicate that helix 5′ in the second cyclin lobe of TFBc and TFIIBc is in contact with the major groove of the BRE. The resolution of these crystal structures is sufficient in both cases to ultimately give a definitive description of how the one-way sign is read.

ACKNOWLEDGMENTS

This work was supported in part by National Institutes of Health grants GM-15225 to P.B.S. and GM-52544 to A.S. F.T.F.T. is the recipient of a Wellcome Trust International Prize Traveling Research Fellowship (049086/Z/96/Z/JMW/LEC).

REFERENCES

Cox J.M., Kays A.R., Sanchez J.F., and Schepartz A. 1998. Preinitiation complex assembly: Potentially a bumpy path. *Curr. Opin. Chem. Biol.* **2:** 11.

Cox J.M., Hayward M.M., Sanchez J.F., Gegnas L.D., van der Zee S., Dennis J.H., Sigler P.B., and Schepartz A. 1997. Bidirectional binding of the TATA box binding protein to the TATA box. *Proc. Natl. Acad. Sci.* **94:** 13475.

DeDecker B.S., O'Brien R., Fleming P.J., Geiger J.H., Jackson S.P., and Sigler P.B. 1996. The crystal structure of a hyperthermophilic archaeal TATA-box binding protein. *J. Mol. Biol.* **264:** 1072.

El Hassan M.A. and Calladine C.R. 1998. Two distinct modes of protein-induced bending in DNA. *J. Mol. Biol.* **282:** 331.

Frey G., Thomm M., Brudigam B., Gohl H.P., and Hausner W. 1990. An archaebacterial cell-free transcription system. The expression of tRNA genes from *Methanococcus vannielii* is mediated by a transcription factor. *Nucleic Acids Res.* **18:** 1361.

Geiger J.H., Hahn S., Lee S., and Sigler P.B. 1996. Crystal structure of the yeast TFIIA/TBP/DNA complex. *Science* **272:** 830.

Guzikevich-Guerstein G. and Shakked Z. 1996. A novel form of the DNA double helix imposed on the TATA-box by the TATA-binding protein. *Nat. Struct. Biol.* **3:** 32.

Hausner W. and Thomm M. 1993. Purification and characterization of a general transcription factor, aTFB, from the archaeon *Methanococcus thermolithotrophicus. J. Biol. Chem.* **268:** 24047.

Hausner W., Wettach J., Hethke C., and Thomm M. 1996. Two transcription factors related with the eucaryal transcription factors TATA-binding protein and transcription factor IIB direct promoter recognition by an archaeal RNA polymerase. *J. Biol. Chem.* **271:** 30144.

Hethke C., Geerling A.C., Hausner W., de Vos W.M., and Thomm M. 1996. A cell-free transcription system for the hyperthermophilic archaeon *Pyrococcus furiosus. Nucleic Acids Res.* **24:** 2369.

Kim J.L., Nikolov D.B., and Burley S.K. 1993. Co-crystal structure of TBP recognizing the minor groove of a TATA element. *Nature* **365:** 520.

Kim Y., Geiger J.H., Hahn S., and Sigler P.B. 1993. Crystal structure of a yeast TBP/TATA-box complex. *Nature* **365:** 512.

Kosa P.F., Ghosh G., DeDecker B.S., and Sigler P.B. 1997. The 2.1-Å crystal structure of an archaeal preinitiation complex: TATA-box binding protein/transcription factor (II)B core/TATA-box. *Proc. Natl. Acad. Sci.* **94:** 6042.

Lagrange T. Kapanidis A.N., Tang H., Reinberg D., and Ebright R.H. 1998. New core promoter element in RNA polymerase II-dependent transcription: Sequence-specific DNA binding by transcription factor IIB. *Genes Dev.* **12:** 34.

Lagrange T., Kim T.-K., Orphanides G., Ebright Y.W., Ebright R.H., and Reinberg D. 1996. High-resolution mapping of nucleoprotein complexes by site-specific protein-DNA photocrosslinking: Organization of the human TBP-TFIIA-TFIIB-DNA quaternary complex. *Proc. Natl. Acad. Sci.* **93:** 10620.

Langer D., Hain J., Thuriaux P., and Zillig W. 1995. Transcription in archaea: Similarity to that in eucarya. *Proc. Natl. Acad. Sci.* **92:** 5768.

Lebrun A., Shakked Z., and Lavery R. 1997. Local DNA stretching mimics the distortion caused by the TATA box-binding protein. *Proc. Natl. Acad. Sci.* **94:** 2993.

Lee S. and Hahn S. 1995. Model for binding of transcription factor TFIIB to the TBP-DNA complex. *Nature* **376:** 609.

Nikolov D.B. and Burley S.K. 1994. 2.1 Å resolution refined structure of a TATA box-binding protein (TBP). *Nat. Struct. Biol.* **1:** 621.

Nikolov D.B., Chen H., Halay E.D., Usheva A.A., Hisatake K., Lee D.K., Roeder R.G., and Burley S.K. 1995. Crystal structure of a TFIIB-TBP-TATA-element ternary complex. *Nature* **377:** 119.

Nikolov D.B., Hu S.-H., Lin J., Gasch A., Hoffmann A., Horikoshi M., Chua N.-H., Roeder R.G., and Burley S.K. 1992. Crystal structure of TFIID TATA-box binding protein. *Nature* **360:** 40.

Qureshi S.A. and Jackson S.P. 1998. Sequence-specific DNA binding by the *S. shibatae* TFIIB homologue, TFB, and its effect on promoter strength. *Mol. Cell.* **1:** 389.

Qureshi S.A., Bell S.D., and Jackson S.P. 1997. Factor requirements for transcription in the archaeon *Sulfolobus shibatae. EMBO J.* **16:** 2927.

Tan S., Hunziker Y., Sargent D.F., and Richmond T.J. 1996. Crystal structure of a yeast TFIIA/TBP/DNA complex. *Nature* **381:** 127.

Zillig W., Palm P., Klenk H.P., Langer D., Hudepohl U., Hain J., Lanzendorfer M., and Holz I. 1993. Transcription in Archaea. In *The biochemistry of Archaea (archaebacteria)* (ed. M. Kates et al.), vol. 26, p. 367. Elsevier Science, Amsterdam.

Transcriptional Regulation by DNA Structural Transitions and Single-stranded DNA-binding Proteins

L.B. ROTHMAN-DENES, X. DAI, E. DAVYDOVA, R. CARTER, AND K. KAZMIERCZAK

Department of Molecular Genetics and Cell Biology, The University of Chicago, Chicago, Illinois 60637

Proteins that bind to single-stranded DNA (SSBs) with high affinity have been purified and characterized from several prokaryotes, eukaryotes, and their viruses (Chase and Williams 1986). These SSBs bind stoichiometrically and cooperatively to single-stranded DNA but without sequence specificity, lowering the melting temperature of DNA. SSBs cover the transient single-stranded regions of DNA that normally arise in vivo as a result of replication, recombination, and repair, thereby removing secondary structures that result from these cellular processes and presenting the DNA in an extended conformation for interaction with proteins involved in DNA metabolism. Several lines of evidence indicate that SSBs are involved in a multitude of protein-protein interactions (Formosa et al. 1983; Falkenberg et al. 1997; Kelman et al. 1998; Kong and Richardson 1998; Sarov-Blat and Livneh 1998). Recently, a number of reports in the literature implicate both nonspecific and site-specific SSBs in transcriptional regulation. Examples of SSBs that regulate transcription are found in both prokaryotes and eukaryotes. In most cases, not many details regarding such activation are known; however, some general characteristics emerge. Site-specific SSBs seem to be implicated in activation and repression in eukaryotes, whereas nonspecific SSBs have been found to have a role in transcription activation in prokaryotes.

SSBs AND TRANSCRIPTIONAL REGULATION

SSBs Involved in Transcriptional Activation

In eukaryotes, a role for several site-specific SSBs in activation is implicated by the requirement for the presence of the SSB's binding site and/or for the protein's synthesis in an appropriate cell-specific manner. In all cases, these SSBs bind preferentially to one strand of their cognate recognition sequence.

An enhancer element (USEIV), present upstream of the sea urchin H1-β histone gene promoter, is responsible for activation of H1-β histone gene expression in a temporally specific manner at the midblastula stage of embryogenesis (Di Liberto et al. 1989). Stage-specific activator protein 1 (SSAP-1), a 43-kD protein that binds to the USEIV, appears just before enhancer activation and undergoes a change in its molecular weight that parallels the increase in H1-β histone expression (De Angelo et al. 1993). SSAP-1 binds to single-stranded and double-stranded DNA in a sequence-specific manner. The DNA-binding activity is localized to a conserved RNA recogni-

tion motif (RRM) (De Angelo et al. 1995), whereas the rest of the protein contains a potent bipartite activation domain that has been shown to interact, in vitro, with TATA-binding protein (TBP), TFIIB, TFII74, and dTAF(II) 110 (DeFalco and Childs 1996). In another example, control of the mouse adipsin gene during adipocyte differentiation requires a sequence that is recognized by a factor which specifically binds to one of its strands. Induction of this factor during differentiation parallels transcription activation (Wilkinson et al. 1990). Similarly, cell-cycle-dependent transcription of the murine *Hft9-a/RanBP1* and *Hft9-c* genes has been correlated with the presence of SSBs that interact with the promoter (Di Matteo et al. 1998).

Further examples include Brn-3a, a member of the POU family of factors that bind to the octamer motif ATGCAAAT (Gerrero et al. 1993), which activates the α-internexin promoter (Budhram-Mahadeo et al. 1995). Activation is not dependent on octamer motifs or the POU domain but on the amino-terminal domain of Brn-3a as well as a sequence not related to the octamer motif (Budhram-Mahadeo et al. 1995). Analysis of the interaction of Brn-3a with its binding site, which confers Brn-3a responsiveness to a heterologous promoter, reveals that Brn-3a interacts strongly with one of the strands of this sequence but poorly with the double-stranded form (Budhram-Mahadeo et al. 1996). In the case of the c-*myc* gene, two SSBs, FBP and hnRNP K, activate transcription by binding to specific sites upstream of the promoter (Takimoto et al. 1993; Duncan et al. 1994; Tomonaga and Levens 1995). A HeLa nuclear extract protein binds preferentially to the noncoding strand of the serum response element required for activation of the platelet-derived growth factor α-chain gene promoter (Wang et al. 1993). In mammary epithelial cell cultures, two proteins that bind to one strand of a sequence present in promoters of many casein genes have been implicated in the induction of the β-casein gene family by insulin, prolactin, and glucocorticoids (Saito and Oka 1996). Full constitutive expression of the thyrotropin receptor gene in thyroid cells requires a DNA element at the promoter that is recognized by an SSB which interacts with one of its strands (Shimura et al. 1995). Finally, MF3 is a sequence-specific SSB from skeletal muscle which forms a stable complex in vitro with one strand of a sequence motif [5′-CATTCCTT-3′] that occurs in several muscle gene promoters (Santoro et al. 1991). Mutations that abolish MF3 binding in vitro prevent activation in vivo.

In contrast, work in our laboratory has shown that nonspecific SSBs are involved in activation of transcription during bacteriophage N4 development. Activation of the bacteriophage N4 early promoters requires *Escherichia coli* single-stranded DNA-binding protein (*Eco* SSB) (Markiewicz et al. 1992). Additionally, the bacteriophage N4-coded P17 protein, which is essential for activation of transcription of the phage middle genes (Zehring et al. 1983), is also a nonspecific SSB (R. Carter and L.B. Rothman-Denes, unpubl.).

SSBs Involved in Transcriptional Repression

SSBs have also been implicated in transcriptional repression. The proximal promoter element MB1 of the mouse myelin basic protein (MBP) gene has a role in cell-type-specific gene expression. Transient expression of protein Myef-2, which binds to one of the strands of the MB1 element, results in down-regulation of transcription of the MBP gene (Haas et al. 1995). SSBs have also been implicated in negative regulation of the lipoprotein lipase gene promoter in HeLa and CHO cells (Tanuma et al. 1995), of the growth hormone gene promoter (Pan et al. 1990), the androgen receptor promoter (Grossman and Tindall 1995), the platelet-derived growth factor α-chain gene promoter (Wang et al. 1994; Liu et al. 1996), the β-casein gene promoter in fibroblasts and myoblasts (Sun et al. 1995) and in mammary epithelial cells (Altiok and Groner 1993), and the smooth muscle α-actin gene promoter in fibroblasts, myoblasts, and visceral smooth muscle cells (Sun et al. 1995; Kimura et al. 1998). SSBs appear to have a role in sterol-mediated repression of transcription of the low-density lipoprotein receptor and of enzymes in the cholesterol biosynthetic pathway through the sterol regulatory element (SRE) (Rajavashisth et al. 1989), and in cell-type specific repression of the *N*-methyl-D-aspartate receptor (NAMDAR1) (Bai et al. 1998). More examples are provided by the family of Y-box proteins, which were initially characterized by their ability to bind to the Y-Box motif (ATTGG) (Didier et al. 1988). These proteins bind to double-stranded DNA, single-stranded DNA and RNA and have therefore been implicated in many cellular processes (Wolffe 1994b). Y-box protein YB-1 is a negative regulator of major histocompatibility complex (MHC) class II genes (Didier et al. 1988) and represses interferon-γ-induced transcription of MHC genes (Ting et al. 1994). Two other Y-box proteins, MY1 and MY1a, interact with a negative *cis*-acting element present in the promoter of the nicotinic acetylcholine receptor δ-subunit gene (Sapru et al. 1996). These proteins bind specifically to one of the strands of their target site and suppress promoter activity.

The involvement of SSBs in transcriptional regulation raises several questions. How do these proteins access their binding sites on double-stranded DNA? How do nonspecific single-stranded DNA-binding proteins achieve specificity of activation? How do they activate or repress transcription?

DNA STRUCTURAL TRANSITIONS LEADING TO SINGLE-STRANDED DNA PROTEIN BINDING

Single-strandedness Induced by Torsional Stress

To activate transcription, SSBs must be able to access single-stranded DNA. Strand separation, which is energetically unfavored, is promoted by the degree of negative supercoiling of the DNA. In vivo superhelical density is determined by the activity of topoisomerases, by the binding of proteins that constrain the writhe and/or twist of DNA, and by the process of transcription, which generates negative supercoiling upstream of actively transcribed promoters (Wang and Lynch 1993). In some of the cases described above, it has been shown that supercoiling of the template is essential for SSB binding (Wang et al. 1993; Michelotti et al. 1996b; Tomonaga and Levens 1996) or activity (Usheva and Shenk 1996).

Noncanonical DNA Structures Induced by Torsional Stress

Template supercoiling also promotes the formation of noncanonical DNA structures such as cruciforms and triple helices (Frank-Kamenetskii 1990). These DNA structures possess regions of single-strandedness that can lead to binding by SSBs.

Our studies on the mechanism of activation at the bacteriophage N4 early promoters provide a direct link between cruciform extrusion and activation by an SSB. Transcription of the coliphage N4 early genes is carried out by a phage-coded, virion-encapsulated DNA-dependent RNA polymerase (vRNAP) that is injected into the cell along with the linear 72-kb double-stranded DNA genome. Surprisingly, vRNAP does not utilize native N4 DNA or other double-stranded linear templates (Falco et al. 1977). However, in contrast to all other RNA polymerases, it transcribes single-stranded, promoter-containing templates accurately and efficiently (Haynes and Rothman-Denes 1985). The three N4 vRNAP promoters share sequence homology from position –18 to +1. This region includes a set of short (five to seven bases) inverted repeats, which are composed of conserved and nonconserved sequences, separated by three bases and centered at –12 (Haynes and Rothman-Denes 1985). Analysis of the transcriptional activity of a large set of mutant promoters indicates that certain conserved sequences as well as the inverted repeats are important for activity (Glucksmann et al. 1992). This striking requirement for promoter recognition suggests the need for DNA secondary structure at the promoter. In vivo, two host factors are required for N4 early transcription: *E. coli* DNA gyrase (Falco et al. 1978) and *Eco* SSB (Markiewicz et al. 1992). In vitro, supercoiled, promoter-containing templates do not support vRNAP activity at physiological superhelical densities unless *Eco* SSB is present. On the basis of these results, we proposed a model whereby the introduction of negative supercoils by DNA gyrase drives hairpin extrusion at the vRNAP promoters. Subsequently, *Eco* SSB binds to this region to yield an "activated promoter," within which vRNAP recognizes the hairpin structure and a subset of the conserved bases present at –18 to +1 on the template strand (Glucksmann et al. 1992).

On the basis of theoretical formulations, extrusion of 5–7-bp stem 3-base loop hairpins requires nonphysiological, high superhelical densities (Vologodskii 1992). To study hairpin extrusion at the vRNAP promoters, topoisomers of 2.2-kb DNA circles containing two vRNAP promoters were generated with a range of superhelical densities (Miller et al. 1996). Circles were treated in the absence of proteins with chemical (chloroacetaldehyde, CAA) and enzymatic (Mung Bean nuclease) single-stranded DNA-specific probes and T7 endonuclease I (T7 Endo I), which recognizes DNA four-way junctions (Dai et al. 1997, 1998). Results of these studies indicated that (1) a structural transition occurs at physiological super-helical densities in an Mg(II)-dependent manner; (2) the pattern of T7 Endo I reactivity indicates cruciform extrusion; (3) Mg(II) and supercoiling are required for both the formation and the maintenance of the cruciform structure; (4) Mg (II) must be stabilizing the cruciform's four-way junction, as it can be replaced by Ca(II), Mn(II), Hexamminecobalt (III), or spermine; and (5) the loop of the template-strand hairpin is highly resistant to single-strand specific probes, whereas the loop of the nontemplate strand hairpin is reactive, indicating that the two hairpins adopt different conformations.

The differential reactivity of the template- and non-template-strand hairpin loops toward single-strand-specific probes suggested the involvement of specific sequences in hairpin extrusion. Mutational analysis confirmed that specific sequences in the template strand, at the stem-loop junction and in the loop (5´-C-GXA-G-3´, where X = G, A, T), are required for extrusion (Dai et al. 1998). These sequences yield the most stable DNA hairpins (Hirao et al. 1994; Dai et al. 1997). Elucidation of the structure of template-strand hairpins by nuclear magnetic resonance (NMR) spectroscopy indicates that the hairpin loops exist as well-ordered, stacked structures (Yoshizawa et al. 1997; M. Kloster and L.B. Rothman-Denes, unpubl.), which explains their unusual stability and their lack of reactivity to single-stranded DNA-specific reagents. This unusual stability must drive the extrusion of these small hairpins at physiological superhelical densities (Dai and Rothman-Denes 1998).

Is hairpin extrusion unique to regulation of N4 vRNAP transcription? Three reports in the literature implicate hairpin extrusion or cruciform formation in transcriptional regulation of other systems. Expression of tyrosine hydroxylase (TH) is specific to catecholaminergic cells. DSE1, a sequence with imperfect dyad symmetry present upstream of the TH promoter, supports promoter activity in TH-expressing cells but is inhibitory in nonexpressing cells (Kim et al. 1998). Hydrolysis with S1 nuclease, sensitivity to OsO4, and reactivity to cruciform-specific antibodies demonstrate that the DSE1 sequence extrudes a

cruciform that interacts with nuclear proteins in super-coiled DNA (Kim et al. 1998). A second example involves the DAX-1 protein, which is an unusual member of the nuclear-receptor superfamily of transcription factors (Zanaria et al. 1994). DAX-1-binding sites are present in the *dax-1* and steroidogenic acute regulatory protein gene (*StAR*) promoters and are required for DAX-1-mediated repression. In vitro, DAX-1 binds to hairpin structures without marked sequence specificity as long as the loop is present and the stem is 10 bp or longer (Zazopoulos et al. 1997). Results of CAA treatment of transfected cells show that the wild-type site, but not mutant sites with base substitutions that disrupt the stem, exists in a hairpin conformation. The DAX-1 protein's in vitro DNA-binding properties, coupled with the requirement for hairpin-forming sequences for DAX-1-mediated repression of the *dax-1* and *StAR* promoters, indicate that hairpin extrusion is required for DAX-1-mediated transcriptional repression (Zazopoulos et al. 1997). Finally, the two strands of the cAMP-responsive enhancer of the enkephalin gene, which contains the CRE-1 and CRE-2 DNA elements, have been shown to interconvert into hairpin structures under certain solution conditions (Mc-Murray et al. 1991, 1994). Two lines of evidence suggest that a structural transition in the enhancer element regulates transcription of the enkephalin gene: CREB binds with high affinity to the hairpin but poorly to the native duplex. In addition, in vivo footprinting and mutational studies suggest that the observed structural transition is required for efficient transcription (Spiro et al. 1993).

Torsional stress also induces the formation of intramolecular triplexes (H-DNA) in vitro in homopurine-homopyrimidine (pur/pyr) mirror repeats (for review, see Frank-Kamenetskii and Mirkin 1995). Sequences that can form H-DNA are abundant in mammalian genomes and often occur in the 5´-flanking regions of genes (Schroth and Ho 1995). For a given pur/pyr mirror repeat, four distinct H-DNA isoforms are possible (for review, see Mirkin 1998). Stabilization of certain H-DNA isoforms requires low pH (Lyamichev et al. 1985; Collier and Wells 1990); however, other isoforms occur at neutral pH in the presence of Mg(II) (Kohwi and Kohwi-Shigematsu 1988). H-DNA was detected in vivo when pur/pyr stretches were cloned into *E. coli* (Kohwi and Kohwi-Shigematsu 1991; Ussery and Sinden 1993), and chromosomal reactivity to a triplex DNA monoclonal antibody was observed in mouse myeloma cells (Lee et al. 1987; Agazie et al. 1994). Evidence implicating pur/pyr stretches in regulation of transcription includes their ability to regulate expression of heterologous promoters (Chen et al. 1993; Santra et al. 1994; Xu and Goodridge 1996), the in vitro formation of supercoil-dependent structures as detected by reactivity to single-stranded DNA-specific probes (Hoffman et al. 1990; Bacolla and Wu 1991; O'Neill et al. 1991; Hollingsworth et al. 1994; Santra et al. 1994; Potaman et al. 1996; Xu and Goodridge 1996; Chen et al. 1997), and the isolation of SSBs that interact preferentially with the pyrimidine- or purine-rich strands of the sequence (Kolluri et al. 1992;

Hollingsworth et al. 1994; Xu and Goodridge 1996; Chen et al. 1997). The 5′-flanking region of the human γ-globin genes has been shown to undergo intramolecular triplex formation in vitro (Ulrich et al. 1992); furthermore, mutations that map to this region and that cause hereditary persistence of fetal hemoglobin alter formation of the structure (Ulrich et al. 1992; Bacolla et al. 1995). Despite the large number of examples where pur/pyr stretches are present in control regions, are S1-sensitive, and undergo a B-DNA to H-DNA transition in vitro, their involvement in either activation or repression of transcription remains conjectural. Indeed, recent reports indicate that some well-characterized pur/pyr sequences in fact do not have a role in transcriptional regulation (Raghu et al. 1994; Nelson et al. 1996; Pahwa et al. 1996; Becker and Maher 1998). The development of methods to detect triplex structures in vivo in mammalian cells is necessary to assess fully the role of triplex DNA in transcription regulation.

SPECIFICITY OF REGULATION BY SINGLE-STRANDED DNA-BINDING PROTEINS

For sequence-specific SSBs, specificity of regulation is provided by access to the cognate sequence in a single-stranded form. In several instances, the target sequences have been shown to be hyperreactive to S1 nuclease (Wang et al. 1993; Liu et al. 1996; Chen et al. 1997; Di Matteo et al. 1998) or potassium permanganate (Duncan et al. 1994; Michelotti et al. 1996b) in vivo, and supercoiling of the template is required for protein binding in vitro (Wang et al. 1993; Tomonaga and Levens 1996).

Studies of the protein FBP reveal an additional requirement for binding to its cognate sequence FUSE, which is a positive regulator of c-*myc* expression in cycling cells (Avigan et al. 1990). FBP binds specifically to the noncoding strand of FUSE, which is present upstream of the c-*myc* promoter (Duncan et al. 1994). FUSE is contiguous with an A+T-rich region that is topologically unwound in supercoiled DNA (Bazar et al. 1995). Analysis of the interaction of FBP with these sequences suggests that the unwound A+T-rich region serves as a nucleation site for FBP-induced melting of FUSE, which is necessary for subsequent interaction of FBP with the FUSE noncoding strand (Bazar et al. 1995).

The achievement of specificity by nonspecific SSBs is less clear and may evoke different mechanisms. In the case of *Eco* SSB, which exhibits no sequence specificity, we have proposed that the specificity of binding at the N4 vRNAP promoters is conferred by the promoter structure generated by template supercoiling. Supercoiled-induced extrusion of the promoter hairpins provides a region of single-strandedness, i.e., the loop of the nontemplate strand hairpin, for *Eco* SSB site-specific invasion (Dai and Rothman-Denes 1998). The recruitment of *Eco* SSB at the N4 vRNAP promoters through a site-specific structure solves the apparent paradox of the use of a nonspecific single-stranded DNA-binding protein for activation of transcription at specific sequences.

MECHANISMS OF REGULATION BY SINGLE-STRANDED DNA-BINDING PROTEINS

A priori, one might expect that SSBs which regulate transcription might simply act as "architectural proteins" since they modify the conformation of DNA (Wolffe 1994a). However, current evidence indicates that SSBs regulate transcription through multiple mechanisms.

Transcriptional Activation

Transcriptional activation through protein-protein interactions. Transcriptional activators bind to specific sites on double-stranded DNA near or far from promoters and stimulate transcription through contacts between their respective "activating regions" and the transcriptional machinery (Ptashne and Gann 1997; Hochschild and Dove 1998), with two known exceptions. The activator of T4 late gene transcription is the Gp45 sliding clamp that loads onto an enhancer (a primer-template junction) (Sanders et al. 1995). The sliding clamp activator subsequently tracks along the DNA toward the promoter where it activates through contacts with σ factor Gp55 and the Gp33 coactivator (Sanders et al. 1997). N4SSB, the activator of *E. coli* RNAP-σ^{70} at the bacteriophage N4 late promoters, activates transcription through direct interactions with the β′-subunit of RNAP, in the absence of DNA binding (Miller et al. 1997). Activators modulate transcription by increasing the binding of RNAP to the promoter or facilitating subsequent steps leading to promoter escape.

In several cases, transcription activation by SSBs occurs through contact between an "activating domain" present in the SSB and a component of the transcriptional machinery, leading to its recruitment to the promoter. Cloning, sequencing, and analysis of the activity of truncated forms of FBP revealed that the 644-amino-acid protein is organized in three domains (Duncan et al. 1994). The central domain contains four highly conserved nucleic-acid-binding KH domains that are required to recognize and unwind the FUSE sequence. The FBP carboxyl terminus functions as a classical *trans*-activation domain which, when tethered to a Gal4 DNA-binding domain, is capable of activating a reporter construct containing Gal4 DNA-binding sites (Duncan et al. 1996). This region contains three copies of the novel sequence motif AW(A/E)(A/E)YY (YM motif). Amino acid substitutions made within the YM motif indicate that it is responsible for transcription activation. Surprisingly, these YM motifs do not act synergistically; a single motif is sufficient for activation. In the full-length protein, the activation domain is masked by the amino terminus of FBP (Duncan et al. 1996). The regulation of the interaction between the amino and carboxyl termini and the identity of FBP's target in the transcription apparatus are unknown.

The CT element, consisting of five copies of the sequence CCCTCCCCA, is present upstream of the c-*myc* gene and acts as a positive regulator of c-*myc* and heterologous promoters in vitro and in vivo in an orientation-independent manner (Takimoto et al. 1993θ). The heterogeneous ribonucleoprotein particle protein K (hnRNP K) binds preferentially to the pyrimidine-rich strand of tandem CT elements spaced by at least three bases (Tomonaga and Levens 1995). hnRNP K contains three KH domains, each capable of binding to the CT element. The role of single-stranded DNA binding in activation is further supported by experiments utilizing a fusion of the VP16 *trans*-activation domain to the KH domains of hnRNP K (Tomonaga and Levens 1996). The VP16-hnRNP K fusion protein activates transcription in vivo from supercoiled but not from linear templates. The amino-terminal 35-amino-acid domain of hnRNP K, which contains 11 acidic residues, contributes to transcriptional activation (Tomonaga and Levens 1995). Coimmunoprecipitation of hnRNAP K with epitope-tagged TBP from nuclear extracts, as well as in vitro binding experiments, indicates that activation by hnRNAP K is mediated through protein-protein interactions with the basal transcription machinery (Michelotti et al. 1996a).

The binding of human Ying-Yang 1 protein (YY1) to its cognate site at +1 of the adenovirus-associated virus type-2 P5 promoter (P5+1) activates transcription (Seto et al. 1991). Initiation of transcription at the P5+1 site requires YY1, TFIIB, RNA polymerase II, and supercoiled template (Usheva and Shenk 1994). TFIIB binds to YY1 and stabilizes YY1's interaction with DNA. Subsequently, YY1 recruits pol II through interaction with the carboxy-terminal domain of the large subunit (Usheva and Shenk 1996). At the P5+1 site, YY1 contacts both strands upstream of +1; however, downstream from +1, YY1 contacts are restricted to the template strand (Houbaviy et al. 1996). Initiation on linear templates requires mismatched sequences at the YY1 site (Usheva and Shenk 1996). These results suggest that supercoiling induces strand separation that permits stable YY1 interaction and recruitment of the transcriptional machinery to the site.

Bacteriophage N4-coded protein P17 is required for transcription of the phage middle genes by N4 RNAPII (Zehring et al. 1983). Although N4 RNAPII shares sequence homology with the phage T7 class of small single-subunit RNA polymerases (S. Willis and L.B. Rothman-Denes, unpubl.), it does not transcribe double-stranded templates. P17 does not interact with double-stranded DNA but binds to single-stranded DNA without specificity. Results of transcription assays and gel-shift experiments indicate that P17 activates N4 RNAPII transcription on single-stranded templates by N4 RNAPII recruitment (R. Carter and L.B. Rothman-Denes, unpubl.). Activation is specific to P17, other SSBs cannot substitute. Although the mechanism of promoter recognition by N4 RNAPII remains to be elucidated, the absolute requirement of P17 for N4 RNAPII transcription in vivo indicates that its single-stranded DNA-binding activity is relevant, and its ability to recruit N4 RNAPII must be an intermediate in the pathway of promoter utilization.

Transcriptional activation through modifications of template structure. As mentioned previously, although physiological superhelical density leads to extrusion of the hairpin required for N4 vRNAP promoter recognition, no vRNAP transcription is detected unless *Eco* SSB is present. Activation is specific to *Eco* SSB; other single-stranded DNA-binding proteins cannot substitute (Markiewicz et al. 1992). The specific requirement for *Eco* SSB is puzzling since SSBs function in replication and recombination by destabilizing secondary structures. However, results of DNase-I-footprinting experiments indicate that the small, stable template strand hairpin persists upon *Eco* SSB binding, whereas the nontemplate strand hairpin is destabilized (Glucksmann-Kuis et al. 1996). Other SSBs tested destabilize the template strand hairpin (Markiewicz et al. 1992; Glucksmann-Kuis et al. 1996), implying that *Eco* SSB activates transcription by maintaining the template strand hairpin that contains the determinants of vRNAP recognition and binding. Three lines of evidence indicate that *Eco* SSB has additional roles. First, *Eco* SSB activates promoters containing inverted repeats that extrude weakly at physiological superhelical densities (with a 4-bp stem or an A:T base pair at the base of the stem), suggesting that *Eco* SSB stabilizes these hairpins (Dai and Rothman-Denes 1998; Dai et al. 1998). Second, promoters that contain four or five G:C base pairs in the hairpin stem display reduced activity on single-stranded templates; although vRNAP binds these promoters efficiently, it is unable to catalyze the formation of the first phosphodiester bond (Glucksmann et al. 1992). However, these promoters are activated by *Eco* SSB to wild-type levels on supercoiled templates (Dai and Rothman-Denes 1998). These results suggest that *Eco* SSB might facilitate promoter clearance, perhaps by destabilizing the promoter hairpin to disrupt initial contacts. Third, vRNAP is able to initiate transcription from promoters present on highly supercoiled templates in the absence of *Eco* SSB (Dai and Rothman-Denes 1998). However, no transcription is detected from a heteroduplex template composed of a wild-type promoter template strand and a nontemplate strand from which the inverted repeat sequences are deleted, i.e., a double-stranded DNA template containing an extruded template strand hairpin (M.A. Glucksmann-Kuis et al., unpubl.). Therefore, we propose that vRNAP requires a single-stranded DNA region, in addition to specific sequences and a hairpin structure, for binding and transcription initiation. How is a single-stranded region generated at this promoter? *Eco* SSB binding to templates of physiological superhelical density might create or stabilize single-strandedness at the promoter, which then allows vRNAP to initiate transcription. Although chemical and nuclease probes do not detect single-stranded bases immediately flanking the promoter hairpin on highly supercoiled templates (Dai et al. 1998), vRNAP may bind to the hairpin and induce a DNA conformational change involving strand opening (facilitated at high superhelical densities), resulting in stable and productive vRNAP-promoter association. These results would indicate that *Eco* SSB first activates N4 vRNAP promoters by acting as an "architectural protein" to provide the initial DNA structure for polymerase binding, and then destabilizes the structure upon promoter clearance.

However, the recent finding that human mitochondrial SSB (*Hsmt*SSB), which shares striking sequence and structural homology with *Eco* SSB (Ragunathan et al. 1997; Yang et al. 1997), fails to activate vRNAP promoters although it does not destabilize the promoter hairpin, suggests that *Eco* SSB-vRNAP interactions may also be required for transcription activation (E. Davydova and L.B. Rothman-Denes, unpubl.).

Binding of an SSB to its target site will provide, a priori, a site of flexibility for bending and twisting of the otherwise rigid double-stranded DNA and therefore allow for optimal protein-protein interactions between activators and the basal transcriptional machinery. Such a scenario has been recently tested by Levens and colleagues (Tomonaga et al. 1998). A CT element, the site of binding of hnRNP K, was placed between the adenovirus TATA box and a GAL4-binding site located 100 bp upstream. Transcriptional activation by a VP16 activator domain fused to the GAL4 DNA-binding domain was dependent on the presence of the wild-type CT between the promoter and the activator binding site. Although not in a natural context, these results reveal an additional possible physiological role for SSBs in transcription activation through changes in DNA structure.

Transcriptional Repression

Although mechanisms of repression by SSBs have not been studied in depth, the arrangement of the SSB-binding sites suggests that repression might be a result of SSB-induced changes in DNA structure obliterating the binding of an activator or a component of the basal transcriptional machinery.

Repression of the vascular smooth muscle α-actin promoter in fibroblasts and myoblasts has been correlated with the interaction of two tissue-specific SSBs, VACssBF1, and VACssBF2, with a 30-bp pur/pyr tract upstream of the promoter (Sun et al. 1995). VACssBF1 binds to the pyrimidine-rich strand, and VACssBF2 binds selectively to the purine-rich strand. The center of this region contains the site of binding of TEF-1, a double-stranded DNA-binding protein with no affinity for single-stranded DNA. Mutations in the TEF-1-binding site abolish promoter activity; in contrast, mutations outside the TEF-1-binding site which impair VACssBF1 and VACssBF2 binding result in transcription activation of the α-actin promoter. These results suggest that the stabilization of single-stranded regions by VACssBF1 and VACssBF2 binding might disrupt the normal double-stranded DNA configuration of the TEF-1 site, preventing TEF-1 binding and leading to repression (Sun et al. 1995).

Human Y-box protein YB-1, a negative regulator of the MHC class II genes, interacts with sequences of the DRA promoter. Recombinant, purified YB-1 specifically binds the sense strand of a DNA fragment containing the DRA X and Y proximal promoter elements (MacDonald et al. 1995). Analysis of the interaction of YB-1 with its binding site, using single-stranded DNA probes spanning either the Y or X elements, reveals that YB-1 binds to the sense strand of the Y element and the opposite strand of the X element. Detectable binding is also observed with the double-stranded Y element which contains the Y-box sequence CCAAT where, upon incubation at 37ºC, YB-1 promotes slow strand opening. The NF-Y/CBF family of proteins also bind to the double-stranded CCAAT sequence. It has been proposed that YB-1 represses transcription by inducing single-strandedness around the CCAAT box sequence and stabilizing its binding through interactions with opposite strands of the Y and X elements. Single-strandedness in this region prevents NF-Y/CBF binding at the MHC DRA promoter, which is the initial step in assembling a transcription-competent protein complex on the promoter proximal elements (MacDonald et al. 1995).

The general transcription cofactor PC4 enhances transcription from a variety of promoters and a wide range of transcriptional activators in vitro (Meisterernst et al. 1991; Ge and Roeder 1994; Kretzschmar et al. 1994). PC4 is a 127-amino-acid protein that contains two domains. The amino-terminal half has low, nonspecific affinity for double-stranded DNA and is required for protein-protein interactions and transcription activation (Kretzschmar et al. 1994). X-ray analysis of the structure of the carboxy-terminal half (P4-CTD) reveals a dimeric fold that provides a surface for two antiparallel single-stranded DNA strands (Brandsen et al. 1997). The P4-CTD, which is highly conserved in yeast (Henry et al. 1996), contains a latent but potent single-stranded DNA-binding activity that is unmasked by phosphorylation (Ge et al. 1994; Kretzschmar et al. 1994; Werten et al. 1998a). The phosphorylated form and the carboxy-terminal half do not stimulate transcription in vitro (Ge et al. 1994; Kretzschmar et al. 1994; Kaiser et al. 1995). PC4 and the PC4-CTD inhibit, rather than activate, transcription in a minimal system containing supercoiled template, recombinant TBP, TFIIB, TFIIEα, TFIIEβ, RAP30, RAP74, and RNAPII (Malik et al. 1998; Werten et al. 1998b; Wu and Chiang 1998). The isolation and characterization of PC4 mutants defective in single-stranded DNA-binding reveal that single-stranded DNA binding is required for repression but not for activation (Werten et al. 1998b). Since PC4 transcriptional repression is relieved by TFIIH, it is difficult to envision a role for PC4 repression at promoters, albeit it has been argued that the counterplay between PC4 and TFIIH could be subjected to regulation (Werten et a. 1998b). As an alternative, the possibility has been raised that PC4 represses transcription by binding to pyrimidine-rich single-stranded DNA structures at nonpromoter regions (Werten et al. 1998b).

CONCLUSIONS AND PROSPECTS

Figure 1 presents an attempt to summarize some of the mechanisms by which DNA structural transitions and SSBs regulate transcription. As mentioned above, the first requirement for regulation of transcription by SSBs is the accessibility of their binding sites, which are normally in a double-stranded conformation. The energetic

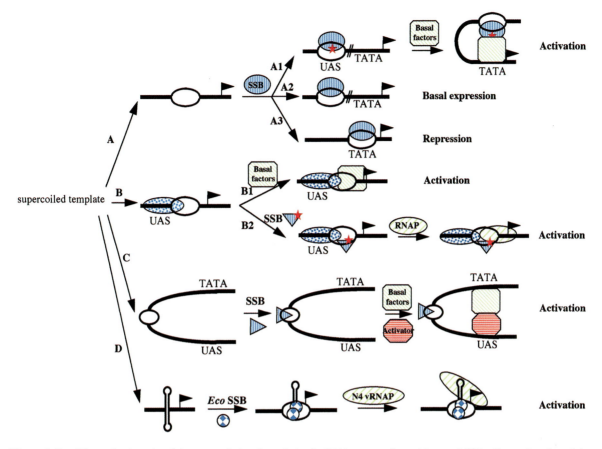

Figure 1. Possible mechanisms involving transcriptional regulation by DNA structural transitions and SSBs. For explanation of the different pathways, see the text. The star in some SSBs represents the transcription activation surface.

barrier to strand separation is provided by local underwinding of the DNA. Conformational transitions to cruciforms or triple helices or binding of SSBs are favored, since they stabilize the underwound region resulting in partial relaxation of superhelical stress. Why should transcription be regulated in response to the local superhelical density? In prokaryotes, the chromosomal superhelical density is partially unrestrained and its level is influenced by the ATP/ADP ratio (Mc Clellan et al. 1990; Hsieh et al. 1991a,b). In the case of coliphage N4, vRNAP-promoter recognition is the first event after injection of the genome into the host. We have suggested that the dependence of N4 vRNAP transcription on template supercoiling allows the phage to monitor the ability of the host cell to support phage growth (Markiewicz et al. 1992). In eukaryotes, nucleosomes restrain negative supercoils, whereas linker DNA appears relaxed under the electron microscope. However, topologically constrained domains exist due to chromosomal attachment to the nuclear matrix (Mirkovitch et al. 1984). In these constrained domains, negative supercoils will accumulate behind the transcribing RNAP (Wu et al. 1988). Such accumulated torsional stress can be relieved by local strand separation which will be stabilized by SSBs; therefore, SSBs will act as sensors of local torsional stress. One might envision that a promoter could undergo further activation

if the SSB has an activating domain. On the other hand, SSB binding could lead to repression if the SSB's binding site overlaps with the site of binding of basal or upstream factors.

In Figure 1, pathway A, unwinding is determined by the intrinsic properties of the sequence. The effect of binding of an SSB on transcription will depend on the location of the unwound region. In pathway A1, the site of SSB binding does not overlap the binding site of the basal transcriptional machinery. Interactions between the SSB's activating domain and a component of the basal transcription machinery will result in transcription activation. The properties of hnRNP K and FBP indicate that they activate the c-*myc* promoter through such a pathway (Duncan et al. 1994, 1996; Tomonaga and Levens 1995) In contrast, when the SSB-binding site overlaps the binding site of an activator (pathway A2) or the basal transcriptional machinery (pathway A3), SSB binding will result in basal expression or repression, respectively. The mode of YB-1 repression of the DRA promoter and of VACssBF1 and VACssBF2 repression of the vascular smooth muscle α-actin promoter in fibroblasts and myoblasts probably follows pathway A3 (MacDonald et al. 1995; Sun et al. 1995).

In pathway B, binding of a protein leads to underwinding of a region adjacent to its binding site. Such an activity is a hallmark of proteins involved in initiation of *E. coli* (dnaA protein) and bacteriophage λ (λ O protein) DNA replication (for review, see Kornberg and Baker 1992). The properties of protein YY1 described previously indicate that it might possess such an underwinding activity in addition to interactions with the single-stranded region. Underwinding might lead to direct transcription activation as in the case of YY1 (pathway B1), or to the recruitment of an SSB that interacts with the basal transcriptional machinery (pathway B2). Such a scenario has been proposed for activation of N4 RNAP II at its promoters (R. Carter and L.B. Rothman-Denes, unpubl.).

In pathway C, the binding of an SSB to a supercoil-driven unwound sequence produces a node of flexibility on the otherwise rigid double-stranded DNA, allowing optimal interactions between upstream activators and their targets in the basal machinery (Tomonaga et al. 1998). In this case, activation does not require the presence of activating domains in the SSBs since they do not contact the basal transcriptional machinery. This is reminiscent of the function of proteins such as integration host factor (IHF) or HMGs that, by binding and bending the DNA at specific sites, provide the correct topology for interactions between activators and the transcriptional machinery (Hoover et al. 1990; Bazett-Jones et al. 1994).

Finally, in pathway D, a threshold superhelical density leads to a DNA structural transition forming a cruciform (as depicted) or H-DNA, which is stabilized by the binding of proteins. The pathway in Figure 1 describes the mechanism of activation of N4 vRNAP promoters, where supercoiling drives extrusion of a cruciform which is invaded by *Eco* SSB, resulting in a template strand hairpin that is recognized by vRNAP (Dai and Rothman-Denes 1998). This pathway should also describe regulation by the DAX-1 protein except that, in that case, the hairpin provides the binding site for the repressor (Zazopoulos et al. 1997). It should also be pointed out that, conversely, hairpin extrusion could inhibit the interaction of sequence-specific DNA-binding proteins with their cognate sites if those sites overlap the extruding sequences. We have shown that replacement of sequences between the –10 and –35 hexamers of the *rrnB* P1 promoter with stable N4 hairpin sequences leads, in vivo, to transcriptional repression of the *rrnB* P1 promoter in a supercoiling-dependent manner (Dai et al. 1997).

We have only summarized a limited number of examples where SSBs or structural transitions are involved in transcription regulation. Many more cases of binding of SSBs to transcriptional control regions or of sequences that can undergo structural transitions (primarily to form H-DNA) at such control regions have been reported in the literature. However, their functional involvement remains to be determined. Further progress requires the development of new methodology to detect structural transitions, and more specifically H-DNA formation, in eukaryotic cells. Moreover, since in many cases the sequences required for SSB binding overlap with the binding site of transcription activators or transcription factors, mutational analysis of the sequences implicated in the structural transition or the SSB-binding site is required. Finally, genetic and structural studies of the SSBs are necessary to define the DNA binding as well as the transcription activation surfaces.

The preliminary nature of these studies notwithstanding, the fact that such a diverse group of SSBs recognize and capitalize from distorsions of the DNA in response to template supercoiling provides a compelling case for the involvement of SSBs in transcription regulation. Why should SSBs regulate transcription? As proposed by Crick (1971), SSBs increase the repertoire of possible DNA-protein interactions at a specific site. In addition, by locking the region of binding in a single-stranded configuration, they drastically change the conformation of the DNA, providing flexibility to allow bending and twisting, which then enables other proteins to interact with each other.

ACKNOWLEDGMENTS

We thank Drs. David Levens, James Maher, and Sergei Mirkin for discussions and critical reading of the text, and Dr. Steve Kowalczykowski for numerous discussions on SSBs. Work in our laboratory was supported by National Institutes of Health grants AI-12575 and GM-54431.

REFERENCES

Agazie Y.M., Lee J.S., and Burkholder G.D. 1994. Characterization of a new monoclonal antibody to triplex DNA and immunofluorescent staining of mammalian chromosomes. *J. Biol. Chem.* **269:** 7019.

Altiok S. and Groner B. 1993. Interaction of two sequence-specific single-stranded DNA-binding proteins with an essential region of the β-casein gene promoter is regulated by lactogenic hormones. *Mol. Cell. Biol.* **13:** 7303.

Avigan M., Strober B., and Levens D. 1990. A far upstream element stimulates c-*myc* expression in undifferentiated leukemia cells. *J. Biol. Chem.* **265:** 18538.

Bacolla A. and Wu F.Y.-H. 1991. Mung bean nuclease cleavage pattern at a polypurine-polypyrimidine sequence upstream from the mouse metallothionein gene. *Nucleic Acids Res.* **19:** 1639.

Bacolla A., Ulrich M.J., Larson J.E., Ley T.J., and Wells R.D. 1995. An intramolecular triplex in the human gamma-globin 5′-flanking region is altered by point mutations associated with hereditary persistence of fetal hemoglobin. *J. Biol. Chem.* **270:** 24556.

Bai G., Norton D.D., Prenger M.S., and Kusiak J.W. 1998. Single-stranded DNA binding proteins and neuron-restrictive silencer factor participate in cell-specific transcriptional control of the *NMDAR1* gene. *J. Biol. Chem.* **273:** 1086.

Bazar L., Meighen D., Harris V., Duncan R., and Levens D. 1995. Targeted melting and binding of a DNA regulatory element by a transactivator of c-*myc*. *J. Biol. Chem.* **270:** 824.

Bazett-Jones D., Leblanc B., Herfort M., and Moss T. 1994. Short-range DNA looping by the *Xenopus* HMG-box transcription factor, xUBF. *Science* **264:** 1134.

Becker N.A. and Maher L.J. 1998. Characterization of a polypurine/polypyrimidine sequence upstream of the mouse metallothionein-I gene. *Nucleic Acids Res.* **26:** 1951.

Brandsen J., Werten S., Van der Vliet P., Meisterernst M., Kroon J., and Gros P. 1997. Crystal structure of the C-terminal domain of the positive cofactor PC4 reveals ssDNA binding modes. *Nat. Struct. Biol.* **4:** 900.

Budhram-Mahadeo V., Morris P.J., Lakin N.D., Dawson S.J., and Latchman D.S. 1996. The different activities of the two activation domains of the Brn-3a transcription factor are dependent on the context of the binding site. *J. Biol. Chem.* **271:** 9108.

Budhram-Mahadeo V., Morris P.J., Lakin N.D., Theil T., Ching G.Y., Lillycrop K.A., Moroy T., Liem R.K., and Latchman D.S. 1995. Activation of the α-internexin promoter by the Brn-3a transcription factor is dependent on the N-terminal region of the protein. *J. Biol. Chem.* **270:** 2853.

Chase J.W. and Williams K.R. 1986. Single-stranded DNA binding proteins required for DNA replication. *Annu. Rev. Biochem.* **55:** 130.

Chen A., Reyes A., and Akeson R. 1993. A homopurine:homopyrimidine sequence derived from the rat neuronal cell adhesion molecule-encoding gene alters expression in transient transfections. *Gene* **128:** 211.

Chen S., Supakar P.C., Vellanoweth R.L., Song C.S., Chatterjee B., and Roy A.K. 1997. Functional role of a conformationally flexible homopurine/homopyrimidine domain of the androgen receptor gene promoter interacting with Sp1 and a pyrimidine single-strand DNA binding protein. *Mol. Endocrinol.* **11:** 3.

Collier D.A. and Wells R.D. 1990. Effect of length, supercoiling, and pH on intramolecular triplex formation. Multiple conformers at pur/pyr mirror repeats. *J. Biol. Chem.* **265:** 10652.

Crick F. 1971. General model for chromosomes of higher organisms. *Nature* **234:** 25.

Dai X., and Rothman-Denes L.B. 1998. Sequence and DNA structural determinants of N4 virion RNA polymerase-promoter recognition. *Genes Dev.* **12:** 2782.

Dai X., Kloster M., and Rothman-Denes L.B. 1998. Sequence-dependent extrusion of a small DNA hairpin at the N4 virion RNA polymerase promoters. *J. Mol. Biol.* **283:** 43.

Dai X., Greizerstein M., Nadas-Chinni K., and Rothman-Denes L.B. 1997. Supercoil-induced extrusion of a regulatory DNA hairpin. *Proc. Natl. Acad. Sci.* **94:** 2174.

DeAngelo D.J., DeFalco J., and Childs G. 1993. Purification and characterization of the stage-specific enhancer-binding protein SSAP-1. *Mol. Cell. Biol.* **13:** 1746.

DeAngelo D.J., DeFalco J., Rybacki L., and Childs G. 1995. The embryonic enhancer-binding protein SSAP contains a novel DNA binding domain which has homology to several RNA-binding motifs. *Mol. Cell. Biol.* **15:** 1254.

DeFalco J. and Childs G. 1996. The embryonic transcription factor stage-specific activator protein contains a potent bipartite activation domain that interacts with several RNA polymerase II basal transcription factors. *Proc. Natl. Acad. Sci.* **93:** 5902.

Didier D., Schiffenbauer J., Woulfe S., Zacheis M., and Schwartz B. 1988. Characterization of the cDNA encoding a protein binding to the major histocompatibility complex class II Y box. *Proc. Natl. Acad. Sci.* **85:** 7322.

Di Liberto M., Lai Z.C., Fei H., and Childs G. 1989. Developmental control of promoter-specific factors responsible for the embryonic activation and inactivation of the sea urchin early histone H3 gene. *Genes Dev.* **3:** 973.

Di Matteo G., Salerno M., Guarguaglini G., Di Fiore B., Palitti F., and Lavia P. 1998. Interaction of single-stranded and double-stranded DNA binding factors and alternative promoter conformation upon transcriptional activation of the *Hft9-a/RanBP1* and *Hft9-c* genes. *J. Biol. Chem.* **273:** 995.

Duncan R., Collins I., Tomonaga T., Zhang T., and Levens D. 1996. A unique transactivation sequence motif is found in the carboxyl-terminal domain of the single-strand-binding protein FBP. *Mol. Cell. Biol.* **16:** 2274.

Duncan R., Bazar L., Michelotti G., Tomonaga T., Krutzsch H., Avigan M., and Levens D. 1994. A sequence-specific, single-strand binding protein activates the far upstream element of c-*myc* and defines a new DNA binding motif. *Genes Dev.* **8:** 465.

Falco S.C., VanderLaan K., and Rothman-Denes L.B. 1977. Virion-associated RNA polymerase required for bacteriophage N4 development. *Proc. Natl. Acad. Sci.* **74:** 520.

Falco S.C., Zivin R., and Rothman-Denes L.B. 1978. Novel template requirements of N4 virion RNA polymerase. *Proc. Natl. Acad. Sci.* **75:** 3220.

Falkenberg M., Bushnell D.A., Elias P., and Lehman I.R. 1997. The UL8 subunit of the heterotrimeric herpes simplex virus type 1 helicase-primase is required for the unwinding of single strand DNA-binding protein (ICP8)-coated DNA substrates. *J. Biol. Chem.* **272:** 22766.

Formosa T., Burke R.L., and Alberts B.M. 1983. Affinity purification of bacteriophage T4 proteins essential for replication and genetic recombination. *Proc. Natl. Acad. Sci.* **80:** 2442.

Frank-Kamenetskii M.D. 1990. DNA supercoiling and unusual DNA structures. In *DNA topology and its biological effects* (ed. N.R. Cozzarelli and J.C. Wang), p. 185. Cold Spring Harbor Laboratory Press, Cold Spring Harbor, New York.

Frank-Kamenetskii M.D. and Mirkin S.M. 1995. Triplex DNA structures. *Annu. Rev. Biochem.* **64:** 65.

Ge H. and Roeder R.G. 1994. Purification, cloning, and characterization of a human coactivator, PC4, that mediates transcriptional activation of class II genes. *Cell* **78:** 513.

Ge H., Zhao Y., Chait B.T., and Roeder R.G. 1994. Phosphorylation negatively regulates the function of coactivator PC4. *Proc. Natl. Acad. Sci.* **20:** 12691.

Gerrero M.R., McEvilly R.J., Turner E., Lin C.R., O'Connell S., Jenne K.J., Hobbs M.V., and Rosenfeld M.G. 1993. Brn-3: A POU-domain protein expressed in the sensory, immune, and endocrine systems that functions on elements distinct from known octamer motifs. *Proc. Natl. Acad. Sci.* **90:** 10841.

Glucksmann M.A., Markiewicz P., Malone C., and Rothman-Denes L.B. 1992. Specific sequences and a hairpin structure in the template strand are required for N4 virion RNA polymerase promoter recognition. *Cell* **70:** 491.

Glucksmann-Kuis M.A., Dai X., Markiewicz P., and Rothman-Denes L.B. 1996. *E. coli* SSB activation of N4 virion RNA polymerase: Specific activation of an essential DNA hairpin required for promoter recognition. *Cell* **84:** 147.

Grossman M.E. and Tindall D.J. 1995. The androgen receptor is transcriptionally suppressed by proteins that bind single-stranded DNA. *J. Biol. Chem.* **270:** 10968.

Haas S., Steplewski A., Siracusa L., Amini S., and Khalili K. 1995. Identification of a sequence-specific single-stranded DNA binding protein that suppresses transcription of the mouse myelin basic protein gene. *J. Biol. Chem.* **270:** 12503.

Haynes L.L. and Rothman-Denes L.B. 1985. N4 virion RNA polymerase sites of transcription initiation. *Cell* **41:** 597.

Henry N.L., Bushnell D.A., and Kornberg R.D. 1996. A yeast transcriptional stimulatory protein similar to human PC4. *J. Biol. Chem.* **271:** 21842.

Hirao I., Kawai G., Yoshizawa S., Nishimura Y., Ishido Y., Watanabe K., and Miura K. 1994. Most compact hairpin-turn structure exerted by a short DNA fragment, d(GCGAAGC) in solution: An extraordinarily stable structure resistant to nucleases and heat. *Nucleic Acids Res.* **22:** 576.

Hochschild A., and Dove S.L. 1998. Protein-protein contacts that activate and repress prokaryotic transcription. *Cell* **92:** 597.

Hoffman E.K., Trusko S.P., Murphy M., and George D.L. 1990. An S1 nuclease-sensitive homopurine/homopyrimidine domain in the c-Ki-ras promoter interacts with a nuclear factor. *Proc. Natl. Acad. Sci.* **87:** 2705.

Hollingsworth M.A., Closken C., Harris A., McDonald C.D., Pahwa G.S., and Maher L.J. 1994. A nuclear factor that binds purine-rich, single-stranded oligonucleotides derived from S1-sensitive elements upstream of the CFTR gene and the MUC1 gene. *Nucleic Acids Res.* **22:** 1138.

Hoover T.R., Santero E., Porter S., and Kustu S. 1990. The integration host factor stimulates interaction of RNA polymerase with NIFA, the transcriptional activator of nitrogen fixation operons. *Cell* **63:** 11.

Houbaviy H., Usheva A., Shenk T., and Burley S. 1996. Cocrystal structure of YY1 bound to the adeno-associated virus P5 initiator. *Proc. Natl. Acad. Sci.* **93:** 13577.

Hsieh L.-S., Burger R., and Drlica K. 1991a. Bacterial DNA supercoiling and the ATP/ADP: Changes associated with transition to anaerobic growth. *J. Mol. Biol.* **219:** 443.

Hsieh L.-S., Rouvière-Yaniv J., and Drlica K. 1991b. Bacterial DNA supercoiling and the ATP/ADP ratio: Changes associated with salt shock. *J. Bacteriol.* **173:** 3914.

Kaiser K., Stelzer G., and Meisterernst M. 1995. The coactivator p15 (PC4) initiates transcriptional activation during TFIIA-TFIID-promoter complex formation. *EMBO J.* **15:** 1933.

Kelman Z., Yuzhakov A., Andjelkovic J., and O'Donnell M. 1998. Devoted to the lagging strand—The subunit of DNA polymerase III holoenzyme contacts SSB to promote processive elongation and sliding clamp assembly. *EMBO J.* **17:** 2436.

Kim E.L., Peng H., Esparza F.M., Maltchenko S.Z., and Stachowiak M.K. 1998. Cruciform-extruding regulatory element controls cell-specific activity of the tyrosine hydroxylase gene promoter. *Nucleic Acids Res.* **26:** 1793.

Kimura K., Saga H., Hayashi K., Obata H., Chimori Y., Ariga H., and Sobue K. 1998. c-Myc gene single-strand binding protein-1, MSSP-1, suppresses transcription of alpha-smooth muscle actin gene in chicken visceral smooth muscle cells. *Nucleic Acids Res.* **26:** 2420.

Kohwi Y. and Kohwi-Shigematsu T. 1988. Magnesium ion dependent triple-helix structure formed by homopurine:homopyrimidine sequences in supercoiled plasmid DNA. *Proc. Natl. Acad. Sci.* **85:** 3781.

———. 1991. Altered gene expression correlates with DNA structure. *Genes Dev.* **5:** 2547.

Kolluri R., Torrey T.A., and Kinniburgh A.J. 1992. A CT promoter element binding protein: Definition of a double-strand and a novel single-strand DNA binding motif. *Nucleic Acids Res.* **20:** 111.

Kong D. and Richardson C.C. 1998. Role of the acidic carboxyl-terminal domain of the single-stranded DNA binding protein of bacteriophage T7 in specific protein-protein interactions. *J. Biol. Chem.* **273:** 6556.

Kornberg A. and Baker T.A. 1992. *DNA replication.* W.H. Freeman, New York.

Kretzschmar M., Kaiser K., Lottspeich F., and Meisterernst M. 1994. A novel mediator of class II gene transcription with homology to viral immediate-early transcriptional regulators. *Cell* **78:** 525.

Lee J.S., Burkholder G.D., Latimer L.J., Haug B.L., and Braun R.P. 1987. A monoclonal antibody to triplex DNA binds to eucaryotic chromosomes. *Nucleic Acids Res.* **15:** 1047.

Liu B., Maul R.S., and Kaetzel D.M.J. 1996. Repression of platelet-derived growth factor A-chain gene transcription by an upstream silencer element. Participation by sequence-specific single-stranded DNA-binding proteins. *J. Biol. Chem.* **271:** 26281.

Lyamichev V.I., Mirkin S.M., and Frank-Kamenetskii M.D. 1985. A pH-dependent structural transition in the homopurine-homopyrimidine tract in superhelical DNA. *J. Biomol. Struct. Dyn.* **3:** 327.

MacDonald G., Itoh-Lindstrom Y., and Ting J. 1995. The transcriptional regulatory protein, YB-1, promotes single-stranded regions in the DRA promoter. *J. Biol. Chem.* **270:** 3527.

Malik S., Guermah M., and Roeder R.G. 1998. A dynamic model for PC4 coactivator function in RNA polymerase II transcription. *Proc. Natl. Acad. Sci.* **95:** 2192.

Markiewicz P., Malone C., Chase J.W., and Rothman-Denes L.B. 1992. E. coli single-stranded DNA binding protein is a supercoiled-template dependent transcriptional activator of N4 virion RNA polymerase. *Genes Dev.* **6:** 2010.

McClellan J.A., Boublikova P., Palecek E., and Lilley D.M.J. 1990. Superhelical torsion in cellular DNA responds directly to environmental and genetic factors. *Proc. Natl. Acad. Sci.* **87:** 8373.

McMurray C.T., Wilson W.D., and Douglass J.O. 1991. Hairpin formation within the enhancer region of the human enkephalin gene. *Proc. Natl. Acad. Sci.* **88:** 666.

McMurray C.T., Juranic N., Chandrasekaran S., Macura S., Li Y., Jones R.L., and Wilson W.D. 1994. Hairpin formation within the human enkephalin enhancer region. 2. Structural studies. *Biochemistry* **33:** 11960.

Meisterernst M., Roy A.L., Lieu H.M., and Roeder R.G. 1991. Activation of class II gene transcription is potentiated by a novel activity. *Cell* **66:** 981.

Michelotti E., Michelotti G., Aronsohn A., and Levens D. 1996a. Heterogeneous nuclear ribonucleoprotein K is a transcription factor. *Mol. Cell. Biol.* **16:** 2350.

Michelotti G., Michelotti E., Pullner A., Duncan R., Eick D., and Levens D. 1996b. Multiple single-stranded *cis* elements are associated with activated chromatin of the human c-*myc* gene in vivo. *Mol. Cell. Biol.* **16:** 2656.

Miller A.A., Wood D., Ebright R.E., and Rothman-Denes L.B. 1997. RNA polymerase β′ subunit: A target of DNA binding-independent activation. *Science* **275:** 1655.

Miller A., Dai X. , Choi M., Glucksmann-Kuis A., and Rothman-Denes L. B. 1996. Single-stranded DNA binding proteins as transcriptional activators. *Methods Enzymol.* **274:** 9.

Mirkin S.M. 1998. Structure and biology of H-DNA. In *Triple helix forming oligonucleotides* (ed. C. Malvy and A. Harel-Bellan). Kluwer Academic, Norwell, Massachusetts. (In press.)

Mirkovitch J., Mirault M.E., and Laemmli U.K. 1984. Organization of the higher order chromatin loop: Specific DNA attachment sites on the nuclear scaffold. *Cell* **39:** 223.

Nelson K.L., Becker N.A., Pahwa G.S., Hollingsworth M.A., and Maher L.J. 1996. Potential for H-DNA in the human MUC1 mucin gene promoter. *J. Biol. Chem.* **271:** 18061.

O'Neill D., Bornschlegel K., Flamm M., Castle M., and Bank A. 1991. A DNA-binding factor in adult hematopoietic cells interacts with a pyrimidine-rich domain upstream from the human delta-globin gene. *Proc. Natl. Acad. Sci.* **88:** 8953.

Pahwa G.S., Maher L.J., and Hollingsworth M.A. 1996. A potential H-DNA element in the MUC1 promoter does not influence transcription. *J. Biol. Chem.* **271:** 26543.

Pan W.T., Liu Q.R., and Bancroft C. 1990. Identification of a growth hormone gene promoter repressor element and its cognate double- and single-stranded DNA-binding proteins. *J. Biol. Chem.* **265:** 7022.

Potaman V.N., Ussery D.W., and Sinden R.R. 1996. Formation of a combined H-DNA/open TATA box structure in the promoter sequence of the human Na, K-ATPase α2 gene. *J. Biol. Chem.* **271:** 13441.

Ptashne M. and Gann A. 1997. Transcriptional activation by recruitment. *Nature* **386:** 569.

Raghu G., Tevosian S., Anant S., Subramanian K.N., George D.L., and Mirkin S.M. 1994. Transcriptional activity of the homopurine-homopyrimidine repeat of the c-Ki-*ras* promoter is independent of its H-forming potential. *Nucleic Acids Res.* **25:** 3271.

Ragunathan S., Ricard C.S., Lohman T.M., and Waksman G. 1997. Crystal structure of the homo-tetrameric DNA binding domain of E. coli single-stranded DNA binding protein determined by multiwavelength x-ray diffraction on the selenomethionyl protein at 2.9 Å resolution. *Proc. Natl. Acad. Sci.* **94:** 6652.

Rajavashisth T.B., Taylor A.K., Andalibi A., Svenson K.L., and Lusis A.J. 1989. Identification of a zinc finger protein that binds the sterol regulatory element. *Science* **245:** 640.

Saito H. and Oka T. 1996. Hormonally regulated double- and single-stranded DNA-binding complexes involved in mouse beta-casein gene transcription. *J. Biol. Chem.* **271:** 8911.

Sanders G.M., Kassavetis G.A., and Geiduschek E.P. 1995. Rules governing the efficiency and polarity of loading a tracking clamp protein onto DNA: Determinants of enhancement in bacteriophage T4 late transcription. *EMBO J.* **15:** 3966.

———. 1997. Dual targets of a transcriptional activator that tracks on DNA. *EMBO J.* **16:** 3124.

Santoro I.M., Yi T.-M., and Walsh K. 1991. Identification of single-stranded DNA binding proteins that interact with muscle gene elements. *Mol. Cell. Biol.* **11:** 1944.

Santra M., Danielson K.G., and Iozzo R.V. 1994. Structural and functional characterization of the human decorin gene promoter. A homopurine-homopyrimidine S1 nuclease-sensitive region is involved in transcriptional control. *J. Biol. Chem.* **269:** 579.

Sapru M.K., Gao J.P., Walke W., Burmeister M., and Goldman D. 1996. Cloning and characterization of a novel transcriptional repressor of the nicotinic acetylcholine receptor delta-subunit gene. *J. Biol. Chem.* **271:** 7203.

Sarov-Blat L.S. and Livneh Z. 1998. The mutagenesis protein MucB interacts with single strand DNA binding protein and induces a major conformational change in its complex with single-stranded DNA. *J. Biol. Chem.* **273:** 5520.

Schroth G.P. and Ho P.S. 1995. Occurrence of potential cruciform and H-DNA forming sequences in genomic DNA. *Nucleic Acids Res.* **23:** 1977.

Seto E., Shi Y., and Shenk T. 1991. YY1 is an initiator sequence-binding protein that directs and activates transcription in vitro. *Nature* **354:** 241.

Shimura H., Shimura Y., Ohmori M., Ikuyama S., and Kohn L.D. 1995. Single-strand DNA- binding proteins and thyroid transcription factor-1 conjointly regulate thyrotropin receptor gene expression. *Mol. Endocrinol.* **9:** 527.

Spiro C., Richard J.P., Chandrasekaran S., Brennan R.G., and McMurray C.T. 1993. Secondary structure creates mismatched base pairs required for high affinity binding of cAMP response-binding protein to the human enkephalin enhancer. *Proc. Natl. Acad. Sci.* **90:** 4606.

Sun S., Stoflet E.S., Cogan J.G., Strauch A.R., and Getz M.J. 1995. Negative regulation of the vascular smooth muscle α-actin gene in fibroblasts and myoblasts: Disruption of enhancer function by sequence-specific single-stranded-DNA-binding proteins. *Mol. Cell. Biol.* **15:** 2429.

Takimoto M., Tomonaga T., Matunis M., Avigan M., Krutsch H., Dreyfuss G., and Levens D. 1993. Specific binding of heterogeneous ribonucleoprotein particle protein K to the human c-*myc* promoter *in vitro*. *J. Biol. Chem.* **268:** 18249.

Tanuma Y., Nakabayashi H., Esumi M., and Endo H. 1995. A silencer element for the lipoprotein lipase gene promoter and cognate double- and single-stranded DNA-binding proteins. *Mol. Cell. Biol.* **15:** 517.

Ting J.P., Painter A., Zeleznik-Le N.J., MacDonald G., Moore T.M., Brown A., and Schwartz B.D. 1994. YB-1 DNA-binding protein represses interferon gamma activation of class II major histocompatibility complex genes. *J. Exp. Med.* **179:** 1605.

Tomonaga T. and Levens D. 1995. Heterogeneous nuclear ribonucleoprotein K is a DNA-binding transactivator. *J. Biol. Chem.* **270:** 4875.

———. 1996. Activating transcription from single-stranded DNA. *Proc. Natl. Acad. Sci.* **93:** 5830.

Tomonaga T., Michelotti G.A., Libutti D., Uy A., Sauer B., and Levens D. 1998. Unrestraining genetic processes with a protein DNA hinge. *Mol. Cell* **1:** 759.

Ulrich M.J., Gray W.J., and Ley T.J. 1992. An intramolecular DNA triplex is disrupted by point mutations associated with hereditary persistence of fetal hemoglobin. *J. Biol. Chem.* **267:** 18649.

Usheva A. and Shenk T. 1994. TATA-binding protein-independent initiation: YY1, TFIIB, and RNA polymerase II direct basal transcription on supercoiled template DNA. *Cell* **76:** 1115.

———. 1996. YY1 transcriptional initiator: Protein interactions and association with a DNA site containing unpaired strands. *Proc. Natl. Acad. Sci.* **93:** 13571.

Ussery D.W. and Sinden R.R. 1993. Environmental influences on the *in vivo* level of intramolecular triplex DNA in *Escherichia coli*. *Biochemistry* **32:** 6206.

Vologodskii A. 1992. *Topology and physics of circular DNA.* CRC Press, Boca Raton, Florida.

Wang J.C. and Lynch A.S. 1993. Transcription and DNA supercoiling. *Curr. Opin. Genet. Dev.* **3:** 764.

Wang Z.-Y., Lin X.-H., Nobukoshi M., and Deuel T.F. 1993. Identification of a single-stranded DNA-binding protein that interacts with an S1 nuclease-sensitive region of the platelet-derived growth factor A-chain gene promoter. *J. Biol. Chem.* **268:** 10681.

Wang Z.Y., Masaharu N., Qiu Q.Q., Takimoto Y., and Deuel T.F. 1994. An S1 nuclease sensitive region in the first intron of the human platelet-derived growth factor A-chain gene contains a negatively acting cell-type specific regulatory element. *Nucleic Acids Res.* **22:** 457.

Werten S., Langen F.W.M., Van Schaik R., Timmers H.T.M., Meistererncst M., and Van der Vliet P.C. 1998a. High affinity DNA binding by the C-terminal domain of the transcriptional coactivator PC4 requires simultaneous interactions with two opposing unpaired strands and results in helix destabilization. *J. Mol. Biol.* **276:** 367.

Werten S., Stelzer G., Goppelt A., Langen F.M., Gros P., Timmers H.T.M., Van der Vliet P.C., and Meistererncst M. 1998b. Interaction of PC4 with melted DNA inhibits transcription. *EMBO J.* **17:** 5103.

Wilkinson W.O., Min H.Y., Claffey K.P., Satterberg B.L., and Spiegelman B.M. 1990. Control of the adipsin gene in adipocyte differentiation. Identification of distinct nuclear factors binding to single-and double-stranded DNA. *J. Biol. Chem.* **265:** 477.

Wolffe A.P. 1994a. Architectural transcription factors. *Science* **264:** 1100.

———. 1994b. Structural and functional properties of the evolutionarily ancient Y-box family of nucleic acid binding proteins. *BioEssays* **16:** 245.

Wu H.-Y., Shyy S., Wang J.C., and Liu L.F. 1988. Transcription generates positive and negative supercoiled domains in the template. *Cell* **53:** 433.

Wu S.-Y. and Chiang C.-M. 1998. Properties of PC4 and an RNA polymerase II complex in directing activated and basal transcription *in vitro*. *J. Biol. Chem.* **273:** 12492.

Xu G. and Goodridge A.G. 1996. Characterization of a polypyrimidine/polypurine tract in the promoter of the gene for chicken malic enzyme. *J. Biol. Chem.* **271:** 16008.

Yang C., Curth U., Urbanke C., and Kang C. 1997. Crystal structure of the human mitochondrial single-stranded DNA binding protein at 2.4 Å resolution. *Nat. Struct. Biol.* **4:** 153.

Yoshizawa S., Kawai G., Watanabe K., Miura K., and Hirao I. 1997. GNA trinucleotide loop sequences producing extraordinarily stable DNA hairpins. *Biochemistry* **36:** 4761.

Zanaria E., Muscatelli F., Bardoni B., Strom T.M., Guioli S., Guo W., Lalli E., Moser C., Walker A.P., McCabe E.R., Meitinger T., Monaco A.P., Sassone-Corsi P., and Camerino G. 1994. An unusual member of the nuclear hormone receptor superfamily responsible for X-linked adrenal hypoplasia congenita. *Nature* **372:** 635.

Zazopoulos E., Lalli E., Stocco D.M., and Sassone-Corsi P. 1997. DNA binding and transcriptional repression by DAX-1 blocks steroidogenesis. *Nature* **390:** 311.

Zehring W.A., Falco S.C., Malone C., and Rothman-Denes L.B. 1983. Bacteriophage N4-induced transcribing activities in *E. coli*. III. A third cistron required for N4 RNA polymerase II activity. *Virology* **126:** 678.

The DPE, a Conserved Downstream Core Promoter Element That Is Functionally Analogous to the TATA Box

T.W. Burke, P.J. Willy, A.K. Kutach, J.E.F. Butler, and J.T. Kadonaga

Department of Biology, 0347, and Center for Molecular Genetics, University of California, San Diego, La Jolla, California 92093-0347

Transcription by RNA polymerase II is a fundamental and important biological process. There are tens of thousands of diverse genes that are transcribed in a highly regulated manner by the RNA polymerase II transcriptional machinery, and thus the mechanisms by which transcription is controlled are necessarily complex (for reviews, see Conaway and Conaway 1993; Smale 1994; Koleske and Young 1995; Zawel and Reinberg 1995; Björklund and Kim 1996; Kaiser and Meisterernst 1996; Orphanides et al. 1996; Reines et al. 1996; Roeder 1996; Svejstrup et al. 1996; Verrijzer and Tjian 1996).

The current data on basal transcription factors indicate that RNA polymerase II along with auxiliary factors, termed transcription factor (TF)IIA, TFIIB, TFIID, TFIIE, TFIIF, and TFIIH, are required for transcription initiation, whereas the polymerase and TFIIF, TFIIS, elongin/SIII, ELL, and P-TEFb are involved in transcriptional elongation. It has also been proposed that some of the RNA polymerase II exists as a component of a large conglomerate termed the holoenzyme, which contains RNA polymerase II (also referred to as the core polymerase) and many other factors (for reviews, see Koleske and Young 1995; Orphanides et al. 1996).

This paper focuses on the core promoter, which is defined herein to be the key DNA sequences that immediately encompass the RNA start site (from about –40 to +40) that are sufficient to direct the accurate initiation of transcription. Two well-studied core promoter elements are the TATA box and the initiator (Inr). The TATA box is an A/T-rich sequence that is present in some, but not all, promoters at about 25–30 nucleotides upstream of the transcription start site. The TATA element is a binding site for the TATA-box-binding polypeptide (TBP) component of the multisubunit TFIID complex (for reviews, see Pugh and Tjian 1992; Smale 1994; Burley and Roeder 1996; Verrijzer and Tjian 1996). The consensus sequence for the TATA box is usually designated as TATAAA, but it has also been shown that there can be considerable variation in the TATA box sequence (see, e.g., Singer et al. 1990; Wiley et al. 1992; Zenzie-Gregory et al. 1993). Immediately upstream of the TATA box, TFIIB contacts the promoter DNA. This TFIIB interaction site, termed the TFIIB recognition element (BRE), has a distinct sequence preference (Lagrange et al. 1998). The extent of similarity between a BRE in a TATA-containing promoter and the consensus sequence for the BRE appears to be parallel to the ability of TFIIB to assemble into a TBP-TFIIB-DNA complex and to promote the transcription process.

The Inr element was identified as a sequence that encompasses the RNA start site that is required to direct accurate transcription in the absence of a TATA box (see, e.g., Smale and Baltimore 1989; Weis and Reinberg 1992; Smale 1994). The consensus for the Inr element is $Py\text{-}Py\text{-}A_{+1}\text{-}N\text{-}^T/_A\text{-}Py\text{-}Py$ (where A_{+1} is the transcription start site) for mammalian genes (Smale and Baltimore 1989; Javahery et al. 1994) and $T\text{-}C\text{-}A_{+1}\text{-}^G/_T\text{-}T\text{-}^T/_C$ for *Drosophila* genes (Hultmark et al. 1986; Purnell et al. 1994; Arkhipova 1995). As with the TATA box element, sequences that resemble the Inr consensus are found in some, but not all, promoters.

IDENTIFICATION OF THE DPE: A DOWNSTREAM CORE PROMOTER ELEMENT

In our analysis of basal transcription by RNA polymerase II, we devised a means for the purification of epitope-tagged TFIID complex (termed eTFIID) from transgenic *Drosophila* (Burke and Kadonaga 1996). TFIID from *Drosophila* appears to consist of TBP and eight TBP-associated factors, which are known as TAFs (see, e.g., Dynlacht et al. 1991; Kokubo et al. 1993; Burke and Kadonaga 1996). Because the binding of TFIID appears to be the first step in the pathway leading to the assembly of the transcription preinitiation complex, we characterized the binding of the purified eTFIID to various TATA-box-containing promoters as well as to TATA-box-deficient (TATA-less) promoters and found that the purified protein bound to both types of promoters. We were particularly interested in the binding of TFIID to the TATA-less promoters and noticed that both of the two TATA-less promoters that we had been studying (*jockey* and *Antennapedia* P2, both from *Drosophila*) had a related segment (7 out of 8 bp identity) at about +30 relative to the start site. We then found that mutation of these sequences in each of the two promoters led to a substantial reduction in the binding of purified TFIID and a 20–100-fold decrease in transcriptional activity. A search of a *Drosophila* promoter database (Arkhipova 1995) for this sequence revealed that it was present in a variety of TATA-less promoters (Fig. 1) (Burke and Kadonaga 1996). On the basis of its properties, we named this conserved core promoter element the DPE, for *downstream promoter element*.

The DPE is a distinct 7-bp element that is located at about +30 (typically, from +28 to +34) relative to the transcription start site. It is present in many TATA-less

Figure 1 section (sequences):

```
                                 Inr
                                +1→
                                 |
297                          TT  TTAGTC  TTAAGCTGAGATCCAAAGAATAA      AGTCGTG  AAACTATT
Antennapedia P2   ATTCACTGGCGT   TCAGTT  GTGAATGAATGGACGTGCCAAAT      AGACGTG  CCGCCGCC
Doc                          G   ACATTC  GGCATTCCACAGTCTTCGGGTGG      AGACGTG  TTTCTTCA
G                                ACAGTC  GCGATCGAACACTCAACGAGTGC      AGACGTG  CCTACGGA
I                                CATTA   CCACTTCAACCTCCGAAGAGATA      AGTCGTG  CCTCTCAG
jockey               AAAAA       TCATTC  ACATGGGAGGATGAGCAATCGAGT     GGACGTG  TTCACAGA

consensus:                       TCA G T                             A GA CGTG
                                     T C                             G  T
```

```
Abdominal-B      CGGTGCAGTGAG   TCAGTG  TGTTGTGTGCCCCAGTCGCGAGCG     GACG  ATCCGTGGAG
brown            CAACGACGTTGC   CCAGTT  GCCGCTCAGTCTTCCACGCGAACA     GTCG  ACGGCGTGTG
caudal           TTGCTCTACGTT   TCAGTA  CGTGTTCGACCTGCATACTGAGTT     GACG  TCGCCACGCA
E74A             CTTGTTCGCCGT   TTAGTT  GTCTTTTGACTGCTGTAACGGACA     GTCG  CAAATTTTGC
E74B             TGACTTTAGATT   TTAGTT  TTGCTGATACCGTAAGAGATAAAT     GACG  TGCCGCGGCG
Fin                      TTCGT  GGATTT  CAATTCGATCGCCGACGTGTGAA      GACG  TTTTTATCGT
glass            TGCAACTGACAA   TTAGTC  GCCGTCTTATGCTCCTCGCCAAAA     GTCG  CTTCTTGCCC
ras2             AGTGTGGATTTC   TCAGTT  AACCGAGAACGGTCACGCTGCTGCT    GTCG  AGGAAGGAAA
singed           TCGCAACGGGTT   TCATTC  CCACTGGAGTGCAGTTCGTGAGCG     GTCG  TTCTCTCCTC

consensus:                      TCA G T                             G A CG
                                    T C                              T
```

Extended DPE Consensus: A GA CGTG   Core DPE Consensus: G A CG
                         G T   T

Figure 1. The DPE is a conserved downstream promoter element that is present in many TATA-less promoters. These and other potential DPE-containing promoters were identified in a search of the *Drosophila* promoter database of Arkhipova (1995) and other DNA sequence databases for promoters containing sequences related to the DPE motifs in the *Drosophila Antennapedia* P2 and *Drosophila jockey* core promoters. In a statistical analysis, Arkhipova (1995) had identified the quadruplet sequences CGTG, ACGY, and CTCG as being overrepresented in the downstream region of TATA-less promoters. The first two of these quadruplets are related to sequences in the extended DPE consensus. This figure is adapted from Burke and Kadonaga (1996). The scope of the DPE consensus sequence has yet to be determined. Hence, it is likely that there are other promoters that possess functionally important DPE sequences that were not identified in this initial search. (Reprinted, with permission, from Burke and Kadonaga 1996.)

promoters and is specifically bound by TFIID, but not by TBP. Binding of purified TFIID to DPE-containing TATA-less promoters requires both the Inr and DPE motifs. These data suggest that the DPE acts in conjunction with the Inr to mediate basal transcription of TATA-less promoters. The TATA, BRE, Inr, and DPE core promoter elements are depicted in Figure 2. In the remainder of this paper, we discuss some of the properties of the DPE. These data suggest that the DPE is a downstream analog of the TATA box.

THE DPE IS A DISTINCT CORE PROMOTER ELEMENT

To determine in a systematic manner the importance of the tentative DPE consensus sequence in basal RNA polymerase II transcription, we constructed and analyzed a series of clustered triple point mutations in the *Drosophila jockey* promoter from +20 to +40 relative to

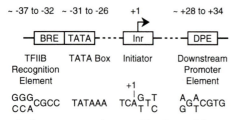

Figure 2. Core promoter elements. The *Drosophila* consensus sequences for the Inr and DPE are indicated.

the RNA start site (Fig. 3). Transcriptional analysis of these mutant promoters with factors derived from either *Drosophila* embryos or human (HeLa) cells revealed that alteration of the core of the DPE consensus, but not the flanking sequences, resulted in a sharp decrease in the efficiency of transcription from the *jockey* promoter. These experiments reveal the importance of the DPE consensus sequence for basal transcription. In addition, the observation that both *Drosophila* and human transcription factors exhibit the same requirement for the DPE sequence suggests that the DPE is used as a core promoter element in both *Drosophila* and humans.

In addition, because the DPE is downstream from the transcription start site, mutation of the DPE sequence results in mutation of the RNA transcript. We therefore tested whether the reduction in transcription that is seen upon mutation of the DPE is due to the generation of an unstable transcript (Burke and Kadonaga 1997). These experiments revealed that the decrease in transcription that occurs upon mutation of the DPE sequence in TATA-less promoters is due to reduced efficiency of the transcription process, rather than to reduced transcript stability.

THE SPACING BETWEEN THE DPE AND THE INR IS IMPORTANT FOR TRANSCRIPTION AND FOR TFIID BINDING

In addition to the generation and analysis of clustered point mutations of the DPE, we tested whether the spacing between the DPE and the Inr is important for pro-

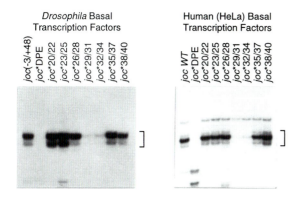

Drosophila Basal Transcription Factors

Human (HeLa) Basal Transcription Factors

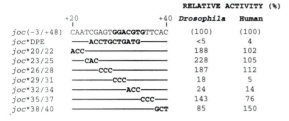

		RELATIVE ACTIVITY (%)	
	+20 +40	*Drosophila*	Human
joc(-3/+48)	CAATCGAGT**GGACGTG**TTCAC	(100)	(100)
*joc**DPE	————**ACCTGCTGATG**————	<5	4
*joc**20/22	ACC———————————————	188	102
*joc**23/25	———CAC————————————	228	105
*joc**26/28	——————CCC—————————	187	112
*joc**29/31	—————————CCC——————	18	5
*joc**32/34	————————————ACC———	24	14
*joc**35/37	———————————————CCC	143	76
*joc**38/40	——————————————————GCT	85	150

Figure 3. The DPE is a distinct core promoter element. Systematic triple point mutagenesis of the *Drosophila jockey* DPE reveals that the conserved DPE consensus is important for basal transcription with either *Drosophila* or HeLa (human) transcription factors. The *jockey* (*joc*) promoter sequence from +20 to +40 relative to the major RNA start site is shown with the corresponding nucleotide substitutions below. Lines indicate unchanged sequences. The DPE consensus, GGACGTG, is highlighted in bold. The transcriptional activity (as quantitated with a phosphorimager) is reported as relative to that of the wild-type *jockey* promoter. (Reprinted, with permission, from Burke and Kadonaga 1996, 1997.)

moter activity. We therefore constructed a series of mutant *jockey* promoters in which the distance between the Inr and the DPE was either increased or decreased by three nucleotide increments (Fig. 4). In vitro transcription analysis of these mutant promoters with either *Drosophila* or HeLa factors revealed that the spacing between the DPE and the Inr is critically important for promoter activity, as a change in spacing of three nucleotides resulted in a 7–20-fold reduction in transcriptional activity. Moreover, as seen with the clustered triple point mutant promoters (Fig. 3), we found that the *Drosophila* and the human factors possess the same requirement for the native spacing between the DPE and the Inr, which further supports the notion that the DPE is conserved from *Drosophila* to humans. We also tested the binding of purified TFIID to the mutant +3 insertion or –3 deletion mutant promoters and observed that the native spacing between the Inr and the DPE is important for TFIID binding to the promoter (Burke and Kadonaga 1997), which is consistent with the transcription data (Fig. 4). Therefore, we conclude that the native spacing between the Inr and the DPE appears to be necessary for TFIID binding and for transcriptional activity of TATA-less, DPE-containing promoters. This strict spacing requirement suggests that the interaction of TFIID with the Inr and DPE elements is somewhat inflexible.

A DPE CAN COMPENSATE FOR DISRUPTION OF A TATA BOX

We investigated whether the DPE in its normal downstream position can function to substitute for the upstream TATA box motif of a TATA-containing promoter. In these experiments, we characterized wild-type and mutant versions of a series of *hunchback* P2-*Antennapedia* P2 (*hb*P2-*Antp*P2) hybrid promoters, which are outlined in Figure 5. The *hb*P2 promoter contains a consensus TATA box (TATAAA) upstream of dual, overlapping consensus Inr sequences at +1 and +5, from which transcription initiates both in vitro and in vivo. The hybrid *hb*P2-*Antp*P2 construction contains the TATA box and Inr of the *hb*P2 promoter fused to the DPE of the *Antp*P2 promoter, with the identical spacing between the DPE and the *hb*P2 transcription start site (+1) as in the wild-type *Antp*P2 promoter. Mutant variants of these promoters with clustered point mutations in the TATA box (*TATA), initiator (*Inr), and/or the DPE (*DPE) were also examined.

This set of promoter constructions was subjected to in vitro transcription analysis (Fig. 5). With the *hb*P2-*Antp*P2 hybrid promoter, the *Antp*P2 DPE compensated for the disruption of the *hb*P2 TATA box (Fig. 5, compare lanes 1 and 2). Notably, the Inr + DPE (Fig. 5, lane 2) promoter and the TATA + Inr (Fig. 5, lane 5) promoter exhibited comparable transcriptional activity. By comparison, a mutant version of the *Antp*P2 DPE was unable to restore activity of the disrupted TATA box (Fig. 5, lanes 5 and 6). These results suggest that the DPE can compensate for the loss of the TATA box.

We also carried out a DNase I footprinting analysis of the *hb*P2-*Antp*P2 constructions to see if the binding of TFIID correlated with the transcriptional activity of the promoters (Burke and Kadonaga 1996). With the *hb*P2-*Antp*P2 and the *hb*P2-*Antp*P2*DPE promoters, both of which contain intact TATA box elements, the DNase I footprints were nearly identical and extended from about –36 to +40 relative to the RNA start site. Thus, in the presence of a TATA box, the DPE had little effect upon the binding of TFIID throughout the promoter region. In contrast, with the TATA-less promoters, the DPE had a significant effect. There was more efficient binding of eTFIID to the *hb*P2-*Antp*P2*TATA promoter than to the *hb*P2-*Antp*P2*TATA*DPE promoter. Hence, the TFIID footprinting results correlate well with the transcription data. It is further notable that there was no apparent DNase I footprint in the region of the mutated TATA box of the *hb*P2-*Antp*P2*TATA promoter. Instead, the footprint of this TATA⁻ Inr⁺ DPE⁺ promoter resembled the footprints of the *Antp*P2 and *joc* TATA-less promoters.

Because a significant decrease in transcription was observed upon mutation of the Inr sequences (Fig. 5, lanes 3 and 7), we tested the binding of TFIID to promoters containing the DPE along with wild-type versus mutant Inr regions (Burke and Kadonaga 1996). DNase I footprinting of the *hb*P2-*Antp*P2 versus the *hb*P2-*Antp*P2*Inr promoters and the *hb*P2-*Antp*P2*TATA versus the *hb*P2-*Antp*P2*TATA*Inr promoters revealed that the Inr se-

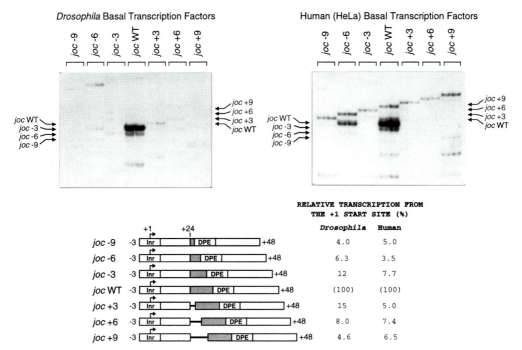

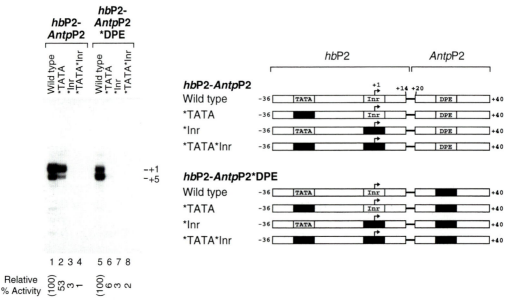

Figure 4. Position of the DPE relative to the RNA start site is important for transcriptional activity with either *Drosophila* or human (HeLa) basal transcription factors. A series of mutant *jockey* promoters was constructed in which the spacing between the DPE and the RNA start site was either increased or decreased by three nucleotide increments. The amount of transcription from the +1 start site (as determined with a phosphorimager) for each mutant relative to that observed with the wild-type promoter is indicated. (Reprinted, with permission, from Burke and Kadonaga 1997.)

quences contribute to the binding of TFIID to the downstream promoter region (comprising the Inr and DPE sequences) of the promoters. Hence, these transcription and footprinting data further support the conclusion that interactions of TFIID with the Inr and DPE sequences are important for basal transcription of TATA-less, DPE-containing promoters.

These data suggest that the DPE at its downstream position functions analogously to a TATA box at its upstream position. First, the DPE and TATA box are both

Figure 5. A DPE can compensate for the loss of a TATA box. The solid black rectangles denote mutant core promoter elements. The term "Wild type" indicates the wild-type sequences for the TATA box and Inr motifs in the *hb*P2 promoter segment. The preferential usage of the +1 start site relative to the +5 start site in the *hb*P2-*Antp*P2 promoter is probably due to the optimal spacing of the DPE for transcription from the +1 site relative to the +5 site (compare lanes 1 and 2 [DPE+] with lane 5 [DPE−]). (Reprinted, with permission, from Burke and Kadonaga 1996.)

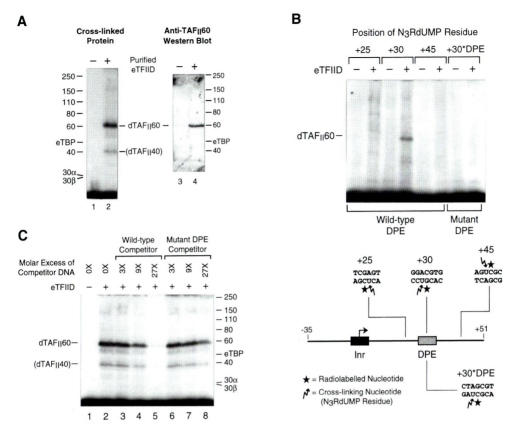

Figure 6. Histone H4-related dTAF$_{II}$60 is cross-linked to the DPE. (*A*) dTAF$_{II}$60 is cross-linked to the *joc* DPE. Photo cross-linking experiments were performed with purified TFIID and the *joc* core promoter. The *joc* promoter photoprobe was a DNA fragment comprising sequences from −35 to +51 relative to the start site, which contained N$_3$RdUMP in the DPE as well as an adjacent radioactively labeled nucleotide (see the +30 probe at the bottom of *B*). TFIID was incubated with the DNA photoprobe, and the samples were subjected to UV irradiation. The complexes were then digested with nucleases to remove excess DNA, and the resulting radioactively labeled proteins were resolved by electrophoresis on a 10% polyacrylamide-SDS gel and detected by autoradiography (*left panel*). The migration of each of the polypeptides that comprise TFIID is indicated. Western blot analysis of the cross-linked protein further confirmed that the major radioactively labeled band corresponds to dTAF$_{II}$60 (*right panel*). Weaker, yet clearly apparent, cross-linking to a species that comigrates with dTAF$_{II}$40 is also indicated. (*B*) Sequence-specific photo cross-linking of the dTAF$_{II}$60 subunit of TFIID to the wild-type DPE, but not to a mutant DPE. Four different *joc* promoter photoprobes, which are indicated at the bottom, were each incubated with TFIID and subjected to cross-linking. The positions of N$_3$RdUMP residues (*arrows*) and radioactively labeled nucleotides (*stars*) are indicated. In the mutant DPE (+30*DPE) probe, the wild-type DPE motif, GGACGTG, was changed to CTAGCGT. (*C*) The interaction of dTAF$_{II}$60 with the DPE photoaffinity probe can be specifically inhibited by competition with a DPE-containing promoter fragment. TFIID was photo cross-linked to the +30 DPE probe (see *B*) in the absence or the presence of the indicated amounts (reported as the molar ratio of competitor DNA to probe DNA) of wild-type DPE or mutant DPE promoter DNA fragments. The wild-type DPE competitor DNA fragment was identical to the photoprobe, except that it was not radioactively labeled and contained a thymidine residue instead of N$_3$RdUMP. The corresponding mutant DPE competitor DNA contained GGACGTG to CTAGCGT substitutions at the DPE. (Reprinted, with permission, from Burke and Kadonaga 1997.)

sequence recognition elements for TFIID. Second, a DPE sequence can compensate for the loss of a TATA box.

THE HISTONE H4-LIKE dTAF$_{II}$60 APPEARS TO INTERACT WITH THE DPE

To determine which components of TFIID were in close proximity to the DPE, we carried out photo cross-linking experiments with purified dTFIID (Burke and Kadonaga 1997). We used the thymidine analog, 5-[*N*-(*p*-azidobenzoyl)-3-amino allyl]deoxyuridine 5′-monophosphate (N$_3$RdUMP; Bartholomew et al. 1990, 1991, 1995), as the cross-linking reagent. N$_3$RdUMP has a photoreactive aryl azide moiety on a 9–10 Å tether that probes the space in the vicinity of the DNA major groove (Bartholomew et al.

1990). Notably, N$_3$RdUMP had been previously used in the analysis of the binding of human TFIID to the TATA-containing, DPE-less adenovirus major late promoter (Oelgeschläger et al. 1996).

In our experiments with the DPE, N$_3$RdUMP was incorporated into a *Drosophila jockey* (*joc*) core promoter probe and then subjected to photo cross-linking with purified dTFIID (Fig. 6). When a N$_3$RdUMP residue was located in the DPE (at +30 relative to the transcription start site), strong cross-linking of an approximately 60-kD polypeptide and weaker cross-linking of an approximately 40-kD polypeptide was observed (Fig. 6A). The apparent molecular masses of these cross-linked polypeptides corresponded to dTAF$_{II}$60 and dTAF$_{II}$40 (which are also known as dTAF$_{II}$62 and dTAF$_{II}$42). The photo cross-

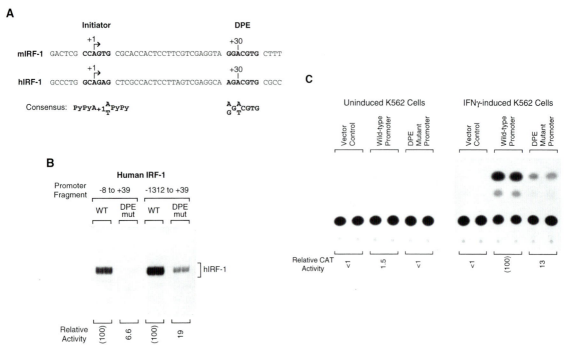

Figure 7. Mammalian DPE-containing, TATA-less promoter. (*A*) The promoter of the IRF-1 gene from mice (m) and humans (h) contains an extended consensus DPE sequence with the correct spacing between the initiator and the DPE. (*B*) The human IRF-1 DPE sequences are functionally important in vitro in the context of a large promoter fragment. Both minimal and larger versions of the human IRF-1 promoters were analyzed. The IRF-1 promoter was transcribed with HeLa (human) factors. The numbers above the autoradiograms indicate the endpoints (relative to the transcription start site) of the promoter constructions that were used. The mutant promoters had clustered point mutations that altered the wild-type DPE consensus from AGACGTG to CTCATGT. The amount of transcription (determined by using a phosphorimager) from each mutant promoter is reported as the percentage of transcription from the corresponding wild-type promoter. (*C*) The DPE in the human IRF-1 promoter is important for IFN-induced transcription in vivo. The human IRF-1 promoter (containing sequences from –1312 to +39 relative to the transcription start site) with a CAT reporter gene was transfected into K562 cells and induced with IFN-γ. CAT activity is reported as relative to that seen with the wild-type promoter. CAT activity was not detectable in the absence of IFN-γ induction. (Reprinted, with permission, from Burke and Kadonaga 1997.)

linking of dTAF$_{II}$60 to the DPE was not seen either when the N$_3$RdUMP residue was located outside of the DPE or when the DPE sequence was altered (Fig. 6B). Moreover, the cross-linking of the photoprobe to dTAF$_{II}$60 and dTAF$_{II}$40 was inhibited by a competitor DNA containing a wild-type DPE, but not competitor DNA containing a mutated DPE (Fig. 6C). In addition, with a photoprobe consisting of the DPE-containing TATA-less *Drosophila Antennapedia* core promoter (instead of the *joc* promoter, as shown in Fig. 6), we observed cross-linking of dTAF$_{II}$60 (strong cross-linking), dTAF$_{II}$40 (weak cross-linking), and an approximately 30-kD species (possibly the histone H2B-like dTAF$_{II}$30α) (Burke and Kadonaga 1997).

These data suggest that dTAF$_{II}$60 and dTAF$_{II}$40 interact with the DPE. Interestingly, dTAF$_{II}$60 and dTAF$_{II}$40 contain histone-fold motifs, which are related to those in the core histones H4 and H3, and can form an α$_2$β$_2$ heterotetramer that resembles the H3-H4 tetramer (Goodrich et al. 1993; Weinzierl et al. 1993; Kokubo et al. 1994; Nakatani et al. 1996; Xie et al. 1996). Thus, a dTAF$_{II}$60-dTAF$_{II}$40 heterotetrameric subcomponent of the TFIID complex may recognize the DPE. It should be noted, however, that binding of TFIID to DPE-containing TATA-less promoters requires both the DPE and Inr motifs, i.e., there appears to be cooperative binding of TFIID

to the Inr and DPE sequences (Burke and Kadonaga 1996). It thus seems improbable that a dTAF$_{II}$60-dTAF$_{II}$40 heterotetramer alone would bind with high affinity to the DPE. Instead, it is more likely that dTAF$_{II}$60 and dTAF$_{II}$40 act in conjunction with other components of TFIID (such as dTAF$_{II}$150 and dTAF$_{II}$250) in the binding to the DPE and Inr motifs.

IDENTIFICATION AND CHARACTERIZATION OF A MAMMALIAN TATA-LESS, DPE-CONTAINING PROMOTER

Since the DPE had been initially studied in *Drosophila*, we wondered whether there might exist TATA-less, DPE-containing promoters in other organisms. We therefore examined the sequences of some mammalian promoters and found that the promoter of the gene encoding interferon regulatory factor-1 (IRF-1) is TATA-less and contains a sequence that conforms to the extended DPE consensus sequence at the appropriate approximately +30 location downstream from the transcription start site (Fig. 6A) (Sims et al. 1993).

We then performed an in vitro transcription analysis of wild-type and mutant versions of the human IRF-1 promoter either as a minimal promoter (from –8 to +39 relative to the start site) or as a larger promoter fragment

TATA-box-containing Promoters

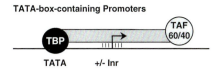

TATA-less, DPE-containing Promoters

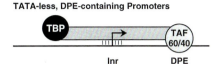

Figure 8. Model for the interaction of TFIID with TATA box-containing versus TATA-less, DPE-containing promoters. This figure depicts two postulated TFIID-promoter interactions and does not exclude other types of functionally important interactions between TFIID and core promoters. The lines between TFIID and the Inr indicate interactions between components of TFIID, such as dTAF$_{II}$150 and/or dTAF$_{II}$250, and the Inr. (Reprinted, with permission, from Burke and Kadonaga 1997.)

(from −1312 to +39 relative to the start site). As shown in Figure 7B, the human IRF-1 promoter requires the DPE sequence for efficient transcription in vitro.

We also analyzed the wild-type and DPE-mutant versions of the IRF-1 promoter by transient transfection analysis (Fig. 7C). These experiments indicated that the DPE motif in the human IRF-1 promoter is important for promoter activity in vivo. We therefore conclude that the IRF-1 promoter is an example of a human TATA-less, DPE-containing promoter. (In addition, these results are consistent with the observation that the HeLa [human] basal transcription factors exhibit the same behavior as the *Drosophila* transcription factors with mutant versions of the *Drosophila jockey* promoter in which the DPE was altered by clustered triple point mutagenesis [see Fig. 3] or by variation of the spacing between the Inr and DPE [see Fig. 4].)

These data hence suggest that the DPE is a core promoter element that is used in both *Drosophila* and humans. This conclusion is in accord with the general observation that basal transcription by RNA polymerase II is conserved throughout eukaryotes and that transcriptional mechanisms in *Drosophila* and humans appear to be closely related. For instance, *Drosophila* promoters can be transcribed with human factors, and vice versa. In addition, each of the *Drosophila* basal transcription factors can be functionally substituted with its human counterpart, and vice versa.

It may be notable that Inr sequences exhibit a much more distinct consensus sequence in *Drosophila* (Hultmark et al. 1986; Purnell et al. 1994; Arkhipova 1995) than in mammals (Smale and Baltimore 1989; Javahery et al. 1994), in which the Inr consensus appears to be more degenerate. By analogy, it is possible that the consensus DPE sequence might be more distinctly apparent in *Drosophila* than in mammals. Hence, although the IRF-1 promoter DPE sequence conforms to the tentative *Drosophila* DPE consensus, it seems likely that there are many mammalian promoters that may or may not fit this consensus, but nevertheless possess fully functional DPE motifs.

THE DPE: A DOWNSTREAM ANALOG OF THE TATA BOX

A simple model for the binding of TFIID to TATA-box-containing promoters and to TATA-less DPE-containing promoters is shown in Figure 8. A summary is as follows.

- The DPE is a distinct core promoter element that is specifically recognized by TFIID. It is present in a subset of promoters: A rough estimate is about 25% of all promoters in *Drosophila*. (By comparison, about half of all *Drosophila* promoters appear to contain a TATA box.) Thus far, functionally active DPE elements have been found only in TATA-less promoters.
- The DPE appears to be conserved from *Drosophila* to humans.
- On the basis of a comparison of promoter sequences, the tentative consensus sequence of the DPE in *Drosophila* is $^A/_G$-G-$^A/_T$-C-G-T-G. This consensus was biased, however, toward the sequences that are present in the *Drosophila jockey* and *Antennapedia* P2 promoters, which were the first two promoters that were found to contain a DPE. Hence, the full range of sequences that comprise functional DPE motifs has yet to be determined.
- Both the Inr and DPE elements are required for binding of TFIID to DPE-containing TATA-less promoters. Thus, there appears to be cooperative binding of TFIID to the Inr and DPE motifs.
- There is, in addition, a stringent requirement for the proper spacing between the Inr and DPE elements. This observation suggests that the interactions of TFIID with the Inr and DPE motifs is somewhat inflexible.
- Interestingly, many long interspersed nuclear elements (LINEs), which include the *jockey*, Doc, G, I, and F elements in *Drosophila*, possess DPE motifs in their core promoters. These non-long terminal repeat (non-LTR) retrotransposons propagate via the use of internal promoters that comprise core promoter elements that are entirely downstream from the transcription start site. Thus, the LINE promoters provide examples in which DPE motifs are used as downstream core promoter elements in vivo.
- The DPE functions as a downstream analog of the TATA box. First, the insertion of a DPE at its downstream position can compensate for the disruption of the TATA box at its upstream position. Second, both elements are bound by TFIID, with TBP binding to the TATA box and dTAF$_{II}$60-dTAF$_{II}$40 interacting with the DPE. Third, analysis of TATA, Inr, and DPE motifs indicated that an Inr in conjunction with either a TATA box or a DPE constitutes an active core promoter. Fourth, many, but not all, naturally occurring promoters appear to be TATA+Inr or Inr+DPE promoters. Therefore, the existence of TATA+Inr and Inr+DPE promoters suggests that it will be important to compare the transcriptional properties of these two types of core promoters, particularly with regard to the biological significance of the presence of one or the other type of these core promoters in a gene.

ACKNOWLEDGMENTS

Research in the laboratory of J.T.K. is supported by grants from the National Institutes of Health and the National Science Foundation.

REFERENCES

Arkhipova I.R. 1995. Promoter elements in *Drosophila melanogaster* revealed by sequence analysis. *Genetics* **139:** 1359.

Bartholomew B., Kassavetis G.A., and Geiduschek E.P. 1991. Two components of *Saccharomyces cerevisiae* transcription factor IIIB (TFIIIB) are stereospecifically located upstream of a tRNA gene and interact with the second-largest subunit of TFIIIC. *Mol. Cell. Biol.* **11:** 5181.

Bartholomew B., Kassavetis G. A., Braun B.R., and Geiduschek E. P. 1990. The subunit structure of *Saccharomyces cerevisiae* transcription factor IIIC probed with a novel photocrosslinking reagent. *EMBO J.* **9:** 2197.

Bartholomew B., Tinker R.L., Kassavetis G.A., and Geiduschek E. P. 1995. Photochemical cross-linking assay for DNA tracking by replication proteins. *Methods Enzymol.* **262:** 476.

Björklund S. and Kim Y.-J. 1996. Mediator of transcriptional regulation. *Trends Biochem. Sci.* **21:** 335.

Burke T.W. and Kadonaga J. T. 1996. *Drosophila* TFIID binds to a conserved downstream basal promoter element that is present in many TATA-box-deficient promoters. *Genes Dev.* **10:** 711.

———. 1997. The downstream core promoter element, DPE, is conserved from *Drosophila* to humans and is recognized by TAF$_{II}$60 of *Drosophila*. *Genes Dev.* **11:** 3020.

Burley S.K. and Roeder R.G. 1996. Biochemistry and structural biology of transcription factor IID (TFIID). *Annu. Rev. Biochem.* **65:** 769.

Conaway R.C. and Conaway J.W. 1993. General initiation factors for RNA polymerase II. *Annu. Rev. Biochem.* **62:** 161.

Dynlacht B.D., Hoey T., and Tjian R. 1991. Isolation of coactivators associated with the TATA-binding protein that mediate transcriptional activation. *Cell* **66:** 563.

Goodrich J.A., Hoey T., Thut C.J., Admon A., and Tjian R. 1993. *Drosophila* TAF$_{II}$40 interacts with both a VP16 activation domain and the basal transcription factor TFIIB. *Cell* **75:** 519.

Hultmark D., Klemenz R., and Gehring W. 1986. Translational and transcriptional control elements in the untranslated leader of the heat shock gene *hsp22*. *Cell* **44:** 429.

Javahery R., Khachi A., Lo K., Zenzie-Gregory B., and Smale S. 1994. DNA sequence requirements for transcriptional initiator activity in mammalian cells. *Mol. Cell. Biol.* **14:** 116.

Kaiser K. and Meisterernst M. 1996. The human general co-factors. *Trends Biochem. Sci.* **21:** 342.

Kokubo T., Gong D.-W., Wootton J.C., Horikoshi M., Roeder R.G., and Nakatani Y. 1994. Molecular cloning of *Drosophila* TFIID subunits. *Nature* **367:** 484.

Kokubo T., Gong D.-W., Yamashita S., Horikoshi M., Roeder R.G., and Nakatani Y. 1993. *Drosophila* 230-kD TFIID subunit, a functional homolog of the human cell cycle gene product, negatively regulates DNA binding of the TATA box-binding subunit of TFIID. *Genes Dev.* **7:** 1033.

Koleske A.J. and Young R.A. 1995. The RNA polymerase II holoenzyme and its implications for gene regulation. *Trends Biochem. Sci.* **20:** 113.

Lagrange T., Kapanidis A.N., Tang H., Reinberg D., and

Ebright R.H. 1998. New core promoter element in RNA polymerase II-dependent transcription: Sequence-specific DNA binding by transcription factor IIB. *Genes Dev.* **12:** 34.

Nakatani Y., Bagby S., and Ikura M. 1996. The histone folds in transcription factor TFIID. *J. Biol. Chem.* **271:** 6575.

Oelgeschläger T., Chiang C.-M., and Roeder R.G. 1996. Topology and reorganization of a human TFIID-promoter complex. *Nature* **382:** 735.

Orphanides G., Lagrange T., and Reinberg D. 1996. The general transcription factors of RNA polymerase II. *Genes Dev.* **10:** 2657.

Pugh B.F. and Tjian R. 1992. Diverse transcriptional functions of the multisubunit eukaryotic TFIID complex. *J. Biol. Chem.* **267:** 679.

Purnell B.A., Emanuel P.A., and Gilmour D. S. 1994. TFIID sequence recognition of the initiator and sequences farther downstream in *Drosophila* class II genes. *Genes Dev.* **8:** 830.

Reines D., Conaway J.W., and Conaway R.C. 1996. The RNA polymerase II general elongation factors. *Trends Biochem. Sci.* **21:** 351.

Roeder R.G. 1996. The role of general initiation factors in transcription by RNA polymerase II. *Trends Biochem. Sci.* **21:** 327.

Sims S.H., Cha Y., Romine M.F., Gao P.-Q., Gottlieb K., and Deisseroth A.B. 1993. A novel interferon-inducible domain: Structural and functional analysis of the human interferon regulatory factor 1 gene promoter. *Mol. Cell. Biol.* **13:** 690.

Singer V.L., Wobbe C.R., and Struhl K. 1990. A wide variety of DNA sequences can functionally replace a yeast TATA element for transcriptional activation. *Genes Dev.* **4:** 636.

Smale S.T. 1994. Core promoter architecture for eukaryotic protein-coding genes. In *Transcription: Mechanisms and regulation* (ed. R.C. Conaway and J.W. Conaway), p. 63. Raven Press, New York.

Smale S.T. and Baltimore D. 1989. The "initiator" as a transcription control element. *Cell* **57:** 103.

Svejstrup J.Q., Vichi P., and Egly J.-M. 1996. The multiple roles of transcription/repair factor TFIIH. *Trends Biochem. Sci.* **21:** 346.

Verrijzer C.P. and Tjian R. 1996. TAFs mediate transcriptional activation and promoter selectivity. *Trends Biochem. Sci.* **21:** 338.

Weinzierl R.O.J., Ruppert S., Dynlacht B.D., Tanese N., and Tjian R. 1993. Cloning and expression of *Drosophila* TAF$_{II}$60 and human TAF$_{II}$70 reveal conserved interactions with other subunits of TFIID. *EMBO J.* **12:** 5303.

Weis L. and Reinberg D. 1992. Transcription by RNA polymerase II: Initiator-directed formation of transcription-competent complexes. *FASEB J.* **6:** 3300.

Wiley S.R., Kraus R.J., and Mertz J.E. 1992. Functional binding of the "TATA" box binding component of transcription factor TFIID to the –30 region of TATA-less promoters. *Proc. Natl. Acad. Sci.* **89:** 5814.

Xie X., Kokubo T., Cohen S.L., Mirza U. A., Hoffmann A., Chait B.T., Roeder R.G., Nakatani Y., and Burley S.K. 1996. Structural similarity between TAFs and the heterotetrameric core of the histone octamer. *Nature* **380:** 316.

Zawel L. and Reinberg D. 1995. Common themes in assembly and function of eukaryotic transcription complexes. *Annu. Rev. Biochem.* **64:** 533.

Zenzie-Gregory B., Khachi A., Garraway I. P., and Smale S.T. 1993. Mechanism of initiator-mediated transcription: Evidence for a functional interaction between the TATA-binding protein and DNA in the absence of a specific recognition sequence. *Mol. Cell. Biol.* **13:** 3841.

The RNA Polymerase II General Transcription Factors: Past, Present, and Future

D. Reinberg,* G. Orphanides, R. Ebright,*† S. Akoulitchev, J. Carcamo, H. Cho,
P. Cortes, R. Drapkin, O. Flores, I. Ha, J.A. Inostroza, S. Kim, T.-K. Kim, P. Kumar,
T. Lagrange, G. LeRoy, H. Lu, D.-M. Ma, E. Maldonado, A. Merino, F. Mermelstein,
I. Olave, M. Sheldon, R. Shiekhattar, N. Stone, X. Sun, L. Weis, K. Yeung, and L. Zawel

*Howard Hughes Medical Institute and Division of Nucleic Acids Enzymology, Department of Biochemistry,
Robert Wood Johnson Medical School, University of Medicine and Dentistry of New Jersey,
Piscataway, New Jersey 0885; †Department of Chemistry and Waksman Institute, Rutgers University,
Piscataway, New Jersey 08854

RNA synthesis is a multistep process. The steps of the transcription cycle were first dissected in studies using the bacterial RNA polymerase (RNAP) (Chamberlin 1974; McClure 1985) and have recently been extended to the eukaryotic RNAP systems. Transcription requires that RNAP associate with the promoter region and form a *stable* initiation complex. In eukaryotes, this is accomplished with the aid of a set of auxiliary factors that are different for each RNAP transcription system (i.e., RNAPI, -II and -III; for review, see Zawel and Reinberg 1995). Transcription by RNAPII minimally requires six factors (TFIIA, TFIIB, TFIID, TFIIE, TFIIF, and TFIIH) termed the general transcription factors (GTFs). Stable association of RNAPII with promoter sequences requires TFIID (or TBP), TFIIB, and TFIIF (for review, see Orphanides et al. 1996). However, the RNAPII transcription system is unique in that after the polymerase has stably associated with promoter sequences, two additional factors, TFIIE and TFIIH, are necessary for transcription (Flores et al. 1990; for review, see Drapkin and Reinberg 1994; Orphanides et al. 1996). This is perhaps related to a unique structure found at the carboxyl terminus of the largest subunit of RNAPII known as the carboxy-terminal domain (CTD) (Koleske and Young 1995; Dahmus 1996).

Studies performed during the last decade have identified each of the GTFs and have provided insights into the part that each plays during the transcription cycle. Each of the GTFs has been purified to apparent homogeneity, and the activities of TFIIA, TFIIB, TFIIE, and TFIIF have been reconstituted from recombinant polypeptides (Fig. 1). In addition, immunopurification procedures have been developed for the isolation of the multisubunit RNAPII, TFIID, and TFIIH factors from human cells (Fig. 1).

Previous studies established that assembly of the transcription complex involves three core promoter elements: (1) the TATA element, located near position –30 in higher eukaryotes, (2) the initiator element encompassing the transcription start site, and (3) the downstream promoter element, located near position +30 (Weis and Reinberg 1992; Smale 1997). TFIID is believed to be responsible for recognition of at least two of these elements. One

subunit of TFIID, the TATA-binding protein (TBP) is responsible for recognition of the TATA element (Burley and Roeder 1996). One or more of the remaining subunits of TFIID, the TBP-associated factors (TAFs), is believed to be responsible for recognition of the downstream promoter element. It is not clear which factor is responsible of recognizing the initiator element.

MATERIALS AND METHODS

All materials used and methods described in the studies presented here have previously been published as is indicated throughout the text.

RESULTS

TFIIB Recognizes a Novel Promoter Element through a Helix-Turn-Helix Motif

Structures have been determined for several polypeptides and polypeptide domains within the transcription complex. These include the TBP core domain (TBPc) alone (Nikolov et al. 1992) and in complex with DNA (J.L. Kim et al. 1993; Y. Kim et al. 1993), the TFIIB core domain alone (TFIIBc) (Bagby et al. 1995) and in a complex with DNA and TBPc (Nikolov et al. 1995), the TFIIB amino-terminal domain alone (Zhu et al. 1996), and TFIIA in a complex with TBP and DNA (Geiger et al. 1996; Tan et al. 1996). The structure of TBP represents a novel protein-fold consisting of a curved, ten-stranded, antiparallel β-sheet, forming a concave lower surface with four α helices on its upper surface (Fig. 2). A loop, or "stirrup," is present at each end of the long axis of TBP and gives the molecule the appearance of a "protein saddle" (Nikolov et al. 1992, 1996; Nikolov and Burley 1994). As predicted from earlier biochemical experiments (Lee et al. 1991; Starr and Hawley 1991), TBP recognizes the minor groove of the 8-bp TATA element. However, these biochemical experiments did not predict the dramatic and unprecedented DNA distortion that accompanies the induced-fit mechanism of TATA element recognition. To facilitate DNA recognition, the minor groove of the TATA element is molded to follow the curved β-sheet on the underside of the TBP saddle. As a

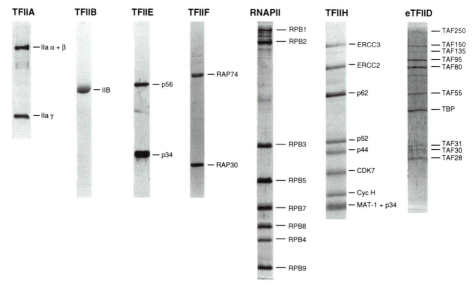

Figure 1. Polypeptide composition of the human general transcription factors. Silver staining of different polyacrylamide gels containing the GTFs. Human native TFIIA is composed of three subunits (Cortes et al. 1992; Coulombe et al. 1992). However, the two larger subunits, α and β, are produced by an unknown protein-processing mechanism (DeJong and Roeder 1993; Ma et al. 1993; Yokomori et al. 1993). The proteins shown in the panel were produced in *E. coli* and the largest subunit is unprocessed. The unprocessed α+β subunits migrate as a 56-kD polypeptide and reconstitute, together with the γ-subunit (14 kD), an active form of the factor (Yokomori et al. 1993; Sun et al. 1994; DeJong et al. 1995). TFIIB is a single polypeptide of 32 kD. TFIIE and TFIIF are composed of two subunits each as indicated on the panel. TFIIA, TFIIB, TFIIE, and TFIIF were produced in bacteria and purified as described by Maldonado et al. (1996). Human RNAPII was purified from HeLa cells as described by Maldonado et al. (1996). The form of the polymerase is the initiation-competent IIA where the CTD is non-phosphorylated. The minor bands between Rpb1 and Rpb2 represent proteolytic fragments derived from Rpb1. The subunits are indicated on the panel. TFIIH was purified from human cells using an affinity purification procedure as described by LeRoy et al. (1998a). The nine subunits composing the factor are indicated on the panel. TFIID was purified from HeLa-LTR cells using an affinity purification procedure initially described by Zhou et al. (1992) and modified as described by Maldonado et al. (1996). TBP and the TBP-associated factors (TAFs) are indicated on the panel.

result, the TATA sequence is partially unwound and is bent in a smooth arc (Fig. 2).

The next factor visualized in the puzzle that makes up the preinitiation complex was TFIIB. The solution structure of TFIIBc demonstrated that the molecule is composed of two domains each containing five α helices with pseudo twofold symmetry (Bagby et al. 1995). The crystallographic structure of a ternary complex of the core domain of TFIIB (TFIIBc), TBPc, and a 16-bp DNA fragment containing the TATA element illustrates how TFIIB recognizes the preformed TBPc-DNA complex (Fig. 2) (Nikolov et al. 1995). The dramatic distortion of the TATA element by TBP allows TFIIB to interact with the phosphodiester backbone of DNA both upstream and downstream from the TATA sequence. TFIIB binds underneath and on one face of the TBP-DNA complex,

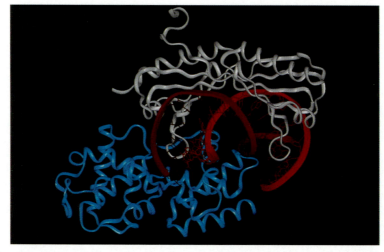

Figure 2. Representation of the X-ray crystallographic structure of the TBP-TFIIB-TATA complex. The figure was prepared starting from the crystallographic structure of the TFIIB-TBP-TATA complex (Nikolov et al. 1995). (*Blue*) TFIIB; (*gray*) TBP; (*red*) DNA.

where it interacts with TBP and DNA (Fig. 2). TFIIBc interacts with the DNA major groove immediately upstream of the TATA element and the DNA minor groove immediately downstream from the TATA element. In the crystallographic structure, details of the interaction between TFIIBc and the DNA major groove upstream of the TATA element are incomplete, since the structure was determined using a DNA fragment containing only three nucleotide pairs upstream of the TATA element (Fig. 2). However, DNA-binding (Lee and Hahn 1995) and protein-DNA photocrosslinking (Lagrange et al. 1996, 1998) experiments revealed an interaction between TFIIB and the DNA major groove upstream of the TATA element and indicated that the interaction is extensive, spanning up to 7–9 nucleotide pairs.

The observation that TFIIB makes extensive interactions with the DNA major groove upstream of the TATA element raised the possibility that the ability of TFIIB to

enter into transcription complexes, and thus the ability of TFIIB to support transcription initiation, may be affected by the DNA sequence upstream of the TATA element. Studies performed by Lagrange et al. (1998) uncovered a sequence element located immediately upstream of the TATA element of the adenovirus major late promoter (AdMLP), having a consensus sequence 5´-G/C-G/C-G/A-C-G-C-C-3´ that affects the ability of TFIIB to enter transcription complexes and support transcription initiation in vitro. Importantly, this sequence element is recognized directly by TFIIB (Fig. 3A; and data not shown) (see Lagrange et al. 1998), involving α helices 4´ and 5´ of TFIIBc (Fig. 3C). The amino acid sequence of helices 4´ and 5´ of TFIIBc matches the sequence profile for the canonical helix-turn-helix (HTH) DNA-binding motif (data not shown, see Fig. 3C). Importantly, the backbone atoms of helices 4´ and 5´ of TFIIBc superimpose on the backbone atoms of the HTHs of bacterial proteins CAP,

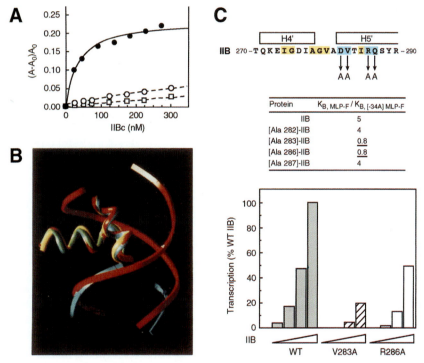

Figure 3. TFIIB is a sequence-specific DNA-binding protein. (*A*) Fluorescence anisotropy analysis of IIB-DNA complex formation. The DNA fragments contain fluorescein. Three different DNA molecules were used in the analysis: wild-type (5´-CT-GAAGGGGGGGCTATAAAAGGGGGT-3´), a fragment where the –34 G-residue was substituted to an A (*open circles*), and a DNA fragment where the G residues at –34 and at –37 were substituted to A (*open squares*). Reactions were performed as described by Lagrange et al. (1998). (*B*) TFIIB recognizes the new sequence element through a helix-turn-helix (HTH) motif: structural evidence. Superimposition of helices 4´ and 5´ of TFIIB (*blue*; Nikolov et al. 1995; coordinates kindly supplied by S. Burley, The Rockefeller University) on the HTHs of CAP (*yellow*; Parkinson et al. 1996), λ repressor (*green*; Beamer and Pabo 1992; coordinates obtained from Brookhaven Protein Data Bank, accession code 1LMB), and Trp repressor (*red*; Otwinowski et al. 1988; coordinates obtained from Brookhaven Protein Data Bank, accession code 1TRO). The three nucleotide pairs upstream of the TATA element present in the crystallographic structure of the TFIIBc-TBPc-TATA complex are shown in ribbon representation and colored *blue*; these nucleotide pairs correspond to positions –34 to –32 of the AdMLP. Nine nucleotide pairs in the Trp repressor–DNA complex are shown in ribbon representation and colored *red*. (*C*) TFIIB recognizes the new sequence element through the HTH motif: functional evidence. (*Top panel*) Amino acid sequences of helices 4´ and 5´ of TFIIB (residues 270–290 of TFIIB; Bagby et al. 1995; Nikolov et al. 1995). Amino acids conserved among known HTHs are highlighted in *yellow*; positions of amino acids that contact DNA base pairs in known structures of HTH-DNA complex structures are marked in *blue*. Substitutions of amino acids of TFIIB that correspond to base-pair-contacting amino acids of HTH DNA-binding proteins are indicated. (*Middle panel*) Effects of substitutions on specificity between G:C and A:T at position –34 of the AdMLP. Data are from fluorescence anisotropy assays of TFIIB-TBPc-DNA complex formation. (*Bottom panel*) Effects on transcription initiation at the AdMLP. Data are from reconstituted transcription assays. TFIIB and TFIIB derivatives were at 1, 2, 4, and 8 nM.

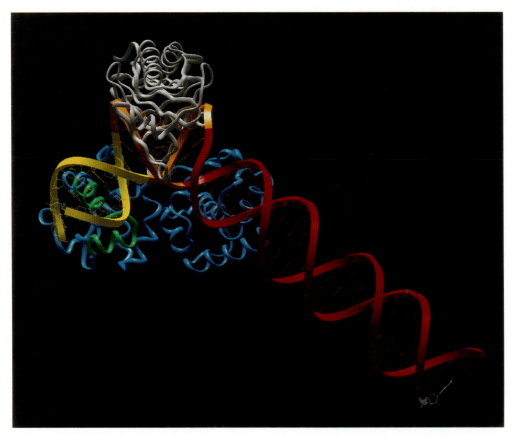

Figure 4. Modeling of the TBP-TFIIB-AdMLP-ternary complex and the recognition of the BRE by TFIIB. A sequence element (TFII*B* *r*ecognition *e*lement, BRE; *yellow*) located immediately upstream of the TATA element (*orange*) of the AdMLP affects the ability of TFIIB to enter transcription complexes and support transcription initiation. The sequence element is recognized by a canonical HTH DNA-binding motif (*green*) within TFIIB. The figure was prepared starting from the crystallographic structure of the TFIIB-TBP-TATA complex (Nikolov et al. 1995) and extending the DNA upstream of the TATA element as B-DNA. TFIIB is *green* and *blue*; TBP is *gray*; DNA is *yellow, orange,* and *red*; the transcription start is *white* (for details, see Lagrange et al. 1998).

λ-repressor, and Trp repressor (Fig. 3B). Moreover, the orientation of helices 4´ and 5´ of TFIIBc, relative to DNA, falls within the range of orientations observed with canonical HTHs (Fig. 3B). The functional significance of the structural similarity between helices 4´ and 5´ of TFIIBc and those of bacterial proteins containing a canonical HTH motif was assessed by constructing alanine substitutions. Amino acids of TFIIB corresponding to the amino acids that in HTH DNA-binding proteins make specificity-determining contacts with DNA base pairs were changed (i.e., Asp-282, Val-283, Arg-286, and Gln-287 on TFIIB, Fig. 3C). The effects on specificity at position 5 of the sequence element, which corresponds to –34 of the AdMLP, were analyzed (Fig. 3C). The Val-283–Ala and Arg-286–Ala substitutions eliminated specificity at position –34; thus, whereas wild-type TFIIB exhibited a fivefold preference for G:C versus A:T at position –34 (Fig. 3C), the substituted TFIIB derivatives exhibited no preference for G:C (Fig. 3C). In addition, these substitutions exhibited decreased affinity for formation of the TFIIB-TBP-DNA complex and decreased effectiveness in supporting transcription initiation at the AdMLP (Fig. 3C). The Val-283 and Arg-286 substitutions exhibit the two defining characteristics of substitutions that permit iden-

tification of amino acids that make specificity-determining contacts with DNA base pairs: (1) elimination of specificity between consensus and nonconsensus base pairs at one position within the DNA site and (2) reduction in affinity for the consensus DNA site ("loss of contact" substitutions; see Ebright 1991). We conclude that Val-283 and Arg-286 of TFIIB determine specificity for G:C versus A:T at position –34, and we propose that these amino acids make direct contact with G:C at position –34 (Fig. 4) (for details, see Lagrange et al. 1998).

These results establish the existence of a fourth core promoter element, the TFIIB-recognition-element (BRE), in addition to the TATA element, the initiator element, and the downstream promoter element, and a second sequence-specific general transcription factor, TFIIB, in addition to TFIID. On the basis of our results, it is possible that TFIIB-BRE interaction will play a part in determining the overall strength of a promoter, the order of preinitiation complex assembly at a promoter, the rate-limiting step in transcription initiation at a promoter, and, given the observation that several transcriptional activators interact with TFIIB (Orphanides et al. 1996), the responsiveness of a promoter to specific transcriptional activators. It is also likely that the IIB-BRE interaction plays a part in determining the upstream-downstream directionality of preinitiation complex assembly and transcription initiation, supplementing the TBP-TATA interaction, which, because of its high degree of twofold symmetry (J.L. Kim et al. 1993; Y. Kim et al. 1993), is insufficient to determine directionality (Xu et al. 1991; Cox et al. 1997). Consistent with this hypothesis are findings obtained with the TFIIB homolog of the archaeon *Sulfolobus shibatae*, TFB, indicating that TFB recognizes a promoter element located immediately upstream of the TATA element (Qureshi and Jackson 1998).

It will be important to investigate whether there exist promoter elements recognized by other GTFs. The possible existence of additional promoter elements is attractive in that it would enable the "encoding" in promoter DNA of a wide range of promoter strengths and promoter characteristics, including responsiveness to specific transcriptional activators and repressors. If documented, the existence of additional promoter elements would require revision of the current view of the promoter DNA as an information-poor scaffold for preinitiation complex assembly to a new view of promoter DNA as an information-rich participant in, and director of, preinitiation complex assembly.

The Trajectory of the DNA in the RNAPII Transcription Preinitiation Complex

Specific and stable association of RNAPII with promoter sequences requires the formation of a complex between RNAPII and TFIIF and the recognition of the TBP-TFIIB-DNA-ternary complex. Human TFIIF is a heterotetrameric factor consisting of 28-kD (RAP30) and 58-kD (RAP74) subunits, and human RNAPII is composed of 12 polypeptides (see Fig. 1). Primary sequence analysis revealed that TFIIF is structurally related to bacterial σ factors: The RAP30 subunit contains two distinct regions with sequence similarity to *Escherichia coli* σ factors (Sopta et al. 1989; Garrett et al. 1992). Moreover, one of the σ homology regions of RAP30 is thought to interact with RNAPII, and TFIIF binds to the same surface on *E. coli* RNA polymerase as *E. coli* σ^{70} (McCracken and Greenblatt 1991). Some of the functions of TFIIF can be accomplished by RAP30 alone. RAP30 can deliver RNAPII to the promoter to support transcription initiation in the absence of RAP74. This is consistent with the ability of RAP30 to interact with TFIIB (Ha et al. 1993), RNAPII (McCracken and Greenblatt 1991), and DNA (Tan et al. 1994).

To define positions of polypeptides relative to promoter DNA within the human TBPolF DNA-protein complex, we performed site-specific protein-DNA photocrosslinking (Kim et al. 1997). We constructed 80 site-specifically derivatized promoter-DNA fragments, each containing a phenyl-azide photoactivatable crosslinking agent incorporated at a single, defined phosphate of the AdMLP (positions –55 to +25; for details, see Lagrange et al. 1996; Kim et al. 1997). For each of the promoter DNA fragments, the TBP-TFIIB-RNAPII-TFIIF-promoter complex (TBPolF) was formed using each of the

DNA fragments, the complex was UV-irradiated, and the polypeptide(s) that crosslinked were identified. Representative data are summarized in both panels of Figure 5 (also see Kim et al. 1997; Ebright, this volume). Three lines of evidence indicate that the crosslinks summarized are specific for the TBPolF complex. (1) The crosslinks are position-dependent. (2) For each polypeptide, except RPB2, the crosslinks are TBP-dependent. (3) Standard photocrosslinking experiments and "in gel" photocrosslinking experiments, with preisolation of the TBPolF complex by nondenaturing gel electrophoresis followed by UV irradiation in situ, yield similar patterns of crosslinking.

Both subunits of TFIIF crosslinked to DNA. The cross-groove pattern of crosslinking indicates that RAP30 interacts with the DNA major groove immediately downstream from the TATA element (between positions –26 and –12, and to a lesser extent between positions –37 and –31, Fig. 5, top). RAP74 interacts with the major groove 10 bp downstream from the TATA element (between positions –16 and –10, Fig. 5, top). These results are in general agreement with results indicating crosslinking of RAP30 at –19 and of RAP74 at –15 and –5 (Robert et al. 1996; Forget et al. 1997). However, Forget et al. (1997) reported that RAP74 interacts with DNA both upstream and downstream from the TATA element. We have not observed crosslinking of RAP74 to sites upstream of the TATA box.

Of the 12 polypeptides composing human RNAPII, only three crosslinked to DNA: RPB1 (the largest subunit, homologous to bacterial RNAP β´-subunit), RPB2 (the second largest subunit, homologous to bacterial RNAP β-subunit), and RPB5 (the fifth largest subunit, shared by the three eukaryotic RNAPs but having no counterpart in bacterial RNAP). RPB1 crosslinks to a single face of the DNA over an extensive region, from positions –53 to +9 (Fig. 5, bottom). RPB2 also crosslinks over an extensive region, from position –49 to +19 (Fig. 5, bottom), and, except between positions –4 and +3, it crosslinks to a single face of the DNA helix, the face opposite that to which RPB1 crosslinks (see Fig. 5, bottom). RPB5 crosslinks between positions +5 and +15 on a single face of the DNA helix, the face to which RPB1 crosslinks (Fig. 5, bottom).

These results establish the existence of an extremely long RNAPII-DNA contact region (~70 bp, ~240 Å) and show that, throughout most of the RNAPII-DNA contact region, RPB1 and RPB2 interact with opposite faces of the DNA helix (Fig. 5, bottom). The RNAPII-DNA contact is continuous except in the vicinity of the TATA element, where a break in the RNAPII-DNA contact permits access to promoter DNA by TBP, TFIIB, and TFIIF (Fig. 5, bottom). The results confirm DNA footprinting experiments showing protection of an extended region (Van Dyke et al. 1988) and establish that protection is due to direct RNAPII-DNA contacts, rather than to RNAPII-induced changes in DNA. We infer that the two largest subunits of RNAPII form opposite walls of an approximately 240-Å channel that interacts with promoter DNA. In view of the fact that 240 Å is approximately 1.7 times the longest dimension of RNAPII (140 x 136 x 100 Å; Darst

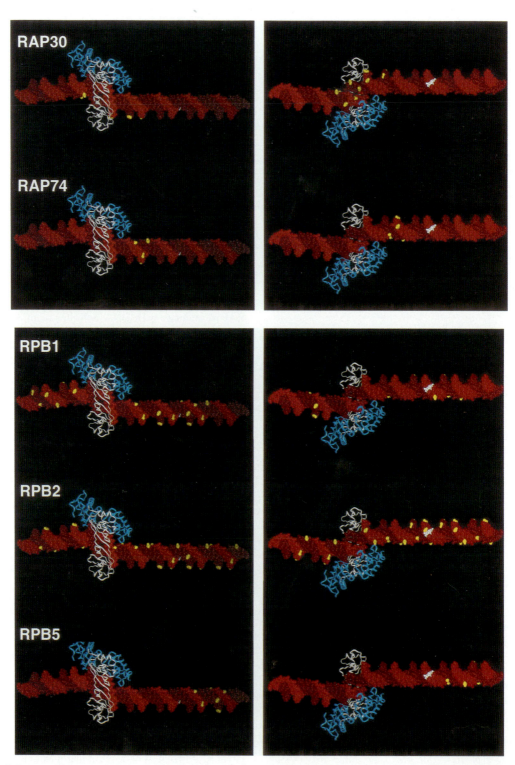

Figure 5. Structural interpretation of protein-DNA photocrosslinking within the TBP-TFIIB-TFIIF-RNAPII-AdMLP complex. (*Top*) Sites at which polypeptides of TFIIF crosslink to DNA (*yellow*). (*Bottom*) Sites at which subunits of RNAPII crosslink to DNA (*yellow*). For each polypeptide, two views are shown of the TBP-TFIIB-promoter complex along the vector of the TBP-induced DNA bend: a "top view," with DNA ends receding from viewer (*left*) and a "bottom view," with DNA ends approaching the viewer (*right*). TBP, *white*; TFIIB, *blue*; DNA nontemplate strand, *light red*; DNA template strand, *dark red*; the transcription start site is *white*. (For details, see Kim et al. [1997] and Ebright [this volume].)

et al. 1991), we further infer that RNAPII must wrap DNA around its surface.

Using electron microscopy, in collaboration with Drs. Yuh-Hwa Wang and Jack Griffith (University of North Carolina, Chapel Hill), we were able to demonstrate that in the TBPolF complex, the DNA is sharply bent. In addition, we were able to demonstrate that in the complex, the contour length of the upstream and downstream DNA segments is about 20% (~170 Å) less than the contour length of the DNA fragment in the absence of protein (see Kim et al. 1997). This indicates that the DNA fragment is compacted by approximately 170 Å or about 50 bp, by its association with the protein mass (for details, see Kim et al. 1997). Compaction of double-stranded DNA of this magnitude under physiological conditions is diagnostic of wrapping of DNA around a protein core. Similar results indicating compaction of DNA were obtained by Forget et al. (1997).

The results discussed above, i.e., the photocrosslinking studies establishing that RNAPII interacts with a segment of promoter DNA that is nearly twice the longest dimension of RNAPII, and the electron microscopy studies establishing that RNAPII compacts promoter DNA by approximately 170 Å, make a compelling case that RNAPII wraps promoter DNA around its circumference.

Our results, together with the high-resolution structure of the TBPc-TFIIBc-TATA complex (Nikolov et al. 1995), the low-resolution structures of RNAPII (Darst et al. 1991) and RNAPII-TFIIB (Leuther et al. 1996), and the proposal that the active center of RNAPII is located between two finger-like projections of RNAPII (Darst et al. 1991; Leuther et al. 1996), permit construction of a specific model for the trajectory of DNA in the TBPolF-promoter complex (Fig. 6) (for details, see Kim et al. 1997; Ebright, this volume). In the model, there are two DNA bends: (1) the approximately 80° TBP-induced DNA bend, phased to maximize contacts to the DNA segment upstream of the TATA element, and (2) a second, approximately 70° DNA bend, centered at or near the transcription start site and phased to maximize contact to the DNA segment at and downstream from the transcription start site (Fig. 6).

The DNA segment corresponding to positions –50 to –6 is placed within a shallow, relatively open, channel (Fig. 6A), consistent with RNAPII-DNA crosslinking at approximately half of the tested phosphates within this DNA segment (Fig. 5, bottom); the DNA segment corresponding to positions –5 to +20 is placed within a deep, nearly completely enclosed, channel (Fig. 6B), consistent with RNAPII-DNA crosslinking at nearly all tested phosphates within this DNA segment (Fig. 5, bottom). This model provides a framework for interpretation of genetic, biochemical, and structural data on RNAPII-dependent transcription initiation and regulation.

On the basis of similarities between RNAPII and other multisubunit RNAPs in three-dimensional structures, and subunit primary structures, we propose that the model described in Figure 6, and further refined in Ebright (this volume), may apply generally to mutisubunit RNA polymerases.

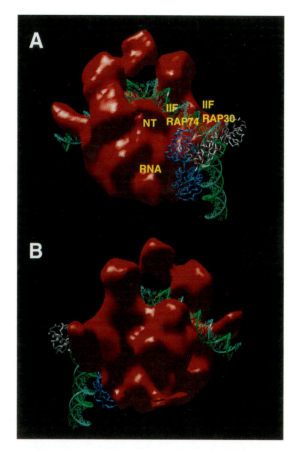

Figure 6. Model for the structure of the TBP-TFIIB-TFIIF-RNAPII-promoter complex. Two views are presented: upstream-end at right in *A*, downstream-end at right in *B*. TBP, *white*; TFIIB, *blue*; positions of TFIIF-RAP30 and TFIIF-RAP74, *yellow*; RNAPII, *red*; DNA template strand, *green*; DNA nontemplate strand, *turquoise*; the transcription start site, *yellow*. NT, proposed binding site for nontemplate strand of the transcription bubble in open and elongation complexes. RNA, proposed emergence point of nascent RNA in elongation complexes. (For details, see Kim et al. [1997] and Ebright [this volume].)

The Stable Association of RNAPII with Promoter Sequences Is Not Sufficient for Transcription

As stated above, after RNAPII has stably associated with promoter sequences (establishment of the TBPolF complex), two other factors, TFIIE and TFIIH, are necessary for *stable* promoter melting (Holstege et al. 1996) and the escape of RNAPII from the promoters (Kumar et al. 1998). TFIIE exists as a heterotetramer of 34-kD and 56-kD subunits (see Fig. 1). The 56-kD subunit contains a sequence with the potential to form a zinc-binding domain and a stretch of amino acids sharing homology with the catalytic loop of a kinase domain (Ohkuma et al. 1991; Peterson et al. 1991; Sumimoto et al. 1991). However, neither purified or recombinant TFIIE possesses kinase activity. TFIIE can enter the preinitiation complex after RNAPII and TFIIF. However, since TFIIE binds to RNAPII in solution (Flores et al. 1989), it is likely that TFIIE joins the preinitiation complex concomitant with

RNAPII and TFIIF. TFIIE recruits TFIIH to the promoter (Flores et al. 1990) and regulates the enzymatic activities of TFIIH (Lu et al. 1992; Drapkin et al. 1994b; Ohkuma and Roeder 1994).

TFIIH, also known as BTF2, was initially purified as a factor required to reconstitute accurate transcription by RNAPII in vitro (Gerard et al. 1991; Flores et al. 1992). BTF2 was originally presumed to be TFIIE (Zheng et al. 1990). However, the availability of antibodies against subunits of TFIIE and TFIIH revealed that BTF2 and TFIIE are distinct (Peterson et al. 1991; Fischer et al. 1992). Moreover, antibodies generated against the 62-kD subunit of BTF2 revealed that BTF2 and TFIIH are the same factor (Fischer et al. 1992; Lu et al. 1992). Purification of TFIIH demonstrated that it is a multisubunit factor composed of nine polypeptides (see Fig. 1) (Conaway and Conaway 1989; Gerard et al. 1991; Flores et al. 1992; Drapkin et al. 1994a; LeRoy et al. 1998a). The subunit composition of TFIIH has proved to be intriguing. cDNA clones encoding the different subunits of TFIIH have been isolated. The largest subunit (p89) was identified as the DNA excision repair protein ERCC3 (*excision repair cross complement*; Schaeffer et al. 1993). Mutations in the ERCC3 gene are responsible for the DNA repair defects observed in patients with xeroderma pigmentosum (XP) group B and Cockayne's syndrome (for review, see Drapkin and Reinberg 1994; Drapkin et al. 1994a; Tanaka and Wood 1994). In addition, ERCC3 was found to contain a DNA-dependent ATPase and a 3′ to 5′ helicase activity. Subsequent studies demonstrated that the second largest subunit of TFIIH (p80) was the DNA excision repair protein ERCC2 (Feaver et al. 1993; Drapkin et al. 1994b; Schaeffer et al. 1994), which corrects the DNA repair defect in cultured cells obtained from patients with XP group D (for review, see Drapkin and Reinberg 1994; Drapkin et al. 1994a; Tanaka and Wood 1994). ERCC2 possesses a 5′ to 3′ ATP-dependent DNA helicase activity. Mutations in yeast TFIIH subunits that are homologous to p62 (TFB1) and p44 (hSSL1) demonstrated that these subunits also perform essential functions in excision repair (Matsui et al. 1995; Wang et al. 1995). Moreover, the 34-kD subunit of TFIIH also has homology with domains of hSSL1, suggesting that it also participates in excision repair (Humbert et al. 1994). The p52 subunit of TFIIH has recently been cloned (Marinoni et al. 1997). This subunit was found to be a homolog of Tfb2, a subunit of yeast TFIIH (Feaver et al. 1997). This subunit is also thought to be involved in DNA repair, because Tfb2 alleles confer a UV-sensitive phenotype, and antibodies raised against the human protein inhibit excision repair in microinjection assays (Feaver et al. 1997; Marinoni et al. 1997). In agreement with these observations, TFIIH is essential for nucleotide excision repair (Aboussekhra et al. 1995; Guzder et al. 1995; Mu et al. 1996).

TFIIH also contains a kinase specific for the CTD of RNAPII (Feaver et al. 1991; Lu et al. 1992; Serizawa et al. 1992). The CTD is composed of multiple, tandemly repeated copies of the heptapeptide Tyr-Ser-Pro-Thr-Ser-Pro-Ser. The kinase activity of mammalian TFIIH resides in the *cdk-activating kinase* (CAK) complex, composed of the catalytic subunit CDK7 and its regulatory subunits cyclin H and MAT1 (Feaver et al. 1994; Roy et al. 1994; Serizawa et al. 1995; Shiekhattar et al. 1995), necessary for the activation of at least three cyclin-dependent kinases that regulate cell cycle progression through G_1 (cdk2 and cdk4) and mitosis (cdc2) (Morgan 1995) in higher eukaryotes. These findings suggested that TFIIH not only participates in transcription and nucleotide excision repair, but may also regulate cell cycle progression.

Studies performed initially in yeast (Svejstrup et al. 1995), and subsequently extended to the mammalian factor (Drapkin et al. 1996; Reardon et al. 1996), demonstrate that TFIIH exists in at least two subcomplexes, a complex containing the core subunits of TFIIH (ERCC3, ERCC2, p62, p50, p44, and p34) and devoid of the kinase complex, referred to as "core-TFIIH," and a complex associated with the kinase complex (CDK7, cyclin H, and MAT-1), referred to as holo-TFIIH (Fig. 7). Core TFIIH appears to be the form involved in excision repair, whereas holo-TFIIH functions in transcription. Interestingly, studies performed in HeLa cells have demonstrated that TFIIH is limiting with respect to CAK and that no more than 20% of the total cellular CAK is associated with the TFIIH complex (Drapkin et al. 1996). Studies with human CAK showed that it exists in three distinct complexes: CAK, a novel ERCC2-CAK complex, and TFIIH (Drapkin et al. 1996; Reardon et al. 1996). Moreover, these studies also demonstrated that the ERCC2-CAK complex, but not CAK, could complement the severely compromised transcription activity of "core-TFIIH" in vitro (Drapkin et al. 1996; Reardon et al. 1996; LeRoy et al. 1998a).

Phosphorylation of CDK7 Positively and Negatively Regulates the Transcription Activity of TFIIH

Genetically, CDK7 homologs have been implicated in transcription in *Saccharomyces cerevisiae* (Cismowski et al. 1995; Valay et al. 1995) and in mitotic regulation in *Schizosaccharomyces pombe* (Buck et al. 1995). Repression of transcription during mitosis has been associated with a number of regulatory mechanisms (Gottesfeld and Forbes 1997). Accumulating evidence indicates that mitotic repression involves direct inactivation of key components of the transcription machinery (Leresche et al. 1996; Segil et al. 1996). To study mitotic repression of transcription and the possible part played by TFIIH, we used HeLa cells arrested with nocodazole. The transcription activity of mitotic nuclear extract prepared from these cells was severely compromised compared to that of the interphase extract (Fig. 8a and data not shown). The addition of purified TFIIH and RNAPII restored the transcriptional activity of the mitotic extract (Fig. 8a). No other combination of GTFs/RNAPII complemented the mitotic extracts (Fig. 8a and data not shown). These results strongly implicate TFIIH and RNAPII in mitotic inhibition of basal transcription. The finding that RNAPII is inactivated in mitotic extracts is consistent with previous studies demonstrating that MPF (composed of cdc2

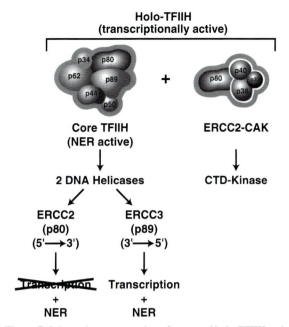

Figure 7. Schematic representation of core- and holo-TFIIH and its associated enzymatic activities. Schematic representation of the composition of core TFIIH, the ERCC2-CAK complex, and holo-TFIIH. Core TFIIH which is composed of six subunits is active in nucleotide excision repair (NER) and contains two helicases. A 5′-3′ is associated with the 80-kD subunit also known as ERCC2 or XPD. This helicase activity is required for NER but dispensable for transcription (Guzder et al. 1994). The second helicase activity (3′-5′) is contained in the 89-kD subunit also known as ERCC3 or XPB. This activity is necessary for NER and transcription (Guzder et al. 1994). The figure also depicts the ERCC2 (XPD)-CAK complex. The CAK complex contains a kinase that is present in the 40-kD polypeptide and is known as CDK7. CDK7 activity is regulated by cyclin H (which is the 38-kD polypeptide) and by MAT1 (which is the 32-kD polypeptide). This ternary complex is known as CAK and its association with core-TFIIH is mediated by the 80-kD polypeptide (ERCC2, XPD) subunit. CAK is active in phosphorylating cdc2, cdk2 and cdk4, and to a lesser extent the CTD of RNAPII. CAK and XPD-CAK complexes exist in HeLa cells. Transcriptionally active TFIIH is contained in holo-TFIIH which results by the association of the ERCC2-CAK complex with core-TFIIH. Holo-TFIIH is apparently the form of the enzyme that phosphorylates the CTD of RNAPII in vivo.

and cyclin B), which is active in mitosis, directly or through a kinase cascade, phosphorylates the CTD of RNAPII resulting in its inactivation (Sherman and O'Farrell 1991; Zawel et al. 1993; Parsons and Spencer 1997).

To expand the results observed with the mitotic-derived TFIIH, we isolated TFIIH from mitotic extracts and analyzed its ability to reconstitute transcription from the AdMLP using a reconstituted transcription system (Fig. 8b). TFIIH from mitotic and interphase extracts was isolated using an immunopurification procedure that yielded a highly purified form of TFIIH (LeRoy et al. 1998a; for details, see Akoulitchev and Reinberg 1998). TFIIH isolated from interphase extracts was transcriptionally active, whereas TFIIH isolated from the mitotic extract was severely compromised in transcription (Fig. 8b). This

was not due to differences in the amounts of TFIIH as determined by Western blot analysis (Fig. 8b).

We next analyzed whether the CAK complex was implicated in mitotic repression. Previously, we (Drapkin et al. 1996) and others (Reardon et al. 1996; Marinoni et al. 1997) have shown that the CAK complex can be dissociated from holo-TFIIH with high ionic strength. The reassociation of CAK with core-TFIIH restores transcriptionally active TFIIH (holo-TFIIH) and requires the formation of the intermediary CAK/ERCC2 complex, which is present in HeLa cell extracts (see Fig. 7) (Drapkin et al. 1996; Reardon et al. 1996). We exploited these features of TFIIH by preparing TFIIH from mitotic extracts and the CAK/ERCC2 complex from interphase extracts (Fig. 8c). Basal transcription reconstituted with mitotic TFIIH showed reduced activity (Fig. 8c, lane 2). No transcription activity was observed when the CAK/ERCC2 complex was added in lieu of TFIIH (lane 1), although it was active in a CTD kinase assay (data not shown). Addition of an excess of purified interphase CAK/ERCC2 to reactions containing mitotic TFIIH led to restoration of basal transcription (Fig. 8c, lanes 3 and 4).

Phosphorylation is an important regulatory modification observed both in mitotic regulation and in the regulation of cyclin-dependent kinases. Earlier reports established that CDK7 is phosphorylated in vivo at two major sites, Ser-164 and Thr-170 (Labbe et al. 1994; Fisher et al. 1995). Phosphorylation on Thr-170 is essential for CDK7 kinase activity in vivo (Labbe et al. 1994), whereas phosphorylation on Ser-164 is not required for activity and is detrimental in stage VI oocytes (Labbe et al. 1994). The Ser-164 site matches the consensus sequence for cdks/MAP kinases (Fig. 8d). We therefore analyzed whether the CDK7 subunit of TFIIH was specifically phosphorylated in mitosis. Interphase and nocodazole-arrested HeLa cells were labeled with [^{32}P]orthophosphate in vivo. CDK7 was immunoprecipitated using ERCC3 antibodies followed by denaturation with SDS and a second immunoprecipitation using anti-CDK7 antibodies (Fig. 8d). The autoradiograph displayed a single phospholabeled polypeptide with increased radioactivity in mitotic-derived CDK7 (Fig. 8d). Importantly, Western blots revealed a shift in the mobility of the mitotic-derived CDK7 (Fig. 8d, bottom panel). In light of these results, we attempted to restore the transcriptional activity of mitotic TFIIH by phosphatase treatment. This approach was possible because the phosphate on Thr-170 in CDK7, which is essential for the activity, is protected from nonspecific dephosphorylation (Labbe et al. 1994). TFIIH immunopurified from the interphase and mitotic extracts was treated with phosphatase and then used in reconstituted TFIIH-dependent transcription. As shown in Figure 8e (top panel), treatment of mitotic TFIIH with phosphatase restored transcriptional activity. Reactivation of TFIIH was dependent on the amount of phosphatase added and was blocked by sodium phosphate, an inhibitor of phosphatases. Western blot analyses of the samples analyzed in transcription show a shift in the mobility of CDK7 associated with mitotic inactive TFIIH (Fig. 8e, bottom panel).

Having established that CDK7 is hyperphosphorylated during mitosis (Fig. 8d,e), we analyzed whether phosphorylation of CDK7 is involved in the regulation of TFIIH. We studied several mutants of CDK7 in vivo. Wild-type CDK7 and individual point mutations in the two sites where phosphorylation occurs in vivo, Ser-164 (S) or Thr-170 (T), were substituted to alanine. The mutant constructs were placed into a mammalian expression vector with an in-frame triple c-Myc tag at the carboxyl terminus (Makela et al. 1995) and were transiently expressed in 293T cells. Following transfection, the cells were treated with nocodazole. c-Myc-TAG affinity-purified complexes were further selected for TFIIH by affinity purification using ERCC3 monoclonal antibodies (Fig. 8f). Interphase and mitotic TFIIHs were then assayed for transcription and kinase activities (Fig. 8f). In agreement with results presented above, mitotic-derived TFIIH was compromised in both transcription and CTD

phosphorylation (lanes 1 and 2) (see also Akoulitchev and Reinberg 1998). Analysis with mutant TFIIH revealed that both the transcriptional and kinase activities are dependent on Thr-170 (lane 4). Importantly, however, the mutation in Ser-164 relieved mitotic repression of TFIIH (lane 3). Inhibitory phosphorylation within the T-loop of a kinase has been described previously (Luo and Lodish 1997). These results strongly suggest that in vivo, both transcriptional and kinase activities of TFIIH depend on the phosphorylation state of its CDK7 subunit.

An earlier study of CDK7 regulation did not detect significant changes in its activity during the cell cycle in the context of the free CAK complex (Tassan et al. 1994; Adamczewski et al. 1996). It now appears that phosphorylation of CDK7 plays a critical regulatory part within the context of TFIIH. Previous data have established that transcription from the AdMLP does not require the kinase activity of CDK7 (Akoulitchev et al. 1995; Makela et al.

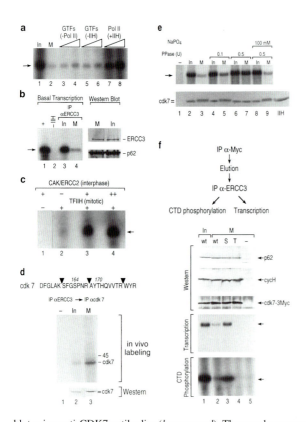

Figure 8. TFIIH activity is negatively regulated upon entry into mitosis. Mitotic and interphase extracts were prepared as described by Akoulitchev and Reinberg (1998). (*a*) Transcriptional activity of mitotic extracts (lane *2*) as compared to the interphase extracts (lane *1*) and complemented with different interphase-derived GTFs and RNAPII (lanes *3–8*). Titration included all GTFs without RNAPII (lanes *3,4*), GTFs and RNAPII without TFIIH (lanes *5,6*), TFIIH and RNAPII (lanes *7,8*). (*b*) Mitotic-derived TFIIH is impaired in its ability to reconstitute transcription. Basal transcription was reconstituted using the AdMLP and purified GTFs (lane *1*). Transcription is dependent on TFIIH (lane *2*). Reactions were reconstituted with immunoaffinity-purified TFIIH from interphase (lane *3*) or mitotic (lane *4*) extracts (for details on the TFIIH preparation used, see Fig. 1 and Akoulitchev and Reinberg [1998]). At the right of the panel, a Western blot is shown containing immunopurified TFIIH from mitotic and interphase extracts used in the reconstituted transcription assay. The subunits analyzed were ERCC3 and p62. (*c*) The ERCC2-CAK complex is impaired in mitotic-derived TFIIH. Basal transcription using the AdMLP was reconstituted using affinity-purified mitotic TFIIH (lanes *2–4*). CAK/ERCC2 purified from interphase extract was added in lieu of TFIIH (lane *1*) or in the presence of mitotic TFIIH (lanes *3,4*). (*d*) The CDK7 subunit of TFIIH is hyperphosphorylated in mitotic extracts. (*Top panel*) Amino acid sequence of a fragment of CDK7 spanning the sites of phosphorylation in vivo (Ser-164 and Thr-170). Arrowheads mark positions of cleavage by trypsin. (*Middle panel*) Cells were grown in the presence of [^{32}P]orthophosphate as described in the text and by Akoulitchev and Reinberg (1998). Cells were lysed and TFIIH was immunoprecipitated using anti-ERCC3 monoclonal antibodies. The precipitate was treated with SDS, followed by a second immunoprecipitation using anti-CDK7 antibodies. The immunoprecipitate was resolved by SDS-PAGE, followed by Western blot using anti-CDK7 antibodies (*lower panel*). The membrane was then analyzed by autoradiography (*top panel*). Double line indicates the resolved shift in the mobility of mitotic CDK7. (*e*) Reactivation of mitotic TFIIH by phosphatase treatment. Basal transcription from the AdMLP was reconstituted with affinity-purified interphase TFIIH (lane *2*), mitotic TFIIH (lane *3*), mitotic or interphase TFIIH pretreated with alkaline phosphatase in the absence or presence of sodium phosphate as indicated on the panel. (Lane *1*) Result using a mock affinity purification procedure. (*Lower panel*) Western blot analysis of CDK7 of the samples analyzed in transcription on the top panel. (*f*) Mutation of CDK7-Ser-164 bypasses mitotic repression. Wild-type and mutant forms of CDK7 were transfected into HeLa cells. (*Top panel*) Procedure used to isolate TFIIH containing the transfected mutated (CDK7-Ser-164 [Ser$_{164}$A (S)] or CDK7-Thr-170 [Thr$_{170}$A (T)] or a control empty vector) CDK7 subunit. The transfected CDK7 contains a c-Myc tagged which allows isolation of the transfected CDK7-containing complexes by immunoprecipitation using anti-Myc antibodies. TFIIH containing the transfected CDK7 subunit was isolated from the pool selected with the myc antibodies using a second immunoprecipitation with anti-ERCC3 monoclonal antibodies. TFIIH was isolated from interphase (lane *1*) and mitotic (lane *2*) cells. The isolated TFIIH was analyzed by Western blot for the presence of the p62, cyclin H, and CDK7-[3xMyc] subunits. The immunoprecipitates were also analyzed for transcription and kinase activities. The lane with the minus (–) symbol denotes the activity present in cells that were transfected with the Myc-Tag, but devoid of the CDK7-coding sequences (empty vector).

1995) but depends on its presence in TFIIH (Drapkin et al. 1996; Reardon et al. 1996; LeRoy et al. 1998a). We have shown that entry into mitosis leads to transcriptional inactivation of TFIIH through hyperphosphorylation of CDK7. It is important to point out that regulatory phosphorylation of CDKs has been shown to induce conformational changes (Russo et al. 1996) and thus may have an important role in regulating the transcriptional activity of TFIIH.

Our studies demonstrate that mitotic repression of basal transcription results from the phosphorylation of TFIIH. It has previously been demonstrated that MPF-mediated phosphorylation of RNAPII (Shermoen and O'Farrell 1991; Zawel et al. 1993; Parsons and Spencer 1997) and phosphorylation of the TAF subunits of TFIID in mitosis (Segil et al. 1996) impair transcription. These studies collectively demonstrate that the cell has developed different mechanisms to silence transcription during mitosis. The multitude of factors affected probably reflects mechanisms to halt transcription at different steps of the transcription cycle.

What Is the Function of TFIIH during the Transcription Cycle?

During its formation, the transcription initiation complex undergoes conformational changes that lead to the establishment of an open complex (Luse et al. 1987; Jiang and Gralla 1993; Holstege et al. 1996). Open complex formation is followed by the formation of the first phosphodiester bond. Following bond formation, RNAP can enter into an abortive mode, characterized by the production of catalytic amounts of short RNA molecules of up to nine nucleotides (McClure 1985; Luse and Jacob 1987; Luse et al. 1987). During the abortive mode of RNA synthesis, the enzyme does not leave the template, but generates numerous dead-end products. Some RNAPII molecules escape the abortive mode and enter into the productive cycle. In this case, RNAP moves away from the promoter enabling a second polymerase molecule to enter into the transcription cycle. This step is defined as promoter clearance.

In the case of RNAPII, transcription requires the hydrolysis of the β–γ bond of ATP or dATP (Bunick et al. 1982). Importantly, it was found that the energy derived from ATP hydrolysis is not required for elongation (Sawadogo and Roeder 1984) but instead for the formation of a stable open complex (Jiang et al. 1993; Holstege et al. 1996) and for a step subsequent to initiation of transcription, most likely promoter clearance (Goodrich and Tjian 1994; Dvir et al. 1996a, 1997). Recent studies have revealed that the ATP-dependent step can be bypassed by supercoiling of the DNA or by partial opening of the DNA around and immediately upstream of the transcription start site (Parvin and Sharp 1993; Pan and Greenblatt 1994; Tantin and Carey 1994; Holstege et al. 1995). This ATP-dependent step is catalyzed by TFIIH: Bypassing the requirement for ATP hydrolysis also bypasses the requirement for TFIIH (Parvin and Sharp 1993; Pan and Greenblatt 1994; Tantin and Carey 1994; Holstege et al. 1995).

To analyze the role of TFIIH in transcription, we developed an assay that measured the different steps of the transcription cycle, i.e., preinitiation complex formation, synthesis of the first phosphodiester bond, abortive synthesis, promoter clearance, and elongation (see Fig. 9a). Transcription complexes were formed on linear DNA molecules attached to magnetic beads as described previously (see Fig. 9a) (Zawel et al. 1995). The DNA used contains five Gal4 DNA-binding sites upstream of the AdMLP and a 50-nucleotide U-less transcription cassette. Subsequent to the formation of transcription complexes, the complexes were extensively washed to remove unbound proteins.

Upon addition of ribonucleoside triphosphates (ATP, CTP, and GTP), and in the presence of all the transcription factors, RNAPII initiated transcription and translocated along the U-less cassette synthesizing RNA molecules up to 50 nucleotides in length, whereupon the enzyme halts due to the lack of UTP. Under these conditions, stalled RNAPII molecules remained bound to the template within active ternary complexes and could be recovered with the beads (Fig. 9b, lane 3). A fraction of the RNAPII molecules that initiated transcription entered into the abortive mode and generated short transcripts that were released and recovered in the aqueous phase (Fig. 9b, lane 4).

When transcription reactions were performed in the absence of TFIIH (or additionally in the absence of TFIIE; data not shown), no full-length RNA was observed (Fig. 9b, compare lanes 2 and 3). Instead, the RNA transcripts recovered with the beads (ternary complexes) were short and were predominantly from 12 to 17 nucleotides in length (Fig. 9b, lane 2). Similar short RNA molecules in ternary complexes were present in the complete reaction but less abundant (compare lane 2 with 3). In addition, in the absence of TFIIH, the amount of released aborted RNA was increased compared to the complete reaction (Fig. 9b, compare lane 1 with 4). All products observed were sensitive to α-amanitin (data not shown).

These results establish a role for TFIIH during the transcription cycle. TFIIH and TFIIE are not required for initiation of transcription. In its absence, initiation of transcription takes place, but a large number of the transcription complexes are aborted. Moreover, RNAPII molecules that escape the abortive mode encounter a block to elongation when RNAPII reaches positions +12 to +17. Interestingly, this block to elongation was also observed, but to a much lesser extent, in reactions performed with all of the transcription factors (Fig. 9b, compare lanes 2 and 3). The accumulation of short RNA molecules (12–17 nucleotides in length) was not an artifact of the assay used, i.e., DNA attached to a solid support and single-round transcription conditions, similar products were observed in solution, under multiple-round transcription conditions (data not shown; see Kumar et al. 1998). Moreover, pausing appears to be sequence-independent, as an almost identical pattern was observed when RNAPII transcribed through G-less or U-less cassettes (data not shown). Additionally, pausing was not

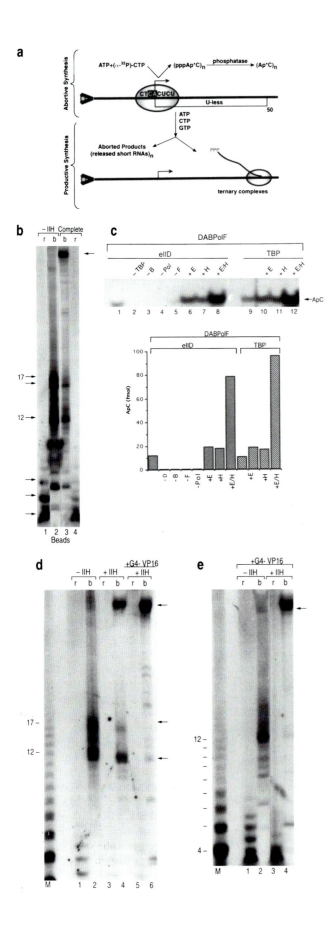

Figure 9. Functional analysis of TFIIH. (*a*) Schematic representation of the DNA template containing the AdMLP and a U-less cassette used in the experiments described below. The DNA sequence of the first 25 nucleotides downstream from the transcription start site is 5′-CTCA(+1)CACACCAACGGGCCCGAAGA-GAGG(+25). Transcription complexes were formed using purified transcription factors, as described by Kumar et al. (1998). The DNA template was biotinylated upstream of the transcription start site and immobilized to streptavidin-coated magnetic beads as described by Zawel et al. (1995). Following formation of the preinitiation complexes, the complexes were washed and transcription was monitored using different assays. For abortive dinucleotide synthesis, the transcription complexes were supplied with ATP and [α-^{32}P]CTP to form the pppApC dinucleotide product. This dinucleotide product was treated with phosphatase to form the ApC. (ApC)$_n$ denotes that the dinucleotide should be produced in catalytic amounts with respect to the amount of active template molecules. Reactions were also analyzed under conditions allowing productive synthesis. In this case, transcription was initiated by the addition of ribonucleoside triphosphates ATP, CTP, and GTP, as indicated in the figure. The partitioning of the expected RNA products is shown. The abortive products, up to nine nucleotides, are expected to be released from the template and recovered in the aqueous phase. The productive products are expected to remain in ternary complexes and therefore recovered with the beads. (*b*) Analysis of the role of TFIIH during transcription. Transcription reactions in the presence and absence of TFIIH, as indicated on the panel, were performed using immobilized DNA template under productive transcription conditions as described in *a* and by Kumar et al. (1998). The products of the reaction were separated as described in panel *a* and analyzed by electrophoresis on denaturing polyacrylamide gels. All lanes in the panel were derived from the same experiment and were exposed for the same period. Arrows on the left denote aborted released products (*lower arrows*), as well as short RNA molecules that were recovered with the beads. The arrow on the right denotes the full-length 50-nucleotide RNA. The numbers represent the size of RNA products. r and b at the top of the panel denote the aborted products that were released (r) and the productive products that remain bound (b) to the DNA. (*c*) Formation of the first phosphodiester bond in an abortive initiation assay. Transcription complexes were provided with ATP and [α-^{32}P]CTP as described in panel *a*. Reactions were performed using TBP or the TFIID complex as indicated on the panel. Dinucleotide synthesis is dependent on TBP (lane *2*), TFIIB (lane *3*), TFIIF (lane *5*), and RNAPII (lane *4*) and is stimulated by TFIIE and TFIIH (lanes *6–8* and *10–12*). The raw data are presented on the top of the panel. The amount of product obtained was quantitated and shown as a bar graph at the bottom of the panel. (*d* and *e*) Effect of Gal4-VP16 on the formation of the stalled complexes. Transcription reactions were performed as described in panel *b*, in the presence and absence of TFIIH and Gal4-VP16, as indicated at the top of the lanes on each panel. Reactions contained TFIID in lieu of TBP and the coactivators TFIIA and PC4. Following transcription, the template-bound (b) and the aqueous (r) phases were separated and the RNA products in these fractions were analyzed on a 28% denaturing polyacrylamide-urea gel. Arrows indicate positions of stalling (12–17 nucleotides) and full-size products.

promoter-specific as a similar pattern was observed with the human HSP70 promoter (data not shown).

Previous studies have indicated that TFIIH and TFIIE are absolutely required for first bond formation (Dvir et al. 1996b; Holstege et al. 1996). In light of the results presented above, we analyzed the formation of the first phosphodiester bond using an abortive initiation assay. Transcription complexes were provided with the two nucleotides required for initiation of transcription, ATP and [α-^{32}P]CTP, and formation of the dinucleotide pppApC was analyzed (see Fig. 9a). Under these conditions, we found that first-bond formation was dependent on TBP (supplied as recombinant or as the TFIID complex), TFIIB, TFIIF, and RNAPII (Fig. 9c) (for details, see Kumar et al. 1998). TFIIE and TFIIH were not required. However, in the presence of both of these factors, first-bond formation was stimulated. We estimate that the dinucleotide was produced at approximately 3% efficiency with respect to the amount of DNA added to the reaction in the absence of TFIIE and/or TFIIH. This low level of first-bond formation was specific, as it was dependent on TBP, TFIIB, and TFIIF, and mutations in the TATA motif eliminated all products (data not shown; for details, see Kumar et al. 1998). In the presence of TFIIE and TFIIH, an approximately fivefold stimulation was observed. The 3% efficiency observed in the absence of TFIIE and/or TFIIH is similar to the proportion of DNA templates utilized in the reconstituted RNAPII transcription system. Therefore, we suggest that the dinucleotide (abortive) product is produced in catalytic amounts only in the presence of TFIIE and TFIIH. In the absence of these factors, the dinucleotide is produced in approximately stoichiometric amounts with respect to the amount of active transcription complex.

Similar conclusions, with respect to factor requirements, were reached by Goodrich and Tjian (1994) using an abortive initiation assay coupled to dinucleotide priming. However, the studies of Dvir et al. (1996b), using a similar approach, suggested that TFIIH is required for first-bond formation. We thought that the discrepancies may reside in the conditions of the transcription assays, and we therefore analyzed the components of the transcription reaction for their ability to support first-bond formation in the absence of TFIIH. Our studies revealed that excess amounts of GTFs capable of binding to DNA, or nonspecific DNA-binding proteins (such as histone H1 or the bacterial Hu protein), inhibited TFIIH-independent first-bond formation (data not shown; see Kumar et al. 1998). Studies by other investigators have demonstrated that TFIIH can overcome repression of transcription mediated by nonspecific DNA-binding proteins (Stelzer et al. 1994). Thus, we suggest that the difference between the studies of Dvir et al. (1996b) and those of Goodrich and Tjian (1994), our observations, and Kumar et al. (1998) resides in the presence of nonspecific DNA-binding proteins or in the concentration of GTFs used. We conclude that TFIIE and TFIIH are not absolutely necessary for first-bond formation. In their presence, however, first-bond formation is strongly stimulated due to the establishment of a *stable* open complex (Jiang et al. 1993, 1996; Holstege et al. 1996, 1997).

Previous studies have shown that TFIIH interacts with activators (Xiao et al. 1994). Therefore, we investigated whether an activator could influence the formation of the paused promoter-proximal complexes. The results presented in Figure 9d illustrate that reactions performed under transcription activation conditions (with TFIID and the coactivators TFIIA and PC4) behave as we have described above; i.e., in the absence of TFIIH, no full-length transcripts were observed, and RNA molecules of 12–17 nucleotides were recovered with the beads (lane 2). In addition, aborted RNA molecules were observed in the aqueous phase (lane 1). The addition of TFIIH resulted in the production of full-length and short transcripts (lane 4), and the amount of aborted RNAs produced was reduced. We observed that addition of Gal4-VP16 to complete transcription reactions resulted in an approximately threefold increase in the production of the full-length RNA (Fig. 9d, compare lane 4 with 6). Importantly, the amount of short RNAs resulting from stalling of the transcription complex was drastically decreased (compare lane 4 with 6). The activator had no appreciable effect on the amount of paused complex produced on reactions lacking TFIIH (Fig. 9e, lane 2). Again, the activator, in the presence of TFIIH, suppressed the formation of promoter-proximal paused complexes and aborted products (Fig. 9e, lanes 3 and 4).

These results establish that an activator can increase the production of the full-length transcript and therefore stimulates the efficiency of transcription. The effect of the activator under the assay conditions was low (threefold stimulation). This is expected, as transcription was performed under single-round conditions and therefore reinitiation (RNAPII loading), which is also regulated by activators (Carcamo et al. 1989), was not observed.

Our findings have implications for previous studies indicating that activators influence the efficiency of elongation in vivo (Yankulov et al. 1994; Blau et al. 1996; for review, see Bentley 1995). These previous studies analyzed the effect of activators on transcription elongation as a function of the distribution of RNAPII on genes using nuclear run-on experiments. Interestingly, a direct correlation between activators capable of stimulating elongation (higher density of RNAPII at the 3´ end of the gene) and their ability to interact with TFIIH was established (Blau et al. 1996). These studies, however, did not distinguish promoter clearance from elongation. Indeed, elongation was defined as "all nucleotide addition steps after initiation," which includes promoter clearance (Blau et al. 1996). In light of our findings demonstrating that TFIIH is required for promoter clearance and that TFIIH does not travel with RNAPII in vitro (Zawel et al. 1995; see below), we believe that those studies measured, in part, the ability of TFIIH to catalyze promoter clearance.

The TFIIH-dependent phenomenon observed in the studies presented here, specifically the accumulation of promoter-proximal paused RNAPII complexes, is reminiscent of studies performed in vivo demonstrating that RNAPIIs that initiate transcription are paused in a region proximal to the promoter. This phenomenon appears to be general, as it has been observed in a number of genes in different species (Lis and Wu 1994; Akhtar et al. 1996). The results establish that TFIIH enters into the transcription complex during the formation of a transcription-competent complex, or prior to the formation of the promoter-proximal complexes. We found that the preformed promoter-proximal paused complex could not be induced to form a full-length transcript by the subsequent addition of TFIIH or other protein fractions (for details, see Kumar et al. 1998). An important question is whether the promoter-proximal paused complex observed in vivo is a consequence of the inability to recruit TFIIH (or an active form of the factor) to the initiation complex, and whether this complex ever engages in productive transcription. We suggest that the concentration of TFIIH in vivo may be limiting and that TFIIH may be recruited to the transcription initiation complex by activators during the formation of the transcription initiation complex.

Recycling of the GTFs during Transcription

The advent of highly purified reconstituted transcription systems has allowed an extensive factor by factor analysis of the initiation/elongation transition. We have analyzed the behavior of human TFIID, TFIIB, TFIIE, TFIIF, and TFIIH. For numerous reasons, understanding the fate of TFIIE and TFIIH has been a subject of particular interest. TFIIH is the only GTF known to contain enzymatic activities (for review, see Drapkin and Reinberg 1994). The discovery that TFIIH participates in nucleotide excision repair led to speculation that TFIIH might travel with RNAPII during elongation facilitating DNA unwinding and or coupling of transcription and NER (Buratowski 1993). Using diverse approaches, including transcription of templates attached to solid sup-

ports coupled to Western blotting and template competition assays, we have constructed an account of the events preceding and immediately following transcription initiation by RNAPII (for details, see Zawel et al. 1995). Our results shed light on which factors compose a transcription-competent initiation complex, which factors remain bound at the promoter following the release of RNAPII, and which factors are released. A summary of our findings is depicted in Figure 10.

We ascertained that five GTFs (TFIID/TBP, IIB, IIF, IIE, IIH) coexist in the complete initiation complex prior to the addition of NTPs. Although it is possible that some of the complexes analyzed were nonviable, the fact that the release of each factor, with the exception of TFIID, was detected functionally and, moreover, the fact that we were able to define a point where release occurs provide further indication that each of the factors is indeed present.

TFIID was found to remain promoter-bound through the transcription cycle. These results are in agreement with several previous studies that utilized fractionated HeLa extracts (Van Dyke et al. 1988; White et al. 1992; Jiang and Gralla 1993), but they contradict observations made using *Drosophila* extracts (Kadonaga 1990). Discrepancies may be attributed to factors such as Mot1 (Auble et al. 1994), which can specifically remove TBP from the DNA.

TFIIB release was detected immediately upon addition of NTPs. The fact that TFIIB is released is somewhat surprising since TFIIB can directly interact with the BRE site as discussed above. Moreover, TFIIB interacts with two components known to stably interact with the promoter, i.e., TFIID, via TBP (Ha et al. 1993), and TAF40 (Goodrich et al. 1993). The release of TFIIB from the transcription complex may be related to

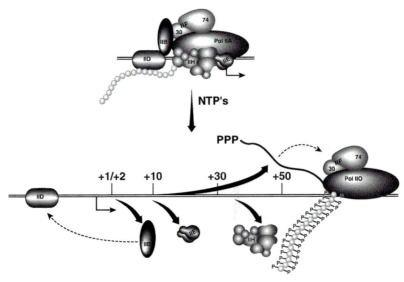

Figure 10. Model depicting the fate of the GTFs during the transition from transcription initiation to elongation. (*Upper panel*) Schematic representation of the preinitiation complex in which it is shown that all the GTFs coexist together with the nonphosphorylated form of RNAPII. (*Lower panel*) Stalled RNAPII in which the CTD is phosphorylated. Following the onset of elongation, the initiation complex is disrupted. TFIID, or TBP, remains bound at the TATA motif. TFIIB is released but reassociates with TFIID (or TBP). TFIIF was released from the initiation complex but reassociates with the stalled polymerase. TFIIE and TFIIH both recycle with TFIIE release occurring first.

changes in the conformation of TFIIB upon initiation of transcription. This hypothesis is consistent with studies of Green and coworkers (Roberts and Green 1994), and Hampsey and coworkers (M. Hampsey, pers. comm.), who independently demonstrated that activators induce a conformational change in TFIIB in vitro and in vivo, respectively. Rather than maintain TFIIB at the promoter, the activator may stimulate TFIIB assembly (Choy and Green 1993).

In an effort to determine whether any GTFs travel with RNAPII during elongation, we developed a procedure for the isolation and analysis of RNAPII elongation complexes (for details, see Zawel et al. 1995). TFIIF was the only initiation factor detected in halted elongation complexes. TFIIF appears to be released from the initiation complex at some point after formation of the first ten phosphodiester bonds. Following its release, TFIIF reassociates with the halted elongation complex. Upon reentry of the polymerase into productive elongation, TFIIF is released. TFIIF is known to tightly interact with RNAPII in solution (Flores et al. 1989; McCracken and Greenblatt 1991). Apparently, this interaction has a dual role. First, TFIIF escorts RNAPII to the assembling initiation complex. The small subunit of TFIIF, RAP30, is independently capable of performing this function (Flores et al. 1991; Killeen et al. 1992). Second, in the event that the polymerase pauses, TFIIF reassociation may serve to facilitate passage through the pause (Price et al. 1989; Bengal et al. 1991). Studies from Burton and coworkers indicate that this property is intrinsic to the large subunit of TFIIF, RAP74 (Chang et al. 1993).

In agreement with these observations, we found that TFIIF was capable of interacting with both the phosphorylated and nonphosphorylated forms of RNAPII (Zawel et al. 1995). This is in contrast to TBP and TFIIE, which interact only with the nonphosphorylated form of RNAPII (Usheva et al. 1992; Maxon et al. 1994). Accordingly, we did not detect TFIIE in the elongation complex, where only phosphorylated RNAPII was detected (see Zawel et al. 1995). Rather, TFIIE appeared to be released before the formation of the tenth phosphodiester bond of the nascent RNA. We cannot at this time determine whether TFIIE is released before, after, or concomitant with TFIIB. Significantly, we have demonstrated that TFIIE release precedes release of TFIIH. This has important mechanistic implications as TFIIE was found to stimulate the TFIIH-CTD kinase activity (Lu et al. 1992; Ohkuma and Roeder 1994) but to negatively regulate the ERCC3-associated DNA helicase activity of TFIIH (Drapkin et al. 1994b). Since the helicase associated with ERCC3 was found to be essential for transcription in yeast (Guzder et al. 1994), we suspect that release of TFIIE is a critical checkpoint during initiation/promoter clearance.

The investigation of whether TFIIH was in the elongation complex was of particular interest to us. The observation that at least two subunits of TFIIH (ERCC2 and ERCC3) contain DNA helicase activity (Schaeffer et al. 1993, 1994; Drapkin et al. 1994a,b) fueled speculation that TFIIH traveled with RNAPII and facilitated DNA

unwinding during elongation. In addition, as stated above, studies have demonstrated that the TFIIH complex participates not only in transcription, but also in DNA excision repair. Since actively transcribed genes are repaired more efficiently than silent genes (Mellon et al. 1986; Hanawalt 1994), it was thought that TFIIH was the component which linked these two processes (Buratowski 1993). Contrary to this prediction, neither the p62, p80/ERCC2, nor p89/ERCC3 subunit of TFIIH was detected in the ternary complex. Toward reconciling our observations, the following points are offered: With regard to the helicase activity aiding elongation, it is known that RNAPII can extensively transcribe double-stranded DNA in the absence of exogenously added helicase (for review, see Kerppola and Kane 1991). In the case of the coupling of transcription with nucleotide excision repair, genetic studies have shown that humans and yeast have specific proteins that couple these processes. CS-A and CS-B (ERCC6) mutants are capable of repairing damaged DNA, but they are defective in the ability to preferentially repair actively transcribed genes (Troelstra et al. 1992). TFIIH can be incorporated into elongation complexes that are stalled at lesions, yet the mechanism by which TFIIH is recruited to RNAPII stalled at a lesion is unclear and requires CS-A and CS-B (Sancar 1996; Sancar et al. 1996; Tantin et al. 1997). Although CS-B (ERCC6) is not a subunit of TFIIH, it has been shown to interact with TFIIH (Iyer et al. 1996) and to associate with stalled ternary complexes (Tantin et al. 1997). Thus, when RNAPII stalls at a lesion, ERCC6 may stimulate formation of repair complexes by recruiting TFIIH. We envision that the release of TFIIH is mechanistically favorable, as it ensures that a limited cellular pool of TFIIH will not be sequestered in elongation complexes and is thus available not only to mediate multiple transcription initiation events, but also to participate in both nucleotide excision repair processes: overall genome repair and preferential repair of actively transcribed genes.

THE NEXT LEVEL: TRANSCRIPTION ON CHROMATIN TEMPLATES

The studies described above have defined the protein machinery that transcribe class II genes and have shed light on their roles in the transcription mechanism. However, in vivo, this machinery encounters its transcription template not as naked DNA, but packaged with histones to form a highly ordered structure known as chromatin. Early attempts to transcribe chromatin templates in vitro using highly purified transcription systems clearly demonstrated that transcription of DNA in its natural chromatin form required additional undefined accessory factors (Kamakaka et al. 1993; Owen-Hughes and Workman 1994; Cairns 1998). These accessory factors must be identified and purified if the transcription process is to be reconstituted using physiologically relevant templates.

With the aim of establishing an in vitro transcription system capable of transcribing chromatin templates, we assembled plasmid chromatin templates using the *Drosophila* S190 assembly extract (Kamakaka et al.

1993) and transcribed them with the reconstituted human RNAPII transcription system we have discussed above. We found that the purified system could not initiate transcription on chromatin templates (data not shown; for details, see Orphanides et al. 1998). Other investigators have shown that activities present in the S190 chromatin assembly extract use the energy of ATP hydrolysis to locally remodel chromatin structure upon binding of a transcriptional activator protein (Pazin et al. 1994; Tsukiyama and Wu 1995). These activities have now been purified (NURF, Tsukiyama and Wu 1995; ACF, Ito et al. 1997; CHRAC, Varga-Weisz et al. 1997). We found that our reconstituted system could initiate transcription on chromatin templates that were remodeled in the S190 extract upon the binding of the model transcriptional activator GAL4-VP16 (data not shown; see Orphanides et al. 1998). Surprisingly, this remodeling was not sufficient for the production of long transcripts. Instead, polymerases that initiated transcription on the remodeled templates encountered a block to transcript elongation at about the +50 position, presumably due to the presence of a nucleosome (Fig. 11a, lane 2). Since no such block was observed in crude HeLa cell nuclear extracts (data not shown), we attempted to purify a factor(s) from the crude extract that could counteract this block to allow RNAPII to synthesize long transcripts. Using this assay, we purified a heterodimeric factor of 140 and 80 kD (Fig. 11b) that we have termed FACT (*facilitate chromatin transcription*) that promotes transcript elongation through nucleosomes (Fig. 11, lane 3) (for details, see Orphanides et al. 1998).

The purification of FACT took us some way toward our goal of reconstituting transcription on chromatin templates. However, in these experiments, transcription initiation was dependent on the remodeling of chromatin templates by activities in the S190 assembly extract. We therefore searched for an analogous activity in HeLa nuclear extracts that would facilitate transcription initiation on chromatin templates. Using an abortive initiation assay that measures the formation of the first phosphodiester bond of the RNA, we purified a protein factor that facilitates transcription initiation on chromatin templates in the presence of the activator GAL4-VP16 (Fig 11c). This factor also possesses ATP-dependent chromatin remodeling and spacing activities, hence the name RSF (*re*modeling and *s*pacing *f*actor) (data not shown; see LeRoy et al. 1998b). Although RNAPII can initiate transcription in the presence of RSF and GAL4-VP16 (Fig. 11d), the synthesis of long transcripts is dependent on FACT (Fig. 11e). We therefore conclude that RSF, FACT, and an activator are the minimal accessory factors required for transcription on chromatin templates.

As we have described, our understanding of the proteins involved in transcription of class II genes has come a long way in the past decade. Initiatives to reconstitute transcription in a more natural chromatin environment have

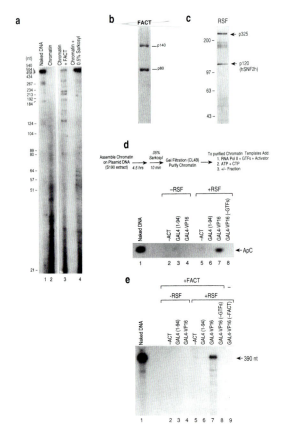

Figure 11. Two novel factors, RSF and FACT, are minimally required for transcription from chromatin templates. (*a*) RNA polymerase II stalls proximal to the promoter on remodeled chromatin templates. Remodeled chromatin (for details, see Orphanides et al. 1998) was incubated with GTFs and RNAPII for 30 min to allow preinitiation complex formation. ATP, UTP, and limiting radioactively labeled CTP was added to allow synthesis of short, 5′ end-labeled RNAs. Excess unlabeled CTP was then added and reactions were incubated for a further 20 min. Naked DNA was used as the template in lane *1*. Reactions containing chromatin templates were scaled-up threefold to aid comparison with the reaction containing naked DNA. FACT (lane *3*) and 0.5% Sarkosyl (lane *4*), which removes histones (Izban and Luse 1991), were added 4 min after the addition of excess CTP. The positions of size markers are indicated (nt). (*b*) Silver staining of an SDS-polyacrylamide gel containing the two subunits of FACT. (*c*) Silver staining of a polyacrylamide-SDS gel containing an aliquot of RSF derived from the last step of purification (Superose 12; for details, see LeRoy et al. 1998b). Peptide sequencing by microcapillary HPLC/ion trap mass spectrometry identified the 135-kD polypeptide as hSNF2h (for details, see LeRoy et al. 1998b). (*d*) Initiation of transcription with the reconstituted system on chromatin templates requires both an activator and RSF. Reaction conditions were reconstituted using the factors depicted in Fig. 1. The presence or absence of RSF and the addition of $GAL4_{1-94}$-VP16 or $GAL4_{1-94}$ are indicated on the figure. GTFs were omitted in lane 8. Lane *1* is transcription initiation reconstituted on naked DNA. (*e*) Productive full-length transcription on chromatin templates requires RSF and FACT. Transcription was reconstituted under the same conditions used for initiation assays except that the reactions were provided with a nucleotide mixture sufficient to generate a 390-nucleotide transcript. RSF, FACT, and $GAL4_{1-94}$-VP16 or $GAL4_{1-94}$ were added as indicated in the figure. GTFs were omitted from the reaction in lane 8. (Lane *1*) Full-length transcription reconstituted on naked DNA.

identified additional chromatin-specific accessory factors with no apparent role on naked DNA. Future studies will be aimed at identifying and purifying other protein factors required for efficient transcription on chromatin templates and at understanding the roles of accessory factors in the regulation of RNAPII transcription.

ACKNOWLEDGMENTS

We thank S. Burley for atomic coordinates. This work was supported by National Institutes of Health grants GM-37120, GM-48518 to D.R., and GM-53665 to R.H.E. and by Howard Hughes Medical Institute investigatorships to D.R and R.H.E.

REFERENCES

Aboussekhra A., Biggerstaff M., Shivji M.K., Vilpo J.A., Moncollin V., Podust V.N., Protic M., Hubscher U., Egly J.M., and Wood R.D. 1995. Mammalian DNA nucleotide excision repair reconstituted with purified protein components. *Cell* **80:** 859.

Adamczewski J.P., Rossignol M., Tassan J.P., Nigg E.A., Moncollin V., and Egly J.M. 1996. MAT1, cdk7 and cyclin H form a kinase complex which is UV light-sensitive upon association with TFIIH. *EMBO J.* **15:** 1877.

Akhtar A., Faye G., and Bentley D.L. 1996. Distinct activated and non-activated RNA polymerase II complexes in yeast. *EMBO J.* **15:** 4654.

Akoulitchev S. and Reinberg D. 1998. Cell cycle regulation of TFIIH transcriptional activity through the CDK7 subunit. *Genes Dev.* **12:** 3541.

Akoulitchev S., Makela T.P., Weinberg R.A., and Reinberg D. 1995. Requirement for TFIIH kinase activity in transcription by RNA polymerase II. *Nature* **377:** 557.

Auble D.T., Hansen K.E., Mueller C.G., Lane W.S., Thorner J., and Hahn S. 1994. Mot1, a global repressor of RNA polymerase II transcription, inhibits TBP binding to DNA by an ATP-dependent mechanism. *Genes Dev.* **8:** 1920.

Bagby S., Kim S., Maldonado E., Tong K.I., Reinberg D., and Ikura M. 1995. Solution structure of the C-terminal core domain of human TFIIB: Similarity to cyclin A and interaction with TATA- binding protein. *Cell* **82:** 857.

Beamer L.J. and Pabo C.O. 1992. Refined 1.8 Å crystal structure of the lambda repressor-operator complex. *J. Mol. Biol.* **227:** 177.

Bengal E., Flores O., Krauskopf A., Reinberg D., and Aloni Y. 1991. Role of the mammalian transcription factors IIF, IIS, and IIX during elongation by RNA polymerase II. *Mol. Cell. Biol.* **11:** 1195.

Bentley D.L. 1995. Regulation of transcriptional elongation by RNA polymerase II. *Curr. Opin. Genet. Dev.* **5:** 210.

Blau J., Xiao H., McCracken S., O'Hare P., Greenblatt J., and Bentley D. 1996. Three functional classes of transcriptional activation domain. *Mol. Cell. Biol.* **16:** 2044.

Buck V., Russell P., and Millar J.B. 1995. Identification of a cdk-activating kinase in fission yeast. *EMBO J.* **14:** 6173.

Bunick D., Zandomeni R., Ackerman S., and Weinmann R. 1982. Mechanism of RNA polymerase II—specific initiation of transcription in vitro: ATP requirement and uncapped runoff transcripts. *Cell* **29:** 877.

Buratowski S. 1993. DNA repair and transcription: The helicase connection (see comments). *Science* **260:** 37.

Burley S.K. and Roeder R.G. 1996. Biochemistry and structural biology of transcription factor IID (TFIID). *Annu. Rev. Biochem.* **65:** 769.

Cairns B.R. 1998. Chromatin remodeling machines: Similar motors, ulterior motives. *Trends Biochem. Sci.* **23:** 20.

Carcamo J., Lobos S., Merino A., Buckbinder L., Weinmann R., Natarajan V., and Reinberg D. 1989. Factors involved in specific transcription by mammalian RNA polymerase II. Role of factors IID and MLTF in transcription from the adenovirus major late and IVa2 promoters. *J. Biol. Chem.* **264:** 7704.

Chamberlin M.J. 1974. The selectivity of transcription. *Annu. Rev. Biochem.* **43:** 721.

Chang C., Kostrub C.F., and Burton Z.F. 1993. RAP30/74 (transcription factor IIF) is required for promoter escape by RNA polymerase II. *J. Biol. Chem.* **268:** 20482.

Choy B. and Green M.R. 1993. Eukaryotic activators function during multiple steps of preinitiation complex assembly. *Nature* **366:** 531.

Cismowski M.J., Laff G.M., Solomon M.J., and Reed S.I. 1995. KIN28 encodes a C-terminal domain kinase that controls mRNA transcription in *Saccharomyces cerevisiae* but lacks cyclin- dependent kinase-activating kinase (CAK) activity. *Mol. Cell. Biol.* **15:** 2983.

Conaway J.W. and Conaway R.C. 1989. A multisubunit transcription factor essential for accurate initiation by RNA polymerase II. *J. Biol. Chem.* **264:** 2357.

Cortes P., Flores O., and Reinberg D. 1992. Factors involved in specific transcription by mammalian RNA polymerase II: Purification and analysis of transcription factor IIA and identification of transcription factor IIJ. *Mol. Cell. Biol.* **12:** 413.

Coulombe B., Killeen M., Liljelund P., Honda B., Xiao H., Ingles C.J., and Greenblatt J. 1992. Identification of three mammalian proteins that bind to the yeast TATA box protein TFIID. *Gene Expr.* **2:** 99.

Cox J., Hayward M., Sanchez J., Gegnas L., van der Zee S., Dennis J., Sigler P., and Schepartz A. 1997. Bidirectional binding of TBP to the TATA box. *Proc. Natl. Acad. Sci.* **94:** 13475.

Dahmus M.E. 1996. Reversible phosphorylation of the C-terminal domain of RNA polymerase II. *J. Biol. Chem.* **271:** 19009.

Darst S.A., Edwards A.M., Kubalek E.W., and Kornberg R.D. 1991. Three-dimensional structure of yeast RNA polymerase II at 16 Å resolution. *Cell* **66:** 121.

DeJong J. and Roeder R.G. 1993. A single cDNA, hTFIIA/α, encodes both the p35 and p19 subunits of human TFIIA. *Genes Dev.* **7:** 2220.

DeJong J., Bernstein R., and Roeder R.G. 1995. Human general transcription factor TFIIA: Characterization of a cDNA encoding the small subunit and requirement for basal and activated transcription. *Proc. Natl. Acad. Sci.* **92:** 3313.

Drapkin R. and Reinberg D. 1994. The multifunctional TFIIH complex and transcriptional control. *Trends Biochem. Sci.* **19:** 504.

Drapkin R., Sancar A., and Reinberg D. 1994. Where transcription meets repair. *Cell* **77:** 9.

Drapkin R., Le Roy G., Cho H., Akoulitchev S., and Reinberg D. 1996. Human cyclin-dependent kinase-activating kinase exists in three distinct complexes. *Proc. Natl. Acad. Sci.* **93:** 6488.

Drapkin R., Reardon J.T., Ansari A., Huang J.C., Zawel L., Ahn K., Sancar A., and Reinberg D. 1994b. Dual role of TFIIH in DNA excision repair and in transcription by RNA polymerase II. *Nature* **368:** 769.

Dvir A., Conaway R.C., and Conaway J.W. 1996a. Promoter escape by RNA polymerase II. A role for an ATP cofactor in suppression of arrest by polymerase at promoter-proximal sites. *J. Biol. Chem.* **271:** 23352.

Dvir A., Tan S., Conaway J.W., and Conaway R.C. 1997. Promoter escape by RNA polymerase II. Formation of an escape-competent transcriptional intermediate is a prerequisite for exit of polymerase from the promoter. *J. Biol. Chem.* **272:** 28175.

Dvir A., Garrett K.P., Chalut C., Egly J.M., Conaway J.W., and Conaway R.C. 1996b. A role for ATP and TFIIH in activation of the RNA polymerase II preinitiation complex prior to transcription initiation. *J. Biol. Chem.* **271:** 7245.

Ebright R. 1991. Identification of amino acid-base pair contacts by genetic methods. *Methods Enzymol.* **208:** 620.

Feaver W.J., Gileadi O., Li Y., and Kornberg R.D. 1991. CTD kinase associated with yeast RNA polymerase II initiation factor b. *Cell* **67:** 1223.

Feaver W.J., Svejstrup J.Q., Henry N.L., and Kornberg R.D. 1994. Relationship of CDK-activating kinase and RNA polymerase II CTD kinase TFIIH/TFIIK. *Cell* **79:** 1103.

Feaver W.J., Henry N.L., Wang Z., Wu X., Svejstrup J.Q., Bushnell D.A., Friedberg E.C., and Kornberg R.D. 1997. Genes for Tfb2, Tfb3, and Tfb4 subunits of yeast transcription/repair factor IIH. Homology to human cyclin-dependent kinase activating kinase and IIH subunits. *J. Biol. Chem.* **272:** 19319.

Feaver W.J., Svejstrup J.Q., Bardwell L., Bardwell A.J., Buratowski S., Gulyas K.D., Donahue T.F., Friedberg E.C., and Kornberg R.D. 1993. Dual roles of a multiprotein complex from *S. cerevisiae* in transcription and DNA repair. *Cell* **75:** 1379.

Fischer L., Gerard M., Chalut C., Lutz Y., Humbert S., Kanno M., Chambon P., and Egly J.M. 1992. Cloning of the 62-kilodalton component of basic transcription factor BTF2. *Science* **257:** 1392.

Fisher R.P., Jin P., Chamberlin H.M., and Morgan D.O. 1995. Alternative mechanisms of CAK assembly require an assembly factor or an activating kinase. *Cell* **83:** 47.

Flores O., Ha I., and Reinberg D. 1990. Factors involved in specific transcription by mammalian RNA polymerase II. Purification and subunit composition of transcription factor IIF. *J. Biol. Chem.* **265:** 5629.

Flores O., Lu H., and Reinberg D. 1992. Factors involved in specific transcription by mammalian RNA polymerase II. Identification and characterization of factor IIH. *J. Biol. Chem.* **267:** 2786.

Flores O., Maldonado E., and Reinberg D. 1989. Factors involved in specific transcription by mammalian RNA polymerase II. Factors IIE and IIF independently interact with RNA polymerase II. *J. Biol. Chem.* **264:** 8913.

Flores O., Lu H., Killeen M., Greenblatt J., Burton Z.F., and Reinberg D. 1991. The small subunit of transcription factor IIF recruits RNA polymerase II into the preinitiation complex. *Proc. Natl. Acad. Sci.* **88:** 9999.

Forget D., Robert F., Grondin G., Burton Z.F., Greenblatt J., and Coulombe B. 1997. RAP74 induces promoter contacts by RNA polymerase II upstream and downstream of a DNA bend centered on the TATA box. *Proc. Natl. Acad. Sci.* **94:** 7150.

Garrett K. P., Serizawa H., Hanley J.P., Bradsher J.N., Tsuboi A., Arai N., Yokota T., Arai K., Conaway R.C., and Conaway J.W. 1992. The carboxyl terminus of RAP30 is similar in sequence to region 4 of bacterial sigma factors and is required for function. *J. Biol. Chem.* **267:** 23942.

Geiger J.H., Hahn S., Lee S., and Sigler P.B. 1996. Crystal structure of the yeast TFIIA/TBP/DNA complex. *Science* **272:** 830.

Gerard M., Fischer L., Moncollin V., Chipoulet J.M., Chambon P. and Egly J.M. 1991. Purification and interaction properties of the human RNA polymerase B(II) general transcription factor BTF2. *J. Biol. Chem.* **266:** 20940.

Goodrich J.A. and Tjian R. 1994. Transcription factors IIE and IIH and ATP hydrolysis direct promoter clearance by RNA polymerase II. *Cell* **77:** 145.

Goodrich J.A., Hoey T., Thut C.J., Admon A., and Tjian R. 1993. *Drosophila* TAFII40 interacts with both a VP16 activation domain and the basal transcription factor TFIIB. *Cell* **75:** 519.

Gottesfeld J.M. and Forbes D.J. 1997. Mitotic repression of the transcriptional machinery. *Trends Biochem. Sci.* **22:** 197.

Guzder S.N., Habraken Y., Sung P., Prakash L., and Prakash S. 1995. Reconstitution of yeast nucleotide excision repair with purified Rad proteins, replication protein A, and transcription factor TFIIH. *J. Biol. Chem.* **270:** 12973.

Guzder S.N., Sung P., Bailly V., Prakash L., and Prakash S. 1994. RAD25 is a DNA helicase required for DNA repair and RNA polymerase II transcription. *Nature* **369:** 578.

Ha I., Roberts S., Maldonado E., Sun X., Kim L.U., Green M., and Reinberg D. 1993. Multiple functional domains of human transcription factor IIB: Distinct interactions with two general transcription factors and RNA polymerase II. *Genes Dev.* **7:** 1021.

Hanawalt P.C. 1994. Transcription-coupled repair and human disease. *Science* **266:** 1957.

Holstege F.C., Fiedler U., and Timmers H.T. 1997. Three transitions in the RNA polymerase II transcription complex during initiation. *EMBO J.* **16:** 7468.

Holstege F.C., van der Vliet P.C., and Timmers H.T. 1996. Opening of an RNA polymerase II promoter occurs in two distinct steps and requires the basal transcription factors IIE and IIH. *EMBO J.* **15:** 1666.

Holstege F.C., Tantin D., Carey M., van der Vliet P.C., and Timmers H.T. 1995. The requirement for the basal transcription factor IIE is determined by the helical stability of promoter DNA. *EMBO J.* **14:** 810.

Humbert S., van Vuuren H., Lutz Y., Hoeijmakers J.H., Egly J.M., and Moncollin V. 1994. p44 and p34 subunits of the BTF2/TFIIH transcription factor have homologies with SSL1, a yeast protein involved in DNA repair. *EMBO J.* **13:** 2393.

Ito T., Bulger M., Pazin M.J., Kobayashi R., and Kadonaga J.T. 1997. ACF, an ISWI-containing and ATP-utilizing chromatin assembly and remodeling factor. *Cell* **90:** 145.

Iyer N., Reagan M.S., Wu K.J., Canagarajah B., and Friedberg E.C. 1996. Interactions involving the human RNA polymerase II transcription/nucleotide excision repair complex TFIIH, the nucleotide excision repair protein XPG, and Cockayne syndrome group B (CSB) protein. *Biochemistry* **35:** 2157.

Izban M.G. and Luse D.S. 1991. Transcription on nucleosomal templates by RNA polymerase II in vitro: Inhibition of elongation with enhancement of sequence-specific pausing. *Genes Dev.* **5:** 683.

Jiang Y. and Gralla J.D. 1993. Uncoupling of initiation and reinitiation rates during HeLa RNA polymerase II transcription in vitro. *Mol. Cell. Biol.* **13:** 4572.

Jiang Y., Smale S.T., and Gralla J.D. 1993. A common ATP requirement for open complex formation and transcription at promoters containing initiator or TATA elements. *J. Biol. Chem.* **268:** 6535.

Jiang Y., Yan M., and Gralla J.D. 1996. A three-step pathway of transcription initiation leading to promoter clearance at an activation RNA polymerase II promoter. *Mol. Cell. Biol.* **16:** 1614.

Kadonaga J.T. 1990. Assembly and disassembly of the *Drosophila* RNA polymerase II complex during transcription. *J. Biol. Chem.* **265:** 2624.

Kamakaka R.T., Bulger M., and Kadonaga J.T. 1993. Potentiation of RNA polymerase II transcription by Gal4-VP16 during but not after DNA replication and chromatin assembly. *Genes Dev.* **7:** 1779.

Kerppola T.K. and Kane C.M. 1991. RNA polymerase: Regulation of transcript elongation and termination. *FASEB J.* **5:** 2833.

Killeen M., Coulombe B., and Greenblatt J. 1992. Recombinant TBP, transcription factor IIB, and RAP30 are sufficient for promoter recognition by mammalian RNA polymerase II. *J. Biol. Chem.* **267:** 9463.

Kim J.L., Nikolov D.B., and Burley S.K. 1993. Co-crystal structure of TBP recognizing the minor groove of a TATA element (see comments). *Nature* **365:** 520.

Kim T.K., Lagrange T., Wang Y.H., Griffith J.D., Reinberg D., and Ebright R.H. 1997. Trajectory of DNA in the RNA polymerase II transcription preinitiation complex. *Proc. Natl. Acad. Sci.* **94:** 12268.

Kim Y., Geiger J.H., Hahn S., and Sigler P.B. 1993. Crystal structure of a yeast TBP/TATA-box complex. *Nature* **365:** 512.

Koleske A.J. and Young R.A. 1995. The RNA polymerase II holoenzyme and its implications for gene regulation. *Trends Biochem. Sci.* **20:** 113.

Kumar K.P., Akoulitchev S., and Reinberg D. 1998. Promoter-proximal stalling results from the inability to recruit transcription factor IIH to the transcription complex and is a regulated event. *Proc. Natl. Acad. Sci.* **95:** 9767.

Labbe J.C., Martinez A.M., Fesquet D., Capony J.P., Darbon

J.M., Derancourt J., Devault A., Morin N., Cavadore J.C., and Doree M. 1994. p40MO15 associates with a p36 subunit and requires both nuclear translocation and Thr176 phosphorylation to generate cdk-activating kinase activity in *Xenopus* oocytes. *EMBO J.* **13:** 5155.

Lagrange T., Kapanidis A.N., Tang H., Reinberg D., and Ebright R.H. 1998. New core promoter element in RNA polymerase II-dependent transcription: Sequence-specific DNA binding by transcription factor IIB. *Genes Dev.* **12:** 34.

Lagrange T., Kim T.K., Orphanides G., Ebright Y.W., Ebright R.H., and Reinberg D. 1996. High- resolution mapping of nucleoprotein complexes by site-specific protein-DNA photocrosslinking: Organization of the human TBP-TFIIA-TFIIB-DNA quaternary complex. *Proc. Natl. Acad. Sci.* **93:** 10620.

Lee D.K., Horikoshi M., and Roeder R.G. 1991. Interaction of TFIID in the minor groove of the TATA element. *Cell* **67:** 1241.

Lee S. and Hahn S. 1995. Model for binding of transcription factor TFIIB to the TBP-DNA complex. *Nature* **376:** 609.

Leresche A., Wolf V.J., and Gottesfeld J.M. 1996. Repression of RNA polymerase II and III transcription during M phase of the cell cycle. *Exp. Cell Res.* **229:** 282.

LeRoy G., Drapkin R., Weis L., and Reinberg D. 1998. Immunoaffinity purification of the human multisubunit transcription factor IIH. *J. Biol. Chem.* **273:** 7134.

LeRoy G., Orphanides G., Lane W.S., and Reinberg D. 1998b. Transcription of chromatin templates minimally requires RNA polymerase II, general transcription factors, FACT and RSF, a novel chromatin remodeling factor. *Science* **282:** 1900.

Leuther K.K., Bushnell D.A., and Kornberg R.D. 1996. Two-dimensional crystallography of TFIIB- and IIE-RNA polymerase II complexes: Implications for start site selection and initiation complex formation. *Cell* **85:** 773.

Lis J.T. and Wu C. 1994. Transcriptional regulation of heat shock genes. In *Transcription: Mechanisms and regulation* (ed. R.C. Conaway and J.W. Conaway), p. 459. Raven Press, New York.

Lu H., Zawel L., Fisher L., Egly J.M., and Reinberg D. 1992. Human general transcription factor IIH phosphorylates the C-terminal domain of RNA polymerase II. *Nature* **358:** 641.

Luo K. and Lodish H.F. 1997. Positive and negative regulation of type II TGF-β receptor signal transduction by autophosphorylation on multiple serine residues. *EMBO J.* **16:** 1970.

Luse D.S. and Jacob G.A. 1987. Abortive initiation by RNA polymerase II in vitro at the adenovirus 2 major late promoter. *J. Biol. Chem.* **262:** 14990.

Luse D.S., Kochel T., Kuempel E.D., Coppola J.A., and Cai H. 1987. Transcription initiation by RNA polymerase II in vitro. At least two nucleotides must be added to form a stable ternary complex. *J. Biol. Chem.* **262:** 289.

Ma D., Watanabe H., Mermelstein F., Admon A., Oguri K., Sun X., Wada T., Imai T., Shiroya T., Reinberg D., and Handa H. 1993. Isolation of a cDNA encoding the largest subunit of TFIIA reveals functions important for activated transcription. *Genes Dev.* **7:** 2246.

Makela T.P., Parvin J.D., Kim J., Huber L.J., Sharp P.A., and Weinberg R.A. 1995. A kinase- deficient transcription factor TFIIH is functional in basal and activated transcription. *Proc. Natl. Acad. Sci.* **92:** 5174.

Maldonado E., Drapkin R., and Reinberg D. 1996. Purification of human RNA polymerase II and general transcription factors. *Methods Enzymol.* **274:** 72.

Marinoni J.C., Roy R., Vermeulen W., Miniou P., Lutz Y., Weeda G., Seroz T., Gomez D.M., Hoeijmakers J.H., and Egly J.M. 1997. Cloning and characterization of p52, the fifth subunit of the core of the transcription/DNA repair factor TFIIH. *EMBO J.* **16:** 1093.

Matsui P., DePaulo J., and Buratowski S. 1995. An interaction between the Tfb1 and Ssl1 subunits of yeast TFIIH correlates with DNA repair activity. *Nucleic Acids Res.* **23:** 767.

Maxon M.E., Goodrich J.A., and Tjian R. 1994. Transcription factor IIE binds preferentially to RNA polymerase IIa and recruits TFIIH: A model for promoter clearance. *Genes Dev.* **8:** 515.

McClure W.R. 1985. Mechanism and control of transcription initiation in prokaryotes. *Annu. Rev. Biochem.* **54:** 171.

McCracken S. and Greenblatt J. 1991. Related RNA polymerase-binding regions in human RAP30/74 and *Escherichia coli* σ 70 (erratum appears in *Science* [1992] **255:** 1195). *Science* **253:** 900.

Mellon I., Bohr V.A., Smith C.A., and Hanawalt P.C. 1986. Preferential DNA repair of an active gene in human cells. *Proc. Natl. Acad. Sci.* **83:** 8878.

Morgan D.O. 1995. Principles of CDK regulation. *Nature* **374:** 131.

Mu D., Hsu D.S., and Sancar A. 1996. Reaction mechanism of human DNA repair excision nuclease. *J. Biol. Chem.* **271:** 8285.

Nikolov D.B. and Burley S.K. 1994. 2.1 Å resolution refined structure of a TATA box-binding protein (TBP). *Nat. Struct. Biol.* **1:** 621.

Nikolov D.B., Chen H., Halay E.D., Hoffman A., Roeder R.G., and Burley S.K. 1996. Crystal structure of a human TATA box-binding protein/TATA element complex. *Proc. Natl. Acad. Sci.* **93:** 4862.

Nikolov D.B., Chen H., Halay E.D., Usheva A.A., Hisatake K., Lee D.K., Roeder R.G., and Burley S.K. 1995. Crystal structure of a TFIIB-TBP-TATA-element ternary complex. *Nature* **377:** 119.

Nikolov D.B., Hu S.H., Lin J., Gasch A., Hoffmann A., Horikoshi M., Chua N.H., Roeder R.G., and Burley S.K. 1992. Crystal structure of TFIID TATA-box binding protein. *Nature* **360:** 40.

Ohkuma Y. and Roeder R.G. 1994. Regulation of TFIIH ATPase and kinase activities by TFIIE during active initiation complex formation. *Nature* **368:** 160.

Ohkuma Y., Sumimoto H., Hoffmann A., Shimasaki S., Horikoshi M., and Roeder R.G. 1991. Structural motifs and potential sigma homologies in the large subunit of human general transcription factor TFIIE. *Nature* **354:** 398.

Orphanides G., Lagrange T., and Reinberg D. 1996. The general transcription factors of RNA polymerase II. *Genes Dev.* **10:** 2657.

Orphanides G., LeRoy G., Chang C.H., Luse D.S., and Reinberg D. 1998. FACT, a factor that facilitates transcript elongation through nucleosomes. *Cell* **92:** 105.

Otwinowski Z., Schevitz R.W., Zhang R.G., Lawson C.L., Joachimiak A., Marmorstein R.Q., Luisi B.F., and Sigler P.B. 1988. Crystal structure of trp repressor/operator complex at atomic resolution (erratum appears in *Nature* [1988] **335:** 837). *Nature* **335:** 321.

Owen-Hughes T. and Workman J.L. 1994. Experimental analysis of chromatin function in transcription control. *Crit. Rev. Eukaryot. Gene Expr.* **4:** 403.

Pan G. and Greenblatt J. 1994. Initiation of transcription by RNA polymerase II is limited by melting of the promoter DNA in the region immediately upstream of the initiation site. *J. Biol. Chem.* **269:** 30101.

Parkinson G., Wilson C., Gunasekera A., Ebright Y. W., Ebright R.E., and Berman H.M. 1996. Structure of the CAP-DNA complex at 2.5 Å resolution: A complete picture of the protein-DNA interface. *J. Mol. Biol.* **260:** 395.

Parsons G.G. and Spencer C.A. 1997. Mitotic repression of RNA polymerase II transcription is accompanied by release of transcription elongation complexes. *Mol. Cell. Biol.* **17:** 5791.

Parvin J.D. and Sharp P.A. 1993. DNA topology and a minimal set of basal factors for transcription by RNA polymerase II. *Cell* **73:** 533.

Pazin M.J., Kamakaka R.T., and Kadonaga J.T. 1994. ATP-dependent nucleosome reconfiguration and transcriptional activation from preassembled chromatin templates. *Science* **266:** 2007.

Peterson M.G., Inostroza J., Maxon M.E., Flores O., Admon A., Reinberg D., and Tjian R. 1991. Structure and functional properties of human general transcription factor IIE. *Nature* **354:** 369.

Price D.H., Sluder A.E., and Greenleaf A.L. 1989. Dynamic interaction between a *Drosophila* transcription factor and RNA polymerase II. *Mol. Cell. Biol.* **9:** 1465.

Qureshi S.A. and Jackson,S.P. 1998. Sequence-specific DNA binding by the *S. shibatae* TFIIB homolog, TFB, and its effect on promoter strength. *Mol. Cell* **1:** 389.

Reardon J.T., Ge H., Gibbs E., Sancar A., Hurwitz J., and Pan Z.Q. 1996. Isolation and characterization of two human transcription factor IIH (TFIIH)-related complexes: ERCC2/CAK and TFIIH (erratum appears in *Proc. Natl. Acad. Sci.* [1996] **93:** 10538). *Proc. Natl. Acad. Sci.* **93:** 6482.

Robert F., Forget D., Li J., Greenblatt J., and Coulombe B. 1996. Localization of subunits of transcription factors IIE and IIF immediately upstream of the transcriptional initiation site of the adenovirus major late promoter. *J. Biol. Chem.* **271:** 8517.

Roberts S.G. and Green M.R. 1994. Activator-induced conformational change in general transcription factor TFIIB. *Nature* **371:** 717.

Roy R., Adamczewski J.P., Seroz T., Vermeulen W., Tassan J.P., Schaeffer L., Nigg E.A., Hoeijmakers J.H., and Egly J.M. 1994. The MO15 cell cycle kinase is associated with the TFIIH transcription-DNA repair factor. *Cell* **79:** 1093.

Russo A.A., Jeffrey P.D., and Pavletich N.P. 1996. Structural basis of cyclin-dependent kinase activation by phosphorylation. *Nat. Struct. Biol.* **3:** 696.

Sancar A. 1996. DNA excision repair (erratum appears in *Annu. Rev. Biochem.* [1997] **66:** VII). *Annu. Rev. Biochem.* **65:** 43.

Sancar G.B., Siede W., and van Zeeland A.A. 1996. Repair and processing of DNA damage: A summary of recent progress. *Mutat. Res.* **362:** 127.

Sawadogo M. and Roeder R.G. 1984. Energy requirement for specific transcription initiation by the human RNA polymerase II system. *J. Biol. Chem.* **259:** 5321.

Schaeffer L., Roy R., Humbert S., Moncollin V., Vermeulen W., Hoeijmakers J.H., Chambon P., and Egly J.M. 1993. DNA repair helicase: A component of BTF2 (TFIIH) basic transcription factor. *Science* **260:** 58.

Schaeffer L., Moncollin V., Roy R., Staub A., Mezzina M., Sarasin A., Weeda G., Hoeijmakers J.H., and Egly J.M. 1994. The ERCC2/DNA repair protein is associated with the class II BTF2/TFIIH transcription factor. *EMBO J.* **13:** 2388.

Segil N., Guermah M., Hoffmann A., Roeder R.G., and Heintz N. 1996. Mitotic regulation of TFIID: Inhibition of activator-dependent transcription and changes in subcellular localization. *Genes Dev.* **10:** 2389.

Serizawa H., Conaway R.C., and Conaway J.W. 1992. A carboxyl-terminal-domain kinase associated with RNA polymerase II transcription factor delta from rat liver. *Proc. Natl. Acad. Sci.* **89:** 7476.

Serizawa H., Makela T.P., Conaway J.W., Conaway R.C., Weinberg R.A., and Young R.A. 1995. Association of Cdk-activating kinase subunits with transcription factor TFIIH. *Nature* **374:** 280.

Sherman A.W. and O'Farrell P.H. 1991. Progression of the cell cycle through mitosis leads to abortion of nascent transcripts. *Cell* **67:** 308.

Shiekhattar R., Mermelstein F., Fisher R.P., Drapkin R., Dynlacht B., Wessling H.C., Morgan D.O., and Reinberg D. 1995. Cdk-activating kinase complex is a component of human transcription factor TFIIH. *Nature* **374:** 283.

Smale S.T. 1997. Transcription initiation from TATA-less promoters within eukaryotic protein-coding genes. *Biochim. Biophys. Acta* **1351:** 73.

Sopta M., Burton Z.F., and Greenblatt J. 1989. Structure and associated DNA-helicase activity of a general transcription initiation factor that binds to RNA polymerase II. *Nature* **341:** 410.

Starr D.B. and Hawley D.K. 1991. TFIID binds in the minor groove of the TATA box. *Cell* **67:** 1231.

Stelzer G., Goppelt A., Lottspeich F., and Meisterernst M. 1994. Repression of basal transcription by HMG2 is counteracted by TFIIH- associated factors in an ATP-dependent process. *Mol. Cell. Biol.* **14:** 4712.

Sumimoto H., Ohkuma Y., Sinn E., Kato H., Shimasaki S., Horikoshi M., and Roeder R.G. 1991. Conserved sequence motifs in the small subunit of human general transcription factor TFIIE. *Nature* **354:** 401.

Sun X., Ma D., Sheldon M., Yeung K., and Reinberg D. 1994. Reconstitution of human TFIIA activity from recombinant polypeptides: A role in TFIID-mediated transcription. *Genes Dev.* **8:** 2336.

Svejstrup J.Q., Wang Z., Feaver W.J., Wu X., Bushnell D.A., Donahue T.F., Friedberg E.C., and Kornberg R.D. 1995. Different forms of TFIIH for transcription and DNA repair: Holo-TFIIH and a nucleotide excision repairosome. *Cell* **80:** 21.

Tan S., Garrett K.P., Conaway R.C., and Conaway J.W. 1994. Cryptic DNA-binding domain in the C terminus of RNA polymerase II general transcription factor RAP30. *Proc. Natl. Acad. Sci.* **91:** 9808.

Tan S., Hunziker Y., Sargent D.F., and Richmond T.J. 1996. Crystal structure of a yeast TFIIA/TBP/DNA complex (see comments). *Nature* **381:** 127.

Tanaka K. and Wood R.D. 1994. Xeroderma pigmentosum and nucleotide excision repair of DNA. *Trends Biochem. Sci.* **19:** 83.

Tantin D. and Carey M. 1994. A heteroduplex template circumvents the energetic requirement for ATP during activated transcription by RNA polymerase II. *J. Biol. Chem.* **269:** 17397.

Tantin D., Kansal A., and Carey M. 1997. Recruitment of the putative transcription-repair coupling factor CSB/ERCC6 to RNA polymerase II elongation complexes. *Mol. Cell. Biol.* **17:** 6803.

Tassan J.P., Schultz S.J., Bartek J., and Nigg E.A. 1994. Cell cycle analysis of the activity, subcellular localization, and subunit composition of human CAK (CDK-activating kinase). *J. Cell. Biol.* **127:** 467.

Troelstra C., van Gool A., de Wit J., Vermeulen W., Bootsma D., and Hoeijmakers J.H. 1992. ERCC6, a member of a subfamily of putative helicases, is involved in Cockayne's syndrome and preferential repair of active genes. *Cell* **71:** 939.

Tsukiyama T. and Wu C. 1995. Purification and properties of an ATP-dependent nucleosome remodeling factor. *Cell* **83:** 1011.

Usheva A., Maldonado E., Goldring A., Lu H., Houbavi C., Reinberg D., and Aloni Y. 1992. Specific interaction between the nonphosphorylated form of RNA polymerase II and the TATA-binding protein. *Cell* **69:** 871.

Valay J.G., Simon M., Dubois M.F., Bensaude O., Facca C., and Faye G. 1995. The KIN28 gene is required both for RNA polymerase II mediated transcription and phosphorylation of the Rpb1p CTD. *J. Mol. Biol.* **249:** 535.

Van Dyke M.W., Roeder R.G., and Sawadogo M. 1988. Physical analysis of transcription preinitiation complex assembly on a class II gene promoter. *Science* **241:** 1335.

Varga-Weisz P.D., Wilm M., Bonte E., Dumas K., Mann M., and Becke, P.B. 1997. Chromatin- remodelling factor CHRAC contains the ATPases ISWI and topoisomerase II (erratum appears in *Nature* [1997] **389:** 1003). *Nature* **388:** 598.

Wang Z., Buratowski S., Svejstrup J.Q., Feaver W.J., Wu X., Kornberg R.D., Donahue T.F., and Friedberg E.C. 1995. The yeast TFB1 and SSL1 genes, which encode subunits of transcription factor IIH, are required for nucleotide excision repair and RNA polymerase II transcription. *Mol. Cell. Biol.* **15:** 2288.

Weis L. and Reinberg D. 1992. Transcription by RNA polymerase II: Initiator-directed formation of transcription-competent complexes. *FASEB J.* **6:** 3300.

White J., Brou C., Wu J., Lutz Y., Moncollin V., and Chambon P. 1992. The acidic transcriptional activator GAL-VP16 acts on preformed template-committed complexes. *EMBO J.* **11:** 2229.

Xiao H., Pearson A., Coulombe B., Truant R., Zhang S., Regier J.L., Triezenberg S.J., Reinberg D., Flores O., Ingles C.J., and Greenblatt J. 1994. Binding of basal transcription factor TFIIH to the acidic activation domains of VP16 and p53. *Mol. Cell. Biol.* **14:** 7013.

Xu L.C., Thali M., and Schaffner W. 1991. Upstream box/TATA box order is the major determinant of the direction of transcription. *Nucleic Acids Res.* **19:** 6699.

Yankulov K., Blau J., Purton T., Roberts S., and Bentley D.L. 1994. Transcriptional elongation by RNA polymerase II is stimulated by transactivators. *Cell* **77:** 749.

Yokomori K., Admon A., Goodrich J.A., Chen J.L., and Tjian R. 1993. *Drosophila* TFIIA-L is processed into two subunits that are associated with the TBP/TAF complex. *Genes Dev.* **7:** 2235.

Zawel L. and Reinberg D. 1993. Initiation of transcription by RNA polymerase II: A multi-step process. *Prog. Nucleic Acid Res. Mol. Biol.* **44:** 67.

———. 1995. Common themes in assembly and function of eukaryotic transcription complexes. *Annu. Rev. Biochem.* **64:** 533.

Zawel L., Kumar K.P., and Reinberg D. 1995. Recycling of the general transcription factors during RNA polymerase II tran-

scription. *Genes Dev.* **9:** 1479.

Zawel L., Lu H., Cisek L.J., Corden J.L., and Reinberg D. 1993. The cycling of RNA polymerase II during transcription. *Cold Spring Harbor Symp. Quant. Biol.* **58:** 187.

Zheng X.M., Black D., Chambon P., and Egly J.M. 1990. Sequencing and expression of complementary DNA for the general transcription factor BTF3. *Nature* **344:** 556.

Zhou Q., Lieberman P.M., Boyer T.G., and Berk A.J. 1992. Holo-TFIID supports transcriptional stimulation by diverse activators and from a TATA-less promoter. *Genes Dev.* **6:** 1964.

Zhu W., Zeng Q., Colangelo C.M., Lewis M., Summers M.F., and Scott R.A. 1996. The N-terminal domain of TFIIB from *Pyrococcus furiosus* forms a zinc ribbon (letter). *Nat. Struct. Biol.* **3:** 122.

Ten Years of TFIIH

F. Coin and J.-M. Egly

Institut de Génétique et de Biologie Moléculaire et Cellulaire, CNRS/INSERM/ULP,
B.P.163, 67404 Illkirch Cedex, C.U. de Strasbourg, France

A complete molecular description of the transcription machinery is now at hand. The essential transcription factors that allow RNA polymerase II (pol II) to transcribe accurately from different promoters have been identified, and when these are added together in an in vitro system, transcription occurs. Crucial questions remain as to how regulatory signals are conveyed from adjacent specific DNA sequences and/or external stimuli to the core promoter and the basal transcription machinery. Three basal transcription factors have been described as potential candidates for transcription regulation: TFIID, which through its connection with cell-type- and gene-specific activators and mediators, selects and regulates the TATA-binding protein (TBP)-associated factors (TAFs) (Hampsey and Reinberg 1997); TFIIB, which stabilizes TBP and is targeted by activators (Roberts et al. 1995), and TFIIH, which possesses several enzymatic activities, which must be regulated. Before investigating how TFIIH could function as a key regulatory point of transcription, it was of interest to analyze TFIIH in the context of the basal transcription apparatus knowing that, besides pol II, it is the only factor with a number of enzymatic activities.

TFIIH is a multisubunit protein complex that was originally identified as a basal transcription factor involved in protein-coding gene transcription (Conaway and Conaway 1989; Gerard et al. 1991). However, the microsequencing of the nine subunits revealed that it participates in two other fundamental cell processes: nucleotide excision repair (NER) and cell cycle regulation (Hoeijmakers et al. 1996; Svejstrup et al. 1996). In addition, TFIIH is connected to human genetic diseases. The discovery that the two largest subunits, XPB and XPD, were two helicases of opposite polarity, first known as components of NER, highlighted the importance of TFIIH in ensuring the integrity of the genome (Feaver et al. 1993; Schaeffer et al. 1993, 1994; Drapkin et al. 1994; Guzder et al. 1994a,b). Defects in this latter function, due to mutations in both helicases, give rise to severe genetic diseases such as xeroderma pigmentosum, trichothiodystrophy and Cockayne syndrome. The molecular bases of these defects are complex and not yet fully understood, but at least some of them are related to defects in the NER pathway (van Vuuren et al. 1994). Two other subunits, a cyclin-dependent kinase cdk7 and cyclin H, form with a third partner MAT1, a cdk7-activating kinase complex (Feaver et al. 1994; Roy et al. 1994b; Serizawa et al. 1995; Shiekhattar et al. 1995; Adamczewski et al. 1996),

which is also involved in the regulation of the cell cycle (Nigg 1996). This complex is capable of phosphorylating cyclin-dependent kinases, transcription factors (Ohkuma and Roeder 1994), the carboxy-terminal domain (CTD) of the largest subunit of pol II (Lu et al. 1992), as well as activators. The remaining polypeptides, p62 (Fischer et al. 1992), p52 (Marinoni et al. 1997), and p34 (Humbert et al. 1994), have no identified motifs or enzymatic functions (Table 1).

This paper updates our knowledge of TFIIH function in transcription. We particularly focus, on the one hand, on the function of both XPB and XPD helicases in the opening of the DNA around the transcription start site and, on the other hand, on the role of the kinase in transcription.

TFIIH IS INVOLVED IN OPENING THE PROMOTER

RNA synthesis requires ATP at a number of transcriptional steps (Bunick et al. 1982; Sawadogo and Roeder 1984), including the formation of an open initiation transcription complex, the synthesis of an RNA transcript (Jiang and Gralla 1995; Jiang et al. 1996), and the phosphorylation of the largest pol II subunit. Energy is indispensable for the first step of transcription initiation. In-

Table 1. TFIIH Composition and Subunit Function

Human	m.w. (kD)	Function	*Saccharomyces cerevisiae*
XPB/ERCC3	89	3´-5´ helicase ATPase; essential for promoter opening	Rad25/Ssl2
XPD/ERCC2	80	5´-3´helicase APTase; CAK anchoring subunit	Rad3
p62	62	n.d.	Tfb1
p52	52	WD domains; interaction with other proteins?	Tfb2
p44	44	DNA binding and stimulation of XPD helicase activity	Ssl1
p34	34	DNA binding?	Tfb4
Cdk7	40	CTD kinase	Kin28
Cyclin H	34	regulation of Cdk7	Ccl1
MAT1	32	ring finger; kinase stimulation	

Cdk7, Cyclin H, MAT1 bracketed as CAK. Kin28, Ccl1 bracketed as TFIIK.

(XPB, XPD) Xeroderma pigmentosum group B and D; (CTD kinase) carboxy-terminal domain kinase; (CAK) CDK-activating kinase

deed, AMP-PNP, an analog of ATP, cannot be used to initiate RNA synthesis, unless there is addition of a dinucleotide representing the first two nucleotides to be transcribed, a point established 15 years ago (Sawadogo and Roeder 1984). A few years later, the Conaways demonstrated that the rat transcription factor δ, which is similar to human TFIIH, possesses an ATPase activity, strongly stimulated in the presence of a DNA fragment that contains the adenovirus major late promoter (AdMLP) TATA box (Conaway and Conaway 1989).

With this in mind, we went on to demonstrate that this DNA-dependent ATPase activity belongs to XPB and XPD, two subunits of TFIIH that possess ATP-dependent helicase activity (Schaeffer et al. 1993, 1994; Roy et al. 1994a). This latter finding made TFIIH the ideal candidate for effecting the promoter melting step in transcription. Footprinting experiments showed that the opening of the DNA around the transcription start site requires ATP as an energy source (Wang et al. 1992). The need for TFIIH for such an opening to allow transcription initiation of AdMLP is dependent on the topology of the promoter: A negatively supercoiled but not linearized DNA template circumvents the requirement for TFIIH (Parvin and Sharp 1993). Transcription from supercoiled templates does not require energy as AMP-PNP can be used instead of ATP (Sawadogo and Roeder 1984). Interestingly, since it was demonstrated that AMP-PNP cannot be used as a cofactor for TFIIH helicase activity, it can be then concluded that TFIIH is the transcription factor using the energy needed to initiate the transcription reaction.

The above data indicated a connection between promoter opening and TFIIH requirement. In agreement with this observation, using a $KMnO_4$-sensitive assay, it was shown that the opening of a 10–20-bp region around the promoter requires the presence of TFIIH in an ATP-dependent manner (Holstege et al. 1996). In consequence, when the AdMLP template is artificially premelted (opened) by introducing a region of unpaired base pairs, TFIIH and ATP are dispensable for RNA synthesis.

Yeast genetics was also used to demonstrate the role of the helicases in this crucial step of transcription initiation. Indeed, *rad25 Arg-392*, which encodes a protein mutated in the ATP-binding site, is defective in pol II transcription, suggesting that Rad25, the yeast counterpart of XPB, functions in DNA duplex opening during transcription (Guzder et al. 1994a) and that a defect of this function affects the viability of the cell. In contrast, yeast strains carrying the *rad3 Arg-48* allele are viable. This shows that XPD, the human counterpart of *rad3*, is likely dispensable for transcription (Sung et al. 1988). This signifies that the XPD helicase activity of TFIIH is not crucial for the synthesis of run-off transcripts but does not exclude a function of XPD in the conformation of the TFIIH complex as well as in its contacts with other transcription factors. In agreement with these observations, being able to reconstitute recombinant TFIIH, we demonstrated that mutation in the ATP-binding site of the XPB subunit abrogated transcription, whereas the same mutation in XPD slightly inhibited transcription (Tirode et al. 1999).

TFIIH HELICASES, TRANSCRIPTION, AND GENETIC DISEASES

Mutations in both XPB and XPD helicases of TFIIH have been found to be responsible for several genetic diseases: xeroderma pigmentosum (XP), Cockayne's syndrome (CS or XP-CS), and trichothiodystrophy (TTD) (van Vuuren et al. 1994). Patients are clinically characterized by the early onset of a severe photosensitivity of the exposed regions of the skin, which can be explained by a deficiency in the NER mechanism. However, some phenotypes, such as neurological and developmental abnormalities as well as keratinoses, cannot be explained simply by deficiencies in NER but are suggested to also involve disruptions in TFIIH-mediated transcription (Bootsma and Hoeijmakers 1993; Vermeulen et al. 1994a). In addition, a striking clinical heterogeneity of these clinical features is observed among XP individuals. This varies from patients with severe neurological problems and a high number of skin tumors to individuals with mild pigmentary abnormalities without signs of neurological defects (Robbins et al. 1991). The most impressive example of the heterogeneity of the XP features is given by the analysis of XPB patients. Only three families of XPB patients have been described so far, two of them displaying a combination of XP and CS syndromes (called XP/CS) and a total absence of DNA repair in vivo and in vitro. The first XP/CS patient (XP11BE), who for some time was the only case described (Weeda et al. 1990), presented numerous skin tumors at an early age as well as developmental and neurological abnormalities. However, two brothers of the second XP/CS family, who are also almost totally defective in NER, exhibit less severe neurological abnormalities and a total absence of skin tumors (Vermeulen et al. 1994b).

The key to understanding the heterogeneity of XPB features may come from the study of the effect of the XPB mutation on the transcription capacities of TFIIH for a particular set of genes. A study from our laboratory (Hwang et al. 1996) draws a parallel between the mutation in XPB found in the XPB11BE patient, the reduction of the helicase activity of XPB, and the decrease in TFIIH transcription, thus at least partially explaining some of the clinical symptoms of XP patients.

The presence of an active XPD helicase has been shown to be dispensable for the survival of the cell but absolutely required for the function of XPD in NER (Sung et al. 1988; Weber et al. 1990). These results do not eliminate further roles for XPD in DNA-protein or protein-protein interactions in the transcription reaction. Highlighting this, a temperature-sensitive mutation of Rad3 (the yeast XPD) has been shown to lead to transcription defects. Extracts from mutated *rad3-ts14* strains support pol II transcription at a reduced rate compared to the wild-type Rad3 extract (Guzder et al. 1994b). This can be correlated with recent findings from our laboratory showing that reconstituted recombinant TFIIH, either lacking XPD or containing mutated XPD in its ATPase-binding site, is much less active than the corresponding wild-type recombinant (Tirode et al. 1999).

XPD helicase activity is regulated through p44 binding (another subunit of TFIIH): Recombinant XPD, carrying mutations in the carboxyl terminus, as found in the majority of XPD patients, failed to interact accurately with p44 (Coin et al. 1998). This explains the inability of these mutated TFIIHs to restore NER in vitro. This study provides a biochemical explanation of the NER phenotype found in XPD patients. Since XPD is also involved in transcription, it would be worthwhile to analyze how a decrease in XPD unwinding activity could affect transcription.

XPB can now be defined as the main helicase involved in promoter opening for in vitro transcription reactions. This does not exclude a role for XPD in the various phases of transcription, including a participation in promoter opening itself, activation through targeting by other factors, and transcription-coupled repair.

THE CAK COMPLEX AND THE PHOSPHORYLATION OF THE RNA POL II CTD

In transcription, phosphorylation of the CTD of the largest subunit of pol II, a highly conserved domain consisting of a heptapeptide sequence (YSPTSPS) tandemly repeated up to 52 times, is believed to be essential in the transition from initiation to elongation. Indeed, CTD phosphorylation is thought to induce the release of the initiation complex from TFIID which remains bound to the promoter. In support of this hypothesis, although only the nonphosphorylated form of pol II (pol IIA) can enter a preinitiation complex, the phosphorylated form (pol II0) is found upon RNA synthesis (Usheva et al. 1992; Maxon et al. 1994). Due to its composition, CTD is the substrate of several protein kinases (Poon and Hunter 1995). Among these, TFIIH is a serious candidate for an in vivo role.

Indeed, we and other investigators have demonstrated that cdk7, a member of the cyclin-dependent kinases (cdk) family, known to have a role in the cell cycle, is a component of TFIIH (Lu et al. 1992; Feaver et al. 1994; Roy et al. 1994b; Serizawa et al. 1995). This group of proteins includes different regulatory components known to coordinate different events of the cell cycle such as cell cycle progression, DNA replication, and transcription. There are three forms of CDK-activating kinase (CAK): free, associated with XPD helicase, and as part of TFIIH (Drapkin et al. 1996; Reardon et al. 1996; Rossignol et al. 1997; Yankulov and Bentley 1997). As a component of TFIIH, the kinase activity of TFIIH is mainly directed toward CTD, but it may also use other substrates, such as the basal transcription factors TBP, TFIIE, and TFIIF. Equally, the transcriptional activators p53 and the retinoic acid receptor α are also known to be phosphorylated by the TFIIH kinase (Ko et al. 1997; Rochette-Egly et al. 1997). CAK has also been found associated with viral proteins such as Tat (Cujec et al. 1997; Garcia-Martinez et al. 1997). As a free complex, CAK has been demonstrated to preferentially phosphorylate components of the cell cycle such as cdk2, cdk1, and cdk4 (for

a review of cyclin-dependent kinases, see Nigg 1995). Although yeast TFIIH has been shown to have the same composition as the mammalian TFIIH (Table 1), there are still some discrepancies concerning the kinase complex. The yeast Kin28/Ccl1 TFIIK complex is not involved in the phosphorylation of the cell cycle kinases and therefore lacks both CAK activity and MAT1 protein (Cismowski et al. 1995; Valay et al. 1995).

The involvement of CAK in the phosphorylation of CTD is supported by several experiments. First, the phosphorylation of pol IIA by TFIIH is strongly stimulated in the presence of the basal transcription machinery and promoter elements. Second, pol II transcription is drastically reduced in a yeast *kin28* temperature-sensitive mutant at the restrictive temperature, and this correlates with a decreased phosphorylation of CTD (Cismowski et al. 1995; Valay et al. 1995). Third, in vivo (Mäkelä et al. 1995) and in vitro (Tirode et al. 1999) studies demonstrated that TFIIH containing a mutated form of *cdk7* cannot phosphorylate pol II even in the presence of basal factors and template. Nevertheless, the *cdk7* mutant was as active as the wild-type protein in supporting TFIIH-dependent basal in vitro transcription in the context of both AdMLP and the yeast Cyc promoter, whereas it cannot support transcription of the dihydrofolate reductase gene. This has to be related to the observation that transcription from promoters such as AdMLP can take place in vitro using a CTD-less form of pol II (Akoulitchev et al. 1995). Together, these data demonstrate that the CTD-kinase activity is not essential for formation of the first phosphodiester bond in RNA synthesis from either promoter (Dahmus 1995; see also Tirode et al. 1999). It is, however, required to stimulate transcription from a special set of promoters, at least in a well-defined in vitro transcription system.

Although not essential for RNA synthesis, TFIIH kinase nevertheless clearly stimulates transcription. Whether this stimulation is due to a combined action with RNA processing remains to be elucidated, given that phosphorylation induces the recruitment of both capping and splicing enzymes on the transcription elongation complex (Cho et al. 1997; Kim et al. 1997; McCracken et al. 1997). This kind of transcriptional activation mediated by cdk7/Kin28 is not unique as it has been shown in yeast that transcription of certain genes can be highly induced even when cells lack Kin28 (Lee and Lis 1998).

Bearing in mind the fact that XPD can be found associated with the CAK complex (Drapkin et al. 1996; Reardon et al. 1996), and also can be dissociated from the "core TFIIH" (Rossignol et al. 1997), we wonder whether CAK activity can be mediated by XPD, which would also function as a bridging factor. Immunopurified TFIIH from an XPD cell line (XP102LO), with a carboxy-terminal mutation in XPD, was shown to contain considerably lower levels of XPD/CAK polypeptides than the corresponding wild type, thus explaining the slight reduction of its transcription activity (F. Coin et al., in prep.). Such a decrease in the overall transcription activity of TFIIH could be due either to the defect of XPD in promoter opening and/or to the defect in CAK that fails to phosphorylate the largest pol II subunit (Drapkin et al. 1996). Accordingly, mutations found in XPD patients, which prevent the integration of XPD in TFIIH, would also modify the incorporation of the CAK complex and result in a decrease in transcription (Tirode et al. 1999).

TFIIH INTERACTING PROTEINS IN TRANSCRIPTION AND NER

The question then arises as to how, and by which factors, TFIIH is regulated in the cell. It is possible that when TFIIH enters either the transcription preinitiation complex or the pol II holoenzyme (Koleske and Young 1994; Ossipow et al. 1995; Maldonado et al. 1996), and/or the DNA damage incision/excision complex, it activates regulatory mechanisms. TFIIH interacts with several basal transcription factors, such as TBP and pol II (Gerard et al. 1991; Drapkin et al. 1994). Equally, TFIIE interacts with TFIIH through the XPB and p62 subunits (Ohkuma and Roeder 1994) and modulates TFIIH kinase, ATPase, and helicase activities.

Several NER factors have been found in the so-called TFIIH⁻ pol II holoenzyme. Their activity has been tested in an NER assay (F. Coin and J.-M. Egly, unpubl.). A homolog of the transcription holoenzymes denoted the repairosome complex, which contains in addition to TFIIH several other NER factors such as Rad1-2-4-10 and 14, has been identified (Svejstrup et al. 1995). This factor was found to lack TFIIK, the yeast homolog of the mammalian CAK, thus demonstrating that the TFIIH kinase is not essential in DNA repair. Using coimmunoprecipitation of in-vitro-translated peptides, Friedberg's group found an interaction between different subunits of TFIIH and the Cockayne syndrome B protein (CSB) as well as the 3′ endonuclease XPG (Iyer et al. 1996). The interaction between TFIIH and CSB may mediate the preferential repair of the transcribed strand of active genes as these regions are known to be repaired faster than the rest of the genome (Mellon et al. 1987). In yeast, Prakash's group has confirmed that TFIIH copurified with Rad2, the yeast counterpart of XPG (Habraken et al. 1996). Together, these results suggest a putative role of XPG and CSB in the transcription process (Tantin et al. 1997; Selby and Sancar 1997).

CONCLUSION AND PERSPECTIVES

Ten years after the discovery of TFIIH, a factor at the crossroads of three different processes—transcription, DNA repair, and cell cycle regulation—the characterization of its various subunits is now complete. To go further in this dissection of TFIIH functions, we need to work with a reconstituted factor. This "artificial" TFIIH will allow us to understand the precise function of each subunit and to identify the consequences of mutations in any subunit on the various processes in which TFIIH is involved. The reconstitution of TFIIH is almost achieved, and preliminary results have highlighted the subunits that are sufficient for basal transcription and repair activity in an in-vitro-reconstituted system (Tirode et al. 1999).

Structural studies, as well as the analysis of TFIIH from XPB and XPD patient cell lines, will be decisive in furthering our understanding of the function of TFIIH. Of course, we also hope for progress in the identification of the relationship of genotype to clinical phenotype.

ACKNOWLEDGMENTS

We thank our group as well as G. Richards for discussion and a critical reading of this manuscript. We are very grateful to the different students and postdocs of the lab (they will recognize themselves), who, in the past, have made important contributions to the study of TFIIH. Our work on TFIIH has been supported for 10 years by grants from the Institut National de la Santé et de la Recherche Médicale (INSERM), Centre National de la Recherche Scientifique (CNRS), Association pour la Recherche contre le Cancer (ARC), and Ligue National contre le Cancer and Université Louis Pasteur.

REFERENCES

Adamczewski J.P., Rossignol M., Tassan J.P., Nigg E.A., Moncollin V., and Egly J.-M. 1996. MAT1, cdk7 and cyclin H form a kinase complex which is UV light-sensitive upon association with TFIIH. *EMBO J.* **15:** 1877.

Akoulitchev S., Mäkelä T.P., Weinberg R.A., and Reinberg D. 1995. Requirement for TFIIH kinase activity in transcription by RNA polymerase II. *Nature* **377:** 557.

Bootsma D. and Hoeijmakers J.H.J. 1993. DNA repair. Engagement with transcription. *Nature* **363:** 114.

Bunick D., Zandomeni R., Ackerman S., and Weinmann R. 1982. Mechanism of RNA polymerase II-specific intitiation by RNA polymerase II *in vitro:* ATP requirement and uncapped runoff transcripts. *Cell* **29:** 877.

Cho E., Tagaki T., Moore C.R., and Buratowski S. 1997. mRNA capping enzyme is recruted to the transcription complex by phosphorylation of the RNA polymerase II carboxy-terminal domain. *Genes Dev.* **11:** 3319.

Cismowski M.J., Laff G.M., Solomon M.J., and Reed S.I. 1995. Kin28 encodes a C-terminal domain kinase that controls mRNA transcription in *Saccharomyces cerevisiae* but lacks cyclin-dependent kinase-activating kinase (CAK) activity. *Mol. Cell. Biol.* **15:** 2983.

Coin F., Marinoni J.C., Rodolfo C., Fribourg S., Pedrini M.A., and Egly J.M. 1998. Mutations in XPD helicase gene result in XP and TTD phenotype, preventing interaction between XPD and the p44 subunit of TFIIH. *Nat. Gen.* **20:** 184.

Conaway R.C. and Conaway J.W. 1989. An RNA polymerase II transcription factor has an associated DNA-dependent ATPase (dATPase) activity strongly stimulated by TATA region of promoters. *Proc. Natl. Acad. Sci.* **86:** 7356.

Cujec T., Okamoto H., Fujinaga K., Meyer J., Chamberlin H., Morgan D.O., and Peterlin B.M. 1997. The HIV transactivator TAT binds to the CDK-activating kinase and activates the phosphorylation of the carboxy-terminal domain of RNA polymerase II. *Genes Dev.* **11:** 2645.

Dahmus M.E. 1995. Phosphorylation of the C-terminal domain of RNA polymerase II. *Biochim. Biophys. Acta* **1261:** 170.

Drapkin R., Le Roy G., Cho H., Akoulitchev S., and Reinberg D. 1996. Human cyclin-dependent kinase-activating kinase exists in three distinct complexes. *Proc. Natl. Acad. Sci.* **93:** 6488.

Drapkin R., Reardon J.T., Ansari A., Huang J.C., Zawel L., Ahn K.J., Sancar A., and Reinberg D. 1994. Dual role of TFIIH in DNA excision repair and in transcription by RNA polymerase II. *Nature* **368:** 769.

Feaver W.J., Svejstrup J.Q., Henry N.L., and Kornberg R.D.

1994. Relationship of CDK-activating kinase and RNA polymerase II CTD kinase TFIIH/TFIIK. *Cell* **79:** 1103.

Feaver W.J., Svejstrup J.Q., Bardwell L., Bardwell A.J., Buratowski S., Gulyas K.D., Donahue T.F., Friedberg E.C., and Kornberg R.D. 1993. Dual roles of a multiprotein complex from *S. cerevisiae* in transcription and DNA repair. *Cell* **75:** 1379.

Fischer L., Gèrard M., Chalut C., Lutz Y., Humbert S., Kanno M., Chambon P., and Egly J.M. 1992. Cloning of the 62-kilodalton component of basic transcription factor BTF2. *Science* **257:** 1392.

Garcia-Martinez L.F., Mavankal G., Neveu J.M., Lane W.S., Ivanov D., and Gaynor R.B. 1997. Purification of a Tat-associated kinase reveals a TFIIH complex that modulates HIV-1 transcription. *EMBO J.* **16:** 2836.

Gerard M., Fischer L., Moncollin V., Chipoulet J.M., Chambon P., and Egly J.M. 1991. Purification and interaction properties of the human RNA polymerase B(II) general transcription factor BTF2. *J. Biol. Chem.* **266:** 20940.

Guzder S.N., Sung P., Bailly V., Prakash L., and Prakash S. 1994a. RAD25 is a DNA helicase required for DNA repair and RNA polymerase II transcription. *Nature* **369:** 578.

Guzder S.N., Qiu H., Sommers C.H., Sung P., Prakash L., and Prakash S. 1994b. DNA repair gene RAD3 of *S. cerevisiae* is essential for transcription by RNA polymerase II. *Nature* **367:** 91.

Habraken Y., Sung P., Prakash S., and Prakash L. 1996. Transcription factor TFIIH and DNA endonuclease Rad2 constitute yeast nucleotide excision repair factor 3: Implications for nucleotide excision repair and Cockayne syndrome. *Proc. Natl. Acad. Sci.* **93:** 10718.

Hampsey M. and Reinberg D. 1997. Transcription: Why are TAFs essential? *Curr. Biol.* **7:** 44.

Hoeijmakers J.H.J., Egly J.M., and Vermeulen W. 1996. TFIIH: A key component in multiple DNA transactions. *Curr. Opin. Genet. Dev.* **6:** 26.

Holstege F.C.P., van der Vliet P.C., and Timmers M.H.T. 1996. Opening of an RNA polymerase II promoter occurs in two distinct steps and requires the basal transcription factors IIE and IIH. *EMBO J.* **15:** 1666.

Humbert S., van Vuuren H., Lutz Y., Hoeijmakers J.H.J., Egly J.M., and Moncollin V. 1994. p44 and p34 subunits of the BTF2/TFIIH transcription factor have homologies with SSL, a yeast protein involved in DNA repair. *EMBO J.* **13:** 2393.

Hwang J.R., Moncollin V., Vermeulen W., Seroz T., van Vuuren H., Hoeijmakers J.H.J., and Egly J.M. 1996. A 3′-5′ XPB helicase defect in repair/transcription factor TFIIH of xeroderma pigmentosum group B affects both DNA repair and transcription. *J. Biol. Chem.* **271:** 15898.

Iyer N., Reagan M.S., Wu K.J., Canagarajah B., and Friedberg E.C. 1996. Interactions involving the human RNA polymerase II transcription factor/nucleotide excision repair complex TFIIH, the nucleotide excision repair protein XPG, and Cockayne syndrome group B (CSB) protein. *Biochemistry* **35:** 2157.

Jiang Y. and Gralla J.D. 1995. Nucleotide requirements for activated RNA polymerase II open complex formation in vitro. *J. Biol. Chem.* **270:** 1277.

Jiang Y., Yan M., and Gralla J.D. 1996. A three-step pathway of transcription initiation leading to promoter clearance at an activated RNA polymerase II promoter. *Mol. Cell. Biol.* **16:** 1614.

Kim E., Du L., Bregman D.B., and Warren S.L. 1997. Splicing factors associate with hyperphosphorylated RNA polymerase II in the absence of pre-mRNA. *J. Cell Biol.* **136:** 19.

Ko L.J., Shieh S.Y., Chen X., Jayaraman L., Tamai K., Taya Y., Prives C., and Pan Z.Q. 1997. p53 is phosphorylated by cdk7-cyclin H in a p36MAT1-dependent manner. *Mol. Cell. Biol.* **17:** 7220.

Koleske A.J. and Young R.A. 1994. An RNA polymerase II holoenzyme responsive to activators. *Nature* **368:** 466.

Lee D.K. and Lis J.T. 1998. Transcription activation independent of TFIIH kinase and the RNA polymerase II mediator in

vivo. *Nature* **393:** 389.

Lu H., Zawel L., Fisher L., Egly J.M., and Reinberg D. 1992. Human general transcription factor IIH phosphorylates the C-terminal domain of RNA polymerase II. *Nature* **358:** 641.

Mäkelä T.P., Parvin J.D., Kim J., Huber L.J., Sharp P.A., and Weinberg R.A. 1995. A kinase-deficient transcription factor TFIIH is functional in basal and activated transcription. *Proc. Natl. Acad. Sci.* **92:** 5174.

Maldonado E., Shiekhattar R., Sheldon M., Cho H., Drapkin R., Rickert P., Lees E., Anderson C.W., Linn S., and Reinberg D. 1996. A human RNA polymerase II complex associated with SRB and DNA-repair proteins. *Nature* **381:** 86.

Marinoni J.C., Roy R., Vermeulen W., Miniou P., Lutz Y., Weeda G., Seroz T., Gomez D.M., Hoeijmakers J.H.J., and Egly J.M. 1997. Cloning and characterization of p52, the fifth subunit of the core of the transcription/DNA repair factor TFIIH. *EMBO J.* **16:** 1093.

Maxon M.E., Goodrich J.A., and Tjian R. 1994. Transcription factor IIE binds preferentially to RNA polymerase IIa and recruits TFIIH: A model for promoter clearance. *Genes Dev.* **8:** 515.

McCracken S., Fong N., Rosonina E., Yankulov K., Brothers G., Siderovski D., Hessel A., Foster S., Shuman S., and Bentley D.L. 1997. 5′-capping enzymes are targeted to pre-mRNA by binding to the phosphorylated carboxy-terminal domain of RNA polymerase II. *Genes Dev.* **11:** 3306.

Mellon I., Spivak G., and Hanawalt P.C. 1987. Selective removal of transcription-blocking DNA damage from the transcribed strand of the mammalian DHFR gene. *Cell* **51:** 241.

Nigg E.A. 1995. Cyclin-dependent protein kinases: Key regulators of the eukaryotic cell cycle. *BioEssays* **17:** 471.

——— 1996. Cyclin-dependant kinase 7: At the cross-roads of transcription, DNA repair and cell cycle control? *Curr. Opin. Cell Biol.* **8:** 312.

Ohkuma Y. and Roeder R.G. 1994. Regulation of TFIIH ATPase and kinase activities by TFIIE during active initiation complex formation. *Nature* **368:** 160.

Ossipow V., Tassan J.P., Nigg E.A., and Schibler U. 1995. A mammalian RNA polymerase II holoenzyme containing all the components required for promoter-specific transcription initiation. *Cell* **83:** 137.

Parvin J.D. and Sharp P.A. 1993. DNA topology and a minimal set of basal factors for transcription by RNA polymerase II. *Cell* **73:** 533.

Poon R.Y.C. and Hunter T. 1995. Innocent bystanders or chosen collaborators? *Curr. Biol.* **5:** 1243.

Reardon J.T., Ge H., Gibbs E., Sancar A., Hurwitz J., and Pan Z.-Q. 1996. Isolation and characterization of two human transcription factor IIH (TFIIH)-related complexes: ERCC2/CAK and TFIIH. *Proc. Natl. Acad. Sci.* **93:** 6482.

Robbins J.H., Brumback R.A., Mendiones M., Barrett S.F., Carl J.R., Cho S., and Denckla M.B., Ganges M.B., Gerber L.H., and Guthrie R.A. 1991. Neurological disease in xeroderma pigmentosum: Documentation of a late onset type of the juvenile onset form. *Brain* **114:** 1335.

Roberts S.G., Choy B., Walker S., Lin Y.S., and Green M.R. 1995. A role for activator-mediated TFIIB recruitment in diverse aspects of transcriptional regulation. *Curr. Biol.* **5:** 508.

Rochette-Egly C., Adam S., Rossignol M., Egly J.-M., and Chambon P. 1997. Stimulation of RARa activation function AF-1 through binding to the general transcription factor TFIIH and phosphorylation by CDK7. *Cell* **90:** 1.

Rossignol M., Kolb-Cheynel I., and Egly J.M. 1997. Substrate specificity of the cdk-activating kinase (CAK) is altered upon assocation with TFIIH. *EMBO J.* **16:** 1628.

Roy R., Schaeffer L., Humbert S., Vermeulen W., Weeda G., and Egly J.M. 1994a. The DNA-dependent ATPase activity associated with the class II transcription factor BTF2/TFIIH. *J. Biol. Chem.* **269:** 9826.

Roy R., Adamczewski J.P., Seroz T., Vermeulen W., Tassan J.P., Schaeffer L., Hoeijmakers J.H.J., and Egly J.M. 1994b. The MO15 cell cycle kinase is associated with the TFIIH transcription-DNA repair factor. *Cell* **79:** 1093.

Sawadogo M. and Roeder R.G. 1984. Energy requirement for specific transcription initiation by the human RNA polymerase II system. *J. Biol. Chem.* **259:** 5321.

Schaeffer L., Moncollin V., Roy R., Staub A., Mezzina M., Sarasin A., Weeda G., Hoeijmakers J.H.J., and Egly J.M. 1994. The ERCC2/DNA repair protein is associated with the class II BTF2/TFIIH transcription factor. *EMBO J.* **13:** 2388.

Schaeffer L., Roy R., Humbert S., Moncollin V., Vermeulen W., Hoeijmakers J.H.J., Chambon P., and Egly J.M. 1993. DNA repair helicase: A component of BTF2 (TFIIH) basic transcription factor. *Science* **260:** 58.

Selby C.P. and Sancar A.. 1997. Cockayne syndrome group B protein enhances elongation by RNA polymerase II. *Proc. Natl. Acad. Sci.* **94:** 11205.

Serizawa H., Mäkelä T.P., Conaway J.W., Conaway R.C., Weinberg R.A., and Young R.A.. 1995. Association of Cdk-activating kinase subunits with transcription factor TFIIH. *Nature* **374:** 280.

Shiekhattar R., Mermelstein F., Fisher R.P., Drapkin R., Dynlacht B., Wessling H.C., Sancar A., Morgan D.O., and Reinberg D.. 1995. Cdk-activating kinase complex is a component of human transcription factor TFIIH. *Nature* **374:** 283.

Sung P., Higgins D., Prakash L., and Prakash S. 1988. Mutation of lysine-48 to arginine in the yeast RAD3 protein abolishes its ATPase and DNA helicase activities but not the ability to bind ATP. *EMBO J.* **7:** 3263.

Svejstrup J.Q., Vichi P., and Egly J.M. 1996. The multiple roles of transcription/repair factor TFIIH. *Trends Biochem. Sci.* **20:** 346.

Svejstrup J.Q., Wang Z., Feaver W.J., Wu X., Bushnell D.A., Donahue T.F., Friedberg E.C., and Kornberg R.D. 1995. Different forms of TFIIH for transcription and DNA repair: Holo-TFIIH and a nucleotide excision repairosome. *Cell* **80:** 21.

Tantin D., Kansal A., and Carey M. 1997. Recruitment of the putative transcription-repair coupling factor CSB/ERCC6 to RNA polymerase II elongation complexes. *Mol. Cell. Biol.* **17:** 6803.

Tirode F., Busso F., Coin F., and Egly J.M. 1999. Reconstitution of the transcription factor TFIIH. Assignment of functions for the three enzymatic subunits XPB, XPD and cdk7. *Mol. Cell*

(in press).

Usheva A., Maldonando E., Goldring A., Lu H., Houbavi C., Reinberg D., and Aloni J. 1992. Specific interaction between the nonphosphorylated form of RNA polymerase II and the TATA-binding protein. *Cell* **69:** 871.

Valay J.G., Simon M., Dubois M.F., Bensaude O., Facca C., and Faye G. 1995. The Kin28 gene is required both for RNA polymerase II mediated transcription and phosphorylation of the Rpb1p CTD. *J. Mol. Biol.* **249:** 535.

van Vuuren A.J., Vermeulen W., Ma L., Weeda G., Appeldoorn E., Jaspers N.G.J., van der Eb A.J., Bootsma D., Hoeijmakers J.H.J., Humbert S., Schaeffer L., and Egly J.M. 1994. Correction of xeroderma pigmentosum repair defect by basal transcription factor BTF2 (TFIIH). *EMBO J.* **13:** 1645.

Vermeulen W., Scott R.J., Potger S., Müller H.J., Cole J., Arlett C.F., Kleijer W.J., Bootsma D., Hoeijmakers J.H.J., and Weeda G.. 1994a. Clinical heterogeneity within xeroderma pigmentosum associated with mutations in the DNA repair and transcription gene ERCC3. *Am. J. Hum. Genet.* **54:** 191.

Vermeulen W., van Vuuren A.J., Chipoulet M., Schaeffer L., Appeldoorn E., Weeda G., Jaspers N.G.J., Priestley A., Arlett C.F., Lehmann A.R., Stefanini M., Mezzina M., Sarasin A., Bootsma D., Egly J.M., and Hoeijmakers J.H.J. 1994b. Three unusual repair deficiencies associated with transcription factor BTF2 (TFIIH). Evidence for the existence of a transcription syndrome. *Cold Spring Harbor Symp. Quant. Biol.* **59:** 317.

Wang W., Carey M., and Gralla J.D. 1992. Polymerase II promoter activation: Closed complex formation and ATP-driven start site opening. *Science* **255:** 450.

Weber C.A., Salazar E.P., Stewart S.A., and Thompson L.H. 1990. ERCC2: cDNA cloning and molecular characterization of human nucleotide excision repair gene with high homology to yeast RAD3. *EMBO J.* **9:** 1437.

Weeda G., van Ham R.C.A., Vermeulen W., Bootsma D., van der Eb A.J., and Hoeijmakers J.H.J. 1990. A presumed DNA helicase encoded by ERCC-3 is involved in the human repair disorders xeroderma pigmentosum and Cockayne's syndrome. *Cell* **62:** 777.

Yankulov K. and Bentley D.L. 1997. Regulation of CDK7 substrate specificity by MAT1 and TFIIH. *EMBO J.* **16:** 1638.

Crossing the Line between RNA Polymerases: Transcription of Human snRNA Genes by RNA Polymerases II and III

R.W. Henry,*† E. Ford,*‡ R. Mital,*§ V. Mittal,*¶ and N. Hernandez*¶
*Cold Spring Harbor Laboratory, Cold Spring Harbor, New York 11724; ‡Genetics Program, State University of New York at Stony Brook, Stony Brook, New York 11794; ¶Howard Hughes Medical Institute, Cold Spring Harbor Laboratory, Cold Spring Harbor, New York 11724

In eukaryotic organisms, transcription is carried out by three RNA polymerases, RNA polymerases I, II, and III. In general, these RNA polymerases recognize promoters with very different structures. For example, RNA polymerase II (pol II) promoters are mostly located upstream of the transcription start site, whereas RNA polymerase III (pol III) promoters are mostly located downstream from the transcription start site. The U1, U2, and U6 small nuclear RNA (snRNA) genes, which encode essential RNA components of small nuclear ribonucleoprotein particles (snRNPs) involved in mRNA splicing, serve as representative models for a group of pol II and pol III genes with unusually similar promoters. The pol II U1 and U2 promoters contain a distal sequence element (DSE) and a proximal sequence element (PSE), whereas the pol III U6 promoter contains a DSE, a PSE, and a TATA box, all located upstream of the transcriptional start site.

The list of genes with U1- or U6-type promoters keeps expanding. Additional pol II members of the family include the genes encoding the other major spliceosomal snRNAs U4 and U5 (for review, see Lobo and Hernandez 1994); those encoding the more recently described snRNAs U11 and U12 (Suter-Crazzolara and Keller 1991; Tarn et al. 1995), which are constituents of minor snRNPs involved in the removal of a small class of introns; and the *Herpesvirus saimiri* U RNA (HSUR) genes, which may be involved in regulation of mRNA stability (Fan et al. 1997 and references therein). Additional examples of pol III family members include the gene encoding the RNA component of human RNase P, required for 5′ processing of tRNAs; the MRP/Th gene, whose RNA product is required for the maturation of the RNA primer for mitochondrial DNA synthesis; and various genes of unknown function such as the 7SK, Y1, and Y3 genes (Lobo and Hernandez 1994). A number of pol III genes contain both U6-type promoter elements as well as the typical gene internal pol III A and B box promoter elements: Genes with such hybrid promoters include the selenocysteine tRNA gene (Carbon and Krol 1991), as well as the neural-cell-specific BC1 RNA gene (Martignetti and Brosius 1995),

and the vault RNA gene (Vilalta et al. 1994), whose functions remain to be determined.

Many of the spliceosomal snRNAs are very abundant, ranging from 100,000 to 1 million molecules per nucleus. As a result, transcription from snRNA-type promoters represents a significant portion of the total amount of transcription initiation events by both pol II and pol III (Dahlberg and Lund 1988). Because the pol II and pol III snRNA promoters are so similar, any difference is likely to be directly relevant to the determination of RNA polymerase choice. These promoters therefore constitute an ideal model system to study what determines the specific recruitment of one type of RNA polymerase. So far, the study of snRNA gene transcription has brought several surprises. Among them was the realization that the TATA-box-binding protein (TBP), long thought to be a factor used exclusively by pol II promoters, was also required for transcription by another RNA polymerase, namely, pol III (Lobo et al. 1991; Margottin et al. 1991; Simmen et al. 1991). TBP was then found to be involved in transcription by all three RNA polymerases, but as part of different complexes: SL1 for pol I, TFIID for pol II, and TFIIIB for pol III (Hernandez 1993). None of these complexes, however, appear to participate in transcription of snRNA genes, although TBP itself is required for transcription from both the pol II and pol III snRNA promoters. Another surprise was our recent finding that the same multisubunit complex SNAP$_c$ binds to and directs transcription from pol II and pol III core snRNA promoters. Thus, the same basal transcription complex can be involved in the nucleation of two classes of initiation complexes, resulting in the recruitment of two different RNA polymerases.

STRUCTURE OF THE HUMAN RNA POLYMERASE II AND III snRNA PROMOTERS

The promoters of most genes can be divided into two distinct functional regions: the core promoter region and the regulatory region. Core promoter regions are sufficient to direct low levels of transcription in vitro and contain the binding sites for the basal transcription factors that promote the assembly of preinitiation complexes. Regulatory regions are responsible for the recruitment of activator or repressor proteins, which modulate the levels of transcription. Figure 1 illustrates the architecture of pol II and pol III snRNA promoters. Both the pol II and pol

Present addresses: †Department of Biochemistry, Michigan State University, E. Lansing, Michigan 48824; §Institute of Molecular Biology, Austrian Academy of Sciences, Billrothstrasse 11, A-5020 Salzburg, Austria.

Human snRNA promoter structure

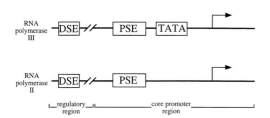

Figure 1. Schematic repesentation of human snRNA promoters. The elements in the regulatory regions and core promoter regions of pol II and pol III snRNA promoters are indicated.

III core promoters contain a PSE, which is interchangeable between the two types of promoters. The pol III core promoter contains, in addition, a TATA box, which in this specific context is responsible for the selective recruitment of pol III. Thus, RNA polymerase specificity is determined by the core snRNA promoters. In contrast, the regulatory region, which contains the DSE and greatly enhances transcription from the core promoter, does not play any part in RNA polymerase selectivity. In fact, like the PSEs, the DSEs of pol II and pol III snRNA promoters are interchangeable (Lobo and Hernandez 1994).

THE PSE RECRUITS THE MULTISUBUNIT BASAL TRANSCRIPTION FACTOR SNAP$_c$

The PSE is a particularly intriguing snRNA promoter element. First, the observation that it can be interchanged between pol II and pol III snRNA promoters suggested early on that it might recruit the same transcription factor in both promoter contexts. Second, it constitutes, by itself, the core pol II snRNA promoter and is thus sufficient to assemble a pol II snRNA-type transcription complex and direct basal transcription. Since transcription from pol II snRNA promoters requires TBP despite the absence of a TATA box, it implies that the PSE is sufficient to recruit TBP, either directly or, more likely, through a PSE-binding factor. Third, in the pol III U6-type promoters, the distance separating the PSE and the TATA box is conserved, suggesting that the PSE and TATA-binding factors might interact. It was therefore of particular interest to identify and purify the PSE binding factor(s).

The identification of factors binding specifically to the PSEs of the human U1, U2, or U6 promoters proved difficult, until Waldschmidt et al. (1991) showed that a fraction derived from a HeLa cell extract contained an activity, which they called PSE-binding protein (PBP), that bound specifically to the PSE of the mouse U6 promoter. Surprisingly, this human factor bound with much lower affinity to the PSEs of the human U1, U2, or U6 promoters. However, the observation that the relative affinity of PBP for the mouse and human U6 PSEs correlated with the relative transcriptional activities of these promoters in HeLa cell extracts strongly suggested that PBP was indeed involved in transcription of the U6 gene (Waldschmidt et al. 1991; Simmen et al. 1992).

We used specific binding to the mouse U6 PSE as an assay to purify a PSE-binding factor from HeLa cell S-100 extracts. The first step was an 18–32% ammonium sulfate precipitation, and the resulting fraction was then chromatographed over a phosphocellulose P-11 column according to the same protocol traditionally used as the first step in the separation of factors involved in pol II and pol III transcription (Matsui et al. 1980; Segall et al. 1980; Samuels et al. 1982). In this protocol, four fractions are generated: The A fraction corresponds to the 100 mM KCl flowthrough, whereas the B, C, and D fractions correspond to successive elutions with 350, 500, and 800–1000 mM KCl, respectively. The TBP-containing complex TFIIIB, required for transcription of pol III genes with gene-internal promoters, elutes in the B fraction, whereas the TBP-containing complexes SL1 and TFIID, required for pol I transcription from rRNA promoters and pol II transcription from mRNA promoters, respectively, elute in the D fraction.

The majority of the PSE-binding activity eluted in the C fraction, which also contained detectable amounts of TBP. The C fraction was further fractionated over a cibacron blue Affigel column and a Mono-Q column. The resulting fractions were highly enriched in PSE-binding activity (we estimate that at this step, the PSE-binding activity had been purified ~2000-fold) and in TBP. Significantly, these fractions were capable of restoring snRNA transcription in HeLa cell extracts that had been depleted of PSE-binding activity with PSE oligonucleotides attached to beads (Sadowski et al. 1993). Curiously, the factor, which we named snRNA activating protein complex (SNAP$_c$), sedimented on a glycerol gradient like a spherical protein of 200 kD (Sadowski et al. 1993). This was considerably larger than the molecular mass of 90 kD estimated for PBP by the same method (Waldschmidt et al. 1991).

The SNAP$_c$-containing Mono-Q fractions were further purified by Mono-S column chromatography followed by glycerol gradient sedimentation. On the Mono-S column, a large fraction of the TBP present in the Mono-Q fraction separated from SNAP$_c$. Nevertheless, substoichiometric amounts of TBP copurified with SNAP$_c$ even after glycerol gradient sedimentation. The glycerol gradient fractions were analyzed for PSE-binding activity, U1 snRNA transcriptional activity, and protein content by SDS-PAGE and silver staining. We observed four major proteins migrating with apparent molecular masses of approximately 200, 50, 45, and 43 kD which copurified with DNA-binding and transcriptional activities (Henry et al. 1995). This composition was in good agreement with that of an independently purified PSE-binding activity called PSE transcription factor or PTF (Yoon et al. 1995). Indeed, the subsequent cloning of cDNAs corresponding to each of these polypeptides confirmed that they corresponded to four SNAP$_c$ subunits, SNAP190 (Wong et al. 1998), SNAP50 (Henry et al. 1996) (also called PTFβ; Bai et al. 1996), SNAP45 (Sadowski et al. 1996) (PTFδ; Yoon and Roeder 1996), and SNAP43 (Henry et al. 1995) (PTFγ; Yoon and Roeder 1996). However, during the purification of SNAP$_c$, we made an addi-

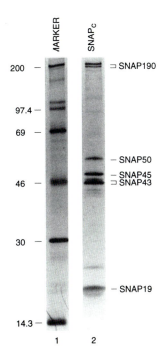

Figure 2. Recombinant core SNAP$_c$ contains five subunits. The proteins constituting SNAP$_c$ were cotranslated in vitro as [^{35}S]methionine-labeled proteins, size-fractionated by 15% SDS-PAGE, and visualized by autoradiography. The identity of each SNAP$_c$ subunit is indicated. The doublets obtained for some subunits may be due to proteolytic breakdown or post-translation modifications.

tional important observation. A small protein of approximately 19 kD also copurified with PSE-binding activity. Initially, we thought that this small protein was a degradation product of the larger proteins; however, recent experiments have demonstrated that it corresponds to a fifth SNAP$_c$ subunit, SNAP19 (Henry et al. 1998). Figure 2 shows the five SNAP$_c$ subunits, SNAP190, SNAP50, SNAP45, SNAP43, and SNAP19, translated in vitro.

SNAP19 IS REQUIRED FOR ASSEMBLY OF SNAP$_c$ FROM IN-VITRO-TRANSLATED SUBUNITS

To determine the architecture of SNAP$_c$ or, more precisely, the network of protein-protein interactions among SNAP$_c$ subunits, we performed coimmunoprecipitation experiments with various subsets of in-vitro-translated SNAP$_c$ subunits. With such experiments, we showed that SNAP43 can associate with SNAP50 in the absence of the other subunits, and similarly that SNAP45 can associate with SNAP190 (Henry et al. 1996; Wong et al. 1998). However, we were unable to obtain complexes containing either three of the four proteins or all four proteins. We then tested the ability of SNAP19 to associate with the other SNAP$_c$ subunits. SNAP19 and each of the other SNAP$_c$ subunits were independently expressed as [^{35}S]methionine-labeled proteins by translation in vitro. SNAP19 was then incubated with each protein in pairwise combinations, and protein-protein associations were

assessed by coimmunoprecipitation of SNAP19 with antibodies directed against each of the other subunits.

As shown in Figure 3A, each SNAP$_c$ subunit was efficiently recognized by its cognate antibody (lane 5 in each panel), and addition of increasing amounts of SNAP19 did not interfere with the ability of the antibody to recognize the primary target (lanes 6–8 in each panel), suggesting that SNAP19 did not mask any epitopes. When incubated with SNAP190, SNAP19 was efficiently coimmunoprecipitated by anti-SNAP190 antibodies (top left panel, lanes 6-8), indicating that these two proteins associate in this assay. This association was specific as SNAP19 was not immunoprecipitated directly by the anti-SNAP190 antibodies (lane 4), and neither SNAP19 (lane 9) nor SNAP190 (lane 10) were immunoprecipitated by preimmune antibodies. However, we did not observe any association of SNAP19 with SNAP45 (top right, lanes 6–8), SNAP43 (bottom left, lanes 6–8), or SNAP50 (bottom right, lanes 6–8). Therefore, in this assay, SNAP19 can associate specifically and exclusively with SNAP190.

We then tested whether the presence of SNAP19 would mediate higher-order complex assembly. Indeed, the SNAP19/SNAP190 protein pair was capable of associating with SNAP43 (data not shown). Since SNAP19 on its own does not associate with SNAP43, this suggests that SNAP43 is involved in weak protein-protein interactions with both SNAP19 and SNAP190, only the sum of which is sufficient to promote coimmunoprecipitation in this assay. As indicated in Figure 3B, this trimeric SNAP19/SNAP190/SNAP43 complex could then associate with SNAP50 (through SNAP43) and SNAP45 (through SNAP190) (Henry et al. 1998), although the stoichiometry of the various subunits within the complex is not known. Thus, SNAP19 is essential for higher-order complex assembly in this assay.

SNAP$_c$ IS REQUIRED FOR TRANSCRIPTION OF HUMAN snRNA GENES BY BOTH RNA POLYMERASES II AND II

The PSE is present in both the pol II and pol III snRNA promoters. An important question was to determine whether SNAP$_c$ was involved in transcription by both RNA polymerases. We first showed that separate depletions of extracts with antibodies directed against SNAP190 (Wong et al. 1998), SNAP50 (Henry et al. 1996), SNAP45 (Sadowski et al. 1996), SNAP43 (Henry et al. 1995), and SNAP19 (Henry et al. 1998) decreased both U1 transcription by pol II and U6 transcription by pol III and that in each case, transcription could be restored by addition of a Mono-Q fraction enriched in SNAP$_c$. These observations suggested that each of these polypeptides was involved in pol II and pol III snRNA gene transcription, but because Mono-Q SNAP$_c$ was not homogeneous, they did not address whether this was as part of the same complex.

To address whether a unique SNAP complex was capable of reconstituting both pol II and pol III snRNA gene transcription, we assembled SNAP$_c$ from recombinant

A

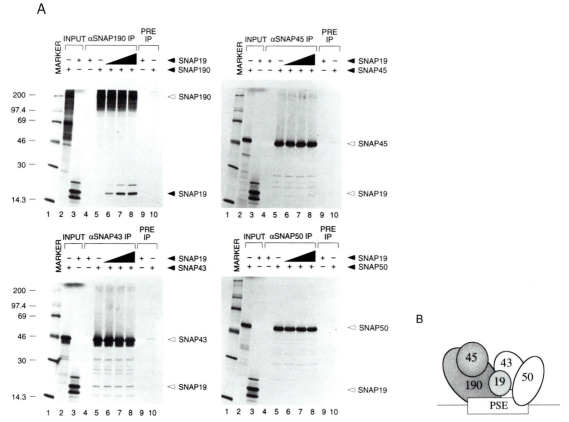

Figure 3. SNAP19 interacts with SNAP190. (*A*) SNAP19 was tested for its ability to interact with SNAP190 (*upper left*), SNAP45 (*upper right*), SNAP43 (*lower left*), and SNAP50 (*lower right*). Each protein was individually translated in vitro and labeled with [³⁵S]methionine; 0, 3, 10, or 30 µl of SNAP19 (lanes *5–8*) were incubated with 10 µl of each of the four other SNAP_c subunits for 1 hr to allow association prior to addition of beads coated with anti-SNAP190, anti-SNAP50, anti-SNAP45, or anti-SNAP43 antibodies as indicated above the panels. (Lane *1*) Protein size standards; (lanes *2* and *3*) 1 µl of the indicated input samples; (lanes *4–8*) immunoprecipitations performed with the specific antibodies indicated above the lanes; (lanes *9* and *10*) immunoprecipitations performed with the relevant preimmune antibodies. The proteins were separated by 15% SDS-PAGE and visualized by autoradiography. (*B*) Schematic representation of SNAP_c. SNAP19 interacts directly with SNAP190. SNAP43 is capable of associating with the SNAP19/SNAP190 complex. SNAP45 and SNAP50 can join the complex through direct interactions with SNAP190 and SNAP43, respectively. The stoichiometry of the various subunits within the complex is not known.

subunits in a baculovirus expression system. As a control, we also coexpressed all SNAP_c subunits except SNAP190. The resulting complexes were then purified by immunoaffinity chromatography with anti-SNAP43 antibodies followed by peptide elution. We first tested the recombinant complexes for specific binding to the PSE in an electrophoretic mobility shift assay (EMSA). As shown in Figure 4, when increasing amounts of recombinant SNAP_c (rSNAP_c) were incubated with a PSE-containing probe (lanes 1–3), but not with a mutant PSE-containing probe (lanes 4–6), a complex was formed. In contrast, proteins present in the control fraction did not bind to either the mutant or wild-type PSE-containing probes (lanes 7–12). The DNA/protein complex formed by rSNAP_c migrated with a mobility similar to that formed by biochemically purified SNAP_c, suggesting that the recombinant complex is similar to the endogenous HeLa cell complex (compare lanes 1–3 with lanes 13–15). Indeed, antibodies directed against each subunit of SNAP_c supershift both the endogenous and recombinant complexes, indicating that rSNAP_c contains all five subunits (Henry et al. 1998).

Thus, we could assemble an rSNAP_c capable of binding DNA in a PSE-dependent manner.

We then determined whether SNAP_c assembled from recombinant subunits was able to reconstitute transcription. For this purpose, a HeLa cell extract was depleted of endogenous SNAP_c with anti-SNAP43 antibodies. As shown in Figure 5 (Henry et al. 1998), this strongly reduced both pol II U1 transcription and pol III U6 transcription, whereas depletion with preimmune antibodies had little effect (compare lanes 2 and 3 in Fig. 5). Upon addition of biochemically purified SNAP_c, U1 and U6 transcriptions were both restored (lanes 8–11). Significantly, U1 and U6 transcriptions were also restored upon addition of increasing amounts of rSNAP_c (lanes 4 and 5) but not of the control fraction (lanes 6 and 7: Note that two different preparations of HeLa cell SNAP_c and rSNAP_c were used in the two panels, such that recovered levels of U1 and U6 transcription cannot be compared directly). The observation that both U1 and U6 transcriptions can be recovered by addition of rSNAP_c strongly suggests that the same SNAP_c recognizes the PSE and

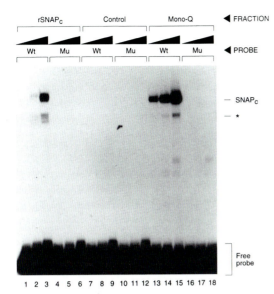

Figure 4. Recombinant SNAP$_c$ binds specifically to the PSE. Insect cells were infected with five recombinant baculoviruses expressing each of the SNAP$_c$ subunits or with viruses expressing all subunits except SNAP190 (control). The resulting complexes were purified by immunoaffinity with antibodies directed against SNAP43 (Henry et al. 1998) and tested for DNA binding in EMSAs as described previously (Sadowski et al. 1993). The reactions contained 1, 3, or 10 μl of rSNAP$_c$ (lanes *1–3* and *4–6*) or control complexes (lanes *7–9* and *10–12*) or biochemically purified SNAP$_c$ (lanes *13–15* and *16–18*), as well as wild-type PSE (lanes *1–3, 7–9, 13–15*) or mutant PSE (lanes *4-6, 10-12, 16-18*) probes. The complex labeled with a star is probably due to proteolysis of some SNAP$_c$ subunits. The protein-DNA complex containing all five subunits is labeled SNAP$_c$.

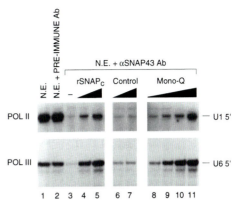

Figure 5. Recombinant SNAP$_c$ functions for both pol II and pol III transcription. Untreated HeLa nuclear extracts (lane *1*) or extracts immunodepleted with rabbit preimmune (lane *2*) or anti-SNAP43 (lanes *3–11*) antibody beads were tested for their ability to support in vitro transcription of U1 snRNA by pol II (*upper panel*) and U6 snRNA by pol III (*lower panel*) as described previously (Sadowski et al. 1993). (Lanes *4* and *5*) 5 and 10 μl of tagged (*top panel*) or untagged (*bottom panel*) rSNAP$_c$ were added; (lanes *6* and *7*) 5 and 10 μl of control fractions were added; (lanes *8–11*) 1, 2, 4, and 8 μl (*top panel*) or 2, 4, 8, and 16 μl (*bottom panel*) of biochemically purified SNAP$_c$ were added. (Reprinted, with permission, from Henry et al. 1998.)

functions as a basal transcription factor in pol II and pol III snRNA promoters.

HOW DOES SNAP$_c$ BIND TO DNA?

UV cross-linking studies performed with PTF/SNAP$_c$ identified a large protein of about 200 kD in close proximity to the DNA (Yoon et al. 1995). Consistent with this observation, the cloning of SNAP190 revealed that the protein contains an unusual Myb domain consisting of four repeats, which we refer to as Ra, Rb, Rc, and Rd (Wong et al. 1998). The Myb domains of typical Myb domain proteins such as c-Myb, A-Myb, and B-Myb (Nomura et al. 1988; Luscher and Eisenman 1990) contain three imperfect tandem repeats called R1, R2, and R3, of which R2 and R3 contact the DNA. Plant Myb domain proteins usually only contain the R2 and R3 repeats (Martin and Paz-Ares 1997). Each repeat consists of three α helices, with the second and third helix forming a variant helix-turn-helix motif (Ogata et al. 1994). In the case of SNAP190, it is not clear which repeats correspond to the DNA-contacting R2 and R3 repeats of c-Myb. However, the Rc and Rd repeats are the most closely related in sequence to the R2 and R3 repeats of c-Myb, and a truncated SNAP190 protein containing only the Rc and Rd repeats can bind to wild-type but not mutant PSE (Wong et al. 1998). Intriguingly, we did not observe DNA binding

by full-length SNAP190 or other SNAP190 truncations. Perhaps other subunits in the complex induce a conformational change in SNAP190 to regulate its ability to bind DNA. Alternatively, or in addition, SNAP$_c$ may be stabilized on the PSE through additional contacts between other SNAP$_c$ subunits and DNA. In particular, UV cross-linking experiments combined with immunoprecipitations indicate that SNAP50 is in close proximity to the DNA within SNAP$_c$, although recombinant SNAP50 on its own does not bind DNA (Henry et al. 1996). Together, these data suggest that the Rc and Rd repeats of SNAP190 as well as the SNAP50 subunit contribute to DNA binding by SNAP$_c$.

THE TATA BOX AND SNAP$_c$ RECRUIT TBP TO HUMAN RNA POLYMERASE III snRNA PROMOTERS

TBP is involved in transcription of genes by pol I, pol II, and pol III, as part of the multisubunit complexes SL1, TFIID, and TFIIIB, respectively. As shown in Table 1, TBP is also required for transcription of snRNA genes by both pol II and pol III, but probably as part of different

Table 1. Requirement for SNAP$_c$, TFIIIC, TFIID, and TFIIB in Transcription of pol II mRNA-type and snRNA-type Promoters, and pol III U6-type and tRNA-type Promoters

	RNA polymerase II		RNA polymerase III	
	mRNA	U1 snRNA	U6 snRNA	tRNA
TBP	TFIID	+	+	TFIIB
SNAP$_c$	–	+	+	–
TFIIIC	–	–	–	+

complexes. Indeed, when an extract is depleted of TBP with anti-TBP antibodies, pol III transcription of both the U6 snRNA gene and the adenovirus 2 (Ad2) VAI gene, a typical pol III gene with a gene-internal promoter, is inhibited. However, whereas transcription from the U6 promoter can be reconstituted by addition of just recombinant TBP, transcription from the VAI promoter can be reconstituted only upon addition of both recombinant TBP and recombinant human BRF (hBRF), a subunit of TFIIIB that is tightly associated with TBP (Mital et al. 1996). These results are confirmed by depletions with anti-hBRF antibodies, which inhibit VAI transcription but not U6 transcription (Mital et al. 1996).

Because other experiments suggested that human U6 transcription does require hBRF in vitro (Wang and Roeder 1995), we wanted to confirm our observation by performing successive depletions, first with antibodies directed against hBRF (α-CSH407) and then with antibodies directed against TBP. As shown in Figure 6, this treatment reduced transcription from both the Ad2 VAI and U6 promoters considerably, as compared to double depletion with preimmune antibodies (compare lanes 4 and 5 with lane 1). Strikingly, and as we had observed before (Mital et al. 1996), U6 transcription, but not VAI transcription, was reconstituted by addition of recombinant TBP (lanes 6–8). In contrast, VAI transcription, but not U6 transcription, was reconstituted by addition of a fraction enriched in TFIIIB (lanes 15–17, 0.38M-TFIIIB fraction). VAI transcription could also be reconstituted by addition of both recombinant TBP and recombinant hBRF (lanes 12–14). In contrast, addition of recombinant hBRF together with recombinant TBP had no positive ef-

fect on U6 transcription beyond that of recombinant TBP alone. In fact, at high concentrations of hBRF, U6 transcription was inhibited (compare lanes 13 and 14 with lane 12). Addition of just hBRF did not restore VAI or U6 transcription (lanes 9–11). Together, these results indicate that U6 transcription does not require the same TFIIIB complex as genes with gene-internal promoter elements. In particular, the U6 promoter does not appear to need hBRF for transcription, at least in vitro.

In yeast, BRF has been shown to have at least two roles: It serves to recruit TBP to pol III promoters through protein-protein interactions with the DNA-bound transcription factor IIIC (TFIIIC), which recognizes the A and B boxes of tRNA promoters, and it contacts pol III. Since hBRF is not required for transcription of the human U6 snRNA gene, both of these functions must be accomplished through another pathway. It is not clear how pol III is recruited to U6-type promoters in mammalian cells, but recruitment of TBP appears to be mediated by both the TATA box and SNAP$_c$. Indeed, we find that full-length TBP, which contains a conserved carboxy-terminal DNA-binding domain and a nonconserved amino-terminal domain, binds very poorly to TATA boxes in general and to the U6 TATA box in particular (Mittal and Hernandez 1997). However, binding of full-length TBP is strongly enhanced in the presence of SNAP$_c$, i.e., full-length TBP and SNAP$_c$ bind cooperatively to their respective binding sites within the U6 promoter. This strongly suggests that TBP interacts with SNAP$_c$ and thus that both protein-DNA interactions with the TATA box and protein-protein interactions with SNAP$_c$ mediate recruitment of TBP to human U6 promoters (Mittal and Hernandez 1997).

Remarkably, cooperative binding is dependent on the nonconserved amino-terminal domain of TBP. Truncated TBP lacking the nonconserved first 96 amino acids binds efficiently to TATA boxes but is unable to recruit SNAP$_c$ to the PSE. These results indicate that the nonconserved amino-terminal domain of TBP performs at least two functions: It down-regulates binding of the protein to TATA boxes and it mediates cooperative binding with SNAP$_c$ (Mittal and Hernandez 1997). This suggests that the TBP amino-terminal domain serves to ensure that the protein not bind to irrelevant A/T-rich sequences but rather be recruited specifically to promoter regions through interactions with other promoter binding factors. It is also worth noting that the "nonconserved" amino-terminal domain of TBP is actually highly conserved among vertebrates. Whereas the run of glutamine residues located in the center of the domain varies in length, the regions before and after the run of glutamine residues are more than 77% identical in TBPs derived from mouse, hamster, *Xenopus*, and viper (Hashimoto et al. 1992; Mittal and Hernandez 1997). Similarly, snRNA promoters have diverged widely between organisms as distant as yeast and humans, but they are quite conserved among vertebrates. Perhaps the conservation of the TBP amino-terminal domain correlates with conservation of the TBP-SNAP$_c$ interaction.

Basal pol II transcription of mRNA-type promoters can be reconstituted in vitro with recombinant TBP; yet in

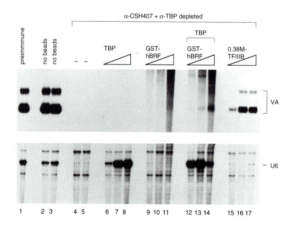

Figure 6. In vitro transcription from the U6 promoter does not require the TBP-hBRF complex. Untreated whole-cell extract (lanes *2* and *3*) or extract double-depleted with preimmune antibody beads (lane *1*) or extract depleted successively with anti-hBRF (α-CSH407) and anti-TBP antibody beads (lanes *4–17*) were tested for their ability to direct Ad2 VAI (*upper panel*) or U6 (*lower panel*) transcription. (Lanes *6–8* and *9–11*) Increasing amounts of recombinant TBP and recombinant GST-hBRF, respectively, were added; (lanes *12–14*) a constant amount of recombinant TBP (corresponding to that added in lane *7*) and increasing amounts of recombinant GST-hBRF were added; (lanes *15–17*) increasing amounts of a fraction enriched in TFIIIB (0.38M-TFIIIB) were added. The bands corresponding to correctly initiated RNA are indicated.

vivo, these types of promoters are thought to recruit the TBP-containing complex TFIID. Similarly, it is well possible that although basal U6 transcription can be reconstituted in vitro with just recombinant TBP, the U6 TATA box is recognized in vivo by a TBP-containing complex. Although the TBP-associated subunits of this putative complex do not appear to be required for basal U6 transcription, they may have a role in mediating the function of activation domains or in directing transcription from chromatin templates.

HOW IS TBP RECRUITED TO HUMAN RNA POLYMERASE II snRNA PROMOTERS?

As for pol III transcription of snRNA genes, depletion of extracts with anti-TBP antibodies reduces pol II transcription of snRNA genes (Sadowski et al. 1993). However, unlike pol III snRNA gene transcription, pol II snRNA gene transcription is restored only to low levels by addition of recombinant TBP (Sadowski et al. 1993). Furthermore, the TBP-containing complexes TFIID and TFIIIB are less active than TBP alone (Sadowski et al. 1993; Yoon and Roeder 1996). Together, these data suggest that pol II transcription of snRNA genes requires a specific TBP-associated factor, which facilitates the recruitment of TBP to pol II snRNA promoters through protein-protein interactions. Intriguingly, high levels of pol II snRNA gene transcription can be reconstituted in TBP-depleted extracts by addition of the $SNAP_c$-enriched Mono-Q fraction, which as mentioned above, also contains significant levels of TBP. It therefore seems likely that a population of TBP present in the Mono-Q fraction is associated with factor(s) specifically required for pol II transcription of snRNA genes.

The composition of this TBP-containing complex remains to be determined, but the observation that low levels of TBP copurify with $SNAP_c$ suggests that it may be loosely associated with $SNAP_c$. Both SNAP43 (Henry et al. 1995; Yoon and Roeder 1996) and SNAP45 (Sadowski et al. 1996; Yoon and Roeder 1996) interact with TBP in vitro, as determined by GST-pull down of recombinant proteins. Furthermore, TBP can also be coimmunoprecipitated from nuclear extracts with antibodies directed against SNAP43 (Henry et al. 1995) and PTFβ/SNAP50 (Bai et al. 1996), although in the latter case, the interaction with TBP is salt-sensitive. A caveat, however, is that we do not know whether these TBP-$SNAP_c$ interactions have a role in pol II or pol III snRNA gene transcription. In any case, since the same $SNAP_c$ binds to pol II and pol III snRNA promoters, it is likely that RNA polymerase specificity is determined by different modes of TBP recruitment.

$SNAP_c$ IS A TARGET FOR THE TRANSCRIPTIONAL ACTIVATOR PROTEIN Oct-1

The DSEs of snRNA-type promoters are composed of several protein-binding sites, one of which is often a binding site for the activator STAF (Schaub et al. 1997)

and another is the octamer motif. The octamer motif recruits the widely expressed activator Oct-1. Oct-1 is a founding member of the POU domain protein family, which contains a large number of transcription factors including Oct-2 and the pituitary transcription factor Pit-1 (Herr et al. 1988). The POU domain is a bipartite DNA-binding domain consisting of two helix-turn-helix-containing DNA-binding structures: an amino-terminal POU-specific (POU_S) domain and a carboxy-terminal POU-homeo (POU_H) domain joined by a flexible linker (Herr and Cleary 1995). The Oct-1 and Oct-2 POU domains have been shown to participate in a number of protein-protein interactions. For example, they can recruit a B-cell-specific cofactor, variously termed OBF-1, OCA-B, or Bob1, to octamer-motif-containing immunoglobulin promoters (Luo et al. 1992; Gstaiger et al. 1995; Luo and Roeder 1995; Strubin et al. 1995). The region of the Oct-1 POU domain contacted by OBF-1 has been mapped in detail by a mutagenesis of all the surface residues in the Oct-1 POU domain and encompasses residues in both the POU_H and the POU_S domains (Babb et al. 1997). In addition, the Oct-1 but not the Oct-2 POU domain can recruit the herpes simplex virus protein VP16 to cis-acting elements in the viral immediate early promoters (Gerster and Roeder 1988; Kristie et al. 1989; Stern et al. 1989), and this interaction is directed by the POU_H domain (Stern et al. 1989).

Early studies on PTF/$SNAP_c$ revealed that the Oct-1 POU domain and PTF bind cooperatively to DNA probes containing a PSE and an octamer motif (Murphy et al. 1992). In contrast, the Pit-1 POU domain, which is only 50% identical to the Oct-1 POU domain, did not display cooperative binding (Mittal et al. 1996). This allowed us to use chimeric POU domains to determine that a single-amino-acid difference, a glutamic acid to arginine change at position 7 within the POU_S domain (E7R mutation), is responsible for the differential abilities of the Oct-1 and Pit-1 POU domains to recruit $SNAP_c$ to the PSE (Mittal et al. 1996). These results suggested that the Oct-1 POU_S domain could recruit $SNAP_c$ to the PSE either by inducing a DNA conformation favorable for $SNAP_c$ binding or by direct protein-protein contacts.

The first suggestion that the Oct-1 POU domain contacts $SNAP_c$ directly came with the cloning of SNAP190. A partial cDNA encoding the carboxy-terminal half of SNAP190 was isolated in a yeast one-hybrid screen designed to identify proteins capable of associating with Oct-1 bound to an octamer motif (Wong et al. 1998). The same screen had earlier resulted in the isolation of cDNAs encoding OBF-1, the B-cell-specific cofactor mentioned above that interacts with octamer-bound Oct-1 (Strubin et al. 1995). The selection of SNAP190-encoding cDNAs in such a screen suggested that the carboxy-terminal half of SNAP190 was capable of interacting with octamer-bound Oct-1, and indeed this could be shown directly with an EMSA (Wong et al. 1998). The region required for the interaction as determined by both EMSAs and in vivo interaction assays was then narrowed down to a small segment, whose interaction with Oct-1 POU was still sensitive to the E7R mutation in Oct-1 POU (Ford et al. 1998).

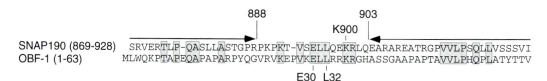

Figure 7. The Oct-1 POU interacting regions in SNAP190 and OBF-1 show sequence similarity. The Oct-1 POU interacting regions of SNAP190 and OBF-1 are shown. Identical amino acids are shaded. The left arrow shows the endpoint of the largest amino-terminal deletion (in the context of a SNAP190 truncation containing carboxy-terminal sequences extending to amino acid 1137) that still interacted with octamer-bound Oct-1 POU in vivo. The right arrow shows the endpoint of the largest carboxy-terminal deletion tested (in a SNAP190 truncation containing amino-terminal sequences extending to amino acid 800) that still interacted with DNA-bound Oct-1 POU in an EMSA (Ford et al. 1998). The borders of the SNAP190 "core" interacting region defined by these deletions (extending from P888 to Q903) are indicated. Mutation of the OBF-1 residues E30 and L32 (Gstaiger et al. 1995) and SNAP190 residue K900 (Ford et al. 1998) debilitates interaction with octamer-bound Oct-1 POU.

Perhaps one of the most convincing methods to demonstrate a direct protein-protein interaction is through the isolation of altered specificity mutants, in which a mutation in one of the partners that disrupts the interaction can be compensated for by a second mutation in the other partner that restores the interaction. We reasoned that the Oct-1 POU$_S$ E7 might be involved in a side chain–side chain interaction with a basic residue in SNAP190 and that the effect of the E7R mutation might be reversed by mutation of an interacting basic residue in SNAP190 to glutamic acid. We therefore mutated basic amino acids in the SNAP190 segment required for interaction with Oct-1 POU to glutamic acids. A single and a double mutation were then introduced into full-length SNAP190, and the mutant SNAP190s were used to assemble mutant SNAP$_c$s with only a single or a double amino acid change. The abilities of these mutant SNAP$_c$s to interact with both wild-type Oct-1 POU and Oct-1 POU E7R were then tested (Ford et al. 1998). These experiments showed first that a single- or double-amino-acid change within the Oct-1 POU interacting region of SNAP190 debilitated or reduced the ability of these two factors to bind cooperatively. Importantly, the loss of cooperative binding correlated with a loss of transcription activation in vitro, indicating that the recruitment of SNAP$_c$ by Oct-1 POU contributes to transcription activation. The results also showed that one SNAP190 mutation, K900E, destroyed the ability of SNAP$_c$ to bind cooperatively with Oct-1 POU but restored the ability of SNAP$_c$ to bind cooperatively with Oct-1 POU E7R. The isolation of this altered specificity mutant showed that cooperative binding of Oct-1 POU and SNAP$_c$ is mediated at least in part by a direct protein-protein interaction probably involving E7 in Oct-1 and K900 in the SNAP190 subunit of SNAP$_c$.

The identification of a small SNAP190 region required for interaction with Oct-1 POU brought a surprise. Although a comparison of full-length SNAP190 with sequences in the databases had failed to detect any related sequence in OBF-1, a direct comparison of this small SNAP190 segment with full-length OBF-1 revealed striking similarities (Ford et al. 1998). As shown in Figure 7, SNAP190 amino acids 888–903, which constitute the core of the Oct-1 POU interacting region (see legend to Fig. 7), and OBF-1 amino acids 22–38, which are con-

tained within the OBF-1 region sufficient for association with octamer-bound Oct-1 POU (OBF-1 amino acids 1–63; Gstaiger et al. 1995), share seven (41%) identical amino acid residues (Ford et al. 1998). Moreover, OBF-1 residues E30 and L32 (Gstaiger et al. 1995), and SNAP190 residue K900 (Ford et al. 1998), which are conserved in both proteins, are in each case essential for interaction with Oct-1. These observations suggest that there are similarities in how OBF-1 and SNAP190 interact with the transcriptional activator Oct-1. This is remarkable, since these two proteins, a cell-specific transcriptional coactivator and a basal transcription factor, respectively, do not share any sequence similarity outside of the Oct-1 interacting regions nor any known functional role.

Together, the results summarized above have important implications for the mechanism of transcription activation. Transcriptional activators are often described as having a DNA-binding domain, whose role is to bring the activator to the correct promoter, and an activation domain, whose role is to activate transcription. Oct-1 provides an example of an activator whose DNA-binding domain does much more than just target the protein to the correct location: In this case, it recruits SNAP$_c$ to the PSE, an event that results in transcription activation. As shown in Figure 8, this is similar to the function of the λ repressor protein cI in the context of the λ P$_{RM}$ promoter. λ cI binds as a dimer upstream of the RNA-polymerase-binding site and activates transcription by recruiting RNA polymerase through direct protein-protein contacts with the σ-subunit (Guarente et al. 1982; Bushman et al. 1989; Li et al. 1994), which like SNAP$_c$ mediates recognition of the core promoter. It is striking that in addition to mediating comparable functions, the Oct-1 POU$_S$ domain and the λ cI repressor share very similar three-dimensional structures (Herr and Cleary 1995 and references therein). It should be noted, however, that Oct-1 possesses, outside of the POU domain, activation domains that specifically enhance snRNA gene transcription (Tanaka et al. 1992; Das et al. 1995). These domains contribute little to the recruitment of SNAP$_c$ to the PSE, which is mediated largely by just the POU domain (Ford and Hernandez 1997), and their mechanism of action is not known. It remains that despite the large complexity of

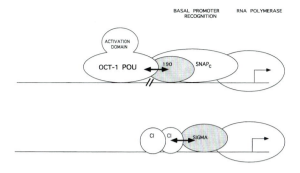

Figure 8. Both Oct-1 POU and λ cI activate transcription by recruiting a core promoter recognition factor. The Oct-1 activator contains an activation domain and a DNA-binding domain. On snRNA promoters, the Oct-1 POU DNA-binding domain activates transcription by recruiting SNAP$_c$ to the core promoter element PSE, and on the λ P$_{RM}$ promoter, the c1 repressor activates transcription by recruiting RNA polymerase through the σ-subunit, which like SNAP$_c$ is involved in core promoter recognition.

the transcription apparatus of higher eukaryotes as compared to that of prokaryotes, some of the same mechanisms are used to achieve transcription activation.

FUTURE DIRECTIONS

The study of snRNA gene transcription in human cells has served as a valuable system for understanding mechanisms of transcription for both pol II and pol III, but many questions remain unanswered. Thus, the same core SNAP$_c$ is required for transcription by both RNA polymerases, and it seems likely that different modes of TBP recruitment by both SNAP$_c$ and promoter sequences set the stage for subsequent recruitment of polymerase-specific factors. However, how TBP is recruited to pol II snRNA promoters, and what the identities of the RNA-polymerase-specific factors are, is presently unknown. Similarly, although the mechanism by which the Oct-1 POU domain recruits SNAP$_c$ and activates transcription is now well understood, we do not know how the Oct-1 activation domains function.

SNAP$_c$ itself may perform a number of functions in addition to nucleating the assembly of PSE-dependent initiation complexes. For example, it has long been known that a "3′ box" located downstream from the pol II snRNA genes works in concert with promoter sequences to direct transcription termination (Lobo and Hernandez 1994). Perhaps by determining the nature of the machinery assembled at the core promoter, SNAP$_c$ influences the processivity and the termination properties of pol II, as well as transcription factor recycling. In addition, by analogy with TFIID, components of SNAP$_c$ may have additional enzymatic capabilities such as kinase and/or acetyltransferase activity. Some of these activities may be important for basal transcription from naked DNA templates, and some may be required for activated transcription or for transcription from chromatin templates. The availability of a fully defined SNAP$_c$ should greatly facilitate addressing these questions.

ACKNOWLEDGMENTS

This work was supported in part by National Institutes of Health grant GM-38810. N.H. is an Associate Investigator with the Howard Hughes Medical Institute.

REFERENCES

Babb R., Cleary M.A., and Herr W. 1997. OCA-B is a functional analog of VP16 but targets a separate surface of the Oct-1 POU domain. *Mol. Cell. Biol.* **17:** 7295.

Bai L., Wang Z., Yoon J.-B., and Roeder R.G. 1996. Cloning and characterization of the β subunit of human proximal sequence element-binding transcription factor and its involvement in transcription of small nuclear RNA genes by RNA polymerases II and III. *Mol. Cell. Biol.* **16:** 5419.

Bushman F.D., Shang C., and Ptashne M. 1989. A single glutamic acid residue plays a key role in the transcriptional activation function of lambda repressor. *Cell* **58:** 1163.

Carbon P. and Krol A. 1991. Transcription of the *Xenopus laevis* selenocysteine tRNA(Ser)Sec gene: A system that combines an internal B box and upstream elements also found in U6 snRNA genes. *EMBO J.* **10:** 599.

Dahlberg J.E. and Lund E. 1988. The genes and transcription of the major small nuclear RNAs. In *Structure and function of major and minor small nuclear ribonucleoprotein particles* (ed. M.L. Birnstiel), p. 38. Springer Verlag, Berlin.

Das G., Hinkley C.S., and Herr W. 1995. Basal promoter elements as a selective determinant of transcriptional activator function. *Nature* **374:** 657.

Fan X.C., Myer V.E., and Steitz J.A. 1997. AU-rich elements target small nuclear RNAs as well as mRNAs for rapid degradation. *Genes Dev.* **11:** 2557.

Ford E. and Hernandez N. 1997. Characterization of a trimeric complex containing Oct-1, SNAP$_c$, and DNA. *J. Biol. Chem.* **272:** 16048.

Ford E., Strubin M., and Hernandez N. 1998. The Oct-1 POU domain activates snRNA gene transcription by contacting a region in the SNAPc largest subunit that bears sequence similarities with the Oct-1 coactivator OBF-1. *Genes Dev.* **12:** 3528.

Gerster T. and Roeder R.G. 1988. A herpesvirus trans-activating protein interacts with transcription factor OTF-1 and other cellular proteins. *Proc. Natl. Acad. Sci.* **85:** 6347.

Gstaiger M., Knoepfel L., Georgiev O., Schaffner W., and Hovens C.M. 1995. A B-cell coactivator of octamer-binding transcription factors. *Nature* **373:** 360.

Guarente L., Nye J.S., Hochschild A., and Ptashne M. 1982. Mutant lambda phage repressor with a specific defect in its positive control function. *Proc. Natl. Acad. Sci.* **79:** 2236.

Hashimoto S., Fujita H., Hasegawa S., Roeder R., and Horikoshi M. 1992. Conserved structural motifs within the N-terminal domain of TFIIDτ from *Xenopus*, mouse and human. *Nucleic Acids Res.* **20:** 3788.

Henry R.W., Sadowski C.L., Kobayashi R., and Hernandez N. 1995. A TBP-TAF complex required for transcription of human snRNA genes by RNA polymerases II and III. *Nature* **374:** 653.

Henry R.W., Ma B., Sadowski C.L., Kobayashi R., and Hernandez N. 1996. Cloning and characterization of SNAP50, a subunit of the snRNA-activating protein complex SNAP$_c$. *EMBO J.* **15:** 7129.

Henry R.W., Mittal V., Ma B., Kobayashi R., and Hernandez N. 1998. Assembly of a functional, core promoter complex (SNAP$_c$) shared by RNA polymerase II and III. *Genes Dev.* **12:** 2664.

Hernandez N. 1993. TBP, a universal eucaryotic transcription factor? *Genes Dev.* **7:** 1291.

Herr W. and Cleary M.A. 1995. The POU domain: Versatility in transcriptional regulation by a flexible two-in-one DNA-binding domain. *Genes Dev.* **9:** 1679.

Herr W., Sturm R.A., Clerc R.G., Corcoran L.M., Baltimore D.,

Sharp P.A., Ingraham H.A., Rosenfeld M.G., Finney M., Ruvkun G., and Horvitz H.R. 1988. The POU domain: A large
conserved region in the mammalian *pit-1, oct-1, oct-2*, and
Caenorhabditis elegans unc-86 gene products. *Genes Dev.* **2:**
1513.

Kristie T.M., LeBowitz J.H., and Sharp P.A. 1989. The octamer-
binding proteins form multi-protein-DNA complexes with the
HSV α-TIF regulatory protein. *EMBO J.* **8:** 4229.

Li M., Moyle H., and Susskind M.M. 1994. Target of the transcriptional activation function of phage λ cI protein. *Science*
263: 75.

Lobo S.M. and Hernandez N. 1994. Transcription of snRNA
genes by RNA polymerases II and III. In *Transcription, mechanisms and regulation* (ed. R.C. Conaway and J.W.
Conaway), p. 127. Raven Press, New York.

Lobo S.M., Lister J., Sullivan M.L., and Hernandez N. 1991.
The cloned RNA polymerase II transcription factor IID selects RNA polymerase III to transcribe the human U6 gene in
vitro. *Genes Dev.* **5:** 1477.

Luo Y. and Roeder R.G. 1995. Cloning, functional characterization and mechanism of action of the B cell-specific transcription coactivator OCA-B. *Mol. Cell. Biol.* **15:** 4115.

Luo Y., Fujii H., Gerster T., and Roeder R.G. 1992. A novel B
cell-derived coactivator potentiates the activation of immunoglobulin promoters by octamer-binding transcription
factors. *Cell* **71:** 231.

Luscher B. and Eisenman R.N. 1990. New light on Myc and
Myb. II. Myb. *Genes Dev.* **4:** 2235.

Margottin F., Dujardin G., Gerard M., Egly J.-M., Huet J., and
Sentenac A. 1991. Participation of the TATA factor in transcription of the yeast U6 gene by RNA polymerase C. *Science*
251: 424.

Martignetti J.A. and Brosius J. 1995. BC1 RNA: Transcriptional
analysis of a neural cell-specific RNA polymerase III transcript. *Mol. Cell. Biol.* **15:** 1642.

Martin C. and Paz-Ares J. 1997. MYB transcription factors in
plants. *Trends Genet.* **13:** 67.

Matsui T., Segall J., Weil P.A., and Roeder R.G. 1980. Multiple
factors required for accurate initiation of transcription by purified RNA polymerase II. *J. Biol. Chem.* **255:** 11992.

Mital R., Kobayashi R., and Hernandez N. 1996. RNA polymerase III transcription from the human U6 and adenovirus
type 2 VAI promoters has different requirements for human
BRF, a subunit of human TFIIIB. *Mol. Cell. Biol.* **16:** 7031.

Mittal V. and Hernandez N. 1997. Role for the amino-terminal
region of human TBP in U6 snRNA transcription. *Science*
275: 1136.

Mittal V., Cleary M.A., Herr W., and Hernandez N. 1996. The
Oct-1 POU-specific domain can stimulate small nuclear RNA
gene transcription by stabilizing the basal transcription complex SNAP$_c$. *Mol. Cell. Biol.* **16:** 1955.

Murphy S., Yoon J.-B., Gerster T., and Roeder R.G. 1992. Oct-
1 and Oct-2 potentiate functional interactions of a transcription factor with the proximal sequence element of small nuclear RNA genes. *Mol. Cell. Biol.* **12:** 3247.

Nomura N., Takahashi M., Matsui M., Ishii S., Date T.,
Sasamoto S., and Ishizaki R. 1988. Isolation of human cDNA
clones of myb-related genes, A-*myb* and B-*myb*. *Nucleic
Acids Res.* **16:** 11075.

Ogata K., Morikawa S., Nakamura H., Sekikawa A., Inoue T.,
Kanai H., Sarai A., Ishii S., and Nishimura Y. 1994. Solution
structure of a specific DNA complex of the Myb DNA-binding domain with cooperative recognition helices. *Cell* **79:**
639.

Sadowski C.L., Henry R.W., Kobayashi R., and Hernandez N.
1996. The SNAP45 subunit of the small nuclear RNA
(snRNA) activating protein complex is required for RNA
polymerase II and III snRNA gene transcription and interacts
with the TATA box binding protein. *Proc. Natl. Acad. Sci.* **93:**
4289.

Sadowski C.L., Henry R.W., Lobo S.M., and Hernandez N.
1993. Targeting TBP to a non-TATA box *cis*-regulatory element: A TBP-containing complex activates transcription from
snRNA promoters through the PSE. *Genes Dev.* **7:** 1535.

Samuels M., Fire A., and Sharp P.A. 1982. Separation and characterization of factors mediating accurate transcription by
RNA polymerase II. *J. Biol. Chem.* **257:** 14419.

Schaub M., Myslinski E., Schuster C., Krol A., and Carbon P.
1997. Staf, a promiscuous activator for enhanced transcription
by RNA polymerases II and III. *EMBO J.* **16:** 173.

Segall J., Matsui T., and Roeder R.G. 1980. Multiple factors are
required for the accurate transcription of purified genes by
RNA polymerase III. *J. Biol. Chem.* **255:** 11986.

Simmen K.A., Waldschmidt R., Bernues J., Parry H.D., Seifart
K.H., and Mattaj I.W. 1992. Proximal sequence element factor binding and species specificity in vertebrate U6 snRNA
promoters. *J. Mol. Biol.* **223:** 873.

Simmen K.A., Bernues J., Parry H.D., Stunnenberg H.G.,
Berkenstam A., Cavallini B., Egly J.-M., and Mattaj I.W.
1991. TFIID is required for in vitro transcription of the human
U6 gene by RNA polymerase III. *EMBO J.* **10:** 1853.

Stern S., Tanaka M., and Herr W. 1989. The Oct-1 homeo-do-
main directs formation of a multiprotein-DNA complex with
the HSV transactivator VP16. *Nature* **341:** 624.

Strubin M., Newell J.W., and Matthias P. 1995. OBF-1, a novel
B cell-specific coactivator that stimulates immunoglobulin
promoter activity through association with octamer-binding
proteins. *Cell* **80:** 497.

Suter-Crazzolara C. and Keller W. 1991. Organization and transient expression of the gene for human U11 snRNA. *Gene
Expr.* **1:** 91.

Tanaka M., Lai J.-S., and Herr W. 1992. Promoter-selective activation domains in Oct-1 and Oct-2 direct differential activation of an snRNA and mRNA promoter. *Cell* **68:** 755.

Tarn W.Y., Yario T.A., and Steitz J.A. 1995. U12 snRNA in
vertebrates: Evolutionary conservation of 5´ sequences implicated in splicing of pre-mRNAs containing a minor class of
introns. *RNA* **6:** 644.

Vilalta A., Kickhoefer V.A., Rome L.H., and Johnson D.L.
1994. The rat vault RNA gene contains a unique RNA polymerase III promoter composed of both external and internal
elements that function synergistically. *J. Biol. Chem.* **269:**
29752.

Waldschmidt R., Wanandi I., and Seifart K.H. 1991. Identification of transcription factors required for the expression of
mammalian U6 genes in vitro. *EMBO J.* **10:** 2595.

Wang Z. and Roeder R.G. 1995. Structure and function of a human transcription factor TFIIIB subunit that is evolutionarily
conserved and contains both TFIIB- and high-mobility-group
protein 2-related domains. *Proc. Natl. Acad. Sci.* **92:** 7026.

Wong M.W., Henry R.W., Ma B., Kobayashi R., Klages N.,
Matthias P., Strubin M., and Hernandez N. 1998. The large
subunit of basal transcription factor SNAP$_c$ is a Myb domain
protein that interacts with Oct-1. *Mol. Cell. Biol.* **18:** 368.

Yoon J.-B. and Roeder R.G. 1996. Cloning of two proximal sequence element-binding transcription factor subunits (γ and δ)
that are required for transcription of small nuclear RNA genes
by RNA polymerases II and III and interact with the TATA-
binding protein. *Mol. Cell. Biol.* **16:** 1.

Yoon J.-B., Murphy S., Bai L., Wang Z., and Roeder R.G..
1995. Proximal sequence element-binding transcription factor
(PTF) is a multisubunit complex required for transcription of
both RNA polymerase II- and RNA polymerase III-dependent
small nuclear RNA genes. *Mol. Cell. Biol.* **15:** 2019.

Transcription Factor IIIB: The Architecture of Its DNA Complex, and Its Roles in Initiation of Transcription by RNA Polymerase III

A. Kumar, A. Grove, G.A. Kassavetis, and E.P. Geiduschek

Department of Biology and Center for Molecular Genetics, University of California, San Diego, La Jolla, California 92093-0634

COMPONENTS OF TRANSCRIPTIONAL INITIATION BY YEAST RNA POLYMERASE III

The core components of the yeast RNA polymerase III (pol III) transcription system are fully enumerated. pol III is brought to its promoters by its central transcription factor (TF)IIIB, which is composed of three subunits: TBP, the ubiquitous component of all eukaryotic nuclear transcription; Brf, the TFII*B*-related and archaeal TF*B*-related component; and B″, a pol-III-specific subunit. All three subunits are required for all transcription by yeast pol III. TFIIIB can find its own way to the promoter in bare DNA through a direct interaction of TBP with a strong TATA box. On genes that lack a strong TATA box, and in vivo, where DNA is packaged into chromatin, TFIIIB is brought to the promoter by TFIIIC, its large (six-subunit) DNA-binding assembly factor. 5S rRNA genes lack a direct DNA-binding site for TFIIIC, the 9-zinc finger TFIIIA serving as a DNA-binding adaptor (Braun et al. 1992; Rowland and Segall 1998). Ten essential genes encode the subunits of these three transcription factors (for review, see Geiduschek and Kassavetis 1995; White 1998); 17 genes (16 essential; 1 not yet tested) encode the subunits of pol III (S. Chédin et al., pers. comm.). Once it has been brought by TFIIIB to the promoter, pol III spontaneously generates an extensive transcription bubble around the transcription start site (Kassavetis et al. 1992).

In this paper, we review recent and current work dealing with the structure and interactions of TFIIIB: The structure-supporting properties of B″ are summarized, and an interpretation of the remarkable stability that B″ confers on the TFIIIB-DNA complex is offered. Extensive protein-protein and protein-DNA interactions of Brf are also specified, and a model based on these interactions is presented. We also summarize new evidence pointing to a post-polymerase-recruitment role for TFIIIB in transcriptional initiation, showing that certain transcription-deficient TFIIIB-DNA complexes bring pol III to the promoter in a state in which it is unable to proceed along the reaction path to initiation of transcription. The existence of closed pol III promoter complexes suggests a similarity between the roles of TFIIIB in transcriptional initiation and the mechanisms of action of certain activators of bacterial transcription, which we discuss in a concluding section.

THE INTERNAL CONNECTIONS OF THE TFIIIB-DNA COMPLEX

B″ Is a Scaffolding Protein

B″ is essential for all yeast pol-III-initiated transcription, in supercoiled as well as linear DNA, in TFIIIC-dependent as well as TFIIIC-independent transcription, and with bare DNA as well as with chromatin templates. Nevertheless, a deletion analysis of B″ shows that this 594-amino-acid protein can be extensively truncated at both ends and that small internal deletions can be made freely without destroying its ability to direct transcription of the yeast U6 snRNA gene (*SNR6*) in the absence of TFIIIC. In fact, no individual segment of this essential protein is essential for U6 RNA synthesis in vitro using supercoiled DNA. This startling property arises from a functional redundancy of two core segments of B″, the presence of either one sufficing for transcription activity in this specific context (Fig. 1) (Kumar et al. 1997).

Separate essential roles for these core segments of B″ are exposed under three circumstances: First, in TFIIIC-dependent transcription (Kumar et al. 1997); second, when B″ with a small internal deletion and Brf with an amino-terminal deletion, which separately retain competence for TFIIIC-independent *SNR6* transcription, are used in combination (Kassavetis et al. 1997); and third, as we show below, when transcription is done in vitro with linear in place of supercoiled DNA (Kassavetis et al. 1998a). These functional complementarities between B″ and Brf and between either protein and the torsional state of DNA support the notion that Brf, B″, and DNA serve as mutually reinforcing and interdependent parts of the TFIIIB-DNA complex.

Brf Makes Multiple Protein-Protein and Protein-DNA Contacts

We have examined an extensive set of fragments of the 596-amino-acid Brf subunit for the ability to assemble a TFIIIB-DNA complex and to recruit pol III to accurately initiating transcription of the *SNR6* gene. Reconstitution of these capabilities of Brf from diverse combinations of inactive and poorly active fragments has also been examined. Yeast Brf is homologous to TFIIB in its amino-terminal half, but distinctive in its carboxy-terminal half (Fig. 1). When Brf is split (at amino acid 283) so as to

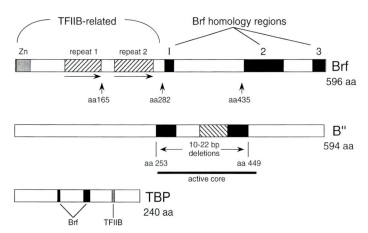

Figure 1. The subunits of *S. cerevisiae* TFIIIB. (*Top line*) Brf (also named yTFIIIB70); the N-proximal TFIIB-related half with its amino-terminal zinc finger and imperfect-repeat segments (amino acids 94–164 and 189–264) and three segments in the C-proximal half with conserved sequence among fungal proteins are indicated; the sequence for fungal homology domain 2 is also conserved in the worm and human. At the indicated sites, Brf can be split into fragments that combine effectively to reconstitute transcriptional activity (see Fig. 2). Fungal homology domain 2 harbors the principal TBP- and B″-binding sites of Brf in the TFIIIB-DNA complex (see Fig. 3). (*Middle line*) B″ (also named yTFIIIB90 and Tfc5); combined amino- and carboxy-terminal truncations define an active core for TFIIIC-dependent transcription. Small internal deletions within this core identify segments (*closed rectangle*) that are only required on an either/or basis in vitro for TFIIIC-independent transcription but are separately required for TFIIIC-dependent transcription. A segment (*hatched area*) covering four 13–22-amino acid deletions generates B″ proteins that are competent for transcription of supercoiled DNA, but not linear DNA, in vitro (see Fig. 4). (*Bottom line*) TBP; surface-exposed amino acids that are implicated in binding Brf (*closed box*) and TFIIB (*hatched*) are indicated (see Figs. 2 and 3) (Shen et al. 1998).

separate these two parts, activity can be reconstituted from the resulting fragments with high efficiency, generating a TFIIIB-DNA complex that is very stable and indistinguishable (by the criterion of in vitro footprinting) from the TFIIIB-DNA complex formed with intact Brf. Proximities of the C and N halves of Brf to DNA as well as interactions with TBP and B″ have been extensively mapped: We used photochemical cross-linking for DNA proximity mapping, and an extensive mutational analysis of the TBP surface (Bryant et al. 1996; Shen et al. 1998) was instrumental for mapping Brf-TBP interactions.

This analysis (Kassavetis et al. 1998b) establishes the following features of the TFIIIB-DNA complex: (1) Both halves of Brf interact with DNA-bound TBP, the C and N halves of Brf binding, respectively, to N- and C-proximal lobes of TBP. (2) The C-proximal half of Brf contributes most of the stability of the TFIIIB-DNA complex through its interactions with TBP and B″. The interacting region of Brf, which has been defined by deletion to 110 amino acids (but which could be smaller), comprises the fungal homology domain 2 (Fig. 1) (Khoo et al. 1994), which is also well conserved in the related proteins from worms and humans (Mital et al. 1996). (3) The N-proximal half of Brf contributes the primary competence for transcription but binds weakly to TBP and B″, generating TFIIIB-DNA complexes that can be detected by photochemical cross-linking, although they are not stable (or barely stable) to gel electrophoresis.

The principal binding sites on TBP for these two Brf fragments have been identified with the aid of specifically designed multiple mutants of TBP (Bryant et al. 1996; Shen et al. 1998) in a photochemical cross-linking experiment that is shown in Figure 2. Each TBP has a triple mutation that allows recognition of a TGTA box as well as the TATA box (Strubin and Struhl 1992). When TFIIIB is reconstituted with this variant TBP (designated TBPm₃), it binds unidirectionally to an *SNR6* promoter with a T*G*TAAATA site (Fig. 2, top) (Whitehall et al. 1995). (Unidirectional DNA binding is essential for un-

ambiguous mapping of protein-DNA proximity by photochemical cross-linking.) One TBP mutant, designated Brf⁻, has three additional mutations in its N-proximal lobe that separately weaken Brf binding and together diminish pol III transcription (Shen et al. 1998). The other TBP mutant, designated IIB⁻, has substitutions in its C-proximal lobe at three amino acids that are critical for TFIIB binding (Bryant et al. 1996), but these mutations do not diminish binding of intact Brf or reduce pol III transcription in vitro (Shen et al. 1998).

In the experiment shown in Figure 2 (bottom), proximity to DNA of B″, the amino-terminal (TFIIB-related) half of Brf and amino acids 435–545 of Brf, is probed at a site (base pair –28, on the nontranscribed strand) on the underside of the saddle-shaped TGTA box–TBPm₃ complex (Fig. 3c). The amino-terminal half of Brf and B″ lie in the vicinity of this DNA site when the TFIIIB-DNA complex is assembled with TBPm₃ (with no other mutations) (Fig. 2, bottom, lane 8). The amino acid 435–545 segment of Brf cross-links poorly at this DNA site, but it is as competent as full-length Brf in assembling B″ into the DNA complex and in bringing B″ into the vicinity of this site for photochemical cross-linking (lane 5 and data not shown; but see Kassavetis et al. 1998b); the assembly of the amino-terminal half of Brf into the DNA complex absolutely requires B″, but this half of Brf (reciprocally) only brings B″ into the vicinity of DNA further upstream of the TATA box (data not shown; see Kassavetis et al. 1998b). When the TFIIIB-DNA complex is assembled with the sextuple TBP-IIB⁻ mutant, cross-linking to the N-half of Brf (amino acids 1–282) drops out (lane 10). When the sextuple TBP-Brf⁻ mutant is used instead to reconstitute TFIIIB, cross-linking to the N-half of Brf is retained, but cross-linking to the amino acid 435–545 fragment is lost (lane 9). In each case, B″ is recruited to the TFIIIB-DNA complex (lanes 9 and 10; cross-linking of Brf [1–282] in lane 9 implies the presence of B″, and an identical experiment with photoreactive side chains upstream of the TATA box demonstrates this explicitly

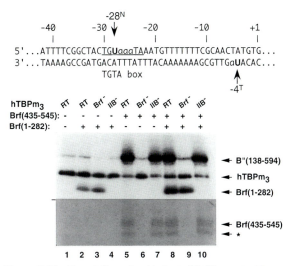

Figure 2. The N- and C-proximal halves of Brf interact with opposite ends of TBP. (*Top*) Photoactive DNA probes. The 62-bp DNA for the experiment shown below (extending from base pairs –58 to +4 of the *SNR6* gene; only base pairs –42 to +1 are shown) has an AT→GC substitution generating a TGTA box (*underlined*) and ABdUMP (5-[*N′*-(*p*-azidobenzoyl)-3-aminoallyl]dUMP; U) replacing dTMP on the nontranscribed strand at base pair –28; dAMP (lowercase a) at base pairs –29, –30, and –31 is ^{32}P-labeled. (*Bottom*) Mapping interactions by photochemical cross-linking. TFIIIB-DNA complexes have been formed with human TBPm$_3$ (the reference type "RT") or with triple mutants of hTBPm$_3$ also containing the amino acid substitutions R231E+R235E+R239S ("Brf$^-$" TBP; defective for binding Brf) or E284R+E286R+L287E ("IIB$^-$" TBP; defective for binding TFIIB and for pol II transcription). The B″ used to assemble TFIIIB-DNA complexes is deleted for amino acids 1–137, but is fully functional for transcription. Brf has been assembled from its amino-terminal 282-amino-acid fragment and the amino-acid 435–545 fragment, which comprises fungal homology domain 2. Cross-linked proteins are identified at the right (the asterisk identifies a fragment of Brf [435–545]). Because the Brf (435–545) fragment cross-links poorly, the photographic exposure of the bottom of the figure (below the mark at the left) has been increased to make the outcome of the analysis easier to see. (Adapted from Kassavetis et al. 1998b.)

Experimental Details: TFIIIB-DNA complexes were formed with 200 fmoles of hTBPm$_3$ (RT, Brf$^-$, or IIB$^-$, as defined above), 300 fmoles each of Brf (1–282) and Brf (435–545), as indicated above each lane, 200 fmoles of B″ (138–594), 8 fmoles of photoactive DNA (–28^N), and 100 ng of poly(dG-dC):poly(dG-dC) as carrier, in a 20-μl volume of buffer containing 40 mM Tris-Cl (pH 8.0), 50 mM NaCl, 7 mM MgCl$_2$, 3 mM 2-mercaptoethanol, 8%(v/v) glycerol, and 60 μg/ml bovine serum albumin. Nuclease treatment, denaturation, and SDS-PAGE allowed ^{32}P-labeled, cross-linked proteins to be resolved and identified by their apparent molecular weights. Samples were processed and analyzed as described and referenced in Kassavetis et al. (1998b).

[data not shown]). These results demonstrate that the N-proximal lobe of TBP interacts with the amino acid 435–545 segment of Brf, that the C-proximal lobe of TBP interacts with the N half of Brf (amino acids 1–282), and that each of these parts of Brf interacts with B″.

The preceding experiment and related analysis specify how the segments of Brf fit into the structure of the TFIIIB-DNA complex (Fig. 3a–c). Brf is intertwined with the TBP-DNA complex and anchored to both lobes of TBP on opposite sides of bent DNA. Brf fills a space

in the TFIIIB-DNA complex that is occupied in a pol II complex by TFIIB and TFIIA: Its N-proximal half is weakly attached to a site on TBP that is also important for binding TFIIB (Fig. 3c,d) and its C-proximal half binds strongly to a site on TBP that overlaps the TFIIA-attachment site (Fig. 3a). Two features of this model merit further comment: (1) The upstream-facing 34-kD subunit of pol III interacts with Brf (Werner et al. 1993), apparently in its TFIIB-homologous amino-terminal half (Khoo et al. 1994). Recent experiments also identify the 17-kD subunit of pol III as interacting with the amino-terminal half of Brf (A. Sentenac; S. Chédin et al.; both pers. comm.). The fact that TFIIB (which interacts specifically with pol II) and Brf occupy similar spaces in the pol II and pol III promoter complexes (Fig. 3c,d) is consistent with the orientation of TBP on DNA determining the same direction of transcription for yeast pol II and pol III (Whitehall et al. 1995). (2) To the extent that the required promoter-marking and polymerase-recruitment capabilities of TFIIIB are incorporated into TBP and Brf, it is remarkable that B″ should be absolutely required for all pol III transcription. This issue is addressed in experiments (described below) that show two surprising results: Promoter-bound Brf (as a Brf-TBP-DNA complex) is nearly ineffectual in bringing pol III to the promoter in the absence of B″, and promoter-bound TFIIIB has a role in transcriptional initiation that transcends passive pol III recruitment.

Multiple Bends Imposed on DNA by TFIIIB

The description of how Brf fits into the TFIIIB-DNA complex (Fig. 3a–c) is drawn on a DNA frame based on the known structures of the TBP-DNA complex (J.L. Kim et al. 1993; Y. Kim et al. 1993; Juo et al. 1996). However, indications that TFIIIB imposes additional distortions on its bound DNA are provided by older experiments that examined the effects on gel electrophoretic mobility of placing intrinsically bent DNA in varying helical phase (Crothers et al. 1991) relative to a 5S RNA gene transcription initiation complex. These experiments indicated that incorporation of B″ into a Brf- and TBP-containing DNA complex further increases DNA distortion and that the direction of the additional DNA bending is such as to move DNA lying upstream and downstream from the TFIIIB-binding site closer together (Braun et al. 1992).

Further evidence for additional bending in the TFIIIB-DNA complex comes from recent experiments examining the effect on assembly of a TFIIIB complex of introducing points of flexibility into DNA with a suboptimal TATA box. The formation of TFIIIB-DNA complexes on DNA is favored when flexibility is introduced between the TATA box and the transcriptional start site (by using hydroxyl radical to remove a single nucleoside). Introducing 4-nucleotide loops (generated by tandem mismatches of identical opposing nucleotides) downstream from a weak TATA box also favors assembly of the TFIIIB-DNA complex. The fact that increasing DNA flexibility downstream from the TATA box facilitates TFIIIB complex formation suggests that TFIIIB assem-

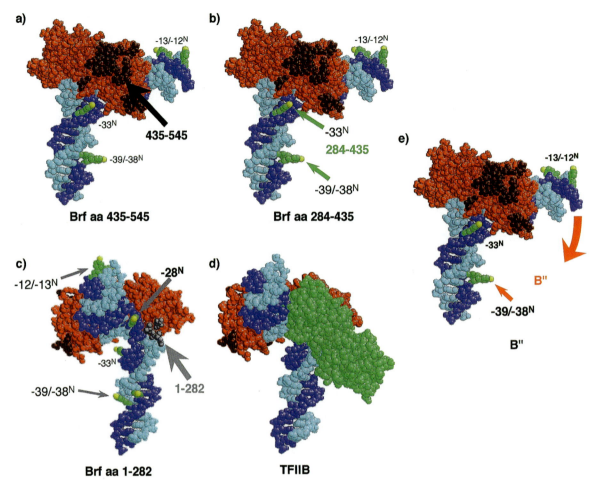

Figure 3. Components and interactions of the TFIIIB-DNA complex. (*a–c*) Locations of three segments of Brf in the complex. Interactions with DNA and TBP are projected onto the structure of the *Arabidopsis thaliana* TBP-DNA complex (J.L. Kim et al. 1993) with TBP (*red*) and DNA (*blue*) (nontranscribed strand [N] darker, and transcribed strand lighter). The photoactive ABdUMP side chains (*green*; with the photochemically generated nitrene in *yellow*) at the indicated locations (+1 is the transcriptional start site) project out of the DNA major groove. (*Maroon*) A surface patch on TBP that is responsible for the principal interaction with Brf; (*gray*) patch responsible for a secondary Brf interaction on the other TBP lobe. (*a*) The amino acid 435–545 Brf segment interacts with TBP as shown (*arrow*) and also anchors B'' into the TFIIIB-DNA complex (not indicated). (*b*) The amino acid 284–435 segment is located in vicinity to DNA upstream of the TATA box (*arrows*) but is also weakly cross-linked from the downstream end of the DNA site (at –13 and –12 of the nontranscribed strand). (*c*) The amino-terminal TFIIB-related half of Brf (amino acids 1–282) interacts relatively weakly with the indicated surface patch on TBP (*heavy arrow*) and is principally accessible for photochemical cross-linking from DNA sites (*smaller arrow* and the two *smallest arrows*), suggesting a general disposition of this segment of Brf in its DNA complex that is TFIIB-like. Note that Brf (1–282) and Brf (435–545) reach the two lobes of TBP on opposite faces of the TBP-DNA complex. (*d*) View of the human TFIIB-*A. thaliana* TBP-DNA complex (Nikolov et al. 1995), in the same aspect as panel *c*, indicating occupancy of the same space by TFIIB and Brf (1–282). (*e*) Entry of B'' into the TFIIIB-DNA complex further bends the DNA, compacting the structure of the TFIIIB-DNA complex (*curved arrow*). The principal site of cross-linking of B'' to DNA is located near the upstream edge of the complex (*short arrow*), although weaker cross-linking is detected along the entire length of the DNA-binding site of TFIIIB. (Adapted from Kassavetis et al. 1998b.)

bly imposes further bending on DNA that is already contorted by TBP (Grove et al. 1999).

TFIIIB Forms a Cage for DNA: An Interpretation of the Extraordinary Stability of TFIIIB-DNA Complexes

The fully assembled yeast TFIIIB-DNA complex is stable to high concentrations of simple electrolytes (e.g., 1 M NaCl) and to displacement by polyanions such a heparin (Kassavetis et al. 1989, 1990). This special stability is specifically contributed by B''; the TBP-Brf-DNA complex is not comparably stable. It is remarkable that this stability of the TFIIIB complex is equally manifested at optimal sites for TFIIIB-DNA binding (such as the TATAAATA site of the *SNR6* gene), at GC-rich sites entirely lacking a TATA box and therefore completely dependent on TFIIIC for TFIIIB-DNA complex assembly, and at sites with entirely diverse TATA-flanking sequence (Joazeiro et al. 1996).

To our knowledge, such very high DNA-sequence-independent stability has no counterpart in protein-DNA interactions: Other especially high-affinity protein-DNA complexes that are also resistant to dissociation by salts and polyanions are known, but changes of sequence at their specific DNA sites invariably diminish complex stability together with affinity. At least a part of the loss of stability arises from the ability of DNA-binding proteins to scan DNA sequence by sliding. When energy barriers to base pair by base pair sliding are relatively low, neighboring sequences are readily accessed, and only the deeper energy well of the highly preferred DNA site diminishes the rate of lateral transfer of the protein ligand to the immediate neighbor DNA site. As a consequence, flanking DNA becomes a conduit for dissociation as well as formation of the specific DNA complex by proteins that can slide along DNA (for review, see von Hippel and Berg 1989).

The structure of the TBP-DNA complex, with its sharp and compound bends (J.L. Kim et al. 1993; Y. Kim et al. 1993; Juo et al. 1996), implies that facile sliding cannot be a significant part of its reaction pathway to formation or dissociation; the energy barrier for a one-base-pair translocation of a fully formed TBP-DNA complex must be quite high. Recent experiments on the effects of DNA flexure at the TATA box on the stability of TBP-DNA complexes reinforce this conclusion (Grove et al. 1998). In these experiments, DNA flexure has been manipulated at the sites of TBP-mediated sharp DNA kinking in two ways: by replacing T-A base pair steps with 5-hydroxymethyluracil(h)-A steps, and by the appropriate placement of 4-nucleotide loops (cf Kahn et al. 1994). The deformability of DNA is known to be sequence-dependent (Travers 1991; Bailly et al. 1996); the T-A step is a site of propensity for DNA kinking (Wolffe and Drew 1995), although it is clearly not the only base pair step at which proteins introduce kinks into DNA (Werner et al. 1996). Experiments with other DNA-bending proteins have suggested that h-for-T substitution at the T-A step increases DNA deformability and particularly favors binding of proteins that sharply bend DNA (Grove et al. 1996, 1997).

These experiments on DNA flexure and TBP binding show a striking effect: The affinity and stability of the TBP-DNA complex can be increased greatly (by two orders of magnitude for the most favorable example) by changing the compliance of DNA at its two sites of kinking by TBP. For example (at 20ºC, in buffer containing $MgCl_2$ and contributing an ionic strength of ~0.12), the mean residence time, $t_{1/2}$, of TBP in a DNA complex is changed from less than 10 minutes to more than 10 hours when h replaces T at just two sites in each DNA strand (Table 1) (Grove et al. 1998).

The conclusion that can be drawn from these observations is that dissociation may be sterically hindered when a protein cannot slide out of a sharply bent and contorted DNA site, when the DNA must unkink as it leaves its protein ligand, and when the DNA is further contorted, as happens when the TBP-DNA complex is converted to the TFIIIB-DNA complex. The resulting steric restrictions on dissociation can be compounded, and dissociation further slowed, by placing protein obstacles along the path of dissociating DNA. Since the obstructing protein does not have to bind directly to DNA (it serves as a cage for the DNA, not its shackles), the stability of the resulting DNA confinement is determined by protein-protein interactions that hold the protein obstruction(s) in place.

The preceding considerations suggest that the special stability of the TBIIIB-DNA complex arises in the following manner: (1) TBP provides the principal distortion of DNA, and B″ imposes an additional bend. (2) B″ also completes a scaffold around the DNA that sterically blocks DNA dissociation. (3) A major contribution to the stability of this scaffold is made by protein-protein interactions. Specific DNA sequence may be recognized by Brf and B″ (compare recent observations on TFIIB and archaeal TFB by Lagrange et al. [1998] and Qureshi and Jackson [1998]), but this is clearly not required to stabilize the scaffold of the TFIIIB-DNA complex (Joazeiro et al. 1996). (4) The protein-protein contacts that retain the scaffold around DNA are stable to dissociation by neutral salts but are sensitive to modest concentrations of chaotropic electrolyte (Kassavetis et al. 1989).

Table 1. Stabilization of TBP-DNA Complexes by a Site-specific Change of the Deformability of DNA

DNA		k_{off} (S^{-1})	k_{on} (M^{-1}s^{-1})	K_d (M)
All-T:	5′-CGTGACTAC**TATAAATA**AATGATCCG-3′	1.1×10^{-3}	2.8×10^5	3.9×10^{-9}
Two-hA:	5′-CGTGACTAC**hAATAAhA**AATGATCCG-3′	1.5×10^{-5}	4.7×10^5	3.2×10^{-11}

The association and dissociation rate constants, k_{on} and k_{off}, respectively, of full-length yeast TBP with DNA, and the derived equilibrium dissociation constant ($K_d^{app} = k_{off}/k_{on}$) were determined for 26-mer duplex DNA containing the TATA box (TATAAATA) of the *S. cerevisiae SNR6* (U6 snRNA) gene. For DNA Two-hA, T-A base pair steps at the two sites of sharp DNA kinking by TBP (J.L. Kim et al. 1993; Y. Kim et al. 1993; Juo et al. 1996) were replaced by 5-hydroxymethyluracil(h)-A steps. (Adapted from Grove et al. 1998.)

Experimental Details: Measured at 20ºC in buffer containing 40 mM Tris-Cl (pH 8.0), 90 mM NaCl, 7 mM $MgCl_2$, 3 mM dithiothreitol, 100 µg/ml bovine serum albumin, and 8%(v/v) glycerol. Rates of association were determined with TBP at varying stoichiometric excess (10–80 nM final concentration after tenfold dilution from stock solutions) added to 1 nM ^{32}P-labeled DNA and incubated for varying lengths of time. Binding was then quenched with 200 ng of poly(dA-dT):poly(dA-dT) per 100 fmoles of TBP, and samples were immediately applied for gel electrophoresis in low-ionic-strength buffer containing $MgCl_2$ (to stabilize TBP-DNA complexes). The fractional formation of competitor-stable TBP complexes was corrected for dissociation in the gel (detectable only for all-T DNA) (Hoopes et al. 1992) yielding an apparent first-order rate constant k_{obs}. The second-order association rate constant k_{on} was determined from the linear regression of k_{obs}^{-1} against [TBP]$^{-1}$. For rates of TBP-DNA complex dissociation in solution, preformed TBP-DNA complexes were challenged with a high excess of poly(dA-dT):poly(dA-dT) for varying times (0– 20 min for all-T DNA and up to 24 hr for Two-hA DNA) before loading for gel electrophoresis. For other details, see Grove et al. (1998).

FUNCTIONS IN TRANSCRIPTIONAL INITIATION

A Postrecruitment Role for TFIIIB in Initiation of Transcription

Deleting the N-proximal one third of Brf (amino acids 1–164) only partly inhibits *SNR6* transcription in supercoiled DNA (Kassavetis et al. 1997); individual small internal deletions of B˝ and more extensive external deletions also do not prevent *SNR6* transcription in supercoiled DNA, as summarized above. For transcription of the *SNR6* gene in linear DNA, the outcome is drastically different: (1) Brf (165–596) is now totally defective for transcription and (2) a survey of B˝ mutants with small internal deletions exposes several, covering an internal approximately 70-amino-acid segment, that are also transcriptionally inactive (Fig. 1). In contrast, under the conditions of these assays, TFIIIB assembled with full-length Brf and B˝ is equally active for transcription of the *SNR6* gene in supercoiled and linear DNAs. The involvement of DNA structure in the transcription defect is emphasized by a further observation: The transcription defect is partly suppressed when the linear DNA is made more deformable. Evidently, a change in DNA that facilitates assembly of the TFIIIB-DNA complex also modifies the impact of these Brf and B˝ deletions on transcription.

To identify the step of transcriptional initiation at which these defective TFIIIB-promoter assemblies fail, we examined pol III recruitment, promoter opening, and abortive initiation. pol III recruitment to linear DNA was readily detected by photochemical cross-linking (Fig. 4) and also by gel retardation.

A control in the experiment shown in Figure 4 yields its own, already referred to, striking outcome: In the absence of B˝, promoter-bound Brf is ineffective in bringing pol III to the promoter (Fig. 4, lane 1). What little cross-linking of pol III subunits occurs (too little to detect in the reproduced figure) appears to reflect an inappropriate orientation or random binding, as judged from the fact that the 82- and 53-kD subunits, which would not be expected to cross-link to DNA upstream of the transcriptional start site (cf Bartholomew et al. 1993), are detected along with the 160-, 128-, and 34-kD subunits. Brf is known to interact with the 34-kD pol III subunit (Werner et al. 1993; Khoo et al. 1994). Since the latter is located at the upstream end of promoter-bound pol III (Bartholomew et al. 1993), and since C34 is one of the pol-III-specific subunits (Brun et al. 1997), it has been assumed that pol III recruitment would take place through this Brf-C34 interaction. A second interaction of the amino-terminal half of Brf, with the pol-III-specific 17-kD subunit (C17), has been identified recently (S. Chédin et al.; A. Sentenac, both pers. comm.). Instead, we see that recruitment does not happen until B˝ joins the TFIIIB-DNA complex (Fig. 4, lane 3). This implies that (1) the pol-III interaction site on Brf is cryptic until B˝ comes along, or (2) the TFIIIB-pol III interaction involves multiple points of contact that are brought into their required register by B˝, or (3) B˝ contributes an essential TFIIIB-pol III contact.

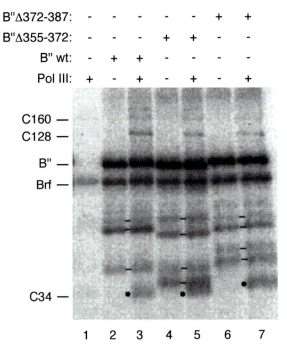

Figure 4. Recruitment of pol III to transcriptionally active and inactive initiation complexes, analyzed by photochemical cross-linking. The 88-bp DNA, extending from base pairs –56 to +32 of an *SNR6*-derived transcription unit has a TGTA box and AB-dUMP(U) substituting for dTMP in the transcribed strand at base pair –4 (Fig. 2, top). TBPm3 has been substituted for wild-type TBP in order to secure unidirectional assembly of TFIIIB and recruitment of pol III over the transcriptional start site at +1 (Kassavetis et al. 1998a). Full-length Brf and TBPm3 are present in all reaction mixtures. The presence of pol III and full-length B˝ or B˝ with deletions covering amino acids 355–372 or 372–387 is indicated above each lane. Proteins that cross-link to DNA are identified at the side. C160 and C128, are the two largest subunits of pol III, and C34 (*closed circle*) is a Brf-interacting subunit. Contaminating B˝ fragments that cross-link to DNA (–) are not material to the outcome of the analysis.

Experimental Details: Protein-DNA complexes were formed in a 20-µl volume of 40 mM Tris-Cl (pH 8.0), 50 mM NaCl, 7 mM MgCl$_2$, 3 mM 2-mercaptoethanol, 6%(v/v) glycerol, 60 µg/ml bovine serum albumin, 5 µg/ml poly(dG-dC):poly(dG-dC), with 8 fmoles of ^{32}P-labeled photoactive DNA, 400 fmoles of TBPm3, 350 fmoles of Brf, 60–90 fmoles of the specified B˝, and 10 fmoles of pol III. TFIIIB-DNA or TBP-Brf-DNA complexes were formed first, and sheared salmon sperm DNA was added to 5 µg/ml for 5 min and was followed by pol III for an additional 10 min, prior to UV irradiation.

In contrast, the transcription-defective TFIIIB-linear DNA complexes assembled with B˝Δ355–372 or B˝Δ372–387 recruit pol III relatively well (although not as well as intact, normal TFIIIB). Only minor differences are discernible between the DNA cross-linking patterns (Fig. 4, compare lanes 5 and 7 with lane 3) or DNase I footprints (not shown; see Kassavetis et al. 1998a) of the transcriptionally competent and inactive TFIIIB-DNA complexes. The failure of the transcriptionally inactive promoter complex materializes at a subsequent step: The transcription bubble fails to open and there is no abortive or productive initiation of transcription (Kassavetis et al. 1998a). These are the properties that define the closed state of the *Esch-*

erichia coli RNA polymerase promoter complex. The combination of linear DNA with, respectively, the truncated Brf or B″ with one of three contiguous small internal deletions generates defective TFIIIB-DNA complexes that recruit pol III to the closed promoter but fail to support a subsequent step of transcriptional initiation.

Why should the success or failure of pol III promoter complexes to initiate transcription revolve around the choice of linear versus supercoiled DNA or around the flexibility of DNA? We propose that DNA supercoiling helps to align parts of the TFIIIB complex for interaction with pol III. We presume that such an alignment secures multipoint contacts with pol III that are required for transcriptional initiation. Because the required alignment is also generated by protein-protein interactions within TFIIIB, the structure and function of the TFIIIB-DNA complex are overspecified, and the involvement of DNA supercoiling is redundant in the context of wild-type TFIIIB. When deletions that eliminate or distort these supporting protein-protein contacts are introduced into Brf or B″, the contribution of DNA supercoiling (or flexure) to alignment becomes determining for transcriptional activity.

The use of multipoint contacts to reinforce polymerase recruitment has been well demonstrated for prokaryotic transcription (Busby and Ebright 1997). On linear DNA, we find that pol III recruitment by the transcriptionally defective TFIIIB[Brf(165–596)]-DNA and TFIIIB[B″ (Δ355–372)]-DNA complexes is diminished, compared to the transcriptionally active complete TFIIIB-linear-DNA complex. This is consistent with the notion that recruitment directly benefits from multiple pol III-TFIIIB contacts. However, the key observation is that the recruited pol III does not open the promoter and is *qualitatively* transcriptionally inactive. Attempts to rescue the defect by lowering ionic strength or changing temperature have been unsuccessful; the failure of transcriptional activity is definitive, and it transcends polymerase recruitment.

The implication of this finding is that a proper alignment of contact sites for pol III on the surface of the TFIIIB-DNA complex facilitates passage of the promoter-bound polymerase along its reaction pathway to transcriptional initiation. One can envisage different possibilities (which are not mutually exclusive) for this involvement: (1) DNA binding aligns an array of sites on a pol-III-interacting surface of TFIIIB. The interaction of these sites with the polymerase favors a subsequent isomerization of the latter; only isomerized pol III is able to proceed to promoter opening and transcriptional initiation. (Since pol III is capable of autonomously initiating transcription at the ends of linear duplex DNA with overhanging 3′ single-stranded ends, this interpretation suggests that an equivalent isomerization of the polymerase may be induced through its interaction with single-stranded DNA ends.) The defective TFIIIB-DNA complexes do not furnish the appropriate contact sites for the required polymerase isomerization and accordingly do not support progression to promoter opening. (2) New contact sites that actively lock the polymerase into its pre-

isomerized configuration may be exposed on the surface of defective TFIIIB-DNA complexes, blocking the reaction path to transcriptional initiation. (3) The preceding proposals assume a static TFIIIB-DNA complex. It is also possible that the TFIIIB-DNA complex changes structure in a way that allows TFIIIB to make different pol III contacts at successive steps of the reaction pathway to transcriptional initiation and that the defective TFIIIB assemblies on linear DNA fail to support this concerted TFIIIB-pol III reaction sequence.

Does the existence of holoenzymes change these considerations? Firm evidence for the existence of a human pol III holoenzyme containing components of TFIIIB and TFIIIC has been presented recently (Wang et al. 1997). A corresponding yeast enzyme has not yet been identified as a biochemically defined entity, but the concentrations of TFIIIB components in the nucleus, which are in the micromolar range (Sethy-Coraci et al. 1998), and the generally macromolecular crowding conditions of nucleoplasm make its existence in vivo probable. However, the question of whether the assembly of (wild type) TFIIIB-pol III complex takes place only on DNA or whether these two multiprotein complexes arrive at their target genes preassembled is essentially independent of the consideration of how the TFIIIB-pol III interaction guides the reaction path to transcriptional initiation.

Genetic analysis of bacterial RNA polymerases has contributed to understanding the reaction pathway to transcriptional initiation by identifying mutations that block progress along that pathway at different steps (Gross et al. 1996; Li et al. 1997; Oguiza and Buck 1997; Severinov and Darst 1997; Wilson and Dombroski 1997). It is likely that a comparable analysis of pol III will contribute to a further dissection of its transcriptional initiation pathway. In fact, an interesting pol III mutation that makes promoter opening cold-sensitive has been identified in the Brf-interacting 34-kD subunit (Brun et al. 1997).

Comparison with Postrecruitment Events in Activation of Transcription by Bacterial RNA Polymerases

The preceding interpretation of our observations about pol III recruitment to transcriptionally inactive promoter complexes provokes comparisons with mechanisms of transcriptional regulation in bacteria.

E. coli RNA polymerase binds to many of its strongest promoters autonomously and avidly, opens the DNA duplex around the transcriptional start site spontaneously, and initiates transcription rapidly without intervention of additional (regulatory) proteins (McClure 1985; Record et al. 1996). At many other promoters, polymerase binding is assisted by one or more transcriptional activators, but the subsequent steps leading to promoter opening and initiation of transcription are largely autonomous. At a third class of promoters, RNA polymerase requires transcriptional activators to assist in some step after binding to the promoter. The best-studied of these promoters readily bind the σ^{54}-RNA polymerase holoenzyme

($E.\sigma^{54}$), forming closed complexes that absolutely require an activator for transcriptional initiation. The σ^{54} subunit may directly dictate this activator control by preventing the polymerase from undergoing a step of protein isomerization that is prerequisite to promoter opening (Record et al. 1996; Craig et al. 1998); the activator is required to free the promoter-bound polymerase from a σ^{54}-imposed block (Wang et al. 1995; Wedel and Kustu 1995; North et al. 1996; Wyman et al. 1997).

The σ^{70}-containing principal *E. coli* RNA polymerase ($E.\sigma^{70}$) and its *Bacillus subtilis* homolog ($E.\sigma^{A}$) also appear to be capable of locking in inactive preinitiation states at some promoters in the absence of a suitable activator protein (Miller et al. 1997; Rowe-Magnus and Spiegelman 1998). Activation of transcription at phage N4 late promoters, which are transcribed by $E.\sigma^{70}$, requires the phage N4 single-stranded DNA-binding protein. Nevertheless, a mutant N4 single-stranded DNA-binding protein that is defective for DNA binding and, as a consequence, a priori incapable of directing recruitment to the promoter, is fully functional as a transcriptional activator (Miller et al. 1997). Although the precise step at which the N4 single-stranded DNA-binding protein generates its activation has not yet been determined, it most probably follows upon polymerase binding. It appears that the *B. subtilis* Spo0A activator protein also is not required to bind $E.\sigma^{A}$ to its Spo0IIG promoter region (Bird et al. 1996) but is required for promoter opening (Rowe-Magnus and Spiegelman 1998). For these two cases, and for other activators that may exert their critical effects on transcription at a postrecruitment step (Caslake et al. 1997; Sanders et al. 1997), the evidence is not as complete as it is for σ^{54}. However, it appears likely that diverse RNA polymerase holoenzymes, and not just $E.\sigma^{54}$, readily bind to, and form closed complexes with, certain of their conjugate promoters but require some effector for executing a subsequent step of transcriptional initiation.

The data that are summarized in the preceding section show that the ability to bind to a promoter in an inactive state is a property that is shared by the bacterial RNA polymerases and pol III. We suggest that TFIIIB serves dual functions for pol III that activators of bacterial transcription also fulfill (in a global sense): TFIIIB clearly and patently recruits pol III to the promoter. It also ensures passage of the polymerase-promoter complex along the subsequent reaction pathway to transcriptional initiation. These recruitment and postrecruitment functions of TFIIIB have been separated by mutation in our experiments.

Recently published observations suggest that pol II may follow the same reaction pathway to transcriptional initiation. In a minimal in vitro transcription system composed of plasmid DNA, TBP, TFIIB, and TFIIF as well as pol II, the transcriptional-positive coactivator PC4 strongly inhibits transcription, but does not prevent formation of a TBP-TFIIB-(sc)DNA-pol II complex (Malik et al. 1998). If it could be shown that transcriptionally silent pol II is correctly placed over the transcriptional start site of these complexes, that would establish a correspondence between the PC4-inhibited TBP-TFIIB promoter complex and polymerase-recruiting but transcrip-tion-defective TFIIIB-DNA complexes that can be generated by mutation.

ACKNOWLEDGMENTS

We are grateful to E. Ramirez and G.A. Letts for skillful assistance with experiments that are reported here. Our research has been supported by a grant from the NIGMS.

REFERENCES

Bailly C., Payet D., Travers A.A., and Waring M.J. 1996. PCR-based development of DNA substrates containing modified bases: An efficient system for investigating the role of the exocyclic groups in chemical and structural recognition by minor groove binding drugs and proteins. *Proc. Natl. Acad. Sci.* **93:** 13623.

Bartholomew B., Durkovich D., Kassavetis G.A., and Geiduschek E.P. 1993. Orientation and topography of RNA polymerase III in transcription complexes. *Mol. Cell. Biol.* **13:** 942.

Bird T.H., Grimsley J.K., Hoch J.A., and Spiegelman G.B. 1996. The *Bacillus subtilis* response regulator Spo0A stimulates transcription of the spoIIG operon through modification of RNA polymerase promoter complexes. *J. Mol. Biol.* **256:** 436.

Braun B.R., Kassavetis G.A., and Geiduschek E.P. 1992. Bending of the *Saccharomyces cerevisiae* 5S rRNA gene in transcription factor complexes. *J. Biol. Chem.* **267:** 22562.

Brun I., Sentenac A., and Werner M. 1997. Dual role of the C34 subunit of RNA polymerase III in transcription initiation. *EMBO. J.* **16:** 5730.

Bryant G.O., Martel L.S., Burley S.K., and Berk A.J. 1996. Radical mutations reveal TATA-box binding protein surfaces required for activated transcription *in vivo*. *Genes Dev.* **10:** 2491.

Busby S. and Ebright R.H. 1997. Transcription activation at class II CAP-dependent promoters. *Mol. Microbiol.* **23:** 853.

Caslake L.F., Ashraf S.I., and Summers A.O. 1997. Mutations in the alpha and sigma-70 subunits of RNA polymerase affect expression of the mer operon. *J. Bacteriol.* **179:** 1787.

Craig M.L., Tsodikov O.V., McQuade K.L., Schlax P.E., Jr., Capp M.W., Saecker R.M., and Record M.T., Jr. 1998. DNA footprints of the two kinetically-significant intermediates in formation of an RNA polymerase-promoter open complex: Evidence that interactions with start site and downstream DNA induce sequential conformational changes in polymerase and DNA. *J. Mol. Biol.* **283:** 741.

Crothers D.M., Gartenberg M.R., and Shrader T.E. 1991. DNA bending in protein-DNA complexes. *Methods Enzymol.* **208:** 118.

Geiduschek E.P. and Kassavetis G.A. 1995. Comparing transcriptional initiation by RNA polymerases I and III. *Curr. Opin. Cell Biol.* **7:** 344.

Gross C.A., Chan C.L., and Lonetto M.A. 1996. A structure/function analysis of *Escherichia coli* RNA polymerase. *Philos. Trans. R. Soc. Lond. B Biol. Sci.* **351:** 475.

Grove A., Galeone A., Mayol L., and Geiduschek E.P. 1996. Localized DNA flexibility contributes to target site selection by DNA-bending proteins. *J. Mol. Biol.* **260:** 120.

Grove A., Figueiredo M., Galeone A., Mayol L., and Geiduschek E.P. 1997. Twin hydroxymethyluracil-A basepair steps define the binding site for the DNA-bending protein TF1. *J. Biol. Chem.* **272:** 13084.

Grove A., Galeone A., Yu E., Mayol L., and Geiduschek E.P. 1998. Affinity, stability and polarity of binding of the TATA binding protein governed by flexure at the TATA box. *J. Mol. Biol.* **282:** 731.

Grove A., Kassavetis G.A., Johnson T.E., and Geiduschek E.P. 1999. The RNA polymerase III-recruiting factor TFIIIB induces a DNA bend between the TATA box and the transcriptional start site. *J. Mol. Biol.* (in press).

Hoopes B.C., LeBlanc J.F., and Hawley, D.K. 1992. Kinetic analysis of yeast TFIID-TATA box complex formation suggests a multi-step pathway. *J. Biol. Chem.* **267:** 11539.

Joazeiro C.A.P., Kassavetis G.A., and Geiduschek E.P. 1996. Alternative outcomes in assembly of promoter complexes: The roles of TBP and a flexible linker in placing TFIIIB on tRNA genes. *Genes Dev.* **10:** 725.

Juo Z.S., Chiu T.K., Leiberman P.M., Baikalov I., Berk A.J., and Dickerson R.E. 1996. How proteins recognize the TATA box. *J. Mol. Biol.* **261:** 239.

Kahn, J.D., Yun, E., and Crothers, D.M. 1994. Detection of localized DNA flexibility. *Nature* **368:** 163.

Kassavetis G.A., Blanco J.A., Johnson T.E., and Geiduschek E.P. 1992. Formation of open and elongating transcription complexes by RNA polymerase III. *J. Mol. Biol.* **226:** 47.

Kassavetis G.A., Braun B.R., Nguyen L.H., and Geiduschek E.P. 1990. *S. cerevisiae* TFIIIB is the transcription initiation factor proper of RNA polymerase III, while TFIIIA and TFIIIC are assembly factors. *Cell* **60:** 235.

Kassavetis G.A., Kumar A., Letts G.A., and Geiduschek E.P. 1998a. A post-recruitment function for the RNA polymerase III transcription initiation factor TFIIIB. *Proc. Natl. Acad. Sci.* **95:** 9196.

Kassavetis G.A., Kumar A., Ramirez E., and Geiduschek E.P. 1998b. The functional and structural organization of Brf, the TFIIIB-related component of the RNA polymerase III transcription initiation complex. *Mol. Cell. Biol.* **18:** 5587.

Kassavetis G.A., Bardeleben C., Kumar A., Ramirez E., and Geiduschek E.P. 1997. Domains of the Brf component of RNA polymerase III transcription factor IIIB (TFIIIB): Functions in assembly of TFIIIB-DNA complexes and recruitment of RNA polymerase to the promoter. *Mol. Cell. Biol.* **17:** 5299.

Kassavetis G.A., Riggs D.L., Negri R., Nguyen L.H., and Geiduschek E.P. 1989. Transcription factor IIIB generates extended DNA interactions in RNA polymerase III transcription complexes on tRNA genes. *Mol. Cell. Biol.* **9:** 2551.

Khoo B., Brophy B., and Jackson S.P. 1994. Conserved functional domains of the RNA polymerase III general transcription factor BRF. *Genes Dev.* **8:** 2879.

Kim J.L., Nikolov D.B., and Burley S.K. 1993. Co-crystal structure of TBP recognizing the minor groove of a TATA element. *Nature* **365:** 520.

Kim Y., Geiger J.H., Hahn S., and Sigler P.B. 1993. Crystal structure of a yeast TBP/TATA-box complex. *Nature* **365:** 512.

Kumar A., Kassavetis G.A., Geiduschek E.P., Hambalko M., and Brent C.J. 1997. Functional dissection of the B″ component of RNA polymerase III transcription factor (TF)IIIB: A scaffolding protein with multiple roles in assembly and initiation of transcription. *Mol. Cell. Biol.* **17:** 1868.

Lagrange T., Kapanidis A.N., Tang H., Reinberg D., and Ebright R.H. 1998. New core promoter element in RNA polymerase II-dependent transcription: Sequence-specific DNA binding by transcription factor IIB. *Genes Dev.* **12:** 34.

Li M., McClure W.R., and Susskind M.M. 1997. Changing the mechanism of transcriptional activation by phage lambda repressor. *Proc. Natl. Acad. Sci.* **94:** 3691.

Malik S., Guermah M., and Roeder R.G. 1998. A dynamic model for PC4 coactivator function in RNA polymerase II transcription. *Proc. Natl. Acad. Sci.* **95:** 2192.

McClure W.R. 1985. Mechanism and control of transcription initiation in prokaryotes. *Annu. Rev. Biochem.* **54:** 171.

Miller A., Wood D., Ebright R.H., and Rothman-Denes L.B. 1997. RNA polymerase beta′ subunit: A target of DNA binding-independent activation. *Science* **275:** 1655.

Mital R., Kobayashi R., and Hernandez N. 1996. RNA polymerase III transcription from the human U6 and adenovirus type 2 VAI promoters has different requirements for human BRF, a subunit of human TFIIIB. *Mol. Cell. Biol.* **16:** 7031.

Nikolov D.B., Chen H., Halay E.D., Usheva A.A., Hisatake K., Lee D.K., Roeder R.G., and Burley S.K. 1995. Crystal structure of a TFIIB-TBP-TATA-element ternary complex. *Nature* **377:** 119.

North A.K., Weiss D.S., Suzuki H., Flashner Y., and Kustu S. 1996. Repressor forms of the enhancer-binding protein NrtC: Some fail in coupling ATP hydrolysis to open complex formation by sigma 54-holoenzyme. *J. Mol. Biol.* **260:** 317.

Oguiza J.A. and Buck M. 1997. DNA-binding domain mutants of sigma-N (sigmaN, sigma54) defective between closed and stable open promoter complex formation. *Mol. Microbiol.* **26:** 655.

Qureshi S.A. and Jackson S.P. 1998. Sequence-specific DNA binding by the *S. shibatae* TFIIB homolog, TFB, and its effect on promoter strength. *Mol. Cell* **1:** 389.

Record M.T., Jr., Reznikoff W.S., Craig M.L., McQuade K.L., and Schlax, P.J. 1996. *Escherichia coli* RNA polymerase (Eσ70), promoters, and the kinetics of the steps of transcription initiation. In Escherichia coli and Salmonella, 2nd edition (ed. F.C. Neidhardt et al.), p. 792. ASM Press, Washington, D.C.

Rowe-Magnus D.A. and Spiegelman G.B. 1998. DNA strand separation during activation of a developmental promoter by the *Bacillus subtilis* response regulator Spo0A. *Proc. Natl. Acad. Sci.* **95:** 5305.

Rowland O. and Segall J. 1998. A hydrophobic segment within the 81-amino-acid domain of TFIIIA from *Saccharomyces cerevisiae* is essential for its transcription factor activity. *Mol. Cell. Biol.* **18:** 420.

Sanders G.M., Kassavetis G.A., and Geiduschek E.P. 1997. Dual targets of a transcriptional activator that tracks on DNA. *EMBO J.* **16:** 3124.

Sethy-Coraci I., Moir R.D., López-de-León A., and Willis I.M. 1998. A differential response of wild type and mutant promoters to TFIIIB70 overexpression in vivo and in vitro. *Nucleic Acids Res.* **26:** 2344.

Severinov K. and Darst S.A. 1997. A mutant RNA polymerase that forms unusual open promoter complexes. *Proc. Natl. Acad. Sci.* **94:** 13481.

Shen Y., Kassavetis G.A., Bryant G.O., and Berk A.J. 1998. Polymerase (Pol) III TATA box-binding protein (TBP)-associated factor Brf binds to a surface on TBP also required for activated Pol II transcription. *Mol. Cell. Biol.* **18:** 1692.

Strubin M. and Struhl K. 1992. Yeast and human TFIID with altered DNA-binding specificity for TATA elements. *Cell* **68:** 721.

Travers A.A. 1991. DNA bending and kinking. *Curr. Opin. Struct. Biol.* **1:** 114.

von Hippel P.H. and Berg O.G. 1989. Facilitated target location in biological systems. *J. Biol. Chem.* **264:** 675.

Wang J.T., Syed A., Hsieh M., and Gralla J.D. 1995. Converting *Escherichia coli* RNA polymerase into an enhancer-responsive enzyme: Role of an NH2-terminal leucine patch in sigma 54. *Science* **270:** 992.

Wang Z., Luo T., and Roeder R.G. 1997. Identification of an autonomously initiating RNA polymerase III holoenzyme containing a novel factor that is selectively inactivated during protein synthesis inhibition. *Genes Dev.* **11:** 2371.

Wedel A. and Kustu S. 1995. The bacterial enhancer-binding protein NTRC is a molecular machine: ATP hydrolysis is coupled to transcriptional activation. *Genes Dev.* **9:** 2042.

Werner M.H., Gronenborn A.M., and Clore G.M. 1996. Intercalation, DNA kinking, and the control of transcription. *Science* **271:** 778.

Werner M., Chaussivert N., Willis I.M., and Sentenac A. 1993. Interaction between a complex of RNA polymerase III subunits and the 70-kDa component of transcription factor IIIB. *J. Biol. Chem.* **268:** 20721.

White R.J. 1998. *RNA polymerase III transcription,* 2nd edition. Springer-Verlag and Landes Bioscience, New York.

Whitehall S.K., Kassavetis G.A., and Geiduschek E.P. 1995. The symmetry of the yeast U6 RNA gene's TATA box and the orientation of the TATA-binding protein in yeast TFIIIB. *Genes Dev.* **9:** 2974.

Wilson C. and Dombroski A.J. 1997. Region 1 of sigma70 is required for efficient isomerization and initiation of transcription by *Escherichia coli* RNA polymerase. *J. Mol. Biol.* **267:** 60.

Wolffe A.P. and Drew H.R. 1995. DNA structure: Implications for chromatin structure and function. In *Chromatin structure and gene expression* (ed. S.C.R. Elgin), p. 27. IRL Press, Oxford, United Kingdom.

Wyman C., Rombel I., North A.K., Bustamante C., and Kustu S. 1997. Unusual oligomerization required for activity of NtrC, a bacterial enhancer-binding protein. *Science* **275:** 1658.

Strength and Regulation without Transcription Factors: Lessons from Bacterial rRNA Promoters

R.L. Gourse, T. Gaal, S.E. Aiyar, M.M. Barker, S.T. Estrem, C.A. Hirvonen, and W. Ross

Department of Bacteriology, University of Wisconsin, Madison, Wisconsin 53706

The determinants responsible for transcription from a particular promoter at a specific time in growth or development are most typically protein factors that interact directly or indirectly with RNA polymerase (RNAP). However, during the last few years, our studies on ribosomal RNA transcription in *Escherichia coli* have led us to appreciate the role of the basal transcription apparatus (i.e., the promoter, RNAP, and RNAP's nucleoside triphosphate substrates) as major determinants of transcription and its regulation. Although there are protein factors that influence *E. coli* rRNA transcription by interacting directly with the basal transcription apparatus, major features of rRNA transcription, i.e., the extraordinary strength of rRNA promoters and the regulation of transcription initiation with the growth rate, are determined (at least in the *rrnB* operon) primarily by features intrinsic to the promoter-RNAP interaction itself and by the concentrations of the initiating nucleoside triphosphates (NTPs) in the cell. These features turn out to be relevant not only to transcription of rRNA, but to transcription of other promoters as well.

This paper focuses on the roles of interactions between the α-subunit of RNAP and the DNA region upstream of the core promoter in bacterial transcription and on the role of RNAP-promoter complex stability in transcription regulation. We make three major points: (1) UP element–α-subunit interactions play an important part in promoter strength in bacteria, and this is especially important in the transcription of rRNA promoters. (2) The interaction between the carboxy-terminal domain of RNAP (αCTD) and DNA is crucial for understanding the mechanisms by which transcription factors interact with the transcription machinery. (3) Initiation complex stability has an important role in the regulation of rRNA promoters by making them sensitive to changing concentrations of the initiating NTP and to the effects of the nucleotide effector ppGpp. We also propose that the extreme stability of certain other promoter-RNAP complexes, compared to the instability of rRNA promoter complexes, at least in part determines the positive response of these other promoters to the effects of ppGpp.

OVERVIEW OF rRNA TRANSCRIPTION IN *E. COLI*

In *E. coli*, the seven rRNA operons can account for well over half the transcription in the cell, producing more RNA than all of the other promoters in the cell combined when cells are dividing rapidly. There are two pro-

moters in each operon, *rrn* P1 and *rrn* P2, with the P1 promoter providing most of the transcription at fast growth rates (Fig. 1). During the past decade or so, we have dissected the *rrnB* P1 promoter using a variety of approaches (Gourse et al. 1986; Dickson et al. 1989; Gaal et al. 1989; Ross et al. 1990, 1993; Leirmo and Gourse 1991; Newlands et al. 1991, 1992, 1993; Bartlett and Gourse 1994; Rao et al. 1994). We discovered that specific recognition of *rrn* P1 promoters by RNAP involves not only the core promoter region (which includes the –10 and –35 hexamers known for many years to interact with the σ-subunit), but also sequences upstream of the –35 hexamer, primarily between about –40 and –60. This upstream region, which we named the UP element, increases transcription dramatically (Ross et al. 1993; Rao et al. 1994; Estrem et al. 1998). Surprisingly, we found that the UP element interacts not with σ, but with the α-subunit of RNAP, specifically with the αCTD (Ross et al. 1993; Blatter et al. 1994). Our studies on UP elements and their interactions with α have led to a new appreciation for the role of this interaction in bacterial transcription in general.

Although we do not focus on the role of transcription factors in this paper, we do not mean to imply that they are irrelevant to rRNA transcription. The transcription factor, FIS (*f*actor for *i*nversion *s*timulation), increases the strength of rRNA promoters. In *rrnB* P1, FIS binds to three sites located between –60 and –150 and increases transcription about fivefold (Ross et al. 1990), and FIS

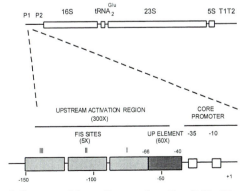

Figure 1. Structure of the *rrnB* operon from *E. coli*. The P1 promoter region is expanded to illustrate the presence of FIS-binding sites (which increase transcription about 5-fold), the UP element (which increases transcription at least 60-fold), and the core promoter region (which includes the –10 and –35 hexamers).

binds upstream of *rrnE* P1 and increases transcription even more (C.A. Hirvonen and R.L. Gourse, unpubl.). In fact, the architecture of the initiation complex at *rrn* P1 promoters is representative of a common motif in bacterial transcription (Bokal et al. 1997). *E. coli* rRNA transcription is also subject to the effects of Nus antitermination factors that bind to the nascent rRNA downstream from the P2 promoter region. These proteins prevent premature transcription termination, presumably in a manner analogous to the way that complexes containing the same factors (and the λN protein) prevent transcription termination in bacteriophage λ (Condon et al. 1995).

The *rrn* P2 promoters, situated approximately 120 bp downstream from the *rrn* P1 promoters in all seven rRNA operons, have not been studied as thoroughly as the P1 promoters. They are clearly less active than the P1 promoters during rapid cell growth, but they are the more significant contributors to rRNA transcription when cells are growing slowly (Sarmientos and Cashel 1983). Like *rrn* P1 promoters, *rrnB* P2 (and probably other *rrn* P2 promoters) contains an UP element, although its effect on transcription is not as large as that of the *rrnB* P1 UP element (Ross et al. 1993, 1998).

Historically, regulation of rRNA transcription has been characterized in three ways: regulation in steady-state growth, inhibition by nutrient starvation, and responses to rapid changes in growth conditions. rRNA synthesis rates (and thus ribosome synthesis rates since rRNA transcription is the rate-limiting step in ribosome synthesis) increase approximately with the square of the steady-state growth rate (μ, doublings per hour), a relationship referred to as growth-rate-dependent control. This relationship was described 40 years ago and has been the subject of intensive investigation ever since (Schaechter et al. 1958; Maaloe and Kjeldgaard 1966; Jinks-Robertson et al. 1983; Gourse et al. 1996; Keener and Nomura 1996). The increase in rRNA synthesis with growth rate allows the cell to meet its increasing demand for protein synthesis at higher growth rates and to conserve biosynthetic energy at lower growth rates. Many models have been proposed to explain the phenomenon, and recently, at least some of the molecular mechanisms responsible have become clearer (Gaal et al. 1997; Bartlett et al. 1998).

Stringent control of rRNA transcription refers most often to the shut-off of rRNA synthesis following amino acid starvation. Like growth-rate-dependent control, this phenomenon has been recognized for many years (Sands and Roberts 1952; Stent and Brenner 1961), and it is part of a larger response of cells to starvation known as the stringent response (Cashel et al. 1996). The inhibition of rRNA transcription is thought to result from direct interactions between RNAP and the product of the *relA* and *spoT* genes, ppGpp.

rRNA synthesis rates respond very rapidly (i.e., within seconds) to changes in nutritional conditions (Gausing 1980). It has not been determined whether the mechanisms responsible for stringent and growth-rate-dependent control are sufficient to explain the responses of rRNA promoters to upshifts and downshifts. Available evidence indicates that the P2 promoters are responsible for the earliest increases in rRNA transcription during upshifts from stationary phase to exponential growth (Sarmientos et al. 1983). It remains to be determined whether all upshifts (e.g., carbon, nitrogen, and phosphate) utilize the same mechanism(s) for controlling rRNA transcription.

ROLE OF αCTD INTERACTIONS WITH UP ELEMENTS IN TRANSCRIPTION OF rRNA AND OTHER BACTERIAL PROMOTERS

rrnB P1 Core Promoter Structure

The *rrn* P1 promoters match the *E. coli* consensus quite well, as might be expected for strong promoters. For example, in *rrnB* P1, the –10 hexamer is a perfect match to the consensus sequence, the –35 hexamer differs from consensus at only one position, and the spacing between the –10 and –35 hexamers is 16 bp, rather than the consensus 17 bp. Mutation of the spacer to 17 bp or the –35 hexamer to consensus increases promoter strength but alters the regulatory properties of the promoter (Dickson et al. 1989). Thus, even at rRNA promoters where high activity is so crucial, the core promoter sequence has evolved not for maximal activity but for proper regulation.

Stimulation of *rrn* P1 Core Promoters by Upstream Sequences

Perhaps to compensate for the less than perfect fit between σ^{70} and the *rrn* P1 core promoters, sequences upstream of the –35 hexamers are key to the extraordinary strength of these promoters. Sequences upstream of –41 in *rrnB* P1 increase promoter activity about 300-fold (Gourse et al. 1986; Rao et al. 1994). Although some of this effect (~5-fold) results from three binding sites for the transcription factor FIS, most (~70-fold) is not dependent on the action of transcription factors (Leirmo and Gourse 1991; Ross et al. 1993; Rao et al. 1994; Estrem et al. 1998). In *rrnD* P1, the factor-independent stimulatory effect of upstream sequences is even greater than in *rrnB* P1, about 90-fold (Ross et al. 1998).

Interactions between the αCTD and Upstream DNA

At *rrnB* P1, we showed that the upstream A+T-rich region from about –40 to –60 (which we named the UP element) increases the rate of transcription initiation by facilitating closed complex formation with RNAP (K_B) and perhaps also by increasing later step(s) in the pathway to open complex formation (k_f) (Rao et al. 1994). The explanation for the effects of the upstream sequences on transcription was surprising but quite simple: There are direct DNA-protein interactions between the UP element and the carboxy-terminal domain of the α-subunit of RNAP (Fig. 2A). We showed that these interactions are responsible for the high activities of *rrnB* P1 (Ross et al. 1993) and *rrnD* P1 (Ross et al. 1998) and probably of the rest of the *rrn* P1 promoters. We showed that α is a DNA-

A.

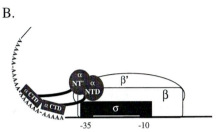

B.

Figure 2. UP elements and phased A-tracts function through interactions with the αCTD. (*A*) Cartoon representation of the interaction between RNAP holoenzyme and the three recognition regions in bacterial promoters: the –10 and –35 hexamers, recognized by the σ-subunit, and the UP element, recognized by the αCTD. (*B*) Cartoon representation of the interaction between RNAP and a promoter with upstream phased A-tracts. We propose that although phased A-tracts lead to macroscopic DNA curvature, stimulation of transcription by A-tracts does not result from the curvature per se, but rather from interactions between the A-tracts and the αCTD.

binding protein both in the context of RNAP holoenzyme and when purified away from the other RNAP subunits, although its ability to interact with UP elements is 100–1000-fold greater as part of holoenzyme (Ross et al. 1993). Even a purified αCTD peptide (Blatter et al. 1994) binds directly and specifically to UP element DNA. Mutant RNAPs lacking the αCTD are unable to make productive interactions with and thus to be stimulated by UP elements.

UP Element Consensus Sequence

Using an in vitro selection modeled after the SELEX method (Tuerk and Gold 1990), followed by an in vivo screen for high transcription activity, we recently determined an UP element consensus sequence (Estrem et al. 1998). This sequence (Fig. 3) stimulates transcription more than 325-fold in vivo and consists of two A+T-rich subsites (proximal, –41 to –44, and distal, –47 to –57; Estrem et al. 1998 and in prep.). Like the *rrnB* P1 UP element, the consensus UP element stimulates transcription when fused to other core promoters (e.g., *lac*), not just *rrnB* P1. The best proximal subsite, consisting primarily of an A-tract from about –41 to –44, is responsible for the majority of the effects of the upstream DNA, increasing transcription in the absence of the distal subsite more than 100-fold (S.T. Estrem et al., in prep.). The best distal subsite also can stimulate transcription independent of the proximal subsite, but it increases transcription only about 16-fold.

The DNA-binding Patch in α

The residues in the αCTD responsible for DNA binding to UP elements were determined using genetic screens for mutations in α that interfered with UP ele-

A.

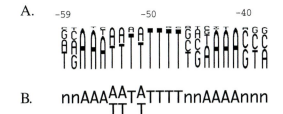

B.

Figure 3. UP element consensus sequence. (*A*) Diagram representing the frequencies of bases present in UP elements obtained from an in vitro selection, followed by an in vivo screen for high activity when upstream sequences were fused to the *rrnB* P1 core promoter. The height of each letter is proportional to the frequency of that base in the selected population. (*B*) UP element consensus sequence derived from *A*. For details, see Estrem et al. (1998).

ment utilization in vivo, followed by alanine scans of the regions defined by the random screen. The effects of the specific residues that we identified in vivo were analyzed in vitro by transcription and footprinting analyses with reconstituted mutant RNAPs (Gaal et al. 1996; Murakami et al. 1996). Seven residues in the αCTD were most critical for DNA interaction and for UP element function: L262, R265, N268, C269, G296, K298, S299. We found that α subunits containing alanine substitutions in any of these seven residues, and in only these residues, failed to complement temperature-sensitive α mutants at the restrictive temperature. Thus, because there is a one-to-one correspondence between the residues required for DNA binding and those required for viability in haploid, we conclude that UP element utilization is essential in *E. coli* (Gaal et al. 1996).

The solution structure of the αCTD was determined by nuclear magnetic resonance (NMR) (Jeon et al. 1995; Gaal et al. 1996) and indicated that the DNA-binding residues form a patch on the αCTD surface (Fig. 4). The identities of the residues in this patch are almost 100% conserved in eubacteria, and thus we predict that the UP element consensus sequence should apply throughout eubacteria. The αCTD contains four standard α helices (α1–4) and a nonstandard α-helix. The critical residues for DNA binding are located in α-helix 1 (262, 265, 268, 269) and in a loop between α helices 3 and 4 (296, 298, 299) which are adjacent on the surface of the domain. Nevertheless, the molecular details of the interaction between the DNA and the αCTD are still unclear. Whether sequence specificity derives from interactions between functional groups on this patch in the αCTD with the minor and/or major grooves of the DNA or with the DNA backbone or both remains to be determined.

UP Element Interactions with αCTD Monomers

Bacterial RNAPs contain two α subunits. We used RNAPs containing one wild-type α-subunit and one α-subunit lacking its αCTD (α heterodimeric RNAPs; S.T. Estrem et al., in prep.) to determine whether both αCTDs are essential for UP element function. We found that promoters containing both UP element subsites (proximal

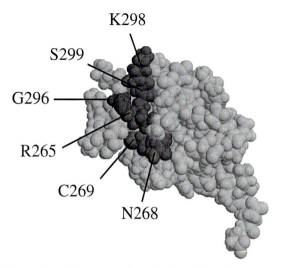

Figure 4. DNA-binding patch on the αCTD. The solution structure is from Jeon et al. (1995). The six residues pictured were identified from a random screen of αCTD mutants resulting in reduced UP element function in vivo, followed by alanine scanning mutagenesis and biochemical analyses of reconstituted mutant RNAPs in vitro. Single alanine substitutions at any of the six pictured residues resulted in a greater than fourfold defect in *rrnB* P1 UP element function in vivo (for details, see Gaal et al. 1996).

and distal) were transcribed only one third as well by the α heterodimeric RNAPs as by the wild-type RNAP, indicating that two-site UP elements require both αCTDs for optimal function in vitro, with one αCTD monomer interacting with each UP element subsite. However, the α heterodimeric RNAPs transcribed promoters containing only the proximal UP element subsite with almost the same efficiency as wild-type RNAP, and the proximal subsite interacted with the remaining αCTD in footprints. On the other hand, the α heterodimeric RNAPs did not transcribe promoters with only the distal UP element subsite as efficiently in vitro as the wild-type RNAP, and the heterodimeric RNAPs failed to protect the distal subsite in these promoters in footprints. In this case, weak protection of the proximal subsite was observed, even though the distal subsite was the one that matched the consensus and even though there was little transcription stimulation.

Our current picture of the interactions between the two UP element subsites and the two α subunits is that each monomer thus makes sequence-specific interactions with a subsite but that the proximal interaction is the preferred one, responsible for most of the stimulation of transcription. The distal interaction can add to the stimulation of transcription, but it is dependent on the presence of the proximal interaction: The second α monomer can only "stretch" to the distal subsite when the proximal subsite is occupied, as if the energy gained from making a sequence-specific interaction with the distal subsite is less than that gained from the alternative sequence nonspecific interaction with the proximal region. This view is consistent with the results of experiments using an αCTD-tethered cleavage reagent (Murakami et al. 1997a).

Previous Evidence for Effects of Upstream Sequences at Other Bacterial Promoters

Although perhaps not widely recognized, it had been reported previously that A+T-rich sequences upstream of the –35 hexamer could strongly affect bacterial promoter strength, independent of the effects of transcription factors (Moran et al. 1981; Banner et al. 1983; Bujard et al. 1987), and footprints had shown that RNAP interacted directly with upstream sequences at certain promoters (Johnson et al. 1983; Busby et al. 1987). Furthermore, it had been shown that A-tracts positioned upstream of the –35 hexamer, in phase with the helical repeat, increased promoter activity in synthetic promoter constructs (see, e.g., Gartenberg and Crothers 1991; for review, see Perez-Martin et al. 1994).

Relationship of Macroscopic DNA Curvature to Transcription Stimulation

Since phased A-tracts result in macroscopic DNA curvature, it was proposed that the observed stimulation in transcription might result from effects of DNA bending (see, e.g., Plaskon and Wartell 1987; Gartenberg and Crothers 1991). As more and more promoters were identified that contained A+T-rich upstream sequences and/or upstream DNA bends, it became more and more accepted that upstream intrinsic curvature stimulates transcription. Although the mechanism by which DNA bending increased transcription was not clear, "structural transmission" effects were proposed in some cases to affect DNA strand opening or to bring DNA sequences upstream of the A-tracts into contact with RNAP (Perez-Martin et al. 1994).

Because of the resemblance between the consensus UP element sequence and phased A-tracts, we decided to evaluate whether αCTD-DNA interactions, rather than DNA curvature per se, might be responsible for the effects of A-tracts on transcription (Aiyar et al. 1998). We confirmed that phased A-tracts, when fused upstream of different promoters, increased transcription in vivo and in vitro. The phased A-tracts stimulated transcription best when the A-tract nearest to the –35 hexamer was centered at about –42, i.e., at a position characteristic of an αCTD-UP element interaction. However, the A-tracts did not stimulate transcription when RNAPs lacked a functional αCTD. Furthermore, the A-tract sequences were protected in footprints with wild-type RNAP but not with the αCTD mutant RNAPs. These data demonstrated that A-tract sequences function as UP elements, increasing transcription by binding to the RNAP αCTD, although the unusual structural features of A-tract DNA could, of course, play a part in α recognition.

Other data also suggested to us that intrinsic curvature per se was not sufficient to explain the effects of A+T-rich upstream sequences on transcription. The amount of stimulation of transcription did not increase in proportion to the number of A-tracts (and thus the bend angle); i.e., four A-tracts stimulated transcription only slightly more than two A-tracts (Aiyar et al. 1998), and even a sequence containing only one A-tract could increase transcription dramatically (S.T. Estrem et al., in prep.). Furthermore, we found that the *rrnB* P1 UP element increased transcription more than sequences containing two or four phased A-tracts, yet displayed little or no macroscopic curvature (Gaal et al. 1994; Aiyar et al. 1998). (Although the *rrnB* P1 promoter region does display intrinsic curvature, the sequences responsible are centered at about –100, upstream of the residues responsible for stimulating transcription [Gourse et al. 1986; Plaskon and Wartell 1987; Gaal et al. 1994].)

Intrinsic curvature may have additional or different roles at certain other promoters. For example, curved DNA can function as a "coactivator," facilitating interactions between RNAP and distantly bound enhancer proteins (Perez-Martin et al. 1994). In addition, we cannot exclude the possibility that there may be cases where upstream A-tracts increase transcription by some kind of "structural transmission" effect such as that proposed for IHF binding at the $ilvP_G$ promoter (Parekh and Hatfield 1996). However, we suggest that in most cases, upstream A-tracts increase transcription through DNA-protein interactions with the αCTD and should be considered UP elements (see Fig. 2B) (Aiyar et al. 1998).

UP Element Distribution in *E. coli* and Other Bacteria

UP elements are not limited to *rrn* P1 promoters. They have also been identified in promoters that utilize σ factors other than σ^{70} (σ^{32}, Newlands et al. 1993; σ^F, Fredrick et al. 1995; M. Kainz and R.L. Gourse, unpubl.) and in promoters from other bacterial species (see, e.g., Fredrick et al. 1995). Some UP elements contain both a distal and proximal subsite, some contain only one subsite (Ross et al. 1998).

The effects of UP element sequences on transcription are approximately proportional to their matches to the consensus sequence, similar to the situation with the –10 and –35 hexamers. Some promoters contain upstream sequences that have no effect on transcription and others increase transcription 2–10-fold, whereas the ones with the closest matches to consensus (e.g., those in *rrn* P1 promoters) increase transcription almost 100-fold (Ross et al. 1998; S.T. Estrem et al., in prep.). Preliminary analyses of the *E. coli* genome sequence (S.T. Estrem et al., unpubl.) suggest that about 2–3% of promoters contain upstream sequences that match the consensus at 11 or more of the 15 crucial positions. This class includes about 19% of promoters transcribing rRNAs and tRNAs and about 2% of promoters transcribing mRNAs. The proportion of promoters containing lower but still significant similarity to consensus is much larger of course. It seems likely that many promoters (conceivably a large fraction of promoters) may derive some contribution of their RNAP recognition from UP element interactions.

Role of αCTD-DNA Interactions in the Action of Transcription Factors

Bacterial transcription factors often function through interactions with either the amino- or carboxy-terminal

domains of α, although not all transcription factors function in this manner (see, e.g., O'Halloran et al. 1989; Li et al. 1994; Miller et al. 1997). In complexes where a transcription factor interacts with DNA well upstream of the –35 hexamer (e.g., centered at –61.5 as in the case of CRP-cAMP at the *lac* promoter), the activator usually makes direct contact with the αCTD (Ebright and Busby 1995). In these promoters, αCTD interacts with the DNA between the –35 hexamer and the factor-binding site. When the activator binds closer to the –35 hexamer, in many cases, αCTD binds upstream of the activator (Murakami et al. 1997b; Belyaeva et al. 1998; Savery et al. 1998). The activator-αCTD interactions in these complexes often involve surface-exposed residues in both proteins directly adjacent to their respective DNA-binding surfaces. However, selections and screens for mutations in αCTD that affect activation have frequently resulted in the identification of residues in the DNA-binding patch, instead of (or in addition to) residues in α involved in protein-protein contacts, suggesting that the DNA-binding residues in α that affect activation do so by stabilizing the activation complex (van Ulsen et al. 1997). Thus, DNA binding by the αCTD is central to the function of factor-dependent as well as factor-independent activation of transcription.

STABILITY OF THE OPEN COMPLEX: IMPLICATIONS FOR TRANSCRIPTIONAL REGULATION

Instability of *rrnB* P1 Complexes with RNAP

The sequence of a promoter determines not only the rate at which it forms an open complex with RNA polymerase, but also the rate at which that complex dissociates. *rrn* P1 promoters form exceedingly unstable binary complexes with RNAP (Table 1). The rate of dissociation of *rrnB* P1 is more than two orders of magnitude faster than the rates of dissociation of more typical *E. coli* promoter complexes under the same solution conditions. We have determined recently that this rapid dissociation rate is crucial for the regulation of *rrn* P1 promoters (Gaal et al. 1997).

Table 1. Complexes between Different *E. coli* Promoters and RNAP Have Very Different Half-lives

Promoter	Half-life
rrnB P1	~1 min
rrnB P1 (C-1T)	~5 min
rrnB P1 (CGC-5→7ATA)	~20 min
rrnE P1	~3 min
argI	>6 hours
hisG	>6 hours
livJ	>6 hours
lysC	>6 hours
pheA	>6 hours
thrABC	>6 hours

Decay rates were measured as described in Gaal et al. (1997) at 30ºC. Complexes with RNAP were formed on supercoiled plasmids in a buffer containing 40 mM Tris-Cl (pH 8.0), 30 mM NaCl, 10 mM MgCl$_2$, and 1 mM DTT.

This rapid rate of dissociation was apparent from early experiments, where it was found that *rrnB* P1 promoter complexes with RNAP could not be footprinted nor could they bind to nitrocellulose filters in the absence of the two initiating NTPs, ATP (+1, the transcription start site) and CTP (+2) (Gourse 1988). It was later found that the stabilization of the interaction between the promoter and RNAP by the initiating NTPs resulted from the formation of a "slipped" complex. In this slipped complex, the priming NTPs, ATP and CTP, moved back to positions –3 and –2 on the template and elongated to form a 5-mer before the RNAP reached the position on the template where the appropriate nucleotide, UTP, was not available (Borukhov et al. 1993). Although there is no evidence that a slipped complex ever occurs in the presence of all four NTPs or in cells, the slipped complex has a much longer lifetime than the *rrnB* P1 open complex, allowing structural characterization by footprinting (Newlands et al. 1991) and analyses of the kinetics of formation of *rrnB* P1 complexes (Leirmo and Gourse 1991; Rao et al. 1994).

The Core Promoter-RNAP Interaction Is Responsible for Growth-rate-dependent Regulation of *rrnB* P1

rrn P1 promoter activity increases in proportion to the growth rate, accounting for growth-rate-dependent regulation of ribosome synthesis (Miura et al. 1981; Gourse et al. 1986). We found that an *rrnB* P1 core promoter, containing sequences only from –41 to +1 with respect to the transcription start site, still displayed the characteristic increase in activity with growth rate, and this regulation occurred even in a strain unable to make ppGpp (Gaal and Gourse 1990; Bartlett and Gourse 1994). Thus, ppGpp, FIS, αCTD interactions with the *rrnB* P1 UP element, the *rrn* P2 promoter, and Nus factors are all dispensable for steady-state regulation of rRNA transcription.

The NTP-sensing Model for Growth-rate-dependent Regulation of *rrn* P1 Promoters

rrn P1 promoters require higher concentrations of the initiating NTP than other promoters for efficient transcription in vitro (Gaal et al. 1997). This requirement derives from the intrinsic instability of the *rrn* P1 promoter complex with RNAP: Solution conditions (e.g., low salt or supercoiling; Ohlsen and Gralla 1992) that increase the stability of the complex decrease the observed K_{NTP} (Gaal et al. 1997). We found that the concentrations of the initiating NTPs (GTP for *rrnD* P1 and ATP for the other six *rrn* P1 promoters) increase with growth rate, and we showed that the initiating NTP stabilizes the unusually unstable *rrn* P1 open complexes. The concentrations of ATP and GTP thereby were proposed to regulate rRNA transcription in vivo (Gaal et al. 1997).

The properties of mutant *rrnB* P1 promoters and mutant RNAPs support this model: The mutants alter the stability characteristics of the promoter-RNAP complexes and thereby alter the regulation of *rrn* P1 transcription in vivo and in vitro. For example, the *rrnB* P1 C-1T mutant promoter (Table 1) makes complexes with RNAP that are about fivefold more stable than the wild-type promoter, and it is therefore transcribed efficiently at lower ATP concentrations in vitro. In vivo, the promoter is transcribed efficiently at all growth rates, as if the concentration of ATP present even at the lowest growth rates is sufficient for stabilization and efficient transcription of the mutant *rrn* P1. Likewise, mutant RNAPs that make promoter complexes less stable than wild-type complexes transcribe *rrnB* P1 and *rrnD* P1 very inefficiently at all growth rates (although FIS can compensate for this deficiency; Bartlett et al. 1998; M. Bartlett and R.L. Gourse, unpubl.), as if the concentrations of ATP and GTP required to stabilize the complex are no longer in the range present in vivo, even at the highest growth rates (Gaal et al. 1997; Bartlett et al. 1998). The control of rRNA transcription by NTP concentration thereby provides a molecular explanation for two long-standing observations in biology: growth-rate-dependent control and homeostatic regulation of ribosome synthesis (Fig. 5).

Role of the Initiating NTP in Stability of the RNAP-Promoter Complex

It seems clear that the instability of the *rrn* P1 open complex is key to rRNA regulation, but we do not yet understand the molecular interactions that determine stability or how the initiating NTP increases that stability. Perhaps the extra hydrogen bonds to the template that result from the presence of the incoming NTP in the active site of the enzyme are sufficient to increase complex stability. Alternatively, the bound initiating NTP could change the conformation of RNAP such that the RNAP contacts with the template are altered, increasing the lifetime of the complex. Interestingly, the presence or absence of the UP element-αCTD interaction did not affect the decay rate of the complex or the apparent K_{NTP} needed for transcrip-

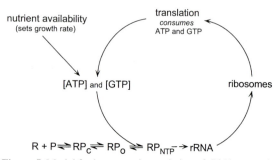

Figure 5. Model for homeostatic regulation of rRNA transcription by the initiating NTP concentration (Gaal et al. 1997). The model proposes that ATP and GTP, whose concentrations vary with growth rate (nutrient availability), regulate rRNA transcription by stabilizing RNAP (R)-*rrn* P1 promoter (P) open complexes (RP_O). rRNA transcription determines the rate of ribosome synthesis and therefore the level of translation. ATP and GTP are consumed during the process of translation, resulting in a feedback signal affecting *rrn* P1 transcription. Initiating NTP pools therefore reflect the balance between protein synthesis rates and the nutritional conditions.

tion in vitro (W. Ross et al., unpubl.). Apparently, UP element contacts with α are important for forming early kinetic intermediates in the pathway to open complex formation, but they do not make an important contribution to the stability of the open complex; i.e., upstream contacts do not affect the stability of the kinetic intermediate that binds the initiating NTP.

ppGpp Destabilizes Open Complexes

We have found that ppGpp destabilizes *rrn* P1 open complexes, thereby reducing transcription in vitro (M.M. Barker et al., unpubl.), and a mutation in the *rrnB* P1 core promoter (CGC-5 to -7ATA) that increases open complex stability reduces the effects of ppGpp on transcription in vitro and in vivo (Table 1) (Josaitis et al. 1995). Therefore, we propose that open complex instability is important not only for growth-rate-dependent control, but also for regulation during the stringent response by making the promoter susceptible to the effects of ppGpp; i.e., the high ppGpp concentrations induced by amino acid starvation make the complex dissociate before transcription can begin. We propose that the larger the stabilizing effects of mutations on stability, the higher the concentration of ppGpp that the promoter can withstand. For example, the 3-bp *rrnB* P1 substitution mutation (CGC-5 to -7ATA) increases the half-life of the RNAP-promoter complex and is almost unaffected by ppGpp after amino acid starvation, whereas the 1-bp substitution mutation (C-1T) increases the stabilty of the complex much less, and as a result, this mutant promoter is still "stringent;" i.e., it is still inhibited by ppGpp after amino acid starvation (Josaitis et al. 1995).

We suggest that the relative stability of most promoter complexes makes them insensitive to the negative effects of even high ppGpp concentrations; i.e., even though all promoter complexes appear to be destabilized by ppGpp in vitro (M.M. Barker et al., unpubl.), if the intrinsic lifetime of a complex is very long relative to the time needed for initiation, then ppGpp will not affect transcription. Consistent with this proposal, we have found that the activities of the non-rRNA promoter complexes listed in Table 1 do not decrease in the presence of ppGpp in vitro or in vivo (M.M. Barker and R.L. Gourse, unpubl.).

ppGpp as a Positive Regulator

Strains lacking ppGpp are polyauxotrophs (Xiao et al. 1991). The mutant RNAPs described above that destabilized *rrn* P1 open complexes, thereby reducing *rrn* P1 transcription, were originally selected for their ability to confer amino acid prototrophy to strains lacking ppGpp (Bartlett et al. 1998). We found that the mutant RNAPs decreased the stabilities of all promoter complexes tested in vitro, not just rRNA promoters, and the mutant RNAPs increased the activities of several amino acid biosynthetic promoters in vivo that formed very stable complexes with RNAP (M.M. Barker and R.L. Gourse, unpubl.). These data are consistent with the proposal that ppGpp is a positive regulator of transcription of amino acid biosynthetic

promoters (see, e.g., Riggs et al. 1986; Shand et al. 1989) and that the mutant RNAPs mimic the effect of ppGpp on these promoters by reducing the stability of the RNAP-promoter complex (Bartlett et al. 1998). We do not yet understand, however, the relationship between stability and transcription of amino acid biosynthetic promoters; i.e., we do not know what step in the transcription mechanism of these promoters might be facilitated by a reduction in open complex stability.

Stability as a Regulatory Determinant

The above discussion emphasizes the importance of promoter-RNAP complex stability as a determinant not only in promoter activity, but also in response to regulatory molecules, in this case the initiating NTP or ppGpp. We conclude that different promoters can be regulated differently depending on their promoter mechanisms. We have proposed that *rrn* P1 promoters are regulated positively by initiating NTP concentrations, because *rrn* P1 open complexes are unstable enough that small increases in stability increase transcription. These same promoters are negatively regulated by ppGpp, because small decreases in stability inhibit transcription. However, the same relative decrease in stability on a promoter forming a more stable complex might lead to positive control: Depending on the promoter mechanism, the same molecular interactions can have opposite outcomes. We do not yet understand the promoter sequences that affect open complex stability. Mutations in the region between the –10 hexamer and the transcription initiation site are crucial (Table 1) (Gaal et al. 1997), but it is unlikely that these sequences alone are sufficient to account for the stabilities of all promoter complexes.

CONCLUDING STATEMENT

We have found that the basic mechanism of a promoter in the absence of activator or repressor proteins can be crucial to its strength and regulation. The role of α interactions in promoter strength and the role of initiating NTP concentrations in regulation of transcription illustrate how the cell originally may have accomplished the major features of rRNA transcription and its regulation without resorting to auxiliary proteins, as if the basic transcription machinery evolved to suit the needs of rRNA promoters. An understanding of recruitment of RNAP through UP element-α interactions and an understanding of the factors that determine open complex stability will be central to an understanding not only of a promoter's activity, but also of its regulation.

ACKNOWLEDGMENTS

Work in our lab is supported by RO1 GM-37048 from the National Institutes of Health and by a Hatch grant from the U.S. Department of Agriculture. S.T.E., M.M.B., and C.A.H. were supported in part by training grants from the National Institutes of Health.

REFERENCES

Aiyar S.E., Gourse R.L., and Ross W. 1998. Upstream A-tracts increase prokaryotic promoter activity through interactions with the RNA polymerase α subunit. *Proc. Natl. Acad. Sci.* (in press).

Banner C.D., Moran C.P., Jr., and Losick R. 1983. Deletion analysis of a complex promoter for a developmentally regulated gene from *Bacillus subtilis. J. Mol. Biol.* **168:** 351.

Bartlett M.S. and Gourse R.L. 1994. Growth rate-dependent control of the *rrnB* P1 core promoter in *Escherichia coli. J. Bacteriol.* **176:** 5560.

Bartlett M.S., Gaal T., Ross W., and Gourse R.L. 1998. RNA polymerase mutants that destabilize RNA polymerase-promoter complexes alter NTP-sensing by *rrn* P1 promoters. *J. Mol. Biol.* **279:** 331.

Belyaeva T.A., Rhodius V.A., Webster C.L., and Busby S.J. 1998. Transcription activation at promoters carrying tandem DNA sites for the *Escherichia coli* cyclic AMP receptor protein: Organisation of the RNA polymerase α subunits. *J. Mol. Biol.* **277:** 789.

Blatter E.E., Ross W., Tang H., Gourse R.L., and Ebright R.H. 1994. Domain organization of RNA polymerase α subunit: C-terminal 85 amino acids constitute a domain capable of dimerization and DNA binding. *Cell* **78:** 889.

Bokal A.J., Ross W., Gaal T., Johnson R.C., and Gourse R.L. 1997. Molecular anatomy of a transcription activation patch: FIS-RNA polymerase interactions at the *Escherichia coli rrnB* P1 promoter. *EMBO J.* **16:** 154.

Borukhov S., Sagitov V., Josaitis C.A., Gourse R.L., and Goldfarb A. 1993. Two modes of transcription initiation *in vitro* at the *rrnB* P1 promoter of *Escherichia coli. J. Biol. Chem.* **268:** 23477.

Bujard H., Brunner M., Deuschle U., Kammerer W., and Knaus R. 1987. Structure-function relationship of *Escherichia coli* promoters. In *RNA polymerase and the regulation of transcription* (ed. W.S. Reznikoff et al.), pp. 95. Elsevier, New York.

Busby S., Spassky A., and Chan B. 1987. RNA polymerase makes important contacts upstream of base pair –49 at the *Escherichia coli* galactose operon P1 promoter. *Gene* **53:** 145.

Cashel M., Gentry D.R., Hernandez V.J., and Vinella D. 1996. The stringent response. In Escherichia coli *and* Salmonella typhimurium: *Cellular and molecular biology,* 2nd edition (ed. F.C. Neidhardt et al.), p. 1458. American Society for Microbiology, Washington, D.C.

Condon C., Squires C., and Squires C.L. 1995. Control of rRNA transcription in *Escherichia coli. Microbiol. Rev.* **59:** 623.

Dickson R.R., Gaal T., deBoer H.A., deHaseth P.L., and Gourse R.L. 1989. Identification of promoter mutants defective in growth-rate-dependent regulation of rRNA transcription in *Escherichia coli. J. Bacteriol.* **171:** 4862.

Ebright R.H. and Busby S. 1995. The *Escherichia coli* RNA polymerase α subunit: Structure and function. *Curr. Opin. Genet. Dev.* **5:** 197.

Estrem S.T., Gaal T., Ross W., and Gourse R.L. 1998. Identification of an UP element consensus sequence for bacterial promoters. *Proc. Natl. Acad. Sci.* **95:** 9761.

Fredrick K., Caramori T., Chen Y.F., Galizzi A., and Helmann J.D. 1995. Promoter architecture in the flagellar regulon of *Bacillus subtilis:* High-level expression of flagellin by the sigma D RNA polymerase requires an upstream promoter element. *Proc. Natl. Acad. Sci.* **92:** 2582.

Gaal T. and Gourse R.L. 1990. Guanosine 3′-diphosphate 5′-diphosphate is not required for growth rate-dependent control of rRNA synthesis in *Escherichia coli. Proc. Natl. Acad. Sci.* **87:** 5533.

Gaal T., Bartlett M.S., Ross W., Turnbough C.L., Jr., and Gourse R.L. 1997. Transcription regulation by initiating NTP concentration: rRNA synthesis in bacteria. *Science* **278:** 2092.

Gaal T., Rao L., Estrem S.T., Yang J., Wartell R.M., and Gourse R.L. 1994. Localization of the intrinsically bent DNA region upstream of the *E. coli rrnB* P1 promoter. *Nucleic Acids Res.* **22:** 2344.

Gaal T., Barkei J., Dickson R.R., deBoer H.A., deHaseth P.L., Alavi H., and Gourse R.L. 1989. Saturation mutagenesis of an *Escherichia coli* rRNA promoter and initial characterization of promoter variants. *J. Bacteriol.* **171:** 4852.

Gaal T., Ross W., Blatter E.E., Tang H., Jia X., Krishnan V.V., Assa-Munt N., Ebright R.H., and Gourse R.L. 1996. DNA-binding determinants of the α subunit of RNA polymerase: Novel DNA-binding domain architecture. *Genes Dev.* **10:** 16.

Gartenberg M.R. and Crothers D.M. 1991. Synthetic DNA bending sequences increase the rate of in vitro transcription initiation at the *Escherichia coli lac* promoter. *J. Mol. Biol.* **219:** 217.

Gausing K. 1980. Regulation of ribosome biosynthesis in *E. coli.* In *Ribosomes: Structure, function and genetics* (ed. G. Chambliss et al.), p. 693. University Park Press, Baltimore.

Gourse R.L. 1988. Visualization and quantitative analysis of complex formation between *E. coli* RNA polymerase and an rRNA promoter *in vitro. Nucleic Acids Res.* **16:** 9789.

Gourse R.L., de Boer H.A., and Nomura M. 1986. DNA determinants of rRNA synthesis in *E. coli:* Growth rate dependent regulation, feedback inhibition, upstream activation, antitermination. *Cell* **44:** 197.

Gourse R.L., Gaal T., Bartlett M.S., Appleman J.A., and Ross W. 1996. rRNA transcription and growth rate-dependent regulation of ribosome synthesis in *Escherichia coli. Annu. Rev. Microbiol.* **50:** 645.

Jeon Y.H., Negishi T., Shirakawa M., Yamazaki T., Fujita N., Ishihama A., and Kyogoku Y. 1995. Solution structure of the activator contact domain of the RNA polymerase α subunit. *Science* **270:** 1495.

Jinks-Robertson S., Gourse R.L., and Nomura M. 1983. Expression of rRNA and tRNA genes in *Escherichia coli:* Evidence for feedback regulation by products of rRNA operons. *Cell* **33:** 865.

Johnson W.C., Moran C.P., Jr., Banner C., Zuber P., and Losick R. 1983. Anatomy of a complex procaryotic promoter under developmental regulation. In *Gene Expression* (ed. D.H. Hamer and M. Rosenberg), p. 235. A.R. Liss New York.

Josaitis C.A., Gaal T., and Gourse R.L. 1995. Stringent control and growth-rate-dependent control have nonidentical promoter sequence requirements. *Proc. Natl. Acad. Sci.* **92:** 1117.

Keener J. and Nomura M. 1996. Regulation of ribosome biosynthesis. In Escherichia coli *and* Salmonella typhimurium: *Cellular and molecular biology,* 2nd edition (ed. F.C. Neidhardt et al.), p. 1417. American Society for Microbiology, Washington, D.C.

Leirmo S. and Gourse R.L. 1991. Factor-independent activation of *Escherichia coli* rRNA transcription. I. Kinetic analysis of the roles of the upstream activator region and supercoiling on transcription of the rrnB P1 promoter *in vitro. J. Mol. Biol.* **220:** 555.

Li M., Moyle H., and Susskind M.M. 1994. Target of the transcriptional activation function of phage lambda cI protein. *Science* **263:** 75.

Maaloe O. and Kjeldgaard N.O. 1966. *Control of macromolecular synthesis: A study of DNA, RNA, and protein synthesis in bacteria.* Benjamin, New York.

Miller A., Wood D., Ebright R.H., and Rothman-Denes L.B. 1997. RNA polymerase β′ subunit: A target of DNA binding-independent activation. *Science* **275:** 1655.

Miura A., Krueger J.H., Itoh S., deBoer H.A., and Nomura M. 1981. Growth-rate-dependent regulation of ribosome synthesis in *E. coli:* Expression of the *lacZ* and *galK* genes fused to ribosomal promoters. *Cell* **25:** 773.

Moran C.P., Jr., Lang N., Banner C.D.B., Haldenwang W.G., and Losick R. 1981. Promoter for a developmentally regulated gene in *Bacillus subtilis. Cell* **25:** 783.

Murakami K., Fujita N., and Ishihama A. 1996. Transcription factor recognition surface on the RNA polymerase α subunit is involved in contact with the DNA enhancer element.

EMBO J. **15:** 4358.

Murakami K., Kimura M., Owens J.T., Meares C.F., and Ishihama A. 1997a. The two α subunits of *Escherichia coli* RNA polymerase are asymmetrically arranged and contact different halves of the DNA upstream element. *Proc. Natl. Acad. Sci.* **94:** 1709.

Murakami K., Owens J.T., Belyaeva T.A., Meares C.F., Busby S.J., and Ishihama A. 1997b. Positioning of two α subunit carboxy-terminal domains of RNA polymerase at promoters by two transcription factors. *Proc. Natl. Acad. Sci.* **94:** 11274.

Newlands J.T., Gaal T., Mecsas J., and Gourse R.L. 1993. Transcription of the *Escherichia coli rrnB* P1 promoter by the heat shock RNA polymerase Eσ³² *in vitro*. *J. Bacteriol.* **175:** 661.

Newlands J.T., Josaitis C.A., Ross W., and Gourse R.L. 1992. Both fis-dependent and factor-independent upstream activation of the *rrnB* P1 promoter are face of the helix dependent. *Nucleic Acids Res.* **20:** 719.

Newlands J.T., Ross W., Gosink K.K., and Gourse R.L. 1991. Factor-independent activation of *Escherichia coli* rRNA transcription. II. Characterization of complexes of rrnB P1 promoters containing or lacking the upstream activator region with *Escherichia coli* RNA polymerase. *J. Mol. Biol.* **220:** 569.

O'Halloran T.V., Frantz B., Shin M.K., Ralston D.M., and Wright J.G. 1989. The MerR heavy metal receptor mediates positive activation in a topologically novel transcription complex. *Cell* **56:** 119.

Ohlsen K.L. and Gralla J.D. 1992. Interrelated effects of DNA supercoiling, ppGpp, and low salt on melting within the *Escherichia coli* ribosomal RNA *rrnB* P1 promoter. *Mol. Microbiol.* **6:** 2243.

Parekh B.S. and Hatfield G.W. 1996. Transcriptional activation by protein-induced DNA bending: Evidence for a DNA structural transmission model. *Proc. Natl. Acad. Sci.* **93:** 1173.

Perez-Martin J., Rojo F., and de Lorenzo V. 1994. Promoters responsive to DNA bending: A common theme in prokaryotic gene expression. *Microbiol. Rev.* **58:** 268.

Plaskon R.R. and Wartell R.M. 1987. Sequence distributions associated with DNA curvature are found upstream of strong *E. coli* promoters. *Nucleic Acids Res.* **15:** 785.

Rao L., Ross W., Appleman J.A., Gaal T., Leirmo S., Schlax P.J., Record M.T., Jr., and Gourse R.L. 1994. Factor independent activation of *rrnB* P1. An "extended" promoter with an upstream element that dramatically increases promoter strength. *J. Mol. Biol.* **235:** 1421.

Riggs D.L., Mueller R.D., Kwan H.S., and Artz S.W. 1986. Promoter domain mediates guanosine tetraphosphate activation of the histidine operon. *Proc. Natl. Acad. Sci.* **83:** 9333.

Ross W., Aiyar S.E., Salomon, J., and Gourse R.L. 1998. *Escherichia coli* promoters with UP elements of different strength:

Modular structure of bacterial promoters. *J. Bacteriol.* **180:** 5375.

Ross W., Thompson J.F., Newlands J.T., and Gourse R.L. 1990. *E.coli* Fis protein activates ribosomal RNA transcription *in vitro* and *in vivo*. *EMBO J.* **9:** 3733.

Ross W., Gosink K.K., Salomon J., Igarashi K., Zou C., Ishihama A., Severinov K., and Gourse R.L. 1993. A third recognition element in bacterial promoters: DNA binding by the α subunit of RNA polymerase. *Science* **262:** 1407.

Sands M.K. and Roberts R.B. 1952. The effects of a tryptophan-histidine deficiency in a mutant of *Escherichia coli*. *J. Bacteriol.* **63:** 505.

Sarmientos P. and Cashel M. 1983. Carbon starvation and growth rate-dependent regulation of the *Escherichia coli* ribosomal RNA promoter: Differential control of dual promoters. *Proc. Natl. Acad. Sci.* **80:** 7010.

Sarmientos P., Contente S., Chinali G., and Cashel M. 1983. Ribosomal RNA operon promoters P1 and P2 show different regulatory responses. In *Gene expression* (ed. D.H. Hamer and M. Rosenberg), p. 65. A.R. Liss, New York.

Savery N.J., Lloyd G.S., Kainz M., Gaal T., Ross W., Ebright R.H., Gourse R.L., and Busby S.J.W. 1998. Transcription activation at class II CRP-dependent promoters: Identification of determinants in the C-terminal domain of the RNA polymerase α subunit. *EMBO J.* **17:** 3439.

Schaechter M., Maaloe O., and Kjeldgaard N.O. 1958. Dependency on medium and temperature of cell size and chemical composition during balanced growth of *Salmonella typhimurium*. *J. Gen. Microbiol.* **19:** 592.

Shand R.F., Blum P.H., Mueller R.D., Riggs D.L., and Artz S.W. 1989. Correlation between histidine operon expression and guanosine 5′-diphosphate-3′-diphosphate levels during amino acid downshift in stringent and relaxed strains of *Salmonella typhimurium*. *J. Bacteriol.* **171:** 737.

Stent G.S. and Brenner S. 1961. A genetic locus for the regulation of ribonucleic acid synthesis. *Proc. Natl. Acad. Sci.* **47:** 2005.

Tuerk C. and Gold L. 1990. Systematic evolution of ligands by exponential enrichment: RNA ligands to bacteriophage T4 DNA polymerase. *Science* **249:** 505.

van Ulsen P., Hillebrand M., Kainz M., Collard R., Zulianello L., van de Putte P., Gourse R.L., and Goosen N. 1997. Function of the C-terminal domain of the α subunit of *Escherichia coli* RNA polymerase in basal expression and integration host factor-mediated activation of the early promoter of bacteriophage Mu. *J. Bacteriol.* **179:** 530.

Xiao H., Kalman M., Ikehara K., Zemel S., Glaser G., and Cashel M. 1991. Residual guanosine 3′,5′-bispyrophosphate synthetic activity of *relA* null mutants can be eliminated by *spoT* null mutations. *J. Biol. Chem.* **266:** 5980.

The Functional and Regulatory Roles of Sigma Factors in Transcription

C.A. Gross,* C. Chan,* A. Dombroski,† T. Gruber,* M. Sharp,‡ J. Tupy,‡ and B. Young‡

*Department of Stomatology and Department of Microbiology and Immunology, and
‡Department of Biochemistry and Biophysics, University of California at San Francisco,
San Francisco, California 94143; †Department of Microbiology and Molecular Genetics,
University of Texas Medical School, Houston, Texas 77030

In prokaryotic cells, a single subunit of RNA polymerase, called sigma, orchestrates the process of transcription initiation. Sigma binds to the multisubunit core RNA polymerase ($\alpha_2\beta\beta'$), creating RNA polymerase holoenzyme ($\alpha_2\beta\beta'\sigma$), which performs transcription initiation. Holoenzyme recognizes the two conserved hexamer sequences that constitute a prokaryotic promoter, exposes the single-stranded DNA template necessary for transcription initiation, and begins synthesizing the nascent RNA chain. When the nascent RNA is five to ten nucleotides long, sigma is released, terminating the initiation phase of transcription. Core RNA polymerase then carries out the elongation and termination phases of transcription.

Sigma was initially implicated in transcription initiation as selectivity factor (Burgess et al. 1969). For many years, the subunit was thought to control traffic, directing core RNA polymerase to promoter sites on the DNA, ensuring biologically accurate transcription. In the past few years, this view has changed dramatically. In bacterial cells, transcription initiation is controlled at three different levels, and sigma is involved in all of them. First, initiation frequency is responsive to the promoter sequence. Because of promoter diversity, transcription initiation frequency can vary over three orders of magnitude (Record et al. 1996). Sigma is intimately involved in the read out of this "hard-wired" information carried in prokaryotic promoter sequences. When sigma binds to core, it not only provides RNA polymerase with a promoter "landing site," but also promotes a major conformational change in the enzyme, converting polymerase to an initiation-competent form. Accumulating evidence indicates that sigma continues to function throughout initiation, communicating information about the promoter and transcription start site to core RNA polymerase. Initiation is also regulated by activators and repressors, which modulate transcription initiation in response to environmental conditions. Some of this information is communicated to RNA polymerase by the interaction of these regulatory proteins with the sigma subunit. Finally, all of the different sigma subunits in the cell (*Escherichia coli* has seven) are utilized in a global regulatory capacity. In addition to the primary sigma used for expression of housekeeping genes, most bacteria have alternative sigmas that are used to regulate expression of genes that respond to altered environmental or developmental signals.

The relative level of gene expression mediated by each sigma is, in part, determined by competition among sigmas for core RNA polymerase.

Given the central role of sigmas in both transcription initiation and global gene regulation, it is not surprising that they have been a focus of intense study. There are actually two different sigma families, which are not homologous to each other. We will not consider members of the σ^{54} family here. We discuss only those sigmas that are homologous to *E. coli* σ^{70} (the "primary sigma" mentioned above). There are more than 100 sigmas known in this protein family; alignment of their sequences revealed four conserved regions, each of which can be further subdivided into smaller regions with high conservation (Helmann and Chamberlin 1988; Lonetto et al. 1992; Wösten 1998). These conserved regions are ubiquitously present in the sigma family of proteins, with the exception of region 1.1, which is found only in primary sigmas (Lonetto et al. 1992; Wilson and Dombroski 1997). Genetic and biochemical analyses have led to functional assignments for some of these regions (Fig. 1). Because most functions of sigma have been found to be linearly arranged along the polypeptide backbone, such approaches have been remarkably successful. However, these successes have ushered in a new phase of research requiring much more sophisticated experiments to understand how this component carries out its role in the initiation process. We review these successes and discuss the roles of sigma in basal transcription initiation, activation, and global gene regulation. We conclude with the challenges posed by future research on sigma.

HOW SIGMA BINDS TO CORE RNA POLYMERASE

Sigma binding to core not only signals the beginning of initiation, but also provokes the conformational changes in both partners that are integral to the initiation process. Moreover, information about the progressive interactions between sigma and DNA is likely to be transmitted to core across the sigma-core interface. Despite its importance, the interaction between sigma and core is poorly understood at present. The information currently available indicates that multiple regions in sigma contact core RNA polymerase. Little is known about the core side of this interface.

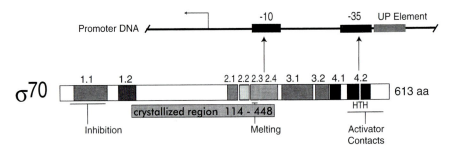

Figure 1. The conserved regions of σ^{70} and their functional assignments. The four conserved regions and the large nonconserved segment in σ^{70} are shown; the proposed roles of some of the different regions and the fragment crystallized by the Darst group are indicated below the linear representation of σ^{70}. A schematic diagram of a typical promoter with conserved regions located at –35 and –10 from the start site of transcription is shown above the linear representation of σ^{70}. Arrows indicate the regions of σ^{70} that contact the conserved promoter regions. The UP element characteristic of very strong promoters is also shown. This element is contacted by the α-subunit of core RNA polymerase.

The first approach to defining the interaction of these proteins examined core binding of various deletion derivatives of sigma. Scott Lesley and Richard Burgess identified region 2.1 as important for binding to core RNA polymerase by using deletion derivatives of both σ^{70} and σ^{32} (Lesley and Burgess 1989; Lesley et al. 1991). Since deletions can disrupt overall structure of a protein, confirmation of such results with point mutations is essential. A single point mutation in region 2.1 of *Bacillus subtilis* σ^E that reduces binding to core RNA polymerase was identified by Charles Moran and colleagues, providing support for the importance of region 2.1 in this reaction (Schuler et al. 1995). One cautionary note about this point mutant is in order. Seth Darst and colleagues recently determined the crystal structure of a proteolytically stable fragment of σ^{70} extending from region 1.2 to 2.4 (Malhotra et al. 1996). The residue in σ^{70} homologous to the *B. subtilis* σ^E mutated residue forms a kink in the extended 2.1 helix. Altering this residue could affect the overall protein fold in this region.

Recent genetic studies suggest that many additional regions of sigma bind to core RNA polymerase. Richard Calendar and colleagues identified point mutations in σ^{32} that result in sigmas with reduced binding to core. Interestingly, these mutations are located in many conserved regions of the polypeptide (regions 2.2, 3.1, 4.1, and 4.2) but not in region 2.1. In addition, one mutation was located in the RpoH box, a region unique to σ^{32} (Joo et al. 1997, 1998). An independent genetic approach, carried out by Meghan Sharp and Carol Gross, identified point mutations with a similar phenotype that are located in σ^{70}. This analysis found mutations in conserved regions 2.1, 2.2, 2.4, 3.1, 4.1, and 4.2 that were important for σ^{70} in binding core polymerase (M. Sharp, unpubl.). Interestingly, the single point mutant identified in region 2.1, when mapped on the crystal structure, is not surface-exposed and thus not to likely to directly contact core polymerase. In addition, using a yeast two-hybrid analysis, Tanja Gruber and Carol Gross have found that region 1.1, which is unique to primary sigmas, also binds to core polymerase (T. Gruber, unpubl.). Thus, conserved regions from the beginning to the end of sigma appear to interact with core.

The binding of both σ^{32} and σ^{70} to core has been studied with protein footprinting. This analysis provides independent evidence that multiple regions of sigma bind to core. This approach assumes that regions of sigma which interact with core should be less susceptible to cleavage by free radicals as holoenzyme than as free sigma. Using this approach with σ^{32}, Cathleen Chan and Carol Gross find protection of the RpoH box and region 3.1 of σ^{32} upon binding to core RNA polymerase, corroborating the mutant analysis of Joo and Calendar (C. Chan, unpubl.). Because of the large size of the molecule, protein footprinting with σ^{70} cannot be carried out to very high resolution. Nonetheless, Hiroki Nagai and Nobuo Shimamoto found that regions of σ^{70} throughout the entire polypeptide chain, including region 1.1, are relatively protected in holoenzyme, corroborating the conclusion of M. Sharp and C.A. Gross (unpubl.) that the σ^{70}-core interface is extensive (Nagai and Shimamoto 1997). Jean-Paul Léonetti and Peter Geiduschek have performed a high-resolution protein footprinting study of the binding of a small, divergent phage-encoded sigma to core. They found that regions 2.1, 2.2, and several others exhibited greater protection in holoenzyme than as free sigma, attesting to the generality of these results for all sigmas of this family (Léonetti et al. 1998).

Taken together, these studies suggest that multiple regions of sigmas contact core RNA polymerase. Although predominantly similar regions in the different sigmas interact with core, each sigma may also use one or more unique regions to form its interface. Even the distantly related promoter specificity factor Mftf1p (found in yeast mitochondrial RNA polymerase) uses amino acids in regions homologous to conserved regions 2 and 3 to interact with its single subunit core polymerase (Cliften et al. 1997). Moreover, a preliminary comparative analysis of the σ^{32} and σ^{70} interface with core indicates that these two sigmas use identical residues within their conserved regions to interact with core (M. Sharp, unpubl.). Perhaps different regions of this interface are important at different times in the transcription cycle. Some mutant sigmas defective in binding to core have altered functional properties; however, no systematic investigation of this proposition has yet been performed.

HOW SIGMA RECOGNIZES PROMOTER DNA

Sigma was first identified by Richard Burgess, Andrew Travers, John Dunn, and Ekhard Bautz as an activity that allowed core RNA polymerase to transcribe biologically relevant DNA with high efficiency (Burgess et al. 1969). The finding that sigma increased the initiation rate, rather than elongation proficiency, immediately suggested that sigma worked by recognizing promoter DNA and depositing RNA polymerase at the transcription start site. The surprising finding that free σ^{70} did not bind to DNA at all, whether or not it had a promoter (Burgess et al. 1969), transiently gave rise to a more complex notion about how sigma might work. Perhaps sigma exposed buried DNA-binding domains in RNA polymerase. This notion soon ran into difficulty when it was realized that core RNA polymerase would have to possess multiple DNA binding domains in order to accommodate the different conserved promoters recognized by the alternative sigma factors discovered by Richard Losick and Jan Pero (Losick and Pero 1981). These alternative sigmas recognized promoters that varied in both conserved hexamers, making this early model seem even more improbable. It soon became clear that sigma, and not core, contained the DNA-binding determinants.

A molecular understanding of the mechanism of promoter recognition began with the selection of mutants in sigma that had altered promoter recognition properties. Two different laboratories (Thomas Gardella and Carey Waldberger with Miriam Susskind and Deborah Siegele and James Hu with Carol Gross) selected sigma variants that compensated for the poor activity of promoters with point mutations in their –10 and –35 conserved regions; these mutants identified two distinct regions in sigma (Gardella et al. 1989; Siegele et al. 1989; Waldberger et al. 1990). At the same time, Peter Zuber and Rich Losick found mutations in an alternative sigma of *B. subtilis* that affected recognition of the –10 region of the promoter (Zuber et al. 1989). These mutations were located at positions homologous to those of the equivalent mutants in σ^{70}, which seemed to indicate that all sigmas of the σ^{70} family recognized promoters in a similar fashion. Since that time, additional mutants in other sigma factors have confirmed the picture that sigmas have two DNA-binding

domains that recognize the two conserved regions of the prokaryotic promoter (Fig. 2). Although these DNA recognition mutants presented a compelling picture of how promoters are recognized, the nagging problem of the inability of free sigma to bind DNA remained. Alicia Dombroski, Tom Record, and Carol Gross solved this contradiction by showing that partial polypeptides of σ^{70}, lacking the amino-terminal conserved region 1.1, were able to bind DNA and faithfully mimic the promoter recognition properties of sigma (Dombroski et al. 1992).

Recognition of the –35 Region of the Promoter

The sigma mutants that specifically suppressed the deleterious effects of base changes in the conserved –35 hexamer were located in the recognition helix of the helix-turn-helix (HTH) motif in region 4.2, implicating this region in –35 recognition (Gardella et al. 1989; Siegele et al. 1989). Mutations at two positions in this helix, RH584 and RC588, altered recognition of two bases in the conserved –35 region. Satisfyingly, these mutant sigmas exhibited the same altered specificity for promoter recognition in an in vitro equilibrium competition binding assay, measuring the ability of partial sigmas (lacking the inhibitory region 1.1) to bind DNA (Dombroski et al. 1992). Jeffrey Owens and Claude Meares confirmed the proximity of region 4.2 to the –35 region of the promoter in the open complex. They observed that iron-EDTA tethered to a cysteine residue in the HTH cleaves promoter DNA in the vicinity of the –35 region of the promoter (Owens et al. 1998). Using UV lasar footprinting, Johannes Geiselman, Henri Buc, and collaborators have been able to show that interaction with the –35 region of the promoter is one of the first specific polymerase-promoter interactions to occur in transcription initiation (Eichenberger et al. 1997). Footprinting of the intermediates in formation of the open complex indicates that polymerase remains bound at this position throughout the initiation process (Schickor et al. 1990; Mecsas et al. 1991). Thus, RNA polymerase recognizes the conserved –35 region of the promoter using the HTH motif in region 4.2 of sigma; this binding persists throughout open complex formation and into the first stages of transcription initiation.

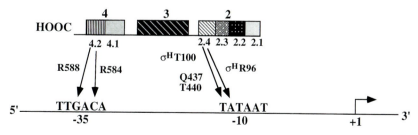

Figure 2. Interactions of sigma with a consensus promoter. The genetic identification of interactions between σ^{70} and a canonical promoter with conserved hexamers centered at –10 and –35 bases upstream of the start point of transcription is indicated. These interactions are deduced from identification and analysis of sigma mutants that specifically suppress particular base changes in the promoter. Except where noted, these interactions were identified in σ^{70}. Interactions of sigma with particular bases in the –35 region of the promoter were reported by Siegele et al. (1989) and Gardella et al. (1989). Interactions of sigma with particular bases in the –10 region of the promoter were reported by Siegele et al. (1989), Zuber et al. (1989), Daniels et al. (1990), and Waldburger et al. (1990). Zuber and Daniels map interactions in the alternative *B. subtilis* sigma factor σ^H. Kenney et al. (1989) report similar interactions in σ^A, the housekeeping sigma of *B. subtilis*.

Recognition of the –10 Region of the Promoter

The interactions of sigma with the –10 region of the promoter are considerably more complex. Since the strand separation process begins within this region, sigma could be involved in sequence-specific interactions with both single-stranded and double-stranded DNA. In addition, nonspecific interactions of sigma with DNA could facilitate strand opening. Below, we recount the observations which establish that the melting transition is driven by sequence-specific interactions with both double-stranded and nontemplate single-stranded DNA and also by nonspecific interactions of sigma with DNA, which recognize and stabilize the unique junction between single-stranded and double-stranded DNA.

The sigma mutants that specifically suppressed the deleterious effects of base changes at the first two positions of the conserved –10 hexamer were located in a helix in region 2.4, implicating this region in –10 recognition. Three mutations in this helix alter recognition of the first base in the –10 hexamer: TI440 and QH437 in σ^{70} and RA100 in σ^H (Gardella et al. 1989; Siegele et al. 1989; Zuber et al. 1989). RA96 in σ^H alters recognition of the second base in the –10 hexamer (Daniels et al. 1990). We believe that it is justifed to extrapolate the RA96 results of σ^H to σ^{70}, even though the consensus distance of this position from the start site differs for the two sigmas. Although the first base of the consensus for σ^H is at –13 and that for σ^{70} is at –12 from the start site, an amino acid in a homologous position recognizes the first base of each consensus, suggesting that the same will be true for recognition of the second position. A comparison of *E. coli* σ^{70} promoters reveals that the distance of the –10 consensus from the start site can vary, reinforcing this suggestion (Lisser and Margalit 1993).

Using an equilibrium competition binding assay, Alicia Dombroski showed that fragments of sigma containing region 2.4 distinguish mutational changes at three positions in the –10 region (positions –12, –11, and –10), indicating that sigma directly recognizes these positions (Dombroski et al. 1993; Dombroski 1997). As these sigma fragments were unable to recognize single-stranded DNA of the correct sequence, these experiments assessed the ability of sigma to specifically recognize double-stranded DNA. They established that sigma discriminates particular base pairs at the first three positions of the –10 consensus sequence (Dombroski 1997).

Other evidence, however, indicated that at least within the context of holoenzyme, sigma also interacts specifically with single-stranded DNA in the –10 region. It was shown that sigma could be cross-linked to DNA between positions –7 and –3 exclusively on the nontemplate strand (Simpson 1979; Buckle and Buc 1994). This provided early evidence for the importance of the nontemplate strand in interacting with sigma. Our current understanding is that sigma makes base-specific contacts with the nontemplate strand in the –10 region of the promoter and then utilizes information from that strand to guide initiation. This model derives primarily from work of Jeffrey Roberts and his collaborators. Christine Roberts found

that when promoters were mismatched at the –12, –11, or –7 positions, only the identity of the nontemplate base influenced the rate of open complex formation and abortive initiation, establishing the importance of the nontemplate strand (Roberts and Roberts 1996). Then, Michael Marr showed that holoenzyme specifically recognized a single-strand oligonucleotide containing the nontemplate sequence of the –10 region. The mutant QH437, which lies in region 2.4 and alters promoter recognition, changes the preference of sigma for nontemplate strand sequences. Therefore, in addition to its role in specific binding of double-stranded DNA, region 2.4 of sigma must also mediate single-strand recognition (Marr and Roberts 1997). In the next section, we consider how core aids sigma with single-strand recognition.

Although region 2.4 recognizes both single-stranded and double-stranded DNA, the recognition rules for these templates appear to differ. If we assume that the defects in open complex formation with mismatched templates reflect the single-strand recognition properties of σ^{70} (as measured by Roberts and Roberts), then G is the only base significantly disfavored in the nontemplate strand at both a –12 and –11 mismatch (Roberts and Roberts 1996). The other nonconsensus bases had either no effect (–12 position) or only small effects (–11 position) on open complex formation. Sigma interaction with double-stranded DNA appears to be more restrictive, as judged from data from the equilibrium filter-binding assays. When mutants in double-stranded promoter sequences were tested, sigma discriminated against two nonconsensus bases at the –12 position and the single base (which was not a G) tested at the –11 position (Dombroski 1997).

Sigma also participates in the strand opening process. While a graduate student with Michael Chamberlin, John Helmann noted a cluster of conserved aromatic residues in region 2.3 of sigma. Since proteins that bind single-stranded DNA or RNA often interact with these nucleic acids by stacking aromatic residues with the nucleotide bases, Helmann and Chamberlin (1988) postulated that region 2.3 of sigma might be directly involved in strand melting. Helmann continued on to prove this proposition using σ^A, the *B. subtilis* housekeeping sigma. By using site-directed mutagenesis to change the appropriate residues in σ^A to alanine, he impaired DNA melting of the promoter. This defect can be overcome by conditions favoring supercoiling or by high temperature, both of which encourage melting (Juang and Helmann 1994). When heteroduplex templates are used to bypass the melting step, these mutants have no defect (Aiyar et al. 1994). As expected, these mutant sigmas have a dominant negative phenotype in vivo (Rong and Helmann 1994). A mutation of a homologous position of *B. subtilis* σ^E, identified by Hal Jones and Charles Moran, has a similar phenotype (Jones and Moran 1992).

The crystal structure of the fragment of σ^{70} solved by Darst and colleagues gives further insight into how region 2.3 might facilitate strand opening (Malhotra et al. 1996). Regions 2.3 and 2.4 form a continuous helix with the amino acids that participate in base-specific recognition and the aromatic residues implicated in strand melting

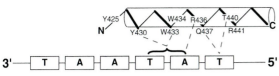

Figure 3. A schematic of the interaction between the regions 2.3 and 2.4 helix and the nontemplate DNA strand in the –10 region of the promoter. A schematic of the α-helix is shown with its solvent-exposed face pointing downward. The nontemplate strand sequence of the –10 consensus element is illustrated schematically below the helix. Dashed lines indicate interactions between specific residues and bases, as determined from genetic or biochemical studies. The specific interactions indicated between residues in sigma and the first two positions of the consensus are based on the genetic studies of Siegele et al. (1989), Zuber et al. (1989), Daniels et al. (1990), and Waldburger et al. (1990). The Y430 and W433 interactions that promote melting are based on studies by Juang and Helmann (1994). These assignments differ somewhat from those indicated in Malhotra et al. (1996) as discussed in the text.

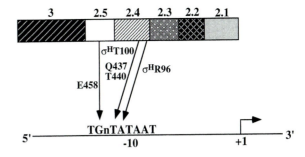

Figure 4. Interactions of sigma with the extended –10 promoter. The schematic shows the interactions between σ^{70} and the extended –10 promoter deduced from identification and analysis of sigma mutants that specifically suppress particular base changes in the promoter. The precise boundaries of region 2.5 have not been determined; in this diagram, it extends from region 2.4 to 3.1.

both positioned on a single face of the helix. Thus, region 2.3 is perfectly placed to facilitate strand opening. A schematic representation of the interaction of this helix with nontemplate strand DNA is shown in Figure 3.

Following open complex formation by holoenzyme, the –12 and –11 positions of the promoter are at a junction between double-stranded and single-stranded DNA. Yuli Guo and Jay Gralla (1988) have shown that σ^{70} holoenzyme interacts preferentially with a fork junction that contains a DNA duplex ending at position –12 and continuing with the single-stranded nontemplate sequence. Holoenzyme binding to this structure is much tighter than its binding to either single-stranded or double-stranded promoter DNA, forming a heparin-resistant complex even at very low temperatures. A structure containing the correct sequence at positions –12 and –11 is preferentially recognized, implicating sigma in the recognition event. The demonstration that σ^{70} fragments containing region 2 specifically recognize bubble constructs with the correct nucleotide pair at position –11 provides evidence consistent with this idea (Dombroski 1997).

In summary, region 2.4 of sigma first recognizes the upstream base pairs in the –10 region of the promoter as double-stranded DNA. Then, regions 2.3 and 2.4 facilitate and stabilize DNA melting. Interaction between region 2.3 and single-stranded DNA promotes open complex formation, and interaction between region 2.4 and the nontemplate strand, as well as preferential binding to a forked DNA structure, stabilizes the open complex.

A Second Kind of –10 Promoter Region

A subset of *E. coli* promoters is dependent on a conserved 5´-TG -3´ located one base upstream of the –10 element of the promoter (Ponnambalam et al. 1986; Keilty and Rosenberg 1987). Interestingly, this "extended –10 promoter" functions without a conserved –35 region. Using the same kind of suppression genetics that identified sigma contacts with the –10 and –35 regions of the promoter, Kerry Barne, Stephen Minchin, Stephen Busby, and their collaborators identified a sigma contact with the

5´-TG -3´ motif (Barne et al. 1997). This region of sigma, located immediately downstream from region 2.4, has been dubbed region 2.5 (Fig. 4). Alicia Dombroski (1997) has provided biochemical evidence for this interaction by showing that a sigma fragment with region 2 recognizes the extended –10 region promoter. Moreover, as is true in vivo, interaction with the 5´-TG -3´ motif is dominant. Even in the absence of the –35 region or other specificity determinants in the –10 region, a promoter with the 5´-TG -3´ motif competed well with the canonical *tac* promoter for recognition by sigma. RNA polymerase-promoter complexes at extended –10 promoters give footprints that are different from those at canonical promoters, indicating that the interactions driving open complex formation differ between the two types of promoters (Chan et al. 1990). Although this type of promoter is relatively rare in *E. coli* (accounting for <10% of its promoters), it accounts for almost 50% of the promoters in some gram-positive organisms (Helmann 1995).

Do the Two DNA-binding Domains in Sigma Interact?

Although sigma contains two independent DNA-binding domains, Alicia Dombroski and Carol Gross have provided some evidence that the activity of these two domains is coordinated at canonical promoters having both –10 and –35 recognition elements (Dombroski et al. 1992). Sigma fragments containing either region 2 or region 4 bind faithfully to their respective segments of the promoter, but they do not sense the presence of the promoter segment recognized by the missing partner. In contrast, when the sigma fragment contains both DNA-binding domains, a single mutation in either conserved hexamer eliminates specific DNA binding to all parts of the promoter. In addition, a sigma fragment that includes both DNA-binding domains is sensitive to the spacing between the –35 and –10 elements of the promoter. These results suggest that regions 2 and 4 of sigma communicate with each other to coordinate recognition of the entire promoter. A mutational change that alters the response of sigma to spacing between promoter elements both in vivo and in vitro has been identified (Dombroski et al. 1996). This mutation is located in the upstream helix of the HTH motif in region 4.2, at a position predicted to be pointing away from the DNA. One possible explanation for the phenotype of this mutant is that it alters communication between the two domains of sigma. A possible precedent for disrupting this communication already exists. Since the 5´-TG -3´ motif of the extended −10 promoter eliminates the requirement for a -35 region, the interaction between region 2.5 of sigma and the 5´-TG-3´ motif may disrupt the communication between regions 2 and 4 of sigma.

Summary

Sigma orchestrates initiation at two different types of promoters, and many of the interactions between sigma and DNA that contribute to promoter recognition and open complex formation are known. At standard promoters, with canonical −10 and −35 conserved regions, RNA polymerase is anchored at the promoter by interactions between the −35 promoter region and the HTH in region 4.2 of sigma. Meanwhile, gymnastics are occurring at the −10 region of the promoter. Following the initial interaction of region 2.4 with the upstream base pairs in the −10 region of the promoter, several activities of sigma facilitate and stabilize strand opening. There is some evidence suggesting that information about the two promoter regions is integrated by interactions within sigma, but this process has not been studied. At alternative "extended −10 promoters," additional contacts between region 2.5 and the extended 5´-TG -3´ motif replace contacts in the −35 region of the promoter. Exactly how these contacts contribute to open complex formation is currently unknown. Comparative studies between open complex formation in canonical promoters having a −10 and −35 region and those relying on the extended −10 element are likely to identify key interactions required for starting transcription.

THE ROLE OF SIGMA IN TRANSCRIPTION INITIATION

Until now, we have concentrated on the specific interactions of sigma with its partners, core RNA polymerase and DNA. In this section, we consider sigma from a more functional point of view and describe how sigma contributes to the process of transcription initiation within the context of holoenzyme.

Transcription Initiation

The prokaryotic transcription initiation cycle has been extensively investigated. RNA polymerase holoenzyme forms a weak complex with double-stranded promoter DNA (the initial closed complex) and isomerizes to one or more additional closed complexes before forming the open complex and initiating transcription. The initial transcribing complex contains sigma and characteristically makes many small abortive RNA products. Eventually, sigma dissociates and an elongation complex is formed. We have a rather good understanding of the process from the point of view of the DNA. The first closed complex is distinguished from the second closed complexes by the size of its footprint: Although the upstream boundary of both complexes is similar, the downstream boundary of the initial closed complex extends at most to about +1. The footprint of the second closed complex is almost as long as that of the open complex. It extends almost to +20, but does not exhibit strand opening. At least two open complexes have been identified; the second requires Mg^{++} for its formation and includes the +1 position in its melted region (Fig. 5) (Record et al. 1996).

Formation of Holoenzyme

When sigma initiates the transcription cycle by binding to core RNA polymerase, a large-scale conformational change occurs in both partners. As one indication of the extent of this change, the location of sigma in holoenzyme cannot be determined by simple comparison of the overall structure of core RNA polymerase with holoenzyme, as determined from two-dimensional electron crystals (Darst et al. 1989; Polyakov et al. 1995). We know that the consequence of these conformational changes is to create a state competent for transcription initiation, but we have only a very general idea of the nature of these changes. For core, one major modification is that a channel, large enough to accommodate duplex DNA, is freely accessible in holoenzyme; the "jaws" of the channel are closed in core. It is unclear whether the "jaws" close only when sigma is lost from the complex or whether that change occurs earlier in the transcription cycle. There are undoubtedly other changes as well, but the current level of resolution at which these structures can be

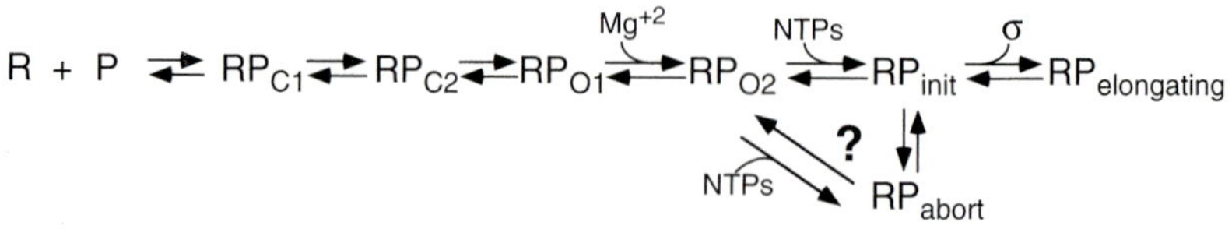

Figure 5. Intermediates in the process of initiation of RNA synthesis for which structural and/or kinetic evidence is available. (R) RNA polymerase holoenzyme; (P) promoter DNA; (RP_{C1}) first closed complex; (RP_{C2}) second closed complex; (RP_{O1}) first open complex; (RP_{O2}) second open complex; (RP_{init}) initiating complex; (RP_{abort}) abortively initiating complex.

studied does not permit their detection. When sigma is bound by core, its DNA recognition properties are altered in at least two respects. First, sigma is now able to bind to DNA and specifically recognize the promoter. Second, sigma acquires the ability to specifically recognize the nontemplate strand of DNA. We have some understanding of the origin of each of these changes.

The ability of sigma to bind DNA and recognize the promoter as part of holoenzyme results primarily from changes in the function of region 1.1. In free sigma, region 1.1 prevents DNA binding, predominantly by binding to the carboxyl terminus of σ^{70} and interfering with the interaction of region 4.2 with DNA, but also by somewhat decreasing the DNA binding of region 2.4 (Dombroski et al. 1992, 1993). Although we have no definite evidence for how holoenzyme antagonizes the inhibitory function of region 1.1, we believe that the core RNA polymerase binds to region 1.1, preventing its inhibitory interaction with the carboxyl terminus of sigma. Tanja Gruber and Carol Gross have direct evidence for the proposed binding, as region 1.1 binds to the amino terminus of β´ in a yeast two-hybrid analysis (T. Gruber, unpubl.). In addition, data on protein accessibility are consistent with this idea: Region 1.1 is less accessible to external reagents in holoenzyme than in free sigma, and region 4.2 is more accessible. Region 1.1 accessibility has been probed in two ways. Marie Strichland and Richard Burgess demonstrated that an epitope in region 1.1 is less accessible to a monoclonal antibody in holoenzyme than in free sigma (Strickland et al. 1988). More recently, Hiroki Nagai and Nobuo Shimamoto (1997) showed that this region is less accessible to protein cleavage reagents in holoenzyme than in free sigma. Conversely, Cathleen Chan and Carol Gross showed that region 4.2 of σ^{32} is more accessible to protein cleavage reagents in holoenzyme than in free sigma (C. Chan, unpubl.).

In the previous section, we reviewed the evidence from the Roberts laboratory that sigma, in the context of the holoenzyme, specifically recognizes the nontemplate strand of the DNA. However, two lines of evidence indicate that free sigma, even when it lacks region 1.1, cannot bind single-stranded DNA specifically. First, using the equilibrium competition binding assay, Alicia Dombroski (1997) could find no evidence that partial sigma polypeptides bind single-stranded DNA specifically. Second, Sandhya Callaci and Tomasz Heyduk have created a sigma with only two tryptophan resides, both located in region 2.3, and then substituted these residues with 5-hydroxy-tryptophan. Using the unique fluorescence of 5-hydroxy-tryptophan, the environment of the region 2.3–2.4 helix can be monitored both in free sigma and in holoenzyme. They find that in free sigma, region 2.3–2.4 is accessible to solvent and does bind single-stranded DNA, but this binding is weak and without specificity. However, when bound to core, solvent accessibility of region 2.3–2.4 decreases and binding specificity increases, now exhibiting a 200-fold preference for nontemplate over template strand. Thus, binding to core alters both the environment of region 2.3—2.4 and its DNA-binding capabilities (Callaci and Heyduk 1998). The analysis of the

σ^{70} interface carried out by Meghan Sharp (unpubl.) indicates that region 2.4 itself and the nearby region 2.2 bind to core. Interaction of either of these regions with core may alter the environment of the 2.3–2.4 helix so that it can specifically recognize the nontemplate strand.

Formation of the Open Complex

Both core RNA polymerase and promoter DNA are likely to be allosteric effectors of sigma function throughout open complex formation, but studies in this area are in a rudimentary state. Although strand separation occurs under conditions where double-stranded DNA is very stable, no input of energy is necessary to drive this reaction, presumably because RNA polymerase induces various distortions in the DNA that lower the activation energy for strand separation (deHaseth et al. 1998). These activities, which are most likely carried out by core RNA polymerase, undoubtedly affect the activities of sigma that facilitate strand opening at the −10 region, but the connection between sigma and core functions is unknown.

At present, there is only one tangible example of how sigma function is modified by interactions that occur during open complex formation. We have just discussed the role of region 1.1 in preventing free sigma from binding to DNA, and how binding to core might antagonize this activity. It turns out that region 1.1 also has a positive role in transcription, facilitating open complex formation. Alicia Dombroski and Christina Wilson showed that holoenzyme lacking region 1.1 initially binds to the promoter at a rate indistinguishable from that of wild-type holoenzyme, but forms open complexes more slowly than its wild-type counterpart (Wilson and Dombroski 1997). When added in *trans* to holoenzyme lacking region 1.1, this region can speed up open complex formation. Thus, the function of region 1.1 is altered once the polymerase-promoter complex is formed. We have an idea about how this could come about. The protein footprinting experiments of Nagai and Shimamoto (1997) indicate that region 1.1 again becomes accessible to cleavage in the open complex. The studies of Tanja Gruber and Carol Gross indicated that region 1.1 binds to the amino terminus of β´ (T. Gruber, unpubl.). But we know that this general region of β´ also binds to DNA (Nudler et al. 1996). Perhaps binding of the amino terminus β´ to DNA and to region 1.1 is mutually exclusive. In this case, after the initial interaction of the amino terminus of β´ with DNA, region 1.1 would be free to rebind to sigma. Even if region 1.1 now rebinds to the same site with which it interacts in free sigma, the context is different. At this point, the carboxyl terminus of sigma is interacting with both core and DNA. Within this context, region 1.1 binding could facilitate open complex formation. It is interesting to note that region 1.1 is unique to the housekeeping sigmas (Lonetto et al. 1992; Wilson and Dombroski 1997). The unique role of these sigmas may necessitate region 1.1 function. Most alternative sigmas recognize a small subset of promoters, which are very similar to each other and generally have longer consensus sequences than the promoters of housekeeping genes. In contrast, the housekeeping sigmas transcribe most of the genome and recognize widely divergent promoters at widely varying rates. Additional interactions of alternative sigmas with their promoters may obviate the need for region 1.1.

Sigma and Phosphodiester Bond Formation

Following open complex formation, RNA synthesis initiates. Although the capacity to catalyze formation of phosphodiester bonds resides in core RNA polymerase, sigma can influence this process since phosphodiester bond formation during the initiation phase of transcription differs in two respects from that during the elongation phase. First, the K_M for binding the initiating nucleotide is very high (Record et al. 1996). Second, the nascent RNA chain spontaneously dissociates during the initiation phase, generating abortive products (Record et al. 1996). Two regions of sigma are currently known to influence the initial transcription properties of the enzyme.

Region 3 is close to the active site of the enzyme, and functional analysis suggests that it modulates initial RNA synthesis. Konstantin Severinov and Alex Goldfarb have shown that region 3 cross-links to the initiating nucleotide, and Jeffrey Owens and Claude Meares have shown that Fe^{++}-EDTA tethered to a cysteine in region 3.2 cleaves the template strand in the vicinity of +1 (Severinov et al. 1994; Owens et al 1998). Not only is region 3 in the right place to influence polymerization, genetic analysis suggests that it does so. James Hernandez and Michael Cashel (1995) have selected mutants in sigma that allow growth in the absence of ppGpp, a nucleotide that influences the initiation properties of RNA polymerase at some promoters. These mutants map solely to region 3.2 and have been shown to alter the abortive initiation process at some promoters (Hernandez et al. 1996). We are left to wonder whether this region of sigma forms part of the binding pocket for the initiating nucleotide and part of any RNA exit channel that exists in holoenzyme.

During elongation, the template strand of DNA has the crucial role, encoding not only the information for the sequence of a protein, but also the signals for modulating elongation by pausing and termination. The role, if any, of the nontemplate strand during elongation has not yet been discovered. In contrast, the nontemplate strand of DNA appears to play a critical part in promoter clearance, and its information is communicated to RNA polymerase through region 2.4 of sigma. Our current evidence suggests that these interactions modulate elongation for as long as sigma remains bound to core RNA polymerase. The heteroduplex experiments of Roberts and Roberts (1996) indicated that any nonconsensus base at the –12, –11, and –7 positions of the nontemplate strand had severe deleterious effects on abortive initiation, even when these bases had no effect on open complex formation. These data indicate that the interactions of this region with sigma do not simply maintain strand opening, they also affect phosphodiester bond formation. Interestingly,

when an additional –10 region is present early in the transcript, sigma continues to affect elongation. Antitermination mediated by the λQ protein requires a pause following production of the first 16 nucleotides of the λP_R´ transcript. This pause occurs when the sigma subunit, which remains bound to its moving RNA polymerase, now interacts with a reiterated –10 region in the nontemplate strand of the DNA (Ring and Roberts 1994; Ring et al. 1996). Similar interactions may account for the very long abortive transcripts produced from some promoters. We have already mentioned that sigma does not interact selectively with the nontemplate strand on its own. Interaction of this region of sigma with core may be necessary to create a single-stranded DNA-binding channel that modulates the initiation process by its interactions with nontemplate DNA. This putative DNA-binding channel could be composed solely of sigma, but more likely, it also has contributions from core RNA polymerase. The information in this strand is then utilized, either by sigma itself or in concert with RNA polymerase, to modulate phosphodiester bond formation catalyzed by holoenzyme.

The role of sigma in initiation ends when it leaves the ternary complex. The forces that eject sigma from transcribing polymerase are currently completely unknown, although probably crucial to our understanding of transcription initiation. We do not favor the idea that sigma is simply left behind as the train leaves the station, as sigma can transiently move with polymerase away from the promoter in at least one situation. If sigma can move with polymerase away from the λP_R´ promoter, we anticipate that it can do so from other promoters as well, implying that an active process removes sigma from transcribing RNA polymerase. It has been proposed that the entry of RNA into the exit channel in core RNA polymerase is the allosteric effector that dissociates sigma.

Summary

Interplay among sigma, core RNA polymerase, and DNA gives rise to sequential changes in each that result in transcription initiation. We are just beginning to understand these interactions. Both core RNA polymerase and DNA modulate the DNA-binding capabilities of sigma in several ways. First, specific recognition of single-stranded nontemplate DNA by sigma requires core RNA polymerase. Second, competitive binding interactions during open complex formation modulate the function of region 1.1 of sigma. Self-inhibition by region 1.1 prevents free sigma from recognizing promoter DNA; inhibition is likely to be antagonized by the competitive binding of this region to the amino terminus of the β´ subunit in holoenzyme. When bound to the promoter, region 1.1 is freed from this interaction, possibly because the amino terminus of β´ binds to DNA, and functions to facilitate open complex formation by an unknown mechanism. Conversely, sigma modulates phosphodiester bond formation by holoenzyme. Sigma may contribute to the binding of the initiating nucleotide, and interactions with the nontemplate strand of sigma alone or in conjunction with core modulate abortive initiation and pausing. The mechanistic basis of these effects is not understood.

THE ROLE OF SIGMA IN ACTIVATION

The basal transcription cycle that we have just described is an important component of bacterial growth, setting the transcription rate of a wide variety of genes in response to their promoter strength. However, efficient growth requires that transcription rates adjust to environmental change. Activation of transcription is a major strategy for accomplishing this. Most transcription activators bind to sites at or near the promoter and make direct contact with one of the subunits of RNA polymerase. Interaction with polymerase enhances either initial binding or subsequent steps on the pathway to transcription initiation, converting a weak promoter to a strong one.

Although any subunit of RNA polymerase can be the target of transcriptional activators, the majority of activators contact either alpha or sigma. Note that these are the only two subunits of polymerase that make specific contacts with the promoter. As we have just discussed extensively, sigma makes contacts with the –35 and –10 regions of the promoter. Alpha contacts the UP element (Ross et al. 1993), a segment of the promoter located at –40 to –60 from the transcription start site (see Fig. 1). A consensus UP element is found in very strong promoters; however, many other promoters have degenerate UP elements. To a large extent, the location of the DNA-binding site of the activator determines the polymerase subunit contacted. When the activator binds at positions at or upstream of –60, it invariably contacts the carboxy-terminal domain of the RNA polymerase α-subunit (αCTD). When the activator binds at or near the –35 region, it usually contacts sigma, but may choose other targets (αCTD, αNTD, and possibly other polymerase subunits) (Ishihama 1993; Busby and Ebright 1994, 1997).

Interestingly, the very first mutant ever identified in sigma affected its activation function. The *alt-1* mutation, identified by Alan Silverstone and John Scaife, affected activated transcription of the arabinose operon and was initially thought to identify an alternative transcription factor (Silverstone et al. 1972). Six years later, Andrew

Travers realized that this mutation was actually located in sigma (Travers et al. 1978). On the basis of mutations in sigma that selectively eliminated function of an activator without altering basal transcription, at least ten activators are thought to contact sigma. These activator-specific mutations cluster in two regions of sigma: a segment at the beginning of the first helix (amino acids 570–580) and a segment downstream from the second helix (amino acids 591–613) of the HTH motif in region 4.2. For some activators, residues in only one of these two regions affect activation, suggesting that these two regions are either distinct in space or perform different functions. A list of activator-sigma interactions suggested by the phenotypes of mutations in sigma is presented in Figure 6.

How might these activators work? The basal promoter function of activator-dependent promoters must be impaired in some way in order to make the promoter responsive to the presence of an activator. A simple way to construct such a promoter is to have weak DNA contacts between the polymerase and promoter, either because of a nonconsensus –10 or –35 region or because of a nonconsensus spacing between these regions. In this case, protein-protein contacts can substitute for the usual DNA-protein contacts found at intrinsically strong promoters, either recruiting RNA polymerase to the promoter or enhancing a subsequent step in transcription. Below, we first discuss some examples of each type and end with a situation that is less resolved but rather interesting. These examples are all drawn from activation of σ^{70} because analysis of these interactions is mechanistically advanced. However, the positions of existing activation-defective mutants in σ^A, the housekeeping sigma of *B. subtilis* (Fig. 6), suggest that this discussion will also apply to housekeeping sigmas of gram-positive organisms. The extent to which the function of alternative sigmas is modulated by activators is currently unclear.

Activation by PhoB is a clear case where sigma-PhoB contacts substitute for sigma-DNA contacts to recruit RNA polymerase to the promoter. The *phoB* promoter lacks a consensus –35 region; instead, the promoter contains a PhoB-binding site at that position. PhoB appears to substitute for a –35 region as RNA polymerase is able to bind to the promoter only when PhoB is present. Con-

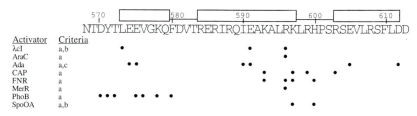

Figure 6. Predicted points of contact between activators and sigma. The sequence of σ^{70} from amino acid 568 to the end of the protein is shown. Three helices (boxed) are shown above the sequence; the first helix and the first half of the second helix constitute the HTH in region 4.2. These helices are predicted from secondary structure analysis and alignment to NarL (Lonetto et al. 1998). The contact points with various activators are shown below the σ^{70} partial sequence. Contacts are based on predictions from mutations in sigma eliminating activation (*a*), second-site suppression of sigma mutants with mutations in activators or vice versa (*b*), or physical interaction between activator and sigma (*c*). All of the data are for activators of *E. coli* σ^{70} except for SpoA, which is for *B. subtilis* σ^A. Data for PhoB are from Kim et al. (1995); for λcI, from Kuldell and Hochschild (1994) and Li et al. (1994); for Ada, from Landini et al. (1998); for AraC, from Hu and Gross (1985); for MerR, from Caslake et al. (1997); for CAP and FNR, from Lonetto et al. (1998); and for SpoA, from Schyns et al. (1997) and Buckner et al. (1998).

sistent with the idea that recognition of the –35 region is unnecessary for the *phoB* promoter, truncations of sigma that remove the recognition helix of the HTH that recognizes the –35 region are still able to transcribe *phoB*, provided the activator is present. The RNA polymerase contact site is identified as sigma because only mutations in sigma confer a defect in PhoB-activated transcription. These mutations are all located in the first helix of this HTH and prevent RNA polymerase from binding to the *phoB* promoter, even when PhoB is present (Makino et al. 1993; Kim et al. 1995).

The bacteriophage T4 protein, AsiA, is a more complex iteration of this same theme. This anti-sigma factor has two roles: to inhibit host transcription and to collaborate with another phage protein, MotA, to activate the T4 middle genes. AsiA accomplishes these two tasks by using protein-protein contacts to control the recruitment of polymerase to the promoter. The groups of Deborah Hinton, Seth Darst, Ed Brody, and Annie Kolb have all contributed to our understanding of this event (Ouhammouch et al. 1994; Hinton et al. 1996; Colland et al. 1998; Severinova et al. 1998). AsiA is a "subtle" anti-sigma; it appropriates, rather than poisons, the activity of σ^{70}. AsiA binds to region 4.2 of σ^{70}. This complex binds to core RNA polymerase but cannot bind to canonical *E. coli* promoters requiring both –35 and –10 recognition. The clue to AsiA action came from the finding that the AsiA-holoenzyme complex transcribes extended –10 promoters. These promoters are not dependent on –35 recognition. Thus, the specific effect of AsiA-σ^{70} contact in region 4.2 is to antagonize recognition of the –35 promoter element, preventing recruitment to canonical *E. coli* promoters. This modified polymerase is now recruited to T4 middle promoters by phage-specific protein-protein contacts. T4 middle promoters are characterized by an extended –10 element and a binding site for the T4 MotA activator at the –35 region of the promoter. Interactions between MotA and the AsiA-holoenzyme selectively recruit polymerase to the T4 middle promoters.

In contrast to these situations, activation by Ada is a clear case where the activator converts a relatively inactive polymerase-promoter complex to a more active one. Ada is a multifunctional protein that reverses alkylation damage to DNA and activates transcription of its own gene. The Ada promoter has an upstream region with an excellent match to the UP element and somewhat nonconsensus –10 and –35 regions. RNA polymerase is brought to the promoter by interaction between α and the strong UP element; Ada is not necessary for this step (Landini and Volkert 1995). However, without Ada, this polymerase-promoter complex is transcriptionally compromised. When Ada is present, it binds upstream of the –35 region of the promoter, at a position partially overlapping that of the UP element, and converts RNA polymerase to a transcriptionally active complex. Paolo Landini, Stephen Busby, and their collaborators have shown that mutations in sigma with the strongest defect in activation affect an acidic amino acid in the first helix; however, mutations affecting acidic residues downstream from the HTH show some defects in activation. Consis-

tent with the idea that Ada interacts with the carboxyl terminus of sigma, in gel-mobility assays, full-length σ^{70}, but not σ^{70} deleted for its last 39 amino acids, supershifts Ada bound to its DNA site (Landini et al. 1998). This is the strongest biochemical evidence to date for an activator-sigma interaction. Currently, it is not clear whether Ada interacts with one or both contact regions in the carboxyl terminus of σ^{70} or how this interaction facilitates initiation.

CAP and FNR are two homologous activators that regulate catabolic genes under aerobic and anaerobic conditions, respectively. These two proteins have several activating surfaces; of these, the AR3 activating surface is believed to interact with a patch of basic residues downstream from the HTH in σ^{70} (Lonetto et al. 1998). Interestingly, the "activation potential" of CAP AR3 depends on the particular sequences at the –35 region, although the preferred nucleotides are not always those in the consensus –35 promoter element (Rhodius et al. 1997). This suggests that interaction with an activator may also function to reposition some portion of the recognition helix on the –35 region of the promoter. This interplay between activator-subunit contacts and DNA-subunit contacts may explain why most activators interact with the two polymerase subunits that make specific DNA contacts.

SIGMA AND GLOBAL REGULATION OF TRANSCRIPTION

From the time that the first sigma was identified, Richard Burgess and Andrew Travers (1970) speculated that alternative sigmas would serve a positive control function, directing RNA polymerase to alternative promoters. That proposition was first demonstrated in *B. subtilis*, where Richard Losick found that a procession of alternative sigma factors executed development (Losick and Pero 1981). Many years later, Alan Grossman and Carol Gross found the first alternative sigma in *E. coli*, which controlled the heat shock response (Grossman et al. 1984). This suggested that alternative sigmas were likely to be widespread, as they spanned the divide between gram-positive and gram-negative bacteria. Today, we know that alternative sigmas control specific regulons in vegetative cells and also control the development of some bacteria and bacteriophage.

The Diversity of Alternative Sigmas

Most alternative sigmas are sufficiently similar to the primary (housekeeping) sigmas that they are readily identified by standard sequence similarity searching methods. However, Michael Lonetto working with Carol Gross and in collaboration with Mark Buttner identified one group of alternative sigmas, called ECF sigmas (because several founding members were involved in *extra*cytoplasmic *function*) that are quite divergent (Lonetto et al. 1994). Indeed, several members of the group had been identified as positive activators before they were reclassified as sigma factors. Although the ECF sigmas align over all four conserved regions, they are quite divergent in the regions 2.3

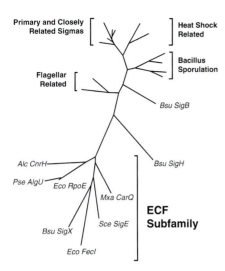

Figure 7. Phylogenetic relationships between the ECF subfamily and other sigma factors. Shown is a minimal unrooted tree, based on distance methods. Primary and closely related sigmas include the primary sigmas from *E. coli, B. subtilis, Anabaena* sp., and *Chlamydia trachomatis*, as well as *E. coli* σ ^{S} and *S. coelicolor* σ^{HrdD}. Heat shock and related sigmas are *E. coli* σ^{32} and *M. xanthus* σ^B. *B. subtilus* sporulation sigmas are σ^K, σ^E, σ^F, and σ^G. Flagellar sigmas are *B. subtilis* σ^D and *S. typhimurium* σ^{FliA}. The distance tree shown was calculated using the Fitch program from the PHYLIP package (Felsenstein 1989). Parsimony trees were constructed by using PAUP (D. Swofford, Illinois Natural History Survey) using a combination of repeated heuristic searches on the full data set, with reweighing of characters based on the 20–30 best trees, and exhaustive searches on subsets of sequences. Region 2, the conserved portion of region 3.2, and region 4 were used in the input to the phylogeny programs.

and 2.4, which are involved in –10 recognition and promoter melting (region 2.3–2.4), and in region 3, which may have a role in initiation. Divergence in these regions made their identification as sigmas difficult, but suggests that a detailed study of their mechanism of initiation would be worthwhile. Interestingly, as shown in the next section, the ECF family is one of the most highly represented types of alternative sigma, with some organisms having as many as 11 sigmas of this type. It is possible that additional sigmas, even more divergent in sequence, remain to be identified.

The evolutionary relationships between primary and alternative sigmas, with emphasis on the ECF family, are presented in Figure 7. The relationship among primary sigmas approximates those based on rRNA comparisons and can be used as an independent measure of species phylogeny. Because the evolution of the alternative sigmas is often driven by the functional and regulatory requirements of the organism, and because alternative sigma genes may exchange between organism, alternative sigmas exhibit groupings that are unrelated to phylogeny (Lonetto et al. 1992, 1994).

Sigma Proliferation and Organismal Complexity

In recent years, several eubacterial genomes have been sequenced, allowing us to determine the number of sigma factors present in these organisms (Table 1). To a certain degree, a correlation can be made between the number of sigma factors found in an organism and the complexity of its "lifestyle," For example, *Mycoplasma genitalium* has one of the smallest genomes for a self-replicating cell-based organism (0.58 Mb), and this bacterium leads a relatively sheltered existence in parasitic association with ciliated epithelial cells of primate genital and respiratory tracts. It has a marked reduction in the number and types of biosynthetic pathways, and gene regulation appears to be less complex than in more metabolically versatile organisms. It is therefore not surprising that this organism can exist with a single sigma factor. On the other extreme, the metabolically and morphologically diverse organism *B. subtilis* has the largest number of sigma factors found in an eubacterium to date (18). It is expected that *B. subtilis* responds to a variety of environmental conditions, and requires extensive transcriptional regulation. As an example, five sigma factors are dedicated mainly to spore formation. It is interesting to compare the sigma factors of the four sequenced proteobacteria: *E. coli, Haemophilus influenzae, Helicobacter pylori,* and *Rickettsia prowazekii.* *E. coli* has the largest genome and the most versatile metabolic capabilities among these four organisms and also contains the largest number of sigma factors (seven). The obligate intracellular parasite *R. prowazekii* only possesses two sigma factors, in accordance with its more limited metabolic diversity. The pathogen *H. pylori* is ex-

Table 1. Sigma Factors Identified in Sequenced Genomes

Organism	Phylum	Genome size (Mb)	Total sigmas	ECF members	Reference
Escherichia coli	γ-proteobacterium	4.60	7	2	Blattner et al. (1997)
Haemophilus influenzae	γ-proteobacterium	1.83	4	2	Fleishmann et al. (1995)
Helicobacter pylori	ε-proteobacterium	1.67	3	0	Tomb et al. (1997)
Rickettsia prowazekii	α-proteobacterium	1.11	2	0	Andersson et al. (1998)
Bacillus subtilis	low-GC gram-positive	4.20	18	7	Kunst et al. (1997)
Mycobacterium tuberculosis	high-GC gram-positive	4.41	13	11	Cole et al. (1998)
Chlamydia trachomatis	chlamydia	1.04	3	0	Stephens et al. (1998)
Synechocystis sp. PCC 6803	cyanobacterium	3.57	9	3	Kaneko et al. (1996a,b)
Aquifex aeolicus	aquificales	1.55	4	0	Deckert et al. (1998)
Borrelia burgdorferi	spirochete	1.44	3	0	Fraser et al. (1997)
Treponema pallidum	spirochete	1.11	5	1	Fraser et al. (1998)
Mycoplasma genitalium	mycoplasma	0.58	1	0	Fraser et al. (1995)
Mycoplasma pneumoniae	mycoplasma	0.81	1	0	Himmelreich et al. (1996)

posed only to the gastric environment and contains about one-third the number of sensor proteins and response regulators found in *E. coli*. Again, the number of sigma factors correlates with the reduced metabolic variety: *H. pylori* manages with three sigma factors.

Regulation by Alternative Sigmas

It is abundantly clear that regulation by alternative sigma factors is widespread. We may wonder why this is so. Although there may be no "reason" for the prevalence of this form of regulation, we note that some consequences of regulating with alternative sigmas cannot be duplicated by using activators that work with the housekeeping sigma. The amount of a particular sigma and its affinity for core RNA polymerase will control the prevalence of that type of holoenzyme in the cell. Competition among sigmas for core RNA polymerase will set the amount of transcription directed by each type of active sigma in the cell. When the housekeeping sigma is partially or completely disabled, most holoenzyme will have the alternative sigma, permitting most transcription to be directed by that sigma. This feature may explain why alternative sigmas often control developmental and heat shock responses. At present, only two instances of proliferation of eukaryotic basal transcription factors for regulatory reasons are known. A cell-type-specific TATA-binding protein has been found in human cells (Dikstein et al. 1996). One group of Archaea has a proliferation of TFIIB subunits, one of which is involved in the heat shock response (C. Daniels, unpubl.).

SUMMARY AND PROSPECTS

Ever since its discovery 30 years ago, the sigma subunit of RNA polymerase has stimulated research into the process of transcription initiation and its regulation in the cell. Each set of questions answered has provoked even more exciting investigations. This continues to be true. We have arrived at a stage where the interactions of sigma with DNA and with core RNA polymerase are known, or will be known shortly. The emerging model is one in which sigma forms a very extensive interface with core RNA polymerase. Many of these interactions are in the vicinity of the DNA-binding domains in sigma, leading us to speculate that the sigma-core interface is used to transmit information about promoter interaction from sigma to core. This, of course, remains to be determined. Understanding the functional significance of the sigma-core and sigma-DNA interactions and how they change during the transcription initiation process are clearly the next priority. Likewise, we are close to identifying the contact points between sigma and many of the activators that regulate transcription, but except for the simplest case of recruitment to the promoter, we do not know how the interaction between activator and sigma accelerates transcription. Finally, the discovery and analysis of alternative sigma factors have provided a wonderful vantage point for understanding how the cell copes with stress and irreversible decisions in development. The availability of genome sequences aids in the discovery of the genes transcribed by each alternative sigma. The mapping of these "regulons" reveals the cellular functions required to cope with a variety of conditions and outlines the overlapping regulatory circuits utilized by the cell to survive. We fully expect that the continued study of sigmas will yield exciting new insights into regulation and cellular function.

REFERENCES

Aiyar S.E., Juang Y.-L., Hellman J.D., and deHaseth P.L. 1994. Mutations in sigma factor that affect the temperature dependence of transcription forrom a promoter, but not from a mismatch bubble in double-stranded DNA. *Biochemistry* **33:** 11501.

Andersson S.G.E., Zomorodipour A., Andersson J.O., Sicheritz-Ponten T., Alsmark U.C.M., Podowski R.M., Naslund A.K., Ericksson A.-S., Winkler, H.H., and Kurland, C.G. 1998. The genome sequence of *Rickettsia prowazekii* and the origin of mitochondria. *Nature* **396:** 133.

Barne K.A., Bown J.A., Busby S.J.W., and Minchin S.D. 1997. Region 2.5 of the *Escherichia coli* RNA polymerase σ⁷⁰ subunit is responsible for the recognition of the 'extended –10' motif at promoters. *EMBO J.* **16:** 4034.

Blattner F.R., Plunkett G., III, Bloch C.A., Perna N.T., Burland V., Riley M., Collado-Vides J., Glasner J.D., Rode C.K., Mayhew G.F., Gregor J., Davis N.W., Kirkpatrick H.A., Goeden M.A., Rose D.J., Mau B., and Shao Y. 1997. The complete genome sequence of *Escherichia coli* K-12. *Science* **277:** 1453.

Buckle M. and Buc H. 1994. On the mechanism of promoter recognition by *E. coli* RNA polymerase. In *Transcription: Mechanism and regulation* (ed. R.C. Conaway and J.W. Conaway), pp. 207. Raven Press, New York.

Buckner C.M., Schyns G., and Moran C.P., Jr. 1998. A region in the *Bacillus subtilis* transcription factor Spo0A that is important for *spoIIG* promoter activation. *J. Bacteriol.* **180:** 3578.

Burgess R.R. and Travers A.A. 1970. *Escherichia coli* RNA polymerase: Purification, subunit structure, and factor requirements. *Fed. Proc.* **29:** 1164.

Burgess R.R., Travers A.A., Dunn J.J., and Bautz E.K.F. 1969. Factor stimulating transcription by RNA polymerase. *Nature* **221:** 43.

Busby S. and Ebright R. 1994. Promoter structure, promoter recognition, and transcription activation in prokaryotes. *Cell* **79:** 743.

———. 1997. Transcription activation at class II CAP-dependent promoters. *Mol. Microbiol.* **23:** 853.

Callaci S. and Heyduk T. 1998. Conformation and DNA binding properties of a single-stranded DNA binding region of σ⁷⁰ subunit from *Escherichia coli* RNA polymerase are modulated by an interaction with the core enzyme. *Biochemistry* **37:** 3312.

Caslake L.F., Ashraf S.I., and Summers A.O. 1997. Mutations in the α and σ-70 subunits of RNA polymerase affect expression of the *mer* operon. *J. Bacteriol.* **179:** 1787.

Chan B., Spassky A., and Busby S. 1990. The organization of open complexes between *Escherichia coli* RNA polymerase and DNA fragments carrying promoters either with or without consensus –35 region sequences. *Biochem. J.* **270:** 141.

Cliften P.F., Park J.-Y., Davis B.P., Jang S.-H., and Jaehning J.A. 1997. Identification of three regions essential for interaction between a σ-like factor and core RNA polymerase. *Genes Dev.* **11:** 2897.

Cole S.T., Brosch R., Parkhill J., Garnier T., Churcher C., Harris D., Gordon S.V., Eiglmeier K., Gas S., Barry C.E., III, Tekaia F., Badcock K., Basham D., Brown D., Chillingworth T., Connor R., Davies R., Devlin K., Feltwell T., Gentles S., Hamlin N., Holroyd S., Hornsby T., Jagels K., Krogh A. et al. 1998. Deciphering the biology of *Mycobacterium tuberculo-*

sis from the complete genome sequence. *Nature* **393:** 537.

Colland F., Orsini G., Brody E.N., Buc H., and Kolb A. 1998. The bacteriophage T4 AsiA protein: A molecular switch for sigma 70-dependent promoters. *Mol. Microbiol.* **27:** 819.

Daniels D., Zuber P., and Losick R. 1990. Two amino acids in an RNA polymerase sigma factor involved in the recognition of adjacent base pairs in the −10 region of a cognate promoter. *Proc. Natl. Acad. Sci.* **87:** 8075.

Darst S.A., Kubalek E.W., and Kornberg R.D. 1989. Three-dimensional structure of *Escherichia coli* RNA polymerase holoenzyme determined by electron crystallography. *Nature* **340:** 730.

Deckert G., Warren P.V., Gaasterland T., Young W.G., Lenox A.L., Graham D.E., Overbeek R., Snead M.A., Keller M., Aujay M., Huber R., Feldman R.A., Short J.M., Olsen G.J., and Swanson R.V. 1998. The complete genome of the hyperthermophilic bacterium *Aquifex aeolicus*. *Nature* **392:** 353.

deHaseth P.L., Zupancic M.L., and Record M.T., Jr. 1998. RNA polymerase-promoter interactions: The comings and goings of RNA polymerase. *J. Bacteriol.* **180:** 3019.

Dikstein R., Zhou S., and Tjian R. 1996. Human TAFII 105 is a cell type-specific TFIID subunit related to hTAFII130. *Cell* **87:** 137.

Dombroski A.J. 1997. Recognition of the -10 promoter sequence by a partial polypeptide of σ⁷⁰ *in vitro*. *J. Biol. Chem.* **272:** 3487.

Dombroski A.J., Walter W.A., and Gross C.A. 1993. Amino-terminal amino acids modulate σ-factor DNA-binding activity. *Genes Dev.* **7:** 2446.

Dombroski A.J., Johnson B.D., Lonetto M., and Gross C.A. 1996. The sigma subunit of *Escherichia coli* RNA polymerase senses promoter spacing. *Proc. Natl. Acad. Sci.* **93:** 8858

Dombroski A.J., Walter W.A., Record, M.T., Jr., Siegele D.A., and Gross C.A. 1992. Polypeptides containing highly conserved regions of transcription initiation factor σ⁷⁰ exhibit specificity of binding to promoter DNA. *Cell* **70:** 501.

Eichenberger P., Déthiollaz S., Buc H., and Geiselmann J. 1997. Structural kinetics of transcription activation at the *malT* promoter of *Escherichia coli* by UV laser footprinting. *Proc. Natl. Acad. Sci.* **94:** 9022.

Felsenstein J. 1989. PHYLIP—Phylogeny inference package (Version 3.2). *Cladistics* **5:** 165.

Fleischmann R.D., Adams M.D., White O., Clayton R.A., Kirkness E.F., Kerlavage A.R., Bult C.J., Tomb J.-F., Dougherty B.A., Merrick J.M., McKenney K., Sutton G., FitzHugh W., Fields C.A., Gocayne J.D., Scott J.D., Shirley R., Liu L.-I., Glodek A., Kelley J.M., Weidman J.F., Phillips C.A., Spriggs T., Hedblom E., and Cotton M.D. et al. 1995. Whole-genome random sequencing and assembly of *Haemophilus influenzae* Rd. *Science* **269:** 496.

Fraser C.M., Gocayne J.D., White O., Adams M.D., Clayton R.A., Fleischmann R.D., Bult C.J., Kerlavage A.R., Sutton G., Kelley J.M., Fritchman J.L., Weidman J.F., Small K.V., Sandusky M., Fuhrmann J.L., Nguyen D.T., Utterback T.R., Saudek D.M., Phillips C.A., Merrick J.M., Tomb J.-F., Dougherty B.A., Bott K.F., Hu P.-C., and Lucier T.S. et al. 1995. The minimal gene complement of *Mycoplasma genitalium*. *Science* **270:** 397.

Fraser C.M., Norris S.J., Weinstock G.M., White O., Sutton G.G., Dodson R., Gwinn M., Hickey E.K., Clayton R., Ketchum K.A., Sodergren E., Hardham J.M., McLeod M.P., Salzberg S., Peterson J., Khalak H., Richardson D., Howell J.K., Chidambaram M., Utterback T., McDonald L., Artiach P., Bowman C., Cotton M.D., and Fujii C. et al. 1998. Complete genome sequence of *Treponema pallidum*, the syphilis spirochete. *Science* **281:** 375.

Fraser C.M., Casjens S., Huang W.M., Sutton G.G., Clayton R.A., Lathigra R., White O., Ketchum K.A., Dodson R., Hickey E.K., Gwinn M., Dougherty B., Tomb J.-F., Fleischmann R.D., Richardson D., Peterson J., Kerlavage A.R., Quackenbush J., Salzberg S., Hanson M., van Vugt R., Palmer N., Adams M.D., Gocayne J.D., and Weidman J. et al. 1997. Genomic sequence of a Lyme disease spirochete, *Borrelia burgdorferi*. *Nature* **190:** 580.

Gardella T., Moyle H., and Susskind M.M. 1989. A mutant *Escherichia coli* σ⁷⁰ subunit of RNA polymerase with altered promoter specificity. *J. Mol. Biol.* **206:** 579.

Grossman A.D., Erickson J.W., and Gross C.A. 1984. The *htpR* gene product of *E. coli* is a sigma factor for heat-shock promoters. *Cell* **38:** 383.

Guo Y. and Gralla J.D. 1998. Promoter opening via a DNA fork junction binding activity. *Proc. Natl. Acad. Sci.* **95:** 11655.

Helmann J.D. 1995. Compilation and analysis of *Bacillus subtilis* σ^A-dependent promoter sequences: Evidence for extended contact between RNA polymerase and upstream promoter DNA. *Nucleic Acids Res.* **23:** 2351.

Helmann J.D. and Chamberlin M.J. 1988. Structure and function of bacterial sigma factors. *Annu. Rev. Biochem.* **57:** 839.

Hernandez V.J. and Cashel M. 1995. Changes in conserved region 3 of *Escherichia coli* σ⁷⁰ mediate ppGpp-dependent functions *in vivo*. *J. Mol. Biol.* **252:** 536.

Hernandez V.J., Hsu L.M., and Cashel M. 1996. Conserved region 3 of *Escherichia coli* final σ⁷⁰ is implicated in the process of abortive transcription. *J. Biol. . Chem.* **27:** 18775.

Himmelreich R., Hilbert H., Plagens H., Pirkl E., Li B.C., and Herrmann R. 1996. Complete sequence analysis of the genome of the bacterium *Mycoplasma pneumoniae*. *Nucleic Acids Res.* **24:** 4420.

Hinton D.M., March-Amegadzie R., Gerber J.S., and Sharma M. 1996. Characterization of pre-transcription complexes made at a bacteriophage T4 middle promoter: Involvement of the T4 MotA activator and the T4 AsiA protein, a sigma 70 binding protein, in the formation of the open complex. *J. Mol. Biol.* **256:** 235.

Hu J.C. and Gross C.A. 1985. Mutations in the sigma subunit of *E. coli* RNA polymerase which affect positive control of transcription. *Mol. Gen. Genet.* **199:** 7.

Ishihama A. 1993. Protein-protein communication within the transcription apparatus. *J. Bacteriol.* **175:** 2483.

Jones C.H. and Moran C.J. 1992. Mutant sigma factor blocks transition between promoter binding and initiation of transcription. *Proc. Natl. Acad. Sci.* **89:** 1958.

Joo D.M., Ng N., and Calendar R. 1997. A σ³² mutant with a single amino acid change in the highly conserved region 2.2 exhibits reduced core RNA polymerase affinity. *Proc. Natl. Acad. Sci.* **94:** 4907.

Joo D.M., Nolte A., Calendar R., Zhou Y.N., and Jin D.J. 1998. Multiple regions on the *Escherichia coli* heat shock transcription factor σ³² determine core RNA polymerase binding specificity. *J. Bacteriol.* **180:** 1095.

Juang Y.-L. and Helmann J.D. 1994. A promoter melting region in the primary σ factor of *Bacillus subtilis*: Identification of functionally important aromatic amino acids. *J. Mol. Biol.* **235:** 1470.

Kaneko T., Sato S., Kotani H., Tanaka A., Asamizu E., Nakamura Y., Miyajima N., Hirosawa M., Sugiura M., Sasamoto S., Kimura, T., Hosouchi T., Matsuno A., Muraki A., Nakazaki N., Naruo K., Okumura S., Shimpo S., Takeuchi C., Wada T., Watanabe A., Yamada M., Yasuda M., and Tabata S. 1996a. Sequence analysis of the genome of the unicellular cyanobacterium *Synechocystis* sp. strain PCC 6803. II. Sequence determination of the entire genome and assignment of potential protein-coding regions. *DNA Res.* **3:** 109.

———. 1996b. Sequence analysis of the genome of the unicellular cyanobacterium *Synechocystis* sp. strain PCC 6803. II. Sequence determination of the entire genome and assignment of potential protein-coding regions. *DNA Res.* (suppl.) **3:** 185.

Keilty S. and Rosenberg M. 1987. Constitutive function of a positively regulated promoter reveals new sequences essential for activity. *J. Biol. Chem.* **262:** 6389.

Kenney T.J., York K., Youngman P., and Moran C.P. 1989. Genetic evidence that RNA polymerase associated with σ^A uses a sporulation-specific promoter in *Bacillus subtilis*. *Proc. Natl. Acad. Sci.* **86:** 9109.

Kim S.-K., Makino K., Amemura M., Nakata A., and Shinagawa H. 1995. Mutational analysis of the role of the first helix of region 4.2 of the σ⁷⁰ subunit of *Escherichia coli* RNA polymerase in transcriptional activation by activator protein

PhoB. *Mol. Gen. Genet.* **248**: 1.

Kuldell N. and Hochschild A. 1994. Amino acid substitutions in the -35 recognition motif of σ^{70} that result in defects in phage λ repressor-stimulated transcription. *J. Bacteriol.* **176**: 2991.

Kunst F., Ogasawara N., Moszer I., Albertini A.M., Alloni G., Azevedo V., Bertero M.G., Bessieres P., Bolotin A., Borchert S., Borriss R., Boursier L., Brans A., Braun M., Brignell S.C., Bron S., Brouillet S., Bruschi C.V., Caldwell B., Capuano V., Carter N.M., Choi S.K., Codani J.J., Connerton I.F., and Cummings N.J. et al. 1997. The complete genome sequence of the gram-positive bacterium *Bacillus subtilis*. *Nature* **390**: 249.

Landini P. and Volkert M.R. 1995. RNA polymerase α subunit binding site in positively controlled promoters: A new model for RNA polymerase-promoter interaction and transcriptional activation in the *Escherichia coli ada* and *aidB* genes. *EMBO J.* **14**: 4329.

Landini P., Brown J.A., Volkert M.R., and Busby S.J.W. 1998. Ada protein-RNA polymerase σ subunit interaction and α subunit-promoter DNA interaction are necessary at different steps in transcription initation at the *Escherichia coli ada* and *aidB* promoters. *J. Biol. Chem.* **273**: 13307.

Léonetti J.-P., Wong K., and Geiduschek E.P. 1998 Core-sigma interaction: Probing the interaction of the bacteriophage T4 gene 55 promoter recognition protein with *E. coli* RNA polymerase core. *EMBO J.* **17**: 1467.

Lesley S.A. and Burgess R.R. 1989. Characterization of the *Escherichia coli* transcription factor σ^{70}: Localization of a region involved in the interaction with core RNA polymerase. *Biochemistry* **28**: 7728.

Lesley S.A., Brow M.D., and Burgess R.R. 1991. Use of *in vitro* protein synthesis from polymerase chain reaction-generated templates to study interaction of *Escherichia coli* transcription factors with core RNA polymerase and for epitope mapping of monoclonal antibodies. *J. Biol. Chem.* **266**: 2632.

Li M., Moyle H., and Susskind M.M. 1994. Target of the transcriptional activation function of phage λ cI protein. *Science* **263**: 75.

Lisser S. and Margalit H. 1993. Compilation of *E. coli* mRNA promoter sequences. *Nucleic Acids Res.* **21**: 1507.

Lonetto M., Gribskov M., and Gross C.A. 1992. The σ^{70} family: Sequence conservation and evolutionary relationships. *J. Bacteriol.* **174**: 3843.

Lonetto M.A., Brown K.L., Rudd K.E., and Buttner M.J. 1994. Analysis of the *Streptomyces coelicolor sigE* gene reveals the existence of a subfamily of eubacterial RNA polymerase σ factors involved in the regulation of extracytoplasmic functions. *Proc. Natl. Acad. Sci.* **91**: 7573.

Lonetto M.A., Rhodius V., Lamberg K., Kiley P., Busby S., and Gross C.A. 1998. Identification of a contact site for different transcription activators in Region 4 of the *Escherichia coli* RNA polymerase σ^{70} subunit. *J. Mol. Biol.* **284**: 1353.

Losick R. and Pero J. 1981. Cascades of sigma factors. *Cell* **25**: 582.

Makino K., Amemura M., Kim S.-K., Nakata A., and Shinagawa H. 1993. Role of the σ^{70} subunit of RNA polymerase in transcriptional activation by activator protein PhoB in *Escherichia coli*. *Genes Dev.* **7**: 149.

Malhotra A., Severinova E., and Darst S.A. 1996. Crystal structure of a σ^{70} subunit fragment from *E. coli* RNA polymerase. *Cell* **87**: 127.

Marr M.T. and Roberts J.W. 1997. Promoter recognition as measured by binding of polymerase to nontemplate strand oligonucleotide. *Science* **276**: 1258.

Mecsas J., Cowing W., and Gross C.A. 1991. Development of RNA polymerase-promoter contacts during open complex formation. *J. Mol. Biol.* **220**: 585.

Nagai H. and Shimamoto N. 1997. Regions of the *Escherichia coli* primary sigma factor σ^{70} that are involved in interaction with RNA polymerase core enzyme. *Genes Cells* **2**: 725.

Nudler E., Avetissova E., Markovtsov V., and Goldfarb A. 1996. Transcription processivity: Protein-DNA interactions holding together the elongation complex. *Science* **273**: 211.

Ouhammouch M., Orsini G., and Brody E.N. 1994. The *asiA*

gene product of bacteriophage T4 is required for middle mode RNA synthesis. *J. Bacteriol.* **176**: 3956.

Owens J.T., Chmura A.J., Murakami K., Fujita N., Ishihama A., and Meares C.F. 1998. Mapping the promoter DNA sites proximal to conserved regions of σ^{70} in an *Escherichia coli* RNA polymerase-*lac*UV5 open promoter complex. *Biochemistry* **37**: 7670.

Polyakov A., Severinova E., and Darst S.A. 1995. Three-dimensional structure of *E. coli* core RNA polymerase: Promoter binding and elongation conformations of the enzyme. *Cell* **83**: 365.

Ponnambalam S., Webster C., Bingham A., and Busby S. 1986. Transcription initiation at the *Escherichia coli* galactose operon promoters in the absence of the normal –35 region sequences. *J. Biol. Chem.* **261**: 16043.

Record M.T., Jr., Reznikoff W.S., Craig M.L., McQuade K.L., and Schlax P.J. 1996. *Escherichia coli* RNA polymerase ($E\sigma^{70}$), promoters, and the kinetics of the steps of transcription initiation. In Escherichia coli *and* Salmonella: *Cellular and molecular biology*, 2nd edition (ed. F.C. Neidhardt et al.), vol. 1, p. 792. ASM Press, Washington, D.C.

Rhodius V.A., West D.M., Webster C.L., Busby S.J.W., and Savery N.J. 1997. Transcription activation at class II CRP-dependent promoters: The role of different activating regions. *Nucleic Acids Res.* **25**: 326.

Ring B.Z. and Roberts J.W. 1994. Function of a nontranscribed DNA strand site in transcription elongation. *Cell* **78**: 317.

Ring B.Z., Yarnell W.S., and Roberts J.W. 1996. Function of *E. coli* RNA polymerase σ factor σ^{70} in promoter-proximal pausing. *Cell* **86**: 485.

Roberts C.W. and Roberts J.W. 1996. Base-specific recognition of the nontemplate strand of promoter DNA by *E. coli* RNA polymerase. *Cell* **86**: 495.

Rong J.C. and Helmann J.D. 1994. Genetic and physiological studies of *Bacillus subtilis* σ^{A} mutants defective in promoter melting. *J. Bacteriol.* **176**: 5218.

Ross W., Goslink K., Salomon J., Igarashi K., Zou C., Ishihama A., Severinov K., and Gourse R. 1993. A third recognition element in bacterial promoters: DNA binding by the α subunit of RNA polymerase. *Science* **262**: 1407.

Schickor P., Metzger W., Werel W., Lederer H., and Heumann H. 1990. Topography of intermediates in transcription initiation of *E. coli*. *EMBO J.* **9**: 2215.

Schuler M.F., Tatti K.M., Wade K.H., and Moran C.P., Jr. 1995. A single amino acid substitution in σ^{E} affects its ability to bind core RNA polymerase. *J. Bacteriol.* **177**: 3687.

Schyns G., Buckner C.M., and Moran C.P., Jr. 1997. Activation of the *Bacillus subtilis spoIIG* promoter requires interaction of Spo0A and the sigma subunit of RNA polymerase. *J. Bacteriol.* **179**: 5605.

Severinov K., Fenyo D., Severinova E., Mustaev A., Chait B.T., Goldfarb A., and Darst S.A. 1994. The σ subunit conserved region 3 is part of "5´-face" of active center of *Escherichia coli* RNA polymerase. *J. Biol. Chem.* **269**: 20826.

Severinova E., Severinov K., and Darst S.A. 1998. Inhibition of *Escherichia coli* RNA polymerase by bacteriophage T4 AsiA. *J. Mol. Biol.* **279**: 9.

Siegele D.A., Hu J.C., Walter W.A., and Gross C.A. 1989. Altered promoter recognition by mutant forms of the σ^{70} subunit of *Escherichia coli* RNA polymerase. *J. Mol. Biol.* **206**: 591.

Silverstone A.E., Gorman M., and Scaife J.G. 1972. ALT: A new factor involved in the synthesis of RNA by *Escherichia coli*. *Mol. Gen. Genet.* **118**: 223.

Simpson R.B. 1979. The molecular topography of RNA polymerase-promoter interaction. *Cell* **18**: 277.

Stephens R.S., Kalman S., Lammel C., Fan J., Marathe R., Aravind L., Mitchell W., Olinger L., Tatusov R.L., Zhao Q., Koonin E.V., and Davis R.W. 1998. Genome sequence of an obligate intracellular pathogen of humans: *Chlamydia trachomatis*. *Science* **282**: 754.

Strickland M.S., Thompson N.E., and Burgess R.R. 1988. Structure and function of the σ-70 subunit of *Escherichia coli* RNA polymerase monoclonal antibodies: Localization of epitopes by peptide mapping and effects on transcription. *Biochemistry* **27**: 5755.

Tomb J.-F., White O., Kerlavage A.R., Clayton R.A., Sutton G.G., Fleischmann R.D., Ketchum K.A., Klenk H.P., Gill S., Dougherty B.A., Nelson K., Quackenbush J., Zhou L., Kirkness E.F., Peterson S., Loftus B., Richardson D., Dodson R., Khalak H.G., Glodek A., McKenney K., Fitzgerald L.M., Lee N., Adams M.D., and Hickey E.K. et al. 1997. The complete genome sequence of the gastric pathogen *Helicobacter pylori*. *Nature* **388:** 539.

Travers A.A., Buckland R., Gorman M., Le Grice S.S.G., and Scaife J.G. 1978. A mutation affecting the subunit of RNA polymerase changes transcriptional specificity. *Nature* **273:** 354.

Waldberger C., Gardella T., Wong R., and Susskind M.M. 1990. Changes in conserved region 2 of *Escherichia coli* σ^{70} affecting promoter recognition. *J. Mol. Biol.* **21:** 267.

Wilson C. and Dombroski A.J. 1997. Region 1 of σ^{70} is required for efficient isomerization and initiation of transcription by *Escherichia coli* RNA polymerase. *J. Mol. Biol.* **267:** 60.

Wösten M.M.S.M. 1998. Eubacterial σ-factors. *FEMS Microbiol. Rev.* **22:** 127.

Zuber P., Healy J., Carter H.L., III, Cutting S., Moran, C.P., Jr., and Losick R. 1989. Mutation changing the specificity of an RNA polymerase σ factor. *J. Mol. Biol.* **206:** 605.

The Bacterial Enhancer-binding Protein NtrC as a Molecular Machine

I. Rombel,* A. North,† I. Hwang,‡ C. Wyman,§ and S. Kustu

Department of Plant and Microbial Biology, University of California, Berkeley, California 94720

NtrC AS A MOLECULAR MACHINE

The bacterial enhancer-binding protein NtrC (nitrogen regulatory protein) C functions as a molecular machine to activate transcription by the σ^{54}-holoenzyme form of RNA polymerase (Weiss et al. 1992; Wedel and Kustu 1995). When phosphorylated on aspartate 54, a reaction that is self-catalyzed, NtrC forms large oligomers that are capable of hydrolyzing ATP and activating transcription (Fig. 1) (Wyman et al. 1997). NtrC contacts σ^{54}-holoenzyme from distant enhancer sites by means of DNA loop formation—although contact appears to be transient—and catalyzes the isomerization of closed promoter complexes to transcriptionally productive open complexes. Although σ^{54}-holoenzyme can bind to promoters in physically detectable closed complexes, it cannot form open complexes in the absence of an activator or energy source because open complex formation is thermodynamically as well as kinetically unfavorable (Wedel and Kustu 1995). Thus, NtrC is viewed as a molecular machine because it must couple the energy available from ATP hydrolysis to the formation of open complexes by σ^{54}-holoenzyme.

In this paper, we address two aspects of NtrC function: the roles of enhancers and our (limited!) understanding of NtrC as a molecular machine. Other aspects of its function, for example, its physiological importance and its characterization as a response regulator of a "two-component" regulatory system, have been reviewed previously (see, e.g., Kustu et al. 1989; Weiss et al. 1992; Porter et al. 1995). Recent experiments indicate that the role of the *glnA* enhancer in facilitating the formation of active NtrC oligomers is at least as important as its recognized role in tethering these oligomers near the promoter. Functional evidence indicates that the active oligomers, which are unusual in that not all dimers are DNA-bound (Wyman et al. 1997), are either octamers or hexamers with special properties. NtrC and other activators of σ^{54}-holoenzyme have a purine nucleotide-binding domain that appears to assume a classic mononucleotide fold (Osuna et al. 1997); recent results allow us to define determinants for ATP binding and hydrolysis and determinants involved in interaction with polymerase, i.e., contact and/or energy-coupling.

METHODS

Overproduction and purification of proteins and assays of their activities. Proteins were overexpressed and purified as described previously (Porter et al. 1993; Flashner et al. 1995; North et al. 1996; North and Kustu 1997; I. Rombel et al., unpubl.). Unless noted otherwise, protein concentrations refer to dimers. The ability of

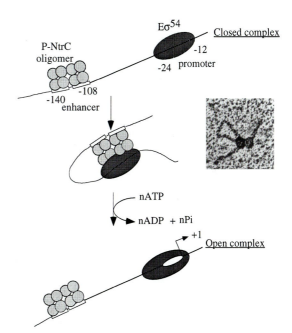

Figure 1. Transcriptional activation by NtrC at the *glnA* promoter of *S. typhimurium*. The *glnA* gene encodes glutamine synthetase. Conserved promoter sequences recognized by σ^{54}-holoenzyme ($E\sigma^{54}$) lie at sites –12 and –24 with respect to the start site of transcription at +1. Boxes represent the two 17-bp NtrC-binding sites that constitute the *glnA* enhancer; they are centered at –108 and –140. NtrC binds to the enhancer, but only the phosphorylated form (P-NtrC) can activate transcription. Active oligomers of P-NtrC must contain not only the two dimers bound to the enhancer, but also an additional dimer(s) bound to these by protein-protein interactions. (*Inset*) Electron micrograph of a looped complex between P-NtrC (*gray*) and the polymerase (*black*); 350 bp of DNA were inserted between enhancer and promoter (Su et al. 1990). Although contact between NtrC and σ^{54}-holoenzyme at the *glnA* promoter occurs by means of random conformational changes in DNA, in many cases, contact between this or other activators and the polymerase is mediated by the DNA-bending protein integration host factor (IHF) (Hoover et al. 1990; Dworkin et al. 1998). Roles of IHF provide interesting parallels for those of the eukaryotic transcription factors LEF1 and SRY (Love et al. 1995; Werner et al. 1996).

Present addresses: *University of Texas Southwestern Medical Center, Department of Internal Medicine and Cardiology, 5323 Harry Hines Boulevard, Dallas, Texas 75235-8573; †Onyx Pharmaceuticals, Richmond, California 94806; ‡Plant Protectants R.U., Korea Research Institute of Bioscience and Biotechnology/KIST Yusung, P.O. Box 115, Taejon, South Korea 305-600; §Department of Cell Biology and Genetics, Erasmus University, Rotterdam, Netherlands.

NtrC proteins to catalyze open complex formation by σ^{54}-holoenzyme was assessed on supercoiled plasmid templates in single-cycle transcription assays. In cases where different forms of NtrC were mixed, conditions were chosen to minimize subunit exchange between dimers, except in control experiments (Porter et al. 1993; Wyman et al. 1997). σ^{54}-holoenzyme (formed by first mixing σ^{54} and RNA polymerase core), DNA template, carbamoyl phosphate (10 mM), and transcription buffer were mixed and then added to the DNA-binding form of the NtrC protein. This mixture was held on ice for 10 minutes and then incubated at 25°C for 10 minutes to allow phosphorylation of NtrC, if this was possible. For data in Figure 6, panel b, the form of NtrC incapable of DNA binding was concomitantly phosphorylated for 10 minutes at 25°C in buffer containing 50 mM Tris-actetate (pH 8.0), 8 mM magnesium acetate, and 10 mM carbamoyl phosphate. Since phosphorylation is transient (Weiss et al. 1992), it must be maintained in situ. After the 10-minute incubation at 25°C, all components were placed on ice, the appropriate amount of the nonbinding form of NtrC was added to the transcription mixture, and formation of open complexes was initiated by addition of ATP (4 mM final concentration) and incubation at 25°C for 5 minutes (total volume 26 μl). Addition of heparin to stop open complex formation and the subsequent synthesis of transcripts were performed essentially as described, as was electrophoresis of transcripts and their quantitation with a PhosphorImager. To achieve complete subunit mixing between homodimeric species of NtrC, the DNA-binding and nonbinding forms of NtrC were mixed and incubated together for 20 minutes at 37°C (Klose et al. 1994). The other components were then added and all subsequent steps were carried out as described above. Transcription buffer contained 50 mM Tris-acetate (pH 8.0), 100 mM potassium acetate, 8 mM magnesium acetate, 27 mM ammonium acetate, 1 mM dithiothreitol (DTT), and 3.5% polyethylene glycol (PEG 6000–8000), all final concentrations in the reaction mixture. The final concentrations of σ^{54} and core RNA polymerase were 50 and 30 nM, respectively. The ATPase activity of NtrC proteins and their ability to bind MgATP in a filter-binding assay were assessed as described previously (Weiss et al. 1991; I. Rombel et al., unpubl.).

Gel filtration chromatography. Sephacryl S-300 HR was packed in an XK16/70 column (Pharmacia Biotech) to a final bed volume of 105 ml. For sieving unphosphorylated or phosphorylated NtrC proteins, the column was equilibrated in B buffer (50 mM Tris-acetate [pH8.3], 50 mM KCl, 5% v/v glycerol, 0.1 mM EDTA, and 1 mM DTT) containing 8 mM $MgCl_2$ or 8 mM $MgCl_2$ + 10 mM carbamoyl phosphate, respectively. It was run at 4°C at 0.25 ml/min, and protein was detected by monitoring absorption at 280 nm. NtrC proteins were phosphorylated for 10 minutes at room temperature immediately prior to loading. A high-molecular-weight gel filtration kit (Pharmacia Biotech) was used to calibrate the column just prior to sieving NtrC proteins.

Scanning force microscopy (SFM). SFM was performed essentially as described previously (Wyman et al. 1997), as was image analysis and determination of the sizes of NtrC complexes at the strong enhancer. NtrC proteins were phosphorylated with the low-molecular-mass donor carbamoyl phosphate or the physiological donor NtrB, which depends on ATP (Weiss et al. 1992). ATP was present in both cases. As an internal control, each experiment included complexes of the relevant unphosphorylated NtrC protein at a single binding site.

RESULTS

The *glnA* Enhancer Nucleates the Formation of Active NtrC Oligomers

To assess the roles of the *glnA* enhancer in promoting the formation of NtrC oligomers and tethering these oligomers in high local concentration near the promoter, we have studied transcriptional activation by mutant forms of NtrC that fail to bind to DNA (North and Kustu 1997). These forms carry three alanine substitutions in the second (recognition) helix of the helix-turn-helix DNA-binding motif and are employed because they are structurally stable. At high concentrations, NtrC^{3ala} proteins can activate transcription directly from solution, commensurate with the view that the essential role of NtrC is to alter the configuration of polymerase-promoter complexes rather than to recruit σ^{54}-holoenzyme to promoters (Weiss et al. 1992).

We previously noted that phosphorylated NtrC^{3ala} and unphosphorylated NtrC^{S160F,3ala}, called a "constitutive form" because it has some ability to activate transcription without being phosphorylated, had different capacities for transcriptional activation (North and Kustu 1997), namely, phosphorylated NtrC^{3ala} activated at markedly lower concentrations than unphosphorylated NtrC^{S160F,3ala} (Fig. 2a). We postulated that this difference was due to a difference in the ability of the two proteins to oligomerize. We now have direct evidence for this. At a monomer concentration of 10 μM, phosphorylated NtrC^{3ala} behaved as a mixture of dimers and higher-order oligomers on a gel filtration column (Fig. 3, trace c), whereas NtrC^{S160F,3ala} was largely dimeric (Fig. 3, trace a). We have extended the comparison to phosphorylated NtrC^{S160F,3ala}. Strikingly, phosphorylation of this protein yielded a preparation that was essentially all large oligomers (Fig. 3, trace b) and that had enormously increased ability to activate transcription (Fig. 2a). The comparison between phosphorylated and unphosphorylated forms of NtrC^{S160F,3ala} shows even more clearly than the original comparison that activation of transcription from solution is limited by the ability of NtrC to oligomerize, rather than to reach polymerase by diffusion. Hence, stimulation of oligomer formation, a recently recognized role for the enhancer, appears to be as fundamental as its widely recognized role in tethering.

In corroboration of the important role of enhancers in nucleating oligomer formation, unphosphorylated NtrC^{S160F} (DNA-binding form), which forms *as many* large oligomers as phosphorylated wild-type NtrC on an ena)

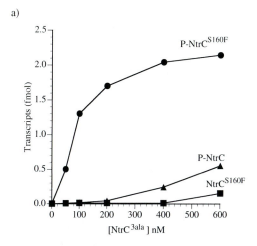

b)

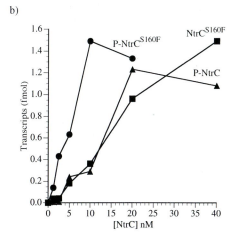

Figure 2. Transcriptional activation at the *glnA* promoter by non-DNA-binding (*a*) or DNA-binding forms of NtrC (*b*). Formation of open complexes was assessed in a single-cycle transcription assay at 37°C as described in Materials and Methods. All forms of NtrC used in panel a carried three alanine substitutions in the DNA-binding motif and were essentially unable to bind to DNA (North and Kustu 1997). Relevant NtrC proteins and their state of phosphorylation are indicated above the curves in both panels. Unphosphorylated NtrC^{3ala} and NtrC were inactive (not shown). The template was supercoiled plasmid pJES534 (1 nM), which carries the strong enhancer at a distance of about 400 bp from the *glnA* promoter (see text; Porter et al. 1993). Note the difference in the *x* axes in the two panels.

hancer (studied by scanning force microscopy; see below; Table 1), also activates transcription as well as wild-type NtrC (Fig. 2b). Phosphorylation of NtrC^{S160F}, which increases the formation of large oligomers at an enhancer above that seen for phosphorylated wild-type NtrC (Table 1), has a corresponding stimulatory effect on transcriptional activation (Fig. 2b).

Functional Characterization of Active NtrC Oligomers

Active NtrC oligomers are unusual in that not all dimers are directly DNA-bound (Wyman et al. 1997). The *glnA* enhancer and most other enhancers for NtrC

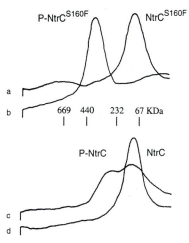

Figure 3. Gel filtration chromatography of non-DNA-binding forms of NtrC. Traces indicate absorbance at 280 nm, and the vertical bars indicate positions of elution of molecular-weight standards used to calibrate the column. Both of the NtrC proteins carried three alanine substitutions in the DNA-binding motif (see text). The proteins (0.5 mg/ml; 100 μg total) and their state of phosphorylation are indicated adjacent to the relevant traces: (*a*) NtrC^{S160F,3ala}; (*b*) P-NtrC^{S160F,3ala}; (*c*) P-NtrC^{3ala}; (*d*) NtrC^{3ala}. Molecular-mass standards were thyroglobulin, 669 kD; ferritin, 440 kD; catalase, 232 kD; and bovine serum albumin, 68 kD.

consist of two binding sites for dimers, and unphosphorylated NtrC is a dimer in solution even at very high concentrations (Fig. 3, trace d). Scanning force micrographs revealed that two dimers of unphosphorylated NtrC were bound to the *glnA* enhancer (Fig. 4, top and Table 1) (Wyman et al. 1997). (The enhancer actually employed, called "the strong enhancer," had the same configuration as the *glnA* enhancer, but the two identical binding sites were of higher affinity than those at *glnA* [Porter et al. 1993, 1995].) When NtrC was phosphorylated, larger oligomers were formed in which one or two additional dimers were bound by protein-protein interactions to those bound to the enhancer (Fig. 4, bottom and Table 1) (Wyman et al. 1997). In vitro complementation between inactive DNA-binding and nonbinding forms of NtrC provided strong functional evidence that the additional dimers were required for transcriptional activation (Wyman et al. 1997). The transcriptionally inactive forms of NtrC carried lesions of two types: (1) amino acid sub-

Table 1. Characterization of NtrC Complexes at the Strong Enhancer[a] by Scanning Force Microscopy

| Protein | Estimated percentage of complexes as | | | |
	one dimer	two dimers	>two dimers	Total[b]
NtrC[c]	22	72	6	455
P-NtrC[c]	8	63	29	383
NtrC^{S160F}	16	50	34	135
P-NtrC^{S160F}	4	35	61	98

[a]610-bp DNA fragment carrying the strong enhancer derived from the *glnA* enhancer (see text).
[b]Total number of nucleoprotein complexes examined.
[c]Data from Wyman et al. (1997).

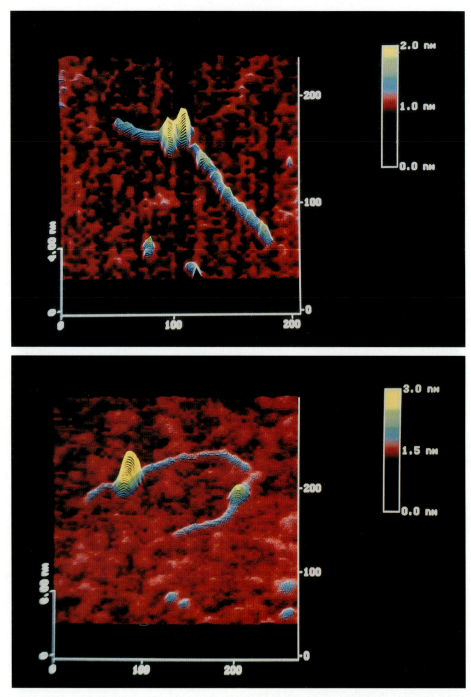

Figure 4. Scanning force micrographs of NtrC (*top*) and phosphorylated NtrC (*bottom*) at the strong enhancer (see text; Wyman et al. 1997). Images are displayed as line plots at a 60° tilt angle to emphasize topography. The *z* dimension (height), which is different in the two panels, is indicated by the color code on the accompanying bar; the mica surface is at half-maximal height. The bottom panel includes a single (unphosphorylated) dimer bound to a single NtrC-binding site that can be used for size comparison.

stitutions at the site of phosphorylation (D54A or D54N) that prevent phosphorylation and (homo) oligomer formation, and (2) the A216V substitution, which leaves the ATPase activity of the phosphorylated protein intact and therefore presumably results in loss of contact with polymerase or energy-coupling. Complementation was performed between NtrC^{D54A} or NtrC^{D54N} bound to the enhancer and phosphorylated NtrC^{A216V,3ala} as the

non-DNA-binding form. The two cooperated to yield activating oligomers.

Variations of the in vitro complementation experiments performed previously allowed us to ask how many dimers of a particular inactive NtrC protein such as NtrC^{A216V} can be assimilated into an active oligomer and whether there are constraints on their positions. The principle employed was that DNA-binding and nonbinding

Transcriptional Activation

Octamer model

Hexamer model

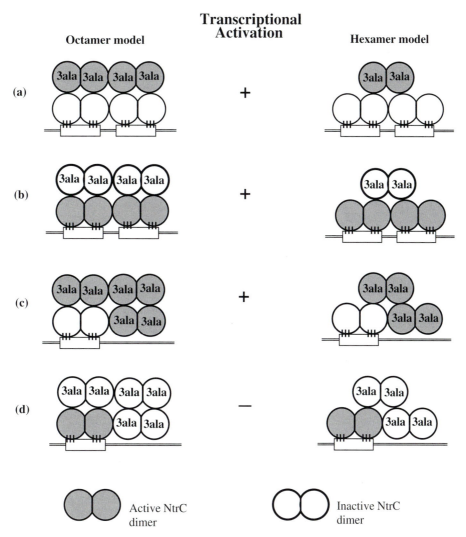

Figure 5. Assimilation of inactive NtrC dimers into active oligomers. DNA-binding forms of NtrC are indicated with three prongs, whereas non-DNA-binding forms, which carry three alanine substitutions in the DNA-binding motif, are smooth. Inactive dimers (*white*) carry the A216V substitution (see text). Results of transcriptional activation assays (Fig. 6) are easily accommodated by a model in which active oligomers are octamers but can also be accommodated if they are hexamers.

forms of NtrC can cooperate to yield oligomers that are tethered to DNA either by pairs of DNA-bound dimers at an enhancer (as described above) or by single DNA-bound dimers at a single binding site (Porter et al. 1993, 1995). For cases in which the non-DNA-binding partner is active, tethered oligomers nevertheless activate transcription better than nontethered oligomers, presumably due to their increased frequency of contacts with σ^{54}-holoenzyme.

Previous studies indicated that one or two inactive dimers of phosphorylated $NtrC^{A216V,3ala}$ can be assimilated into an active oligomer with two enhancer-bound dimers of $NtrC^{D54A}$ or $NtrC^{D54N}$ (Wyman et al. 1997). (The number of dimers of $NtrC^{A216V,3ala}$ depends on whether the active entity is a hexamer or an octamer [see Fig. 5b].) The same was true with $NtrC^{D54E}$ as the DNA-bound partner (Fig. 6a). At high concentrations, the latter retains weak ability to activate transcription by itself, presumably because it carries residual negative charge at the

site of phosphorylation and has slight capacity to form homo-oligomers (Klose et al. 1993). Not surprisingly, phosphorylated $NtrC^{A216V,3ala}$ could also be assimilated into an active oligomer with two DNA-bound dimers of phosphorylated NtrC itself (Fig. 6b). Conversely, two dimers of phosphorylated $NtrC^{A216V}$ (DNA-binding form) could be assimilated into an active oligomer with either phosphorylated $NtrC^{3ala}$ or $NtrC^{D54E,3ala}$ (Fig. 6b and a, respectively), and thus inactive forms of NtrC bearing the A216V lesion can serve as either the DNA-bound or the non-DNA-bound partner (Fig. 5a and b, respectively). The two alternative explanations for this positional flexibility are (1) the interaction of mutant forms of NtrC carrying the A216V substitution with other forms of NtrC in active oligomers compensates directly for their defects, i.e., allows the A216V mutant forms themselves to interact productively with σ^{54}-holoenzyme, and (2) not all dimers of an active oligomer must interact with the polymerase and there is no absolute requirement as to

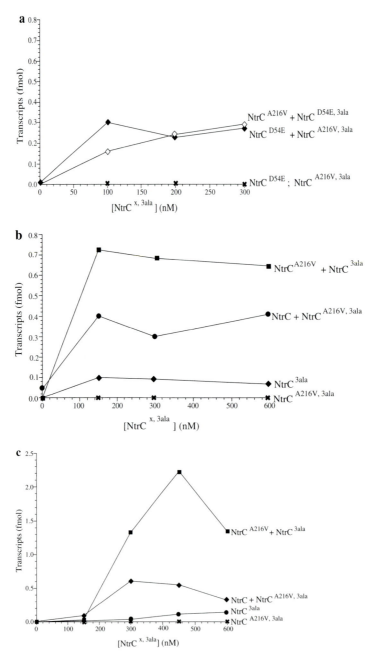

Figure 6. In vitro complementation by mutant forms of NtrC at the strong enhancer (*a* and *b*; see text) or a single strong binding site (*c*). Formation of open complexes at the *glnA* promoter was assessed in a single-cycle transcription assay as described in Materials and Methods. Mixtures of DNA-binding and nonbinding (3ala) forms of NtrC or individual NtrC proteins are indicated beside the curves. (*a*) The template was supercoiled plasmid pJES534 (0.5 nM), which carries the strong enhancer at a distance of about 400 bp from the *glnA* promoter (Porter et al. 1993). The DNA-binding form of NtrC^{A216V} or NtrC^{D54E} was held at 4 nM, whereas the concentration of NtrC^{D54E,3ala} or NtrC^{A216V,3ala} was varied as indicated. (*b*) The template was the same as in panel *a*. The DNA-binding form of NtrC^{A216V} or NtrC was held at 4 nM, whereas the concentration of NtrC^{3ala} or NtrC^{A216V,3ala} was varied as indicated. (*c*) The template was supercoiled plasmid pJES640 (10 nM), which carries a single symmetrical NtrC-binding site about 400 bp upstream of the *glnA* promoter (Porter et al. 1993). The DNA-binding form of NtrC^{A216V} or NtrC was held constant at 5 nM, whereas the concentration of NtrC^{3ala} or NtrC^{A216V,3ala} was varied as indicated. Maximum template utilization was 2.5% (0.35 fmole/13 fmoles total), 5% (0.7 fmole/13 fmoles total), and 1% (2.5 fmoles/250 fmoles total) in panels *a*, *b*, and *c*, respectively.

whether it is the DNA-bound or nonbound dimers that do so. As expected, synergistic effects between DNA-bound and nonbound forms of NtrC were lost if subunit exchange between dimers was allowed (not shown; Wyman et al. 1997) because heterodimers of two such forms of NtrC essentially fail to bind to DNA (Klose et al. 1994).

Just as two inactive dimers of phosphorylated NtrC^{A216V} could tether an active oligomer to an enhancer, a single phosphorylated dimer could tether an active oligomer to a single binding site (Fig. 6c). Presumably, either two or three active dimers of phosphorylated NtrC[3ala] were also present in such active oligomers, depending on whether the active entity is a hexamer or an octamer (Fig. 5c). However, in this case, the converse arrangement failed. If a single active dimer of phosphorylated NtrC served as tether, phosphorylated NtrC[A216V,3ala] could not serve as its partner (Figs. 5d and 6c). The easiest explanation for this result, and therefore the one we prefer, is a quantitative one: that the active oligomer is an octamer and one or two—but not three—inactive dimers carrying the A216V lesion can be assimilated (Fig. 5d vs. other panels). An alternative explanation is that active oligomers are hexamers and that two inactive dimers carrying the A216V substitution can be assimilated in some positions but not others. The matter is of interest because other members of the purine nucleotide-binding protein family to which NtrC belongs (see below) are known to form hexamers (Abrahams et al. 1994; Hingorani et al. 1997), but as far as we are aware, none are known to form octamers. Neither gel filtration nor scanning force microscopy (SFM) allowed us to distinguish between hexamers and octamers of NtrC structurally, but ultracentrifugation studies may do so (Farez-Vidal et al. 1996; B.T. Nixon, pers. comm.).

Dissecting the NtrC Machine: Determinants Involved in Binding and Hydrolysis of ATP and Interaction with σ54-Holoenzyme

The central domain of NtrC is directly responsible for transcriptional activation (Fig. 7), and a homolog of this domain is present in all activators of σ54-holoenzyme. Recent secondary structure predictions coupled with the use of recognition algorithms for protein folds indicated that the central activation domain of NtrC and the corresponding domain of other activators adopt a mononucleotide-binding fold similar to those of the eukaryotic signaling protein p21[ras] and the G domain of the bacterial polypeptide elongation factor EF-Tu (Osuna et al. 1997).

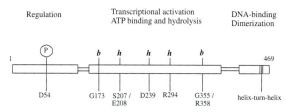

Figure 7. Domain structure of NtrC (not to scale). NtrC is composed of three functional domains: an amino-terminal regulatory domain (~120 amino acids) that is phosphorylated at aspartate 54 (D54), a central catalytic domain (~240 amino acids) that contains determinants for ATP binding and hydrolysis, as well as for transcriptional activation, and a carboxy-terminal domain (~90 amino acids) that contains a helix-turn-helix DNA-binding motif and dimerization determinants. Residues within the central catalytic domain of NtrC that appear to be required for MgATP binding are indicated by *b*, whereas residues that appear to be required for ATP hydrolysis but not binding are indicated by *h* (I. Rombel et al., unpubl.). Residues G173, S207/E208, D239, R294, and G355/R358 lie in conserved regions 1, 3, 4, 6, and 7, respectively, for activators of σ54-holoenzyme (Osuna et al. 1997).

Table 2. ATPase Activities and Apparent Equilibrium Dissociation Constants for MgATP of Wild-type and Mutant Forms of NtrC

Protein	ATPase activity[a] (%)	K_d for MgATP[b] (μM)
Wild type	100[c]	135
NtrC[G173N]	≤2	≥10000
NtrC[S207F]	≤2	50
NtrC[E208Q]	12	105
NtrC[A216V]	100	100
NtrC[G219K]	56	40
NtrC[A220T]	80	100
NtrC[D239N]	≤2	20
NtrC[D239C]	≤2	25
NtrC[D239A]	≤2	30
NtrC[R294C]	≤2	190
NtrC[G355V]	≤2	825
NtrC[R358C]	≤2	≥10000
NtrC[R358H]	≤2	≥10000

[a]From Weiss et al. (1991), North et al. (1996), and I. Rombel et al. (unpubl.). Proteins were phosphorylated.
[b]Binding by unphosphorylated proteins was assessed as described in Materials and Methods and titration data were plotted to determine K_d. K_d values for ATP in the absence of Mg++ were severalfold higher for all proteins except those carrying substitutions for D239 (I. Rombel et al., unpubl.). The absence of divalent cation affected each of the latter three proteins differently.
[c]100% corresponds to about 500 pmoles/15′/10 μl at a concentration of 1 μM protein.

Lesions to loss of transcriptional activation occur throughout the central domain of NtrC over a span of almost 200 residues (Weiss et al. 1991; North et al. 1996). Many, but not all, of these lesions, decrease ATP hydrolysis (summarized in Table 2). Recent filter-binding assays have allowed us to distinguish between regions of the central domain involved in ATP binding and those involved in ATP hydrolysis per se (I. Rombel et al., unpubl.). All of these regions are highly conserved among activators of σ54-holoenzyme. Regions implicated in nucleotide binding were the Walker A motif or P-loop and the region around residues G355/R358, which may interact with the nucleotide base (Table 2; Fig. 7). (These have also been referred to as conserved regions 1 and 7 among activators of σ54-holoenzyme [Osuna et al. 1997].) Residues implicated in nucleotide hydrolysis were D239, which is the conserved aspartate in the putative Walker B motif (conserved region 4), a residue that appears to be involved in coordinating the divalent cation (I. Rombel et al., unpubl.), R294 (in conserved region 6), which may be a catalytic residue (see Discussion), and residues S207 and E208, which have been proposed to lie in a region analogous to the Switch I effector region of p21[ras] and other purine nucleotide-binding proteins (conserved region 3; Osuna et al. 1997). Note that residues such as A216, which was discussed above, also lie in the putative Switch I region. The fact that unphosphorylated, dimeric NtrC appears to bind nucleotide as well as the phosphorylated form(s) (I. Rombel et al., unpubl.) provides evidence that oligomerization controls nucleotide hydrolysis directly.

DISCUSSION

In addition to their better-known role in tethering NtrC in high concentration near promoters, enhancers have a fundamental role in nucleating the formation of active

NtrC oligomers (Austin and Dixon 1992; Porter et al. 1995; Wyman et al. 1997). In fact, it is the latter that appears to limit transcriptional activation from solution (Figs. 2 and 3) (North and Kustu 1997), and enhancers can be viewed as allosteric effectors that stimulate oligomer formation at low concentrations of phosphorylated NtrC (Lefstin and Yamamoto 1998). Functional tests of the assimilation of inactive dimers of NtrC into active oligomers are easily rationalized if the oligomers are octamers but can also be rationalized if they are hexamers with subtle positional requirements for the presence of active dimers (see Fig. 5). Phosphorylation and oligomer formation by NtrC appear to control nucleotide hydrolysis per se rather than nucleotide binding (Farez-Vidal et al. 1996; I. Rombel et al., unpubl.). Whether they are also required for contact with σ^{54}-holoenzyme is not known, but this does not appear to be the case for the homologous activator DctD (Lee and Hoover 1995; Wang et al. 1997).

As is the case for well-studied members of the purine nucleotide-binding protein family to which they belong, NtrC and other activators of σ^{54}-holoenzyme have recognizable Walker A and B motifs that are involved in nucleotide binding and hydrolysis, respectively (Tables 2 and 3; Fig. 7). It is noteworthy that both motifs end in unusual residues. The Walker A motif, which has been shown to form a loop that wraps around the β phosphate group of the bound nucleotide in a number of proteins (see, e.g., Abrahams et al. 1994; Wittinghofer 1994), ends in a conserved lysine residue that is usually followed by a threonine or serine (Walker et al. 1982). In activators of σ^{54}-holoenzyme, the conserved lysine is followed by a glutamate or aspartate (Osuna et al. 1997), the consequences of which are not clear. In a subset of purine nucleotide-binding proteins with a nucleic-acid-stimulated ATPase activity (DExx proteins), the conserved aspartate of the Walker B motif, which is involved in coordination of the divalent cation essential for nucleotide hydrolysis (see, e.g., Wittinghofer 1994), is followed by a glutamate (Hodgman 1988; Koonin 1993). This is the case for NtrC and other activators of σ^{54}-holoenzyme, and it is known that enhancers can greatly stimulate the ATPase activity of phosphorylated NtrC (Austin and Dixon 1992; Flashner et al. 1995; Porter et al. 1995; North et al. 1996). Unlike other DExx proteins, NtrC does not function as a helicase (Wedel et al. 1990) or site-specific helicase (Wedel and Kustu 1995), rather, it appears to cause a change in the conformation of σ^{54}-holoenzyme, which allows the polymerase to act as a site-specific helicase, i.e., to form

open complexes (Wang et al. 1995). NtrC and other activators of σ^{54}-holoenzyme lack motifs downstream from the Walker B motif that are characteristic of helicases.

In addition to the Walker A motif, the region around residues 355/358 of NtrC also appears to be involved in nucleotide binding. Given that ribose is not likely to be a strong binding determinant (I. Rombel, et al., unpubl.), this region may be involved in binding the nucleotide base. Residue R294 has no role in nucleotide binding but appears to be essential for nucleotide hydrolysis. Although we cannot exclude the possibility that R294 is essential for appropriate oligomerization of NtrC, we postulate that it may be a catalytic residue because a number of other purine nucleotide-binding proteins have catalytic arginines (Bourne et al. 1991; Coleman et al. 1994; Sondek et al. 1994; Mittal et al. 1996). Replacement of the catalytic arginine in $G_{i\alpha 1}$ with cysteine, the same replacement we studied, greatly impaired GTPase activity without affecting GTP binding (Coleman et al. 1994). Moreover, this replacement resulted in a substantial decrease in the affinity for GDP·AlF4⁻, an analog of the transition state for GTP hydrolysis.

Finally, the region between residues 207 and 220 of NtrC, which lies between the Walker A and B motifs, appears to be the Switch I effector region. Switch I, which undergoes a large conformational change upon nucleotide hydrolysis, can play a critical part in biological output (see, e.g., Bourne et al. 1991; Story and Steitz 1992; Kim et al. 1993; Hilgenfeld 1995). Several amino acid substitutions at positions 216–220 cause profound defects in transcriptional activation. However, there is little, if any, decrease in ATP binding or hydrolysis (Table 2) (Weiss et al. 1991; North et al. 1996; I. Rombel et al., unpubl.). Hence, the substitutions at positions 216–220 apparently affect only contact with σ^{54}-holoenzyme or coupling of energy to open complex formation, functions of NtrC for which we unfortunately do not have direct biochemical assays. Although amino acid substitutions at positions 207 and 208 do decrease ATP hydrolysis (Table 2) (Weiss et al. 1992), we have noted that the decrease in ATPase activity caused by the sterically conservative E208Q substitution is not sufficient to account for the profound decrease in transcriptional activation by NtrC^{E208Q} (North et al. 1996). This protein must also be defective in contact with polymerase or energy-coupling. Strikingly, one substitution for the serine residue that corresponds to S207 of NtrC in the homologous activator DctD impaired the ability of the mutant protein (DctD^{S212I}) to cross-link to σ^{54}-holoenzyme (Wang et al. 1997).

Table 3. Motifs in the Central Domain of NtrC

ATP binding	ATP hydrolysis	ATP hydrolysis	ATP hydrolysis	ATP binding
G173	S207/E208	D239	R294	G355/R358
Walker A binds β-phosphate	Switch I effector	Walker B/Switch II coordinates Mg⁺⁺	catalytic residue?	binds nucleotide base
GESGTG^{173}KE	ES^{207}ELFGHEKGAFTGA	GGTLFLD^{239}EIG	FR^{294}EDLfhRLNV	WPG^{355}NVR^{358}qLEN

For the NtrC proteins of *Salmonella typhimurium, Escherichia coli, Klebsiella pneumoniae, Proteus vulgaris, Rhizobium meliloti, Bradyrhizobium parasponiae*, and *Agrobacterium tumefaciens* (Flashner et al. 1995; Osuna et al. 1997). The motifs listed are also referred to as conserved regions 1, 3, 4, 6, and 7, respectively, among activators of σ^{54}-holoenzyme.

Recent experiments indicate that responses of NtrC to phosphorylation and enhancers—both of which stimulate its ATPase activity—can be altered by amino acid substitutions within the Switch I region (D. Yan and S. Kustu, unpubl.). Phosphorylation and binding to an enhancer can be thought of as analogous to G-activating proteins (GAPS) for small G proteins, and like GAPS, they appear to be perceived through the Switch I region of NtrC. Although σ^{54}-holoenzyme is the "target" of NtrC and also appears to interact with the Switch I region, it has not been shown to affect the ATPase activity of NtrC (A. North and S. Kustu, unpubl.).

Our present efforts are directed at obtaining biochemical assays for contact between NtrC and σ^{54}-holoenzyme and assays for the predicted conformational changes that should occur in NtrC upon binding and hydrolysis of nucleoside triphosphate (see, e.g., Story and Steitz 1992; Abrahams et al. 1994). Although contact between NtrC and polymerase has been observed in both electron and scanning force micrographs (Su et al. 1990; Rippe et al. 1997), this contact appears to be transient; the dissociation constant (K_d) has been estimated at 10 μM or higher (T. Heyduk, pers. comm.). Contact between the homologous activator DctD and σ^{54}-holoenzyme has been demonstrated by cross-linking (Lee and Hoover 1995). Although ATP hydrolysis and transcriptional activation by DctD require that it be phosphorylated, contact with polymerase does not. On the basis of the mechanisms of energy-coupling in purine nucleotide-binding proteins such as myosin and the F1-ATPase, the founding member of the family, cyclic conformational changes in NtrC are likely to provide the basis for coupling the energy available from nucleotide hydrolysis to a change in the conformation of σ^{54}-holoenzyme (Wang et al. 1995; Wedel and Kustu 1995).

CONCLUDING REMARKS

Like σ^{70}, the most abundant sigma factor in bacteria, σ^{54} is required to transcribe genes whose products have diverse physiological functions. This raises the question of why bacteria might employ one of these sigma factors in a particular case rather than the other. We have speculated that the use of σ^{54} may confer one notable advantage—the capacity to vary transcriptional efficiency at a given promoter over a wide range. As noted in the introduction, σ^{54}-holoenzyme itself is transcriptionally silent, although it can form physically detectable closed complexes at a number of promoters (Popham et al. 1989). Isomerization to open complexes depends on an activator and the energy available from hydrolysis of ATP. When activated, transcription by σ^{54}-holoenzyme can reach high levels (see, e.g., MacNeil et al. 1981; He et al. 1997). Hence, genes transcribed by this form of polymerase can be silent or very highly expressed, depending on the physiological or environmental conditions. This wide range of control of the rate of transcription—not usually accessible to σ^{70}-holoenzyme at a single promoter—is achieved by coupling energy to the process of open complex formation.

That the potential advantage discussed above, or some other aspect of σ^{54} function, is fundamental to the process of transcription in bacteria is indicated by the depth and breadth of distribution of σ^{54} in the bacterial domain (Kaufman and Nixon 1996; Deckert et al. 1998). σ^{54} is one of only three alternative sigma factors encoded by *Aquifex aeolicus*, a member of the most deeply divergent bacterial lineage known. As expected, a number of activators (five) of σ^{54}-holoenzyme are also encoded on the relatively small *Aquifex* genome. Notably, σ^{54} is also one of only two alternative sigma factors encoded by *Chlamydia trachomatis* (Stephens et al. 1998), a member of another distinct bacterial lineage. σ^{54} occurs widely among members of the large gram-negative lineage proteobacteria, where gene products under its control participate in processes as diverse as biological nitrogen fixation, utilization of hydrogen as an energy source, pathogenesis, and development (Kustu et al. 1989; Brun and Shapiro 1992; Keseler and Kaiser 1997; Lenz et al. 1997; Wu et al. 1997; Arora et al. 1998). σ^{54} appears to be essential in the sporulating proteobacterium *Myxococcus xanthus* (Keseler and Kaiser 1997), in which some 14 activators of σ^{54}-holoenzyme are in process of being characterized (Kaufman and Nixon 1996; L. Gorski and D. Kaiser, in prep.).

ACKNOWLEDGMENTS

We thank Norm Pace, Petra and Volker Wendisch, and Dalai Yan for critical reading of the manuscript. This work was supported by National Institutes of Health grant GM-38361 to S.K.

REFERENCES

Abrahams J.P., Leslie A.G., Lutter R., and Walker J.E. 1994. Structure at 2.8 Å resolution of F_1-ATPase from bovine heart mitochondria. *Nature* **370:** 621.

Arora S.K., Ritchings B.W., Almira E.C., Lory S., and Ramphal R. 1998. The *Pseudomonas aeruginosa* flagellar cap protein, FliD, is responsible for mucin adhesion. *Infect. Immun.* **66:** 1000.

Austin S. and Dixon R. 1992. The prokaryotic enhancer binding protein NtrC has an ATPase activity which is phosphorylation and DNA dependent. *EMBO J.* **11:** 2219.

Bourne H.R., Sanders D.A., and McCormick F. 1991. The GTPase superfamily: Conserved structure and molecular mechanism. *Nature* **349:** 117.

Brun Y.V. and Shapiro L. 1992. A temporally controlled sigma-factor is required for polar morphogenesis and normal cell division in *Caulobacter*. *Genes Dev.* **6:** 2395.

Coleman D.E., Berghuis A.M., Lee E., Linder M.E., Gilman A.G., and Sprang S.R. 1994. Structures of active conformation of $G_{i\alpha 1}$ and the mechanism of GTP hydrolysis. *Science* **265:** 1405.

Deckert G., Warren P.V., Gaasterland T., Young W.G., Lenox A.L., Graham D.E., Overbeek R., Snead M.A., Keller M., Aujay M., Huber R., Feldman R.A., Short J.M., Olsen G.J., and Swanson R.V. 1998. The complete genome of the hyperthermophilic bacterium *Aquifex aeolicus*. *Nature* **392:** 353.

Dworkin J., Ninfa A.J., and Model P. 1998. A protein-induced DNA bend increases the specificity of a prokaryotic enhancer-binding protein. *Genes Dev.* **12:** 894.

Farez-Vidal M.E., Wilson T.J., Davidson B.E., Howlett G.J., Austin S., and Dixon R.A. 1996. Effector-induced self-association and conformational changes in the enhancer-binding protein NtrC. *Mol. Microbiol.* **5:** 779.

Flashner Y., Weiss D.S., Keener J., and Kustu S. 1995. Constitutive forms of the enhancer-binding protein NtrC: Evidence that essential oligomerization determinants lie in the central activation domain. *J. Mol. Biol.* **249:** 700.

He L., Soupene E., and Kustu S. 1997. NtrC is required for control of *Klebsiella pneumoniae* NifL activity. *J. Bacteriol.* **179:** 7446.

Hilgenfeld R. 1995. Regulatory GTPases. *Curr. Opin. Struct. Biol.* **5:** 810.

Hingorani M., Washington M.T., Moore K., and Patel S. 1997. The dTTPase mechanism of T7 DNA helicase resembles the binding change mechanism of the F_1-ATPase. *Proc. Natl. Acad. Sci.* **94:** 5012.

Hodgman T.C. 1988. A new superfamily of replicative proteins. *Nature* **333:** 22.

Hoover T.R., Santero E., Porter S., and Kustu S. 1990. The integration host factor stimulates interaction of RNA polymerase with NIFA, the transcriptional activator for nitrogen fixation operons. *Cell* **63:** 11.

Kaufman R.I. and Nixon B.T. 1996. Use of PCR to isolate genes encoding σ54-dependent activators from diverse bacteria. *J. Bacteriol.* **178:** 3967.

Keseler I.M. and Kaiser D. 1997. σ^{54}, a vital protein for *Myxococcus xanthus. Proc. Natl. Acad. Sci.* **94:** 1979.

Kim S.-H., Privé G.G., and Milburn M.V. 1993. Conformational switch and structural basis for oncogenic mutations of *Ras* proteins. In *Handbook of experimental pharmacology* (ed. B.F. Dickey and L. Birnbaumer), p 177. Springer-Verlag, Berlin.

Klose K.E., Weiss D.S., and Kustu S. 1993. Glutamate at the site of phosphorylation of nitrogen-regulatory protein NtrC mimics aspartyl-phosphate and activates the protein. *J. Mol. Biol.* **232:** 67.

Klose K.E., North A.K., Stedman K.M., and Kustu S. 1994. The major dimerization determinants of the nitrogen regulatory protein NtrC from enteric bacteria lie in its carboxy-terminal domain. *J. Mol. Biol.* **241:** 233.

Koonin E.V. 1993. A common set of conserved motifs in a vast variety of putative nucleic acid-dependent ATPases including MCM proteins involved in the initiation of eukaroyotic DNA replication. *Nucleic Acids Res.* **21:** 2541.

Kustu S., Santero E., Keener J., Popham D., and Weiss D. 1989. Expression of σ^{54} (*ntrA*)-dependendent genes is probably united by a common mechanism. *Microbiol. Rev.* **53:** 367.

Lee J.H. and Hoover T.R. 1995. Protein crosslinking studies suggest that *Rhizobium meliloti* C4-dicarboxylic acid transport protein D, a σ^{54}-dependent transcriptional activator, interacts with σ^{54} and the beta subunit of RNA polymerase. *Proc. Natl. Acad. Sci.* **92:** 9702.

Lefstin J.S. and Yamamoto K.R. 1998. Allosteric effects of DNA on transcriptional regulators. *Nature* **392:** 885.

Lenz O., Strack A., Tran-Betcke A., and Friedrich B. 1997. A hydrogen-sensing system in transcriptional regulation of hydrogenase gene expression in Alcaligenes species. *J. Bacteriol.* **179:** 1655.

Love J.J., Li X., Case D.A., Giese K., Grosschedl R., and Wright P.E. 1995. Structural basis for DNA bending by the architectural transcription factor LEF-1. *Nature* **376:** 791.

MacNeil D., Zhu J., and Brill W.J. 1981. Regulation of nitrogen fixation in *Klebsiella pneumoniae:* Isolation and characterization of strains with nif-lac fusions. *J. Bacteriol.* **145:** 348.

Mittal R., Ahmadian M.R., Goody R.S., and Wittinghofer A. 1996. Formation of a transition-state analog of the Ras GTPase reaction by Ras·GDP, tetrafluoroaluminate, and GTPase-activating proteins. *Science* **273:** 115.

North A.K. and Kustu S. 1997. Mutant forms of the enhancer-binding protein NtrC can activate transcription from solution. *J. Mol. Biol.* **267:** 17.

North A.K., Weiss D.S., Suzuki H., Flashner Y., and Kustu S. 1996. Repressor forms of the enhancer-binding protein NtrC: Some fail in coupling ATP hydrolysis to open complex formation by σ54-holoenzyme. *J. Mol. Biol.* **260:** 317.

Osuna J., Soberón X., and Morett E. 1997. A proposed architecture for the central domain of the bacterial enhancer-binding proteins based on secondary structure prediction and fold recognition. *Protein Sci.* **6:** 543.

Popham D.L., Szeto D., Keener J., and Kustu S. 1989. Function of a bacterial activator protein that binds to transcriptional enhancers. *Science* **243:** 629.

Porter S.C., North A.K., and Kustu S. 1995. Mechanism of transcriptional activation by NtrC. In *Two-component signal transduction* (ed. J.A. Hoch and T.J. Silhavy), p 147. American Society for Microbiology, Washington, D.C.

Porter S.C., North A.K., Wedel A.B., and Kustu S. 1993. Oligomerization of NtrC at the *glnA* enhancer is required for transcriptional activation. *Genes Dev.* **7:** 2258.

Rippe K., Guthold M., von Hippel P.H., and Bustamante C. 1997. Transcriptional activation via DNA-looping: Visualization of intermediates in the activation pathway of *E. coli* RNA polymerase x σ54 holoenzyme by scanning force microscopy. *J. Mol. Biol.* **270:** 125.

Sondek J., Lambright D.G., Noel J.P., Hamm H.E., and Sigler P.B. 1994. GTPase mechanism of G proteins from the 1.7-Å crystal structure of transducin α-GDP-AlF4. *Nature* **372:** 276.

Stephens R.S., Kalman S., Fenner C., and Davis R. 1998. Chlamydia Genome Project Website (http://chlamydia-www.berkeley.edu:4231/).

Story R.M. and Steitz T.A. 1992. Structure of the recA protein-ADP complex. *Nature* **355:** 374.

Su W., Porter S., Kustu S., and Echols H. 1990. DNA-looping and enhancer activity: Association between DNA-bound NtrC activator and RNA polymerase at the bacterial *glnA* promoter. *Proc. Natl. Acad. Sci.* **87:** 5504.

Walker J.E., Saraste M., Runswick M.J., and Gay N.J. 1982. Distantly related sequences in the α - and β-subunits of ATP synthase, myosin, kinases and other ATP-requiring enzymes and a common nucleotide binding fold. *EMBO J.* **1:** 945.

Wang J.T., Syed A., Hsieh M., and Gralla J.D. 1995. Converting *Escherichia coli* RNA polymerase into an enhancer-responsive enzyme: Role of an NH$_2$-terminal leucine patch in σ54. *Science* **270:** 992.

Wang Y.-K., Lee J.H., Brewer J.M., and Hoover T.R. 1997. A conserved region in the σ54-dependent activator DctD is involved in both binding to RNA polymerase and coupling ATP hydrolysis to activation. *Mol. Microbiol.* **26:** 373.

Wedel A., and Kustu S. 1995. The bacterial enhancer-binding protein NtrC is a molecular machine: ATP hydrolysis is coupled to transcriptional activation. *Genes Dev.* **9:** 2042.

Wedel A., Weiss D., Popham D., Dröge P., and Kustu S. 1990. A bacterial enhancer functions to tether a transcriptional activator near a promoter. *Science* **248:** 486.

Weiss D.S., Batut J., Klose K.E., Keener J., and Kustu S. 1991. The phosphorylated form of the enhancer-binding protein NtrC has an ATPase activity that is essential for activation of transcription. *Cell* **67:** 155.

Weiss D.S., Klose K.E., Hoover T.R., North A.K., Porter S.C., Wedel A.B., and Kustu S. 1992. Prokaryotic transcriptional enhancers. In *Transcriptional regulation* (ed. S.L. McKnight and K.R. Yamamoto), p 667. Cold Spring Harbor Laboratory Press, Cold Spring Harbor, New York.

Werner M.H., Huth J.R., Gronenborn A.M., and Clore G.M. 1996. Molecular determinants of mammalian sex. *Trends Biochem. Sci.* **21:** 302.

Wittinghofer A. 1994. The structure of transducin G alpha t: More to view than just ras. *Cell* **76:** 201.

Wu J., Ohta N., Benson A.K., Ninfa A.J., and Newton A. 1997. Purification, characterization, and reconstitution of DNA-dependent RNA polymerases from *Caulobacter crescentus. J. Biol. Chem.* **272:** 21558.

Wyman C., Rombel I., North A.K., Bustamante C., and Kustu S. 1997. Unusual oligomerization required for activity of NtrC, a bacterial enhancer-binding protein. *Science* **275:** 1658.

Gene Transcription by Recruitment

Z. ZAMAN, A.Z. ANSARI, L. GAUDREAU, J. NEVADO, AND M. PTASHNE
Program in Molecular Biology, Memorial Sloan-Kettering Cancer Center, New York, New York 10021

An argument has been made that at many but not all genes, transcriptional activators work by "recruitment" (for review, see Ptashne and Gann 1997). The role of a specific activator, in this mechanism, is simply to "locate" the transcriptional machinery at a specific gene. According to this idea, repression could be effected by any of several different mechanisms. In the following sections, we review recent experiments from our laboratory that address various issues concerning the mechanism of transcriptional activation and repression.

GENE ACTIVATION BY RECRUITMENT

A key prediction of the recruitment model is that because the sole function of an activator is to bring the transcriptional machinery to the DNA, activators can be dispensed with provided the machinery can be localized to the promoter by some other means. We have demonstrated this effect in experiments performed both in vitro and in vivo.

In Vitro

As reported by Gaudreau et al. (1998), we found that in a cell-free system containing the yeast RNA polymerase II holoenzyme (purified according to Koleske et al. 1996) and supplemented with yeast TFIIE and TBP, transcription was efficiently activated by DNA-bound Gal4+Hap4 (Hap4 fused to the DNA-binding domain of Gal4), provided the holoenzyme was present at a low concentration. At higher concentrations of holoenzyme, however, the fully "activated" levels of transcription were attained in the absence of activator, and the activator had no further effect. This experiment is analogous to that performed by Meyer (Meyer and Ptashne 1980; Meyer et al. 1980) using *Escherichia coli* RNA polymerase, in which it was shown that the levels of transcription elicited by the DNA-bound activator λ repressor were also reached simply by increasing the concentration of polymerase in the absence of activator.

In Vivo

"Activator bypass" experiments performed in yeast demonstrate that high levels of transcription can be achieved by recruiting the transcriptional machinery to DNA in the absence of any classical activator. An early demonstration of this effect was gleaned from a study of the Gal11P mutation. This mutation changes a single amino acid in the holoenzyme component Gal11 and thereby confers upon an otherwise inert DNA-tethered peptide (the dimerization region of Gal4) the ability to activate transcription to high levels (Himmelfarb et al. 1990). A variety of experiments indicated that activation is a result of an interaction between the DNA-tethered peptide and the mutant holoenzyme (Fig. 1A) and that the only requirement for the activation was simple binding energy (Barberis et al. 1995; Farrell et al. 1996). More re-

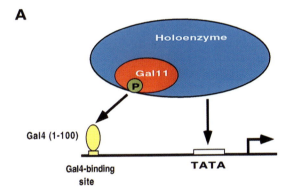

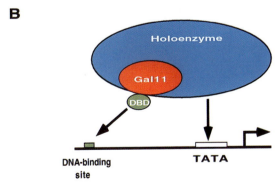

Figure 1. Recruitment of the holoenzyme in "activator bypass" experiments is sufficient to activate gene transcription. (*A*) Dimerization domain of Gal4 binds to the mutant holoenzyme component Gal11P and thereby recruits the holoenzyme to the promoter. (*B*) DNA-binding domain (DBD) fused directly to Gal11 recruits the holoenzyme to a promoter bearing cognate DNA-binding sites. We assume that whatever required components are missing from the holoenzyme (e.g., TBP and TF11E), those components bind cooperatively with the holoenzyme to the promoter.

cently, we have shown (Gaudreau et al. 1998) that as predicted, the tethered Gal4 dimerization region activates transcription in vitro when presented with a holoenzyme (supplemented with TBP and TFIIE) isolated from Gal11P cells but not when presented with holoenzyme isolated from wild-type cells.

As an extension of the results with Gal11P, we found that fusion proteins bearing Gal11 attached to a DNA-binding domain (e.g., the DNA-binding domain of LexA) can, when bound to DNA, activate transcription in vivo to very high levels. We call such a fusion protein (e.g., LexA+Gal11) a "nonclassical" activator because it lacks any natural activating region, and we presume it works by inserting the Gal11 moiety into the holoenzyme and recruiting that complex to DNA (Fig. 1B) (Ptashne and Gann 1997). Fusions bearing other holoenzyme components (e.g., LexA+Srbs) can also activate (Laurent et al. 1991; Farrell et al. 1996; Hengartner et al. 1998; Keaveney and Struhl 1998), but until recently, the behavior of these fusions was not studied systematically. We have now, in collaboration with the Struhl laboratory, studied the activities of a group of nonclassical activators at several yeast promoters.

In brief, our results (L. Gaudreau et al., in prep.) are as follows. The ability of the typical fusion protein to activate transcription can be strongly influenced by three factors: the location of the DNA-binding sites in relation to the TATA box, the promoter sequence between the TATA box and activator binding sites, and the sequence of the gene downstream from the TATA box. In all cases tested, classical activators worked impervious to these factors, as for the most part did fusions bearing Gal11. One possible reason for the failure of any given configuration to work would be, of course, that under those conditions, the nonclassical activator simply failed to recruit any part of the machinery. This explanation is evidently incorrect: In every case tested, the nonclassical activator was found to work highly synergistically with a classical activator bound to sites nearby. Moreover, in no case have we been able to detect any positive effect of a classical activating region without a DNA-binding moiety in such a synergy experiment. We therefore suggest that consistent with other evidence (see below), classical activating regions touch more than one—perhaps many—sites on the transcriptional machinery, and those multiple interactions ensure recruitment, or stabilization, of all necessary components regardless of the promoter variables we have tested in our experiments. In contrast, nonclassical activators, we surmise, interact with the holoenzyme in a much more restricted fashion, and so their activities are more sensitive to promoter configuration.

We do not know why DNA-tethered Gal11 (in contrast to other tethered transcriptional machinery components) works so well under a wide array of conditions. One possibility might be that different forms of the holoenzyme work at certain different promoters but that all of these forms include Gal11. Indeed, of the two forms of the holoenzyme complex that have been isolated from yeast (Kim et al. 1994; Koleske and Young 1994; Chang and

Jaehning 1997; Shi et al. 1997), both contain Gal11. The second form of holoenzyme isolated by Shi et al. (1997) is devoid of any Srb proteins but contains Gal11, Cdc73, and Paf1. It is thus conceivable that Gal11 might be able to associate with different forms of holoenzymes, thus ensuring that DNA-tethered Gal11 would efficiently activate transcription at a wide variety of genes.

We have also found that in transient transfection assays in mammalian cells, the typical nonclassical activator (bearing a mammalian or yeast transcriptional machinery component fused to a DNA-binding domain) works poorly or not at all on a reporter gene, but in every case this nonclassical activator works highly synergistically with a classical activator bound nearby (J. Nevado et al., in prep.). As in the yeast experiments, the positive effects of the classical activator, in every case tested, required DNA binding of that activator.

We have also used nonclassical activators to ask whether a classical activating region is required for the action of two complexes, Swi/Snf and Ada/Gcn5, both of which are believed to help overcome the inhibitory effects of nucleosomes (Struhl 1996; Kadonaga 1998). Gaudreau et al. (1997) found that activation by DNA-tethered Gal11 and Srb2, as well as that by classical activators, was sufficient to remodel nucleosomes at the *PHO5* promoter concomitant with gene activation, but Swi/Snf was not required for those effects. However, a promoter independent of Swi/Snf and Ada/Gcn5 can be made dependent on both by modifying the strengths and/or positions of the activator-binding sites (Burns and Peterson 1997; Gaudreau et al. 1997; Pollard and Peterson 1997; Gregory et al. 1998; L. Gaudreau and M. Ptashne, unpubl.). At such promoters, both functions can be brought into play in activator-bypass experiments using DNA-tethered Gal11 (Gaudreau et al. 1997; Gregory et al. 1998; L. Gaudreau and M. Ptashne, unpubl.). Thus, neither Swi/Snf nor Ada/Gcn5 need be targeted independently of the holoenzyme by activating regions at these promoters.

TARGETS OF CLASSICAL ACTIVATING REGIONS

We noted above the idea that classical activating regions may see multiple targets in the transcriptional machinery. Previous work has shown that classical activating regions bind in vitro to many components of the transcriptional machinery, including TBP, TAFs, TFIIB, and TFIIH (Stringer et al. 1990; Ingles et al. 1991; Lin et al. 1991; Goodrich et al. 1993; Roberts et al. 1993; Xiao et al. 1994; Zhu et al. 1994; Kobayashi et al. 1995). In one study, we found that Gal4 specifically interacts with TBP and TFIIB and that the strength of the interactions with several Gal4 activation domain mutants correlated well with the ability of Gal4 or its mutants to activate transcription in vivo (Wu et al. 1996). In the following section, we summarize our recent experiments which suggest that two yeast holoenzyme components—Srb4 and Srb10—are also targets of Gal4.

Srb4 and Srb10 as Targets of Gal4

Previous genetic evidence indicated that nested truncations of the carboxy-terminal heptapeptide repeats of the largest subunit of the polymerase proportionately impaired the ability of Gal4 to activate transcription in response to galactose (Allison and Ingles 1989; Scafe et al. 1990). This region of the polymerase does not bind the activation domain of Gal4 directly (A.Z. Ansari and M. Ptashne, unpubl.), but it does stably associate with the "mediator/Srb complex." As the length of the CTD proportionally affects both the binding of the Srb complex and the ability of Gal4 to activate transcription, we suspected that this complex could be a direct target of Gal4.

Label-transfer photo-affinity cross-linking experiments performed with the transcriptionally competent holoenzyme identified TBP, Srb4, and Srb10 as proteins that were in close proximity to the activation domain of Gal4 (Koh et al. 1998; A.Z. Ansari and M. Ptashne, unpubl.). These experiments were performed with all components in solution as described in Koh et al. (1998). Applying a battery of methods that detect direct protein-protein interactions, namely, affinity chromatography, immunoprecipitation, surface plasmon resonance (SPR), and GST pulldowns, we confirmed that TBP and Srb4 individually interact with Gal4 (for experimental details, see Koh et al. 1998). Identical analyses with purified Srb10 have now shown that it binds Gal4 directly and that the strength of Srb10-Gal4 interaction is similar in magnitude to that observed for interaction between TBP and Gal4 (A.Z. Ansari and M. Ptashne, unpubl.). Thus, Gal4 interacts with at least three different proteins bearing little sequence similarity to each other.

Further analysis led to the identification of Srb4 mutants that could complement defects in the activation domain of Gal4. These Srb4 "up-mutants" were mapped to the regions on the protein that are bound by the Gal4 activation domain (Koh et al. 1998). Although these Srb4 up-mutants strongly complement the defects in the Gal4 activation domain, they also weakly complement a few other defective activating regions (Koh et al. 1998; A.Z. Ansari et al., unpubl.). Srb4 is now known to interact with the few other activating regions (S. Koh, pers. comm.). Taken together, these results suggest that Srb4 is a common target for a variety of transcriptional activators.

To test the relevance of Gal4-Srb10 interactions in vivo, we deleted Srb10 from yeast cells and found that Gal4 was severely impaired in its ability to activate transcription (Liao et al. 1995). LexA+Gal11 activated transcription about twice as efficiently as did DNA-tethered Gal4 in this Srb10 deleted strain (A.Z. Ansari et al., unpubl.). This difference suggests that in addition to its general role in holoenzyme function, Srb10 may have a specific role in responding to activators such as Gal4. Srb10 is a cyclin-dependent kinase that has a role in both positive and negative regulation of transcription (Liao et al. 1995). When bound to Srb11, its cyclin partner, Srb10 specifically phosphorylates the carboxy-terminal heptapeptide repeats of the largest subunit of the RNA poly-

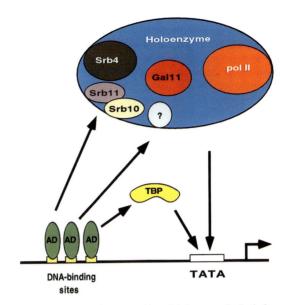

Figure 2. Activators interact with multiple targets in the holoenzyme. Multiple activators bound at a UAS simultaneously interact with several components of the holoenzyme and cooperatively recruit it to the promoter.

merase (Liao et al. 1995). The human homologs of Srb10/11, Cdk8/CycC, respectively, show similar substrate specificity and have recently been found to be part of a complex that associates with viral activators E1A and VP16 (Gold et al. 1996; Rickert et al. 1996). This suggests that some targets of activators may be conserved from yeast to humans.

These and other results support the idea that the activating region of Gal4 can interact with multiple targets in the holoenzyme. Activators bound at an upstream activating sequence (UAS), which typically contains multiple activator binding sites, would according to the recruitment model simultaneously interact with different surfaces of the holoenyzme and thereby work synergistically (Fig. 2). Furthermore, the ability to interact with several surfaces of the holoenzyme is consistent with the observation that a classical activator is able to activate transcription from several different positions with respect to the TATA box.

TWO MODES OF REPRESSION

Several modes of transcriptional repression would be consistent with the recruitment model for gene transcription. These would include, for example, inactivation of the recruited transcriptional machinery at the promoter, masking of the activating region of an activator, or the presentation of a barrier to recruitment in the form of nucleosomes. In the following section, we describe repression of transcription activated by two different kinds of DNA-bound activators: "classical" activators such as the yeast proteins Gcn4 and Gal4, and nonclassical activators bearing a holoenzyme component fused to a DNA-binding domain (e.g., LexA+Gal11). We subject each form of activated transcription to repression mediated by DNA-tethered Tup1, a yeast transcriptional repressor (Johnson 1995), and to teleomeric repression as described by Aparicio and Gottschling (1994). Our results are consistent with the idea that these two modes of transcriptional repression work by different molecular mechanisms.

Tup1-mediated Repression

In a series of experiments studying transcription in vivo elicited by various classical and nonclassical activators, we found that in every case, Tup1, tethered upstream of the activator, had a greater negative effect on activation by the nonclassical activator (Z. Zaman and M. Ptashne, in prep.). For example, using a yeast gene reporter template bearing Gal4 DNA-binding sites upstream of LexA DNA-binding sites, we tested repression by DNA-bound Tup1 of transcription elicited by LexA+Gal11 or LexA+Gal4 . We found that even though both fusions activated transcription to similar levels, LexA+Gal11 was repressed some 20-fold, whereas LexA-Gal4 was repressed only 5-fold. Similar experiments using LexA fusions of other holoenzyme components (e.g., Srb2) and classic activators (e.g., Gcn4) indicated that without exception, the holoenzyme components were far more sensitive to repression by DNA-bound Tup1 than were classical activators. Other investigators have reported that deletion of Srb10 or 11 decreases repression by Tup1 (Kuchin et al 1995; Wahi and Johnson 1995; Kuchin and Carlson 1998), and we have found that deletion of either of those components virtually abolishes activation elicited by nonclassical activators and diminishes that elicited by classical activators.

These results, taken with the observations described above that suggest that Srb10 is a target of classical activating regions, indicate first that Tup1 works by inactivating the recruited holoenzyme, and second that classical activating regions and Tup1 interact with Srb10. The model remains only suggestive, however, in view of the paucity of evidence showing a direct and relevant interaction between Tup1 and Srb10. These results do not support the idea suggested by Saha et al. (1993) that repressors, including Tup1, work by interacting with activating regions (because the nonclassical activators have no such regions), but they do not exclude this as a possible additional mechanism. In other experiments, we have failed to find any interaction between Tup1 and classical activating regions as assayed in two-hybrid experiments in vivo and binding experiments in vitro (Z. Zaman et al., unpubl.).

Heterochromatin-mediated Repression

Yeast telomers are believed to form heterochromatic structures that repress transcription of nearby genes (Grunstein 1998). We have found that using the same activators as those used to study Tup1 repression, transcription elicited by classical activators is somewhat more sensitive to telomeric repression than is transcription elicited

by nonclassical activators. This result is in striking contrast to those obtained using Tup1 and suggests that the molecular mechanism of repression mediated by telomeres is different from that mediated by Tup1.

To perform this experiment, we used a *URA3*-based reporter bearing Gal4 DNA-binding sites integrated in yeast such that it was subject to telomeric repression (Aparicio and Gottschling 1994). Cell growth on media containing 5-fluoroacetic acid is inhibited if cells express *URA3*, and this represents a very sensitive assay for telomeric repression of the *URA3* gene. We tested the ability of a classical activator Gal4 and Gal4+Gal11 to overcome telomeric repression in this system. We observed that in contrast to results seen with DNA-tethered Tup1, Gal4-activated transcription was far more sensitive to telomeric repression than was that activated by Gal4+Gal11. High levels of the protein Sir3 are known to increase telomeric repression of classical activation (Renauld et al. 1993); we found that overexpression of Sir3 led to increased repression of Gal4-mediated activation, whereas there was little if any additional effect on Gal4+Gal11-mediated activation (Z. Zaman and M. Ptashne, unpubl.). Taken together, our results suggest that Tup1-mediated repression and telomeric repression represent two distinct modes of gene regulation, particularly distinguishable by activator bypass experiments.

ACKNOWLEDGMENTS

We thank members of the Ptashne laboratory for discussions and comments on the manuscript, S.S. Koh and R. Young for discussions and for communicating unpublished results. Research described herein was supported in part by the National Institutes of Health to M.P. A.Z.A. was supported by the Helen Hay Whitney Foundation during the initial stages of this work, L.G. was supported in part by a fellowship from the M.R.C. of Canada, J.N. was a recipient of a fellowship from the Spanish Ministry of Education (F.P.I), and Z.Z. was supported by EMBO and a Clinical Scholars training fellowship from the Winston Foundation.

REFERENCES

Allison L.A. and Ingles C.J. 1989. Mutations in RNA polymerase II enhance or supress mutations in GAL4. *Proc. Natl. Acad. Sci.* **86:** 2794.

Aparicio O.M. and Gottschling D.E. 1994. Overcoming telomeric silencing: A *trans*-activator competes to establish gene expression in a cell cycle-dependent way. *Genes Dev.* **8:** 1133.

Barberis A., Pearlberg, J., Simkovich N., Farrell S., Reinagel P., Bamdad C., Sigal G., and Ptashne, M. 1995. Contact with a component of the polymerase II holoenzyme suffices for gene activation. *Cell* **81:** 359.

Burns L.G. and Peterson C.L. 1997. The yeast SWI-SNF complex facilitates binding of a transcriptional activator to nucleosomal sites in vivo. *Mol. Cell. Biol.* **17:** 4811.

Chang M. and Jaehning J.A. 1997. A multiplicity of mediators: Alternative forms of transcription complexes communicate with transcriptional regulators. *Nucleic Acids Res.* **25:** 4861.

Farrell S., Simkovich, N., Wu Y., Barberis A., and Ptashne M. 1996. Gene activation by recruitment of the RNA polymerase II holoenzyme. *Genes Dev.* **10:** 2359.

Gaudreau L., Adam M, and Ptashne M. 1998. Activation of transcription in vitro by recruitment of the yeast RNA polymerase II holoenzyme. *Mol. Cell* **1**: 913.

Gaudreau L., Schmid A., Blaschke, D., Ptashne M., and Hörz W. 1997. RNA polymerase II holoenzyme recruitment is sufficient to remodel chromatin at the yeast *PHO5* promoter. *Cell* **89**: 55.

Gold M.O., Tassan J.P. Nigg E.A., Rice A.R., and Hermann C.H. 1996. Viral transactivators E1A and VP16 interact with a large complex that is associated with CTD kinase activity and contains CDK8. *Nucleic Acids Res.* **24**: 3771.

Goodrich J.A., Hoey T., Thut C.J., Admon A., and Tjian R. 1993. *Drosophila* TAFII40 interacts with both VP16 activation domain and the basal transcription factor TFIIB. *Cell* **75**: 519.

Gregory P.D., Schmid A., Zavari M., Lui L., Berger S.L., and Hörz W. 1998. Absence of Gcn5 HAT activity defines a novel state in the opening of chromatin at the PHO5 promoter in yeast. *Mol. Cell* **1**: 4503.

Grunstein M. 1998. Yeast heterochromatin: Regulation of its assembly and inheritance by histones. *Cell* **93**: 325.

Hengartner C.J., Myer V.E., Liao S.M., Wilson C.J., Koh S.S., and Young R.A. 1998. Temporal regulation of RNA polymerase II by Srb10 and Kin28 cyclin-dependent kinases. *Mol. Cell* **2**: 43.

Himmelfarb H.J., Pearlberg J., Last D.H., and Ptashne M. 1990. GAL11P: A yeast mutation that potentiates the effect of weak GAL4-derived activators. *Cell* **63**: 1299.

Ingles C.J., Sjales M., Cress W.D., Triezenberg S.J., and Greenblatt J. 1991. Reduced binding of TFIID to transcriptionally compromised mutants of VP16. *Nature* **351**: 588.

Johnson A.D. 1995. The price of repression. *Cell* **81**: 655.

Kadonaga J.T. 1998. Eukaryotic transcription: An interlaced network of transcription factors and chromatin-modifying machines. *Cell* **92**: 307.

Keaveney M. and Struhl K. 1998. Activator-mediated recruitment of the RNA polymerase II machinery is the predominant mechanism for transcriptional activation in yeast. *Mol. Cell* **1**: 917.

Kim Y.J., Bjorklund S., Li Y., Sayre M.H., and Kornberg R.D. 1994. A multiprotein mediator of transcriptional activation and its interaction with the C-terminal repeat domain of RNA polymerase II. *Cell* **77**: 599.

Kobayashi N., Boyer T.G., and Berk A.J. 1995. A class of activation domain interacts directly with TFIIA and stimulates TFIIA-TFIID-promoter complex assembly. *Mol. Cell. Biol.* **15**: 6465.

Koh S.S., Ansari A.Z., Ptashne M., and Young R.A. 1998. An activator target in the RNA polymerase II holoenzyme. *Mol. Cell* **1**: 895.

Koleske A.J. and Young R.A. 1994. An RNA polymerase II holoenzyme responsive to activators. *Nature* **368**: 466.

Koleske A.J., Chao D.M., and Young R.A. 1996. Purification of yeast RNA polymerase II holoenzymes. *Methods Enzymol.* **273**: 176.

Kuchin S. and Carlson M. 1998. Functional relationships of Srb10-Srb11 kinase, carboxy-terminal domain kinase CTDK-I, and transcriptional corepressor Ssn6-Tup1. *Mol. Cell. Biol.* **18**: 1163.

Kuchin S., Yeghiayan P., and Carlson M. 1995. Cyclin-dependent protein kinase and cyclin homologs SSN3 and SSN8 contribute to transcriptional control in yeast. *Proc. Natl. Acad. Sci.* **92**: 4006.

Laurent B.C., Treitel M.A., and Carlson M. 1991. Functional in-

terdependence of the yeast SNF2, SNF5, and SNF6 proteins in transcriptional activation. *Proc. Natl. Acad. Sci.* **88**: 2687.

Liao S.M., Zhang J., Jeffery D.A., Koleske AJ., Thompson C.M., Chao D.M., Viljoen M., van Vuuren H.J.J., and Young R.A. 1995. A kinase-cyclin pair in the RNA polymerase II holoenzyme. *Nature* **374**: 193.

Lin Y.-S., Ha I., Maldonado E., Reinberg D., and Green M. 1991. Binding of general transcription factor TFIIB to an acidic activating region. *Nature* **353**: 569.

Meyer B.J. and Ptashne M. 1980. Gene regulation at the right operator (OR) of bacteriophage lambda. III. Lambda repressor directly activates gene transcription. *J. Mol. Biol.* **139**: 195.

Meyer B.J., Maurer R., and Ptashne M. 1980. Gene regulation at the right operator (OR) of bacteriophage lambda. II. OR1, OR2, and OR3: Their roles in mediating the effects. *J. Mol. Biol.* **139**: 163.

Pollard K.J. and Peterson C.L. 1997. Role for ADA/GCN5 products in antagonizing chromatin-mediated transcriptional repression. *Mol. Cell. Biol.* **17**: 6212.

Ptashne M. and Gann A. 1997. Transcriptional activation by recruitment. *Nature* **386**: 569.

Renauld H., Aparicio O.M., Zierath P.D., Billington B.L., Chhablani S.K., and Gottschling D.E. 1993. Silent domains are assembled continuously from the telomere and are defined by promoter distance and strength, and by SIR3 dosage. *Genes Dev.* **7**: 1133.

Rickert P., Seghezzi W., Shanahan F., Cho H., and Lees E. 1996. Cyclin C/CDK8 is a novel CTD kinase associated with RNA polymerase II. *Oncogene* **12**: 2631.

Roberts S.G.E.., Ha I., Maldonado E., Reinberg D., and Green M.R. 1993. Interaction between an acidic activator and transcription factor TFIIB is required for transcriptional activation. *Nature* **363**: 741.

Saha S., Brickman J.M., Lehming N., and Ptashne M. 1993. New eukaryotic trancriptional repressors. *Nature* **363**: 648.

Scafe C., Chao D., Lopes J., Hirsch J.P., Henry S., and Young R.A. 1990. RNA polymerase II C-terminal repeat influences response to transcriptional enhancer signals. *Nature* **347**: 491.

Shi X., Chang M., Wolf A.J., Chang C., Frazer-Abel A.A., Wade P.A., Burton Z.F., and Jaehning J.A. 1997. Cdc37p and Paf1p are found in a novel RNA polymerase II-containing complex distinct from the Srbp-containing complex. *Mol. Cell. Biol.* **17**: 1160.

Stringer K.F., Ingles C.J., and Greenblatt J. 1990. Direct and selective binding of an acidic transcriptional activator domain to the TATA -box factor TFIID. *Nature* **345**: 783.

Struhl K. 1996. Chromatin structure and RNA polymerase II connection: Implication for transcription. *Cell* **84**: 179.

Wahi M. and Johnson A.D. 1995. Identification of genes required for α2 repression in *Saccharomyces cerevisiae*. *Genetics* **140**: 79.

Wu Y., Reece R.J., and Ptashne M. 1996. Quantitation of putative activator-target affinities predicts transcriptional activating potentials. *EMBO J.* **15**: 3951.

Xiao H., Pearson A., Coulombe B., Truant R., Zhang S., Regier R.L., Triezenberg S.J., Reinberg D., Flores O., Ingles C.J., and Greenblatt J. 1994. Binding of basal transcription factor TFIIH to the acidic activation domain of VP16 and p53. *Mol. Cell. Biol.* **14**: 7013.

Zhu H., Joliot V., and Prywes R. 1994. Role of transcription factor TFIIF in serum response factor-activated transcription. *J. Biol. Chem.* **269**: 3489.

Use of Artificial Activators to Define a Role for Protein-Protein and Protein-DNA Contacts in Transcriptional Activation

S.L. DOVE AND A. HOCHSCHILD

Department of Microbiology and Molecular Genetics, Harvard Medical School, Boston, Massachusetts 02115

DNA-bound regulatory proteins interact with RNA polymerase (RNAP) to control gene expression at the transcriptional level. The interactions of a number of prokaryotic activators with RNAP have been extensively studied, but the analysis of natural activators has not revealed whether contact between a DNA-bound protein and RNAP is sufficient to activate transcription. This paper reviews work done to address this question. Our strategy was to replace the natural interaction between a DNA-bound activator and RNAP with heterologous protein-protein interactions.

RNAP in *Escherichia coli* consists of an enzymatic core composed of subunits α, β, and β' in the stoichiometry $\alpha_2\beta\beta'$ and one of several alternative σ factors that directs binding to a specific promoter class (Burgess 1976; Helmann and Chamberlin 1988). Most genes in *E. coli* are transcribed by RNAP holoenzyme containing σ^{70}. A typical σ^{70}-dependent promoter is defined by two elements known as the –10 and –35 hexamers (generally separated by 16–18 bp), both of which are contacted directly by σ in the holoenzyme (Fig. 1A) (for review, see Lonetto et al. 1992). It should, however, be noted that σ^{70} does not bind to the promoter on its own (Dombroski et al. 1992); i.e., the preassembled holoenzyme is the DNA-binding species.

CONTACT BETWEEN A DNA-BOUND PROTEIN AND A HETEROLOGOUS PROTEIN DOMAIN FUSED TO THE RNA POLYMERASE α SUBUNIT CAN ACTIVATE TRANSCRIPTION

Many natural activators apparently contact the carboxy-terminal domain (CTD) of the α-subunit of RNAP (Ishihama 1992; Ebright and Busby 1995). The α-subunit, which initiates the assembly of RNAP holoenzyme by forming a dimer (for review, see Ishihama 1981), consists of two independently folded domains separated by a flexible linker region (Blatter et al. 1994; Jeon et al. 1997). The amino-terminal domain (NTD) mediates formation of the α dimer, which serves as a scaffold upon which the other subunits assemble (Kimura et al. 1994). The CTD, on the other hand, is a DNA-binding domain that stimulates transcription from certain promoters by binding specifically to a DNA sequence (called the UP element) that is located upstream of the –35 region (Fig. 1B) (Ross et al. 1993; Blatter et al. 1994; Jeon et al. 1995; Gaal et al. 1996). At other promoters, the α-CTD mediates the effects of transcriptional activators (for reviews, see Ishihama 1993; Ebright and Busby 1995), and finally, at some promoters, the α-CTD has no essential role (Igarashi et al. 1991; Igarashi and Ishihama 1991).

An intensively studied example of an activator that interacts with the α-CTD is the *E. coli* cAMP receptor protein (CRP, also known as CAP). At the familiar *lac* promoter (*plac*), CRP binds to a site centered at position –61.5 and contacts an α-CTD (for review, see Ebright and Busby 1995), thereby stabilizing the association of at least one α-CTD with the DNA between the CRP-binding site and the promoter –35 region (Fig. 1C) (Kolb et al. 1993). This suggests that one effect of the interaction between CRP and the α-CTD is the stabilization of the binding of RNAP to the promoter (Busby and Ebright 1994; Hochschild and Dove 1998). However, it is not clear whether the stabilization provided by such a protein-protein contact is sufficient to activate transcription. To answer this question, we replaced the α-CTD with a heterologous protein domain that does not ordinarily mediate transcriptional activation, namely, the oligomerization domain of the bacteriophage λ cI protein (λcI).

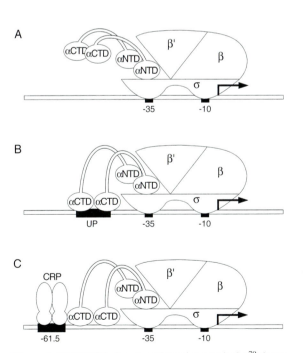

Figure 1. (*A*) RNAP holoenzyme bound at a typical σ^{70}-dependent promoter. (*B*) RNAP holoenzyme bound at a promoter with an associated UP element. (*C*) Protein-protein interaction between CRP and the α-CTD at the *lac* promoter. (Adapted, with permission, from Hochschild and Dove 1998 [copyright Cell Press].)

A

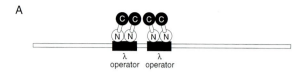

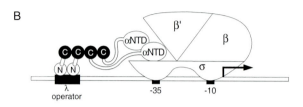

B

C

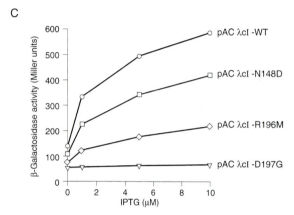

Figure 2. Contact between λcI-CTD dimers activates transcription. (*A*) Domain structure of λcI and interaction of adjacently bound λcI dimers. (*B*) Contact between DNA-bound λcI dimer and λcI-CTD fused to α-NTD. Shown is the artificial *plac* derivative bearing a λ operator centered at position –62. (*C*) Effects of λcI and λcI mutants on transcription in vivo in the presence of the α-cI chimera. Cells harboring the *plac*-based reporter construct depicted in *B* and compatible plasmids directing expression of the α-cI chimera and the indicated λcI variant were assayed for β-galactosidase activity. Both λcI and the α-cI chimera were expressed under the control of IPTG-inducible promoters, and β-galactosidase was assayed as a function of IPTG concentration. The indicated λcI cooperativity mutants have been characterized previously (Whipple et al. 1994); the substitution D197G abolishes cooperativity, as assayed in vivo, whereas the substitutions N148D and R196M confer weak and intermediate cooperativity defects, respectively. (Adapted, with permission, from Dove et al. 1997 [copyright Macmillan].)

λcI (λ repressor) is a two-domain protein that binds as a dimer to its specific recognition site (operator) (for review, see Ptashne 1992). The λcI-NTD contacts the DNA, whereas the λcI-CTD mediates formation of the λcI dimer as well as a higher-order dimer-dimer interaction that results in cooperative binding of the dimers to pairs of operator sites (Fig. 2A) (for review, see Sauer et al. 1990). Having replaced the α-CTD with the λcI-CTD, we demonstrated that contact between a DNA-bound λcI dimer and the λcI-CTDs tethered to RNAP resulted in transcriptional activation from a *plac* derivative bearing a λ operator in place of the CRP-binding site (centered 62 bp upstream from the start point of transcription) (Fig.

2B,C) (Dove et al. 1997). Although λcI has a natural activating region in its NTD that can contact the σ-subunit of RNAP when λcI is bound directly adjacent to the promoter –35 region (at position –42) (for review, see Hochschild and Dove 1998), λcI cannot activate transcription when bound at position –62, and this activating region does not have a role in the activation observed in the presence of the α-cI chimera (Dove et al. 1997). We note that this experiment was performed with cells that contain both wild-type α (encoded by the chromosome) and the chimera (encoded by a plasmid). Thus, the magnitude of the observed stimulatory effect is an underestimate of the effect that would be observed with a homogeneous preparation of RNAP containing the α-cI chimera.

We took advantage of a set of λcI mutants altered specifically in their abilities to participate in the dimer-dimer interaction responsible for cooperativity (Whipple et al. 1994, 1998) to demonstrate that the strength of the protein-protein interaction of the λcI-CTDs determines the magnitude of gene activation in this artificial system. We showed that a λcI mutant unable to participate in this protein-protein interaction was unable to activate transcription in the presence of the α-cI chimera, but that mutants with specific but less severe defects elicited intermediate levels of activation (Fig. 2C) (Dove et al. 1997).

This set of experiments demonstrates that an arbitrarily selected protein domain with no determinants for binding to DNA can mediate transcriptional activation when tethered to the α-NTD simply by providing a contact surface for a suitably positioned DNA-binding protein, the strength of the protein-protein contact presumably determining the degree to which the binding of RNAP is stabilized. In principle, any sufficiently strong interaction between a DNA-bound protein and any subunit of RNAP should stabilize the binding of RNAP and would therefore be expected to activate transcription from a suitable promoter. It is not clear how strongly natural activators interact with RNAP; in the case of CRP at *plac*, the dissociation constant has been estimated to be in the millimolar range (R.H. Ebright, pers. comm.). The dissociation constant for the interaction of λcI dimers is approximately 10^{-6} M (Sauer 1979), suggesting that interaction energies in this range are sufficient. Furthermore, most natural activators appear to interact with either the α-CTD or the σ-subunit, both of which can contact the DNA directly (Busby and Ebright 1994). Under these circumstances, even weaker protein-protein interactions might be expected to mediate a significant amount of stimulation through the stabilization of suboptimal contacts between an RNAP subdomain and the DNA (Busby and Ebright 1994; Hochschild and Dove 1998).

TWO ACTIVATING REGIONS FUNCTION SYNERGISTICALLY

As mentioned above, λcI can activate transcription when bound at position –42, using an activating region in its NTD to contact the σ-subunit. We took advantage of the fact that a DNA-bound λcI dimer has the potential to

A

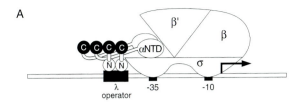

B

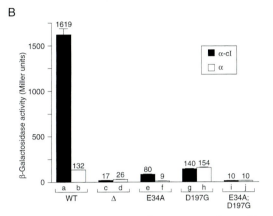

Figure 3. (*A*) Contacts between λcI dimer bound at position –42 and RNAP. The DNA-bound λcI dimer uses the natural activating region in its NTD to contact the σ-subunit and its CTD to contact the CTD of the α-cI chimera. (*B*) Effects of wild-type and mutant λcI proteins bound at position –42 on transcription in vivo in the presence or absence of the α-cI chimera. Cells harboring the modified P_{RM}-*lacZ* reporter construct depicted in *A* and containing either the α-cI chimera (*filled bars*) or wild-type α only (*open bars*) and the indicated λcI variant were assayed for β-galactosidase activity. λcI-E34A is a positive control mutant that is completely unable to activate transcription from λP_{RM} (Bushman et al. 1989). λcI-D197G is a cooperativity mutant that is unable to participate in the dimer-dimer interaction that mediates cooperative binding to DNA (Whipple et al. 1994). (Adapted, with permission, from Dove et al. 1997 [copyright Macmillan].)

participate in two distinct protein-protein interactions to show that the natural activating region and the CTD can function synergistically to activate transcription when λcI is bound at position –42 (Dove et al. 1997). The test promoter used for this experiment was a derivative of λP_{RM} (the promoter normally activated by λcI) bearing a single λ operator site centered at position –42 (Fig. 3A). Whereas λcI stimulated transcription from this test promoter less than 10-fold in the absence of the α-cI chimera (Fig. 3B, compare d and b), the magnitude of the stimulation increased to approximately 100-fold in the presence of the chimera (Fig. 3B, compare c and a). We demonstrated that this high level of λcI-stimulated transcription depends on both the natural activating surface (located in the NTD) and the artificial activating surface (located in the CTD) by using two λcI mutants. One (a classical positive control mutant) is specifically defective for activation through the natural contact with the σ-subunit and the other (a cooperativity mutant) is specifically defective for activation through the dimer-dimer interaction with the α-cI chimera. Each mutant stimulated transcription by less than tenfold (Fig. 3B, compare c with e

and g, and d with h), and the positive control mutant only stimulated transcription in the presence of the α-cI chimera (Fig. 3B, compare e with f). These findings demonstrate that for an activator capable of contacting two components of RNAP, the effect of these two contacts on transcription can be synergistic. In a striking parallel, an experiment performed in yeast cells similarly demonstrates a synergistic effect of two activating regions (one natural and the other artificial) present on a single DNA-bound activator (Farrell et al. 1996).

This experiment done with the λ P_{RM} derivative also demonstrates that the stimulatory effect of the engineered interaction between λcI and the α-cI chimera does not depend on the *lac* promoter, in particular, nor on the precise geometry of the interacting components.

Natural activators have also been described that can interact simultaneously with more than one target surface on RNAP. These include CRP, the related global regulator FNR, and the bacteriophage Mu Mor protein (for review, see Rhodius and Busby 1998; see also Artsimovitch et al. 1996). In the case of CRP, two activating regions have been identified, AR1 and AR2, which contact the α-CTD and the α-NTD, respectively. At *plac* and other similarly arranged promoters, only AR1 can be utilized, but when the CRP-binding site overlaps the –35 region (such as at the *gal* promoter, which bears a single site centered at position –41.5), both AR1 and AR2 are utilized simultaneously (Niu et al. 1996). Similarly, the *E. coli* FNR protein bears two activating regions, one of which corresponds to AR1 of CRP and targets the α-CTD and the other which evidently targets the σ-subunit (Li et al. 1998).

A PROTEIN-PROTEIN CONTACT MODEL FOR TRANSCRIPTIONAL ACTIVATION

Our findings with the α-cI chimera suggest that any protein-protein interaction could, in principle, trigger transcriptional activation provided the relevant protein domains are fused to the α-NTD and to λcI (or another DNA-binding protein). They further suggest that contact between a DNA-bound protein and any accessible surface of RNAP could activate transcription. Summarized below are the results of experiments done to test the generality of our findings.

Contact between Protein Domains X and Y Fused, Respectively, to the α-NTD and λcI Can Activate Transcription

To confirm that other protein domains known to interact can mediate transcriptional activation, we took advantage of a well-characterized interaction involving two yeast proteins, the GAL4 protein (a transcriptional regulator) and a mutant form of the GAL11 protein (a component of the RNA polymerase II holoenzyme). Although wild-type GAL11 does not interact with GAL4, this GAL11 mutant (called GAL11^P) bears a single-amino-acid substitution that results in an apparently fortuitous interaction with the dimerization domain of GAL4 which

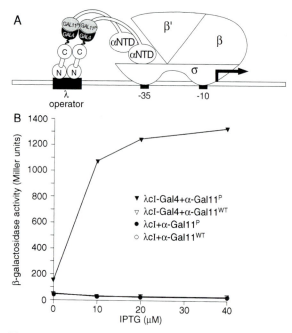

Figure 4. Transcriptional activation by the λcI-GAL4 fusion protein in the presence of α-GAL11^P fusion protein. (*A*) Contact between GAL4 moiety of λcI-GAL4 fusion protein and GAL11^P moiety fused to α-NTD. (*B*) Effect of λcI-GAL4 on transcription in vivo in the presence of α-GAL11^P fusion protein. Cells harboring the modified *plac-lacZ* reporter construct (with the λ operator centered at position –62) and compatible plasmids directing expression of the indicated pair of proteins under the control of IPTG-inducible promoters were assayed for β-galactosidase. (Reprinted, with permission, from Dove and Hochschild 1998.)

can trigger transcriptional activation in yeast (Himmelfarb et al. 1990; Barberis et al. 1995). We found that when fused to the carboxyl terminus of λcI, the GAL4 dimerization domain strongly activated transcription (by a factor of ~45) in cells containing an α-GAL11^P chimera, but not in cells containing an α-GAL11^{wt} chimera, indicating that this protein-protein interaction can also trigger transcriptional activation in *E. coli* (Fig. 4) (Dove and Hochschild 1998). Interestingly, the equilibrium dissociation constant for the GAL4/GAL11^P interaction has been reported to be 10^{-7} M (Farrell et al. 1996), whereas the dissociation constant for the λcI-CTD tetramerization interaction is approximately 10^{-6} M (Sauer 1979), and the former interaction mediates correspondingly greater activation (~45-fold as compared with ~15-fold).

Contact between Protein Domains X and Y Fused, Respectively, to the ω-subunit and λcI Can Activate Transcription

To test the hypothesis that contact with any subunit of RNAP can activate transcription, and to ask specifically if a component of RNAP other than the α-subunit can mediate the effects of artificial activators, we took advantage of a small protein known as the ω-subunit of RNAP, which unlike the α-subunit is not essential for cell growth (Gentry and Burgess 1989). The ω-subunit is found

tightly associated with *E. coli* RNAP, but its function is unknown (Gentry et al. 1991). Importantly, for our purposes, the growth of cells that bear a chromosomal deletion of the ω gene (*rpoZ*) is indistinguishable from that of wild-type cells (Gentry et al. 1991).

We fused to the ω-subunit the GAL11^P fragment that we had previously fused to α, creating an ω-GAL11^P chimera. We found that the λcI-GAL4 fusion protein activated transcription approximately 30-fold in cells containing the ω-GAL11^P chimera but not in cells containing an otherwise identical ω-GAL11^{wt} chimera (Fig. 5) (Dove and Hochschild 1998). The experiments performed with α chimeras were done using cells that contained both wild-type α (since the α-CTD is required for viability) and a particular chimera. For these experiments, in contrast, we could delete *rpoZ* from our reporter strain, thus ensuring that wild-type ω protein would not compete with the ω chimera for the binding site on RNAP.

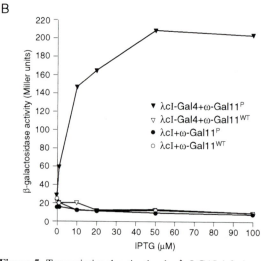

Figure 5. Transcriptional activation by λcI-GAL4 fusion protein in the presence of ω-GAL11^P fusion protein. (*A*) Contact between GAL4 moiety of λcI-GAL4 and GAL11^P moiety fused to ω. (*B*) Effect of λcI-GAL4 on transcription in vivo in the presence of ω-GAL11^P fusion protein. Cells harboring the modified *plac-lacZ* reporter construct (with the λ operator centered at position –62) and compatible plasmids directing expression of the indicated pair of proteins under the control of IPTG-inducible promoters were assayed for β-galactosidase. (Reprinted, with permission, from Dove and Hochschild 1998.)

DIRECT FUSION OF DNA-BINDING PROTEINS TO THE ω SUBUNIT OR THE α-NTD CAN ACTIVATE TRANSCRIPTION

The examples of transcriptional activation we have described presumably reflect a strengthened association of RNAP with the promoter. The binding of RNAP should also be stabilized by the direct fusion of a DNA-binding protein to one of the RNAP subunits provided the DNA-binding domain can interact with a cognate recognition site provided upstream of the promoter. We first tested this hypothesis by linking the ω-subunit directly to λcI, in effect replacing the GAL4/GAL11ᵖ bridge with a covalent link. To this end, we fused ω to the carboxyl terminus of λcI (Fig. 6A). We found that the resulting λcI-ω chimera stimulated transcription approximately 70-fold in the Δ*rpoZ* derivative of our standard reporter strain (with the λ operator centered at position –62) (Fig. 6B) (Dove and Hochschild 1998). This stimulation was entirely abrogated when the cells bore the *rpoZ⁺* allele (i.e., contained wild-type ω protein). Furthermore, the introduction of a DNA-binding mutation into the cI moiety of the chimeric gene abolished the activation.

This experiment is, however, subject to one mechanistic uncertainty. We do not know whether the λcI-ω chimera is stably associated with RNAP holoenzyme prior to promoter binding, or alternatively, whether the chimera binds to the λ operator and subsequently helps recruit RNAP to the associated promoter. To resolve this uncertainty, we tested the effect of fusing a DNA-binding protein directly to the α-NTD, which is an integral and obligate component of the enzymatic core. In this case, we could not, however, use λcI because fusion of its amino terminus to another protein might disrupt essential contacts between its amino-terminal arm and the central portion of the operator (Beamer and Pabo 1992). Instead, we took advantage of CRP, which is also a two-domain protein that binds its recognition site as a dimer but contacts the DNA using its CTD (Schultz et al. 1991). We constructed an α-CRP chimera and tested its ability to activate transcription from a promoter bearing a single CRP recognition site centered at position –61.5 (essentially the wild-type *lac* promoter) (Fig. 7). There was, however, one complication in the design of this experiment: Since CRP is a transcriptional activator that ordinarily activates transcription from *plac* and other similarly arranged promoters by contacting the α-CTD (see above), it was necessary to inactivate this positive control function. In fact, we inactivated both AR1 and AR2 by the introduction of two previously characterized amino acid substitutions into the CRP moiety of the fusion protein. The resulting α-CRP chimera activated transcription approximately 10-fold (S.L. Dove and A. Hochschild, in prep.), and this activation was abolished by the introduction of a DNA-binding mutation into the *crp* moiety of the chimeric gene. Experiments performed in vitro with reconstituted RNAP containing the α-CRP chimera but no wild-type α suggest that the magnitude of the stimulatory effect is actually much larger (~100-fold; S.L. Dove and A. Hochschild, in prep.).

This activation is presumably analogous to the activation that is mediated by an UP element, which functions by binding the α-CTD (Blatter et al. 1994; Jeon et al. 1995; Gaal et al. 1996). Kinetic experiments have indicated that at least one effect of the UP element is to stabilize the initial binding of RNAP (Rao et al. 1994), a mechanism that we presume is primarily responsible for the activation observed in our experiments.

We have discussed experiments which indicate that direct or indirect contacts between protein domains fused to RNAP and the DNA can activate transcription. Parallel

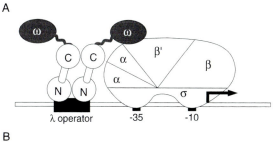

A

B

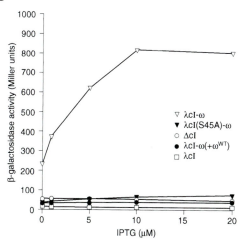

Figure 6. Transcriptional activation by λcI-ω fusion protein. (*A*) Covalent link to the ω-subunit connects DNA-bound λcI dimer with RNAP. (*B*) Effect of λcI-ω fusion protein on transcription in vivo from modified *lac* promoter. Cells harboring the modifed *plac-lacZ* reporter construct (with the λ operator centered at position –62) and a plasmid directing expression of λcI or the indicated λcI-ω fusion protein under the control of an IPTG-inducible promoter were assayed for β-galactosidase. The λcI(S45A)-ω variant bears an amino acid substitution in the λcI DNA-binding domain that disrupts DNA binding. The cells bore a deletion of the gene encoding wild-type ω (*rpoZ*), except in the case of the (+ωᵂᵀ) control. (Reprinted, with permission, from Dove and Hochschild 1998.)

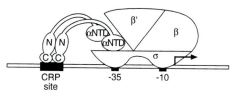

Figure 7. α-CRP chimera incorporated into RNAP can bind to promoter-associated CRP-recognition site.

findings in yeast (for reviews, see Struhl 1996; Ptashne and Gann 1997) similarly show that transcription can be activated by direct or indirect connections between DNA-bound protein domains and either the RNAP II holoenzyme (Himmelfarb et al. 1990; Barberis et al. 1995; Farrell et al. 1996; Gaudreau et al. 1997) or the TATA-binding protein (TBP) (Chatterjee and Struhl 1995; Klages and Strubin 1995; Xiao et al. 1995; see also Apone et al. 1996; Gonzalez-Couto et al. 1997; Lee and Struhl 1997; Keaveney and Struhl 1998). These experiments with artificial activators imply that strengthening the association of essential components of the transcription machinery with promoter DNA can suffice to activate transcription in both prokaryotes and eukaryotes (Struhl 1996; Ptashne and Gann 1997; Hochschild and Dove 1998; see also Gaudreau et al. 1998; Keaveney and Struhl 1998).

TRANSCRIPTIONAL REPRESSION BY TETHERING RNAP TO THE PROMOTER

The initiation of transcription is a multistep process, any one of which can be rate limiting (for review, see Gussin 1996). In vitro studies have led to the following general scheme. RNAP first recognizes and binds to double-stranded promoter DNA, forming what is known as the closed complex. The closed complex must then undergo an isomerization process, leading to the formation of the transcriptionally active open complex in which the DNA strands are locally melted. RNAP can then direct the synthesis of short abortive products, but for full-length transcripts to be generated, RNAP must escape from the promoter (a step referred to as promoter clearance). The artificial activators we have described presumably function, at least in part, by stabilizing closed complex formation. In principle, however, the stabilization of RNAP binding could have a repressive rather than a stimulatory effect on transcription, for example, if overstabilization of the open complex inhibits promoter clearance. Recent studies of the bacteriophage ϕ29 p4 protein, which functions as both an activator and a repressor of transcription, in fact suggest that a stabilizing protein-protein interaction can either activate or repress transcription depending on the characteristics of the target promoter (Monsalve et al. 1998). Protein p4 interacts with the α-CTD, and both activation and repression depend on this interaction (for review, see Rojo et al. 1998). Monsalve et al. (1998) have shown that this protein-protein interaction can stabilize closed complex formation at a promoter that is inefficiently recognized by RNAP or impede promoter clearance at a promoter that is efficiently recognized, presumably through overstabilization of the open complex.

It will be interesting to determine whether transcriptional repression can be achieved through arbitrarily selected protein-protein or protein-DNA interactions. The simplest test would take advantage of our demonstration that transcriptional activation can be achieved through the covalent linkage of a DNA-binding protein to the α-

subunit (which is an integral component of core RNAP). Under these circumstances, the escape of RNAP from a test promoter presumably requires that the contacts between the DNA-binding protein and its cognate recognition site be broken. Thus, it will be informative to test the effect of systematically increasing the affinity of such a protein-DNA interaction.

As is suggested by the analysis of protein p4, the nature of the promoter may also help to determine whether the effect of a tethered DNA-binding protein is to activate or to repress transcription. Thus, it will also be important to test a variety of different promoters to determine if the α-CRP chimera (or others) can repress as well as activate transcription. Finally, it may be possible to potentiate both stimulatory and repressive effects by providing for additional stabilizing protein-protein interactions.

MECHANISTIC CONSIDERATIONS

The presumption that the artificial activators we have described function by stabilizing the binding of RNAP to the promoter implies that RNAP binding must be a rate-limiting step in the initiation process at those promoters that are susceptible to this form of activation. Thus, if promoter occupancy is not limiting for a given promoter in vivo, then we would predict that this form of activation would be ineffective. Activatable promoters have been described whose activities are not limited by promoter occupancy (see, e.g., Kustu et al. 1991; Summers 1992). A well-characterized example is provided by the *Salmonella glnA* promoter (North and Kustu 1997), which is recognized by a form of RNAP in which σ^{70} is replaced by the alternative sigma factor σ^{54}. In the absence of the activator (NTRC), this promoter binds RNAP in a stable but transcriptionally inactive closed complex. The role of the activator is to catalyze the isomerization of this closed complex to a transcriptionally active open complex, a requirement that presumably could not be bypassed by the use of an artificial activator of the sort we have described.

Future work will be directed toward the mechanistic questions we have raised. Kinetic analysis will allow us to test the hypothesis that the artificial activators we have described stabilize closed complex formation, as well as reveal whether other steps in the initiation process can be affected. The use of a variety of different promoters, with distinct kinetic properties, will allow us to test the hypothesis that these activators work only on promoters for which RNAP binding is limiting and to examine how the promoter affects both activator and repressor function.

ACKNOWLEDGMENTS

This work was supported by National Institutes of Health grant GM-44025 (A.H.) and by an established investigatorship from the American Heart Association (A.H.).

REFERENCES

Apone L.M., Virbasius C.M., Reese J.C., and Green M.R. 1996. Yeast TAF(II)90 is required for cell-cycle progression through G$_2$/M but not for general transcription activation. *Genes Dev.* **10:** 2368.

Artsimovitch I., Murakami K., Ishihama A., and Howe M.M. 1996. Transcription activation by the bacteriophage Mu Mor protein requires the C-terminal regions of both α and σ^{70} subunits of *Escherichia coli* RNA polymerase. *J. Biol. Chem.* **271:** 32343.

Barberis A., Pearlberg J., Simkovich N., Farrell S., Reinagel P., Bamdad C., Sigal G., and Ptashne M. 1995. Contact with a component of the polymerase II holoenzyme suffices for gene activation. *Cell* **81:** 359.

Beamer L.J. and Pabo C.O. 1992. Refined 1.8 Å crystal structure of the λ repressor-operator complex. *J. Mol. Biol.* **227:** 177.

Blatter E.E., Ross W., Tang H., Gourse R.L., and Ebright R.H. 1994. Domain organization of RNA polymerase α subunit: C-terminal 85 amino acids constitute a domain capable of dimerization and DNA binding. *Cell* **78:** 889.

Burgess R.R. 1976. Purification and physical properties of *E. coli* RNA polymerase In *RNA polymerase* (ed. R. Losick and M. Chamberlin), p. 69. Cold Spring Harbor Laboratory, Cold Spring Harbor, New York.

Busby S. and Ebright R.H. 1994. Promoter structure, promoter recognition, and transcription activation in prokaryotes. *Cell* **79:** 743.

Bushman F.D., Shang C., and Ptashne M. 1989. A single glutamic acid residue plays a key role in the transcriptional activation function of λ repressor. *Cell* **58:** 1163.

Chatterjee S. and Struhl K. 1995. Connecting a promoter-bound protein to TBP bypasses the need for a transcriptional activation domain. *Nature* **374:** 820.

Dombroski A.J., Walter W.A., Record M.T., Jr., Siegele D.A., and Gross C.A. 1992. Polypeptides containing highly conserved regions of transcription initiation factor σ 70 exhibit specificity of binding to promoter DNA. *Cell* **70:** 501.

Dove S.L. and Hochschild A. 1998. Conversion of the ω subunit of *Escherichia coli* RNA polymerase into a transcriptional activator or an activation target. *Genes Dev.* **12:** 745.

Dove S.L., Joung J.K., and Hochschild A. 1997. Activation of prokaryotic transcription through arbitrary protein-protein contacts. *Nature* **386:** 627.

Ebright R.H. and Busby S. 1995. The *Escherichia coli* RNA polymerase α subunit: Structure and function. *Curr. Opin. Genet. Dev.* **5:** 197.

Farrell S., Simkovich N., Wu Y., Barberis A., and Ptashne M. 1996. Gene activation by recruitment of the RNA polymerase II holoenzyme. *Genes Dev.* **10:** 2359.

Gaal T., Ross W., Blatter E.E., Tang H., Jia X., Krishnan V.V., Assa-Munt N., Ebright, R.H., and Gourse R.L. 1996. DNA-binding determinants of the α subunit of RNA polymerase: Novel DNA-binding domain architecture. *Genes Dev.* **10:** 16.

Gaudreau L., Adam M., and Ptashne M. 1998. Activation of transcription in vitro by recruitment of the yeast RNA polymerase II holoenzyme. *Mol. Cell* **1:** 913.

Gaudreau L., Schmid A., Blaschke D., Ptashne M., and Horz W. 1997. RNA polymerase II holoenzyme recruitment is sufficient to remodel chromatin at the yeast PHO5 promoter. *Cell* **89:** 55.

Gentry D.R. and Burgess R.R. 1989. rpoZ, encoding the omega subunit of *Escherichia coli* RNA polymerase, is in the same operon as spoT. *J. Bacteriol.* **171:** 1271.

Gentry D., Xiao H., Burgess R., and Cashel M. 1991. The ω subunit of *Escherichia coli* K-12 RNA polymerase is not required for stringent RNA control in vivo. *J. Bacteriol.* **173:** 3901.

Gonzalez-Couto E., Klages N., and Strubin M. 1997. Synergistic and promoter-selective activation of transcription by recruitment of transcription factors TFIID and TFIIB. *Proc. Natl. Acad. Sci.* **94:** 8036.

Gussin G.N. 1996. Kinetic analysis of RNA polymerase-promoter interactions. *Methods Enzymol.* **273:** 45.

Helmann J.D. and Chamberlin M.J. 1988. Structure and function of bacterial σ factors. *Annu. Rev. Biochem.* **57:** 839.

Himmelfarb H.J., Pearlberg J., Last D.H., and Ptashne M. 1990. GAL11P: A yeast mutation that potentiates the effect of weak GAL4-derived activators. *Cell* **63:** 1299.

Hochschild A. and Dove S.L. 1998. Protein-protein contacts that activate and repress prokaryotic transcription. *Cell* **92:** 597.

Igarashi K. and Ishihama A. 1991. Bipartite functional map of the *E. coli* RNA polymerase α subunit: Involvement of the C-terminal region in transcription activation by cAMP-CRP. *Cell* **65:** 1015.

Igarashi K., Hanamura A., Makino K., Aiba H., Mizuno T., Nakata A., and Ishihama A. 1991. Functional map of the α subunit of *Escherichia coli* RNA polymerase: Two modes of transcription activation by positive factors. *Proc. Natl. Acad. Sci.* **88:** 8958.

Ishihama A. 1981. Subunit of assembly of *Escherichia coli* RNA polymerase. *Adv. Biophys.* **14:** 1.

———. 1992. Role of the RNA polymerase α subunit in transcription activation. *Mol. Microbiol.* **6:** 3283.

———. 1993. Protein-protein communication within the transcription apparatus. *J. Bacteriol.* **175:** 2483.

Jeon Y.H., Yamazaki T., Otomo T., Ishihama A., and Kyogoku Y. 1997. Flexible linker in the RNA polymerase α subunit facilitates the independent motion of the C-terminal activator contact domain. *J. Mol. Biol.* **267:** 953.

Jeon Y.H., Negishi T., Shirakawa M., Yamazaki T., Fujita N., Ishihama A., and Kyogoku Y. 1995. Solution structure of the activator contact domain of the RNA polymerase α subunit. *Science* **270:** 1495.

Keaveney M. and Struhl K. 1998. Activator-mediated recruitment of the RNA polymerase II machinery is the predominant mechanism for transcriptional activation in yeast. *Mol. Cell* **1:** 917.

Kimura M., Fujita N., and Ishihama A. 1994. Functional map of the α subunit of *Escherichia coli* RNA polymerase. Deletion analysis of the amino-terminal assembly domain. *J. Mol. Biol.* **242:** 107.

Klages N. and Strubin M. 1995. Stimulation of RNA polymerase II transcription initiation by recruitment of TBP in vivo. *Nature* **374:** 822.

Kolb A., Igarashi K., Ishihama A., Lavigne M., Buckle M., and Buc H. 1993. *E. coli* RNA polymerase, deleted in the C-terminal part of its α-subunit, interacts differently with the cAMP-CRP complex at the lacP1 and at the galP1 promoter. *Nucleic Acids Res.* **21:** 319.

Kustu S., North A.K., and Weiss D.S. 1991. Prokaryotic transcriptional enhancers and enhancer-binding proteins. *Trends Biochem. Sci.* **16:** 397.

Lee M. and Struhl K. 1997. A severely defective TATA-binding protein-TFIIB interaction does not preclude transcriptional activation in vivo. *Mol. Cell. Biol.* **17:** 1336.

Li B., Wing H., Lee D., Wu H.C., and Busby S. 1998. Transcription activation by *Escherichia coli* FNR protein: Similarities to, and differences from, the CRP paradigm. *Nucleic Acids Res.* **26:** 2075.

Lonetto M., Gribskov M., and Gross C.A. 1992. The σ 70 family: Sequence conservation and evolutionary relationships. *J. Bacteriol.* **174:** 3843.

Monsalve M., Calles B., Mencia M., Salas M., and Rojo F. 1998. Transcription activation or repression by phage φ29 protein p4 depends on the strength of the RNA polymerase-promoter interactions. *Mol. Cell* **1:** 99.

Niu W., Kim Y., Tau G., Heyduk T., and Ebright R.H. 1996. Transcription activation at class II CAP-dependent promoters: Two interactions between CAP and RNA polymerase. *Cell* **87:** 1123.

North A.K. and Kustu S. 1997. Mutant forms of the enhancer-binding protein NtrC can activate transcription from solution. *J. Mol. Biol.* **267:** 17.

Ptashne M. 1992. *A genetic switch*, 2nd edition. Cell Press and Blackwell Scientific, Cambridge, Massachusetts.

Ptashne M. and Gann A. 1997. Transcriptional activation by recruitment. *Nature* **386:** 569.

Rao L., Ross W., Appleman J.A., Gaal T., Leirmo S., Schlax P.J., Record M.T., Jr., and Gourse R L. 1994. Factor independent activation of rrnB P1. An "extended" promoter with an upstream element that dramatically increases promoter strength. *J. Mol. Biol.* **235:** 1421.

Rhodius V.A. and Busby S.J.W. 1998. Positive activation of gene expression. *Curr. Opin. Microbiol.* **1:** 152.

Rojo F., Mencia M., Monsalve M., and Salas M. 1998. Transcription activation and repression by interaction of a regulator with the α subunit of RNA polymerase: The model of phage ϕ29 protein p4. *Prog. Nucleic Acid Res. Mol. Biol.* **60:** 29.

Ross W., Gosink K.K., Salomon J., Igarashi K., Zou C., Ishihama A., Severinov K., and Gourse R.L. 1993. A third recognition element in bacterial promoters: DNA binding by the α subunit of RNA polymerase. *Science* **262:** 1407.

Sauer R.T. 1979. "Molecular characterization of the lambda repressor and its gene cI." Ph.D. thesis, Harvard University, Cambridge, Massachusetts.

Sauer R.T., Jordan S.R., and Pabo C.O. 1990. λ Repressor: A model system for understanding protein-DNA interactions and protein stability. *Adv. Protein Chem.* **40:** 1.

Schultz S.C., Shields G.C., and Steitz T.A. 1991. Crystal structure of a CAP-DNA complex: The DNA is bent by 90 degrees. *Science* **253:** 1001.

Struhl K. 1996. Chromatin structure and RNA polymerase II connection: Implications for transcription. *Cell* **84:** 179.

Summers A.O. 1992. Untwist and shout: A heavy metal-responsive transcriptional regulator. *J. Bacteriol.* **174:** 3097.

Whipple F.W., Hou E.F., and Hochschild A. 1998. Amino acid-amino acid contacts at the cooperativity interface of the bacteriophage λ and P22 repressors. *Genes Dev.* **12:** 2791.

Whipple F.W., Kuldell N.H., Cheatham L.A., and Hochschild A. 1994. Specificity determinants for the interaction of λ repressor and P22 repressor dimers. *Genes Dev.* **8:** 1212.

Xiao H., Friesen J.D., and Lis J.T. 1995. Recruiting TATA-binding protein to a promoter: Transcriptional activation without an upstream activator. *Mol. Cell. Biol.* **15:** 5757.

Activation and the Role of Reinitiation in the Control of Transcription by RNA Polymerase II

S. Hahn

Howard Hughes Medical Institute and Fred Hutchinson Cancer Research Center, Seattle, Washington 98109-1024

Given our current understanding of the pathway for transcription by RNA polymerase II (pol II), there are a few general models to explain how activators can promote high levels of transcription. One possibility is that activators promote rapid reinitiation of transcription, by a pathway that bypasses a rate-limiting step used in formation of the first preinitiation complex upon gene induction. Possible mechanisms for the pathway of reinitiation and its stimulation by activators are reviewed. Activation-specific defects observed using certain basal factor mutations may be explained by defects in reinitiation of transcription.

The papers in this Symposium volume illustrate the great advances in our understanding of the mechanisms for pol II transcription and its control during the past 20 years. Although there is still a great deal to be learned, most of the components of the transcription machinery have been identified, and a general outline for how these components assemble at promoters and initiate transcription is known (Fig. 1) (Orphanides et al. 1996; Roeder 1996; Nikolov and Burley 1997; Hampsey 1998). The first step in transcription initiation is recruitment of the transcription machinery to the promoter to form a preinitiation complex (PIC). The transcription machinery consists of more than 50 polypeptides, many of which can assemble at the promoter in a stepwise fashion in vitro (Buratowski et al. 1989). However, genetic and biochemical data suggest that a large subset of these factors are preassembled in the pol II holoenzyme complex and associate at the promoter in a single step (Koleske and Young 1995; Chao et al. 1996; Maldonado et al. 1996; Greenblatt 1997; Myers et al. 1998). A conservative model, based on available data, is that TFIID and TFIIA assemble at the promoter first with the holoenzyme using

these factors on DNA as a platform for assembly. However, cooperative interactions between all of these factors may make assembly of the machinery at the promoter a concerted process (Struhl 1996). Once the necessary factors are assembled at the promoter, the PIC undergoes an ATP-dependent transition to the open complex, which likely involves at least one DNA helicase activity associated with the TFIIH factor (Bunick et al. 1982; Sawadogo and Roeder 1984; Wang et al. 1992; Goodrich and Tjian 1994; Pan and Greenblatt 1994; Holstege et al. 1995, 1997; Divr et al. 1996). Next, polymerase initiates RNA synthesis and can produce multiple short abortive RNA products or enter the productive elongation mode whereby it releases contacts with the promoter and synthesizes mRNA (Luse and Jacob 1987; Goodrich and Tjian 1994; Divr et al. 1997; Holstege et al. 1997). Finally, polymerase and other components of the transcription machinery that dissociate from the promoter during the initiation and elongation steps must be recruited again so that multiple cycles of transcription (reinitiation) can occur. In principle, any of these steps could be rate-limiting under certain in vivo or in vitro conditions and may be a possible target of transcription control by activators. Below, I review our current understanding of the reinitiation pathway for pol II and the evidence for the importance of reinitiation in gene regulation.

POSSIBLE MECHANISMS OF TRANSCRIPTION ACTIVATION

Much controversy currently exists as to the target(s) of transcription activator proteins. However, from the above transcription pathway, there are only a few general mechanisms by which activators can increase the

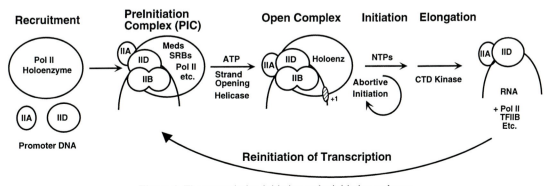

Figure 1. The transcription initiation and reinitiation pathway.

rate of RNA synthesis. These mechanisms are not mutually exclusive.

Recruitment

The first (and currently most popular) model for gene activation is the recruitment model (Ptashne and Gann 1997). In this model, the DNA-binding domain of an activator interacts specifically with its target site in promoters, whereas the activation domain of the activator binds to a component of the transcription machinery, thus recruiting it to the promoter. There is good evidence that this is a viable mechanism of activation. For example, fusion of a DNA-binding domain to many components of the transcription machinery in yeast can partially bypass the requirement for an activator (Keaveney and Struhl 1998 and references therein). Also consistent with this model are the numerous interactions observed in vitro and in vivo between activation domains and components of the transcription machinery (for review, see Burley and Roeder 1996; Ptashne and Gann 1997; Koh et al. 1998). However, DNA-binding domain fusion to subunits of the transcription machinery often gives transcription levels significantly below that promoted by strong natural activators. This suggests that recruitment is not the only mechanism of activator action.

Antirepression

A second possible mode of activator action is antirepression. One well-known example is occlusion of promoter accessibility by histones and other chromatin-associated proteins (see Paranjape et al. 1994). Recent biochemical results suggest that accessibility in chromatin can be regulated by the combined action of histone acetylases and deacetylases (and perhaps kinases) and by chromatin remodeling activities such as SWI/SNF and NURF (Kingston et al. 1996; Grunstein 1997; Pazin and Kadonaga 1997; Struhl 1998). In addition, the findings that known transcription coactivators such as p300/CBP have histone acetylase activity support a strong connection between antirepression by histones and activation of transcription (Bannister and Kouzarides 1996; Ogryzko et al. 1996).

Another possible antirepression mechanism involves repressors specific for subunits of the pol II transcription machinery (Auble et al. 1994; Poon et al. 1994; Catron et al. 1995; Kaiser and Meisterernst 1996; Gadbois et al. 1997; Kim et al. 1997; Prelich 1997; Yeung et al. 1997; Li and Manley 1998). For example, the factors NC2 (DR1/DRAP1) and MOT1 are known to directly target TBP (TATA-binding protein) (Inostroza et al. 1992; Auble et al. 1994). An effective mechanism of activation would bypass the ability of these activators to block formation of active transcription complexes. For example, MOT1 has been shown to block basal but not activated transcription (Auble et al. 1994). Thus, in at least one case, an activator can overcome the effect of the specific transcription repressor.

PIC Activity

A third possible mechanism for increasing the rate of initiation is to increase the activity of the preinitiation complexes. Indirect evidence suggests that preinitiation complexes may form in both productive (active) and nonproductive (inactive) states (Wang et al. 1992; Hahn 1993; Herschlag and Johnson 1993; Chi et al. 1995). The ratio of these two states may be influenced by the specific repressors described above (Antirepression) or frequent nonproductive PIC formation may be inherent in assembly of the transcription machinery. In any case, the activators could influence the ratio of active to inactive complexes, leading to more frequent initiation. The initiation pathway also involves at least several steps after PIC formation such as open complex formation, helicase activity, and the catalytic step of first phosphodiester bond formation (Orphanides et al. 1996). In principle, if any of these steps were rate-limiting, a rate increase caused by the activator would lead to a faster rate of initiation.

Elongation Rate

Another step that could potentially be affected by activators is an increase in the rate of promoter clearance and elongation of polymerase (Yankulov et al. 1994; Krumm et al. 1995; Blau et al. 1996; Brown et al. 1998). One possibility is that the activator may block or reduce the time spent in production of short abortive products. This would lead to faster promoter clearance allowing faster binding and initiation by the next polymerase and other factors. Additionally, it is known that polymerase tends to pause a short distance downstream from the start site (~25–30 bp) at many promoters (O'Brien and Lis 1991; Bentley 1995; Shilatifard et al. 1997). Certain classes of activators have been found to reduce this promoter proximal pausing, which may result from the loading of elongation factors or a direct interaction with polymerase or other factors that travel with the polymerase.

Reinitiation

Finally, activators could stimulate transcription by influencing the rate at which multiple cycles of transcription occur. Since highly expressed genes are transcribed at a rapid rate (see below), it has been proposed that the bulk of mRNA synthesized from strong promoters arises from the reinitiation pathway (Zawel et al. 1995; Struhl 1996). One specific effect of an activator could involve stabilization of factors such as TFIID and TFIIA, believed to be left at the promoter after initiation. This may involve stabilization via cooperative protein-protein interactions (Lieberman and Berk 1991; Chi et al. 1995; Chi and Carey 1996) or an enzymatic activity such as an acetylase or kinase that contributes to stabilization. Another possibility is that the factors left at the promoter are targets for repressors which would either remove them (i.e., MOT1) or block binding of the next polymerase (i.e., NC2). Activators could, in theory, favor the recruitment of holoenzyme over disassembly or inactivation of partial PICs by making the rate of holoenzyme recruitment faster than the rate of factor inactivation or dissociation. Finally, in the context of chromatin, activators are probably required to keep the promoter in an accessible state for multiple cycles of transcription, likely through mechanisms such as those outlined above (Antirepression).

EVIDENCE FOR A SEPARATE PATHWAY FOR REINITIATION DISTINCT FROM INITIATION

For genes that are expressed at a high rate, it seems logical to not completely disassemble the transcription machinery at the promoter after each round of initiation (for one exception to this model, see Kadonaga 1990). Initiation rates at highly expressed pol II promoters are thought to approach the rate of initiation from much simpler systems such as *Escherichia coli* and T7 RNA polymerases. In yeast, initiation rates at an enhanced HIS3 promoter can occur about once every 6–8 seconds or about once every 15 seconds at the natural yeast DED1 promoter (Iyer and Struhl 1996). In *Drosophila*, hsp70 initiation has been estimated to occur approximately once every 4 seconds (O'Brien and Lis 1993). This high rate of initiation is similar to that of eukaryotic polymerase I (~1 RNA/5 sec) (Reeder and Lang 1997), *E. coli* polymerase (~1 RNA/2–3 sec) (Kennell and Riezman 1977), and T7 polymerase (~1 RNA/1.2 sec) (Martin and Coleman 1987). Once the activator is bound to DNA, nucleosome modification and rearrangement and the binding of factors to DNA such as TFIID are potentially slow steps in initial formation of PICs. It is plausible that these and/or other slow steps would be bypassed if continued high-level transcription is required.

Since it is currently difficult if not impossible to separate initiation from reinitiation using in vivo models, most experiments probing reinitiation are carried out in vitro. One strict view of reinitiation in vitro is that multiple cycles of initiation must occur from the same promoter that led to first round transcription. Currently, only one assay can directly test for this, the colliding polymerase assay (Szentirmay and Sawadogo 1994). In this assay, elongation of polymerase is stopped at the end of a G-free cassette by omission of GTP, or more recently, a roadblock to elongation consisting of a psoralen-cross-linked oligonucleotide (Szentirmay et al. 1998). This block in elongation leads to pile up of polymerases which initiated subsequently to the first round of transcription and gives rise to RNA products progressively shorter by about 30 bases. However, most experiments probing reinitiation rely on either kinetic analysis of RNA synthesis and/or detergents or competitor DNAs to block reinitiation of transcription (Hawley and Roeder 1985, 1987; White et al. 1992; Sheridan et al. 1997; Yean and Gralla 1997; Kraus and Kadonaga 1998; Sandaltzopoulos and Becker 1998). In nearly all in vitro transcription systems (with a few reported exceptions; Holstege et al. 1997; Sandaltzopoulos and Becker 1998), the utilization of promoter templates is very low (typically 3% to <0.1%)

(Luse and Roeder 1980; Manley et al. 1980; Dignam et al. 1983; Hawley and Roeder 1987; Lue and Kornberg 1987; Wootner et al. 1991). Because of low template utilization, these indirect methods cannot distinguish multiple cycles of initiation at the same promoter from new initiations occurring at different promoters. However, both indirect methods have consistently shown that multiple cycles of transcription occur faster than initiation in the first round of transcription (see below). These facts may be reconciled by the recent finding that a large fraction of yeast PICs are inactive in RNA synthesis and polymerase and other factors can dissociate from these inactive complexes, leaving TFIID and TFIIA at the promoter (J. Ranish and S. Hahn, unpubl.). Thus, even if initiations are not occurring from the same promoter in multiple cycles, they may likely use the prebound TFIID and TFIIA and possibly other factors as a platform for rapid PIC assembly.

One argument for a separate pathway of transcription for initiation and reinitiation is that promoters used in the first cycle of transcription are preferentially used in subsequent cycles (Hawley and Roeder 1987). This finding suggests that one or more factors remain committed to promoters after initiation by polymerase.

A second argument for a separate pathway is that multiple cycles of initiation occur much more rapidly than formation of the first functional preinitiation complex (Hawley and Roeder 1987; White et al. 1992; Yean and Gralla 1997). The rate of initial preinitiation complex formation is typically measured by mixing activator and extracts or factors in the absence of nucleotides. After variable assembly times, nucleotides are added and the number of functional preinitiation complexes are determined by single-round transcription assays. This is usually measured by a short pulse of NTPs (~2 min) or by adding the detergent Sarkosyl a short time after NTP addition to block additional rounds of initiation. Alternatively, the rate of functional PIC formation has been assayed by ATP-dependent opening of the DNA strands around the transcription start site by permanganate probing (Jiang and Gralla 1993). All three methods show that initial PIC formation is slow, with a typical half-time of approximately 10–20 minutes. The rate in subsequent cycles of initiation is typically measured by determining the difference in RNA synthesis in single-round transcription assays with the level synthesized 30–60 minutes after NTP addition. Measurements of RNA synthesized from preformed PICs at different times after NTP addition show an initial burst of RNA synthesis from preformed complexes followed by a slower rate of RNA synthesis which occurs linearly for 30–60 minutes (Fig. 2) (Yean and Gralla 1997). This slower rate is the rate of RNA synthesis during multiple cycles of transcription. In other experiments, the rate of reinitiation has been measured by the rate of trapping open complexes with α-amanitin during ongoing transcription as probed by permanganate reactivity (Jiang and Gralla 1993). Measured by these assays, in experiments with nonchromatin templates, this rate is typically five- to sevenfold greater than the rate of initial PIC formation. This difference in rate of transcrip-

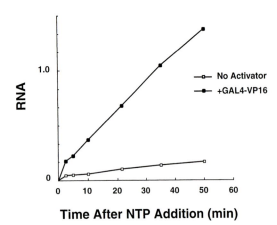

Figure 2. Kinetics of transcription from preformed preinitiation complexes. Yeast transcription complexes were formed in the presence or absence of the activator GAL4-VP16 at the yeast HIS4 core promoter; 0.4 mM NTPs were added at time 0, and the reaction was allowed to proceed for various times indicated, after which the RNA products were assayed by primer extension. The number of functional initiation complexes is given by the size of the initial burst of transcription, and the rate of multiple cycles of transcription is calculated from the initial burst and the slope of product formation after the burst. In this example, GAL4-VP16 increases both the number of functional preinitiation complexes and the rate of multiple transcription cycles.

tion complex formation between the first and subsequent cycles leads to the conclusion that one or more rate-limiting steps have been bypassed in later cycles of transcription.

FACTORS REMAINING AT THE PROMOTER AFTER INITIATION

One step in preinitiation complex formation that can be rate-limiting in vitro is the binding of TFIID and TFIIA to DNA which, along with activator, form a complex capable of recruiting the other components of the transcription machinery. In experiments using partially purified factors, preincubation of TFIID, TFIIA, and activator with promoter greatly increased the rate of preinitiation complex formation upon addition of other factors, compared to the rate of PIC formation when all factors were added at once (Wang et al. 1992). This initial step is possibly slow because of a conformational change in TFIID-DNA contacts, which has been observed in the presence of TFIIA and activator and likely involves wrapping of promoter DNA by the TAF (TBP-associated factors) subunits of TFIID (Horikoshi et al. 1988; Lieberman and Berk 1991; Chi et al. 1995; Chi and Carey 1996).

In experiments utilizing DNase footprinting or templates immobilized to beads, TBP, TAFs, TFIIA, and activator remain bound to the promoter after NTP addition to preformed PICs (Van Dyke et al. 1988; Roberts et al. 1995; Zawel et al. 1995; Sandaltzopoulos and Becker 1998; J. Ranish and S. Hahn, unpubl.). This observation would explain how subsequent cycles of initiation could occur rapidly, bypassing the slow step of TFIID and TFIIA binding. In addition, TBP and the TAFs are

thought to make sequence-specific interactions with the promoter, so it is logical that they would remain behind at the promoter after dissociation of the other components.

This simple model for the reinitiation pathway predicts that there would be a special role for the factors remaining behind at the promoter (TBP, TAFs, TFIIA, and possibly others) as well as the TATA box and other sequence-specific promoter elements that contribute to the stability or activity of this partial PIC. Recent experiments by Yean and Gralla (1997) have shown an important role for the TATA element in multiple cycles of initiation in vitro. These authors measured transcription kinetics from preformed PICs using a series of promoters that contained mutations in either the activator binding site, the TATA box, or the initiator (Inr) element. Mutations in the TATA box had the most striking effects on multiple cycles of initiation (about fivefold decrease) and also decreased the number of active PIC complexes formed (about sixfold decrease). In contrast, the Inr element had only a small or no effect on the number of cycles of transcription. It was proposed that mutation of the TATA led to greater dissociation rates of TFIID from promoters that had initiated, leading to a slower rate of forming the next PIC. Alternatively, the TATA element could influence the activity of TFIID left behind at the promoter in recruiting the next round of polymerase and other factors. In accordance with either model, transcription from the DHFR promoter, a naturally TATA-less promoter, had a rate of reinitiation nearly equal to the rate of formation of the first PIC.

It has also been proposed that TAFs have a role in secondary initiation events (Oelgeschlager et al. 1998). This proposal stems from the observation that TAF-depleted extracts have lower absolute levels of transcription under multiround conditions as compared to normal extracts. In contrast, in single-round assays, the TAF-depleted extracts sometimes give higher levels of transcription than normal extracts. More experiments will be needed, however, to establish conclusively if one or more TAFs have a specific role in reinitiation.

POSSIBLE ACTIVATION-SPECIFIC DEFECTS DUE TO A DEFECT IN REINITIATION

Since the bulk of transcription in vivo from strong activated promoters may derive mainly from reinitiation, it is possible that some mutants defective in activated transcription may result from specific defects in reinitiation. Several TBP mutants tested in vivo are potential candidates for specific reinitiation defects. First, TBP mutants have been isolated in the DNA-binding surface of TBP which have a preferential defect in activated transcription (Kim et al. 1994; Arndt et al. 1995; Lee and Struhl 1995; Stargell and Struhl 1996). This phenotype could result from an unstable complex of TBP-TAFs-TFIIA left behind at the promoter after initiation that rapidly dissociates. Second, a TBP mutant defective in activated transcription was described that was defective in interaction with TFIIA (Stargell and Struhl 1995). Again, this may result in instability of the TFIID-TFIIA-DNA complex leading to a specific reinitiation defect. Finally, extensive mutagenesis of TBP surface residues found surface regions of TBP that were important for activated transcription (Bryant et al. 1996). One region determined by this mutagenesis was that known to interact with TFIIA. A second region of TBP defined for activated transcription may interact with TAFs or another unidentified factor.

How could any of these mutants have a specific effect on activated but not basal transcription? First, it is not clear what basal transcription measured in vivo really is. One proposal is that basal transcription does not occur in vivo at all and that low levels of transcription seen in the absence of the normal activator are promoted by some unknown weak activator. In any case, when transcription initiation rates are low, there is likely a different rate-limiting step in initiation compared to when the initiation rate is high. For example, stability of TFIID binding in the absence of activator may be much less important than the accessibility of promoter DNA or the rate of holoenzyme binding or open complex formation. Thus, defects in TBP binding or stability may have only minor effects under such basal conditions but have a dramatic effect when high levels of transcription are induced and a stable TFIID-DNA complex is required for utilization of the reinitiation pathway.

EFFECT OF ACTIVATORS ON REINITIATION

There have been conflicting reports on the effects of activators on reinitiation of transcription with effects ranging from none to greater than 100-fold. As discussed below, this may be the result of the particular activator or promoter used or the assays employed to distinguish single from multiround transcription.

In an elegant series of experiments, Crabtree and colleagues (Ho et al. 1996) probed for the requirement of the transcription activation domain in vivo at an already induced promoter. This was done using a reversible linkage between the activation domain and the DNA-binding domain. When the linkage between these two domains was blocked by a competitor molecule, gene expression decreased markedly with initial kinetics consistent with the expected dissociation of the two domains. The conclusion drawn by these authors was that the activation domain is required for reinitiation of transcription in vivo. However, it could not be proven conclusively from these in vivo experiments whether the transcription observed resulted from the reinitiation pathway or involved complete dissociation of PIC components after each round of synthesis. Assuming that the transcription seen in vivo was due to reinitiation, then as discussed above, the activator could have at least several nonexclusive roles that would require its continuous presence.

Using human transcription extracts with nonchromatin assembled templates, little or no effect of activator on reinitiation has been observed. In one study by Yean and Gralla (1997), neither GC boxes (presumably bound by SP1) nor the AH activation domain had a significant effect (less than twofold) on the rate of multiple cycles of initiation in vitro. In this system, the main effect of the activator was to increase the number of functional preinitiation complexes. Likewise, in another study using VP16, this activator had a very small effect on multiple cycles of initiation, but had a five- to sixfold effect on recruitment of functional preinitiation complexes (White et al. 1992). In contrast, in the yeast system, VP16 increases both the number of functional PICs and the rate of multiple cycles of transcription by about threefold (Fig. 2) (J. Ranish and S. Hahn, unpubl.).

Using chromatin templates, several activators have been reported to have dramatic effects specifically on reinitiation of transcription. In one study, estrogen receptor (ER) increased transcription only 3-fold under single-round transcription conditions, but stimulated transcription 80-fold under multiround conditions (Kraus and Kadonaga 1998). These data, along with other experiments, suggested that ER was promoting high levels of reinitiation (~30 rounds of transcription) at a small fraction of chromatin-associated templates. In another study, the human immunodeficiency virus type-1 (HIV-1) enhancer was found to have a small effect on single-round transcription but a greater than 100-fold effect under multiround conditions (Sheridan et al. 1997). Again, the conclusion was that the enhancer was stimulating a large number of initiations (~100 cycles) at a small fraction of promoters. One model based on these studies is that the remodeled chromatin on these templates in association with the activator somehow enhances the process of reinitiation. One potential problem with this interpretation, however, is the data pertaining to the quantitation of single-round versus multiround transcription. In the experiments using chromatin templates, 0.1% Sarkosyl was used to block transcription from all but preformed PICs to measure the level of single-round transcription. However, it is possible that this concentration of Sarkosyl inactivates elongation factors necessary to traverse the chromatin template and thus gives an artificially low number for single-round transcription. Further experiments will be required to resolve this issue and the effect of these activators on multiround transcription on chromatin templates.

In one other study using chromatin templates, the effect of heat shock factor (HSF) activator was measured using templates that had prebound TFIID and TFIIA (Sandaltzopoulos and Becker 1998). HSF was found to have no effect on transcription under single-round conditions but a strong effect under multiround conditions. This effect may be due to a specific effect of HSF on reinitiation. However, it may also be due to the fact that TFIID and TFIIA were already preassembled at the promoter.

OTHER FACTORS INVOLVED IN REINITIATION OF TRANSCRIPTION

In principle, other factors involved in recycling and reassembly of the holoenzyme would also have an effect on reinitiation of transcription. These factors would include the pol II carboxy-terminal domain (CTD) phosphatase

(Archambault et al. 1997), as it is known that the CTD enters the PIC in an unphosphorylated form and becomes phosphorylated during elongation. Two other factors were observed to have effects on reinitiation. One of these is SII, a pol II elongation factor (Szentirmay and Sawadogo 1991, 1993). A strong requirement for SII in reinitiation was seen using G-free cassette templates, but this requirement has not been seen using one other promoter not fused to a G-free cassette (Szentirmay et al. 1998). Another factor affecting multiple cycles of transcription is TFIIF (Lei et al. 1998). In this study, mutations were made in the large TFIIF subunit that severely limited multiround transcription but had little effect on single-round transcription. These authors speculated that this effect was due to slow dephosphorylation of the pol II CTD since TFIIF can stimulate the activity of the CTD phosphatase (Archambault et al. 1997), although direct experiments to test this model have not been done.

CONCLUSIONS

Reinitiation of transcription is one of several steps in the transcription pathway that may be enhanced by transcription activators. Biochemical data suggest that reinitiation bypasses one or more rate-limiting steps necessary for formation of transcription complexes upon initial gene induction. At strong promoters, it has been postulated that the bulk of transcription in vivo derives from the reinitiation pathway. Reinitiation is thus likely a critical step in maintaining high levels of gene expression from induced or constitutive promoters. Several TBP mutations that have been isolated as having activation-specific defects may be specifically defective in the reinitiation pathway. Future work on this pathway will further our understanding of the mechanisms needed for control of gene expression after initial induction of transcription and the role of specific promoter elements in control of transcription.

ACKNOWLEDGMENTS

I thank Jeff Ranish, Natalya Yudkovsky, and Ron Reeder for their comments on the manuscript and for many helpful discussions. This work was supported by grant GM-42551 from the National Institutes of Health. S.H. is an associate investigator of the Howard Hughes Medical Institute.

REFERENCES

Archambault J., Chambers R.S., Kobor M.S., Ho Y., Cartier M., Bolotin D., Andrews B., Kane C. M., and Greenblatt J. 1997. An essential component of a C-terminal domain phosphatase that interacts with transcription factor IIF in *Saccharomyces cerevisiae*. *Proc. Natl. Acad. Sci.* **94:** 14300.

Arndt K.M., Ricupero-Hovasse S., and Winston F. 1995. TBP mutants defective in activated transcription in vivo. *EMBO J.* **14:** 1490.

Auble D.T., Hansen K.E., Mueller C.G.F., Lane W.S., Thorner J., and Hahn S. 1994. Mot1, a global repressor of RNA polymerase II transcription, inhibits TBP binding to DNA by an ATP-dependent mechanism. *Genes Dev.* **8:** 1920.

Bannister A.J. and Kouzarides T. 1996. The CBP co-activator is a histone acetyltransferase. *Nature* **384:** 641.

Bentley D.L. 1995. Regulation of transcriptional elongation by RNA polymerase II. *Curr. Opin. Genet. Dev.* **5:** 210.

Blau J., Xiao H., McCracken S., O'Hare P., Greenblatt J., and Bentley D. 1996. Three functional classes of transcriptional activation domain. *Mol. Cell. Biol.* **16:** 2044.

Brown S.A., Weirich C.S., Newton E.M., and Kingston R.E. 1998. Transcriptional activation domains stimulate initiation and elongation at different times and via different residues. *EMBO J.* **11:** 3146.

Bryant G.O., Martel L.S., Burley S.K., and Berk A.J. 1996. Radical mutations reveal TATA-box binding protein surfaces required for activated transcription in vivo. *Genes Dev.* **10:** 2491.

Bunick D., Zandomeni R., Ackerman S., and Weinmann R. 1982. Mechanism of RNA polymerase II-specific initiation of transcription in vitro: ATP requirement and uncapped runoff transcripts. *Cell* **29:** 877.

Buratowski S., Hahn S., Guarente L., and Sharp P.A. 1989. Five intermediate complexes in transcription initiation by RNA polymerase II. *Cell* **56:** 549.

Burley S.K. and Roeder R.G. 1996. Biochemistry and structural biology of transcription factor IID (TFIID). *Annu. Rev. Biochem.* **65:** 769.

Catron K.M., Zhang H., Marshall S.C., Inostroza J.A., Wilson J.M., and Abate C. 1995. Transcriptional repression by Msx-1 does not require homeodomain DNA-binding sites. *Mol. Cell. Biol.* **15:** 861.

Chao D.M., Gadbois E.L., Murray P.J., Anderson S.F., Sonu M.S., Parvin J.D., and Young R.A. 1996. A mammalian SRB protein associated with an RNA polymerase II holoenzyme. *Nature* **380:** 82.

Chi T. and Carey M. 1996. Assembly of the isomerized TFIIA-TFIID-TATA ternary complex is necessary and sufficient for gene activation. *Genes Dev.* **10:** 2540.

Chi T., Lieberman P., Ellwood K., and Carey M. 1995. A general mechanism for transcriptional synergy by eukaryotic activators. *Nature* **377:** 254.

Dignam J.D., Lebovitz R.M., and Roeder R.G. 1983. Accurate transcription initiation by RNA polymerase II in a soluble extract from isolated mammalian nuclei. *Nucleic Acids Res.* **11:** 1475.

Divr A., Tan S., Conaway J.W., and Conaway R.C. 1997. Promoter escape by RNA polymerase II. *J. Biol. Chem.* **272:** 28175.

Divr A., Garrett K.P., Chalut C., Egly J.-M., Conaway J.W., and Conaway R.C. 1996. A role for ATP and TFIIH in activation of the RNA pol II preinitiation complex prior to transcription initiation. *J. Biol. Chem.* **271:** 7245.

Gadbois E.L., Chao D.M., Reese J.C., Green M.R., and Young R.A. 1997. Functional antagonism between RNA polymerase II holoenzyme and global negative regulator NC2 in vivo. *Proc. Natl. Acad. Sci.* **94:** 3145.

Goodrich J.A. and Tjian R. 1994. Transcription factors IIE and IIH and ATP hydrolysis direct promoter clearance by RNA polymerase II. *Cell* **77:** 145.

Greenblatt J. 1997. RNA polymerase II holoenzyme and transcriptional regulation. *Curr. Opin. Cell Biol.* **9:** 310.

Grunstein M. 1997. Histone acetylation in chromatin structure and transcription. *Nature* **389:** 349.

Hahn S. 1993. Transcription: Efficiency in activation. *Nature* **363:** 672.

Hampsey M. 1998. Molecular genetics of the RNA polymerase II general transcriptional machinery. *Microbiol. Mol. Biol. Rev.* **62:** 465.

Hawley D.K. and Roeder R.G. 1985. Separation and partial characterization of three functional steps in transcription initiation by human RNA polymerase II. *J. Biol. Chem.* **260:** 8163.

———. 1987. Functional steps in transcription initiation and reinitiation from the major late promoter in a HeLa nuclear extract. *J. Biol. Chem.* **262:** 3452.

Herschlag D. and Johnson F.B. 1993. Synergism in transcriptional activation: A kinetic view. *Genes Dev.* **7:** 173.

Ho S.N., Biggar S.R., Spencer D.M., Schreiber S.L., and Crabtree G.R. 1996. Dimeric ligands define a role for transcriptional activation domains in reinitiation. *Nature* **382:** 822.

Holstege F.C.P., Fiedler U., and Timmers H.T.M. 1997. Three transitions in the RNA polymerase II transcription complex during initiation. *EMBO J.* **16:** 7468.

Holstege F.C.P., Tantin D., Carey M., van der Vliet P.C., and Timmers H.T.M. 1995. The requirement for the basal transcription factor IIE is determined by the helical stability of promoter DNA. *EMBO J.* **14:** 810.

Horikoshi M., Hai T., Lin Y.S., Green M.R., and Roeder R.G. 1988. Transcription factor ATF interacts with the TATA factor to facilitate establishment of a preinitiation complex. *Cell* **54:** 1033.

Inostroza J.A., Mermelstein F.H., Ha I., Lane W.S., and Reinberg D. 1992. Dr1, a TATA-binding protein-associated phosphoprotein and inhibitor of class II gene transcription. *Cell* **70:** 477.

Iyer V. and Struhl K. 1996. Absolute mRNA levels and transcriptional initiation rates in *Saccharomyces cerevisiae. Proc. Natl. Acad. Sci.* **93:** 5208.

Jiang Y. and Gralla J.D. 1993. Uncoupling of initiation and reinitiation rates during HeLa RNA polymerase II transcription in vitro. *Mol. Cell. Biol.* **13:** 4572.

Kadonaga J.T. 1990. Assembly and disassembly of the *Drosophila* RNA polymerase II complex during transcription. *J. Biol. Chem.* **265:** 2624.

Kaiser K. and Meisterernst M. 1996. The human general co-factors. *Trends Biochem. Sci.* **21:** 342.

Keaveney M. and Struhl K. 1998. Activator-mediated recruitment of the RNA polymerase II machinery is the predominant mechanism for transcriptional activation in yeast. *Mol. Cell* **1:** 917.

Kennell D. and Riezman H. 1977. Transcription and translation initiation frequencies of the *E. coli lac* operon. *J. Mol. Biol.* **114:** 1.

Kim S., Na J.G., Hampsey M., and Reinberg D. 1997. The Dr1/DRAP1 heterodimer is a global repressor of transcription in vivo. *Proc. Natl. Acad. Sci.* **94:** 820.

Kim T.K., Hashimoto S., Kelleher R.J., Flanagan P.M., Kornberg R.D., Horikoshi M., and Roeder R.G. 1994. Effects of activation-defective TBP mutations on transcription initiation in yeast. *Nature* **369:** 252.

Kingston R.E., Bunker C.A., and Imbalzano A.N. 1996. Repression and activation by multiprotein complexes that alter chromatin structure. *Genes Dev.* **10:** 905.

Koh S.S., Ansari A.Z., Ptashne M., and Young R.A. 1998. An activator target in the RNA polymerase II holoenzyme. *Mol. Cell* **1:** 895.

Koleske A.J. and Young R.A. 1995. The RNA polymerase II holoenzyme and its implications for gene regulation. *Trends Biochem. Sci.* **20:** 113.

Kraus W.L. and Kadonaga J.T. 1998. p300 and estrogen receptor cooperatively activate transcription via differential enhancement of initiation and reinitiation. *Genes Dev.* **12:** 331.

Krumm A., Hickey L.B., and Groudine M. 1995. Promoter-proximal pausing of RNA polymerase II defines a general rate-limiting step after transcription initiation. *Genes Dev.* **9:** 559.

Lee M. and Struhl K. 1995. Mutations on the DNA-binding surface of TATA-binding protein can specifically impair the response to acidic activators in vivo. *Mol. Cell. Biol.* **15:** 5461.

Lei L., Ren D., Finkelstein A., and Burton Z.F. 1998. Functions of the N- and C-terminal domains of human RAP74 in transcriptional initiation, elongation, and recycling of RNA polymerase II. *Mol. Cell. Biol.* **18:** 2130.

Li C. and Manley J.L. 1998. Even-skipped represses transcription by binding TATA binding protein and blocking the TFIID-TATA box interaction. *Mol. Cell. Biol.* **18:** 3771.

Lieberman P.M. and Berk A.J. 1991. The Zta *trans*-activator protein stabilizes TFIID association with promoter DNA by direct protein-protein interaction. *Genes Dev.* **5:** 2441.

Lue N.F. and Kornberg R.D. 1987. Accurate initiation at RNA polymerase II promoters in extracts from *Saccharomyces cerevisiae. Proc. Natl. Acad. Sci.* **84:** 8839.

Luse D.S. and Jacob G.A. 1987. Abortive initiation by RNA polymerase II in vitro at the adenovirus 2 major late promoter. *J. Biol. Chem.* **262:** 14990.

Luse D. and Roeder R.G. 1980. Accurate transcription initiation on a purified mouse β-globin DNA fragment in a cell-free system. *Cell* **20:** 691.

Maldonado E., Shiekhattar R., Sheldon M., Cho H., Drapkin R., Rickert P., Lees E., Anderson C.W., Linn S., and Reinberg D. 1996. A human RNA polymerase II complex associated with SRB and DNA-repair proteins. *Nature* **381:** 86.

Manley J.L., Fire A., Cano A., Sharp P.A., and Gefter M.L. 1980. DNA-dependent transcription of adenovirus genes in a soluble whole-cell extract. *Proc. Natl. Acad. Sci.* **77:** 3855.

Martin C.T. and Coleman J.E. 1987. Kinetic analysis of T7 RNA polymerase-promoter interactions with small synthetic promoters. *Biochemistry* **26:** 2690.

Myers L.C., Gustafsson C.M., Bushnell D.A., Lui M., Erdjument-Bromage H., Tempst P., and Kornberg R.D. 1998. The Med proteins of yeast and their function through the RNA polymerase II carboxy-terminal domain. *Genes Dev.* **12:** 45.

Nikolov D.B. and Burley S.K. 1997. RNA polymerase II transcription initiation: A structural view. *Proc. Natl. Acad. Sci.* **94:** 15.

O'Brien T. and Lis J.T. 1991. RNA polymerase II pauses at the 5′ end of the transcriptionally induced *Drosophila* hsp70 gene. *Mol. Cell. Biol.* **11:** 5285.

———. 1993. Rapid changes in *Drosophila* transcription after an instantaneous heat shock. *Mol. Cell. Biol.* **13:** 3456.

Oelgeschlager T., Tao Y., Kang Y.K., and Roeder R.G. 1998. Transcription activation via enhanced preinitiation complex assembly in a human cell-free system lacking TAFIIs. *Mol. Cell* **1:** 925.

Ogryzko V.V., Schiltz R.L., Russanova V., Howard B.H., and Nakatani Y. 1996. The transcriptional coactivators p300 and CBP are histone acetyltransferases. *Cell* **87:** 953.

Orphanides G., Lagrange T., and Reinberg D. 1996. The general transcription factors of RNA polymerase II. *Genes Dev.* **10:** 2657.

Pan G. and Greenblatt J. 1994. Initiation of transcription by RNA polymerase II is limited by melting of the promoter DNA in the region immediately upstream of the initiation site. *J. Biol. Chem.* **269:** 30101.

Paranjape S.M., Kamakaka R.T., and Kadonaga J.T. 1994. Role of chromatin structure in the regulation of transcription by RNA polymerase II. *Annu. Rev. Biochem.* **63:** 265.

Pazin M.J. and Kadonaga J.T. 1997. SWI2/SNF2 and related proteins: ATP-driven motors that disrupt protein-DNA interactions? *Cell* **88:** 737.

Poon D., Campbell A.M., Bai Y., and Weil P.A. 1994. Yeast TAF170 is encoded by MOT1 and exists in a TATA box-binding protein (TBP)-TBP assoicated factor complex distinct from transcription factor TFIID. *J. Biol. Chem.* **269:** 23135.

Prelich G. 1997. *Saccharomyces cerevisiae* BUR6 encodes a DRAP1/NC2α homolog that has both positive and negative roles in transcription in vivo. *Mol. Cell. Biol.* **17:** 2057.

Ptashne M. and Gann A. 1997. Transcriptional activation by recruitment. *Nature* **386:** 569.

Reeder R.H. and Lang W.H. 1997. Terminating transcription in eukaryotes: Lessons learned from RNA polymerase I. *Trends Biochem. Sci.* **22:** 473.

Roberts S.G., Choy B., Walker S.S., Lin Y.S., and Green M.R. 1995. A role for activator-mediated TFIIB recruitment in diverse aspects of transcriptional regulation. *Curr. Biol.* **5:** 508.

Roeder R.G. 1996. The role of general initiation factors in transcription by RNA polymerase II. *Trends Biochem. Sci.* **21:** 327.

Sandaltzopoulos R. and Becker P.B. 1998. Heat shock factor increases the reinitiation rate from potentiated chromatin templates. *Mol. Cell. Biol.* **18:** 361.

Sawadogo M. and Roeder R.G. 1984. Energy requirement for specific transcription initiation by the human RNA polymerase II system. *J. Biol. Chem.* **259:** 5321.

Sheridan P.L., Mayall T.P., Verdin E., and Jones K.A. 1997. Histone acetyltransferases regulate HIV-1 enhancer activity in vitro. *Genes Dev.* **11:** 3327.

Shilatifard A., Conaway J.W., and Conaway R.C. 1997. Mechanism and regulation of transcriptional elongation and termination by RNA polymerase II. *Curr. Opin. Genet. Dev.* **7:** 199.

Stargell L.A. and Struhl K. 1995. The TBP-TFIIA interaction in the response to acidic activators in vivo. *Science* **269:** 75.

———. 1996. A new class of activation-defective TATA-binding protein mutants: Evidence for two steps of transcriptional activation in vivo. *Mol. Cell. Biol.* **16:** 4456.

Struhl K. 1996. Chromatin structure and RNA polymerase II connection: Implications for transcription. *Cell* **84:** 179.

———. 1998. Histone acetylation and transcriptional regulatory mechanisms. *Genes Dev.* **12:** 599.

Szentirmay M.N. and Sawadogo M. 1991. Transcription factor requirement for multiple rounds of initiation by human RNA polymerase II. *Proc. Natl. Acad. Sci.* **88:** 10691.

———. 1993. Synthesis of reinitiated transcripts by mammalian RNA polymerase II is controlled by elongation factor SII. *EMBO J.* **12:** 4677.

———. 1994. Sarkosyl block of transcription reinitiation by RNA polymerase II as visualized by the colliding polymerases reinitiation assay. *Nucleic Acids Res.* **22:** 5341.

Szentirmay M.N., Musso M., Van Dyke M.W., and Sawadogo M. 1998. Multiple rounds of transcription by RNA Pol II at covalently cross-linked templates. *Nucleic Acids Res.* **26:** 2754.

Van Dyke M.W., Roeder R.G., and Sawadogo M. 1988. Physical analysis of transcription preinitiation complex assembly on a class II gene promoter. *Science* **241:** 1335.

Wang W., Gralla J.D., and Carey M. 1992. The acidic activator GAL4-AH can stimulate polymerase II transcription by promoting assembly of a closed complex requiring TFIID and TFIIA. *Genes Dev.* **6:** 1716.

White J., Brou C., Wu J., Lutz Y., Moncollin V., and Chambon P. 1992. The acidic transcriptional activator GAL-VP16 acts on preformed template-committed complexes. *EMBO J.* **11:** 2229.

Wootner M., Wade P.A., Bonner J., and Jaehning J.A. 1991. Transcriptional activation in an improved whole-cell extract from *Saccharomyces cerevisiae*. *Mol. Cell. Biol.* **11:** 4555.

Yankulov K., Blau J., Purton T., Roberts S., and Bentley D.L. 1994. Transcriptional elongation by RNA polymerase II is stimulated by transactivators. *Cell* **77:** 749.

Yean D. and Gralla J. 1997. Transcription reinitiation rate: A special role for the TATA box. *Mol. Cell. Biol.* **17:** 3809.

Yeung K., Kim S., and Reinberg D. 1997. Functional dissection of a human Dr1-DRAP1 repressor complex. *Mol. Cell. Biol.* **17:** 36.

Zawel L., Kuman K.P., and Reinberg D. 1995. Recycling of the general transcription factors during RNA polymerase II transcription. *Genes Dev.* **9:** 1479.

Cofactor Requirements for Transcriptional Activation by Sp1

A.M. Näär, S. Ryu, and R. Tjian

Howard Hughes Medical Institute, University of California, Berkeley, California 94720

The promoter selectivity factor Sp1 is required for efficient transcriptional activation of many cellular and viral genes. Our laboratory has employed in vitro biochemical strategies to investigate the mechanisms of Sp1 regulation of gene expression. Our recent reconstitution experiments revealed a requirement for multiple cofactor activities that interface with the basal transcription machinery to support activation by Sp1. Previous studies demonstrated that the activation domains of Sp1 and other transcriptional regulators can bind directly to TBP-associated factors (TAF$_{II}$s). These factors appear to serve as coactivators mediating recruitment of the transcriptional machinery to the promoter by activators. Further refinement of the in vitro transcription system resulted in the identification of additional cofactors required for Sp1 activation, including a novel factor, CRSP, whose function remains to be established.

Although most of these studies were performed with single activators on simple promoters, Sp1 often cooperates with other transcriptional activators on complex promoters in vivo. To investigate cofactor requirements for Sp1 function in the context of more complex promoters, we attempted to recapitulate in vitro the synergistic activation by Sp1 and the sterol-regulated factor SREBP-1a at the LDL receptor (LDLR) promoter. Using a reconstituted human transcription system, we found that chromatin, TAF$_{II}$s, a CRSP-containing fraction, and a novel SREBP-binding coactivator activity, which includes CBP, are all required to mediate efficient activation by Sp1 and SREBP-1a. The development of a chromatin-mediated transcription system will allow us to conduct detailed studies of cofactor function in coordinating the action of Sp1 with other activators at complex promoters.

The efforts by many researchers during the last two decades have unveiled a highly complex regulatory machinery required to direct mammalian organogenesis and maintenance of homeostatic functions. Intricate spatial and temporal patterns of gene expression are governed by the cooperative actions of a host of general and gene selective transcription factors that function to enhance or repress gene activity in response to genetic programs and environmental cues. Much of our knowledge about these processes comes from biochemical dissection of the transcriptional apparatus, which has provided insights into the mechanistic principles controlling gene expression. In particular, this approach has been instrumental to investigation of the requirements for activators to stimulate transcription of their target genes. The development of reconstituted transcription reactions in vitro has resulted in the discovery of several classes of cofactors that functionally connect activators with the general transcriptional machinery.

This paper presents a perspective using Sp1 as a model activator to introduce the role of cofactors in mediating transcriptional activation, followed by a discourse on recent advances in our understanding of the requirements for Sp1 to function cooperatively with other activators on complex promoters in the context of chromatin, and concludes with a discussion of future investigations.

PROPERTIES OF THE SEQUENCE-SPECIFIC TRANSCRIPTION FACTOR, SP1

The realization in the late 1970s that purified RNA polymerase II was not sufficient to direct promoter-selective initiation stimulated a search for factors that functioned to discriminate between promoters in a sequence-specific manner. Using an in vitro transcription approach, specificity protein 1 (Sp1) was discovered based on its ability to selectively activate transcription of the viral SV40 promoter (Dynan and Tjian 1983a,b). The Sp1 protein can be structurally divided into a carboxy-terminal zinc finger DNA-binding domain preceded by a glutamine-rich *trans*-activation domain (Fig. 1). Sp1 thus conforms to the modular organization of separable activation and DNA-binding domains that is the hallmark of many eukaryotic transcription factors. The Sp1 DNA-binding domain was found to interact specifically with *cis*-acting sequences just upstream of the promoter elements required for basal transcription (Fig. 1) (Kadonaga et al. 1987, 1988). This arrangement of Sp1 cognate DNA elements in relation to general promoter sequences such as the TATA box and initiator is now recognized as a common feature of many genes and established Sp1 as a sequence-specific promoter proximal transcriptional activator. Constitutive housekeeping genes frequently harbor several strong binding sites for Sp1, whereas Sp1-binding sites in the promoter region of inducible genes are generally much weaker. Recent studies revealed that Sp1 often cooperates with other signal-responsive upstream enhancer binding factors to activate inducible genes.

IDENTIFYING TARGETS OF SP1 THAT MEDIATE ACTIVATION OF TRANSCRIPTION IN VITRO

The development of reconstituted in vitro transcription reactions revealed that accurate initiation of transcription by RNA polymerase II requires the coordinate action of several general transcription factors (GTFs), including

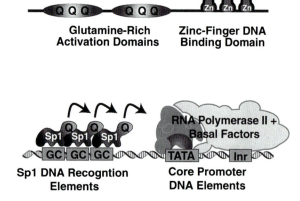

Glutamine-Rich Activation Domains   **Zinc-Finger DNA Binding Domain**

Sp1 DNA Recogntion Elements   **Core Promoter DNA Elements**

Figure 1. Structural and functional domains of the human transcription factor Sp1. (*Top panel*) Relative arrangement of the glutamine-rich activation domains and zinc finger DNA-binding domain of Sp1. (*Bottom panel*) Sp1 tethered to the DNA template by its triple zinc fingers interacting with GC box elements. The binding of Sp1 to the template positions this regulatory protein to somehow help signal the preinitiation complex containing RNA polymerase II and basal factors to assemble at the core promoter and trigger transcription.

TFIIA, TFIIB, TFIID, TFIIE, TFIIF, and TFIIH. Since TFIID was found to interact with the TATA box present in the promoter region of many mRNA encoding genes and was suggested to nucleate the formation of a preinitiation complex with the other GTFs, it was proposed to serve as a direct target for activators (Fig. 2) (Sawadogo and Roeder 1985; Horikoshi et al. 1988). The purification of yeast TFIID as a single small TATA-binding protein (TBP) and subsequent cloning of the gene encoding TBP allowed testing of this hypothesis. Although basal levels of transcription could be reconstituted with recombinant TBP from yeast or higher eukaryotes, only partially purified TFIID from *Drosophila* or human sources was able

to mediate transcriptional stimulation by Sp1 and other sequence-specific activators. This result indicated that other factors present in the TFIID fraction were critical to direct activated transcription. The fact that *Drosophila* and human TFIID fractions capable of supporting activated transcription were found to purify as large complexes (>700 kD) suggested that the putative TBP-associated factors (TAFs) were important mediators of activating signals (Goodrich and Tjian 1994). However, isolation of these factors was initially hampered by the difficulty in purifying large quantities of TFIID by conventional chromatographic methods.

TBP-ASSOCIATED FACTORS

A breakthrough came with the generation of antiserum directed against TBP that was capable of immunoprecipitating the native TFIID complex. This allowed purification of sufficient quantities of TFIID to determine its subunit composition. A series of polypeptides were found to associate with TBP; removal of these from TBP using heat or urea abolished activator response in transcription, whereas readdition of the eluted TAFs to transcription reactions containing TBP restored activation (Dynlacht et al. 1991; Tanese et al. 1991). Immunoprecipitation using antibodies directed against either TBP or TAFs identified complexes with identical subunit compositions, establishing TAFs as stoichiometric components of TFIID. Together, these findings indicated the importance of TAFs to TFIID function and prompted the cloning of TAF genes. The isolation of *Drosophila* cDNAs encoding TAFs allowed generation of recombinant TAF proteins which were then tested for binding to activation domains. The *Drosophila* TAF$_{II}$110 was found to interact with the glutamine-rich activation domain of Sp1 but not with other activation domains tested (Hoey et al. 1993). Interestingly, acidic activators such as VP16 and p53 were subsequently shown to bind to another set of TAFs,

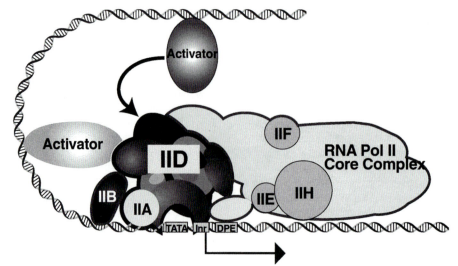

Figure 2. TFIID complex as a potential target for activators. Template-bound transcriptional activators must convey the activating signal to the general transcription apparatus to initiate RNA synthesis. TFIID was proposed to constitute a direct target for activators because it is thought to act at the initial core promoter recognition step in the assembly cascade resulting in the formation of a preinitiation transcription complex.

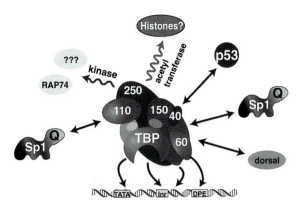

TAFs can serve as:

- ◆ **Core promoter recognition factors**
- ◆ **Activator Targets**
- ◆ **Enzymes that modify proteins**

Figure 3. TAFs perform multiple functions and have been found to serve as targets for multiple activators. For example, Sp1 interacts with *Drosophila* TAF$_{II}$110 (and the human TAF$_{II}$130 homolog) as well as with the dTAF$_{II}$150 subunit; the transcriptional regulator p53 binds to dTAF$_{II}$40 (or hTAF$_{II}$32), and the *Drosophila* transcription factor *Dorsal* can bind to dTAF$_{II}$60. TFIID also contains multiple core promoter recognition factors, including TBP and several TAFs. Thus, the dTAF$_{II}$150 and dTAF$_{II}$60 subunits not only may function as activator targets, but are also thought to recognize DNA sequences within the initiator (Inr) core element and downstream promoter elements (DPE), respectively. TFIID has also been demonstrated to harbor two distinct enzymatic functions. The TAF$_{II}$250 protein was found to mediate kinase activity, with the RAP74 subunit of TFIIF as one candidate substrate. Additionally, TAF$_{II}$250 has been shown to contain acetyltransferase activity, modifying histones and possibly other proteins.

the *Drosophila* TAF$_{II}$40 and TAF$_{II}$60 or the human counterparts TAF$_{II}$32/70. These results suggested that different classes of activation domains can target different TAFs and led to a model where the TAF components of TFIID function as coactivators, serving as bridges between activation domains and the transcriptional apparatus (Fig. 3) (Goodrich and Tjian 1994). Support for this model came from studies showing that partial TFIID complexes assembled in vitro with recombinant TAFs and TBP could at least partially substitute for TFIID to support activated levels of transcription (Chen et al. 1994). Furthermore, synergy between different activators in vitro correlated with targeting of multiple TAFs within TFIID to effect cooperative recruitment of TFIID to the promoter (Sauer and Tjian 1997).

The highly complex regulatory requirements, however, for accurate temporal and spatial gene expression during organogenesis as well as for exquisitely sensitive responses to hormonal and environmental signals suggest the need for a more diverse repertoire of transcriptional targets. Interestingly, a number of tissue-specific TFIID and TFIID-like complexes have recently been identified. A novel human TAF (hTAF$_{II}$105) is highly enriched in a TFIID complex derived from B cells (Dikstein et al. 1996b) and could function to direct activation by B-cell-

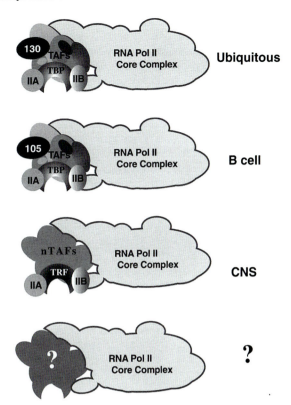

Figure 4. Multiple TFIIDs and TRF complexes. In addition to the prototypic TFIID complex composed of the universal TBP subunit and multiple TAF$_{II}$s that appears to be ubiquitous in all eukaryotic cells, recent studies have uncovered both cell-type-selective TAFs (i.e., B-cell TAF$_{II}$105) and cell-type-specific TBP-like molecules (i.e., neuronal TRF-1). Preliminary biochemical analysis of TRF-1 suggests that the TBP-like molecule and another relative TRF-2 both can interact with TFIIA and TFIIB. Moreover, TRF-1 can bind in a sequence-specific manner to TATA elements and mediate basal levels of transcription in vitro. Like TBP, TRF-1 appears to be part of a large multisubunit complex; however, immunoblot analysis using antisera directed against TAF$_{II}$ subunits suggested that the TRF-1 complex composition may be distinct from that of TFIID. Recent studies also reveal that a subset of TAFs (i.e., histone-like TAFs) can be associated with distinct multisubunit complexes (i.e., SAGA and P/CAF), suggesting that both TAFs and TBP can functionally participate in different transcription and/or chromatin modifying complexes.

restricted activators (Fig. 4). Two forms of TFIID were isolated from human HeLa cervical carcinoma cells differing in their TAF compositions and showing activator specificity (Jacq et al. 1994). A TBP-free TAF-containing complex (TFTC) has also been reported recently (Wieczorek et al. 1998). Intriguingly, recent findings suggest that TFIID shares some of its TAFs with novel yeast and human coactivators, the SAGA and PCAF complexes (Grant et al. 1998; Ogryzko et al. 1998). In addition to TAFs, the yeast complex and its human counterpart contain "adaptor" proteins (ADA2/3), the GCN5/PCAF histone acetyltransferase, and several Spt gene products which have been genetically linked to TBP in yeast. These multifactorial coactivators may be recruited to specific promoters by activators interacting with the TAF subunits, or via other contacts, and could influence select steps in the activation process, such as access to chromatin-restricted promoters or the subsequent recruitment of the general transcription factors.

In addition to cell-type-specific TAFs and other TAF-containing putative coactivator complexes, recent evidence also revealed the existence of novel cell-type-specific TBP-related molecules. For example, the sequence of a *Drosophila* gene associated with a *shaker* neurological phenotype was found to exhibit high homology with TBP and was termed TBP-related factor (TRF) (Crowley et al. 1993). The TRF gene is expressed primarily in the central nervous system (CNS) and the gene product was shown to bind TATA sequences in conjunction with TFIIA and TFIIB and support basal transcription in vitro (Hansen et al. 1997). Native TRF was found to purify as a large complex; however, it did not interact with the known TAFs. These results suggest the presence of a CNS-restricted TFIID-like complex that may mediate *trans*-activation by neuron-enriched activators via a specific set of TRF-associated factors (Fig. 4).

TAFs not only may serve as targets for activation domains, but they appear to contribute other functions as well. TFIID was found to be required for Sp1 function on TATA-less promoters (Pugh and Tjian 1990; Smale et al. 1990), and recent evidence indicates the involvement of specific TAFs in the recognition of the initiator sequence (Verrijzer and Tjian 1996; Kaufmann et al. 1998) and downstream promoter elements (DPE) (Fig. 3) (Burke and Kadonaga 1997). Furthermore, the human and *Drosophila* TAF$_{II}$250 harbors intrinsic kinase (Dikstein et al. 1996a; O'Brien and Tjian 1998) and histone acetyltransferase activity (Fig. 3) (Mizzen et al. 1996), although the functional contribution of these activities to transcription has not been worked out in detail.

OTHER COFACTORS AND TARGETS OF ACTIVATORS

Direct recruitment of TFIID or TFIID-like complexes is only one of several possible rate-limiting steps that may be affected by transcriptional regulators. The subsequent steps in the formation of a preinitiation complex, open complex formation and promoter clearance, as well as transcript elongation and posttranscriptional events could serve as tentative points of regulatory intervention by activators. Additionally, the restrictions imposed by chromatin in vivo need to be overcome by activators to induce gene expression. It is thus not surprising that recent findings suggest that activation domains may target other cofactors or components of the transcriptional machinery, as well as chromatin-modifying complexes. For example, the acidic activation domains of VP16 and p53 have been shown to interact with multiple targets, including GTFs and several classes of cofactors (Triezenberg 1995; Levine 1997).

The CBP/p300 coactivators can form complexes with other coactivators, including PCAF and SRC-1/GRIP-1/NCoAs, all of which contain histone acetyltransferase activity. CBP/p300 complexes can be recruited by many classes of activators and have been proposed to mitigate the repressive effects of chromatin. CBP was also found to interact with RNA polymerase II to mediate CREB activation (Struhl 1998). Another class of coactivators (TRAP/DRIP complexes) have been described that bind in a ligand-dependent manner to the thyroid hormone and vitamin D receptors, respectively, and which appear to be required for full transcriptional activation by these receptors in vitro, although the mechanism of coactivation is unknown (Fondell et al. 1996; Rachez et al. 1998).

Efficient transcriptional activation may also require other cofactors not directly targeted by activation domains to facilitate steps in the transcriptional process. For example, the USA (upstream stimulatory activity) cofactor was initially identified as a crude fraction important for transcriptional stimulation by Sp1 and several other classes of activators (Meisterernst et al. 1991). However, only the VP16 activation domain appears to interact directly with a component of the USA fraction (PC4) (Ge and Roeder 1994). Although the identity and action of some of the activities found in the USA fraction remain to be established, most USA components do not appear to be directly recruited via activation domains and may function in a more general fashion to suppress basal or nonspecific transcription and indirectly potentiate activation by activators.

ATP-dependent chromatin remodeling complexes, such as SWI/SNF complexes, NURF, CHRAC, ACF, and RSC, are thought to function as cofactors, which may be required for activators and the general transcription machinery to access their target sequences in a chromatin-restricted environment (Cairns 1998). However, no evidence is presently available indicating that these activities are directly recruited by activators (or repressors). Hence, this class of cofactors may act in concert with other chromatin-modifying activities to enhance binding of activators and GTFs. Whether ATP-dependent nucleosomal remodeling is required for Sp1 activation on chromatin templates is currently under investigation.

The mediator complex exemplifies another class of cofactors that like TAFs may be more integral to the general transcription machinery. The mediator has been described in yeast as a large complex that associates with the carboxy-terminal repeat domain (CTD) of RNA polymerase II and is required for activated transcription both in vitro and in vivo (Björklund and Kim 1996). The mediator is composed of more than 30 polypeptides, including SRBs and other proteins, such as RGR1 and GAL11, which have been genetically linked to transcriptional regulation in yeast. Recent findings suggest that RNA polymerase II/mediator complexes could be recruited by the VP16 activation domain (Hengartner et al. 1995), although the functional relevance of this interaction is still unclear.

CRSP, A NOVEL SP1 COFACTOR

In our effort to identify, purify, and characterize all of the cofactors necessary and sufficient to mediate Sp1-dependent activation of transcription in vitro, we have re-

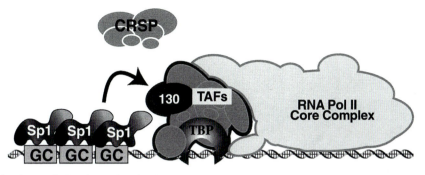

Figure 5. A novel cofactor (CRSP) is required for full transcriptional activation by Sp1. Although Sp1 activation is dependent on TFIID and select subunits of the complex, such as TAF$_{II}$130 and TAF$_{II}$150, recent biochemical reconstitution of in vitro transcription revealed the requirement for a novel cofactor, CRSP. This cofactor is thought to be a multisubunit complex, composed of several previously unidentified polypeptides. The mechanism of action of CRSP is not known and it is possible that this cofactor operates either at the preinitiation step or at postinitiation events to enhance the synthesis of RNA during Sp1-mediated activation of transcription.

cently developed an assay that allowed us to detect a novel activity that potentiates activation by Sp1 (S. Ryu et al., unpubl.). This *cofactor required for Sp1* activation (CRSP) consists of a multisubunit complex containing between six and eight polypeptides that range from 200 to 30 kD. Preliminary analysis revealed that CRSP is distinct from other cofactors that have been isolated (i.e., PC4, CBP, SAGA, etc.) and that activation by Sp1 requires both TBP/TAFs and CRSP (Fig. 5). The biochemical properties of the CRSP subunits and their mode of action are currently under investigation.

COFACTOR REQUIREMENTS FOR SYNERGISTIC ACTIVATION BY SP1 AND SREBP-1A AT THE LDL RECEPTOR PROMOTER

The molecular mechanisms of Sp1 activation have been studied extensively (Pascal and Tjian 1991; Hoey et al. 1993; Chen et al. 1994), but the regulation of inducible promoters by Sp1 working in concert with other enhancer-binding factors remains largely unexplored. Many of these inducible genes appear to be governed by gene-selective activators that work synergistically with Sp1 to direct transcription (Krey et al. 1995; Look et al. 1995; Merika and Orkin 1995; Sheridan et al. 1995; Pazin et al. 1996). However, the cofactor requirements and molecular interactions that allow combinatorial regulation by multiple transcriptional activators are poorly defined. The regulation of the low-density lipoprotein (LDL) receptor gene by cholesterol provides a useful model system to investigate the molecular mechanisms of synergistic activation by Sp1 together with other enhancer-binding factors. The LDL receptor (LDLR) promoter sequences mediating cholesterol regulation consist of a DNA element recognized by the sterol-responsive element-binding proteins (SREBP-1 and 2) flanked by Sp1-binding sites. SREBPs belong to the basic helix-loop-helix class of transcription factors, with the unusual property of being anchored to the endoplasmic reticulum (ER) membrane until released by cholesterol-regulated proteolysis (Brown and Goldstein 1997).

Transient transfection experiments showed that SREBPs and Sp1 can function synergistically to activate transcription of the LDLR gene (Fig. 6A) (Sanchez et al. 1995; Athanikar et al. 1997). To investigate the molecular requirements for cooperative activation by SREBP and Sp1, we have recently recapitulated SREBP/Sp1 synergy in an in vitro transcription system (Näär et al. 1998). A portion of the evidence favoring a requirement for novel cofactors in the synergistic activation of Sp1/SREBP-1a in the context of chromatin is presented below.

SYNERGISTIC ACTIVATION BY SP1 AND SREBP-1A IN VITRO DEPENDS ON CHROMATIN

Activation by SREBP-1a and Sp1 on the LDLR-derived template was evaluated in a highly purified human in vitro transcription reaction consisting of recombinant as well as purified and partially purified RNA polymerase II general transcription factors and cofactors. Although this system supports high levels of activated transcription on naked DNA templates, no synergy of Sp1 and SREBP-1a was observed (Fig. 6B, lanes 1–4).

One difference between the in vitro and in vivo systems that could influence the coordinate action of multiple activators is the likely packaging of the in vivo template into chromatin. To investigate this possibility, the LDLR template was assembled into chromatin in vitro using a *Drosophila* chromatin assembly system (Kamakaka et al. 1993) followed by testing of SREBP-1a/Sp1 activation on chromatin with the purified human transcription system. A high degree of synergy was observed (Fig. 6B, lanes 5–8), suggesting that it is possible to reconstitute transcription reactions in vitro using a chromatin template and purified factors that recapitulate synergistic activation by SREBP-1a and Sp1 on the LDLR-derived template. The purified transcription system also allowed investigation of various cofactor requirements for Sp1/SREBP-1a activation not possible to address using crude transcription systems on chromatin templates.

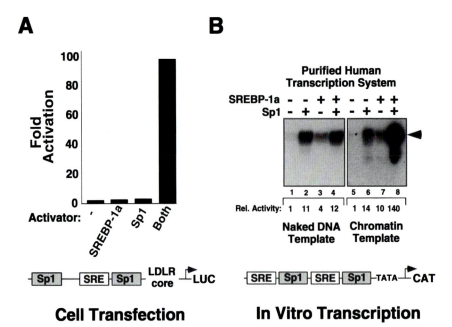

A — **Cell Transfection**

B — **In Vitro Transcription**

Figure 6. Sp1/SREBP-1a-dependent transcriptional synergy in a purified in vitro transcription system requires the template to be assembled into chromatin. (A) Sp1 and SREBP-1a cooperate to activate transcription at the LDL receptor promoter in vivo. Transient transfection of *Drosophila* Schneider cells with vectors directing expression of human Sp1 and SREBP (SREBP-1a[aa 1-487]) or vector alone (none) together with a reporter construct containing the human LDL receptor promoter and enhancer sequences fused to the luciferase reporter gene. The organization of the LDLR promoter is depicted below the bar graph. (B) Sp1 and SREBP-1a require chromatin for synergistic activation of the LDLR-derived template. In vitro transcription results with no activator added (lanes *1,5*), Sp1 (lanes *2,6*), SREBP-1a (lanes *3,7*), or both activators added (lanes *4,8*). The activity relative to basal transcription (lane *1,5* arbitrarily set to 1) is shown below the transcription panels (Rel. Activity). In vitro transcription reactions were performed using purified activators (1–5 nM), a highly purified human transcription system, and an LDLR promoter-derived reporter plasmid (0.6 nM) known to support SREBP-1a/Sp1 synergy in vivo. The LDLR promoter chromatin template was assembled using a *Drosophila* chromatin assembly system. Transcription products were analyzed by primer extension and denaturing PAGE.

TAFS ARE IMPORTANT FOR ACTIVATION BY SP1/SREBP-1A ON CHROMATIN

Previous studies identified the TAFs as potential coactivators for mediating activation in *Drosophila* and human reconstituted transcription reactions on naked DNA templates (Tjian and Maniatis 1994). Experiments were therefore conducted to investigate whether the TAF subunits of TFIID are also required for activation on chromatin templates. Substituting recombinant human TBP for antibody affinity-purified human TFIID resulted in the loss of activation observed with Sp1 alone, as well as synergistic activation by Sp1 and SREBP-1a (Fig. 7). This finding suggests that TAFs serve important functions in directing transcriptional responses to activators on chromatin templates. High levels of activation with naked DNA and chromatin templates also require a fraction containing the CRSP cofactor important for Sp1 activation on naked DNA templates (data not shown). However, since Sp1/SREBP-1a-dependent synergy only occurs on nucleosomal LDLR templates, additional chromatin-specific cofactors may also be necessary to observe cooperative activation on chromatin templates.

THE COACTIVATOR CBP IS IMPLICATED IN SREBP-1A/SP1-DEPENDENT CHROMATIN-SPECIFIC SYNERGY

Previous results from our laboratory demonstrated that the coactivator CBP could interact with the SREBP activation domain and stimulate SREBP transcriptional activation in vivo (Oliner et al. 1996). The reconstitution of SREBP-1a and Sp1 synergy on the LDLR template in a defined in vitro transcription system allowed testing of

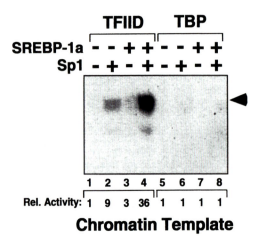

Figure 7. Sp1/SREBP-1a activation at the LDLR chromatin template requires TAFs. In vitro transcription reactions using the purified transcription system and chromatin templates with TFIID or TBP are presented. Activation in the absence of activators (lanes *1,5*), or in the presence of Sp1 (lanes *2*,*6*), SREBP-1a (lanes *3* and *7*), or both activators (lanes *4,8*), together with either immunopurified human TFIID (lanes *1–4*) or recombinant human TBP (lanes *5–8*) is shown here.

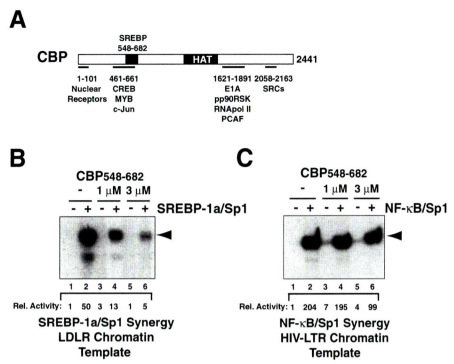

Figure 8. The SREBP-binding domain of CBP inhibits SREBP-1a/Sp1 chromatin-specific transcriptional synergy in a dominant negative fashion. (*A*) Depiction of the domains of CBP involved in protein-protein interactions and HAT activity. The CBP domain (amino acids 548–682) interacting with the activation domain of SREBP-1a is highlighted. (*B*) Addition of CBP548-682 polypeptide to SREBP-1a/Sp1-driven transcription reactions with the LDLR-promoter chromatin template results in a dose-dependent inhibition of transcription. (*C*) Minimal effects on NF-κB/Sp1-dependent transcription from HIV-LTR chromatin template by addition of CBP548-682 protein.

the potential involvement of CBP in chromatin-dependent synergy in vitro. Mapping studies identified a 135-amino-acid region of CBP as sufficient for binding to the first 50 amino acids of SREBP-1a (Fig. 8A). When adding increasing concentrations of this CBP polypeptide to in vitro transcription reactions reconstituted with the LDLR chromatin template, the SREBP-1a/Sp1-mediated transcriptional activation was strongly inhibited in a dose-dependent manner (Fig. 8B). The chromatin-dependent synergistic activation by NF-κB and Sp1 on the HIV-LTR was used as a control for the specificity of the inhibition because NF-κB interacts with a different portion of CBP. Addition of CBP548-682 did not significantly affect NF-κB/Sp1 synergy on the HIV-LTR chromatin template, attesting to the specificity of the peptide (Fig. 8C). The CBP peptide also failed to significantly inhibit either basal or activated transcription on the naked DNA template (data not shown).

RECONSTITUTION OF SREBP-1A/SP1 ACTIVATION ON CHROMATIN REQUIRES A CBP-CONTAINING ACTIVITY

The dominant negative inhibition by the CBP peptide indicated that the chromatin transcription reactions contained a CBP-like activity. When reconstituting transcription reactions with chromatin assembled using low levels of chromatin assembly extract, activation by SREBP-1a/Sp1 was substantially reduced (data not shown), suggesting that the chromatin assembly extract

provides one or more limiting cofactor(s) important for SREBP-1a/Sp1 activation. On the basis of this assay, a human CBP-containing cofactor (PC0.5M) was partially purified that could substitute for a limiting cofactor(s) in the S-190 extract (data not shown). Furthermore, depletion of the PC0.5M cofactor fraction using GST-SREBP-1a(aa1-50) resulted in greatly reduced cofactor activity, suggesting that the SREBP-1a activation domain interacts with components of a coactivator activity (data not shown).

Since the activation domain of SREBP-1a binds to CBP and could deplete a coactivator activity, it could potentially be used as an affinity resin to purify the CBP-containing coactivator. HeLa cell nuclear extract or the PC0.5M fraction was initially applied to the GST-SREBP(aa1-50) affinity resin or a control resin consisting of the glutamine-rich activation domain A of Sp1 fused to GST, and the specifically bound fractions were analyzed by SDS-PAGE followed by silver staining. The data indicated that several proteins (m.w. 280–30 kD), including CBP, bound strongly to the GST-SREBP(aa1-50) affinity resin, whereas no specific interaction with the GST-Sp1A control beads was detected (Fig. 9A). These results together suggest that SREBP-1a recruits an activation domain-specific coactivator.

When adding the highly purified SREBP-binding proteins to transcription reactions with chromatin template, strong stimulation of SREBP-1a/Sp1-dependent activation was observed, suggesting that the proteins eluted from the SREBP resin (Fig. 9B) constitute a SREBP-

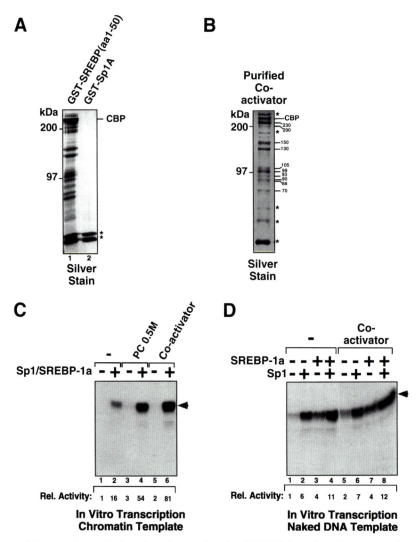

Figure 9. A CBP-containing coactivator binds to the activation domain of SREBP-1a and mediates chromatin-specific activation by SREBP-1a/Sp1. (*A*) Several nuclear proteins, including CBP, interact selectively with the SREBP-1a activation domain. HeLa cell nuclear extract was purified using affinity resins containing the activation domains of SREBP-1a (GST-SREBP-1a[aa1-50]) (lane *1*) or Sp1 (GST-Sp1A[aa83-262]) (lane *2*). The bound fractions were analyzed by SDS-PAGE/silver staining. The asterisks indicate the positions of nonspecific proteins. (*B*) The CBP-containing coactivator eluted from the SREBP-1a activation domain affinity column. The cofactor present in the PC0.5M fraction was further purified using the GST-SREBP-1a(aa1-50) resin. Bound proteins were eluted using buffer containing 0.1% deoxycholate or the CBP548-682 peptide. The eluate was analyzed using 7% SDS-PA gel/silver staining. Molecular-weight standards are indicated to the left of the panel. The migration of CBP and specific SREBP-binding proteins is indicated to the right. The asterisks denote nonspecific proteins that are not part of the coactivator activity. (*C*) The CBP-containing coactivator is required for full activation by SREBP-1a/Sp1 in S-190-limited chromatin transcription reactions. This panel shows the effect of adding either the PC0.5M cofactor fraction (lanes *3,4*) or the affinity-purified coactivator (lanes *5,6*) to S-190 limited chromatin transcription reactions in the absence (lanes *1,3,5*) or presence (lanes *2,4,6*) of SREBP-1a/Sp1. (*D*) The purified coactivator exert little effect on transcription with the naked DNA template. Transcription reactions performed with naked DNA templates in the absence (lanes *1–4*) or presence (lanes *5–8*) of affinity-purified CBP-containing coactivator. SREBP-1a (lanes *3,4,7,8*) and Sp1 (lanes *2,4,6,8*) were added as indicated.

binding coactivator which can substitute for the crude PC0.5M cofactor fraction (Fig. 9C). Additionally, when testing the eluted SREBP-binding proteins in transcription reactions with naked DNA templates, little effect on basal or activated transcription was observed (Fig. 9D). These findings suggest that a coactivator capable of selective binding to the activation domain of SREBP-1a participates in mediating SREBP-1a/Sp1 activation in the context of chromatin.

DISCUSSION

Sp1/SREBP-1a-dependent transcriptional synergy was reconstituted at the LDLR promoter in vitro. The results indicate that the coordinate activation by Sp1 and SREBP-1a requires the LDLR template to be assembled into chromatin. Using a purified human transcription system, multiple distinct coactivators were found to be necessary to mediate chromatin-dependent synergy by

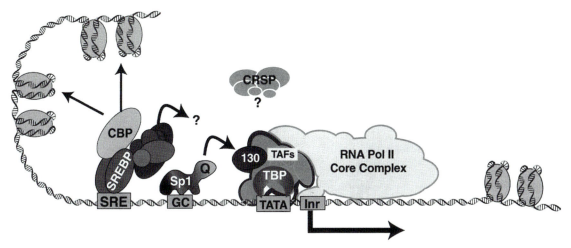

Figure 10. Synergistic activation by SREBP-1a and Sp1 on the LDLR chromatin template requires multiple cofactors. The recruitment of a CBP-containing coactivator by SREBP could serve to alleviate chromatin-mediated repression of transcription by facilitating access of activators and the transcriptional machinery to a nucleosomal template. The HAT activity of CBP may participate in remodeling of the nucleosomal structure, whereas the function of the other components of the SREBP-binding coactivator remains to be established. Sp1 has been shown to recruit TFIID via $TAF_{II}130$, which could help nucleate formation of the preinitiation complex.

SREBP-1a/Sp1, including TAFs, and a CBP-containing multiprotein coactivator that directly binds to the activation domain of SREBP-1a (Fig. 10). The TAF requirement for high levels of activation by Sp1 and SREBP-1a may be explained at least partly by the well-documented interactions between Sp1 and $hTAF_{II}130$ in recruiting TFIID to the promoter (Chen et al. 1994; Gill et al. 1994). Since SREBP-1a was not found to directly contact TFIID, or other components of the core transcriptional machinery (data not shown), cooperativity cannot be explained by direct targeting of multiple GTFs. Indeed, SREBP-1a/Sp1 cooperativity may be a function of recruitment of both a CBP-containing coactivator and TFIID to the promoter (Fig. 10). This dual requirement of a CBP-containing coactivator and TAFs for activation of transcription by SREBP-1a and Sp1 is reminiscent of the cofactor interactions utilized by the cAMP/PKA-activated transcription factor CREB. Both recruitment of TFIID and CBP are required for CREB activation; thus, it appears that CREB embodies functions of both Sp1 and SREBP-1a (Chrivia et al. 1993; Ferreri et al. 1994; Nakajima et al. 1997a). Although CREB appears to activate transcription efficiently on naked DNA templates and interacts with a CBP/RNA RNA polymerase II complex (Kee et al. 1996; Nakajima et al. 1997b), our results suggest that the CBP-containing coactivator recruited by SREBP-1a performs a chromatin-dependent function required for SREBP-1a/Sp1 synergy. Chromatin-mediated regulation of transcription may be a strategy utilized by many classes of transcription factors responding to various signaling pathways. For example, NF-κB was found to interact with p300/CBP (Perkins et al. 1997) and was recently shown to synergize with Sp1 on the HIV-LTR in a chromatin-dependent manner (Pazin et al. 1996; Sheridan et al. 1997). Interestingly, p300/CBP recruited by NF-κB is associated with a cyclin-dependent kinase activity (Perkins et al. 1997), suggesting that multiple biochemical activities may be contained within p300/CBP complexes.

FUTURE PROSPECTS

Chromatin-dependent Regulation by Sp1

The CBP-containing coactivator harbors histone acetyltransferase (HAT) activity and appears to be necessary for full activation by SREBP-1a/Sp1 on chromatin templates, while having little effect when using naked DNA templates. It is tempting to speculate that acetylation of histones by CBP may facilitate access of SREBP-1a and Sp1 to nucleosomal templates and aid in the recruitment of the preinitiation complex to the core promoter. Studies of the role of CBP and HAT activity in the chromatin-dependent synergistic activation by SREBP-1a/Sp1 and NF-κB/Sp1 will likely require the use of purified chromatin templates, purified or recombinant histone deacetylases, and the development of HAT-specific inhibitors. Other enzymatic activities, such as kinases, could also be associated with the CBP-containing coactivator and might modify histones or components of the transcriptional apparatus.

We are currently cloning and characterizing the genes encoding the polypeptides in the CBP-containing coactivator to elucidate their functional contribution in mediating SREBP-1a/Sp1 synergistic activation on chromatin templates. The active coactivator fraction contains CBP and at least ten other stoichiometric polypeptides ranging from 240 to 30 kD. Peptide sequencing revealed that most of these proteins are novel human gene products.

ATP-dependent nucleosomal remodeling activities such as SWI/SNFs, NURF, CHRAC, ACF, and RSC (Cairns 1998) may have important roles in cooperative activation by Sp1 with other natural activators, such as SREBP-1a and NF-κB, on chromatin templates. This class of cofactors may not be directly recruited to the promoter by activation domains but may function in a more general capacity to facilitate binding of activators and the core transcriptional machinery to chromatin templates. Future studies will be aimed at elucidating the role of

ATP-dependent remodeling factors in the synergistic activation by Sp1 with SREBP-1a and other activators on chromatin templates.

The eventual development of a purified chromatin transcription system should provide us with valuable tools for future detailed molecular studies of chromatin-dependent mechanisms of transcription regulation of natural genes by multiple activators.

ACKNOWLEDGMENTS

We thank Shane Albright for help in preparing the figures and members of the Tjian laboratory for critical reading of the manuscript.

REFERENCES

Athanikar J.N., Sanchez H.B., and Osborne T.F. 1997. Promoter selective transcriptional synergy mediated by sterol regulatory element binding protein and Sp1: A critical role for the Btd domain of Sp1. *Mol. Cell. Biol.* **17:** 5193.

Björklund S. and Kim Y.J. 1996. Mediator of transcriptional regulation. *Trends Biochem. Sci.* **21:** 335.

Brown M.S. and Goldstein J.L. 1997. The SREBP pathway: Regulation of cholesterol metabolism by proteolysis of a membrane-bound transcription factor. *Cell* **89:** 331.

Burke T.W. and Kadonaga J.T. 1997. The downstream core promoter element, DPE, is conserved from *Drosophila* to humans and is recognized by TAF$_{II}$60 of *Drosophila*. *Genes Dev.* **11:** 3020.

Cairns B.R. 1998. Chromatin remodeling machines: Similar motors, ulterior motives. *Trends Biochem. Sci.* **23:** 20.

Chen J.L., Attardi L.D., Verrijzer C.P., Yokomori K., and Tjian R. 1994. Assembly of recombinant TFIID reveals differential coactivator requirements for distinct transcriptional activators. *Cell* **79:** 93.

Chrivia J.C., Kwok R.P., Lamb N., Hagiwara M., Montminy,M.R., and Goodman R.H. 1993. Phosphorylated CREB binds specifically to the nuclear protein CBP. *Nature* **365:** 855.

Crowley T.E., Hoey T., Liu J.K., Jan Y.N., Jan L.Y., and Tjian R. 1993. A new factor related to TATA-binding protein has highly restricted expression patterns in *Drosophila*. *Nature* **361:** 557.

Dikstein R., Ruppert S., and Tjian R. 1996a. TAFII250 is a bipartite protein kinase that phosphorylates the base transcription factor RAP74. *Cell* **84:** 781.

Dikstein R., Zhou S., and Tjian R. 1996b. Human TAF$_{II}$105 is a cell type-specific TFIID subunit related to hTAFII130. *Cell* **87:** 137.

Dynan W.S. and Tjian R. 1983a. Isolation of transcription factors that discriminate between different promoters recognized by RNA polymerase II. *Cell* **32:** 669.

———. 1983b. The promoter-specific transcription factor Sp1 binds to upstream sequences in the SV40 early promoter. *Cell* **35:** 79.

Dynlacht B.D., Hoey T., and Tjian R. 1991. Isolation of coactivators associated with the TATA-binding protein that mediate transcriptional activation. *Cell* **66:** 563.

Ferreri K., Gill G., and Montminy M. 1994. The cAMP-regulated transcription factor CREB interacts with a component of the TFIID complex. *Proc. Natl. Acad . Sci.* **91:** 1210.

Fondell J.D., Ge H., and Roeder R.G. 1996. Ligand induction of a transcriptionally active thyroid hormone receptor coactivator complex. *Proc. Natl. Acad . Sci.* **93:** 8329.

Ge H. and Roeder R.G. 1994. Purification, cloning, and characterization of a human coactivator, PC4, that mediates transcriptional activation of class II genes. *Cell* **78:** 513.

Gill G., Pascal E., Tseng Z.H., and Tjian R. 1994. A glutamine-rich hydrophobic patch in transcription factor Sp1 contacts

the dTAF$_{II}$110 component of the *Drosophila* TFIID complex and mediates transcriptional activation. *Proc. Natl. Acad. Sci.* **91:** 192.

Goodrich J.A. and Tjian R. 1994. TBP-TAF complexes: Selectivity factors for eukaryotic transcription. *Curr. Opin. Cell Biol.* **6:** 403.

Grant P.A., Schieltz D., Pray-Grant M.G., Steger D.J., Reese J.C., Yates III J.R., and Workman J.L. 1998. A subset of TAFIIs are integral components of the SAGA complex required for nucleosomal acetylation and transcriptional stimulation. *Cell* **94:** 45.

Hansen S.K., Takada S., Jacobson R.H.J., Lis J.T., and Tjian R. 1997. Transcription properties of a cell type-specific TATA-binding protein, TRF. *Cell* **91:** 71.

Hengartner C.J., Thompson C.M., Zhang J., Chao D.M., Liao S.M., Koleske A.J., Okamura S., and Young R.A. 1995. Association of an activator with an RNA polymerase II holoenzyme. *Genes Dev.* **9:** 897.

Hoey T., Weinzierl R.O., Gill G., Chen J.L., Dynlacht B.D., and Tjian R. 1993. Molecular cloning and functional analysis of *Drosophila* TAF110 reveal properties expected of coactivators. *Cell* **72:** 247.

Horikoshi M., Carey M.F., Kakidani H., and Roeder R.G. 1988. Mechanism of action of a yeast activator: Direct effect of GAL4 derivatives on mammalian TFIID-promoter interactions. *Cell* **54:** 665.

Jacq X., Brou C., Lutz Y., Davidson I., Chambon P., and Tora L. 1994. Human TAF$_{II}$30 is present in a distinct TFIID complex and is required for transcriptional activation by the estrogen receptor. *Cell* **79:** 107.

Kadonaga J.T., Carner K.R., Masiarz F.R., and Tjian R. 1987. Isolation of cDNA encoding transcription factor Sp1 and functional analysis of the DNA binding domain. *Cell* **51:** 1079.

Kadonaga J.T., Courey A.J., Ladika J., and Tjian R. 1988. Distinct regions of Sp1 modulate DNA binding and transcriptional activation. *Science* **242:** 1566.

Kamakaka R.T., Bulger M., and Kadonaga J.T. 1993. Potentiation of RNA polymerase II transcription by Gal4-VP16 during but not after DNA replication and chromatin assembly. *Genes Dev.* **7:** 1779.

Kaufmann J., Ahrens K., Koop R., Smale S.T., and Muller R. 1998. CIF150, a human cofactor for transcription factor IID-dependent initiator function. *Mol. Cell . Biol.* **18:** 233.

Kee B.L., Arias J., and Montminy M.R. 1996. Adaptor-mediated recruitment of RNA polymerase II to a signal-dependent activator. *J. Biol. Chem.* **271:** 2373.

Kre G., Mahfoudi A., and Wahli W. 1995. Functional interactions of peroxisome proliferator-activated receptor, retinoid-X receptor, and Sp1 in the transcriptional regulation of the acyl-coenzyme-A oxidase promoter. *Mol. Endocrinol.* **9:** 219.

Levine A.J. 1997. p53, the cellular gatekeeper for growth and division. *Cell* **88:** 323.

Look D.C., Pelletier M.R., Tidwell R.M., Roswit W.T., and Holtzman M.J. 1995. Stat1 depends on transcriptional synergy with Sp1. *J. Biol. Chem.* **270:** 30264.

Meisterernst M., Roy A.L., Lieu H.M., and Roeder R.G. 1991. Activation of class II gene transcription by regulatory factors is potentiated by a novel activity. *Cell* **66:** 981.

Merika M. and Orkin S.H. 1995. Functional synergy and physical interactions of the erythroid transcription factor GATA-1 with the Kruppel family proteins Sp1 and EKLF. *Mol. Cell. Biol.* **15:** 2437.

Mizzen C.A., Yang X.J., Kokubo T., Brownell J.E., Bannister A.J., Owen-Hughes T., Workman J., Wang L., Berger S.L., Kouzarides T., Nakatani Y., and Allis C.D. 1996. The TAF(II)250 subunit of TFIID has histone acetyltransferase activity. *Cell* **87:** 1261.

Näär A.M., Beaurang P., Robinson K., Oliner J.D., Avizonis D., Zwicker J., Scheek S., Kadonaga J.T., and Tjian R. 1998. Chromatin, TAFs, and a novel multiprotein co-activator are required for synergistic activation by Sp1 and SREBP-1a in vitro. *Genes Dev.* **12:** 3020.

Nakajima T., Uchida C., Anderson S.F., Parvin J.D., and Montminy M. 1997a. Analysis of a cAMP-responsive activator reveals a two-component mechanism for transcriptional induction via signal-dependent factors. *Genes Dev.* **11:** 738.

Nakajima T., Uchida C., Anderson S.F., Lee C.G., Hurwitz J., Parvin J.D., and Montminy M. 1997b. RNA helicase A mediates association of CBP with RNA polymerase II. *Cell* **90:** 1107.

O'Brien T. and Tjian R. 1998. Functional analysis of the human TAF$_{II}$250 N-terminal kinase domain. *Mol.Cell* **1:** 905.

Ogryzko V.V., Kotani T., Zhang X., Schiltz R.L., Howard T., Yang X.-J., Howard B.H., Qin J., and Nakatani Y. 1998. Histone-like TAFs within the PCAF histone acetylase complex. *Cell* **94:** 35.

Oliner J.D., Andresen J.M., Hansen S.K., Zhou S., and Tjian R. 1996. SREBP transcriptional activity is mediated through an interaction with the CREB-binding protein. *Genes Dev.* **10:** 2903.

Pascal E. and Tjian R. 1991. Different activation domains of Sp1 govern formation of multimers and mediate transcriptional synergism. *Genes Dev.* **5:** 1646.

Pazin M.J., Sheridan P.L., Cannon K., Cao Z., Keck J.G., Kadonaga J.T., and Jones K.A. 1996. NF-κB-mediated chromatin reconfiguration and transcriptional activation of the HIV-1 enhancer in vitro. *Genes Dev.* **10:** 37.

Perkins N.D., Felzien L.K., Betts J.C., Leung K., Beach D.H., and Nabel G.J. 1997. Regulation of NF-κB by cyclin-dependent kinases associated with the p300 coactivator. *Science* **275:** 523.

Pugh B.F. and Tjian R.1990. Mechanism of transcriptional activation by Sp1: Evidence for coactivators. *Cell* **61:** 1187.

Rachez C., Suldan Z., Ward J., Chang C.-P.W., Burakov D., Erdjument-Bromage H., Tempst P., and Freedman L.P. 1998. A novel protein complex that interacts with the vitamin D3 receptor in a ligand-dependent manner and enhances VDR transactivation in a cell-free system. *Genes Dev.* **12:** 1787.

Sanchez H.B., Yieh L., and Osborne T.F. 1995. Cooperation by sterol regulatory element-binding protein and Sp1 in sterol regulation of low density lipoprotein receptor gene. *J. Biol. Chem.* **270:** 1161.

Sauer F. and Tjian R. 1997. Mechanisms of transcriptional activation: Differences and similarities between yeast, *Drosophila*, and man. *Curr. Opin. Genet. Dev.* **7:** 176.

Sawadogo M. and Roeder R.G. 1985. Interaction of a gene-specific transcription factor with the adenovirus major late promoter upstream of the TATA box region. *Cell* **43:** 165.

Sheridan P.L., Mayall T.P., Verdin E., and Jones K.A. 1997. Histone acetyltransferases regulate HIV-1 enhancer activity in vitro. *Genes Dev.* **11:** 3327.

Sheridan P.L., Sheline C.T., Cannon K., Voz M.L., Pazin M.J., Kadonaga J.T., and Jones K.A. 1995. Activation of the HIV-1 enhancer by the LEF-1 HMG protein on nucleosome-assembled DNA in vitro. *Genes Dev.* **9:** 2090.

Smale S.T., Schmidt M.C., Berk A.J., and Baltimore D. 1990. Transcriptional activation by Sp1 as directed through TATA or initiator: Specific requirement for mammalian transcription factor IID. *Proc. Natl. Acad . Sci.* **87:** 4509.

Struhl K. 1998. Histone acetylation and transcriptional regulatory mechanisms. *Genes Dev.* **12:** 599.

Tanese N., Pugh B.F., and Tjian R. 1991. Coactivators for a proline-rich activator purified from the multisubunit human TFIID complex. *Genes Dev.* **5:** 2212.

Tjian R. and Maniatis T. 1994. Transcriptional activation: A complex puzzle with few easy pieces. *Cell* **77:** 5.

Triezenberg S.J. 1995. Structure and function of transcriptional activation domains. *Curr. Opin. Genet. Dev.* **5:** 190.

Verrijzer C.P. and Tjian R. 1996. TAFs mediate transcriptional activation and promoter selectivity. *Trends Biochem. Sci.* **21:** 338.

Wieczorek E., Brand M., Jacq X., and Tora L. 1998. Function of TAF(II)-containing complex without TBP in transcription by RNA polymerase II (comments). *Nature* **393:** 187.

Role of General and Gene-specific Cofactors in the Regulation of Eukaryotic Transcription

R.G. ROEDER

Laboratory of Biochemistry and Molecular Biology, The Rockefeller University, New York, New York 10021

HISTORICAL PERSPECTIVES AND BASIC PRINCIPLES

Prokaryotic and eukaryotic genes are activated and regulated at the level of transcription through two generally distinct classes of DNA elements and cognate transcription factors: (1) common core promoter elements (located at or near the transcription initiation site) and interacting RNA polymerases and associated accessory factors (comprising the general transcriptional machinery) that mediate basal transcription, and (2) gene-specific regulatory elements (usually located distal to the core promoter elements) and interacting gene-specific factors, either constitutive or induced, that modulate the functions of the general transcription machinery at corresponding core promoters (Fig. 1). The intrinsic ability of the general transcriptional machinery to mediate accurate transcription from core promoters, especially those containing consensus elements, has provided detailed information on basic aspects of promoter recognition and transcription initiation (see below). Equally important to understand, however, are the mechanisms by which regulatory factors modulate the functions of the general transcriptional machinery on corresponding target genes. Especially relevant questions are (1) whether different regulatory factors recognize the same or distinct general factor targets, (2) whether the interactions are direct or indirect (i.e., involve other cofactors), (3) how the interactions modulate general factor functions, and (4) the basis for the synergistic functions of multiply bound regulatory factors.

The most common regulatory paradigm in bacteria involves interactions of gene-specific activators with a common RNA polymerase holoenzyme ($\alpha_2\beta\beta'\sigma$) (for review, see Busby and Ebright 1994), although gene-specific regulation through structural variations in the general transcription machinery is now evident from the presence of distinct σ factors that directly recognize different core promoters (for review, see Gross et al. 1992). This latter principle was first established in eukaryotes through the discovery of multiple forms (I, II, III) of nuclear RNA polymerase (Roeder and Rutter 1969) with distinct subunit structures and with distinct functions, respectively, in the transcription of large ribosomal RNA genes (class I), all protein-coding and some small nuclear RNA genes (class II), and most small structural RNA genes (class III) (for review, see Roeder 1976; Sentenac 1985). It was then extended and dramatically reinforced

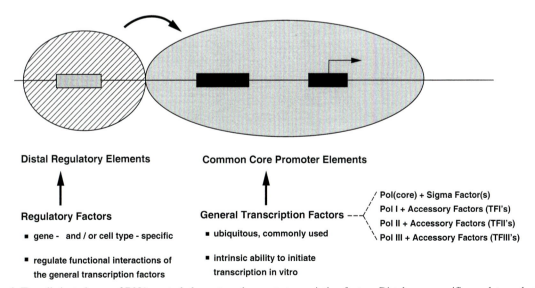

Distal Regulatory Elements   **Common Core Promoter Elements**

Regulatory Factors   **General Transcription Factors** ---<
- gene - and / or cell type - specific   - ubiquitous, commonly used
- regulate functional interactions of the general transcription factors   - intrinsic ability to initiate transcription in vitro

- Pol(core) + Sigma Factor(s)
- Pol I + Accessory Factors (TFI's)
- Pol II + Accessory Factors (TFII's)
- Pol III + Accessory Factors (TFIII's)

Figure 1. Two distinct classes of DNA control elements and cognate transcription factors. Distal gene-specific regulatory elements are recognized by various gene- and cell-specific factors (*hatched area*) that directly (arrow) or indirectly (Fig. 3) regulate the interactions or functions of RNA polymerases and general transcription factors (*gray area*) at common core promoter elements. Bacteria contain a single core RNA polymerase [Pol (core)] and a number of σ factors with different promoter specificities, whereas eukaryotic cells contain three different nuclear RNA polymerases (Pol I, Pol II, and Pol III) that act with three distinct classes of accessory factors (TFIs, TFIIs, and TFIIIs) at distinct core promoter elements.

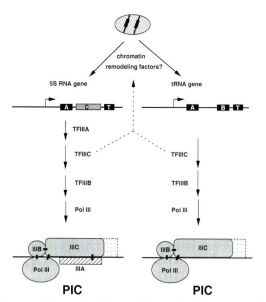

Figure 2. Pathway for PIC assembly on class III genes and the role of a gene-specific regulatory factor in recruitment of general initiation factors. In the simple case of a tRNA gene, common multisubunit factors TFIIIC, TFIIIB, and Pol III (*gray areas*) suffice for PIC formation, nucleated by direct TFIIIC binding to A and B boxes, and for transcription. In the specialized case of the 5S RNA gene, which utilizes the same common factors, the essential gene-specific regulatory factor TFIIIA (*hatched area*) binds directly to the A and C boxes and facilitates the stable binding of TFIIIC (through both DNA and TFIIIA contacts), which in turn promotes assembly of TFIIIB and Pol III into a functional PIC as in the case of the tRNA gene. Thus, from a mechanistic view, the major role of the gene-specific activator is to facilitate recruitment of common factors and concomitant PIC assembly. The short solid bars in the PICs indicate contacts between specific subunits in the different factors (for review of human factors, see Z. Wang et al. 1997; for yeast factors, see Chédin et al., this volume). Factor interactions over the terminator (T) regions are indicated by the dashed line enclosures and may reflect either extended TFIIIC interactions induced by other factors involved in termination and reinitiation or direct interactions of such factors (Wang and Roeder 1998). Potential chromatin remodeling activities for class III genes include histone acetyltransferase activities intrinsic to several subunits within TFIIIC (Kundu et al. 1999). For additional references, see text.

through the discovery—somewhat surprising in view of the structural complexity (up to 16 subunits) of the catalytically active core RNA polymerases—of essential RNA polymerase-specific accessory factors (Fig. 1) (Parker and Roeder 1977; Ng et al. 1979; Weil et al. 1979; Matsui et al. 1980; Segall et al. 1980). Studies during the following 15 years have led to a more complete description (including cognate cDNA cloning) of these common multisubunit accessory factors. They include TFIIIC and TFIIIB for RNA polymerase III (for review, see Geiduschek and Kassavetis 1992; Willis 1993; Z. Wang et al. 1997); TFIIA, TFIIB, TFIID, TFIIE, TFIIF, and TFIIH for RNA polymerase II (for review, see Orphanides et al. 1996; Roeder 1996; Hampsey 1998); and several factors for RNA polymerase I (for review, see Grummt 1998; Nomura 1998).

Early mechanistic studies, in turn, led to the identification of primary core promoter recognition factors and or-

dered pathways of preinitiation complex (PIC) assembly, first for class III genes (Lassar et al. 1983; for review, see Geiduschek and Kassavetis 1992), then for class II genes (Parker and Topol 1984; Sawadogo and Roeder 1985; Nakajima et al. 1988; Van Dyke et al. 1988; Buratowski et al. 1989; for review, see Roeder 1996; Orphanides et al. 1996), and finally for class I genes (for review, see Schnapp and Grummt 1991; Nomura 1998). The basic assembly pathways for class III genes (tRNA, VA RNA) containing the most common class III core promoter elements (A and B boxes) and for class II genes containing the most common class II core promoter element (TATA box) are shown in the right-hand parts of Figures 2 and 3, respectively. They emphasize, respectively, primary interactions of TFIIIC with A and B boxes and primary interactions of TFIID with the TATA box, forming complexes that nucleate the assembly of the remaining cognate factors and RNA polymerases into functional preinitiation complexes. Apart from the initial biochemical analyses, various biophysical and chemical studies also have contributed greatly to our appreciation of the structure and assembly of the cognate PICs (for review, see Geiduschek and Kassavetis 1992; Burley and Roeder 1996; Kim et al. 1997; Nikolov and Burley 1997). More recent studies have suggested the possibility of simplified PIC assembly pathways involving preassembled RNA polymerase–general factor complexes (for review, see Koleske and Young 1995; Greenblatt 1997; Z. Wang et al. 1997; Seither et al. 1998). However, the utilization of such pathways in vivo remains to be established and, in any case, the various DNA-protein and protein-protein interactions deduced from studies with isolated factors (Figs. 2 and 3) would almost certainly be relevant to the alternate assembly pathways.

Early studies with cloned genes and purified RNA polymerases and cognate accessory factors also set the stage for an analysis, in eukaryotes, of the fundamental but then unknown question of the nature of gene-specific activation mechanisms and possible similarities to the now well-established bacterial paradigm of direct contacts between gene-specific DNA-bound activators and core RNA polymerase or accessory factor (σ) subunits (for review, see Busby and Ebright 1994). That the situation in eukaryotes is formally analogous to that in bacteria was initially established by studies of class III genes. Whereas tRNA genes are activated optimally by RNA polymerase III and the common accessory factors TFIIIC and TFIIIB, beginning with promoter recognition by TFIIIC, activation of the 5S RNA gene requires, in addition, TFIIIA (Fig. 2). TFIIIA was the first of many eukaryotic gene-specific transcriptional activators to be identified as such, purified, and cloned (Engelke et al. 1980; Ginsburg et al. 1984). It was shown both to bind in a site-specific manner to the 5S promoter (A and C boxes) and to facilitate recruitment of common factors (initially TFIIIC and, consequently, TFIIIB and Pol III) whose interactions are otherwise limiting because of a variant core promoter structure (Engelke et al. 1980; Lassar et al. 1983; Bieker et al. 1985; Setzer and Brown 1985; Kassavetis et al. 1990). These early mechanistic studies, in-

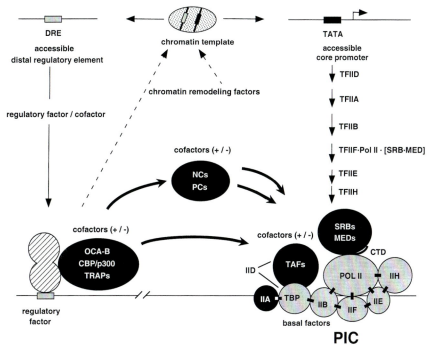

Figure 3. A comprehensive scheme summarizing (1) the stepwise PIC assembly pathway for general transcriptional factors (*gray areas*) on a simple TATA-containing class II core promoter and (2) three classes of cofactors (*black areas*) that variously modulate the function of DNA-binding regulatory factors (*hatched area*) on corresponding target genes. The stepwise PIC assembly pathway indicated is that elaborated with purified factors and begins with recognition of the TATA element by the TBP component of TFIID. Interactions between specific basal factors are indicated by short solid bars. The assembly and/or function of the PIC may be regulated in several ways: by direct interactions of DNA-bound regulatory factors, by direct interactions of certain cofactors (e.g., NC2) acting alone, or by direct interactions of cofactors that are recruited to the promoter and modulated by interactions with DNA-bound regulatory factors. In the latter case, the cofactors act as adapters to transmit signals from DNA-bound regulatory factors to general factors. The cofactors (see text for definitions) include a group associated with the basal transcriptional machinery (TFIIA, the TAF$_{II}$ components of TFIID, and the SRB and MED components of the Pol II holoenzyme); a group associated with gene-specific regulatory factors (OCA-B, CBP/p300, and TRAPs, as well as others not indicated); and a miscellaneous group of potentially more general positive and negative cofactors (PCs and NCs). All three groups, which are not necessarily mutually exclusive, include both positively and negatively acting cofactors, as discussed in the text. These cofactors also include targeted histone acetyltransferases (e.g., CBP/p300) and deacetylases that, along with distinct ATP-dependent factors such as NURF and SWI-SNF complexes, are involved in chromatin remodeling (*dashed arrows*) prior to, or concomitant with, regulatory factor binding and PIC assembly (for review, see Struhl 1998; Workman and Kingston 1998). For additional references and discussion, see text. The conventions used here for regulatory factors (*hatched area*), cofactors (*black areas*), and basal factors (*gray area*) are maintained in other figures.

cluding the first demonstration of accurate eukaryotic gene-specific transcription in a cell-free system reconstituted with isolated components of the general transcriptional machinery (Parker et al. 1976; Parker and Roeder 1977), not only stimulated subsequent studies of class II and class I genes, but, indeed, also proved to be especially relevant in light of current views on activator functions through general factor recruitment on class II genes (Ptashne and Gann 1997). However, as will be seen below, regulation of class II genes involves complexities far beyond those of prokaryotic and eukaryotic class I and III genes.

Subsequent studies of regulation of the vastly more diverse group of class II genes have led to the identification and cloning, beginning with the glucocorticoid receptor (Payvar et al. 1981; Miesfeld et al. 1984) and Sp1 (Dynan and Tjian 1983; Kadonaga et al. 1987), of a large number of site-specific DNA-binding proteins that serve as transcriptional activators and repressors. The earliest in vitro mechanistic studies of class II gene activators also implicated TFIID, the primary core promoter recognition factor, as an activator target (Sawadago and Roeder 1985;

Abmayr et al. 1988; Horikoshi et al. 1988; Workman et al. 1988) and further demonstrated a mutually exclusive binding of TFIID and nucleosomal histones to the core promoter as a point of regulation (Workman and Roeder 1987). This early work also set the stage for more detailed studies of the underlying mechanisms involving TFIID and for identification of both other activator targets and a broad group of "cofactors." As discussed here, cofactors are operationally defined as factors that, although not essential for low levels of accurate core promoter transcription by RNA polymerase II and a minimal set of accessory factor components in vitro (defined as basal transcription), are required for optimal levels of induction or repression (from the basal level of transcription) by gene-specific DNA-binding regulatory proteins. Implicit in this definition is the fact that high levels of induction in vitro may depend in part on the further imposition of natural constraints (see below) that repress or restrict the intrinsic activity of positively acting basal factors on core promoter elements. These constraints include specific negative cofactors, chromatin structure, limiting general factor concentrations, weak core promoter-binding sites, and competing templates (Roeder 1991).

Given the above definition, and for convenience, the cofactors can be placed into several groups (Fig. 3): (1) those whose primary interactions are with specific DNA-bound activators (or repressors), such as the early described B-cell-specific OCA-B (Luo et al. 1992) and the ubiquitous CBP (Chrivia et al. 1993); (2) those that either are part of or are intimately associated with the general transcriptional machinery, including TFIIA (Meisterernst et al. 1991; for review, see Hampsey 1998), TATA-binding protein (TBP)-associated components of TFIID (Hoffmann et al. 1990; Pugh and Tjian 1990; for review, see Burley and Roeder 1996; Verrijzer and Tjian 1996), and the SRB/MED components associated with RNA polymerase II (Thompson et al. 1993; Kim et al. 1994; Koleske and Young 1994; for review, see Koleske and Young 1995; Hampsey 1998; Myers et al. 1998); and (3) those that appear to be associated (in the absence of promoters) neither with activators nor with general factors and to function as more general cofactors, including human *u*pstream *s*timulatory *a*ctivity (USA)-derived positive and negative cofactors (Meisterernst and Roeder 1991; Meisterernst et al. 1991; for review, see Kaiser and Meisterernst 1996) and certain nucleosome remodeling factors (for review, see Workman and Kingston 1998). It is to be emphasized that each of these three groups, which may not be mutually exclusive, contains both positively acting and negatively acting cofactors. These cofactors, which in several cases act as adapters between DNA-binding regulatory factors and the general transcriptional machinery, add an important new layer of complexity to eukaryotic transcriptional regulation and further distinguish most eukaryotic and prokaryotic regulatory pathways. Selected examples of such coactivators, from studies in our laboratory, are discussed below. Other important findings related to activation mechanisms that are not discussed here include the identification of a number of distinct components of the general transcription machinery as targets both for the same activators and for different activators (for review, see Wu et al. 1996; Koh et al. 1998) and the identification of concerted activator interactions with multiple components of the general transcriptional machinery as a basis for transcriptional synergy between multiply bound activators (for review, see Chi et al. 1995; Ptashne and Gann 1997).

GENERAL POSITIVE AND NEGATIVE COFACTORS

In human-cell-free systems reconstituted with natural general initiation factors (including TFIID) and RNA polymerase II, high levels of induction by natural activators (Sp1, USF) were found to depend on the USA fraction and to reflect the presence (in USA) of both a negative cofactor (NC) activity(ies) that selectively represses basal activity and a positive cofactor (PC) activity(ies) that preferentially enhances activator function (Meisterernst and Roeder 1991; Meisterernst et al. 1991). These and earlier studies showing activator-reversible, chro-

matin-assembly-mediated inhibition of TFIID binding and function (Workman and Roeder 1987; Workman et al. 1988) suggested the simple model for promoter regulation shown in Figure 4. In this model, a given promoter DNA template has an intrinsic activity with basal factors that can be repressed either by general negative cofactors or by nucleosomal histones (within chromatin) to a more physiological ground state and elevated to a higher (activated) level, either from the intrinsic basal activity state or from the ground state, through the combined action of activators and positive cofactors (coactivators). Activation from the ground state can be viewed as a two-step process, involving both antirepression and net activation stages that may involve different coactivators and mechanisms.

Currently resolved positive cofactors that can show independent activities in vitro, especially with model activators containing various metazoan activation domains fused to the DNA-binding domain of yeast GAL4, include (1) PC1, at least one component of which is poly-ADP-ribose polymerase (Meisterernst et al. 1997); (2)

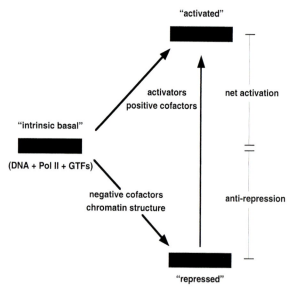

Figure 4. States of promoter activity defined by in vitro transcription assays. A given DNA template exhibits an intrinsic basal transcription activity with purified Pol II and purified general transcription factors (GTFs). This activity can be repressed by general negative cofactors (such as NC2 or MOT1) and by preassembly of the DNA template into a chromatin structure, generating a more physiological repressed or ground state. An activated state is reached in response to gene-specific (DNA-bound) activators and various positive cofactors, either from the intrinsic basal state or from the repressed state. The overall level of induction in response to activators will thus depend on the extent to which the basal activity is repressed. Activation from the repressed state can be viewed as a two step process involving an antirepression step that restores activity to the intrinsic basal level and a net-activation step that leads to an overall net increase in activity above the basal level; and the two steps may involve different positive cofactors. Thus, the in vitro observations indicate that activation in vivo is not simply an antirepression phenomenon. The model also applies to the action of gene-specific DNA-bound repressors (and corepressors) acting directly on general factors or on a chromatin template. For references, see text. (Reprinted, with permission, from Burley and Roeder 1998 [copyright Cell Press].)

PC2, an approximately 500-kD complex of unidentified polypeptides (Kretzschmar et al. 1994b); (3) PC3/Dr2, identified as topoisomerase I (Kretzschmar et al. 1993; Merino et al. 1993); (4) PC4, a 15-kD polypeptide whose activity is regulated by phosphorylation (Ge and Roeder 1994; Ge et al. 1994; Kretzschmar et al. 1994a) and that appears to have a yeast homolog (Henry et al. 1996; Knaus et al. 1996); (5) two closely related polypeptides derived from the natural PC4 fraction, one (p52) that shows a broad specificity and overall activity similar to that of the structurally unrelated PC4 and another (p75) that is less active but shows some activator selectivity (Ge et al.1998); (6) less well characterized PC5 (Halle et al. 1995) and PC6 (Kaiser and Meisterernst 1996) activities; and (7) HMG2 (Shykind et al. 1995). The PC1, PC2, PC3, PC4, and p52/p75 activities, as well as NC1 (below), are derived from the conventional USA fraction, whereas the other positive cofactors described here, as well as NC2 (below), are derived from different nuclear extract chromatographic fractions. The mechanism of action of these PCs is best understood for PC4, which shows interactions both on and off the promoter with a variety of activation domains and with both TFIIA and RNA polymerase II (Ge and Roeder 1994; Malik et al. 1998). Consistent with the observations of multiple interactions, PC4 has been reported to act both during and after assembly of the TFIIA·TFIID·DNA complex (Kaiser et al. 1995). Thus, PC4 acts at least in part as an adapter between activators and the general transcription machinery (Fig. 3), facilitating recruitment of the latter (Kaiser et al. 1995) by a mechanism(s) that also may involve stabilization of an early preinitiation complex and reversible formation of an inactive intermediate (Malik et al. 1998).

The general paradigm of PC4 as an adapter between activators and the general transcriptional machinery is consistent with mechanisms implicated in the action of other coactivators (for review, see Guarente 1995). With the exception of p52, which appears to interact both with a VP16 activation domain and with general transcription factors, a similar mechanism has not yet been demonstrated for any of the other PCs. However, PC1, PC3/Dr2, and HMG2, as well as PC4, have nonspecific DNA-binding activities and function at high cofactor:DNA ratios, suggesting that they

may act in part as architectural proteins that stabilize the PIC in response to (or in conjunction with) activators (Kretzschmar et al. 1993; Shykind et al. 1995, 1997). These observations also raise the possibility that some of the coactivators, especially those (e.g., PC1/poly-ADP ribose polymerase and PC3/topoisomerase I) that have other well-defined functions in topological or structural modifications of DNA or chromatin may link these functions to the activation of specific genes in vivo. A somewhat surprising and potentially related finding is that the apparently structurally distinct PCs can function independently of each other in vitro, especially on model promoters in response to multiply bound activators containing metazoan activation domains fused to the yeast GAL4 DNA-binding domain. This suggests that they may act by different mechanisms and affect different steps in PIC assembly and function. Consistent with this possibility, cooperative functions (e.g., between PC2 and PC4) can be observed in some situations with natural activators and promoters (Guermah et al. 1998; Luo et al. 1998; see also below). Thus, major unanswered questions concern the polypeptide compositions of the incompletely purified PCs (especially the very potent PC2), the relative importance and possible synergistic functions of the various PCs under more physiological conditions, the mechanism(s) of action of the PCs and their potential interactions (including their presence within larger complexes), and the potential relationships of the PCs to other more genetically defined coactivators such as the yeast and human SRB/MED complexes (below).

The earlier-described negative cofactors (for review, see Lee and Young 1998) include human NC1 (Meisterernst et al. 1991), human NC2/Dr1–DRAP1 (Meisterernst and Roeder 1991; Inostroza et al. 1992), human Dr2/PC3/topoisomerase I (Kretzschmar et al. 1993; Merino et al. 1993), and yeast ADI/MOT1 (Auble and Hahn 1993). NC1 and NC2 were found to inhibit basal transcription through direct interactions with TBP (TATA-binding polypeptide) that prevent (or reverse) TBP-TFIIB interactions but are competitive with TFIIA-TBP interactions (Fig. 5), whereas MOT1 acts to inhibit basal transcription through an ATP-dependent reaction that reverses TBP binding to the promoters. Genetic anal-

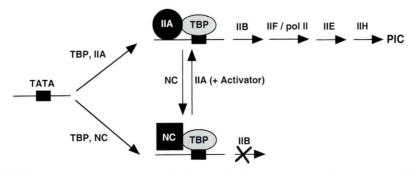

Figure 5. Action of TBP-interacting negative cofactors. As discussed in the text, negative cofactors such as NC1 and NC2 interact with DNA-bound TBP to form complexes that prevent TFIIB binding and subsequent PIC assembly. These interactions are competitive with TFIIA-TBP interactions that, at high TFIIA concentrations, reverse the inhibition and facilitate PIC assembly. The selective effects of NCs on basal transcription suggest that activators (and positive cofactors) help TFIIA overcome the inhibition. For references and other negative cofactors, see text.

yses of yeast MOT1 and subsequently identified yeast homologs of the two human NC2 subunits, as well as the cloning of a human homolog of yeast MOT1, have provided strong support for the general relevance and in vivo functions as global repressors of the biochemically identified negative cofactors (for review, see Lee and Young 1998; Hampsey 1998). These negative cofactors thus provide a potentially general mechanism for the selective and readily reversible suppression of basal activity through TBP/TFIID interactions, perhaps on genes otherwise poised for activation (Hoffmann et al. 1997). However, a recent study has shown that a DNA-bound repressor can interact with and enhance the function of NC2, thus suggesting gene-specific repressor functions involving NC2 as an adapter (Ikeda et al. 1998). At the same time, it also is apparent, most notably from studies of PC3/Dr2/topoisomerase I (Kretzschmar et al. 1993; Merino et al. 1993) and PC4 (Ge and Roeder 1994; Kretzschmar et al. 1994a; Guermah et al. 1998; Malik et al. 1998), that some of the human cofactors may act both to repress basal transcription and to mediate activator function (see also Meisterernst et al. 1991). This phenomenon may be a more general property of cofactors composed of either single polypeptides or groups of polypeptides (below).

BASAL-FACTOR-ASSOCIATED COFACTORS

TFIID-associated TAF$_{II}$s

As mentioned above, the earliest studies of in vitro activation mechanisms for class II genes implicated TFIID, the primary core promoter recognition factor, as a potential target, showing both quantitative (Sawadogo and Roeder 1985; Abmayr et al. 1988; Workman et al. 1988; Lieberman and Berk 1994) and qualitative (Horikoshi et al. 1988; Lieberman and Berk 1994; Chi and Carey 1996) effects of activators on TFIID binding that were correlated with enhanced recruitment or function of other components of the preinitiation complex (Fig. 6) (for review, see Roeder 1996). TFIID is now known to consist of the TATA-binding polypeptide (TBP) and a large number of TBP-associated factors (TAF$_{II}$s) (for review, see Burley and Roeder 1996; Tansey and Herr 1997; Hampsey 1998; Lee and Young 1998). A role for TAF$_{II}$s as coactivators was first suggested by the observations that whereas TFIID could mediate both basal and activator-dependent transcription in assay systems reconstituted with purified factors (Sawadogo and Roeder 1985; Nakajima et al. 1988), only basal activity was observed when TBP replaced TFIID in these systems (Hoffmann et al. 1990; Pugh and Tjian 1990). Further physical and functional analyses of the *Drosophila* and human TFIID complexes (Dynlacht et al. 1991; Tanese et al. 1991; Zhou et al. 1992; Chiang et al. 1993), including the cloning of TAF$_{II}$ cDNAs and the demonstration of specific activator-TAF$_{II}$ interactions that correlated with activator function in vitro (Hoey et al. 1993; for review, see Burley and Roeder 1996; Verrijzer and Tjian 1996), suggested that TAF$_{II}$s are generally required for activator function and led to more refined models emphasizing either specific activator-TAF interactions leading to TFIID recruitment (Sauer et al.

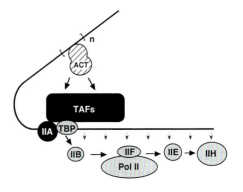

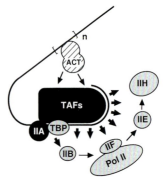

Figure 6. Models for activator function via quantitative versus qualitative effects on TFIID binding to the promoter. In the upper model, emphasizing increased recruitment of TFIID, activator interactions with TBP or TAF$_{II}$s increase promoter occupancy by TFIID, which in turn leads to enhanced PIC formation through the basal interactions established for TBP, TFIIB, Pol II/TFIIF, TFIIE, and TFIIH (Figs. 3 and 8). In the lower model, stressing isomerization of the TFIID-promoter complex, activator interactions result in a topological change in the TFIID-promoter complex that effects a more efficient recruitment of Pol II and the other general factors; this topological change, which likely involves DNA wrapping, results in more extensive DNA-TFIID interactions (more than an 80-bp region) that may create a more stable docking platform for the other factors. Apart from activator interactions with TBP and specific TAF$_{II}$s, the TFIID recruitment and isomerization steps may also involve activator interactions with TFIIA. For references, see text.

1995) or activator interactions leading to TFIID isomerization (Chi and Carey 1996; Oelgeschläger et al. 1996) as the basis for activation (Fig. 6) (for review, see Roeder 1996; Verrijzer and Tjian 1996). Activator-TBP interactions (Stringer et al. 1990) have also been implicated in activation (Wu et al. 1996 and references therein).

More recently, the proposal of a general TAF requirement for activation has been challenged by studies in yeast indicating that TAF$_{II}$s are dispensable for activation of most genes (Moqtaderi et al. 1996; Walker et al. 1996). Significantly, however, in a more recent investigation of human TAF$_{II}$ function in nuclear extracts, which contain a more natural complement of nuclear factors than purified systems, we have observed dramatic TBP-dependent activation of transcription by GAL4-VP16 and GAL4-CTF fusion proteins at the level of PIC formation in the absence of TFIID-specific TAF$_{II}$s (Oelgeschläger et. al. 1998). In the same study, single-round transcription assays also revealed promoter-specific inhibitory effects of TAF$_{II}$s on PIC formation (see also below) that appear to be compensated for by positive effects of TAF$_{II}$s on reinitation under standard (multiple-round) assay conditions. In this regard, it is important to stress that the absolute levels of activity observed in the presence of the activators in standard assays were comparable for equivalent inputs of TBP and TFIID, further indicating that the observed TAF-independent activation was not due to simple antirepression effects from potentially lower basal activities with TBP. Although this analysis has not yet been extended to a broad group of natural activators and promoters, it is important to note that the same activators (GAL4-VP16 and GAL4-CTF) which did not require TFIID-specific TAF$_{II}$s in the nuclear extract assay do require these TAF$_{II}$s when assayed in the standard purified reconstituted system.

Although there are several reasons why TAF$_{II}$s may be essential for activator function in purified reconstituted systems but not in unfractionated HeLa nuclear extracts or in yeast (Oelgeschläger et al. 1998), one simple explanation is that TAF$_{II}$s are functionally redundant with other coactivators that are present in intact cells and nuclear extracts but absent in the purified systems. Indeed, our initial studies have indicated the function of SRB7 (or associated components) in nuclear extracts (Oelgeschläger et al. 1998) as well as the function of an SRB/MED-containing complex in a more purified system (below). Finally, although these studies suggest that human TAF$_{II}$s indeed may not be generally required for activation, in accord with the yeast studies, they do not preclude the possibility that the human TAF$_{II}$s may be required for the function of some activators through the previously proposed coactivator mechanisms involving direct activator-TAF$_{II}$ interactions that result in increased recruitment of TFIID (Abmayr et al. 1988; Workman et al. 1988; Lieberman and Berk 1994; Chi et al. 1995; Sauer et al. 1995) and/or in topological rearrangements (isomerization) of TFIID-promoter complexes (Horikoshi et al. 1988; Lieberman and Berk 1994; Chi and Carey 1996; Oelgeschläger et al. 1996) that each in turn lead to increased recruitment or function of other general factors (Fig. 6). An alternative possibility not involving direct activator-TFIID interactions is that intrinsic interactions of TFIID with broad (80 bp) regions of the promoter (Sawadogo and Roeder 1985; Horikoshi et al. 1988; Nakajima et al. 1988), including DNA wrapping around TFIID (Oelgeschläger et al. 1996), may result in more stable binding of other general transcription factors that directly interact with (and are recruited by) activators; in this scenario, the various interactions could be concerted and mutually stabilized (see also Fig. 8 below). These and other considerations, including the greater stability of TFIID (TBP-TAF$_{II}$ interactions) and the much greater complexity of expressed genes and regulatory programs (e.g., in development) in metazoans relative to yeast, make it probable that TAF$_{II}$s will have more essential functions both as coactivators and as core promoter selectivity factors (below) in metazoans (see also Tansey and Herr 1997).

In addition to possible coactivator functions, TAF$_{II}$s also show core promoter-selective functions. This is most evident from studies showing an absolute TAF requirement for basal transcription from promoters that contain alternate core promoter elements, namely, initiator (Inr) and/or DPE elements (for review, see Smale 1997), in the absence of TATA elements (Martinez et al. 1994; Burke and Kadonaga 1996). Interestingly, basal transcription from TATA-less Inr-containing promoters in human systems also requires additional factors (TICs, for TAF- and Inr-dependent cofactors) that are not required for basal transcription from TATA-containing promoters, but are required for TAF$_{II}$-dependent Inr functions (Fig. 7) (Martinez et al. 1998b). Hence, for such promoters, TAF$_{II}$s and TICs must also be regarded as basal factors. Both TAF$_{II}$s and a subset of TICs are also required for functional synergy between TATA and Inr or DPE elements, indicating core promoter-selective functions on composite core pro-

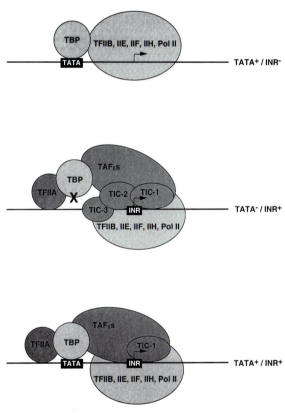

Figure 7. TAF$_{II}$s, TICs, and TFIIA have core promoter-selective functions in basal transcription. In purified systems, TBP, TFIIB, TFIIE, TFIIF, TFIIH, and Pol II suffice for basal transcription from conventional TATA-containing promoters (Orphanides et al. 1996; Roeder 1996). In contrast, basal transcription from TATA-less promoters containing pyrimidine-rich Inr elements requires additional factors (darker shaded symbols) that include TFIIA, TAF$_{II}$s, and three novel, partially purified factors designated TIC-1, TIC-2, and TIC-3 (Martinez et al. 1998b and references therein). On composite core promoters containing both TATA and Inr elements, the synergistic function of Inr and TATA elements (Smale 1997) requires TFIIA, TAF$_{II}$s, and TIC-1, but not TIC-2 and TIC-3. The $\times$ in the center panel indicates that the TATA-binding function of TBP is not required for TFIID function on TATA-less Inr-containing core promoters (Martinez et al. 1995). The specific factor(s) that directly contacts the Inr element, presumably a TAF or TIC component, is unknown. For additional references, see text.

moters (Fig. 7) (Kaufmann and Smale 1994; Martinez et al. 1994, 1995, 1998b; Verrijzer et al. 1995; Burke and Kadonaga 1996). Core-promoter-selective effects of metazoan $TAF_{II}s$ are also evident from the ability of $TAF_{II}s$ either to inhibit (e.g., on weak TATA-containing promoters lacking Inr or DPE elements) or to enhance (e.g., on strong TATA- and Inr-containing promoters) the binding and/or function of TBP in basal transcription (Nakatani et al. 1990; Verrijzer et al. 1995; Emami et al. 1997; Guermah et al. 1998; Oelgeschläger et al. 1998). Recent genetic studies in yeast suggesting that TAF_{II} requirements for specific genes are dictated by (as yet undetermined) core promoter elements, rather than by upstream activating sequences (Shen and Green 1997), are consistent with our earlier in vitro studies showing core promoter-selective functions of metazoan $TAF_{II}s$.

At least some of the positive core-promoter-selective functions of $TAF_{II}s$ appear to be mediated by direct TAF_{II}-DNA contacts (Kaufmann and Smale 1994; Purnell et al. 1994; Verrijzer et al 1994; Oelgeschläger et al. 1996; Burke and Kadonaga 1997). In contrast, the inhibitory effects of $TAF_{II}s$ on TBP binding and function may be due in large part to interactions of the amino terminus of the largest TAF_{II} (*Drosophila* $TAF_{II}230$/human $TAF_{II}250$/ yeast $TAF_{II}145$) with TBP (Kokubo et al. 1994). Interestingly, this appears to involve interactions with two distinct regions of TBP that show competitive interactions, respectively, with the activator VP16 (Nishikawa et al. 1997) and with TFIIA (Kokubo et al. 1998). More compelling evidence for this inhibitory effect is provided by a recent high-resolution nuclear magnetic resonance (NMR) structure showing that a part of the $TAF_{II}230$ inhibitory domain mimics the (unwound) TATA element in binding to the underside of TBP (Liu et al. 1998). Hence, the reversal of this promoter-specific TAF_{II}-mediated constraint to TBP function may be a significant point of control either for specific activators and coactivators or for specific core promoter elements.

In the context of TFIID, $TAF_{II}s$ thus appear to have coactivator functions mainly on the basis of in vitro studies with metazoan factors, whereas core-promoter-selective functions are evident both from in vivo studies in yeast (above) and from in vitro (above) and in vivo (Suzuki-Yagawa et al. 1997; E.H. Wang et al. 1997) studies in metazoans. Interestingly, recent studies (Grant et al. 1998; Martinez et al. 1998a; Ogryzko et al. 1998) have shown that a subset of $TAF_{II}s$, most notably those with histone folds (for review, see Burley and Roeder 1996), are also present in distinct coactivator complexes that contain histone acetyltransferases. This may reflect specialized functions of these $TAF_{II}s$, either in complex formation or in subsequent DNA or chromatin interactions (Hoffmann et al. 1997), that are shared by both TFIID and TAF-containing HAT complexes.

TFIIA

TFIIA was originally categorized as a general (basal) factor from studies in partially purified systems in which the presence of TBP-interacting negative cofactors such as NC1 or NC2 may have necessitated the action of TFIIA to reverse their inhibitory effects on basal transcription (for review, see Orphanides et al. 1996). This effect is mediated through competitive interactions between TFIIA and NCs with a common site on TBP (see Fig. 5) (Meisterernst and Roeder 1991; Meisterernst et al. 1991; Kim et al. 1995). As mentioned above, TFIIA may likewise have a role in countering inhibitory interactions of the amino terminus of a specific TAF (yTAF$_{II}$145/ dTAF$_{II}$230/hTAF$_{II}$250) on TBP binding to some promoters (Kokubo et al. 1998). In more purified systems, TFIIA (like $TAF_{II}s$) is often required for optimal activator-dependent transcription but not for basal transcription from standard TATA-containing promoters, thus indicating that on such promoters TFIIA acts more like a conventional coactivator than an essential basal factor (Meisterernst et al. 1991; for review, see Roeder 1996). This notion is supported by recent reports of activator interactions with TFIIA that enhance TFIID binding to target promoters (Ozer et al. 1994; Kobayashi et al. 1995) and of general coactivator (PC4) interactions with TFIIA that may enhance TFIID binding (Ge and Roeder 1994; Kretzschmar et al. 1994a; Kaiser et al. 1995). The activator/coactivator interactions with TFIIA could stimulate the intrinsic ability of TFIIA to counter the action of TBP-interacting negative cofactors (above) or to enhance TBP/TFIID binding (for review, see Orphanides et al. 1996; Roeder 1996) and isomerization of TFIID-promoter complexes (Chi and Carey, 1996; Oelgeschläger et al. 1996) through either TBP or $TAF_{II}110/135$ interactions (for review, see Burley and Roeder 1996).

Our more recent studies have shown that like TAFs, TFIIA is also required for basal transcription from TATA-less Inr-containing promoters in purified systems, indicating that it should also be considered a core promoter-specific basal factor (Fig. 7) (Martinez et al. 1998b). Thus, depending on promoter context and the presence of various negative cofactors, TFIIA may either act as a coactivator that directly (through TFIID interactions) or indirectly (through effects on negative cofactors) enhances TBP/TFIID functions or as a true basal factor that presumably enhances TFIID and/or TIC functions through Inr elements. Biochemical and genetic analyses of TFIIA have shown that the direct coactivator function can be separated from both the antirepression function and the TBP-binding function (Kang et al. 1995; Ma et al. 1996; Ozer et al. 1996).

SRB and MED Components

Recent genetic and biochemical studies in yeast have identified a group of factors, including both *s*uppressors of *R*NA polymerase *B*(II) mutations (SRBs) and other associated factors (MED proteins, GAL11, SIN4, RGR1, PGD1, and ROX3), that interact reversibly with RNA polymerase II (to form a holoenzyme) and serve as important cofactors (positive and negative) for various activators and repressors (Thompson et al. 1993; Kim et al. 1994; Koleske and Young 1994; for review, see Koleske and Young 1995; Hampsey 1998; Myers et al. 1998).

Subsequent studies in mammalian cells identified various RNA polymerase II complexes that contained human homologs of yeast SRB7, SRB10 (a cyclin-dependent kinase) and SRB11 (an SRB10-associated cyclin), as well as various general transcription factors and other cofactors (for review, see Orphanides et al. 1996; Greenblatt 1997; Hampsey 1998). Human homologs of MED6 and MED7 have also been reported (Lee et al. 1997; Myers et al. 1998). These results, and our recent demonstration (above) of TAF-independent transcriptional activation in human cell-free systems, raised the possibility that a human complex analogous to the yeast mediator complex might also be active in systems not requiring human TAF$_{II}$s. Indeed, human SRB7 immunodepletion studies have implicated hSRB7 and/or associated proteins in activator-dependent transcription in HeLa nuclear extracts (Oelgeschläger et al. 1998).

In more recent studies (Gu et al. 1999), we have used cell lines expressing epitope-tagged hSRB7, hSRB10 (CDK8), and hSRB11 (cyclin C) to affinity-purify an *S*RB and *MED* *c*ofactor *c*omplex (SMCC) that has an approximate size of 1.5 MD and, on the basis of peptide sequence and immunoblot analyses, contains human homologs of several yeast mediator components (SRB7, SRB10, SRB11, MED6, MED7, and RGR1), the yeast negative regulator NUT2 (Tabtiang and Herskowitz 1998), and the yeast positive regulator SOH1 (Fan et al. 1996). This complex lacks RNA polymerase II, general transcription factors, and other factors (CBP, BRCA-1, and SWI2) variably detected in mammalian holoenzyme preparations (for review, see Chang and Jaehning 1997; Greenblatt 1997; Hampsey 1998). Somewhat surprisingly, in a cell-free system reconstituted with RNA polymerase II, general initiation factors, and PC4, SMCC was found to inhibit activator-mediated transcription. This result is consistent with reports that SRB10 and SRB11 act mainly as negative regulators in vivo (for review, see Carlson 1997) and with a recent report that the SRB10 kinase can inhibit holoenzyme/mediator function in vitro by phosphorylation of the RNA polymerase II CTD prior to PIC formation (Hengartner et al. 1998). However, repression by SMCC occurred in the human reconstituted system independently of the presence of the CTD, was not observed when PC4 was replaced by PC2 (which also serves as an effective coactivator), and was correlated with an ability of SMCC to very actively phosphorylate PC4. Since PC4 phosphorylation is known to inactivate the coactivator function (Ge et al. 1994; Kretzschmar et al. 1994a), these results suggest both that the SMCC-mediated repression may involve PC4 phosphorylation and that this may serve as a novel pathway for repression in the cell. However, given that a number of yeast mediator components other than SRB10 are involved in repression (for review, see Carlson 1997; Hampsey 1998; Myers et al. 1998), other components of SMCC (e.g., RGR1 or NUT2) may also exert repressive effects by as yet unknown mechanisms. Although the observation of both activated transcription and SMCC-mediated repression in the absence of the CTD was somewhat surprising in view of the CTD requirements for yeast mediator function

(Kim et al. 1994; Meyers et al. 1998), the finding of intragenic (as well as extragenic) suppressors of RPB1 CTD truncations argues for other RNA polymerase II–mediator interactions as well (Nonet and Young 1989). The observations of genetic interactions between SOH1 and RPB1 and RPB2 in yeast (Fan et al. 1996) suggest that the human SOH1 in SMCC may have such a function.

Consistent with documented coactivator functions for the yeast mediator complex, and the presence in SMCC of components (e.g., MED6 and SOH1) implicated in gene activation in yeast, we also have shown that SMCC possesses coactivator functions (Gu et al. 1999). In our highly purified system, maximal SMCC coactivator activity was elicited either at limiting concentrations of TFIIH or in its absence. Under these conditions, SMCC exhibited both an intrinsic (albeit weak) coactivator function in the absence of any other general coactivators (specifically PC4 or PC2) and a marked synergism with PC4 in mediating activator function, but had little effect on basal transcription in the presence or absence of PC4. Thus, SMCC shows some of the coactivator properties attributed to the yeast SRB/mediator complex, although the latter was shown to have a significant effect on basal transcription as well as activator function (Kim et al. 1994). Our results further suggest that SMCC has a bona fide coactivator activity that, in some cases, may operate independently of TFIIH, with which it shares similarities that include the presence of a cyclin-kinase pair and the ability to phosphorylate the RNA polymerase II CTD. Consistent with this possibility, mutations in the TFIIH kinase function do not appear to impair activator-dependent transcription of certain genes either in vitro (Mäkelä et al. 1995) or in yeast (Lee and Lis 1998; McNeil et al. 1998) or *Drosophila* (Larochelle et al. 1998). At the same time, our in vitro results do not exclude the possibility that with other general coactivator(s) in place of or in addition to PC4, or with other physiological constraints on specific TFIIH (or SMCC) inhibitory activities, SMCC may also operate as a coactivator in the presence of or in conjunction with TFIIH. The inhibition observed when both TFIIH and SMCC are present in our highly purified reconstituted system could represent, for example, dosage-dependent synergistic effects of associated kinases on phosphorylation and inactivation of PC4 or another essential factor that may not occur under more physiological conditions.

These analyses thus provide evidence for a novel human complex, with a small subset of components homologous to yeast SRB/mediator components, that shows intrinsic repression and coactivator functions in vitro. As originally proposed for the yeast SRB/mediator complex, SMCC most likely reflects a complex involved in processing both positive and negative signals from corresponding transcriptional activators and repressors (see Fig. 3). On the basis of studies of the yeast SRB/mediator (Hengartner et al. 1995; Koh et al. 1998; Song and Carlson 1998), it is anticipated that its mechanism of action will involve, at least in part, interactions with specific DNA-bound activators (or repressors) and consequent re-

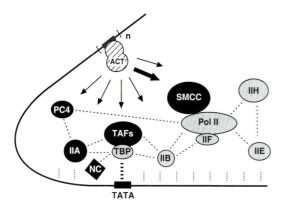

Figure 8. Model for transcriptional activation emphasizing primary activator interactions with the human SRB- and MED-containing cofactor complex (SMCC). The SMCC interactions (*thick arrow*) may facilitate recruitment of RNA polymerase II and interacting factors and may act in concert with secondary activator interactions (*thin arrows*) involving one or more other factors (e.g., PC4, TFIIA, TFIID, TFIIB, and TFIIH). Multiple interactions may account for synergy between multiply bound activators. Activator interactions may facilitate not only the formation (through recruitment), but also the subsequent function, of the PIC, and in either case may involve not only interactions with positive factors, but also an inhibition of negative regulatory cofactors, such as NCs or SMCC components, that may otherwise restrict the function of the general transcriptional machinery. Similarly, DNA-bound repressors (not shown) may act either by directly inhibiting the function of the general transcriptional machinery or by stimulating the function of negative regulatory cofactors. (*Dashed lines*) Specific protein-protein and protein-DNA interactions; (*small dotted lines*) largely nonspecific DNA-protein interactions within the PIC. For references and further discussion, see text.

cruitment (or inhibition) of RNA polymerase II and interacting general factors. Consistent with this idea, very recent studies (cited in proof) have shown direct interactions of SMCC with activators (p53, VP16) whose functions are dependent on SMCC in assays reconstituted with purified factors (Gu et al. 1999; M. Ito et al., unpubl.). Thus, the model in Figure 8 emphasizes dominant interactions (thick arrow) with RNA polymerase-II-associated SMCC factors as a basis for activator function. It also indicates the possibility of other activator interactions that may act in concert with activator-SMCC interactions, perhaps in an activator-specific manner, to facilitate PIC formation or function (for discussion, see Wu et al. 1996; Koh et al. 1998). Similarly, interactions of DNA-bound repressors or corepressors with SMCC components implicated in repression may also facilitate gene-specific repression, although effects of yeast mediator mutations on derepression of reporter genes lacking upstream activation sequences indicate more global negative regulatory functions for some yeast mediator components (for review, see Carlson 1997). The human SMCC complex also contains a number of components that, based on peptide sequence and yeast database analyses, have no yeast counterparts (Gu et al. 1999). Hence, SMCC asppears to be only distantly related to the yeast mediator and may be expected to have novel functions

that could include interactions with metazoan-specific regulatory factors. Future studies of human SMCC must determine the specificity and mechanism of action of this complex in conjunction with more natural DNA-binding activators and repressors and natural templates.

ACTIVATOR-ASSOCIATED OR ACTIVATOR-RECRUITED COACTIVATORS

Oct-1-interacting Coactivators OCA-B and OCA-S

The octamer element ATTTGCAT is a key control element in a number of genes that are regulated in different ways through this element (for review, see Luo et al. 1992). These include the immunoglobulin (Ig) promoters that are activated in B lymphoid cells and the histone H2B promoters that are activated in S phase of the cell cycle (Fig. 9). B cells contain a B-cell-enriched octamer-binding protein (Oct-2) that had been assumed to be the cell-specific regulatory factor for Ig promoters (for re-

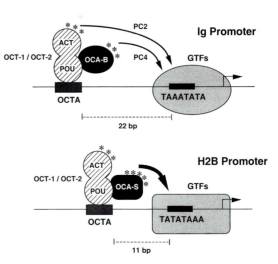

Figure 9. Models for cell-specific coactivator functions in the differential activation of Ig and H2B promoters through a common upstream regulatory element. The Ig and H2B promoters are activated in B cells and in S phase of the cell cycle, respectively, through a common octamer element (OCTA) that is recognized either by the ubiquitous Oct-1 or by the related B-cell-enriched Oct-2. However, activation of the Ig promoter is critically dependent on the B-cell-specific coactivator OCA-B, whereas activation of the H2B promoter is mediated by OCA-S. OCA-B contains both a POU-interaction domain that is involved in promoter recruitment and an activation domain (*asterisks*) that acts synergistically with an Oct-1 (or Oct-2) activation domain (*asterisks*). In a purified system, optimal Oct-1 and OCA-B function requires both PC2 and PC4, the latter showing direct interaction with the OCA-B activation domain. OCA-S functions optimally with just the POU domain of Oct-1, suggesting that it contains a more potent activation domain (*asterisks*), although a direct physical interaction with the POU domain (as indicated) has not yet been shown. The selective functions of OCA-B and OCA-S with a common activator on the Ig and H2B promoters appear to depend on core promoter sequences, most notably consensus (H2B) versus nonconsensus (Ig) TATA elements, rather than the characteristic distances between the octamer and TATA elements. The different shapes of the PICs on the Ig and H2B core promoters indicate presumed, but as yet unknown, differences in composition and/or structure. For references, see text.

view, see Staudt and Lenardo 1991). However, our early biochemical work showed that the major determinant for B-cell-specific transcription of Ig promoters in vitro is a B-cell-specific coactivator, designated OCA-B, that can act in conjunction either with the ubiquitous octamer-binding activator Oct-1 (to which it is preferentially bound in B cells) or with Oct-2 (Luo et al. 1992). This finding provided the first example of a dominant tissue-specific promoter regulatory mechanism for which the major determinant is a non-DNA-binding coactivator, rather than a tissue-specific DNA-binding protein, and this mechanism has proved to be a more general paradigm for cell-specific regulatory mechanisms.

Further mechanistic studies have shown that OCA-B is recruited to the promoter through interactions of a specific OCA-B domain with the DNA-binding domain (POU domain) of Oct-1 and that activation domains on Oct-1 and OCA-B act synergistically to activate Ig promoters (Luo et al. 1998 and references therein) (Fig. 9). We have further shown that optimal Oct-1 and OCA-B function in a purified cell-free system requires both PC2 and PC4 and that the activation domain of OCA-B appears to act through direct interactions with PC4 or a functional equivalent (Luo et al. 1998). The relevance of OCA-B to Ig heavy chain gene activation in vivo has been demonstrated by targeted gene disruption in mice, although the requirement for OCA-B function is restricted to the antigen-dependent phase of B-cell differentiation (for references, see Luo et al. 1998). This could represent either a true stage-specific requirement that reflects variable enhancer usage during B-cell differentiation (Ong et al. 1998 and references therein) or the presence of a compensatory coactivator (possibly related to OCA-B) during the antigen-independent B-cell differentiation state when OCA-B is not required.

In the case of the H2B promoter, our earlier studies showed that activation through the octamer element requires Oct-1 but is unaffected by OCA-B (Luo et al. 1992; Luo and Roeder 1995), whereas more recent studies have identified a novel activity (OCA-S) that activates the H2B promoter in conjunction with Oct-1 but is without effect on Ig promoter function through the octamer element (Y. Luo and R.G. Roeder, unpubl.). Although the composition of OCA-S is not firmly established, it appears to be a multicomponent complex (Y. Luo and R.G. Roeder, unpubl.) and to show full activity on the H2B promoter in conjunction with just the DNA-binding (POU) domain of Oct-1 (Luo and Roeder 1995). Hence, if OCA-S is indeed recruited to the promoter through Oct-1, it presumably has a strong activation domain(s) that operates independently of the defined Oct-1 activation domains (Fig. 9).

Although the identification and characterization of OCA-B and OCA-S have provided major insights into the biological and molecular basis of the differential cell-specific regulation of the Ig and H2B promoters, the fact that the cognate octamer regulatory elements, and thus the corresponding Oct-1 complexes, are identical fails to explain the molecular basis for differential coactivator function on the two promoters. The fixed 11-bp distance between the

H2B octamer and TATA elements and the larger, variable ($\geq$20 bp) distances between the Ig octamer and TATA elements (Fig. 9) suggested a topological basis for this discrimination. However, our preliminary studies have indicated that the basis for coactivator specificity lies mainly in the Ig and H2B core promoter regions, and most likely in the indicated divergent TATA elements (Y. Luo et al., unpubl.). Whereas the H2B promoter contains a strong consensus TATA, Ig promoters in general contain weak nonconsensus TATA elements. This interesting observation reemphasizes the importance of understanding the structure and function of core promoter elements and corresponding preinitiation complexes, especially in light of the emerging view of TAF$_{II}$ functions in core promoter selectivity (above). We speculate that differences, perhaps subtle, in the structure, rate of formation, or stability of these complexes (Fig. 9) may determine whether Oct-1–coactivator interactions are productive. Previous studies also have emphasized the importance of specific core promoter elements for the function of specific upstream regulatory elements and associated factors (for review, see Martinez et al. 1995; Smale 1997).

Thyroid Hormone Receptor-associated Polypeptides

Nuclear hormone receptors comprise a superfamily whose individual members bind to and, in a ligand-dependent manner, activate transcription of specific genes that regulate diverse physiological processes such as cell growth, differentiation, and homeostasis (for review, see Mangelsdorf and Evans 1995). These receptors typically contain an amino-terminal activation domain, a central DNA-binding domain, a carboxy-terminal ligand-binding domain, and a carboxy-terminal activation domain whose function is dependent on conformational changes associated with ligand binding. Despite the specificity intrinsic to individual receptors and cognate ligands and DNA-binding sites, as well as the complexity of the general transcriptional machinery that serves as the ultimate target of nuclear receptors, other essential coactivators have been identified by a variety of experimental approaches and corresponding functions demonstrated by in vitro and/or in vivo assays (for review, see Glass et al. 1997).

Our earlier studies showed that the thryoid hormone receptor (TR) and its DNA-binding partner RXR have minimal activity, even in the presence of ligand (T_3), in a cell-free system composed of general transcription factors, the USA coactivator fraction, and a DNA template with TR-binding sites. However, it was possible to isolate *TR-associated proteins* (TRAPs) by affinity purification of an epitope (FLAG)-tagged receptor from cells grown in the presence of ligand (Fondell et al. 1996). This TR-TRAP complex (see Fig. 3), containing nine to ten major TRAPs ranging in size from 80 to 240 kD, was shown to markedly enhance transcription from the receptor (relative to TR and T_3 alone) in a manner dependent on RXR and the general coactivator USA. At the same time, studies from other laboratories identified a number of distinct coactivators with broad receptor specificities (for review, see Glass et al. 1997; Voegel et al. 1998; Yuan et al. 1998). The best characterized of these include (1) CBP and p300, which also serve as coactivators for a diverse group of activators (Goldman et al. 1997), and (2) the SRC family of coactivators (Oñate et al. 1995), which appear to function mainly with receptors. These coactivators showed ligand-dependent receptor interactions and in vivo functions with a broad group of nuclear hormone receptors that included TR but, on the basis of peptide sequence and immunoreactivity assays, are distinct from the TRAPs and not present in our in vitro assays (Fondell et al. 1999).

The diversity of the coactivators implicated in nuclear receptor function was initially surprising, and seemingly contradictory. However, the TRAP functions were demonstrated in vitro on naked DNA templates, whereas p300/CBP and SRC family members, and the interacting PCAF, were shown to regulate receptor function in vivo, presumably on natural chromatin templates, and to have histone acetyltransferase (HAT) activities that, on related coactivators, have been implicated in chromatin remodeling associated with gene activity (for review, see Struhl 1998). These considerations, along with demonstrations of ligand-independent interactions of nuclear receptors with various corepressors (Heinzel et al. 1997; Nagy et al. 1997), led us to propose a multistep model that is compatible with all of the available data. Major features of this model, presented in Figure 10, include (1) ligand-independent binding of nuclear receptors (to target DNA sites within chromatin) along with corepressors (SMRT/ NCoR, SIN3, and histone deacetylases) that help maintain a repressed state by deacetylation of nucleosomal histones; (2) ligand-mediated dissociation of (co)repressors plus concomitant binding of coactivators (SRC-1 related factors, p300/CBP, and PCAF) that contain, or interact with factors that contain, HAT activity, with resulting acetylation of nucleosomal histones or possibly other factors (below); and (3) binding of TRAPs (or a receptor-TRAP complex), perhaps with displacement of other coactivators (or a receptor-coactivator complex), and subsequent interactions with general initiation factors or coactivators. This latter step could also involve direct ligand-independent interactions of nuclear receptors with general initiation factors (for review, see Voegel et al. 1998).

In more recent studies of the structure and function of the TRAPs, we have verified by molecular cloning and subsequent characterization that they are distinct from other nuclear receptor coactivators, that they reside in a single large complex in the absence of TR, and that they are anchored to the receptor through ligand-dependent interactions between the TRAP220 subunit and the receptor carboxyl terminus (Fig. 10) (Yuan et al. 1998). Interestingly, these interactions appear to be mediated through LXXLL motifs (in TRAP220) that have been implicated in ligand-dependent interactions of other coactivators with nuclear hormone receptors (for review, see Voegel et al. 1998). In vivo functions of TRAP220 have been verified by both transfection analyses and the use of dominant negative forms of TRAP220 containing the LXXLL motifs (Yuan et al. 1998).

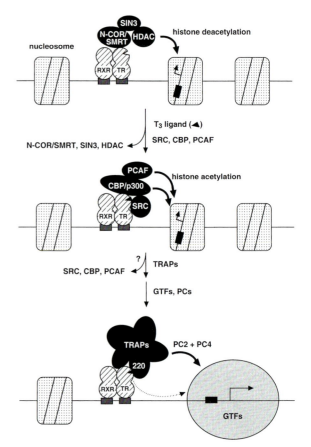

Figure 10. Model for a multistep pathway involving distinct sets of cofactors in target gene regulation by thyroid hormone receptor. The model depicts (1) a repressed state in which unliganded thyroid hormone receptor (TR), bound in association with RXR, interacts with a corepressor (N-COR or SMRT) that, through an intermediate (SIN3), recruits a histone deacetylase (HDAC) that acts on histones within adjacent nucleosomes; (2) an intermediate state, induced by ligand binding to TR and dissociation of corepressors, in which TR-bound coactivators (CBP/p300, PCAF, and SRC) with histone acetyltransferase activities acetylate histones in adjacent nucleosomes; and (3) an activated state in which TR-bound TRAPs have replaced (or bound simultaneously with) the other TR-bound coactivators and, in conjunction with PC2 and PC4, facilitate formation of a functional PIC containing the general transcriptional machinery (Fondell et al. 1996; Yuan et al. 1998). (*Dashed line*) A low, ligand-dependent activity of TR alone in conjunction with PC2 and PC4. The ligand-dependent interaction of TR with the TRAP complex involves direct interactions of TRAP220 with TR. The fact that TRAP220 interacts in a ligand-dependent manner with a number of nuclear hormone receptors indicates a broader function of the TRAP complex (Yuan et al. 1998). For other references, see text. (Reprinted, with permission, from Fondell et al. 1999.)

Although the receptor binding and coactivator functions of the TRAPs were originally described for TR, recent studies have also shown ligand-dependent interactions of TRAP220 not only with TR, but also with RARα, RXRα, VDR, PPARα, PPARγ, and ERα (Yuan et al. 1998). This suggests a much broader specificity for TRAPs, as reported for other coactivators (above), and studies with the dominant negative form of TRAP220 are consistent with this view. In view of the target gene DNA-binding and activation specificity intrinsic to the various receptors and cognate ligands, the question arises as to the reason for the unexpected complexity of nuclear receptor coactivators (most notably the 9–10 subunit TRAP complex). This most likely relates to the extreme diversity of physiological processes that involve nuclear receptor function and the evolution of secondary regulatory mechanisms that are mediated through other (parallel) signaling pathways that affect the abundance or activity of these cofactors. Thus, cell- or cell-state-specific variations in coactivators, as originally observed for the B-cell-specific OCA-B (Luo et al. 1992), could help explain various cell and promoter-specific effects of nuclear receptor ligand functions.

Finally, the complexity of receptor-interacting coactivators, most notably the 9–10-subunit TRAP complex, may also reflect a functional redundancy with other coactivators, such as those associated with the general transcriptional machinery (see Fig. 3). Indeed, recent studies suggest that the TAF_{II} components of TFIID are dispensable for TR-TRAP function in vitro (Fondell et al. 1999). However, TR-TRAP function still requires USA components, and, in this more physiological assay for activator function, the USA-derived PC2 and PC4 components show synergistic effects. In this regard, it will be important to understand not only the mechanisms involved in the transitions between the various corepressor and coactivator complexes in the model of Figure 10, but also the actual mechanisms by which the cofactors modulate the structure and function of chromatin and of the general transcriptional machinery.

In summary, the identification of the TRAP complex provides an example of a very complex and novel coactivator that is recruited to the promoter through primary, and apparently rather stable, interactions with a conventional DNA-binding activator. The continued dissection of its structure and function should provide new insights not only into specific aspects of nuclear hormone receptor function, but also into transcriptional regulation in general, since it seems probable that some components of the TRAP complex, or even the TRAP complex itself, will be involved in the function of other activators. Indeed, very recent studies (cited in proof) have indicated a close relationship (if not near identity) between recently isolated TR-TRAP complexes (containing an additional complement of small polypeptides) and the SMCC complex, as well as TR-TRAP complex-mediated activation by Gal4-VP16 and Gal4-p53 and SMCC-mediated activation by TR (Gu et al. 1998; M. Ito et al., unpubl.). This surprising but highly significant finding represents a satisfying convergence of different coactivator studies and emphasizes an apparently major role for the above-described SMCC/TRAP complexes in activator function, as well as a mechanism involving stable activator-coactivator interactions prior to DNA binding.

CBP/p300 as a Coactivator for p53

CBP is one of the earliest-described coactivators that is recruited to target promoters through interactions with DNA-binding activators (Chrivia et al. 1993) and, with its close relative p300, is now recognized as a coactivator for a large and diverse group of activators (for review, see Goldman et al. 1997). As first indicated for the yeast coactivator GCN5 (Brownell et al. 1996; Kuo et al. 1998), CBP/p300 has a HAT activity (Ogryzko et al. 1996) that is now known to be essential for its transcription function (Martinez-Balbas et al. 1998). Along with observations that transcriptional activation of specific genes is correlated with hyperacetylation of associated histones (for review, see Turner and O'Neill 1995), these results have led to the view that the coactivator function of CBP/p300 involves targeted acetylation of histones and that these chromatin modifications allow functional interactions of other transcriptional factors (for review, see Struhl 1998). However, other functions of CBP/p300, perhaps as components of an RNA poly-

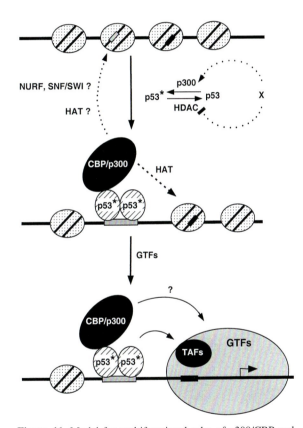

Figure 11. Model for multifunctional roles of p300/CBP and p53 in p53-mediated activation of target genes. In response to a signal (x) such as DNA damage, the latent form of p53 interacts with and is acetylated by p300, with a concomitant increase in DNA-binding activity. Activated p53 (p53*) binds stably to target gene sites, which may depend on chromatin remodeling by NURF or SWI/SNF complex activities (or possibly the p300/CBP HAT activity), and bound p300/CBP then modifies adjacent nucleosomes through its HAT activity. This in turn facilitates interactions of p53* (e.g., with specific TAF_{II}s) that enhance formation of a functional preinitiation complex. In this model, p53 and p300 both have dual and interrelated functions in the overall activation pathway: p300 to activate p53 and to remodel chromatin, and p53 to recruit both p300 and components of the PIC. As indicated by the arrow with the question mark, CBP/p300 may also function through interactions with components (notably Pol II) of the general transcription machinery. For references, see text.

merase II holoenzyme, are also indicated (Nakajima et al. 1997).

Our own studies of CBP/p300 have focused on activation by the tumor suppressor p53, a DNA-binding transcriptional activator that was found by us and others to interact physically with CBP/p300 in vitro and to function synergistically with CBP/p300 in vivo (for review, see Gu and Roeder 1997). We also found that p53 acts synergistically with CBP when the latter is tethered to the promoter as a GAL4-CBP fusion protein (Gu et al. 1997). This result indicated that p53 does not function exclusively to recruit CBP/p300 to the promoter, consistent with previous studies indicating a (secondary) function of p53 involving interactions of its activation domain with TAF_{II} components of TFIID (for review, see Levine 1997; Gu et al. 1997). Since p53 has a latent DNA-binding activity that is both activated in response to cellular stress signals and important for activation of target genes (for review, see Levine 1997), these studies suggested a model involving binding of activated p53 to target gene sites within a chromatin template, perhaps with the assistance of nucleosome remodeling factors such as NURF or SWI-SNF complexes (for review, see Workman and Kingston 1998), followed by CBP/p300 recruitment, CBP/p300-mediated acetylation of adjacent nucleosomal histones, and recruitment of TFIID through $p53$-TAF_{II} contacts (Fig. 11).

Previous studies have shown that the DNA-binding activity of p53 is regulated (inhibited) by a highly basic carboxy-terminal regulatory domain and that modification of this region by in vitro phosphorylation, antibody binding, or deletion can activate the (latent) DNA-binding activity (for review, see Levine 1997). In more recent studies of this phenomenon, in conjunction with coactivator interactions, we discovered that p300 could acetylate human p53 both in vitro and in vivo and that the acetylation site is present in the carboxy-terminal regulatory domain and conserved in p53 from other species. Most remarkably, this acetylation was found to stimulate dramatically (>20-fold) the DNA-binding activity of an otherwise inactive form of p53 (Gu and Roeder 1997). More recently, we have found that the acetylated form of p53 can be deacetylated by an affinity-purified histone deacetylase complex (W. Gu and R.G. Roeder, unpubl.). Thus, these observations indicated a novel pathway for p53 activation and suggested a modified model for p53-mediated activation of target genes. In this model (Fig. 11), stress signals, such as DNA damage, somehow activate p300/CBP and/or inhibit histone deacetylases acting on acetylated p53, leading to an increase in the steady-state level of active p53 that in turn binds to DNA sites in association with p300/CBP. Subsequent secondary functions of p300 (e.g., histone acetylation) and p53 (e.g., TAF_{II} interactions) then lead to target gene activation. Confirmation of this model awaits further tests, both in vivo and in vitro, with chromatin templates, mutant versus wild-type forms of p53 and p300, and additional general coactivators. Given our newer results (above) on TAF_{II}-independent activation and the human SRB- and MED-containing complex (SMCC), it will be important to determine which of these components are essential and whether there is a functional redundancy in coactivators indicative of different p53-mediated gene activation pathways.

Although these analyses have contributed significantly to our understanding of p53 function, they also have provided the first example of an acetylation-mediated change in the function of a nonhistone transcriptional regulatory protein. As such, they have significant implications regarding potentially more diverse molecular mechanisms for the expanding group of acetyltransferase-containing coactivators whose primary (or exclusive) targets have been presumed to be histones. We predict that other cellular regulatory proteins will prove to be functionally modified in a similar manner. Indeed, recent studies have demonstrated p300-mediated acetylation of nuclear hormone receptors (W. Gu et al., unpubl.), although functional relevance has not yet been demonstrated.

CONCLUSIONS AND PERSPECTIVES

Since the discovery of multiple eukaryotic nuclear RNA polymerases almost 30 years ago, various biochemical and genetic analyses have given us a deeper understanding of the structure, function, regulation, and mechanism of action of these RNA polymerases, their cognate accessory factors, and an ever-increasing group of gene-specific regulatory factors. Allowing for the much greater complexity of both the RNA polymerases (12 subunits for RNA polymerase II) and their accessory factors (~32 polypeptides for the class II factors), which are formally equivalent to bacterial core RNA polymerases and singly associated σ factors, many of the fundamental principles of regulatory factor-RNA polymerase/accessory factor interactions seen in bacteria also apply in the case of eukaryotes. At the same time, intense activity in the past few years has revealed a surprisingly complex array of cofactors, principally for class II genes, that are needed for effective communication between gene-specific regulatory factors and the general transcriptional machinery. These have functions both in regulating the accessibility of DNA control elements within chromatin and in modulating the binding and function of the general transcriptional machinery. Whereas biochemistry heretofore has had the leading role in the identification and mechanistic analyses of these factors, genetics has had an increasingly important role in these efforts as well, and in providing essential in vivo checks on functions deduced from in vitro analyses. Although the true diversity, mechanism of action, genetic specificity, and potential redundancy of such cofactors remain to be established, it is clear that they represent an important layer of complexity for eukaryotic gene control that is not seen in bacteria and that must be investigated, case by case, if the physiological regulation of specific eukaryotic genes is to be understood and selectively manipulated.

ACKNOWLEDGMENTS

I thank members of my laboratory for their contributions to the work summarized here, for their permission to cite unpublished work, and for their critical comments on the manuscript. I am especially indebted to Zhengxin Wang for preparation of the figures. Work from my laboratory was supported by grants from the National Institutes of Health.

REFERENCES

Abmayr S.M. , Workman J.L. and Roeder R.G. 1988. The pseudorabies immediate early protein stimulates in vitro transcription by facilitating TFIID:promoter interactions. *Genes Dev.* **2:** 542.

Auble D.T. and Hahn S. 1993. An ATP-dependent inhibitor of TBP binding to DNA. *Genes Dev.* **7:** 844.

Bieker J.J., Martin P.L., and Roeder R.G. 1985. Formation of a rate-limiting intermediate in 5S RNA gene transcription. *Cell* **40:** 119.

Brownell J.E., Zhou J., Ranalli T., Kobayashi R., Edmondson D.G., Roth S.Y., and Allis C.D. 1996. *Tetrahymena* histone acetyltransferase A: A homolog to yeast Gen5p linking histone acetylation to gene activation. *Cell* **84:** 843.

Buratowski S., Hahn S., Guarente L., and Sharp P.A. 1989. Five intermediate complexes in transcription initiation by RNA polymerase II. *Cell* **56:** 549.

Burke T.W. and Kadonaga J.T. 1996. *Drosophila* TFIID binds to a conserved downstream basal promoter element that is present in many TATA-box deficient promoters. *Genes Dev.* **10:** 711.

———. 1997. The downstream core promoter element, DPE, is conserved from *Drosophila* to humans and is recognized by $TAF_{II}60$ of Drosophila. *Genes Dev.* **11:** 3020.

Burley S.K. and Roeder R.G. 1996. Biochemistry and structural biology of transcription factor IID (TFIID). *Annu. Rev. Biochem.* **65:** 769.

———. 1998. TATA box mimicry by TFIID: Autoinhibition of pol II transcription. *Cell* **94:** 551.

Busby S. and Ebright R.H. 1994. Promoter structure, promoter recognition, and transcription activation in prokaryotes. *Cell* **79:** 743.

Carlson M. 1997. Genetics of transcriptional regulation in yeast: Connections to the RNA polymerase II CTD. *Annu. Rev. Cell Dev. Biol.* **13:** 1.

Chang M. and Jaehning J.A. 1997. A multiplicity of mediators: Alternative forms of transcription complexes communicate with transcriptional regulators. *Nucleic Acids Res.* **25:** 4861.

Chi T. and Carey M. 1996. Assembly of the isomerized TFIIA-TFIID-TATA ternary complex is necessary and sufficient for gene activation. *Genes Dev.* **10:** 2540.

Chi T., Lieberman P., Ellwood K., and Carey M. 1995. A general mechanism for transcriptional synergy by eukaryotic activators. *Nature* **377:** 254.

Chiang C.-M., Ge H., Wang Z., Hoffmann A., and Roeder R.G. 1993. Unique TATA-binding protein-containing complexes and cofactors involved in transcription by RNA polymerase II and III. *EMBO J.* **12:** 2749.

Chrivia J.C., Kwok R.P., Lamb N., Hagiwara M., Montminy M.R., and Goodman R.H. 1993. Phosphorylated CREB binds specifically to the nuclear protein CBP. *Nature* **365:** 855.

Dynan W.S. and Tjian R. 1983. The promoter-specific transcription factor Sp1 binds to upstream sequences in the SV40 early promoter. *Cell* **35:** 79.

Dynlacht B.D., Hoey T., and Tjian R. 1991. Isolation of coactivators associated with the TATA-binding protein that mediate transcriptional activation. *Cell* **66:** 563.

Emami K.H., Jain A., and Smale S.T. 1997. Mechanism of synergy between TATA and initiator: Synergistic binding of TFIID following a putative TFIIA-induced isomerization. *Genes Dev.* **11:** 3007.

Engelke D.R., Ng S.-Y., Shastry B.S., and Roeder R.G. 1980. Specific interaction of a purified transcription factor with an internal control region of 5S RNA genes. *Cell* **19:** 717.

Fan H.-Y., Cheng K.K., and Klein H.L. 1996. Mutations in the RNA polymerase II transcription machinery suppress the hyperrecombination mutant hpr1 of *Saccharomyces cerevisiae*. *Genetics* **142:** 749.

Fondell J.D., Ge H., and Roeder R.G. 1996. Ligand induction of a transcriptionally active thyroid hormone receptor coactivator complex. *Proc. Natl. Acad. Sci.* **93:** 8329.

Fondell J.D., Guermah M., Malik S., and Roeder R.G. 1999. Thyroid hormone receptor associated proteins (TRAPs) and general positive cofactors mediate thyroid hormone receptor function in the absence of $TAF_{II}s$. *Proc. Natl. Acad. Sci.* (in press).

Ge H. and Roeder R.G. 1994. Purification, molecular cloning and functional characterization of a human coactivator PC4 that mediates transcriptional activation of class II genes. *Cell* **78:** 513.

Ge H., Si Y., and Roeder R.G. 1998. Isolation of cDNAs encoding novel transcription coactivators p52 and p75 reveals an alternate regulator mechanism of transcriptional activation. *EMBO J.* **17:** 6723.

Ge H., Zhao Y.-M., Chait B.T., and Roeder R.G. 1994. Phosphorylation negatively regulates the function of coactivator PC4. *Proc. Natl. Acad. Sci.* **91:** 12691.

Geiduschek P.E. and Kassavetis G.A. 1992. RNA polymerase III transcription complexes. In *Transcription regulation* (ed. S.L. McKnight and K.R. Yamamoto), p. 247. Cold Spring Harbor Laboratory Press, Cold Spring Harbor, New York.

Ginsberg A.M., King B.O., and Roeder R.G. 1984. *Xenopus* 5S gene transcription factor TFIIIA: Characterization of a cDNA clone and measurement of RNA levels throughout development. *Cell* **39:** 479.

Glass C.K., Rose D.W., and Rosenfeld M.G. 1997. Nuclear receptor coactivators. *Curr. Opin. Cell Biol.* **9:** 222.

Goldman P.S., Tran V.K., and Goodman R.H. 1997. The multifunctional role of the co-activator CBP in transcriptional regulation. *Recent Prog. Horm. Res.* **52:** 103.

Grant P.A., Schieltz D., Pray-Grant M.G., Steger D.J., Joseph C.R., Yates J.R., III, and Workman J.L. 1998. A subset of $TAF_{II}s$ are integral components of the SAGA complex required for nucleosome acetylation and transcriptional stimulation. *Cell* **94:** 45.

Greenblatt J. 1997. RNA polymerase II holoenzyme and transcriptional regulation. *Curr. Opin. Cell Biol.* **9:** 310.

Gross C.A., Lonetto M., and Losick R. 1992. Bacterial sigma factors. In *Transcriptional regulation*. (ed. S.L. McKnight and K.R. Yamamoto), p. 129, Cold Spring Harbor Laboratory Press, Cold Spring Harbor, New York.

Grummt I. 1998. Initiation of murine rDNA transcription. In *Transcription of eukaryotic ribosomal RNA genes by RNA polymerase I* (ed. M.R. Paule), p. 135, Springer-Verlag, Amsterdam.

Gu W. and Roeder R.G. 1997. Activation of p53 sequence-specific DNA binding by acetylation of its C-terminal domain. *Cell* **90:** 595.

Gu W., Shi X.-L., and Roeder R.G. 1997. Synergistic activation of transcription by CBP and p53. *Nature* **387:** 819.

Gu W., Malik S., Ito M., Yuan C.-X., Fondell J.D., Zhang X., Martinez E., Qin J., and Roeder R.G. 1999. A novel human SRB/MED-containing cofactor complex (SMCC) involved in transcription regulation. *Mol. Cell* (in press).

Guarente L. 1995. Transcriptional coactivators in yeast and beyond. *Trends Biochem. Sci.* **20:** 517.

Guermah M., Malik S., and Roeder R.G. 1998. Involvement of TFIID and USA components in transcriptional activation of the HIV promoter by NFκB and Sp1. *Mol. Cell. Biol.* **18:** 3234.

Halle J.P., Stelzer G., Goppelt A., and Meisterernst M. 1995. Activation of transcription by recombinant upstream stimulatory factor 1 is mediated by a novel positive cofactor. *J. Biol. Chem.* **270:** 21307.

Hampsey M. 1998. Molecular genetics of the RNA polymerase II general transcriptional machinery. *Microbiol. Rev.* **62:** 465.

Heinzel T., Lavinsky R.M., Mullen T.-M., Söderström M., Laherty C.D., Torchia J., Yang W.-M., Brard G., Ngo S.D., Davie J.R., Seto E., Eisenman R.N., Rose D.W., Glass C.K., and Rosenfeld M.G. 1997. A complex containing N-CoR, mSin3 and histone deacetylase mediates transcriptional repression. *Nature* **387:** 43.

Hengartner C.J., Myer V.E., Liao S.-M., Wilson C.J., Koh S.S., and Young R.A. 1998. Temporal regulation of RNA polymerase II by Srb10 and Kin 28 cyclin dependent kinases. *Mol. Cell* **2:** 45.

Hengartner C.J., Thompson C.M., Zhang J.H., Chao D.M., Liao S.M., Koleske A.J., Okamura S., and Young R.A. 1995. Association of an activator with an RNA polymerase II holoenzyme. *Genes Dev.* **9:** 897.

Henry N.L., Bushnell D.A., and Kornberg R.D. 1996. A yeast transcriptional stimulatory protein similar to human PC4. *J. Biol. Chem.* **271:** 21842.

Hoey T., Weinzierl R.O., Gill G., Chen J.L., Dynlacht B.D., and Tjian R. 1993. Molecular cloning and functional analysis of *Drosophila* TAF110 reveal properties expected of coactivators. *Cell* **72:** 247.

Hoffmann A., Oelgeschläger T., and Roeder R.G. 1997. Considerations of transcriptional control mechanisms: Do TFIID-core promoter complexes recapitulate nucleosome-like functions? *Proc. Natl. Acad. Sci.* **94:** 8928.

Hoffmann A., Sinn E., Yamamoto T., Wang J., Roy A., Horikoshi M., and Roeder R.G. 1990. Highly conserved core domain and unique N terminus with presumptive regulatory motifs in a human TATA factor (TFIID). *Nature* **346:** 387.

Horikoshi M., Hai T., Lin Y.-S., Green M.R., and Roeder R.G. 1988. Transcription factor ATF interacts with the TATA factor to facilitate establishment of a preinitiation complex. *Cell* **54:** 1033.

Ikeda K., Halle J.P., Stelzer G., Meisterernst M., and Kawakami K. 1998. Involvement of negative cofactor NC2 in active repression by zinc finger-homeodomain transcription factor AREB6. *Mol. Cell. Biol.* **18:** 10.

Inostroza J.A., Mermelstein F.H., Ha I., Lane W.S., and Reinberg D. 1992. Dr1, a TATA-binding protein-associated phosphoprotein and inhibitor of class II gene transcription. *Cell* **70:** 477.

Kadonaga J.T., Carner K.C., Masiarz F.R., and Tjian R. 1987. Isolation of cDNA encoding transcription factor Sp1 and functional analysis of the DNA binding domain. *Cell* **51:** 1079.

Kaiser K. and Meisterernst M. 1996. The human general cofactors. *Trends Biochem. Sci.* **21:** 342.

Kaiser K., Stelzer G., and Meisterernst M. 1995. The coactivator p15 (PC4) initiates transcriptional activation during TFIIA-TFIID-promoter complex formation. *EMBO J.* **14:** 3520.

Kang J.J., Auble D.T., Ranish J.A., and Hahn S. 1995. Analysis of the yeast transcription factor TFIIA: Distinct functional regions and a polymerase II-specific role in basal and activated transcription. *Mol. Cell. Biol.* **15:** 1234.

Kassavetis G.A., Braun B.R., Nguyen L.H., and Geiduschek E. P. 1990. *S. cerevisiae* TFIIIB is the transcription initiation factor proper of RNA polymerase III, while TFIIIA and TFIIIC are assembly factors. *Cell* **60:** 235.

Kaufmann J. and Smale S.T. 1994. Direct recognition of initiator elements by a component of the transcription factor IID complex. *Genes Dev.* **8:** 821.

Kim T.K., Zhao Y., Ge H., Bernstein R., and Roeder R.G. 1995. TATA-binding protein residues implicated in a functional interplay between negative cofactor NC2 (Dr1) and general factors TFIIA and TFIIB. *J. Biol. Chem.* **270:** 10976.

Kim T.K., Lagrange T., Wang Y.H., Griffith J.D., Reinberg D., and Ebright R.H. 1997. Trajectory of DNA in the RNA polymerase II transcription preinitiation complex. *Proc. Natl. Acad. Sci.* **94:** 12268.

Kim Y., Bjorklund J.S., Li Y., Sayre M.H., and Kornberg R.D. 1994. A multiprotein mediator of transcriptional activation and its interaction with the C-terminal repeat domain of RNA polymerase II. *Cell* **77:** 599.

Knaus R., Pollock R., and Guarente L. 1996. Yeast SUB1 is a suppressor of TFIIB mutations and has homology to the human co-activator PC4. *EMBO J.* **15:** 1933.

Kobayashi N., Boyer T.G., and Berk A.J. 1995. A class of activation domains interacts directly with TFIIA and stimulates TFIIA-TFIID-promoter complex assembly. *Mol. Cell. Biol.* **15:** 6465.

Koh S.S., Ansari A.Z., Ptashne M., and Young R.A. 1998. An activator target in the RNA polymerase II holoenzyme. *Mol. Cell* **1:** 895.

Kokubo T., Swanson M.J., Nishikawa J., Hinnebusch A., and Nakatani Y. 1998. The yeast TAF145 inhibitory domain and TFIIA competitively bind to TATA-binding protein. *Mol. Cell. Biol.* **18:** 1003.

Kokubo T., Yamashita S., Horikoshi M., Roeder R.G., and Nakatani Y. 1994. Interaction between the N-terminal domain of the 230 kDa subunit and the TATA box-binding subunit of TFIID negatively regulates TATA-box binding. *Proc. Natl. Acad. Sci.* **91:** 3520.

Koleske A.J. and Young R.A. 1994. An RNA polymerase II holoenzyme responsive to activators. *Nature* **368:** 466.

———. 1995. The RNA polymerase II holoenzyme and its implications for gene regulation. *Trends Biochem. Sci.* **20:** 113.

Kretzschmar M., Meisterernst M., and Roeder R.G. 1993. Identification of human DNA topoisomerase I as a cofactor for activator-dependent transcription by RNA polymerase II. *Proc. Natl. Acad. Sci.* **90:** 1508.

Kretzschmar M., Kaiser K., Lottspeich F., and Meisterernst M. 1994a. A novel mediator of class II gene transcription with homology to viral immediate-early transcriptional regulators. *Cell* **78:** 525.

Kretzschmar M., Stelzer G., Roeder R.G., and Meisterernst M. 1994b. RNA polymerase II cofactor PC2 facilitates activation of transcription by GAL4-AH in vitro. *Mol. Cell. Biol.* **14:** 3927.

Kundu T., Wang Z., and Roeder R.G. 1999. Human TFIIIC relieves chromatin-mediated repression of RNA polymerase III transcription and contains an intrinsic histone acetyltransferase activity. *Mol. Cell. Biol.* (in press).

Kuo M.-H., Zhou J., Jambeck P., Churchill M.E.A., and Allis C.D. 1998. Histone acetyltransferase activity of yeast Gcn5p is required for the activation of target genes in vivo. *Genes Dev.* **12:** 627.

Larochelle S., Pandur J., Fisher R.P., Salz H.K., and Suter, B. 1998. Cdk 7 is essential for mitosis and for in vivo Cdk-activating kinase activity. *Genes Dev.* **12:** 370.

Lassar A.B., Martin P.L., and Roeder R.G. 1983. Transcription of class III genes: Formation of preinitiation complexes. *Science* **222:** 740.

Lee D.-K. and Lis J.T. 1998. Transcriptional activation independent of TFIIH kinase and the RNA polymerase II mediator in vivo. *Nature* **393:** 389.

Lee T.I. and Young R.A. 1998. Regulation of gene expression by TBP-associated proteins. *Genes Dev.* **12:** 1398.

Lee Y.C., Min S., Gim B.S., and Kim Y.-J. 1997. A transcriptional mediator protein that is required for activation of many RNA polymerase promoters and is conserved from yeast to humans. *Mol. Cell. Biol.* **17:** 4622.

Levine A.J. 1997. The cellular gatekeeper for growth and division. *Cell* **88:** 323.

Lieberman P.M. and Berk A.J. 1994. A mechanism for TAFs in transcriptional activation: Activation domain enhancement of TFIID-TFIIA-promoter DNA complex formation. *Genes Dev.* **8:** 995.

Liu D., Ishima R., Tong K.I., Bagby S., Kokubo T., Muhandiram D.R., Kay L.E., Nakatani Y., and Ikura M. 1998. Solution structure of TBP-TAF$_{II}$230 complex: protein mimicry of the minor groove surface of the TATA box unwound by TBP. *Cell* **94:** 573.

Luo Y. and Roeder R.G. 1995. Cloning, functional characterization, and mechanism of action of the B-cell-specific transcriptional coactivator OCA-B. *Mol. Cell. Biol.* **15:** 4115.

Luo Y., Fujii H., Gerster T., and Roeder R.G. 1992. A novel B cell-derived coactivator potentiates the activation of immunoglobulin promoters by octamer-binding transcription factors. *Cell* **71:** 231.

Luo Y., Stevens S., Xiao H., and Roeder R.G. 1998. Coactivation by OCA-B: Definition of critical regions and synergism with general cofactors. *Mol. Cell. Biol.* **7:** 3803.

Ma D., Olave I., Merino A., and Reinberg D. 1996. Separation of the transcriptional coactivator and antirepression functions of transcription factor IIA. *Proc. Natl. Acad. Sci.* **93:** 6583.

Mäkelä T.P., Parvin J.D., Kim J., Huber L. J., Sharp P.A., and Weinberg R.A. 1995. A kinase-deficient transcription factor TFIIH is functional in basal and activated transcription. *Proc. Natl. Acad. Sci.* **92:** 5174.

Malik S., Guermah M., and Roeder, R.G. 1998. A dynamic model for PC4 coactivator function in RNA polymerase II transcription. *Proc. Natl. Acad. Sci.* **95:** 2192.

Mangelsdorf D.J. and Evans R.M. 1995. The RXR heterodimers and orphan receptors. *Cell* **83:** 841.

Martinez E., Chiang C.-M., Ge H., and Roeder R.G. 1994. TATA-binding protein-associated factor(s) in TFIID function through the initiator to direct basal transcription from a TATA-less class II promoter. *EMBO J* **13:** 3115.

Martinez E., Kundu T.K., Fu J., and Roeder R.G. 1998a. A human SPT3-TAF$_{II}$31-GCN5-L acetylase complex distinct from TFIID. *J. Biol. Chem.* **373:** 23781.

Martinez E., Ge H., Tao Y., Yuan C.-X., and Roeder R.G. 1998b. Novel cofactors (TICs) and TFIIA mediate functional core promoter selectivity by the human TAF$_{II}$150-containing TFIID complex. *Mol. Cell. Biol.* **18:** 6571.

Martinez E., Zhou Q., L'Etoile N.D., Oelgeschläger T., Berk A.J., and Roeder R.G. 1995. Core promoter-specific function of a mutant transcription factor TFIID defective in TATA box binding. *Proc. Natl. Acad. Sci.* **92:** 11864.

Martinez-Balbas M., Bannister A.J., Martin K., Haus-Seuffert P., Meisterernst M., and Kouzarides T. 1998. The acetyltransferase activity of CBP stimulates transcription. *EMBO J.* **17:** 2886.

Matsui T., Segall J., Weil P.A., and Roeder R.G. 1980. Multiple factors required for accurate initiation of transcription by purified RNA polymerase II. *J. Biol. Chem.* **255:** 11992.

McNeil J.B., Agah H., and Bentley D. 1998. Activated transcription independent of the RNA polymerase II holoenzyme in budding yeast. *Genes Dev.* **12:** 12510.

Meisterernst M. and Roeder R.G. 1991. A family of proteins that interact with TFIID and regulate promoter activity. *Cell* **67:** 557.

Meisterernst M., Stelzer G., and Roeder R.G. 1997. Poly (ADP-ribose) polymerase enhances activator-dependent transcription in vitro. *Proc. Natl. Acad. Sci.* **94:** 2261.

Meisterernst M., Roy A., Lieu M., and Roeder R.G. 1991. Activation of class II gene transcription by regulatory factors is potentiated by a novel activity. *Cell* **66:** 981.

Merino A., Madden K.R., Lane W.S., Champoux J.J., and Reinberg D. 1993. DNA topoisomerase I is involved in both repression and activation of transcription. *Nature* **365:** 227.

Miesfeld R., Okret S., Wilkstrom A.-C., Wrange O., Gustafsson J.-A., and Yamamoto K. 1984. Characterization of a steroid receptor gene and mRNA in wild-type and mutant cells. *Nature* **312:** 779.

Moqtaderi Z., Bai Y., Poon D., Weil P.A., and Struhl K. 1996. TBP-associated factors are not generally required for transcriptional activation in yeast. *Nature* **383:** 188.

Myers L.C., Gustafsson C.M., Bushnell D.A., Lui M., Erdjument-Bromage H., Tempst P., and Kornberg R.D. 1998. The Med proteins of yeast and their function through the RNA polymerase II carboxy-terminal domain. *Genes Dev.* **12:** 45.

Nagy L., Kao H.-Y., Chakravarti D., Lin R.J., Hassig C.A., Ayer D.E., Schreiber S.L., and Evans R.M. 1997. Nuclear receptor repression mediated by a complex containing SMRT, mSin3A, and histone deacetylase. *Cell* **89:** 373.

Nakajima N., Horikoshi M., and Roeder R.G. 1988. Factors involved in specific transcription by mammalian RNA polymerase II: Purification, genetic specificity and TATA box-promoter interactions of TFIID. *Mol. Cell. Biol.* **8:** 4028.

Nakajima T., Uchida C., Anderson S., Parvin J., and Montiminy M. 1997. Analysis of a cAMP-responsive activator reveals a

two-component mechanism for transcription induction via signal-dependent factors. *Genes Dev.* **11:** 738.

Nakatani Y., Horikoshi M., Brenner M., Yamamoto T., Besnard F., Roeder R.G., and Freese E. 1990. A downstream initiation element required for efficient TATA box binding and in vitro function of TFIID. *Nature* **348:** 86.

Ng S.-Y., Parker C.S., and Roeder R.G. 1979. Transcription of cloned *Xenopus* 5S RNA genes by *X. laevis* RNA polymerase III in reconstituted systems. *Proc. Natl. Acad. Sci.* **76:** 136.

Nikolov D.B. and Burley S.K. 1997. RNA polymerase II transcription initiation: A structural view. *Proc. Natl. Acad. Sci* **94:** 15.

Nishikawa J., Kokubo T., Horikoshi M., Roeder R.G., and Nakatani Y. 1997. *Drosophila* TAF$_{II}$230 and the transcriptional activator VP16 bind competitively to the TATA box-binding domain of the TATA box-binding protein. *Proc. Natl. Acad. Sci.* **94:** 85.

Nomura M. 1998. Transcription factors used by *Saccharomyces cerevisiae* RNA polymerase I and the mechanism of initiation. In *Transcription of eukaryotic ribosomal RNA genes by RNA polymerase I* (ed. M.R. Paule), p. 155. Springer-Verlag, Amsterdam.

Nonet M.L. and Young R.A. 1989. Intragenic and extragenic suppressors of mutations in the heptapeptide repeat domain of *Saccharomyces cerevisiae* RNA polymerase II. *Genetics* **123:** 715.

Oelgeschläger T., Chiang C.-M., and Roeder R.G. 1996. Topology and reorganization of a human TFIID-promoter complex. *Nature* **382:** 735.

Oelgeschläger T., Tao Y., Kang Y.K., and Roeder R.G. 1998. Transcriptional activation via enhanced preinitiation complex assembly in a human cell free system lacking TAF$_{II}$s. *Mol. Cell* **1:** 925.

Ogryzko V.V., Schiltz R.L., Russanova V., Howard B.H., and Nakatani Y. 1996. The transcriptional coactivators p300 and CBP are histone acetyltransferases. *Cell* **87:** 953.

Ogryzko V.V., Kotani T., Zhang X., Schiltz R.L., Howard T., Yang X.-J., Howard B.H., Qin J., and Nakatani Y. 1998. Histone-like TAFs within the PCAF histone acetylase complex. *Cell* **94:** 35.

Oñate S.A., Tsai S.Y., Tsai M.-J., and O'Malley B.W. 1995. Sequence and characterization of a coactivator for the steroid hormone receptor superfamily. *Science* **270:** 1354.

Ong J., Stevens S., Roeder R.G., and Eckhardt L.A. 1998. 3′ IgH enhancer elements shift synergistic interactions during B-cell development. *J. Immunol.* **160:** 4896.

Orphanides G., LaGrange T., and Reinberg D. 1996. The general initiation factors of RNA polymerase II. *Genes Dev.* **10:** 2657.

Ozer J., Bolden A.H., and Lieberman P.M. 1996. Transcription factor IIA mutations show activator-specific defects and reveal a IIA function distinct from stimulation of TBP-DNA binding. *J. Biol. Chem.* **271:** 11182.

Ozer J., Moore P.A., Bolden A.H., Lee A., Rosen C.A., and Lieberman P.M. 1994. Molecular cloning of the small (γ) subunit of human TFIIA reveals functions critical for activated transcription. *Genes Dev.* **8:** 2324.

Parker C.S. and Roeder R.G. 1977. Selective and accurate transcription of the *Xenopus laevis* 5S RNA genes in isolated chromatin by purifed RNA polymerase III. *Proc. Natl. Acad. Sci.* **74:** 44.

Parker C.S. and Topol J. 1984. A *Drosophila* RNA polymerase II transcription factor contains a promoter-region-specific DNA-binding activity. *Cell* **36:** 357.

Parker C.S., Ng S.-Y., and Roeder R.G. 1976. Selective transcription of the 5S RNA genes in isolated chromatin by RNA polymerase III. In *Molecular mechanisms in the control of gene expression* (ed. D.P. Nierlich et al.), p. 223. Academic Press, New York.

Payvar F., Wränge O., Carlstedt-Duke J., Okret S., Gustafsson J.-A., and Yamamoto K.R. 1981. Purified glucocorticoid receptors bind selectively in vitro to a cloned DNA fragment whose transcription is regulated by glucocorticoids in vivo. *Proc. Natl. Acad. Sci.* **78:** 6628.

Because binding of TFIID to promoter DNA is the initial step in PIC assembly, it may be an important determinant for the rate and efficiency of this process. As discussed above, in vitro studies have shown that activators can cause qualitative and/or quantitative changes in the TFIID-promoter interaction (for recent reviews, see Burley and Roeder 1996; Pugh 1996; Ranish and Hahn 1996). Furthermore, in vivo DNA footprinting/cross-linking experiments suggest that TFIID binding may be rate-limiting for transcription on some but not all promoters (Wu 1984; Selleck and Majors 1987; Giardina and Lis 1993; Chen et al. 1994a).

TFIID binding could be limiting in the cell for several reasons. First, the stoichiometric amount of TFIID may be limiting relative to the number of active promoters. In support of this view, transcriptional activity can be limited by the intracellular TFIID concentration (Colgan and Manley 1992; Klein and Struhl 1994; for review, see Struhl 1995). Although TBP is relatively abundant in the cell (Lee and Young 1998), the effective concentration of TFIID may be low because TBP is also present in complexes other than TFIID (for review, see Hernandez 1993; Lee and Young 1998). Second, in the cell, DNA is packaged into chromatin, and nucleosomes can exclude the binding of TBP to promoter DNA (see, e.g., Workman and Roeder 1987; Imbalzano et al. 1994). Finally, the binding of TFIID on some promoters may involve a rate-limiting isomerization step (see, e.g., Chi and Carey 1996). Both the initial binding and the isomerization can be stimulated by activators. Interestingly, for some promoters, the TFIID-promoter complex displays an extensive footprint even in the absence of activators (Nakajima et al. 1988), suggesting differences in rate-limiting steps among promoters.

In vitro studies indicate that following initiation, a subset of GTFs including TFIID remain bound to the promoter (Roberts et al. 1995; Zawel et al. 1995). Whether TFIID remains bound may depend on the promoter sequence and the type of activators. For some promoters, each transcription cycle may require re-binding of TFIID.

The TAF$_{II}$ Components of TFIID

TFIID was originally purified as an activity required to reconstitute a RNA polymerase II in vitro basal transcription reaction. The high molecular weight of this partially purified TFIID immediately suggested that it was a multisubunit complex (Nakajima et al. 1988). The purification and cloning of TBP led to the subsequent identification and cloning of other TFIID subunits in the higher eukaryotes, humans, and *Drosophila*. These higher eukaryotic TFIIDs were shown to consist of TBP and at least 8–12 tightly bound subunits, the TBP-associated factors, TAF$_{II}$s (for review, see Burley and Roeder 1996).

In most in vitro transcription systems, TBP (which lacks TAF$_{II}$s) could support an activator-independent "basal" transcription reaction, but it was unable to respond to an activator, whereas TFIID (which contains TAF$_{II}$s) could support both basal and activated transcription. Thus, in vitro, one or more TAF$_{II}$s appeared to have

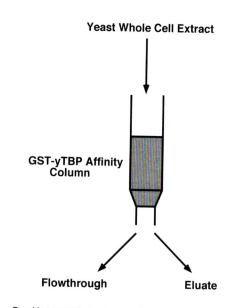

Figure 1. Experimental strategy to isolate yeast TAF$_{II}$s. For detailed experimental conditions, see Reese et al. (1994).

an obligatory "coactivator" activity. On the basis of this observation, protein-protein interaction studies, and TFIID reconstitution experiments, it was proposed that TAF$_{II}$s are the obligatory targets of activators (for review, see Burley and Roeder 1996; Goodrich and Tjian 1994).

The general transcription machinery is highly conserved from yeast to humans. Moreover, some yeast activators can function in mammalian cells, and likewise, some mammalian activators work in yeast (for review, see Ptashne 1988). It was therefore surprising that early studies indicated that TFIID from higher eukaryotes differed considerably from that of yeast: Purified yeast TFIID is a single polypeptide, TBP, that is highly homologous to its human counterpart (Buratowski et al. 1988).

The considerable advantages of yeast as an experimental system to study in vivo function provided a strong incentive to identify yeast TAF$_{II}$s. To isolate yeast TAF$_{II}$s, we performed protein affinity chromatography using yeast TBP (yTBP) as the immobilized ligand to isolate from a yeast whole-cell extract a protein complex required specifically for activated transcription by RNA polymerase II (see Fig. 1) (Reese et al. 1994). A GST-yTBP column bound a complex comprising approximately 11 major polypeptides that shared several important similarities with higher eukaryotic TFIID, including large size, in vitro coactivator activity, and specificity for RNA polymerase II transcription. Most importantly, microsequence and cloning revealed that this yTAF$_{II}$ complex contained homologs of known subunits of higher eukaryotic TFIID. Subsequently, yTAF$_{II}$s were also identified by coimmunoprecipitation with an α-TBP antibody (Poon et al. 1995) and through database searches (Moqtaderi et al. 1996a).

Figure 2 summarizes our current knowledge of yTAF$_{II}$s. The major conclusions from this figure are that

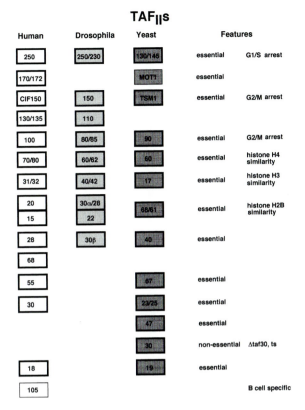

Figure 2. Yeast TAF$_{II}$s and their metazoan homologs.

in almost all instances, a yTAF$_{II}$ has a higher eukaryotic TAF$_{II}$ homolog. Thus, like all other components of the transcription machinery, TAF$_{II}$s have also been highly conserved. Second, with the exception of TAF30/TFG3/ANC1, all yTAF$_{II}$s are essential for viability. Thus, each essential TAF$_{II}$ must perform at least one obligate, nonredundant function.

A further similarity between TFIID in higher eukaryotes and yeast concerns the TBP-TAF$_{II}$ interface. In higher eukaryotes, TBP interacts with hTAF$_{II}$250 (Takada et al. 1992; Hisatake et al. 1993; Kokubo et al. 1993; Ruppert et al. 1993; Zhou et al. 1993), whereas in the yTAF$_{II}$ complex, it is the homologous yTAF$_{II}$145 with which TBP interacts (Reese et al. 1994). Presumably, the relatively weak affinity of TBP for yTAF$_{II}$145 is relevant to the reason yTAF$_{II}$s proved to be technically difficult to identify.

The elaborate nature of the yTAF$_{II}$ complex (~12 subunits) (Reese et al. 1994) raised the possibility that TAF$_{II}$s may have functions other than as general coactivators. It has been postulated that the multiple TAF$_{II}$s of higher eukaryotic TFIID provide interaction sites for the different types of transcriptional activation domains (Tjian and Maniatis 1994; Burley and Roeder 1996). However, the complexity of the yTAF$_{II}$ complex is essentially equivalent to that of higher eukaryotes; yet yeast appear to contain only a single class of activation domain, acidic and nonacidic activation domains are inactive in yeast (Ptashne 1988; Ponticelli et al. 1995). Thus, it seemed unlikely that the complexity of the yTAF$_{II}$ com-

plex was solely to accommodate different classes of yeast activation domains.

yTAF$_{II}$s Are Not General Coactivators In Vivo

Originally, TAF$_{II}$ function was analyzed in vitro using transcription systems reconstituted from purified components (Verrijzer and Tjian 1996). These studies led to the so-called "coactivator hypothesis," which posited that TAF$_{II}$s are the obligatory targets of activators and that different activator-coactivator combinations selectively regulate transcription.

The function of TAF$_{II}$s had not been systematically investigated in vivo. To do so, we used two independent strategies to functionally inactivate yTAF$_{II}$s: temperature-sensitive mutations and conditional depletion (summarized in Fig. 3). We found that following inactivation or depletion of six different yTAF$_{II}$s, including yTAF$_{II}$145, the core yTAF$_{II}$ that contacts TBP did not compromise transcription of a variety of yeast genes driven by diverse activators (Apone et al. 1996; Walker et al. 1996). Likewise, following temperature-sensitive inactivation of yTAF$_{II}$145, TSM1, or yTAF$_{II}$90, no significant decrease was seen in total poly(A)$^+$ RNA synthesis

A. Temperature sensitive inactivation

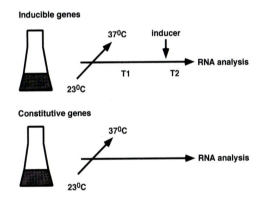

B. Conditional depletion

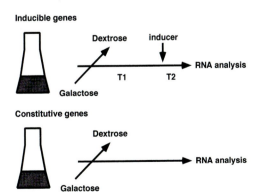

Figure 3. Experimental strategies for analyzing yTAF$_{II}$ functions in vivo. For detailed experimental procedures, see Walker et al. (1996).

(Walker et al. 1997). Although inactivation of all yTAF$_{II}$s behaved similarly, the results with yTAF$_{II}$145 were most persuasive: yTAF$_{II}$145 is the only yTAF$_{II}$ known to contact TBP directly, and its higher eukaryotic homolog, TAF$_{II}$250, is always required to reconstitute TFIID activity in vitro (Chen et al. 1994b). From these results, we conclude that activated transcription can occur in the absence of multiple yTAF$_{II}$s. Kevin Struhl and colleagues reached a similar conclusion using a third independent strategy to inactivate TAF$_{II}$s (Moqtaderi et al. 1996b).

An implication of these results was that yTAF$_{II}$s are not essential targets of activators and that in these instances, other pathways for transcription activation must be operative. As discussed above, other PIC components, such as TFIIB, TBP, or components of the RNA polymerase II holoenzyme complex, may serve as the activator targets (Zawel and Reinberg 1992; Tjian and Maniatis 1994). The ability of proteins containing a DNA-binding domain fused to a PIC component (e.g., TBP, TFIIB, and GAL11) to activate transcription also argues that activator-TAF$_{II}$ interactions are not obligatory.

The yeast results were not inconsistent with the limited in vivo analysis of TAF$_{II}$ function in higher eukaryotes. For example, mammalian cell lines harboring a temperature-sensitive TAF$_{II}$250 allele do not have a global defect in RNA polymerase II transcription under nonpermissive conditions (Hirschhorn et al. 1984; Liu et al. 1985). Furthermore, inactivation of TAF$_{II}$250 did not prevent transcriptional activation of the c-*fos* gene, a highly inducible eukaryotic promoter (Wang and Tjian 1994). Although transcription of the cyclin A gene was reportedly compromised following TAF$_{II}$250 inactivation (Wang and Tjian 1994), these cells were arrested in G$_1$; cyclin A is transcribed only in S phase (Henglein et al. 1994). Thus, the apparent transcription defect may have been an indirect effect of the cell cycle arrest and not a direct consequence of TAF$_{II}$250 inactivation. Consistent with this possibility, nuclear run-off experiments indicate that transcription of the cyclin A promoter is affected relatively late after TAF$_{II}$250 inactivation, suggesting that it is not a primary effect (Suzuki-Yagawa et al. 1997).

TAF$_{II}$s and the Cell Cycle

Although inactivation of yTAF$_{II}$s does not affect transcription of many genes, most yTAF$_{II}$s are required for viability, and inactivation of some yTAF$_{II}$s results in distinct cell cycle phenotypes (Apone et al. 1996; Walker et al. 1996). In particular, inactivation of yTAF$_{II}$145 leads to a G$_1$/S arrest (Walker et al. 1996), analogous to the results with its mammalian homolog, TAF$_{II}$250 (Talavera and Basilico 1977). Conversely, following inactivation of yTAF$_{II}$90 and yTAF$_{II}$150 (TSM1), cells arrest in G$_2$/M. These observations raise the possibility that TAF$_{II}$s may have a specialized role in the transcriptional control of the cell cycle and cellular growth state and further suggest that individual yTAF$_{II}$s mediate cell cycle progression through unique activities or targets. Consistent with this idea, yTAF$_{II}$45 is required for transcription of genes such

as G$_1$ and B-type cyclins, whose protein products are required for G$_1$/S progression (Walker et al. 1997).

A further connection between TAF$_{II}$s and the cell cycle is that TAF$_{II}$s appear to be regulated by the cellular growth state. For example, as yeast cell growth slows down in response to high density, the levels of several yTAF$_{II}$s and TBP decrease dramatically, whereas other GTFs are unaffected (Walker et al. 1997). Another example is when cells enter mitosis, RNA polymerase-II-directed transcription is shut off (Prescott and Bender 1962). Experiments in mammalian cells suggest that the shut-off mechanism involves mitotic-specific phosphorylation of TAF$_{II}$s and TBP (Segil et al. 1996).

Finally, a role for TAF$_{II}$s in cell cycle control is also suggested by a phylogenetic comparison of transcription and cell cycle machineries (summarized in Fig. 4). The most primitive organism with a eukaryotic-like transcription apparatus are the Archaea, which contain homologs of eukaryotic RNA polymerase II, and several GTFs including TBP and TFIIB (for review, see Klenk and Doolittle 1994), but lack TAF$_{II}$ homologs (Bult et al. 1996). Thus, TAF$_{II}$s are present only in organisms with a eukaryotic cell cycle.

	Eubacteria	Archea	Eukaryotes
Eukaryotic-like GTFs (RNA pol II, TBP, TFIIB)	NO	YES	YES
TAF$_{II}$s	NO	NO	YES
Cell Cycle	NO	NO	YES

Figure 4. Phylogenetic comparison of transcription machinery and cell cycle.

yTAF$_{II}$145 Is a Core Promoter Selectivity Factor

The fact that yTAF$_{II}$s were highly conserved, tightly associated with known transcription components, essential for viability but not required for general transcription was enigmatic. What then was the essential function provided by yTAF$_{II}$s? The possibility that we favored was that rather than providing a general transcription function, yTAF$_{II}$ activity was promoter-specific. To test this idea, we used differential display, a systematic mRNA screening approach to identify genes transcriptionally dependent on yTAF$_{II}$145 (Shen and Green 1997). This analysis confirmed that yTAF$_{II}$145 was dispensable for transcription of the vast majority of yeast genes. However, a small minority of genes, most notably RPS genes, were found to be highly dependent on yTAF$_{II}$145. Unexpectedly, for a comparable number of genes, transcription paradoxically increased following yTAF$_{II}$145 inactivation, suggesting that yTAF$_{II}$145 could also selectively repress transcription. yTAF$_{II}$145-dependent genes were also identified based on the characteristic cell cycle phenotype of TAF$_{II}$145 mutants (see below): Transcription of G$_1$ and certain B-type cyclin genes was also yTAF$_{II}$145-dependent (Walker et al. 1997).

The results of the above studies indicated that there were two classes of genes: The vast majority whose transcription did not require yTAF$_{II}$145, and a minority whose transcription is yTAF$_{II}$145-dependent. To determine the portion of the gene that conferred yTAF$_{II}$145 dependence, we used an experimental strategy outlined in Figure 5 (Shen and Green 1997). In brief, we constructed a series of chimeric promoters by fusing different combinations of UASs, which harbor the activator-binding sites, with core promoters. We unexpectedly found that the portion of these genes that rendered them yTAF$_{II}$145-dependent was the core promoter, not the UAS. In fact, a yTAF$_{II}$145-dependent promoter retained the yTAF$_{II}$145 requirement even when its transcription was artificially driven in the absence of an activator. Taken together, these results indicate that yTAF$_{II}$145 functions in recognition and selection of core promoters by a mechanism not involving upstream activators.

The in vivo results with yTAF$_{II}$145 highlight distinctions among different core promoters and emphasize how in addition to the activators, the core promoter can contribute to transcriptional regulation. In this regard, a variety of studies have demonstrated differential responsiveness of various core promoters to upstream activators (Simon et al. 1988; Taylor and Kingston 1990; Das et al. 1995; Emami et al. 1995), perhaps reflecting differences in the rate-limiting step(s) for transcription activation.

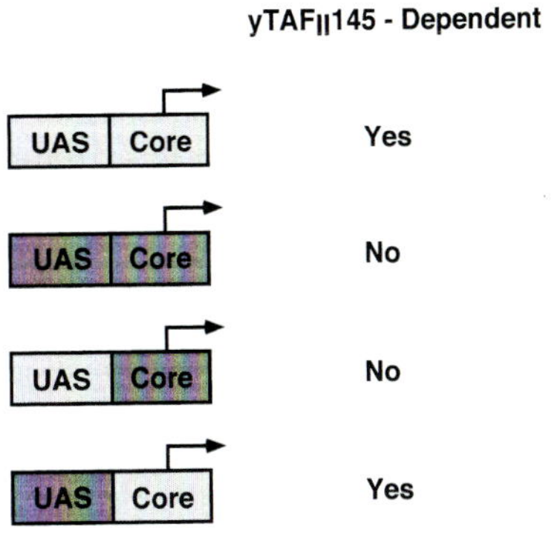

TAF$_{II}$s Contact the Core Promoter In Vitro

Although the in vivo results described above clearly indicate that the core promoter can dictate TAF$_{II}$ dependence, the basis for recognition and selection remains to be elucidated. For the reasons described below, we favor a model involving direct contacts between TAF$_{II}$s and the core promoter.

Several independent lines of evidence are indicative of interactions between TAF$_{II}$s and the core promoter. First, whereas TBP gives rise to a discrete footprint that covers only the TATA box (Burley and Roeder 1996), the DNase I footprint of TFIID on some core promoters is substantially larger (Nakajima et al. 1988; Zhou et al. 1992; Chiang et al. 1993; Kaufmann and Smale 1994; Purnell et al. 1994; Sypes and Gilmour 1994). This difference in footprinting is highly suggestive of TAF$_{II}$-DNA contacts. The occurrence of this extended footprint on only some promoters further suggests differential affinity of TAF$_{II}$s for various core promoters. Second, site-specific DNA cross-linking studies have directly demonstrated TAF$_{II}$-core promoter contacts. For example, Oelgeschlager et al. (1996) have shown that several TAF$_{II}$s are engaged in position-dependent contacts with the core promoter. Burke and Kadonaga (1997) have shown specific photocross-linking of dTAF$_{II}$60 and dTAF$_{II}$40 to the downstream promoter element (DPE). Third, in addition to the TATA box, several of the other core promoter elements appear to function through interactions with TAF$_{II}$s (Nakatani et al. 1990; Kaufmann and Smale 1994; Purnell et al. 1994; Verrijzer et al. 1994; Martinez et al. 1995; Burke and Kadonaga 1997; Kaufmann et al. 1998). Fourth, binding-

Figure 5. Mapping yTAF$_{II}$145-dependent promoter elements. For detailed experimental procedures and results, see Shen and Green (1997).

site selection experiments indicate that TFIID interacts with the initiator in a sequence-specific fashion. Remarkably, this selected sequence matches a consensus deduced by comparison of bone fide *Drosophila* promoters (Purnell et al. 1994). Finally, the in vitro DNA-binding specificity of TBP-TAF$_{II}$ complexes differs from that of TBP alone (Verrijzer et al. 1995), which is most readily explained by TAF$_{II}$-DNA contacts.

Taken together, these data indicate that TAF$_{II}$s contact the core promoter and suggest that these contacts affect the affinity and specificity of TFIID binding. To date, however, there has been no rigorous demonstration that the affinity of TFIID for the promoter is greater than that of TBP alone, particularly in vivo.

Potential Mechanisms for Core Promoter Selectivity by TAF$_{II}$s

An important future goal will be to understand how TAF$_{II}$s act in vivo as core promoter selectivity factors. We suggest several reasonable possibilities. First, the TAF$_{II}$-core promoter contacts described above may affect the affinity and/or specificity of TFIID for certain core promoters. This may be particularly important in vivo where the amount of TFIID may be limiting relative to approximately 4500 actively transcribed genes (Velculescu et al. 1997). Second, TAF$_{II}$-core promoter contact may introduce allosteric effects in TFIID that modulate its activity and hence transcription. DNA-binding-induced conformational changes have been observed with several transcriptional activators, in particular nuclear hormone receptors (Lefstin and Yamamoto 1998). Third, differences in core promoters' chromatin structure may affect TFIID binding, and TAF$_{II}$s may be differentially required. In this regard, hTAF$_{II}$250 and its yeast homolog, yTAF$_{II}$145, have intrinsic histone acetyltransferase (HAT) enzyme activity (Mizzen et al. 1996), which could act to overcome a repressive nucleosomal structure.

Multiple TBP- and TAF$_{II}$-containing Complexes

The differential requirement of individual TAF$_{II}$s for transcription of specific genes raises the possibility of multiple TAF$_{II}$-containing complexes. Several lines of evidence indicate that there are distinct complexes comprising different combinations of TFIID components. First, there are tissue-specific TAF$_{II}$s (Dikstein et al. 1996) and TBPs (Hansen et al. 1997). The notion that TBP may not be required for transcription of all class II genes is also supported by identification of a TAF$_{II}$-containing complex that lacks TBP and can support RNA polymerase II transcription (Wieczorek et al. 1998). Second, TFIID complexes with different compositions of TAF$_{II}$s have been reported. For example, hTAF$_{II}$30 is present in a complex called TFIIDβ but not in another complex called TFIIDα (Jacq et al. 1994; Mengus et al. 1995); hTAF$_{II}$170 is present in a complex called B-TFIID but not in the D-TFIID complex (Timmers and Sharp 1991; Timmers et al. 1992). Interestingly, hTAF$_{II}$ 170/172 (Chicca et al. 1998) is a homolog of yeast MOT1, a global repressor of RNA polymerase II transcription (Auble et al. 1994). The presence of hTAF$_{II}$170 in a subpopulation of TFIID suggests a mechanism for transcriptional repression of certain promoters. Finally, TAF$_{II}$s are present in complexes other than TFIID. For instance, yeast TAF30 is present in TFIIF and the SWI/SNF chromatin remodeling complex (Cairns et al. 1996). Recent studies have shown that a subset of yTAF$_{II}$s are components of the SAGA (*Spt-Ada-Gcn5-acetyltransferase*) complex and are required for both integrity and complete HAT activity (Grant et al. 1998). The functions of TAF$_{II}$s present in complexes other than TFIID remained to be defined. In any case, these observations emphasize that the role of yTAF$_{II}$s in the TFIID complex represents only a part of total TAF$_{II}$ function.

PERSPECTIVE

To date, the analysis of TAFII function has been most extensive for a single yTAF$_{II}$, yTAF$_{II}$145. It remains to be determined whether other TAF$_{II}$s, like yTAF$_{II}$145, are required for transcription of a particular subset of genes and, if so, whether they will function through the activator or the core promoter. It is possible that every gene requires at least one TAF$_{II}$ for expression. Alternatively, some genes may be transcribed in the complete absence of any functional TAF$_{II}$s. The ability to perform whole-genome analysis in yeast using DNA chip microarrays will expedite these studies.

The findings with yTAF$_{II}$s highlight the potential for obtaining significantly different results using in vivo and various in vitro transcription systems. Although the explanation for the differences in TAF$_{II}$ requirements in vivo and in vitro remain to be elucidated, it is worth pointing out the substantial differences in the manner in which these experiments were carried out. One of the most striking differences is that in vivo, a single gene competes for transcription components with many other active genes. For example, in a yeast cell, there are approximately 4500 actively transcribed genes (Velculescu et al. 1997). In contrast, an in vitro transcription assay contains a single (or at most two) DNA template typically added at a concentration to optimize transcription. Sec-

ond, a typical in vitro transcription reaction is performed using naked (not chromatin) DNA templates, generally with artificial (not natural) promoters and activators. Finally, many biochemical reconstitution experiments use only a subset of transcription components; for example, SRBs are required for transcription in vivo but are dispensable in certain reconstituted systems. These differences may account for, at least in part, the apparent discrepancy in TAF$_{II}$ requirements in vivo and in vitro.

In vitro approaches will ultimately be required to elucidate the detailed mechanisms of TAF$_{II}$ action. The challenge is to develop a system that faithfully recapitulates the in vivo situation. In this regard, several in vitro transcription studies have achieved activated transcription in the absence of TAF$_{II}$s. For example, in a coupled chromatin assembly/transcription system, TBP alone supported activated transcription (Workman et al. 1991). In vitro transcription systems involving yeast holoenzyme complexes have observed activated transcription in the absence of yTAF$_{II}$s (Y.-J. Kim et al. 1994; Koleske and Young 1994). The recent report of a mammalian in vitro system, which supports activated transcription in a TAF$_{II}$-independent fashion (Oelgeschlager et al. 1998), is a significant advance. The next step will be to develop an in vitro system that mimics the TAF$_{II}$-dependent and TAF$_{II}$-independent transcription of specific genes observed in vivo.

ACKNOWLEDGMENTS

We thank Judy Mondor for secretarial assistance. W.-C.S. is a special fellow of the Leukemia Society of America, and M.R.G. is an investigator of the Howard Hughes Medical Institute. C.-M.A.V. was supported by a fellowship from the National Institutes of Health, M.M. was supported by a fellowship from the European Molecular Biology Organization. This work was supported, in part, by a grant from the National Institutes of Health to M.R.G.

REFERENCES

Apone L.M., Virbasius C.A., Reese J.C., and Green M.R. 1996. Yeast TAF$_{II}$ 90 is required for cell-cycle progression through G$_2$/M but not for general transcription activation. *Genes Dev.* **10:** 2368.

Auble D.T., Hansen K.E., Mueller C.G., Lane W.S., Thorner J., and Hahn S. 1994. Mot1, a global repressor of RNA polymerase II transcription, inhibits TBP binding to DNA by an ATP-dependent mechanism. *Genes Dev.* **8:** 1920.

Bult C.J., White O., Olsen G.J., Zhou L., Fleischmann R.D., Sutton G.G., Blake J.A., FitzGerald L.M., Clayton R.A., Gocayne J.D., Kerlavage A.R., Doughtery B.A., Tomb J.F., Adams M.D., Reich C.I., Overbeek R., Kirkness E.F., Weinstock K.G., Merrick J.M., Glodek A., Scott J.L., Geoghagen N.S.M., and Venter J.C. 1996. Complete genome sequence of the methanogenic archaeon, *Methanococcus jannaschii*. *Science* **273:** 1058.

Buratowski S., Hahn, S., Sharp, P.A., and Guarente L., 1988. Function of a yeast TATA element-binding protein in a mammalian transcription system. *Nature* **334:** 37.

Burke T.W. and Kadonaga J.T. 1997. The downstream core promoter element DPE, is conserved from *Drosophila* to humans and is recognized by TAF$_{II}$60 of *Drosophila*. *Genes Dev.* **11:** 3020.

Burley S.K. and Roeder R.G. 1996. Biochemistry and structural biology of transcription factor IID (TFIID). *Annu. Rev. Biochem.* **65:** 769.

Cairns B.R., Henry N.L., and Kornberg R.D. 1996. TFG3/TAF30/ANC1, a component of the yeast SWI/SNF complex that is similar to the leukemogenic proteins ENL and AF-9. *Mol. Cell. Biol.* **16:** 3308.

Carcamo J., Lobos S., Merino A., Buckbinder L., Weinmann R., Natarajan V., and Reinberg D. 1989. Factors involved in specific transcription by mammalian RNA polymerase II. Role of factors IID and MLTF in transcription from the adenovirus major late and IVa2 promoters. *J. Biol. Chem.* **264:** 7708.

Chao D.M., Gadbois E.L., Murray P.J., Anderson S.F., Sonu M.S., Parvin J.D., and Young R.A. 1996. A mammalian SRB protein associated with an RNA polymerase II holoenzyme. *Nature* **380:** 82.

Chen J., Ding M., and Pederson D.S. 1994a. Binding of TFIID to the CYC1 TATA boxes in yeast occurs independently of upstream activating sequences. *Proc. Natl. Acad. Sci.* **91:** 11909.

Chen J.-L., Attardi L.D., Verrijzer C.P., Yokomori K., and Tjian R. 1994b. Assembly of recombinant TFIID reveals differential requirements for distinct transcriptional activators. *Cell* **79:** 93.

Chi T. and Carey M. 1996. Assembly of the isomerized TFIIA-TFIID-TATA ternary complex is necessary and sufficient for gene activation. *Genes Dev.* **10:** 2540.

Chi T., Lieberman P., Ellwood K., and Carey M. 1995. A general mechanism for transcriptional synergy by eukaryotic activators. *Nature* **377:** 254.

Chiang C.M., Ge H., Wang Z., Hoffmann A., and Roeder R.G. 1993. Unique TATA-binding protein-containing complex and cofactors involved in transcription by RNA polymerase II and III. *EMBO J.* **12:** 2749.

Chicca J.J., II, Auble D T., and Pug, B F. 1998. Cloning and biochemical characterization of TAF-172, a human homolog of yeast Mot1. *Mol. Cell. Biol.* **18:** 1701.

Choy B. and Green M.R. 1993. Eukaryotic activators function during multiple steps of preinitiation complex assembly. *Nature* **366:** 531.

Colgan J., and Manley J.L. 1992. TFIID can be rate limiting in vivo for TATA-containing, but not TATA-lacking, RNA polymerase II promoters. *Genes Dev.* **6:** 304.

Das G., Hinkley C.S., and Herr W. 1995. Basal promoter elements as a selective determinant of transcriptional activator function. *Nature* **374:** 657.

Dikstein R., Zhou S., and Tjian R. 1996. Human TAF$_{II}$105 is a cell type-specific TFIID subunit related to hTAF$_{II}$130. *Cell* **87:** 137.

Dynlacht B.D., Hoey T., and Tjian R. 1991. Isolation of coactivators associated with the TATA-binding protein that mediate transcriptional activation. *Cell* **55:** 563.

Emami K.H., Navarre W.W., and Smale S.T. 1995. Core promoter specificities of the Sp1 and VP16 transcriptional activation domains. *Mol. Cell. Biol.* **15:** 5906.

Gaudreau L., Adam M., and Ptashne M. 1998. Activation of transcription in vitro by recruitment of the yeast RNA polymerase II holoenzyme. *Mol. Cell* **1:** 913.

Giardina C. and Lis J. T. 1993. DNA melting on yeast RNA polymerase II promoters. *Science* **261:** 759.

Goodrich J.A. and Tjian R. 1994. TBP-TAF complexes: Selectivity factors for eukaryotic transcription. *Curr. Opin. Cell Biol.* **6:** 403.

Grant P.A., Schieltz D., Pray-Grant M.G., Steger D.J., Reese J.C., Yates J.R., III, and Workman J.L. 1998. Identification of novel components within the SAGA histone acetyltransferase complex required for the acetylation of nucleosomal histone. *Cell* **94:** 45.

Hai T.W., Horikoshi M., Roeder R.G., and Green M.R. 1988. Analysis of the transcription factor ATF in the assembly of a functional preinitiation complex. *Cell* **54:** 1043.

Hansen S.K., Takada S., Jacobson R.H., Lis J.T., and Tjian R. 1997. Transcription properties of a cell type-specific TATA-binding protein, TRF. *Cell* **91:** 71.

Heller H. and Bengal E. 1998. TFIID (TBP) stabilizes the binding of MyoD to its DNA site at the promoter and MyoD facilitates the association of TFIIB with the preinitiation complex. *Nucleic Acids Res.* **26:** 2112.

Henglein B., Chenivesse X., Wang J., Eick D., and Brechot C. 1994. Structure and cell cycle-regulated transcription of the human cyclin A gene. *Proc. Natl. Acad. Sci.* **91:** 5490.

Hernandez N. 1993. TBP, a universal eukaryotic transcription factor? *Genes Dev.* **7:** 1291.

Hirschhorn R.R., Aller P., Yuan Z.-A., Gibson C.W., and Baserga R. 1984. Cell-cycle specific cDNAs from mammalian cells temperature sensitive for growth. *Proc. Natl. Acad. Sci.* **81:** 6004.

Hisatake K., Hasegawa S., Takada R., Nakatani Y., Horikoshi M., and Roeder R.G. 1993. The p250 subunit of native TATA box-binding factor TFIID is the cell-cycle regulatory protein CCG1. *Nature* **362:** 179.

Imbalzano A.N., Kwon H., Green M.R., and Kingston R.E. 1994. Facilitated binding of TATA-binding protein to nucleosomal DNA. *Nature* **370:** 481.

Jacq X., Brou C., Lutz Y., Davidson I., Chambon P., and Tora L. 1994. Human TAF$_{II}$30 is present in a distinct TFIID complex and is required for transcriptional activation by the estrogen receptor. *Cell* **79:** 107.

Johnson F.B. and Krasnow M.A. 1992. Differential regulation of transcription preinitiation complex assembly by activator and repressor homeo domain proteins. *Genes Dev.* **6:** 2177.

Katagiri F., Yamazaki K., Horikoshi M., Roeder R.G., and Chua N.H. 1990. A plant DNA-binding protein increases the number of active preinitiation complexes in a human in vitro transcription system. *Genes Dev.* **4:** 1899.

Kaufmann J. and Smale S.T. 1994. Direct recognition of initiator elements by a component of the transcription factor IID complex. *Genes Dev.* **8:** 821.

Kaufmann J., Ahrens K., Koop R., Smale S.T., and Muller R. 1998. CIF150, a human cofactor for TFIID-dependent initiator function. *Mol. Cell. Biol.* **18:** 233.

Kim T.K., Hashimoto S., Kelleher R.J., III, Flanagan P.M., Kornberg R.D. Horikoshi M., and Roeder R.G. 1994. Effects of activation-defective TBP mutations on transcription initiation in yeast. *Nature* **369:** 252.

Kim Y.-J., Bjorklund S., Li Y., Sayre M.H., and Kornberg R.D. 1994. A multiprotein mediator of transcriptional activation and its interaction with the C-terminal repeat domain of RNA polymerase II. *Cell* **77:** 599.

Klein C. and Struhl K. 1994. Increased recruitment of TATA-binding protein to the promoter by transcriptional activation domains in vivo. *Science* **206:** 280.

Klenk H.-P. and Doolittle W.F. 1994. Archaea and eukaryotes versus bacteria? *Curr. Biol.* **4:** 920.

Kobayashi N., Boyer T.G., and Berk A.J. 1995. A class of activation domains interacts directly with TFIIA and stimulates TFIIA-TFIID-promoter complex assembly. *Mol. Cell. Biol.* **15:** 6465.

Koh S.S., Ansari A.Z., Ptashne M., and Young R.A. 1998. An activator target in the RNA polymerase II holoenzyme. *Mol. Cell* **1:** 895.

Kokubo T., Gong D.-W., Yamashita S., Horikoshi M., Roeder R.G., and Nakatani Y. 1993. *Drosophila* 230-kD TFIID subunit, a functional homolog of the human cell cycle gene product, negatively regulates DNA binding of the TATA box-binding subunit of TFIID. *Genes Dev.* **7:** 1033.

Koleske A.J. and Young R.A. 1994. An RNA polymerase II holoenzyme responsive to activators. *Nature* **368:** 466.

———. 1995. The RNA polymerase II holoenzyme and its implications for gene regulation. *Trends Biochem. Sci.* **20:** 113.

Lee T.I. and Young R.A. 1998. Regulation of gene expression by TBP-associated proteins. *Genes Dev.* **12:** 1398.

Lefstin J. and Yamamoto K. 1998. Allosteric effects of DNA on transcriptional regulators. *Nature* **392:** 885.

Lieberman P.M. and Berk A.J. 1994. A mechanism for TAFs in transcriptional activation: Activation domain enhancement of TFIID-TFIIA-promoter DNA complex formation. *Genes Dev.* **8:** 995.

Lin Y.S and Green M.R. 1991. Mechanism of action of an acidic transcriptional activator in vitro. *Cell* **64:** 971.

Lin Y.S., Ha I., Maldonado E., Reinberg D., and Green M.R. 1991. Binding of general transcription factor TFIIB to an acidic activating region. *Nature* **353:** 569.

Liu H.T., Gibson C.W., Hirschhorn R.R., Rittling S., Baserga R., and Mercer W.E. 1985. Expression of thymidine kinase and dihydrofolate reductase gene in mammalian ts mutants of the cell cycle. *J. Biol. Chem.* **260:** 3269.

Maldonado E., Sheikhattar R., Sheldon M., Cho H., Drapkin R., Rickert P., Lees E., Anderson C.W., Linn S., and Reinberg D. 1996. A human RNA polymerase II complex associated with SRB and DNA repair proteins. *Nature* **381:** 86.

Martinez E., Zhou Q., L'Etoile N.D., Oelgeschlager T., Berk A.J., and Roeder R.G. 1995. Core promoter-specific function of a mutant transcription factor TFIID defective in TATA-box binding . *Proc. Natl. Acad. Sci.* **92:** 11864.

Mengus G., May M., Jacq X., Staub A., Tora L., Chambon P., and Davidson I. 1995. Cloning and characterization of hTAF$_{II}$18, hTAF$_{II}$20 and hTAF$_{II}$28: Three subunits of human transcription factor TFIID. *EMBO J.* **14:** 1520.

Mitchell P.J. and Tjian R. 1989. Transcriptional regulation in mammalian cells by sequences-specific DNA binding proteins. *Science* **245:** 371.

Mizzen C.A., Yang X.-J., Kokubo T., Brownell J.E., Bannister A.J., Owen-Hughes T., Workman J., Wang L., Berger S.L., Kouzarides T., Nakatani Y., and Allis C. D. 1996. The TAF$_{II}$250 subunit of TFIID has histone acetyltransferase activity. *Cell* **87:** 1261.

Moqtaderi Z., Yale J.D., Struhl K., and Buratowski S. 1996a. Yeast homologues of higher eukaryotic TFIID subunits. *Proc. Natl. Acad. Sci.* **93:** 14654.

Moqtaderi Z., Bai Y., Poon D., Weil P.A., and Struhl K. 1996b. TBP-associated factors are not generally required for transcriptional activation in yeast. *Nature* **383:** 188.

Nakajima N., Horikoshi M., and Roeder R.G. 1988. Factors involved in specific transcription by mammalian RNA polymerase II: Purification, genetic specificity, and TATA box-promoter interactions of TFIID. *Mol. Cell. Biol.* **8:** 4028.

Nakatani Y., Horikoshi M., Brenner M., Yamamoto T., Besnard F., Roeder R.G., and Freese E. 1990. A downstream initiation element required for efficient TATA box binding and in vitro function of TFIID. *Nature* **348:** 86.

Oelgeschlager T., Chiang C.-M., and Roeder R.G. 1996. Topology and reorganization of a human TFIID-promoter complex. *Nature* **382:** 735.

Oelgeschlager T., Tao Y. Kang Y.K., and Roeder R.G. 1998. Transcription activation via enhanced preinitiation complex assembly in a human cell-free system lacking TAF$_{II}$s. *Mol. Cell* **1:** 925.

Orphanides G., Lagrange T., and Reinberg D. 1996. The general transcription factors of RNA polymerase II. *Genes Dev.* **10:** 2657.

Ponticelli A.S., Pardee T.S., and Struhl K. 1995. The glutamine-rich activation domains of human Sp1 do not stimulate transcription in *Saccharomyces cerevisiae*. *Mol. Cell. Biol.* **15:** 983.

Poon D., Bai Y., Campbell A.M., Bjorklund S., Kim Y.J., Zhou S., Kornberg R.D., and Weil P.A. 1995. Identification and characterization of a TFIID-like multiprotein complex from *Saccharomyces cerevisiae*. *Proc. Natl. Acad. Sci.* **92:** 8224.

Prescott D.M. and Bender M.A. 1962. Synthesis of RNA and protein during mitosis in mammalian tissue culture cells. *Exp. Cell Res.* **26:** 260.

Ptashne M. 1988. How eukaryotic transcriptional activators work. *Nature* **335:** 683.

Ptashne M. and Gann A. 1990. Activators and targets. *Nature* **346:** 329.

———. 1997. Transcriptional activation by recruitment. *Nature* **386:** 569.

Pugh B.F. 1996. Mechanisms of transcription complex assembly. *Curr. Opin. Cell Biol.* **8:** 303.

Purnell B.A., Emanuel P.A., and Gilmour D.S. 1994. TFIID sequence recognition of the initiator and sequences further

downstream in *Drosophila* class II genes. *Genes Dev.* **8:** 830.

Ranish J.A. and Hahn S. 1996. Transcription: Basal factors and activation. *Curr. Opin. Genet. Dev.* **6:** 151.

Reese J.C., Apone L., Walker S.S., Griffin L.A., and Green M.R. 1994. Yeast TAF$_{II}$s in a multisubunit complex required for activated transcription. *Nature* **371:** 523.

Roberts S.G.E., Choy B., Walker S.S., Lin S.S., and Green M.R. 1995. A role for activator-mediated TFIIB recruitment in diverse aspects of transcriptional regulation. *Curr. Biol.* **5:** 508.

Roberts S.G.E., Ha I., Maldonado E., Reinberg D., and Green M.R. 1993. Interaction between an acidic activator and transcription factor TFIIB is required for transcriptional activation. *Nature* **363:** 741.

Roeder R.G. 1996. The role of general initiation factors in transcription by RNA polymerase II. *Trends Biochem. Sci.* **21:** 327.

Ruppert S., Wang E.H., and Tjian R. 1993. Cloning and expression of human TAF$_{II}$ 250: A TBP-associated factor implicated in cell-cycle regulation. *Nature* **362:** 175.

Segil N., Guermah M., Hoffmann A., Roeder R.G., and Heintz N. 1996. Mitotic regulation of TFIID: Inhibition of activator-dependent transcription and changes in subcellular localization. *Genes Dev.* **10:** 2389.

Selleck S.B., and Majors J. 1987. In vivo DNA-binding properties of a yeast transcription activator protein. *Mol. Cell. Biol.* **7:** 3260.

Shen W.-C. and Green M.R. 1997. Yeast TAF$_{II}$145 functions as a core promoter selectivity factor, not a general coactivator. *Cell* **90:** 615.

Simon M.C., Fisch T.M., Benecke B.J., Nevins J.R., and Heintz N. 1988. Definition of multiple, functionally distinct TATA element, one of which is a target in the *hsp70* promoter for E1A regulation. *Cell* **52:** 723.

Stringer K.F., Ingles C.J., and Greenblatt J. 1990. Direct and selective binding of an acidic transcriptional activation domain to the TATA-box factor TFIID. *Nature* **345:** 783.

Struhl K. 1995. Yeast transcriptional regulatory mechanisms. *Annu. Rev. Genet.* **29:** 651.

Suzuki-Yagawa Y., Guermah M., and Roeder R.G. 1997. The ts13 mutation in the TAF(II)250 subunit (CCG1) of TFIID directly affects transcription of D-type cyclin genes in cells arrested in G1 at the nonpermissive temperature. *Mol. Cell. Biol.* **17:** 3284.

Sypes M.A. and Gilmour D.S. 1994. Protein/DNA crosslinking of a TFIID complex reveals novel interactions downstream of the transcription start. *Nucleic Acids Res.* **22:** 807.

Takada R., Nakatani Y., Hoffmann A., Kokubo T., Hasegawa S., Roeder R.G., and Horikoshi M. 1992. Identification of human TFIID components and direct interaction between a 250-kDa polypeptide and the TATA box-binding protein (TFIIDτ). *Proc. Natl. Acad. Sci.* **89:** 11809.

Talavera A. and Basilico C. 1977. Temperature sensitive mutants of BHK cells affected in cell cycle progression. *J. Cell. Physiol.* **92:** 425.

Tanese N., Pugh B.F., and Tjian R. 1991. Coactivators for a proline-rich activator purified from the multisubunit human TFIID complex. *Genes Dev.* **5:** 2212.

Taylor I.C.A. and Kingston R.E. 1990. Factor substitution in human HSP70 gene promoter: TATA-dependent and TATA-independent interactions. *Mol. Cell. Biol.* **10:** 165.

Timmers H.T. and Sharp P.A. 1991. The mammalian TFIID protein is present in two functionally distinct complexes. *Genes Dev.* **5:** 1946.

Timmers H.T., Meyers R.E., and Sharp P.A. 1992. Composition of transcription factor B-TFIID. *Proc. Natl. Acad. Sci.* **89:** 8140.

Tjian R. 1996. The biochemistry of transcription in eukaryotes: A paradigm for multisubunit regulatory complexes. *Philos. Trans. R. Soc. Lond. B. Biol. Sci.* **351:** 491.

Tjian R. and Maniatis T. 1994. Transcriptional activation: A complex puzzle with few easy pieces. *Cell* **77:** 5.

Velculescu V.E., Zhang L., Zhou W., Vogelstein J., Basrai M.A., Bassett D.E., Hieter P., Vogelstein B., and Kinzler K.W. 1997. Characterization of the yeast transcriptome. *Cell*

88: 243.

Verrijzer C.P and Tjian R. 1996. TAFs mediate transcriptional activation and promoter selectivity. *Trends Biochem. Sci.* **21:** 338.

Verrijzer C.P., Chen J.-L., Yokomori K., and Tjian R. 1995. Binding of TAFs to core elements directs promoter selectivity by RNA polymerase II. *Cell* **81:** 1115.

Verrijzer C.P., Yokomori K., Chen J.-L., and Tjian R. 1994. *Drosophila* TAFII150: Similarity to yeast gene TSM-1 and specific binding to core promoter DNA. *Science* **264:** 933.

Walker S.S., Reese J.C., Apone L.M., and Green M.R. 1996. Transcription activation in cells lacking $TAF_{II}s$. *Nature* **383:** 185.

Walker S.S., Shen W.-C., Reese J.C., Apone L.M., and Green M.R. 1997. Yeast $TAF_{II}145$ required for transcription of G1/S cyclin genes and regulated by the cellular growth state. *Cell* **90:** 607.

Wampler S.L. and Kadonaga J.T. 1992. Functional analysis of *Drosophila* transcription factor IIB. *Genes Dev.* **6:** 1542.

Wang E.H. and Tjian R. 1994. Promoter-selective defect in cell cycle mutant ts13 rescued by $hTAF_{II}250$. *Science* **263:** 811.

Wang W., Carey M., and Gralla J.D. 1992a. Polymerase II promoter activation: Closed complex formation and ATP-driven start site opening. *Science* **255:** 450.

Wang W., Gralla J.D., and Carey M. 1992b. The acidic activator GAL4-AH can stimulate polymerase II transcription by promoting assembly of a closed complex requiring TFIIA and TFIID. *Genes Dev.* **6:** 1716.

White J., Brou C., Wu J., Lutz Y., Moncollin V., and Chambon P.

1992. The acidic transcriptional activator GAL-VP16 acts on preformed template-committed complexes. *EMBO J.* **11:** 2229.

Wieczorek E., Brand M., Jacq X., and Tora L. 1998. Function of TAFII-containing complex without TBP in transcription by RNA polymerase II. *Nature* **393:** 187.

Workman J.L. and Roeder R.G. 1987. Binding of transcription factor TFIID to the major late promoter during in vitro nucleosome assembly potentiates subsequent initiation by RNA polymerase II. *Cell* **51:** 613.

Workman J.L., Roeder R.G., and Kingston R.E. 1990. An upstream transcription factor, USF (MLTF), facilitates the formation of preinitiation complexes during in vitro chromatin assembly. *EMBO J.* **9:** 1299.

Workman J.L., Taylor I.C., and Kingston R.E. 1991. Activation domains of stably bound GAL4 derivatives alleviate repression of promoter by nucleosomes. *Cell* **64:** 533.

Wu C. 1984. Activation protein factor binds in vitro to upstream control sequences in heat shock gene chromatin. *Nature* **311:** 81.

Zawel L. and Reinberg D. 1992. Advances in RNA polymerase II transcription. *Curr. Opin. Cell. Biol.* **4:** 488.

Zawel L., Kumar K.P., and Reinberg D. 1995. Recycling of the general transcription factors during RNA polymerase II transcription. *Genes Dev.* **9:** 1479.

Zhou Q., Boyer T.G., and Berk A.J. 1993. Factors (TAFs) required for activated transcription interact with TATA box-binding protein conserved core domain. *Genes Dev.* **7:** 180.

Zhou Q., Liberman P.M., Boyer T.G., and Berk A.J. 1992. Holo-TFIID supports transcriptional stimulation by diverse activators and from a TATA-less promoter. *Genes Dev.* **6:** 1964.

Mechanism and Regulation of Yeast RNA Polymerase II Transcription

R.D. KORNBERG

Department of Structural Biology, Stanford School of Medicine, Stanford California 94305

The mRNA synthetic machinery of eukaryotes comprises general and gene-specific components. The general components, which include RNA polymerase II and general transcription factors (GTFs), recognize and initiate transcription of a core, or minimal promoter, usually consisting of a TATA box and transcription start site. The gene-specific components include activator and repressor proteins, which interact with DNA elements termed enhancers and operators, respectively. Research summarized here is directed toward the mechanisms of both transcription initiation and its regulation.

Cell-free systems supporting basal (unregulated) transcription have been resolved to homogeneity, revealing a requirement for five GTFs, termed TFIIB, -E, -F, -H, and TATA-binding protein (TBP) (Conaway and Conaway 1993, 1997). Additional factors, such as TFIIA, have also been identified, but these have little influence on the basal reaction performed with pure transcription proteins. The 15 essential subunits of the GTFs have been conserved across species from yeast to humans. The challenge of understanding the transcription initiation mechanism is to explain why so many accessory proteins are required and to define their roles. This challenge is met in the discussion to follow through a combination of biochemical and structural data obtained in a yeast transcription system.

Regulation requires communication between the general and gene-specific components of the transcription machinery. Regulatory information must be transduced from enhancers or operators to promoters. Enhancer-promoter communication was at first thought to be direct. Evidence was obtained for activator-GTF interaction both in vitro and in vivo (Stringer et al. 1990; Ingles et al. 1991; Truant et al. 1993; Roberts and Green 1994). Studies of regulation in cell extracts, however, indicated that additional factors were required. Attention in *Drosophila* and human systems focused on TAF$_{II}$s, a set of about eight proteins that interact TBP. A TAF$_{II}$-TBP complex was shown to support activated transcription in vitro (Chen et al. 1994). In the yeast system, an altogether different result was obtained. A 20-protein complex termed Mediator, unrelated to TAF$_{II}$s, was isolated as a required component for transcriptional activation reconstituted with essentially pure proteins (Kim et al. 1994). Mediator interacts with RNA polymerase II to form a "holoenzyme" complex (Thompson et al. 1993; Kim et al. 1994).

The question of whether TAF$_{II}$s play a part in enhancer-promoter communication in vivo similar to that indicated by the studies in vitro was addressed by genetic means, with surprising results: Deletion of TAF$_{II}$ genes or destruction of the proteins in yeast had no effect on the transcription of any inducible gene tested (Moqtaderi et al. 1996; Walker et al. 1996). At a small number of promoters where TAF$_{II}$s were important for transcription, the sequences involved were in the immediate vicinity of the TATA box and did not include upstream (enhancer) elements (Shen and Green 1997). Evidently, TAF$_{II}$s augment the sequence specificity of TBP, rather than enabling a response to upstream elements.

A similar analysis must be performed for Mediator to assess its involvement in regulation in vivo. A Mediator protein mutation has already been described that abolishes both basal and activated transcription of a wide range of promoters in yeast (Thompson and Young 1995). Other Mediator protein mutations have been shown to diminish activation and repression at subsets of yeast promoters (Lee et al. 1997; Myers et al. 1999). Here, we review recent studies of Mediator and discuss the implications for the mechanism of transcription control.

CHARACTERIZATION OF GENERAL TRANSCRIPTION FACTORS

Binding and catalytic activities of the GTFs have been characterized by various means, including gel shift, biosensor chip, and X-ray crystallographic analyses (Bushnell et al. 1996; Roeder 1996 and references therein). The results may be summarized as follows. TBP binds on the outer surface of a bend in TATA DNA, creating a context for interaction of TFIIB, which binds on the inner surface of the bend, primarily with DNA as well. TFIIB is required for association of the TBP-DNA complex with an RNA polymerase–TFIIF complex and has therefore been suggested to have a "bridging" role. TFIIB-polymerase interaction determines the location of the transcription start site. TFIIE binds to the complex of polymerase with the other factors and recruits TFIIH, which possesses two catalytic activities: a DNA-dependent ATPase/helicase, believed to melt DNA around the transcription start site, facilitating the initiation of transcription, and a cyclin-dependent protein kinase, responsible for hyperphosphorylation of the polymerase carboxy-terminal domain (CTD) occurring at the initiation of transcription, which appears to cause the displacement of Mediator (Svejstrup et al. 1997).

STRUCTURE OF RNA POLYMERASE II

The three-dimensional structure of RNA polymerase II has been determined at 16-Å resolution by electron microscopy and image processing of two-dimensional crystals in negative stain (Darst et al. 1991). A cleft appropriate in size for binding DNA was observed (Fig. 1). Electron crystallographic analysis of an RNA polymerase II transcription complex, containing DNA and RNA, supported this role of the cleft (C. Poglitsch et al., in prep.). Extension of the analysis also revealed the likely mechanism for entry of DNA in the cleft. Two conformations of the polymerase were detected, differing by the presence (closed conformation) or absence (open conformation) of a domain of protein density in the mouth of the cleft (Asturias et al. 1997, 1998). The results were consistent with binding/release of DNA in the open conformation, as required for transcription initiation and termination, and retention of DNA in the closed conformation, presumably to enhance the processivity of RNA chain elongation.

STRUCTURE OF RNA POLYMERASE II: GENERAL TRANSCRIPTION FACTOR COMPLEXES

Difference electron crystallographic analysis between RNA polymerase II–general transcription factor complexes and the polymerase alone has revealed the sites of interaction of TFIIB and TFIIE with the enzyme (Fig. 1) (Leuther et al. 1996). The results were paradoxical: TFIIB bound at a location remote from the active center cleft, seemingly inconsistent with its role in start site determination, and TFIIE interacted with the polymerase domain defining the closed conformation, although it also has a role in initiation, presumed to involve the open con-

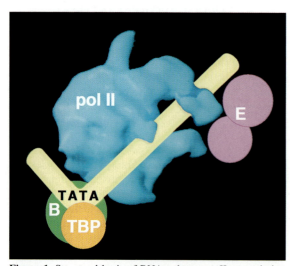

Figure 1. Structural basis of RNA polymerase II transcription start site determination. The three-dimensional structure of RNA polymerase II at 16-Å resolution in the open conformation is shown in blue. TBP, TFIIB, and TFIIE (*orange, green,* and *purple*, respectively) are drawn symbolically, to scale, in locations revealed by X-ray and electron crystallography. The yellow cylinder symbolizes duplex DNA, also to scale.

formation of the enzyme. The paradoxes were resolved by a proposed mechanism of transcription initiation. The proposed mechanism is described below in four stages, although the actual number of steps in the initiation process is unknown.

MECHANISM OF TRANSCRIPTION INITIATION

1. The distance from TFIIB to the active center cleft revealed by electron crystallography is about 110 Å, corresponding to 32 bp of DNA. The coincidence of this distance with a spacing of about 30 bp between the TATA box and transcription start site of almost all RNA polymerase II promoters suggests a simple stereochemical basis for start site selection (Fig. 1): TFIIB escorts a TBP-TATA box complex to the polymerase and orients the complex so that a straight path of the DNA across the surface of the enzyme places the start site in the active center. There appears to be a channel in the polymerase structure defining such a path (Fig. 1), although the association of DNA with this channel remains to be demonstrated.

2. We suppose that entry of DNA in the active center cleft provokes a conformational change from the open to the closed state of the polymerase. This change creates the site for binding TFIIE. The presence of TFIIE therefore signals the acquisition of the closed conformation. It marks the completion of the polymerase-DNA interaction.

3. TFIIH does not contact RNA polymerase II directly, but rather it binds to TFIIE and enters the initiation complex by virtue of this interaction. TFIIE staging the entry of TFIIH makes sense in light of the catalytic activities of TFIIH. These activities terminate the regulatory phase of the initiation process and trigger the transcription reaction, so they should only come into play following the completion of polymerase-DNA interaction.

4. The mechanism of start site determination by TFIIB proposed here implies the nonspecific interaction of polymerase II with a considerable length of DNA. The possibility arises of interaction with nonpromoter DNA. The proposed DNA channel would be occluded, inhibiting transcription. Such inhibition may be prevented by TFIIF, shown previously to disrupt nonspecific polymerase-DNA interaction (Conaway and Conaway 1997). We propose that TFIIF controls access to the DNA channel, allowing entry of DNA only in a complex with TBP and TFIIB. A location of TFIIF at some point along the channel, between the TFIIB-binding site and the active center cleft, would be anticipated (Fig. 2). Indeed, cross-linking experiments have demonstrated the proximity of TFIIF to the DNA between the TATA box and transcription start site (Ebright, this volume and references therein).

The resulting picture of the RNA polymerase II initiation complex (Fig. 2) may be viewed as a solution, in outline, of the transcription initiation problem. It assigns a

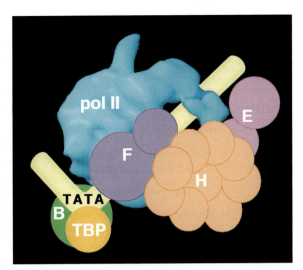

Figure 2. Topography of the RNA polymerase II transcription initiation complex. RNA polymerase II is shown in the closed conformation. The symbolism is the same as in Fig. 1, with the addition of TFIIF and TFIIH (*pale purple* and *pale orange*, respectively), drawn symbolically, to scale, in locations based on the role for TFIIF suggested in the text, and on evidence for TFIIH-CTD and TFIIH-TFIIE interaction.

simple, essential role to all five general transcription factors. It embodies results of biochemical studies (Buratowski et al. 1989 and references cited above) and structural information on all components except TFIIF and TFIIH, but it is nonetheless hypothetical. Further studies will be needed to test the validity of the proposed functional roles and structural interactions.

MEDIATOR

Genes for all 20 subunits of yeast Mediator have been identified (Myers et al. 1998 and refererences therein). They fall into three groups: *SRB* genes, identified from a genetic screen for CTD-interacting proteins; a second group of genes, identified from screens for mutations affecting repression, as well as activation of transcription; and *MED* genes, encoding Mediator subunits not revealed by previous screens. Of the 11 *SRB* gene products described, only 5, Srb2, -4, -5, -6, and -7, are required for function of Mediator in vitro. The second group of previously identified genes includes *SIN4*, *RGR1*, *GAL11*, and *HRS1*, which have been characterized genetically as "global repressors," and are involved in regulation of overlapping if not identical sets of genes. Of the 20 Mediator genes, 10 are essential for viability of yeast.

Consistent with the genetic findings, members of the *SIN4/RGR1* group appear to be clustered in a module, or subassembly, of the Mediator complex. One line of evidence for such a module is the loss of all its members but no other Mediator subunits from an *rgr1* truncation mutant (Li et al. 1995). Organization within the module is revealed by the loss of subsets of members from various deletion mutants (Myers et al. 1999). Recent biochemical and genetic studies of the deletion mutants have con-

firmed the importance of the module for transcriptional activation (Myers et al. 1999). These studies together with previous genetic evidence establish a dual role for the *SIN4/RGR1* module in activation and repression.

MECHANISM OF TRANSCRIPTIONAL ACTIVATION

It has often been suggested that transcriptional activation results from "recruitment" of transcription factors to promoters by activator-factor interaction (Ptashne and Gann 1997). This mechanism was first proposed for RNA polymerase II transcription on the basis of evidence for direct activator-TBP interaction in vitro. Subsequent reports of activator-TFIIB and activator-TFIIH interaction raised the possibility of activation by recruitment of these factors as well. Most recently, activator-Mediator interaction was taken to indicate recruitment of the RNA polymerase II holoenzyme. Support for these ideas came from effects of DNA-binding domain–TBP and DNA-binding domain–Mediator fusions upon transcription in vivo.

In addition or as an alternative to recruitment, an activator may influence a step subsequent to initiation complex formation, such as promoter opening or the transition to transcription elongation. Such a downstream effect would be more compatible with the dual role of the *SIN4/RGR1* module in activation and repression. Action of the module would only be required at a single point in the initiation pathway. An activator would spur progress of the reaction past this point, whereas a repressor would restrain or prevent it.

REFERENCES

Asturias F., Chang W., Li Y., and Kornberg R. 1998. Electron crystallography of yeast RNA polymerase II preserved in vitreous ice. *Ultramicroscopy* **70:** 133.

Asturias F., Meredith G., Poglitsch C., and Kornberg R. 1997. Two conformations of RNA polymerase II revealed by electron crystallography. *J. Mol. Biol.* **272:** 536.

Buratowski S., Hahn S., Guarente L., and Sharp P.A. 1989. Five intermediate complexes in transcription initiation by RNA polymease II. *Cell* **56:** 549.

Bushnell D.A., Bamdad C., and Kornberg R.D. 1996. A minimal set of RNA polymerase II transcription protein interactions. *J. Biol. Chem.* **271:** 20170.

Chen J.-L., Attardi L.D., Verrijzer C.P., Yokomori K., and Tjian R. 1994. Assembly of recombinant TFIID reveals differential coactivator requirements for distinct transcriptional activator. *Cell* **79:** 93.

Conaway R. and Conaway J.W. 1993. General initiation factors for RNA polymerase II. *Annu. Rev. Biochem.* **62:** 161.

———. 1997. General transcription factors for RNA polymerase II. *Prog. Nucleic Acid Res. Mol. Biol.* **56:** 327.

Darst S.A., Edwards A.M., Kubalek E.W., and Kornberg R.D. 1991. Three-dimensional structure of yeast RNA polymerase II at 16Å resolution. *Cell* **66:** 121.

Ingles C., Shales M., Cress W., Triezenberg S., and Greenblatt J. 1991. Reduced binding of TFIID to transcriptionally compromised mutants of VP16. *Nature* **351:** 588.

Kim Y.J., Bjorklund S., Li Y., Sayre M.H., and Kornberg R.D. 1994. A multiprotein mediator of transcriptional activation and its interaction with the C-terminal repeat domain of RNA polymerase II. *Cell* **77:** 599.

Lee Y., Min S., Gim B., and Kim Y. 1997. A transcriptional mediator protein that is required for activation of many RNA polymerase II promoters and is conserved from yeast to humans. *Mol. Cell. Biol.* **17:** 4622.

Leuther K.K., Bushnell D.A., and Kornberg R.D. 1996. Two-dimensional crystallography of transcription factor IIB- and IIE-RNA polymerase II complexes: Implications for start site selection and initiation complex formation. *Cell* **85:** 773.

Li Y., Bjorklund S., Jiang Y.W., Kim Y.-J., Lane W.S., Stillman D.J., and Kornberg R.D. 1995. Yeast global transcriptional regulators Sin4 and Rgr1 are components of mediator complex/RNA polymerase II holoenzyme. *Proc. Natl. Acad. Sci.* **92:** 10864.

Moqtaderi Z., Bai Y., Poon D., Weil P. A., and Struhl K. 1996. TBP-associated factors are not generally required for transcriptional activation in yeast. *Nature* **383:** 188.

Myers L.C., Gustafsson C.M., Hayashibara K.C., Brown P.O., and Kornberg R.D. 1999. Mediator protein mutations that selectively abolish activated transcription. *Proc. Natl. Acad. Sci.* **96:** (in press).

Myers L.C., Gustafsson C.M., Bushnell D.A., Lui M., Erdjument-Bromage H., Tempst P., and Kornberg R.D. 1998. The Med proteins of yeast and their function through the RNA polymerase II C-terminal domain. *Genes Dev.* **12:** 45.

Ptashne M. and Gann A. 1997. Transcriptional activation by recruitment. *Nature* **386:** 569.

Roberts S. and Green M. 1994. Activator-induced conformational change in general transcription factor TFIIB. *Nature* **371:** 717.

Roeder R. 1996. The role of general initiation factors in transcription by RNA polymerase II. *Trends Biochem. Sci.* **21:** 327.

Shen W.-C. and Green M.R. 1997. Yeast TAF$_{II}$145 functions as a core promoter selectivity factor, not a general coactivator. *Cell* **90:** 615.

Stringer K.F., Ingles C.J., and Greenblatt J. 1990. Direct and selective binding of an acidic transcriptional activation domain to the TATA-box factor TFIID. *Nature* **345:** 783.

Svejstrup J., Li Y., Fellows J., Gnatt A., Bjorklund S., and Kornberg R. 1997. Evidence for a mediator cycle at the initiation of transcription. *Proc. Natl. Acad. Sci.* **94:** 6075.

Thompson C.M. and Young R.A. 1995. General requirement for RNA polymerase II holoenzymes *in vivo. Proc. Natl. Acad. Sci.* **92:** 4587.

Thompson C.M., Koleske A.J., Chao D.M., and Young R.A. 1993. A multisubunit complex associated with the RNA polymerase II CTD and TATA-binding protein in yeast. *Cell* **73:** 1361.

Truant R., Xiao H., Ingles C.J., and Greenblatt J. 1993. Direct interaction between the transcriptional activation domain of human p53 and the TATA box-binding protein. *J. Biol. Chem.* **268:** 2284.

Walker S.S., Reese J.C., Apone L.M., and Green M.R. 1996. Transcription activation in cells lacking TAF$_{II}$s. *Nature* **383:** 185.

Functional and Structural Analysis of the Subunits of Human Transcription Factor TFIID

I. Davidson, C. Romier, A.-C. Lavigne, C. Birck,* G. Mengus,† O. Poch, and D. Moras
Institut de Génétique et de Biologie Moléculaire et Cellulaire, CNRS/INSERM/ULP,
163-67404 Illkirch Cédex, C.U. de Strasbourg France

Transcription initiation of protein-coding genes in eukaryotes requires the assembly of a macromolecular complex containing the general transcription factors TFIIA, TFIIB, TFIID, TFIIE, TFIIF, and TFIIH (for review, see Orphanides et al. 1996; Roeder 1996), together with RNA polymerase II. TFIID is a multisubunit complex composed of the TATA-binding protein (TBP) and TBP-associated factors (TAF$_{II}$s). TAF$_{II}$s were first characterized in the TFIID from *Drosophila* embryos, and then from human HeLa cells, and more recently in yeast (Dynlacht et al. 1991; Pugh and Tjian 1991; Tanese et al. 1991; Takada et al. 1992; Zhou et al. 1992; Brou et al. 1993; Chiang et al. 1993; Reese et al. 1994; Poon et al. 1995). They range in size from 250 kD in metazoans (145 kD in yeast) to 15 kD. The cDNAs for 11 human (h)TAF$_{II}$s have been cloned, and homologs for all but one (hTAF$_{II}$135) exist in yeast (Moqtaderi et al. 1996a). Genes encoding homologs of hTAF$_{II}$55, hTAF$_{II}$30, and hTAF$_{II}$18 exist in *Drosophila*, but the corresponding proteins have not been described as *Drosophila* (d)TFIID subunits and therefore may not be tightly associated with TBP. Similarly, the existence of a homolog of dTAF$_{II}$150 in hTFIID is as yet subject to some controversy. Homologous TAF$_{II}$s from different species contain highly conserved regions, often corresponding structured domains, along with nonconserved often highly charged regions. The structured domains, for example, the histone-fold motif, likely mediate many of the conserved TAF-TAF interactions. Although originally described as TFIID components, some TAF$_{II}$s have also been found in other complexes such as TFTC, the SAGA complex in yeast, and the PCAF complex in human cells (Wieczorek et al. 1998; see other relevant chapters in this volume).

The identification and cloning of TAF$_{II}$s have led to significant advances in our understanding of their function. Both biochemical and genetic studies have shown that TAF$_{II}$s participate in promoter recognition (Verrijzer et al. 1995; Verrijzer and Tjian 1996; Shen and Green 1997). TAF$_{II}$250 has also been shown to possess kinase and histone acetyltransferase activities (Dikstein et al. 1996; Mizzen et al. 1996). However, initial interest in TAF$_{II}$s came from the observation that recombinant TBP could support basal, but not activated, transcription in re-

constituted systems in vitro from animal cell extracts, whereas both functions were supported by native TFIID (Dynlacht et al. 1991; Zhou et al. 1992; Brou et al. 1993; Chiang et al. 1993). Subsequently, many specific activator-TAF$_{II}$ interactions have been described, and multiple activator-TAF$_{II}$ interactions can result in transcriptional synergy in vitro (Goodrich et al. 1993; Hoey et al. 1993; Chen et al. 1994; Jacq et al. 1994; Sauer et al. 1995 and references therein). Such experiments led to the idea that TAF$_{II}$s can act as coactivators in vitro, linking the gene-specific regulators to the basal transcription machinery by direct protein-protein interactions.

RESULTS AND DISCUSSION

TAF$_{II}$28 and TAF$_{II}$135 Function as Specific Coactivators in Transfected Mammalian Cells

The ability of TAF$_{II}$s to function as coactivators in vitro has been extensively documented. With this in mind, we asked whether hTAF$_{II}$s could also act as specific coactivators in transfected mammalian cells. To address this, we considered two simple possibilities. Either a given TAF$_{II}$ is functionally limiting and its overexpression will increase the activity of activators that normally function via this TAF$_{II}$ or the TAF$_{II}$ is already at an optimal concentration and its overexpression will result in titration or "squelching" of a downstream or possibly upstream target protein, resulting in diminished activation by a given *trans*-activator. To test these possibilities, we looked at the effect of overexpression of hTAF$_{II}$s on activation by different activators. To simplify the assay, we first used chimeric activators composed of the DNA-binding domain of the yeast activator GAL4 (G4) fused to various activation domains along with a reporter gene under the control of a minimal G4-responsive promoter.

Previously, it had been shown that a chimeric G4-activator (G4-RXRβ[DE]) comprising the ligand-binding domain (DE region) with the ligand-inducible activation function-2 (AF-2) of the nuclear receptor (NR) for 9-*cis* retinoic acid (retinoid X receptor [RXR]) β only minimally activated transcription in transfected Cos cells (Nagpal et al. 1993) from a minimal promoter. However, when a vector expressing hTAF$_{II}$28 was cotransfected with G4-RXRβ(DE), a strong ligand-dependent transcriptional activation was observed (Fig. 1, lanes 1–5). Coexpression of hTAF$_{II}$28 also increased transcriptional activation by five- to sevenfold for the AF-2s of the es-

Present addresses: *The EMBL, Meyerhofstrasse 1, 69012, Heidelberg, Germany; †Institut de Pharmacologie et de Biologie Structurale, CNRS, UPR 9062, 205 route de Narbonne, 31077 Toulouse, France.

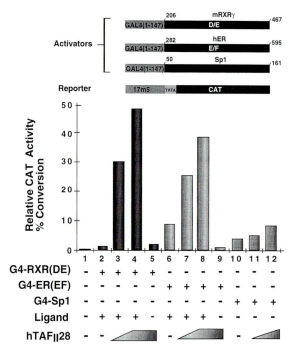

Figure 1. hTAF$_{II}$28 potentiates activation in transfected mammalian cells. The effects of coexpression of hTAF$_{II}$28 on activation by the chimeric G4-RXR, ER, and Sp1 activators in transfected Cos cells are shown graphically. The structures of the chimeric activators and the GAL4-responsive CAT reporter are diagrammed above the graph. The numbers represent the amino acid coordinates of each fragment in the native proteins. 17m5 shows the presence of five palindromic 17-mer GAL4-binding sites. The transfected plasmids used in each lane are shown below the graph. Cos cells were transfected by the calcium phosphate coprecipitation method. Transfections contained 1 μg of the CAT reporter and 2 μg of the RSV-luciferase reporter as internal control, 0.25 μg of the G4-ER or G4-Sp1 activator, or 1 μg of the G4-RXR activator expression vectors along with 0, 0.5, or 2 μg of the hTAF$_{II}$28 expression vector; 100 nM 9-cis retinoic acid or 50 nM β-estradiol was added as indicated (ligand +/–). After correction for transfection efficiency using the internal luciferase control, CAT assays were performed. The percentage of acetylated chloramphenicol was calculated by quantitative phosphorimager analysis of thin-layer chromatography plates and is presented graphically.

trogen receptor (ER) and vitamin D3 receptor (VDR) (G4-ER[EF], in Fig. 1, lanes 6–9, and G4-VDR[DE] in Fig. 4B, lanes 2–4). Smaller, but significant, increases were also seen with other NRs, but no significant effect on *trans*-activation by chimeric activators that do not belong to the nuclear receptor superfamily (see, e.g., G4-Sp1 in Fig. 1, lanes 10–12) was observed. Expression of hTAF$_{II}$28 also potentiated activation by full-length RXR homodimers bound to RXR-responsive elements, whereas expression of other hTAF$_{II}$s had no significant effect on activation by either the G4-RXR chimera or the full-length RXR (May et al. 1996). These results show that hTAF$_{II}$28 can act as a specific coactivator that will potentiate activation by several NRs in transfected mammalian cells.

The ability of hTAF$_{II}$28 to act as a transcriptional coactivator required interactions with TBP. Mutants of hTAF$_{II}$28 that impaired interactions with TBP led to a

loss of coactivator activity in transfected cells (May et al. 1996). In the course of such experiments, we precisely mapped amino acids in an α-helix of TAF$_{II}$28 required for the interaction with TBP (see also below).

Although hTAF$_{II}$28 acts as a coactivator for several nuclear receptors, no direct hTAF$_{II}$28-NR interactions could be observed either in vitro or in yeast two-hybrid assays. Furthermore, mutants of hTAF$_{II}$28, which do not interact with TBP and do not act as coactivators themselves, act as dominant negative repressors of the ER or RXR AF-2 activation induced by wild-type TAF$_{II}$28. This suggests that the hTAF$_{II}$28 interaction with the NRs is mediated indirectly by a titratable factor. Obvious candidates are the transcriptional intermediary factors (TIFs) which themselves associate with the liganded nuclear receptors (summarized in Fig. 2) (for a discussion of TIFs, see Chambon 1996; LeDouarin et al. 1997; Voegel et al. 1998 and references therein).

In agreement with the idea that TAF$_{II}$28 is functionally limiting in Cos cells, no TAF$_{II}$28 could be detected in TFIID from untransfected Cos cells using a mixture of monoclonal and polyclonal antibodies against hTAF$_{II}$28, whereas transfected hTAF$_{II}$28 did stably associate with Cos cell TFIID. These results suggest that TAF$_{II}$28 is depleted in Cos cells and that association of a fraction of the overexpressed hTAF$_{II}$28 with the Cos cell TFIID raises the concentration of TAF$_{II}$28-containing TFIID, resulting in increased activation (summarized in Fig. 2). In contrast, in HeLa cells, where hTAF$_{II}$28 can be readily detected in the TFIID, overexpression of hTAF$_{II}$28 led to a reduction of activation by the ER and VDR AF-2s (May et al. 1996).

The activity of a second type of activator is also potentiated by expression of hTAF$_{II}$28, but in this case, direct activator-TAF$_{II}$ interactions are invloved. The viral activator Tax forms a ternary complex by interacting with both hTAF$_{II}$28 and TBP (Caron et al. 1997). Overexpression of either TBP or hTAF$_{II}$28 potentiates activation by Tax, but coexpression of both proteins has a much stronger effect (summarized in Fig. 3).

Transcriptional activation in transfected mammalian cells can also be potentiated by the expression of hTAF$_{II}$135. Overexpression of this hTAF$_{II}$ strongly stimulates ligand-dependent activation by the AF-2s of the retinoic acid receptor (RAR) and the VDR (Fig. 4A, lanes, 4–9, and Fig. 4B, lanes 5–6). Activation by the VDR AF-2 was also potentiated by coexpression of hTAF$_{II}$28 (Fig. 4B, lanes 3–4), and coexpression of both TAF$_{II}$s resulted in an additive effect (Fig. 4B, lanes 7–8). In contrast to hTAF$_{II}$28, hTAF$_{II}$135 expression did not affect activation by the ER or RXR AF-2s (see, e.g., Fig. 4A, lanes 10–12). Therefore, hTAF$_{II}$135 is a specific coactivator for a limited subset of NRs.

To investigate the molecular mechanism by which these hTAF$_{II}$s enhance NR AF-2 activity, we used a promoter in which the TATA element had been mutated to TGTA. This TGTA element is not efficiently recognized by wild-type TBP, but it is recognized by the altered specificity mutant TBP spm3 (Strubin and Struhl 1992). A G4-responsive luciferase reporter gene under the con-

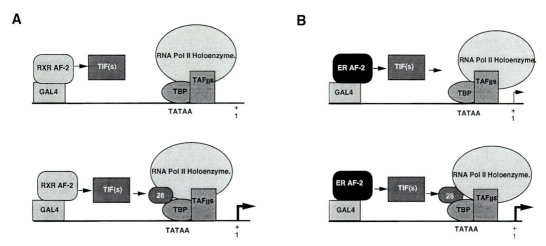

Figure 2. Model for hTAF$_{II}$28 coactivator function. (*A*) (*Upper panel*) The chimeric G4-RXR activator does not activate transcription from the minimal GAL4-responsive promoter in transfected Cos cells where TAF$_{II}$28 has not been detected in the endogenous TFIID. (*Lower panel*) Transfected hTAF$_{II}$28 associates with endogenous Cos cell TFIID possibly via direct interactions with TBP and allows activation by G4-RXR. Activation requires interactions with transcriptional intermediary factors (TIFs). (*B*) Coexpression of hTAF$_{II}$28 potentiates activation by the G4-ER. As in *A*, the effect of hTAF$_{II}$28 requires its association with Cos cell TFIID and probably is mediated through interactions with TIFs.

trol of a promoter with a mutated TGTA element is inactive even in the presence of coexpressed TBP spm3 (Fig. 4C, lane 1). The RAR AF-2 alone did not activate tran-

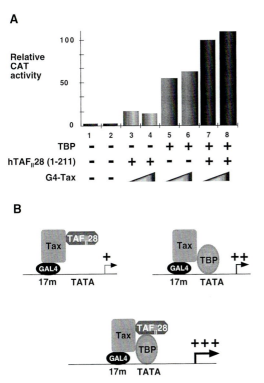

Figure 3. Expression of hTAF$_{II}$28 potentiates activation by the viral activator Tax. (*A*) Effect of coexpression of hTAF$_{II}$28 on activation by the chimeric G4-Tax activator in transfected HeLa cells. The transfected expression vectors are shown below the graph. CAT assays were performed and quantified as described by Caron et al. (1997). (*B*) Results of in vitro Tax-TBP and Tax-hTAF$_{II}$28 interaction assays, along with the effect of these interactions on activation in transfected HeLa cells as shown in panel *A*.

scription from this promoter, whereas coexpression with TBP spm3 resulted in a stimulation of transcription (Fig. 4C, lanes 2 and 5). Coexpression of hTAF$_{II}$135 along with TBP spm3 resulted in a strong increase in activation by the RAR AF-2 compared to that observed with TBP spm3 and the RAR AF-2 alone (compare lanes 5 and 7–8 in Fig. 4). In contrast, expression of hTAF$_{II}$135 in the presence or absence of TBP spm3 did not promote activation by the ER (Fig. 4C, lanes 12–14). Activation by the ER could, however, be potentiated by coexpression of TBP spm3 and hTAF$_{II}$28 (Fig. 4C, lanes 11 and 16–17). These results show that both hTAF$_{II}$135 and hTAF$_{II}$28 can functionally cooperate with coexpressed TBP spm3 specifically to enhance activation by NR AF-2s.

In the above experiments, it is particularly interesting to note that coexpression of hTAF$_{II}$135 or hTAF$_{II}$28 enhanced RAR or ER AF-2 activity even in the absence of TBP spm3 (Fig. 4C, lanes 6 and 15). Therefore, expression of these TAF$_{II}$s facilitates NR AF-2-dependent formation of initiation complexes containing endogenous TFIID. This suggests that expression of TAF$_{II}$135 and TAF$_{II}$28 may enhance the recruitment of endogenous TFIID by NR AF-2s, thereby compensating for the low affinity of this TFIID for the TGTA element (Mengus et al. 1997).

The above results, as well as the wealth of in vitro experiments, suggest that transcriptional activator proteins can recruit TFIID to promoters through specific direct or indirect interactions with TAF$_{II}$s. In this context, it is surprising that preliminary genetic experiments in yeast suggest that inactivation of TAF$_{II}$s does not globally effect transcription and *trans*-activation. In such experiments, TAF$_{II}$s have been found to rather play an essential part in cell cycle progression (Apone et al. 1996; Moqtaderi et al. 1996b; Walker et al. 1996). This phenotype reflects the requirement of TAF$_{II}$s, for example, yTAF$_{II}$145 or yTAF$_{II}$90, for transcription of some cyclin genes. The requirement of yTAF$_{II}$145 for transcription of cyclin genes

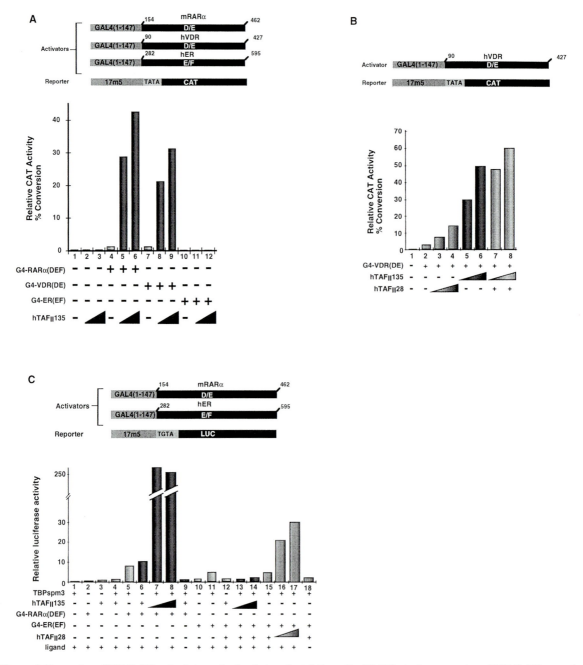

Figure 4. Expression of hTAF$_{II}$135 potentiates activation in transfected Cos cells. (*A*) Effect of coexpression of hTAF$_{II}$135 on activation by the G4-RAR, G4-VDR, and G4-ER activators in transfected Cos cells. Transfections and quantification of CAT assays were performed as described in Fig. 1. As indicated below each lane, transfections contained 0.25 μg of the chimeric activators whose structures are shown above the graph along with 0, 3, or 7 μg of the hTAF$_{II}$135 expression vector. All transfections contained the appropriate ligands. (*B*) Both hTAF$_{II}$28 and hTAF$_{II}$135 enhance VDR AF-2. Transfections contained 0, 2, or 4 μg of the hTAF$_{II}$28 expression vector and 0, 3, or 7 μg of the hTAF$_{II}$135 expression vector as indicated. Transfections in lanes *7* and *8* contained the suboptimal 2 μg of the hTAF$_{II}$28 expression vector. (*C*) hTAF$_{II}$135 and hTAF$_{II}$28 functionally cooperate with TBP spm3 to promote activation of transcription from a reporter containing a mutated TGTA rather than a TATA element. The structure of the G4-responsive reporter containing a mutated TGTA element is indicated above the graph. Transfections were performed as described using the expression vectors shown below each lane. Luciferase values were corrected for transfection efficiency using a cotransfected β-galactosidase expression vector as internal control and are displayed graphically. Where indicated, transfections contained 3 or 7, or 2 or 4 μg of the hTAF$_{II}$135 or hTAF$_{II}$28 expression vectors, respectively, along with 1 μg of the TBP spm3 expression vector.

involves core promoter recognition, rather than the response to an upstream activator (Shen and Green 1997; Walker et al. 1997). Interestingly, the cell cycle phenotype observed in yTAF$_{II}$145 mutants is reminiscent of

what is observed in mammalian cells bearing a temperature-sensitive mutant of its homolog TAF$_{II}$250 (Sekiguchi et al. 1991; Suzuki-Yagawa et al. 1997). Therefore, although the results of the transfection and in vitro experiments are suggestive, more definitive answers concerning the role of TAF$_{II}$s in mammalian cells will require genetic analyses.

Structure of the hTAF$_{II}$28-hTAF$_{II}$18 Heterodimer: A Novel Histone-fold Pair in the TFIID Complex

A better understanding of the function of the TFIID complex obviously requires a detailed knowledge of its molecular organization. To this end, we and other investigators have identified a plethora of TAF-TAF and TAF-TBP interactions (Mengus et al. 1995; Dubrovskaya et al. 1996; Lavigne et al. 1996 and references therein). TAF$_{II}$250 was the first to be shown to interact with TBP (Takada et al. 1992; Ruppert et al. 1993), and biochemical studies have shown that the amino-terminal domain of yeast or metazoan TAF$_{II}$250 interacts with a region of the concave underside of the saddle-shaped TBP. TAF$_{II}$250 competes with activators such as VP16 for binding this region. Binding of TATA-containing DNA is also competed by TAF$_{II}$250, resulting in an inhibition of basal transcription in vitro (Kokubo et al. 1994a; Bai et al. 1997; Nishikawa et al. 1997). Aside from TAF$_{II}$250, hTAF$_{II}$80, hTAF$_{II}$28, hTAF$_{II}$20, and hTAF$_{II}$18 (or in some cases their dTAF$_{II}$ homologs) have also been shown to interact with TBP, although their sites of interaction have not as yet been determined (Kokubo et al. 1994b; Mengus et al. 1995; Hoffmann and Roeder 1996).

Of the many TAF-TAF interactions, only that between dTAF$_{II}$62 and dTAF$_{II}$42 has been described at the molecular level. These TAF$_{II}$s show sequence homology with histones H3 and H4, and X-ray crystallography has shown that they interact via a histone-fold motif and form an H3/H4-like heterotetramer (Xie et al. 1996). From sequence homology, it is probable that dTAF$_{II}$28/20 (hTAF$_{II}$20) also contains a histone-fold motif analogous to H2B (Hoffmann et al. 1996). These observations, together with other biochemical data, suggest the existence of a nucleosome-like octamer core in the TFIID complex around which the other TAF$_{II}$s associate (Burley and Roeder 1996; Oelgeschlager et al. 1996).

To further understand the molecular organization of TFIID, we chose to characterize the complex formed by hTAF$_{II}$28 and hTAF$_{II}$18. We had previously shown that these two TAF$_{II}$s interact directly with each other in vitro, and the regions of each TAF$_{II}$ involved in these interactions had been characterized (Mengus et al. 1995). The structure of the hTAF$_{II}$28/hTAF$_{II}$18 complex was solved at 2.6 Å resolution by X-ray crystallography (Birck et al. 1998). This structure shows that the conserved carboxy-terminal domain of hTAF$_{II}$28 interacts with the central region of hTAF$_{II}$18 via a canonical histone-fold motif consisting of a long central α-helix (termed α2) flanked on each side by a loop and a shorter α-helix (L1, L2 and α1, α3; Fig. 5). hTAF$_{II}$28 contains an additional amino-terminal αN-helix analogous to the αN-helix of histone H3 (Luger et al. 1997). In contrast, proteolytic cleavage during crystallization removed the putative hTAF$_{II}$18 L2 and α3-helix, and these regions

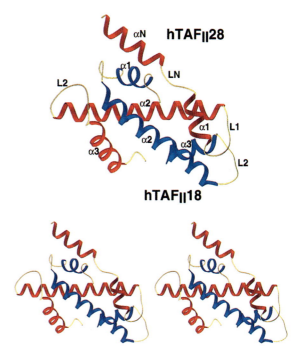

Figure 5. Structure of the hTAF$_{II}$28/hTAF$_{II}$18 heterodimer determined by X-ray crystallography. (*Top*) Ribbon representation of the crystal structure of the hTAF$_{II}$28/hTAF$_{II}$18 heterodimer at 2.6 Å resolution, together with the modeled L2 and α3-helix of hTAF$_{II}$18. (*Red*) α helices of hTAF$_{II}$28; (*blue*) hTAF$_{II}$18; (*yellow*) loops; their identities (αN, α1, LN, L1, etc.) are indicated. (*Bottom*) Left-right stereoview of the hTAF$_{II}$28/hTAF$_{II}$18 heterodimer.

were modeled on the basis of secondary structure prediction and homology with the known structure of dTAF$_{II}$62 (Birck et al. 1998).

Extensive intermolecular contacts occur in heterodimers, allowing a tight interaction of these proteins. About half of these predominantly hydrophobic intermolecular contacts involve the two α2 helices, with a remarkable hydrophobic stacking between F64 of hTAF$_{II}$18 and F161 of hTAF$_{II}$28 at the crossover of the two α2 helices. Strong intermolecular interactions also occur between the hTAF$_{II}$18 L1 and the hTAF$_{II}$28 L2 and the hTAF$_{II}$28 αN-helix and the hTAF$_{II}$18 α1-helix via a network of hydrogen bonds, some of which are water-mediated, and hydrophobic stacking. Some of the intermolecular and intramolecular interactions observed in the hTAF$_{II}$18/hTAF$_{II}$28 complex are reminiscent of those observed in the histone H3-H4 dimer.

The existence of a histone-fold in these TAF$_{II}$s was not predictable from primary sequence analysis, but manual alignment of the sequences encoding their histone-fold motifs with those of other histone proteins did reveal homology. For TAF$_{II}$18, the best homology is found with histone H4 itself and with hTAF$_{II}$80 and dTAF$_{II}$62, which are structurally assigned to the H4 family (Table 1). Most of the sequence similarities noted between TAF$_{II}$18 and the H4 family are, however, clustered in the α2 and α3 helices, the α1-helix being poorly conserved. Consequently,

Table 1. Percentage of Homology between hTAF$_{II}$28, hTAF$_{II}$18, and Other Histone-fold-containing Proteins

	hH3	dTAF$_{II}$42	hTAF$_{II}$31	hH4	dTAF$_{II}$62	hTAF$_{II}$80	hH2A	hH2B	hNC2β
hTAF$_{II}$28	35	37	39	48	49	41	45	37	47
hTAF$_{II}$18	36	36	38	43	35	43	36	34	36

database searches with only the residues of these helices did detect homology with hTAF$_{II}$80, suggesting that TAF$_{II}$18 can be assigned as a member of the H4 family.

The homology of TAF$_{II}$18 with H4 and the presence of the αN-helix in hTAF$_{II}$28 suggest it to be an H3-like protein. Surprisingly, however, for the hTAF$_{II}$28 α1-helix, the best homology is found with histone H4; for the α2 and α3 helices, comparable homology with both the H3 and H4 families is seen (Table 1). Furthermore, the homology between TAF$_{II}$28 and the H2B-like transcriptional repressor NC2β/DR1 (Goppelt et al. 1996; Mermelstein et al. 1996) is as high as with H4. Therefore, the histone-fold motif of hTAF$_{II}$28 is atypical and cannot be unambiguously assigned to the existing histone families.

Further database searches revealed the existence of at least seven true TAF$_{II}$28 homologs and four TAF$_{II}$18 homologs in diverse organisms from the yeast *Caenorhabditis elegans* and *Drosophila* to humans. The residues re-

quired for the formation of the histone-fold are well conserved in all of these homologs. The sequence comparisons, however, do show in yTAF$_{II}$40 (the yeast TAF$_{II}$28 homolog) the presence of a large insertion domain in the L2. Such insertions have not been previously described within a histone-fold motif.

Database searches using sequence profiles derived from the true TAF$_{II}$28 homologs detected four fungal SPT3 proteins revealing a novel homology between TAF$_{II}$28 and the carboxy-terminal region of SPT3, in addition to the known homology between hTAF$_{II}$18 and the amino terminus of SPT3 (Mengus et al. 1995). Five regions of homology exist between hTAF$_{II}$18 and hTAF$_{II}$28, two of which are conserved between TAF$_{II}$18 and the amino-terminal domain of SPT3, and three between TAF$_{II}$28 and the carboxy-terminal domain of SPT3 (Birck et al. 1998). In each of these regions, the conservation of numerous amino acids involved in intra- and in-

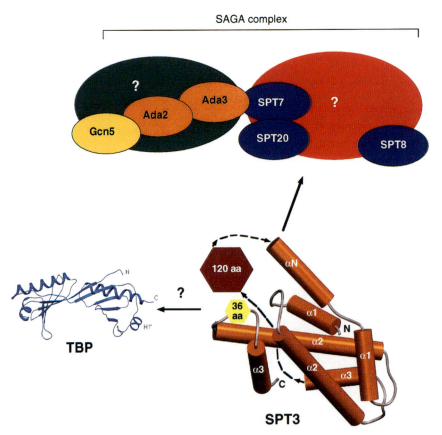

Figure 6. Molecular model of yeast SPT3 and its interactions in the SAGA complex. A molecular model of the yeast protein SPT3 is proposed based on sequence homology with the histone-fold regions of hTAF$_{II}$18 and hTAF$_{II}$28 in the amino- and carboxy-terminal regions of SPT3, respectively. Modeling of the α helices and loops is based on the coordinates of the hTAF$_{II}$18/hTAF$_{II}$28 heterodimer. The presence of the 120-amino-acid region between the amino- and carboxy-terminal histone folds and the 36-amino-acid insertion in the carboxy-terminal L2 loop are indicated. The interactions of SPT3 with other components of the SAGA complex are schematized. The question marks within the red and green ovals indicate the existence of further unidentified components of the ADA and SPT complexes.

termolecular interactions required for the formation of the histone-like dimer strongly suggests that SPT3 contains two complementary histone folds homologous to those of $hTAF_{II}28$ and $hTAF_{II}18$. In SPT3, the presence of these two histone-fold motifs in the same polypeptide and the presence of a 120-residue linker domain between them clearly permit formation of the histone pair through intramolecular interactions, rather than the classical intermolecular interactions (see Fig. 6). Such a potential structure has not been previously noted in eukaryotic proteins. Alternatively, it is also possible that the SPT3 histone folds mediate interactions between SPT3 and other components of the SAGA complex.

In yeast, SPT3 is known to interact physically and functionally, with TBP playing an important part in the transcription of a subset of genes (Eisenmann et al. 1992; Collar 1996; Madison and Winston 1997). In yeast, SPT3 is part of a large complex (the SAGA complex) containing the ADA coactivators and the GCN5 histone acetyltransferase (Fig. 6) (Grant et al. 1997; Roberts and Winston 1997). A complex (PCAF complex) with a similar subunit composition including a homolog of SPT3 has also been described in human cells (see relevant chapter in this volume). Yeast genetic studies have established a functional link between SPT3 and TBP. The G174E mutation in α-helix H1´ of TBP is suppressed by mutation E240K in SPT3, and the E240K mutation is itself suppressed by several mutations in TBP, the majority of which map to α-helix H1´ (Eisenmann et al. 1992). E240 is located in the carboxy-terminal α2-helix, suggesting that this helix and

α-helix H1´ of TBP mediate the putative interactions between these proteins. Moreover, mutation K74N in the amino-terminal region of ySPT3 is also a mild suppressor of the G174E mutation in TBP. K74 is highly conserved in both the SPT3 and $hTAF_{II}18$ proteins, further highlighting similarities in the way in which SPT3 and the $hTAF_{II}28/hTAF_{II}18$ heterodimer interact with TBP.

The above genetics are reminiscent of our previous data showing that mutation of three of the exposed glutamic acid residues in the $hTAF_{II}28$ α2-helix abolish physical and functional interactions with TBP in transfected mammalian cells (May et al. 1996). Together, these data indicate an important role of this α2-helix of both $hTAF_{II}28$ and SPT3 in interactions with TBP, showing that the homology between these helices is not only structural, but also functional. Further biochemical and functional studies have allowed us to identify critical residues in $TAF_{II}28$ and TBP required for their interactions (A.C. Lavigne et al., in prep.). These results support the idea that the helix H1´ of TBP is involved in interactions with $TAF_{II}28$ as well as SPT3.

Understanding the structure of the $TAF_{II}28/TAF_{II}18$ heterodimer provides a basis for studying the interactions of these TAF_{II}s with other TFIID components (summarized in Fig. 7). Interactions with other TAF_{II}s may involve the exposed residues of their αN, α1, and α3 helices and the corresponding LN, L1, and L2. Nevertheless, as the $hTAF_{II}28/hTAF_{II}18$ histone-fold could not be detected in classical sequence alignments, it is possible that other TAF_{II}s may also interact via as yet

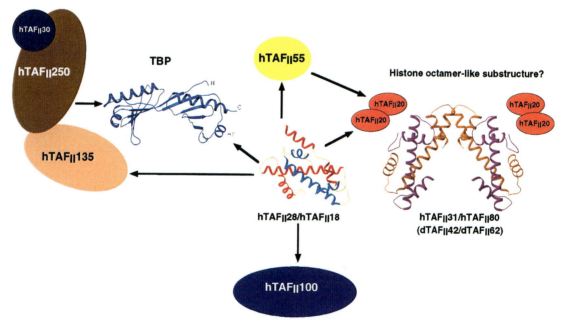

Figure 7. Summary of the structures and interactions within the TFIID complex. A selection of interactions among TFIID subunits is shown schematically. TBP, $hTAF_{II}28$, $hTAF_{II}18$, $dTAF_{II}42$, and $dTAF_{II}62$ are depicted by ribbon representations of their known structures, other TAF_{II}s are shown as colored forms. TBP and $dTAF_{II}42/dTAF_{II}62$ coordinates come from Nikolov et al. (1996) and Xie et al. (1996). Known interactions are indicated by arrows or by the overlap of the colored forms. The putative histone octamer-like substructure composed of the $hTAF_{II}31(dTAF_{II}42)/hTAF_{II}80(dTAF_{II}62)$ heterotetramer (represented as a ribbon structure) and two $hTAF_{II}20$ homodimers is also depicted. For the sake of simplicity, only a subset of the known interactions are depicted, but the additional interactions, $hTAF_{II}20$-$hTAF_{II}135$, $hTAF_{II}20$-TBP, and $hTAF_{II}80$-TBP, are noteworthy as they provide a basis for direct tethering of the putative histone octamer-like substructure to TBP and other TAF_{II}s.

undetected histone folds and that this may be a commonly used motif for TAF-TAF interactions.

Three situations exist where one of the partners of the hTAF$_{II}$28/hTAF$_{II}$18 pair is missing. In Cos cell TFIID, we have previously shown that TAF$_{II}$28 is depleted, in the TFIIDα subcomplex lacking hTAF$_{II}$30, TAF$_{II}$18 is depleted (Jacq et al. 1994; Mengus et al. 1995), and in dTFIID, there is a TAF$_{II}$28 homolog (dTAF$_{II}$30β), but the presence of the TAF$_{II}$18 homolog has not been detected. Consequently, in each situation, TAF$_{II}$18 and TAF$_{II}$28 may form a heterodimer with another TAF$_{II}$ or form homodimers to reconstitute a histone-fold pair. It is therefore conceivable that alternate TAF-TAF interactions may occur via histone folds.

What is the significance of the presence of multiple histone-like pairs in the TFIID, SAGA, and PCAF transcription factor complexes? It has been suggested that these complexes contain a histone-like octamer substructure. Although our finding that hTAF$_{II}$28, hTAF$_{II}$18, and SPT3 are novel histone-fold-containing proteins does not rule out this possibility, it does indicate that this simple model is insufficient to account for all of the interactions among histone-fold-containing proteins observed in these complexes. Furthermore, the amino acids required for classical histone-DNA interactions are not conserved in the histone-like TAF$_{II}$s or SPT3. Alternatively, the presence of multiple histone pairs within these complexes may simply reflect the fact that this interaction motif facilitates compact and tight protein-protein interactions necessary for the formation of such stable complexes.

SUMMARY

The past few years have brought many new insights concerning the structure and function of TAF$_{II}$ proteins. In the future, further biochemical and structural studies will no doubt lead to a greater understanding of the molecular organization of TFIID complexes. A better understanding of the function of metazoan, in particular, mammalian, TAF$_{II}$s in cell cycle progression and gene activation will, however, require the use of novel genetic techniques in addition to the biochemical analyses.

ACKNOWLEDGMENTS

We thank Y.G. Gangloff and L. Perletti for critical reading of the manuscript. C.R. is supported by an EMBO fellowship, G.M by a fellowship from the Ligue National contre le Cancer, and AC.L. by fellowships from the Ligue National contre le Cancer and the Association pour la Recherche contre le Cancer (A.R.C). This work was supported by grants from the CNRS, the INSERM, the Hôpital Universitaire de Strasbourg, the Ministère de la Recherche et de la Technologie, the A.R.C., and the Ligue National contre le Cancer.

REFERENCES

Apone L.M., Virbasius C.M., Reese J.C., and Green M.R. 1996. Yeast TAF$_{II}$90 is required for cell-cycle progression through G2/M but not for general transcription activation. *Genes Dev.* **10:** 2368.

Bai Y., Perez G.M., Beechem J.M., and Weil P.A. 1997. Structure-function analysis of TAF130: Identification and characterization of a high-affinity TATA binding protein interaction domain in the N terminus of yeast TAF$_{II}$130. *Mol. Cell. Biol.* **17:** 3081.

Birck C., Poch O., Romier C., Ruff M., Mengus G., Lavigne A.C., Davidson I., and Moras D. 1998. Human TAF$_{II}$28 and TAF$_{II}$18 interact through a histone fold encoded by atypical evolutionary conserved sequence motifs also found in the SPT3 family. *Cell* **94:** 239.

Brou C., Chaudhary S., Davidson I., Lutz Y., Wu J., Egly J.M., Tora L., and Chambon, P. 1993. Distinct TFIID complexes mediate the effect of different transcriptional activators. *EMBO J.* **12:** 489.

Burley S.K. and Roeder R.G. 1996. Biochemistry and structural biology of transcription factor IID (TFIID). *Annu. Rev. Biochem.* **65:** 769.

Caron C., Mengus G., Dubrowskaya V., Roisin A., Davidson I., and Jalinot P. 1997. Human TAF$_{II}$28 interacts with the human T cell leukemia virus type I Tax transactivator and promotes its transcriptional activity. *Proc. Natl. Acad. Sci.* **94:** 3662.

Chambon P. 1996. A decade of molecular biology of retinoic acid receptors. *FASEB J.* **10:** 940.

Chen J.L., Attardi L.D., Verrijzer C.P., Yokomori K., and Tjian R. 1994. Assembly of recombinant TFIID reveals differential coactivator requirements for distinct transcriptional activators. *Cell* **79:** 93.

Chiang C.M., Ge H., Wang Z., Hoffmann A., and Roeder R.G. 1993. Unique TATA-binding protein-containing complexes and cofactors involved in transcription by RNA polymerases II and III. *EMBO J.* **12:** 2749.

Collart M.A. 1996. The NOT, SPT3, and MOT1 genes functionally interact to regulate transcription at core promoters. *Mol. Cell. Biol.* **16:** 6668.

Dikstein R., Ruppert S., and Tjian R. 1996. TAF250 is a bipartite protein kinase that phosphorylates the base transcription factor RAP74. *Cell* **84:** 781.

Dubrovskaya V., Lavigne A.C., Davidson I., Acker J., Staub A., and Tora L. 1996. Distinct domains of hTAF$_{II}$100 are required for functional interaction with transcription factor TFIIFβ (RAP30) and incorporation into the TFIID complex. *EMBO J.* **15:** 3702.

Dynlacht B.D., Hoey T., and Tjian R. 1991. Isolation of coactivators associated with the TATA-binding protein that mediate transcriptional activation. *Cell* **66:** 563.

Eisenmann D.M., Arndt K.M., Ricupero S.L., Rooney J.W., and Winston F. 1992. SPT3 interacts with TFIID to allow normal transcription in *Saccharomyces cerevisiae. Genes Dev.* **6:** 1319.

Goodrich J.A., Hoey T., Thut C.J., Admon A., and Tjian R. 1993. *Drosophila* TAF$_{II}$40 interacts with both a VP16 activation domain and the basal transcription factor TFIIB. *Cell* **75:** 519.

Goppelt A., Stelzer G., Lottspeich F., and Meisterernst M. 1996. A mechanism for repression of class II gene transcription through specific binding of NC2 to TBP-promoter complexes via heterodimeric histone fold domains. *EMBO J.* **15:** 3105.

Grant P.A., Duggan L., Cote J., Roberts S.M., Brownell J.E., Candau R., Ohba R., Owen-Hughes T., Allis C.D., Winston F., Berger S.L., and Workman J.L. 1997. Yeast Gcn5 functions in two multisubunit complexes to acetylate nucleosomal histones: Characterization of an Ada complex and the SAGA (Spt/Ada) complex. *Genes Dev.* **11:** 1640.

Hoey T., Weinzierl R.O., Gill G., Chen J.L., Dynlacht B.D., and Tjian R. 1993. Molecular cloning and functional analysis of *Drosophila* TAF110 reveal properties expected of coactivators. *Cell* **72:** 247.

Hoffmann A. and Roeder R.G. 1996. Cloning and characterization of human TAF20/15. Multiple interactions suggest a central role in TFIID complex formation. *J. Biol. Chem.* **271:** 18194.

Hoffmann A., Chiang C.M., Oelgeschlager T., Xie X., Burley S.K., Nakatani Y., and Roeder R.G. 1996. A histone octamer-like structure within TFIID (comments). *Nature* **380:** 356.

Jacq X., Brou C., Lutz Y., Davidson I., Chambon P., and Tora L. 1994. Human TAF$_{II}$30 is present in a distinct TFIID complex and is required for transcriptional activation by the estrogen receptor. *Cell* **79:** 107.

Kokubo T., Yamashita S., Horikoshi M., Roeder R.G., and Nakatani Y. 1994a. Interaction between the N-terminal domain of the 230-kDa subunit and the TATA box-binding subunit of TFIID negatively regulates TATA-box binding. *Proc. Natl. Acad. Sci.* **91:** 3520.

Kokubo T., Gong D.W., Wootton J.C., Horikoshi M., Roeder R.G., and Nakatani, Y. 1994b. Molecular cloning of *Drosophila* TFIID subunits. *Nature* **367:** 484.

Lavigne A.C., Mengus G., May M., Dubrovskaya V., Tora L., Chambon P., and Davidson I. 1996. Multiple interactions between hTAF$_{II}$55 and other TFIID subunits. Requirements for the formation of stable ternary complexes between hTAF$_{II}$55 and the TATA-binding protein. *J. Biol. Chem.* **271:** 19774.

LeDouarin B., Nielsen A.L., You J., Chambon P., and Losson R. 1997. TIF1α: A chromatin-specific mediator for the ligand-dependent activation function AF-2 of nuclear receptors? *Biochem. Soc. Trans.* **25:** 605.

Luger K., Mader A.W., Richmond R.K., Sargent D.F., and Richmond T.J. 1997. Crystal structure of the nucleosome core particle at 2.8 Å resolution. *Nature* **389:** 251.

Madison J.M. and Winston F. 1997. Evidence that Spt3 functionally interacts with Mot1, TFIIA, and TATA-binding protein to confer promoter-specific transcriptional control in *Saccharomyces cerevisiae*. *Mol. Cell. Biol.* **17:** 287.

May M., Mengus G., Lavigne A.C., Chambon P., and Davidson I. 1996. Human TAF$_{II}$28 promotes transcriptional stimulation by activation function 2 of the retinoid X receptors. *EMBO J.* **15:** 3093.

Mengus G., May M., Carre L., Chambon P., and Davidson I. 1997. Human TAF$_{II}$135 potentiates transcriptional activation by the AF-2s of the retinoic acid, vitamin D3, and thyroid hormone receptors in mammalian cells. *Gene Dev.* **11:** 1381.

Mengus G., May,M., Jacq X., Staub A., Tora L., Chambon P., and Davidson I. 1995. Cloning and characterization of hTAF$_{II}$18, hTAF$_{II}$20 and hTAF$_{II}$28: Three subunits of the human transcription factor TFIID. *EMBO J.* **14:** 1520.

Mermelstein F., Yeung K., Cao J., Inostroza J.A., Erdjument-Bromage H., Eagelson K., Landsman D., Levitt P., Tempst P., and Reinberg D. 1996. Requirement of a corepressor for Dr1-mediated repression of transcription. *Genes Dev.* **10:** 1033.

Mizzen C.A., Yang X.J., Kokubo T., Brownell J.E., Bannister A.J., Owen-Hughes T., Workman J., Wang L., Berger S.L., Kouzarides T., Nakatani Y., and Allis C.D. 1996. The TAF$_{II}$250 subunit of TFIID has histone acetyltransferase activity. *Cell* **87:** 1261.

Moqtaderi Z., Yale J.D., Struhl K., and Buratowski S. 1996a. Yeast homologues of higher eukaryotic TFIID subunits. *Proc. Natl. Acad. Sci.* **93:** 14654.

Moqtaderi Z., Bai Y., Poon D., Weil P.A., and Struhl K. 1996b. TBP-associated factors are not generally required for transcriptional activation in yeast. *Nature* **383:** 188.

Nagpal S., Friant S., Nakshatri H., and Chambon P. 1993. RARs and RXRs: Evidence for two autonomous transactivation functions (AF-1 and AF-2) and heterodimerization in vivo. *EMBO J.* **12:** 2349.

Nikolov D.B., Chen H., Halay E.D., Hoffmann A., Roeder R.G., and Burley S.K. 1996. Crystal structure of a human TATA box-binding protein/TATA element complex. *Proc. Natl. Acad. Sci.* **93:** 4862.

Nishikawa J., Kokubo T., Horikoshi M., Roeder R.G., and Nakatani Y. 1997. *Drosophila* TAF$_{II}$230 and the transcriptional activator VP16 bind competitively to the TATA box-binding domain of the TATA box-binding protein. *Proc. Natl. Acad. Sci.* **94:** 85.

Oelgeschlager T., Chiang C.M., and Roeder R.G. 1996. Topology and reorganization of a human TFIID-promoter complex. *Nature* **382:** 735.

Orphanides G., Lagrange T., and Reinberg D. 1996. The general transcription factors of RNA polymerase II. *Genes Dev.* **10:** 2657.

Poon D., Bai Y., Campbell A.M., Bjorklund S., Kim Y.J., Zhou S., Kornberg R.D., and Weil, P.A. 1995. Identification and characterization of a TFIID-like multiprotein complex from *Saccharomyces cerevisiae*. *Proc. Natl. Acad. Sci.* **92:** 8224.

Pugh B.F. and Tjian R. 1991. Transcription from a TATA-less promoter requires a multisubunit TFIID complex. *Genes Dev.* **5:** 1935.

Reese J.C., Apone L., Walker S.S., Griffin L.A., and Green M.R. 1994. Yeast TAF$_{II}$S in a multisubunit complex required for activated transcription. *Nature* **371:** 523.

Roberts S.M. and Winston F. 1997. Essential functional interactions of SAGA, a *Saccharomyces cerevisiae* complex of Spt, Ada, and Gcn5 proteins, with the Snf/Swi and Srb/mediator complexes. *Genetics* **147:** 451.

Roeder R.G. 1996. Nuclear RNA polymerases: Role of general initiation factors and cofactors in eukaryotic transcription. *Methods Enzymol.* **273:** 165.

Ruppert S., Wang E.H., and Tjian R. 1993. Cloning and expression of human TAF$_{II}$250: A TBP-associated factor implicated in cell-cycle regulation. *Nature* **362:** 175.

Sauer F., Hansen S.K., and Tjian R. 1995. Multiple TAF$_{II}$s directing synergistic activation of transcription. *Science* **270:** 1783.

Sekiguchi T., Nohiro Y,. Nakamura Y., and Nishimoto. T. 1991. The human CCG1 gene essential for progression of the G1 phase, encodes a 210-kD nuclear DNA binding protein. *Mol. Cell. Biol.* **11:** 3317.

Shen W.C. and Green M.R. 1997. Yeast TAF(II)145 functions as a core promoter selectivity factor, not a general coactivator. *Cell* **90:** 615.

Strubin M. and Struhl K. 1992. Yeast and human TFIID with altered DNA-binding specificity for TATA elements. *Cell* **68:** 721.

Suzuki-Yagawa Y., Guermah M., and Roeder R.G. 1997. The ts13 mutation in the TAF$_{II}$250 subunit (CCG1) of TFIID directly affects transcription of D-type cyclin genes in cells arrested in G1 at the nonpermissive temperature. *Mol. Cell. Biol.* **17:** 3284.

Takada R., Nakatani Y., Hoffmann A., Kokubo T., Hasegawa S., Roeder R.G., and Horikoshi M. 1992. Identification of human TFIID components and direct interaction between a 250-kDa polypeptide and the TATA box-binding protein (TFIIDτ). *Proc. Natl. Acad. Sci.* **89:** 11809.

Tanese N., Pugh B.F., and Tjian R. 1991. Coactivators for a proline-rich activator purified from the multisubunit human TFIID complex. *Genes Dev.* **5:** 2212.

Verrijzer C.P. and Tjian R. 1996. TAFs mediate transcriptional activation and promoter selectivity. *Trends Biochem. Sci.* **21:** 338.

Verrijzer C.P., Chen J.L., Yokomori K., and Tjian R. 1995. Binding of TAFs to core elements directs promoter selectivity by RNA polymerase II. *Cell* **81:** 1115.

Voegel J.J., Heine M.J., Tini M., Vivat V., Chambon P., and Gronemeyer H. 1998. The coactivator TIF2 contains three nuclear receptor-binding motifs and mediates transactivation through CBP binding-dependent and -independent pathways. *EMBO J.* **17:** 507.

Walker S.S., Reese J.C., Apone L.M., and Green M.R. 1996. Transcription activation in cells lacking TAF$_{II}$S. *Nature* **383:** 185.

Walker S.S., Shen W.C., Reese J.C., Apone L.M., and Green M.R. 1997. Yeast TAF$_{II}$145 required for transcription of G1/S cyclin genes and regulated by the cellular growth state. *Cell* **90:** 607.

Wieczorek E., Brand M., Jacq X., and Tora L. 1998. Function of TAF$_{II}$-containing complex without TBP in transcription by RNA polymerase II. *Nature* **393:** 187.

Xie X., Kokubo T., Cohen S.L., Mirza U.A., Hoffmann A., Chait B.T., Roeder R.G., Nakatani Y., and Burley S.K. 1996. Structural similarity between TAFs and the heterotetrameric core of the histone octamer. *Nature* **380:** 316.

Zhou Q., Lieberman P.M., Boyer T.G., and Berk A.J. 1992. Holo-TFIID supports transcriptional stimulation by diverse activators and from a TATA-less promoter. *Genes Dev.* **6:** 1964.

Mechanisms of Viral Activators

A.J. BERK,* T.G. BOYER,* A.N. KAPANIDIS,† R.H. EBRIGHT† N.N. KOBAYASHI,* P.J. HORN,‡
S.M. SULLIVAN,‡ R. KOOP,*§ M.A. SURBY,* AND S.J. TRIEZENBERG‡

*Molecular Biology Institute, University of California, Los Angeles, California 90095-1570; †Howard Hughes Medical Institute,
Waksman Institute, and Department of Chemistry, Rutgers University, Piscataway, New Jersey 08854-8020; ‡Department of
Biochemistry, Michigan State University, East Lansing, Michigan 48824-1319

Viral regulatory proteins were among the first eukaryotic transcriptional activators identified (Berk et al. 1979; Jones and Shenk 1979; Post et al. 1981; Campbell et al. 1984). They generally induce a very high rate of viral gene transcription and thus are excellent models for analyzing mechanisms of transcriptional activation. Our studies have focused primarily on the adenovirus large E1A, Epstein-Barr virus Zebra, and herpes simplex virus VP16 proteins.

ADENOVIRUS LARGE E1A PROTEIN ACTIVATES TRANSCRIPTION FROM MULTIPLE PROMOTERS, PROBABLY BY INFLUENCING THE ACTIVITY OF A GENERAL TRANSCRIPTION FACTOR(S)

To search for an example of an activator of transcription in animal cells, we analyzed viral transcription in cells infected with adenovirus mutants blocked in the early phase of infection as defined by their inability to induce viral DNA synthesis. By analogy with bacteriophages with genomes of a similar size (~40 kb) such as T7 and λ, we anticipated that one or more adenovirus early proteins might regulate the expression of other early viral genes required for viral DNA replication. First, nuclease protection methods were developed to simplify the identification, characterization, and quantitation of specific RNAs (Berk and Sharp 1977, 1978a). This led to the unexpected discovery of RNA splicing in early viral mRNAs (Berk and Sharp 1978a,b; Berget et al. 1978). The hope that an early viral gene product would control the transcription of a set of viral genes was fulfilled by the finding that the adenovirus 5 (Ad5) mutant *hr1* (Harrison et al. 1977) expresses greatly reduced levels of mRNAs processed from each of the five early transcription units (Berk et al. 1979). The critical mutation in *hr1* was mapped to the E1A transcription unit at the left end of the viral genome (Frost and Williams 1978). Similar results were observed for mutant *dl*312, a nearly complete deletion of E1A (Jones and Shenk 1979). Moreover, the E1A function stimulated expression of viral nuclear, unspliced primary transcripts, leading to the conclusion that it activates transcription from viral promoters (Berk et al. 1979). This conclusion was confirmed by in vitro nuclear run-on assays (Nevins 1981).

Two alternatively spliced mRNAs of 12S and 13S are processed from the E1A transcription unit during the early phase of adenovirus infection (Fig. 1, left) (Berk 1986). All five host-range mutants isolated that had similar defects in early transcription had mutations which affect the 46-amino-acid region of the large E1A protein encoded in the portion of the 13S E1A mRNA not included in the 12S E1A mRNA (Ricciardi et al. 1981; Glenn and Ricciardi 1985). This, plus experiments with constructed viral mutants that express only the 12S or 13S E1A mRNAs (Montell et al. 1982, 1984; Winberg and Shenk 1984), led to the conclusion that the large E1A protein, as opposed to the smaller E1A protein translated from the 12S mRNA, is primarily responsible for activating transcription from early viral promoters. The unique region of large E1A required for its activation function is highly conserved between different human and simian adenoviruses and consequently is called conserved region 3 (CR3; Kimelman et al. 1985).

The large E1A protein was also found to activate transcription from cellular promoters cloned onto the viral genome (Gaynor et al. 1984; Hearing and Shenk 1985). In the experiment of Figure 1, an adenovirus recombinant was constructed in which the E1A region was substituted with the rat preproinsulin I gene (Fig. 1, right). This recombinant was coinfected into HeLa cells with Ad5 mutants bearing a nearly complete deletion of E1A (*dl*312), or an Ad5 variant with the wild-type E1A region (*dl*309) but otherwise isogenic with *dl*312 (Jones and Shenk 1979), or mutants with defects in E1A 5′ splice sites that prevent RNA splicing of the 12S (pm975) or 13S (*dl*1500) E1A mRNAs (Fig. 1, left) (Montell et al. 1982, 1984). Only viruses that express the 13S E1A mRNA were able to stimulate expression of the preproinsulin gene (Fig. 1, right). Later studies from the Michael Green laboratory demonstrated that the amino-terminal 39 amino acids of E1A CR3 function as a potent activation domain when fused to a DNA-binding domain (Lillie and Green 1989; Martin et al. 1990). This region includes four cysteines whose sulfhydryl groups chelate a Zn^{++} ion, forming a zinc finger of unknown structure (Culp et al. 1988). Consequently, large E1A residues 140–178 form a zinc finger structure critical for activation of early viral promoters as well as other mammalian promoters cloned onto the adenovirus genome.

The *cis*-acting region of the Ad5 E1B promoter that responds to activation by large E1A was mapped between –127 and +5 of the transcription initiation site, since this

---

§*Present address:* Institute for Molecular Biology and Tumor Research, Philipps University, D-35033 Marburg, Germany.

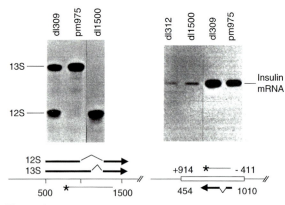

Figure 1. Adenovirus large E1A encoded by the 13S E1A mRNA activates transcription from promoters on the viral chromosome. (*Left*) S1 protection assays of adenovirus E1A mRNAs using the 5′-end labeled probe indicated at the bottom along with a diagram of the major early E1A mRNAs. Numbers refer to base pairs from the left end of the Ad5 genome. E1A mRNAs expressed from Ad5 variant *dl*309 and mutants pm975 and *dl*1500 protect the fragments shown in the autoradiogram at the top. Mutant pm975 contains a T→G transversion at base pair 975, altering the invariant GT at the 5′ end of the 12S mRNA intron to GG, thereby blocking splicing of the 12S mRNA. This mutation does not alter the large E1A protein encoded by the 13S mRNA because a glycine GG<u>T</u> codon is changed to a synonymous GG<u>G</u> codon. Mutant *dl*1500 contains a 5-bp deletion of the 13S mRNA 5′ splice site and consequently does not express the 13S mRNA. Wild-type 12S and 13S mRNAs are expressed from the Ad5 variant *dl*309 used to construct mutants pm975 and *dl*1500. (*Right*) S1 protection assay of preproinsulin mRNA expressed from an Ad5 recombinant containing the rat preproinsulin gene substituted for the E1A region, as diagrammed at the bottom. HeLa cells were coinfected with the preproinsulin recombinant and one Ad5 mutant, as indicated at the top of each lane. The preproinsulin recombinant contained rat genomic DNA from –411 to +914 relative to the rat preproinsulin 1 gene transcription start site (Lomedico et al. 1980) in place of Ad5 sequence from 454 to 1010 (*bottom*). Mutant *dl*312 contains a deletion of nearly the entire E1A region. The 3′-end labeled DNA probe used is diagrammed at the bottom. Ad5 mutants expressing the 13S mRNA encoding large E1A (*dl*309 and pm975) activated preproinsulin transcription, whereas mutant *dl*1500 expressing the 12S mRNA only did not. (Adapted from Montell et al. 1984.)

region was shown to support activation when fused to a CAT reporter gene in a transient transfection assay (Dery et al. 1987). The critical large E1A-responsive region was mapped by analyzing a systematic set of constructed viral mutants (Wu et al. 1987). Transcription from mutants with deletions from –127 to –65, –55, –45, and –35, just upstream of the TATA box, was activated to the same extent as the wild-type promoter, even though deletion of an SP1 site from –48 to –39 (Fig. 2) diminished the overall level of transcription by five- to tenfold (Wu et al. 1987; Wu and Berk 1988). Further deletion to –25, including the TATA box, eliminated E1B transcription entirely. Importantly, as shown in Figure 2, mutations that eliminated the TATA box specifically, such as LS-35/-25, resulted in a low level of transcription that was largely unaffected by large E1A. (The low level of transcription from mutant LS-35/-26 and LS-30/-23 [Wu and Berk 1988] apparently is directed to the normal start site by the E1B consensus initiator sequence [Smale et al., this vol-

ume].) These results indicated that it is the E1B TATA box which responds to large E1A activation. The finding that the E1B TATA box alone responds led us to suggest that large E1A influences the activity of a general transcription factor.

AN E1A CR3-TBP INTERACTION PROBABLY DOES NOT CONTRIBUTE TO ACTIVATION

The TATA box was known to be the binding site for the general transcription factor TFIID (Davison et al. 1983; Parker and Topol 1984; Sawadogo and Roeder 1985). Because we found that a TATA box responded to large E1A activation and because TFIID had not been purified and characterized at that time, we joined the effort to define this central general transcription factor. Efforts to purify TFIID from nuclear extracts of HeLa cells were frustrated by the very low concentrations of the protein and the broad elution profile of TFIID activity from ion exchange columns. The breakthrough in this area came when Buratowski et al. (1988) reported that TFIID activity in extracts of *Saccharomyces cerevisiae* could be assayed in conjunction with general transcription factors partially purified from HeLa cells and that the yeast activity eluted off of ion exchange columns in sharp peaks. In retrospect, we now know that this is because the TATA-box-binding protein (TBP) is found free in yeast extracts, whereas it is bound tightly to TBP-associated factors (TAFs) forming the multisubunit TFIID protein in extracts from metazoan cells (Verrijzer and Tjian 1996). Many of the TAFs are extensively phosphorylated (Zhou et al. 1992). Variations in the phosphorylation of TAFs in different TFIID molecules may result in the broad elution profile of TFIID from ion exchange columns, complicating the purification of TFIID by standard methods. We (Schmidt et al. 1989) and others (Hernandez 1993) quickly purified yeast TBP and isolated the yeast gene based on its partial amino acid sequence. Polymerase chain reaction (PCR)-based methods then allowed us (Kao et al. 1990) and others to isolate a cDNA for human TBP as well as for TBPs from multiple other species (Burley and Roeder 1996).

The isolation of a cDNA for human TBP made it possible to test directly the hypothesis that large E1A influences TFIID activity. Initial studies revealed that isolated large E1A and fusion proteins containing E1A CR3 bound directly to purified human TBP (Horikoshi et al. 1991; Lee et al. 1991). This led us to hypothesize that the E1A CR3-TBP interaction might be the basis of the large E1A activation mechanism. Further studies with a set of E1A CR3 mutants (Webster and Ricciardi 1991) were consistent with this hypothesis: A subset of mutants in the zinc finger activation domain region that are defective for activation in vivo exhibited reduced binding to TBP in vitro using a coimmunoprecipitation assay (Geisberg et al. 1994).

To assess further the role of the E1A CR3-TBP interaction in transcription activation, we performed fluorescence anisotropy experiments (Heyduk et al. 1993, 1996) to quantify equilibrium binding constants for E1A-TBP

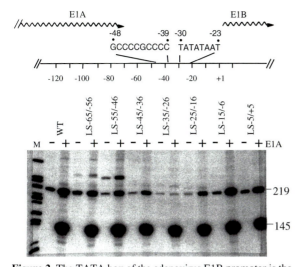

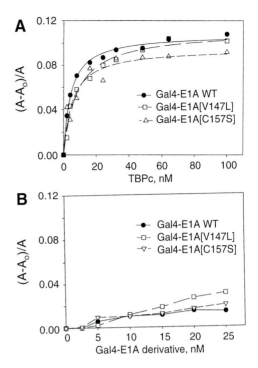

Figure 2. The TATA box of the adenovirus E1B promoter is the *cis*-acting sequence that responds to large E1A. (*Top*) Diagram of the 69-bp intergenic region between the Ad5 E1A and E1B transcription units. Numbers refer to base pairs relative to the E1B transcription initiation site. The major E1B promoter elements are shown, an SP1-binding site, and a TATA box. (*Bottom*) S1 protection assays using a probe 5'-end labeled at +219. Cytoplasmic RNA analyzed was isolated from HeLa cells infected with Ad5 mutants containing mutations in the E1B promoter region as shown at the top of pairs of lanes in the autoradiogram. LS-65/-56 refers to a linker-scanning mutation which alters most of the base pairs in the region from –65 to –56, and so on, for each of the LS mutants. Each of the E1B promoter mutants (including the mutant with a wild-type E1B promoter region [WT]) were constructed in a *dl*1500 background so that large E1A protein was not expressed in cells infected with the mutant alone. In lanes marked +E1A, large E1A was provided by coinfecting cells with mutant S13B which has a wild-type E1A region and a 5-bp deletion in the E1B-coding region ending at +145. E1B mRNA expressed from the E1B promoter mutants generated a 219-base S1-protected fragment; E1B mRNA expressed from S13B generated a 145-base protected fragment. Transcription from each of the E1B promoter mutants was activated by large E1A, with the notable exception of the TATA box mutant LS-35/-26. (Adapted from Wu et al. 1987; Wu and Berk 1988.)

Figure 3. Fluorescence anisotropy analysis of the E1A-TBP interaction. (*A*) Fluorescein-labeled (Gal4-E1A)-DNA complexes (10 nM Gal4-E1A plus 0.5 nM DNA fragment FGal4 [F-CAATCGGAGGACTGTCCTCCGTAGG]) were titrated at 25°C by addition of 0–100 nM TBPc in 20 mM HEPES (pH 7.9), 60 mM KCl, 10 mM MgCl$_2$, 0.2 mM EDTA, 20 µM ZnCl$_2$, 10 mM β-mercaptoethanol, 0.1 mM PMSF, 100 µg/ml BSA, 2.5 µg/ml poly([d(G-C)]:poly[(d[G-C]), and 12% glycerol. (*B*) Fluorescein-labeled TBPc-DNA complexes (30 nM TBPc plus 1 nM DNA fragment MLP24FL of Heyduk et al. [1996]) were titrated at 25°C by addition of 0–25 nM Gal4-E1A in 20 mM Tris-HCl (pH 8.0), 20 mM HEPES, 60 mM KCl, 10 mM MgCl$_2$, 8 mM (NH4)$_2$SO$_4$, 0.05 mM EDTA, 20 µM ZnCl$_2$, 0.5 mM DTT, 0.1 mM PMSF, 2.4% PEG-8000, 5 µg/ml poly([d(G-C)]:poly[d(G-C)]), and 5% glycerol. For both *A* and *B*, fluorescence anisotropy values were determined using a PanVera Beacon fluorescence polarization instrument, and equilibrium binding constants were calculated using nonlinear regression.

and E1A-[TBP-TATA DNA] interactions (Fig. 3) (A. Kapanidis et al., unpubl.). The assays in Figure 3A were designed to reanalyze whether potency in transcription activation and equilibrium binding constants for the E1A-TBP interaction are correlated. A Gal4-E1A fusion protein consisting of the Gal4 DNA-binding domain (residues 1–147) followed by wild-type E1A residues 121–223 (including CR3) activates transcription from a template with Gal4-binding sites, both in vivo (Lillie and Green 1989) and in vitro (Boyer and Berk 1993). The human TBP core domain (TBPc), when assembled with TAFs into a TFIID complex, supports activation by Gal4-E1A (Zhou et al. 1993). We prepared the Gal4-E1A fusion protein and confirmed its activity in in vitro transcription reactions. In addition, a 25-bp 5'-fluorescein-labeled DNA fragment having the DNA site for Gal4 was prepared ("FGal4"). (Gal4-E1A)-FGal4 complexes then were formed using saturating concentrations of Gal4-E1A. Next, the (Gal4-E1A)-FGal4 complexes were titrated with increasing concentrations of TBPc, and formation of

TBPc-(Gal4-E1A)-FGal4 complex was assessed by increase in fluorescence anisotropy. The results (Fig. 3A) confirm that E1A-TBP interaction occurs (Kb = 2.2 ± 0.2 × 10^8 M^{-1} for wild-type Gal4-E1A) but, importantly, do not confirm the existence of a correlation between potency in transcription activation and equilibrium binding constants for the E1A-TBP interaction. Thus, V147L and C137S mutants of Gal4-E1A, which result in large defects in transcription activation in vivo and in vitro (Webster and Ricciardi 1991; T.G. Boyer and A.J. Berk, unpubl.) result in little or no defects in E1A-TBP interaction (Kb = 1.2 ± 0.2 × 10^8 M^{-1} and 1.8 ± 0.7 × 10^8 M^{-1}).

The assays in Figure 3B were designed to determine whether the E1A-TBP interaction can occur while TBP is bound to a TATA element (an assumption of all simple models for transcription activation mediated by an E1A-TBP interaction). For these experiments, a preformed, saturated TBP-TATA complex (prepared using a 24-bp 5'-fluorescein-labeled DNA fragment having the adenovirus major late promoter TATA element; Heyduk et al.

1996) was titrated with wild-type Gal4-E1A. No significant change in fluorescence anisotropy was observed in the 0–25 nM concentration range, suggesting that E1A does not interact with the TBP-TATA DNA complex in this concentration range. (Presumably, the observed E1A-TBP interaction in the absence of the TATA element involves determinants of TBP that are within, or that overlap, the DNA-binding surface of TBP.)

Taken together, the absence of a verifiable correlation between potency in transcription activation and binding constants for the E1A-TBP interaction (Fig. 3A) and the absence of significant E1A-[TBP-TATA] interaction (Fig. 3B) argue against the hypothesis that transcription activation is mediated by an E1A-TBP interaction.

A second argument against the model that a direct interaction between E1A CR3 and TBP participates in the mechanism of large E1A activation came from a systematic study of mutations in TBP surface residues (Bryant et al. 1996). TBP mutants were designed based on the structure of the conserved core domain bound to TATA box DNA (Burley and Roeder 1996). Single-amino-acid changes were introduced into most of the surface residues of the TBP core domain that do not make direct interactions with DNA. The mutant TBPs also contained three mutations affecting the DNA-binding surface which allow the molecule to bind to a TGTAAA box. This makes it possible to assay mutant TBP function in vivo by using a reporter gene with a TGTAAA box that has very low affinity for the endogenous wild-type TBP (Strubin and Struhl 1992).

Each of the mutant TBPs was assayed in a transient transfection assay in HeLa cells for its ability to support activation in response to Gal4-E1A and Gal4-VP16, a fusion of the Gal4 DNA-binding domain (residues 1–147) to the 78 carboxy-terminal residues of the herpes simplex virus type-1 (HSV-1) VP16 activator protein (Triezenberg et al. 1988). The VP16 activation domain is chemically very different from that of the E1A CR3 activation domain. The E1A activation domain is not acidic overall, and activation function is quite sensitive to single-amino-acid substitutions; conservative substitutions at 9 (Webster and Ricciardi 1991) of the 39 residues in the activation domain (Martin et al. 1990) reduce activation to less than 20% of wild type. In contrast, the VP16 activation domain is overall extremely acidic and quite insensitive to single alanine substitutions (Regier et al. 1993). Despite this dissimilarity in chemistry, each of the 89 mutations in TBP surface residues had a similar effect on activation by both Gal4-E1A and Gal4-VP16 (Fig. 4).

Since mutations in TBP surface residues had similar effects on activation by E1A CR3 and VP16, we conclude that TBP makes similar molecular interactions during the process of activation by both activators. Mutations that significantly decreased function in response to both activators were in TBP surface residues that contact TFIIB (E284 and L287; Burley and Roeder 1996), residues that contact TFIIA (A184, N189, E191, R205; Geiger et al. 1996; Tan et al. 1996), and residues on the convex surface of TBP opposite the DNA-binding surface (G175, C176, R231, R235, R239, F250, P272) that we hypothesize influence interactions with TAF subunits in TFIID. It is very unlikely that the chemically different E1A CR3 and VP16 activation domains interact identically with the surface of TBP. Consequently, our systematic mutational analysis of TBP surface residues did not support the significance of an interaction between E1A CR3 and TBP for the mechanism of large E1A activation.

If CR3 does not interact with TBP during the process of activation, what cellular proteins does it interact with? In recent work, we have found that CR3 interacts with a specific subunit of a low-abundance multiprotein complex. This subunit is conserved between nematodes and mammals but is absent from the *S. cerevisiae* genome. Moreover, the extensively purified multiprotein complex greatly increases Gal4-E1A activation in in vitro transcription assays. This work will be reported in detail elsewhere (T.G. Boyer et al., in prep.) so as to not preclude publication in a primary journal.

PURIFIED HELA CELL TFIID CONTAINS 11 MAJOR TAFS

Work from the Tjian laboratory was the first to report that TFIID in extracts from metazoan cells consists of a TBP subunit bound stably to multiple additional TAF subunits (Dynlacht et al. 1991). We contributed to the study of TFIID by developing a simple method for the rapid purification of the multisubunit protein approximately 25,000-fold to near homogeneity from a HeLa cell nuclear extract. Epitope-tagged TBP was used to purify the multisubunit TFIID protein according to the method of Field et al. (1988). HeLa cells were stably transformed with a retrovirus vector that expresses human TBP with an amino-terminal nine-amino-acid epitope from influenza virus hemagglutinin (HA1) that is bound by monoclonal antibody 12CA5 (Zhou et al. 1992). One transformed clone was selected in which about 80% of the steady-state TBP is epitope-tagged. The overall steady-state concentration of TBP (epitope-tagged plus endogenous TBP) in extracts from these LTRα3 cells is comparable to the wild-type TBP concentration in the parental HeLa cells (Zhou et al. 1992). We postulate that this occurs because expression of the epitope-tagged TBP from the moderately strong murine leukemia virus long terminal repeat (MLV LTR) promoter in the vector results in the synthesis of about eightfold more epitope-tagged TBP than is expressed from the endogenous TBP gene, that the tagged and endogenous TBPs compete for assembly into complexes with endogenous TAFs, and that the excess tagged and endogenous TBP molecules are rapidly degraded.

TFIID was purified from LTRα3 cells by preparation of nuclear extract and fractionation on phosphocellulose to prepare the high-salt D-fraction that contains the bulk of TFIID activity (Dignam et al. 1983). TBP complexes present in this fraction were bound to a 12CA5 matrix, washed with buffer containing 1 M KCl to remove proteins that were not tightly bound to the TFIID complex, and then eluted in a low-salt buffer containing a high concentration of the synthetic 9-mer HA1 peptide (Field et al.

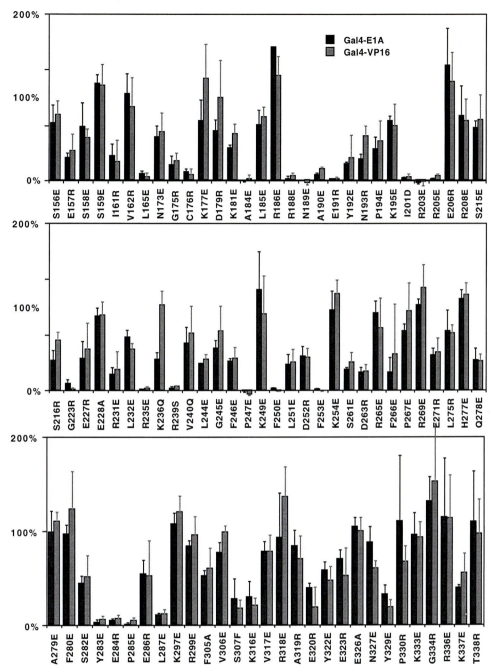

Figure 4. TBP mutants with single changes in surface residues respond similarly to activation by Gal4-E1A and by Gal4-VP16. Transient transfection assays were performed with four plasmids: (1) A luciferase reporter gene with five Gal4-binding sites upstream of the c-*fos* TATA box region in which the TATAAAGG box was mutated to TGTAAAGG; (2) an expression vector for human TBP containing three mutations affecting the TBP DNA-binding surface that greatly increase the affinity for the TGTAAAGG box (altered binding specificity TBP) plus a single additional mutation on the surface of TBP that does not interact with DNA; an expression vector for Gal4(1–147)-E1A (121–223) or Gal4(1–147)-VP16(413–490); a β-galactosidase expression vector using the SV40 early promoter-enhancer for normalizing transfection efficiency. When an expression vector for altered binding site TBP with otherwise wild-type surface residues was used ("wild-type altered binding site TBP"), luciferase expression was 10–15-fold higher than when no altered binding site TBP expression vector was used. The activities of TBP mutants with surface residue mutations as shown are expressed as the percentage of this increase in luciferase relative to wild-type altered binding site TBP. (*Black bars*) Activities in transfections with Gal4-E1A; (*gray bars*) activities in transfections with Gal4-VP16. Standard deviations determined from at least three independent assays are shown. (Reprinted, with permission, from Bryant et al. 1996.)

1988; Zhou et al. 1992). The affinity of 12CA5 for HA1 is optimal for binding and elution of HA1 epitope-tagged proteins. TBP tagged with an HA1 mutant sequence with fivefold lower affinity for 12CA5 was not quantitatively bound to the matrix. Substitution of 12CA5 with monoclonal antibody HA11 (Santa Cruz Biochemicals), which has higher affinity for HA1 peptide than does 12CA5, results in quantitative binding but inefficient elution with HA1 peptide.

HA1 peptide-eluted protein was fractionated by gel filtration on a Sephacryl-S400 column. The bulk of total protein and epitope-tagged TBP eluted in a single peak after the column void volume and well ahead of thyroglobulin (670 kD) (Fig. 5). Epitope-tagged TBP migrates just below the 43-kD marker, as determined by immunoblotting with 12CA5. Analysis of TFIID assembled with an amino-terminally deleted form of TBP revealed that a TAF comigrates with epitope-tagged, full-length TBP and consequently is not clearly visualized in this gel (Zhou et al. 1993). The band at 28-kD resolved into two bands on gels of higher polyacrylamide concentration. Consequently, 11 major TAFs were observed in what appears to be a single, major multisubunit TFIID protein in the phosphocellulose D-fraction prepared from HeLa cells (Fig. 5). Additional minor polypeptides coeluted with the peak of TFIID. These may be additional TAFs present on one or more minor fractions of TFIID. Alternatively, they may result from proteolytic cleavage of higher-molecular-weight TAFs in vivo or during the purification procedure.

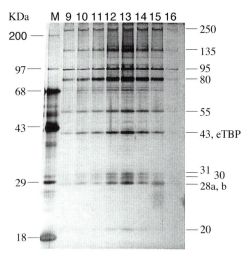

Figure 5. Silver-stained 10% polyacrylamide gel of peak fractions of epitope-tagged TFIID from a Sephacryl-S400 gel filtration column. Epitope-tagged TFIID was isolated from LTRα3 cells expressing human TBP with an amino-terminal nine-amino-acid HA1 epitope. LTRα3 cell nuclear extract was fractionated by phosphocellulose. The 0.5–1.0 M KCl fraction was bound to a column of anti-HA1 monoclonal antibody 12CA5 and eluted with HA1 peptide as described previously (Zhou et al. 1992). The eluted material was subjected to gel filtration on Sephacryl-S400. Fraction numbers are indicated at the top. (M) Polypeptide molecular weight markers. TAFs are indicated at the right according to their approximate molecular weights as enumerated in the review by Burley and Roeder (1996).

A SUBSET OF ACTIVATION DOMAINS PROMOTE ASSEMBLY OF TFIID AND TFIIA ON PROMOTER DNA

TFIID purified as described above was used in experiments to analyze the influence of viral activators on TFIID binding to promoter DNA. DNA protein complexes formed with purified TFIID and labeled DNA probes containing several activator binding sites (~225 bp) did not migrate significantly into low-percentage polyacrylamide gels. However, they did migrate into 1% agarose gels; 5 mM Mg^{++} in the gel and electrophoresis buffers stabilized the complexes so that they resolved into a band observed after autoradiography (Fig. 6A) (Lieberman and Berk 1994). The mobilities of these DNA protein complexes in agarose gels were similar whether they were formed with TFIID (~750 kD) or with TBP (38 kD), although treatment with polyclonal antibody to TAF250 caused a supershift of the TFIID-containing, but not TBP-containing, complexes (Lieberman and Berk 1994). Consequently, we believe that the decreased mobility of TFIID-containing complexes in agarose gels results principally from the bend introduced into DNA by TBP binding (Burley and Roeder 1996).

TFIID binding to promoter DNA was assayed by Mg^{++} agarose gel electrophoretic mobility shift assay (EMSA) at various time intervals after addition of protein to a probe containing a TATA box and seven upstream binding sites for the Epstein-Barr virus activator designated Zebra (also called Zta), which has a basic zipper DNA-binding domain at its carboxyl terminus and an activation domain at its amino terminus (Lieberman and Berk 1990; Chi and Carey 1993). This template supports strong activation by recombinant Zebra in in vitro transcription reactions with HeLa nuclear extract (Lieberman and Berk 1990). Under the conditions used, TFIID plus TFIIA bound to the probe with slow kinetics (Fig. 6A). However, when sufficient Zebra was added to the binding reaction to saturate the Zebra sites on the probe, the kinetics of TFIID binding increased dramatically; the total amount of probe bound at equilibrium also increased. This activation of TFIID binding required both the Zebra activation domain and the presence of TFIIA in the binding reaction (Lieberman and Berk 1994). Thus, Zebra greatly stimulated the assembly of a TFIID-TFIIA promoter DNA complex (DA complex).

The conformation of the TFIID-promoter DNA complex was also modified by Zebra as shown by DNase I footprinting assays performed at higher concentrations of TFIID and TFIIA that saturated the probe in the absence of the activator (Fig. 6B). TFIID alone protected the TATA box region of the probe only, generating a footprint equivalent to that produced by TBP. Addition of TFIIA extended the footprint in the upstream direction, as expected from the position of TFIIA binding relative to TBP (Geiger et al. 1996; Tan et al. 1996), but did not modify the footprint further. However, when Zebra was bound to the probe simultaneously with TFIID, protections and hypersensitive sites were observed in the region downstream from the TATA box that were accentuated by the additional binding of TFIIA. Hypersensitive sites surrounding the transcription start site at –5, +2, and +12 were especially prominent. These modifications of the footprint were not observed in complexes formed with Zebra, TFIIA, and TBP, nor in complexes with a deleted form of Zebra lacking its activation domain (Lieberman

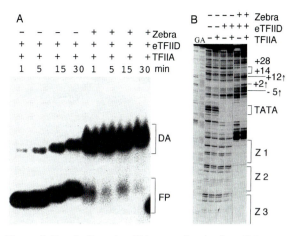

Figure 6. Epstein-Barr virus Zebra protein stimulates DA complex assembly and interactions of TAFs with DNA downstream from the TATA box. (*A*) Agarose gel shift assays using a probe with seven Zebra-binding sites upstream of the Ad2 E4 TATA box (–170 to +54 relative to the transcription initiation site). Purified epitope-tagged TFIID, TFIIA, and sufficient Zebra to saturate the Zebra-binding sites were added to two binding reactions as shown at the top. Equal aliquots of the binding reactions were removed at the indicated times after addition of protein to probe DNA and immediately layered into wells of a running agarose gel. An autoradiogram of the dried gel is shown. (*B*) DNase I footprinting assays. The indicated purified proteins were added in sufficient quantities to saturate their binding sites on the labeled DNA. (Lane G) Probe fragments generated by the Maxam-Gilbert G reaction; (lane A) fragments generated by the G+A reaction. The positions of the TATA box and the first three Zebra-binding sites are shown. Zebra plus TFIID produced a region of partial protection from +14 to +28 indicated by the bracket at the right, and hypersensitive sites indicated by up arrows following the sequence position number. These footprint changes downstream from the TATA box were not produced by Zebra alone (not shown). (Adapted from Lieberman and Berk 1994.)

and Berk 1994). Consequently, we interpret the downstream protection and hypersensitive sites to be due to Zebra activation-domain-induced interactions of promoter DNA with TAF subunits of TFIID. Because they occur in the region of the transcription start site, they are located precisely where they could influence initiation by RNA polymerase II.

To test the generality of the ability of activation domains to promote the assembly of a DA complex on promoter DNA, and to induce TAF-DNA interactions near the transcription start site, we analyzed these activities on probes with appropriate binding sites for Gal4-E1A and Gal4-VP16 (Kobayashi et al. 1995). Gal4-VP16 displayed these activities, whereas Gal4-E1A did not. Furthermore, the VP16 activation domain can be split approximately in half, with each half exhibiting significant activation domain activity (Regier et al. 1993). The VP16 carboxy-terminal activation subdomain promoted DA complex formation, whereas the amino-terminal half did not. Consequently, DA complex assembly activity is associated with a subset of activation domains.

To test the significance of DA complex assembly activity for the mechanism of activation, we studied VP16 carboxy-terminal subdomain mutants (Kobayashi et al. 1998). A set of Gal4-VP16C mutants containing single,

double, or triple mutations in the 38-amino-acid VP16C subdomain were selected for study that exhibited a range of activation function from near that of Gal4-VP16C wild-type to about 1% of that activity. The mutants were expressed in and purified from *Escherichia coli* and analyzed for their ability to promote DA complex assembly using Mg^{++} agarose EMSA as in Figure 6. Figure 7 shows plots of the activities of these mutants for transcriptional activation in a transient transfection assay and for DA complex assembly activity. An excellent correlation was observed between DA complex assembly activity and in vivo activation function. The same mutants also were assayed for their ability to bind in vitro to several proposed targets of the VP16C activation domain: TBP, TFIIB, *Drosophila* TAF40, and the p62 subunit of TFIIH (Kobayashi et al. 1998). No comparable correlation was observed between activation function and the interaction of VP16C mutants with any of these proposed targets, as measured by coprecipitation and "GST pull-down" assays, with the possible exception of some correspondence with TFIIH p62 subunit binding.

The results summarized in Figure 7 argue strongly that the Gal4-VP16C DA-complex assembly activity is an important aspect of its activation mechanism. How might this activity stimulate activation in vivo? A complete preinitiation complex composed of TFIID, A, B, E, F, H, and RNA polymerase II and capable of initiating transcription can assemble on a DA complex (Fig. 8) (Van Dyke et al. 1988; Buratowski et al. 1989). In vitro, assembly of the DA complex is a rate-limiting step in transcription initiation (Wang et al. 1992). Consequently, the ability of Gal4-VP16C to stimulate the rate of DA complex assembly overcomes a rate-limiting step in initiation. Second, the VP16C DA complex assembly activity may counteract the function of the transcription inhibitors NC2 (also called Dr1-DRAP1; Kim et al. 1995; Kim et al. 1997) and Mot1 (Auble et al. 1997). Inhibition of in vitro transcription by these factors is prevented by association of TFIIA with preinitiation complexes.

SUMMARY

Adenovirus large E1A, Epstein-Barr virus Zebra, and herpes simplex virus VP16 were studied as models of animal cell transcriptional activators. Large E1A can activate transcription from a TATA box, a result that leads us to suggest that it interacts with a general transcription factor. Initial studies showed that large E1A binds directly to the TBP subunit of TFIID. However, analysis of multiple E1A and TBP mutants failed to support the significance of this in vitro interaction for the mechanism of activation. Recent studies to be reported elsewhere indicate that conserved region 3 of large E1A, which is required for its activation function, binds to one subunit of a multisubunit protein that stimulates in vitro transcription in response to large E1A and other activators.

A method was developed for the rapid purification of TFIID approximately 25,000-fold to near homogeneity from a cell line engineered to express an epitope-tagged form of TBP. Purified TFIID contains 11 major TAFs

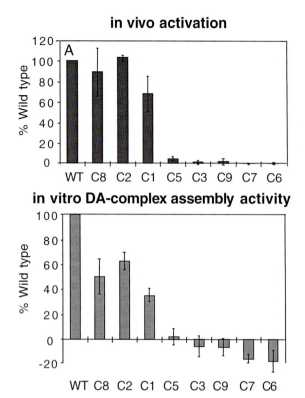

in vivo activation

in vitro DA-complex assembly activity

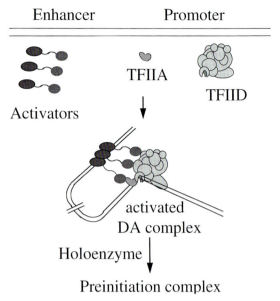

Enhancer   Promoter

Activators

TFIIA

TFIID

activated
DA complex

Holoenzyme

Preinitiation complex

Figure 8. A class of activation domains stimulate DA complex assembly. The DA complexes formed may be activated through interactions between one or more TFIID-TAFs and DNA near the transcription initiation site. Subsequent binding of the remaining general transcription factors and RNA polymerase II, possibly in the form of a preassembled holoenzyme, generates a complete preinitiation complex that initiates transcription.

Figure 7. Correlation between DA complex assembly activity and in vivo activation function for mutants of the VP16 carboxy-terminal activation subdomain. (*Top*) Transient transfection assays in COS cells were performed with an expression vector for Gal4(1–147)-VP16C(VP16 amino acids 453–490), or Gal4-VP16C mutants and a reporter gene containing five Gal4-binding sites upstream of the Ad5 TATA box controlling expression of luciferase. VP16C mutants designated C1, C2, etc., each contained one, two, or three mutations in the VP16C sequence. Activity is shown as percentage of luciferase activity induced compared to that observed for the Gal4 fusion to the wild-type VP16C sequence (WT). (*Bottom*) DA complex assembly activity was measured by the agarose gel-shift assay shown in Fig. 6. A probe containing five Gal4-binding sites upstream of the Ad5 E4 TATA box and transcription initiation site was incubated with sufficient recombinant Gal4-VP16CWT or Gal4-VP16C mutant to saturate the Gal4-binding sites on the probe, and sub-saturating amounts of purified eTFIID and purified recombinant human TFIIA, and the products of the 30-min binding reaction were subjected to electrophoresis as in Fig. 6A. Percent wild-type DA complex assembly activity was calculated as (fraction of probe shifted by Gal4-VP16C mutant+TFIID+TFIIA) – (fraction of probe shifted by TFIID+TFIIA alone) / (fraction of probe shifted by Gal4-VP16CWT+TFIID+TFIIA) – (fraction of probe shifted by TFIID+TFIIA alone). For mutants with negative values, a smaller percentage of the probe was shifted when the Gal4-VP16C mutant was added to the reaction than in its absence. (Adapted from Kobayashi et al. 1998.)

ranging in mass from approximately 250 to 20 kD. Zta and VP16, but not large E1A, greatly stimulate the rate and extent of assembly of a TFIID-TFIIA complex on promoter DNA (DA complex). For VP16, this is a function of the carboxy-terminal activation subdomain. An excellent correlation was found between the ability of VP16C mutants to stimulate DA complex assembly and their ability to activate transcription in vivo. Consequently, for a subset of activation domains, DA complex

assembly activity is an important component of the overall mechanism of activation.

ACKNOWLEDGMENTS

This research was supported by grants CA-25235 from the National Cancer Institute to A.J.B., GM-53665 from the NIGMS, and a Howard Hughes Medical Institute investigatorship to R.H.E., and AI-27323 from the NIAID to S.J.T. A.J.B. thanks all members of the Berk lab past and present for their science and friendship.

REFERENCES

Auble D.T., Wang D., Post K.W., and Hahn S. 1997. Molecular analysis of the SNF2/SWI2 protein family member MOT1, an ATP-driven enzyme that dissociates TATA-binding protein from DNA. *Mol. Cell. Biol.* **8:** 4842.

Berk, A.J. 1986. Adenovirus promoters and E1A transactivation. *Annu. Rev. Genet.* **20:** 45.

Berk A.J. and Sharp P.A. 1977. Sizing and mapping of early adenovirus mRNAs by gel electrophoresis of S1 endonuclease-digested hybrids. *Cell* **12:** 721.

———. 1978a. Spliced early messenger RNAs of simian virus 40. *Proc. Natl. Acad. Sci.* **75:** 1274.

———. 1978b. Structure of the adenovirus 2 early mRNAs. *Cell* **14:** 695.

Berk A.J., Lee F., Harrison T., Williams J., and Sharp P.A. 1979. Pre-early adenovirus 5 gene products regulate transcription and processing of early viral messenger RNAs. *Cell* **17:** 935.

Berget S.M., Berk A.J., Harrison T., and Sharp, P.A. 1978. Spliced segments at the 5´ termini of adenovirus 2 late mRNA: A role for heterogeneous nuclear RNA in mammalian cells. *Cold Spring Harbor Symp. Quant. Biol.* **42:** 523.

Boyer T.G. and Berk A.J. 1993. Functional interactions of adenovirus E1A with holo-TFIID. *Genes Dev.* **7:** 1810.

Bryant G.O., Martel L.S., Burley S.K., and Berk A.J. 1996. Radical mutagenesis reveals TATA-box binding protein surfaces required for activated transcription in vivo. *Genes Dev.* **10:** 2491.

Buratowski S., Hahn S., Guarente L., and Sharp, PA. 1989. Five intermediate complexes in transcription initiation by RNA polymerase II. *Cell* **56:** 549.

Buratowski S., Hahn S., Sharp P.A., and Guarente L. 1988. Function of a yeast TATA element-binding protein in a mammalian transcription system. *Nature* **334:** 37.

Burley S.K. and Roeder R.G. 1996. Biochemistry and structural biology of transcription factor IID. *Annu. Rev. Biochem.* **65:** 769.

Campbell M.E., Palfreyman J.W., and Preston, C.M. 1984. Identification of herpes simplex virus DNA sequences which encode a *trans*-acting polypeptide responsible for stimulation of immediate early transcription. *J. Mol. Biol.* **180:** 1.

Chi T. and Carey M. 1993. The ZEBRA activation domain: Modular organization and mechanism of action. *Mol. Cell. Biol.* **11:** 7045.

Culp J.S., Webster L.C., Friedman D.J., Smith C.L., Huang W.J., Wu F.Y., Rosenberg M., and Ricciardi R.P. 1988. The 289-amino acid E1A protein of adenovirus binds zinc in a region that is important for trans-activation. *Proc. Natl. Acad. Sci.* **85:** 6450.

Davison B.L., Egly J.M., Mulvihill E.R., and Chambon P. 1983. Formation of stable preinitiation complexes between eukaryotic class B transcription factors and promoter sequences. *Nature* **301:** 680.

Dignam J.D., Martin P.L., Shastry B.S., and Roeder R.G. 1983. Eukaryotic gene transcription with purified components. *Methods Enzymol.* **101:** 582.

Dery C.V., Herrmann C.H., and Mathews M.B. 1987. Response of individual adenovirus promoters to the products of the E1A gene. *Oncogene* **2:** 15.

Dynlacht B.D., Hoey T., and Tjian R. 1991. Isolation of coactivators associated with the TATA-binding protein that mediate transcriptional activation. *Cell* **66:** 563.

Field J., Nikawa J., Broek D., MacDonald B., Rodgers L., Wilson I.A., Lerner R.A., and Wigler M. 1988. Purification of a RAS-responsive adenylyl cyclase complex from *Saccharomyces cerevisiae* by use of an epitope addition method. *Mol. Cell. Biol.* **8:** 2159.

Frost E. and Williams J. 1978. Mapping temperature-sensitive and host-range mutations of adenovirus type 5 by marker rescue. *Virology* **91:** 39.

Gaynor R.B., Hillman D., and Berk A.J. 1984. Adenovirus early region lA protein activates transcription of a non-viral gene introduced into mammalian cells by infection or transfection. *Proc. Natl. Acad. Sci.* **81:** 1193.

Geiger J.H., Hahn S., Lee S., and Sigler P.B. 1996. Crystal structure of the yeast TFIIA/TBP/DNA complex. *Science* **272:** 830.

Geisberg J.V., Lee W.S., Berk A.J., and Ricciardi R.P. 1994. The zinc finger region of the E1A transactivating domain complexes with the TATA box-binding protein. *Proc. Natl. Acad. Sci.* **91:** 2488.

Glenn G.M. and Ricciardi R.P. 1985. Adenovirus 5 early region 1A host range mutants hr3, hr4, and hr5 contain point mutations which generate single amino acid substitutions. *J. Virol.* **56:** 66.

Harrison T., Graham F., and Williams, J. 1977. Host-range mutants of adenovirus type 5 defective for growth in HeLa cells. *Virology* **77:** 319.

Hearing P. and Shenk T. 1985. Sequence-independent autoregulation of the adenovirus type 5 E1A transcription unit. *Mol. Cell. Biol.* **11:** 3214.

Hernandez N. 1993. TBP, a universal eukaryotic transcription factor? *Genes Dev.* **7:** 1291.

Heyduk T., Ma Y., Tang H., and Ebright, R.H. 1996. Fluorescence anisotropy: Rapid, quantitative assay for protein-DNA and protein-protein interaction. *Methods Enzymol.* **274:** 492.

Heyduk T., Lee J.C., Ebright Y.W., Blatter E.E., Zhou, Y., and Ebright, RH. 1993. CAP interacts with RNA polymerase in solution in the absence of promoter DNA. *Nature* **364:** 548.

Horikoshi N., Maguire K., Kralli A., Maldonado E., Reinberg D., and Weinmann R. 1991. Direct interaction between adenovirus E1A protein and the TATA box binding transcription factor IID. *Proc. Natl. Acad. Sci.* **88:** 5124.

Jones N. and Shenk T. 1979. An adenovirus type 5 early gene function regulates expression of other early viral genes. *Proc. Natl. Acad. Sci.* **76:** 3665.

Kao C.C., Lieberman P.M., Schmidt M.C., Zhou Q., Pei R., and Berk A.J. 1990. Cloning of a transcriptionally active human TATA binding factor. *Science* **248:** 1646.

Kim S., Na J.G., Hampsey M., and Reinberg D. 1997. The Dr1/DRAP1 heterodimer is a global repressor of transcription in vivo. *Proc. Natl. Acad. Sci.* **94:** 820.

Kim T.K., Zhao Y., Ge H., Bernstein R., and Roeder R.G. 1995. TATA-binding protein residues implicated in a functional interplay between negative cofactor NC2 Dr1 and general factors TFIIA and TFIIB. *J. Biol. Chem.* **270:** 10976.

Kimelman D., Miller J.S., Porter D., and Roberts B.E. 1985. E1a regions of the human adenoviruses and of the highly oncogenic simian adenovirus 7 are closely related. *J. Virol.* **53:** 399.

Kobayashi N., Boyer T. G., and Berk A.J. 1995. Functional interactions between a class of activators and TFIIA. *Mol. Cell. Biol.* **15:** 6465.

Kobayashi N., Horn P.J., Sullivan S.M., Triezenberg S.J., Boyer T.G., and Berk A.J. 1998. DA-complex assembly activity required for VP16C transcriptional activation. *Mol. Cell. Biol.* **18:** 4023.

Lee W.S., Kao C.C., Bryant G.O., Liu X., and Berk A.J. 1991. Adenovirus E1A activation domain binds the basic repeat in the TATA box transcription factor. *Cell* **67:** 365.

Lieberman P.M. and Berk A.J. 1990. *In vitro* transcriptional activation, dimerization and DNA binding specificity of the Epstein-Barr virus Zta protein. *J. Virol.* **64:** 2560.

———. 1994. A mechanism for TAFs in transcriptional activation: Activation domain enhancement of TFIID-TFIIA-promoter DNA complex formation. *Genes Dev.* **8:** 995.

Lillie J.W. and Green M.R. 1989. Transcription activation by the adenovirus E1a protein. *Nature* **338:** 39.

Lomedico P.T., Rosenthal N., Kolodner R., Efstratiadis A., and Gilbert W. 1980. The structure of rat preproinsulin genes. *Ann. N.Y. Acad. Sci.* **343:** 425.

Martin K.J., Lillie J.W., and Green M.R. 1990. Evidence for interaction of different eukaryotic transcriptional activators with distinct cellular targets. *Nature* **346:** 147.

Montell C., Courtois G., Eng C., and Berk A. 1984. Complete transformation by adenovirus 2 requires both E1A proteins. *Cell* **36:** 951.

Montell C., Fisher E.F., Caruthers M.H., and Berk A.J. 1982. Resolving the functions of overlapping viral genes by site-specific mutagenesis at a mRNA splice site. *Nature* **295:** 380.

Nevins J.R. 1981. Mechanism of activation of early viral transcription by the adenovirus E1A gene product. *Cell* **26:** 213.

Parker C.S. and Topol J. 1984. A *Drosophila* RNA polymerase II transcription factor contains a promoter-region-specific DNA-binding activity. *Cell* **36:** 357.

Post L.E., Mackem S., and Roizman B. 1981. Regulation of alpha genes of herpes simplex virus: Expression of chimeric genes produced by fusion of thymidine kinase with alpha gene promoters. *Cell* **24:** 555.

Regier J.L., Shen F., and Triezenberg S.J. 1993. Pattern of aromatic and hydrophobic amino acids critical for one of two subdomains of the VP16 transcriptional activator. *Proc. Natl. Acad. Sci.* **90:** 883.

Ricciardi R.P., Jones R.L., Cepko, C.L., Sharp P.A., and Roberts B.E. 1981. Expression of early adenovirus genes requires a viral encoded acidic polypeptide. *Proc. Natl. Acad. Sci.* **78:** 6121.

Sawadogo M. and Roeder R.G. 1985. Interaction of a gene-specific transcription factor with the adenovirus major late promoter upstream of the TATA box region. *Cell* **43:** 165.

Schmidt M.C., Kao C.C., Pei R., and Berk, A.J. 1989. Yeast TATA-box transcription factor gene. *Proc. Natl. Acad. Sci.* **86:** 7785.

Strubin M. and Struhl K. 1992. Yeast and human TFIID with altered DNA-binding specificity for TATA elements. *Cell* **68:** 721.

Tan S., Hunziker Y., Sargent D.F., and Richmond T.J. 1996. Crystal structure of a yeast TFIIA/TBP/DNA complex. *Nature* **381:** 127.

Triezenberg S.J., Kingsbury R.C., and McKnight S.L. 1988. Functional dissection of VP16, the *trans*-activator of herpes simplex virus immediate early gene expression. *Genes Dev.* **2:** 718.

Van Dyke M.W., Roeder R.G., and Sawadogo M. 1988. Physical analysis of transcription preinitiation complex assembly on a class II gene promoter. *Science* **241:** 1335.

Verrijzer C.P. and Tjian R. 1996. TAFs mediate transcriptional activation and promoter selectivity. *Trends Biochem. Sci.* **21:** 338.

Wang W., Carey M., and Gralla J.D. 1992. Polymerase II promoter activation: Closed complex formation and ATP-driven start site opening. *Science* **255:** 450.

Webster L.C. and Ricciardi R.P. 1991. *Trans*-dominant mutants of E1A provide genetic evidence that the zinc finger of the *trans*-activating domain binds a transcription factor. *Mol. Cell. Biol.* **11:** 4287.

Winberg G. and Shenk T. 1984. Dissection of overlapping functions within the adenovirus type 5 E1A gene. *EMBO J.* **8:** 1907.

Wu L. and Berk A.J. 1988. Transcriptional activation by the pseudorabies virus immediate early protein requires the TATA box element in the adenovirus 2 ElB promoter. *Virology* **167:** 318.

Wu L., Rosser D.S.E., Schmidt M., and Berk A.J. 1987. TATA-box implicated in ElA transcription activation of a simple adenovirus 2 promoter. *Nature* **326:** 512.

Zhou Q., Boyer T.G., and Berk A.J. 1993. Factors (TAFs) required for activated transcription interact with TATA-box binding protein. *Genes Dev.* **7:** 180.

Zhou Q., Lieberman P.M., Boyer T.G., and Berk A.J. 1992. Holo TFIID supports transcriptional stimulation by diverse activators and from a TATA-less promoter. *Genes Dev.* **6:** 1964.

Cooperative Assembly of RNA Polymerase II Transcription Complexes

K. Ellwood,* T. Chi,† W. Huang,* K. Mitsouras,* and M. Carey*

*Department of Biological Chemistry, University of California School of Medicine, Los Angeles, California 90095–1737

Synergy is a regulatory phenomenon that controls the expression of many eukaryotic genes. There are two interrelated manifestations of synergy: the greater than additive effect of increasing numbers of sites on promoter activity and the cooperative or sigmoidal transcriptional response to increasing concentrations of activator. The former effect is the basis of combinatorial control, and the latter allows genes to be turned on and off over small changes in activator concentration, providing a sensitive molecular switch.

Our laboratory has been pursuing the biochemical mechanism of synergistic gene activation, and we review our key experiments in these areas. We show that transcriptional synergy can be recapitulated in cell-free extracts using model reporter templates bearing multiple activator binding sites. Both the levels of transcription and the amount of open complex assembly mediated by either the Epstein-Barr virus (EBV) activator ZEBRA or GAL4-VP16 are synergistic with respect to the number of upstream promoter sites. This cooperative assembly of the transcription complex correlates well with recruitment of purified TFIIA and TFIID to a core promoter. Other events that may be relevant to synergistic transcription are also discussed, including putative conformational changes in the general machinery and isomerization of the closed to the open complex.

THE USE OF MODEL SYSTEMS FOR STUDYING MAMMALIAN GENE ACTIVATION

A typical RNA polymerase II (pol II) promoter contains upstream regulatory elements, which bind activators and repressors, and a core region, encompassing the TATA box, initiator, and downstream sequence elements (Ernst and Smale 1995). The organization of these sites is called promoter architecture. Promoter architecture can be designed to either restrict or optimize the way proteins bind to DNA, leading to a wide range of promoter responses. For this reason, the architecture is hypothesized to be a central determinant in regulation of the timing and levels of gene expression in response to cellular signals.

Our laboratory has been investigating the biochemical mechanism of gene activation through the use of model systems composed of either GAL4-VP16 or the EBV ZEBRA protein, and idealized reporter templates bearing either one or multimerized high-affinity binding sites for

the activators (Carey et al. 1990a,b, 1992; Lehman et al. 1998). Our philosophy has been to employ the idealized templates to systematically probe promoter architecture with the goal of understanding which variables are important for activity and what ranges they function within. We found that by varying the number, affinity, and positioning of activator binding sites along with the core promoter and its affinity for the general machinery, we could tune promoter activity over a 100-fold range in vitro. Although this type of study has been popular in the gene regulation field, we have coupled it with detailed biochemistry to generate insights into the mechanisms of gene regulation. We review our studies using the model systems as they form the framework for our recent efforts to understand the principles governing differential regulation of the genes constituting the EBV lytic cycle (Miller 1990). Our focus is on the mechanism underlying synergistic transcription complex assembly.

Figure 1 is a cartoon summarizing our view that the transcription complex is a network of interconnected protein-protein and protein-DNA interactions. Every interaction has a free energy that is related to the affinity constant K by the Gibbs equation $K = e^{-\Delta G/RT}$. Because the components are all linked, any change in the affinity or energy of one interaction affects the overall free energy of the complex and by necessity influences the binding of other components of the complex. This simple idea forms the foundation for understanding synergy and its effect on promoter activity.

RESULTS AND DISCUSSION

The Mechanism of Synergy

The model reporter templates were first employed to show that ZEBRA and GAL4-VP16 would activate transcription synergistically in HeLa extracts and by cotransfection into Cos cells (Carey et al. 1990b, 1992). GAL4-VP16 and ZEBRA manifest both forms of synergy on the model reporter templates, the effect of sites and the effect of concentration. Our hypothesis is that both forms of synergy are due to the simultaneous interactions of multiple activators with the transcriptional machinery (Fig. 1). As discussed above, an exponential relationship exists between interaction-free energy and affinity. Therefore, doubling or tripling the number of activators that are simultaneously contacting the general machinery would, in principle, negating entropic effects, double or triple the total free energy of the interaction. According to the

†*Present address:* Howard Hughes Medical Institute, Beckman Center, Stanford University, Stanford, California 94305.

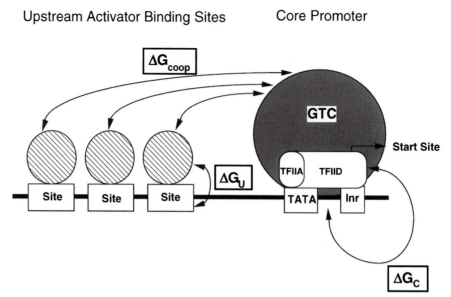

Figure 1. Energetics of transcription complex assembly. A typical promoter contains upstream activator binding sites (U) and a core promoter (C). Activators stimulate transcription by binding to the upstream promoter and recruiting limiting components of the general pol II machinery (the general transcription complex: GTC) to DNA. We hypothesize that the final transcription complex has an affinity or free energy that reflects the protein-DNA interactions between the activators and its sites (ΔG_U), the general machinery and the core (ΔG_C), and a protein-protein interaction energy between activators and the general machinery (Δg_{coop}) which allows preinitiation complex assembly. We imagine that there is a minimum free energy (ΔG_T) or barrier required to assemble this complex and that exceeding that barrier results in cooperative transcription complex assembly. Because of the logarithmic relationship between energy and the affinity, small changes in the energy can dramatically influence the affinity and thus transcription. Because the protein-protein and protein-DNA interactions form a network, their energies are linked to the final free energy of transcription complex assembly. Thus, changing one energy must influence the overall free energy of the complex and the manner in which the energy is distributed. In this fashion, a strong core promoter should allow binding of TFIID and this in turn should promote cooperative binding of ZEBRA to the upstream sites.

Gibbs equation, the linear increase in free energy would generate an exponential increase in the affinity (K) of the pol II holoenzyme for multiple versus single activators. In our model, the exponential increase in affinity would, when the transcriptional machinery was present at subsaturating concentrations, generate a synergistic increase in transcription complex recruitment to the promoter. This model would explain the first manifestation of synergy, the greater than additive effects of increasing the number of activators. However, it does not directly address the reason that there is a sigmoidal response of a gene to increasing activator concentration. This latter effect can in part be explained by imagining that the ability of the general machinery to bind multiple activators means that it could serve as a multimerization interface. In such a scenario, the interaction of multiple activators with the machinery would lead to a cooperative effect on binding of the activators to DNA. Such an effect could explain the sigmoidal transcriptional response to activator concentration.

We addressed the first aspect of the hypothesis in two studies where in vitro transcription was employed to compare templates bearing increasing numbers of sites for their response to recombinant ZEBRA or GAL4-VP16 in HeLa nuclear extracts. We found that when GAL4-VP16 or ZEBRA was raised to sufficiently high levels, the activator would saturate the templates by DNase I and gel-shift assays. Under such conditions, transcription was nevertheless synergistic, when comparing templates bearing multiple sites to those containing a single site. This result supported the idea that the synergy occurred after binding of the activators and could not have been influenced by cooperative binding of the activators to the templates (Carey et al. 1990b, 1992).

As discussed above, this result does not preclude cooperative binding of the activators when they are present at subsaturating conditions, it simply demonstrates that a significant component of synergy can be observed when activators are saturating. The data support the hypothesis that the activators are simultaneously interacting with the general machinery to recruit it to a promoter. Because the upstream and core promoters are physically and therefore energetically linked through activator-general factor interactions as illustrated in Figure 1, the interactions should exert reciprocal effects on binding of the interacting proteins to DNA when the proteins are limiting (Carey et al. 1990b). This type of mechanism would generate additional specificity in gene activation by allowing transcription complex assembly to be a concerted reaction, possibly requiring a threshold free energy. Only when the appropriate energy was achieved could the transcription complex assemble in a concerted fashion onto DNA. The threshold free energy could be varied in different ways by chromatin or other factors related to the physiological state of the cell.

In an effort to address this second aspect of the model, we systematically combined strong upstream promoters with weak core promoters and weak upstream promoters with strong cores. We asked whether the strong core could compensate for a weak upstream promoter and vice versa (Lehman et al. 1998). Strong upstream promoters were defined by the presence of either high affinity or multiple binding sites for the activator ZEBRA. Strong core promoters were defined as those exhibiting high basal transcription levels in vitro and a high affinity for TFIID and TFIIA. We predicted, based on the hypothesis of Figure 1, that if the proteins bound to the core and upstream promoters were physically linked, they should exert reciprocal cooperative effects on each others binding to DNA. This binding effect in turn would be manifested as an effect on the transcription.

Our experiment revealed that low-affinity activator binding sites supported similar levels of activation from strong core promoters as weak core promoters linked to strong upstream promoters. We are attempting to model this effect more precisely using statistical thermodynamics (M. Carey et al., in prep.). This reciprocity, or ability of one portion of the promoter to compensate for the other, established that the two segments of the promoter were energetically linked and implied, albeit indirectly, that proteins binding the core and upstream promoters (activators and the general machinery) bound their sites cooperatively. We are currently attempting to demonstrate this effect more directly by DNase I footprinting.

Activator-mediated Transcription Complex Assembly

To address the issue of whether transcription levels correlated directly with transcription complex assembly, we studied formation of pol II open complexes on the reporter templates in the HeLa extracts. The last or penultimate step in transcription preinitiation complex assembly is melting of the DNA to form the open complex. We reasoned that if we could identify and quantitate these complexes, they could be used as a measure of complete transcription complex assembly. To identify the open complexes, we employed a potassium permanganate modification assay. Thymidine residues in melted DNA are sensitive to potassium permanganate modification, which generates thymidine glycols. The modifications can be identified by primer extension analysis as *Taq* DNA polymerase stalls at the modified residues. The intensity of the primer extension serves as a measure of the amount of open complex.

We found, when comparing different templates, that the relative amount of activator-stimulated permanganate-sensitive open complexes on the different templates in Figure 2B closely parallels the relative amount of transcription in Figure 2A (Wang et al. 1992a; Chi and Carey 1993). This correlation suggests that transcription complex assembly is the primary determinant of the amount of transcription.

To demonstrate that the permanganate sensitivity is actually measuring pol II complex assembly, Figure 2C shows that open complex formation like pol II initiation requires the ATP β-γ phosphoanhydride bond and that ad-dition of nucleoside triphosphates causes the complexes to disappear, consistent with the idea that pol II elongates out of the start site. The complexes are clearly due to pol II because α-amanitin inhibits the elongation.

An important conclusion drawn from these studies is that if there are postinitiation mechanisms controlling elongation or reinitiation in the in vitro systems, these mechanisms must have effects approximately proportional to the original amount of open complex. This observation does not exclude the idea that differences in the amount of reinitiation or rate of promoter escape, for example, might generate large differences in overall transcription levels on different promoters. However, such effects must be proportional to the amount of open complex.

Promoter Melting and Activation

To understand how the promoter melting step was linked to activation, we asked whether a premelted start site would bypass the requirement for an activator. We created the premelted templates in a two-step procedure. First, we mutated the 10-bp region encompassing the E4 start site and mixed the mutated template (Tantin and Carey 1994) with the wild-type template. After denaturation and renaturation, the strands reassociate following a binomial distribution to generate wild-type and mutant parental templates and two sibling heteroduplexes. The heteroduplexes are purified using a special gel system and then subjected to in vitro transcription assays. Importantly, we found that reactions on the heteroduplex template were stimulated by activators to the same extent as on duplex DNAs. The data implied that the activator must primarily influence early steps in transcription and that the early steps must be completed to allow pol II to join the complex. We discuss the steps affected by activator below. First, however, we discuss the mechanics of open complex assembly, which also proved to be very interesting on these heteroduplex templates.

Studies by the Sharp laboratory had shown that supercoiled DNA templates could bypass the requirements for TFIIH and the ATP β-γ bond (Parvin and Sharp 1993; Timmers 1994). This finding led to the hypothesis that TFIIH might mediate promoter melting directly. We predicted that if TFIIH did indeed mediate ATP-dependent DNA melting that heteroduplex templates should bypass the requirement for a β-γ bond. Indeed, we found that such templates could bypass the β-γ bond requirement; i.e., reactions with heteroduplex but not wild-type duplex DNA templates could utilize the ATP analogs AMP-PNP and ATPγS to initiate and elongate transcription (Tantin and Carey 1994). In reconstituted systems, the heteroduplex also directly bypassed the requirement for TFIIE and TFIIH for basal transcription and for TFIIH alone for activated transcription (Holstege et al. 1995; Tantin et al. 1996). The experiment described above does not exclude the possibility that TFIIH is required at subsequent steps. Indeed, TFIIH is likely involved in allowing pol II to escape from the promoter (Dvir et al. 1996, 1997). Further-

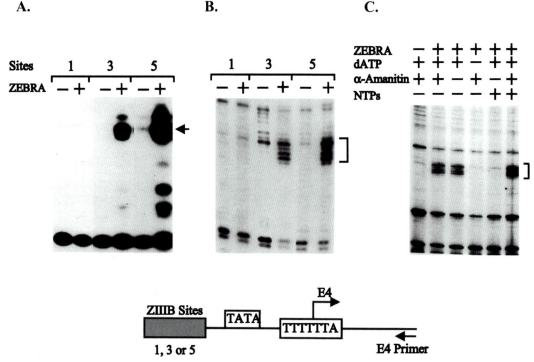

Figure 2. Comparison of open complex with transcription. (*A*) Results of a transcription reaction performed in a HeLa nuclear extract comparing the levels of ZEBRA-stimulated transcription on model templates bearing 1, 3, or 5 sites upstream of the adenovirus E4 TATA. Transcription was measured by primer extension. (*B*) Results of a parallel permanganate open complex assay performed in the same nuclear extract on the same templates in the presence of ATP. (*C*) Validity of the permanganate assay is established by showing that open complexes are stimulated by ZEBRA, require ATP for their formation, disappear in the presence of nucleotides, and are due to pol II as shown by the sensitivity of the transcription to α-amanitin.

more, activator-TFIIH interactions may facilitate this escape step. However, in our biochemical system, we only measured large transcripts, and effects on promoter-proximal pausing would not have been detected.

Synergistic Recruitment of the "DA Complex"

To identify the early step(s) in transcription complex assembly that is directly affected by activators, we employed a partially purified, reconstituted transcription system containing crude TFIIA, TFIID, a TFIIE/IIF/IIH fraction, recombinant TFIIB, and pure pol II. This system supported activation by GAL4 derivatives and by ZEBRA to the same extent in the crude HeLa extracts. Kinetic "order-of-addition" experiments were then used to study the rate-determining step in transcription complex assembly. The open complex was used as an endpoint for a transcriptionally active complex. Assembly of an activator-TFIID-TFIIA complex (the DA complex) was shown to be rate-limiting for transcription and open complex assembly (Wang et al. 1992b).

The idea that activators affect DA complex assembly was later directly established by Lieberman and Berk (1994) and colleagues, who used DNase I footprinting and gel-shift studies to show that immunopurified TFIID and TFIIA could be recruited to a core promoter by ZEBRA. We reasoned that if such a step was important for gene activation, it should be incorporated into the mech-

anism of synergy. In collaboration with Lieberman, we employed gel-shift and footprinting experiments along with purified TFIID and recombinant human TFIIA to show that GAL4-VP16 and ZEBRA synergistically recruited the DA complex to templates bearing multiple sites, but not single sites (Chi et al. 1995). The key footprinting data are shown in Figure 3. Whereas ZEBRA is saturating on templates bearing 1, 3, and 7 sites, the concentrations of TFIID were set at subsaturating levels and did not generate a strong footprint over the core TATA box. Together, however, ZEBRA strongly recruits DA to the TATA box on templates bearing 3 and 7 sites, generating a 25-bp footprint over TATA and a greater than 50-bp series of downstream enhancements and protections called the extended footprint (see below). The recruitment was dependent on TBP-associated factors (TAFs), although our study did not determine if the activators directly interacted with TAFs. A similar result was observed with GAL4-VP16.

Previous studies by Green and colleagues had shown that GAL4-VP16 interacted directly with TFIIB. This data prompted us to determine the effect of TFIIB on DA recruitment in the presence and absence of activators (Lin and Green 1991; Lin et al. 1991; Roberts et al. 1995). We found that inclusion of TFIIB in the reaction led to enhanced recruitment of the DA complex. Green and colleagues found that GAL4-VP16 could recruit TFIIB to either TFIID or TATA-box-binding polypeptide (TBP)

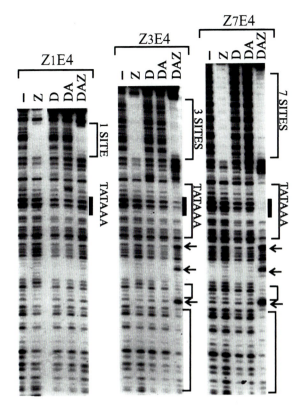

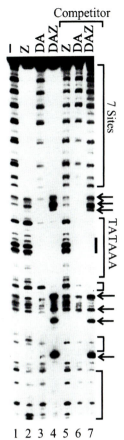

Figure 3. Synergistic recruitment of TFIIA and TFIID. The binding and DNase I footprinting reactions contained templates bearing 1, 3, or 7 sites, recombinant ZEBRA (Z), a mixture of the two recombinant TFIIA subunits (A), and immunopurified TFIID (D). The main point is that ZEBRA is present at saturating concentrations and generates footprints of the expected sizes on all three templates. In contrast, TFIID and TFIIA are subsaturating. On templates bearing 7 and 3 sites, ZEBRA strongly recruits the DA complex to the TATA generating a 25-bp footprint over TATA and a 50-bp or so series of enhancements and protections referred to as the extended footprint. Despite the fact that ZEBRA binds tightly to the 1-site template, it fails to recruit the DA complex. The data agree well with the results of the transcription and open complex assays in Fig. 2. We propose that transcriptional synergy is first manifested during assembly of the DA complex, the nucleating step in transcription complex assembly. (Reprinted, with permission, from Chi et al. 1995 [copyright Macmillan].)

1 2 3 4 5 6 7

Figure 4. ZEBRA is required for inducing, but not maintaining, isomerization of the DA complex. A DNA fragment bearing 7 high-affinity ZEBRA-binding sites upstream of the adenovirus E4 core promoter (Z_7E4T) was incubated with recombinant ZEBRA (Z), template saturating concentrations of TFIID (D), and recombinant TFIIA (A). After preincubation, a molar excess of a competitor oligonucleotide was added as indicated; the oligonucleotide bore a high-affinity ZEBRA-binding site. The mixtures were then subjected to DNase I footprinting analysis, and the digestion products were fractionated on a sequencing gel. The brackets indicate protected regions and the arrows are enhancements. The black rectangle denotes the location of the TATA box. The main point is that ZEBRA induces an isomerization characterized by the extended footprint and that once formed, the complex is stable even after ZEBRA removal. Subsequent experiments showed that this complex can support modest levels of activated transcription albeit not full levels. (Reprinted, with permission, from Chi and Carey 1996.)

(Choy et al. 1993; Roberts et al. 1993, 1995). In agreement with these previous studies, we found that the ability of the activator-TFIIB interaction to enhance recruitment of the DA complex could also be observed with TBP (Chi et al. 1995).

The recruitment of the DA complex also led to a TAF-dependent isomerization event characterized by a 75-bp TFIID footprint, much larger than the 25-bp footprint generated by TBP alone over a TATA box. This effect was first noticed by Roeder and colleagues almost a decade ago, although its significance had not been fully established (Horikoshi et al. 1988). Figure 4 shows an example of a ZEBRA-induced extended footprint. TFIID alone generates a 25-bp footprint over TATA. Prebinding TFIID followed by addition of ZEBRA, however, led to formation of the extended footprint. By saturating the promoter with DA

complex in the absence of activator, and then removing the activator by competitor oligos, we found that the isomerized DA complex had a higher affinity for TFIIB consistent with its higher transcriptional activity (Chi and Carey 1996). This isomerized complex was found to support moderate levels of transcription in a TFIID-depleted nuclear extract (i.e., above basal but not fully activated) after ZEBRA removal. These and other data in the field led our group and others to propose a two-step model for gene activation where the activator recruits the DA complex in one step and TFIIB in the other (Fig. 5).

Although the view derived from studies in the model systems was informative from a mechanistic standpoint, it did not consider the nuances imposed by assembly of

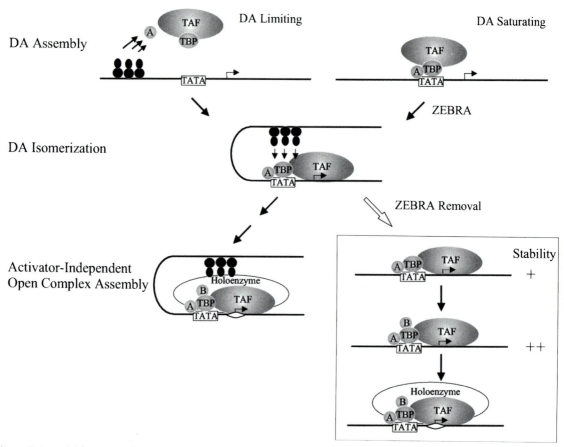

Figure 5. A model for gene activation and synergy. This figure summarizes several salient features of our mechanistic studies over the last 7 years. Multiple activators (in *black*) bind upstream of the TATA box and initiation site. These synergistically interact with TFIID and TFIIA as denoted by the arrows, resulting in DA complex assembly with a concomitant isomerization, which generates an extended footprint, apparently due to TAFs, covering the start site (arrow) and downstream regions. The presence of TFIIB enhances the synergistic action of ZEBRA and the stability of the complex. The isomerized complexes then serve as a platform for the entry of holoenzyme (or the other general factors); the start site is subsequently melted in the presence of ATP to form the open complex before initiation. However, ZEBRA is dispensable for these latter steps: The isomerized DA complex, once formed, is relatively stable in the absence of ZEBRA and able to bind TFIIB and the holoenzyme to generate the final open complex. The activator is, however, required for the highest levels of transcription, indicating that it may be necessary either for reinitiation or for efficient recruitment of the holoenzyme (or both). (Reprinted, with permission, from Chi and Carey 1996.)

transcription complexes on natural promoters. Studies by Maniatis and Grosschedl showed that natural promoters often employ cooperative activator-activator interactions to assemble multiple molecules of the same activator or diverse activator combinations into nucleoprotein structures called enhanceosomes (Giese et al. 1992; Giese and Grosschedl 1993; Tjian and Maniatis 1994; Grosschedl 1995; Thanos and Maniatis 1995). Enhanceosome assembly generally requires architectural proteins that bend and twist the DNA to facilitate the cooperativity (Carey 1998). Furthermore, the cooperativity is dependent on the proper stereo-specific positioning of activators. The proper positioning of the activators is thought to create a unique interface for interaction with the general machinery and hence represents a means of ensuring the specificity of the transcriptional response. Only when the correct interface is displayed can the activators recruit TFIID, TFIIA, and the remaining general factors (Carey 1998). We conclude by summarizing the system we are currently using to study how enhanceosome assembly and recruitment of the general machinery correlates with gene activation.

Figure 6. Several ZEBRA-responsive promoters. The line drawings illustrate the widely divergent promoter organization of different ZEBRA-responsive EBV genes. We have shown 6 of more than 30 different genes containing multiple known ZEBRA-responsive elements of ZREs. We have identified at least 20 different ZEBRA sites by footprinting and sequence searches. These sites fall within a 20-fold range of affinities, although the cooperative effects of HMG-1 and 2 have not been taken into account in this analysis. Our goal is to correlate systematically the levels and timing of gene activation in vivo with the ability of these promoters to assemble transcription complexes. In addition to ZEBRA sites and sites for another EBV regulator called Rta, sites for the cellular Sp1 protein seem to predominate in computer searches of EBV promoters, although it is important to consider that the EBV genome is unusually GC-rich and these sites may not be physiological.

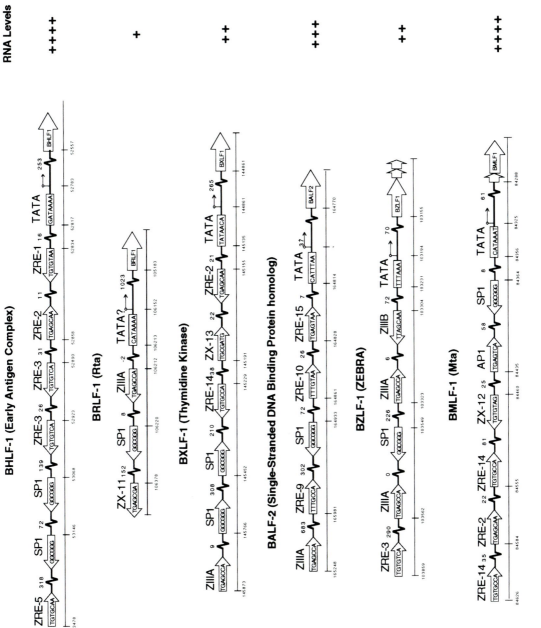

Figure 6. (*See facing page for legend.*)

The Mechanism of Synergy on a Natural ZEBRA-responsive Promoter

We originally chose ZEBRA as a model activator because it differentially controls the transcription of many different genes involved in the EBV lytic cycle. We have performed extensive footprinting analysis and homology searches of known sites to identify more than three dozen known and potentially ZEBRA-responsive genes, a subset of which are shown in Figure 6. The wide range of transcriptional responses elicited by the lytic promoters represents an excellent opportunity to apply the principles derived from the model system toward understanding a regulatory hierarchy.

It is believed that promoter architecture, the positioning of the activator binding sites relative to each other and to the core promoter, represents a mechanism for controlling enhanceosome assembly and therefore the level and timing of the transcriptional response. There are two key characteristics of enhanceosomes, the cooperative binding of activators to their sites and the requirement for architectural proteins to mold the DNA to permit the cooperative protein-protein interactions. There are many classes of architectural factors (Werner and Burley 1997). HMGI(Y) and HMG-1(2) represent prototypes for families that bend the DNA. These proteins have been shown to have both specific and global effects on pol II transcription.

Because we could find no binding sites for known sequence-specific bending proteins, we investigated the ability of the relatively nonspecific HMG-1 and HMG-2 to assist ZEBRA binding to the upstream promoter. We found that both HMG-1 and Sp-1 had strong stimulatory effects on ZEBRA DNA binding and both stimulated in in vitro transcription and cotransfection assays.

Recruitment of the DA complex by ZEBRA is also highly cooperative. When subsaturating amounts of either ZEBRA or TFIIA and TFIID are incubated with the template, little protection was observed as determined by DNase I footprinting. However, together, we found that ZEBRA recruited the DA complex to the GATAA box of the BHLF-1 promoter but the DA complex also has a reciprocal effect on binding of ZEBRA. The original data will be shown elsewhere (K. Ellwood and M. Carey, in prep.). This reciprocity was predicted based on the energetic link between the upstream and core promoters. Previous efforts on the model templates had revealed weak effects. The reciprocity phenomenon was probably most evident on the natural promoter due to the lower affinity upstream and core promoters and the unique arrangement of binding sites within the promoter. The binding effect required the activation domain of ZEBRA as a truncated version of ZEBRA lacking the amino-terminal nonacidic activation domain failed to recruit the complex.

Recent studies have established the view that many of the general factors are assembled into a holoenzyme. Although holoenzymes isolated from yeast and mammalian extracts vary considerably in terms of composition, the yeast holoenzyme isolated by Young and colleagues contains TFIIB and is complementable by TFIID and TFIIE

(Koleske and Young 1994). This enzyme has also shown to interact directly with activators in affinity chromatography experiments (Hengartner et al. 1995). The unique properties of the holoenzyme suggested that it could participate in the second step of the two-step recruitment model. Although a complex containing TFIIB and complementable solely by TFIIA and TFIID had not yet been isolated from mammalian extracts, the rationale that such a complex exists is compelling. First, it would agree with biochemical data suggesting that binding of the DA complex and recruitment of TFIIB are two biochemically separable steps. Second, most transcription occurs in multiple rounds. the first round is slow and takes much longer than reinitiation (Jiang and Gralla 1993). Biochemical data suggest that TFIID and probably TFIIA stay bound to the core promoter during elongation but the remaining factors dissociate from the complex (Zawel et al. 1995). Thus, a TFIIB-containing holoenzyme lacking TFIID and TFIIA would make sense from a regulatory standpoint because it could support the rapid reinitiation observed in vitro.

In an effort to isolate such a holoenzyme from HeLa extracts, we employed GST-VP16 and GST-ZEBRA affinity chromatography. We then subjected the eluate to a second round of affinity chromatography and assayed it for transcriptional activity. The eluate displayed a low basal activity and the ability to respond weakly to activators, but activity was greatly stimulated by TFIIA and TFIID. To determine the composition and whether the various factors constituting the putative holoenzyme existed in a complex, we subjected our affinity eluate to gel filtration. We found by immunoblotting that TFIIB, TFIIE, TFIIF, TFIIH, pol II, SWI/SNF, p300, and other components comigrated as a large more than 2-MD complex on the gel filtration column (W. Huang and M. Carey, in prep.).

Our future studies will be to compare cooperative binding of ZEBRA, DA complex assembly, and holoenzyme recruitment on a wide variety of ZEBRA-responsive promoters to determine if aspects of transcription complex assembly in vitro can be correlated with gene expression in vivo.

ACKNOWLEDGMENTS

This work described herein was supported by grants from the National Institutes of Health (GM-46424 and GM-057283) and the American Cancer Society.

REFERENCES

Carey M. 1998. The enhanceosome and transcriptional synergy. *Cell* **92:** 5.
Carey M., Leatherwood J., and Ptashne M. 1990a. A potent GAL4 derivative activates transcription at a distance in vitro. *Science* **247:** 710.
Carey M., Lin Y.S., Green M.R., and Ptashne M. 1990b. A mechanism for synergistic activation of a mammalian gene by GAL4 derivatives. *Nature* **345:** 361.
Carey M., Kolman J., Katz D.A., Gradoville L., Barberis L., and Miller G. 1992. Transcriptional synergy by the Epstein-Barr virus transactivator ZEBRA. *J. Virol.* **66:** 4803.
Chi T. and Carey M. 1993. The ZEBRA activation domain:

Modular organization and mechanism of action. *Mol. Cell. Biol.* **13:** 7045.

———. 1996. Assembly of the isomerized TFIIA—TFIID—TATA ternary complex is necessary and sufficient for gene activation. *Genes Dev.* **10:** 2540.

Chi T., Lieberman P., Ellwood K., and Carey M. 1995. A general mechanism for transcriptional synergy by eukaryotic activators. *Nature* **377:** 254.

Choy B., Roberts S.G., Griffin L.A., and Green M.R. 1993. How eukaryotic transcription activators increase assembly of preinitiation complexes. *Cold Spring Harbor Symp. Quant. Biol.* **58:** 199.

Dvir A., Conaway R.C., and Conaway J.W. 1996. Promoter escape by RNA polymerase II. A role for an ATP cofactor in suppression of arrest by polymerase at promoter-proximal sites. *J. Biol. Chem.* **271:** 23352.

Dvir A., Tan S., Conaway J.W., and Conaway R.C. 1997. Promoter escape by RNA polymerase II. Formation of an escape-competent transcriptional intermediate is a prerequisite for exit of polymerase from the promoter. *J. Biol. Chem.* **272:** 28175.

Ernst P. and Smale S.T. 1995. Combinatorial regulation of transcription. I. General aspects of transcriptional control. *Immunity* **2:** 311.

Giese K. and Grosschedl R. 1993. LEF-1 contains an activation domain that stimulates transcription only in a specific context of factor-binding sites. *EMBO J.* **12:** 4667.

Giese K., Cox J.. and Grosschedl R. 1992. The HMG domain of lymphoid enhancer factor 1 bends DNA and facilitates assembly of functional nucleoprotein structures. *Cell* **69:** 185.

Grosschedl R. 1995. Higher-order nucleoprotein complexes in transcription: Analogies with site-specific recombination. *Curr. Opin. Cell Biol.* **7:** 362.

Hengartner C.J., Thompson C.M., Zhang J., Chao D.M., Liao S.M., Koleske A.J., Okamura S., and Young R.A. 1995. Association of an activator with an RNA polymerase II holoenzyme. *Genes Dev.* **9:** 897.

Holstege F.C., Tantin D., Carey M., van der Vliet P.C., and Timmers H.T. 1995. The requirement for the basal transcription factor IIE is determined by the helical stability of promoter DNA. *EMBO J.* **14:** 810.

Horikoshi M., Carey M.F., Kakidani H., and Roeder R.G. 1988. Mechanism of action of a yeast activator: Direct effect of GAL4 derivatives on mammalian TFIID-promoter interactions. *Cell* **54:** 665.

Jiang Y. and Gralla J.D. 1993. Uncoupling of initiation and reinitiation rates during HeLa RNA polymerase II transcription in vitro. *Mol. Cell. Biol.* **13:** 4572.

Koleske A.J. and Young R.A. 1994. An RNA polymerase II holoenzyme responsive to activators (comments). *Nature* **368:** 466.

Lehman A.M., Ellwood K.B., Middleton B.E., and Carey M. 1998. Compensatory energetic relationships between upstream activators and the RNA polymerase II general transcription machinery. *J. Biol. Chem.* **273:** 932.

Lieberman P.M. and Berk A.J. 1994. A mechanism for TAFs in transcriptional activation: Activation domain enhancement of TFIID-TFIIA–promoter DNA complex formation. *Genes Dev.* **8:** 995.

Lin Y.S. and Green M.R. 1991. Mechanism of action of an acidic transcriptional activator in vitro. *Cell* **64:** 971.

Lin Y.S., Ha I., Maldonado E., Reinberg D., and Green M.R. 1991. Binding of general transcription factor TFIIB to an acidic activating region. *Nature* **353:** 569.

Miller G. 1990. The switch between latency and replication of Epstein-Barr virus. *J. Infect. Dis.* **161:** 833.

Parvin J.D. and Sharp P.A. 1993. DNA topology and a minimal set of basal factors for transcription by RNA polymerase II. *Cell* **73:** 533.

Roberts S.G., Choy B., Walker S.S., Lin Y.S., and Green M.R. 1995. A role for activator-mediated TFIIB recruitment in diverse aspects of transcriptional regulation. *Curr. Biol.* **5:** 508.

Roberts S.G., Ha I., Maldonado E., Reinberg D., and Green M.R. 1993. Interaction between an acidic activator and transcription factor TFIIB is required for transcriptional activation. *Nature* **363:** 741.

Tantin D. and Carey M. 1994. A heteroduplex template circumvents the energetic requirement for ATP during activated transcription by RNA polymerase II. *J. Biol. Chem.* **269:** 17397.

Tantin D., Chi T., Hori R., Pyo S., and Carey M. 1996. Biochemical mechanism of transcriptional activation by GAL4-VP16. *Methods Enzymol.* **274:** 133.

Thanos D. and Maniatis T. 1995. Virus induction of human IFNβ gene expression requires the assembly of an enhanceosome. *Cell* **83:** 1091.

Timmers H.T. 1994. Transcription initiation by RNA polymerase II does not require hydrolysis of the β-γ phosphoanhydride bond of ATP. *EMBO J.* **13:** 391.

Tjian R. and Maniatis T. 1994. Transcriptional activation: A complex puzzle with few easy pieces. *Cell* **77:** 5.

Wang W., Carey M., and Gralla J.D. 1992a. Polymerase II promoter activation: Closed complex formation and ATP-driven start site opening. *Science* **255:** 450.

Wang W., Gralla J.D., and Carey M. 1992b. The acidic activator GAL4-AH can stimulate polymerase II transcription by promoting assembly of a closed complex requiring TFIID and TFIIA. *Genes Dev.* **6:** 1716.

Werner M.H. and Burley S.K. 1997. Architectural transcription factors: Proteins that remodel DNA. *Cell* **88:** 733.

Zawel L., Kumar K.P., and Reinberg D. 1995. Recycling of the general transcription factors during RNA polymerase II transcription. *Genes Dev.* **9:** 1479.

Transcription Regulation, Initiation, and "DNA Scrunching" by T7 RNA Polymerase

G.M.T. Cheetham,*† D. Jeruzalmi,† and T.A. Steitz*†‡

Howard Hughes Medical Institute and *Departments of Molecular Biophysics and Biochemistry† and Chemistry,‡
Yale University, New Haven, Connecticut 06520-8114*

What are the structural features of a DNA-dependent RNA polymerase that account for its many additional functional properties when compared to a DNA polymerase? Besides forming phosphodiester bonds during the synthesis of new RNA chains by a mechanism that is the same as that for DNA polymerases (Steitz 1993), RNA polymerases must first recognize a specific duplex DNA sequence and melt out the promoter to form an initiation bubble. Unlike DNA polymerases, they can initiate RNA synthesis primed by a single nucleotide and have activity that is subject to regulation by other proteins. Finally, RNA polymerases can cycle abortively (Carpousis and Gralla 1980) at the initiation of transcription, making short RNA transcripts before entering the elongation phase that results in complete mRNA transcripts. Since T7 DNA polymerase is homologous to the very well studied DNA polymerase I (pol I) family of DNA polymerases (Davanloo et al. 1984), comparison of the structures of its substrate complexes with those of the DNA polymerases provides considerable insight into the structural basis of their functional differences.

Most of the genes in the T7 virus genome are transcribed by the T7 RNA polymerase, which can be inhibited by one of these gene products, the T7 lysozyme (Moffatt and Studier 1987). The T7 RNA polymerase is a single polypeptide chain with a molecular mass of approximately 99 kD that is homologous to other phage and mitochondrial RNA polymerases (Cermakian et al. 1996). The T7 RNA polymerase residues that form a highly conserved polymerase catalytic active site include Asp-537, Asp-812, His-811, and Lys-631 (Osumi-Davis et al. 1992).

STRUCTURE OF T7 RNA POLYMERASE COMPLEXED WITH A TRANSCRIPTIONAL INHIBITOR, T7 LYSOZYME

Zhang and Studier (1997) have shown that the binding of the T7 lysozyme to the T7 RNA polymerase inhibits the ability of the polymerase to elongate RNA transcripts beyond about 15 nucleotides. In an effort to provide a framework for understanding the structural basis of this regulation by the phage lysozyme, the structure of the polymerase complexed with lysozyme has been determined and refined at 2.8 Å resolution (Jeruzalmi and Steitz 1998). The structure determination of this complex proved to be far more challenging than most structures due to the large size of the complex, the modest resolu-

tion of the diffraction pattern, and the significant variation in the structures of different copies of the polymerase in its two crystal forms. Our structure of the T7 RNA polymerase shows that significant errors were made in the earlier structure determination of this enzyme by B.C. Wang and colleagues (Souza et al. 1993). The large amino-terminal domain (which is unique to the RNA polymerase in the pol I family) was misconnected and largely traced backward, whereas significant translocations of the sequence relative to the structure were made in the "fingers" and "thumb" domains.

The lysozyme is found bound to the side of the polymerase that is opposite the active site region (Fig. 1). It is interacting at the junction between the "fingers," "palm," and amino-terminal domains as if to lock their relative positioning in place. It does not appear that the lysozyme is repressing elongation synthesis through a direct inter-

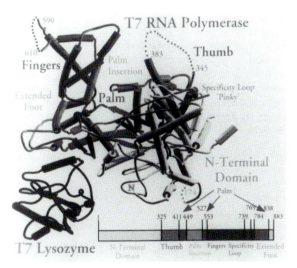

Figure 1. Structure of the T7 RNA polymerase-T7 lysozyme (PL) complex. Schematic representation of the PL complex, in which α helices are depicted as tubes and β strands are shown as arrows. This representation is shaded by domain, subdomain, or module, with the amino-terminal domain (8–325), *light gray,* the thumb (326–411), the palm (412–440, 528–553, 875–883), the palm insertion module (450–527), the fingers (554–739, 769–784), the pinky specificity loop (740–769), and the extended foot module (838–879), *darker gray,* and T7 lysozyme with bound mercury atom, *dark gray.* Domain boundaries are depicted within the primary sequence of T7 RNA polymerase, represented as a bar. (Reprinted, with permission, from Jeruzalmi and Steitz 1998.)

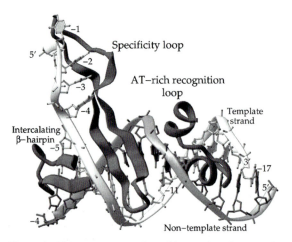

Figure 2. Ribbons representation of interactions between the upstream binding domain of the T7 promoter and the T7 RNA polymerase. Sequence-specific recognition is accomplished in the major groove by an extended β-hairpin "specificity" loop. Recognition of the AT-rich promoter sequence in the minor groove is accomplished by a loop emanating from the amino-terminal domain.

STRUCTURE OF T7 RNA POLYMERASE BOUND TO PROMOTER DNA

The sequences of the T7 promoters used by T7 RNA polymerase are highly conserved over a 20-bp stretch (Oakley and Coleman 1977; Rosa 1979; Carter and McAllister 1981; Dunn and Studier 1981; McAllister and Morris 1981 and references therein). The sequence between –6 and –10 is absolutely conserved and constitutes a major recognition site in phage polymerases. Biochemical and mutagenic experiments were interpreted to mean that side chains from the pinky are responsible for recognizing a specific promoter DNA sequence (Rong et al. 1998).

The T7 RNA polymerase was cocrystallized with a 17-bp duplex DNA promoter that consists of the class III promoter sequence from –1 to –17. This crystal structure has been refined at 2.4 Å resolution (G.M.T. Cheetham et al., in prep.). The T7 promoter sequences are recognized by the pinky antiparallel β-ribbon, which binds in the major groove to the absolutely conserved promoter sequence, and also by a loop that binds into the minor groove (Fig. 2). This recognition of A-T base pairs in the promoter at positions –17 to –13 result in a minor groove that is more shallow and wider than regular B-form DNA. The inherent flexibility that A-T base pair stretches support (Steitz 1990) may thus be an important factor in the initial electrostatic nonspecific promoter recognition during DNA translocation by T7 RNA polymerase. Furthermore, extrapolation of the upstream binding domain of the T7 promoter DNA helix observed in this complex strongly suggests that the position +2 through +5 region may be located very close to residues 590–610 at the top the fingers domain prior to isomerization to an open conformation (Fig. 3). The extended β-hairpin formed by these residues may be important for opening or stability of the downstream end of the transcription bubble (this is supported by preliminary experimental evidence showing the nontemplate strand interacting with the fingers domain) (G.M.T. Cheetham and T.A. Steitz, unpubl.). In the structure presented here, however, the duplex DNA is de-

action with the polymerase active site, but rather may be serving to inhibit some conformational transition required for processive elongation synthesis.

Comparison of the fingers and palm domains of the RNA polymerase with the homologous domains found in the Klenow fragment from *Escherichia coli* DNA polymerase I (Ollis et al. 1985) shows several important insertions in the RNA polymerase (Souza 1996; Jeruzalmi and Steitz 1998). The fingers of the RNA polymerase contain insertions that make them taller than in the Klenow fragment and also a major insertion, termed the "pinky" (or little finger) that consists largely of an antiparallel β-ribbon and has an important role in promoter sequence recognition (Raskin et al. 1992). The palm domain has a large insertion that resides at the back of the cleft between the fingers and thumb domain. The function of this insertion is at present unknown.

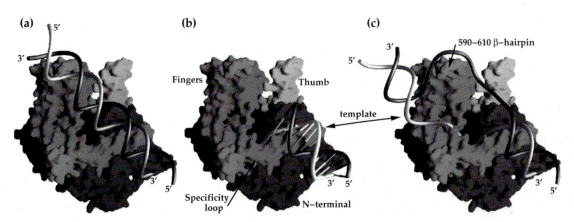

Figure 3. Schematic representations of the T7 RNA polymerase 17-bp T7 promoter complex refined at 2.4 Å resolution (*b*), and models for the "closed" (*a*) and "open" (*c*) DNA complexes. Nucleic acid strands are depicted as "worms," calculated from backbone P positions, with the template colored *dark gray* and the nontemplate strand *white*, whereas the protein is in surface representation.

natured at base pair –4 by a protein loop emanating from the amino-terminal domain that intercalates residue Val-237 between the base pairs at –4 and –5. Although the three nucleotides closest to the 3′ end of the nontemplate strand are not visible in the experimental electron density map, the four single-stranded nucleotides at the 5′ end of the template strand plunge into the active site hole of the polymerase.

ON THE NATURE OF TRANSCRIPTION INITIATION AND ABORTIVE SYNTHESIS

The cocrystal structure provides some insight into the nature of abortive RNA synthesis that occurs at the initiation of transcription and suggests that this process requires accumulation of the DNA template strand at the polymerase active site. Ikeda and Richardson (1986) determined the extent of promoter protection from methidiumpropyl-EDTA·Fe(II)-produced hydroxyl radical hydrolysis in the presence of T7 RNA polymerase at various early stages in the elongation reaction. The complex formed between the promoter and the enzyme in the absence of nucleotides showed a 17-bp-long protection that was from approximately –4 to –20. Formation of an open promoter and synthesis of a trinucleotide extended the protection from hydrolysis to include base pairs further downstream, a total of 29 bp. Synthesis of a six-nucleotide RNA extended this protection by an additional 2–4 bp downstream. Although protection of the downstream DNA sequences extended as the open complex was formed and RNA was synthesized, the extent of protection of the upstream sequences did not change. Upon formation of a 15-nucleotide RNA, both the upstream and the downstream now changed and the polymerase protected a total of 24 bp.

At least two possible models explain the ability of the polymerase to expand its downstream protection of the DNA while maintaining the upstream protection: protein inchworming or DNA scrunching. In one model, the duplex recognition portion of the enzyme and the polymerase portion of the enzyme would reside on different domains and could move independently of each other as RNA synthesis proceeded (the protein inchworming model). The polymerase would thus "walk down" the DNA in much the same way as an inchworm would walk down a branch by alternately extending and compressing. An alternative model, however, would keep the enzyme structure constant but require that the template and nontemplate DNA strands be accommodated progressively in the polymerase active site as synthesis proceeds. In this model, the single-stranded DNA would be compacted or "bunched" near the synthesis site (the DNA scrunching model; Fig. 4). We have demonstrated that the amino-terminal domain and fingers domain are responsible for recognition of the upstream binding domain of the promoter and some interactions with the incoming ribonucleotide triphosphate, whereas the catalytic active site resides on the palm domain. Despite this, however, our crystal structure and models built with the RNA primer and template would suggest that the latter model is correct for the T7 RNA polymerase (G.M.T. Cheetham et al., in prep.).

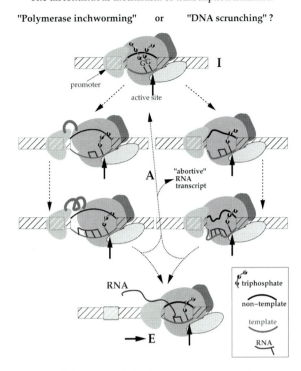

The discontinuous mechanism of transcription initiation

"Polymerase inchworming" or "DNA scrunching" ?

Figure 4. Schematic and simplified representation of two possible mechanisms for accommodating newly synthesized RNA and DNA strands during transcription initiation and abortive RNA synthesis. In the absence of DNA translocation, while strong promoter contacts are retained, the transcription bubble must expand, facilitated by relative movement of polymerase domains, or the DNA strands must "scrunch" in the region of the catalytic active site. A corollary of the latter model is that the distance between the active site and promoter contacts remains constant.

Because of the homology between the T7 RNA and DNA polymerases, it is possible to construct an accurate model of the primer-template for the +1 to +3 region, including the priming ribonucleotide triphosphate and incoming ribonucleotide triphosphate (Fig. 5), by homology modeling using the DNA polymerase ternary structure, determined by Doublié et al. (1998). The RNA polymerase can be oriented on the DNA polymerase by superposition of the closely similar palm domain structures. Then assuming that the synthesis of product RNA off the template proceeds in a nearly identical fashion in both the RNA and DNA polymerases, the DNA promoter from the RNA polymerase can be positioned on the DNA polymerase or, vice versa, the primer-template and deoxynucleotide triphosphate from the DNA polymerase can be positioned on the RNA polymerase. Having done that, it is clear that the primer-template helix and the helix corresponding to the duplex promoter DNA are neither parallel to each other nor do they overlap.

After positioning the 3′ nucleotide of the primer strand and its corresponding template nucleotide from the DNA

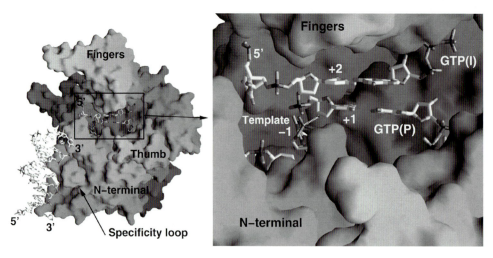

Figure 5. Model of the T7 RNA polymerase initiation complex derived from a cocrystal structure and homology modeling. The distance between the experimentally determined position for the 5′ phosphate of the −1 template position and the 3′ hydroxyl of the modeled +1 template position is approximately 1.5 Å. Incoming rGTP molecules at the initiating and priming positions base pair with template positions +2 and +1, respectively.

polymerase complex into the active site of the T7 RNA polymerase by homology modeling, many aspects of substrate specificity can be understood (G.M.T. Cheetham et al., in prep.). Perhaps surprisingly, the 5′ phosphate of the −1 nucleotide of the promoter template bound to the RNA polymerase lies within 1.5 Å of the 3′ hydroxyl of the homology-modeled template +1 nucleotide, suggesting that the homology model is very good and that the +1 to +3 template nucleotides do, indeed, bind very similarly in these two enzymes. The 2′ hydroxyl group of the priming nucleotide and that of the incoming nucleotide are seen to make hydrogen-binding interactions with the RNA polymerase, which are not observed, of course, in the DNA polymerase.

Two major reasons that account for the ability of the RNA polymerase to initiate at a single nucleotide, whereas DNA polymerase cannot, become apparent (G.M.T. Cheetham et al., in prep.). First and perhaps most importantly, the RNA polymerase exactly positions the template strand in the active site, which provides an important part of the binding sites for the bases of the incoming nucleotide triphosphates at the +1 and +2 positions. Single-stranded DNA bound at the DNA polymerase active sites would presumably not have such precise positioning. Furthermore, there is an additional interaction between the 2′ hydroxyl of the priming nucleotide and the RNA polymerase that would further stabilize the positioning of this nucleotide.

Extending the RNA primer on the DNA template by further homology modeling with the DNA polymerase suggests that only 2 or 3 bp can be made between the newly synthesized RNA and the DNA before the RNA must peel off (G.M.T. Cheetham et al., in prep.). Furthermore, there appears to be no room for more than three of four nucleotides of RNA in the pocket before it must emerge. This is because the cleft into which the duplex product of DNA synthesis is observed to bind in both the *Taq* polymerase (Eom et al. 1996) and T7 DNA poly-

merase (Doublié et al. 1998) is blocked in the RNA enzyme by the large amino-terminal domain. Hybrid DNA-RNA intermediates have been demonstrated to be important for keeping the RNA 3′ terminus in the active site during elongation transcription by the *E. coli* RNA polymerase (Nudler et al. 1997). Furthermore, topological considerations of the RNA winding around the DNA require the RNA to peel off the DNA after only a couple of base pairs. Left unclear by these modeling studies is exactly what happens to the presumably accumulating template strand as RNA synthesis proceeds. It seems abundantly clear, however, that the synthesis active site and the promoter-binding region of the enzyme cannot move relative to each other in an inchworming fashion. The pinky antiparallel β-ribbon that is recognizing the promoter DNA comes from the fingers domain, which is also providing part of the binding site for the incoming nucleoside triphosphate. These two sites cannot be moving significantly relative to each other during synthesis.

Proteolysis (Muller et al. 1988) and mutagenic (He et al. 1997) and photocrosslinking (Sastry and Ross 1998) experiments have suggested a major contact site between residues 144 and 168, located in the amino-terminal domain of T7 RNA polymerase, and the 5′ terminus of the transcribed RNA. On the basis of our corrected model for the amino-terminal domain, these residues are located on a concave surface that faces the catalytic active site and modeled ribonucleotides, at a distance of approximately 25 Å. The position of the putative RNA-binding site, with respect to the active site and incoming ribonucleotide triphosphates, is significantly different from that previously proposed (Sastry and Ross 1998). The RNA-binding site is also adjacent to the AT-rich binding and pinky specificity determining loops. We therefore conclude that binding to this part of the amino-terminal domain may contribute to release of the strong polymerase-promoter contacts and trigger the transition to processive elongation RNA synthesis.

Future studies of T7 RNA polymerase will address numerous questions. What happens to the DNA that accumulates in the bubble during the initial stages of abortive RNA synthesis, and furthermore, what is the nature of the conversion from abortive to elongation synthesis? How is the RNA peeled off the DNA template and how long is the binding site for the transcript for the enzyme? Finally, how does the binding of T7 phage lysozyme prevent the extension of this transcript beyond about 15 nucleotides?

ACKNOWLEDGMENTS

Figures were prepared with RIBBONS (Carson 1991), BOBSCRIPT (Esnouf 1997), and GRASP (Nichols et al. 1993).

REFERENCES

Carpousis A.J. and Gralla J.D. 1980. Cycling of ribonucleic acid polymerase to produce oligonucleotides during initiation in vitro at the lac UV5 promoter. *Biochemistry* **19:** 3245.

Carson M. 1991. Ribbons 2.0. *J. Appl. Crystallogr.* **A24:** 958.

Carter A.D. and McAllister W.T. 1981. Sequences of three class II promoters for the bacteriophage T7 RNA polymerase. *J. Mol. Biol.* **153:** 825.

Cermakian N., Ikeda T.M., Cedergren R., and Gray M.W. 1996. Sequences homologous to yeast mitochondrial and bacteriophage T3 and T7 RNA polymerases are widespread through the eukaryotic lineage. *Nucleic Acids Res.* **24:** 648.

Davanloo P., Rosenberg A.H., Dunn J.J., and Studier F.W. 1984. Cloning and expression of the gene for bacteriophage T7 RNA polymerase. *Proc. Natl. Acad. Sci.* **81:** 2035.

Doublié S., Tabor S., Long A.M., Richardson C.C., and Ellenberger T. 1998. Crystal structure of a bacteriophage T7 DNA replication complex at 2.2 Å resolution. *Nature* **391:** 251.

Dunn J. and Studier F.W. 1981. Nucleotide sequence from the genetic left end of bacteriophage T7 DNA to the beginning of gene 4. *J. Mol. Biol.* **148:** 303.

Eom S.H., Wang J., and Steitz T.A. 1996. Structure of Taq polymerase with DNA at the polymerase active site. *Nature* **382:** 278.

Esnouf R. 1997. An extensively modified version of MolScript that includes greatly enhanced coloring capabilities. *J. Mol. Graphics* **15:** 133.

He B., Rong M., Durbin R.K., and McAllister W.T. 1997. A mutant T7 RNA polymerase that is defective in RNA binding and blocked in the early stages of transcription. *J. Mol. Biol.* **265:** 275.

Ikeda R.A. and Richardson C.C. 1986. Interactions of the RNA polymerase of bacteriophage T7 with its promoter during binding and initiation of transcription. *Proc. Natl. Acad. Sci.* **83:** 3614.

Jeruzalmi D. and Steitz T.A. 1998. Structure of the T7 RNA polymerase complexed to the transcriptional inhibitor T7 lysozyme. *EMBO J.* **17:** 4101.

McAllister W.T. and Morris C. 1981. Utilization of bacteriophage T7 late promoters in recombinant plasmids during infection. *J. Mol. Biol.* **153:** 527.

Moffatt B.A. and Studier F.W. 1987. T7 lysozyme inhibits transcription by T7 RNA polymerase. *Cell* **49:** 221.

Muller D.K., Martin C.T., and Coleman J.E. 1988. Processivity of proteolytically modified forms of T7 RNA polymerase. *Biochemistry* **27:** 5763.

Nichols A., Bharadwaj R., and Honig B. 1993. Graphical representation and analysis of surface properties. *Biophys. J.* **64:** A116.

Nudler E., Mustaev A., Lukhtanov E., and Goldfarb A. 1997. The RNA-DNA hybrids maintains the register of transcription by preventing backtracking of RNA polymerase. *Cell* **89:** 33.

Oakley J.L. and Coleman J.E. 1977. Structure of a promoter for T7 RNA polymerase. *Proc. Natl. Acad. Sci.* **74:** 4266.

Ollis D.L., Brick P., Hamlin R., Xuong N.G., and Steitz T.A. 1985. Structure of large fragment of *Escherichia coli* DNA polymerase I complexed with dTMP. *Nature* **313:** 762.

Osumi-Davis P.A., Aguilera M.C., Woody R.W., and Woody A.Y.M. 1992. Asp 537, Asp812 are essential and Lys631, His811 are catalytically significant in bacteriophage T7 RNA polymerase activity. *J. Mol. Biol.* **226:** 37.

Raskin C.A., Diaz G., Joho K., and McAllister W.T. 1992. Substitution of a single bacteriophage T3 residue in bacteriophage T7 RNA polymerase at position 748 results in a switch in promoter specificity. *J. Mol. Biol.* **228:** 506.

Rong M., He B., McAllister W.T., and Durbin R.K. 1998. Promoter specificity determinants of T7 RNA polymerase. *Proc. Natl. Acad. Sci.* **95:** 515.

Rosa M.D. 1979. Four T7 RNA polymerase promoters contain an identical 23 bp sequence. *Cell* **16:** 815.

Sastry S. and Ross B.M. 1998. RNA-binding site in T7 RNA polymerase. *Proc. Natl. Acad. Sci.* **95:** 9111.

Souza R. 1996. Structural and mechanistic relationships between nucleic acid polymerases. *Trends Biochem. Sci.* **21:** 186.

Souza R., Chung Y.J., Rose J.P., and Wang, B.-C. 1993. Crystal structure of bacteriophage T7 RNA polymerase at 3.3 Å resolution. *Nature* **364:** 595.

Steitz T.A. 1990. Structural studies of protein-nucleic acid interactions: The sources of sequence specific binding. *Q. Rev. Biophys.* **23:** 205.

———. 1993. DNA- and RNA-dependent DNA polymerases. *Curr. Opin. Struct. Biol.* **3:** 31.

Zhang X. and Studier F.W. 1997. Mechanism of inhibition of bacteriophage T7 RNA polymerase by T7 lysozyme. *J. Mol. Biol.* **269:** 10.

Structural Studies of *Escherichia coli* RNA Polymerase

S.A. Darst, A. Polyakov, C. Richter, and G. Zhang

The Rockefeller University, New York, New York 10021

In some bacteriophages, a single polypeptide, such as the 110-kD T7 RNA polymerase (RNAP), will suffice for the transcription of a handful of bacteriophage genes. All transcription in eubacteria is performed by one core RNAP, which typically comprises four subunits with a total molecular mass of approximately 400 kD. *Escherichia coli* RNAP comprises an essential catalytic core of two α subunits (each 36.5 kD), one β-subunit (150.6 kD), and one β'-subunit (155.2 kD). Transcription in eukaryotes is performed by three distinct enzymes, which typically comprise more than a dozen subunits and have total molecular masses of approximately 500 kD.

High-resolution crystal structures of the T7 RNAP and closely related molecules such as DNA polymerase I Klenow fragment and reverse transcriptase have appeared, sometimes with templates, substrates, or inhibitors bound (Doublie et al. 1998; Jeruzalmi and Steitz 1998; Kiefer et al. 1998). All of these RNAPs catalyze exactly the same chemical reaction, and the detailed catalytic mechanism used by each enzyme is almost certainly the same. The similarities end there, however. There is very little, if any, evolutionary relationship between the single-subunit family of polymerases and the multisubunit cellular RNAPs, and the essential catalytic center, composed of one polypeptide in the case of the single-subunit enzymes, appears to be built up from widely separated portions of the two large β and β' subunits that are positioned close together in the three-dimensional structure of the multisubunit enzyme assembly (Mustaev et al. 1997). However, with the cloning and sequencing of the eukaryotic RNAP subunits, sequence homologies between the essential bacterial core subunits and subunits that make up more than two-thirds the mass of the eukaryotic enzymes were revealed (Archambault and Friesen 1993). These sequence homologies point to structural and functional homologies, making *E. coli* RNAP an excellent model system for understanding the multisubunit cellular RNAPs in general.

In both prokaryotes and eukaryotes, the core RNAP is catalytically competent to synthesize processively RNA chains from a DNA template, but it is unable to recognize promoters and initiate specific RNA synthesis from double-stranded templates. Additional protein factors are required for specific transcription initiation. In prokaryotes, this function is provided by an additional subunit called σ, which binds to the core RNAP to form what is called the holoenzyme, which can specifically locate promoters and form open complexes (Burgess et al. 1969; Travers and Burgess 1969). After RNA chain initiation and the first translocation event of the enzyme away from the promoter, the σ-subunit is released and the core enzyme completes the cycle.

Over the years, a wealth of biochemical, biophysical, and genetic information has accumulated on *E. coli* RNAP and its complexes with nucleic acids and accessory factors, and the functional aspects of the *E. coli* transcription cycle are by far the best understood. Recent years have seen an explosion of fundamental information regarding RNAP structure and function from biochemical and genetic studies of the *E. coli* system (Gross et al. 1996). Nevertheless, the enzyme itself, in terms of its structure/function relationship, remains somewhat of a black box. An essential step toward understanding the mechanism of transcription and its regulation is to determine the three-dimensional structures of RNAP and its complexes with DNA, RNA, and regulatory factors. For this purpose, we use a combination of approaches, including electron microscopy and image processing to determine low-resolution structures of intact RNAPs and transcription complexes, biophysical and biochemical methods to dissect the functional and structural architecture of the individual subunits, and X-ray crystallography to determine high-resolution structures of RNAP components and accessory factors.

RESULTS

Structures of Multisubunit Cellular RNA Polymerases by Electron Microscopy and Image Processing

Low-resolution structures of RNA polymerases from negatively stained crystals. Structures to date of the multisubunit cellular RNAPs have come from electron microscopy and image processing of negatively stained crystals, resulting in an image of the depression or cast that the protein leaves in a heavy-metal embedding medium such as uranyl acetate. Low-resolution, three-dimensional structures of *E. coli* RNAP holoenzyme (Darst et al. 1989), and yeast RNAPs II (Darst et al. 1991) and I (Schultz et al. 1993), were determined by electron microscopy of negatively stained two-dimensional crystals tilted at various angles to the incident electron beam (Amos et al. 1982). Our subsequent structure of *E. coli* core RNAP, which lacks the promoter-specific σ-subunit, revealed dramatic conformational changes compared with the *E. coli* RNAP holoenzyme but resembled yeast RNAP II (Polyakov et al. 1995). Although each structure contains a thumb-like projection surrounding a groove or channel about 25 Å in diameter, which is the appropriate size to accommodate double-stranded nucleic acid, the thumb of *E. coli* RNAP holoenzyme defines a deep but open groove on the surface of the molecule, whereas the thumb of *E. coli* core RNAP and yeast RNAP II forms part of a ring of protein density that completely surrounds the channel. This may define "promoter binding" and "elongation" conformations of RNAP, as both *E. coli* core and yeast RNAP II are capable of processive elongation of RNA chains but are incapable of specific promoter recognition without additional factors. The open groove of holoenzyme in the promoter-binding conformation would facilitate the recognition and loading of double-stranded promoter DNA within the groove. Closing of the thumb to completely surround the DNA upon transition to the elongation conformation would stabilize the elongating complex on the template, facilitating highly processive elongation of the nascent RNA chain.

Structural studies of E. coli *core RNA polymerase by cryo-electron microscopy of frozen-hydrated crystals.*

The low-resolution structure of *E. coli* core RNAP described above was determined by electron microscopy analysis of negatively stained, flattened, helical crystals (Polyakov et al. 1995). Since the crystals were preserved in negative stain, only the surface topography (i.e., an envelope of the surface) was revealed. In addition, since in the original crystal the rows of molecules wrapped around the diameter of the tube with helical symmetry, the flattening of the tubes compressed the molecules together, possibly distorting them somewhat.

We are therefore investigating the structure of *E. coli* core RNAP by cryo-electron microscopy (cryo-EM) and image processing of the tubular, helical crystals preserved by flash-freezing and embedding them in a thin layer of amorphous ice. This method of preservation and structure determination (Dubochet et al. 1988) will ultimately yield a more detailed structure of the polymerase due to the preservation in a completely native state in the absence of stain. With this preservation method, the helical symmetry of the crystals is maintained, allowing the application of powerful approaches developed over the years for reconstructing three-dimensional structures from images of helical assemblies (DeRosier and Moore 1970; Toyoshima and Unwin 1990; Unwin 1993; Beroukhim and Unwin 1995). Although the specimen is potentially preserved to higher resolution, because of the absence of stain and the increased sensitivity of the sample to radiation damage, images are taken under minimal dose conditions. This results in images with low ratios of signal to noise, which must be overcome by collecting and averaging together many images, usually ten or so. Our analysis of the data collected thus far indicates that a 12-Å-resolution structure should be obtainable in the near future. Multiple views of a preliminary structure, resulting from the averaging of data from two images and extending to a resolution of 16 Å, are shown in Figure 1.

Overall, the structure looks similar to the *E. coli* core RNAP structure from negative stain and bears an even more striking resemblance to the yeast RNAP II structure. Even this preliminary map has two remarkable features. First, as seen in the other RNAP structures determined in negative stain, there is a well-defined channel completely surrounded by protein density. The channel is about 40-Å diameter at one end but constricts to about 20-Å diameter at the other end (Fig. 1d). Second, two prominent domains of the RNAP are revealed in the cryo-EM structure that were not visible in negative stain (denoted by * and ** in Fig. 1). We believe that these domains were not visible in the negative stain structure due to disorder caused either by a staining artifact or as an effect of the flattening of the tubular crystals.

From an analysis of the data, it is clear that a structure of *E. coli* core RNAP from ice-embedded helical crystals to approximately 12-Å resolution is well within reach. This will represent the most detailed structure of a multisubunit, cellular RNAP yet. The analysis described above has many other significant advantages:

1. The structure will be free of any potential staining artifacts and represent the native structure of the RNAP molecule. The preliminary analysis (Fig. 1) already reveals several domains that were not visible in negative stain due either to staining artifacts or to effects of flattening the cylindrical tubes.
2. Since a single image of the helical crystal affords a large number of different views of the RNAP molecule, the resulting structure will not suffer from the "missing cone" effect arising from missing data due to the difficulty in collecting very high tilts from two-dimensional crystals (Amos et al. 1982). The resolution of the structure will thus be isotropic in three-dimensional space.
3. The structure determined from the ice-embedded crystals represents the true distribution of electron scattering mass within the specimen. This means that the internal density distribution of the protein can be visualized, not just an envelope of the protein surface.

We believe that these advantages and the resulting improved structure will allow us to discern the subunit organization of the RNAP. Further important questions, such as which of the densities correspond to the α, β, and β′ subunits of core RNAP, where functional sites and domains within the large subunits are located, and where regulatory factors interact with the RNAP, are being addressed with further experiments. For instance, we have located the density within the cryo-EM map corresponding to β dispensable region II (a 110-amino-acid region centered at approximately position 998; Borukhov et al. 1991) as the domain labeled ** in Figure 1 (C. Richter et al., unpubl.). We have also developed a general strategy for locating the binding sites of regulatory factors on the RNAP and have recently used it to visualize the transcript elongation factor GreB bound to core RNAP (Polyakov et al. 1998).

X-ray Crystallography of E. coli RNA Polymerase Components

Although significant progress has been made in the analysis of the *E. coli* RNAP structure by electron microscopy and image processing, this method is unlikely to

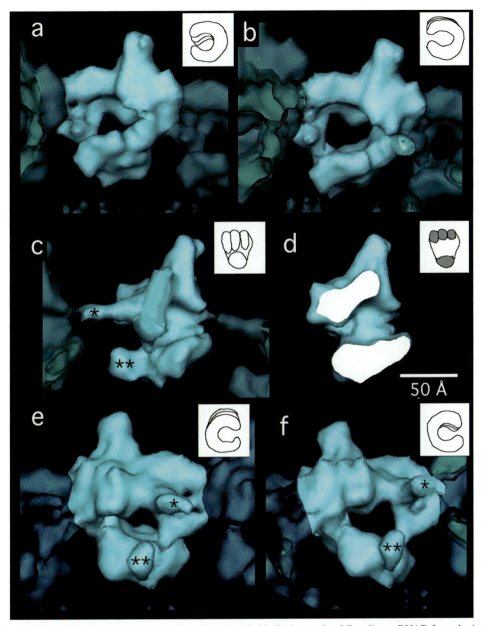

Figure 1. Views of the preliminary reconstruction from ice-embedded helical crystals of *E. coli* core RNAP. In each view, a single-core RNAP molecule is highlighted. A drawing of a "hand" in the upper right corner of each view is shown in an orientation similar to that of the RNAP molecule. (*c,e,f*) * and ** denote domains that were not visible in the earlier structure determined in negative stain. (*a*) View parallel with the helix axis; (*b*) view from the wide end of the channel, roughly down its axis; (*c*) view perpendicular to the channel; (*d*) similar view as *c* except the protein density surrounding the channel nearest the viewer has been cut away; (*e*) view parallel with the helix axis but from the opposite side as *a*; (*f*) view from the narrow end of the channel, roughly down its axis.

ultimately provide high-resolution information. For this we turn to X-ray crystallography. Although we lack three-dimensional crystals of *E. coli* RNAP suitable for X-ray analysis, we can still obtain useful structural information by examining individual components of the RNAP. Despite extensive efforts on the part of many laboratories (especially for the α and σ subunits), the individual RNAP subunits have not yielded to crystallization. Nevertheless, our work on the β (Severinov et al. 1995; Wang et al. 1997), β´ (Severinov et al. 1996), and σ70 subunits (Severinova et al. 1996), and the work of others on the α-sub-

unit (Igarashi and Ishihama 1991; Blatter et al. 1994), reveals that the individual subunits comprise structural and functional subdomains that can be amenable to high-resolution structural studies. A case in point is our analysis of the domain architecture of the σ^{70}-subunit (Severinova et al. 1996), and subsequent crystal structure determination of a domain that contains σ homology region 2 (Lonetto et al. 1992), which has been implicated in binding of σ to core RNAP, and recognition and melting of the –10 promoter consensus element (Malhotra et al. 1996; Darst et al. 1997). Below we describe our latest results on the crystal structure

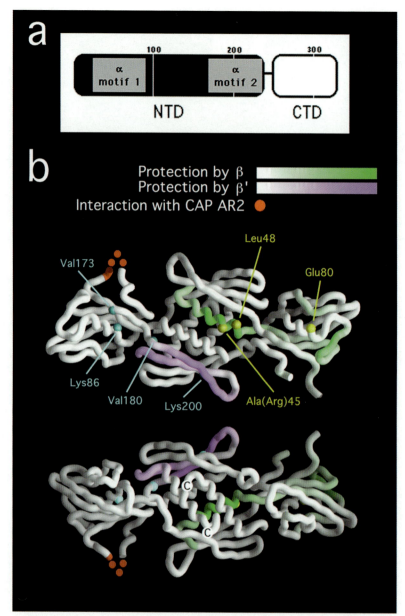

Figure 2. (*a*) Schematic diagram showing the domain structure of the *E. coli* RNAP α-subunit. (*Black box*) NTD crystallized in this study (α residues 1–235); (*gray boxes*) regions conserved in sequence between α homologs of prokaryotic, archaebacterial, chloroplast, and eukaryotic RNAPs. (*b*) (*top*) Backbone representation of the αNTD dimer viewed along the dimer twofold axis. Backbone residues are color-coded according to the hydroxyl-radical footprinting data of Heyduk et al. (1996) so that regions protected from hydroxyl-radical cleavage by β or β′ are colored *green* or *magenta*, respectively. For clarity, the regions protected by β are shown on one αNTD monomer, and the regions protected by β′ on the other. The actual distribution of determinants across the αNTD dimer is unknown. Shown in *yellow* or *light blue* are the α-carbon positions of mutations that cause defects in β or β′ binding, respectively (Kimura et al. 1994; Kimura and Ishihama 1995a,b). (*Red*) Region of αNTD found to interact with CAP-AR2 at class II CAP sites (αNTD residues 162–165) (Niu et al. 1996). (*Bottom*) View along the dimer twofold axis from the opposite direction as the middle view. The carboxyl termini of the two αNTD monomers are denoted. (Reprinted, with permission, from Zhang and Darst 1998 [copyright AAAS].)

of the *E. coli* RNAP α-subunit amino-terminal domain, which represents the first high-resolution structure of a core RNAP component required for RNAP assembly and basal transcription.

In addition to having key roles in transcription initiation, the α-subunit initiates RNAP assembly (Zillig et al. 1976; Ishihama 1981) by dimerizing into a platform with which the large β and β′ subunits interact. The α-subunit comprises two independent domains, the amino-terminal domain (NTD; residues 8–235) and carboxy-terminal domain (CTD; residues 249–329), connected by a flexible, 14-residue linker (Igarashi and Ishihama 1991; Blatter et al. 1994; Jeon et al. 1997). The αCTD is required for the interaction with upstream promoter elements (Ross et al. 1993) and is the target for a wide array of transcription activators (Ishihama 1993). The solution structure of αCTD consists of a compact fold of four short α helices (Jeon et al. 1995). The αNTD is necessary and sufficient in vivo and in vitro for RNAP assembly and basal transcription (Hayward et al. 1991; Igarashi and Ishihama 1991). The regions of conserved sequence between α homologs of prokaryotic, archaebacterial, chloroplast, and eukaryotic RNAPs (Fig. 2a, α motifs 1 and 2) are contained within the NTD (Ebright and Busby 1995), as are the determinants for α interaction with the RNAP β and β′ subunits (Igarashi et al. 1990, 1991; Igarashi and Ishihama 1991; Blatter et al. 1994; Kimura et al. 1994; Kimura and Ishihama 1995a,b; Heyduk et al. 1996). We have determined the X-ray crystal structure of αNTD to a resolution of 2.5 Å (Zhang and Darst 1998).

The αNTD monomer comprises two distinct, flexibly linked domains, only one of which participates in dimerization (Fig. 2b). In the αNTD dimer, a pair of helices from one monomer interact with the cognate helices of the other to form an extensive hydrophobic core. All of the determinants for interactions with the other RNAP subunits lie on one face of the αNTD dimer. On the opposite face, sites known to interact with the other RNAP subunits are not found, and located on this face are the carboxyl termini of the two αNTD monomers (Fig. 2b). Thus, the αCTDs and the β and β′ subunits are located on opposite faces of the αNTD structure.

Although the αCTD is the target for a wide array of transcription activators (Ishihama 1993), at least one interaction between an activator (catabolite activator protein, or CAP) and αNTD, which is essential for activation at class II CAP-dependent promoters, has been identified. The protein-protein interactions between CAP and αNTD occur between the basic activating region 2 (AR2) of CAP and a stretch of four acidic residues of αNTD (Niu et al. 1996). This region of the αNTD structure comprises a highly exposed loop (Fig. 2b, shown in red on only one αNTD monomer). A short stretch of residues in this region is disordered in the crystal structure.

The largest subunits of prokaryotic RNAPs (β′ and β) exhibit strong sequence conservation with homologs in eukaryotic RNAPs (Archambault and Friesen 1993). Less obvious evolutionary relationships have been proposed between α and two families of eukaryotic RNAP subunits related to *Saccharomyces cerevisiae* Rpb3 and Rpb11, and a number of recent studies suggest that an Rpb3-Rpb11 heterodimer serves as the eukaryotic analog of the prokaryotic α_2 homodimer (Azuma et al. 1993; Lalo et al. 1993; Pati 1994; Ulmasov et al. 1996; Svetlov et al. 1998). Sequence alignments, combined with secondary structure predictions and the known structure of αNTD, provide strong support for these proposals (Zhang and Darst 1998).

DISCUSSION

Progress toward a High-resolution Model of *E. coli* RNA Polymerase by Combining X-ray Crystallography and Cryo-EM Results

A major step toward interpreting the low-resolution structure of *E. coli* RNAP would be to relate the cryo-

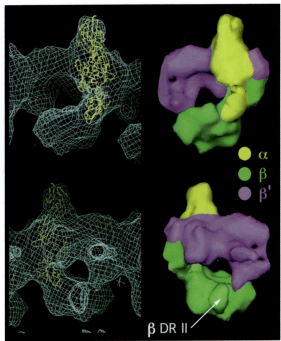

Figure 3. (*Top and bottom*) Two views (corresponding to Fig. 1a and e) of the 16-Å-resolution structure of *E. coli* core RNAP from cryo-EM of helical crystals. (*Left*) The cryo-EM density map is shown as a *light blue* net. (*Yellow*) α-carbon backbone of the αNTD structure (Zhang and Darst 1998), which is superimposed inside the cryo-EM structure. (*Right*) Model for the organization of the RNAP subunits with respect to the cryo-EM structure, which is based on the superimposition of the αNTD structure, and the identification of β dispensable region II (DRII) is illustrated by the color coding (α in *yellow*, β in *green*, β′ in *magenta*).

EM-derived density map (Fig. 1) with specific subunits and functional sites within the enzyme. The 2.5-Å-resolution crystal structure of the *E. coli* RNAP αNTD represents the first high-resolution structure of a core RNAP component that is essential for RNAP assembly and basal transcription. The αNTD dimer has a very distinctive shape (Fig. 2b), and upon examination of the cryo-EM structure, a feature in the density map that matches the unusual shape of the αNTD becomes apparent. The backbone of the αNTD structure fits extremely well into this part of the cryo-EM structure (Fig. 3). With this fit and our identification of β dispensable region II mentioned above, we can generate an outline of the subunit organization (color coding in Fig. 3). In this picture, we have assumed that the αCTDs are disordered and invisible in the cryo-EM density map, which remains to be confirmed.

This fit of the αNTD into the cryo-EM map reveals the location of the loop on αNTD that interacts with CAP-AR2 at class II CAP-dependent promoters (Niu et al. 1996; Zhang and Darst 1998). In the model of the αNTD fit into the cryo-EM map, this loop is somewhat buried on one αNTD monomer but highly exposed on the other (shown in red in Fig. 4a). At class II CAP-dependent promoters (where the CAP site is centered at –41.5 bp with respect to the transcription start site), AR2 of the downstream subunit of the CAP dimer interacts with αNTD, whereas AR1 of the upstream CAP subunit interacts with

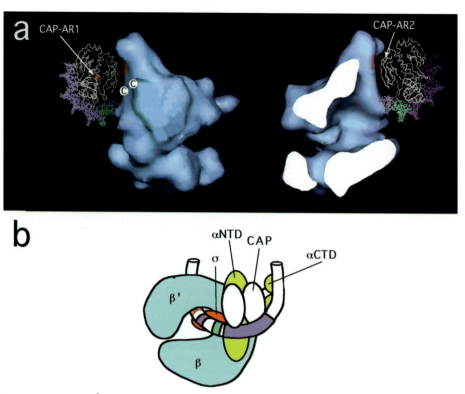

Figure 4. (*a*) Two views of the 16-Å-resolution structure of *E. coli* core RNAP from cryo-EM of helical crystals. The right view is similar (but not identical) to the view of Fig. 1d, with some of the protein density surrounding the RNAP channel cut away. The left view is rotated approximately 180° about the vertical axis. The red blotch on the surface of the RNAP density denotes the location of the exposed αNTD loop that interacts with CAP-AR2 at class II CAP-dependent promoters (Niu et al. 1996). Next to the RNAP density is the structure of CAP bound to its DNA site (Schultz et al. 1991) positioned such that AR2 of the downstream CAP subunit is next to the exposed loop of αNTD. (*White*) α-carbon backbone of the CAP dimer. (*Orange*) Residues comprising AR1 of the upstream CAP subunit, which interacts with αCTD. (*Red*) Residues comprising AR2 of the downstream CAP subunit, which interacts with αNTD. The bent DNA from the CAP-DNA crystal structure is shown in *violet,* except that the region corresponding to the –35 consensus promoter element is *light green*. (*b*) Schematic model incorporating the model shown in *a*, except also including the inferred location of the σ[70]-subunit (*red*) and the αCTDs, which interact with the DNA upstream of the CAP site. The protein components are labeled. The color coding for the DNA sites are as follows: (*red line*) Transcription start site; (*magenta*) –10 promoter element; (*light green*) –35 promoter element; (*violet*) CAP site.

αCTD (Niu et al. 1996). We can combine all of these pieces of information, along with the location of AR2 on the crystal structure of CAP bound to its DNA site (Schultz et al. 1991), to generate a model of the interactions between RNAP and CAP at a class II CAP-dependent site (Fig. 4a). It should be noted that we have modeled CAP interacting with the core RNAP structure, whereas CAP really interacts with holoenzyme. During the transition from the open promoter complex of the RNAP holoenzyme to the committed elongation complex of core RNAP, the RNAP likely attains the "closed channel" conformation of core RNAP despite the presence of σ (Polyakov et al. 1995). Evidence for this comes from hydroxyl-radical footprinting data of RNAP kinetically trapped in different stages of the transcription cycle (Metzger et al. 1989; Schickor et al. 1990; Mecsas et al. 1991). In this regard, it is interesting to note the position of the –35 promoter consensus element in the model (illustrated in light green in Fig. 4a). The major groove of the –35 consensus element is exposed and facing the RNAP, suggesting a location for homology region 4.2 of the σ[70]-sub-

unit, with which it interacts (Gardella et al. 1989; Waldburger et al. 1990). From Figure 4a, it is clear that if the DNA were extended from the downstream end of the CAP-DNA complex, it would direct the DNA through the RNAP channel. If the approximate dimensions of B-form DNA were maintained, the –10 element would lie in the center of the channel, whereas the +1 site (at the RNAP catalytic center) would lie just where the channel constricts to its narrowest diameter of about 20 Å. These observations make the model consistent with the complete protection of both DNA strands from hydroxyl-radical cleavage within the –10 element, and with the fact that the +1 site is near the downstream boundary of the elongating RNAP footprint (Metzger et al. 1989; Schickor et al. 1990; Mecsas et al. 1991). All of these observations are combined in a schematic model of all the RNAP holoenzyme subunits (including σ and the αCTDs), CAP, and DNA containing a promoter and a CAP site (Fig. 4b). Our current efforts are focused toward obtaining experimental evidence for the proposed location of the αNTD within the cryo-EM structure to support this model.

ACKNOWLEDGMENTS

We thank present and past members of the Darst lab, including E.A. Campbell, N. Loizos, A. Malhotra, K. Severinov, E. Severinova, and Y. Wang, for intellectual contributions to this work. A.P. was supported by funds from the Norman and Rosita Winston Foundation and an NRSA award (GM-17708-01). G.Z. was also supported by an NRSA award (GM-19441-01). S.A.D. is a Pew Scholar in the Biomedical Sciences. This work was supported in part by The Rockefeller University, and by grants to S.A.D. from the Lucille P. Markey Charitable Trust, the Human Frontier Science Project, the Irma T. Hirschl Trust, the Pew Foundation, the March of Dimes, and the National Institutes of Health (GM-53759).

REFERENCES

Amos L.A., Henderson R., and Unwin P.N.T. 1982. Three-dimensional structure determination by electron microscopy of two-dimensional crystals. *Prog. Biophys. Mol. Biol.* **39:** 183.

Archambault J. and Friesen J.D. 1993. Genetics of RNA polymerases I, II, and III. *Microbiol. Rev.* **57:** 703.

Azuma Y., Yamagishi M., and Ishihama A. 1993. Subunits of the *Schizosaccharomyces pombe* RNA polymerase II: Enzyme purification and structure of the subunit 3 gene. *Nucleic Acids Res.* **21:** 3749.

Beroukhim R. and Unwin N. 1995. Three-dimensional location of the main immunogenic region of the acetylcholine receptor. *Neuron* **15:** 323.

Blatter E., Ross W., Tang H., Gourse R., and Ebright R. 1994. Domain organization of RNA polymerase alpha subunit: C-terminal 85 amino acids constitute a domain capable of dimerization and DNA binding. *Cell* **78:** 889.

Borukhov S., Severinov K., Kashlev M., Lebedev A., Bass I., Rowland G.C., Lim P.-P., Glass R.E., Nikiforov V., and Goldfarb A. 1991. Mapping of trypsin cleavage and antibody-binding sites and delineation of a dispensable domain in the β subunit of *Escherichia coli* RNA polymerase. *J. Biol. Chem.* **266:** 23921.

Burgess R.R., Travers A.A., Dunn J.J., and Bautz E.K.F. 1969. Factor stimulating transcription by RNA polymerase. *Nature* **221:** 43.

Darst S.A., Kubalek E.W., and Kornberg R.D. 1989. Three-dimensional structure of *Escherichia coli* RNA polymerase holoenzyme determined by electron crystallography. *Nature* **340:** 730.

Darst S.A., Edwards A.M., Kubalek E.W., and Kornberg R.D. 1991. Three-dimensional structure of yeast RNA polymerase II at 16 Å resolution. *Cell* **66:** 121.

Darst S.A., Roberts J.W., Malhotra A., Marr M., Severinov K., and Severinova, E. 1997. Pribnow box recognition and melting by *Escherichia coli* RNA polymerase. *Nucleic Acids Mol. Biol.* **11:** 27.

DeRosier D.J. and Moore P.B. 1970. Reconstruction of three-dimensional images from electron micrographs of structures with helical symmetry. *J. Mol. Biol.* **52:** 355.

Doublie S., Tabor S., Long A.M., Richardson C.C., and Ellenberger T. 1998. Crystal structure of a bacteriophage T7 DNA replication complex at 2.2 Å resolution. *Nature* **391:** 251.

Dubochet J., Adrian M., Chang J.-J., Homo J.-C., Lepault J., McDowall A.W., and Schultz P. 1988. Cryo-electron microscopy of vitrified specimens. *Q. Rev. Biophys.* **21:** 129.

Ebright R.H. and Busby S. 1995. *Escherichia coli* RNA polymerase α subunit: Structure and function. *Curr. Opin. Genet. Dev.* **5:** 197.

Gardella T., Moyle, T., and Susskind, M.M. 1989. A mutant *Escherichia coli* σ[70] subunit of RNA polymerase with altered promoter specificity. *J. Mol. Biol.* **206:** 579.

Gross C.A., Chan C.L., and Lonetto M.A. 1996. A structure/function analysis of *Escherichia coli* RNA polymerase. *Philos. Trans. R. Soc. Lond. B Biol. Sci.* **351:** 475.

Hayward R.S., Igarashi K., and Ishihama A. 1991. Functional specialization within the alpha-subunit of *Escherichia coli* RNA polymerase. *J. Mol. Biol.* **221:** 23.

Heyduk T., Heyduk E., Severinov K., Tang H., and Ebright R.H. 1996. Rapid epitope mapping by hydroxyl-radical protein footprinting: Determinants of RNA polymerase alpha subunit for interaction with beta, beta´, and sigma subunits. *Proc. Natl. Acad. Sci.* **93:** 10162.

Igarashi K. and Ishihama A. 1991. Bipartite functional map of the *E. coli* RNA polymerase α subunit: Involvement of the C-terminal region in transcription activation by cAMP-CRP. *Cell* **65:** 1015.

Igarashi K., Fujita N., and Ishihama A. 1990. Sequence analysis of two temperature-sensitive mutations in the alpha subunit gene (rpoA) of *Escherichia coli* RNA polymerase. *Nucleic Acids Res.* **18:** 5945.

———. 1991. Identification of a subunit assembly domain in the alpha subunit of *Escherichia coli* RNA polymerase. *J. Mol. Biol.* **218:** 1.

Ishihama A. 1981. Subunit assembly of RNA polymerase. *Adv. Biophys.* **14:** 1.

———. 1993. Protein-protein communication within the transcription apparatus. *J. Bacteriol.* **175:** 2483.

Jeon Y.H., Yamazaki T., Otomo T., Ishihama A., and Kyogoku Y. 1997. Flexible linker in the RNA polymerase α subunit facilitates the independent motion of the C-terminal activator contact domain. *J. Mol. Biol.* **267:** 953.

Jeon Y.H., Negishi T., Shirakawa M., Yamazaki T., Fujita N., Ishihama A., and Kyogoku Y. 1995. Solution structure of the activator contact domain of the RNA polymerase α subunit. *Science* **270:** 1495.

Jeruzalmi D. and Steitz T.A. 1998. Structure of T7 RNA polymerase complexed to the transcriptional inhibitor T7 lysozyme. *EMBO J.* **17:** 4101.

Kiefer J.R., Mao C., Braman J.C., and Beese L.S. 1998. Visualizing DNA replication in a catalytically active *Bacillus* DNA polymerase crystal. *Nature* **391:** 304.

Kimura M. and Ishihama A. 1995a. Functional map of the alpha subunit of *Escherichia coli* RNA polymerase: Insertion analysis of the amino-terminal assembly domain. *J. Mol. Biol.* **248:** 756.

———. 1995b. Functional map of the alpha subunit of *Escherichia coli* RNA polymerase: Amino acid substitution within the amino-terminal assembly domain. *J. Mol. Biol.* **254:** 342.

Kimura M., Fujita N., and Ishihama A. 1994. Functional map of the alpha subunit of *Escherichia coli* RNA polymerase. Deletion analysis of the amino-terminal assembly domain. *J. Mol. Biol.* **242:** 107.

Lalo D., Carles C., Sentenac A., and Thuriaux P. 1993. Interactions between three common subunits of yeast RNA polymerases I and III. *Proc. Natl. Acad. Sci.* **90:** 5524.

Lonetto M., Gribskov M., and Gross C.A. 1992. The σ[70] family: Sequence conservation and evolutionary relationships. *J. Bacteriol.* **174:** 3843.

Malhotra A., Severinova E., and Darst S.A. 1996. Crystal structure of a σ[70] subunit fragment from *Escherichia coli* RNA polymerase. *Cell* **87:** 127.

Mecsas J., Cowing D.W., and Gross C.A. 1991. Development of RNA polymerase-promoter contacts during open complex formation. *J. Mol. Biol.* **220:** 585.

Metzger W., Schickor P., and Heumann H. 1989. A cinematographic view of *Escherichia coli* RNA polymerase translocation. *EMBO J.* **8:** 2745.

Mustaev A., Kozlov M., Markovtsov V., Zaychikov E., Denissova L., and Goldfarb A. 1997. Modular organization of the catalytic center of RNA polymerase. *Proc. Natl. Acad. Sci.* **94:** 6641.

Niu W., Kim Y., Tau G., Heyduk T., and Ebright R.H. 1996. Transcription activation at class II CAP-dependent promoters: Two interactions between CAP and RNA polymerase. *Cell* **87:** 1123.

Pati U.K. 1994. Human RNA polymerase II subunit hRPB14 is homologous to yeast RNA polymerase I, II, and III subunits (AC19 and RPB11) and is similar to a portion of the bacterial RNA polymerase α subunit. *Gene* **145:** 289.

Polyakov A., Severinova E., and Darst S.A. 1995. Three-dimensional structure of *Escherichia coli* core RNA polymerase: Promoter binding and elongation conformations of the enzyme. *Cell* **83:** 365.

Polyakov A., Richter C., Malhotra A., Koulich D., Borukhov S., and Darst S.A. 1998. Visualization of the binding site for the transcript cleavage factor GreB on *Escherichia coli* RNA polymerase. *J. Mol. Biol.* **281:** 465.

Ross W., Gosink K., Salomon J., Igarashi K., Zou C., Ishihama A., Severinov K., and Gourse R.L. 1993. A third recognition element in bacterial promoters: DNA binding by the alpha subunit of RNA polymerase. *Science* **262:** 1407.

Schickor P., Metzger W., Wladyslaw, W., Lederer H., and Heumann H. 1990. Topography of intermediates in transcription initiation of *E. coli*. *EMBO J.* **9:** 2215.

Schultz P., Celia H., Riva M., Sentenac A., and Oudet P. 1993. Three-dimensional model of yeast RNA polymerase I determined by electron microscopy of two-dimensional crystals. *EMBO J.* **12:** 2601.

Schultz S.C., Shields G.C., and Steitz T.A. 1991. Crystal structure of a CAP-DNA complex: The DNA is bent by 90 degrees. *Science* **253:** 1001.

Severinov K., Mustaev A., Kukarin A., Muzzin O., Bass A., Darst S.A., and Goldfarb A. 1996. Structural modules of the large subunits of RNA polymerase: Introducing archaebacterial and chloroplast split sites in the β and β′ subunits of *Escherichia coli* RNA polymerase. *J. Biol. Chem.* **271:** 27969.

Severinov K., Mustaev A., Severinova E., Bass I., Landick R., Nikiforov V., Goldfarb A., and Darst S.A. 1995. Assembly of functional *Escherichia coli* RNA polymerase using beta subunit fragments. *Proc. Natl. Acad. Sci.* **92:** 4591.

Severinova E., Severinov K., Fenyö D., Marr M., Brody E.N., Roberts J.W., Chait B.T., and Darst S.A. 1996. Domain organization of the *Escherichia coli* RNA polymerase σ[70] subunit. *J. Mol. Biol.* **263:** 637.

Svetlov V., Nolan K., and Burgess R.R. 1998. Rpb3: Stoichiometry and sequence determinants of the assembly into yeast RNA polymerase II *in vivo*. *J. Biol. Chem.* **273:** 10827.

Toyoshima C. and Unwin N. 1990. Three-dimensional structure of the acetylcholine receptor by cryoelectron microscopy and helical image reconstruction. *J. Cell Biol.* **111:** 2623.

Travers A.A. and Burgess R.R. 1969. Cyclic re-use of the RNA polymerase sigma factor. *Nature* **222:** 537.

Ulmasov T., Larkin R.M., and Guilfoyle T.J. 1996. Association between 36- and 13.6-kDa α-like subunits of *Arabidopsis thaliana* RNA polymerase II. *J. Biol. Chem.* **271:** 5085.

Unwin N. 1993. Nicotinic acetylcholine receptor at 9 Å resolution. *J. Mol. Biol.* **229:** 1101.

Waldburger C., Gardella T., Wong R., and Susskind M.M. 1990. Changes in conserved region 2 of *Escherichia coli* σ[70] affecting promoter recognition. *J. Mol. Biol.* **215:** 267.

Wang Y., Severinov K., Loizos N., Fenyö D., Heyduk E., Heyduk T., Chait B.T., and Darst S.A. 1997. Determinants for *Escherichia coli* RNA polymerase assembly within the β subunit. *J. Mol. Biol.* **270:** 648.

Zhang G. and Darst S.A. 1998. Structure of the *Escherichia coli* RNA polymerase α subunit N-terminal domain. *Science* **281:** 262.

Zillig W., Palm P., and Heil A. 1976. Function and reassembly of subunits of DNA-dependent RNA polymerase. In *RNA polymerase* (ed. R. Losick and M. Chamberlin), p. 101. Cold Spring Harbor Laboratory Press, Cold Spring Harbor, New York.

Interaction of *Escherichia coli* σ^{70} with Core RNA Polymerase

R.R. Burgess,* T.M. Arthur,*† and B.C. Pietz*†

*McArdle Laboratory for Cancer Research and †Department of Bacteriology, University of Wisconsin-Madison, Madison Wisconsin 53706

The RNA polymerase of *Escherichia coli* is a large, multisubunit enzyme existing in two forms. The core enzyme, consisting of subunits β and β′ and an α-subunit dimer (Burgess 1969a,b), carries out processive transcription elongation followed by termination. When one of a variety of sigma (σ) factors is added to core, the holoenzyme is formed. The σ-subunit confers promoter-specific DNA binding and transcription initiation capabilities to the enzyme (Burgess and Travers 1970; Helmann and Chamberlin 1988; Gross et al. 1992, 1996). σ^{70} of *E. coli* was the first σ factor to be described and characterized (Burgess et al. 1969). Since then, numerous σ factors have been discovered throughout the eubacterial kingdom, including six alternative σ factors in *E. coli*. Each σ directs transcription initiation from a specific set of promoters to transcribe genes usually with related functions. This control of transcription is mediated partially through the competition of the individual σ factors for the core enzyme and is a major part of global gene regulation in bacteria (Zhou and Gross 1992). Elucidation of the structural characteristics of the core RNA polymerase σ factor-binding interaction will be very beneficial in fully understanding this aspect of regulation.

As the number of identified σ factors increased, it became apparent that they shared several regions of amino acid sequence similarity (Gribskov and Burgess 1986; Helmann and Chamberlin 1988; Lonetto et al. 1992). Work has been ongoing to assign specific functions to these conserved regions (Gardella et al. 1989; Lesley and Burgess 1989; Siegele et al. 1989; Dombrowski et al. 1993; Waldburger and Susskind 1994). Deletion analysis of σ^{70} identified a segment of the protein that overlaps region 2.1 (residues 361–390) as being necessary and sufficient for core binding (Lesley and Burgess 1989). A mutation in a homologous region of *Bacillus subtilis* σ^{E} has also been shown to affect core binding (Shuler et al. 1995). However, recent findings of core-binding mutations in other conserved and nonconserved regions of σ^{32} (Zhou et al. 1992; Joo et al. 1997, 1998) and σ^{70}(C. Gross, pers. comm.) have led to the idea of multiple binding sites for the σ-subunit on the core enzyme .

To date, little is known about the location of the σ-binding sites on the core subunits. Two studies have identified deletions in the β or β′ subunits that produce subunits still capable of forming core enzyme structures but not holo. First, a β-subunit truncation, missing about 200 amino acids of the carboxyl terminus, was shown by glycerol gradient centrifugation to migrate with the other core subunits but was never seen in the σ^{70}-containing fractions (Glass et al. 1986). Second, when immunoprecipitation assays were performed using reconstituted RNA polymerase containing β′ deletion mutants missing 201–477, the core subunits were recovered in the same fraction but lacked σ^{70} (Luo et al. 1996). This idea that σ^{70} binding is affected by perturbations of the carboxyl terminus of β and the amino terminus of β′ is consistent with experiments showing that these two subunit termini are physically close together and can be fused through a flexible linker and still form a functional enzyme (Severinov et al. 1997). Recent protein-protein footprinting data have identified a similar region on β′ and two new sites on β for possible interactions with the σ^{70} subunit (Owens et al. 1998).

The β and β′ subunits each contains regions that have high sequence homology with the two largest subunits of eukaryal polymerases (Allison et al. 1985; Sweetser et al. 1987; Jokerst et al. 1989). Some of these conserved regions may act as interaction domains. We define an interaction domain as the minimal region of a protein that can independently fold to form the secondary and tertiary structures required to interact with another protein, DNA, RNA, or ligand. Interaction domains will always be larger than the actual binding site, the amino acids being in direct contact with the binding partner. Therefore, additional work will be needed to identify the critical residues involved in making binding contacts. Severinov et al. (1992, 1995, 1996) demonstrated the domain-like properties of β and β′ by reconstitution of functional RNA polymerase from fragmented β and β′ subunits. This indicates that the properties of the protein do not require the entire intact length of the subunit but rather can be generated with smaller domain modules.

In this study, we set out to map the protein-protein interaction domains on both β and β′ required for the binding of σ^{70}. Chemical cleavage and polymerase chain reaction (PCR) methods were used to generate fragments of β and β′ for use in mapping the interaction domains. Using far-Western blotting and nickel nitrilotriacetic acid (Ni^{++}-NTA) coimmobilization assays (Blanar and Rutter 1992; Kaelin et al. 1992; Wang et al. 1997), we were able to map a strong specific binding site for σ^{70} to the amino terminus of β′. This paper shows that this binding site is located within a span of residues (260–309) that overlaps conserved region B of β′ (Jokerst et al. 1989).

This work is part of a large body of research on *E. coli* RNA polymerase and σ factors that has been carried out by this laboratory and its collaborators from 1965 to the present. Highlights of this research are summarized in Table 1.

Table 1. Highlights of *E. coli* σ/Core Research
from the Burgess Group

1965–1990	Purification of *E. coli* RNA polymerase (Burgess 1969a,b, 1971; Burgess and Jendrisak 1975; Gross et al. 1976; Hager et al. 1990)
1966–1975	Subunit structure of *E. coli* core RNA polymerase (Burgess 1969a,b, 1971)
1968–1969	Discovery of σ^{70} (Burgess et al. 1969; Travers and Burgess 1969; Burgess and Travers 1970)
1971–1979	Purification of σ^{70} (Burgess and Travers 1971; Lowe et al. 1979)
1978–1985	Physical chemistry of RNA polymerase/DNA interactions (deHaseth et al. 1978; Taylor and Burgess 1979; Strauss et al. 1980a,b, 1981; Shaner et al. 1982; Roe et al. 1984, 1985)
1979–1980	Renaturation of σ^{70} eluted out of SDS gel (Hager and Burgess 1980)
1977–1984	Temperature-sensitive mutants of σ^{70} (Gross et al. 1978, 1984; Burgess et al. 1979; Liebke et al. 1980; Lowe et al. 1981; Grossman et al. 1983)
1977–1978	Mapping the σ^{70} gene (Gross et al. 1978)
1978–1979	Cloning the σ^{70} gene (Gross et al. 1979)
1980–1981	Sequencing the σ^{70} gene (Burton et al. 1981)
1983–1985	Sequencing the σ^{70} operon from *E. coli* and *S. typhimurium* (Burton et al. 1983; Erickson et al. 1985)
1984–1985	σ^{70} operon regulation (Grossman et al. 1984, 1985; Taylor et al. 1984)
1982–1983	Overproduction of σ^{70} (Gribskov and Burgess 1983)
1985–1986	Computer analysis of σ sequences (Gribskov and Burgess 1986)
1986–1993	Cloning, sequencing, and purification of the ω subunit of RNA polymerase (Gentry and Burgess 1986, 1989, 1990, 1993; Gentry et al. 1991)
1984–on	Isolation of monoclonal antibodies to *E. coli* transcription machinery (Lesley et al. 1987; Thompson et al. 1992)
1988–on	Epitope mapping of monoclonal antibodies (Strickland et al. 1988; Lesley et al. 1991; Rao et al. 1996; Breyer et al. 1997)
1989–90	Monoclonal antibodies inhibition of σ factors in S-30 extracts (Jovanovich et al. 1989a,b; Lesley et al. 1990)
1990–on	Immunoaffinity purification of RNA polymerase and σ factors (Thompson et al. 1992; Burgess and Knuth 1996)
1989–on	Determining core-binding site on σ^{70} (Lesley and Burgess 1989; McMahan and Burgess 1994; this paper)
1993–1996	Overproduction and purification of σ^{70} using Sarkosyl (Nguyen et al. 1993a; Burgess 1996; Burgess and Knuth 1996)
1993–1997	Overproduction, purification, and characterization of σ^{S} (Nguyen et al. 1993b; Gentry et al. 1993; Nguyen and Burgess 1996, 1997)
1996–on	Novel uses of His-tag technology in polymerase studies (McMahan and Burgess 1996; Rao et al. 1996; Arthur and Burgess 1998; this paper)

METHODS

Plasmids. Plasmids were constructed according to the method of Arthur and Burgess (1998). These include overexpression vectors for carboxy-terminal hexahistidine (His$_6$)-tagged β and β′ and amino-terminally His$_6$-tagged β and β′ (Wang et al. 1995). Vectors expressing unmodified truncated fragments of β′ were obtained by PCR cloning of the desired fragment and placing the fragment into either pET-21a or pET-24a (Novagen) (Studier

et al. 1990). Tagged truncated fragments of β′ and σ were created by amplifying the specified regions via PCR and inserting into an expression vector pET-21a derivative that had been modified to fuse an amino- or carboxy-terminal His$_6$ and, in some cases, a heart muscle kinase (HMK) recognition site to the expressed protein. All products created by PCR were sequenced to ensure that no mutations had been introduced. To use σ^{70} or the β′$_{1-309}$ fragment as radioactive probes, vectors were created by placing these genes into a derivative of pET-28b vector that contained the amino-terminal His$_6$ and HMK recognition site fusion that adds a total of 13 extra amino acids (MHHHHHHARRASV) to the amino terminus. A similar vector, pET-33b(+), is available (Novagen, Inc.) that allows expression of proteins with His$_6$ and HMK recognition site tags.

Expression and purification of proteins. Plasmids were transformed into BL21(DE3) (Novagen) for expression (Studier et al. 1990). The cells were grown in 1-liter cultures at 37°C in LB medium with either 100 µg/ml ampicillin or 50 µg/ml kanamycin. The cultures were grown to an A$_{600}$ between 0.6 and 0.8 and then induced with 1 mM isopropyl β-D-thiogalactoside (IPTG); 3 hours after induction, the cells were harvested by centrifugation at 8000g for 15 minutes and frozen at –20°C until use.

The cells were thawed and resuspended in 10 ml of lysis buffer (40 mM Tris-HCl at pH 7.9, 0.3 M KCl, 10 mM EDTA, and 0.1 mM PMSF) and lysozyme was added to 100 µg/ml. The cells were incubated on ice for 15 minutes and then sonicated three times in 60-second bursts. The recombinant protein in the form of inclusion bodies was separated from the soluble lysate by centrifugation at 27,000g for 15 minutes. The inclusion body pellet was resuspended, by sonication, in 10 ml of lysis buffer + 2% (w/v) sodium deoxycholate (DOC). The mixture was centrifuged at 27,000g for 15 minutes and, the supernatant was discarded. The DOC-washed inclusion bodies were resuspended in 10 ml of deionized water and centrifuged at 27,000g for 15 minutes. The water wash was repeated, and the inclusion bodies were aliquoted into 1-mg pellets and frozen at –20°C until use.

σ^{70} inclusion bodies were solubilized, refolded, and purified according to a variation of the procedure of Gribskov and Burgess (1986). The inclusion bodies were solubilized by resuspension in 6 M guanidine-HCl. The proteins were allowed to refold by diluting the denaturant 64-fold with Buffer A (50 mM Tris-HCl, 0.5 mM EDTA, and 5% [v/v] glycerol) in twofold steps over 2 hours. DE52 resin (1 g; Whatman) was added and mixed with slow stirring for 24 hours at 4°C. The resin was then collected in a 10-ml column and washed, and the protein was eluted with a gradient from 0.1 to 1 M NaCl in Buffer A. The σ^{70} fractions were pooled and dialyzed overnight against 1 liter of storage buffer (50 mM Tris-HCl, 0.5 mM EDTA, 0.1 M NaCl, 0.1 mM DTT, and 50% (v/v) glycerol) and stored at –20°C.

Whole-cell lysate. Cells containing truncated β′ or σ^{70} expression plasmids were grown to an A$_{600}$ of 0.6–0.8

and induced with 1 mM IPTG. The cells were grown for an additional 30 minutes. A 200-µl sample was removed and sonicated 3 × 30 seconds. Glycerol (20 µl) and 20 µl of SDS-sample buffer were added and heated for 2 minutes at 95°C and then stored at –20°C until analysis by SDS-PAGE.

Protein cleavage. β and β´ inclusion bodies were subjected to chemical cleavage (see below) and then purified by nickel affinity chromatography as follows: The cleavage reaction was loaded onto 300 µl of Ni^{++}-NTA resin (Qiagen) in a BioRad minicolumn. The resin had been preequilibrated with Buffer B (20 mM Tris-HCl at pH 7.9, 500 mM NaCl, 5 mM imidazole, 0.1% [v/v] Tween-20, and 10% [v/v] glycerol) + 8 M urea. The protein-bound resin was washed with 10 column volumes of Buffer B + 8 M urea followed by 10 column volumes of Buffer B to allow refolding. The resin was then washed with 500 µl of Buffer B + 40 mM imidazole. The protein was eluted with 500 µl of Buffer B + 200 mM imidazole. The eluted fractions were stored at –20°C.

NTCB cleavage (Jacobson et al. 1973): Inclusion body protein (1 mg) was resuspended in 1 ml of Buffer B + 8 M urea. DTT was added to fivefold molar excess over the thiol groups in the protein. The mixture was incubated for 15 minutes at 37°C to reduce any disulfide bonds. 2-nitro-5-thiocyanobenzoic acid (NTCB) was added to fivefold molar excess over total sulfhydryl groups. The pH was adjusted to 9.5 with NaOH. The reaction mixture was incubated for 2 hours at room temperature. The cleavage mixture was diluted 1:10 in Buffer B + 8 M urea and loaded onto a Ni^{++}-NTA column as described above.

Hydroxylamine cleavage (Bornstein and Bolian 1970): Inclusion body protein (1 mg) was resuspended in 1 ml of Buffer B + 8 M urea. Solubilized protein (500 µl) was added to 500 µl of hydroxylamine cleavage solution (0.4 M CHES at pH 9.5, 4 M hydroxylamine-HCl) and incubated 2 hours at 42°C. β-mercaptoethanol was added to 0.1 M and incubated 10 minutes at 37°C. The mixture was diluted 1:10 in Buffer B + 8 M urea and loaded onto a Ni^{++}-NTA column as described above.

Far-Western blot analysis. *Dot blot or slot blot:* Inclusion body proteins resuspended in Buffer B + 8 M urea were spotted directly onto a 0.05-µm pore size nitrocellulose membrane (Schleicher & Schuell) using a Schleicher & Schuell "MINIFOLD" dot-blot or slot-blot apparatus. The wells were washed three times with Buffer B. The nitrocellulose was blocked by incubation in HYB buffer (20 mM HEPES at pH 7.2, 200 mM KCl, 2 mM MgCl$_2$, 0.1 mM ZnCl$_2$, 1 mM DTT, 0.5% [v/v] Tween-20, 1% (w/v) nonfat dry milk) for 16 hours at 4°C.

Gel blot: Protein cleavage fragments or whole-cell lysates were separated by SDS-PAGE. The proteins were electrophoretically transferred onto nitrocellulose. The nitrocellulose was blocked by incubating in HYB buffer for 16 hours at 4°C.

Labeling: Labeling of σ^{70} or β´$_{1-309}$ was done in a 100-µl reaction volume; 50 µl of 2× kinase buffer (40 mM Tris-HCl at pH 7.4, 200 mM NaCl, 24 mM MgCl$_2$, 2 mM DTT, and 50% [v/v] glycerol) was added to 50 µg of protein. cAMP-dependent kinase-catalytic subunit (240 units; Promega) was added, and the total volume was brought up to 99 µl with deionized water; 1 µl of [γ-^{32}P]ATP (0.15 mCi/µl) was added. The mixture was incubated at room temperature for 30 minutes. The reaction mixture was then loaded onto a Biospin-P6 column (BioRad) preequilibrated with 1× kinase buffer and spun at 1100g for 4 minutes. The flow-through was collected and stored at –20°C.

Probing: The blocked nitrocellulose was incubated in 10 ml of HYB buffer with 4 × 10^5 cpm/ml ^{32}P-labeled σ^{70} or β´$_{1-309}$ for 3 hours at room temperature. The blot was washed three times with 10 ml of HYB buffer for 3 minutes each. The blot was then dried and exposed to film or PhosphorImager (Molecular Dynamics).

Coimmobilization assay. His$_6$-tagged, truncated β´ (1 mg) was solubilized in 1 ml of Buffer C (20 mM Tris-HCl at pH 7.9, 200 mM NaCl, 5 mM imidazole, 0.1% [v/v] Tween-20, and 10% (v/v) glycerol) + 8 M urea. Protein solution (20 µg) was loaded onto 150 µl of Ni^{++}-NTA resin. The column was washed with 15 column volumes of Buffer C + 8 M urea, followed by a 15 column-volume wash with Buffer C to allow refolding. Then, 30 µg of native σ^{70} was loaded onto the column. The column was washed with 20 volumes of Buffer C. The bound proteins were eluted with 300 µl of Buffer C + 250 mM imidazole. Samples from the σ^{70} flow-through, wash, and elution fractions were analyzed by SDS-PAGE.

RESULTS

σ^{70} Interacts Strongly with the β´-subunit and Weakly with the β-subunit in Far-Western Blot Analysis

Previously, we developed a method that could be used to map antibody epitopes using His$_6$-tagged purification of partially cleaved proteins (Rao et al. 1996). A schematic of this method is shown in Figure 1. To apply this method for mapping the protein-protein interactions of σ^{70} on β or β´, we needed to determine if σ^{70} could bind to individual β and β´ subunits outside of the core enzyme complex. Far-Western assays of dot blots were used to assess the binding. Inclusion body proteins of β and β´ were separately solubilized in urea and spotted onto nitrocellulose. Bovine serum albumin (BSA) was spotted as a control for nonspecific binding. The nitrocellulose was blocked and the denaturant was washed away. The blot was then probed with ^{32}P-labeled σ^{70}. Both β and β´ subunits bound σ^{70}; the BSA control did not (data not shown). Thus, both β and β´ subunits can individually bind σ^{70}, although the signal is much stronger for β´ than for β. We performed an additional test to assess the specificity of the far-Western analysis using σ^{70} as a probe. A cell lysate from a log-phase culture was separated by SDS-PAGE, blotted onto nitrocellulose, and probed with σ^{70}. The only strong signal produced had the same mobility as that of β or β´ (data not shown). The absence of

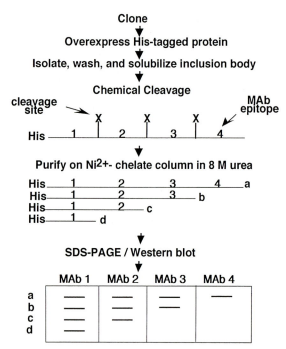

Figure 1. Epitope mapping schematic.

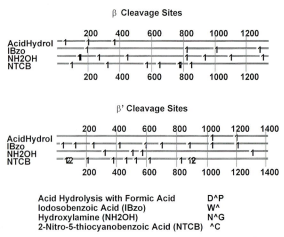

Figure 2. Chemical cleavage site map. Specific chemical cleavage sites on β and β′ were determined from the amino acid sequence using the MacVector computer program (Oxford Molecular Group).

other strong signals indicates that σ^{70} is not binding nonspecifically to β and/or β′. Minor bands were observed as expected since other proteins have been shown to interact with σ^{70} (activators, anti-σ, etc.) (Ishihama 1993; Jishage and Ishihama 1998).

A Strong, Specific Binding Site for σ^{70} Is Located Near the Amino Terminus of the β′-subunit

To map the σ^{70} interaction sites on β and β′, we performed far-Western analysis of chemical cleavage products of the two large subunits. The amino acid sequences of both were analyzed using MacVector software (Oxford Molecular Group) to identify specific chemical cleavage sites (Fig. 2). On the basis of this analysis, cleavage reagents were chosen that produced an array of products following partial digestion that provide the highest resolution for mapping. Both amino- and carboxy-terminal His_6-tagged constructs of β and β′ were subjected to cleavage under denaturing conditions. The products of the cleavage reaction were purified under denaturing conditions using Ni^{++}-NTA resin to isolate cleavage fragments containing an His_6 tag. When the cleavage fragments were fractionated by SDS-PAGE, they produced an ordered fragment ladder of descending-sized fragments with a common end (either amino or carboxyl terminus depending on the placement of the His_6 tag) (see schematic in Fig. 3). These purified fragments were then identified on the basis of their mobility in SDS-PAGE, and their exact size was determined on the basis of the cleavage site that produced them (Fig. 4A) (Arthur and Burgess 1998). The use of both amino- and carboxy-terminally His_6-tagged fragments allows the positive identification of both the amino and carboxyl termini of the interaction domain. The σ^{70} probe will only bind the

fragments that have an intact interaction domain. The amino-terminally His_6-tagged β′ ladders produced by hydroxylamine and NTCB cleavage both contained several fragments that retained the ability to bind σ^{70} (Fig. 4B, lanes 3 and 7). Thus, a large portion of the carboxyl terminus of β′ can be removed without affecting σ^{70} binding. The smallest fragment to bind σ^{70} was the 1–309-amino-acid fragment of β′ in the hydroxylamine ladder (Fig 4B, lane 3). In the carboxy-terminally His_6-tagged ladders, only full-length β′ bound σ^{70} (Fig. 4B, lanes 4 and 8). These results indicated that a strong specific binding site is located within amino acids 1–309 of β′ (β'_{1-309}). The β fragment ladders failed to produce signals strong enough to map the interaction domain effectively. For the remainder of this study, we focus only on mapping the σ^{70}-β′ interaction domain.

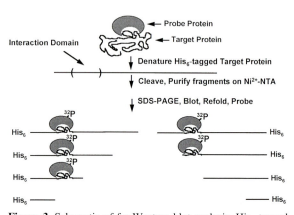

Figure 3. Schematic of far-Western blot analysis. His_6-tagged target protein was cleaved, and the fragments were purified on Ni^{++}-NTA, fractionated on SDS-PAGE, and blotted onto nitrocellulose. The denaturant was washed away from the blotted protein fragments, and interaction domains on fragments were allowed to refold. These interaction domains can be identified by probing with a radioactively labeled protein. The interaction domains were mapped by identifying fragments that have part of their interaction domain missing and can no longer bind the labeled protein probe.

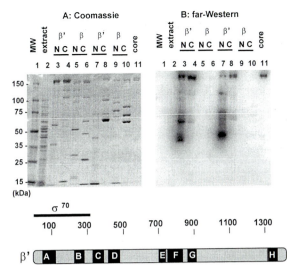

Figure 4. Ordered fragment ladder far-Western. β and β′ subunits were cleaved with hydroxylamine and NTCB. The Ni^{++}-NTA-purified fragments were separated on identically loaded 8–16% SDS-PAGE gels (Novex). One gel (*A*) was stained with Coomassie blue and the other (*B*) was transferred to nitrocellulose and probed with ^{32}P-σ^{70}. (*1*) Markers; (*2*) *E. coli* lysate; (*3*) N-His$_6$-β′/cut with hydroxylamine; (*4*) C-His$_6$-β′/hydroxylamine; (*5*) N-His$_6$-β/hydroxylamine; (*6*) C-His$_6$-β/hydroxylamine; (*7*) N-His$_6$-β′/NTCB; (*8*) C-His$_6$-β′/NTCB; (*9*) N-His$_6$-β/NTCB; (*10*) C-His$_6$-β/NTCB; (*11*) purified core polymerase. Schematic at bottom illustrates where the σ^{70} interaction domain is located with respect to the conserved regions (*lettered boxes*) of β′ (Jokerst et al. 1989).

Interaction Domain Narrowed to Amino Acids 60–309 of β′ by Far-Western Analysis with Truncated Fragments

In trying to define this binding site more precisely, we made various truncated fragments using PCR. Using the β′$_{1-309}$ fragment as a starting point, we made constructs that were truncated at either the amino or carboxyl terminus. DNA coding for the truncated fragments was cloned into overexpression plasmids. When cells containing these plasmids had been grown to an A$_{600}$ of 0.6, expression was induced. The cells were only allowed to grow for 30 minutes after induction. A whole-cell lysate from each culture was made and used for far-Western assays. Short expression times kept the expression level of the induced protein comparable to that of the other proteins in the lysate. The use of the whole-cell lysate in far-Western assays was an internal control to ensure that binding was specific for the protein of interest. This also meant that the various proteins would not have to be purified and could be expressed without purification tags. When constructs were made where the carboxyl terminus of β′$_{1-309}$ was truncated beyond amino acid 300, the binding of σ^{70} was lost. However, the amino terminus of the same fragment could be truncated up to 60 amino acids without diminishing the signal. β′$_{100-309}$ still showed binding, but at a lower level, and β′$_{150-309}$ did not bind σ^{70} (Arthur and Burgess 1998). These results narrowed the σ^{70}-binding site to β′$_{60-309}$. Western blot experiments using anti-β′

monoclonal antibodies were done to ensure that the protein fragments were being efficiently transferred to the nitrocellulose and that they were fragments of β′ (data not shown).

Coimmobilization Assays Further Narrow Interaction Domain to β′$_{260-309}$

Ni^{++}-NTA-coimmobilization assays were used to confirm and extend the results that had been produced using far-Western blotting. The proteins to be assayed for binding σ^{70} were fused to His$_6$ purification tags and overexpressed in the form of inclusion bodies. The inclusion body protein was solubilized with 8 M urea and loaded onto Ni^{++}-NTA resin. The denaturant was washed away, allowing the proteins to refold while still remaining bound to the resin. Native σ^{70} was then loaded onto the column. The column was washed and the bound proteins were then eluted with imidazole. Any truncated protein that contained the interaction domain for σ^{70} would cause σ^{70} to be bound and to be in the eluted fraction. The results of these binding experiments are consistent with the far-Western blotting experiments with respect to defining the carboxy-terminal boundary of the domain. β′$_{1-309}$ bound σ^{70}, whereas β′$_{1-300}$ and β′$_{1-280}$ did not bind σ^{70} (Arthur and Burgess 1998).

For the amino-terminal boundary, the results showed that more of the amino terminus could be removed without affecting σ^{70} binding than was seen by the far-Western assay. Several amino-terminally truncated fragments, all having amino acid 309 as the carboxy-terminal boundary followed by a His$_6$ tag, were constructed and used in coimmobilization assays. Truncations to residues 33, 60, 100, 178, and 200 still produced fragments capable of binding σ^{70} (data not shown). β′$_{260-309}$, the smallest fragment we could make and still manipulate efficiently in our assay, retained the ability to bind σ^{70}. To find the amino terminus of the interaction domain, truncations greater than residue 240 were made from full-length β′. A truncation of the first 260 residues of β′ (β′$_{260-C}$) bound σ^{70}, whereas β′$_{270-C}$ showed diminished binding and β′$_{280-C}$ showed no detectable binding of σ^{70} (Arthur and Burgess 1998). Taken together, these results indicate that a strong σ^{70}-binding site on the core polymerase is located within the amino acid 260–309 region of β′. Results of both far-Western and coimmobilization assays for a variety of truncated β′ fragments are summarized in Figure 5.

Mapping the β′$_{1-309}$ Interaction on σ^{70}

Having identified this major interaction between β′ and full-length σ^{70}, we began studies to localize the region on σ^{70} that is the interaction partner with β′$_{1-309}$. To accomplish this, we have used probing techniques similar to those described above. These experiments are essentially the opposite of the β′ mapping. In this case, we probed cloned His$_6$-tagged σ^{70} truncations with a ^{32}P-labeled β′$_{1-309}$. The His$_6$-tagged σ^{70} fragments were made by cloning due to lack of suitable chemical and protease cleavage sites in this protein.

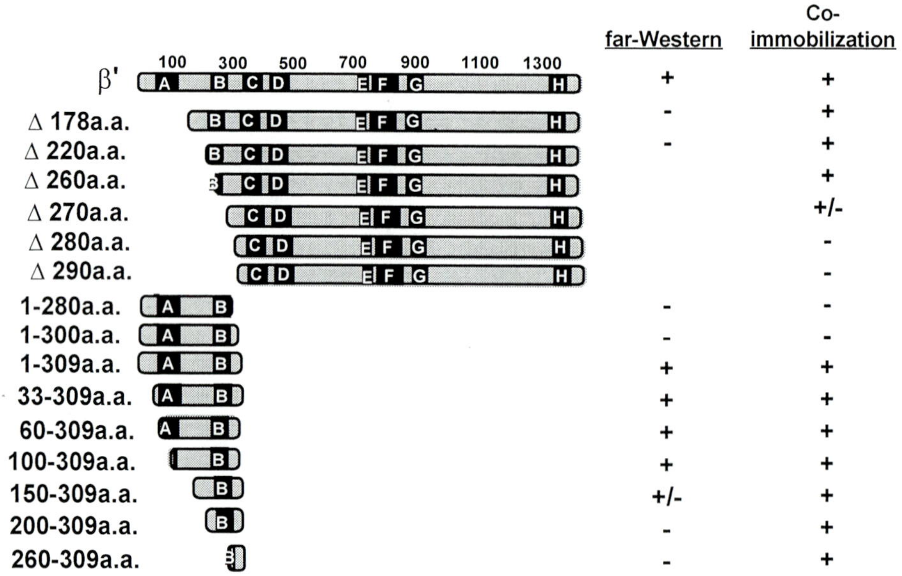

Figure 5. Summary of binding assay results. Results from far-Western blotting and Ni^{++}-NTA coimmobilization assays indicate that a σ^{70}-binding site is located within residues 260–309 of β'. (+) Detectable binding; (+/–) inconclusive; (–) no detectable binding.

The σ^{70} fragments were overexpressed using the T7 expression system and purified as described above. In the far-Western experiments, the fragments were run on 20% SDS-PAGE. The gels were blotted onto nitrocellulose membrane and probed as was described above with ^{32}P-labeled β'_{1-309}. In the far-Slot blots, the fragments denatured in 8 M urea were applied individually to different wells of a slot-blot apparatus onto nitrocellulose. These blots were probed identically to the far-Western blots.

The far-Western blots were not as sensitive as the far-Slot blots. Not as much protein can be loaded onto an SDS-PAGE gel as can be applied to a blotting apparatus. Interestingly, the carboxy-terminally His$_6$-tagged fragments were less reactive to the β'_{1-309} probe since only the full-length σ^{70}-C-His$_6$ could be detected in the far-Western blots. Smaller carboxy-terminally His$_6$-tagged fragments could be detected by loading more protein in the far-Slot blots.

A summary of the β'_{1-309} probing data is depicted in Figure 7. It appears that only fragments containing region 2.1 are able to react positively in these assays. The $\sigma^{70}_{250-390}$ fragment has not yet been tested in these assays, but one would predict that it would contain the domain necessary to interact with β'_{1-309}. Far-Western and coimmobilization assays are currently in progress to confirm this prediction and further narrow the interaction domain on σ^{70}.

DISCUSSION

What Is Known About Regions of Core Involved in σ Binding

To date, several biochemical and genetic studies have contributed to what is known about the putative core-binding domains on σ; however, much less is known about the sites on core that bind σ. In the holoenzyme as-

sembly pathway, β' is added to the $\alpha_2\beta$ complex, and then σ is added to form the holoenzyme (Ishihama 1981). This would suggest that either the major σ-binding site is located on β' or it is formed in cooperation with α and/or β upon β' assembly into the core enzyme. The isolation of σ^{70}-β' complexes provides evidence for the former (Luo et al. 1996). In this paper, we have localized a strong binding site for σ^{70} on β', as well as identified low-level binding affinity for σ^{70} to β. Thus, we conclude that β' provides the major binding interaction for σ^{70} in the holoenzyme, whereas β adds a secondary binding interaction. Multiple core-binding sites on σ have been suggested in light of σ mutations apparently affecting core binding that map outside of conserved region 2.1 (Zhou et al. 1992; Joo et al, 1997, 1998; C. Gross, pers. comm.).

Locating a Major σ^{70}-binding Site on β'

A primary finding of this work is that a strong binding site for σ^{70} is located within residues 260–309 of β'. A deletion of residues 201–477 of β' has been reported previously to produce a mutant protein that could still form core but not holoenzyme (Luo et al. 1996). The problem with such deletion studies is that one cannot conclude that the binding site is located in the region deleted, but merely that the region, when deleted, prevents correct formation of the interaction domain. Results obtained from protein-protein hydroxyl radical footprinting experiments indicated that a similar region of β' (residues 228–461) was physically close to σ^{70} (Owens et al. 1998). There is some difficulty in interpreting these results since the assay indicates physical proximity (within ~10 Å) of the proteins, but does not prove protein-protein contact. From our findings, it can be concluded that a major σ^{70}-binding site is located within these regions.

The σ^{70} interaction domain on β' that we have identified here (see Fig. 6) contains several residues located in conserved region B (Jokerst et al. 1989). This region has no known function. Secondary structural predictions derived from the PHD program (Rost et al. 1994) for residues 260–309 indicate one α-helix from residue 264 to residue 283 connected by a loop to a second helix from residue 292 to residue 309. These predicted helices are also predicted to form coiled coils (Lupas et al. 1991). This is of particular interest since similar predictions were made for residues 355–391 of σ^{70}. These residues overlap conserved region 2.1. The crystal structure of the protease-resistant fragment of σ^{70} confirmed the prediction that the helix-containing region 2.1 forms a coiled coil with conserved region 1.2 (Malhotra et al. 1996). Coiled coils have been shown to be involved in many protein-protein interactions (Landschulz et al. 1988; Gentz et al. 1989; O'Shea et al. 1989), suggesting that $\beta'_{260\text{-}309}$ may be interacting with region 2.1 of σ^{70}.

Previously in our lab, we developed a method for quickly mapping epitopes for monoclonal antibodies (Rao et al. 1996). We describe here an application of that method to identify domains involved in other protein-protein interactions. Ordered fragment ladder far-Western blotting was used to map the σ^{70}-binding site on β' to within $\beta'_{60\text{-}309}$. This method relies on the fact that after the removal of the denaturant, some fraction of the blotted protein is able to refold and produce the proper conformation for binding of the probe. The specificity of the assay was demonstrated by probing whole-cell lysates and identifying β' as the major binding interaction. The combination of specific chemical cleavage of proteins and far-Western analysis provided a very rapid and effective way to localize this protein-protein interaction. Cloning and screening individual truncated fragments was necessary only after the interaction domain had been targeted. Making truncations of β' all along its length would have been a long and tedious process. The protein cleavage and Ni^{++} column purification procedures can be done in 1 day, thus, making the assay more expedient and less tedious. Once a fragment ladder sample is prepared, it can be used for many experiments.

To confirm and extend the results obtained with far-Western analysis, Ni^{++}-coimmobilization assays were

performed. These experiments also demonstrated that fragments from the amino terminus to residue 309 could still bind σ^{70}, whereas removal of just nine carboxy-terminal residues to amino acid 300 would abolish binding. The results obtained from the amino-terminally truncated fragments in these assays gave better resolution of the interaction domain location than was obtained from far-Western assays. Up to 260 residues could be removed from the amino terminus without affecting σ^{70} binding. When 270 residues were removed, binding of σ^{70} was diminished but not abolished, suggesting that either part of the binding site had been removed or the binding site was intact but hindered from proper refolding due to the loss of upstream residues. To ensure that the binding site was what we were actually mapping and not just a region required for proper folding of the actual binding site, protein fragments were made from residues 260–309 of β' and shown to be sufficient for binding. We believe that the difference in the identified interaction domain size between the far-Western assay ($\beta'_{60\text{-}309}$) and the coimmobilization assay ($\beta'_{260\text{-}309}$) is consistent with the properties of each assay. The far-Western assay requires the interaction domain to refold and properly present the binding site while some portion of the protein is attached to the nitrocellulose membrane. We believe that proteins bound to nitrocellulose are more conformationally restricted than proteins bound only at one terminus as in the Ni^{++}-NTA coimmobilization assay. Therefore, more of the protein length is required to keep the interaction domain away from the membrane surface. The combination of mapping methods provides a rapid, high-resolution procedure for identification of protein interaction domains.

$\beta'_{1\text{-}309}$ Binds Region 2 of σ^{70} and to Most of the Other Bacterial and Phage σ Factors Tested

Results shown in Figure 7 suggest that the region from amino acids 250–390 of σ^{70} is responsible for the binding of labeled $\beta'_{1\text{-}309}$ probe in a far-Western assay. We are

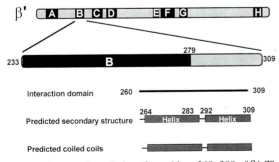

Figure 6. Structural predictions for residues 260–309 of β'. The secondary structural prediction, using the PHD prediction program (Rost et al. 1994), of the identified interaction domain is predicted to contain two α helices. These regions are also predicted to form coiled coils (Lupas et al. 1991).

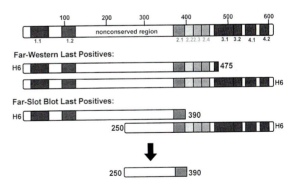

Figure 7. Summary of $\beta'_{1\text{-}309}$ interaction mapping on σ^{70}. Preliminary far-Western and far-Slot blot results indicate that the interaction between $\beta'_{1\text{-}309}$ and σ^{70} is localized to a region adjacent to and including region 2.1 (see text for details). Shown are a representation of the smallest amino- and carboxy-terminal His_6-tagged fragments that were detected (the last positive) when probed with ^{32}P-labeled $\beta'_{1\text{-}309}$ in either far-Western or far-Slot blot assays. The bottom fragment represents the minimal region contained in all of the positive fragments.

presently working to narrow this interaction domain further and to show that $\sigma^{70}_{250-390}$ fragment or even a smaller fragment can bind β'_{1-309} by itself.

We have work in progress to map the binding sites on core for the other six known *E. coli* σ factors. It has been hypothesized that the core-binding region on the σ factors must be highly conserved since they all must bind core polymerase (Helmann and Chamberlin 1988; Lonetto et al. 1992; Gross et al. 1996). This has led to region 2 as the primary candidate since it is the most highly conserved region. One can hypothesize that all of the σ factors in the cell may bind to the same site or sites on core polymerase. Preliminary evidence being gathered in our lab suggests that a labeled β'_{1-309} probe does in fact interact with other *E. coli* σ factors as well as with σ factors from other bacteria and phage. We are now in a position to test this hypothesis and determine if the other σ factors can also bind $\beta'_{260-309}$.

The identification of interaction domains on β' and σ^{70} puts us in an excellent position to carry out biophysical studies of complexes between these two interaction domains.

General Utility of Ordered Fragment Ladder Far-Westerns in Interaction Domain Mapping

Far-Western analysis (separating a mixture of proteins by SDS-PAGE, transferring them to a nitrocellulose membrane, and probing with a nonantibody protein probe) has been used for the last 5–8 years to study protein-protein interactions. Recently, our lab has developed a powerful variation on far-Western analysis. This "ordered fragment ladder far-Western" method represents a major advance in mapping protein-protein interaction domains that should prove to be very useful to many researchers studying multisubunit protein complexes.

Basically, all that is required is an overexpression clone of a protein to be studied that has a terminal His6-tag and a suitable labeled probe protein. Then (1) purify the His6-tag protein, (2) cleave it at specific sites with appropriate chemicals or proteases, (3) purify His6-tag-containing fragments on a nickel-chelate column under denaturing conditions, (4) separate fragments by SDS-PAGE to form an ordered fragment ladder, (5) transfer peptides to nitrocellulose membrane and allow refolding of peptides, (6) incubate blocked membrane with ^{32}P-labeled probe protein, and (7) wash briefly and expose to film or a PhosphorImager.

This method and variations on it can be used in epitope mapping of monoclonal antibodies (see Fig. 1); mapping of protein-protein interaction domains (see Fig. 3); mapping sites of modification, such as phosphorylation or cross-linking tags; and mapping of DNA- or RNA-binding sites. When used for interaction domain mapping with a ^{32}P-labeled protein probe, relatively weak interactions can be detected because a probe can be made that is more than 10^5 dpm/μg and the final wash after incubation of the blot with the probe and before exposure to PhosphorImaging can be as little as 5–10 minutes. It should be noted, however, that this method does not map the contact site, but rather the whole region needed to form the structural domain in which the interaction site resides. It will not give positive results (1) if the binding is too weak, relative to background binding; (2) if the interaction domain involves parts of two different-sized polypeptides or perhaps even two distal regions on the same polypeptide; or (3) if the domain is inefficient in refolding or, in particular, if there is a strong nitrocellulose membrane binding site in the middle of the interaction domain that prevents refolding on the membrane. As a result, only positive results are meaningful. Nevertheless, it is a rapid means of locating a region containing an interaction domain. It can direct one to focus more tedious mapping approaches, such as cloning individual truncated fragments or making multiple mutations, on a relatively small region of the target polypeptide.

Other σ-core-related Studies

In addition to the research results presented above, we have used immunoaffinity chromatography with a monoclonal antibody to β', NT73, to isolate core and all forms of holoenzyme from cells grown under different growth conditions, and we are measuring levels of σ factors in cells and bound to core (Thompson et al. 1992; D. Jensen and R. Burgess, in prep.).

We have probed changes in σ^{70} conformation of free σ^{70}, σ^{70} bound to core, or σ^{70} in an open promoter complex by measuring changes in accessibility to a number of proteases (S. McMahan and R. Burgess, in prep.). Our results indicate that there are protease-sensitive sites between regions 1.1 and 1.2, in the highly acidic region around residue 190, at the end of region 2.4, in regions 3.1 and 3.2, just before region 4.1, and just after region 4.2 that change significantly.

Finally, we have analyzed the function of σ^{70} region 2.1 by site-directed mutagenesis. Mutations in amino acid positions 383, 385, and 386 were shown to be nonfunctional in vivo and defective in vitro. Results indicate that although these mutations do not have a large effect on σ^{70} binding to core polymerase under the conditions tested, they result in moderate effects on open promoter complex formation and severe defects in productive transcription (L. Rao and R. Burgess, in prep.).

ACKNOWLEDGMENTS

We thank Robert Landick for generously providing plasmids pRL663 and pRL706. We also thank N. Thompson, V. Svetlov, K. Nolan, L. Rao, S. McMahan, and D. Jensen for technical input and discussion. This work was supported by National Institutes of Health grant GM-28575 to R.B. and by National Institutes of Health Biotechnology training grant GM-08349 fellowships to T.A and B.P.

REFERENCES

Allison L.A., Moyle M., Shales M., and Ingles C.J. 1985. Extensive homology among the largest subunits of eukaryotic and prokaryotic RNA polymerases. *Cell* **42:** 599.

Arthur T.M. and Burgess R.R. 1998. Localization of a σ^{70} binding site on the N-terminus of the E. coli RNA polymerase β´ subunit. *J. Biol. Chem.* **273:** 31381.

Blanar M.A. and Rutter W.J. 1992. Interaction cloning: Identification of a helix-loop-helix zipper protein that interacts with c-Fos. *Science* **256:** 1014.

Bornstein P. and Bolian G. 1970. Cleavage at Asn-Gly bonds with hydroxylamine. *Methods Enzymol.* **47:** 132.

Breyer M.J., Thompson N.E., and Burgess R.R. 1997. Identification of the epitope for a highly cross-reactive monoclonal antibody on the major sigma factor of bacterial RNA polymerase, *J. Bacteriol.* **179:** 1404.

Burgess R.R. 1969a. A new method for the large scale purification of E. coli deoxyribonucleic acid-dependent ribonucleic acid polymerase. *J. Biol. Chem.* **244:** 6160.

———. 1969b. Separation and characterization of the subunits of ribonucleic acid polymerase. *J. Biol. Chem.* **244:** 6168.

———. 1971. RNA polymerase. *Annu. Rev. Biochem.* **40:** 711.

———. 1996. Purification of overproduced E. coli RNA polymerase sigma factors by solubilizing inclusion bodies and refolding from Sarkosyl. *Methods Enzymol.* **273:** 145.

Burgess R.R. and Jendrisak J.J. 1975. A procedure for the rapid, large-scale purification of E. coli DNA-dependent RNA polymerase involving Polymin P precipitation and DNA cellulose chromatography. *Biochemistry* **14:** 4634.

Burgess R.R. and Knuth M.K. 1996. Purification of a recombinant protein overproduced in E. coli. In *Strategies for protein purification and characterization: A laboratory manual* (ed. D. Marshak et al.), p. 205, Cold Spring Harbor Laboratory Press, Cold Spring Harbor, New York.

Burgess R.R. and Travers A.A. 1970. E. coli RNA polymerase: Purification, subunit structure, and factor requirements. *Fed. Proc.* **29:** 1164.

———. 1971. Purification of the RNA polymerase sigma factor. *Methods Enzymol.* **21:** 500.

Burgess R.R., Gross C.A., Walter W., and Lowe P.A. 1979. Altered chemical properties in three mutants of E. coli RNA polymerase sigma subunit. *Mol. Gen. Genet.* **175:** 251.

Burgess R.R., Travers A.A., Dunn J.J., and Bautz E.K.F. 1969. Factor stimulating transcription by RNA polymerase. *Nature* **221:** 43.

Burton Z.F., Gross C.A., Watanabe K.K., and Burgess R.R. 1983. The operon that encodes the sigma subunit of RNA polymerase also encodes ribosomal protein S21 and DNA primase in E. coli K12. *Cell* **32:** 335.

Burton Z., Burgess R.R., Lin J., Moore D., Holder S., and Gross C.A. 1981. The nucleotide sequence of the cloned *rpoD* gene for the RNA polymerase sigma subunit from E. coli K12. *Nucleic Acids Res.* **9:** 2889.

deHaseth P.L., Lohman T.M., Burgess R.R., and Record M.T., Jr. 1978. Nonspecific interactions of E. coli RNA polymerase with native and denatured DNA: Differences in the binding behavior of core and holoenzyme. *Biochemistry* **17:** 1612.

Dombrowski A.J., Walter W.A., and Gross C.A. 1993. Amino-terminal amino acids modulate sigma-factor DNA binding activity. *Genes Dev.* **7:** 2446.

Erickson B.D., Burton Z.F., Watanabe K.K., and Burgess R.R. 1985. Nucleotide sequence of the *rpsU-dnaG-rpoD* operon from *Salmonella typhimurium* and a comparison of this sequence with the homologous operon of *E. coli. Gene* **40:** 67.

Gardella T., Moyle H., and Susskind M.M. 1989. A mutant E. coli σ^{70} subunit of RNA polymerase with altered promoter specificity. *J. Mol. Biol.* **206:** 579.

Gentry D.R. and Burgess R.R. 1986. The cloning and sequence of the gene encoding the omega subunit of E. coli RNA polymerase. *Gene* **48:** 33.

———. 1989. rpoZ, encoding the omega subunit of E. coli RNA polymerase, is in the same operon as spoT. *J. Bacteriol.* **171:** 1271.

———. 1990. Overproduction and purification of the omega subunit of E. coli RNA polymerase. *Protein Expr. Purif.* **1:** 81.

———. 1993. Crosslinking of E. coli RNA polymerase subunits:

Identification of β´ as the binding site of omega. *Biochemistry* **32:** 11224.

Gentry D., Xiao H., Burgess R.R., and Cashel M. 1991. The omega subunit of E. coli K-12 RNA polymerase is not required for stringent RNA control in vivo. *J. Bacteriol.* **173:** 3901.

Gentry D.R., Hernandez V.J., Nguyen L.H., Jensen D.B., and Cashel M. 1993. Synthesis of the stationary-phase σ factor, σ^S, is positively regulated by ppGpp. *J. Bacteriol.* **175:** 7982.

Gentz R., Rauscher F.J., III., Abate C., and Curran T. 1989. Parallel association of Fos and Jun leucine zippers juxtaposes DNA binding domains. *Science* **243:** 1695.

Glass R.E., Honda A., and Ishihama A. 1986. Genetic studies on the β subunit of E. coli RNA polymerase IX. The role of the C-terminus in enzyme assembly. *Mol. Gen. Genet.* **203:** 492.

Gribskov M. and Burgess R.R. 1983. Overexpression and purification of the sigma subunit of E. coli RNA polymerase. *Gene* **26:** 109.

———. 1986. Sigma factors from E. coli, B. subtilis, phage SPO1, and phage T4 are homologous proteins. *Nucleic Acids Res.* **14:** 6745.

Gross C.A., Chan C.L., and Lonetto M.A. 1996. A structure/function analysis of E. coli RNA polymerase. *Philos. Trans. R. Soc. Lond. B Biol. Sci.* **351:** 475.

Gross C.A., Lonetto M. and Losick R. 1992. Bacterial sigma factors. In: *Transcriptional regulation* (ed. S. McKnight and K. Yamamoto), p.129, Cold Spring Harbor Laboratory Press, Cold Spring Harbor, New York.

Gross C., Engbaek F., Flammang T., and Burgess R. 1976. Rapid micromethod for the purification of E. coli RNA polymerase and the preparation of bacterial extracts active in RNA synthesis. *J. Bacteriol.* **128:** 382.

Gross C.A., Blattner F.R., Taylor W.E., Lowe P.A., and Burgess R.R. 1979. Isolation and characterization of transducing phage coding for sigma subunit of E. coli RNA polymerase. *Proc. Natl. Acad. Sci.* **76:** 5789.

Gross C.A., Grossman A.D., Liebke H., Walter W., and Burgess R.R. 1984. Effects of the mutant sigma allele *rpoD800* on the synthesis of specific macromolecular components of the E. coli K12 cell. *J. Mol. Biol.* **172:** 283.

Gross C., Hoffman J., Ward C., Hager D., Burdick G., Berger H., and Burgess R. 1978. Mutation affecting thermostability of sigma subunit of E. coli RNA polymerase lies near the *dnaG* locus at about 66 min on the E. coli genetic map. *Proc. Natl. Acad. Sc.* **75:** 427.

Grossman A.D., Burgess R.R., Walter W., and Gross C.A. 1983. Mutations in the lon gene of E. coli K12 phenotypically suppress a mutation in the sigma subunit of RNA polymerase. *Cell* **32:** 151.

Grossman A.D., Ullman A., Burgess R.R., and Gross C.A. 1984. Regulation of cyclic AMP synthesis in E. coli K-12: Effects of the *rpoD800* sigma mutation, glucose, and chloramphenicol. *J. Bacteriol.* **158:** 110.

Grossman A.D., Taylor W.E., Burton Z.F., Burgess R.R., and Gross C.A. 1985. Stringent response in E. coli induces expression of heat shock proteins. *J. Mol. Biol.* **186:** 357.

Hager D.A. and Burgess R.R. 1980. Elution of proteins from sodium dodecyl sulfate-polyacrylamide gels, removal of SDS, and renaturation of enzymatic activity: Results with sigma subunit of E. coli RNA polymerase, wheat germ DNA topoisomerase, and other enzymes. *Anal. Biochem.* **109:** 76.

Hager D., Jin D.J., and Burgess R.R. 1990. Use of Mono Q high resolution ion exchange chromatography to obtain highly pure and active E. coli RNA polymerase. *Biochemistry* **29:** 7890.

Helmann J.D. and Chamberlin M.J. 1988. Structure and function of bacterial sigma factors. *Annu. Rev. Biochem.* **57:** 839.

Ishihama A. 1981. Subunit assembly of E. coli RNA polymerase. *Adv. Biophys.* **14:** 1.

———. 1993. Protein-protein communication within the transcription apparatus. *J. Bacteriol.* **175:** 2483.

Jacobson G.R., Schaffer M.H., Stark G.R., and Vanaman T.C. 1973. Specific chemical cleavage in high yield at the amino peptide bonds of cysteine and cystine residues. *J. Biol. Chem.* **248:** 6583.

Jishage M. and Ishihama A. 1998. A stationary phase protein in *E. coli* with binding activity to the major σ subunit of RNA polymerase. *Proc. Natl. Acad. Sci.* **95:** 4953.

Jokerst R.S., Weeks J.R., Zehring W.A., and Greenleaf A.L. 1989. Analysis of the gene encoding the largest subunit of RNA polymerase II in *Drosophila*. *Mol. Gen. Genet.* **215:** 266.

Joo D.M., Ng N., and Calendar R. 1997. A σ^{70} mutant with a single amino acid change in the highly conserved region 2.2 exhibits reduced core RNA polymerase affinity. *Proc. Natl. Acad. Sci.* **94:** 4907.

Joo D.M., Nolte A., Calendar R., Zhou Y.N., and Jin D.J. 1998. Multiple regions on the *E. coli* heat shock transcription factor σ^{32} determine core RNA polymerase binding specificity. *J. Bacteriol.* **180:** 1095.

Jovanovich S.B., Lesley S.A., and Burgess R.R. 1989a. In vitro use of monoclonal antibodies in *E. coli* S-30 extracts to determine the RNA polymerase sigma subunit required by a promoter. *J. Biol. Chem.* **264:** 3794.

Jovanovich S.B., Record M.T., Jr., and Burgess R.R. 1989b. In an *E. coli* coupled transcription-translation system, expression of the osmoregulated gene *proU* is stimulated at elevated potassium concentrations and by an extract from cells grown at high osmolality. *J. Biol. Chem.* **264:** 7821.

Kaelin W.G., Jr., Krek W., Sellers W.R., DeCaprio J.A., Ajechenbaum F., Fuchs C.S., Chittenden T., Li Y., Farnham P. J., Blanar M.A., Livingston D. M., and Flemington E.K. 1992. Expression cloning of a cDNA encoding a retinoblastoma-binding protein with E2F-like properties. *Cell* **70:** 351.

Landschulz W.H., Johnson P.F., and McKnight S.L. 1988. The leucine zipper: A hypothetical structure common to a new class of DNA binding proteins. *Science* **240:** 1759.

Lesley S.A. and Burgess R.R. 1989. Characterization of the *E. coli* transcription factor σ^{70}: Localization of a region involved in the interaction with core RNA polymerase. *Biochemistry* **28:** 7728.

Lesley S.A., Brow M.A.D., and Burgess R.R. 1991. Use of in vitro protein synthesis from PCR-generated templates to study interaction of *E. coli* transcription factors with core RNA polymerase and for epitope mapping of monoclonal antibodies. *J. Biol. Chem.* **266:** 2632.

Lesley S.A., Thompson N.E., and Burgess R.R. 1987. Studies of the role of the *E. coli* heat shock regulatory protein σ^{32} by the use of monoclonal antibodies. *J. Biol. Chem.* **262:** 5404.

Lesley S.A., Jovanovich S.B., Tse-Dinh Y.-C., and Burgess R.R. 1990. Identification of a heat shock promoter in the *topA* gene of *E. coli*. *J. Bacteriol.* **172:** 6871.

Liebke H., Gross C., Walter W., and Burgess R. 1980. A new mutation *rpoD800*, affecting the sigma subunit of *E. coli* RNA polymerase is allelic to two other sigma mutants. *Mol. Gen. Genet.* **177:** 277.

Lonetto M., Gribskov M., and Gross C.A. 1992. The σ^{70} family: Sequence conservation and evolutionary relationships. *J. Bacteriol.* **174:** 3843.

Lowe P.A., Hager D.A., and Burgess R.R. 1979. Purification and properties of the sigma subunit of *E. coli* DNA-dependent RNA polymerase. *Biochemistry* **18:** 1344.

Lowe P.A., Aebi U., Gross C., and Burgess R.R. 1981. In vitro thermal inactivation of a temperature-sensitive sigma subunit mutant (*rpoD800*) of *E. coli* RNA polymerase proceeds by aggregation. *J. Biol. Chem.* **256:** 2010.

Luo J., Sharif K.A., Jin R., Fujita N., Ishihama A., and Krakow J.S. 1996. Molecular anatomy of the β′ subunit of the *E. coli* RNA polymerase: Identification of regions involved in polymerase assembly. *Genes Cells* **1:** 819.

Lupas A., Van Dyke M., and Stock J. 1991. Predicting coiled coils from protein sequences. *Science* **252:** 1162.

Malhotra A., Severinova E., and Darst S.A. 1996. Crystal structure of a σ^{70} subunit fragment from *E. coli* RNA polymerase. *Cell* **87:** 127.

McMahan S.A. and Burgess R.R. 1994. Use of aryl azide crosslinkers to investigate protein-protein interactions: An

optimization of important conditions as applied to *E. coli* RNA polymerase and localization of a σ^{70}-α crosslink to the C-terminal region of α. *Biochemistry* **33:** 12092.

———. 1996. Single-step synthesis and characterization of biotinylated nitrilotriacetic acid, a unique reagent for the detection of histidine-tagged proteins immobilized on nitrocellulose. *Anal. Biochem.* **236:** 101.

Nguyen L.H. and Burgess R.R. 1996. Overproduction and purification of σ^S, the *E. coli* stationary phase specific sigma transcription factor. *Protein Expression Purif.* **8:** 17.

———. 1997. Comparative analysis of the interactions of *E. coli* σ^S and σ^{70} RNA polymerase holoenzyme with the stationary-phase-specific *bolAp1* promoter. *Biochemistry* **36:** 1748.

Nguyen L.H., Jensen D.B., and Burgess R.R. 1993a. Overproduction and purification of σ^{32}, the *E. coli* heat shock transcription factor. *Protein Exp. Purif.* **4:** 425.

Nguyen L.H., Jensen D.B., Thompson N.E., Gentry D.R., and Burgess R.R. 1993b. In vitro functional characterization of overproduced *E. coli katF/rpoS* gene product. *Biochemistry* **32:** 11112.

O'Shea E.K., Rutkowski R., and Kim P.S. 1989. Evidence that the leucine zipper is a coiled coil. *Science* **243:** 538.

Owens J.T., Miyake R., Murakami K., Chmura A.J., Fujita N., Ishihama A., and Meares C.F. 1998. Mapping the σ^{70} subunit contact sites on *E. coli* RNA polymerase with a σ^{70}-conjugated chemical protease. *Proc. Natl. Acad. Sci.* **95:** 6021.

Roe J.-H., Burgess R.R., and Record M.T., Jr. 1984. Kinetics and mechanism of the interaction of *E. coli* RNA polymerase with the λ P$_R$ promoter. *J. Mol. Biol.* **176:** 495.

———. 1985. Temperature dependence of the rate constants of the *E. coli* RNA polymerase-lambda P$_R$ promoter interaction: Assignment of the kinetic steps corresponding to protein conformational change and DNA opening. *J. Mol. Biol.* **184:** 441.

Rao L., Jones D.P., Nguyen L.H., McMahan S.A., and Burgess R.R. 1996. Epitope mapping using histidine-tagged protein fragments: Application to *E. coli* RNA polymerase σ^{70}. *Anal. Biochem.* **241:** 173.

Rost B., Sander C., and Schneider R. 1994. PHD—An automatic mail server for protein secondary structure prediction. *CABIOS* **10:** 53.

Severinov K., Mooney R., Darst S.A., and Landick R. 1997. Tethering of the large subunits of *E. coli* RNA polymerase. *J. Biol. Chem.* **272:** 24137.

Severinov K., Mustaev A., Kaslev M., Borukhov S., Nikiforov V., and Goldfarb A. 1992. Dissection of the β subunit in the *E. coli* RNA polymerase into domains by proteolytic cleavage. *J. Biol. Chem.* **267:** 12813.

Severinov K., Mustaev A., Kukarin A., Muzzin O., Bass I., Darst S.A., and Goldfarb A. 1996. Structural modules of the large subunits of RNA polymerase: Introducing achaebacterial and chloroplast split sites in the β and β′ subunits of *E. coli* RNA polymerase. *J. Biol. Chem.* **271:** 27969.

Severinov K., Mustaev A., Severinova E., Bass I., Kashlev M., Landick R., Nikiforov V., Goldfarb A., and Darst S. 1995. Assembly of functional *E. coli* RNA polymerase containing β subunit fragments. *Proc. Natl. Acad. Sci.* **92:** 4591.

Shaner S.L., Piatt D.M., Wensley C.G., Yu H., Burgess R.R., and Record M.T., Jr. 1982. Aggregation equilibria of *E. coli* RNA polymerase: Evidence for anion-linked conformational transitions in the protomers of core and holoenzyme. *Biochemistry* **21:** 5539.

Shuler M.F., Tatti K.M., Wade K.H., and Moran C.P., Jr. 1995. A single amino acid substitution in σ^E affects its ability to bind to core RNA polymerase. *J. Bacteriol.* **177:** 3687.

Siegele D.A., Hu J.C., Walter W.A., and Gross C.A. 1989. Altered promoter recognition by mutant forms of the σ^{70} subunit of *E. coli* RNA polymerase. *J. Mol. Biol.* **206:** 591.

Strauss H.S., Burgess R.R., and Record M.T., Jr. 1980a. Binding of *E. coli* RNA polymerase holoenzyme to a bacteriophage T7 promoter-containing fragment: Selectivity exists over a wide range of solution conditions. *Biochemistry* **19:** 3496.

———. 1980b. Binding of *E. coli* RNA polymerase holoenzyme

to a bacteriophage T7 promoter-containing fragment: Evaluation of promoter binding constants as a function of solution conditions. *Biochemistry* **19:** 3504.

Strauss H.S., Boston R.S., Record M.T., Jr., and Burgess R.R. 1981. Variables affecting the selectivity and efficiency of retention of DNA fragments by *E. coli* RNA polymerase in the nitrocellulose-filter-binding assay. *Gene* **13:** 75.

Strickland M.S., Thompson N.E., and Burgess R.R. 1988. Structure and function of the σ^{70} subunit of *E. coli* RNA polymerase. Monoclonal antibodies: Localization of epitopes by peptide mapping and effects on transcription. *Biochemistry* **27:** 5755.

Studier F.W., Rosenberg A.H., Dunn J.J., and Dubendorff J.W. 1990. Use of T7 RNA polymerase to direct expression of cloned genes. *Methods Enzymol.* **185:** 60.

Sweetser D., Nonet M., and Young R.A. 1987. Prokaryotic and eukaryotic RNA polymerases have homologous core subunits. *Proc. Natl. Acad. Sci.* **84:** 1192.

Taylor W.E. and Burgess R.R. 1979. *E. coli* RNA polymerase binding and initiation of transcription on fragments of lambda rifd 18 DNA containing promoters for lambda genes and for rrnB, tufB, rplC,A, rplJ,L, and rpoB,C genes. *Gene* **6:** 331.

Taylor W.E., Straus D.B., Grossman A.D., Burton Z.F., Gross C.A., and Burgess R.R. 1984. Transcription from a heat-inducible promoter causes heat shock regulation of the sigma

subunit of *E. coli* RNA polymerase. *Cell* **38:** 371.

Thompson N.E., Hager D.A., and Burgess R.R. 1992. Isolation and characterization of a polyol-responsive monoclonal antibody useful for gentle purification of *E. coli* RNA polymerase. *Biochemistry* **31:** 7003.

Travers A.A. and Burgess R.R. 1969. Cyclic re-use of the RNA polymerase sigma factor. *Nature* **222:** 537.

Waldburger C. and Susskind M.M. 1994. Probing the informational content of *E. coli* σ^{70} region 2.3 by combinatorial cassette mutagenesis. *J. Mol. Biol.* **235:** 1489.

Wang D., Meier T.I., Chan C.L., Feng G., Lee D.N., and Landick R.L. 1995. Discontinuous movements of DNA and RNA in RNA polymerase accompany formation of a paused transcription complex. *Cell* **81:** 341.

Wang Y., Severinov K., Loizos N., Fenyo D., Heyduk E., Heyduk T., Chait B.T., and Darst S.A. 1997. Determinants for *E. coli* RNA polymerase assembly within the β subunit. *J. Mol. Biol.* **270:** 648.

Zhou Y. and Gross C.A. 1992. How a mutation in the gene encoding σ^{70} suppresses the defective heat shock response caused by a mutation in the gene encoding σ^{32}. *J. Bacteriol.* **174:** 7128.

Zhou Y.N., Walter W.A., and Gross C.A. 1992. A mutant σ^{32} with a small deletion in conserved region 3 of σ has reduced affinity for core RNA polymerase. *J. Bacteriol.* **174:** 5005.

The Transition from Initiation to Elongation by RNA Polymerase II

D.S. LUSE* AND I. SAMKURASHVILI*†

*Department of Molecular Biology, Lerner Research Institute, Cleveland Clinic Foundation,
Cleveland, Ohio 44195; †Department of Molecular Genetics, Biochemistry and Microbiology,
University of Cincinnati College of Medicine, Cincinnati, Ohio 45267

In recent years, the importance of the control of transcript elongation in the overall process of gene regulation has been increasingly appreciated (for review, see Uptain et al. 1997). In particular, regulated pausing of newly initiated RNA polymerases in the promoter-proximal region of transcription units has frequently been observed. The prototype for this sort of transcriptional control is provided by the *Drosophila hsp70* gene. RNA polymerases can access the hsp70 promoter at normal temperatures, but the newly initiated polymerases pause 21–35 bases downstream. Upon heat shock, these paused polymerases are released into productive elongation (for review, see Lis and Wu 1993). This control mechanism is not confined to heat shock genes but has been observed in a variety of other transcription units as well (see, e.g., Rougvie and Lis 1990; Krumm et al. 1992; Rasmussen and Lis 1995; for review, see Uptain et al. 1997).

Although many examples of promoter-proximal control have been described, the molecular basis for this type of regulation is not yet well understood. Perhaps newly initiated RNA polymerase is especially susceptible to pausing. It would then be relatively easy for regulatory molecules to exaggerate this tendency to pause and thereby stop transcript elongation completely. RNA polymerases do pass through an abortive initiation process just after the start of transcription, but abortive initiation by RNA polymerase II (pol II) results in termination, not pausing (Luse and Jacob 1987; Jacob et al. 1991), and in any event, this initial transcriptional stage is completed after about ten bonds have been made (Luse and Jacob 1987; Jacob et al. 1991). A potential clue to the mechanism of promoter-proximal pausing is provided by the resemblance of this event to transcriptional arrest. Arrested polymerases stop transcription, despite the presence of NTP substrates, but they do not terminate (Izban and Luse 1992b). Rapid resumption of transcription by arrested polymerases requires cleavage of a 5–17-nucleotide segment from the 3′ end of the nascent RNA, a reaction that is greatly stimulated by the transcript elongation factor TFIIS (also called SII) for pol II (Izban and Luse 1992b; Reines et al. 1992), or GreB for bacterial RNA polymerase (Borukhov et al. 1993).

Arrest has been explained by postulating an alternative pathway to normal transcript elongation. In this alternative mode, RNA polymerases can translocate upstream instead of forming the next phosphodiester bond. The translocated complexes remain stably clamped to the

template during the sliding process (Reeder and Hawley 1996; Komissarova and Kashlev 1997b; Landick 1997; Nudler et al. 1997; see also Guajardo and Sousa 1997). Since the transcription bubble and the RNA:DNA hybrid must also translocate upstream, the 3′ end of the nascent RNA will be removed from the polymerase's catalytic site. Thus, cleavage of the transcript at the new, upstream location of the active center is necessary to realign the 3′ end of the RNA with the catalytic site and allow RNA synthesis to continue (Rudd et al. 1994; Orlova et al. 1995). At template locations that encode long runs of U residues in the transcript, the RNA:DNA hybrid is exceptionally unstable, and upstream translocation may be particularly favored (Reeder and Hawley 1996; Nudler et al. 1997).

Despite the ability to restart transcription after a long inactive period, RNA polymerases paused in the promoter-proximal region are not identical in properties to the well-characterized arrested pol II complexes. RNA polymerases paused early in transcription on the *Drosophila hsp70* gene may be released into elongation with the detergent Sarkosyl (Rougvie and Lis 1988), which does not cause arrested polymerases to resume transcription and which inhibits the action of TFIIS (Izban and Luse 1992b). In addition, the long stretches of T residues on the nontemplate strand that are usually seen in arrest sites have not been identified as a sequence feature of those initially transcribed regions in which regulated pausing takes place (see, e.g., Rasmussen and Lis 1993). However, studies of prokaryotic RNA polymerase suggest a structural similarity between arrested polymerases and those paused early in RNA synthesis. It has been shown that *Escherichia coli* RNA polymerases stalled by NTP starvation during the initial phases of transcription (i.e., before the addition of approximately the 28th base to the nascent RNA) have a strong tendency to translocate upstream as assayed by exonuclease III (Exo III) footprinting (Nudler et al. 1994; Komissarova and Kashlev 1997a; see also Krummel and Chamberlin 1992) even though prokaryotic polymerases do not arrest in promoter-proximal regions when provided with high levels of NTPs. It is important to note that some pol II complexes stalled by NTP starvation at DNA sequences that do not normally cause arrest also display upstream-translocated footprints (Samkurashvili and Luse 1996). A subset of these upstream-translocated pol II complexes will not resume transcription when NTPs are added back

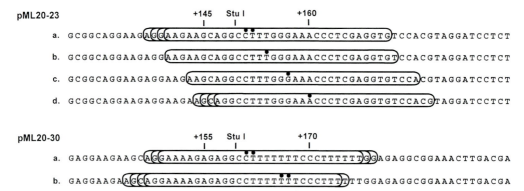

Figure 1. Schematic summary of the Exo III protection experiments on the pML-23 and -30 constructs. Only the nontemplate strand of DNA is shown. The elongation complexes are represented by the ovals. The positions of transcript 3′ ends are designated by dots. (Reprinted, with permission, from Samkurashvili and Luse 1996 [copyright American Society for Biochemistry and Molecular Biology].)

to the reaction (Samkurashvili and Luse 1996). It seems reasonable to suppose that pol II transcription complexes in the early stages of elongation might be particularly likely to stop transcription in response to regulatory intervention if pol II, like bacterial RNA polymerase, tends to translocate upstream when stalled in the promoter-proximal region. It therefore becomes important to examine the translocation behavior of pol II during the initial phase of transcript elongation.

We have approached this question by determining the Exo III footprints for a series of pol II transcription complexes stalled because of NTP limitation between positions +20 and +51. None of these complexes were arrested; i.e., all of them could elongate their RNA chains in 5-minute chase reactions. However, the Exo III footprints of complexes with 20–25-nucleotide RNAs were

identical and resembled those of arrested complexes, whereas the footprint of a +27 complex was displaced forward by 17 bases. In contrast, transcription of a template in which the initial transcribed sequence was duplicated beginning at +98 showed coordinate advance of the polymerase footprint with RNA synthesis during transcription from +122 to +130. Thus, the pol II transcription complex, like *E. coli* RNA polymerase, has a strong tendency to translocate upstream during the early stages of transcript elongation.

MATERIALS AND METHODS

Plasmids. The pML20-40 series plasmids, described in detail elsewhere (Samkurashvili and Luse 1998), are based on pML20-23 (see Figs. 1 and 2), which contains

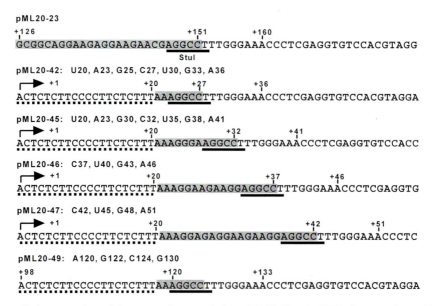

Figure 2. Sequence of relevant portions of the nontemplate strand of the pML20-40 series DNAs in comparison with pML20-23. The arrows indicate the transcription start site (+1). The first 20 nucleotides of the initially transcribed regions from pML20-42, -45, -46, and -47 are designated by dashed underlining. The same sequence is also duplicated in pML20-49 starting at position +98. Gray boxes represent T-free segments, each of which has a *Stu*I site (*solid underline*) at its downstream end. Note that the sequences downstream from the *Stu*I sites are identical in all constructs. (Reprinted, with permission, from Samkurashvili and Luse 1998 [copyright American Society for Microbiology].)

the adenovirus 2 major late promoter cloned into pUC18 (Samkurashvili and Luse 1996). The initial transcribed regions of pML20-42, pML20-45, pML20-46, and pML20-47 are shown in Figure 2. The pML20-42 construct was made by deleting the segment of pML20-23 between +22 and the StuI site, which begins at +147; pML20-45, -46, and -47 were produced from pML20-42 by the addition of 5, 10, or 15 bp of sequence downstream from position +22 as shown in Figure 2. In pML20-49, the initial transcribed region of pML20-42 was duplicated beginning at +98 (Fig. 2). The segment of pML20-49 from +20 to +98 is G-free on the nontemplate strand.

Assembly, purification, and analysis of ternary transcription complexes.

Template preparation and assembly of preinitiation complexes are described in detail elsewhere (Samkurashvili and Luse 1996, 1998). All transcription reactions were performed on DNA templates labeled on one end only. Plasmid DNA was linearized with either SstI for nontemplate strand labeling or with PstI for labeling of the template strand. On all templates, preinitiation complexes were advanced to position +20 (U20 complexes) by incubation with 2 mM ApC, 10 μM dATP, 20 μM UTP, and 1 μM [α-^{32}P]CTP at 30°C for 5 minutes, followed by another 5-minute incubation after the addition of CTP to 20 μM. The stalled ternary complexes were further purified by Sarkosyl rinsing: Sarkosyl was added to 1% and incubated for 5 minutes at 30°C, followed by Bio-Gel A1.5m gel filtration. Sarkosyl-rinsed U20 complexes were incubated at 37°C with 8 mM MgCl$_2$ and either 20 μM ATP, GTP, and CTP for 5 minutes (for all templates except pML20-49) or 20 μM ATP, UTP, and CTP for 10 minutes (for pML20-49), followed by another round of gel filtration. The last step was omitted for production of the U20, A23, G25, and C27 complexes on the pML20-42 template.

After the Sarkosyl rinsing/gel filtration step, all complexes were incubated with restriction enzymes StuI and either HindIII (for SstI end-labeled templates) or EcoRI (for PstI end-labeled templates); all digestions used 0.2 units of enzyme/μl for 10 minutes at 37°C. After restriction digestion, some complexes were further incubated with a subset of the NTPs to walk them to their final positions for footprinting. Exo III digestion was then carried out for 6 minutes at 37°C with either 2 units/μl or 5 units/μl final concentration. Sample preparation and electrophoresis were performed as described previously (Samkurashvili and Luse 1996, 1998).

RESULTS

Experimental Design and Initial Studies

For our initial studies (Samkurashvili and Luse 1996) of pol II translocation as a function of transcript elongation, we generated transcription complexes stalled far downstream (~150 bases) from transcription start. We use the term "stalled" to describe RNA polymerases that have stopped transcription because the next NTP required for chain elongation is missing from the reaction mixture.

Stalled polymerases are not generally arrested; i.e., they resume transcription when the necessary NTPs are supplied. Our overall strategy (for detailed description, see Samkurashvili and Luse 1996, 1998) involved assembling preinitiation complexes on single end-labeled DNA templates by incubation in HeLa nuclear extracts. After a gel filtration step to remove contaminating NTPs, the RNA polymerases were advanced to +20 with a subset of the NTPs and then extensively purified by the addition of Sarkosyl followed by another round of gel filtration to remove both the detergent and the proteins stripped from the complexes. The resulting "Sarkosyl-rinsed" transcription complexes were fully active for further RNA synthesis and lacked any detectable transcript elongation factors. These complexes could be walked to desired downstream locations by a combination of incubation with subsets of the NTPs and additional rounds of gel filtration as necessary. The positions of the RNA polymerases could then be determined from the upstream and downstream boundaries of protection against Exo III digestion which they provided. The plasmid templates for the footprinting study were designed so that the point on the template where the RNA polymerase would be finally stalled before Exo III digestion was at or near a cleavage site for the restriction enzyme StuI. The complexes were treated with StuI before digestion with Exo III. The StuI cleavage step was necessary because only about 1% of the templates are occupied by pol II in a typical in vitro transcription reaction using HeLa nuclear extracts. Unless the background created by incomplete Exo III digestion of the large excess of nontranscribed DNAs was eliminated, it was not possible to detect the Exo III protection conferred by the transcription complexes (Samkurashvili and Luse 1996).

Our first Exo III assays were done with the pML20-23 template. In this construct, the adenovirus 2 major late TATA element and initiator direct transcription. Downstream from +20, the point at which complexes were stalled for Sarkosyl rinsing, we inserted a DNA fragment that contained a 130-bp T-free (nontemplate strand) cassette, followed by a StuI site and a segment of triplet repeat DNA (nontemplate strand sequence of ...TTTGGGAAACCC...; see Fig. 1). Polymerases could be stalled at the beginning of this segment or after the T, G, or A triplets. In all cases, the resulting transcription complexes were fully active for further RNA synthesis. When the locations of these stalled polymerases were mapped by Exo III footprinting, the results shown in Figure 1 were obtained (Samkurashvili and Luse 1996). Exonuclease digestion revealed a transcription complex that protected 30–35 nucleotides of template, with the 3′ end of the nascent RNA located just upstream of the center of protection. Significantly, the protection pattern advanced downstream in near synchrony with transcript elongation, as one would expect for transcriptionally competent complexes. Thus, we had identified a DNA segment which apparently provides no barrier to normal translocation by pol II during RNA synthesis.

We also examined the footprints of transcription complexes on the pML20-30 template, in which the DNA downstream from the *Stu*I site in pML20-23 was replaced with a segment of human DNA bearing a strong arrest site from the histone H3.3 gene (see Fig. 1) (Samkurashvili and Luse 1996). When polymerases were walked to the beginning of the arrest site (where they remained transcriptionally competent), they gave a template protection pattern similar to that seen on pML20-23 (Fig. 1, line *a* under pML20-30). In contrast, when transcription was allowed to continue and the polymerases arrested 4–5 bases further downstream, the protection pattern of the arrested complexes was displaced upstream from that of the stalled complexes (Fig. 1, line *b* under pML20-30).

Exonuclease Protection Conferred by RNA Polymerase II Complexes Stalled at Sequential Sites in the Initially Transcribed Region

To study transcription complexes paused early in the elongation process, we constructed a new set of four templates, pML20-42, -45, -46, and -47, which allowed us to stall pol II at many locations from 20 to 51 bases downstream from transcription start (see Fig. 2). All of the pML20-40 series constructs are based on pML20-23, on which we had observed the polymerase footprint translocate in synchrony with transcription (Fig. 1) (Samkurashvili and Luse 1996). In pML20-23, the *Stu*I site begins at +147 (see Fig. 2). To construct pML20-42, the DNA between +22 and +147 in pML20-23 was deleted; for pML20-45, -46, and -47, an additional 5, 10, or 15 bases upstream of the *Stu*I site were added back into pML20-42. The original sequence upstream of the *Stu*I site in pML20-23 was generally preserved in constructing pML20-45, -46, and -47 (Fig. 2), but some base changes were necessary to allow convenient assembly of transcription complexes at desired template locations.

Exo III studies were initially performed on complexes stalled at positions +20, +23, +25, and +27 on the pML20-42 template. Each complex was fully elongation competent when challenged with all four NTPs (data not shown; see Samkurashvili and Luse 1998). The downstream, or front-edge, boundaries of template protection for the U20, A23, G25, and C27 complexes were determined using Exo III digestion after *Stu*I cleavage, as described above. The results are shown in Figure 3A. Most of the template DNA was cleaved by *Stu*I (lanes 1 and 2), consistent with transcription of only a few percent of the templates by pol II. When U20 complexes were treated with Exo III, the *Stu*I-resistant DNA was truncated to a band whose downstream edge mapped to +28 (dotted band, lanes 3 and 4). This band was absent when the complexes were chased before Exo III digestion (lanes 15 and 16), consistent with our assignment of the band as the front edge of the U20 transcription complex. The location of the U20 complex front edge, only seven nucleotides downstream from the last transcribed base, was surprising. This configuration is typical for arrested complexes, in which the RNA polymerase has undergone upstream translocation (Samkurashvili and Luse 1996; see also Gu et al. 1993), even though the U20 complex shows no evidence of arrest. Furthermore, the front edges of the A23

and G25 complexes were identical to the U20 front edge (Fig. 3A, lanes 6,7 and 9,10), whereas the leading edge of the C27 complex was displaced 17 bases down the template, to +45 (lanes 12 and 13).

As with the leading edges, the upstream boundaries of the U20, A23, and G25 complexes were essentially identical, whereas the C27 boundary was displaced well downstream (Fig. 3B). The rear-edge boundary was reproducibly more diffuse for several of the complexes (compare with Fig. 3A); we also observed this effect with the upstream boundaries of a number of the complexes that we studied previously (Samkurashvili and Luse 1996). The bands corresponding to the upstream boundaries were absent when the complexes were chased to run-off before Exo III digestion (Fig. 3B, lanes 15 and 16).

The results from Figure 3 are summarized schematically in Figure 6. The overall dimensions of the complexes, as judged from the length of template protected, did not change substantially among the four stalled complexes, consistent with the results obtained with stalled complexes on pML20-23 (Fig. 1). However, the strongly discontinuous advance of the Exo III footprint in complexes U20-C27 is in sharp contrast to the observations with pML20-23. The results in Figure 3 are very similar to those reported for a series of *E. coli* RNA polymerase complexes stalled at positions +25 to +30 (Nudler et al. 1994). It is particularly striking that the Exo III footprint of the bacterial polymerase advanced by nine bases as a result of adding only three more bases to the transcript by a complex at +27 (Nudler et al. 1994). Thus, a major structural transition may occur for both bacterial RNA polymerase and mammalian pol II during RNA synthesis about 25 bases downstream from transcription start. We return to this possibility in the Discussion.

The unexpected results with the Exo III footprints of the U20-C27 complexes prompted us to ask whether pol II complexes stalled somewhat further downstream from +1 would also show anomalous footprints. The design of the pML20-42 template allowed us to generate complexes paused at positions +30, +33, and +36 (U30, G33, and A36 complexes). As with the U20-C27 complexes, the U30-A36 complexes were fully elongation competent (Samkurashvili and Luse 1998). The front-edge boundaries of the C27, U30, G33, and A36 complexes, shown in Figure 4A, were unanticipated in two respects. First, the leading edges for the U30, G33, and A36 complexes (solid dots) were not substantially displaced downstream in comparison with the C27 complex downstream boundary. Additionally, the boundaries for the U30, G33, and A36 complexes were clearly partitioned between a set of bands at +44 to +49 and a band at +28 (open dot); this band appeared as a minor part of the C27 front-edge pattern as well (lanes 3 and 4, Fig. 4A; see also lane 13 of Fig. 3A). Both the +44 to +49 and the +28 bands were absent in the chase control (lanes 15–16). Thus, there appeared to be two populations of RNA polymerases within each of the stalled complexes. The more upstream conformation was the more abundant complex for G33 and A36, and possibly for U30 as well, even though the last

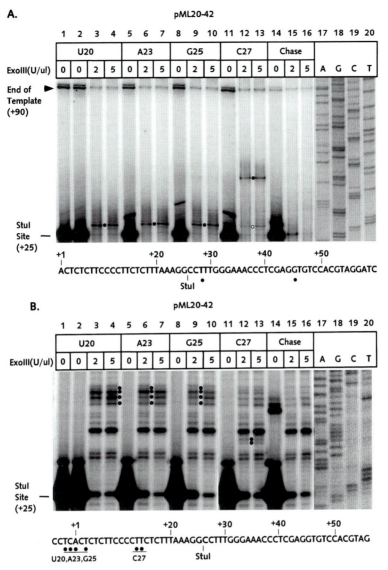

Figure 3. Exo III footprints of U20-C27 complexes on the pML20-42 template. A segment of nontemplate DNA strand sequence is shown at the bottom of each figure. (*A*) The front-edge boundaries of pol II elongation complexes were determined with DNA labeled at the 5´ end of nontemplate strand. DNAs were resolved on a 6% polyacrylamide gel. Exo III reactions on U20 complexes chased to the end of the template with all four NTPs are shown in lanes *14–16*. Solid dots mark the position of the major downstream protection boundary for each complex; the residual protection at +28 in lanes 12 and 13 is indicated by the open dot. The numbers in the left margin indicate the position of bands relative to the transcription start site. The actual lengths of the DNA fragments equal the number indicated plus the 70-bp upstream fragment. Exact DNA length markers (lanes *17–20*) were generated by primer extension from the same DNA template used for transcription. (*B*) The rear-edge boundaries of RNA polymerase elongation complexes were obtained as in *A* except that DNA was labeled at the 5´ end of template strand. DNAs were resolved on a 10% polyacrylamide gel. The actual length of the *Stu*I-cut DNA fragment is 75 nucleotides. (Reprinted, with permision, from Samkurashvili and Luse 1998 [copyright American Society for Microbiology].)

transcribed base is downstream from +28 for each of these complexes. In our earlier studies, the presence of two Exo III footprints for a single transcription complex correlated with a subset of arrested complexes (Samkurashvili and Luse 1996). However, the U30, G33, and A36 complexes did not show a significant population that failed to chase (Samkurashvili and Luse 1998). The upstream Exo III boundaries of the C27, U30, G33, and A36 complexes are shown in Figure 4B. Unlike the front-edge boundaries, the rear edges were displaced downstream with RNA synthesis, although the extent of this displacement did not match

the increase in transcript length. Surprisingly, we failed to detect partitioning between two distinct conformations as we did for the front edges.

The experimental results from Figure 4 are summarized in Figure 6. For clarity, only the more downstream of the U30-A36 front edges are shown. It seems very unlikely that the transcription complex could adopt a conformation in which it protects less than 10 bp of template (which would be the case for the A36 complex with a leading edge at +27 and a trailing edge +17 to +21). In this context, it is important to note that our footprinting

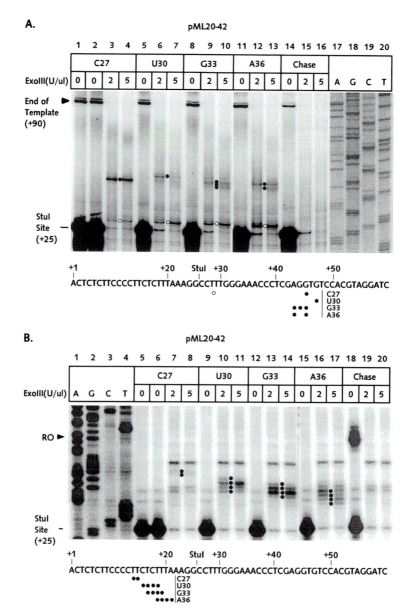

Figure 4. The Exo III footprints of pol II elongation complexes between positions +27 and +36 on the pML20-42 template. The sequence of the nontemplate DNA strand is shown at the bottom of each figure. Numbers in the left margins of both panels indicate distances downstream from transcription start. U20 complexes were supplied with ATP, GTP, and CTP to stall polymerase at position +27; the C27 complexes were then gel-filtered, treated with *Stu*I and advanced to the indicated positions by incubating with a subset of NTPs. (*A*) Front-edge boundaries were obtained with DNAs labeled at the 5′ end of the nontemplate strand. DNAs were resolved on a 6% polyacrylamide gel. Exo III reactions on C27 complexes chased to the end of the template with all four NTPs are shown in lanes *14–16*. The front edges of the more downstream conformations (see text) are marked by solid dots and the front edges of the more upstream conformations are indicated by open dots. Exact DNA markers are shown in lanes *17–20*. (*B*) The rear edge boundaries of RNA polymerase elongation complexes were obtained as in *A*, except that the DNA was labeled at the 5′ end of template strand. DNAs were resolved on a 10% polyacrylamide gel. (Reprinted, with permission, from Samkurashvili and Luse 1998 [copyright American Society for Microbiology].)

experiment involves 6 minutes of exposure to the exonuclease. If a particular stalled complex is in true equilibrium between two template locations, progressive Exo III digestion will tend to emphasize the more upstream of the possible front edges and the more downstream of the possible rear edges, thus leading to an apparently shortened footprint. This model does not indicate why we failed to detect a second, more upstream conformation in the rear

edge determination (Fig. 4B). At present, we do not have a good explanation for this discrepancy, but it is worth noting that the upstream Exo III boundaries for the U30-A36 complexes are quite diffuse. It would be very difficult to detect a second boundary if the signal from this edge were also spread out over many bands.

To generate complexes stalled downstream from position +36, we used the pML20-45, -46, and -47 templates (see Fig. 2). We determined front and back edges of Exo III protection for a series of 12 complexes from C32 to A51. The results of these experiments are summarized in Figure 6 (primary data not shown; see Samkurashvili and Luse 1998). Most of these complexes showed the predicted Exo III footprint for stalled, transcriptionally competent complexes; i.e., from 30 to 35 bp of DNA were protected with the last transcribed base located just upstream of the center of the footprint (Samkurashvili and Luse 1996). However, the U35 complex on pML40-45 and the U40 complex on pML40-46 did not conform to this pattern. Thus, tight linkage of footprint translocation with RNA synthesis was not achieved even with complexes containing RNAs as long as 40 nucleotides. Synchrony between RNA synthesis and footprint translocation was fully observed only with the set of four complexes from the pML20-47 template. Note that the two complexes with anomalous footprints, U35 and U40, each has three U residues at the 3′ ends of their nascent RNAs.

Exonuclease Protection Patterns Conferred by RNA Polymerase II Complexes Stalled Far Downstream from Transcription Start

Footprint translocation and transcription were tightly linked during far-downstream transcription on pML20-23, but this linkage was lost during transcription of similar DNA sequences in a promoter-proximal location (Fig. 6). This difference could have resulted from the difference in transcript lengths in the two sets of complexes. However, the otherwise analogous complexes on pML20-23 and the pML20-40 series templates did not have identical upstream DNA contacts, because of the initial G-free cassette in the pML20-40 series plasmids. To assess the role of transcript length without the complication of differences in template sequence, we constructed the pML20-49 template. In this DNA, the segment of pML20-42 from +1 to +90 was duplicated beginning at +98 in pML20-49. The duplicated regions are separated by a G-free cassette. The sequence of pML20-49 from +95 downstream is shown in Figure 2.

We assembled U20 complexes on the pML20-49 template, rinsed the complexes with Sarkosyl, and advanced the RNA polymerases through the G-free cassette to +120 with ATP, UTP, and CTP. After a second round of gel filtration, some of the A120 complexes were walked to G122, C124, or G130. The front- and rear-edge boundaries of these complexes as determined with Exo III are shown in Figure 5, A and B, and summarized in Figure 6. Despite considerable effort, we failed to detect upstream or downstream edges for the A120 complex, although this complex was stable and fully transcriptionally competent (data not shown). The C124 complex has the identical underlying template sequence to the C27 complex (see Fig. 6), and the front-edge footprints of both complexes were essentially identical. However, there is a dramatic difference in the footprints of the analogous pair of G25 and G122 complexes. The G25 footprint is translocated far upstream, such that the front edge is only 2 bp from the last transcribed base, whereas the front edge of

the G122 complex is 18–20 bp downstream from the last transcribed base, as is typical of transcriptionally competent complexes. Finally, the G130 complex also has a typical stalled-complex footprint, whereas the analogous G33 complex has an abnormally short footprint shifted upstream relative to the last transcribed base. Most significantly, the G130 footprint showed no evidence for a second, upstream conformation, in contrast to the G33 complex. Since the template and transcript sequences are identical for at least 26 bases upstream of the last transcribed base for the G25-G122, C27-C124, and G33-G130 complex pairs, we conclude that the upstream translocation of the Exo III footprint for the G25 and G33 complexes cannot be attributed to the sequence context and therefore most probably results from proximity to the transcription start site.

DISCUSSION

We found that pol II transcription complexes pass through a major structural transition about 25 bases downstream from the transcription start site. In the case of several of the complexes we studied, for example, those with 20-, 23-, or 25-nucleotide nascent RNAs, the predominant template location as judged from Exo III footprinting was far upstream of the expected position. This is a characteristic of arrested transcription complexes (Gu et al. 1993; Wang et al. 1995; Samkurashvili and Luse 1996; Komissarova and Kashlev 1997a,b; Nudler et al. 1997). However, all of the complexes we tested were elongation competent.

To justify this somewhat paradoxical finding, it is useful to briefly review our current understanding of the molecular mechanisms involved in normal transcript elongation and in transcriptional arrest. Based primarily on studies of *E. coli* RNA polymerase, a "sliding clamp" model has been proposed which suggests that RNA polymerase is able to translocate upstream along the DNA template as an alternative to making additional phosphodiester bonds (for review, see Landick 1997). Upstream movement carries the transcription bubble, the RNA:DNA hybrid, and the active site away from the 3′ end of the nascent RNA, resulting in a ternary complex that cannot continue transcription (Nudler et al. 1997; Komissarova and Kashlev 1997a,b). Upstream translocation is thought to be driven primarily by the presence of a relatively weak RNA:DNA hybrid at the 3′ end of the transcript (Reeder and Hawley, 1996; Nudler et al. 1997). In particular, those complexes with the weakest possible hybrid at the 3′ end (U:A) should be especially prone to upstream translocation.

This model envisions arrested complexes as stably occupying upstream template locations, and it also suggests the existence of elongation-competent complexes in equilibrium between two or more conformers, with only the most downstream conformation capable of productive elongation (Komissarova and Kashlev 1997a). Many of our findings support this latter aspect of the sliding clamp model. Partitioning of the pol II elongation complex between two locations is clearly illustrated by the com-

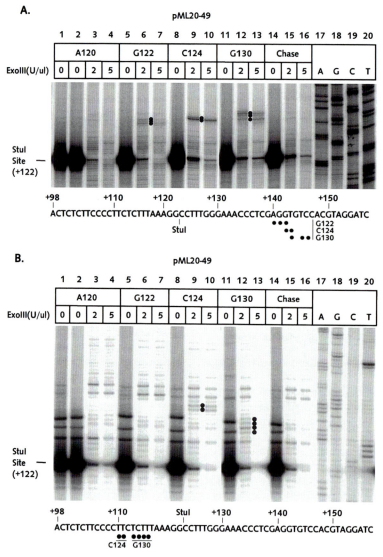

Figure 5. Exo III footprints of pol II elongation complexes stalled far from the transcription start site on the pML20-49 template. The sequence of the nontemplate DNA strand is shown at the bottom of each figure. Numbers in the left margins of both panels indicate distances downstream from transcription start. (*A*) Front-edge boundaries were determined with DNAs labeled at the 5′ end of the nontemplate strand; DNAs were resolved on a 6% polyacrylamide gel. Exo III reactions on A120 complexes chased to the end of the template with all four NTPs are shown in lanes *14–16*. Dots mark the positions of the major boundaries; note that no boundary could be detected above the chase background for the A120 complex. Exact DNA length markers are shown in lanes *17–20*. (*B*) Rear-edge boundaries of RNA polymerase elongation complexes were obtained as in *A* except DNA was labeled at the 5′ end of template strand. DNAs were resolved on a 10% polyacrylamide gel. Boundaries were detected only for the C124 and G130 complexes. (Reprinted, with permisison, from Samkurashvili and Luse 1998 [copyright American Society for Microbiology].)

plexes stalled at positions 27, 30, 33, and 36 on the pML20-42 template (Fig. 4A). In the C27 complex, the predominant footprint was that expected for a transcriptionally competent polymerase (Samkurashvili and Luse 1996), with the transcript 3′ end centrally located. However, as transcription progressed to +30, +33, and +36, two distinct footprints became evident. The more upstream of these had the same front edge as complexes U20, A23, and G25. Thus, we suppose that complexes U30, G33, and A36 are in equilibrium between a relatively stable upstream conformation and a transcriptionally competent downstream conformation. During the 5-minute chase period, essentially 100% of the complexes

must occupy the downstream conformation for at least the brief time required to make an additional bond, since all of the complexes could resume transcription in 5 minutes when supplied with excess NTPs. Based on the relative proportion of the two footprints, the upstream conformer represents the predominant structure for the U30, G33, and A36 complexes. The most dramatic example of the existence of multiple conformers was provided by the A120 complex on the pML20-49 template, which was fully active in transcription but in many experiments gave no detectable Exo III protection boundaries. We interpret the failure to detect any A120 footprint as evidence for many conformations that were relatively stable during the

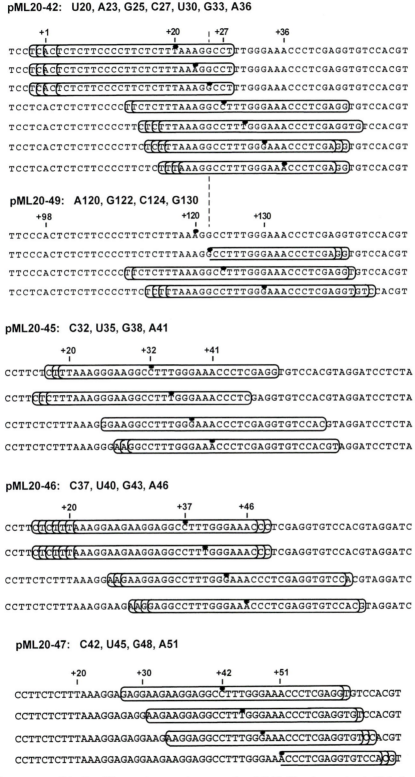

Figure 6. Schematic summary of the Exo III protection experiments on the pML20-40 series constructs. Only the nontemplate strand of DNA is shown. The elongation complexes are represented by the ovals. The positions of transcript 3′ ends are designated by dots. Note that only the more downstream conformation is shown for complexes U30, G33, and A36. The pML20-42 and 20-49 sequences are aligned to facilitate comparison of the footprints of pairs of analogous complexes; that is, complexes for which the transcript and template sequences are identical in the region near the 3′ end of the nascent RNA. (Reprinted, with permission, from Samkurashvili and Luse 1998 [copyright American Society for Microbiology].)

course of experiment. With the footprint signal spread out over many bands, we presume that it became undetectable above the background.

Some aspects of our results are not easily explained by the sliding clamp model as described above. First, although we clearly observed upstream conformers for many of our complexes, upstream translocation in most of these cases could not have been driven by weak RNA:DNA hybrids. Complexes A23 and G25, for example, both show a single, apparently very stable template location well upstream of the expected position, but the 3′ ends of the RNAs in these complexes are not U-rich. What drives upstream translocation for these complexes, and why are they not arrested?

The predominance of the upstream conformation in complexes U20, A23, and G25 must be primarily due to the length of the transcript and not to template or transcript sequence. This is demonstrated by comparison with analogous complexes assembled on the pML20-49 template. Note that the RNA:DNA hybrids are identical in the G25-G122, C27-C124, and G33-G130 pairs (see Fig. 6). However, the complexes stalled on the pML20-49 template did not show any upstream-displaced conformers. Since both sets of complexes were prepared in an identical manner, including Sarkosyl rinsing at position +20, the only difference is the lengths of the nascent RNAs.

It is not immediately obvious why promoter-proximal stalling should strongly favor sliding back along the template. However, a number of observations have suggested that a structural transition should take place in ternary transcription complexes approximately 30 bases downstream from transcription start. First, as noted above, Exo III footprinting studies of *E. coli* RNA polymerase showed an abrupt shift away from the upstream-translocated footprint in a comparison of complexes with 27- or 30-nucleotide nascent RNAs (Nudler et al. 1994). In an earlier study in which *E. coli* ternary complexes were resolved by electrophoresis, Straney and Crothers (1985) demonstrated that complexes with up to 25-nucleotide RNAs and complexes with more than 35-nucleotide RNAs have distinct electrophoretic mobilities. This observation was echoed by work from our laboratory, which showed that pol II ternary complexes with 15- or 35-nucleotide RNAs could be resolved electrophoretically (Linn and Luse 1991). It is interesting that the first of the pol II complexes analyzed in the present study which showed a stable downstream conformer is C27. The length of RNA which early pol II transcription complexes will protect against attack by ribonuclease is about 25 bases (D. McKean and D. Luse, unpubl.; see also Gu et al. 1996). Thus, it is tempting to speculate that filling an RNA-binding site or channel is necessary to begin to lock the RNA polymerase into a stable elongation configuration. If this idea is correct, the association of more than 25 bases of RNA with the polymerase must be required to complete the conversion to the elongation-committed form, since a tight linkage between transcript elongation and downstream translocation of the Exo III footprint is not achieved until the nascent RNA is more than 40 bases long (see Fig. 6).

It is important to note that although transcript length is an overriding feature in the positioning of early elongation complexes on the template, transcript/template sequence also plays a part. For example, the G38 complex on the pML20-45 template was stalled at almost the same distance downstream from +1 as the C37 complex on the pML20-46 template, but the DNA protection pattern was quite different between these two complexes (Fig. 6). The U35 complex on the pML20-45 template and the U40 complex on the pML20-46 template both had nascent RNAs which end in three U residues, so it is not surprising that the footprints for these complexes were displaced upstream. However, complexes U30 on pML20-42 template and U45 on pML20-47 template, which also had three U residues at the end of their RNAs, gave footprints that were not displaced upstream. Thus, the exact sequence context is crucial in determining whether upstream displacement will occur.

Complexes such as G33 and A36, which showed a partitioning between upstream and downstream conformers in footprinting, were entirely transcriptionally active when chased. This presumably reflects a very high probability that all of these complexes can occupy the downstream, transcriptionally competent state for at least the time required to make another bond (probably about 0.2 second; see Izban and Luse 1992a) during the 5 minutes of incubation with high levels of NTPs. It is more difficult to understand the transcriptional activity of complexes such as U20, A23, and G25. The only detectable footprints for these complexes show a conformation in which the polymerase has translocated far upstream. This is the same pattern we (Samkurashvili and Luse 1996) and other workers (Gu et al. 1993; Komissarova and Kashlev 1997b; Nudler et al. 1997) have observed with arrested complexes. If the G25 complex, for example, can occupy a downstream conformation for a sufficient time to resume transcription during a 5-minute chase, why do arrested complexes fail to resume elongation under the same conditions? This difference presumably reflects the importance of U-rich 3′ ends in the arrest process. We suppose that all complexes which show only upstream conformers by Exo III analysis have some probability of sliding downstream toward the transcriptionally competent configuration. However, the active site in arrested complexes may be prevented from reaching the 3′ end of the RNA because this segment of RNA is U-rich in arrested complexes. The very weak U:A hybrid would destabilize the complex to an increasing extent as downstream translocation continued, making it quite unlikely that the active site would actually reach the 3′ end. In complexes such as G25, this barrier would not exist. Thus, two complexes that both show footprints translocated far upstream of the expected location might nevertheless have very different abilities to resume transcript elongation.

Arrested elongation complexes may cleave RNA as far upstream as 17 bases from the 3′ end when exposed to TFIIS or pyrophosphate (Izban and Luse 1993b; Rudd et al. 1994). The size of the cleavage products correlates with the extent of upstream translocation of the elongation complex, because the catalytic site of RNA polymerase is the cleaving agent (Rudd et al. 1994). The increment of TFIIS-mediated cleavage in complexes stalled early in elongation has been addressed in a limited way in earlier work from this laboratory (Izban and Luse 1992b, 1993a). In particular, TFIIS-mediated cleavage in a 20-mer complex (with a sequence very similar to the U20 complex tested here) liberated primarily dinucleotides and a much lower level of large cleavage fragments (Izban and Luse 1993a). This result appears to disagree with the footprint of the U20 complex determined in the present study, which shows the U20 complex exclusively in the upstream conformation. To explain this apparent contradiction, note that during upstream translocation, RNA polymerase should pass through intermediate steps between the most downstream and upstream conformers. If the residence time of RNA polymerase at these intermediate positions is usually sufficient for cleavage to occur in the presence of TFIIS, one would generally observe processive cleavage of the transcript from the 3′ end with the release of dinucleotide fragments. As noted above, the weak RNA:DNA hybrids in arrested complexes might cause the most downstream of the translocation intermediates in such complexes to be extremely short-lived. Cleavage would therefore be improbable until RNA polymerase reached more upstream locations. It is important to emphasize that different kinetics do not affect the equilibrium distribution between upstream and downstream configurations. As a result, the experimentally observed footprints would look similar for truly arrested and upstream translocated, but elongation-competent, stalled complexes.

In summary, pol II elongation complexes stalled early in elongation (i.e., before the addition of 40–45 bases to the nascent RNA) have a different structure from pol II complexes stalled at more downstream locations. Complexes in the promoter-proximal region adopt a stable template location far upstream of the expected position for transcriptionally competent complexes. Since upstream translocation is almost certainly a requirement for arrest to occur, complexes that have already taken this step may be more sensitive to regulatory factors which themselves are not sufficient to force normally elongating polymerases into arrest. Nucleosomes are clearly among the factors that might interfere with pol II in the promoter-proximal region and modulate transcript elongation (Izban and Luse 1992a; Brown et al. 1996; Chang and Luse 1997). Further studies on transcription complexes at the initiation-elongation transition will be needed to explore the physiological relevance of this transition to transcriptional regulation.

ACKNOWLEDGMENTS

We thank Drs. David Setzer, Pieter deHaseth, Richard Gronostajski, Donna Driscoll, and Richard Padgett for advice and encouragement during the course of these studies. This research was supported by grant GM-29487 from the National Institutes of Health.

REFERENCES

Borukhov S., Sagitov V., and Goldfarb A. 1993. Transcript cleavage factors from *E. coli*. *Cell* **72:** 459.

Brown S.A., Imbalzano A.N., and Kingston R.E. 1996. Activator-dependent regulation of transcriptional pausing on nucleosomal templates. *Genes Dev.* **10:** 1479.

Chang C.H. and Luse D.S. 1997. The H3/H4 tetramer blocks transcript elongation by RNA polymerase II in vitro. *J. Biol. Chem.* **272:** 23427.

Gu W.G., Wind M., and Reines D. 1996. Increased accommodation of nascent RNA in a product site on RNA polymerase II during arrest. *Proc. Natl. Acad. Sci.* **93:** 6935.

Gu W., Powell W., Mote J., Jr., and Reines D. 1993. Nascent RNA cleavage by arrested RNA polymerase II does not require upstream translocation of the elongation complex on DNA. *J. Biol. Chem.* **268:** 25604.

Guajardo R. and Sousa R. 1997. A model for the mechanism of polymerase translocation. *J. Mol. Biol.* **265:** 8.

Izban M.G. and Luse D.S. 1992a. Factor-stimulated RNA polymerase II transcribes at physiological elongation rates on naked DNA but very poorly on chromatin templates. *J. Biol. Chem.* **267:** 13647.

———. 1992b. The RNA polymerase II ternary complex cleaves the nascent transcript in a 3′→5′ direction in the presence of elongation factor SII. *Genes Dev.* **6:** 1342.

———. 1993a. SII-facilitated transcript cleavage in RNA polymerase II complexes stalled early after initiation occurs in primarily dinucleotide increments. *J. Biol. Chem.* **268:** 12864.

———. 1993b. The increment of SII-facilitated transcript cleavage varies dramatically between elongation competent and incompetent RNA polymerase II ternary complexes. *J. Biol. Chem.* **268:** 12874.

Jacob G.A., Luse S.W., and Luse D.S. 1991. Abortive initiation is increased only for the weakest members of a set of down mutants of the adenovirus 2 major late promoter. *J. Biol. Chem.* **266:** 22537.

Komissarova N. and Kashlev M. 1997a. RNA polymerase switches between inactivated and activated states by translocating back and forth along the DNA and the RNA. *J. Biol. Chem.* **272:** 15329.

———. 1997b. Transcriptional arrest: *Escherichia coli* RNA polymerase translocates backward, leaving the 3′ end of the RNA intact and extruded. *Proc. Natl. Acad. Sci.* **94:** 1755.

Krumm A., Meulia T., Brunvand M., and Groudine M. 1992. The block to transcriptional elongation within the human c-*myc* gene is determined in the promoter-proximal region. *Genes Dev.* **6:** 2201.

Krummel B. and Chamberlin M.J. 1992. Structural analysis of ternary complexes of *Escherichia coli* RNA polymerase-deoxyribonuclease I footprinting of defined complexes. *J. Mol. Biol.* **225:** 239.

Landick R. 1997. RNA polymerase slides home: Pause and termination site recognition. *Cell* **88:** 741.

Linn S.C. and Luse D.S. 1991. RNA polymerase II elongation complexes paused after the synthesis of 15- or 35-base transcripts have different structures. *Mol. Cell. Biol.* **11:** 1508.

Lis J. and Wu C. 1993. Protein traffic on the heat shock promoter: Parking, stalling, and trucking along. *Cell* **74:** 1.

Luse D.S. and Jacob G.A. 1987. Abortive initiation by RNA polymerase II in vitro at the adenovirus 2 major late promoter. *J. Biol. Chem.* **262:** 14990.

Nudler E., Goldfarb A., and Kashlev M. 1994. Discontinuous mechanism of transcription elongation. *Science* **265:** 793.

Nudler E., Mustaev A., Lukhtanov E., and Goldfarb A. 1997. The RNA-DNA hybrid maintains the register of transcription by preventing backtracking of RNA polymerase. *Cell* **89:** 33.

Orlova M., Newlands J., Das A., Goldfarb A., and Borukhov S. 1995. Intrinsic transcript cleavage activity of RNA polymerase. *Proc. Natl. Acad. Sci.* **92:** 4596.

Rasmussen E.B. and Lis J.T. 1993. In vivo transcriptional pausing and cap formation on three *Drosophila* heat shock genes.

Proc. Natl. Acad. Sci. **90:** 7923.

———. 1995. Short transcripts of the ternary complex provide insight into RNA polymerase II elongational pausing. *J. Mol. Biol.* **252:** 522.

Reeder T.C. and Hawley D.K. 1996. Promoter proximal sequences modulate RNA polymerase II elongation by a novel mechanism. *Cell* **87:** 767.

Reines D., Ghanouni P., Li Q.Q., and Mote J. 1992. The RNA polymerase II elongation complex. Factor-dependent transcription elongation involves nascent RNA cleavage. *J. Biol. Chem.* **267:** 15516.

Rougvie A.E. and Lis J.T. 1988. The RNA polymerase II molecule at the 5′ end of the uninduced *hsp70* gene of *D. melanogaster* is transcriptionally engaged. *Cell* **54:** 795.

———. 1990. Postinitiation transcriptional control in *Drosophila melanogaster. Mol. Cell. Biol.* **10:** 6041.

Rudd M.D., Izban M.G., and Luse D.S. 1994. The active site of RNA polymerase II participates in transcript cleavage within arrested ternary complexes. *Proc. Natl. Acad. Sci.* **91:** 8057.

Samkurashvili I. and Luse D.S. 1996. Translocation and transcriptional arrest during transcript elongation by RNA polymerase II. *J. Biol. Chem.* **271:** 23495.

———. 1998. Structural changes in the RNA polymerase II transcription complex during transition from initiation to elongation. *Mol. Cell. Biol.* **18:** 5343.

Straney D.C. and Crothers D.M. 1985. Intermediates in transcription initiation from the *E. coli* lac UV5 promoter. *Cell* **43:** 449.

Uptain S.M., Kane C.M., and Chamberlin M.J. 1997. Basic mechanisms of transcript elongation and its regulation. *Annu. Rev. Biochem.* **66:** 117.

Wang D., Meier T.I., Chan C.L., Feng G., Lee D.N., and Landick R. 1995. Discontinuous movements of DNA and RNA in RNA polymerase accompany formation of a paused transcription complex. *Cell* **81:** 341.

Role of RNA Polymerase II Carboxy-terminal Domain in Coordinating Transcription with RNA Processing

S. McCracken,* E. Rosonina,* N. Fong,† M. Sikes,‡ A. Beyer,‡ K. O'Hare,§
S. Shuman,¶ and D. Bentley†

*Banting and Best Department of Medical Research and Department of Molecular and Medical Genetics,
University of Toronto, Ontario, Canada, M5G 1L6; †Department of Biochemistry and Molecular Genetics,
UCHSC B121, Denver, Colorado 80262; ‡Department of Microbiology, University of Virginia, Charlottesville,
Virginia 22901; §Department of Biochemistry, Imperial College, London SW7 2AZ, United Kingdom;
¶Memorial Sloan-Kettering Cancer Center, New York, New York 10021

The largest subunit of RNA polymerase II (pol II) has an unusual structure at its carboxyl terminus comprising a repeated heptad motif. This carboxy-terminal domain (CTD) has up to 52 copies of the heptad whose consensus sequence (YSPTSPS) is completely conserved between fungi and vertebrates (Allison et al. 1985; Corden et al. 1985). Although the largest subunits of pol I, pol II, and pol III are members of the same family which includes the B′ and A′ subunits of eubacterial and archaeal RNA polymerases, the CTD is unique to pol II. The function of the CTD has been the object of intense speculation (Corden and Ingles 1992). It is required for cell viability and for transcriptional activation of some genes in budding yeast and mammalian cells (Nonet et al. 1987; Bartolomei et al. 1988; Scafe et al. 1990; Gerber et al. 1995), but paradoxically, it is dispensable for transcription from the adenovirus major late promoter in vitro. Recent observations have unexpectedly implicated the CTD not only in synthesis of the primary transcript, but also in the RNA processing events that are uniquely directed to pol II transcripts.

METHODS

Plasmids. pGal5 HIV-2 CAT, pGal5 HIV-2 CATΔ, pSVβ128, pSPVA, and expression vectors for Gal4-fusion proteins were described previously (Blau et al. 1996; McCracken et al. 1997a,b). HA-WT and HA-Δ5 expression vectors for α-amanitin-resistant pol II were a gift from J. Corden (Gerber et al. 1995). pcDNA3T7 and pcDNA3T7CTD expression vectors for T7 RNA polymerase (amino acids 2–883) were constructed with amino-terminal SV40 T-antigen nuclear localization signals in the vector pcDNA3 (Invitrogen). An alanine to glycine substitution at position 7 of T7 RNA polymerase was introduced accidentally in these constructs. pcDNA3T7CTD contains the complete murine pol II CTD fused to amino acid 2. The pW and pG reporters for T7 and pol II transcription are described in Figure 5.

Transfections. 293 cells were transiently transfected with 5 μg of reporter plasmid, 0.5 or 1 μg of activator expression plasmid, 0.5 μg of pSPVA, and 0.5 μg of pol II expression vector by calcium phosphate precipitation. α-amanitin (2.5 μg/ml) was added 12–15 hours after trans-

fection, and cells were collected at 60–65 hours post-transfection. RNA was prepared by guanidine isothiocyanate/acid phenol extraction and treated with DNase I. Nuclear and cytoplasmic RNAs were prepared after lysis in 10 mM HEPES at pH 7.6, 10 mM NaCl, 3 mM CaCl₂, and 0.5% Nonidet P40 (NP-40).

Affinity chromatography. Expression of GST-CTD fusion proteins and their phosphorylation in HeLa nuclear extract was described previously (McCracken et al. 1997b). Binding reactions were carried out in 20 mM HEPES at pH 7.9, 100 mM NaCl, 0.1 mM EDTA, 1 mM dithiothreitol (DTT), 20% glycerol, and 0.1% NP-40. In some cases, 0.5% nonfat milk was included in the binding reactions. Beads were washed in binding buffer and eluted in the same buffer with 0.6 M or 1 M NaCl.

Immunoprecipitation. Precleared nuclear extract (4.5 mg) was mixed for 2 hours at 4°C with 20 μl of protein A–Sepharose beads containing 5 μg of affinity-purified rabbit antibody. Beads were blocked in 1% nonfat milk prior to use. The beads were washed four times with 100 μl of 20 mM HEPES at pH 7.9, 100 mM NaCl, 0.1 mM EDTA, 1 mM DTT, 0.1% NP-40, and 20% glycerol, resuspended in Laemmli sample buffer, and analyzed by Western blot.

Recombinant yeast capping enzymes. Recombinant yeast guanylyltransferase (Ceg1) and methyltransferase (Abd1) and corresponding polyclonal antibodies were described previously (Schwer and Shuman 1994, 1996; Wang and Shuman 1997).

RESULTS AND DISCUSSION

The CTD and Activation of Transcriptional Elongation

Transcription by pol II is controlled by regulatory mechanisms whose complexity is unequaled among eukaryotic RNA polymerases. In addition to control of RNA chain initiation, pol II is also subject to control of postinitiation events: promoter clearance and subsequent chain elongation (Krumm et al. 1993; Bentley 1995). Different sequence-specific activator proteins preferentially stimulate either initiation, postinitiation, or both initiation and postinitiation steps in the transcription cycle. We and others have investigated the role of CTD in the response to different functional classes of activators. The herpes simplex virus VP16 protein and the cellular activators E2F and p53 belong to a class of activators that stimulates both initiation and elongation. The SV40 enhancer also stimulates both of these steps in transcription. Nuclear run-on analysis of transcription activated by the SV40 enhancer or VP16 demonstrates a high polymerase density throughout the length of the gene (Yankulov et al. 1994; Krumm et al. 1995; Blau et al. 1996). In contrast, a second class of activators gives rise to transcription complexes that elongate poorly as indicated by a drop in polymerase density between the 5′ and 3′ ends of the gene. These activators, which include Sp1 and CTF, stimulate initiation but not elongation (Blau et al. 1996). Mutation of four phenylalanine residues to alanine (442, 473, 475, 479) within the activation domain (residues 411–490) of VP16 (mutant SW6; Walker et al. 1993) converts it to the second class of activator that stimulates predominantly initiation (Blau et al. 1996). A third class of activator is represented by the human immunodeficiency virus (HIV) Tat proteins that enhance elongation but have little effect on initiation (Kao et al. 1987; Marciniak and Sharp 1991).

The effect of the CTD on the function of different classes of activators has been investigated using the strategy of Gerber et al. (1995). Cells were transiently transfected with expression vectors for α-amanitin-resistant mutants of the mouse pol II large subunit bearing either a full-length CTD (WT, 52 heptad repeats) or a truncated CTD with only five heptad repeats (marked ΔCTD in Fig. 1). After synthesis of the resistant pol II subunit for a brief period, α-amanitin was added to inhibit the host cell's pol II. Subsequent transcription in the presence of the toxin is carried out by pol II molecules that have incorporated the resistant large subunit. This technique was used to show that activation by the SV40 enhancer, Gal4-VP16, and HIV Tat was severely inhibited by truncation of the CTD to five heptad repeats, whereas activation by Sp1 was almost unaffected (Gerber et al. 1995; Okamoto et al. 1996).

In the experiment shown in Figure 1, we investigated the effect of CTD truncation on activation by wild-type and mutant VP16. The Gal5 HIV-2 CAT reporter gene with the HIV-2 basal promoter and TAR region and five upstream Gal4-binding sites was transiently transfected into 293 cells with expression vectors for the wild-type Gal4-VP16 or Gal4-SW6 mutant and the wild-type or ΔCTD α-amanitin-resistant large subunit. Total RNA from α amanitin-treated cells was analyzed by RNase protection which distinguished correctly initiated RNAs that read through the TAR sequence (RT) from transcripts which read around from upstream of the start site (RA). CTD truncation significantly inhibited activation by wild-type Gal4-VP16 (Fig. 1, compare lanes 3 and 6) but had little effect on transcription activated by the Gal4-SW6 mutant (Fig. 1, compare lanes 1 and 4). HIV Tat synergizes with Gal4-SW6 to enhance production of full-length transcripts in the presence of wild-type pol II (Fig.

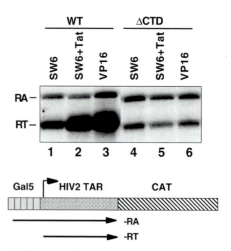

Figure 1. CTD truncation specifically inhibits *trans*-activation by HIV Tat and wild-type Gal4-VP16 but not the Gal4-SW6 mutant. RNase protection analysis of transcripts from α-amanitin-treated 293 cells transiently transfected with the Gal5-HIV2 CAT reporter gene, expression vectors for Gal4-SW6, Gal4-VP16, and HIV1 Tat and either HA-WT (WT) or HA-Δ5 (ΔCTD) expression vectors for α-amanitin-resistant pol II large subunit (Gerber et al. 1995). Note that transcription with ΔCTD pol II preferentially inhibited activation by Tat and Gal4-VP16 but not Gal4-SW6 and that it caused enrichment for readaround (RA) relative to readthrough (RT) transcripts.

1, compare lanes 1 and 2) (Blau et al. 1996). This effect of Tat was abolished when the CTD was truncated (Fig. 1, lanes 2 and 5). The abrogation of Tat function by CTD truncation in other assay systems has been reported previously (Chun and Jeang 1996; Okamoto et al. 1996; Yang et al. 1996). Together, these results demonstrate a striking correlation between those activators that stimulate transcriptional elongation and those that are inhibited by truncation of the CTD. The CTD therefore appears to be required for the response to activators that stimulate transcriptional elongation but not for the response to activators that stimulate only initiation.

Several lines of evidence suggest that modification of the CTD by phosphorylation is required for stimulation of transcriptional elongation by activators. The CTD undergoes a cycle of hyperphosphorylation and dephosphorylation which is linked to the cycle of initiation, elongation, and termination of transcription (Dahmus 1996). Elegant in vivo experiments showed that CTD phosphorylation correlates with the transition from pausing to productive elongation which occurs when the *Drosophila Hsp70* gene is induced by heat shock (O'Brien et al. 1994). It is difficult to identify with certainty the kinase(s) responsible for CTD phosphorylation in vivo; however, two strong candidates are TFIIH and P-TEFb. TFIIH is an integral part of the preinitiation complex at the promoter and contains a tripartite cyclin-dependent kinase (cdk7-cyclinH-MAT1) that phosphorylates the CTD in vitro (Roy et al. 1994). Circumstantial evidence exists for a role of TFIIH in regulating transcriptional elongation. In affinity chromatography experiments, TFIIH binds a subset of activation domains, which coincides well with those that stimulate elongation (Xiao et al.

1994; Blau et al. 1996), including HIV Tat. Tat actually stimulates the CTD kinase activity of TFIIH (Parada and Roeder 1996; Garcia-Martinez et al. 1997). Transcriptional elongation was inhibited by conditional inactivation of Kin28, the TFIIH kinase subunit in budding yeast (Akhtar et al. 1996; McNeil et al. 1998) and by microinjection of anti-TFIIH antibodies into *Xenopus* oocytes (Yankulov et al. 1996). A second cdk cyclin pair, cdk9-cyclin T, has been directly implicated in stimulation of pol II elongation. This kinase forms part of a larger complex called P-TEFb that was originally purified as a general elongation factor (Marshall et al. 1996). P-TEFb directly contacts Tat through cyclin T and acts as a cofactor for stimulation of elongation by Tat (Mancebo et al. 1997; Zhu et al. 1997; Wei et al. 1998). Phosphorylation of the CTD by P-TEFb almost certainly contributes to its elongation function, but it remains possible that other substrates are also involved. Whether CTD phosphorylation by P-TEFb and TFIIH has overlapping or distinct and perhaps complementary functions remains to be determined.

How CTD phosphorylation enhances transcriptional elongation remains an intriguing puzzle. Phosphorylation could act directly by affecting the elongation properties of the polymerase or it could act indirectly by affecting recruitment of other factors that positively or negatively affect elongation. We have investigated the effect of CTD deletion on transcriptional elongation (Fig. 2). The Gal5 HIV-2 CAT reporter gene was transiently transfected into 293 cells with expression vectors for activator proteins and α-amanitin-resistant pol II large subunits. Transcripts were analyzed by RNase protection to detect readaround (RA) and readthrough (RT) transcripts as well as correctly initiated transcripts that terminated prematurely

(TM). Prematurely terminated transcripts with 3′ ends in the TAR sequence are thought to be stabilized by virtue of their secondary structure. To reveal differences in the fraction of prematurely terminated transcripts made with different activators, the amounts of RNA analyzed were normalized to give approximately equal signals. With wild-type pol II, the fraction of readthrough versus terminated transcripts varied depending on the activation domain as expected. Prematurely terminated RNAs accumulated in the presence of the Gal4 DNA-binding domain (1–147), Gal4-Sp1, and Gal4-SW6, indicating poor elongation efficiency or processivity (Fig. 2, lanes 1–3). In contrast, activation by Gal4-VP16 or Gal4-SW6+Tat results in relatively few prematurely terminated RNAs, indicating efficient elongation (Fig. 2, lanes 4 and 5). CTD truncation (ΔCTD) specifically reduced accumulation of prematurely terminated transcripts in the presence of Gal4(1–147), Gal4-Sp1, and Gal4-SW6 (Fig. 2, compare lanes 1–3 with 6–8). This result suggests that the CTD exerts a negative effect on elongation which is relieved by truncation of this domain or by the action of activators like Tat and VP16. We cannot eliminate the possibility, however, that the short transcripts made by CTD truncated polymerase are specifically destabilized relative to those made by pol II with a full-length pol II. CTD phosphorylation stimulated by Tat and possibly also VP16 may therefore antagonize some negative effector of elongation. Candidate negative regulators of elongation include NTEF (Marshall and Price 1992) and DRB sensitivity-inducing factor (DSIF), the human homolog of budding yeast Spt4 and 5 (Wada et al. 1998).

The CTD and mRNA Processing

Transcription by ΔCTD pol II consistently resulted in a relative increase in abundance of the RNAs which transcribed through the promoter region (marked RA in Figs. 1 and 2). Transcripts that extend through the promoter could be synthesized by polymerases which read around the entire plasmid without terminating downstream from the CAT gene. Termination of pol II transcription requires a wild-type poly(A) site, although the basis for poly(A)-dependent termination is unknown (Whitelaw and Proudfoot 1986; Logan et al. 1987; Connelly and Manley 1988). We therefore tested whether transcripts made by the CTD-deleted polymerase were properly cleaved at the poly(A) site. RNase protection mapping of cleavage at the SV40 late poly(A) site showed that transcripts made by ΔCTD pol II accumulated as uncleaved precursors (Fig. 3, lane 3). Analysis of nuclear (N) and cytoplasmic (C) RNA fractions revealed that uncleaved precursors made by ΔCTD pol II were exclusively nuclear as were precursors made by wild-type pol II (Fig. 3). Failure to process ΔCTD transcripts is therefore not because they escape retention in the nucleus.

The CTD requirement for efficient cleavage of the primary transcript at the poly(A) site implicates this domain in coupling polyadenylation with pol II transcription. In addition, these observations show that the CTD is required for termination by pol II downstream from a wild-

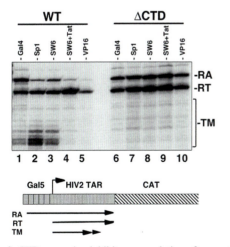

Figure 2. CTD truncation inhibits accumulation of prematurely terminated transcripts. RNase protection of Gal5 HIV-2 CAT transcripts from α-amanitin-treated 293 cells as in Fig. 1. The amounts of RNA analyzed were normalized to give approximately equal signals. The amounts of total RNA analyzed in lanes *1–10* are 18.0, 1.9, 4.3, 1.1, 0.2, 9.2, 9.1, 23.0, 5.4, and 1.7 µg, respectively. Note that transcription by ΔCTD pol II gives fewer prematurely terminated RNAs (TM) than wild-type pol II (compare lanes *1–3* with *6–8*).

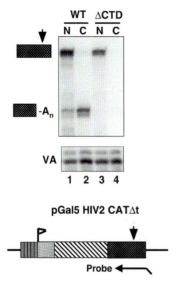

Figure 3. Transcription by CTD-truncated pol II inhibits cleavage at the SV40 late poly(A) site. RNase protection of equal amounts of nuclear (N) and cytoplasmic (C) RNA from α-amanitin-treated 293 cells transiently transfected with the Gal5HIV-2 CATΔt reporter (McCracken et al. 1997b), Gal4-VP16 activator, and HA-WT (WT) or HA-Δ5 (ΔCTD) expression vectors. Note that CTD truncation (lanes *3* and *4*) results in accumulation of uncleaved precursors in the nucleus and little or no detectable mature 3′ ends, whereas wild-type pol II transcripts are cleaved and exported to the cytoplasm. Note that sixfold more cell equivalents of nuclear RNA were analyzed relative to cytoplasmic RNA.

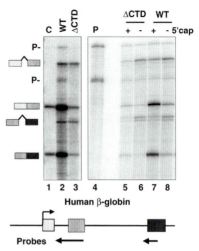

Figure 4. Inhibition of splicing and 5′ capping by CTD truncation. RNase protection of human β-globin transcripts from the reporter plasmid, pSVβ128 (McCracken et al. 1997b) in transiently transfected α-amanitin-treated 293 cells. Two antisense probes (P) for introns 1 and 2 were used. Transcription by ΔCTD pol II (lane *3*) inhibited splicing of both introns relative to full-length α-amanitin-resistant pol II (WT, lane *2*). Note that the background transcripts in control cells (C) treated with α-amanitin are efficiently spliced (lane *1*). CTD truncation reduced the fraction of capped globin transcripts from 74% (lanes *7,8*) to 39% (lanes *5,6*). Note that splicing of capped transcripts made by ΔCTD pol II (32% of RNAs with unspliced intron 1, lane *5*) is less efficient than that of capped transcripts made by wild-type pol II (8% with unspliced intron 1, lane *7*).

type poly(A) site. The termination defect may be simply a secondary effect of failure to cleave the RNA at the poly(A) site. Alternatively, the CTD may be required downstream from the poly(A) site for polymerase to terminate. The latter possibility is suggested by the apparent "runaway" elongation properties of ΔCTD pol II (see Fig. 2, lanes 6–10).

Capping, splicing, and cleavage/polyadenylation do not proceed independently of one another. The 5′ cap stimulates splicing of the first intron (Izaurralde et al. 1994) and 3′ processing (Cooke and Alwine 1996); 3′ processing in turn stimulates splicing of the last intron (Niwa and Berget 1991) and conversely intron sequences can stimulate 3′ processing (Niwa et al. 1990; Chiou et al. 1991). We therefore investigated whether the polyadenylation defect is associated with any defects in splicing or capping when the CTD is truncated. Splicing of human β-globin RNA transcribed by either wild-type or ΔCTD pol II was analyzed by RNase protection using antisense probes that span the 3′ splice sites of both introns (see diagram in Fig. 4). This experiment showed that splicing of introns 1 and 2 was indeed significantly inhibited when the gene was transcribed by ΔCTD relative to wild-type pol II (Fig. 4, compare lanes 2 and 3). In control cells (C) treated with α-amanitin, a background of spliced β-globin transcripts was detected showing that α-amanitin per se does not inhibit splicing. In related studies, Du and Warren (1997) and Yuryev et al. (1996) found that excess CTD in *trans* inhibits splicing in vivo and in vitro.

To measure the extent of mRNA capping, total RNA from transfected cells expressing wild-type and ΔCTD pol II was separated into capped and uncapped fractions using beads coated with the cap-binding protein GST-eIF4E (Edery et al. 1995; McCracken et al. 1997b). RNase protection analysis of the capped and uncapped fractions showed that CTD truncation reduced the fraction of β-globin transcripts with a 5′ cap from 74% to 39% (Fig. 4, compare lanes 5, 6 with 7, 8). In conclusion, CTD truncation inhibits all three of the major mRNA processing steps: capping, splicing, and cleavage/polyadenylation.

The CTD Fused to T7 RNA Polymerase Does Not Support RNA Processing

Since the CTD is necessary for efficient processing of pol II transcripts, we asked whether it might also be sufficient to permit processing of transcripts made by a foreign RNA polymerase. For these experiments, we expressed a chimera of the bacteriophage T7 RNA polymerase with full-length mouse pol II CTD. T7 RNA polymerase (amino acids 2–883) and a chimera containing the CTD fused to the amino terminus of the phage polymerase were expressed with an amino-terminal SV40 large T nuclear localization signal. As templates, we used plasmids with upstream LexA- and Gal4-binding sites, T7 promoter, CAT gene with SV40 t intron, and wild-type (pW) or mutant (pM) SV40 early poly(A) sites followed by a T7 ter-

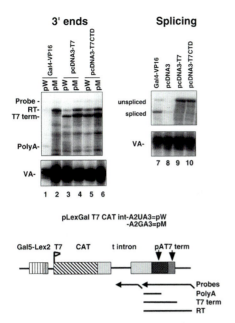

Figure 5. CTD fusion to T7 RNA polymerase is not sufficient to confer 3′ processing or splicing. RNase protection of transcripts from 293 cells transfected with either the pW or pM reporter gene. These reporters have the CAT gene with T7 promoter, SV40 t intron, wild-type or mutant SV40 early poly(A) site, and T7 terminator (see diagram). Gal4-VP16 was used to activate pol II transcription in lanes *1, 2*, and *7*. pcDNA3 vector (Invitrogen) or expression vectors for T7 or T7-CTD RNA polymerase were cotransfected to activate transcription from the T7 promoter. The T7-CTD construct fused the full-length mouse CTD to the amino terminus of T7 RNA polymerase. pol II transcripts of pW but not pM were cleaved at the poly(A) site (lanes *1,2*; note low transfection efficiency in lane *1*). There was no cleavage of T7 or T7-CTD transcripts at the poly(A) site of the pW reporter (compare lanes *3,5* with *4,6*). Splicing of the SV40 t intron was analyzed for transcripts of the pW reporter (lanes *7–10*). Neither T7 nor T7-CTD transcripts (lanes *9,10*) were efficiently spliced relative to pol II transcripts (lanes *7,8*).

minator element. The reporter genes could be transcribed either by T7 RNA polymerase or by pol II which initiates at a cryptic promoter 6–11 bases upstream of the T7 start site (Sandig et al. 1993). Both forms of T7 RNA polymerase produced abundant transcripts, most of which terminated at the T7 terminator (Fig. 5, lanes 3–6). The T7 transcripts were not cleaved at the poly(A) site or spliced regardless of the presence of the CTD (Fig. 5, lanes 3–6 and 9, 10). In contrast, Gal4-VP16-activated pol II transcripts from the same template were efficiently spliced and polyadenylated (Fig. 5, lanes, 1, 7, and 8). In the context of a fusion with T7 RNA polymerase, the CTD is therefore not sufficient to permit either splicing or 3′ processing. Other components of the pol II ternary complex in addition to the CTD might be necessary to recruit processing factors to nascent transcripts. Alternatively, there are many trivial explanations for this negative result. The CTD fused to T7 may not be accessible or it may not be phosphorylated appropriately to permit recruitment of processing factors. It is also possible that the rapid elongation rate of T7 RNA polymerase relative to pol II is not compatible with cotranscriptional processing.

Direct versus Indirect Effects of the CTD on mRNA Processing

Because capping, splicing, and 3′ processing are interdependent, it is not obvious which effects of CTD truncation are direct and which are indirect. Capping occurs when the nascent RNA is only about 25 bases long (Coppola et al. 1983; Rasmussen and Lis 1993), well before other processing steps have taken place, and is presumably unaffected by subsequent processing steps. The effect of CTD truncation on capping is therefore likely to be direct. CTD truncation inhibited cleavage at the poly(A) site in reporter genes with and without introns (see Fig. 3), eliminating the possibility that reduced splicing causes inhibition of 3′ processing. On the other hand, our results do not exclude the possibility that reduced splicing of 3′ introns is an indirect effect of inhibiting cleavage at the poly(A) site. In theory, inhibition of polyadenylation and splicing when the CTD is truncated could be a consequence of reduced capping. In fact, among transcripts made by wild-type pol II, the uncapped fraction is indeed enriched for unspliced and nonpolyadenylated precursors (McCracken et al. 1997b). In the experiment shown in Figure 4, 39% of β-globin intron 1 transcripts in the uncapped fraction were unspliced versus only 8% in the capped fraction (Fig. 4, lanes 7 and 8). However, among those transcripts made by ΔCTD pol II that have a 5′ cap, splicing was still reduced relative to capped transcripts made by wild-type pol II; 32% of capped β-globin transcripts made by ΔCTD pol II retained unspliced intron 1 versus only 8% for wild-type pol II (Fig. 4, compare lanes 5 and 7). The effect of CTD truncation on splicing therefore cannot be fully explained as a secondary effect of reduced capping. Capped transcripts made by ΔCTD pol II were also polyadenylated less efficiently than capped RNAs made by wild-type pol II (see Fig. 3 in McCracken et al. 1997b). In conclusion, abrogation of capping and polyadenylation, which occur when the primary transcript is made by CTD-truncated pol II, cannot easily be explained as secondary consequences of defects in other RNA processing steps. To what extent splicing is affected independently of the reduction in cleavage/polyadenylation remains to be resolved.

Interactions between Capping Enzymes and the CTD

We have looked for protein-protein contacts between the CTD and processing factors that might explain the CTD dependence of RNA processing in vivo. The reduced fraction of transcripts with a 5′ cap when a gene is transcribed by ΔCTD pol II could be due to enhanced decapping or, more likely, to reduced capping of primary transcripts. We therefore investigated the possibility that capping is enhanced by recruitment of the relevant enzymes to the CTD. In affinity chromatography experiments, the bifunctional mammalian triphosphatase-guanylyltransferase capping enzyme bound specifically to the CTD. Binding of this capping enzyme was absolutely specific for the phosphorylated form of the CTD. Even very low levels of CTD phosphorylation at Ser-5 by recombinant cdk7-cyclin H were sufficient to permit binding (McCracken et al. 1997b). The carboxy-terminal guanylyltransferase domain of the mouse capping enzyme is sufficient for CTD binding, and the amino-terminal triphosphatase domain does not bind (Ho et al. 1998). The recombinant budding yeast guanylyltransferase, Ceg1, also bound directly to the phosphorylated CTD. Ceg1 is tightly associated with the triphosphatase, Cet1 (Itoh et al. 1987), which is encoded by a separate gene. By virtue of this interaction, Ceg1 is probably responsible for recruiting Cet1 to the CTD. Following the triphosphatase and guanylyltransferase reactions, the cap structure is completed by the 7-methyltransferase which acts in concert with guanylyltransferase to drive the overall equilibrium in the forward direction (Shuman 1995). In yeast, the 7-methyltransferase, Abd1, does not physically associate with the other capping enzymes (Mao et al. 1995), but like Ceg1, it binds directly to the phosphorylated CTD. The interactions of recombinant Abd1 and Ceg1 with the phospho-CTD were compared by binding to resins with different concentrations of immobilized GST-phospho-CTD ligand. The results in Figure 6 show that Abd1 binds at lower ligand concentration, indicating that it has a higher affinity for the phosphorylated CTD than Ceg1. The direct binding of both Ceg1 and Abd1 to the phosphorylated CTD suggests a way in which these two enzymes could work together in a concerted reaction without actually forming a stable complex with each other. The recruitment of all three capping enzymes to the phosphorylated pol II CTD shortly after initiation may enhance the efficiency of the capping reaction and provides an explanation for why this process is dependent on the CTD.

Interactions between Polyadenylation Factors and the CTD

The polyadenylation factors, cleavage stimulation factor (CstF), and cleavage polyadenylation specificity factor (CPSF), in crude nuclear extracts both bound specifi-

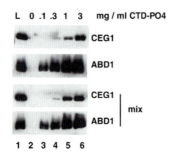

Figure 6. Binding of recombinant *Saccharomyces cerevisiae* capping enzymes to phosphorylated CTD. Guanylyltransferase, Ceg1, and 7-methyltransferase, Abd1, at 4.4 μg/ml and bound individually or in combination to glutathione Sepharose beads containing 0, 0.1, 0.3, 1.0, and 3.0 mg/ml phosphorylated GST-CTD. The GST fusion to full-length mouse CTD was phosphorylated in crude HeLa nuclear extract to about three phosphates/mole. The beads were eluted with 1 M NaCl; 0.5% of the load (L) and 2.0% of the eluates were analyzed by Western blotting with the respective antisera. Note that Abd1 binds at a lower concentration of immobilized GST-CTD ligand, consistent with a higher binding affinity than Ceg1. There was no evidence for cooperative binding when the two proteins were mixed.

cally to a GST-CTD affinity column (Fig. 7A, lanes 1–3). When the high-salt eluate of the GST-CTD column was rechromatographed on a second GST-CTD column, the CstF and CPSF were bound a second time (Fig. 7A, lanes 4–6). Poly(A) polymerase, on the other hand, did not bind to the CTD (McCracken et al. 1997a). In contrast to the capping enzymes, modest phosphorylation of the CTD did not greatly affect binding of these two polyadenylation factors (McCracken et al. 1997a). To try to identify which polypeptides are responsible for contacting the CTD, in-vitro-translated subunits of CstF were tested for binding to GST-CTD. The in-vitro-translated 50-kD subunit of CstF bound specifically to the full-length CTD (1–52) but not to the amino-terminal 15 heptad repeats (1–15; Fig. 7B, lanes 2 and 3), even though the latter was present at a higher concentration on the column (Fig. 7B, lanes 4 and 5). CPSF coimmunoprecipitates with CstF from HeLa nuclear extract (Fig. 7C, lane 3); therefore, it is possible that CPSF is recruited to the CTD indirectly by virtue of its association with CstF.

CPSF and CstF copurify with a high-molecular-weight pol II holoenzyme prepared by affinity chromatography on a GST-TFIIS column (Pan et al. 1997). To verify the significance of the association between CstF and pol II in vivo, we investigated their localization in *Drosophila* salivary gland polytene chromosomes by immunofluorescence. For these experiments, we used a monoclonal antibody (H14) specific for the CTD phosphorylated at Ser-5 of the heptad repeat (Bregman et al. 1995; Patturajan et al. 1998) and a rabbit polyclonal antibody directed against the carboxy-terminal 17 residues of human CstF p77, which is homologous to the *Drosophila* suppressor of forked su(f) gene product (predicted molecular mass of 84 kD). Within a 15-amino-acid segment of the antigenic peptide, the *Drosophila* and human homologs are identical at 12 positions. In Western blots of *Drosophila* extract, the anti-CstF p77 antibody cross-reacts with a strong band that approximately comigrates with HeLa CstF p77 (Fig. 8A, lanes 1 and 2). We tested whether the cross-reacting *Drosophila* protein was the su(f) gene product by Western blotting of extracts from two strains (S2/NN and 3DES/NN) that are heterozygous for complementing lethal alleles of su(f) that alter the protein's carboxyl terminus (Simonelig et al. 1996). The cross-reacting protein band present in the wild-type was absent from both mutants (Fig. 8A, compare lane 3 with lanes 4 and 5) proving that it, in fact, corresponds to the (su)f gene product. Coimmunostaining of polytene chromosomes with the anti-phospho-CTD and anti-CstF p77 antibodies showed a small subset of bands that stain strongly for both proteins (Fig. 8B). Moreover, the distribution of immunofluorescence for both antibodies appears to be similar at each stained band. These results are therefore consistent with the idea that protein-protein contacts occur between transcriptionally active phosphorylated pol II and CstF in vivo. We hypothesize that interactions between CstF and the CTD contribute to coupling 3′ processing with pol II transcription in vivo. Recently, the CTD was also shown to be important for cleavage at the poly(A) site in vitro in the absence of transcription (Hirose and Manley 1998). It remains to be tested whether mutants that disrupt

Figure 7. Polyadenylation factors CPSF and CstF bind to the CTD. (*A*) HeLa nuclear extract (0.55 mg) was chromatographed on microcolumns (20 μl) of GST or the GST-CTD1–52 fusion with full-length mouse CTD. The HeLa load (6.2%, lane *1*) and high-salt eluates (31%, lanes *2,3*) were analyzed by Western blot with antibodies against CstF p77 and CPSF p160. The GST-CTD1–52 eluate was rechromatographed on GST or GST-CTD1–52 columns; 7.5% of the load (lane *4*) and 37.5% of the eluates (lanes *5, 6*) were analyzed. CstF and CPSF in this partially purified fraction re-bound to the CTD (lane *6*). (*B*) In-vitro-translated [³⁵S]methionine-labeled CstF p50 binds specifically to full-length GST-CTD1–52 but not to a fusion with the amino-terminal 15 heptad repeats (lanes *2,3*); 10% of the load (lane *1*) and 50% of the high-salt eluates (lanes *2,3*) were analyzed. Coomassie-stained GST-CTD fusion proteins from equal amounts of beads used for the binding reactions are shown in lanes *4* and *5*. (*C*) Polyadenylation factors CPSF and CstF coimmunoprecipitate. HeLa extract (lane *1*) was immunoprecipitated with rabbit anti-GST control or rabbit anti-CstF p77 (lanes *2,3*). The precipitates were immunoblotted with anti-CstF p77 and anti-CPSF p160 antibodies.

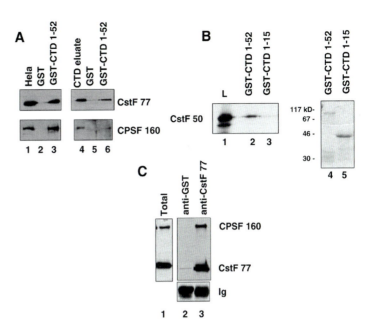

CstF-CTD interaction actually inhibit cleavage/polyadenylation in vitro or in vivo.

The "mRNA Factory"

In summary, the CTD has important roles in two very different aspects of mRNA synthesis: the regulation of transcription in response to activators and repressors and in processing of the primary transcript by capping, splicing, and cleavage/polyadenylation. The molecular basis of these diverse functions appears to lie in the ability of the CTD to serve as a scaffold for assembly of several protein complexes with pol II. It makes direct protein-protein contacts with the mediator complex, with capping and polyadenylation factors, and possibly also with splicing factors (Yuryev et al. 1996). The CTD may be viewed as an interface where multiprotein complexes involved in transcription, RNA processing, and perhaps even other nuclear events can communicate with one another. It is possible that processing factors influence pol II transcription and conversely that transcription factors influence RNA processing through cross-talk mediated by protein contacts with the CTD. We suggest that protein contacts with the CTD are central to assembly of "mRNA factory" complexes that carry out both synthesis and processing of the pre-mRNA.

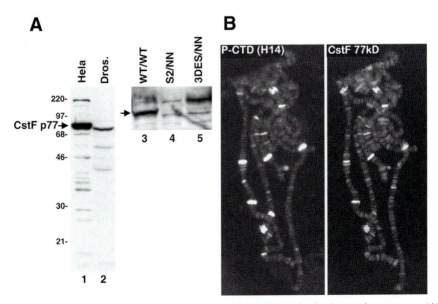

Figure 8. Colocalization of CstF with phosphorylated pol II on *Drosophila* salivary gland polytene chromosomes. (*A*) Cross-reaction of rabbit anti-human CstF p77 antibody with the homologous *Drosophila* su(f) gene product. Western blots of HeLa nuclear extract (lane *1*) and extracts from wild-type (lanes *2,3*) and heterozygous double-mutant su(f) flies (lanes *4,5*). The lack of a comigrating band in the su(f) mutant extracts shows that the antibody recognizes the su(f) gene product. (*B*) Polytene chromosomes were double stained with a mouse monoclonal antibody H14 that recognizes phosphorylated CTD and rabbit anti-CstF p77.

ACKNOWLEDGMENT

This work was supported by a grant from the Medical Research Council of Canada.

REFERENCES

Akhtar A., Faye G., and Bentley D. 1996. Distinct activated and non-activated RNA polymerase II complexes in yeast. *EMBO J.* **15:** 4654.

Allison L.A., Moyle M., Shales M., and Ingles C.J. 1985. Extensive homology among the largest subunits of eukaryotic and prokaryotic RNA polymerases. *Cell* **42:** 599.

Bartolomei M.S., Halden N.F., Cullen C.R., and Corden J.L. 1988. Genetic analysis of the repetitive carboxyl-terminal domain of the largest subunit of mouse RNA polymerase II. *Mol. Cell. Biol.* **8:** 330.

Bentley D. 1995. Regulation of transcriptional elongation by RNA polymerase II. *Curr. Opin. Genet. Dev.* **5:** 210.

Blau J., Xiao H., McCracken S., O'Hare P., Greenblatt J., and Bentley D. 1996. Three functional classes of transcriptional activation domain. *Mol. Cell. Biol.* **16:** 2044.

Bregman D.B., Du L., van der Zee S., and Warren S.L. 1995. Transcription-dependent redistribution of the large subunit of RNA polymerase II to discrete nuclear domains. *J. Cell Biol.* **129:** 287.

Chiou H.C., Dabrowski C., and Alwine J.C. 1991. Simian virus 40 late mRNA leader sequences involved in augmenting mRNA accumulation via multiple mechanisms, including increased polyadenylation efficiency. *J. Virol.* **65:** 6677.

Chun R.F. and Jeang K.T. 1996. Requirements for RNA polymerase II carboxyl-terminal domain for activated transcription of human retroviruses human T-cell lymphotropic virus I and HIV 1. *J. Biol. Chem.* **271:** 27888.

Connelly S. and Manley J.L. 1988. A functional mRNA polyadenylation signal is required for transcription termination by RNA polymerase II. *Genes Dev.* **2:** 440.

Cooke C. and Alwine J.C. 1996. The cap and the 3´ splice site similarly affect polyadenylation efficiency. *Mol. Cell. Biol.* **16:** 2579.

Coppola J.A., Field A.S., and Luse D.S. 1983. Promoter-proximal pausing by RNA polymerase II in vitro: Transcripts shorter than 20 nucleotides are not capped. *Proc. Natl. Acad. Sci.* **80:** 1251.

Corden J.L. and Ingles C.J. 1992. Carboxy-terminal domain of the largest subunit of eukaryotic RNA polymerase II. In *Transcriptional regulation* (ed. S.L. McKnight and K.R. Yamamoto), p. 81. Cold Spring Harbor Laboratory Press, Cold Spring Harbor, New York.

Corden J.L., Cadena D.L., Ahearn J.M., Jr., and Dahmus M.E. 1985. A unique structure at the carboxyl terminus of the largest subunit of eukaryotic RNA polymerase II. *Proc. Natl. Acad. Sci.* **82:** 7934.

Dahmus M.E. 1996. Reversible phosphorylation of the C-terminal domain of RNA polymerase II. *J. Biol. Chem.* **271:** 19009.

Du L. and Warren S.L. 1997. Functional interaction between the carboxy-terminal domain of RNA and pre-mRNA splicing. *J. Cell Biol.* **136:** 5.

Edery I., Chu L.L., Sonenberg N., and Pelletier J. 1995. An efficient strategy to isolate full-length cDNAs based on an mRNA cap retention procedure (CAPture). *Mol. Cell. Biol.* **15:** 3363.

Garcia-Martinez L.F., Mavankal G., Neveu J.M., Lane W.S., Ivanov D., and Gaynor R.B. 1997. Purification of a Tat-associated kinase reveals a TFIIH complex that modulates HIV-1 transcription. *EMBO J.* **16:** 2836.

Gerber H.P., Hagmann M., Seipel K., Georgiev O., West M.A., Litingtung Y., Schaffner W., and Corden J.L. 1995. RNA polymerase II C-terminal domain required for enhancer-driven transcription. *Nature* **374:** 660.

Hirose Y. and Manley J.L. 1998. RNA polymerase II is an essential mRNA polyadenylation factor. *Nature* **395:** 93.

Ho C.K., Sriskanda V., McCracken S., Bentley D., Schwer B., and Shuman S. 1998. The guanylyltransferase domain of mammalian mRNA capping enzyme binds to the phosphorylated carboxyl-terminal domain of RNA polymerase II. *J. Biol. Chem.* **273:** 9577.

Itoh N., Yamada H., Kaziro Y., and Mizumoto K. 1987. Messenger RNA guanylyltransferase from *Saccharomyces cerevisiae*. Large scale purification, subunit functions, and subcellular localization. *J. Biol. Chem.* **262:** 1989.

Izaurralde E., Lewis J., McGuigan C., Jankowska M., Darzynkiewicz E., and Mattaj I.W. 1994. A nuclear cap binding protein complex involved in pre-mRNA splicing. *Cell* **78:** 657.

Kao S.Y., Calman A.F., Luciw P.A., and Peterlin B.M. 1987. Anti-termination of transcription within the long terminal repeat of HIV-1 by *tat* gene product. *Nature* **330:** 489.

Krumm A., Hickey L., and Groudine M. 1995. Promoter-proximal pausing of RNA polymerase II defines a general rate-limiting step after transcription initiation. *Genes Dev.* **9:** 559.

Krumm A., Meulia T., and Groudine M. 1993. Common mechanisms for the control of eukaryotic transcriptional elongation. *BioEssays* **15:** 659.

Logan J., Falck-Pedersen E., Darnell J.E., Jr., and Shenk T. 1987. A poly(A) addition site and a downstream termination region are required for efficient cessation of transcription by RNA polymerase II in the mouse beta maj-globin gene. *Proc. Natl. Acad. Sci.* **84:** 8306.

Mancebo H.S.Y., Lee G., Flygare J., Tomassini J., Luu P., Zhu Y.R., Peno J.M., Blau C., Hazuda D., Price D., and Flores O. 1997. P-TEFb kinase is required for HIV Tat transcriptional activation in vivo and in vitro. *Genes Dev.* **11:** 2633.

Mao X., Schwer B., and Shuman S. 1995. Yeast mRNA cap methyltransferase is a 50-kilodalton protein encoded by an essential gene. *Mol. Cell. Biol.* **15:** 4167.

Marciniak R.A. and Sharp P.A. 1991. HIV-1 Tat protein promotes formation of more-processive elongation complexes. *EMBO J.* **10:** 4189.

Marshall N.F. and Price D.H. 1992. Control of formation of two distinct classes of RNA polymerase II elongation complexes. *Mol. Cell. Biol.* **12:** 2078.

Marshall N.F., Peng J.M., Xie Z., and Price D.H. 1996. Control of RNA polymerase II elongation potential by a novel carboxyl-terminal domain kinase. *J. Biol. Chem.* **271:** 27176.

McCracken S., Fong N., Yankulov K., Ballantyne S., Pan G.H., Greenblatt J., Patterson S.D., Wickens M., and Bentley D.L. 1997a. The C-terminal domain of RNA polymerase II couples messenger RNA processing to transcription. *Nature* **385:** 357.

McCracken S., Fong N., Rosonina E., Yankulov K., Brothers G., Siderovski D., Hessel A., Foster S., Shuman S., and Bentley D. 1997b. 5´-Capping enzymes are targeted to pre-mRNA by binding to the phosphorylated carboxy-terminal domain of RNA polymerase II. *Genes Dev.* **11:** 3306.

McNeil J.B., Agah H., and Bentley D. 1998. Activated transcription independent of the RNA polymerase II holoenzyme in budding yeast. *Genes Dev.* **12:** 2510.

Niwa M. and Berget S.M. 1991. Mutation of the AAUAAA polyadenylation signal depresses in vitro splicing of proximal but not distal introns. *Genes Dev.* **5:** 2086.

Niwa M., Rose S.D., and Berget S.M. 1990. In vitro polyadenylation is stimulated by the presence of an upstream intron. *Genes Dev.* **4:** 1552.

Nonet M., Sweetser D., and Young R.A. 1987. Functional redundancy and structural polymorphism in the large subunit of RNA polymerase II. *Cell* **50:** 909.

O'Brien T., Hardin S., Greenleaf A., and Lis J.T. 1994. Phosphorylation of RNA polymerase II C-terminal domain and transcriptional elongation. *Nature* **370:** 75.

Okamoto H., Sheline C.T., Corden J.L., Jones K.A., and Peterlin B.M. 1996. Trans-activation by human immunodeficiency virus tat protein requires the C-terminal domain of RNA polymerase II. *Proc. Natl. Acad. Sci.* **93:** 11575.

Pan G., Aso T., and Greenblatt J. 1997. Interaction of elongation factors TFIIS and elongin A with a human RNA polymerase II holoenzyme capable of promoter-specific initiation and responsive to transcriptional activators. *J. Biol. Chem.* **272:** 24563.

Parada C.A. and Roeder R.G. 1996. Enhanced processivity of RNA polymerase II triggered by Tat-induced phosphorylation of its carboxy-terminal domain. *Nature* **384:** 375.

Patturajan M., Schulte R.J., Sefton B.M., Berezney R., Vincent M., Bensaude O., Warren S.L., and Corden J.L. 1998. Growth-related changes in phosphorylation of yeast RNA polymerase II. *J. Biol. Chem.* **273:** 4689.

Rasmussen E.B. and Lis J.T. 1993. In-vivo transcriptional pausing and cap formation on 3 *Drosophila* heat-shock genes. *Proc. Natl. Acad. Sci.* **90:** 7923.

Roy R., Adamczewski J.P., Seroz T., Vermeulen W., Tassan J.P., Schaeffer L., Nigg E.A., Hoeijmakers J., and Egly J.M. 1994. The MO15 cell-cycle kinase is associated with the TFIIH transcription DNA-repair factor. *Cell* **79:** 1093.

Sandig V., Lieber A., Bahring S., and Strauss M. 1993. A phage T7 class-III promoter functions as a polymerase II promoter in mammalian cells. *Gene* **131:** 255.

Scafe C., Chao D., Lopes J., Hirsch J.P., Henry S., and Young R.A. 1990. RNA polymerase II C-terminal repeat influences response to transcriptional enhancer signals. *Nature* **347:** 491.

Schwer B. and Shuman S. 1994. Mutational analysis of yeast mRNA capping enzyme. *Proc. Natl. Acad. Sci.* **91:** 4328.

———. 1996. Conditional inactivation of mRNA capping enzyme affects yeast pre-mRNA splicing in vivo. *RNA* **2:** 574.

Shuman S. 1995. Capping enzyme in eukaryotic mRNA synthesis. *Prog. Nucleic Acid Res. Mol. Biol.* **50:** 101.

Simonelig M., Elliott K., Mitchelson A., and O'Hare K. 1996. Interallelic complementation at the suppressor of forked locus of *Drosophila* reveals complementation between suppressor of forked proteins mutated in different regions. *Genetics* **142:** 1225.

Wada T., Takagi T., Yamaguchi Y., Ferdous A., Imai T., Hirose S., Sugimoto S., Yano K., Hartzog G.A., Winston F., Buratowski S., and Handa H. 1998. DSIF, a novel transcription elongation factor that regulates RNA polymerase II processivity, is composed of human Spt4 and Spt5 homologs. *Genes Dev.* **12:** 343.

Walker S., Greaves R., and O'Hare P. 1993. Transcriptional activation by the acidic domain of Vmw65 requires the integrity of the domain and involves additional determinants distinct from those necessary for TFIIB binding. *Mol. Cell. Biol.;* **13:** 5233.

Wang S.P. and Shuman S. 1997. Structure-function analysis of the mRNA cap methyltransferase of *Saccharomyces cerevisiae.* *J.Biol. Chem.* **272:** 14683.

Wei P., Garber M.E., Fang S.M., Fischer W.H., and Jones K.A. 1998. A novel CDK9-associated C-type cyclin interacts directly with HIV-1 Tat and mediates its high-affinity, loop-specific binding to TAR RNA. *Cell* **92:** 451.

Whitelaw E. and Proudfoot N. 1986. Alpha-thalassaemia caused by a poly(A) site mutation reveals that transcriptional termination is linked to 3′ end processing in the human alpha 2 globin gene. *EMBO J.* **5:** 2915.

Xiao H., Pearson A., Coulombe B., Truant R., Zhang S., Regier J.L., Triezenberg S.J., Reinberg D., Flores O., Ingles C.J., and Greenblatt J. 1994. Binding of basal transcription factor TFIIH to the acidic activation domains of VP16 and p53. *Mol. Cell. Biol.* **14:** 7013.

Yang X.Z., Herrmann C.H., and Rice A.P. 1996. The human immunodeficiency virus tat proteins specifically associate with tak in vivo and require the carboxy-terminal domain of RNA polymerase II for function. *J. Virol.* **70:** 4576.

Yankulov K., Blau J., Purton T., Roberts S., and Bentley D. 1994. Transcriptional elongation by RNA polymerase II is stimulated by transactivators. *Cell* **77:** 749.

Yankulov K.Y., Pandes M., McCracken S., Bouchard D., and Bentley D.L. 1996. TFIIH functions in regulating transcriptional elongation by RNA-polymerase II in *Xenopus* oocytes. *Mol. Cell. Biol.* **16:** 3291.

Yuryev A., Patturajan M., Litingtung Y., Joshi R., Gentile C., Gebara M., and Corden J. 1996. The CTD of RNA polymerase II interacts with a novel set of SR-like proteins. *Proc. Natl. Acad. Sci.* **93:** 6975.

Zhu Y.R., Peery T., Peng T.M., Ramanathan Y., Marshall N., Marshall T., Amendt B., Mathews M.B., and Price D.H. 1997. Transcription elongation factor P-TEFb is required for HIV-1 Tat transactivation in vitro. *Genes Dev.* **11:** 2622.

Fractions to Functions: RNA Polymerase II Thirty Years Later

N.A. WOYCHIK

Department of Molecular Genetics and Microbiology, University of Medicine and Dentistry of New Jersey,
Robert Wood Johnson Medical School, Piscataway, New Jersey 08854

Thirty years ago, Roeder and Rutter (1969) demonstrated the existence of multiple RNA polymerases in eukaryotes. Extending these pivotal findings, RNA polymerase II (pol II) was systematically purified from multiple organisms to reveal a striking pattern of similarity in subunit number and mass from yeasts to metazoans (Sentenac 1985). However, emphasis soon shifted toward defining other novel regulatory features in higher organisms—DNA elements and factors that exert their influence at promoters. Intensive studies defined a surprising array of gene-specific factors, general transcription factors, and regulatory elements. As the bits of the puzzle were found and pieced together, the events occurring at promoters in vivo began to come into view and were further sharpened with the intersection of biochemical and genetic studies that uncovered the existence of the pol II holoenzyme.

Now that most of the general players that participate in regulated gene expression have been identified, each is being subjected to even more intense dissection, including pol II. The goal is to understand what signals are transmitted to the transcriptional machinery, how the relay of signals reach their targets, what holoenzyme components are targets of these signals, and the subsequent molecular consequences. This paper focuses on these issues from the perspective of pol II.

RNA POLYMERASE II THEN AND NOW

How does the pol II holoenzyme compare to the classically defined pol II? Fractions possessing pol II activity were identified by two properties: their ability support promoter-independent transcription on nonspecific templates, and their sensitivity to transcription inhibition in the presence of the toxin α-amanitin (Sentenac 1985). Using these criteria to purify the enzyme, or by simply cloning genes encoding subunit orthologs, most organisms appear to have a 12-subunit pol II (Young 1991; Sentenac et al. 1992; Woychik and Young 1994). However, this traditionally defined 12 subunit enzyme alone will not support basal or activated transcription in vitro. Basal transcription occurs only upon addition of the general transcription factors (TFIIB, TFIID, TFIIE, TFIIF, TFIIH) to purified pol II (Buratowski 1994; Orphanides et al. 1996; Roeder 1996). Expansion of the basal complex to include the multitude of proteins comprising the RNA polymerase holoenzyme now supports activated transcription (Koleske and Young 1995; Bjorklund and Kim 1996; Greenblatt 1997).

The pol II holoenzyme was defined on the basis of its ability to support activated transcription, but the jury is still out on exactly which factors and how many additional proteins associate with pol II throughout different stages of the transcription cycle. Clearly, pol II subunits function in concert with many other proteins throughout the transcription cycle—the holoenzyme and mediator components (Koleske and Young 1995; Bjorklund and Kim 1996; Greenblatt 1997), elongation factors (Reines et al. 1996; Uptain et al. 1997; Shilatifard 1998), and possibly RNA processing factors (Neugebauer and Roth 1997). Therefore, the classically defined pol II needs to be associated with the relevant accessory proteins in order to function as it would be expected to in vivo, i.e., have the ability to specifically initiate at promoter sequences, respond to activators, and elongate at a rapid pace without pausing.

SOMETHING OLD, SOMETHING NEW

Like many of the components of the transcriptional machinery and other molecular machines, the sequence, structure, and function of RNA polymerase subunits are well conserved among eukaryotes (Young 1991; Sentenac et al. 1992; Woychik and Young 1994) and between eukaryotes and archaebacteria (Baumann et al. 1995; Keeling and Doolittle 1995; Langer et al. 1995; Olsen and Woese 1997). In contrast to the majority of transcription factors and other pol II holoenzyme components, the three largest subunits also have counterparts in eubacteria (Woychik and Young 1994; K. Linask et al., in prep.). The specific functions of the remaining nine subunits were not apparent by their sequence. Notably absent were also the eukaryotic counterparts for the family of σ factors. Instead, higher organisms apparently employ multiple mechanisms that enlist a variety of proteins to direct the initiation complex to different classes of promoters. To date, the only general transcription factors found in organisms other than eukaryotes are the archaebacterial counterparts of the TATA-box-binding protein (TBP) and TFIIB (Baumann et al. 1995; Keeling and Doolittle 1995; Langer et al. 1995; Olsen and Woese 1997).

We and others have demonstrated that RNA polymerase subunits have a striking degree of functional conservation (Table 1). Replacement of many of the 12 yeast *Saccharomyces cerevisiae* subunits (designated RPB1-RPB12) with their orthologs from a spectrum of organisms revealed that all of them partially or fully support normal yeast cell growth when added intact or as chimeras. Interestingly, these substitution experiments seem to work best when subunits from more complex organisms are substituted in place of their less evolved counterparts. Thus, human RPB6 and RPB10 function well in yeast, but their archaebacterial counterparts do not appear to support any growth in *S. cerevisiae* (McKune and Woychik 1994a). However, both human and yeast TBPs can function in place of their archaebacterial counterpart in an archaeal cell-free transcription system (Wettach et al. 1995). Therefore, it appears that as one moves up the evolutionary ladder, the subunits acquire new features but retain the old. This acquisition of function is sometimes reflected by an increase in the length of the subunit sequence, e.g., archaebacterial RPB5 (Klenk et al. 1992) and RPB6 (McKune and Woychik 1994a) are only about one-third the size of their more evolved relatives. However, there is also an example of a human subunit, hsRPB4, that is substantially shorter than its yeast cognate (Khazak et al. 1998).

STOICHIOMETRY AND SUBUNIT INTERACTIONS

Independent of the genetic and biochemical approaches used to evaluate subunit function, the foundation for understanding subunit functions was laid with the identification of contacts among a subset of subunits. These interactions were assessed by a variety of in vivo and in vitro techniques such as two-hybrid/interaction trap screens, far-Western analysis, genetic suppression, and coprecipitation using GST fusion proteins (Table 2). These data corroborated earlier biochemical data suggesting an interaction between RPB4 and RPB7. They also confirmed speculation that RPB3 and RPB11 interact and do so in a manner dependent on the presence of a domain (α-motif) analogous to an assembly domain in α. These mapped contacts have also been useful in strengthening functional data, especially when multiple subunits are assigned similar functions (such as RPB1, RPB5, and RPB3 in activation). The usefulness of these mapped interactions will continue to be appreciated as the number of functional assignments grow.

Initial estimates of subunit stoichiometry were based on quantification and normalization of immunoprecipitated bands from methionine radioactively labeled cells. The average pol II molecule appeared to be composed of one copy each of RPB1, 2, 6, 8, and 10 and two copies of RPB3, 5, and 9. In contrast, approximately half of the pol II molecules have RPB4 and RPB7, consistent with biochemical reports demonstrating that they form a dissociable subcomplex (Dezelee et al. 1976; Ruet et al. 1980; Edwards et al. 1991). Since then, their interaction has been well documented in vitro and in vivo (Khazak et al. 1995, 1998; Larkin and Guilfoyle 1998) and even exploited to identify human RPB4 (which eluded sequence-dependent identification due to its smaller size and less-conserved sequence) (Khazak et al. 1998). Furthermore, the substoichiometric association of RPB4 and RPB7 with pol II is consistent with reports suggesting a preferential association of RPB4 with the enzyme only under stress conditions (Choder and Young 1993). Two additional subunits, one each comigrating with RPB9 and RPB10, were identified after the initial stoichiometric assignments. Therefore, the published values for RPB9-RPB12 require further scrutiny.

Initial experiments suggested that RPB3 (like α, its tentatively assigned relative in eubacteria) is present in two copies per RNA polymerase molecule, but this assignment does not hold up after more rigorous analysis. Several independent investigators have not been able to detect a high-affinity interaction (such as that seen with α subunits or between RPB3 and RPB11) between RPB3 homodimers (Ulmasov et al. 1996; Larkin and Guilfoyle 1997; Svetlov et al. 1998; Yasui et al. 1998). In vitro reconstitution experiments implementing His6-tagged and untagged forms of RPB3 and RPB11 revealed that only stable RPB3-RPB11 heterodimers, and no stable homodimers of either RPB3 or RPB11, were recovered (Larkin and Guilfoyle 1997). This work was further strengthened by in vivo studies demonstrating that upon immunopurification of pol II from cells expressing equal pools of

Table 1. Functional Substitution of Yeast *S. cerevisiae* RNA Polymerase II Subunits In Vivo

Yeast subunit	Origin of cognate	Functional substitution	Reference
RPB1	mouse	yes	Singleton and Wilcox (1998)
RPB2	–	n.d.[a]	–
RPB3	–	n.d.	–
RPB4	human	yes-partial	Khazak et al. (1998)
RPB5	human	no	McKune et al. (1995); Shpakovski et al. (1995)
		yes-chimera	T. Miyao and N. Woychik (in prep.)
RPB6	human	yes	McKune and Woychik (1994b); Shpakovski et al. (1995)
	Schizosaccharomyces pombe	yes	Shpakovski (1994)
RPB7	human	yes-partial	Khazak et al. (1995)
RPB8	human	yes	McKune et al. (1995); Shpakovski et al. (1995)
RPB9	human	yes-partial	McKune et al. (1995)
RPB10	human	yes	McKune et al. (1995); Shpakovski et al. (1995)
RPB11	human	yes	K. Linask et al. (in prep.)
RPB12	human	yes	Shpakovski et al. (1995)

[a]n.d. indicates not determined.

Table 2. RNA Polymerase II Subunit Interactions

Test subunit(s)[a]	Interacting subunit(s)	Organism	Method(s) used	Reference
RPB1-3, RPB5-8, RPB10-12	multiple combinations	*S. pombe*	cross-linking	Ishiguro et al. (1998)
RPB3	RPB1, RPB2, RPB3, RPB5, RPB11	*S. pombe*	far-Western and GST-pull down	Yasui et al. (1998)
RPB4	RPB7	human	interaction trap	Khazak et al. (1995)
RPB7	RPB4	human		Khazak et al. (1998)
RPB11	RPB3	human	two-hybrid	Fanciulli et al. (1998)
All 12	56 interactions	human	GST-pull down	Acker et al. (1997)
RPB4	RPB7	*Arabidopsis*	renaturation of recombinant proteins	Larkin and Guilfoyle (1998)
RPB7	RPB4	*Arabidopsis*		
All 10	RPB2-RPB3-RPB11 core subcomplex	*S. pombe*	association after denaturation with 6 M urea	Kimura et al. (1997)
RPB3	RPB11	*Arabidopsis*	renaturation of recombinant proteins	Larkin and Guilfoyle (1997)
RPB11	RPB3	*Arabidopsis*		
RPB11	RPB3	*Arabidopsis*	two-hybrid and/or immuno-precipitation	Ulmasov et al. (1996)
RPB3	RPB11	*Arabidopsis*		
RPB5	RPB1, RPB2, RPB3	*S. pombe*	far-Western	Miyao et al. (1996, 1998)
RPB1	RPB6	*S. cerevisiae*	genetic suppression	Archambault et al. (1990, 1992a)
RPB1	RPB2	*S. cerevisiae*	genetic suppression	Martin et al. (1990)

[a]Nomenclature for yeast *S. cerevisiae* subunits is used for consistency.

His$_6$-tagged and untagged RPB3, only tagged RPB3 is present in pol II after Ni-NTA column treatment—never both tagged and untagged (Svetlov et al. 1998). Finally, quantification of Coomassie-stained RNA polymerase suggested a stoichiometry of one (Kolodziej et al. 1990). In total, these results strongly suggest that only one RPB3 subunit is assembled with pol II. The initial experiment, which measured relative intensities of subunit bands immunoprecipitated from ^{35}S-labeled cells (Kolodziej et al. 1990), yielded misleading results due to the limitations of the experiment. Since the antibody used for immunoprecipitation was specific for the epitope on the tagged RPB3 subunit antibody, the amount of RPB3 measured represented *all* RPB3 molecules in the cell, not just those assembled with pol II. Thus, in hindsight, it appears that only a portion of RPB3 in the cell is assembled with the enzyme.

The RPB3/RPB11 heterodimer has been suggested to be related to the eubacterial α$_2$-homodimer. Indeed, these two subunits do interact in vivo and in vitro (Ulmasov et al. 1996; Larkin and Guilfoyle 1997; Svetlov et al. 1998; Yasui et al. 1998): The α-motif in RPB3 and RPB11 is required for their association with each other (Larkin and Guilfoyle 1997), and the α-motif is required for the association of RPB3 with pol II (Svetlov et al. 1998). Although a mutation in the RPB3 α-motif results in an assembly defect (a phenotype seen with many α mutations in eubacteria), it is not known if a deletion of this region in RPB3 or RPB11 is lethal (the predicted result if this association triggers enzyme assembly as in α). In addition, the characterization of assembly defects in the RPB3 α-motif point mutant only followed the presence of the three largest subunits (Kolodziej and Young 1991) and did not assess whether the mutation affects RPB3's association with RPB11 in vivo. Therefore, although strongly suggestive, the unequivocal assignment of the RPB3/11 pair as the ortholog of the α$_2$-homodimer has yet to be established. Since RPB11 is approximately one-third the

size of α, it is unlikely that the functional parallels between eubacterial α$_2$-homodimers and eukaryotic RPB3/11 heterodimers will be striking.

RNA POLYMERASE II STRUCTURE

In the past 10 years, several laboratories have made exciting progress toward determining high-resolution structures of the complex, multisubunit eubacterial and eukaryotic RNA polymerases. Initial studies reported three-dimensional structures obtained by electron microscopy of two-dimensional crystals on positively charged lipid layers (Darst et al. 1989; Edwards et al. 1990). Using this approach, resolution of the yeast *S. cerevisiae* enzyme was further improved (from ~30 Å to ~16 Å resolution) using a mutant pol II lacking the substoichiometrically represented subunits RPB4 and RPB7 (pol II/Δ4/7) (Darst et al. 1991a,b). These studies and others revealed a common structural feature, a 25-Å-diameter cleft, in *Escherichia coli* RNA polymerase, pol I (Schultz et al. 1990), and pol II/Δ4/7. Refinement of the *E. coli* structure revealed more striking similarities between the eubacterial and eukaryotic enzymes; not only do they share a 25-Å-diameter cleft, they also have an arm of protein density surrounding the cleft (Polyakov et al. 1995). These two features not only are shared among RNA polymerases, but are present in a variety of DNA polymerases (Davies et al. 1994; Sawaya et al. 1994; Kim et al. 1995; Korolev et al. 1995). Whereas the 25-Å-diameter channel was suggested to bind duplex DNA, the arm of protein density was proposed to clamp DNA in the cleft and enhance processivity of elongation (Beese et al. 1993; Sawaya et al. 1994; Sousa et al. 1994; Polyakov et al. 1995). Structural studies of *E. coli* RNA polymerase revealed different conformations of this arm (Polyakov et al. 1995)—open and closed—suggesting that other RNA polymerases also exist in analogous states.

To complement the structural studies of pol II/Δ4/7

Table 3. RNA Polymerase II Subunit Features and Functions

Subunit[a]			References
RPB1[b]	*Features:*[c]	β′ ortholog	Young (1991)
		conserved regions designated A–H	Young (1991)
		target for α-amanitin resistance mutants	Bartolomei and Corden (1995)
		in vivo degradation triggered by α-amanitin binding	Nguyen et al. (1996)
		carboxy-terminal repeat YSPTSPS	Corden (1990); Chao and Young (1991)
		phosphorylated	Woychik and Young (1994)
		glycosylated	Cervoni et al. (1997)
		photocrosslinks to DNA	Kim et al.(1997)
		binds zinc	Treich et al. (1991)
		acidic activation domain in region H	Xiao et al. (1994)
		region H interacts with TBP and TFIIB	Xiao et al. (1994)
		target for ubiquitin-mediated degradation	Huibregtse et al. (1997)
	Functions:	start site selection	Berroteran et al. (1994)
		DNA binding	Young (1991)
		activation	Scafe et al. (1990); Liao et al. (1991)
		genetic interaction with TFIIS	Archambault et al. (1992b)
		in vitro studies also implicate TFIIS interaction	Wu et al. (1996)
RPB2	*Features:*	β ortholog	Young (1991)
		conserved regions designated A-I	Young (1991)
		glycosylated	Cervoni et al. (1997)
		photocrosslinks to DNA	Kim et al. (1997)
		binds zinc	Treich et al. (1991)
	Functions:	active site within domain H	Riva et al. (1990); Treich et al. (1992)
		start site selection	Hekmatpanah and Young (1991)
		select mutants display 6-azauracil sensitivity/slower elongation rates	Powell and Reines (1996)
RPB3	*Features:*	weak similarity to α-sequence/function	Woychik and Young (1994)
		α-motif	Woychik and Young (1994)
		binds zinc	Treich et al. (1991)
	Functions:	assembly	Kolodziej and Young (1991)
		activation	K. Linask et al. (in prep.)
RPB4	*Features:*	σ[s]-like	Khazak et al. (1995)
		substoichiometric complex with RPB7	Young (1991)
	Functions:	required for stress response	Choder and Young (1993)
		preferentially assembles with enzyme during stationary phase	Choder and Young (1993)
RPB5	*Features:*	interacts with hepatitis virus X protein and TFIIB	Cheong et al. (1995); Lin et al. (1997)
		photocrosslinks to DNA	Kim et al. (1997)
		glycosylated	Cervoni et al. (1997)
	Functions:	activation	Miyao and Woychik (1998)
RPB6	*Features:*	phosphorylated	Woychik and Young (1994)
		glycosylated	Cervoni et al. (1997)
	Functions:	assembly and large subunit stability	Nouraini et al. (1996)
RPB7	*Features:*	substoichiometric complex with RPB4	Young (1991)
	Functions:	unknown	
RPB8	*Features:*	structure determined by NMR	Krapp et al. (1998)
	Functions:	unknown	
RPB9	*Features:*	binds zinc	Treich et al. (1991)
		zinc-binding regions form zinc ribbon	Jeon et al. (1994)
		carboxy-terminal zinc-binding domain determined by NMR	Wang et al. (1998)
	Functions:	elongation through DNA arrest sites	Awrey et al. (1997)
		start site selection	Furter-Graves et al. (1994); Hull et al. (1995)
RPB10	*Features:*	binds zinc	Carles et al. (1991)
	Functions:	unknown	
RPB11	*Features:*	α-motif	Woychik and Young (1994)
	Functions:	unknown	
RPB12	*Features:*	binds zinc	Carles et al. (1991)
	Functions:	unknown	

[a]Nomenclature for yeast *S. cerevisiae* subunits is used for consistency.
[b]Excludes details on the carboxy-terminal domain.
[c]Features/functions listed represent those known at the time of manuscript preparation. Since none of the subunits have been subjected to exhaustive functional studies, the lists likely represent a small subset of the actual array of functions possessed by the subunits.

alone, a 15.7-Å-resolution structure of the enzyme bound to two basal transcription factors, TFIIE and TFIIB, was also determined (Leuther et al. 1996). In addition, the isolation of crystals of a paused pol II/Δ4/7 elongation complex capable of diffraction to 3.5-Å resolution was reported (Gnatt et al. 1997). The work on pol II/Δ4/7 was further enriched by resolution of the structure of the complete 12-subunit *S. cerevisiae* RNA polymerase by the same two-dimensional method (Meredith et al. 1996; Asturias et al. 1997; Jensen et al. 1998) and by electron crystallography of the enzyme preserved in vitreous ice (Asturias et al. 1998). These studies revealed that although the overall structures of the wild-type and mutant enzymes were similar, crystals from the wild-type enzyme favored the closed conformation (Meredith et al. 1996; Asturias et al. 1997; Jensen et al. 1998). Thus, 4/7 likely stabilizes the polymerase-DNA complex since polymerase is thought to first bind promoter DNA in an open conformation and convert to the more stable closed form before the onset of RNA synthesis.

In addition to studying the multisubunit pol II, structural information about single subunits or subunit domains is beginning to add a new dimension to our perception of the enzyme. Nuclear magnetic resonance (NMR) is being employed to determine the structure of the small-molecular-mass subunits, whose sequences alone have provided few hints to their functions. The first high-resolution structure obtained by this method was for the common subunit RPB8 (Krapp et al. 1998), and more structures are certain to follow. These individual structures will complement the evolving picture of the entire enzyme and will likely be useful in identifying locations of subunits within the overall structure of pol II.

Two domains of individual subunits, the carboxy-terminal domain (CTD) of RPB1 and the zinc-binding region of RPB9, have also been subjected to more detailed inspection. The work with the CTD involved identifying its position in the overall structure of pol II by either antibody localization or difference map comparison of structures with and without the domain (Meredith et al. 1996). These results revealed that the CTD is projecting into solution and is conformationally mobile, consistent with its role in transcriptional activation as a link to multiple protein complexes. Finally, the RPB9 subunit contains two zinc-binding regions ($C-X_2-C-X_{18}-C-X_2-C$ and $C-X_2-C-X_{24}-C-X_2-C$) that had first been proposed to be zinc ribbons similar to those in TFIIS and TFIIB (Jeon et al. 1994; Zhu et al. 1996). Recent NMR analysis of the zinc-binding region from an RPB9 counterpart in archaebacteria (Wang et al. 1998) solidified these suggestions but revealed some differences between its structure and those of TFIIS and TFIIB.

UNMASKING SUBUNIT FUNCTIONS

Although valuable functional information can be extracted by interaction and structural analysis, these approaches are more fruitful when coupled with genetic suppression and biochemical analysis of mutants. Al-

though the initial thrust of functional studies of RNA polymerase were directed toward the RPB1-CTD, more information about other domains of RPB1 and the remaining 11 subunits is coming to light. One disadvantage of using an exclusively biochemical approach is that the functions unveiled are entirely dependent on the inherent limitations of the assays employed. Since there is a built-in bias to look for traditional transcriptional defects and select assays based on current favored models, we risk overlooking certain functions. Genetic suppressors can often illuminate a connection with a system not previously considered to be linked to the gene at hand. Suppression experiments also can help extract clues when none were revealed after a battery of in vitro assays. Ideally, the genetic and biochemical information intersect, unearthing a stronger model, better mechanism, or new function. Using the wealth of tools at hand, the goal of a detailed structure-function map for each subunit of pol II should be attainable. A compilation of the functions associated with RNA polymerase subunits (excluding the CTD, which is covered in detail elsewhere) is presented in Table 3. This information, intertwined with data obtained for the other components of the transcriptional machinery, will provide invaluable contributions to our understanding of the cross-talk between regulatory proteins and RNA polymerase.

REFERENCES

Acker J., de Graaff M., Cheynel I., Khazak V., Kedinger C., and Vigneron M. 1997. Interactions between the human RNA polymerase II subunits. *J. Biol. Chem.* **272:** 16815.

Archambault J., Schappert K.T., and Friesen J.D. 1990. A suppressor of an RNA polymerase II mutation of *Saccharomyces cerevisiae* encodes a subunit common to RNA polymerases I, II, and III. *Mol. Cell. Biol.* **10::** 6123.

Archambault J., Drebot M.A., Stone J.C., and Friesen J.D. 1992a. Isolation and phenotypic analysis of conditional-lethal, linker-insertion mutations in the gene encoding the largest subunit of RNA polymerase II in *Saccharomyces cerevisiae. Mol. Gen. Genet.* **232:** 408.

Archambault J., Lacroute F., Ruet A., and Friesen J.D. 1992b. Genetic interaction between transcription elongation factor TFIIS and RNA polymerase II. *Mol. Cell. Biol.* **12:** 4142.

Asturias F.J., Chang W., Li Y., and Kornberg R.D. 1998. Electron crystallography of yeast RNA polymerase II preserved in vitreous ice. *Ultramicroscopy* **70:** 133.

Asturias F.J., Meredith G.D., Poglitsch C.L., and Kornberg R.D. 1997. Two conformations of RNA polymerase II revealed by electron crystallography. *J. Mol. Biol.* **272:** 536.

Awrey D.E., Weilbaecher R.G., Hemming S.A., Orlicky S.M., Kane C.M., and Edwards A.M. 1997. Transcription elongation through DNA arrest sites. A multistep process involving both RNA polymerase II subunit RPB9 and TFIIS. *J. Biol. Chem.* **272:** 14747.

Bartolomei M.S. and Corden J.L. 1995. Clustered α-amanitin resistance mutations in mouse. *Mol. Gen. Genet.* **246:** 778.

Baumann P., Qureshi S.A., and Jackson S.P. 1995. Transcription: New insights from studies on Archaea. *Trends Genet.* **11:** 279.

Beese L.S., Friedman J.M., and Steitz T.A. 1993. Crystal structures of the Klenow fragment of DNA polymerase I complexed with deoxynucleoside triphosphate and pyrophosphate. *Biochemistry* **32:** 14095.

Berroteran R.W., Ware D.E., and Hampsey M. 1994. The *sua8* suppressors of *Saccharomyces cerevisiae* encode replacements of conserved residues within the largest subunit of RNA polymerase II and affect transcription start site selection similarly to *sua7* (TFIIB) mutations. *Mol. Cell. Biol.* **14**: 226.

Bjorklund S. and Kim Y.J. 1996. Mediator of transcriptional regulation. *Trends Biochem. Sci.* **21**: 335.

Buratowski S. 1994. The basics of basal transcription by RNA polymerase II. *Cell* **77**: 1.

Carles C., Treich I., Bouet F., Riva M., and Sentenac A. 1991. Two additional common subunits, ABC10 alpha and ABC10 beta, are shared by yeast RNA polymerases. *J. Biol. Chem.* **266**: 24092.

Cervoni L., Turano C., Ferraro A., Ciavatta P., Marmocchi F., and Eufemi M. 1997. Glycosylation of RNA polymerase II from wheat germ. *FEBS Lett.* **417**: 227.

Chao D.M. and Young R.A. 1991. Tailored tails and transcription initiation: The carboxyl terminal domain of RNA polymerase II. *Gene Expr.* **1**: 1.

Cheong J.H., Yi M., Lin Y., and Murakami S. 1995. Human RPB5, a subunit shared by eukaryotic nuclear RNA polymerases, binds human hepatitis B virus X protein and may play a role in X transactivation. *EMBO J.* **14**: 143.

Choder M. and Young R.A. 1993. A portion of RNA polymerase II molecules has a component essential for stress responses and stress survival. *Mol. Cell. Biol.* **13**: 6984.

Corden J.L. 1990. Tails of RNA polymerase II. *Trends Biochem. Sci.* **15**: 383.

Darst S.A., Kubalek E.W., and Kornberg R.D. 1989. Three-dimensional structure of *Escherichia coli* RNA polymerase holoenzyme determined by electron crystallography. *Nature* **340**: 730.

Darst S.A., Edwards A.M., Kubalek E.W., and Kornberg R.D. 1991a. Three-dimensional structure of yeast RNA polymerase II at 16 Å resolution. *Cell* **66**: 121.

Darst S.A., Kubalek E.W., Edwards A.M., and Kornberg R.D. 1991b. Two-dimensional and epitaxial crystallization of a mutant form of yeast RNA polymerase II. *J. Mol. Biol.* **221**: 347.

Davies J.F., II, Almassy R.J., Hostomska Z., Ferre R.A., and Hostomsky Z. 1994. 2.3 Å crystal structure of the catalytic domain of DNA polymerase beta. *Cell* **76**: 1123.

Dezelee S., Wyers F., Sentenac A., and Fromageot P. 1976. Two forms of RNA polymerase B in yeast. Proteolytic conversion *in vitro* of enzyme BI into BII. *Eur. J. Biochem.* **65**: 543.

Edwards A.M., Kane C.M., Young R.A., and Kornberg R.D. 1991. Two dissociable subunits of yeast RNA polymerase II stimulate the initiation of transcription at a promoter *in vitro*. *J. Biol. Chem.* **266**: 71.

Edwards A.M., Darst S.A., Feaver W.J., Thompson N.E., Burgess R.R., and Kornberg R.D. 1990. Purification and lipid-layer crystallization of yeast RNA polymerase II. *Proc. Natl. Acad. Sci.* **87**: 2122.

Fanciulli M., Bruno T., Di Padova M., De Angelis R., Lovari S., Floridi A., and Passananti C. 1998. The interacting RNA polymerase II subunits, hRPB11 and hRPB3, are coordinately expressed in adult human tissues and down-regulated by doxorubicin. *FEBS Lett.* **427**: 236.

Furter-Graves E.M., Hall B.D., and Furter R. 1994. Role of a small RNA pol II subunit in TATA to transcription start site spacing. *Nucleic Acids Res.* **22**: 4932.

Gnatt A., Fu J., and Kornberg R.D. 1997. Formation and crystallization of yeast RNA polymerase II elongation complexes. *J. Biol. Chem.* **272**: 30799.

Greenblatt J. 1997. RNA polymerase II holoenzyme and transcriptional regulation. *Curr. Opin. Cell Biol.* **9**: 310.

Hekmatpanah D.S. and Young R.A. 1991. Mutations in a conserved region of RNA polymerase II influence the accuracy of mRNA start site selection. *Mol. Cell. Biol.* **11**: 5781.

Huibregtse J.M., Yang J.C., and Beaudenon S.L. 1997. The large subunit of RNA polymerase II is a substrate of the Rsp5 ubiquitin-protein ligase. *Proc. Natl. Acad. Sci.* **94**: 3656.

Hull M.W., McKune K., and Woychik N.A. 1995. RNA polymerase II subunit RPB9 is required for accurate start site selection. *Genes Dev.* **9**: 481.

Ishiguro A., Kimura M., Yasui K., Iwata A., Ueda S., and Ishi-hama A. 1998. Two large subunits of the fission yeast RNA polymerase II provide platforms for the assembly of small subunits. *J. Mol. Biol.* **279**: 703.

Jensen G.J., Meredith G., Bushnell D.A., and Kornberg R.D. 1998. Structure of wild-type yeast RNA polymerase II and location of Rpb4 and Rpb7. *EMBO J.* **17**: 2353.

Jeon C., Yoon H., and Agarwal K. 1994. The transcription factor TFIIS zinc ribbon dipeptide Asp-Glu is critical for stimulation of elongation and RNA cleavage by RNA polymerase II. *Proc. Natl. Acad. Sci.* **91**: 9106.

Keeling P.J. and Doolittle W.F. 1995. Archaea: Narrowing the gap between prokaryotes and eukaryotes. *Proc. Natl. Acad. Sci.* **92**: 5761.

Khazak V., Sadhale P.P., Woychik N.A., Brent R., and Golemis E.A. 1995. Human RNA polymerase II subunit hsRPB7 functions in yeast and influences stress survival and cell morphology. *Mol. Biol. Cell* **6**: 759.

Khazak V., Estojak J., Cho H., Majors J., Sonoda G., Testa J.R., and Golemis E.A. 1998. Analysis of the interaction of the novel RNA polymerase II (pol II) subunit hsRPB4 with its partner hsRPB7 and with pol II. *Mol. Cell. Biol.* **18**: 1935.

Kim T.K., Lagrange T., Wang Y.H., Griffith J.D., Reinberg D., and Ebright R.H. 1997. Trajectory of DNA in the RNA polymerase II transcription preinitiation complex. *Proc. Natl. Acad. Sci.* **94**: 12268.

Kim Y., Eom S.H., Wang J., Lee D.S., Suh S.W., and Steitz T.A. 1995. Crystal structure of *Thermus aquaticus* DNA polymerase. *Nature* **376**: 612.

Kimura M., Ishiguro A., and Ishihama A. 1997. RNA polymerase II subunits 2, 3, and 11 form a core subassembly with DNA binding activity. *J. Biol. Chem.* **272**: 25851.

Klenk H.P., Palm P., Lottspeich F., and Zillig W. 1992. Component H of the DNA-dependent RNA polymerases of Archaea is homologous to a subunit shared by the three eucaryal nuclear RNA polymerases. *Proc. Natl. Acad. Sci.* **89**: 407.

Koleske A.J. and Young R.A. 1995. The RNA polymerase II holoenzyme and its implications for gene regulation. *Trends Biochem. Sci.* **20**: 113.

Kolodziej P.A. and Young R.A. 1991. Mutations in the three largest subunits of yeast RNA polymerase II that affect enzyme assembly. *Mol. Cell. Biol.* **11**: 4669.

Kolodziej P.A., Woychik N., Liao S.M., and Young R.A. 1990. RNA polymerase II subunit composition, stoichiometry, and phosphorylation. *Mol. Cell. Biol.* **10**: 1915.

Korolev S., Nayal M., Barnes W.M., Di Cera E., and Waksman G. 1995. Crystal structure of the large fragment of *Thermus aquaticus* DNA polymerase I at 2.5-Å resolution: Structural basis for thermostability. *Proc. Natl. Acad. Sci.* **92**: 9264.

Krapp S., Kelly G., Reischl J., Weinzierl R.O., and Matthews S. 1998. Eukaryotic RNA polymerase subunit RPB8 is a new relative of the OB family. *Nat. Struct. Biol.* **5**: 110.

Langer D., Hain J., Thuriaux P., and Zillig W. 1995. Transcription in Archaea: Similarity to that in eucarya. *Proc. Natl. Acad. Sci.* **92**: 5768.

Larkin R.M. and Guilfoyle T.J. 1997. Reconstitution of yeast and *Arabidopsis* RNA polymerase α-like subunit heterodimers. *J. Biol. Chem.* **272**: 12824.

———. 1998. Two small subunits in *Arabidopsis* RNA polymerase II are related to yeast RPB4 and RPB7 and interact with one another. *J. Biol. Chem.* **273**: 5631.

Leuther K.K., Bushnell D.A., and Kornberg R.D. 1996. Two-dimensional crystallography of TFIIB- and IIE-RNA polymerase II complexes: Implications for start site selection and initiation complex formation. *Cell* **85**: 773.

Liao S.M., Taylor I.C., Kingston R.E., and Young R.A. 1991. RNA polymerase II carboxy-terminal domain contributes to the response to multiple acidic activators in vitro. *Genes Dev.* **5**: 2431.

Lin Y., Nomura T., Cheong J., Dorjsuren D., Iida K., and Murakami S. 1997. Hepatitis B virus X protein is a transcriptional modulator that communicates with transcription factor IIB and the RNA polymerase II subunit 5. *J. Biol. Chem.* **272**: 7132.

Martin C., Okamura S., and Young R. 1990. Genetic exploration of interactive domains in RNA polymerase II subunits. *Mol. Cell. Biol.* **10:** 1908.

McKune K. and Woychik N. 1994a. Halobacterial S9 operon contains two genes encoding proteins homologous to subunits shared by eukaryotic RNA polymerases I, II, and III. *J. Bacteriol.* **176:** 4754.

———. 1994b. Functional substitution of an essential yeast RNA polymerase subunit by its highly conserved mammalian counterpart. *Mol. Cell. Biol.* **14:** 4155.

McKune K., Moore P., Hull M.W., and Woychik N.A. 1995. Six human RNA polymerase subunits functionally substitute for their yeast counterparts *in vivo. Mol. Cell. Biol.* **15:** 6895.

Meredith G.D., Chang W.H., Li Y., Bushnell D.A., Darst S.A., and Kornberg R.D. 1996. The C-terminal domain revealed in the structure of RNA polymerase II. *J. Mol. Biol.* **258:** 413.

Miyao T. and Woychik N. 1998. RNA polymerase II subunit RPB5 plays a role in transcriptional activation. *Proc. Natl. Acad. Sci.* **95:** 15281.

Miyao T., Honda A., Qu Z., and Ishihama A. 1998. Mapping of Rpb3 and Rpb5 contact sites on two large subunits, Rpb1 and Rpb2, of the RNA polymerase II from fission yeast. *Mol. Gen. Genet.* **259:** 123.

Miyao T., Yasui K., Sakurai H., Yamagishi M., and Ishihama A. 1996. Molecular assembly of RNA polymerase II from the fission yeast *Schizosaccharomyces pombe:* Subunit-subunit contact network involving Rpb5. *Genes Cells* **1:** 843.

Neugebauer K.M. and Roth M.B. 1997. Transcription units as RNA processing units. *Genes Dev.* **11:** 3279.

Nguyen V.T., Giannoni F., Dubois M.F., Seo S.J., Vigneron M., Kedinger C., and Bensaude O. 1996. In vivo degradation of RNA polymerase II largest subunit triggered by α-amanitin. *Nucleic Acids Res.* **24:** 2924.

Nouraini S., Archambault J., and Friesen J.D. 1996. Rpo26p, a subunit common to yeast RNA polymerases, is essential for the assembly of RNA polymerases I and II and for the stability of the largest subunits of these enzymes. *Mol. Cell. Biol.* **16:** 5985.

Olsen G.J. and Woese C.R. 1997. Archaeal genomics: An overview. *Cell* **89:** 991.

Orphanides G., Lagrange T., and Reinberg D. 1996. The general transcription factors of RNA polymerase II. *Genes Dev.* **10:** 2657.

Polyakov A., Severinova E., and Darst S.A. 1995. Three-dimensional structure of *E. coli* core RNA polymerase: Promoter binding and elongation conformations of the enzyme. *Cell* **83:** 365.

Powell W. and Reines D. 1996. Mutations in the second largest subunit of RNA polymerase II cause 6-azauracil sensitivity in yeast and increased transcriptional arrest *in vitro. J. Biol. Chem.* **271:** 6866.

Reines D., Conaway J.W., and Conaway R.C. 1996. The RNA polymerase II general elongation factors. *Trends Biochem. Sci.* **21:** 351.

Riva M., Carles C., Sentenac A., Grachev M.A., Mustaev A.A., and Zaychikov E.F. 1990. Mapping the active site of yeast RNA polymerase B (II). *J. Biol. Chem.* **265:** 16498.

Roeder R.G. 1996. The role of general initiation factors in transcription by RNA polymerase II. *Trends Biochem. Sci.* **21:** 327.

Roeder R.G. and Rutter W.J. 1969. Multiple forms of DNA-dependent RNA polymerase in eukaryotic organisms. *Nature* **224:** 234.

Ruet A., Sentenac A., Fromageot P., Winsor B., and Lacroute F. 1980. A mutation of the B220 subunit gene affects the structural and functional properties of yeast RNA polymerase B *in vitro. J. Biol. Chem.* **255:** 6450.

Sawaya M.R., Pelletier H., Kumar A., Wilson S.H., and Kraut J. 1994. Crystal structure of rat DNA polymerase beta: Evidence for a common polymerase mechanism. *Science* **264:** 1930.

Scafe C., Chao D., Lopes J., Hirsch J.P., Henry S., and Young R.A. 1990. RNA polymerase II C-terminal repeat influences response to transcriptional enhancer signals. *Nature* **347:** 491.

Schultz P., Celia H., Riva M., Darst S.A., Colin P., Kornberg R.D., Sentenac A., and Oudet P. 1990. Structural study of the

yeast RNA polymerase A. Electron microscopy of lipid-bound molecules and two-dimensional crystals. *J. Mol. Biol.* **216:** 353.

Sentenac A. 1985. Eukaryotic RNA polymerases. *CRC Crit. Rev. Biochem.* **18:** 31.

Sentenac A., Riva M., Thuriaux P., Buhler J.-M., Treich I., Carles C., Werner M., Ruet A., Huet J., Mann C., Chiannilkulchai N., Stettler S., and Mariotte S. 1992. Yeast RNA polymerase subunits and genes. In *Transcriptional regulation* (ed. S.L. McKnight and K.R. Yamamoto), p. 27. Cold Spring Harbor Laboratory Press, Cold Spring Harbor, New York.

Shilatifard A. 1998. The RNA polymerase II general elongation complex. *Biol. Chem.* **379:** 27.

Shpakovski G.V. 1994. The fission yeast *Schizosaccharomyces pombe* rpb6 gene encodes the common phosphorylated subunit of RNA polymerase and complements a mutation in the corresponding gene of *Saccharomyces cerevisiae. Gene* **147:** 63.

Shpakovski G.V., Acker J., Wintzerith M., Lacroix J.F., Thuriaux P., and Vigneron M. 1995. Four subunits that are shared by the three classes of RNA polymerase are functionally interchangeable between *Homo sapiens* and *Saccharomyces cerevisiae. Mol. Cell. Biol.* **15:** 4702.

Singleton T.L. and Wilcox E. 1998. The largest subunit of mouse RNA polymerase II (RPB1) functionally substituted for its yeast counterpart in vivo. *Gene* **209:** 131.

Sousa R., Rose J., and Wang B.C. 1994. The thumb's knuckle. Flexibility in the thumb subdomain of T7 RNA polymerase is revealed by the structure of a chimeric T7/T3 RNA polymerase. *J. Mol. Biol.* **244:** 6.

Svetlov V., Nolan K., and Burgess R.R. 1998. Rpb3, stoichiometry and sequence determinants of the assembly into yeast RNA polymerase II in vivo. *J. Biol. Chem.* **273:** 10827.

Treich I., Riva M., and Sentenac A. 1991. Zinc-binding subunits of yeast RNA polymerases. *J. Biol. Chem.* **266:** 21971.

Treich I., Carles C., Sentenac A., and Riva M. 1992. Determination of lysine residues affinity labeled in the active site of yeast RNA polymerase II(B) by mutagenesis. *Nucleic Acids Res.* **20:** 4721.

Ulmasov T., Larkin R.M., and Guilfoyle T.J. 1996. Association between 36- and 13.6-kDa α-like subunits of *Arabidopsis thaliana* RNA polymerase II. *J. Biol. Chem.* **271:** 5085.

Uptain S.M., Kane C.M., and Chamberlin M.J. 1997. Basic mechanisms of transcript elongation and its regulation. *Annu. Rev. Biochem.* **66:** 117.

Wang B., Jones D.N., Kaine B.P., and Weiss M.A. 1998. High-resolution structure of an archaeal zinc ribbon defines a general architectural motif in eukaryotic RNA polymerases. *Structure* **6:** 555.

Wettach J., Gohl H.P., Tschochner H., and Thomm M. 1995. Functional interaction of yeast and human TATA-binding proteins with an archaeal RNA polymerase and promoter. *Proc. Natl. Acad. Sci.* **92:** 472.

Woychik N.A. and Young R.A. 1994. Exploration of RNA polymerase II structure and function. In *Transcription: Mechanisms and regulation* (ed. R.C. Conaway and J.W. Conaway), p. 227. Raven Press, New York.

Wu J., Awrey D.E., Edwards A.M., Archambault J., and Friesen J.D. 1996. In vitro characterization of mutant yeast RNA polymerase II with reduced binding for elongation factor TFIIS. *Proc. Natl. Acad. Sci.* **93:** 11552.

Xiao H., Friesen J.D., and Lis J.T. 1994. A highly conserved domain of RNA polymerase II shares a functional element with acidic activation domains of upstream transcription factors. *Mol. Cell. Biol.* **14:** 7507.

Yasui K., Ishiguro A., and Ishihama A. 1998. Location of subunit-subunit contact sites on RNA polymerase II subunit 3 from the fission yeast *Schizosaccharomyces pombe. Biochemistry* **37:** 5542.

Young R.A. 1991. RNA polymerase II. *Annu. Rev. Biochem.* **60:** 689.

Zhu W., Zeng Q., Colangelo C.M., Lewis M., Summers M.F., and Scott R.A. 1996. The N-terminal domain of TFIIB from *Pyrococcus furiosus* forms a zinc ribbon. *Nat. Struct. Biol.* **3:** 122.

Antitermination by Bacteriophage λ Q Protein

J.W. ROBERTS, W. YARNELL, E. BARTLETT, J. GUO, M. MARR, D.C. KO,*
H. SUN, AND C.W. ROBERTS
*Section of Biochemistry, Molecular and Cell Biology, Biotechnology Building, Cornell University,
Ithaca, New York 14853*

The antitermination regulators of bacteriophage λ, proteins encoded by its genes *Q* and *N*, provided the first examples of transcription factors that modify the elongation behavior of RNA polymerase. The category now includes the NusA, NusG, and Gre proteins in bacteria, as well as RNA polymerase II (pol II) factors such as Tat, elongin, SII, and pTEF. Transcription antitermination was recognized originally through biochemical identification of bacteriophage λ terminators that require the bacterial termination factor Rho, along with genetic evidence that regulated expression downstream from the terminators derives from promoters located upstream (Roberts 1969). The existence of pol II elongation regulators, including the likely function of carboxy-terminal domain (CTD) phosphorylation in elongation, indicates the generality of such modification.

Why should there exist factors that antiterminate or more generally promote elongation? Clearly, they provide a mode of regulation. But beyond this, such factors may have evolved as part of the apparatus that ensures persistence of the elongation complex through long transcription units. Transcripts extend as long as hundreds of kilobases in eukaryotes, requiring hours of stability of the functional elongation complex; even the 26-kb phage λ late transcript, perhaps a record for a prokaryotic messenger, requires about 10 minutes. The process of initiation is a pathway that begins with reversible association of RNA polymerase and promoter DNA in the closed complex, and proceeds to a highly stable elongation complex that is disrupted only by termination sequences or enzymes (Rho); antiterminators then may seal the complex even against these influences.

Q and N engage the RNA polymerase very differently, but to the same end: They antiterminate at both simple (intrinsic) and Rho-dependent terminators, and they enhance elongation in some fashion that appears in vitro as decreased pausing (Roberts 1996). It has always seemed likely that antipausing and antitermination are related, although the relationship has been unclear. A simple notion is that pausing is necessary for termination because transcript release requires time and competes with elongation, and the only function of Q is to inhibit this pause. Recent experiments suggest that this is not true: Q (W. Yarnell and J.W. Roberts, unpubl.) and probably N (Rees et al. 1997) directly stabilize the complex even in slow elongation or static complexes, so that a kinetic mechanism need not be invoked. Nevertheless, it seems most likely that the same mechanistic change underlies both this stabilization and the inhibition of pausing.

Several recent developments have particularly enhanced interest in the Q antiterminator of λ. First, we realized that engagement of Q occurs while the σ^{70} initiation factor is still present, and in fact requires σ^{70} (Ring et al. 1996), suggesting that modification occurs before or while the mature elongation complex is formed; thus, understanding Q may illuminate the transition between initiation and elongation. Engagement of Q is strongly reminiscent of regulatory events that affect promoter-proximal paused pol II (Lis, this volume), acting to release it from the initiation process and to potentiate efficient elongation. Second, refinement of the model of mature elongation complexes and the forces that construct them gives a better framework by which to understand how both termination and antitermination may operate (Nudler et al. 1996, 1997; Reeder and Hawley 1996; Komissarova and Kashlev 1997). *Escherichia coli* RNA polymerase and Q provide a simple and defined system with which to study the structure and properties of the transcription elongation complex and particularly elements that stabilize and destabilize it.

GENERAL FEATURES OF Q ACTION

The 22.5-kD phage λ Q protein is required for high-level expression of phage late genes. It acts only on RNA polymerase initiated at a single phage promoter with associated *qut* (Q utilization) sequences, and its primary target is a terminator about 200 nucleotides downstream from the promoter, or somewhat closer for related phages (Roberts 1996). Phage λ relatives (phage 82, φ80, phage 21, P22) have homologous arrangements, but Q proteins of different specificity, and thus provide useful variants.

The Q polypeptide alone can provide stable modification of RNA polymerase in vitro, in contrast to the large assembly of proteins that N protein requires; the complete reaction requires DNA containing the *qut* site, holoenzyme, Q, and NTPs and ions required for transcription. The NusA elongation factor does considerably stimulate the activity of Q-modified RNA polymerase at terminators (Yang et al. 1987; Liu et al. 1996), and it affects the process of modification at the promoter (Yarnell and Roberts 1992). However, NusA is not an essential component of the modification, and its relationship to the downstream complex is unknown.

Present address: Department of Developmental Biology, Stanford University, Stanford, California 94305.

THE *qut* SITE AND THE ENGAGEMENT OF Q INTO THE ELONGATION COMPLEX

The complexity of the *qut* site reflects the engagement process; *qut* consists of the promoter –35 and –10 elements, a QBE (Q-binding element) between and partially overlapping these promoter elements, and sequences in the early transcribed region. The QBE binds Q during engagement with RNA polymerase (Yarnell and Roberts 1992) (as well as in free DNA in vitro; Bartlett 1998), whereas the essential early transcribed sequences turned out to have a surprising and novel role: They are a repeat of parts of the –10 promoter consensus element, and their role is to bind σ^{70} after it releases the promoter but is still bound to the core RNA polymerase (Ring et al. 1996). This novel $E\sigma^{70}$ reaction traps an early elongation complex containing 16 or 17 nucleotides of RNA (Figs. 1 and 2), which then escapes into normal synthesis after a pause of about 30 seconds (Grayhack et al. 1985; Kainz and Roberts 1992). When Q is added in vitro, the pause is greatly shortened and the escaped RNA polymerase carries the Q polypeptide as a subunit (Fig. 1) (Grayhack et al. 1985; W. Yarnell and J.W. Roberts, unpubl.). In effect, Q replaces σ^{70} as a subunit, although it is not known if these events are coordinated or separate.

The requirement for σ^{70} in Q function allowed a genetic screen for σ^{70} mutations defective in pausing (Ko et al. 1998.). These mutations define a region of σ^{70} that is required to stabilize both the early pause and the open promoter complex, i.e., the conformation of holoenzyme that is bound tightly to the bases of the –10 region nontemplate strand. The mutations are on an exposed face of the highly conserved region 2.2, which is thought to be involved in σ^{70}-core interactions (Malhotra et al. 1996) .

Engagement of Q at the *qut* site represents a pathway of sequential interactions that must involve surfaces which the enzyme normally exposes during the transitions from promoter binding to mature elongation complex. Thus, Q does not bind open complex, at which the QBE must be covered by the enzyme, but can bind once the $E\sigma^{70}$ advances to the pause and the QBE is exposed. More strikingly, Q cannot bind when the enzyme is farther downstream (Grayhack et al. 1985; W. Yarnell and J.W. Roberts, unpubl.), even though once engaged it remains in a stable complex with the transcribing enzyme. The permissive boundary has not been exactly defined, but it presumably requires $E\sigma^{70}$ to be close to the QBE. The resistance of fully competent elongation complexes to Q suggests that Q binds a surface that is available during the transition period represented by the pause but is cryptic or altered during elongation. Understanding how Q binds core may reveal important elements of this transition.

STRUCTURE OF THE PAUSED COMPLEX

Several lines of evidence indicate that σ^{70} is required for the early pause and for Q function (Ring et al. 1996). First, the pause-inducing sequence of λ, and its counterparts in the cousin phages 80, 21, and 82, has a clear relation to the –10 consensus; furthermore, mutations that impair pausing also change these consensus elements.

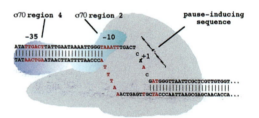

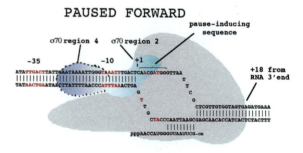

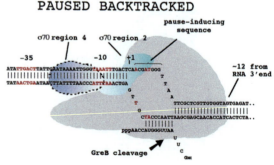

Figure 2. Open complex, and two states of the early σ^{70}-dependent paused complex, at the λ late gene promoter.

Figure 1. Pathway of phage λ gene *Q* antiterminator engagement with RNA polymerase.

Second, initiation from premelted bubbles, which core carries out efficiently and accurately, yields no pausing at +16/+17 unless σ^{70} also is present. Third, direct assay shows σ^{70} present in paused complex, but much reduced in complexes of DNA with mutant pause-inducing sequences that are stopped at the same site (and which also are refractory to modification by Q in vivo and in vitro). Finally, the length of the pause varies among mutants and variants of σ factor (M. Marr et al., unpubl.).

The realization that sequences required for the +16/17 pause are in the region of the melted transcription bubble suggested that only one DNA strand might be active, and experiments with heteroduplexes showed that the nontemplate strand is the important one (Ring and Roberts 1994). However, since the pause-inducing sequence is primarily a reiteration of the –10 promoter consensus sequence, this implies that promoter recognition at –10 also primarily involves the nontemplate strand. The importance is the implication that development of the open promoter complex might reflect primarily establishment of the nontemplate strand base-specific interactions with σ^{70} in holoenzyme (Roberts and Roberts 1996). In fact, the essential promoter-σ^{70} interactions can be measured with a 19-nucleotide oligonucleotide containing the nontemplate strand –10 promoter consensus (Marr and Roberts 1997). Thus, region 2.4 of σ^{70}, which recognizes –10, is shown in the figures bound to the nontemplate strand.

An important element of the current view of an elongation complex is the cleavage reaction that removes several or more nucleotides of RNA from the 3´ end of the transcript (Surratt et al. 1991). This cleavage is greatly stimulated by GreA and GreB proteins in bacteria (Borukhov et al. 1993) (and by the SII elongation factor in eukaryotes; Izban and Luse 1992; Reines 1992) but is catalyzed by the active center of RNA polymerase itself (Orlova et al. 1995). Cleavage occurs when the complex "backtracks," reversing the transit of the transcription bubble in concert with rewinding of the DNA/RNA hybrid toward the 5´ end of the RNA (Reeder and Hawley 1996; Komissarova and Kashlev 1997; Nudler et al. 1997). The consequence is that the 3´ end of the transcript becomes unpaired, and an internal phosphodiester bond becomes placed in the catalytic center, allowing internal

cleavage. Backtracking is promoted in stalled complexes in vitro, and presumably in vivo as well, when the energetics of nucleic acid and protein interactions favor the more backward positions. Pausing may reflect backtracking or may mark sites where backtracking is favored.

Since the +16/17 σ^{70}-dependent paused complex examined in vitro is very sensitive to GreB-stimulated removal of its four 3´ nucleotides of RNA (McDowell 1994), it likely is frequently in the configuration shown in Figure 2 ("Paused Backtracked"). The downstream boundary of RNA polymerase on DNA, determined by exonuclease III digestion to be 11 bp downstream from the nucleotide encoding the RNA 3´ end (W. Yarnell and J.W. Roberts, unpubl.), is appropriate to an elongation complex that has retracted at the front by 4–6 bp (Nudler et al. 1997). The bubble has to unwind by a further 4–5 bp in order to template the RNA, presumably while the back end is still restrained by σ^{70} region 2.4 binding to the pause-inducing sequence in the nontemplate DNA strand (Fig. 2, "Paused Forward"). It is thought that there is flexibility in either the protein or the template in order to explain the existence of these variant structures, as has been considered for the initiation process (Chamberlin 1994; Mustaev et al. 1994).

It is not known how the paused complex may be partitioned between forward and backtracked positions in vivo. Since Gre proteins act in vivo, there may be a quick resynthesis that maintains most RNA polymerase in the forward position; this would depend on the relative rates of the reactions involved.

Paused complex formation and the subsequent bubble retraction to the backtracked position may be exactly analogous to abortive initiation in open complex. Abortive products also are sensitive to Gre-induced cleavage (Hsu et al. 1995), but their fate in the absence of cleavage is to be released, presumably because there is no 5´ RNA extension to stabilize the transcript either as RNA/DNA hybrid or through binding to RNA polymerase.

The structure shown in Figure 2 is confirmed by comparison with its counterpart in phage 82, for which the pause occurs at +25, consistent with an 8-bp rightward displacement of the pause-inducing sequence (Fig. 3) (W.

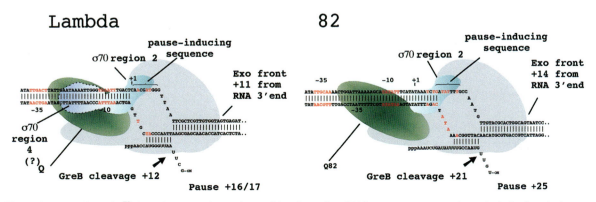

Figure 3. Comparison of σ^{70}-dependent paused complexes of the phages λ and 82 late gene promoters, shown in the backtracked configuration.

Yarnell and J.W. Roberts, unpubl.). As for λ, the 82 paused complex is retracted at the front by 4 nucleotides in vitro (GreB-induced cleavage to +21), and the downstream exonuclease III (Exo III) boundary of DNA digestion is 14 bp downstream from the RNA 3′ end.

Exo III digestion from upstream (digestion of the bottom strand, labeled at its 5′ end) further confirms the structures shown. Thus, the λ paused complex stopped by nucleotide deprivation at +15 shows a strong barrier at −11 that is missing or much weaker in complexes mutant in the pausing sequence, and thus lacking σ^{70}, or allowing σ^{70} to be easily displaced (Yarnell and Roberts 1992). As conjectured in Figure 2, this boundary could be region 2 of σ^{70}, which likely contacts four to five nucleotide pairs to the left of the −10 consensus as duplex DNA. A similar boundary is present in the 82 complex, displaced as expected to −3. It is interesting that the strength of the complex, measured as the time required for Exo III to breach these barriers, is related to the half-life of the pause, among different σs (W. Yarnell and J.W. Roberts, unpubl.): σ^{70}, *Bacillus subtilis* σ^{A}, and σ^{S}, distinct proteins that recognize essentially the same −10 consensus. Thus, the strength of the pause appears to be determined by the same forces that impede Exo III digestion, likely dominated by the σ-nontemplate strand DNA contact.

The paused complex is restrained primarily by σ^{70}, but other elements determine its half-life. Primarily, the G/C base pairs at +7, +8, and +9 just downstream from the −10 consensus match (AACGAT for λ pR′) contribute to the pause (Yang 1988; Ring and Roberts 1994), although it is not known how. They appear to have a preferential nontemplate strand effect (Ring and Roberts 1994), suggesting interaction with RNA polymerase; instead, or in addition, they might act by stabilizing the RNA/DNA hybrid in the backtracked position (Fig. 2). G/C richness in this segment is common to the *qut* sequences of λ, 82, 80, and 21 (Roberts 1992). A second element is the sequence at the front of the transcription bubble, consistent with the involvement of energetics of base pairing in bubble movement (Goliger and Roberts 1989; W. Yarnell and J.W. Roberts, unpubl.).

INTERACTION OF Q WITH RNA POLYMERASE AT THE PAUSE

Sequences of qut^{λ} between −35 and −10, and partly overlapping −10, make up the QBE (Yarnell and Roberts 1992; Bartlett 1998), to which Q binds either in free DNA or in the paused complex (Fig. 4), probably as a dimer (Bartlett 1998). Footprint analysis shows that Q binds the QBE in a transcription complex stopped at the pause site on wild-type DNA, whereas no binding is detected in a complex stopped at the same site on pause-defective mutant DNA (Yarnell and Roberts 1992). Thus, it appears that the presence of σ^{70} is required to reveal this binding site in a transcription complex, whereas in the absence of σ^{70}, some part of RNA polymerase obscures the QBE. Possibly, Q must bind σ^{70} in order to contact the QBE in the context of a transcription complex. Both yeast two-hybrid and mutational analysis suggest that Q^{λ} does bind

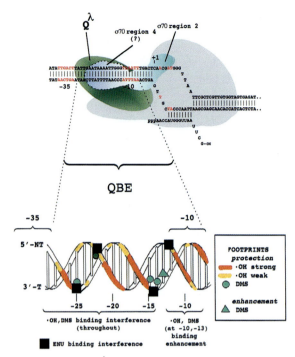

Figure 4. Model of Q^{λ} interaction with the early paused complex.

region 4 of σ^{70}, as indicated by the placement of region 4 in Figure 4 (H.Sun and J.W. Roberts; D.C. Ko and J.W. Roberts; both unpubl.).

Q also must interact with core subunits, because Q^{82} is bound to the elongating complex in the absence of σ^{70} (W. Yarnell and J.W. Roberts, unpubl.), although there is no information about which subunits are involved. A mutation in *rpoB* encoding subunit-β that affects function of Q of the λ relative HK022 could indicate an important region, possibly a contact site (Atkinson and Gottesman 1992).

DNA protection experiments provide physical evidence for Q interaction with the paused complex. Thus, addition of Q^{λ} or Q^{82} to its receptive complex imposes a barrier to Exo III digestion from upstream at about −30 (Yarnell and Roberts 1992). This site corresponds well to the extent of the *qut* sequences inferred from deletion analysis and from footprinting of free DNA. Notably, this same boundary is imposed by Q^{82}, even though the paused complex is about eight nucleotides farther downstream (W. Yarnell and J.W. Roberts, unpubl.).

The MPE footprint of Q-bound paused complex reveals not only protection from −30 to −10, as in free DNA, but also distinct changes from about −2 to +5, including enhancement of reactivity between −2 and +1 (Yarnell and Roberts 1992). Since Q appears not to contact this region in purified DNA, the changes presumably reflect Q-induced modification of holoenzyme subunits. In addition, since the pause-inducing −10 repeat contacted by σ^{70} occurs in just this region (from +1 to +7), σ^{70} is a strong candidate to be involved in the changed interactions.

How does Q actually alter the catalytic properties of the paused complex, and how is this related to antitermination downstream? There would appear to be a relationship: Antipausing occurs both at the early pause site and during transcription downstream (Grayhack et al. 1985; Yang and Roberts 1989). However, there are distinct differences between the early paused and the mature elongating enzyme, such as the presence of σ^{70} and the shortness of the transcript. One potential activity of Q at the pause site that would not apply downstream is "anti-activation." Q could act through the same pathway by which activators like λ repressor and CAP stimulate the rate of open complex formation (Hawley and McClure 1982; Niu et al. 1996) and thus stabilize the holoenzyme–single-stranded DNA interaction, but in the opposite sense: Q might overcome the σ^{70}-dependent pause by destabilizing the interaction of holoenzyme with the nontemplate DNA strand of the pause-inducing sequence. On the other hand, elements that act along with σ^{70} to constrain the early pause also are potential targets: RNA-protein interactions, DNA winding or unwinding energies at the transcription bubble boundaries, RNA/DNA hybrid strength, or single-stranded DNA–protein interactions within the transcription bubble.

The effect of Q^λ on elongation at the pause site can be quantified by measuring the rate of single nucleotide addition to washed, immobilized complexes (McDowell 1994). Q^λ stimulates the rate of addition of GMP to the +16 RNA about three- to fivefold over a range of GTP from 5 μM to 1 mM, saturating at about 50 μM and a rate of 3 per minute. Q has no effect on the rate for the nonpausing mutant, which is two to three times the maximum rate produced by Q on wild-type DNA. Neither the pause nor the effect of Q^λ on the pause appears as a simple change in the kinetic parameters for single nucleotide addition. These experiments also show that the Q effect does not require transcript cleavage (e.g., from some trace Gre protein or by the nonstimulated endogenous reaction) to some earlier position of the transcript, because only GTP is provided in the reaction. However, these reactions may be artifactual in the sense that Gre proteins are present in vivo, and the natural dynamics of cleavage and resynthesis of backtracked paused complexes might change the actual rates. Thus, the maximum rate of elongation from +16 to +17 in vitro of 3–6 per minute may be dominated by the time required for the backtracked complex to isomerize back to the forward position.

We know little about the detailed pathway of the Q-mediated release of the paused complex through which Q eventually replaces σ^{70}. A potential clue involves Gre proteins and backtracked complexes again. Besides speeding modified complexes downstream, a curious effect of adding Q at the pause is to induce arrest in a region centered about ten nucleotides downstream from the pause site (Grayhack et al. 1985; C.W. Roberts and J.W. Roberts; W. Yarnell and J.W. Roberts; both unpubl.). This is illustrated in Figure 5 for Q^{21}, which provides a more extreme example than Q^λ or Q^{82}: Most of the modified complexes are in this region unless they are released by GreB, which increases Q^{21}-dependent readthrough

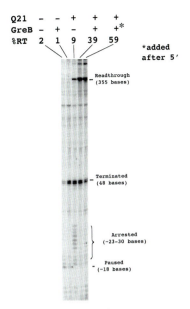

Figure 5. GreB dependence of Q^{21} activity. Single-round synthesis was performed under standard reaction conditions from a linear template containing the phage 21 late gene promoter and terminator. Proteins were added as shown (100 nM GreB; 300 nM Q^{21}).

from 9% to about 50%. Paused complexes that escape without addition of Q do not arrest in this region. These results suggest that interaction between Q and the QBE persists, or can be remade, while the catalytic center has advanced at least a turn of helix downstream from the pause site. Such arrested complexes could reflect failure of Q to release from the QBE after it has bound RNA polymerase during elongation and might represent a trapped intermediate; further study of their formation and properties might help illuminate the process of modification.

DOWNSTREAM EFFECTS OF Q MODIFICATION

Of course, the important regulatory event is antitermination. Bypass of terminators by modified RNA polymerase is very efficient, as measured by the fraction of modified complexes that pass an otherwise stringent terminator. However, the process of modification itself is less efficient, usually on the order of 50%. At least part of the reason for this inefficiency in vitro is the failure of all RNA polymerase to retain σ^{70} and to pause; thus, about 20% of complexes immobilized and washed at the pause site of wild-type DNA do not pause upon addition of substrate (McDowell 1994).

To illustrate the efficiency of antitermination by modified complexes, we show a reaction in which two terminators are transcribed in succession, so that only modified complexes meet the second terminator (t82) (W. Yarnell and J.W. Roberts, unpubl.). As shown in Figure 6, well over 95% of these complexes end up as readthrough. The second panel is a control to demonstrate that t82 is active: A point mutation in the first terminator (t500M) allows a

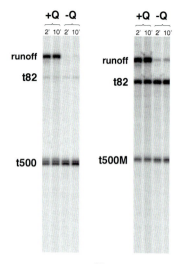

Figure 6. Demonstration that Q^{82}-modified RNA polymerase antiterminates with very high efficiency. The experiment is described in the text.

fraction of the enzyme through, and in the absence of Q, this stops efficiently at t82.

The focus of current interest is to understand how Q-mediated changes in elongation properties prevent termination and to discover what this reveals about termination itself. It appears that modification by Q stabilizes RNA polymerase at termination release sites, even in transcription made indefinitely slow by low substrate concentration. This effect is illustrated in Figure 7 (W. Yarnell and J.W. Roberts, unpubl.), which also shows the antipausing effect of Q and demonstrates the persistence of the modification in washed complexes. For this experiment, immobilized complexes either modified by Q^{82} or not were stopped by nucleotide deprivation at nucleotide 77, and then elongated slowly in 50 nM CTP, 1 μM ATP, GTP, and UTP; both released (S) and bound (P) RNAs are displayed.

First, note that Q-modified complexes advance faster in the region between nucleotide 76 and the t82 release site, yet both modified and unmodified enzymes detect the same collection of pause sites. (Only about half of the complexes are modified, but the most advanced complexes should be entirely modified.) Second, despite an average elongation rate of well less than 1 nucleotide per second, antitermination is very efficient in the advancing wave of Q-modified complexes. Some of the advanced Q-modified enzyme that eventually reads through appears to pause (and possibly backtrack) at t82; this result is consistent with the interpretation that Q inhibits but does not eliminate pausing and does strongly inhibit transcript release. Note that about twice as much transcript is released at 24 minutes in the –Q as in the +Q reactions, consistent with a 50% modification of complexes by Q.

Despite these suggestions that the basic mechanism of antitermination is to stabilize complexes against release, rather than simply speeding enzyme through a rate-limiting step in release, it is impossible to make rigorous conclusions without knowing the exact effect of the modification on the rate of elongation at the site of release. The average rate of Q-mediated elongation cannot be used, because it is just at the release site, where the elongation complex must have unique properties, that the number is likely to vary significantly from any average. Instead, it is necessary to measure the stability of the complex in a more direct way. Our current efforts use an oligonucleotide-mediated release reaction (W. Yarnell and J.W. Roberts, unpubl.) that simulates termination in order to examine the effect of modification by Q.

ACKNOWLEDGMENT

We thank the National Institutes of Health for support.

REFERENCES

Atkinson B.L. and Gottesman M.E. 1992. The *Escherichia coli* Rpo B60 mutation blocks coliphage Hk022 Q-function. *J. Mol. Biol.* **227:** 29.

Bartlett E. 1998. "Characterization of the lambda Q binding site." Ph.D. thesis, Cornell University, Ithaca, New York.

Borukhov S., Sagitov V., and Goldfarb A. 1993. Transcript cleavage factors from *Escherichia coli. Cell* **72:** 459.

Chamberlin M. 1994. New models for the mechanism of transcription elongation and its regulation. *Harvey Lect.* **88:** 1.

Goliger J.A. and Roberts J.W. 1989. Sequences required for antitermination by phage 82 Q protein. *J. Mol. Biol.* **210:** 461.

Grayhack E.J., Yang X., Lau L.F., and Roberts J.W. 1985. Phage λ gene Q antiterminator recognizes RNA polymerase near the promoter and accelerates it through a pause site. *Cell* **42:** 259.

Hawley D. and McClure W. 1982. Mechanism of activation of transcription initiation from the λPRM promoter. *J. Mol. Biol.* **157:** 493.

Hsu L.M., Vo N.V., and Chamberlin M.J. 1995. *Escherichia coli* transcript cleavage factors GreA and GreB stimulate promoter escape and gene expression in vivo and in vitro. *Proc.*

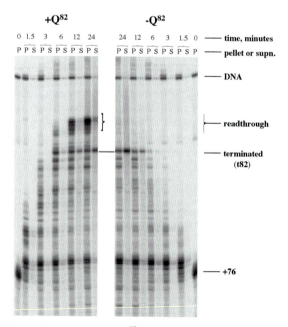

Figure 7. Progression of Q^{82}-modified and unmodified RNA polymerase through DNA containing a terminator, at very low NTP concentration (50 nM CTP; 1 μM ATP, GTP, CTP).

Natl. Acad. Sci. **92:** 11588.

Izban M.G. and Luse D.S. 1992. The RNA polymerase II ternary complex cleaves the nascent transcript in a 3′-5′ direction in the presence of elongation factor SII. *Genes Dev.* **6:** 1342.

Kainz M. and Roberts J.W. 1992. Structure of transcription elongation complexes *in vivo*. *Science* **255:** 838.

Ko D.C., Marr T., Guo J., and Roberts J.W. 1998. A surface of *E. coli* σ^{70} required for promoter function and antitermination by phage Lambda Q protein. *Genes Dev.* **12:** 3276.

Komissarova N. and Kashlev M. 1997. Transcriptional arrest: *Escherichia coli* RNA translocates backward, leaving the 3′ end of the RNA intact and extruded. *Proc. Natl. Acad. Sci.* **94:** 1755.

Liu K., Zhang Y., Severinov K., Das A., and Hanna M. 1996. Role of *Escherichia coli* RNA polymerase α subunit in modulation of pausing, termination and anti-termination by the transcription elongation factor NusA. *EMBO J.* **15:** 150.

Malhotra A., Severinova E., and Darst S.A. 1996. Crystal structure of a σ70 subunit fragment from *E. coli* RNA polymerase. *Cell* **87:** 127.

Marr M.T. and Roberts J.W. 1997. Promoter recognition as measured by binding of polymerase to nontemplate strand oligonucleotide. *Science* **276:** 1258.

McDowell J.C. 1994. "Relation of transcription termination and antitermination to the kinetics of transcription elongation." Ph.D. thesis, Cornell University, Ithaca, New York.

Mustaev A., Zaychikov E., Severinov K., Kashlev M., Polyakov N.-V., and Goldfarb A. 1994. Topology of the RNA polymerase active center probed by rifampicin-nucleotide compounds. *Proc. Natl. Acad. Sci.* **91:** 12036.

Niu W., Kim Y., Tau G., Heyduk T., and Ebright R. 1996. Transcription activation at class II CAP-dependent promoters: Two interactions between CAP and RNA polymerase. *Cell* **87:** 1123.

Nudler E., Avetissova E., Markovtsov V., and Goldfarb A. 1996. Transcription processivity: Protein-DNA interactions holding together the elongation complex. *Science* **273:** 211.

Nudler E., Mustaev A., Lukhtanov E., and Goldfarb A. 1997. The RNA-DNA hybrid maintains the register of transcription preventing backtracking of RNA polymerase. *Cell* **89:** 33.

Orlova M., Newlands J., Das A., Goldfarb A., and Borukhov S. 1995. Intrinsic transcript cleavage activity of RNA polymerase. *Proc. Natl. Acad. Sci.* **92:** 4596.

Reeder T.C. and Hawley D.K. 1996. Promoter proximal sequences modulate RNA polymerase II by a novel mechanism. *Cell* **87:** 767.

Rees W.A., Weitzel S.E., Das A., and von Hippel P.H. 1997. Regulation of the elongation -termination decision at intrinsic terminators by antitermination protein N of phage λ. *J. Mol. Biol.* **273:** 797.

Reines D. 1992. Elongation factor-dependent shortening by template-engaged RNA polymerase II. *J. Biol. Chem.* **267:** 3795.

Ring B.Z. and Roberts J.W. 1994. Function of a nontranscribed DNA strand site in transcription elongation. *Cell* **78:** 317.

Ring B.Z., Yarnell W.S., and Roberts J.W. 1996. Function of *E. coli* RNA polymerase Sigma factor sigma 70 in promoter-proximal pausing. *Cell* **86:** 485.

Roberts C.W. and Roberts J.W. 1996. Base-specific recognition of the nontemplate strand of promoter DNA by *E. coli* RNA polymerase. *Cell* **86:** 495.

Roberts J.W. 1969. Termination factor for RNA synthesis. *Nature* **224:** 1168.

——. 1992. Antitermination and the control of transcription elongation. In *Transcriptional regulation*, Book I (ed. S.L. McKnight and K.R. Yamamoto), p. 389. Cold Spring Harbor Laboratory Press, Cold Spring Harbor, New York.

——. 1996. Transcription termination and its control. In *Regulation of gene expression in* Escherichia coli (ed. E.C.C. Lin and A.S. Lynch), p. 27. R.G. Landes, Austin, Texas.

Surratt C.K., Milan S.C., and Chamberlin M.J. 1991. Spontaneous cleavage of RNA in ternary complexes of *Escherichia coli* RNA polymerase and its significance for the mechanism of transcription. *Proc. Natl. Acad. Sci.* **88:** 7983.

Yang X. 1988. "Transcription antitermination mediated by lambdoid phage Q proteins." Ph.D. thesis, Cornell University, Ithaca, New York.

Yang X. and Roberts J.W. 1989. Gene Q antiterminator proteins of *Escherichia coli* phages 82 and lambda suppress pausing by RNA polymerase at a rho-dependent terminator and at other sites. *Proc. Natl. Acad. Sci.* **86:** 5301.

Yang X., Hart C.M., Grayhack E.J., and Roberts J.W. 1987. Transcription antitermination by phage λ gene Q protein requires a DNA segment spanning the RNA start site. *Genes Dev.* **1:** 217.

Yarnell W.S. and Roberts J.W. 1992. The phage λ gene Q transcription antiterminator binds DNA in the late gene promoter as it modifies RNA polymerase. *Cell* **69:** 1181.

Structure and Mechanism in Transcriptional Antitermination by the Bacteriophage λ N Protein

J. GREENBLATT,*† T.-F. MAH,*† P. LEGAULT,*†‡ J. MOGRIDGE,*† J. LI,* AND L.E. KAY†‡

*Banting and Best Department of Medical Research; †Department of Molecular and Medical Genetics;
‡Protein Engineering Network of Centers of Excellence and Departments of Biochemistry and Chemistry,
University of Toronto, Toronto, Ontario, Canada M5G 1L6

Each of the two early operons and the late operon of bacteriophage λ contain transcriptional terminators (Fig. 1a). As a consequence, the pattern of transcription during a lytic infection by λ can be viewed as a cascade of transcriptional antitermination mechanisms (for review, see Das 1993; Greenblatt et al. 1993; Friedman and Court 1995; Richardson and Greenblatt 1996). These antitermination mechanisms are "processive" in the sense that they can influence termination through multiple terminators over many kilobases of DNA and many minutes of transcription time. The *N* gene is the first gene to be transcribed in the leftward early p_L operon and encodes an antitermination factor. Once N protein is made, it modifies *Escherichia coli* RNA polymerase transcribing the early p_L and p_R operons so that the RNA polymerase transcribes efficiently through all the intrinsic and Rho-dependent terminators of both operons. This leads to the ex-

pression of genes involved in DNA replication (*O, P*) and genetic recombination (*exo, int, xis*). Some time later, the *Q* gene, which is located far downstream from the rightward early promoter p_R, is transcribed as a consequence of antitermination by N. Its product, the Q protein, modifies the *E. coli* RNA polymerase molecules initiating constitutively at the late promoter p'_R so that the RNA polymerase passes through the downstream terminator t'_R and transcribes all of the late lysis and morphogenesis genes of the phage. The N and Q proteins are operon-specific activator proteins and recognize their target operons because they have specific sites, *nut* ("N utilization") in the case of N and *qut* ("Q utilization") in the case of Q. Similar antitermination mechanisms exist in related lambdoid bacteriophages like phage 21 and the *Salmonella* phage P22, but the λ N protein is unable to function with the *nut* sites of phages 21 and P22.

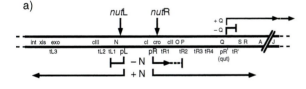

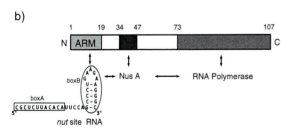

Figure 1. Transcriptional regulation by the N protein of bacteriophage λ. (*a*) Partial genetic and transcriptional map of λ showing the early promoters p_L and p_R, the late promoter p'_R, various terminators (t_{R1}, t_{R2}, ...), the N utilization sites *nutL* and *nutR*, the Q utilization site *qutR*, and the effects of the N and Q antiterminator proteins on transcription. The *int*, *xis*, and *exo* genes are involved in genetic recombination, the *c*I, *c*II, and *c*III genes in the control of lysogeny, the *O* and *P* genes in DNA replication, the *S* and *R* genes in bacterial cell lysis, and the genes *A* through *J* in phage particle morphogenesis. (*b*) Functional domains of the N protein and their interactions with the *nut* site RNA, NusA, and RNA polymerase. Amino acids 1–19 constitute an arginine-rich motif (ARM).

RNA LOOPING

The *nutL* and *nutR* sites are located in transcribed regions downstream from p_L and p_R, respectively. Observations made initially by Friedman and colleagues that mutations allowing ribosomes to translate across a *nut* site prevent antitermination by N (Olson et al. 1982, 1984; Warren and Das 1984; Zuber et al. 1987) provided the first evidence that *nut* sites might function as RNA rather than DNA. This evidence led us to suggest that N and bacterial antitermination cofactors might remain associated with the *nut* site RNA and RNA polymerase during chain elongation, leading to the formation of an RNA loop (see Fig. 2) (Greenblatt 1984). This RNA looping model for antitermination was supported subsequently by several observations: First, we showed that the *nut* site in the nascent transcript is protected from chemical modification and ribonuclease digestion by the presence of N and its bacterial cofactors during transcription in vitro (Nodwell and Greenblatt 1991); second, we found that protection of the *nut* site RNA is weakened by mutations in RNA polymerase that interfere with antitermination by N (Nodwell and Greenblatt 1991); third, Whalen and Das (1990) showed that removal of the *nut*-site-containing DNA fragment from a transcription template by digestion with a restriction enzyme once RNA polymerase had moved past the *nut* site does not interfere with antitermi-

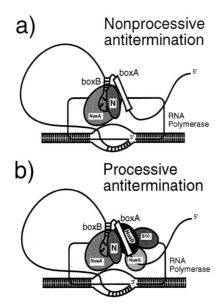

Figure 2. RNA looping models for (*a*) nonprocessive antitermination at a terminator located just downstream from a *nut* site and (*b*) processive antitermination mediated by N over kilobases of DNA. See text for details.

nation by N. According to this model, the *nut* site is effectively a mobile RNA enhancer. A similar RNA looping model involving a mobile RNA enhancer was proposed later to account for transcriptional antitermination by the Tat *trans*-activator proteins of the immunodeficiency viruses (Sharp and Marciniak 1989).

ASSEMBLY OF THE N-MODIFIED ELONGATION COMPLEX

The analysis of mutations in *E. coli* genes, the *nus* genes, that prevent the growth of λ because they interfere with antitermination by N (for review, see Friedman et al. 1984) led to the identification of three *E. coli* proteins, NusA, NusB, and ribosomal protein S10 (NusE), that participate in antitermination by N (Friedman and Baron 1974; Keppel et al. 1974; Friedman et al. 1976, 1981; Das and Wolska 1984; Goda and Greenblatt 1985). Subsequently, we reconstituted processive antitermination by N in a fully purified system containing seven proteins (RNA polymerase, N, Rho factor, NusA, NusB, S10, and NusG), and this led to the identification of NusG as a fourth *E. coli* protein involved in antitermination by N (Li et al. 1992). NusG also interacts with termination factor Rho (Li et al. 1993) and is important for termination mediated by Rho in some circumstances (Sullivan and Gottesman 1992; Li et al. 1993; Nehrke and Platt 1994; Burns and Richardson 1995). Although a mutation in the *nusG* gene suppresses the deleterious effects of the *nusA1* and *nusE71* mutations on antitermination by N (Sullivan et al. 1992), NusG appears to be nonessential for antitermination by N in vivo (Sullivan and Gottesman 1992). Therefore, yet another unidentified bacterial protein may

participate in antitermination by N and make NusG redundant for antitermination in vivo.

Purification of elongation complexes from transcription reactions containing N, RNA polymerase, and all four *E. coli* Nus proteins enabled us to show that N and the four Nus factors become associated stably with the elongation complex if the DNA template contains a *nut* site (see Fig. 2b) (Horwitz et al. 1987; Mason and Greenblatt 1991). This entire assembly process is apparently DNA-independent: We used gel mobility shift assays to show that N, RNA polymerase, and the four Nus factors can also be made to associate stably with *nut* site RNA in the absence of DNA (Mogridge et al. 1995, 1998a). Stable association of RNA polymerase with *nut* site RNA requires N and all four Nus proteins (Mogridge et al. 1998a), which probably explains why all four Nus factors are also required for processive antitermination by N in vitro (Mason et al. 1992a; DeVito and Das 1994). Among the many protein-protein interactions in this complex, three of the Nus factors, NusA, NusG, and S10, interact directly with RNA polymerase (Greenblatt and Li 1981b; Mason and Greenblatt 1991; Li et al. 1992). In the absence of NusB, NusG, and S10, we found that association of RNA polymerase with the N-NusA-*nut* complex is unstable (Mogridge et al. 1995, 1998a). In this situation (see Fig. 2a), antitermination by N is nonprocessive but still occurs at a terminator located just downstream from a *nut* site (Whalen et al. 1988; Mason et al. 1992a). In these circumstances, it is likely that antitermination fails further downstream because the weakly bound N-NusA-*nut* complex dissociates from RNA polymerase as the RNA loop grows large and its localizing effect is lost (Nodwell and Greenblatt 1991; Van Gilst et al. 1997). In the nonprocessive N-modified transcription complex shown in Figure 2a, N and NusA interact with each other (Greenblatt and Li 1981a), with RNA polymerase (Greenblatt and Li 1981b; Mogridge et al. 1998b), and with the *nut* site RNA (Chattopadhyay et al. 1995; Mogridge et al. 1995; Tan and Frankel 1995; Cilley and Williamson 1997; Su et al. 1997b; Van Gilst et al. 1997). In the stable, processive complex shown in Figure 2b, there are additional protein-protein (Mason and Greenblatt 1991; Li et al. 1992; Mason et al. 1992b) and protein-RNA (Nodwell and Greenblatt 1993; Mogridge et al. 1998a) interactions that are likely to account for the increased stability of the complex.

The ability of N to antiterminate at intrinsic and Rho-dependent terminators in the absence of NusB, NusG, and S10 (Whalen et al. 1988; Mason et al. 1992a) implies that the basic antitermination mechanism resides in N and NusA. On its own, NusA increases the termination efficiency at intrinsic terminators (Greenblatt et al. 1981; Grayhack and Roberts 1982; Schmidt and Chamberlin 1987). Moreover, high concentrations of N can suppress termination in the absence of NusA or a *nut* site (Rees et al. 1996). This and other observations (Mogridge et al. 1998b) have implied that N can antiterminate by directly contacting RNA polymerase. The *nut* site RNA and bacterial cofactors may participate in antitermination mostly because they localize N in the vicinity of the RNA polymerase molecules that are transcribing the λ p_L and p_R operons.

RECOGNITION OF *NUT* (*BOXA+BOXB*) BY N AND BACTERIAL PROTEINS

The *nut* sites have at least two important elements, *boxA* (Friedman and Olson 1983) and *boxB* (Fig. 1b) (Salstrom and Szybalski 1978; Doelling and Franklin 1990; Chattopadhyay et al. 1995). The *boxA* elements, located upstream of *boxB*, are very similar in the *nut* sites of various lambdoid phages and are closely related to *boxA* antiterminator elements in the *E. coli rrn* operons, which encode ribosomal RNA (see Fig. 3d) (Li et al. 1984). We used gel mobility shift assays to show that the *rrn* and λ *boxA* elements are recognized by NusB and S10 (Nodwell and Greenblatt 1993; Mogridge et al. 1998a), which themselves form a heterodimer (Mason et al. 1992b). Moreover, there is evidence that NusB is required for *rrn boxA*-dependent antitermination in vitro (Squires et al. 1993) and is titrated by overproduction of *boxA*-containing RNA in vivo (Friedman et al. 1990). One possibility is that *boxA* is recognized by NusB and that the interaction of NusB with *boxA* is somehow stabilized by S10.

In the case of *rrn boxA*, NusB and S10 are sufficient for weak but detectable binding in a gel mobility shift assay (Nodwell and Greenblatt 1993). This is not the case for λ

boxA, for which *boxA*-dependent protein binding, dependent on the sequence GCUCUU in *boxA*, requires NusB, S10, and NusG as well as the preassembly on a complete *nut* site of N, NusA, and RNA polymerase (Mogridge et al. 1998a). This implies that λ *boxA* is a mutant version of *rrn boxA* in which the binding of NusB and S10 depends on N and *boxB*. Stabilization of the N-modified complex (Fig. 2) by NusB, NusG, and S10 requires the *boxA* sequence (Mogridge et al. 1998a).

The *boxB* elements located downstream from *boxA* differ considerably in their nucleotide sequences among the various lambdoid phages, but they are all, at least theoretically, capable of forming small RNA hairpins (Franklin 1985). The construction and assay in vivo of chimeric N proteins revealed that their phage specificity resides in short arginine-rich motifs (ARMs) located near the amino termini of the various N proteins (Lazinski et al. 1989). Gel mobility shift assays and ribonuclease protection assays were then used to show that the N proteins bind specifically to *boxB* RNA in vitro (Chattopadhyay et al. 1995; Mogridge et al. 1995; Tan and Frankel 1995). Moreover, the in vitro binding specificity of the various ARMs could explain the *nut* site specificity of the N proteins in vivo: For example, the λ N protein binds to λ *boxB* but not phage 21 or P22 *boxB*, whereas the ARM of

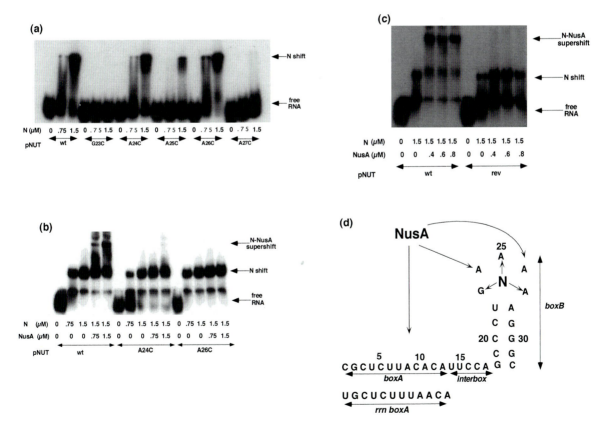

Figure 3. Effects of *boxA* and *boxB* loop mutations on the binding of N and NusA to *nutR* site RNA in gel mobility shift assays. (*a*) Binding of the indicated concentrations of N to wild-type *nut* site RNA and to *nut* site RNAs containing the indicated mutations in the *boxB* loop. (*b*) Effects of the indicated *boxB* loop mutations on the binding of NusA to the N-*nut* site complex. (*c*) Effect of reversing the *boxA-interbox* sequence upstream of *boxB* on the binding of NusA to the N-*nut* site complex. (*d*) Sequence of the *nutR* RNA, including a comparison of the λ *boxA* sequence with the *boxA* sequences found in the *E. coli rrn* operons. (Data in panels *a–c* are reprinted, with permission, from Mogridge et al. 1995.)

phage 22 N binds equally well to *boxB* RNA from phage λ or phage P22 (Tan and Frankel 1995). This corresponds to observations that the λ N protein cannot function with the *nut* sites of phages 21 and P22 (Friedman et al. 1973; Hilliker and Botstein 1976) unless N is overproduced (Schauer et al. 1987; Franklin and Doelling 1989), whereas the P22 N protein has specificity but can function to some extent with the *nut* sites of heterologous phages (Hilliker and Botstein 1976; Lazinski et al. 1989).

The λ *boxB* element is a 15-nucleotide hairpin with a 5-bp stem and a five nucleotide loop (Fig. 1b). The analysis of mutations in *boxB* has revealed the nucleotide sequence requirements for binding and antitermination by N. Although base pairing in the stem is important for antitermination, the identities of all but the top U-A base pair seem to have little importance. In contrast, the identities of all five loop nucleotides are critical for antitermination in vivo and in vitro (Doelling and Franklin 1990; Chattopadhyay et al. 1995). It was therefore of great interest when most studies found that only the identities of nucleotides 1, 3, and 5 of the loop, and not the identities of nucleotides 2 and 4, were important for the binding of N (Fig. 3a) (Chattopadhyay et al. 1995; Mogridge et al. 1995; Tan and Frankel 1995). This implied that nucleotides 2 and 4 are required for some other aspect of antitermination by N.

STRUCTURE OF A COMPLEX CONTAINING THE ARGININE-RICH MOTIF OF N AND *BOXB* RNA

High-resolution structures have now been determined for four ARMs bound to their target RNAs. Although ARMs all consist of a high density of arginine residues in a short stretch of only 10–20 amino acids, these structures have revealed that there is substantial flexibility in the ways in which ARMs recognize RNA. The ARMs of human immunodeficiency virus type-1 (HIV-1) Rev and bovine immunodeficiency virus (BIV) Tat make nonspecific and base-specific contacts with both RNA strands in widened major grooves of double-stranded regions of the HIV-1 Rev response element (RRE) and BIV TAR element, respectively (Puglisi et al. 1995; Ye et al. 1995, 1996; Battiste et al. 1996). Nevertheless, Rev recognizes the RNA as a regular α-helix, whereas BIV Tat forms a β-hairpin.

The N proteins of phages λ and P22 recognize their cognate *boxB* RNA hairpins in yet another way (Fig. 4a,b). We (Legault et al. 1998) and Cai et al. (1998) used heteronuclear magnetic resonance to produce structures of the ARMs of these N proteins bound to their cognate *boxB* RNAs. Both bind on the major groove face, but neither N protein penetrates deeply into the major groove. Rather, each N peptide forms a bent α-helix that interacts almost entirely with the loop and only one strand, the 5′ strand, of the regular A-form hairpin stem. These structures are consistent with the observation that λ N protein protects only the loop and the 5′ strand of the stem from ribonuclease digestion (Chattopadhyay et al. 1995). The different modes of RNA recognition by the ARMs of Tat,

Rev, and the N proteins may be partly explained by the flexibility of RNA structure and the ability of arginine side chains to form a variety of hydrophobic, hydrogen-bonding and electrostatic interactions with RNA.

The λ and P22 N peptides interact in very similar ways with their *boxB* hairpin stems (Fig. 4). In each case, for example, the methyl group of an alanine residue (Ala-3 in λ N protein) makes hydrophobic contacts with riboses and bases of the first two nucleotides of the 5′ stem, and an arginine side chain (Arg-7 in λ N protein) enters the major groove, where it makes hydrophobic contacts with nucleotides in the stem and probably forms a hydrogen bond with the first nucleotide of the loop. This may explain why all tested mutations at Arg-7 of the λ N protein abolish antitermination (Franklin 1993). However, the striking bend in the α-helix of the λ N peptide (Su et al. 1997a; Legault et al. 1998) is displaced toward the amino terminus by about one α-helical turn in the P22 peptide (Cai et al. 1998), and the two N peptides interact very differently with their hairpin loops (Fig. 4).

Interestingly, the five nucleotide loops of the λ and P22 *boxB* elements adopt a fold identical to that of the GAAA

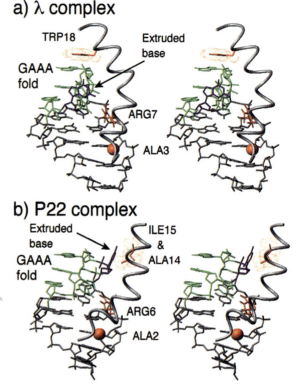

Figure 4. Stereoviews comparing the structures of (*a*) the phage λ ARM peptide (amino acids 1–22) and (*b*) the phage P22 ARM peptide (amino acids 14–28; pdb file 1A4T) bound to their cognate *boxB* RNAs (Cai et al. 1998; Legault et al. 1998). (*Green*) Stacked nucleotides of the GNRA folds of the *boxB* loops; (*purple*) nucleotides that extrude from the GNRA folds, loop nucleotide 4 in λ, and loop nucleotide 3 in P22; (*gray*) *boxB* hairpin stems and the α-helical peptides. Individual representative structures were generated using the program MOLMOL (Koradi et al. 1996) and were oriented relative to each other by superimposing the bases of the GAAA folds and the U-A closing base pairs. See text for other details.

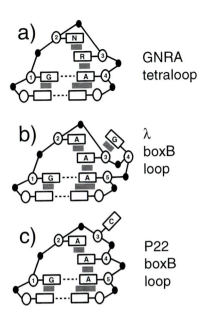

Figure 5. GNRA folds in the λ and P22 *boxB* loops. (*a*) Structural features of the GAAA tetraloop (Heus and Pardi 1991) and summarization of the base requirements for a GNRA fold; (*b*) structural features of the λ *boxB* GAAGA pentaloop (Legault et al. 1998). Nucleotide 4 is extruded from the GNRA fold and is required for the interaction of NusA with the complex. The base of extruded nucleotide 4 interacts with the ribose of loop nucleotide 3. (*c*) Structural features of the P22 *boxB* GACAA pentaloop (Cai et al. 1998). Nucleotide 3 is extruded from the GNRA fold and interacts with the P22 N peptide.

tetraloop, a member of the family of GNRA tetraloops (Heus and Pardi 1991). In this type of GNRA fold (Fig. 5a), the first and last nucleotides of the loop form a sheared G-A base pair, there is a large change in direction of the phosphate backbone between the first and second nucleotides, and the second, third, and fourth nucleotides stack sequentially on the 3′ stem.

Quite remarkably, the five nucleotide loops of the λ and P22 *boxB* elements use different means to create four nucleotide GNRA folds in five nucleotide loops (Fig. 5b,c). In the case of the GAAGA *boxB* loop of λ *nutL*, nucleotides 1, 2, 3, and 5 form a GAAA fold which excludes nucleotide 4 (Fig. 5b). Moreover, nucleotides 1, 3, and 5, which are critical for the formation of the GNRA fold, are precisely those residues in the loop which are critical for the binding of the λ N protein (see Fig. 3a) (Mogridge et al. 1995). This suggests that the λ N protein recognizes the shape of the GNRA fold, and indeed, the indole ring of Trp-18 of λ N is stacked on the top nucleotide of the fold (i.e., nucleotide 2 of the loop) (Figs. 4a and 5b) (Su et al. 1997b; Legault et al. 1998). The excluded nucleotide 4 does not interact with N and can be deleted without the loss of binding (Legault et al. 1998). Therefore, the λ N protein recognizes a genuine GNRA tetraloop.

In striking contrast, nucleotides 1, 2, 4, and 5 of the *boxB* loop of P22 form a GAAA fold which excludes nucleotide 3 (Cai et al. 1998) (Fig. 5c). In this case, the excluded nucleotide 3, a cytidine, makes hydrophobic con-

tacts with several residues of the P22 N peptide (Fig. 4b). Since P22 N binds equally well to the P22 *boxB* element, whose loop is GACAA, and the *boxB* element of λ *nutL*, whose loop is GAAGA (Tan and Frankel 1995), the λ *boxB* loop most likely can form two different kinds of GAAA tetraloops, excluding nucleotide 4 when λ N is bound and nucleotide 3 when P22 N is bound. Viewed this way, it is easy to understand how the P22 N protein can function to some extent with either *boxB* of P22 or *boxB* of λ (Hilliker and Botstein 1976). More generally, it seems plausible that any loop of the form $G(N_x)RA$ or $GNR(N_x)A$ ($x = 1, 2, 3...$) might be capable of forming a GNRA fold. In that case, GNRA folds may be very common in ribosomal RNAs and elsewhere.

DOMAINS OF N

Nuclear magnetic resonance (NMR) and circular dichroic (CD) experiments have shown that the 107-amino-acid λ N protein is disordered (Van Gilst et al. 1997; Legault et al. 1998). This lack of secondary and tertiary structure and the large number of basic residues in N (22% arginine plus lysine) probably explain why N has a half-life of only 2 minutes in vivo (Konrad 1968; Rabovsky and Konrad 1970; Schwartz 1970; Greenblatt 1973). We and other investigators have used various fragments of N in protein binding, gel mobility shift, and antitermination assays and found that N has at least three functionally important regions (see Fig. 1): The amino-terminal ARM (amino acids 1–19) interacts with *nut* site RNA and generates the operon specificity in this control system (Lazinski et al. 1989; Tan and Frankel 1995; Su et al. 1997a; Mogridge et al. 1998b); a central region of N (amino acids 34–47) interacts directly with NusA (Mogridge et al. 1998b); and a carboxy-terminal region (amino acids 73–107) interacts with RNA polymerase (Mogridge et al. 1998b). There may also be an interaction with RNA polymerase of the amino-terminal half of N (amino acids 1–47). In analogy with the nomenclature used for proteins that activate transcriptional initiation, the middle and carboxy-terminal regions of N behave as activation domains for antitermination in transcription assays (Fig. 6b) (Mogridge et al. 1998b). NMR experiments have shown that only the amino-terminal ARM of N, and not its downstream antitermination regions, becomes folded when N interacts with *boxB* RNA (Van Gilst and von Hippel 1997; Van Gilst et al. 1997; Mogridge et al. 1998b). Therefore, N is a multidomain disordered protein whose domains must fold independently when they encounter their targets.

INTERACTION OF THE N-*NUT* COMPLEX WITH NUSA

Transcriptional antitermination requires that the N-*nut* site complex communicate via RNA looping with elongating *E. coli* RNA polymerase, as illustrated in Figure 2. A key aspect of this link to RNA polymerase is provided by NusA, an *E. coli* elongation factor that can interact di-

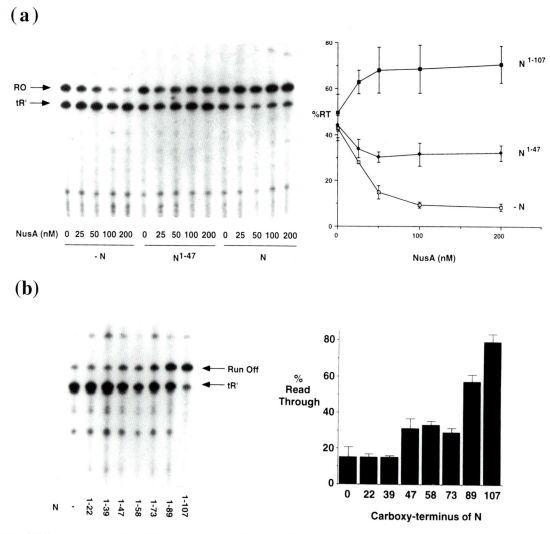

Figure 6. Effects on transcription of the antitermination domains of the λ N protein. (*a*) $N^{1–47}$ supports an intermediate level of antitermination. (*b*) The carboxy-terminal region of N is also important for antitermination. Transcription reactions containing 25 nM RNA polymerase, 100 nM NusA unless otherwise indicated, and 100 nM full-length N or various deletions mutants of N, as indicated, were analyzed on a denaturing polyacrylamide gel and autoradiographed. The DNA fragment used as a template contained p_L, *nutL*, and t'_R. The positions of terminated (t'_R) and runoff (RO) transcripts are indicated. The experiments were quantitated and expressed as percent terminator read-through (RT). (Reprinted, with permission, from Mogridge et al. 1998b [copyright Cell Press].)

rectly with both RNA polymerase (Greenblatt and Li 1981b) and a central region of N (Greenblatt and Li 1981a). If these direct protein-protein interactions were strong enough, there would probably be no *nut* site specificity for antitermination mediated by N, and indeed, *nut* site specificity is lost when N is overproduced (Schauer et al. 1987; Franklin and Doelling 1989; Lazinski et al. 1989). That a *nut* site is normally required for antitermination implies that the N-NusA and NusA-RNA polymerase interactions are too weak to drive the assembly process at physiological concentrations of N (about 10^{-7} M; Greenblatt 1973; Greenblatt et al. 1980). Since we have estimated that the affinity of N for NusA is about 10^6 M^{-1} (Formosa et al. 1992 and unpubl.) and have found that the affinity of N for a NusA-RNA polymerase complex is about 5–10 times lower than its affinity for free NusA (J. Li and J. Greenblatt, unpubl.), *nut*-site-independent

dent interaction of N with RNA polymerase should occur only 1–2% of the time in vivo and have little biological effect. Strengthening of the N-NusA interaction by RNA looping (Fig. 2a) and by direct interactions between NusA and the *nut* site RNA would then make antitermination *nut*-site-specific.

The interaction of NusA with the N-*nut* site complex can be assessed by gel mobility shift assays utilizing ^{32}P-labeled *nut*-site-containing RNA, as shown in Figure 3b (Mogridge et al. 1995). NusA does not bind the RNA detectably on its own but does cause a supershift in the presence of N. Most importantly, NusA does not bind to the N-*nut* site complex when nucleotide 2 or 4 of the *boxB* loop is changed from G or A to C (Fig. 3b) (Mogridge et al. 1995). These loop nucleotides have little effect on the binding of N but are required for antitermination by N (Fig. 3a) (Doelling and Franklin 1989; Chattopadhyay et al. 1995; Mogridge et al. 1995). When nucleotide 4, which extrudes from the GNRA fold of the *boxB* loop (Fig. 5b), is deleted, there is very little effect on the binding of N, but NusA no longer binds to the N-*nut* site complex (Legault et al. 1998). Therefore, nucleotide 4 of the loop has a key role in enabling the antitermination factor N to communicate with the *E. coli* transcription apparatus via NusA. In analogy with the ways in which protein modules like the SH2, SH3, PTB, PDZ, and WW domains organize signaling complexes (Pawson and Scott 1997), *boxB* serves as an adaptor module that helps couple N to NusA.

The *nut* site RNA upstream of *boxB* also has a strong effect on the binding of NusA to the N-*nut* site complex. As shown in Figure 3c (Mogridge et al. 1995), the reversal of the upstream *boxA* and *interbox* sequences prevents the binding of NusA. Further analysis with single and multiple point mutations in the *boxA* and *interbox* regions (T.-F. Mah and J. Greenblatt, unpubl.) has shown that the identities of nucleotides in the 3′ half of *boxA* are most important for the binding of NusA.

The data in Figure 3, b and c, imply that NusA might have multiple direct interactions with the *nut* site RNA. Sequence comparisons have revealed that NusA has an S1 domain (Fig. 7c), which is a presumptive RNA-binding domain whose prototypes are found in bacterial ribosomal protein S1. The S1 domain is a member of the family of OB folds (Bycroft et al. 1997), which bind oligonucleotides and oligosaccharides. As well, NusA has two KH domains, which are also thought to bind RNA (Fig. 7c) (Musco et al. 1996, 1997). The *nusA1* mu-

tation prevents antitermination by N in vivo and in vitro (Friedman and Baron 1974; Das and Wolska 1984; Goda and Greenblatt 1985; Horwitz et al. 1987) without affecting the binding of N to NusA (Greenblatt and Li 1981a). Instead, the *nusA1* mutation prevents the binding of NusA to the N-*nut* site complex (Fig. 7a) (Mogridge et al. 1995), presumably because the NusA1 mutant protein has a defect in RNA binding. Consistent with this idea, the *nusA1* mutation is located in the S1 domain of NusA (Craven and Friedman 1991; Ito et al. 1991). This and other observations (Y.N. Yu et al., in prep.) indicate that the S1 domain of NusA has an important role in binding the *nut* site RNA. Since we have also found that ribosomal protein S1 can itself bind quite selectively to RNAs that contain the *rrn* or λ *boxA* sequence (Mogridge and Greenblatt 1998), one possibility is that the S1 domain of NusA interacts with the *boxA* region of the λ *nut* site RNA. The particular region of NusA that recognizes the loop of *boxB* has not yet been identified.

In any case, the interaction of NusA with an N-*nut* site complex is likely to be at least tripartite (Fig. 8). First, there is a direct interaction of NusA with amino acids 34–47 of N (Mogridge et al. 1998b). Second, there may be an interaction of NusA with extruded nucleotide 4 of the *boxB* loop and/or the minor groove face of the *boxB* A-form stem, which is not occupied by N (Mogridge et al. 1995; Legault et al. 1998). Third, there may be a direct interaction of NusA with *boxA*. It would not be surprising if these interactions had a substantial effect on the conformation of NusA and altered its interaction with RNA polymerase.

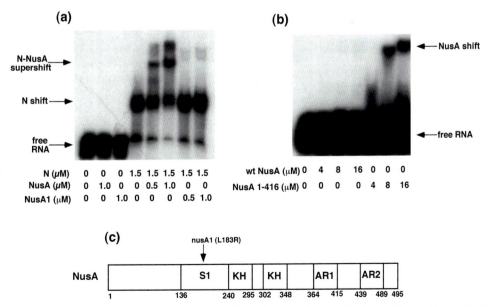

Figure 7. Interaction of NusA with *nut* site RNA. (*a*) Gel mobility shift assays showing the effect of the *nusA1* mutation in the S1 domain of NusA on the binding of NusA to the N-*nut* site complex. Reactions contained ^{32}P-labeled *nut* site RNA and the indicated concentration of N and NusA (Mogridge et al. 1995). (*b*) Gel mobility shift assays showing the effect of deleting the carboxy-terminal 79 amino acids of NusA on the binding of NusA to ^{32}P-labeled *nut* site RNA. (*c*) Diagram showing the S1 and KH domains of NusA and the position of the *nusA1* mutation. AR1 and AR2 are acidic, degenerate, repeated sequences in NusA.

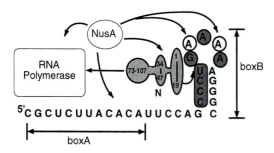

Figure 8. Diagram emphasizing the role of NusA in stabilizing an N-modified transcription complex. NusA interacts with *boxA*, the *boxB* loop, a central region of N, and RNA polymerase. It may facilitate the interaction of the carboxyl terminus of N with RNA polymerase.

MECHANISM OF ANTITERMINATION

The presence of S1 and KH domains in NusA and the inability of NusA to bind RNA on its own imply that there might be occlusion of one or more of the RNA-binding domains in NusA. As shown in Figure 7b, the deletion of 79 amino acids from the carboxyl terminus of NusA does allow NusA to bind *nut*-site-containing RNA weakly in a gel mobility shift assay. This implies that the carboxy-terminal region of NusA has a role in inhibiting the binding of NusA to RNA. Perhaps the interaction of full-length NusA with RNA is only possible in the presence of N or in the context of a transcription complex.

NusA has been shown to bind to the α-subunit of *E. coli* RNA polymerase (Liu et al. 1996). As well, cross-linking experiments using NusA derivatized with a photoactivatable cross-linking reagent have shown that NusA is close to the large subunits, β and β′, of RNA polymerase (J. Li and J. Greenblatt, unpubl.). In analogy to the way in which interaction of σ[70] with RNA polymerase relieves the inhibitory effect of the amino terminus of σ[70] on promoter-specific DNA binding (Dombroski et al. 1992, 1993), it is possible that the interaction of NusA with RNA polymerase relieves the inhibitory effect of the carboxyl terminus of NusA and allows NusA to bind RNA. Nuclease protection experiments and protein-RNA cross-linking experiments (Landick and Yanofsky 1987; Liu and Hanna 1995) have shown that NusA interacts with or is close to RNA nucleotides upstream of the 3′ end of the nascent transcript in a transcription complex. Interaction of NusA with the nascent transcript may be important for NusA to enhance pausing and termination by RNA polymerase.

Amino acids 34–47 of N suffice to bind NusA (see Figs. 1b and 8) (Mogridge et al. 1998b). The presence of N, as well as NusA, in a transcription complex may therefore redirect one or more of the RNA-binding domains of NusA away from the 3′ region of the transcript and onto the *nut* site RNA. This phenomenon, if it occurs, is likely to contribute to the ability of N to inhibit pausing and termination by RNA polymerase, both of which are stimulated by NusA (for review, see Richardson and Greenblatt 1996) and likely to involve an interaction of NusA with the nascent transcript. Amino acids 1–47 of N can bind *boxB* and NusA (see Fig. 1b) and suffice to partly

suppress termination of transcription in the presence of NusA (Fig. 6a) (Mogridge et al. 1998b). In contrast, amino acids 1–39 of N can bind *boxB* but not NusA and have no effect on termination at an intrinsic terminator (Fig. 6b). Perhaps amino acids 39–47 of N influence antitermination because they bind NusA and serve to position an RNA-binding domain of NusA near the *nut* site RNA, rather than near the 3′ end of the transcript. Subsequent occupation of the S1 and KH domains of NusA by *nut* site RNA would then prevent the binding to NusA of newly synthesized RNA and therefore prevent NusA from contributing to the termination of transcription.

N has little effect on transcription in the absence of NusA in standard reactions containing 10^{-7} M N protein (Whalen et al. 1988; Mason et al. 1992a). At higher concentrations of N and under conditions of low ionic strength, N can suppress termination in the absence of NusA (Rees et al. 1996), and this effect depends on amino acids 73–107 of N (Mogridge et al. 1998b). Amino acids 73–107 of N, which bind to RNA polymerase, also have a major effect on the efficiency of antitermination in the presence of NusA when the concentration of N is 10^{-7} M (Fig. 6) (Mogridge et al. 1998b). In these circumstances, it seems likely that the interaction of RNA polymerase-bound NusA molecules with N and the *nut* site RNA would position amino acids 73–107 of N near a critical surface of RNA polymerase (Fig. 8). In this view, the interaction of the N-*nut* site complex with NusA would suppress termination both by interfering with the ability of NusA to enhance termination and by positioning amino acids 73–107 of N near RNA polymerase (see Figs. 1b and 8).

How amino acids 73–107 of N suppress termination is not yet clear. One possibility, which we favor, is that N might prevent the reverse translocation by RNA polymerase which can occur at pause sites and terminators (Nudler et al. 1997). Reverse translocation necessarily involves the extrusion from the RNA-DNA hybrid in the transcription bubble of the 3′ end of the nascent transcript. In principle, N could inhibit reverse translocation simply by occupying the space that would otherwise be occupied by the extruded 3′ end of the transcript.

ACKNOWLEDGMENTS

This research was supported by grants from the Natural Sciences and Engineering Research Council of Canada (L.E.K.) and the Medical Research Council of Canada (J.G. and L.E.K.). P.L. is a Terry Fox research fellow of the National Cancer Institute of Canada supported with funds provided by the Terry Fox Run. J.G. is a Distinguished Scientist of the Medical Research Council of Canada. J.G. and L.E.K. are International Research Scholars of the Howard Hughes Medical Institute.

REFERENCES

Battiste J.L., Mao H., Rao S.N., Tan R., Muhandiram D.R., Kay L.E., Frankel A.D., and Williamson J.R. 1996. α Helix-RNA major groove recognition in a HIV-1 Rev peptide-RRE RNA complex. *Science* **273:** 1547.

Burns C. and Richardson J.P. 1995. NusG is required to overcome a kinetic limitation to Rho function at an intragenic terminator. *Proc. Natl. Acad. Sci.* **92:** 4738.

Bycroft M., Hubbard T.J., Proctor M., Freund S.M., and Murzin A.G. 1997. The solution structure of the S1 RNA binding domain: A member of an ancient nucleic acid-binding fold. *Cell* **88:** 235.

Cai Z., Gorin A., Frederick R., Ye X., Hu W., Majumdar A., Kettani A., and Patel D.J. 1998. Solution structure of P22 transcriptional antitermination N peptide-*boxB* RNA complex. *Nat. Struct. Biol.* **5:** 203.

Chattopadhyay S., Garcia-Mena J., DeVito J., Wolska K., and Das A. 1995. Bipartite function of a small RNA hairpin in transcription antitermination in bacteriophage λ. *Proc. Natl. Acad. Sci.* **92:** 4061.

Cilley C.D. and Williamson J.R. 1997. Analysis of bacteriophage N protein and peptide binding to *boxB* using polyacrylamide gel coelectrophoresis (PACE). *RNA* **3:** 57.

Craven M.G. and Friedman D.I. 1991. Analysis of the *Escherichia coli NusA10(Cs)* allele: Relating nucleotide changes to phenotypes. *J. Bacteriol.* **173:** 1485.

Das A. 1993. Control of transcription termination by RNA-binding proteins. *Annu. Rev. Biochem.* **62:** 893.

Das A. and Wolska K. 1984. Transcription antitermination *in vitro* by lambda N gene product: Requirement for a phage *nut* site and the product of host *nusA, nusB* and *nusE* genes. *Cell* **38:** 165.

DeVito J. and Das A. 1994. Control of transcription processivity in phage λ: Nus factors strengthen the termination-resistant state of RNA polymerase induced by N antitermination. *Proc. Natl. Acad. Sci.* **91:** 8660.

Doelling J.H. and Franklin N.C. 1989. Effects of all single base substitutions in the loop of *boxB* on antitermination by bacteriophage λ's N protein. *Nucleic Acids Res.* **17:** 5565.

Dombroski A.J., Walter W.A., and Gross C.A. 1993. Amino-terminal amino acids modulate sigma-factor DNA-binding activity. *Genes Dev.* **7:** 2446.

Dombroski A.J., Walter W.A., Record M.T., Jr., Siegele D.A., and Gross C.A. 1992. Polypeptides containing highly conserved regions of transcription initiation factor σ70 exhibit specificity of binding to promoter DNA. *Cell* **70:** 501.

Formosa T., Barry J., Alberts B.M., and Greenblatt J. 1992. Using protein affinity chromatography to probe the structure of protein machines. *Methods Enzymol.* **208:** 24.

Franklin N.C. 1985. Conservation of genome form but not sequence in the transcription antitermination determinants of bacteriophage λ, φ21 and P22. *J. Mol. Biol.* **181:** 75.

———. 1993. Clustered arginine residues of bacteriophage λ N protein are essential to antitermination of transcription, but their locale cannot compensate for *boxB* loop defects. *J. Mol. Biol.* **231:** 343.

Franklin N.C. and Doelling J.H. 1989. Over-expression of "N" antitermination proteins of bacteriophage λ, φ21, and P22 causes loss of N specificity. *J. Bacteriol.* **171:** 2513.

Friedman D.I. and Baron L.S. 1974. Genetic characterization of a bacterial locus involved in the activity of the N function of phage lambda. *Virology* **58:** 141.

Friedman D.I. and Court D.L. 1995. Transcription antitermination: The lambda paradigm updated. *Mol. Microbiol.* **18:** 191.

Friedman D.I. and Olson E.R. 1983. Evidence that a nucleotide sequence, "*boxA*", is involved in the action of the NusA protein. *Cell* **34:** 143.

Friedman D.I., Baumann M., and Baron L.S. 1976. Cooperative effects of bacterial mutations affecting lambda N gene expression. I. Isolation and characterization of a *nusB* mutant. *Virology* **73:** 119.

Friedman D.I., Wilgus G.S., and Mural R.J. 1973. Gene N regulator function of phage lambda *imm21*: Evidence that a site of N action differs from a site of N recognition. *J. Mol. Biol.* **81:** 505.

Friedman D.I., Olson E.R., Johnson L.L., Alessi D., and Craven M.G. 1990. Transcription-dependent competition for a host factor: The function and optimal sequence of the phage λ *boxA* transcription antitermination signal. *Genes Dev.* **4:** 2210.

Friedman D.I., Sauer A.T., Baumann M.R., Baron L.S., and Adhya S.L. 1981. Evidence that ribosomal protein S10 participates in the control of transcription termination. *Proc. Natl. Acad. Sci.* **78:** 1115.

Friedman D.I., Olson E.R., Georgopoulos C., Tilly K., Herskowitz I., and Banuett F. 1984. Interactions of bacteriophage and host macromolecules in the growth of bacteriophage λ. *Microbiol. Rev.* **48:** 299.

Goda Y. and Greenblatt J. 1985. Efficient modification of *E. coli* RNA polymerase *in vitro* by the N gene transcription antitermination protein of bacteriophage lambda. *Nucleic Acids Res.* **13:** 2569.

Grayhack E.J. and Roberts J.W. 1982. The phage λ gene Q product: Activity of a transcription antiterminator *in vitro*. *Cell* **30:** 637.

Greenblatt J. 1973. Regulation of the expression of the N gene of bacteriophage lambda. *Proc. Natl. Acad. Sci.* **70:** 421.

———. 1984. Regulation of transcription termination in *E. coli*. *Can. J. Biochem. Cell. Biol.* **62:** 79.

Greenblatt J. and Li J. 1981a. The *nusA* gene product of *Escherichia coli:* Its identification and a demonstration that it interacts with the gene N transcription antitermination protein of bacteriophage lambda. *J. Mol. Biol.* **147:** 11.

———. 1981b. Interaction of the sigma factor and the *nusA* gene protein of *E. coli* with RNA polymerase in the initiation-termination cycle of transcription. *Cell* **24:** 421.

Greenblatt J., Malnoe P., and Li J. 1980. Purification of the gene N transcription antitermination protein of bacteriophage lambda. *J. Biol. Chem.* **255:** 1465.

Greenblatt J., McLimont M., and Hanly S. 1981. Termination of transcription by the *nusA* gene protein of *Escherichia coli*. *Nature* **292:** 215.

Greenblatt J., Nodwell R., and Mason S.W. 1993. Transcriptional antitermination. *Nature* **364:** 401.

Heus H.A. and Pardi A. 1991. Structural features that give rise to the unusual stability of RNA hairpins containing GNRA loops. *Science* **253:** 191.

Hilliker S. and Botstein D. 1976. Specificity of genetic elements controlling regulation of early functions in temperate bacteriophages. *J. Mol. Biol.* **106:** 537.

Horwitz R.W., Li J., and Greenblatt J. 1987. An elongation control particle containing the N gene transcription antitermination protein of bacteriophage lambda. *Cell* **51:** 631.

Ito K., Egawa K., and Nakamura Y. 1991. Genetic interaction between the β´ subunit of RNA polymerase and the arginine-rich domain of *Escherichia coli* NusA protein. *J. Bacteriol.* **173:** 1492.

Keppel F., Georgopoulos C., and Eisen H. 1974. Host interference with expression of the lambda N gene product. *Biochimie* **56:** 1503.

Konrad M.W. 1968. Dependence of "early" lambda bacteriophage RNA synthesis on bacteriophage-directed protein synthesis. *Proc. Natl. Acad. Sci.* **59:** 171.

Koradi R., Billeter M., and Wüthrich K. 1996. MOLMOL: A program for display and analysis of macromolecular structures. *J. Mol. Graphics* **14:** 51.

Landick R. and Yanofsky C. 1987. Isolation and structural analysis of the *Escherichia coli trp* leader paused transcription complex. *J. Mol. Biol.* **196:** 363.

Lazinksi D., Grzadzielska E., and Das A. 1989. Sequence-specific recognition of RNA hairpins by bacteriophage antiterminators requires a conserved arginine-rich motif. *Cell* **59:** 207.

Legault P., Li J., Mogridge J., Kay L.E., and Greenblatt J. 1998. NMR structure of the bacteriophage λ N peptide/*boxB* RNA complex: Recognition of a GNRA fold by an arginine-rich motif. *Cell* **93:** 289.

Li J., Mason S.W., and Greenblatt J. 1993. Elongation factor NusG interacts with termination factor ρ to regulate termination and antitermination of transcription. *Genes Dev.* **7:** 161.

Li J., Horwitz R., McCracken S., and Greenblatt J. 1992. NusG, a new *Escherichia coli* elongation factor involved in transcription antitermination by the N protein of phage lambda. *J. Biol. Chem.* **267:** 6012.

Li S.C., Squires C.L., and Squires C. 1984. Antitermination of *E. coli* rRNA transcription is caused by a control region segment containing lambda *nut*-like sequences. *Cell* **38:** 851.

Liu K. and Hanna M.M. 1995. NusA interferes with interaction between the nascent RNA and the C-terminal domain of the alpha subunit of RNA polymerase in *Escherichia coli* transcription complexes. *Proc. Natl. Acad. Sci.* **92:** 5012.

Liu K., Zhang Y., Severinov K., Das A., and Hanna M.M. 1996. Role of *Escherichia coli* alpha subunit in modulation of pausing, termination and anti-termination by the transcription elongation factor NusA. *EMBO J.* **15:** 150.

Mason S.W. and Greenblatt J. 1991. Assembly of transcription elongation complexes containing the N protein of phage λ and the *Escherichia coli* elongation factors NusA, NusB, NusG, and S10. *Genes Dev.* **5:** 1504.

Mason S.W., Li J., and Greenblatt J. 1992a. Host factor requirements for processive antitermination of transcription and suppression of pausing by the N protein of bacteriophage λ. *J. Biol. Chem.* **267:** 19418.

———. 1992b. A direct interaction between two *Escherichia coli* antitermination factors, NusB and ribosomal protein S10. *J. Mol. Biol.* **223:** 55.

Mogridge J. and Greenblatt J. 1998. Specific binding of *Escherichia coli* ribosomal protein S1 to *boxA* transcriptional antiterminator RNA. *J. Bacteriol.* **180:** 2248.

Mogridge J., Mah T.-F., and Greenblatt J. 1995. A protein-RNA interaction network facilitates the template-independent cooperative assembly on RNA polymerase of a stable antitermination complex containing the λ N protein. *Genes Dev.* **9:** 2831.

———. 1998a. Involvement of *boxA* nucleotides in the formation of a stable ribonucleoprotein complex containing the bacteriophage λ N protein. *J. Biol. Chem.* **273:** 4143.

Mogridge J., Legault P., Li J., van Oene M., Kay L.E., and Greenblatt J. 1998b. Independent ligand induced folding of the RNA binding domain and two functionally distinct antitermination regions in the disordered N protein of bacteriophage λ. *Mol. Cell* **1:** 265.

Musco G., Kharrat A., Stier G., Faternali F., Gibson T.J., Nilges M., and Pastore A. 1997. The solution structure of the first KH domain of FMR1, the protein responsible for the fragile X syndrome. *Nat. Struct. Biol.* **4:** 712.

Musco G., Stier G., Joseph C., Castiglione Morelli M.A., Nilges M., Gibson T.J., and Pastore A. 1996. Three-dimensional structure and stability of the KH domain: Molecular insights into the fragile X syndrome. *Cell* **85:** 237.

Nehrke K.W. and Platt T. 1994. A quaternary transcription termination complex. Reciprocal stabilization by Rho factor and NusG protein. *J. Mol. Biol.* **243:** 830.

Nodwell J.R. and Greenblatt J. 1991. The *nut* site of bacteriophage λ is made of RNA and is bound by transcription antitermination factors on the surface of RNA polymerase. *Genes Dev.* **5:** 2141.

———. 1993. Recognition of *boxA* antiterminator RNA by the *E. coli* antitermination factors, NusB and ribosomal protein S10. *Cell* **72:** 261.

Nudler E., Mustaev A., Lukhtanov E., and Goldfarb A. 1997. The RNA-DNA hybrid maintains the register of transcription by preventing backtracking of RNA polymerase. *Cell* **89:** 33.

Olson E.R., Flamm E.L., and Friedman D.I. 1982. Analysis of *nutR:* A region of phage lambda required for antitermination of transcription. *Cell* **31:** 61.

Olson E.R., Tomich C.-S.C., and Friedman D.I. 1984. The nusA recognition site: Alteration in its sequence or position relative to upstream translation interferes with the action of the N antitermination function of phage λ. *J. Mol. Biol.* **180:** 1053.

Pawson T. and Scott J.D. 1997. Signaling through scaffold, anchoring, and adaptor proteins. *Science* **278:** 2075.

Puglisi J.D., Chen L., Blanchard S., and Frankel A.D. 1995. Solution structure of a bovine immunodeficiency virus Tat-TAR peptide-RNA complex. *Science* **270:** 1200.

Rabovsky D. and Konrad M. 1970. Gene expression in bacteriophage lambda. I. The kinetics of the requirement for *N* gene product. *Virology* **40:** 10.

Rees W.A., Weitzel S.E., Yager T.D., Das A., and von Hippel P.H. 1996. Bacteriophage λ N protein can induce transcription antitermination *in vitro*. *Proc. Natl. Acad. Sci.* **93:** 342.

Richardson J.P. and Greenblatt J. 1996. Control of RNA chain elongation and termination. In Escherichia coli *and* Salmonella typhimurium: *Cellular and molecular biology*, 2nd edition (ed. F.C. Neidhardt et al.), p. 822. American Society of Microbiology, Washington, D.C.

Salstrom J.S. and Szybalski W. 1978. Coliphage λ *nutL⁻:* A unique class of mutants defective in the site of gene *N* product utilization for antitermination of leftward transcription. *J. Mol. Biol.* **124:** 195.

Schauer A.T., Carver D.L., Bigelow B., Baron L.S., and Friedman D.I. 1987. λ N antitermination system: Functional analysis of phage interactions with the host nusA protein. *J. Mol. Biol.* **194:** 679.

Schmidt M.C. and Chamberlin M.J. 1987. *NusA* protein of *Escherichia coli* is an efficient transcription termination factor for certain termination sites. *J. Mol. Biol.* **195:** 809.

Schwartz M. 1970. On the function of the *N* cistron in phage lambda. *Virology* **40:** 23.

Sharp P.A. and Marciniak R.A. 1989. HIV TAR: An RNA enhancer? *Cell* **59:** 229.

Squires C.L., Greenblatt J., Li J., Condon C., and Squires C. 1993. Ribosomal RNA antitermination *in vitro:* Requirement for Nus factors and one or more unidentified cellular components. *Proc. Natl. Acad. Sci.* **90:** 970.

Su L., Radek J.T., Hallenga K., Hermanto P., Chan G., Labeots L.A., and Weiss M.A. 1997a. RNA recognition by a bent α-helix regulates transcriptional antitermination in phage λ. *Biochemistry* **36:** 12722.

Su L., Radek J.T., Labeots L.A., Hallenga K., Hermanto P., Chen H., Nakagawa S., Zhao M., Kates S., and Weiss M.A. 1997b. An RNA enhancer in a phage transcriptional antitermination complex functions as a structural switch. *Genes Dev.* **11:** 2214.

Sullivan S.L. and Gottesman M.E. 1992. Requirement for *E. coli* NusG protein in factor-dependent transcription termination. *Cell* **68:** 989.

Sullivan S.L., Ward D.F., and Gottesman M.E. 1992. Effect of *Escherichia coli nusG* function on λ N-mediated transcription antitermination. *J. Bacteriol.* **174:** 1339.

Tan R. and Frankel A.D. 1995. Structural variety of arginine-rich RNA-binding peptides. *Proc. Natl. Acad. Sci.* **92:** 5282.

Van Gilst M.R. and von Hippel P.H. 1997. Assembly of the N-dependent antitermination complex of phage lambda: NusA and RNA bind independently to different unfolded domains of the N protein. *J. Mol. Biol.* **274:** 160.

Van Gilst M.R., Rees W.A., Das A., and von Hippel P.H. 1997. Complexes of N antitermination protein of phage λ with specific and nonspecific target sites on the nascent transcript. *Biochemistry* **36:** 1514.

Warren F. and Das A. 1984. Formation of termination-resistant transcription complex at phage lambda *nut* locus: Effects of altered translation and a ribosomal mutation. *Proc. Natl. Acad. Sci.* **81:** 3612.

Whalen W. and Das A. 1990. Action of an RNA site at a distance: Role of the *nut* genetic signal in transcription antitermination by phage λ *N* gene product. *New Biol.* **2:** 975.

Whalen W., Ghosh B., and Das A. 1988. NusA protein is necessary and sufficient *in vitro* for phage λ *N* gene product to suppress a ρ-independent terminator placed downstream of *nutL*. *Proc. Natl. Acad. Sci.* **85:** 2494.

Ye X., Kumar R.A., and Patel D.J. 1995. Molecular recognition in the bovine immunodeficiency virus Tat peptide-TAR RNA complex. *Chem. Biol.* **2:** 827.

Ye X., Gorin A., Ellington A.D., and Patel D.J. 1996. Deep penetration of an α helix into a widened RNA major groove in the HIV-1 Rev peptide-RNA aptamer complex. *Nat. Struct. Biol.* **3:** 1026.

Zuber M., Patterson T.A., and Court D.L. 1987. Analysis of *nutR*, a site required for transcription antitermination in phage lambda. *Proc. Natl. Acad. Sci.* **84:** 4514.

Mechanistic Model of the Elongation Complex of *Escherichia coli* RNA Polymerase

N. Korzheva,*† A. Mustaev,* E. Nudler,‡ V. Nikiforov,¶ and A. Goldfarb*

*Public Health Research Institute, New York, New York 10016; ‡Department of Biochemistry,
New York University Medical Center, New York, New York 10016; ¶Institute of Molecular Genetics,
Russian Academy of Sciences, Moscow, Russia

RNA polymerase (RNAP) is the central enzyme of gene expression and the focus of genetic regulation. To understand regulation in molecular detail, one needs to relate the few RNAP basic functions to a myriad of influences that affect its activity. In other words, RNAP can be viewed as a core molecular machine that is target to regulatory signals.

During elongation, RNAP functions as part of the ternary elongation complex (TEC) in association with the DNA template and RNA product. A meaningful model of TEC must address two principal questions. First, it must be determined how nucleic acid–protein interactions ensure two intuitively contradicting features: tight association of the three components and the ease with which RNAP translocates along DNA and RNA. Second, the model should explain termination in response to a specific signal encoded in the DNA sequence.

During the past decade, in the course of our and others' research, a succession of working models of TEC was put forward often using analogies and imagery of mechanical devices (Yager and von Hippel 1987: Chamberlin 1995; Nudler et al. 1997; Gelles and Landick 1998; Mooney et al. 1998). Our current model is presented in Figure 1. The model envisages TEC as a DNA-RNA framework around which the RNAP core enzyme is assembled. Five distinct parts of the polynucleotide scaffold are distinguished (Fig. 1A): the upstream and downstream DNA duplexes, the 9-bp DNA-RNA heteroduplex (hybrid), the single-stranded region of the nontemplate DNA in the transcription bubble, and the emerging single-stranded RNA upstream of the hybrid. In the case of the "backtracked" complex (Reeder and Hawley 1996; Komissarova and Kashlev 1997a,b; Nudler et al. 1997), an additional segment of single-stranded RNA appears downstream from the bubble. It is assumed that the architecture of protein in the moving TEC provides groves and channels to maintain the polynucleotide scaffold in a constant configuration.

The model envisages crucial sites of nucleic acid–protein interaction along the trajectory of template DNA in TEC (Fig. 1B). The DNA-binding site (DBS) holds on to the downstream DNA duplex and acts as a sliding clamp essential for TEC stability (Nudler et al. 1996). Next is the hybrid-binding site (HBS) that accommodates the RNA-DNA heteroduplex. The forward edge of HBS is merged with the catalytic center of RNAP, whose distinctive feature shown in Figure 1B is the coordinated Mg++ ion (Zaychikov et al. 1996). During RNAP backtracking, the catalytic center disengages from the RNA 3´ terminus, and the protein slides upstream, "unzipping" the hybrid and releasing the downstream segment of single-stranded RNA, whereas the catalytic center remains aligned with the front edge of the hybrid (Komissarova and Kashlev 1997a,b; Nudler et al. 1997). On this basis, we call the downstream part of HBS the "front zip-lock" (FZ). Here, we argue that a similar mechanism operates at the upstream edge of HBS, introducing the notion of the "rear zip-lock" (RZ) that maintains the stability of the

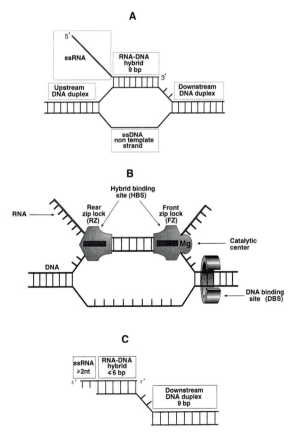

Figure 1. Architecture of TEC. (*A*) Full DNA-RNA scaffold; (*B*) model of TEC inferred from this work; (*C*) minimal RNA-DNA scaffold.

---

†*E-mail address:* kornata@phri.nyu.edu

complex and ensures displacement of the RNA product from the template during elongation. Together, FZ and RZ are responsible for keeping the length of the hybrid at 8 or 9 bp.

The progress in unraveling the mysteries of transcription depends on the availability of adequate experimental techniques. In the absence of a crystal structure of RNAP, we have to rely on indirect approaches. We have developed an arsenal of chemical and discriminatory biochemical assays that are discussed here in the context of presentation of our model of TEC.

MATERIALS AND METHODS

RNA polymerase and DNA template. RNAP with a histidine tag in the β′-subunit was purified according to the modified procedure of Kashlev et al. (1993; as described in Kashlev et al. 1996). The 270-bp *Eco*RI-*Hin*dIII DNA fragment carrying the T7A1 promoter was excised from the pENtR2 plasmid as described previously (Nudler et al. 1994).

In vitro transcription, pyrophosphorolysis, and product analysis. Transcriptional assays using RNAP immobilized on Ni^{++}-NTA agarose beads (Qiagen) were carried out as described by Nudler et al. (1994). Basic transcription buffer (TB) contained 40 mM Tris-HCl at pH 7.9, 40 mM KCl, and 5 mM MgCl$_2$. Radioactively labeled stable complexes with 11-nucleotide RNA were synthesized from a primer (10 μM) in the presence of 0.6 μM [α-^{32}P]ATP and 25 μM GTP at 37°C for 5 minutes. The complexes were then "walked" to the desired position (Nudler et al. 1994). Pyrophosphorolysis was performed at 37°C for 5 minutes at pyrophosphate concentrations specified in the figure legends. All kinetics were measured at 20°C in TB containing 1 M KCl. Cleavage of the RNA by RNase T1 was performed as described by Komissarova and Kashlev (1997a).

RNA electrophoresis was performed using 23% polyacrylamide–8 M urea denaturing gel with 20:3 acrylamide and *N,N′*-methylene bisacrylamide proportions. To quantify results, the gels were scanned with a phosphoimager (Molecular Dynamics).

Oligonucleotides were obtained from Oligos Etc., Inc. and Operon Technologies, Inc. The synthesis of the photoreactive trinucleotide primer will be described elsewhere. Cross-linking was activated by irradiation of the samples at 254 nm for 3 minutes.

THE MINIMAL TEMPLATE SYSTEM

To delineate crucial sites of DNA interactions, we determined the minimal DNA template sufficient to maintain stable and functional TEC (Nudler et al. 1996). In these experiments, the criterion of TEC stability was its ability to resume elongation after prolonged exposure to high ionic strength. Resistance to high salt reflects strong nonionic interactions that hold TEC together and are the

hallmark of elongating RNAP. In contrast, the RNAP initiation complex at the promoter or the paused complex committed to termination is salt-sensitive (Krummel and Chamberlin 1989; Arndt and Chamberlin 1990).

The experimental system for minimal template determination was made possible by two technical advances that greatly increased the discriminating power of in vitro transcription assays. The first technique known as "RNAP walking," first developed by Kashlev et al (1993), utilizes RNAP with six extra histidine residues fused to the carboxyl terminus of the β′-subunit. The enzyme adsorbs to Ni^{++} agarose beads through the His$_6$ tag. This allows assays to be performed in the solid phase, which in turn permits rapid exchange of reaction components by centrifugation and washing of the beads. With the use of this technique, His$_6$-tagged RNAP can be walked along the template in controlled steps, so that homogeneous preparations of defined elongation intermediates can be obtained. The second technique of "template switching" (Nudler et al. 1996) makes use of the ability of RNAP to transpose from one DNA template fragment to another without the loss of the transcript. After the switch, transcription continues on the secondary template (Fig. 2A).

By walking RNAP along secondary templates of varying configurations, it was possible to establish the minimal size and the optimal parameters of the DNA component in TEC (Nudler et al. 1996). As is evident from Figure 2B, the minimal TEC template is represented by secondary template number 7. Three elements of the minimal template can be distinguished: (1) 9-bp DNA duplex spanning positions from +3 to +11 relative to the RNA 3′ terminus, (2) two nucleotides of nonpaired template DNA at positions +1 and +2, and (3) six nucleotides of template DNA at positions –1 to –6 which may participate in the DNA-RNA hybrid. Note that TEC in this assay always contained an RNA transcript 55 nucleotides long. Deviations from these template parameters decreased the salt stability of TEC. It is important to note that some deviations from the optimal configuration led to a less dramatic effect than others. For example, in the case of template 10, significant residual salt stability was retained in comparison to other control templates (e.g., templates 9 and 11). The case of template 10 indicates that when as little as four nucleotides of DNA are available to form DNA-RNA hybrid, TEC is still fairly stable, arguing against major direct contribution of extended DNA-RNA heteroduplex to the forces that hold RNA in TEC.

THE MINIMAL PRODUCT SYSTEM

A similar question can be asked with regard to RNA in TEC: What is the minimal size of the transcript that is retained in the salt-stable complex? An adequate experimental system to address this question should allow analysis of TECs with short RNAs. The initiation complex at the promoter is not suitable for this purpose for several reasons: (1) It contains the initiation factor σ, (2) it does not hold the RNA product in the complex but makes short RNA catalytically in repeated cycles of

A

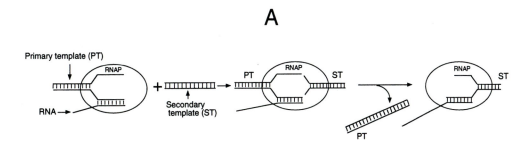

B

#	Secondary template Sequence	Ternary complex	Half-life (min) 50 mM KCl	500 mM KCl
1	n1/t1 `AGCCGATAACAATTTCACACAGGA` / `TCGCCTATTGTTAAAGTGTGTCCT`	+55	>120	80
2	t1 `TCGCCTATTGTTAAAGTGTGTCCT`	+55	>120	<1
3	n2/t1 `TAACAATTT A` / `TCGCCTATTGTTAAAGTGTGTCCT`	+55	>120	2
4	n3/t1 `ACAATTTCA` / `TCGCCTATTGTTAAAGTGTGTCCT`	+55	>120	60
5	n4/t1 `AATTTCACA` / `TCGCCTATTGTTAAAGTGTGTCCT`	+55	>120	8
6	n5/t1 `TTTCACACA` / `TCGCCTATTGTTAAAGTGTGTCCT`	+55	>120	<1
7	n3/t1 `ACAATTTCA` / `TCGCCTATTGTTAAAGT`	+55	>120	60
8	n4/t3 `TAACAATTTCA` / `TCGCCTATTGTTAAA`	+55	>120	1.5
9	n3/t4 `ACAATTTCA` / `TCATTGTTAAAGTGTGTCCT`	+51	30	<1
10	n3/t5 `ACAATTTCA` / `TCGCATTGTTAAAGTGTGTCCT`	+53	>120	10
11	t2 `TCGCCTATTGTTAAAGT`	+55	>120	<1

Figure 2. Determination of the minimal DNA template for the transcription complex. (*A*) Scheme of template-to-template switching during RNA synthesis; (*B*) stability for the switched transcription complexes assembled on different secondary templates. Arrows indicate the position of the catalytic center in the "switched" elongation complex (for details, see Nudler et al. 1996).

abortive initiation, and (3) it is highly salt-sensitive, indicating that elongation-specific nonionic interactions (the subject of our interest) are not present (Mooney et al. 1998). The transition from initiation to elongation is a complex metamorphosis that simultaneously involves the release of σ, the stabilization of the transcript, and the emergence of salt stability; it occurs when the transcript reaches 10–11 nucleotides in length (Grachev and Zaychikov et al. 1980). What we need is a system to study true σ-less ternary complexes with RNA shorter than 10 nucleotides.

To this end, we employed the reaction of processive pyrophosphorolysis (Rozovskaya et al. 1982), whereby a true elongation complex carrying a long transcript was driven in the reverse direction, resulting in progressive shortening of RNA (Fig. 3A). The beads carrying the initial elongation complex were extensively washed to remove traces of σ, exposed to pyrophosphate, and then to 1 M NaCl for different intervals before the retention of radioactive RNA was determined. The results indicate that

products less than 6 nucleotides were not retained at all; TEC with RNAs of 8 nucleotides and longer were highly stable, whereas the 7-nucleotide RNA-carrying TEC displayed small but significant residual salt stability (Fig. 3B,C). The complexes with 7-, 8-, and 9-nucleotide products of pyrophosphorolysis were normally competent in elongation. To rule out sequence-specific effects, this experiment was repeated on a different sequence site. The 26-nucleotide RNA TEC was first treated with RNase T1, which cleaves RNA at 16 nucleotides from the 3′ terminus (Komissarova and Kashlev 1997a), and then subjected to pyrophosphorolysis. Again, the minimal stable product was 8-nucleotide RNA, whereas the 7-nucleotide transcript displayed low-level residual stability (Fig. 3D).

The minimal size of the product providing for salt-stable TEC is thus 8 nucleotides. RNA nucleotides at positions –7 and –8 appear to be the critical incremental element, since the 6-nucleotide RNA was not retained in the complex. Similar conclusions were reached by Sidorenkov et al. (1998) using TEC artificially reconstituted from

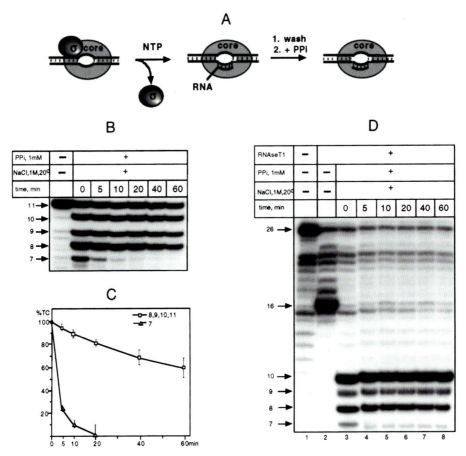

Figure 3. Salt stability of TEC in the Minimal Product System. (*A*) Route to the TEC with short RNA; (*B*) time course of the decay of TECs generated by pyrophosphorolysis of TEC carried 11-nucleotide RNA; (*C*) graphic representation of the data from *B*. Each of the 8-, 9-, 10-, and 11-nucleotide products displayed stability within the limits represented by vertical bars. (*D*) Stability of TEC produced by RNase T1 cleavage of 26 TECs and subsequent pyrophosphorolysis. (Lane *1*) Original 26-nucleotide TEC; (lane *2*) the 16-nucleotide TEC is the RNase T1 cleavage product of the 26-nucleotide TEC; (lanes *3–8*) time course for the decay of TECs obtained by pyrophosphorolysis in lane *2*.

core RNAP, DNA fragments, and RNA primer. However, in their system, the principal drop of TEC stability was between positions –6 and –7, and the 7-nucleotide product was nearly as stable as the 9-nucleotide RNA.

Superimposition of the minimal RNA product and the minimal DNA template yields the minimal nucleic acid scaffold shown in Figure 1C. It should be noted that we arrived at this structure by inference from minimization of either DNA or RNA in the context of the other component being full-sized. It remains to be tested whether the minimal scaffold in Figure 1C is actually sufficient to maintain stable TEC.

THE DNA-RNA HYBRID

The concept of the DNA-RNA heteroduplex has been hotly debated in the transcription field for nearly a decade (Yager and von Hippel 1987, 1991; Chamberlin 1995). The question of the existence and extent of the hybrid was linked to the issue of its functional role, or absence thereof, in holding the RNA product in TEC. In one view,

RNA is held by pairing to DNA in an extended heteroduplex, from 9 to 12 bp in length, which is thus thought to be a key determinant of TEC stability (Yager and von Hippel 1987, 1991; Sidorenkov et al. 1998). The other view postulates that TEC is maintained primarily through crucial interactions between protein and nascent RNA (Landick 1997; Nudler et al. 1998). In the extreme opinion, the extended hybrid is not merely unessential, but it is not at all present in TEC (Rice et al. 1991; Chamberlin 1995; Uptain et al. 1997).

In an attempt to directly determine the length of the hybrid, we (Nudler at al. 1997) utilized a cross-linkable probe incorporated into the pyrimidine ring of a single nucleotide in nascent RNA so that it did not impair the functioning of TEC. As the polymerase was walked down the template during elongation, the cross-linkable group traveled through TEC in the reverse direction and could be chemically activated at each step to probe its immediate vicinity. Sequence mapping of the resulting RNA-DNA cross-links demonstrated that the probe cross-linked exclusively to the template base across the hybrid. The results of this experiment showed that the

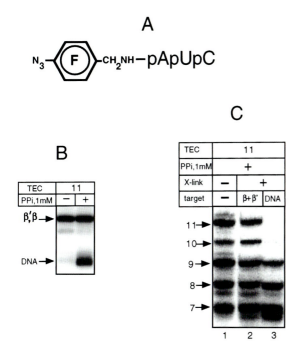

Figure 4. Mapping of the RNA:DNA hybrid in TEC in the Minimal Product System. (*A*) Structure of the photocross-linkable primer used for cross-linking experiments. (*B*) SDS-PAGE separation of the products cross-linked to DNA and RNAP subunits in complexes. Cross-link in 11-nucleotide TEC with or without pyrophosphate treatment preceding UV irradiation. (*C*) RNA products cross-linked to RNAP subunits and DNA. (Lane *1*) Total RNA products after pyrophosphorolysis of 11-nucleotide TEC before cross-linking; (lane *2*) RNAs cross-linked to protein recovered by acid hydrolysis; (lane *3*) cross-linked products recovered from DNA.

RNA probe was consistently cross-linking to DNA at positions from –2 to –8, but not –10 relative to the RNA 3′ terminus. The reactivity of position –9 was not determined in these experiments because of sequence constraints.

We report here further details of the hybrid parameters in a Minimal Product System. A photocross-linkable probe was attached to the 5′-terminal phosphate (Fig. 4A), and a set of TECs was then obtained by pyrophosphorolysis. The cross-linking was induced in TECs carrying 11-, 10-, 9-, 8-, and 7-nucleotide transcripts. In all cases, strong cross-linking to the protein β and β′ subunits was observed, but only with 7-, 8-, and 9-nucleotide RNA cross-linking to DNA (Fig. 4B,C). This result shows that RNA nucleotides at –7, –8, and –9 are aligned to the DNA template, indicating that the length of the hybrid in this system is 9 bp.

ROLE OF HETERODUPLEX IN TEC STABILITY

A 9-bp hybrid is thus present in TEC, but does it contribute significantly to TEC's salt stability? From the pa-

rameters of the minimal DNA template (see Fig. 1C), it appears that a salt-stable TEC is possible when the hybrid is only 6 bp long (template 7 in Fig. 2B). Thus, in the Minimal Template System, RNA does not need to be base-paired to DNA at positions –7, –8, –9 in order to be stably held in TEC.

To assess the significance of base pairing in the Minimal Product System, we prepared a series of complexes in which the transcript sequence was only partially complementary to the template. This was accomplished by using partially noncomplementary primers with two mismatches at their 5′ termini to initiate transcription, followed by 3′ shortening of the transcripts by pyrophosphorolysis (Fig. 5). It is clear that noncomplementary RNA nucleotides at the –8 and –9 positions (curve 9, Fig. 5B) stabilize the transcript more than 30-fold as compared to the fully complementary 7-nucleotide RNA control (curve 7, Fig. 5A). Furthermore, when the residual stability of TEC carrying the 8-nucleotide RNA with mismatches at –7, –8 (curve 8, Fig. 5B) is compared with that of the 6-nucleotide TEC, the stabilization due to unpaired nucleotides is the most dramatic: The retention of 6-nucleotide RNA is beyond the power of our detection method. These observations are in apparent conflict with the conclusions of Sidorenkov et al. (1998), who did not see any stabilizing effect of nonpaired RNA at positions –8, –9. However, in their experiments, the base-line stability of the complementary 7-nucleotide transcript was only half that of the complementary 9-nucleotide transcript, so their result is not informative. Thus, in the crucial positions –7, –8, –9, we included single-stranded RNA rather than the DNA-RNA hybrid into the minimal scaffold shown in Figure 1C.

It should be noted, however, that in the Minimal Product System, TEC stabilization effected by mismatched nucleotides at –8, –9 was not complete: Fully hybridized 9-nucleotide RNA was held approximately three times stronger (Fig. 5), reflecting some incremental role of base pairing. This result is in disagreement with the conclusion derived from the Minimal Template System (Fig. 2, template 7) that DNA at positions –7, –8, and –9 is not needed at all for salt-stable TEC. There are two possible explanations for this discrepancy that reflect the principal difference between the two systems: In the Minimal Template System, TEC carries excessive RNA, whereas in the Minimal Product System, TEC contains excessive DNA.

One explanation is that the full-sized transcript may be additionally stabilized through upstream interactions up to position –14 (Nudler et al. 1996). In the case of shortened transcripts in the Minimal Product System, base pairing at –7, –8, and –9 could compensate for the loss of the upstream RNA-protein contacts.

The other explanation is that the upstream DNA duplex may exert a destabilizing effect on the complex that is somehow countered by RNA base pairing at –7, –8, and –9. In the Minimal Template System, such effect—and the requirement for base pairing—would not be detected because the upstream duplex is not there.

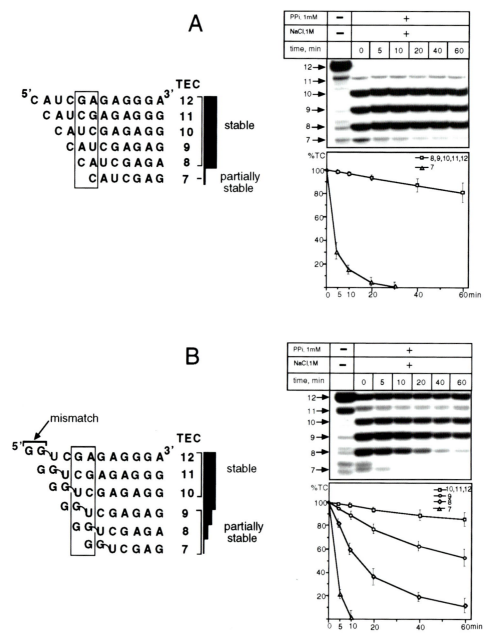

Figure 5. Stability of TECs containing matching (*A*) and miss-matching (*B*) residues at RNA 5′ terminus. The ladders of RNAs generated by pyrophosphorolysis are listed on the left. Positions –7, –8 are boxed.

THE UPSTREAM DNA DUPLEX

To test the latter possibility, the Minimal Product System was set up with a mismatched bubble, in which the upstream DNA duplex had been destroyed by five non-complementary nucleotides (Fig. 6A). This led to strong stabilization of the pyrophosphate-shortened transcripts so that the minimal salt-stable product was 7 nucleotides long, whereas the 6-nucleotide RNA displayed substantial intermediate salt stability (Fig. 6B).

The extended RNA upstream of position –7 thus appears to be needed for salt stability of TEC only in the context of the intact upstream DNA duplex. This points to

some sort of competitive relationship between RNA at positions –8, –9 and the upstream DNA segment in the Minimal Product System.

This result can be explained through the mechanism of a collapsing bubble, a modification of an early suggestion by Yager and von Hippel (1991). In our model, the interactions with the minimal scaffold structure shown in Figure 1C provide full TEC stability. However, in the context of the full-sized template, such stable structure is sterically incompatible with a less than perfect 12-bp single-stranded DNA bubble. The bubble, in turn, needs to be stabilized by the "R-loop" provided by the 9-bp heteroduplex. In the Minimal Template System, there is no

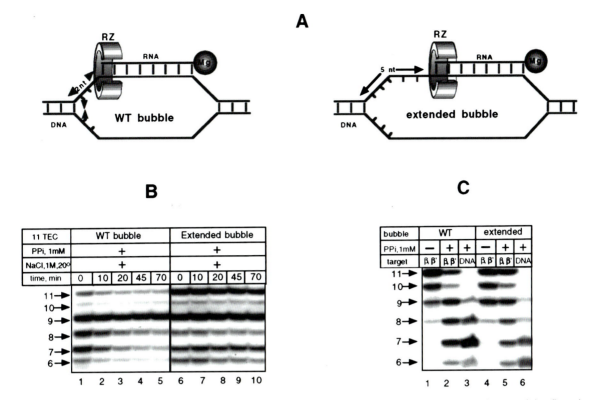

Figure 6. Mapping of the RNA-DNA hybrid and stability for TECs obtained on a 66-bp synthetic DNA template containing five mismatched thymidine residues in the nontemplate strand and the control (WT) synthetic template. (*A*) Scheme of TEC with wild-type and extended bubble. (*B*) Kinetics of TEC stability. (*C*) Cross-linked RNAs recovered from protein (lanes *1,2,4,5*) or DNA (lanes *3,6*) in TEC with (lanes *2,3,5,6*) and without (lanes *1,4*) pyrophosphorolysis.

bubble to collapse, so extended heteroduplex is not required, and RNA:DNA heteroduplex as short as 6 bp is sufficient to maintain the stability of TEC (Fig. 2B). In our scheme (Fig. 1B), the 9-bp heteroduplex is an additional essential component of the complete "natural" TEC, whereas it is dispensable in the Minimal Template System.

It is important to stress that in this scenario, the role of the hybrid is not an alternative to the notion of crucial protein-RNA interaction, an RNA sliding clamp that prevents dissociation of the transcript, allows lateral translocation of RNA through protein, and does not depend on the presence of complementary DNA. From the data on minimal transcripts, such a site should cover RNA positions from –6 to –9, thus accommodating the upstream part of heteroduplex in normal TEC.

ZIP LOCKS

A sliding clamp translocating along a duplex chain can in principle be configured as a simple ring or a zip-lock that destroys the duplex on one side and at the same time holds the chains together. That the front edge of DNA-RNA hybrid is passing through a zip-lock (FZ in Fig. 1B) follows from the phenomenon of backtracking (Komissarova and Kashlev 1997a,b; Nudler et al. 1997). To test whether a zip-lock mechanism operates at the rear end of the hybrid, we determined the extent of RNA-DNA base

pairing on the extended bubble, in which the upstream DNA duplex was destroyed by five mismatched bases incorporated into a synthetic DNA template. The rationale was that a zip-lock would channel nascent RNA away from the template even in the absence of competing DNA duplex. To this end, cross-linking from the 5′ terminus of short transcripts in the Minimal Product System was determined (Fig. 6C). It turned out that regardless of the state of the upstream DNA duplex, the edge of RNA-DNA hybridization was invariably at position –9. This result argues that it is the protein architecture rather than the dynamic equilibrium of duplexes that ensures displacement of the product from the template, in support of the zip-lock notion.

TERMINATION

Factor-independent termination signals in *Escherichia coli* are composed of a characteristic hairpin in RNA with a stem located seven nucleotides upstream of the termination point, followed by a stretch of T residues in the 3′ terminus (Uptain et al. 1997). The hairpin alone is not sufficient to cause termination but induces dramatic salt destabilization of TEC (Arndt and Chamberlin 1990; Nudler et al. 1995). In the context of our model, the RNA hairpin can be imagined to act by invading the zip-lock area and displacing RNA from crucial interactions maintaining the salt stability of TEC.

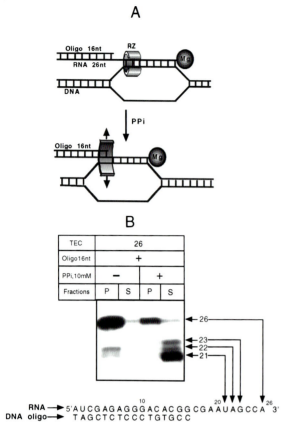

A

B

TEC	26			
Oligo16nt	+			
PPi,10mM	−		+	
Fractions	P	S	P	S

← 26
← 23
← 22
← 21

RNA ⟶ 5'AUCGAGAGGGACACGGCGAAUAGCCA 3'
DNA oligo⟶ TAGCTCTCCCTGTGCC

Figure 7. Pyrophosphorolysis promoted disintegration of the TEC by oligonucleotide complementary to RNA product. (*A*) Scheme of the experiment. (*B*) Electrophoretic analysis of RNAs released from 26-nucleotide TEC after pyrophosphorolysis. Fraction P–radioactive material associated with the pellet of Ni-agarose beads. Fraction S–radioactive material released in solution.

We mimicked this effect in the Minimal Product System by an antisense oligonucleotide hybridized to an upstream segment of nascent RNA (Fig. 7A). When the 3′ terminus of the transcript was progressively shortened by pyrophosphorolysis, the preannealed oligonucleotide began to exert a strong destabilizing effect as soon as its downstream edge invaded position −8 and reached maximum at −6 (Fig. 7B). Thus, the interactions of the transcript that ensures TEC salt stability within the putative RZ site appear to be the primary target of the termination signal. A similar effect of antisense oligonucleotides was reported by Artsimovitch and Landick (1998).

From these results, a model of termination emerges which envisages two possible causes of TEC dissociation: (1) destruction of crucial protein-RNA interactions within the rear zip-lock and (2) destabilization of the heteroduplex leading to the collapse of the DNA bubble. Both mechanisms can be effected by the stem of the termination hairpin acting at the rear edge of the hybrid at positions −7 to −9. The two mechanisms are not mutually exclusive. In fact, our results dissected two components of TEC stability using the two discriminative assay systems: In the Minimal Template System, there is no con-

tribution of the upstream part of the hybrid, whereas in the Minimal Product System, both the strength of the full-sized heteroduplex and RNA protein interactions seem to be important. Thus, the two views on the nature of forces holding the RNA transcript in TEC (Yager and von Hippel 1991; Chamberlin 1995) can peacefully coexist in the present model.

ACKNOWLEDGMENTS

Research in the laboratory of A.G. was supported by National Institutes of Health grant GM-30717. Research in the laboratory of E.N. was supported by Searle Scholar Award. Research in the laboratory of V.N. was supported by research grants 96-04-49019 and 96-1598076 from the Russian Foundation for Basic Research.

REFERENCES

Arndt K.M. and Chamberlin M.J. 1990. RNA chain elongation by *Escherichia coli* RNA polymerase. Factors affecting the stability of elongating ternary complexes. *J. Mol. Biol.* **213:** 79.

Artsimovitch I. and Landick R. 1998. Interaction of a nascent RNA structure with RNA polymerase is required for hairpin-dependent transcriptional pausing but not for transcript release. *Genes Dev.* **12:** 3110.

Chamberlin M.J. 1995. New models for mechanism of transcription elongation and its regulation. *Harvey Lect.* **88:** 1.

Gelles J. and Landick R. 1998. RNA polymerase as a molecular motor. *Cell* **93:**13.

Grachev M.A. and Zaychikov E.F. 1980. Initiation by *Escherichia coli* RNA-polymerase: Transformation of abortive to productive complex. *FEBS Lett.* **115(1):** 23.

Kashlev M., Martin E., Polyakov A., Severinov K., Nikiforov V., and Goldfarb A. 1993. Histidine-tagged RNA polymerase: Dissection of the transcription cycle using immobilized enzyme. *Gene* **130:** 9.

Kashlev M., Nudler E., Severinov K., Borukhov S., Komissarova N., and Goldfarb A. 1996. Histidine-tagged RNA polymerase of *Escherichia coli* and transcription in solid phase. *Methods Enzymol.* **274:** 326.

Komissarova N. and Kashlev M. 1997a. Arrest of transcription: *E. coli* RNA polymerase translocates backward leaving the 3′ end of the RNA intact and extruded. *Proc. Natl. Acad. Sci.* **94:** 1755.

———. 1997b. RNA polymerase switches between inactivated and activated states by translocating back and forth along the DNA and the RNA. *J. Biol. Chem.* **272:** 15329.

Krummel B. and Chamberlin M.J. 1989. RNA chain initiation by *Escherichia coli* RNA polymerase. Structural transitions of the enzyme in early ternary complexes. *Biochemistry* **28:** 7829.

Landick R. 1997. RNA polymerase slides home: Pause and termination site recognition. *Cell* **88:** 741.

Mooney R.A., Artsimovitch I., and Landick R. 1998. Information processing by RNA polymerase: Recognition of regulatory signals during RNA chain elongation. *J. Bacteriol.* **180:** 3265.

Nudler E., Goldfarb A., and Kashlev M. 1994. Discontinuous mechanism of transcription elongation. *Science* **265:** 793.

Nudler E., Avetissova E., Markovtsov V., and Goldfarb A. 1996. Transcription processivity: Protein-DNA interaction holding together the elongation complex. *Science* **273:** 211.

Nudler E., Kashlev M., Nikiforov V., and Goldfarb A. 1995. Coupling between transcription termination and RNA polymerase inchworming. *Cell* **81:** 351.

Nudler E., Mustaev A., Lukhtanov E., and Goldfarb A. 1997. The RNA-DNA hybrid maintains the register of transcription by preventing backtracking of RNA polymerase. *Cell* **89:** 33.

Nudler E., Gusarov I., Avetissova E., Kozlov M., and Goldfarb

A. 1998. Spatial organization of transcription elongation complex in *Escherichia coli*. *Science* **281:** 424.

Reeder T.C., and Hawley D.K. 1996. Promoter proximal sequences modulate RNA polymerase II elongation by a novel mechanism. *Cell* **87:** 767.

Rice G.A, Kane C., and Chamberlin M. 1991. Footprinting analysis of mammalian RNA polymerase II along its transcript: An alternative view of transcription elongation. *Proc. Natl. Acad. Sci.* **88:** 4245.

Rozovskaya T.A., Chenchick F.A., and Beabealashvilli R.S. 1982. Processive pyrophosphorolysis of RNA by *Escherichia coli* RNA polymerase. *FEBS Lett.* **137:** 100.

Sidorenkov I., Komissarova N., and Kashlev M. 1998. Crucial role of the DNA:RNA hybrid in the processivity of transcription. *Mol. Cell* **2:** 55.

Uptain S.M., Kane C., and Chamberlin M. 1997. Basic mechanisms of transcription elongation and its regulation. *Annu. Rev. Biochem.* **66:** 117.

Yager T.D. and von Hippel P. 1987. Transcript elongation and termination. In Escherichia coli *and* Salmonella typhimurium: *Cellular and molecular biology* (ed. F.C. Neidhardt), p. 1241. American Society for Microbiology, Washington, D.C.

———. 1991. A thermodynamic analysis of RNA transcript elongation and termination in *Escherichia coli*. *Biochemistry* **30:** 1097.

Zaychikov E., Martin E., Denissova L., Kozlov M., Markovtsov V., Kashlev M., Heumann H., Nikiforov V., Goldfarb A., and Mustaev A. 1996. Mapping catalytic residues in the RNA polymerase active center. *Science* **273:** 107.

Promoter-associated Pausing in Promoter Architecture and Postinitiation Transcriptional Regulation

J. LIS

Section of Biochemistry, Molecular and Cell Biology, Biotechnology Building, Cornell University, Ithaca, New York 14853

EUKARYOTIC TRANSCRIPTION PROVIDES A RICH REPERTOIRE OF REGULATORY TARGETS: MULTIPLE FACTORS AND DISCRETE STEPS

The transcriptional regulation of a eukaryotic gene is specified by the interplay of specific regulatory factors, general transcription factors (GTFs), RNA polymerase II (pol II), DNA sequence elements, and the chromatin structure of the promoter. The thousands of genes in a eukaryotic cell share a general transcriptional machinery and a common transcriptional pathway (Orphanides et al. 1996). Nonetheless, variations in regulatory mechanisms are becoming increasingly apparent.

Hundreds of specific upstream activators and repressors have been identified. These proteins bind near a particular target gene through specific interactions either with DNA sequences or with other proteins that are themselves targeted to specific DNA sequences. The combination of factors that can interact with a specific gene collaborates with the GTFs and the core promoter to set the regulatory mechanisms of the gene. The copious number and combinations of specific factors provide for a rich repertoire of regulation. In addition to this variety of upstream factors, the GTFs also show variety in that variant GTFs or alternative complexes containing GTFs add additional sophistication to the regulatory process (Grant et al. 1997; Hansen et al. 1997).

The regulation of mRNA-encoding genes generally occurs at steps early in the transcription cycle and can be specified by core promoter DNA (–40 to +40), which contains the TATA box, the transcription start, and the downstream promoter element (Burke and Kadonaga 1997), and a few hundred base pairs of upstream sequences, which contain gene-specific activator and repressor-binding sites. This expression can be further influenced in some genes by enhancer regions that can reside thousands of base pairs from a transcriptional start site. Fusing the RNA leader of a particular gene (about +40) to a reporter gene is often sufficient to recapitulate the expression level and regulatory properties of the normal gene. This indicates that transcription beyond about +40 of the transcription unit is usually not a critical part of the regulation. Nonetheless, for pol II to progress to this point and form a competent elongational complex requires numerous distinct steps and a large battery of interacting general factors. Any of these steps and general factors can potentially be rate limiting and a target of regulation by specific activators and repressors. Some of the prominent steps in producing an elongationally competent pol II complex are outlined in Figure 1.

1. *Opening chromatin.* Chromatin structure can create an early barrier to gene expression and in some cases can prevent access of transcription factors to the promoter (Taylor et al. 1991). The ability of GTFs to gain access to their target sequence is likely to be facilitated not only by sequence-specific DNA-binding factors, but also by chromatin remodeling machines (Wu 1997).

2. *Binding TFIID.* A critical foundation of the promoter is the GTF complex that contains TBP and can interact with the TATA box or start site region or both (Purnell et al. 1994).

3. *Recruiting the preinitiation complex.* Other factors and pol II are then recruited via ordered assembly (Buratowski et al. 1989) or as a holoenzyme that contains many of the GTFs, mediator, and the core pol II (Thompson et al. 1993; Kim et al. 1994).

4. *Forming the open complex.* The bound pol II progresses from a closed to an open complex (Holstege et al. 1997).

5. *Initiation.* pol II forms the first phosphodiester bond to initiate transcription. pol II at this stage can synthesize and release short 7–14-nucleotide transcripts reiteratively (Holstege et al. 1997).

6. *Promoter clearance.* pol II passes beyond this phase of synthesizing short abortive transcripts as it acquires a longer, stably associated RNA (Goodrich and Tjian 1994).

7. *Promoter escape.* pol II with relatively short, stably associated RNAs of 18–45 nucleotides can pause. The production of a fully competent elongational complex is defined as escape.

The steps outlined in Figure 1 provide a relatively low-resolution view of mechanism. Each of these steps may be further divided. For example, the recruiting of a preinitiation complex can occur via one step (Thompson et al. 1993) or multiple distinct steps (Buratowski et al. 1989). When considering molecular mechanisms of activator or repressor function, the step or steps affected may need to be considered at these higher levels of molecular detail.

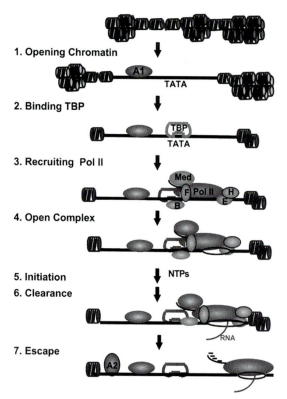

1. Opening Chromatin

2. Binding TBP

TATA

3. Recruiting Pol II

TATA

4. Open Complex

Med

F Pol II H

B E

5. Initiation

6. Clearance

NTPs

RNA

7. Escape

A2

Figure 1. Steps in early transcription. Each step is described in the text. The labels identify various factors and complexes: (TBP) TATA-binding protein; (Med) mediator complex of the pol II holoenzyme; (F, E, H, and B) corresponding TFII general transcription factors; (pol II) RNA polymerase II. (A1 and A2) Specific activators shown here to be acting early and late in the process, although in principle they could act at any regulated step.

A CASE FOR REGULATION AT THE LEVEL OF PROMOTER ESCAPE: *DROSOPHILA HSP70*

The *Drosophila hsp70* gene is rapidly and vigorously activated by heat shock. An instantaneous heat shock triggers a 200-fold increase in the level of transcription in 3 minutes (O'Brien and Lis 1993). This activation is orchestrated by the DNA sequence-specific activator, heat shock factor (HSF), which upon heat shock rapidly trimerizes and binds to heat shock loci (Westwood et al. 1991). The framework for this rapid activation was first suggested by measurements demonstrating the DNase-I-hypersensitive structure of the *hsp70* promoter (Wu 1980). This open chromatin configuration could provide HSF and the general transcription machinery rapid access to specific sequences of the promoter of heat shock genes. This promoter appears to be further primed for transcription in that pol II and TBP are already an integral part of the *hsp70* promoter even before heat shock activation.

Four distinct classes (I–IV) of measurements performed directly on *Drosophila* cells or intact nuclei support the existence of pol II on the 5′ end of *uninduced hsp70* genes (Fig. 2). These assays quantify the amount of this polymerase, delimit its precise location, and define some of its features. Approximately one pol II is transcriptionally engaged, but paused, on each *hsp70* gene (Rougvie and Lis 1988). This pol II is paused at sites covering the interval from +21 to +35, with two peaks of pausing within this interval that are separated by a turn of the DNA helix (Rasmussen and Lis 1993). This pol II is largely hypophosphorylated (O'Brien et al. 1994), and its associated short RNA is uncapped when pol II is at the

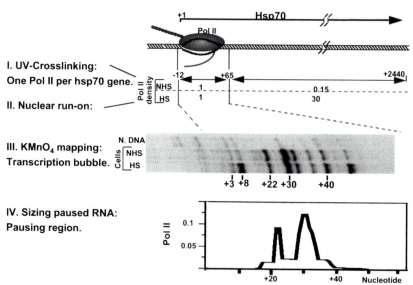

I. UV-Crosslinking:
One Pol II per hsp70 gene.

II. Nuclear run-on:

III. KMnO₄ mapping:
Transcription bubble.

IV. Sizing paused RNA:
Pausing region.

Figure 2. Summary of four different classes of experiments describing the paused pol II on the *Drosophila hsp70* gene. (I) UV cross-linking and immunoprecipitation analyses first revealed the high density of pol II on the 5′ end relative to the body of the *hsp70* gene in uninduced NHS cells (Gilmour and Lis 1986). The corresponding levels in induced HS cells were also derived from these experiments and are illustrated below the *hsp70* map. (II) Nuclear run-on assays demonstrated that the density of transcriptionally engaged pol II is virtually identical to that seen by cross-linking (Rougvie and Lis 1988). (III) Potassium permanganate treatments of intact cells identified sites on the *hsp70* gene that are hyperreactive relative to that seen in naked DNA (N.DNA). These cover the region of the expected transcription bubble created by the paused polymerase (Giardina et al. 1992). (IV) Distribution of pause sites was determined on the uninduced *hsp70* by sizing RNAs associated with the paused polymerase (Rasmussen and Lis 1993). The graph represents relative densities (normalized to a total area of one paused pol II) at different positions in the pause region.

start of the paused interval but is largely capped at the distal portion of the pause region (Rasmussen and Lis 1993).

Cross-linking

The initial evidence for pol II association with the uninduced *hsp70* gene was obtained more than a decade ago from in vivo cross-linking and immunoprecipitation (Gilmour and Lis 1985, 1986). When developing our UV cross-linking approaches to measure protein density on a specific DNA sequence in vivo, we were surprised to find a high density of pol II on the 5′ end of the uninduced *Drosophila hsp70* gene, because it had been assumed that all transcription was regulated at the level of pol II recruitment or "initiation." We had expected that a low level of polymerase would exist on the gene prior to heat shock and the level would increase 200-fold upon activation, mirroring the change in transcription as seen in the analysis of pulse-labeled transcripts (Lis et al. 1981). Although the 3′ half of the gene showed the expected increase in polymerase density, the promoter region (–12 to +65) had a density of polymerase before heat shock that was equivalent to one pol II per gene. This absolute estimate was calculated from the cross-linking of polymerase to the heat-shock-induced gene, where rates of RNA synthesis and direct electron microscopy visualization of growing RNA chains in Miller spreads indicate 30 RNA polymerases are on the fully activated *hsp70* gene (O'Brien and Lis 1993). The pol II seen by UV cross-linking was shown not to be simply an artifact of recruiting pol II during the UV irradiation of cells, since the same density is observed with a 10-minute irradiation with mercury lamps as is seen by a 60-μsec flash with a xenon flash lamp (Gilmour and Lis 1986) . More recently, we have used UV cross-linking and antibodies specific to the hypo- or hyperphosphorylated pol II to show that the paused pol II is hypophosphorylated (O'Brien et al. 1994).

Nuclear Run-on

A pol II that cross-links to the 5′ end of the *hsp70* gene could in principle be at any of a number of discrete steps in the process of early transcription (Fig. 1). Nuclear run-on assays demonstrated that this pol II has initiated transcription but is elongationally paused. In isolated nuclei from uninduced cells, the promoter-associated pol II is capable of transcribing under conditions that prevent transcription initiation. A short run-on reaction performed in the presence of Sarkosyl or high-salt concentrations shows a large burst of transcription that is restricted to the 5′ end of the gene (Rougvie and Lis 1988), and comparison of the amount of this run-on product with that from induced nuclei indicated that there is one engaged pol II per *hsp70* gene. Longer run-on reactions demonstrate that this polymerase can progress through the body of the gene. Interestingly, if nuclei are very carefully prepared from uninduced cells and no Sarkosyl or high salt is added, this promoter-associated polymerase transcribes very inefficiently in a run-on reaction. From these results, we hypothesized that a transcriptionally engaged pol II resides on the 5′ end of the *hsp70* gene and that it is normally paused in vivo (Rougvie and Lis 1988).

Permanganate "Bubble Mapping"

The evidence that the promoter-associated pol II is in a paused configuration was derived initially from analysis of run-on reactions performed with isolated nuclei. To examine cells directly, we (Giardina et al. 1992) used the reagent potassium permanganate, which preferentially modifies T residues in single-stranded DNA. We reasoned that a single-stranded DNA bubble should lie in the wake of a paused pol II and be detected by permanganate. After a brief 30-minute permanganate treatment of cells, DNA was purified, and sites of modification were cleaved. Ligation-mediated PCR revealed the sites of modification and mapped the single-stranded region to an interval consistent with that expected from the location of pause sites mapped by sizing RNAs (see below and Fig. 2). Thus, analysis of intact *Drosophila* cells with a brief (30-min) chemical treatment provides evidence for pausing in vivo.

Sizing Paused RNAs

The precise location of the pause or pauses could be determined by sizing the short, rare RNAs associated with the paused polymerase. To achieve this, we developed new sensitive approaches for purifying and assaying rare RNAs. In the first strategy (Rasmussen and Lis 1993), radioactively labeled, chain-terminated RNAs were generated by performing run-on reactions in the presence of various combinations of radioactively labeled nucleoside triphosphates and chain-terminating nucleoside triphosphates. The short RNAs were hybridized to completion with excess biotin-labeled oligonucleotide, and the biotin oligonucleotide complexes were recovered with avidin-coated magnetic beads. After extensive washing, the labeled RNAs were sized by gel electrophoresis, and the pause sites were deduced from the sizes obtained with different combinations of labeled and chain-terminating NTPs. The pause sites reside between positions +21 and +35, with two peaks of pausing separated by approximately one turn of the DNA helix. This suggests a sidedness to pausing where the pol II may interact with factors that inhibit its progress; however, other explanations are possible. Interestingly, the isolated RNAs contain a mixture of capped and uncapped species (Rasmussen and Lis 1993). RNAs near the start of the pause region are largely uncapped, whereas those at the end are fully capped. This unexpected extra information from these assays defines the point during RNA synthesis where capping occurs in vivo, and this result agrees well with in vitro studies of capping with vaccinia virus, which also show that capping occurs early in transcription (Hagler and Shuman 1992).

The second strategy for sizing paused RNAs again made use of biotin oligonucleotides to select unlabeled RNAs extracted from nuclei. These RNAs were then amplified and labeled by ligation-mediated polymerase chain reaction (Rasmussen and Lis 1995). Analysis of the size of amplified fragments allowed the derivation of pause sites. These results are in excellent agreement with the first strategy, and this approach is more sensitive and could be applied to genes that have relatively low levels of paused pol II (Rasmussen and Lis 1995).

TBP AND GAGA FACTOR ALSO OCCUPY THE UNINDUCED *HSP70* PROMOTER

A by-product of transcription bubble mapping with permanganate was the ability to detect TBP protection of the TATA sequences. TATA sequences are generally in non-B-form DNA and are hypersensitive to modification with potassium permanganate. In cells, however, the *hsp70* (and *hsp26*) TATA sequences are relatively protected from permanganate modification (Giardina et al. 1992). Since this pattern of protection appears to be identical to that generated with purified cloned promoter DNA and purified recombinant TBP, we conclude that TBP is bound to the TATA box of the uninduced *hsp70* promoter.

GAGA factor (GAF) appears to have an important role in heat shock gene expression. Mutations in the strong binding sites (multiple GA repeats) for GAF impair the function of the *hsp70* and *hsp26* promoters. GAF is not a traditional transcription activator in vitro, but it appears to have a role in overcoming repression imposed by histones (Wilkins and Lis 1997). Since GAF is constitutively present in nuclei and is a prime candidate for having a role in establishing the potentiated promoter, we examined GAF's occupancy of various promoters in cells directly by UV cross-linking and immunoprecipitation (O'Brien et al. 1995). These studies demonstrate that GAF is present on the uninduced *hsp70* and *hsp26* promoter regions.

The resulting image of the architecture of the uninduced *hsp70* promoter is depicted in Figure 3. An open chromatin structure extends over all critical features of the promoter. Within this region are a single paused pol II

that is distributed over the region +21 to +35, GAF interacting with the GAGA sequences and TBP occupying the TATA box. Together, these various features of the *hsp70* promoter constitute what we have called the *potentiated* promoter (Lis and Wu 1995). Additional GTFs could also be present at the potentiated promoter, and additional cross-linking and immunofluorescence experiments are required to determined which are present, where precisely they are located, and how their distribution changes upon activation.

DNA SEQUENCES CRITICAL IN ESTABLISHING AND ACTIVATING THE POTENTIATED PROMOTER

The functional elements of the *hsp70* promoter have been identified by analysis of transgenic fly lines containing a variety of promoter mutations. Initially, these analyses demonstrated the critical role of heat shock elements (the targets to which HSF binds) for heat-induced activation (Xiao and Lis 1988). Although mutations in the elements to which HSF binds affect heat shock gene activation, they have little effect on establishing the paused pol II (Lee et al. 1992). The full activation of the heat shock promoter was also found to be dependent on GAGA elements, the targets to which GAF binds (Glaser et al. 1990; Lee et al. 1992). In addition, these GAGA elements were found to be critical for the formation of the nuclease hypersensitivity of the *hsp26* and *hsp70* promoters (Lu et al. 1992; Shopland et al. 1995). Interestingly, the GAGA elements were also found to be important for establishing a paused pol II and bound TBP (Shopland et al. 1995). Therefore, GAF protein and GAGA elements are critical for generating the potentiated promoter. Full heat-induced activation also requires GAF either to form this promoter structure or to participate directly in the activation.

Sequences of the core promoter region were also found to be required for generating paused pol II (Lee et al.

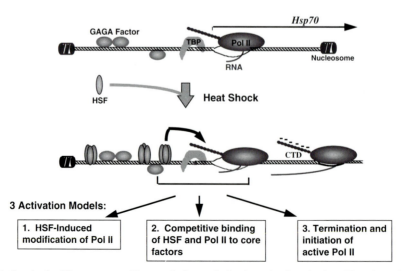

Figure 3. Model depicting the *hsp70* promoter architecture before and after heat shock activation. The minus signs associated with the CTD (C-terminal domain of the largest subunit of pol II) indicate the hyperphosphorylated state. The three models of activation and other features of this sketch are described in the text.

1992). Although deletions of the regions downstream from +30 cause only a modest (1.5-fold) reduction in paused pol II and a 2-fold reduction on the activation of transcription following heat shock, a 3′ deletion that extends to position +23 results in a greater reduction in the level of paused polymerase on the *hsp70* gene (3.5-fold) and also further impairs its inducibility in response to heat shock. Deletions entering farther into the core promoter show even greater reduction in pausing (to background levels) and further reduce heat-inducible transcription. Therefore, both GAGA sequences and core promoter sequences extending through the pause region are important in establishing the potentiated *hsp70* promoter and its subsequent activation.

Interestingly, although the mutagenesis of the *hsp70* promoter is far from exhaustive, none of the mutations in the *hsp70* promoter leads to an elevated constitutive expression of *hsp70*. Such mutations might be expected if specific sequences elements bound factors that acted as blockades to pol II. In addition, no mutation has been found that exhibits normal pausing and disrupts the inducibility of *hsp70*. In contrast, we find that the mutations that reduce pausing also reduce the inducibility of the gene. Thus, the degree of pausing and the strength of the activated promoter appear to be coupled.

WHY BUILD A POTENTIATED PROMOTER? SPEED AND ACCESS

Stress rapidly induces genes that have paused pol II (Stewart et al. 1990; O'Brien and Lis 1993), and it is tempting to speculate that the open promoter with a transcriptionally paused pol II may provide such a gene with the ability to be very rapidly activated. The first wave of polymerase is already engaged in transcription before induction, and the recruitment of additional pol II may be facilitated by this promoter architecture which is open and includes pre-bound TBP. Kinetic analyses using UV-flash cross-linking and nuclear run-on assays directly revealed the dynamics of pol II movement and accumulation on the *hsp70* gene in *Drosophila* cells (O'Brien and Lis 1993). The first wave of pol II moves detectably beyond the pause region of the *hsp70* gene within 70 seconds following an instantaneous heat shock. Within 3 minutes, the density of pol II on *hsp70* is near its fully induced level. This rapid activation appears to be similar to that observed for c-*fos* (Stewart et al. 1990), which also has a promoter-paused pol II (Plet et al. 1995; and see below).

The open structure of the heat shock promoters is also critical for the rapid recruitment of HSF. In vitro, Kingston and colleagues (Taylor et al. 1991) have shown that HSF binds much less well to heat shock elements that are in nucleosomes than in naked DNA. Transgenic lines that carry *hsp70* gene promoter mutations provided an opportunity to assess the effects of promoter architecture on HSF binding in vivo. Immunofluorescence assays of *Drosophila* polytene chromosomes can be used to localize sites of HSF binding. Indeed, transgenic lines carrying functional *hsp70* genes create new bands of HSF at the sites of insertions (Shopland et al. 1995). Surpris-

ingly, mutations that disrupt the leader sequences underlying the paused polymerase, but do not disrupt heat shock elements, reduce (to undetectable levels) the heat-shock-induced recruitment of HSF to the mutant transgene as seen by immunofluorescence and footprinting (Shopland et al. 1995). These mutants prevent the formation of paused polymerase and alter the architecture of the uninduced promoter. Likewise, mutations in the major GAGA element of the *hsp70* transgenes also prevent pausing and reduce the induced HSF binding to this transgene (Shopland et al. 1995). Binding of TBP to the *hsp70* promoter in vivo is also affected in both of these classes of mutation. Therefore, it appears that the overall promoter architecture is dependent on the interplay of TBP, GAF, and the paused polymerase and that this in turn determines whether the heat shock promoter is accessible to HSF.

MECHANISM OF TRANSCRIPTION ACTIVATION: POL II RECRUITMENT IS NOT SUFFICIENT

Heat shock triggers the trimerization and highly specific binding of HSF to chromosomal sites (Westwood et al. 1991). Although many chromosomal sites recruit HSF, the most prominent sites of HSF localization are the loci containing heat shock genes. HSF binds tightly to the multiple heat shock elements of these genes and triggers the 200-fold increase in transcription of major heat shock genes. To account for the basal level of *hsp70* expression in uninduced cells, pol II must escape the pause mode and enter into productive elongation once every 10 minutes. After heat shock and HSF binding to the promoter, pol II must escape to productive elongation once every 4 seconds (this follows from the density of polymerase being one pol II per 80 bp and the elongation rate being 1.2 kb/min [O'Brien and Lis 1993]). The mechanism by which HSF stimulates this dramatic and rapid increase in transcription is not understood, but at least three models should be considered. These models differ in detail, but they all have the common feature of being clearly distinct from regulatory models where activators simply recruit pol II to the promoter directly or recruit GTFs that in turn recruit pol II (Ptashne and Gann 1997). pol II is already recruited to promoters that display pausing, and this recruited pol II encounters what appears to be a rate-limiting step early in its elongation.

Modification of the Paused Pol II Complex

Paused pol II is impaired in its elongational competence relative to polymerases that have progressed beyond the pause region. Perhaps the paused pol II is modified to an elongationally competent form in response to HSF activation. A modification could alter pol II's properties or associations, allowing it to progress to an elongationally competent form. The modification could potentially be of a component of this paused structure other than pol II itself, such as chromatin structure, since studies by Kingston's group of pausing on human *hsp70* support a role of chromatin modification in HSF activation

(Brown and Kingston 1997). In contrast, studies by Gilmour and colleagues of *Drosophila hsp70* promoter pausing in vitro show that the pause can form in nuclear extracts in the absence of assembled chromatin (Li et al. 1996) and even on templates too short to support a downstream nucleosome (Benjamin and Gilmour 1998). Perhaps the resolution of these apparently contradictory conclusions is that the pause can be specified in the absence of chromatin, but the level of pausing is enhanced (and regulated) by chromatin, and the degree of this enhancement may vary for different genes or systems.

One modification of pol II that strongly influences its properties is the phosphorylation of the carboxy-terminal domain (CTD) of the largest subunit. The form of pol II that can enter a promoter is hypophosphorylated, whereas the elongational form is hyperphosphorylated (Lu et al. 1991; Dahmus 1994).

Greenleaf and colleagues (Weeks et al. 1993) examined the chromosomal distribution of hypophosphorylated and hyperphosphorylated epitopes of the pol II CTD by immunofluorescence of polytene chromosomes. Many chromosomal sites are sharply labeled by antibody to the hypophosphorylated form of pol II, whereas the hyperphosphorylated form of pol II is associated with numerous diffuse puffs and interbands, many of which show little overlap with sites of the hypophosphorylated polymerase. Sites containing inserted *Drosophila hsp70* transgenes (Lis et al. 1983) show new (relative to wild type) sharp bands of hypophosphorylated pol II in uninduced animals (Weeks et al. 1993). Presumably, these are the paused polymerases associated with the *hsp70* genes. Upon heat shock, the large puffs generated at these sites are labeled with antibodies to both forms of pol II.

At higher resolution, O'Brien et al. (1994) examined the distribution of the different forms of pol II on several genes using UV cross-linking and immunoprecipitation. These studies show that the paused pol II at the start of uninduced heat shock genes is indeed hypophosphorylated, whereas the pol II population on the body of the induced gene is composed of polymerases that contain both the hyper- and hypophosphorylated epitopes. The transcribing polymerases contain some heptapepide repeats of the CTD that are and some that are not phosphorylated. Therefore, the paused pol II lacks phosphorylation and is like the form that enters the promoter (Dahmus 1994). In contrast, the activated body of the gene is covered with phosphorylated pol II.

It is tempting to consider that the escape of paused pol II may depend on a phosphorylation by a kinase that is either recruited or activated by HSF. The general factor TFIIH is composed of multiple subunits that include an essential helicase activity and an essential cyclin-dependent kinase (*CDK7/KIN28*), which appears to be responsible for a significant portion of phosphorylated pol II CTD in yeast (Cismowski et al. 1995; Valay et al. 1995). Indeed, TFIIH is recruited to *Drosophila* heat shock loci during heat shock; however, it is not clear whether this is a requirement for the essential helicase or kinase activities or both (B. Schwartz and J.T. Lis, unpubl.). TFIIH kinase becomes largely insoluble in extracts upon (maximal) heat shock temperature treatments of HeLa cells (Dubois et al. 1997), indicating that heat shock promoters may be built in ways that make use of specialized conditions or requirements of heat shock. These findings leave open the possible participation of another kinase that may be recruited or stimulated by HSF.

Competition of HSF with pol II for the Core Promoter

Promoter-associated pausing may be a consequence of pol II's affinity for a strong core promoter. Mutations in the core promoter reduce both pausing and heat-induced activation on *hsp70* (Lee et al. 1992). The heat shock promoters are extremely strong, presumably a consequence of their open chromatin configuration and their extensive core promoter-TFIID interactions (Purnell et al. 1994), and they allow for rapid pol II entry. Some of the same interactions that assist pol II binding to the promoter may persist and slow its escape. pol II may be able to initiate and begin transcription, but it is tethered and lacks the ability to break away into a fully elongational mode.

Perhaps the binding of HSF disrupts pol II/core promoter interactions by direct competition with pol II for binding to components of the core promoter. In this regard, it is of interest that HSF is an acidic activator that binds very tightly to TBP (Mason and Lis 1997) and that pol II itself possesses a strong acidic activation domain in the H-homology region (adjacent to the CTD) of the largest subunit of pol II (Xiao et al. 1994). This region can bind competitively with HSF for a region on TBP, and a single point mutation in TBP (L114K) disrupts binding to TBP of both HSF and the H region of pol II (Mason and Lis 1997). We do not know if the H region of pol II contacts TBP in vivo. The possibility of such an interaction is not inconsistent with the intriguing observation that mutations in the critical phenylalanines of the H domain of the largest subunit of yeast pol II influence transcription start site selection in vivo, allowing pol II to reach start sites farther downstream (H. Xiao and E. Guzman, unpubl.).

HSF-induced Replacement of Paused Polymerase with Different Elongationally Competent Polymerases

One question that is extremely difficult to resolve unambiguously is whether the paused pol II is actually the polymerase that enters into productive elongation. Models have been proposed which suggest that the upstream activators of genes act to trigger new pol II initiations that, through the participation of upstream activators, are elongationally competent and displace the paused pol II which is elongationally incompetent (Cullen 1993; Krumm et al. 1993). In such a model, the paused polymerases may have a role in maintaining an open promoter but do not contribute directly to the population of elongating pol II. The main problem with this class of model comes from a quantitative consideration of the levels of pausing on the active gene and its implications concerning the mechanics of the process.

The fact that pausing persists on the activated *hsp70* gene makes this third model harder to rationalize. The *hsp70*-paused polymerase is evident in run-on assays in cells induced at submaximal heat shock temperatures that produce a low enough density of transcribing polymerase to allow detection of the paused polymerase (O'Brien and Lis 1991). With a fully induced *hsp70* gene, we have also demonstrated with the high-resolution potassium permanganate assay that pausing occurs at nearly the same steady-state level as seen in uninduced cells (see Fig. 2) (Giardina et al. 1992). Since the level of pause region occupancy is one pol II per gene, a fully induced gene that fires an elongationally competent pol II every 4 seconds must instantly reestablish a paused pol II. This new paused polymerase would remain for 4 seconds only to be displaced by a newly initiated and elongationally competent pol II. The process would then have to be repeated with each round of transcription to account for the steady-state occupancy of the pause site during heat shock. This seems to be mechanistically clumsy and improbable, although it is difficult to rule out. This model is easier to accept if the elongationally competent polymerases can also encounter the slow pausing step of early elongation, but because the pol IIs have been modified, they spend only 4 seconds rather than 10 minutes at this rate-limiting step.

Resolving the fate of the paused polymerase is technically difficult, and to date, the attempts have not been completely satisfying. On the one hand, the paused pol II is engaged in transcription, and, in nuclear run-on assays, it is capable of elongating deep into the body of the *hsp70* gene (Rougvie and Lis 1988). These results demonstrate that the paused polymerase can transcribe. Moreover, a fraction of the paused polymerase is associated with RNAs that are capped and appear to be ready for elongation (Rasmussen and Lis 1995). On the other hand, in an attempt to examine transcripts after a 1-minute instantaneous heat shock, we have observed that a significant fraction of short transcripts were no longer chased to longer transcripts when run-on reactions were done in the presence of Sarkosyl (Rasmussen and Lis 1995). We used the shorthand of referring to these as terminated transcripts in that paper. If indeed they are truly terminated, then under these conditions, some paused polymerases do not give rise to elongating complexes. These experiments, however, do not distinguish between terminated RNAs and RNAs associated with arrested polymerases. It is possible that at least some paused pol II molecules progress to an intermediate state (indeed, most of the transcripts that are unable to be elongated are of a length between the two peaks of normally paused RNAs) that is not capable of transcribing, and only upon full modification can they escape to productive elongation. Additionally, these experiments need to be re-examined in a manner that provides a more natural heat shock protocol, where the temperature is not raised instantaneously. Finally, this particular assay, although very sensitive, needs to be developed into a quantitative assay, to allow a strict accounting of RNAs that are paused, arrested, elongating, and terminated.

The three models described are not mutually exclusive.

First, pol II may require more than one molecular event to escape the pause; modification (model 1) and competition (model 2) could both participate as means of triggering pol II escape. Second, the modification of polymerase during activation in model 1 may occur not only late in the process, when pol II is paused, but also at various early steps in transcription as in model 3. This modification (whether early or late) could reduce the length of time it takes for pol II to escape the pause region.

PROMOTER-ASSOCIATED PAUSING IS SOMEWHAT GENERAL: C-*MYC*, C-*FOS*, AND OTHERS

The heat shock gene promoters and regulatory regions appear to use much of the same transcriptional machinery as other genes. The upstream and core promoter elements of *hsp70* and the associated protein factors can function with elements and factors of other genes (Lis and Wu 1995). Enhancers of developmentally regulated genes can drive expression of an *hsp70* promoter, and heat shock elements can often drive the expression of other core promoters. These hybrid combinations may not always work as efficiently as native *hsp70*, but at some level, the regulatory and core machinery of a variety of genes can communicate.

Although this paper has as its focus the pausing on major heat shock genes, a variety of other *Drosophila* genes, such as β1-*tubulin* and *Gapdh-1* and -2, show some level of paused pol II. These genes display higher densities of pol II at their 5′ ends than in the body of the gene, as judged from in vivo UV cross-linking/immunoprecipitation experiments, and this extra 5′ pol II (over the density found on the body of the gene), like the paused pol II of heat shock genes, is stimulated to transcribe in the presence of Sarkosyl or high salt (Rougvie and Lis 1990). The levels of pausing on other genes is not as high as on the *hsp70* gene, where we estimate one pol II molecule per promoter. These constitutively expressed genes presumably load and fire paused pol II at rates that lead to a lower steady-state level than seen for the *hsp70* gene. Alternatively, perhaps these constitutive genes are governed by a stochastic process such that only a fraction of the cells have active genes and paused pol II.

Immunofluorescence studies of Greenleaf and colleagues using antibody to hypophosphorylated CTD of pol II showed that there are many sites on polytene chromosomes (>100) that are labeled (Weeks et al. 1993). The *hsp70* loci and new *hsp70* loci in transgenic lines contain the hypophosphorylated form of pol II. It is tempting to speculate that these other sites also represent paused polymerases like those seen on *hsp70*. The test of this idea requires additional experimentation.

Elongational control of c-*myc* has been known for some time. Although initial in vitro and *Xenopus* injection studies localized the regulation at the first exon/intron border in the region surrounding +400, further analysis of nuclei with run-on assays and in cells with potassium permanganate localized the block to c-*myc* elongation to a promoter pause site at an interval centered at +30 (Krumm et al. 1992; Strobl and Eick 1992). This paused pol II is remarkably similar to that on *Drosophila hsp70* in terms of its location and properties (Krumm et al 1995; Albert et al. 1997). The c-*fos* gene also appears to have a paused pol II that is very similar to that of c-*myc* and *hsp70* (Plet et al. 1995). Some level of pausing is quite common in mammalian genes composed of a variety of upstream and core promoter elements (Krumm et al. 1995; Blau et al. 1996). This generality of pausing indicates that promoter-associated pausing is an integral feature of promoters. Perhaps the early phase of elongation is a generally slow process, and in some genes, the property is exploited as a major point of regulation.

PROMOTER ARCHITECTURE IN YEAST: IS THERE PAUSING?

The power of yeast genetics makes yeast a particularly attractive system for investigating promoter-associated pausing. But does this form of pausing exist in yeast? Our attempts to demonstrate pausing in yeast have not yielded a definitive clear example of pausing. We have examined the *HSP82, SSA4* (*hsp70*), and *GAL1* and *GAL10* genes for transcription bubbles associated with pausing by potassium permanganate assays both before and after induction of the genes (Giardina and Lis 1993, 1995 and unpubl.). Unlike *Drosophila* heat shock genes, there is no evidence of polymerase on the uninduced genes by this assay or by the nuclear run-on assay (D. Lee and J. Lis, unpubl.). In addition, in contrast with *Drosophila hsp70*, very little, if any, TBP is associated with the uninduced gene. The reason for the absence of obvious pausing in the yeast heat shock homologs of *Drosophila* genes is not clear. Perhaps yeast does not have to build a potentiated promoter because the chromatin is generally more accessible to transcription factors. The lack of a bona fide linker histone could lead to a generally less compact and more accessible chromatin.

After activation of these heat shock and *GAL* genes, TBP and pol II are recruited, and pol-II-dependent permanganate-hypersensitive regions are clearly visible on these promoters. These have the appearance of the transcription bubble seen with paused pol II on *Drosophila* heat shock genes. This region begins at the same position relative to the TATA box in both *Drosophila* and yeast. In the case of *Drosophila*, this region extends to the pause site; however, in yeast, it extends to the transcription start site which is further downstream and less precisely positioned in yeast than in higher eukaryotes (Giardina and Lis 1993, 1995). Perhaps pol II enters and begins its interaction with the promoters of both higher eukaryotes and yeast by a similar mechanism, positioned by TBP/TATA box complex. Such a model is in agreement with the distance (equivalent to 32 bp of B-form DNA) between the yeast pol II active site and the site of TFIIB binding derived from two-dimensional crystallography of TFIIB-pol II complexes of Kornberg and colleagues (Leuther et al. 1996). Subsequent biochemical events, such as initiation, are clearly different in yeast relative to higher eukaryotes, and these may account for differences

in the observed melting relative to the sites of initiation. There is additional permanganate hypersensitivity after the transcription start site in the activated yeast genes we have examined, but this hypersensitivity is less prominent than that seen upstream of the start site. Other reports of pausing in yeast have been made, but they have suffered from technical problems. Thus, the issue of whether pausing exists in yeast remains an open question. Although yeast would be an attractive system to study paused pol II, the recent increase in sequence information and development of genetic tools in higher eukaryotes should also allow the rigorous genetic and biochemical dissection of pausing in vivo in more complex systems as well.

ACKNOWLEDGMENTS

I thank the following past and present lab members for making this review possible by their critical contributions to the analysis of paused polymerase and heat shock promoter architecture and function (in historical order): Dave Gilmour, Nancy Costlow Lee, Ann Rougvie, Jeff Simon, Bob Glaser, Janis Werner, Ed Wong, Xiao Hua, Hyun-sook Lee, Tom O'Brien, Eric Rasmussen, Olga Perisic, Mary Fernandez, Charlie Giardina, Merce Perez-Riba, Lindsay Shopland, Chris Wilkins, Kazunori Hirayoshi, Adam Law, Paul Mason, Ernie Guzman, Janine Lin, Dong-ki Lee, and Brian Schwartz.

REFERENCES

Albert T., Mautner J., Funk J.O., Hoertnagel K., Pullner A., and Eick D. 1997. Nucleosomal structures of c-*myc* promoters with transcriptionally engaged RNA polymerase II. *Mol. Cell. Biol.* **17:** 4363.

Benjamin L.R. and Gilmour D.S. 1998. Nucleosomes are not necessary for promoter-proximal pausing in vitro on the *Drosophila* hsp70 promoter. *Nucleic Acids Res.* **26:** 1051.

Blau J., Xiao H., McCracken S., O'Hare P., Greenblatt J., and Bentley D. 1996. Three functional classes of transcriptional activation domains. *Mol. Cell. Biol.* **16:** 2044.

Brown S.A. and Kingston R.E. 1997. Disruption of downstream chromatin directed by a transcriptional activator. *Genes Dev.* **11:** 3116.

Buratowski S., Hahn S., Guarente L., and Sharp P.A. 1989. Five intermediate complexes in transcription initiation by RNA polymerase II. *Cell* **56:** 549.

Burke T.W. and Kadonaga J.T. 1997. The downstream core promoter element, DPE, is conserved from *Drosophila* to humans and is recognized by TAF$_{II}$60 of *Drosophila*. *Genes Dev.* **11:** 3020.

Cismowski M.J., Laff G.M., Solomon M.J., and Reed S.I. 1995. KIN28 encodes a C-terminal domain kinase that controls mRNA transcription in *Saccharomyces cerevisiae* but lacks cyclin-dependent kinase-activating kinase (CAK) activity. *Mol. Cell. Biol.* **15:** 2983.

Cullen B.R. 1993. Does HIV-1 Tat induce a change in viral initiation rights? *Cell* **73:** 417.

Dahmus M.E. 1994. The role of multisite phosphorylation in the regulation of RNA polymerase II activity. *Prog. Nucleic Acid Res. Mol. Biol.* **48:** 143.

Dubois M.F., Vincent M., Vigneron M., Adamczewski J., Egly J.M., and Bensaude O. 1997. Heat-shock inactivation of the TFIIH-associated kinase and change in the phosphorylation sites on the C-terminal domain of RNA polymerase II. *Nucleic Acids Res.* **25:** 694.

Giardina C. and Lis J.T. 1993. DNA melting on yeast RNA

polymerase II promoters. *Science* **261:** 759.

——. 1995. Dynamic protein-DNA architecture of a yeast heat shock promoter. *Mol. Cell. Biol.* **15:** 2737.

Giardina C., Perez Riba M., and Lis J.T. 1992. Promoter melting and TFIID complexes on *Drosophila* genes in vivo. *Genes Dev.* **6:** 2190.

Gilmour D.S. and Lis J.T. 1985. In vivo interactions of RNA polymerase II with genes of *Drosophila melanogaster*. *Mol. Cell. Biol.* **5:** 2009.

——. 1986. RNA polymerase II interacts with the promoter region of the noninduced hsp-70 gene in *Drosophila melanogaster* cells. *Mol. Cell. Biol.* **6:** 3984.

Glaser R.L., Thomas G.H., Siegfried E., Elgin S.C.R., and Lis J.T. 1990. Optimal heat-induced expression of the *Drosophila hsp26* gene requires a promoter sequence containing (CT)$_n$·(GA)$_n$ repeats. *J. Mol. Biol.* **211:** 751.

Goodrich J.A. and Tjian R. 1994. Transcription factors IIE and IIH and ATP hydrolysis direct promoter clearance by RNA polymerase II. *Cell* **77:** 145.

Grant P.A., Duggan L., Cote J., Roberts S.M., Brownell J.E., Candau R., Ohba R., Owen-Hughes T., Allis C.D., Winston F., Berger S.L., and Workman J.L. 1997. Yeast Gcn5 functions in two multisubunit complexes to acetylate nucleosomal histones: Characterization of an Ada complex and the SAGA (Spt/Ada) complex. *Genes Dev.* **11:** 1640.

Hagler J. and Shuman S. 1992. A freeze-frame view of eukaryotic transcription during elongation and capping of nascent mRNA. *Science* **255:** 983.

Hansen S.K., Takada S., Jacobson R.H., Lis J.T., and Tjian R. 1997. Transcription properties of a cell type-specific TATA-binding protein, TRF. *Cell* **91:** 71.

Holstege F.C.P., Fiedler U., and Timmers H.T.M. 1997. Three transitions in the RNA polymerase II transcription complex during initiation. *EMBO J.* **16:** 7468.

Kim Y.-J., Bjorklund S., Li Y., Sayre M.H., and Kornberg R.D. 1994. A multiprotein mediator of transcriptional activation and its interaction with the C-terminal repeat domain of RNA polymerase II. *Cell* **77:** 599.

Krumm A., Hickey L.B., and Groudine M. 1995. Promoter-proximal pausing of RNA polymerase II defines a general rate-limiting step after transcription initiation. *Genes Dev.* **9:** 559.

Krumm A., Meulia T., and Groudine M. 1993. Common mechanisms for the control of eukaryotic transcriptional elongation. *BioEssays* **15:** 659.

Krumm A., Meulia T., Brunvand M., and Groudine M. 1992. The block to transcriptional elongation within the human c-*myc* gene is determined in the promoter-proximal region. *Genes Dev.* **6:** 2201.

Lee H.S., Kraus K.W., Wolfner M.F., and Lis J.T. 1992. DNA sequence requirements for generating paused polymerase at the start of Hsp70. *Genes Dev.* **6:** 284.

Leuther K.K., Bushnell D.A., and Kornberg R.D. 1996. Two-dimensional crystallography of TFIIB- and IIE-RNA polymerase II complexes: Implications for start site selection and initiation complex formation. *Cell* **85:** 773.

Li B., Weber J.A., Chen Y., Greenleaf A.L., and Gilmour D.S. 1996. Analyses of promoter-proximal pausing by RNA polymerase II on the *hsp70* heat shock gene promoter in a *Drosophila* nuclear extract. *Mol. Cell. Biol.* **16:** 5433.

Lis J. and Wu C. 1995. Promoter potentiation and activation: Chromatin structure and transcriptional induction of heat shock genes. In *Chromatin structure and gene expression* (ed. S.C.R. Elgin), p. 71. IRL Press, Oxford, United Kingdom.

Lis J.T., Simon J., and Sutton C. 1983. New heat shock puffs and β-galactosidase activity resulting from transformation of *Drosophila* with hsp70-lacZ hybrid gene. *Cell* **35:** 403.

Lis J.T., Neckameyer W., Dubensky R., and Costlow N. 1981. Cloning and characterization of nine heat-shock-induced mRNAs of *Drosophila melanogaster*. *Gene* **15:** 67.

Lu H., Flores O., Weinmann R., and Reinberg D. 1991. The nonphosphorylated form of RNA polymerase II preferentially associates with the preinitiation complex. *Proc. Natl. Acad. Sci.* **88:** 10004.

Lu Q., Wallruth L.L., Allan B.D., Glaser R.L., Lis J.T., and Elgin S.C.R. 1992. Promoter sequence containing (CT)$_n$·(GA)$_n$ repeats is critical for the formation of the DNase I hypersensitive sites in the *Drosophila hsp26* gene. *J. Mol. Biol.* **225:** 985.

Mason P.B., Jr. and Lis J.T. 1997. Cooperative and competitive protein interactions at the hsp70 promoter. *J. Biol. Chem.* **272:** 33227.

O'Brien T. and Lis J.T. 1991. RNA polymerase II pauses at the 5′ end of the transcriptionally induced *Drosophila* hsp70 gene. *Mol. Cell. Biol.* **11:** 5285.

——. 1993. Rapid changes in *Drosophila* transcription after an instantaneous heat shock. *Mol. Cell. Biol.* **13:** 3456.

O'Brien T., Hardin S., Greenleaf A., and Lis J.T. 1994. Phosphorylation of RNA polymerase II C-terminal domain and transcriptional elongation. *Nature* **370:** 75.

O'Brien T., Wilkins R.C., Giardina C., and Lis J.T. 1995. Distribution of GAGA protein on *Drosophila* genes in vivo. *Genes Dev.* **9:** 1098.

Orphanides G., Lagrange T., and Reinberg D. 1996. The general transcription factors of RNA polymerase II. *Genes Dev.* **10:** 2657.

Plet A., Eick D., and Blanchard J.M. 1995. Elongation and premature termination of transcripts initiated from c-*fos* and c-*myc* promoters show dissimilar patterns. *Oncogene* **10:** 319.

Ptashne M. and Gann A. 1997. Transcriptional activation by recruitment. *Nature* **386:** 569.

Purnell B.A., Emanuel P.A., and Gilmour D.S. 1994. TFIID sequence recognition of the initiator and sequences farther downstream in *Drosophila* class II genes. *Genes Dev.* **8:** 830.

Rasmussen E.B. and Lis J.T. 1993. *In vivo* transcriptional pausing and cap formation on three *Drosophila* heat shock genes. *Proc. Natl. Acad. Sci.* **90:** 7923.

——. 1995. Short transcripts of the ternary complex provide insight into RNA polymerase II elongational pausing. *J. Mol. Biol.* **252:** 522.

Rougvie A.E. and Lis J.T. 1988. The RNA polymerase II molecule at the 5′ end of the uninduced *hsp70* gene of *Drosophila melanogaster* is transcriptionally engaged. *Cell* **54:** 795.

——. 1990. Postinitiation transcriptional control in *Drosophila melanogaster*. *Mol. Cell. Biol.* **10:** 6041.

Shopland L.S., Hirayoshi K., Fernandes M., and Lis J.T. 1995. HSF access to heat shock elements in vivo depends critically on promoter architecture defined by GAGA factor, TFIID, and RNA polymerase II binding sites. *Genes Dev.* **9:** 2756.

Stewart A.F., Herrera R.E., and Nordheim A. 1990. Rapid induction of c-*fos* transcription reveals quantitative linkage of RNA polymerase II and DNA topoisomerase I enzyme activities. *Cell* **60:** 141.

Strobl L.J. and Eick D. 1992. Hold back of RNA polymerase II at the transcription start site mediates down-regulation of c-*myc* in vivo. *EMBO J.* **11:** 3307.

Taylor I.C.A., Workman J.L., Schuetz T.J., and Kingston R.E. 1991. Facilitated binding of GAL4 and heat shock factor to nucleosomal templates: Differential function of DNA-binding domains. *Genes Dev.* **5:** 1285.

Thompson C.M., Koleske A.J., Chao D.M., and Young R.A. 1993. A multisubunit complex associated with the RNA polymerase II CTD and TATA-binding protein in yeast. *Cell* **73:** 1361.

Valay J.G., Simon M., Dubois M.F., Bensaude O., Facca C., and Faye G. 1995. The *KIN28* gene is required both for RNA polymerase II mediated transcription and phosphorylation of the Rpb1p CTD. *J. Mol. Biol.* **249:** 535.

Weeks J.R., Hardin S.E., Shen J., Lee J.M., and Greenleaf A.L. 1993. Locus-specific variation in phosphorylation state of RNA polymerase II in vivo: Correlations with gene activity and transcript processing. *Genes Dev.* **7:** 2329.

Westwood J.T., Clos J., and Wu C. 1991. Stress-induced oligomerization and chromosomal relocalization of heat-shock factor. *Nature* **353:** 822.

Wilkins R.C. and Lis J.T. 1997. Dynamics of potentiation and activation: GAGA factor and its role in heat shock gene regulation. *Nucleic Acids Res.* **25:** 3963.

Wu C. 1980. The 5′ end of *Drosophila* heat shock genes in chromatin are hypersensitive to DNase I. *Nature* **286:** 854.

———. 1997. Chromatin remodeling and the control of gene expression. *J. Biol. Chem.* **272:** 28171.

Xiao H. and Lis J.T. 1988. Germline transformation used to define key features of heat-shock response elements. *Science* **239:** 1139.

Xiao H., Friesen J.D., and Lis J.T. 1994. A highly conserved domain of RNA polymerase II shares a functional element with acidic activation domains of upstream transcription factors. *Mol. Cell. Biol.* **14:** 7507.

Mechanism of Promoter Escape by RNA Polymerase II

J.W. Conaway,*†‡ A. Dvir,** R.J. Moreland,†‡ Q. Yan,†‡ B.J. Elmendorf,†‡
S. Tan,§ and R.C. Conaway†

*Howard Hughes Medical Institute and †Program in Molecular and Cell Biology, Oklahoma Medical Research
Foundation, Oklahoma City, Oklahoma 73104; ‡Department of Biochemistry and Molecular Biology,
University of Oklahoma Health Sciences Center, Oklahoma City, Oklahoma 73190; **Department of
Biological Sciences, Oakland University, Rochester, Michigan 48309-4401; §Department of Pathology,
Stanford University School of Medicine, Stanford, California 94305

Transcription initiation by RNA polymerase II (pol II) is an elaborate biochemical process requiring minimally the five general initiation factors TFIIB, TFIID (or TBP), TFIIE, TFIIF, and TFIIH and an ATP cofactor (for review, see Conaway and Conaway 1993; Roeder 1996). Biochemical studies of transcription by pol II in this minimal enzyme system have revealed that initiation is a multistage process beginning with assembly of pol II and all five general initiation factors into a stable preinitiation complex at the promoter and culminating with ATP-dependent formation of the open complex, synthesis of the first few phosphodiester bonds of nascent transcripts, and escape of polymerase from the promoter. Furthermore, substantial progress defining the functions of the general initiation factors in early stages of initiation has been achieved. These studies have established roles for TFIID, TFIIB, and TFIIF in selective binding of pol II to its promoters and for TFIIF, TFIIE, and the DNA helicase activity of TFIIH in ATP-dependent formation of the open complex.

In contrast to the abundance of information on the functions of the general initiation factors in preinitiation stages of transcription, relatively little is known about their contributions to early elongation and promoter escape, prior to their dissociation from elongating pol II. Our laboratory is engaged in biochemical studies investigating the contributions of the general initiation factors to early elongation and promoter escape by pol II (Dvir et al. 1996a; 1997a,b). Here, we describe evidence that efficient promoter escape requires conversion of the early elongation complex to an "escape-competent" intermediate in a step that exhibits a transient requirement for an ATP cofactor, the general initiation factors TFIIE and TFIIH, and template DNA extending 40–50 bp downstream from the transcriptional start site. Failure of the early pol II elongation complex to undergo conversion to the escape-competent intermediate results either in abortive transcription or in arrest by a large fraction of polymerases at promoter-proximal sites. Thus, efficient promoter escape by pol II requires that early elongation intermediates undergo a critical ATP-dependent structural transition that is likely driven by interaction of polymerase and/or one or more of the general initiation factors with template DNA extending 40–50 bp downstream from the transcriptional start site.

METHODS

Preparation of pol II and the general initiation factors. pol II (Serizawa et al. 1992) and TFIIH (rat δ; Conaway and Conaway 1989; Conaway et al. 1992) were purified from rat liver nuclear extracts as described previously. Recombinant yeast TBP (AcA 44 fraction; J.W. Conaway et al. 1991) and TFIIB (rat α; Tsuboi et al. 1992) were expressed in *Escherichia coli* and purified as described previously. Recombinant TFIIE was prepared as described by Peterson et al. (1991), except that the 56-kD subunit was expressed in *E. coli* strain BL21(DE3)-pLysS. Recombinant TFIIF was purified as described by Tan et al. (1994) from *E. coli* strain JM109(DE3) coinfected with M13mpET-RAP30 and M13mpET-RAP74.

Assay of transcription by pol II and the general initiation factors. Transcription reactions were performed essentially as described by Dvir et al. (1996a, 1997a,b). Preinitiation complexes were assembled at the AdML promoter at 28°C by a 45-minute incubation of 35-μl reaction mixtures containing 20 mM HEPES-NaOH (pH 7.9), 20 mM Tris-HCl (pH 7.9), 60 mM KCl, 4 mM MgCl$_2$, 0.1 mM EDTA, 1 mM DTT, 0.5 mg/ml BSA, 2% (w/v) polyvinyl alcohol, 7% (v/v) glycerol, 6 units of RNAasin, ~50 ng of yeast TBP, ~10 ng of TFIIB, ~20 ng of TFIIF, ~20 ng of TFIIE, 0.1 unit of pol II, and the amounts of TFIIH and AdML template DNA indicated in the figure legends. Transcription reactions were performed at 28°C with the concentrations of nucleotides indicated in the figure legends. Transcription reactions measuring synthesis of trinucleotide or 3′-*O*-MeG-terminated transcripts were stopped by addition of 6 μl of a solution of 100 mM EDTA and 0.5 mg/ml proteinase K to 15 μl of reaction mixture. After 15 minutes at room temperature, 25 μl of a solution of 10 M urea, 0.025% bromophenol blue, and 0.025% xylene cyanol FF were added to stopped reaction mixtures. The samples were vortexed for 10 seconds, heated at 70°C for 5 minutes, and analyzed by electrophoresis through 25% (w/v) acrylamide, 3% (w/v) bisacrylamide, 7 M urea gels as described by Jacob et al. (1991) Transcription reactions measuring full-length runoff transcripts were stopped by addition of an equal volume of a solution of 200 mM Tris-HCl (pH 7.6), 300 mM NaCl, 25 mM EDTA, 2% SDS, and 0.5 mg/ml proteinase K. After 15 minutes at room temperature, transcripts were precipitated with ethanol, resuspended in 25 μl of a solution of 10 M urea, 0.025% bromophenol blue, and 0.025% xylene cyanol FF, heated at 70°C for 5 minutes, and anlayzed by electrophoresis through 6% (w/v) acrylamide, 0.8% (w/v) bisacrylamide, 7 M urea gels. Gels were imaged by autoradiography or on a Molecular Dynamics PhosphorImager.

RESULTS

A Transient Requirement for an ATP Cofactor for Efficient Promoter Escape by pol II

Transcription initiation by pol II and the general initiation factors requires an ATP cofactor at least in part to drive formation of the open complex by the TFIIH DNA helicase prior to synthesis of the first phosphodiester bond of nascent transcripts (Wang et al. 1992; Dvir et al. 1996b; Holstege et al. 1996; Yan and Gralla 1997). ATPγS is a potent inhibitor of open complex formation (Dvir et al. 1996b; Holstege et al. 1996). To investigate the possibility that an ATP cofactor is also required subsequent to open complex formation, during early elongation or promoter escape, we tested the effect of ATPγS on extension of short approximately 5–8-nucleotide transcripts by pol II elongation complexes initiated from the adenovirus major late promoter (AdMLP) derivative shown in Figure 1A. These and subsequent experiments were performed using a transcription system reconstituted with recombinant TBP, TFIIB, TFIIE, and TFIIF and highly purified pol II and TFIIH.

Paused elongation complexes containing transcripts with a maximum length of approximately 5–8 nucleotides were synthesized in the presence of 200 μM of the initiating dinucleotide CpU, 5 μM ATP, 10 nM UTP, and 0.5 μM [α-^{32}P]CTP (Fig. 1B, lane 1). Synthesis of these transcripts was inhibited by 1 μg/ml α-amanitin (data not shown). A significant fraction of the transcripts synthesized under these conditions were abortive tri- or tetranucleotide transcripts, which could not be chased into longer transcripts. Stably initiated transcripts were then chased into longer transcripts in the presence of ATP (Fig. 1, lanes 2–6) or ATPγS (Fig. 1, lanes 7–11) by addition of 200 μM CTP, 100 μM UTP, and 100 μM of the RNA chain-terminating nucleotide 3′-O-methylguanosine 5′-triphosphate (3′-O-MeGTP), which prevents most transcription beyond the first G in the AdMLP transcript. Because ATP is not incorporated into CpU-initiated transcripts between positions 4 and 17, any differences in the efficiency of elongation in the presence or absence of ATP must elect a role for ATP other than as a substrate for RNA synthesis by pol II.

In the presence of ATP, most of the stably initiated transcripts were rapidly chased into 18-nucleotide 3′-O-MeG-terminated RNAs. In the presence of ATPγS, most of the short transcripts were rapidly chased into 9–13-nucleotide transcripts, and synthesis of the full-length 3′-O-MeG-terminated transcript was dramatically reduced, reaching a maximum level within 10 minutes of addition of the chase ribonucleoside triphosphates.

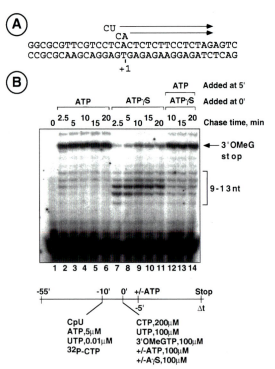

Figure 1. ATPγS inhibits promoter escape by pol II. (A) Sequence of the transcriptional start site and early transcribed region of the AdMLP. +1 indicates the position of transcription initiation from the AdMLP in vivo. CU and CA indicate the positions of transcription initiation primed by CpU and CpA, respectively. (B) CpU-initiated ~5–8-nucleotide transcripts were synthesized by pol II and the general initiation factors as described under Methods by a 10-min pulse in the presence of 200 μM CpU, 5 μM ATP, 10 nM UTP, and 0.5 μM [α-^{32}P]CTP. Transcription reactions were carried out with ~150 ng of TFIIH (rat δ, DEAE 5-PW fraction; Conaway et al. 1992) and ~20 ng of the EcoRI to NdeI fragment from pDN-AdML. Paused pol II elongation complexes were chased for the times indicated in the figure in presence of 200 μM CTP, 100 μM UTP, 100 μM 3′-O-MeGTP and ATP or ATPγS. (Lane 1) Reaction mixture stopped before the chase; (lanes 2–11) chase in the presence of either 100 μM ATP or 100 μM ATPγS; (lanes 12–14) after chase for 5 min in the presence of 100 μM ATPγS, 500 μM ATP was added to reaction mixtures for the times indicated in the figure. (AγS) ATPγS.

The approximately 9–13-nucleotide transcripts synthesized in the presence of ATPγS could be either terminated and released or associated with arrested, but potentially active, elongation complexes. To address this question, we asked whether the approximately 9–13-nucleotide transcripts synthesized in the presence of ATPγS could be chased into longer transcripts following addition of ATP. As shown in Figure 1B, lanes 12–14, a significant fraction of these transcripts could be chased into 3′-O-MeG-terminated transcripts when ATP was added to reaction mixtures at a fivefold molar excess over ATPγS, indicating that they were contained in arrested, but potentially active, elongation complexes. The ability of ATPγS to inhibit elongation by promoter-proximally paused transcription complexes did not depend on the particular initiating dinucleotide, since ATPγS also inhibited extension of short transcripts initiated with ATP and with the dinucleotides UpC and CpA (Dvir et al. 1996a). In addition, we observed that ATPγS inhibited elongation by promoter-proximally paused transcription complexes even when the labeling phase of the reaction was limited to 1 minutes, thereby reducing the length of the pause (Dvir et al. 1996a).

To rule out the possibility that arrest by early pol II elongation complexes might be induced artificially by ATPγS, two different approaches were taken. First, promoter-proximally paused elongation complexes containing short CpU-initiated transcripts were purified by gel filtration to remove ATP and unincorporated nucleotides. Transcripts associated with purified elongation complexes were then chased into longer transcripts by addition of UTP, CTP, and 3′-O-MeGTP in either the presence or absence of ATP. As shown in Figure 2A, in the presence of ATP, nearly all of the short transcripts were chased into 3′-O-MeG-terminated transcripts. In the absence of ATP, however, only a small fraction of the short transcripts could be extended into 16-nucleotide, U-terminated transcripts; the remaining transcripts were arrested or terminated following synthesis of about 10–14 nucleotide transcripts.

Second, promoter-proximally paused elongation complexes were treated with immobilized hexokinase to remove ATP. Hexokinase hydrolyzes ATP in the presence of glucose to give ADP and glucose-6-phosphate. Consistent with the results of Figure 2A, nearly all of the short transcripts were chased into 3′-O-MeG-terminated transcripts in the presence of ATP, whereas in the absence of added ATP, only a small fraction of short transcripts were successfully extended into the 16-nucleotide, U-terminated product (Fig. 2B).

Our results thus argue that an ATP cofactor is required for efficient promoter escape by pol II. Consistent with previous results indicating that ATP is not required for elongation by pol II under most conditions (Bunick et al. 1982; Ernst et al. 1983; Sawadogo and Roeder 1984; Conaway and Conaway 1988), the role of the ATP cofactor appears to be limited to very early elongation. Once early elongation complexes have synthesized approximately 9–10-nucleotide transcripts, they are no longer susceptible to ATPγS-induced promoter-proximal arrest and do not require an ATP cofactor for further elongation (Sawadogo and Roeder 1984; Conaway and Conaway 1988; Dvir et al. 1996a). These findings, together with our observation that, in the presence of ATPγS or in the absence of ATP, elongation complexes containing short approximately 5–8-nucleotide transcripts can extend those transcripts several nucleotides before arresting, suggest that efficient promoter escape by pol II has only a transient requirement for ATP.

A Transient Requirement for TFIIE and TFIIH for Efficient Promoter Escape by pol II

As part of our effort to understand how ATP is used to promote efficient escape of pol II from the promoter, we sought to determine which components of the transcription system mediate ATP-dependent suppression of arrest

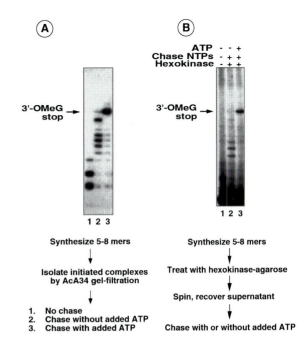

Figure 2. A transient requirement for an ATP cofactor for efficient promoter escape by pol II. (*A*) Depletion of ATP from early pol II elongation complexes by AcA 34 gel filtration. CpU-initiated ~5–8-nucleotide transcripts were synthesized by pol II and the general initiation factors as described under Methods by a 10-min pulse in the presence of 200 μM CpU, 5 μM ATP, 10 nM UTP, and 0.5 μM [α-^{32}P]CTP. Transcription reactions were carried out with ~150 ng of TFIIH (rat δ; DEAE 5-PW fraction; Conaway et al. 1992) and ~20 ng of the *Eco*RI to *Nde*I fragment from pDN-AdML. pol II elongation complexes contained in five reaction volumes were applied at room temperature to a 2.5-ml AcA 34 column packed in a pasteur pipette. The column was preequilibrated and eluted in transcription reaction salts. Fractions of 100 μl were collected. Void volume fractions containing pol II elongation complexes were pooled and divided into three equal portions. One portion was not chased. The other two portions were chased with 200 μM CTP, 100 μM UTP, and 100 μM 3′-O-MeGTP, in the presence or absence of 100 μM ATP. (*B*) Depletion of ATP by treatment of early pol II elongation complexes with immobilized hexokinase. CpU-initiated ~5–8-nucleotide transcripts were synthesized by pol II and the general initiation factors as described under Methods and in panel *A*; 15 μl of hexokinase-agarose (Sigma) (~2.5 units of hexokinase) equilibrated 1:1 with transcription reaction salts and 2 μl of 100 μM dextrose were then added to each 30-μl transcription reaction and incubated for 15 min at 28°C. The hexokinase-agarose was removed by centrifugation for 30 sec in a microcentrifuge, and the supernatants were either not chased or chased with 200 μM CTP, 100 μM UTP, and 100 μM 3′-O-MeGTP, in the presence or absence of 100 μM ATP. To reduce trace amounts of contaminating ATP in the chase ribonucleoside triphosphates, CTP, UTP, and 3′-O-MeGTP were also treated with immobilized hexokinase prior to use. (NTPs) Ribonucleoside triphosphates.

by early pol II elongation complexes. Because TFIIH is the only component of the transcription system known to possess ATPase activity (Conaway and Conaway 1989; Feaver et al. 1991; R.C. Conaway et al. 1991; Schaeffer et al. 1993) and because TFIIE and TFIIH are both required for ATP-dependent formation of the open complex by the TFIIH DNA helicase (Pan and Greenblatt

1994; Holstege et al. 1995, 1996), we sought to determine whether these initiation factors function in ATP-dependent suppression of promoter-proximal arrest. To distinguish the requirements for TFIIE and TFIIH in ATP-dependent formation of the open complex from their potential requirements in promoter escape, we took advantage of the artificial AdMLP derivative Ad(-9/-1) (Fig. 3A), which contains a premelted transcriptional start site and does not require ATP, TFIIE, or TFIIH for synthesis of the first phosphodiester bond of nascent transcripts (Pan and Greenblatt 1994; Tantin and Carey 1994; Holstege et al. 1995, 1996).

In the experiment of Figure 3B, transcription initiation from the premelted Ad(-9/-1) promoter was reconstituted with pol II and various combinations of the general initiation factors. In agreement with previous studies (Pan and Greenblatt 1994; Holstege et al. 1996), transcription initiation by pol II at the Ad(-9/-1) promoter was dependent only on TBP, TFIIB, and TFIIF and was inhibited by α-amanitin at 1 μg/ml (Fig. 3) (Dvir et al. 1997a). pol II early elongation complexes suffered arrest at promoter-proximal sites when transcription was carried out in the absence of either TFIIE, TFIIH, or ATP. In contrast, a substantial fraction of early elongation complexes suc-

cessfully synthesized 3′-O-MeG-terminated transcripts when reactions also included TFIIE, TFIIH, and ATP.

Further investigation of the mechanism of TFIIH action in suppression of arrest revealed that TFIIH can prevent arrest even when added to transcription reactions after elongation complexes have synthesized approximately 5–8-nucleotide transcripts. In these experiments, pol II elongation complexes containing CpU-initiated, approximately 5–8-nucleotide transcripts were synthesized from the Ad(-9/-1) promoter in the absence of TFIIH and purified by gel filtration. Transcripts associated with purified elongation complexes were then chased into longer transcripts by addition of 100 μM UTP, 200 μM CTP, and 100 μM 3′-O-MeGTP, in the presence or absence of various concentrations of TFIIH and either 100 μM ATP or 100 μM ATPγS. As shown in Figure 3C, in the presence of ATP but in the absence of TFIIH, a substantial fraction of early elongation complexes suffered arrest after synthesizing about 9–13 nucleotide transcripts. Addition of TFIIH to early elongation complexes significantly increased the fraction of approximately 5–8-nucleotide transcripts that could be chased into 18-nucleotide 3′-O-MeG-terminated transcripts. Notably, the collection of transcripts synthesized from the duplex AdMLP in the

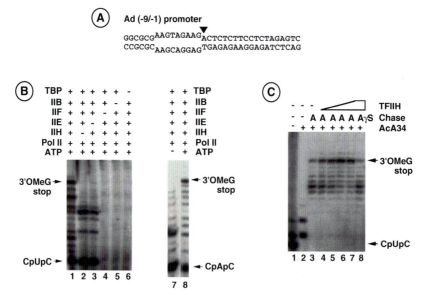

Figure 3. A transient requirement for TFIIE and TFIIH for efficient promoter escape by pol II. (*A*) Structure of the transcriptional start site and early transcribed region of the premelted AdMLP derivative Ad(-9/-1). (*Inverted closed triangle*) Position of transcription initiation from the wild-type AdMLP in vivo. (*B*) CpU- or CpA-initiated transcripts were synthesized by pol II from the Ad(-9/-1) promoter in 30-min reactions performed as described under Methods in the presence of the indicated combinations of general initiation factors, 5 μM ATP, 5 μM UTP, 100 μM 3′-O-MeGTP, 0.5 μM [α-32P]CTP, and either 200 μM CpU or 200 μM CpA. (IIB) TFIIB; (IIE) TFIIE; (IIF) TFIIF; (IIH) TFIIH; (pol II) RNA polymerase II. Transcription reactions were carried out with ~10 ng of TFIIH (rat δ, SP-5PW fraction; Conaway et al. 1992) and ~10 ng of the M13mp19-AdML-derived *Kpn*I to *Ava*II fragment containing the premelted Ad(-9/-1) promoter; the premelted template was prepared as described by Dvir et al. (1997). (*C*) CpU-initiated ~5–8-nucleotide transcripts were synthesized by pol II from the Ad(-9/-1) promoter as described under Methods by a 15-min pulse in the presence of TBP, TFIIB, TFIIE, and TFIIF and 200 μM CpU, 5 μM ATP, 10 nM UTP, and 0.5 μM [α-32P]CTP. Synthesis of CpU-initiated ~5–8-nucleotide transcripts was carried out in the absence of TFIIH with ~10 ng of the M13mp19-AdML-derived *Kpn*I to *Ava*II fragment containing the premelted Ad(-9/-1) promoter. pol II elongation complexes were purified by AcA 34 gel filtration as described in *A* of Fig. 2. Paused pol II elongation complexes were then chased in the presence of 100 μM ATP or 100 μM ATPγS and varying concentrations of TFIIH. (Lane *1*) Transcripts before AcA 34 gel filtration; (lane *2*) transcripts associated with isolated elongation complexes before the chase; (lanes *3–8*) transcripts resulting from chase of isolated elongation complexes. The chase reactions in lanes *4–8* contained ~1.5 ng (lane *4*), ~6 ng (lane *5*), ~15 ng (lane *6*), or ~30 ng (lanes *7* and *8*) of TFIIH (rat δ, SP 5-PW fraction; Conaway et al. 1992). TFIIH was added to reaction mixtures immediately before addition of chase ribonucleoside triphosphates. (A) ATP; (AγS) ATPγS. (Reprinted, with permission, from Dvir et al. 1996a.)

presence of ATPγs and from the Ad(-9/-1) promoter in either the absence of TFIIH or the presence of ATPγS is very similar (Fig. 3C, lanes 3 and 8 and data not shown), suggesting that TFIIE, TFIIH, and ATP function together in the same step in promoter escape on both templates.

Efficient Promoter Escape by pol II Exhibits a Transient Requirement for Template DNA Extending 40–50 bp Downstream from the Transcriptional Start Site

In the course of experiments investigating the mechanism of promoter escape, we discovered that synthesis of transcripts longer than about 10 nucleotides depends on the presence of template DNA extending 40–50 bp downstream from the transcriptional start site. In these experiments, we took advantage of the plasmid pDN-AdML, which contains AdML core promoter sequences from –50 to +10 inserted between the KpnI and XbaI sites in the polylinker of pUC-18 (Conaway and Conaway 1988). The EcoRI to NdeI fragment from pDN-AdML was used as a DNA template in transcription reactions. The NdeI site is located approximately 250 bp downstream from the AdML transcriptional start site. As illustrated in Fig-

ure 4A, pDN-AdML can be cleaved by the restriction enzymes PstI, SphI, HindIII, and HaeIII, which cut the template strand at sites 23, 29, 39, and 48 nucleotides downstream from the AdML transcriptional start site. By digesting the template with these restriction enzymes before transcription initiation or after synthesis of short transcripts, we could assess the requirements for downstream template DNA during initiation, promoter escape, and subsequent elongation by pol II.

To determine how much downstream template DNA is required for synthesis of the first phosphodiester bond of nascent transcripts, we carried out CpU-primed abortive initiation assays. To determine how much downstream template DNA is required for very early elongation and promoter escape, we carried out transcription in the presence of CpU, ATP, UTP, [α-^{32}P]CTP, and the RNA chain-terminating nucleotide 3´-O-MeGTP.

In the experiment of Figure 4B, preinitiation complexes were assembled on the EcoRI to NdeI fragment of pDN-AdML and treated with either HaeII, HindIII, SphI, or PstI. The reaction mixtures were then divided into two equal portions and assayed for synthesis of CpU-primed trinucleotide transcripts or synthesis of CpU-primed 3´-O-MeG-terminated 18-nucleotide transcripts.

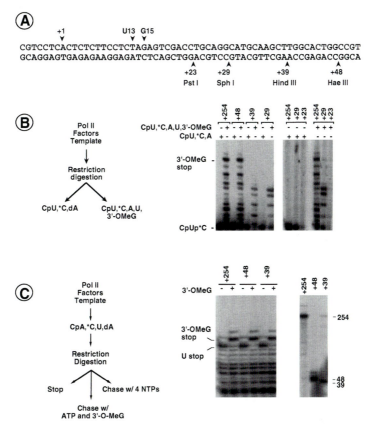

Figure 4. Efficient promoter escape by pol II exhibits a transient requirement for template DNA extending 40–50 bp downstream from the transcriptional start site. (A) Location of restriction sites downstream from the AdMLP in pDN-AdML. The transcriptional start site is indicated by +1. The sites of restriction enzyme cleavage are indicated below the pDN-AdML template sequence. The U residue (U13) and G residue (G15) at positions +13 and +15 of the AdML transcript are indicated above the pDN-AdML template sequence. (B) Preinitiation complexes containing pol II and the general initiation factors were assembled as described under Methods. Transcription reactions were carried out with ~10 ng of TFIIH (rat δ, SP 5-PW fraction; Conaway et al. 1992) and ~10 ng of the EcoRI to NdeI fragment from pDN-AdML. After assembly of preinitiation complexes, DNA templates were digested for 30 min at 28°C.

Synthesis of CpU-primed trinucleotide transcripts was unaffected by digestion of preinitiation complexes with *Hae*III and only modestly reduced by digestion with either *Hin*dIII or *Sph*I. In contrast, synthesis of trinucleotide transcripts was almost completely inhibited by digestion of preinitiation complexes with *Pst*I. Thus, synthesis of the first phosphodiester bond of nascent transcripts does not require template DNA downstream from +29 but is strongly dependent on the presence of template DNA between +23 and +28, even though the preinitiation complex does not protect DNA in this region from digestion by restriction enzymes.

Like synthesis of CpU-primed trinucleotide transcripts, synthesis of CpU-primed 3′-*O*-MeG-terminated 18-nucleotide transcripts was unaffected by digestion of preinitiation complexes with *Hae*III prior to initiation. In contrast, little or no 3′-*O*-MeG-terminated 18-nucleotide transcripts were synthesized when preinitiation complexes were digested with *Hin*dIII or *Sph*I. Under these conditions, CpU-primed transcripts reached a maximum length of only about 10–12 nucleotides. Thus, promoter escape by very early pol II elongation complexes is strongly dependent on the presence of template DNA extending 40–50 bp downstream from the AdML transcriptional start site, even though this region of the DNA template is not essential for assembly of the preinitiation complex and synthesis of the first few phosphodiester bonds of transcripts.

Although digestion of preinitiation complexes with *Hin*dIII prior to transcription initiation is sufficient to inhibit synthesis of transcripts longer than about 10–12 nucleotides, digestion of early elongation complexes with *Hin*dIII after synthesis of approximately 14-nucleotide transcripts does not prevent their further extension. In the experiment of Figure 4C, preinitiation complexes were assembled on the *Eco*RI to *Nde*I fragment from pDN-AdML. Transcription was carried out in the presence of the initiating dinucleotide Cpa, UTP, [α-³²P]CTP, and dATP to satisfy the ATP cofactor requirement for formation of the open complex and promoter escape. Under these conditions, pol II will synthesize transcripts paused at U13, immediately preceding the first A residue that must be incorporated into the transcript. Following treatment of elongation complexes with either *Hae*III or *Hin*dIII, transcription intermediates were assayed for their abilities to extend the CpA-primed 14-nucleotide transcripts to 3′-*O*-MeG-terminated 16-nucleotide transcripts or to full-length runoff transcripts. As shown in the middle panel of Figure 4C, digestion of elongation intermediates with either *Hae*III or *Hin*dIII had no effect on the efficiency with which CpA-primed, 14-nucleotide transcripts were chased into 3′-*O*-MeG-terminated 16-nucleotide transcripts. In addition, full-length runoff transcripts of the expected length were synthesized when nascent transcripts were chased into longer RNAs in the presence of all four ribonucleoside triphosphates (Fig. 4C, right panel). Similar results were obtained when transcription was initiated with the dinucleotide CpU (data not shown).

DISCUSSION

Our laboratory is engaged in biochemical studies investigating the mechanism of promoter escape by pol II in a minimal transcription system reconstituted with the five general initiation factors TBP, TFIIB, TFIIE, TFIIF, and TFIIH. In this paper, we describe findings arguing that promoter escape requires conversion of the early pol II elongation complex to an "escape-competent" intermediate in a step that exhibits a transient requirement for an ATP cofactor, the general initiation factors TFIIE and TFIIH, and template DNA extending 40–50 bp downstream from the transcriptional start site. Our findings can be summarized as follows.

- When deprived of ATP, early pol II elongation complexes that have initiated transcription in the presence of the five general initiation factors and synthesized transcripts shorter than about 10 nucleotides are prone to arrest at promoter-proximal sites after addition of just a few nucleotides to the nascent transcript. Addition of ATP to early pol II elongation complexes prior to arrest is sufficient to prevent arrest, and a fraction of elongation complexes that have arrested can be reactivated by addition of ATP.

- Early pol II elongation complexes that have initiated transcription from the premelted Ad(-9/-1) promoter in the presence of general initiation factors TBP, TFIIB, and TFIIF, but in the absence of TFIIE, TFIIH, or both factors, are prone to arrest at promoter-proximal sites similar to those at elongation complexes deprived of ATP. Arrest by early pol II elongation complexes that have initiated transcription from the Ad(-9/-1) promoter in the absence of just TFIIH can be suppressed even when TFIIH is added to transcription reactions after early elongation complexes have synthesized the first few phosphodiester bonds of nascent transcripts, indicating that TFIIH can act directly on the pol II elongation complex without first entering the preinitiation complex. Thus, TFIIE and TFIIH have dual roles in transcription: They are essential for ATP-dependent formation of the open complex prior to synthesis of the first phosphodiester bond of transcripts (Pan and Greenblatt 1994; Holstege et al. 1995, 1996; Yan and Gralla 1997), and they are needed for ATP-dependent suppression of arrest by early pol II elongation complexes prior to their escape from the promoter. TFIIE, TFIIH, and ATP do not, however, appear to be essential for escape of pol II from the promoter, since a fraction of polymerases are capable of exiting the promoter in their absence.

Exactly how TFIIE and TFIIH function together to expedite efficient promoter escape by pol II is presently unknown. It is possible that the TFIIH DNA helicase may have a role in this process. TFIIH is the only general initiation factor known to possess detectable ATPase activity (Conaway and Conaway 1989; R.C. Conaway et al. 1991; Feaver et al. 1991; Schaeffer et al. 1993), and efficient promoter escape by pol II is blocked by ATPγS, an inhibitor of the TFIIH DNA helicase (Serizawa et al. 1993b; Holstege et al.

1996, 1997), but not by H-8, an inhibitor of the TFIIH CTD kinase (Serizawa et al. 1993a). On the basis of evidence indicating that TFIIE is capable of stabilizing TFIIH with both polymerase and the preinitiation complex (Conaway and Conaway 1990; Gerard et al. 1991), TFIIE may prevent premature arrest by pol II at least in part by stabilizing the interaction of TFIIH with the early elongation complex.

• In the absence of template DNA extending 40–50 bp downstream from the transcriptional start site, early pol II elongation complexes that have initiated transcription in the presence of the five general initiation factors and synthesized transcripts shorter than about 10 nucleotides are prone to arrest at promoter-proximal sites similar to those of elongation complexes deprived of ATP, TFIIE, or TFIIH. Template DNA downstream from +40 is dispensable for assembly of the preinitiation complex, for initiation and synthesis of the first approximately 10–12 phosphodiester bonds of nascent transcripts, and for further extension of transcripts longer than about 13 nucleotides. In contrast, pol II does not normally require an extended region of downstream template DNA for efficient elongation and is able to transcribe to the extreme 3′ end of most DNA templates during synthesis of runoff transcripts. Thus, our findings argue that efficient promoter escape requires that the pol II elongation complex undergoes a critical structural transition in a step that depends on an ATP cofactor, the general initiation factors TFIIE and TFIIH, and interaction of one or more components of the transcriptional machinery with template DNA extending 40–50 bp downstream from the transcriptional start site.

ACKNOWLEDGMENTS

This work was supported in part by National Institutes of Health grant GM-41628 and by funds provided to the Oklahoma Medical Research Foundation by the H.A. and Mary K. Chapman Charitable Trust. J.W.C. is an investigator of the Howard Hughes Medical Institute.

REFERENCES

Bunick D., Zandomeni R., Ackerman S., and Weinmann R. 1982. Mechanism of RNA polymerase II-specific initiation of transcription in vitro: ATP requirement and uncapped runoff transcripts. Cell 29: 877.

Conaway J.W., Bradsher J.N., and Conaway R.C. 1992. Mechanism of assembly of the RNA polymerase II preinitiation complex. Transcription factors delta and epsilon promote stable binding of the transcription apparatus to the initiator element. J. Biol. Chem. 267: 10142.

Conaway J.W., Hanley J.P., Garrett K.P., and Conaway R.C. 1991. Transcription initiated by RNA polymerase II and transcription factors from liver. Structure and action of transcription factors epsilon and tau. J. Biol. Chem. 266: 7804.

Conaway R.C. and Conaway J.W. 1988. ATP activates transcription initiation from promoters by RNA polymerase II in a reversible step prior to RNA synthesis. J. Biol. Chem. 263: 2962.

———. 1989. An RNA polymerase II transcription factor has an associated DNA- dependent ATPase (dATPase) activity

strongly stimulated by the TATA region of promoters. Proc. Natl. Acad. Sci. 86: 7356.

———. 1990. Transcription initiated by RNA polymerase II and purified transcription factors from liver. Transcription factors alpha, beta gamma, and delta promote formation of intermediates in assembly of the functional preinitiation complex. J. Biol. Chem. 265: 7559.

———. 1993. General initiation factors for RNA polymerase II. Annu. Rev. Biochem. 62: 161.

Conaway R.C., Garrett K.P., Hanley J.P., and Conaway J.W. 1991. Mechanism of promoter selection by RNA polymerase II: Mammalian transcription factors alpha and beta gamma promote entry of polymerase into the preinitiation complex. Proc. Natl. Acad. Sci. 88: 6205.

Dvir A., Conaway R.C., and Conaway J.W. 1996a. Promoter escape by RNA polymerase II: A role for an ATP cofactor in suppression of arrest by polymerase at promoter-proximal sites. J. Biol. Chem. 271: 23352.

———. 1997a. A role for TFIIH in controlling the activity of early RNA polymerase II elongation complexes. Proc. Natl. Acad. Sci. 94: 9006.

Dvir A., Garrett K.P., Chalut C., Egly J.M., Conaway J.W., and Conaway, R.C. 1996b. A role for ATP and TFIIH in activation of the RNA polymerase II preinitiation complex prior to transcription initiation. J. Biol. Chem. 271 7245.

Dvir A., Tan S., Conaway J.W., and Conaway R.C. 1997b. Promoter escape by RNA polymerase II. Formation of an escape competent intermediate is a prerequisite for exit of polymerase from the promoter. J. Biol. Chem. 272: 28175.

Ernst H., Filipowicz W., and Shatkin A.J. 1983. Initiation by RNA polymerase II and formation of runoff transcripts containing unblocked and unmethylated 5′ termini. Mol. Cell. Biol. 3: 2172.

Feaver W.J., Gileadi O., Li, Y., and Kornberg R.D. 1991. CTD kinase associated with yeast RNA polymerase II initiation factor b. Cell 67: 1223.

Gerard M., Fischer L., Moncollin V., Chipoulet J.M., Chambon P., and Egly J.M. 1991. Purification and interaction properties of the human RNA polymerase B(II) general transcription factor BTF2. J. Biol. Chem. 266: 20940.

Holstege F.C.P., Fiedler U., and Timmers H.T.M. 1997. Three transitions in the RNA polymerase II transcription complex during initiation. EMBO J. 16: 7468.

Holstege F.C.P., van der Vliet P.C., and Timmers H.T.M. 1996. Opening of an RNA polymerase II promoter occurs in two distinct steps and requires the basal transcription factors IIE and IIH. EMBO J. 15: 1666.

Holstege F., Tantin D., Carey M., van der Vliet P.C., and Timmers H.T.M. 1995. The requirement for basal transcription factor IIE is determined by the helical stability of promoter DNA. EMBO J. 14: 810.

Jacob G.A., Luse S.W., and Luse D.S. 1991. Abortive initiation is increased only for the weakest members of a set of down mutants of the adenovirus 2 major late promoter. J. Biol. Chem. 266: 22537.

Pan G. and Greenblatt J. 1994. Initiation of transcription by RNA polymerase II is limited by melting of the promoter DNA in the region immediately upstream of the initiation site. J. Biol. Chem. 269: 30101.

Peterson M.G., Inostroza J., Maxon M.E., Flores O., Admon A., Reinberg D., and Tjian R. 1991. Structure and functional properties of human general transcription factor IIE. Nature 354: 369.

Roeder R.G. 1996. The role of general initiation factors in transcription by RNA polymerase II. Trends. Biochem. Sci. 21: 327.

Sawadogo M. and Roeder R.G. 1984. Energy requirement for specific transcription initiation by the human RNA polymerase II system. J. Biol. Chem. 259: 5321.

Schaeffer L., Roy R., Humbert S., Moncollin V., Vermeulen W., Hoeijmakers J.H.J., Chambon P., and Egly J.M. 1993. DNA repair helicase: A component of BTF2/TFIIH basic transcription factor. Science 260: 58.

Serizawa H., Conaway J.W., and Conaway R.C. 1993a. Phosphorylation of C-terminal domain of RNA polymerase II is not required in basal transcription. *Nature* **363:** 371.

Serizawa H., Conaway R.C., and Conaway J.W. 1992. A carboxyl-terminal-domain kinase associated with RNA polymerase II transcription factor delta from rat liver. *Proc. Natl. Acad. Sci.* **89:** 7476.

——. 1993b. Multifunctional RNA polymerase II initiation factor delta from rat liver: Relationship between carboxyl-terminal domain kinase, ATPase, and DNA helicase activities. *J. Biol. Chem.* **268:** 17300.

Tan S., Conaway R.C., and Conaway J.W. 1994. A bacteriophage vector suitable for site-directed mutagenesis and high level expression of multisubunit proteins in *E. coli. BioTechniques* **16:** 824.

Tantin D. and Carey M. 1994. A heteroduplex template circumvents the energetic requirement for ATP during activated transcription by RNA polymerase II. *J. Biol. Chem.* **269:** 17397.

Tsuboi A., Conger K., Garrett K.P., Conaway R.C., Conaway J.W., and Arai N. 1992. RNA polymerase II initiation factor alpha from rat liver is almost identical to human TFIIB. *Nucleic Acids Res.* **20:** 3250.

Wang W., Carey M., and Gralla J.D. 1992. Polymerase II promoter activation: Closed complex formation and ATP-driven start site opening. *Science* **255:** 450.

Yan M. and Gralla J.D. 1997. Multiple ATP-dependent steps in RNA polymerase II promoter melting and initiation. *EMBO J.* **16:** 7457.

RNA Polymerase II Elongation Control

J. Peng, M. Liu, J. Marion, Y. Zhu, and D.H. Price
Department of Biochemistry, University of Iowa, Iowa City, Iowa 52242

The expression of many eukaryotic genes is controlled in part at the level of transcription elongation (for review, see Bentley 1995; Shilatifard 1998; Yamaguchi et al. 1997). On the basis of results obtained from a *Drosophila* in vitro transcription system, a model (Fig. 1) for the general control of elongation was described (Kephart et al. 1992; Marshall and Price 1992) that was consistent with data obtained in vitro and in vivo from many studies. According to the model, all RNA polymerase II (pol II) molecules that initiate from a promoter enter abortive elongation and are destined to produce only short transcripts due to the action of negative transcription elongation factors (N-TEF). Escape from this negative control is accomplished through the action of positive transcription elongation factors (P-TEF) that allow productive elongation. Similar to antitermination mechanisms described in prokaryotes (Greenblatt et al. 1993; Roberts 1993), control of the elongation phase of transcription by pol II is accomplished by the regulated elimination of early transcriptional blocks.

Significant progress has been made in identifying the factors involved in pol II elongation control. Two components of N-TEF have been found that limit the elongation potential of polymerases that initiate from a promoter. The best characterized is *Drosophila* factor 2, an ATP-dependent pol II termination factor (Xie and Price 1996). Factor 2 exhibits a double-stranded DNA-dependent ATPase activity and has been hypothesized to bind to the template upstream of a stalled polymerase molecule and, in the presence of ATP, locate the polymerase and cause transcript release (Xie and Price 1997). The second factor, DSIF, was identified in the Handa laboratory as a negative regulator of pol II processivity in a human in vitro transcription system (Wada et al. 1998). DSIF is composed of the human homologs of the yeast SPT4 and SPT5 proteins (Hartzog et al. 1998; Wada et al. 1998). P-TEFb is an essential component in the positive regulation of elongation (Marshall and Price 1995). The purified *Drosophila* factor acts after initiation (Marshall and Price 1995) and possesses kinase activity (Marshall et al. 1996). Cloning of the subunits of both *Drosophila* and human P-TEFb indicated that the factor is a cyclin-dependent kinase composed of CDK9 and cyclin T (Zhu et al. 1997; Peng et al. 1998a,b). After pol II has made the transition into productive elongation, it is affected by elongation factors, such as S-II (Guo and Price 1993), TFIIF (Kephart et al. 1994), ELL (Shilatifard et al. 1996), and elongin (Aso et al. 1995), that increase the efficiency of transcription elongation (Reines et al.1996). Although the effects of the latter factors are important, the transition from abortive to productive elongation is the major regulated step (Wright 1993).

The carboxy-terminal domain (CTD) of the large subunit of pol II is phosphorylated during the transcription cycle at a time coincident with elongation regulation, (for review, see Dahmus 1994, 1995). Recently, the CTD was shown to be required for elongation control (Chun and Jeang 1996; Marshall et al. 1996). A number of kinases can phosphorylate the CTD (Dahmus 1995; Kang and Dahmus 1995), but only P-TEFb has been shown to modify the functional properties of pol II (Marshall et al. 1996). P-TEFb has been shown to phosphorylate the CTD of pol II in an early elongation complex at a time it is known to functionally modify the elongation properties of the polymerase (Marshall et al. 1996). Both the transcriptional activity and the CTD kinase activity of P-TEFb are inhibited by DRB (Marshall et al. 1996), a nucleoside analog that blocks the production of mRNA-length transcripts in vivo.

Here, we extend our understanding of elongation control by developing a human transcription system that allows reconstitution of the process on elongation complexes. Using extensively washed early elongation complexes formed using a HeLa cell nuclear extract (HNE), we demonstrate that P-TEFb can phosphorylate the CTD of human pol II and that even in the presence of crude HNE the phosphorylation of the CTD is carried out by P-TEFb. As indicated by the generation of DRB-sensitive runoff transcripts, elongation control could be reconstituted on the washed complexes by readdition of HNE. This indicates that all factors associated with the polymerase during initiation could be removed and then added back during elongation. Furthermore, much of the effect of P-TEFb could be achieved by prephosphorylating the polymerase in early elongation complexes and then adding HNE in the absence of further P-TEFb activity. Using recombinant factors, we found that both factor

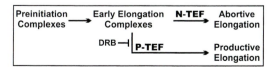

Figure 1. RNA polymerase II elongation control. The model indicates that early elongation complexes that arise from initiation from a promoter enter abortive elongation because of the default action of negative transcription elongation factors (N-TEF). The terminal fate of these complexes can be rescued by the action of positive transcription elongation factors (P-TEF) that allow the transition into productive elongation.

2 and TFIIF had appropriate effects on the isolated early elongation complexes. However, no effect of P-TEFb was seen. This result demonstrates that other factors are necessary for reconstituting elongation control.

EXPERIMENTAL PROCEDURES

Generation of the immobilized template. The template with the 560-nucleotide runoff was synthesized by polymerase chain reaction (PCR) using a vector containing a full cytomegalovirus (CMV) promoter (Peng et al. 1998b), an upstream biotinylated primer (5′-TGTAACT GAGCTAACATAACC), and a downstream primer (5′-GATGAGATGTGACGAACG). The primers were separated from the template by chromatography on Sephadex G-200. The template was coupled to streptavidin-conjugated Dynabeads (Dynal) as described by Marshall and Price (1992).

In vitro transcription. Transcriptions were essentially performed by a pulse-chase protocol (Marshall and Price 1992). Preinitiation complexes were formed by incubating an immobilized template with HNE for 10 minutes in 55 mM HMK (20 mM HEPES, 7 mM MgCl$_2$, and 55 mM KCl) plus 50 μM DRB. Transcription was initiated by the addition of a pulse solution, which contained 5 μCi of [α-^{32}P]CTP and brought the reaction mixture to 600 μM ATP, GTP, and UTP and 7 mM MgCl$_2$. After 30 seconds, the reaction was stopped by the addition of EDTA to 20 mM. These early elongation complexes were washed three times with 1 M HMKS (20 mM HEPES, 7 mM MgCl$_2$, 1 M KCl, and 1% Sarkosyl) and then once with 55 mM HMKB (20 mM HEPES, 7 mM MgCl$_2$, 55 mM KCL, and 200 μg/ml BSA). The washed early elongation complexes were resuspended into 55 mM HMKB and chased with 600 μM of each NTP for 5 minutes under various conditions.

Kinase assay. The assay was performed as described recently (Peng et al. 1998b), except that the substrate was the extensively washed early elongation complexes as described above.

Purification of human recombinant TFIIF. Purification was performed as described by Wang et al. (1994) with modification. RAP74 with a carboxy-terminal His tag was generated in DE3 cells using pET23d-RAP74 NspV. After induction, cells were lysed in 100 mM HGKEDP (25 mM HEPES at pH 7.6, 15% glycerol, 100 mM KCl, 0.1 mM EDTA, 1 mM DTT, and PMSF) plus 1% Triton X-100. The lysate was centrifuged at 100,000g for 1 hour. The pellet was dissolved in a denaturing buffer (25 mM HEPES at pH 7.6, 0.2 mM EDTA, 0.2 M EGTA, 10 mM DTT, and 6 M urea) followed by chromatography on a nickel column. RAP30 was obtained by a similar method using pET11d-RAP30 without using a nickel column. TFIIF was reconstituted from purified RAP74 and RAP30 using the dialysis method and further purified on Mono-S and Mono-Q columns.

RESULTS

Our goal is to reconstitute pol II elongation control using a completely defined transcription system. Since a complex mixture of factors is needed for initiation of transcription, we first wanted to determine if the process could be reconstituted on highly purified early elongation complexes derived by initiation from the CMV promoter using a crude HNE. A template containing the full CMV promoter was immobilized to paramagnetic beads and incubated with HNE. Transcription was initiated with a 30-second pulse containing limiting levels of [^{32}P]CTP, and the resulting early elongation complexes were washed extensively with a buffer containing 1 M KCl and 1% Sarkosyl and then returned to a low-salt buffer. Nucleotides were added to the isolated complexes and RNA was isolated at the indicated times of chasing (Fig. 2). Transcripts in the starting early elongation complexes were less than 30 nucleotides in length. Polymerases moved steadily down the template until reaching the 560-nucleotide runoff. As expected, the rate of movement indicated that general elongation factors had been removed by the stringent washing.

To determine if the isolated early elongation complexes could act as substrate for the kinase activity of P-TEFb, complexes were formed using a cold pulse followed by the extensive washing procedure used above. Increasing amounts of recombinant human P-TEFb composed of CDK9 and cyclin T2a (Peng et al. 1998b) were incubated with the isolated complexes for 5 minutes in the presence of 100 μM [γ-^{32}P]ATP. Phosphorylation of the large subunit of pol II was monitored by the shift in mobility seen in a silver-stained protein gel (Fig. 3A) and by the incorporation of label into the same band by autoradiography of the dried gel (Fig. 3B). Within the limits of detection on the silver-stained gel, none of the polymerases in the isolated early elongation complexes were phosphorylated to the IIo form during initiation. However, there was a quantitative shift of all polymerases to the IIo form during incubation with levels of P-TEFb at or above 0.5 pmole.

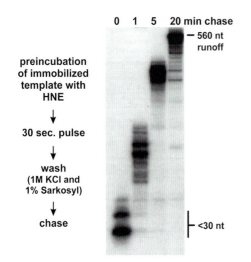

Figure 2. Transcription on immobilized template. Early elongation complexes were generated by 30-sec pulse, washed extensively, and chased for the indicated periods of time.

We next wanted to determine if the elongation properties of the isolated early elongation complexes could be converted back to those found for complexes normally initiated and elongated in the presence of a crude extract. Addition of whole HNE to the isolated complexes followed by a 5-minute chase gave rise to a pattern of transcripts similar to that seen in a normal pulse/chase experiment (Fig. 4). Most transcripts stopped prematurely, and the small amount of runoff transcript was sensitive to DRB. P-TEFb is limiting in the HNE, and addition of an extract supplemented with recombinant P-TEFb caused a higher level of DRB-sensitive runoff transcript. These results indicate that elongation control can be reconstituted on isolated early elongation complexes. Components of N-TEF must be able to reassociate with the stripped elongation complexes and cause the polymerase to pause, and P-TEFb must be able to act to suppress the action of those factors.

We were interested in determining whether phosphorylation of the polymerase by P-TEFb *before* the addition of the extract would also suppress the action of the negative factors. Using conditions similar to those shown in Figure 3, the isolated early elongation complexes were phosphorylated with P-TEFb and then washed again with high salt and Sarkosyl to remove the P-TEFb. The same pattern of transcripts was seen during a chase in the presence of extract supplemented with recombinant P-TEFb regardless of the phosphorylation state of the polymerase at the beginning of the chase. However, the effect of adding DRB during the chase was greatly reduced when phosphorylated complexes were used. Evidently, prephosphorylation of the polymerase allows significant protection from effect of the negative factors. DRB does have some effect when added during the chase, and this could be due to further phosphorylation of the polymerase during the chase or due to phosphorylation of some other factor(s) in the extract added back.

To add support to the idea that P-TEFb phosphorylates pol II in early elongation complexes, we took advantage of the immobilized template add-back system. Early elongation complexes were generated and washed as before, and HNE supplemented with recombinant P-TEFb was added back. Phosphorylation of the polymerase was monitored by the incorporation of label from [γ-^{32}P]ATP into the large subunit. Time points from 30 seconds to 2 minutes were taken +/– DRB. After the brief period of kinase action, the complexes were again washed with high salt and Sarkosyl, and the labeled protein was analyzed by autoradiography of the dried gel (Fig. 5). The polymerase in early elongation complexes was phosphorylated in a DRB-sensitive manner, indicating that P-TEFb was responsible.

In an initial attempt to develop a defined elongation control system, we examined the effect of adding recombinant P-TEFb, factor 2 (Liu et al. 1998), and TFIIF individually or in combination to the isolated early elongation complexes. Two chase time points of 1 and 5 minutes were taken for each factor combination (Fig. 6). P-TEFb had no effect on the elongation properties even though enough was added to completely phosphorylate the polymerase within a few minutes. Factor 2 had the expected

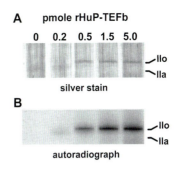

Figure 3. Phosphorylation of CTD by P-TEFb in early elongation complexes. The washed early elongation complexes were incubated for 5 min in the presence of 100 μM ATP and various amounts of recombinant human P-TEFb (CDK9/T2a; see Peng et al. 1998b). The treated early elongation complexes were eluted from the beads with SDS and analyzed on a 6–15% gradient gel followed by silver staining (*A*) and autoradiography (*B*).

effect of causing only short transcripts to appear. These short transcripts were released from the elongation complexes due to the termination activity of the factor (data not shown). TFIIF also had the expected effect of dramatically increasing the elongation rate. If the 5-minute point with no addition is compared to the 1-minute point with TFIIF, it is clear that a greater than fivefold increase in rate was observed. P-TEFb was not able to reverse the effect of factor 2 or TFIIF. When added in about tenfold molar excess together with factor 2, TFIIF was able to partially suppress the effect of factor 2. Most of the polymerases moved with a rate similar to that seen with TFIIF alone and only a small fraction stopped earlier due to factor 2. It is not clear if TFIIF directly blocks factor 2 action by binding to paused polymerases and interfering with the interaction of factor 2 and the polymerase or if just the

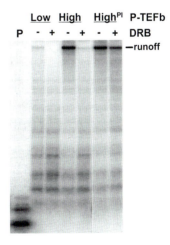

Figure 4. Reconstitution of DRB-sensitive transcription with isolated early elongation complexes. The early elongation complexes were formed, washed, and chased in the presence of HNE or HNE supplemented with 2.2 pmoles of P-TEFb. DRB was added at 50 μM. The superscript "PI" indicated that the early elongation complexes were preincubated with 1 pmole P-TEFb and 600 μM ATP for 5 min to allow phosphorylation of the CTD of pol II. The phosphorylated early elongation complexes were washed again to remove P-TEFb before the chase.

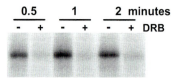

Figure 5. DRB-sensitive phosphorylation of CTD in early elongation complexes in the presence of HNE. The washed early elongation complexes were incubated for the indicated time in the presence of 10 μM ATP and HNE with 2.2 pmoles of P-TEFb. The treated early elongation complexes were washed again and then analyzed as in Fig. 3B.

increase in rate caused factor 2 to be unable to catch the moving polymerase. Again, P-TEFb had very little effect on the action of TFIIF or factor 2 even when all three were present together. All of these results together indicate that other negative factors are missing that can be supplied by the crude HNE.

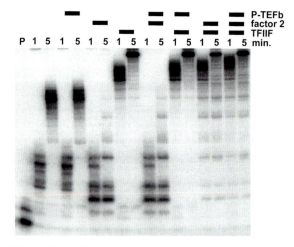

Figure 6. The effect of recombinant P-TEFb, factor 2, and TFIIF on transcription elongation. The early elongation complexes were generated, washed, and chased for 1 or 5 min under various combinations of approximately 1.5 pmoles of P-TEFb, 0.2 pmole of factor 2 (Liu et al. 1998), and 2 pmoles of TFIIF.

DISCUSSION

We have developed a system for reconstituting DRB-sensitive transcription using an add-back assay with isolated early elongation complexes. Crude HNE added to stringently washed early elongation complexes was able to recreate both the negative and positive aspects of elongation control. This indicates that very little communication between initiation events and the elongation control phase of transcription is needed. It should be possible to use this add-back assay to identify required elongation control components and to study their mechanisms of action.

The function of P-TEFb was examined with two experiments that both supported the idea that the CTD was the target of kinase activity of P-TEFb. Under conditions that reconstituted DRB-sensitive runoff transcripts, the polymerases in early elongation complexes were phosphorylated by P-TEFb. In addition, prephosphorylation of the polymerase before addition of HNE containing other required elongation control components allowed most of the polymerases to escape from the effect of the negative factors. This supports the idea that phosphorylation of the CTD somehow weakens the interaction between the negative factors and the CTD or in some other way decreases the effectiveness of the negative factors.

Using recombinant factors, we examined the functional interactions among P-TEFb, factor 2, and TFIIF. Although P-TEFb was able to phosphorylate the polymerase in early elongation complexes, under the conditions used, we did not detect any effect of P-TEFb on the elongation stimulatory activity of TFIIF or on the termination activity of factor 2. Clearly, other components of N-TEF are needed to reconstitute DRB-sensitive transcription. TFIIF dramatically suppressed the termination activity of factor 2. The mechanism of this suppression has not been determined, but it is possible that TFIIF interacts with paused polymerases and directly inhibits the interaction of factor 2 with the polymerase. It is also possible that TFIIF causes a conformation change in the

paused polymerases (Price et al. 1989) and that the altered conformation does not interact productively with factor 2. If the mechanism of factor 2 action involves translocation along the DNA in an ATP-dependent manner (Xie and Price 1997, 1998), it may be that factor 2 simply cannot catch up with the rapidly moving, TFIIF-stimulated, elongation complex.

Our current understanding of the factors involved in eukaryotic elongation control is summarized in Figure 7. The process can be divided into two phases. Initiation takes place before the first phase and includes the abortive initiation and promoter clearance steps. Phase one begins with the arrest of early elongation complexes at promoter-proximal positions through the action of N-TEF. So far, DSIF (Wada et al. 1998) is the only known component of N-TEF that is involved in the arrest, but it is likely that others are also involved. Factor 2 and P-TEFb then functionally compete for control of the early elongation complexes by dictating the subsequent path followed by the transcription complex. If the ATP-dependent termination activity of factor 2 dominates, then the polymerase performs the final act of abortive elongation. If the kinase activity of P-TEFb is able to phosphorylate the CTD of the polymerase, the effect of N-TEF is reversed and the polymerases leave the first phase and enter productive elongation. The mechanism of suppression of the negative factors is not clear. Surprisingly, removal of the CTD did not significantly reduce the effect of N-TEF (Marshall et al. 1996). Although our results here provide further support for the idea that the effect of P-TEFb is through phosphorylation of the CTD, it is possible that phosphorylation of other components of the elongation complex contributes to the action of P-TEFb. The development of a defined elongation control system will allow the identification of other possible phosphorylation targets. The second phase of elongation occurs after the ef-

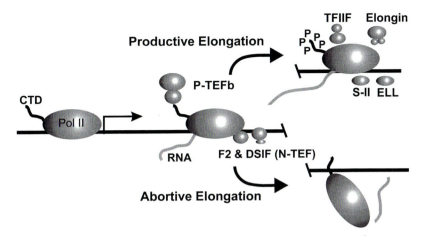

Figure 7. Factors involved in elongation control. The diagram indicates the action of N-TEF in stopping pol II after initiation from a promoter and the functional competition between P-TEFb and factor 2 for the paused complexes. Factor 2 causes termination and P-TEFb phosphorylates the CTD and allows the polymerase to enter productive elongation. The efficiency of elongation of productive complexes is enhanced by the action of the general elongation factors TFIIF, elongin, S-II, and ELL.

fect of the negative factors has been suppressed. Other general elongation factors such as TFIIF, elongin, S-II, and ELL act on the polymerase to ensure efficient passage through intrinsic elongation blocks in the template and allow the polymerase to transcribe RNAs millions of nucleotides in length.

Why is there an elongation control process in eukaryotes? One possibility is that eukaryotes need a mechanism to generally limit mRNA production in a postinitiation process. This would allow reduction in the energy-dependent production of proteins when they are not needed without disrupting the initiation process. The structure of the extended promoter including enhanceosomes (Carey 1998) and surrounding chromatin would not need to be changed, allowing for rapid reversal. Another attractive possibility is that the early block to elongation and subsequent release of the block provides a window of opportunity for the loading of factors into the elongation complex. Similar to the model of antitermination in prokaryotes (Greenblatt et al. 1993; Roberts 1993), the arrested polymerase may allow the formation of an elongation control particle. This transcription machine would be able to efficiently transcribe long stretches of DNA and would be able to revert back to a nonprocessive form after the mature 3′ end of the mRNA was reached. An important component of the elongation control particle in eukaryotes could be the RNA processing machinery. Evidence is building that links the phosphorylation of the CTD to the association of enzymes and factors involved in 5′ capping and 3′ polyadenylation (Cho et al. 1997; McCracken et al. 1997a,b; Tanner et al. 1997). Perhaps the assembly of the eukaryotic elongation control particle is regulated by the activity of P-TEFb to guarantee that transcription of long mRNAs is only started if the processing machinery is appropriately associated. The transcription system we have described here should be useful in examining the coupling of transcription with RNA processing.

ACKNOWLEDGMENTS

We thank L. Lei and Z. Burton for providing the TFIIF expression vectors. This work was supported by National Institutes of Health grant GM-35500.

REFERENCES

Aso T., Lane W.S., Conaway J.W., and Conaway R.C. 1995. Elongin (SIII): A multisubunit regulator of elongation by RNA polymerase II. *Science* **269:** 1439.

Bentley D.L. 1995. Regulation of transcriptional elongation by RNA polymerase II. *Curr. Opin. Genet. Dev.* **5:** 210.

Carey M. 1998. The enhanceosome and transcriptional synergy. *Cell* **92:** 5.

Cho E.J., Takagi T., Moore C.R., and Buratowski S. 1997. mRNA capping enzyme is recruited to the transcription complex by phosphorylation of the RNA polymerase II carboxy-terminal domain. *Genes Dev.* **11:** 3319.

Chun R.F. and Jeang K.T. 1996. Requirements for RNA polymerase II carboxyl-terminal domain for transcription of human retroviruses, human T-cell lymphotrophic virus, and HIV-1. *J. Biol. Chem.* **271:** 27888.

Dahmus M.E. 1994. The role of multisite phosphorylation in the regulation of RNA polymerase II activity. *Prog. Nucleic Acid Res. Mol. Biol.* **48:** 143.

——. 1995. Phosphorylation of the C-terminal domain of RNA polymerase II. *Biochim. Biophys. Acta* **1261:** 171.

Greenblatt J., Nodwell J.R., and Mason S.W. 1993. Transcriptional antitermination. *Nature* **364:** 401.

Guo H. and Price D.H. 1993. Mechanism of DmS-II-mediated pause suppression by *Drosophila* RNA polymerase II. *J. Biol. Chem.* **268:** 18762.

Hartzog G.A., Wada T., Handa H., and Winston F. 1998. Evidence that Spt4, Spt5, and Spt6 control transcription elongation by RNA polymerase II in *Saccharomyces cerevisiae*. *Genes Dev.* **12:** 357.

Kang M.E. and Dahmus M.E. 1995. The unique C-terminal domain of RNA polymerase II and its role in transcription. *Adv. Enzymol. Relat. Areas Mol. Biol.* **71:** 41.

Kephart D.D., Marshall N.F., and Price D.H. 1992. Stability of *Drosophila* RNA polymerase II elongation complexes in vitro. *Mol. Cell. Biol.* **12:** 2067.

Kephart D.D., Wang B.Q., Burton Z.F., and Price D.H. 1994. Functional analysis of *Drosophila* factor 5 (TFIIF), a general transcription factor. *J. Biol. Chem.* **269:** 13536.

Liu M., Xie Z., and Price D.H. 1998. A human RNA polymerase II transcription termination factor is a SWI2/SNF2 family member. *J. Biol. Chem.* **273:** 25541.

Marshall N.F. and Price D.H. 1992. Control of formation of two distinct classes of RNA polymerase II elongation complexes. *Mol. Cell. Biol.* **12:** 2078.

——. 1995. Purification of P-TEFb, a transcription factor required for the transition into productive elongation. *J. Biol. Chem.* **270:** 12335.

Marshall N.F., Peng J.M., Xie Z., and Price D.H. 1996. Control of RNA polymerase II elongation potential by a novel carboxyl-terminal domain kinase. *J. Biol. Chem.* **271:** 27176.

McCracken S., Fong N., Yankulov K., Ballantyne S., Pan G., Greenblatt J., Patterson S.D., Wickens M., and Bentley D.L. 1997a. The C-terminal domain of RNA polymerase II couples mRNA processing to transcription. *Nature* **385:** 357.

McCracken S., Fong N., Rosonina E., Yankulov K., Brothers G., Siderovski D., Hessel A., Foster S., Shuman S., and Bentley D.L. 1997b 5´-Capping enzymes are targeted to pre-mRNA by binding to the phosphorylated carboxy-terminal domain of RNA polymerase II. *Genes Dev.* **11:** 3306.

Peng J.M., Marshall N., and Price D.H. 1998a. Identification of a cyclin subunit required for function of *Drosophila* P-TEFb. *J. Biol. Chem.* **273:** 13855.

Peng J. M., Zhu Y., Milton J.T., and Price D.H. 1998b. Identification of multiple cyclin subunits of human P-TEFb. *Genes Dev.* **12:** 755.

Price D.H., Sluder A.E., and Greenleaf A.L. 1989. Dynamic interaction between a *Drosophila* transcription factor and RNA polymerase II. *Mol. Cell. Biol.* **9:** 1465.

Reines D., Conaway J.W., and Conaway R.C. 1996. The RNA polymerase II general elongation factors. *Trends Biochem. Sci.* **21:** 351.

Roberts J.W. 1993. RNA and protein elements of *E. coli* and lambda transcription antitermination complexes. *Cell* **72:** 653.

Shilatifard A. 1998. The RNA polymerase II general elongation complex. *Biol. Chem.* **379:** 27.

Shilatifard A., Lane W.S., Jackson K.W., Conaway R.C., and Conaway J.W. 1996. An RNA polymerase II elongation factor encoded by the human ELL gene. *Science* **271:** 1873.

Tanner S., Stagljar I., Georgiev O., Schaffner W., and Bourquin J.P. 1997. A novel SR-related protein specifically interacts with the carboxy-terminal domain (CTD) of RNA polymerase II through a conserved interaction domain. *Biol. Chem.* **378:** 565.

Wada T., Takagi T., Yamaguchi K., Ferdous A., Imai T., Hirose S., Sugimoto S., Yano K., Hartzog G.A., Buratowski S., Winston F., and Handa H. 1998. DSIF, a novel transcription elongation factor that regulates RNA polymerase II processivity, is composed of human Spt4 and Spt5 homologs. *Genes Dev.* **12:** 343.

Wang B.Q., Lei L., and Burton Z.F. 1994. Importance of codon preference for production of human RAP74 and reconstitution of the RAP30/74 complex. *Protein Expr. Purif.* **5:** 476.

Wright S. 1993. Regulation of eukaryotic gene expression by transcriptional attenuation. *Mol. Biol. Cell* **4:** 661.

Xie Z. and Price D.H. 1996. Purification of an RNA polymerase II transcript release factor from *Drosophila. J. Biol. Chem.* **271:** 11043.

——. 1997. *Drosophila* factor 2, an RNA polymerase II transcript release factor, has DNA-dependent ATPase activity. *J. Biol. Chem.* **272:** 31902.

——. 1998. Unusual nucleic acid binding properties of factor 2 an RNA polymerase II transcript release factor. *J. Biol. Chem.* **273:** 3771.

Yamaguchi K., Wada T., and Handa H. 1998. Interplay between positive and negative elongation factors: Drawing a new view of DRB. *Genes Cells* **3:** 9.

Zhu Y.R., Peery T., Peng J.M., Ramanathan Y., Marshall N., Marshall T., Amendt B., Mathews M.B., and Price D.H. 1997. Transcription elongation factor P-TEFb is required for HIV-1 Tat transactivation in vitro. *Genes Dev.* **11:** 2622.

HIV-1 Tat Interacts with Cyclin T1 to Direct the P-TEFb CTD Kinase Complex to TAR RNA

M.E. GARBER, P. WEI, AND K.A. JONES

Regulatory Biology Laboratory, The Salk Institute for Biological Studies, La Jolla, California 92037-1099

Studies of viral transcription factors have yielded many important insights into the different mechanisms that regulate RNA polymerase II (RNAPII) transcription. The human immunodeficiency virus type-1 (HIV-1) proviral promoter is induced strongly in cytokine-activated T cells and macrophages and provides a useful system to study the processes that control transcription initiation and elongation in a chromatin environment (Fig. 1). Because the virus has, on average, only 48 hours in an activated T cell to complete its entire virus life cycle, it must transition quickly to an active transcription program following integration into the host-cell genome (for review, see Emerman and Malim 1998). To respond to this challenge, HIV-1 evolved a compact and powerful enhancer which is coupled to promoter and downstream control sequences that sharply restrict transcription elongation by RNAPII. This restriction establishes a requirement for the virus-encoded regulatory protein, Tat, which is targeted to the nucleus and acts as an HIV-1 promoter-specific transcription elongation factor (for review, see Jones and

Peterlin 1994; Jones 1997; Cullen 1998; Emerman and Malim 1998). Tat functions through direct binding to the viral transcript at a specific site in the TAR RNA stem-loop structure. Taken together, the genetic studies of HIV-1 Tat *trans*-activation provide compelling evidence that the efficiency of RNAPII elongation can be regulated through the transient binding of a transcription elongation factor to the nascent RNA transcript. To better understand this process, efforts have focused on identifying the block(s) to productive transcription elongation at the HIV-1 promoter in the absence of Tat and on understanding how Tat functions through TAR RNA to enhance transcription elongation.

Genetic studies with chimeric activators have shown that Tat can work independently of TAR if tethered through a heterologous RNA-binding domain (Selby and Peterlin 1990; Southgate et al. 1990; Southgate and Green 1991), indicating that the specific mechanism of binding to RNA is not a critical factor in Tat *trans*-activation. Moreover, activation of transcription through binding to RNA has also been observed with chimeric proteins that contain activation domains from unrelated transcription factors (e.g., HSV-VP16; Tiley et al. 1992; Luo et al. 1993; Madore and Cullen 1993). Nevertheless, only Tat can stimulate transcription through TAR RNA. The transcriptional activation domain of Tat contains a cysteine-rich region and a short hydrophobic "core" motif that abuts a 9-amino-acid arginine-rich RNA-recognition motif (ARM). Purified Tat protein binds to TAR RNA through the ARM, and a key arginine residue (R52) directly contacts the bulge of the hairpin (for review, see Gait and Karn 1993). However, free (uncomplexed) Tat is unable to recognize residues in the loop of the TAR hairpin that are essential for *trans*-activation, and consequently it has been suggested that the upper stem and loop of TAR RNA may be recognized by a host-cell factor.

The identification of a nuclear kinase (the *T*at-*a*ssociated *k*inase, or TAK) as the transcriptional cofactor for Tat first came from biochemical studies of nuclear proteins that interact specifically with the HIV-1 Tat *trans*-activation domain in protein affinity selection (GST-Tat "pull-down") experiments (Herrmann and Rice 1993, 1995). Single-amino-acid substitutions in the *trans*-activation domains of either HIV-1, HIV-2, or EIAV Tat were found to block the binding of Tat to nuclear TAK in a manner that correlated exactly with Tat function in vivo. The catalytic subunit of TAK was subsequently identified as CDK9 (Yang et al. 1997; Zhu et al. 1997), a cdc2-related kinase (originally called PITALRE; Grana

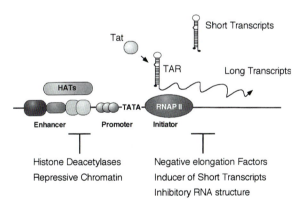

Figure 1. Transcriptional regulation of the HIV-1 promoter. Shown is the minimal region of the HIV-1 enhancer (–85 to –300) that functions in concert with the promoter to counteract the repressive effects of chromatin structure and histone deacetylase complexes. The enhancer is proposed to recruit histone acetyltransferases (HATs) as part of a transcriptional coactivator complex. HIV-1 Tat acts subsequent to initiation and facilitates an early step in elongation through binding to TAR RNA. A strong block to RNAPII elongation is established by an inhibitory RNA structure that forms in the nascent transcript prior to the formation of the TAR structure (Palangat et al. 1998). Additional inhibitors of RNAPII elongation might include CDK inhibitors and CTD phosphatases associated with the RNAPII initiation complex, as well as the "inducer of short transcripts" element at the HIV-1 promoter (Pessler and Hernandez 1998).

et al. 1994; Garriga et al. 1996) that functions in vitro as a CTD (carboxy-terminal domain) kinase and as a key component of the *positive transcription elongation factor*, P-TEFb (Kephart et al. 1992; Marshall and Price 1992, 1995; Marshall et al. 1996; Yang et al. 1996). In support of a central role for CDK9 in *trans*-activation, it was shown that Tat activity requires the CTD (Chun and Jeang 1996; Okamoto et al. 1996; Parada and Roeder 1996; Yang et al. 1996) and is blocked by specific inhibitors of CDK9 (Mancebo et al. 1997). In contrast, short (attenuated) transcripts that are synthesized at the HIV-1 promoter in the absence of Tat do not require the CTD and are not inhibited by compounds that block CDK9 or other CTD kinases (Cujec et al. 1997b; Pessler and Hernandez 1998). Factor-independent pausing at the HIV-1 promoter is observed with purified RNAPII in vitro and is mediated by an alternative RNA structure that forms at the 5′ end of the transcript and competes for folding of the TAR hairpin (Palangat et al. 1998).

Although CDK9 is part of the P-TEFb complex that interacts with HIV-1 Tat in nuclear extracts, Tat does not bind directly to recombinant CDK9 in vitro (M.E. Garber and K.A. Jones, unpubl.). We recently identified an 87-kD protein that is present with CDK9 in P-TEFb complexes isolated from nuclear extracts by GST-Tat affinity chromatography. Analysis of the cDNA encoding p87 revealed that it is a novel human cyclin, called CycT1, which is a member of the family of C-type cyclins (Wei et al. 1998). Cyclin T1 was also isolated independently as one of three possible cyclin partners for CDK9 (Peng et al. 1998). We showed that HIV-1 Tat binds directly through its *trans*-activation domain to CycT1, and that the Tat-CycT1 complex binds cooperatively to TAR RNA in a manner that requires sequences in both the stem and loop of the hairpin structure (Wei et al. 1998). Within the complex, binding to the TAR bulge region is mediated by the ARM of Tat, whereas CycT1 (and potentially also Tat) must recognize bases in the upper stem and loop of the RNA. In addition to CDK9 and CycT1, the P-TEFb complex contains several other unidentified subunits that might also be required for Tat *trans*-activation (Mancebo et al. 1997; Zhu et al. 1997; Zhou et al. 1998).

In this paper, we show that sequences at the amino terminus of Tat, which overlap the *trans*-activation domain, significantly block binding to TAR RNA in vitro. Moreover, we find that the activation domain also contributes to the RNA-binding specificity of Tat. This is most evident in studies using HIV-2 RNA, which forms a duplicated TAR hairpin structure that is bound more strongly by Tat than HIV-1 TAR. Binding of Tat to HIV-2 TAR RNA is reduced by mutations affecting the loop of both hairpin structures, and loop-sensitive binding to TAR is lost upon truncation of the *trans*-activation domain. These findings indicate that Tat possesses an intrinsic ability to recognize sequences in the loop region of the HIV-2 RNA structure and that significant determinants of RNA-binding specificity may lie outside of the ARM. HIV-1 Tat and CycT1 bind cooperatively to TAR RNA with an affinity comparable to that observed with the free Tat ARM, but with much greater loop sequence speci-

ficity. Finally, we show that addition of Tat to HeLa nuclear extracts induces the binding of CDK9, CycT1, and other P-TEFb subunits to TAR RNA. Thus, through its ability to interact with CycT1, Tat is able to recruit the entire P-TEFb complex to the RNA. Possible roles for binding of the P-TEFb CTD kinase to RNA in the regulation of RNAPII elongation are discussed.

MATERIALS AND METHODS

DNA constructs. Bacterial expression vectors for glutathione-*S*-transferase (GST) Tat fusions were obtained from Dr. Andrew Rice through the AIDS Research and Reference Program (NIH): HIV-1 Tat (amino acids 1–86; two exon), HIV-2 Tat (amino acids 1–99; one exon), HIV-2 Tat (amino acids 1–99; Δ8-47), and HIV-2 Tat (amino acids 1–130; two exon). All other Tat constructs were cloned by standard PCR procedures into the *Bam*HI/*Eco*RI sites of pGEX-2T. For all Tat clones, the parent vectors HIV-1 Tat (HXB2 isolate, amino acids 1–86) and HIV-2 Tat (ROD isolate, amino acids 1–130) were used as templates. The GST expression vector for human cyclin T1 was described previously (Wei et al. 1998). HIV-1 TAR RNA probes were transcribed in vitro from the plasmids pH96 wild-type (+1 to +80), pH 96 loop substitution mutant (+30/+33; +1 to +80), pTAR-1S wild type (+17 to +43), and pTAR-1S loop substitution mutant (+29/+34; +17 to +34) as described by Wei et al. (1998). pHIV-2 TAR wild type and loop mutant (LM) RNA, which substitutes the loop sequence UGGG with GUUU in both loops, was used to generate the HIV-2 TAR RNA probes (+1 to +123). The single and double bulge deletion HIV-2 TAR RNA constructs were described previously (Rhim and Rice 1994).

Preparation of TAR RNA and gel mobility shift experiments. TAR synthesis was performed in a 0.08-ml reaction volume at 37°C for 1 hour in RNA-binding buffer: 40 mM Tris-Cl, pH 8.0, 2 mM spermidine, 20 mM DTT, 6 mM MgCl$_2$, [^{32}P]UTP (30 μCi, 800 Ci/mmole, 20 μCi/μl, Amersham), 20 μM rUTP, 0.5 mM each of rATP, rGTP, rCTP, 1 pmole linear DNA template, 0.8 units/μl T7/Sp6 RNA polymerase (Ambion), and 100 units RNasin (USB). After digestion of the DNA template with 2 units of DNase I (Promega) per microgram of DNA, reaction mixtures were extracted once with phenol:chloroform, once with chloroform, ammonium acetate added to 2.5 M and precipitated with 2.5 volumes of ethanol. The RNA pellet was dissolved in 0.1 M NaCl and applied to a G-50 spin column (Boehringer). TAR RNA used for the gel shift probe was heated to 88°C in the presence of 1 mM MgCl$_2$ and cooled to room temperature over 15 minutes. RNA gel shift experiments with Tat and CycT1 were carried out as described by Wei et al. (1998). The gel shift reactions also included 1 μg of total HeLa competitor RNA, and the CycT1-Tat:TAR complexes were separated on native polyacrylamide gels.

Preparation of recombinant Tat protein. Bacteria (BL21 DE3) were grown to a density of 0.9 A$_{600}$ and induced with 0.1 mM IPTG for 7 hours at 30°C. Cells were resuspended in lysis buffer (1× PBS containing 10 mM DTT, 1 mM PMSF, 2 mg/ml benzamidine, 2× protease inhibitors, 5 mM EDTA, and 200 µg/ml lysozyme), sonicated for 30 seconds, and frozen and thawed once. The cell lysate was clarified by centrifugation in an SW41 rotor for 30 minutes at 30,000 rpm. The supernatant was bound to 500 µl of glutathione beads for 1 hour at 4°C, washed three times in 1× PBS containing 10 mM DTT and 0.1 mM PMSF, rinsed once in Tat elution buffer (50 mM Tris-Cl, pH 7.5, 100 mM NaCl, 10 mM DTT, and 0.1 mM PMSF) and eluted in Tat elution buffer containing 10 mM glutathione. Glycerol was added to 10% (final concentration) and the Tat protein was stored at –80°C prior to use.

TAR RNA affinity resins and in vitro kinase experiments. RNA-binding reactions contained binding buffer (25 mM Tris-HCl, pH 8.0, 11% glycerol, 78 mM KCl, 17 mM NaCl, 0.02% NP-40, 5.2 mM MgCl$_2$, 3 mM DTT, 0.5 mM EDTA), 500 µg of HeLa nuclear extract, and 200 ng of GST-cleaved Tat in a 120-µl reaction volume. The mixture was preincubated at 30°C for 30 minutes and chilled on ice prior to the addition of 10 µg of randomly biotinylated TAR RNA bound to 20 µl of streptavidin beads (Pierce). The reactions were rotated with the TAR beads overnight at 4°C; approximately 10% of the TAR RNA was released from the beads following the overnight incubation. Beads were washed twice at 4°C in 120 µl of wash buffer (25 mM Tris-Cl, pH 7.9, 6.2 mM MgCl$_2$, 0.5 mM EDTA, and 10% glycerol) containing 75 mM KCl, 0.05% NP-40, and 5 mM DTT. The proteins were eluted at 4°C in 120 µl of 2× wash buffer containing 1 M KCl, 0.05% NP-40, and 5 mM DTT. Input and flowthough fractions were diluted to 600 µl in EBC (50 mM Tris-Cl, pH 8.0, 120 mM NaCl, 0.5% NP-40, and 5 mM DTT) containing 0.25% gelatin. The 1.0 M eluate fractions were diluted to 600 µl in EBC containing 0.25% gelatin and no NaCl. Input, flowthough, and eluate fractions from the TAR beads were immunoprecipitated with a polyclonal antibody specific for HIV-1 or HIV-2 Tat (obtained from Dr. Bryan Cullen through the AIDS Research and Reference Program, NIH). Immunoprecipitations, washes, and kinase activity were performed as described previously (Yang et al. 1996).

Protein:protein interaction experiments. Aliquots (2 µg) of (GST-cleaved) HIV-1 Tat (amino acids 1–86) or activation domain HIV-1 Tat (amino acids 1–48) proteins were incubated with 3 µg of GST-CycT1 pre-bound to glutathione-*S*-Sepharose beads (15 µl) in 500 µl of binding buffer (40 mM HEPES, pH 8.0, 0.5% NP-40, and 10 mM DTT) containing 120 mM NaCl, for 2 hours at 4°C. Beads were washed extensively (3 × 15 minutes at 4°C) in 400 µl of binding buffer containing 500 mM NaCl prior to the addition of SDS sample buffer. One-half of the protein sample was analyzed by 12% SDS-PAGE, and the proteins were visualized by Western blot with anti-HA (BMB) and anti-GST (Santa Cruz) monoclonal antibodies to detect HIV-1 Tat and CycT1, respectively.

RESULTS AND DISCUSSION

We showed previously that the interaction of HIV-1 Tat with CycT1 greatly enhances binding to TAR RNA (Wei et al. 1998). Interestingly, the CycT1-Tat complex is unable to recognize loop mutant TAR RNAs, even though these same mutant RNAs are bound by Tat in the absence of CycT1. Thus, within the CycT1-Tat complex, the ARM of Tat is prevented from binding to RNA in the absence of specific bases in the loop of the RNA. These findings suggest that the CycT1-Tat complex forms an induced fit with TAR and that the structure of the complex may be critical for binding to RNA. We were interested in understanding how the interaction of Tat with CycT1 could dramatically enhance binding to TAR RNA, given that the free Tat protein (i.e., the Tat ARM) had previously been shown to bind TAR RNA with high affinity. To address this question, we compared the binding of full-length and activation domain-truncated HIV-1 and HIV-2 Tat proteins to wild-type and mutant HIV-1 and HIV-2 TAR RNAs in vitro.

Truncation of the Tat *Trans*-activation Domain Greatly Enhances Binding to TAR RNA In Vitro

Synthetic peptides or short protein fragments that contain the 9-amino-acid arginine-rich motif (ARM) of HIV-1 Tat binds strongly to TAR RNA in vitro (K_d approximately 10 nM; for review, see Gait and Karn 1993). In contrast, we find that full-length Tat binds weakly to TAR RNA in the absence of CycT1 (Fig. 2). To determine whether binding of HIV-1 Tat to TAR RNA is blocked by the amino terminus of the protein, various truncated HIV-1 and HIV-2 Tat proteins were generated, and their ability to bind to wild-type TAR RNA was tested in gel mobility shift experiments. Removal of the Tat transcription activation domain enhanced binding to TAR by approximately 20-fold in these experiments (Fig. 2, compare lanes 2 and 5). Both one-exon and two-exon forms of HIV-2 Tat were found to bind weakly to TAR RNA (lanes 2, 6, and 8). RNA-binding activity increased approximately 4-fold upon truncation of the amino terminus of Tat (amino acids 8–47; lane 3) and was enhanced further upon removal of the Cys-rich subdomain (amino acids 50–65; lane 4). Similar results were obtained with mutants of HIV-1 Tat, which binds more weakly to HIV-1 than HIV-2 TAR RNA (data not shown). The full-length Tat proteins used in these experiments were active in their ability to support TAR-dependent Tat *trans*-activation in vitro, and at least 30% of the Tat in these preparations bound specifically to CycT1 (data not shown). We conclude from this experiment that residues amino-terminal to the ARM strongly interfere with the binding of Tat to TAR RNA in vitro. The interfering region(s) does not correspond precisely with the activation domain of Tat, since point mutations in the activation domain that abrogate transcriptional activation (e.g., C22G, K41A) do not enhance binding of Tat to TAR RNA (data not shown). The inhibition of TAR RNA-binding activity by the amino terminus of Tat might serve to ensure that Tat will

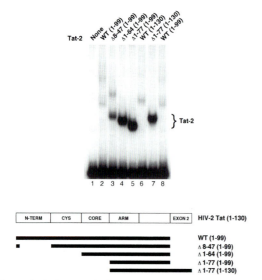

Figure 2. Amino-terminal sequences overlapping the *trans*-activation domain interfere with binding of HIV-1 Tat to TAR RNA in vitro. RNA-binding reactions contained 5 ng of wild-type (HIV-2) TAR RNA and 150 ng of wild-type or truncated GST-Tat-2 proteins, as indicated above each lane. Complexes were subjected to nondenaturing gel electrophoresis (6% polyacrylamide gel) at room temperature. The full-length GST-Tat-2 (aa 1–99) proteins used in lanes 2 and 8 were derived from independent protein preparations.

not bind RNA in the absence of CycT1 and the TAK/P-TEFb complex.

These results suggest that Tat may resemble regulated DNA-binding transcription factors such as Ets-1 and p53, which are prevented from binding to their target sites by the conformation of the native protein in the absence of coactivators (Wasylyk et al. 1992; Jonsen et al. 1996; for reviews on p53, see Ko and Prives 1996; Levine 1997). Autoinhibition of Ets-1 DNA-binding activity is alleviated through its ability to bind cooperatively with other transcription factors to composite Ets-responsive enhancers, whereas protein modifications have been proposed to regulate p53. These studies highlight the importance of protein folding on the control of domain accessibility and function for tightly regulated transcription factors. Such a mode of regulation might serve as a common mechanism to prevent a factor from binding to its target site or activating transcription in the absence of a specific transcriptional coactivator or an appropriate inducing signal.

The HIV Tat Proteins Bind to HIV-2 TAR RNA in a Loop-dependent Manner In Vitro

To determine whether residues in the activation domain might also influence the RNA-binding specificity of Tat, we compared the ability of the full-length and activation domain-deleted Tat proteins to bind to various mutant HIV-1 and HIV-2 TAR RNAs in gel mobility shift experiments (Fig. 3). To assess the RNA-binding specificity of the full-length (FL) and truncated (ARM) Tat proteins on the same gel, it was necessary to use levels of

full-length Tat that were sixfold higher than the Tat ARM protein to compensate for the difference in RNA-binding activity (Fig. 3A, compare lanes 1 and 9). Both full-length HIV-1 and HIV-2 Tat proteins bound more strongly to HIV-2 than to HIV-1 TAR RNA (Fig. 3A, compare lanes 1 and 7) and formed multiple complexes with HIV-2 TAR RNA. Binding of the full-length Tat proteins was significantly reduced, but not eliminated, by substitutions affecting the loops of the HIV-2 TAR RNA structure (Fig. 3A, compare lanes 1 and 2). Only a rapidly

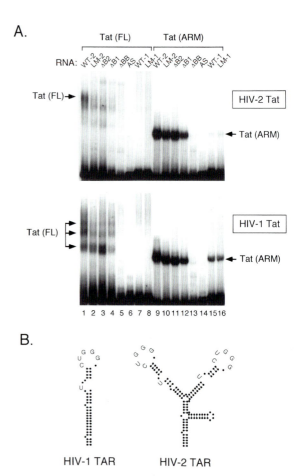

Figure 3. Full-length, but not amino-terminal truncated, Tat proteins bind to HIV-2 TAR RNA in a loop-dependent manner in vitro. (*A*) Binding of HIV-1 or HIV-2 Tat proteins to wild-type or mutant HIV-1 or HIV-2 TAR RNAs was analyzed by gel mobility shift experiments. Binding reactions contained 1 μg of full-length HIV-1 or HIV-2 GST-Tat proteins, as indicated, or 150 ng of the HIV-1 or HIV-2 GST-Tat ARM (aa 48–86) or GST-Tat-2 ARM (aa 78–99). RNA:protein complexes are indicated with arrowheads. HIV-2 TAR RNA sequences were either wild-type (WT-2) or contained a 6-base substitution of residues in both loops (LM-2), a 3′ bulge deletion (ΔB2), a 5′ bulge deletion (ΔB1), or both 5′ and 3′ deletion of both bulges (ΔBB), or antisense (AS). HIV-1 TAR probes were either wild-type (WT-1) or substituted at all positions in the loop (+29/+34; LM-1). Native gel electrophoresis was carried out at room temperature. (*B*) Predicted secondary structure of HIV-1 and HIV-2 TAR RNAs. Both of these TAR RNA structures support efficient *trans*-activation by HIV-1 Tat, whereas the HIV-2 Tat protein activates transcription well only through the HIV-2 TAR RNA structure.

migrating complex that formed with HIV-1 Tat was unaffected by the TAR-2 loop mutation. Short deletions that remove the unpaired bulge from either the first (5′) or second (3′) hairpin of the HIV-2 TAR structure also reduced binding of the full-length Tat proteins, and no binding was observed to the double bulge mutant or antisense RNA. Competition experiments indicate that Tat binds with eight- to tenfold higher affinity to wild-type HIV-2 TAR RNA than to the loop mutant RNA (data not shown). In addition, we have shown that HIV-2 TAR loop sequences are weakly protected by HIV-2 Tat in RNase footprint experiments and that this binding is greatly enhanced in the presence of CycT1 (Wei et al. 1998). Modest effects of loop sequences on the HIV-1 Tat:TAR interaction have also been observed previously (Bohjanen et al. 1996).

These findings indicate that HIV-1 Tat possesses an inherent ability to recognize TAR loop sequences, albeit with much less precision than is observed in the presence of CycT1. Ethylation-interference footprinting experiments reveal that the loop-dependent full-length Tat-2:TAR-2 complexes contained Tat proteins bound to both stems of the structure, whereas the loop-insensitive complexes formed with the Tat ARM protein contained Tat bound to only one of the stems (data not shown). Therefore, it is possible that sequences in the loop of the RNA structure may be necessary for Tat to bind to both stems of the HIV-2 TAR structure.

Truncation of the Tat *Trans*-activation Domain Reduces TAR RNA-binding Specificity

The observation that Tat binds more avidly to TAR-2 RNA than to TAR-1 RNA was unexpected, because previous RNA-binding studies have indicated that these two RNAs are recognized in a nearly equivalent manner by Tat in vitro (Rhim and Rice 1993), and loop-dependent binding of Tat to TAR-2 RNA has not been reported previously. In considering possible reasons for this discrepancy, we noted that many TAR RNA-binding studies used either truncated Tat proteins or peptides containing the ARM. Consequently, we examined whether the RNA-binding specificity of the truncated Tat ARM protein might differ from that of full-length Tat. Interestingly, binding of the Tat ARM proteins was not significantly affected by HIV-2 TAR loop mutations (Fig. 3A, compare lanes 9 and 10), nor was binding reduced by mutations affecting either of the bulges of HIV-2 TAR RNA. Binding of the Tat ARM was eliminated by removal of both bulges, and no binding was observed to antisense RNA. Moreover, the HIV-1 Tat ARM bound strongly to both HIV-1 and HIV-2 TAR RNAs (Fig. 3A, compare lanes 9 and 15), whereas full-length HIV-1 Tat bound preferentially to HIV-2 TAR RNA. As expected, the HIV-2 Tat ARM protein did not bind HIV-1 TAR RNA. We conclude that the Tat ARM binds TAR with much greater affinity but reduced specificity than the full-length Tat protein. In particular, the activation domain appears to be required in order for Tat to discriminate features of the upper stem and loop of HIV-2 TAR RNA.

Thus, residues in the activation domain may contribute to TAR RNA recognition in addition to binding CycT1. These findings are consistent with structural studies, which show that Tat is oriented on TAR-1 RNA in a manner that would position the *trans*-activation domain in the vicinity of the loop (Wang et al. 1996; Huq and Rana 1997).

The finding that the *trans*-activation domain enhances the specificity of the binding to HIV-2 TAR RNA supports previous observations that the HIV-1 Tat "core" region alters the specificity of the Tat:TAR interaction in vitro (Churcher et al. 1993; Aboul-ela et al. 1995). Although we have not examined the specific contribution of core residues to TAR RNA-binding activity, our results indicate that the amino terminus of Tat significantly reduces TAR RNA-binding activity within the context of the full-length protein. In contrast, Churcher et al. (1993) reported that ARM (and ARM+core) peptides have a lower affinity for TAR than the full-length Tat-1 protein, although the latter study compared the activity of peptides with that of recombinant protein. To date, it has been shown that the cysteine residues and amino acid K41 in the core form part of the interaction surface between Tat and CycT1, and further studies will be required to determine the role of the other essential residues in the Tat *trans*-activation domain.

Tat Induces Binding of Nuclear CDK9/P-TEFb Complexes to TAR RNA

An important question raised by these findings is whether the interaction between Tat and CycT1 induces binding of the entire P-TEFb complex to TAR RNA. To determine whether TAK/P-TEFb remains associated with Tat upon binding to TAR RNA, crude HeLa nuclear extracts were incubated with Tat under conditions that support Tat-mediated transcription in vitro, and then incubated with TAR RNA-coupled streptavidin beads. TAR RNA-binding proteins were eluted with high salt, and the Tat-associated proteins were analyzed by immunoprecipitation experiments, using antisera specific for Tat or CDK9. Proteins were visualized following incubation of the beads with [γ-^{32}P]ATP, as described by Yang et al. (1996). We found that Tat induces the binding of nuclear CDK9 to TAR RNA-coupled streptavidin beads (Fig. 4A, compare lanes 1 and 3 for the HIV-1 Tat experiment, and lanes 8 and 10 for the HIV-2 Tat experiment). In contrast, CDK9 did not associate with beads lacking RNA, or with beads containing equivalent amounts of the HIV-1 or HIV-2 loop mutant RNAs (Fig. 4A, compare lanes 1 and 4 for the HIV-1 Tat experiment, and lanes 8 and 11 for the HIV-2 Tat experiment). The identity of the autophosphorylated 42-kD protein as CDK9 was confirmed by Western blot (data not shown). The Tat:TAK (P-TEFb) complex bound on TAR RNA could be dissociated with a high-ionic-strength buffer and was present in the eluate fraction beads containing either HIV-1 or HIV-2 TAR RNAs (Fig. 4A, lanes 6 and 13), and not in the eluate fractions from loop mutant TAR RNA beads (lanes 7 and 14) or beads lacking any RNA (lanes 5 and 12). As observed previously, CDK9 also phosphorylates HIV-2, but not

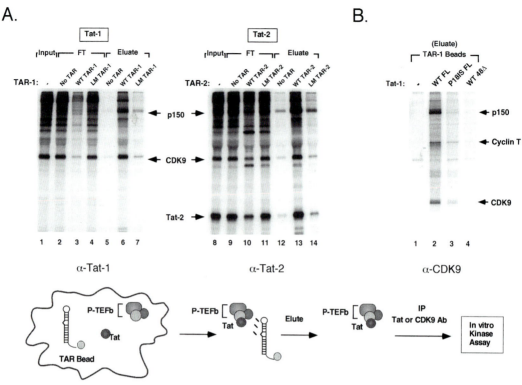

Figure 4. Tat induces loop-specific binding of P-TEFb to TAR RNA in nuclear extracts. (*A*) HIV-1 Tat (lanes *1–7*) or HIV-2 Tat (lanes *8–14*) proteins were incubated with HeLa nuclear extract in the presence of TAR RNA-coupled streptavidin beads. TAR RNA-binding proteins were eluted with high salt, immunoprecipitated with anti-Tat antibodies, and analyzed for CDK9/ P-TEFb activity by incubation of the beads with [γ-^{32}P]ATP. Lanes *1* and *8* show the input CDK9 activity and lanes *2* and *9* show the flowthough (FT) fractions from streptavidin beads that lack TAR. Lanes *3* and *10* show the flowthrough (FT) fractions from beads containing WT HIV-1 and HIV-2 TAR RNA, respectively, and lanes *4* and *11* show the flowthrough fractions from loop mutant TAR RNA-coupled beads. Lanes *5–7* and *12–14* show the eluted TAR RNA-binding proteins. Labeled proteins were separated on a 10% SDS-PAGE gel and visualized by autoradiography. (*B*) Tat is required for CDK9/P-TEFb to bind to TAR RNA. HeLa nuclear extract was incubated in the absence (lane *1*) or presence (lanes *2–4*) of wild-type or mutant HIV-1 Tat proteins (aa 1–86), as indicated above each lane, and eluates from wild-type TAR RNA beads were examined by immunoprecipitation with antisera specific to CDK9 and visualized using the in vitro kinase assay. A schematic of the procedure used to examine binding of P-TEFb to TAR RNA is shown at the bottom of the figure.

HIV-1, Tat protein. In addition, we observe several other high-molecular-weight proteins in the eluate fractions from the TAR RNA beads (e.g., a p150 protein is indicated in the figure), which we also find in immunoprecipitates with anti-CDK9 or anti-CycT1 and therefore may be other subunits of P-TEFb (for review, see Jones 1997).

To assess whether Tat was required for the P-TEFb to associate with TAR RNA, HeLa nuclear extracts were incubated with TAR-1 RNA beads in the presence or absence of Tat, and the presence of the 42-kD CDK9 was examined following immunoprecipitation with anti-CDK9 antibodies. Importantly, binding of CDK9 and the other P-TEFb subunits to HIV-1 TAR beads required the HIV-1 Tat protein (Fig. 4B, compare lanes 1 and 2). Most importantly, binding of P-TEFb to TAR in nuclear extracts requires the activation domain as well as the ARM of Tat, and no binding of P-TEFb components was observed using loop mutant TAR RNAs. We conclude that binding of nuclear CDK9/P-TEFb to TAR RNA requires the *trans*-activation domain as well as the ARM of Tat. Thus, the interaction of HIV-1 Tat with CycT1 does not

displace binding to CDK9, and instead Tat functions to recruit the entire P-TEFb complex to TAR RNA.

Cooperative Binding of HIV-1 Tat and CycT1 to TAR-1 RNA In Vitro

The TAR RNA-binding properties of the nuclear P-TEFb-Tat complex reflect the specificity of the recombinant CycT1-Tat complex (Wei et al. 1998), indicating that the association of Tat with CycT1 is sufficient for TAR RNA recognition. At limiting protein concentrations, we find that the full-length HIV-1 Tat is unable to bind HIV-1 TAR RNA (Fig. 5, lane 3), whereas the truncated HIV-1 Tat ARM protein binds readily to both wild-type and loop mutant HIV-1 TAR RNAs (lanes 5,6). In the presence of CycT1, full-length Tat binds with high affinity to TAR RNA (lane 7), whereas the truncated Tat ARM protein is unaffected, since it is unable to interact with CycT1 (lane 9). As reported previously, CycT1 does not bind detectably to TAR RNA in the absence of Tat (Wei et al. 1998). Thus, under these conditions, binding of Tat and CycT1 to TAR RNA is completely coopera-

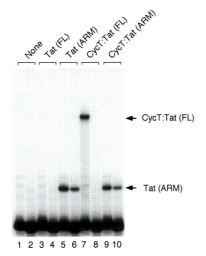

Figure 5. Cooperative binding of HIV-1 Tat and recombinant CycT1 to TAR RNA in gel mobility shift experiments. HIV-1 TAR RNA probes contained the upper half of the stem-loop structure (i.e., the minimal functional TAR RNA; +17 to +43) and were either the wild-type sequence (odd-numbered lanes), or contained a 6-base substitution of residues in the loop (+29/+34; even-numbered lanes). Where indicated, binding reactions contained either 30 ng of (GST-cleaved) full-length HIV-1 Tat (FL) or 12 ng of (GST-cleaved) truncated HIV-1 Tat that lacks the activation domain but contains the arginine-rich motif (ARM; aa 48–86), and 100 ng of (GST-cleaved) CycT1 (aa 1–303). Complexes were subject to native gel electrophoresis at 4°C. RNA:protein complexes are indicated by the arrow. The CycT1 did not bind TAR RNA in the absence of HIV-1 Tat in this experiment (data not shown), as we have reported previously (Wei et al. 1998). Thus, under these conditions, the binding of both Tat and CycT1 to TAR is highly cooperative.

tive, and, unlike the Tat ARM protein, the CycT1-Tat complex does not bind to loop mutant TAR RNA (lane 8). We have shown previously that none of these protein:RNA complexes form on bulge mutant TAR RNAs and that the Tat ARM is critical for binding of the CycT1-Tat complex to the RNA. Consequently, all of the amino acids in HIV-1 Tat that have been shown to be required for *trans*-activation in vivo are required for Tat to form a stable complex with CycT1 on TAR RNA in vitro.

These findings raised the question of whether the Tat *trans*-activation domain is more active than the full-length Tat protein in its ability to interact with recombinant CycT1. To address this question, GST-CycT1 was coupled to glutathione-*S*-Sepharose beads and incubated with (GST-cleaved; HA-tagged) full-length Tat (amino acids 1–86) or activation domain HIV-1 Tat (amino acids 1–48) proteins, in the presence and absence of HIV-1 TAR RNA. The bound Tat proteins were detected by immunoblot with HA-monoclonal antibody, and the CycT1 proteins were visualized with an anti-GST monoclonal antibody. As shown in Figure 6, both full-length (amino acids 1–86) and activation domain (amino acids 1–48) Tat proteins bound to the GST-CycT1 resin (lanes 1 and 3). Binding to CycT1 in this assay was specific because Tat proteins that carry an inactivating point mutation (C22G) in the *trans*-activation domain failed to bind CycT1 (Fig. 6, lanes 2 and 4). The interaction between

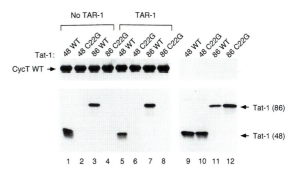

Figure 6. Specific binding of full-length HIV-1 Tat (aa 1–86) or activation domain HIV-1 Tat (aa 1–48) to CycT1 in vitro. The ability of wild-type or mutant Tat proteins to bind to GST-CycT1-coupled beads was analyzed by Western blot. GST-CycT1 protein (50 pmoles) was coupled to beads and incubated with the HA-tagged Tat-1 *trans*-activation domain (aa 1–48; 180 pmoles), or HA-tagged full-length Tat-1 (aa 1–86; 180 pmoles), as indicated above each lane. Where indicated, reactions contained wild-type HIV-1 TAR RNA (+1 to +80; 315 pmoles), which was added simultaneously with Tat to the CycT1 beads (lanes *5–8*). Proteins were visualized by Western blot using monoclonal antisera specific to GST and HA. Lanes *9–12* contained aliquots corresponding to 10% of the input protein, for the full-length Tat-1 aa 1–86 (labeled 86; lanes *11* and *12*) or the Tat-1 *trans*-activation domain aa 1–48 (labeled 48; lanes *9* and *10*) proteins. Reactions contained either the wild-type (WT) proteins, or proteins containing an inactivating mutation within the *trans*-activation domain (C22G).

Tat and CycT1 was unaffected by either the Tat ARM or TAR RNA (Fig. 6, compare lanes 1–4 with lanes 5–8). These data suggest that TAR RNA does not enhance the interaction between Tat and CycT1, although we cannot eliminate the possibility that the binding of Tat to the beads was saturating under these conditions even though the beads contained a two- to threefold molar excess of CycT1 protein relative to Tat.

A Model for TAR RNA Recognition and the Control of HIV-1 Transcription by Tat and the P-TEFb CTD Kinase Complex

The HIV-1 Tat activation domain, when juxtaposed with the ARM, strongly inhibits binding to TAR. The native Tat protein is unfolded and unstructured (Bayer et al. 1995) and binds poorly to TAR RNA. The first step in Tat *trans*-activation involves binding to the CycT1 subunit of P-TEFb, which greatly enhances binding to TAR and imposes a stringent requirement for specific sequences in the loop of the RNA (Fig. 7). Binding of Tat to CycT1 is proposed to impart a stable structure to the Tat activation domain and to relieve the inhibition of RNA-binding imposed by residues at the amino terminus of Tat. Residues in CycT1, and possibly also Tat, are likely to contact the loop of the nascent TAR RNA structure and may rearrange the conformation of the RNA to induce an optimal fit. Thus, the cooperative binding of Tat and CycT1 might be explained through a mechanism that involves the simultaneous refolding of both proteins and TAR RNA during complex formation.

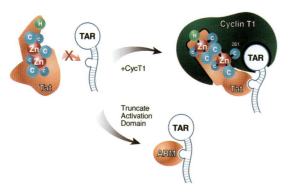

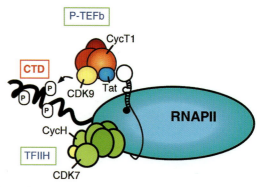

Figure 7. The interaction of HIV-1 Tat with CycT1 is critical for high-affinity, loop-specific binding to TAR RNA. The full-length HIV-1 Tat protein binds only weakly to TAR RNA in vitro. Tat binds in a highly cooperative manner with CycT1 to TAR RNA in vitro, and sequences in both the loop and bulge of TAR are critical. Alternatively, high-affinity binding of Tat to TAR RNA is also observed upon truncation of the *trans*-activation domain, which frees the arginine-rich motif (ARM) of Tat to bind to the bulge of the RNA structure. Binding of Tat to CycT1 is proposed to induce a conformational change in Tat that promotes binding to TAR RNA and might also alter the structure of the cyclin. Tat binds CycT1 through its *trans*-activation domain, which includes a hydrophobic "core" motif as well as a Cys-rich region that coordinates zinc. The interaction between Tat and CycT1 is proposed to form a structure that induces a conformational rearrangement of the TAR RNA through a mechanism of "induced fit."

Figure 8. Transient binding of the Tat-P-TEFb complex to nascent HIV-1 TAR RNA is proposed to regulate phosphorylation of the RNAPII CTD and to facilitate the transition to productive transcription elongation.

The Cys-rich subdomain of Tat has been shown to bind two atoms of zinc per monomer of Tat, potentially coordinated through five cysteine residues (Frankel et al. 1988a,b; Huang and Wang 1996). Six of the seven cysteine residues (and a histidine) in the Cys-rich subdomain of Tat are critical for *trans*-activation, although the arrangement of these closely spaced residues (CxxCx-CxxCxxC) does not match any of the well-defined zinc-binding structures. The C22G substitution destroys binding of Tat to recombinant CycT1 in vitro and prevents association of Tat with TAK/P-TEFb in nuclear extracts. We have recently shown that zinc is required for the CycT1-Tat interaction, and that a cysteine at residue 261 in CycT1 is required to bind Tat (Garber et al. 1998). These findings raise the interesting possibility that the Tat and CycT1 proteins coordinate zinc as part of their interaction surface. As indicated in Figure 7, high-affinity binding of Tat to TAR RNA can also be achieved by truncation of the Tat *trans*-activation domain, although the specificity of the Tat ARM:TAR complex is much reduced compared to that of the CycT1-Tat:TAR complex. It will be important to examine the relative affinity and stability of these two TAR RNA complexes to better assess the changes in Tat and CycT1 that accompany binding to RNA.

It has not yet been established whether P-TEFb is a component of the RNAPII initiation complex, but, if it is, the interaction with Tat may occur on the surface of RNAPII, or prior to binding of P-TEFb to RNAPII. Tat has been reported to be present in the RNAPII holoenzyme (Keen et al. 1996; Cujec et al. 1997a) or the

RNAPII preinitiation complex (Garcia-Martinez et al. 1997); however, it is not clear whether Tat needs to bind RNAPII prior to initiation in order to function. It will be interesting to learn whether the activation domain of Tat is needed for its association with RNAPII, as would be predicted if the interaction was mediated indirectly through P-TEFb.

The evidence available to date indicates that Tat forms a complex with CycT1 in order to direct binding of the TAK/P-TEFb complex to the nascent TAR RNA structure (Fig. 8). This interaction is required for the progression to efficient elongation, and is marked by phosphorylation of the RNAPII CTD. Two possible roles have been suggested for binding to TAR RNA: (1) recruitment of P-TEFb to the HIV-1 promoter and (2) activation of the CDK9 kinase activity, leading to the phosphorylation of the CTD and any other substrates that may be important for transcription elongation. These two possible roles for TAR RNA are not mutually exclusive, and binding may serve both to position the kinase at the promoter and in a location important for regulating transcription elongation, and to induce (or derepress) CDK9 activity. Tat and TAR RNA do not affect the high level of phosphorylation of the CTD observed with recombinant CDK9-CycT1 complexes in vitro (T.P. Mayall and K.A. Jones, unpubl.), possibly because this system lacks a specific CDK9 inhibitor(s). The interaction of Tat-TAK/P-TEFb with TAR is transient, and at later stages in elongation Tat associates with RNAPII elongation complexes in a TAR-independent manner (Keen et al. 1997). The events that cause the disruption of the TAK/P-TEFb-Tat:TAR complex are unknown, although CDK9-mediated phosphorylation of the CTD, CycT1, or other P-TEFb subunits may play a role in this process.

PERSPECTIVES

The specific interaction between HIV-1 Tat and CycT1, and the subsequent binding of the P-TEFb complex to TAR RNA, is a critical step in Tat *trans*-activation. With the identification of CycT1 as the RNA-binding cofactor for Tat, it may now be possible to find new antiviral compounds that block the interaction of Tat with CycT1 or interfere with the binding of the P-TEFb-Tat complex to TAR RNA, without affecting CDK9 activity or the normal cellular function of P-TEFb. Recent studies have shown that chimeric proteins which tether either CycT1 or CDK9 to a heterologous RNA element are sufficient to activate transcription in a manner that requires the catalytic activity of CDK9 (Fujinaga et al. 1998; Gold et al. 1998), indicating that the ultimate goal of Tat is to direct CDK9 to the nascent transcript. To fully understand the mechanism of Tat *trans*-activation, it will be critical to learn the sequence of events that unfold subsequent to binding of the Tat-P-TEFb complex to TAR RNA and to understand the processes that regulate CDK9 kinase activity and the timing of CTD phosphorylation during transcription. The other subunits of the P-TEFb complex remain to be identified, and it will be interesting to learn their role and possible impact on Tat *trans*-activation. Similarly, it will be critical to understand the function of other host cell factors that regulate early RNAPII elongation, including the human homologs of Spt5 and Spt6 (Wada et al. 1998; Wu-Baer et al. 1998), as well as any other negative factors that may block phosphorylation of the CTD during transcription initiation.

It is interesting to consider further that the balance between the positive and negative controls of HIV-1 transcription might be regulated differently in various cells, with important consequences for HIV-1 infection and pathogenesis. For example, TAK/P-TEFb activity is greatly enhanced upon cytokine activation of primary T cells and monocytes, concomitant with induction of HIV-1 transcription (Yang et al. 1997). T-cell activation increases the steady-state levels of CycT1 protein, apparently through a translational control mechanism, which leads to greatly enhanced CDK9 activity in these cells (Herrmann et al. 1998). Variations in CycT1 levels and corresponding changes in CDK9 activity are likely to influence Tat activity directly. In addition, transcription factors that inhibit Tat activity could play an important role in controlling HIV-1 infection. A stable reservoir of latent HIV-infected CD4$^+$ memory T cells has been identified that may serve to repopulate infection throughout the body, and this reservoir appears to persist for the life of the T cell (for review, see Finzi and Siliciano 1998; Ho 1998). Although the events that lead to viral latency remain obscure, the observation that a defective Tat protein can induce postintegration latency of HIV-1 in cell culture (Emiliani et al. 1998) raises the possibility that cellular conditions which down-regulate Tat *trans*-activation might induce viral latency. Thus, as CycT1, CDK9, and other cellular factors critical for Tat *trans*-activation are identified and characterized, it will be important to establish whether the regulation of these proteins contributes to the dynamics of HIV-1 infection in different cells of the immune system.

ACKNOWLEDGMENTS

We thank Andrew Rice (Baylor University College of Medicine) for many of the Tat and TAR constructs that were used in this study. Our research is funded by grants from the National Institutes of Health (AI-44615), the California Universitywide AIDS Research Program, and the Leila Y. and G. Harold Mathers Foundation.

REFERENCES

Aboul-ela F., Karn J., and Varani G. 1995. The structure of the human immunodeficiency virus type-1 TAR RNA reveals principles of RNA recognition by Tat protein. *J. Mol. Biol.* **253:** 313.

Bayer P., Kraft M., Ejchart A., Westendorp M., Frank R., and Rosch P. 1995. Structural studies of HIV-1 Tat protein. *J. Mol. Biol.* **247:** 529.

Bohjanen P.R., Colvin R., Puttaraju M., Been M., and Garcia-Blanco M.A. 1996. A small circular TAR RNA decoy specifically inhibits Tat-activated HIV-1 transcription. *Nucleic Acids Res.* **24:** 3733.

Chun R.F. and Jeang K.T. 1996. Requirements for RNA polymerase II carboxyl-terminal domain for activated transcription of human retroviruses human T-cell lymphotrophic virus I and HIV-1. *J. Biol. Chem.* **271:** 27888.

Churcher M.J., Lamont C., Hamy F., Dingwall C., Green S.M., Lowe A.D., Butler J.G., Gait M.J., and Karn J. 1993. High affinity binding of TAR RNA by the human immunodeficiency virus type-1 Tat protein requires base-pairs in the RNA stem and amino acids flanking the basic region. *J. Mol. Biol.* **230:** 90.

Cujec T.P., Cho H., Maldonado E., Meyer J., Reinberg D., and Peterlin B.M. 1997a. The human immunodeficiency virus transactivator Tat interacts with the RNA polymerase II holoenzyme. *Mol. Cell. Biol.* **17:** 1817.

Cujec T.P., Okamoto H., Fujinaga K., Meyer J., Chamberlin H., Morgan D.O., and Peterlin B.M. 1997b. The HIV transactivator Tat binds to the CDK-activating kinase and activates the phosphorylation of the carboxy-terminal domain of RNA polymerase II. *Genes Dev.* **11:** 2645.

Cullen B.R. 1998. HIV-1 auxiliary proteins: Making connections in a dying cell. *Cell* **93:** 685.

Emerman M. and Malim M. 1998. HIV-1 regulatory/accessory genes: Keys to unraveling viral and host cell biology. *Science* **280:** 1880.

Emiliani S., Fischle W., Ott M., Van Lint C., Amella C.A., and Verdin E. 1998. Mutations in the tat gene are responsible for human immunodeficiency virus type 1 postintegration latency in the U1 cell line. *J. Virol.* **72:** 1666.

Finzi D. and Siliciano R.F. 1998. Viral dynamics in HIV-1 infection. *Cell* **93:** 665.

Frankel A., Bredt D., and Pabo C. 1988a. Tat protein from human immunodeficiency virus forms a metal-linked dimer. *Science* **240:** 70.

Frankel A., Chen L., Cotter R., and Pabo C. 1988b. Dimerization of the tat protein from human immunodeficiency virus: A cysteine-rich peptide mimics the normal metal-linked dimer interface. *Proc. Natl. Acad. Sci.* **85:** 6297.

Fujinaga K., Cujec T., Peng J., Garriga J., Price D., Grana X., and Peterlin B.M. 1998. The ability of positive transcription elongation factor factor B to transactivate human immunodeficiency virus transcription depends on a functional kinase domain, cyclin T1, and Tat. *J. Virol.* **72:** 7154.

Gait M. and Karn J. 1993. RNA recognition by the human immunodeficiency virus Tat and Rev proteins. *Trends Biochem. Sci.* **18:** 255.

Garber M.E., Wei P., KewalRamani V., Mayall T.P., Herrmann C.H., Rice A.P., Littman D.R., and Jones K.A. 1998. The interaction between HIV-1 Tat and human Cyclin T1 requires zinc and a critical cysteine residue that is not conserved in the murine Cyclin T1 protein. *Genes Dev.* **12:** 3512.

Garcia-Martinez L.F., Ivanov D., and Gaynor R.B. 1997. Association of Tat with purified HIV-1 and HIV-2 transcription preinitiation complexes. *J. Biol. Chem.* **272:** 6951.

Garriga J., Segura E., Mayol X., Grubmeyer C., and Grana X. 1996. Phosphorylation site specificity of the CDC2-related kinase PITALRE. *Biochem. J.* **320**: 983.

Gold M.O., Yang X., Herrmann C.H., and Rice A.P. 1998. PITALRE, the catalytic subunit of TAK, is required for human immunodeficiency virus Tat transactivation in vivo. *J. Virol.* **72**: 4448.

Grana X., De Luca A., Sang N., Fu Y., Claudio P., Rosenblatt J., Morgan D., and Giordano A. 1994. PITALRE, a nuclear CDC2-related protein kinase that phosphorylates the retinoblastoma protein in vitro. *Proc. Natl. Acad. Sci.* **91**: 3834.

Herrmann C.H. and Rice A.P. 1993. Specific interaction of the human immunodeficiency virus Tat proteins with a cellular protein kinase. *Virology* **197**: 601.

———. 1995. Lentivirus Tat proteins specifically associate with a cellular protein kinase, TAK, that hyperphosphorylates the carboxyl-terminal domain of the large subunit of RNA polymerase II: Candidate for a Tat cofactor. *J. Virol.* **69**: 1612.

Herrmann C.H., Carroll R.G., Wei P., Jones K.A., and Rice A.P. 1998. Tat-associated kinase, TAK, activity is regulated by distinct mechanisms in peripheral blood lymphocytes and promonocytic cell lines. *J. Virol.* **72**: 9881.

Ho D.D. 1998. Toward HIV eradication or remission: The tasks ahead. *Science* **280**: 1866.

Huang H.-W. and Wang K.-T. 1996. Structural characterization of the metal binding site in the cysteine-rich region of HIV-1 Tat protein. *Biochem. Biophys. Res. Commun.* **227**: 615.

Huq I. and Rana T. 1997. Probing the proximity of the core domain of an HIV-1 tat fragment in a tat-TAR complex by affinity cleaving. *Biochemistry* **36**: 12592.

Jones K.A. 1997. Taking a new TAK on Tat transactivation. *Genes Dev.* **11**: 2593.

Jones K.A. and Peterlin B.M. 1994. Control of RNA initiation and elongation at the HIV-1 promoter. *Annu. Rev. Biochem.* **63**: 717.

Jonsen M., Petersen J., Xu Q.P., and Graves B.J. 1996. Characterization of the cooperative function of inhibitory sequences in Ets-1. *Mol. Cell. Biol.* **16**: 2065.

Keen N.J., Churcher M.J., and Karn J. 1997. Transfer of Tat and release of TAR RNA during the activation of the human immunodeficiency virus type-1 transcription elongation complex. *EMBO J.* **16**: 5260.

Keen N., Gait M., and Karn J. 1996. Human immunodeficiency virus type-1 Tat is an integral component of the activated transcription-elongation complex. *Proc. Natl. Acad. Sci.* **93**: 2505.

Kephart D.D., Marshall N.F., and Price D.H. 1992. Stability of *Drosophila* RNA polymerase II elongation complexes in vitro. *Mol. Cell. Biol.* **12**: 2067.

Ko J. and Prives C. 1996. p53: Puzzle and paradigm. *Genes Dev.* **10**: 1054.

Levine A. 1997. p53, the cellular gatekeeper for growth and division. *Cell* **88**: 323.

Luo Y., Madore S., Parslow T., Cullen B.R., and Peterlin B.M. 1993. Functional analysis of interactions between Tat and the *trans*-activation response element of human immunodeficiency virus type 1 in cells. *J. Virol.* **67**: 5617.

Madore S. and Cullen B.R. 1993. Genetic analysis of the cofactor requirement for human immunodeficiency virus type 1 Tat function. *J. Virol.* **67**: 3703.

Mancebo H., Lee G., Flygare J., Tomassini J., Luu P., Zhu Y., Blau C., Hazuda D., Price D.H., and Flores O. 1997. P-TEFb kinase is required for HIV Tat transcriptional activation in vivo and in vitro. *Genes Dev.* **11**: 2633.

Marshall N. and Price D.H. 1992. Control of formation of two distinct classes of RNA polymerase II elongation complexes. *Mol. Cell. Biol.* **12**: 2078.

———. 1995. Purification of P-TEFb, a transcription factor required for the transition into productive elongation. *J. Biol. Chem.* **270**: 12335.

Marshall N., Peng J., Xie Z., and Price D.H. 1996. Control of RNA polymerase II elongation potential by a novel carboxyl-terminal domain kinase. *J. Biol. Chem.* **271**: 27176.

Okamoto H., Sheline C.T., Corden J.L., Jones K.A., and Peterlin B.M. 1996. Trans-activation by human immunodeficiency virus Tat protein requires the C-terminal domain of RNA polymerase II. *Proc. Natl. Acad. Sci.* **93**: 11575.

Palangat M., Meier T.I., Keene R.G., and Landick R. 1998. Transcriptional pausing at +62 of the HIV-1 nascent RNA modulates formation of the TAR structure. *Mol. Cell* **1**: 1033.

Parada C.A. and Roeder R.G. 1996. Enhanced processivity of RNA polymerase II triggered by Tat-induced phosphorylation of its carboxy-terminal domain. *Nature* **384**: 375.

Peng J., Zhu Y., Milton J., and Price D.H. 1998. Identification of multiple cyclin subunits of human P-TEFb. *Genes Dev.* **12**: 755.

Pessler F. and Hernandez N. 1998. The HIV-1 inducer of short transcripts activates the synthesis of 5,6-dichloro-1-β-D-benzimidazole-resistant short transcripts in vitro. *J. Biol. Chem.* **273**: 5375.

Rhim H. and Rice A.P. 1993. TAR RNA binding properties and relative transactivation activities of human immunodeficiency virus type 1 and 2 Tat proteins. *J. Virol.* **67**: 1110.

———. 1994. Functional significance of the dinucleotide bulge in stem-loop1 and stem-loop 2 of HIV-2 TAR RNA. *Virology* **202**: 202.

Selby M.J. and Peterlin B.M. 1990. Trans-activation by HIV-1 Tat via a heterologous RNA binding protein. *Cell* **62**: 769.

Southgate C. and Green M.R. 1991. The HIV-1 Tat protein activates transcription from an upstream DNA-binding site: implications for Tat function. *Genes Dev.* **5**: 2496.

Southgate C., Zapp M.L., and Green M.R. 1990. Activation of transcription by HIV-1 Tat protein tethered to nascent RNA through another protein. *Nature* **345**: 640.

Tiley L.S., Madore S.J., Malim M.H., and Cullen B.R. 1992. The VP16 transcription activation domain is functional when targeted to a promoter-proximal RNA sequence. *Genes Dev.* **6**: 2077.

Wada T., Takagi, T., Yamaguchi Y., Ferdous A., Imai T., Hirose S., Sugimoto S., Yano K., Hartzog G.A., Winston F., Buratowski S., and Handa H. 1998. DSIF, a novel transcription elongation factor that regulates RNA polymerase II processivity, is composed of human Spt4 and Spt5 homologs. *Genes Dev.* **12**: 343.

Wang Z., Wang X., and Rana T.M. 1996. Protein orientation in the Tat-TAR complex determined by psoralen photocrosslinking. *J. Biol. Chem.* **271**: 16995.

Wasylyk C., Kerckaert J.P., and Wasylyk B. 1992. A novel modulator domain of Ets transcription factors. *Genes Dev.* **6**: 965.

Wei P., Garber M.E., Fang S.M., Fischer W.H., and Jones K.A. 1998. A novel CDK9-associated C-type cyclin interacts directly with HIV-1 Tat and mediates its high-affinity, loop-specific binding to TAR RNA. *Cell* **92**: 451.

Wu-Baer F., Lane W.S., and Gaynor R.B. 1998. Role of the human homolog of the yeast transcription factor SPT5 in HIV-1 Tat-activation. *J. Mol. Biol.* **277**: 179.

Yang X., Herrmann C.H., and Rice A.P. 1996. The human immunodeficiency virus Tat proteins specifically associate with TAK in vivo and require the carboxyl-terminal domain of RNA polymerase II for function. *J. Virol.* **70**: 4576.

Yang X., Gold M., Tang D., Lewis D., Aguilar-Cordova E., Rice A.P., and Herrmann C.H. 1997. TAK, an HIV Tat-associated kinase, is a member of the cyclin-dependent family of protein kinases and is induced by activation of peripheral blood lymphocytes and differentiation of promonocytic cell lines. *Proc. Natl. Acad. Sci.* **94**: 12331.

Zhou Q., Chen D., Pierstorff E., and Luo K. 1998. Transcription elongation factor P-TEFb mediates Tat activation of HIV-1 transcription at multiple stages. *EMBO J.* **17**: 3681.

Zhu Y., Pe'ery T., Peng J., Ramanathan Y., Marshall N., Marshall T., Amendt B., Mathews M., and Price D. 1997. Transcription elongation factor P-TEFb is required for HIV-1 Tat transactivation in vitro. *Genes Dev.* **11**: 2622.

The Yeast RNA Polymerase III Transcription Machinery: A Paradigm for Eukaryotic Gene Activation

S. Chédin, M.L. Ferri, G. Peyroche, J.C. Andrau, S. Jourdain, O. Lefebvre, M. Werner, C. Carles, and A. Sentenac

Service de Biochimie et Génétique Moléculaire, CEA/Saclay, F-91191 Gif-sur-Yvette Cedex, France

Since the discovery of an RNA polymerase activity in rat liver nuclei four decades ago, each successive phase in the development of this field has revealed additional subtlety and complexity of the transcription apparatus. The three forms of nuclear RNA polymerase have been purified from a variety of eukaryotic organisms, and in vitro transcription studies have conclusively established their gene specificity: RNA polymerase I (pol I) synthesizes ribosomal RNA precursors, whereas RNA polymerase II (pol II) transcribes protein-encoding genes and RNA polymerase III (pol III) synthesizes small RNAs (mostly 5S RNA and tRNA) (Sentenac 1985). The availability of cell-free systems capable of accurately transcribing exogenous DNA opened the way to the identification of the DNA elements, promoters, enhancers, and other regulatory signals that direct and regulate accurate initiation of transcription, and this, in turn, encouraged a general search for specific DNA-binding proteins. The harvest was plentiful. In yeast, genetic and biochemical approaches have already identified 80 dedicated transcription factors (Svetlov and Cooper 1995), but analysis of the complete genome sequence suggests that even as many as 6% of all yeast genes may be involved in transcription regulation. The identification of the general transcription factors (GTFs) for all three reconstituted transcription systems was less straightforward and constitutes in itself a remarkable achievement. The entire set of GTFs comprises approximately 50 polypeptides, two thirds of which are devoted to pol II (Orphanides et al. 1996). The in vitro sequential assembly of the GTFs into preinitiation complexes gave insights into the function and interaction properties of the different components. However, this stepwise assembly model was challenged by the discovery of preassembled pol II holoenzymes containing both several GTFs and other novel factors (Koleske and Young 1995). The primary step of gene activation could be considered as the recruitment of a holoenzyme at the promoter under the combinatorial control of gene-specific transcription factors.

Although most studies have been focused on the pol II transcription system in view of its pivotal role in gene expression, the pol III transcription system attracted much interest as a paradigm to study eukaryotic gene activation. Retrospectively, it is interesting to note several breakthroughs that derived from the analysis of the pol III system. The first demonstration of specifically initiated transcription in vitro of a pol III gene (Parker and Roeder 1997; Wu 1978) encouraged the development of a pol II

transcription system and initiated promoter analysis. The characterization and mutagenesis of promoter elements were first done with simple genes such as 5S RNA and tRNA genes (Bogenhagen et al. 1980; Koski et al. 1980; Kurjan et al. 1980; Galli et al. 1981). The first eukaryotic transcription factor to be purified and cloned was TFIIIA (Engelke et al. 1980; Ginsberg et al. 1984), the first factor on a long list of zinc finger proteins (Miller et al. 1985). The discovery that the TATA-binding protein (TBP) was required for transcription of some class III genes (Lobo et al. 1991; Margottin et al. 1991; Simmen et al. 1991) initiated the recognition of its universal role in gene transcription and reflections on the concerted evolution of the three transcription systems. The first multistep assembly of preinitiation complexes (Lassar et al. 1983) and the first topography of fully active transcription complexes by protein-DNA cross-linking (Bartholomew et al. 1991, 1993; Braun et al. 1992) were achieved on tRNA and 5S RNA genes. The description of the cascade of protein-protein interactions that leads to the recruitment of RNA polymerase is also more advanced in the case of pol III. These observations were undoubtedly facilitated by the relative simplicity of the class III system, especially in yeast, which lends itself readily to biochemical and genetic studies.

TFIIIC, A FLEXIBLE, MULTIFUNCTIONAL FACTOR

pol III transcribes genes encoding highly conserved small RNAs, many of which participate in translation (tRNA, 5S rRNA) or are required for mRNA or rRNA processing (U6 RNA, RPR1 RNA, plant U3 RNA, MRP RNA), for protein trafficking (7SL RNA), for viral regulation (EBER RNA, adenovirus-associated VA RNA), or are involved in some hitherto unknown functions (7SK RNA, *Alu*I, and other reiterated gene families). These genes show a marked diversity of promoter organization. Genes such as the 5S rRNA and tRNA genes have internal control regions within the transcribed DNA sequences; others, such as the vertebrate U6 small nuclear RNA (snRNA) gene, have a more conventional class-II-like upstream promoter with a functional TATA box; still others have a mixed promoter structure.

Intrigued by the intragenic promoter of tRNA genes, and encouraged by the apparent simplicity of these transcription units, we began the fractionation of yeast extracts to identify the DNA-binding factors required for tRNA gene activation. The tRNA promoters are made of two different sequence elements, the A and B blocks, and in the 273 tRNA genes identified in the yeast genome, the distance between these promoter elements can vary widely from 31 bp to 93 bp. Therefore, we expected to isolate two distinct DNA-binding proteins on the basis of their DNA footprinting properties. This was not the case. The purified factor, called TFIIIC or Tau, was an exceptionally large multisubunit protein of about 600 kD that interacted with the split promoter of a variety of tRNA genes (Camier et al. 1985; Gabrielsen et al. 1989). Although the two DNA-binding activities could not be separated by conventional means, electron microscopy of TFIIIC-DNA complexes (Schultz et al. 1989) and selective proteolysis (Marzouki et al. 1986) demonstrated the existence of two linked DNA-binding domains, τA and τB, of about 300 kD each, each interacting with one promoter element. Binding of τB to the B block is predominant and enhances the ability of τA to bind to the A block. When challenged with a series of stretched or internally deleted tRNA$_3^{Leu}$ genes having variable A to B block distances, the TFIIIC-DNA interaction displayed a remarkable adaptability to the variable A-B distances and to their relative helical orientation (Baker et al. 1987; Fabrizio et al. 1987).

The primary step in tRNA gene activation is the binding of TFIIIC to the intragenic promoter. The main function of TFIIIC is then to assemble TFIIIB upstream of the transcription start site (Kassavetis et al. 1990). While studying the U6 RNA gene which has an atypical promoter configuration, with a distal B block downstream from the terminator signal, we found no requirement for TFIIIC using a purified transcription system, whereas the B block, and by inference TFIIIC, was necessary in crude extracts (Moenne et al. 1990; Gerlach et al. 1995). We surmised that TFIIIC may have the additional role of counteracting chromatin repression. Indeed, TFIIIC could reactivate the U6 RNA gene with preassembled nucleosomes (Burnol et al. 1993). Chromatin disruption in vivo, induced by histone H4 depletion, stimulated the transcription of various promoter-deficient genes, but not of wild-type U6 RNA genes. Therefore, there is a competition between chromatin assembly and transcription complex formation, but this competition is much in favor of the transcription factors (Marsolier et al. 1995). It will be interesting to investigate the interaction of TFIIIC with chromatin templates. Our preliminary results give no suggestion that TFIIIC is endowed with histone acetyltransferase activity, at least not for free histones.

It is remarkable that TFIIIC possesses an assemblage of different functions that are assigned to separate proteins in class II genes. It plays the part of enhancer-binding proteins (B-block interaction at a variable distance from the start site), of proximal element-binding factors (through A-block binding), of an activator (through TFIIIB binding; see below), and of an antirepressor (relieving chromatin repression). All of these functions are combined into a flexible multisubunit factor for optimal efficiency of the pol III system.

To dissect the structure and the function of yeast TFIIIC, we have undertaken the identification and molecular

Table 1. Components of the Yeast Class III Transcription Apparatus

Polypeptides	M_r ($\times 10^{-3}$)	Null gene	Phenotype	References
RNA polymerase III				
C160	162	RPO31 (RPC160)	lethal	1–3
C128	129	RET1	lethal	4
C82	74	RPC82	lethal	2, 5
C53	47	RPC53	lethal	2, 6
AC40	38	RPC40	lethal	2, 7
C37	32	RPC37	not done	35
C34	36	RPC34	lethal	2, 8
C31	28	RPC31	lethal	2, 9
ABC27	25	RPB5	lethal	10
C25	24	RPC25	lethal	11
ABC23	18	RPO26	lethal	12
AC19	16	RPC19	lethal	13
C17	19	RPC17	lethal	36
ABC14.5	17	RPB8	lethal	10
C11	13	RPC11	lethal	37
ABC10α	8	RPC10	lethal	14
ABC10β	8	RPB10	lethal	15
Transcription factors				
TFIIIA	50	TFC2	lethal	16
TFIIIB				
TFIIIB90/B″	68	TFC5	lethal	17–19
TFIIIB70/BRF1	67	PCF4, (BRF1 TDS4)	lethal	20–22
TBP	27	SPT15	lethal	23–27
TFIIIC				
τ138	132	TFC3	lethal	28
τ131	120	TFC4 (PCF1)	lethal	29–30
τ95	74	TFC1	lethal	31, 32
τ91	75	TFC6	lethal	33
τ60	68	TFC8	lethal	38
τ55	49	TFC7	lethal	34

(1) Allison et al. 1985; (2) Riva et al. 1986; (3) Gudenus et al. 1988; (4) James et al. 1991; (5) Chiannilkulchai et al. 1992; (6) Mann et al. 1992; (7) Mann et al. 1987; (8) Stettler et al. 1992; (9) Mosrin et al. 1990; (10) Woychik et al. 1990; (11) Sadhale and Woychik 1994; (12) Archambault et al. 1990; (13) Dequard-Chablat et al. 1991; (14) Treich et al. 1992; (15) Woychik et al. 1993; (16) Archambault et al. 1992; (17) Kassavetis et al. 1995; (18) Roberts et al. 1996; (19) Rüth et al. 1996; (20) Buratowski and Zhou 1992; (21) Colbert and Hahn 1992; (22) López-De-León et al. 1992; (23) Cavallini et al. 1989; (24) Eisenmann et al. 1989; (25) Hahn et al. 1989; (26) Horikoshi et al. 1989; (27) Schmidt et al. 1989; (28) Lefebvre et al. 1992; (29) Marck et al. 1993; (30) Rameau et al. 1994; (31) Swanson et al. 1991; (32) Parsons and Weil 1992; (33) Arrebola et al. 1998; (34) Manaud et al. 1998; (35) A. Flores et al., in prep.; (36) M.L. Ferri et al., in prep.; (37) Chédin et al. 1998; (38) E. Deprez et al., in prep.

cloning of each subunit. The subunit structure of TFIIIC was established by identifying the polypeptides specifically complexed to a tRNA gene (Gabrielsen et al. 1989; Bartholomew et al. 1990). The polypeptides were characterized by microsequencing and gene cloning, and their TFIIIC subunit status was confirmed by a combination of biochemical and genetic methods. Yeast TFIIIC comprises six polypeptides that were all found to be essential for yeast cell viability (Table 1). Figure 1 summarizes the genetic and protein-protein interactions that were found

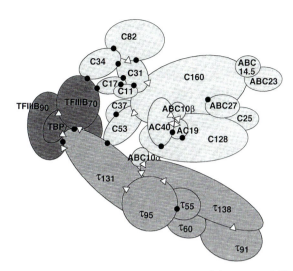

Figure 1. Interactions among components of the yeast pol III transcription complex. Suppression studies using multicopy plasmids revealed genetic interactions. The arrowheads point toward the component harboring the conditional mutation. Polypeptide pairs that reacted positively in two-hybrid assays are joined by a black circle. (Adapted from A. Flores et al. [in prep.] and M.L. Ferri et al. [in prep.].)

to link these subunits together, in good agreement with the TFIIIC-DNA cross-linking data of P. Geiduschek and his colleagues (Geiduschek and Kassavetis 1992).

THE TFIIIC-TFIIIB CONNECTION

TFIIIB is the general initiation factor for yeast class III genes (Kassavetis et al. 1990). Once assembled upstream of the start site, it is sufficient to recruit pol III and direct accurate transcription initiation in the absence of TFIIIC (Kassavetis et al. 1990; Margottin et al. 1991). The spontaneous dissociation of TFIIIB activity into separate components delayed their identification, requiring a combination of biochemical and genetic approaches. TBP is part of TFIIIB, which contains two additional components, BRF1/TFIIIB70 and B´´/TFIIIB90 (see Table 1). These components make TFIIIB appear to be analogous to TFIID, which consists of TBP and TBP-associated factors (TAFs), but the situation is more complex. First, there is no evidence for a stable TFIIIB entity, when not in a DNA-bound state (Kassavetis et al. 1991; Huet et al. 1994). Second, one TFIIIB component, TFIIIB70, is structurally and functionally related to TFIIB, which is not a TAF. TFIIIB therefore combines the function of TFIID and of TFIIB.

A puzzling behavior of TFIIIB was its incapacity to bind by itself to TATA-less promoters (the usual class III promoters). TFIIIB assembly on the DNA requires prior binding of TFIIIC (Kassavetis et al. 1989). This prompted us to investigate the protein-protein interactions in which TFIIIB participates. Contrary to our expectations, the major TBP-binding domain of TFIIIB70 was not included in the TFIIB-related amino-terminal part of the protein but instead lies in the carboxy-terminal half (Khoo et al. 1994; Chaussivert et al. 1995). Furthermore, it was the

TFIIB-related repeats that were found to interact with TFIIIC, at the level of its τ131 subunit (Chaussivert et al. 1995). These interaction data fit well with recent functional studies with truncated TFIIIB70 polypeptides (Kassavetis et al. 1997). There is no reason to believe, however, that the TFIIB-related, amino-terminal half of TFIIIB70 is positioned differently than is TFIIB in TBP·TFIIB·DNA complexes (Nikolov et al. 1995).

Another remarkable property of TFIIIB is the extreme stability of TFIIIB·DNA complexes once formed (Kassavetis et al. 1990, 1997). It has been suggested that TFIIIB could hide DNA-binding domains to avoid random dispersion of this factor on irrelevant DNA sites (Geiduschek and Kassavetis 1992). As cryptic DNA-binding sites were found in transcription factors like σ^{70} (Dombroski et al. 1992) and RAP30 (Tan et al. 1994), we tested for the presence of a cryptic DNA-binding domain in TFIIIB70 by limited proteolysis. We found that the 87 carboxy-terminal residues form a protease-resistant domain that binds DNA tightly and is important for the formation, stability, and function of preinitiation complexes (Huet et al. 1997). This finding suggests a functional analogy between the carboxy-terminal half of TFIIIB70 and RAP30, the small subunit of TFIIIF. RAP30 was shown to bind both TFIIB and pol II (Killeen and Greenblatt 1992; Ha et al. 1993) and to have a carboxy-terminal cryptic DNA-binding domain (Tan et al. 1994). The carboxy-terminal half of TFIIIB70 shows no clear sequence similarity to RAP30, but it provides a similar link between the TFIIB-like domain, DNA, and RNA polymerase (see below).

On the TFIIIC side, τ131 is probably responsible for the complete assembly of TFIIIB on the DNA (see Fig. 1). This conclusion was first supported by protein-DNA cross-linking data that mapped τ131 within the TFIIIB-binding region (Bartholomew et al. 1991). τ131 contains 11 tetratricopeptide motives (TPRs) distributed in three blocks of 5, 4, and 2 TPRs, along its sequence. Two-hybrid experiments showed that TFIIIB70 and τ131 interact primarily through the 165 amino-terminal amino acids of τ131 that harbor the first TPR (Chaussivert et al. 1995). Similarly, τ131 was found to interact with B´´/TFIIIB90 (Rüth et al. 1996) and more unexpectedly with at least one pol III subunit (Fig. 1) (A. Flores et al., in prep.). Its function is therefore critical for preinitiation complex formation. Drastic conformational changes are likely to occur in τ131 as the efficiency of its photocross-linking to DNA greatly varies during the TFIIIB assembly process (Kassavetis et al. 1992). Our two-hybrid analysis also suggests the existence of reversible intramolecular interactions within τ131 (Chaussivert et al. 1995). Several lines of evidence suggest that the second TPR (TPR2) is involved in intramolecular interactions that shield TFIIIB-binding sites: A mutation in TPR2 favored TFIIIB recruitment (Rameau et al. 1994), and we found that a precise deletion of TPR2 greatly enhanced τ131-TFIIIB90 binding (Rüth et al. 1996). τ131 probably flips between two states that expose or mask TFIIIB-binding sites. Such conformational changes may be important to trigger the assembly of TFIIIB only when TFIIIC is properly bound to

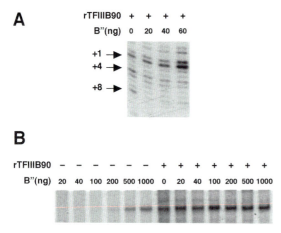

Figure 2. Start-site selection depends on a B″ component distinct from TFIIIB90. (*A*) Start-site selection is influenced by a B″ component. The *SUP4* tRNA gene was transcribed in the presence of affinity-purified TFIIIC, recombinant TFIIIB (rTBP, rTFIIIB90, and rTFIIIB70), highly purified pol III, and increasing amounts of partially purified B″. B″ fraction favored initiation at the +1 start site as seen by primer extension of the RNA transcripts. (*B*) Transcription of the *SUP4* tRNA gene as a function of B″ concentration. Note that in the absence of rTFI-IIB90, amounts of B″ sufficient to alter the start site usage do not support transcription.

DNA and to favor TFIIIC displacement when pol III transcribes through the A and B blocks. τ131 might therefore be an important regulatory target on TFIIIC, inasmuch as this polypeptide is phosphorylated in vivo (see Fig. 3).

Compared to the set of pol II general transcription factors (GTFs), composed of about 30 polypeptides, the set of 10 pol III GTFs thus far identified is modest in number (Table 1). There are hints, however, that additional components exist (Dieci et al. 1993) which influence transcription efficiency. The partially purified B″ fraction which brings TFI-IIB90 is more active on the TATA-less *SUP4* tRNA gene than recombinant B″/TFIIIB90 (Rüth et al. 1996). We observed that trace amounts of the B″ fraction that are unable per se to support transcription of the *SUP4* tRNA gene stimulated transcription driven by recombinant TFIIIB90 and influenced start site selection as seen by primer extension analysis (Fig. 2). This effect on start site selection is likely to be related to the flexible linkage between TFIIIC and TFIIIB that allows selection of alternative sites in vitro (Joazeiro et al. 1996). We suggest that the B″ fraction contains an auxiliary protein that stabilizes the preinitiation complex at the best position on TATA-less genes.

RNA POLYMERASE III, A MINIHOLOENZYME

During the course of these studies on pol III factors, we were pursuing in parallel the analysis of pol III subunits as the ultimate sensors of the molecular interactions that trigger transcriptional initiation. pol III is the more complex of the three forms of nuclear RNA polymerase. Seventeen polypeptides remain physically associated with the enzyme during purification (Fig. 3). Five polypep-

tides are shared with pol I and pol II, and two additional polypeptides are shared with pol I only. All of the pol III subunits, but one, have now been cloned, and as in the case of TFIIIC and TFIIIB components, all of them were found to be essential for yeast cell growth (see Table 1). The higher complexity of pol III as compared to pol II is due to the presence of a set of pol-III-specific polypeptides (i.e., unrelated to any pol I or pol II subunits). We first thought that some of them may be structurally related to pol II factors. This was not the case. Therefore, we considered the possibility that these specific subunits had been recruited by pol III during evolution to interact with the class III transcription factors. Indeed, antibodies directed to two of them, C34 and C53, were found to block the specific transcription of a tRNA gene without much affecting RNA synthesis in a nonspecific assay (Huet et al. 1985). C53 is one of the three pol III subunits that are phosphorylated in vivo, together with the common subunits AC40, ABC23, and AC19 (Fig. 3). Therefore, C53 could well be a specific target for regulation of class III genes at the polymerase level.

We concentrated on a triad of subunits C34, C31, and C82 that spontaneously dissociates from the enzyme under various conditions. In particular, a mutation in the zinc finger domain at the amino terminus of the largest subunit C160 caused the loss of this triad during enzyme

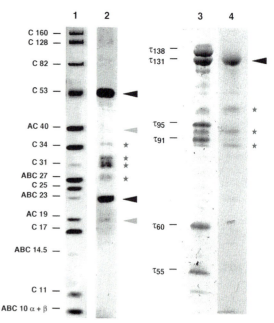

Figure 3. Subunits of pol III and TFIIIC phosphorylated in vivo. pol III and TFIIIC were purified from cells exponentially growing in the presence of ^{32}Pi or ^{33}Pi and analyzed by SDS-PAGE. (Lane *1*) pol III subunits were silver-stained and are identified on the left-hand side; (lane *2*) autoradiograph of the same gel. The black and gray arrows indicate, respectively, strong and weak radioactive signals coinciding with subunits C53, AC40, ABC23, and AC19. The stars point to radioactive bands disappearing after DNase treatment and resistant to proteinase K digestion. (Lane *3*) TFIIIC subunits stained with Coomassie blue; (lane *4*) autoradiograph of the same gel. τ131 is ^{33}P-labeled in vivo (*black arrowhead*). The stars indicate radioactive bands unrelated to TFIIIC subunits.

purification (Werner et al. 1992). First, we found in two-hybrid experiments that C34, C31, and C82 are able to interact with one another in vivo. More importantly, C34 was found to interact with TFIIIB70, thus providing the first link between TFIIIB and the pol III enzyme (Werner et al. 1993). This result was in keeping with DNA cross-linking experiments showing that among all RNA polymerase subunits cross-linked, C34 localized the furthest upstream on the promoter (Bartholomew et al. 1993). We also obtained genetic and biochemical evidence that C31 was essential for specific transcription at the initiation step (Thuillier et al. 1995), but we could not demonstrate an interaction of that subunit with TFIIIB. Recently, however, we found that C31 interacts with a newly characterized subunit, C17, that interacts with TFIIIB70 (M.L. Ferri et al., in prep.). These observations imply that all of these subunits must be located upstream, in the vicinity of TFIIIB, as indicated in the scheme (Fig. 1).

The prevailing model for transcriptional activation posits that activators work simply by recruitment of the transcriptional machinery. Knowing in a broad outline the protein-protein interactions leading to pol III recruitment, we set about testing this model in vivo. We attempted to reactivate a promoter-deficient U6 RNA gene harboring Gal4 sites by fusing the Gal4 DNA-binding domain separately to each of 15 different components of the pol III system. We found that expression of several fusion polypeptides belonging to TFIIIC or TFIIIB reactivated the gene. In contrast, none of the eight polymerase subunit fusions were able to trigger transcription (Marsolier et al. 1994). We concluded that recruitment (i.e., increasing the local concentration) of the polymerase was not sufficient. A mutational analysis of the C34 subunit that acts as a bridge between TFIIIB and the enzyme sheds new light on the recruitment issue. With a mutant form of pol III defective in C34·TFIIIB70 interaction, the transcriptional defect could be compensated in vitro by simply increasing the polymerase concentration, as predicted by the recruitment model. In contrast, with a different mutation also affecting C34·TFIIIB70 interaction, the in vitro transcription defect was barely suppressed by increasing the enzyme concentration: The mutant enzyme was defective in open complex formation (Brun et al. 1997). This unexpected finding indicated that a proper C34·TFIIIB70 interaction is important not only for enzyme recruitment, but also at a later stage, possibly to shift the polymerase into an initiation-competent configuration. Artificial recruitment of pol III via Gal4 sites was unable to meet this additional requirement. We noted previously that the high in vitro transcription efficiency of pol III is due to its rapid recycling on the same template, after termination (Dieci and Sentenac 1996). A model was proposed wherein the polymerase is directly transferred from the termination site to the promoter-TFIIIB complex. It is possible that some TFIIIB·pol III interactions may be specifically involved in the efficient recycling of pol III.

In the subunit interaction map (see Fig. 1), a new small subunit, C11, interacts with C17 and C31, based on two-hybrid experiments. This polypeptide, however, does not seem to be involved in enzyme recruitment by the preinitiation complex, since an enzyme devoid of C11 initiates properly. Our recent results indicate instead that C11 is directly or indirectly responsible for the RNA cleavage activity exhibited by pol III and for efficient termination (Chédin et al. 1998). An exonuclease activity has been found for various RNA polymerases engaged in ternary complexes. Retraction occurs through 3′→5′ hydrolytic cleavage of the nascent transcript. At physiological pH, the intrinsic nuclease activity of RNA polymerase is greatly stimulated by effectors such as TFIIS for pol II and GreA and B for *Escherichia coli* RNA polymerase (Uptain et al. 1997). No such cofactor was found for pol III (Whitehall et al. 1994). One could wonder why pol III has retained as an integral subunit a component that activates the nuclease activity. There is strong evidence that nucleolytic cleavage is essential to traverse arrest sites of various sorts. It may also serve a proofreading function to remove misincorporated nucleotides (Thomas et al. 1998). These functions are likely to be important for RNA polymerases that transcribe very long genes like pol II. A notable difference between pol II and pol III is the large difference in their transcript size. pol III transcribes very short genes and has to terminate efficiently, in contrast to pol II. We suggest that the intrinsic retracting activity of pol III facilitates the termination process by slowing down the enzyme at the terminator (see the model Fig. 4). In keeping with this contention, pol III was found to catalyze multiple cleavage-synthesis cycles at terminators (Bobkova and Hall 1997). The role of C11 subunit may thus be to switch the enzyme from the elongation mode to the retracting mode, rather than being directly responsible for the hydrolytic activity per se. C11 probably acts in concert with the two large subunits (Shaaban et al. 1996; Thuillier et al. 1996; E. Bobkova et al., in prep.).

The question arises as to whether the pol III transcription machinery exists as a holoenzyme. The structural complexity of pol III with its 17 subunits, its intrinsic (cofactor independent) nuclease activity, its ability to terminate transcription efficiently, again without additional cofactor, and the presence of enzyme-specific subunits for preinitiation complex recognition, all suggest that pol III exists as a stable miniholoenzyme for optimal transcription efficiency. The possibility that pol III like pol II and pol I is fully assembled, as a ready-made functional unit, with the general transcription factors TFIIIB and TFIIIC, has been raised recently by the identification of a human pol III holoenzyme (Wang et al. 1997). Our attempts to identify a functional form of pol III holoenzyme in yeast have met with partial success. The yeast pol III enzyme purified by immunoadsorption from crude extracts does contain detectable amounts of TBP, TFIIIB70, and TFIIIC (represented by τ55 subunit) as evidenced by Western blotting (see Fig. 5A,B). However, this potential holoenzyme form is but a very minor part of the total enzyme fraction. It contains functional amounts of TFIIIC but must be supplemented by TFIIIB components for transcription activity (Fig. 5C). One possibility is that the yeast pol III holoenzyme is unstable at high dilutions, as

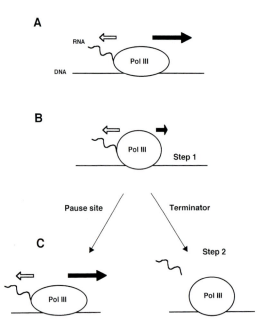

Figure 4. A model for pol III transcription termination. (*A*) Elongating pol III may exist under two different conformations: an elongation conformation (*closed arrow*) and a cleavage conformation (*open arrow*). (*A*) During processive elongation, the probability for the enzyme to be in the elongation mode is much higher than for being in the cleavage mode. As a result, a large majority of pol III elongates. (*B*) At specific sites of the template (transitory pause sites or terminators), the probability for the pol III to be in the elongation mode is strongly reduced. The enzyme oscillates between the cleavage and the elongation modes and performs multiple cleavage-synthesis cycles. The slowing down of the enzyme at the terminator is the first step of the pol III termination process. (*C*) pol III can resume elongation through pause sites by allowing a proportion of enzyme molecules to statistically read through the pause site at each cleavage-synthesis cycle. At the terminator, in a second step, probably related to the poly(U) 3´ end of the transcript in the ternary complex, the RNA is released and the pol III can reinitiate.

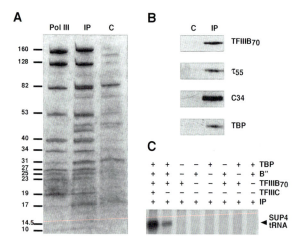

Figure 5 A subpopulation of pol III is associated with transcription factors. (*A*) Epitope-tagged pol III was immunopurified in one step from a concentrated crude extract (IP) and analyzed by SDS-PAGE alongside with purified pol III (Pol III) and a mock control using a cell extract with untagged pol III (C). Proteins were stained with Coomassie blue. (*B*) The same samples (IP and C) were analyzed by Western blotting for the presence of TFIIIB70, TFIIIC subunit τ55, pol III subunit C34, and TBP. All four components copurified with the polymerase. (*C*) The immunopurified pol III is transcriptionally active on Sup4 tRNA gene. It contains TFIIIC activity, but needs to be supplemented by TFIIIB components.

is probably the case of TFIIIB (Sethy-Coraci et al. 1998), and is mainly dissociated into subcomplexes in crude extracts. Alternatively, the holoenzyme could be a minor form of pol III in the nucleus. In fact, what is known of the mode of interaction of TFIIIB with the DNA appears to preclude a one-step recruitment of a TFIIIB-containing holoenzyme. Once formed, TFIIIB·DNA complexes are unusually stable, probably because TFIIIB forms a multicomponent ring around the DNA, and can direct multiple rounds of transcription (Kassavetis et al. 1990). The stepwise assembly of TFIIIB on the DNA, described in vitro (Kassavetis et al. 1992), is therefore likely to reflect the in vivo situation. There might exist an activity that catalyzes the disassembly of TFIIIB·DNA complexes, but it has not yet been found. The fact remains that the form of pol III immunopurified from crude extracts contains several other polypeptides that do not correspond to known components of the pol III transcription system (see Fig. 5A). The identification of these polypeptides might reveal yet unknown links between pol III transcription and other aspects of class III RNA metabolism.

CONCLUDING REMARKS

The pol III transcription apparatus appears to be relatively simple as compared to the versatile pol II system. Nevertheless, a collection of 26 polypeptides participates in transcription complex formation on tRNA genes, amounting to a total mass of approximately 1500 kD. Most remarkably, all of these polypeptides were individually found to be essential for cell viability (Table 1). There must be only a minimal level of functional redundancy among them. Although additional, dispensable components may exist, the main actors have been identified. The mapping of their interconnections is under way (Fig. 1) and should be completed. Already, protein-protein interaction studies have shed light on the activation pathway that flows from TFIIIC to TFIIIB toward some specific pol III subunits. This flow of information should not be considered as the mere stepwise recruitment of individual components culminating with enzyme docking on the DNA. There is strong evidence in favor of major conformational changes occurring at each step, at the level of TFIIIC (in τ131), of TFIIIB (in TFIIIB70), and of the RNA polymerase (via C34·TFIIIB70 interaction). A forthcoming challenge will be to provide a dynamic model of gene activation, answering such questions as: What happens to τ131 and to TFIIIB70 during TFIIIB assembly on the DNA? What happens to the enzyme when C34 and C17 are contacted by TFIIIB? What is the role for cryptic protein-protein or protein-DNA interaction domains during transcription complex assembly and during promoter clearance?

When comparing the yeast and the human pol III systems, it is remarkable that the most conserved components—other than those devoted to the basic transcription process—are those engaged in protein-protein interactions critical for transcription complex assembly. τ131, TFIIIB (TBP, TFIIIB70/BRF1, and TFIIIB90/B''), the triad of pol-III-specific subunits (C34, C31, C82), all have structural and functional homologs in human cells (Wang and Roeder 1995, 1997; Teichmann et al. 1997; Arrebola et al. 1998). This is apparently not the case for the components that anchor TFIIIC on the DNA (L'Etoile et al. 1994; Lagna et al. 1994; Arrebola et al. 1998). If alternative solutions could be found when a mere physical interaction was needed, more complex interactions involving concomitant structural changes were those more likely to be conserved during evolution.

Looking at TFIIIC and TFIIIB function, one could draw direct, functional equivalences in the operational modes of gene activation used by the pol III and pol II systems. TFIIIC acts as an enhanceosome-activator multiprotein complex; TFIIIB corresponds to a combination of TFIID, TFIIB, and possibly RAP30. What plays the part of pol-III-specific subunits in the pol II system among the mediator components or the general class II transcription factors? Some pol III subunits mediate the productive recruitment of the enzyme, but the essential role of the others (they are all essential) is still unknown and deserves to be investigated.

ACKNOWLEDGMENTS

We especially acknowledge the many contributions to this work from several members or previous members of the laboratory: M. Riva, P. Thuriaux, C. Conesa, J. Huet, A. Ruet, C. Marck, R. Arrebola, E. Deprez, H. Dumay, B. Buffin-Meyer, A. Flores, I. Brun, G. Dieci, N. Chaussivert, N. Manaud, J. Rüth, E. Favry, and C. Boschiero. Many thanks to Ben Hall for improving this manuscript and for communicating unpublished results. This work was supported by contract BIO4-CT95-0009 from the European Union and by a MENESRIP contract ACC-SV1.

REFERENCES

Allison L.A., Moyle M., Shales M., and Ingles C.J. 1985. Extensive homology among the largest subunits of eukaryotic and prokaryotic RNA polymerases. *Cell* **42:** 599.

Archambault J., Schappert K.T., and Friesen J.D. 1990. A suppressor of an RNA polymerase II mutation of *Saccharomyces cerevisiae* encodes a subunit common to RNA polymerases I, II, and III. *Mol. Cell. Biol.* **10:** 6123.

Archambault J., Milne C.A., Schappert K.T., Baum B., Friesen J.D., and Segall J. 1992. The deduced sequence of the transcription factor TFIIIA from *Saccharomyces cerevisiae* reveals extensive divergence from *Xenopus* TFIIIA. *J. Biol. Chem.* **267:** 3282.

Arrebola R., Manaud N., Rozenfeld S., Marsolier M.C., Lefebvre O., Carles C., Thuriaux P., Conesa C., and Sentenac A. 1998. τ91, an essential subunit of yeast TFIIIC, cooperates with τ138 in DNA binding. *Mol. Cell. Biol.* **18:** 1.

Baker R.E., Camier S., Sentenac A., and Hall B.D. 1987. Gene size differentially affects the binding of yeast transcription factor τ to two intragenic regions. *Proc. Natl. Acad. Sci.* **84:** 8768.

Bartholomew B., Kassavetis G.A., and Geiduschek E.P. 1991. Two components of *Saccharomyces cerevisiae* transcription factor IIIB (TFIIIB) are stereospecifically located upstream of a tRNA gene and interact with the second-largest subunit of TFIIIC. *Mol. Cell. Biol.* **11:** 5181.

Bartholomew B., Durkovich D., Kassavetis G.A., and Geiduschek E.P. 1993. Orientation and topography of RNA polymerase III in transcription complexes. *Mol. Cell. Biol.* **13:** 942.

Bartholomew B., Kassavetis G.A., Braun B.R., and Geiduschek E.P. 1990. The subunit structure of *Saccharomyces cerevisiae* transcription factor IIIC probed with a novel photocrosslinking reagent. *EMBO J.* **9:** 2197.

Bobkova E. and Hall B. 1997. Substrate specificty of the RNase activity of yeast RNA polymerase III. *J. Biol. Chem.* **272:** 22832.

Bogenhagen D.F., Sakonju S., and Brown D.D. 1980. A control region in the center of the 5S RNA gene directs specific initiation of transcription. II. The 3' border of the region. *Cell* **19:** 27.

Braun B.R., Bartholomew B., Kassavetis G.A., and Geiduschek E.P. 1992. Topography of transcription factor complexes on the *Saccharomyces cerevisiae* 5S RNA gene. *J. Mol. Biol.* **228:** 1063.

Brun I., Sentenac A., and Werner M. 1997. Dual role of the C34 subunit of RNA polymerase III in transcription initiation. *EMBO J.* **16:** 5730.

Buratowski S. and Zhou H. 1992. A suppressor of TBP mutations encodes an RNA polymerase III transcription factor with homology to TFIIB. *Cell* **71:** 221.

Burnol A.-F., Margottin F., Huet J., Almouzni G., Prioleau M.-N., Méchali M., and Sentenac A. 1993. TFIIIC relieves repression of U6 snRNA transcription by chromatin. *Nature* **362:** 475.

Camier S., Gabrielsen O.S., Baker R.E., and Sentenac A. 1985. A split binding site for transcription factor τ on the tRNA$_3^{Glu}$ gene. *EMBO J.* **4:** 491.

Cavallini B., Faus I., Matthes H., Chipoulet J.M., Winsor B., Egly J.M., and Chambon P. 1989. Cloning of the gene encoding the yeast protein BTF1Y, which can substitute for the human TATA box-binding factor. *Proc. Natl. Acad. Sci.* **86:** 9803.

Chaussivert N., Conesa C., Shaaban S., and Sentenac A. 1995. Complex interactions between yeast TFIIIB and TFIIIC. *J. Biol. Chem.* **270:** 15353.

Chédin S., Riva M., Schultz P., Sentenac A., and Carles C. 1998. The RNA cleavage activity of RNA polymerase III is mediated by an essential TFIIS-like subunit and is important for transcription terminaion. *Genes Dev.* (in press).

Chiannilkulchai N., Stalder R., Riva M., Carles C., Werner M., and Sentenac A. 1992. RPC82 encodes the highly conserved, third-largest subunit of RNA polymerase C (III) from *Saccharomyces cerevisiae*. *Mol. Cell. Biol.* **12:** 4433.

Colbert T. and Hahn S. 1992. A yeast TFIIB-related factor involved in RNA polymerase III transcription. *Genes Dev.* **6:** 1940.

Dequard-Chablat M., Riva M., Carles C., and Sentenac A. 1991. RPC19, the gene for a subunit common to yeast RNA polymerases A (I) and C (III). *J. Biol. Chem.* **266:** 15300.

Dieci G. and Sentenac A. 1996. Facilitated recycling pathway for RNA polymerase III. *Cell* **84:** 245.

Dieci G., Duimio L., Coda-Zabetta F., Sprague K.U., and Ottonello S. 1993. A novel RNA polymerase III transcription factor fraction that is not required for template commitment. *J. Biol. Chem.* **268:** 11199.

Dombroski A.J., Walter W.A., Record M.T.J., Siegele D.A., and Gross C.A. 1992. Polypeptides containing highly conserved regions of transcription initiation factor σ70 exhibit specificity of binding to promoter DNA. *Cell* **70:** 501.

Eisenmann D.M., Dollard C., and Winston F. 1989. SPT15, the gene encoding the yeast TATA binding factor TFIID, is required for normal transcription initiation in vivo. *Cell* **58:** 1183.

Engelke D.R., Ng S.Y., Shastry B.S., and Roeder R.G. 1980. Specific interaction of a purified transcription factor with an internal control region of 5S RNA genes. *Cell* **19:** 717.

Fabrizio P., Coppo A., Fruscoloni P., Benedetti P., Di Segni G., and Tocchini-Valentini G. 1987. Comparative mutational analysis of wild-type and streched tRNA3Leu gene promoters. *Proc. Natl. Acad. Sci.* **84:** 8763.

Gabrielsen O.S., Marzouki N., Ruet A., Sentenac A., and Fromageot P. 1989. Two polypeptide chains in yeast transcription factor τ interact with DNA. *J. Biol. Chem.* **264:** 7505.

Galli G., Hofstetter H., and Birnstiel M.L. 1981. Two conserved sequence blocks within eukaryotic tRNA genes are major promoter elements. *Nature* **294:** 626.

Geiduschek E.P. and Kassavetis G.A. 1992. RNA polymerase III transcription complexes. In *Transcriptional regulation* (ed. S.L. McKnight and K.R. Yamamoto), p. 247. Cold Spring Harbor Laboratory Press, Cold Spring Harbor, New York.

Gerlach V., Whitehall S.K., Geiduschek E.P., and Brow D.A. 1995. TFIIIB placement on a yeast U6 RNA gene in vivo is directed primarily by TFIIIC rather than by sequence-specific DNA contacts. *Mol. Cell. Biol.* **15:** 1455.

Ginsberg A.M., King B.O., and Roeder R.G. 1984. Xenopus 5S gene transcription factor, TFIIIA: Characterization of a cDNA clone and measurement of RNA levels throughout development. *Cell* **39:** 479.

Gudenus R., Mariotte S., Moenne A., Ruet A., Memet S., Buhler J.M., Sentenac A., and Thuriaux P. 1988. Conditional mutants of RPC160, the gene encoding the largest subunit of RNA polymerase C in *Saccharomyces cerevisiae*. *Genetics* **119:** 517.

Ha I., Roberts S., Maldonado E., Sun X., Kim L.U., Green M.R., and Reinberg D. 1993. Multiple functional domains of human transcription factor IIB: Distinct interactions with two general transcription factors and RNA polymerase II. *Genes Dev.* **7:** 1021.

Hahn S., Buratowski S., Sharp P.A., and Guarente L. 1989. Yeast TATA-binding protein TFIID binds to TATA elements with both consensus and nonconsensus DNA sequences. *Proc. Natl. Acad. Sci.* **86:** 5718.

Horikoshi M., Wang C.K., Fujii H., Cromlish J.A., Weil P.A., and Roeder R.G. 1989. Cloning and structure of a yeast gene encoding a general transcription factor TFIID that binds to the TATA box. *Nature* **341:** 299.

Huet J., Conesa C., Carles C., and Sentenac A. 1997. A cryptic DNA binding domain at the COOH terminus of TFIIIB70 affects formation, stability, and function of preinitiation complexes. *J. Biol. Chem.* **272:** 18341.

Huet J., Riva M., Sentenac A., and Fromageot P. 1985. Yeast RNA polymerase C and its subunits. Specific antibodies as structural and functional probes. *J. Biol. Chem.* **260:** 15304.

Huet J., Conesa C., Manaud N., Chaussivert N., and Sentenac A. 1994. Interactions between yeast TFIIIB components. *Nucleic Acids Res.* **22:** 3433.

James P., Whelen S., and Hall B.D. 1991. The RET1 gene of yeast encodes the second-largest subunit of RNA polymerase III. Structural analysis of the wild-type and ret1-1 mutant alleles. *J. Biol. Chem.* **266:** 5616.

Joazeiro C.A.P., Kassavetis G.A., and Geiduschek E.P. 1996. Alternative outcomes in assembly of promoter complexes: The roles of TBP and a flexible linker in placing TFIIIB on tRNA genes. *Genes Dev.* **10:** 725.

Kassavetis G.A., Braun B.R., Nguyen L.H., and Geiduschek E.P. 1990. *S. cerevisiae* TFIIIB is the transcription initiation factor proper of RNA polymerase III, while TFIIIA and TFIIIC are assembly factors. *Cell* **60:** 235.

Kassavetis G.A., Bardeleben C., Kumar A., Ramirez E., and Geiduschek E.P. 1997. Domains of the Brf component of RNA polymerase III transcription factor IIIB (TFIIIB): Functions in assembly of TFIIIB-DNA complexes and recruitment of RNA polymerase to the promoter. *Mol. Cell. Biol.* **17:** 5299.

Kassavetis G.A., Bartholomew B., Blanco J.A., Johnson T.E., and Geiduschek E.P. 1991. Two essential components of the *Saccharomyces cerevisiae* transcription factor TFIIIB: Transcription and DNA-binding properties. *Proc. Natl. Acad. Sci.* **88:** 7308.

Kassavetis G.A., Riggs D.L., Negri R., Nguyen L.H., and Geiduschek E.P. 1989. Transcription factor IIIB generates extended DNA interactions in RNA polymerase III transcription complexes on tRNA genes. *Mol. Cell. Biol.* **9:** 2551.

Kassavetis G.A., Nguyen S.T., Kobayashi R., Kumar A., Geiduschek E.P., and Pisano M. 1995. Cloning, expression, and function of TFC5, the gene encoding the B" component of the *Saccharomyces cerevisiae* RNA polymerase III transcription factor TFIIIB. *Proc. Natl. Acad. Sci.* **92:** 9786.

Kassavetis G.A., Joazeiro C.A.P., Pisano M., Geiduschek E.P., Colbert T., Hahn S., and Blanco J.A. 1992. The role of the TATA-binding protein in the assembly and function of the multisubunit yeast RNA polymerase III transcription factor, TFIIIB. *Cell* **71:** 1055.

Khoo B., Brophy B., and Jackson S.P. 1994. Conserved functional domains of the RNA polymerase III general transcription factor BRF. *Genes Dev.* **8:** 2879.

Killeen M.T. and Greenblatt J.F. 1992. The general transcription factor RAP30 binds to RNA polymerase II and prevents it from binding nonspecifically to DNA. *Mol. Cell. Biol.* **12:** 30.

Koleske A.J. and Young R.A. 1995. The RNA polymerase II holoenzyme and its implications for gene regulation. *Trends Biochem. Sci.* **20:** 113.

Koski R.A., Clarkson S.G., Kurjan J., Hall B.D., and Smith M. 1980. Mutations of the yeast SUP4 tRNA^{Tyr} locus: Transcription of the mutant genes in vitro. *Cell* **22:** 415.

Kurjan J., Hall B.D., Gillam S., and Smith M. 1980. Mutations at the yeast SUP4 tRNATyr locus: DNA sequence changes in mutants lacking suppressor activity. *Cell* **20:** 701.

Lagna G., Kovelman R., Sukegawa J., and Roeder R.G. 1994. Cloning and characterization of an evolutionarily divergent DNA-binding subunit of mammalian TFIIIC. *Mol. Cell. Biol.* **14:** 3053.

Lassar A.B., Martin P.L., and Roeder R.G. 1983. Transcription of class III genes: Formation of preinitiation complexes. *Science* **222:** 740.

Lefebvre O., Carles C., Conesa C., Swanson R.N., Bouet F., Riva M., and Sentenac A. 1992. TFC3: Gene encoding the B-block binding subunit of the yeast transcription factor IIIC. *Proc. Natl. Acad. Sci.* **89:** 10512.

L'Etoile N.D., Fahnestock M.L., Shen Y., Aebersold R., and Berk A.J. 1994. Human transcription factor IIIC box B binding subunit. *Proc. Natl. Acad. Sci.* **91:** 1652.

Lobo S.M., Lister J., Sullivan M.L., and Hernandez N. 1991. The cloned RNA polymerase II transcription factor IID selects RNA polymerase III to transcribe the human U6 gene in vitro. *Genes Dev.* **5:** 1477.

López-De-León A., Librizzi M., Puglia K., and Willis I.M. 1992. PCF4 encodes an RNA polymerase III transcription factor with homology to TFIIB. *Cell* **71:** 211.

Manaud N., Arrebola R., Buffin-Meyer B., Lefebvre O., Voss H., Riva M., Conesa C., and Sentenac A. 1998. A chimeric subunit of yeast transcription factor IIIC forms a subcomplex with tau95. *Mol. Cell. Biol.* **18:** 3191.

Mann C., Buhler J.M., Treich I., and Sentenac A. 1987. RPC40, a unique gene for a subunit shared between yeast RNA polymerases A and C. *Cell* **48:** 627.

Mann C., Micouin J.Y., Chiannilkulchai N., Treich I., Buhler J.M., and Sentenac A. 1992. RPC53 encodes a subunit of *Saccharomyces cerevisiae* RNA polymerase C (III) whose inactivation leads to a predominantly G1 arrest. *Mol. Cell. Biol.* **12:** 4314.

Marck C., Lefebvre O., Carles C., Riva M., Chaussivert N., Ruet A., and Sentenac A. 1993. The TFIIIB-assembling subunit of yeast transcription factor TFIIIC has both tetratricopeptide repeats and basic helix-loop-helix motifs. *Proc. Natl. Acad. Sci.* **90:** 4027.

Margottin F., Dujardin G., Gérard M., Egly J.M., Huet J., and Sentenac A. 1991. Participation of the TATA factor in transcription of the yeast U6 gene by RNA polymerase C. *Science*

251: 424.

Marsolier M.-C., Chaussivert N., Lefebvre O., Conesa C., Werner M., and Sentenac A. 1994. Directing transcription of an RNA polymerase III gene via GAL4 sites. *Proc. Natl. Acad. Sci.* **91:** 11938.

Marsolier M.-C., Tanaka S., Livingstone-Zatchej M., Grunstein M., Thoma F., and Sentenac A. 1995. Reciprocal interferences between nucleosomal organization and transcriptional activity of the yeast *SNR6* gene. *Genes Dev.* **9:** 410.

Marzouki N., Camier S., Ruet A., Moenne A., and Sentenac A. 1986. Selective proteolysis defines two DNA binding domains in yeast transcription factor τ. *Nature* **323:** 176.

Miller J., McLachlan A.D., and Klug A. 1985. Repetitive zinc-binding domains in the protein transcription factor IIIA from *Xenopus* oocytes. *EMBO J.* **4:** 1609.

Moenne A., Camier S., Anderson G., Margottin F., Beggs J., and Sentenac A. 1990. The U6 gene of *Saccharomyces cerevisiae* is transcribed by RNA polymerase C (III) *in vivo* and *in vitro*. *EMBO J.* **9:** 271.

Mosrin C., Riva M., Beltrame M., Cassar E., Sentenac A., and Thuriaux P. 1990. The RPC31 gene of *Saccharomyces cerevisiae* encodes a subunit of RNA polymerase C (III) with an acidic tail. *Mol. Cell. Biol.* **10:** 4737.

Nikolov D.B., Chen H., Halay E.D., Usheva A.A., Hisatake K., Lee D.K., Roeder R.G., and Burley S.K. 1995. Crystal structure of a TFIIB-TBP-TATA-element ternary complex. *Nature* **377:** 119.

Orphanides G., Lagrange T., and Reinberg D. 1996. The general transcription factors of RNA polymerase II. *Genes Dev.* **10:** 2657.

Parker C.S. and Roeder R.G. 1977. Selective and accurate transcription of the *Xenopus laevis* 5S RNA genes in isolated chromatin by purified RNA polymerase III. *Proc. Natl. Acad. Sci.* **74:** 44.

Parsons M.C. and Weil P.A. 1992. Cloning of TFC1, the *Saccharomyces cerevisiae* gene encoding the 95-kDa subunit of transcription factor TFIIIC. *J. Biol. Chem.* **267:** 2894.

Rameau R., Puglia K., Crowe A., Sethy I., and Willis I.M. 1994. A mutation in the second largest subunit of TFIIIC increases a rate-limiting step in transcription by RNA polymerase III. *Mol. Cell. Biol.* **14:** 822.

Riva M., Memet S., Micouin J.Y., Huet J., Treich I., Dassa J., Young R., Buhler J.M., Sentenac A., and Fromageot P. 1986. Isolation of structural genes for yeast RNA polymerases by immunological screening. *Proc. Natl. Acad. Sci.* **83:** 1554.

Roberts S., Miller S.J., Lane W.S., Lee S., and Hahn S. 1996. Cloning and functional characterization of the gene encoding TFIIIB90 subunit of RNA polymerase III transcription factor TFIIIB. *J. Biol. Chem.* **271:** 14903.

Rüth J., Conesa C., Dieci G., Lefebvre O., Düsterhöft A., Ottonello S., and Sentenac A. 1996. A suppressor of mutations in the class III transcription system encodes a component of yeast TFIIIB. *EMBO J.* **15:** 1941.

Sadhale P.P. and Woychik N.A. 1994. C25, an essential RNA polymerase III subunit related to the RNA polymerase II subunit RPB7. *Mol. Cell. Biol.* **14:** 6164.

Schmidt M.C., Kao C.C., Pei R., and Berk A.J. 1989. Yeast TATA-box transcription factor gene. *Proc. Natl. Acad. Sci.* **86:** 7785.

Schultz P., Marzouki N., Marck C., Ruet A., Oudet P., and Sentenac A. 1989. The two DNA-binding domains of yeast transcription factor τ as observed by scanning transmission electron microscopy. *EMBO J.* **8:** 3815.

Sentenac A. 1985. Eukaryotic RNA polymerases. *CRC Crit. Rev. Biochem.* **18:** 31.

Sethy-Coraci I., Moir R.D., López-De-León A., and Willis I.M. 1998. A differential response of wild type and mutant promoters to TFIIIB70 overexpression in vivo and in vitro. *Nucleic Acids Res.* **26:** 2344.

Shaaban S., Bobkova E., Chudzig D., and Hall B. 1996. In vitro analysis of elongation and termination by mutant RNA polymerases with altered termination behavior. *Mol. Cell. Biol.*

16: 6468.

Simmen K.A., Bernués J., Parry H.D., Stunnenberg H.G., Berkenstam A., Cavallini B., Egly J.-M., and Mattaj I.W. 1991. TFIID is required for *in vitro* transcription of the human U6 gene by RNA polymerase III. *EMBO J.* **10:** 1853.

Stettler S., Mariotte S., Riva M., Sentenac A., and Thuriaux P. 1992. An essential and specific subunit of RNA polymerase III (C) is encoded by gene RPC34 in *Saccharomyces cerevisiae*. *J. Biol. Chem.* **267:** 21390.

Svetlov V.V. and Cooper T.G. 1995. Review: Compilation and characteristics of dedicated transcription factors in *Saccharomyces cerevisiae*. *Yeast* **11:** 1439.

Swanson R.N., Conesa C., Lefebvre O., Carles C., Ruet A., Quemeneur E., Gagnon J., and Sentenac A. 1991. Isolation of TFC1, a gene encoding one of two DNA-binding subunits of yeast transcription factor tau (TFIIIC). *Proc. Natl. Acad. Sci.* **88:** 4887.

Tan S., Garrett K.P., Conoway R.C., and Conoway J.W. 1994. Cryptic DNA-binding domain in the C terminus of RNA polymerase II general transcription factor RAP30. *Proc. Natl. Acad. Sci.* **91:** 9808.

Teichmann M., Dieci G., Huet J., Rüth J., Sentenac A., and Seifart K. 1997. Functional interchangeability of TFIIIB components from yeast and human cells *in vitro*. *EMBO J.* **16:** 4708.

Thomas M.J., Platas A.A., and Hawley D.K. 1998. Transcriptional fidelity and proofreading by RNA polymerase II. *Cell* **93:** 627.

Thuillier V., Brun I., Sentenac A., and Werner M. 1996. Mutations in the alpha-amanitin conserved domain of the largest subunit of yeast RNA polymerase III affect pausing, RNA cleavage and transcriptional transitions. *EMBO J.* **15:** 618.

Thuillier V., Stettler S., Sentenac A., Thuriaux P., and Werner M. 1995. A mutation in the C31 subunit of *Saccharomyces cerevisiae* RNA polymerase III affects transcription initiation. *EMBO J.* **14:** 351.

Treich I., Carles C., Riva M., and Sentenac A. 1992. RPC10 encodes a new mini subunit shared by yeast nuclear RNA polymerases. *Gene Expr.* **2:** 31.

Uptain S., Kane C., and Chamberlin M. 1997. Basic mechanisms of transcript elongation and its regulation. *Annu. Rev. Biochem.* **66:** 117.

Wang Z. and Roeder R.G. 1995. Structure and function of a human transcription factor TFIIIB subunit that is evolutionarily conserved and contains both TFIIB- and high-mobility-group protein 2-related domains. *Proc. Natl Acad. Sci.* **92:** 7026.

———. 1997. Three human RNA polymerase III-specific subunits form a subcomplex with a selective function in specific transcription initiatiton. *Genes Dev.* **11:** 1315.

Wang Z., Luo T., and Roeder R.G. 1997. Identification of an autonomously initiating RNA polymerase III holoenzyme containing a novel factor that is selectively inactivated during protein synthesis inhibition. *Genes Dev.* **11:** 2371.

Werner M., Chaussivert N., Willis I.M., and Sentenac A. 1993. Interaction between a complex of RNA polymerase III subunits and the 70-kDa component of transcription factor IIIB. *J. Biol. Chem.* **268:** 20721.

Werner M., Hermann-Le Denmat S., Treich I., Sentenac A., and Thuriaux P. 1992. Effect of mutations in a zinc binding domain of yeast RNA polymerase C (III) on enzyme function and subunit association. *Mol. Cell. Biol.* **12:** 1087.

Whitehall S.K., Bardeleben C., and Kassavetis G.A. 1994. Hydrolytic cleavage of nascent RNA in RNA polymerase III ternary transcription complexes. *J. Biol. Chem.* **269:** 2299.

Woychik N.A., Liao S.M., Kolodziej P.A., and Young R.A. 1990. Subunits shared by eukaryotic nuclear RNA polymerases. *Genes Dev.* **4:** 313.

Woychik N.A., McCune K., Lane W.S., and Young R.A. 1993. Yeast RNA polymerase II subunit RPB11 is related to a subunit shared by RNA polymerase I and III. *Gene Expr.* **3:** 77.

Wu G.J. 1978. Adenovirus DNA-directed transcription of 5.5S RNA in vitro. *Proc. Natl. Acad. Sci.* **75:** 2175.

The Regulation of Gene Activity by Histones and the Histone Deacetylase RPD3

N. Suka, A.A. Carmen, S.E. Rundlett, and M. Grunstein
Department of Biological Chemistry, UCLA School of Medicine and the Molecular Biology Institute,
University of California, Los Angeles, California 90095

In yeast, upstream activator sequences (UASs) are usually in nuclease-hypersensitive regions believed to be devoid of nucleosomes even prior to the binding of the activator proteins. This may result from the constitutive binding of factors that exclude core particles. This exclusion positions adjacent nucleosomes that repress nearby TATA elements and basal transcription. The mechanism by which repression occurs and is alleviated has recently received considerable attention and involves, in certain cases, the deacetylation and acetylation of the histone amino termini. This undoubtedly results in a structural change in chromatin that enables gene regulation. However, genes are differentially sensitive to regulation by histones. It is probable that certain genes utilize proteins that function independently of histones to mediate repression and activation, whereas others use histone modification as the dominant form of regulation.

UAS ELEMENTS ARE GENERALLY NOT COMPLEXED WITH NUCLEOSOMES, UNLIKE TATA ELEMENTS

In more complex eukaryotes, most chromatin is found in highly folded 30-nm fibers complexed with histone H1. However, this may not be the case for a promoter in a cell line in which it is induced. For example, the terminally repressed β-globin promoter in a mouse nonerythroid cell is fully complexed with nucleosomes. In contrast, the inducible promoter in the erythroid cell contains a 700-bp region at the promoter that is sensitive to the chemical nuclease MPE-Fe(II) (because its nucleosomes there are disrupted) (Benezra et al. 1986). This is true even before the gene is activated. A parallel to this latter state exists in yeast. Yeast chromatin is largely unfolded (10-nm fiber), and no protein in yeast has been shown to have histone H1 function. This may help explain why most yeast genes are inducible in every cell cycle. Moreover, most yeast UAS elements at which activators bind are uncomplexed with nucleosomes even prior to gene activation. For example, the UAS element of the *GAL1* gene is sensitive to MPE-Fe(II) even before *GAL1* is activated by *GAL4* (Lohr 1984; Fedor et al. 1988). A similar situation exists at other yeast genes examined which include *PHO5* (Almer and Hörz 1986); *SUC2* (Matallana et al. 1992) and *URA3* (Tanaka et al. 1996). Thus, it is likely that proteins other than the activator may keep the UAS element free for binding to the activator. At *GAL1*, one such protein is GRF2 whose binding at the *GAL1*

UAS may exclude a nucleosomal core particle and thereby position adjacent nucleosomes (Chasman et al. 1990).

Although the UAS element is often nucleosome-free, this is not the case for the TATA element. Since nucleosomes are packed so closely in yeast (~10–15 bp apart), TATA elements are either in nucleosomal structures or at the nucleosome edge at a number of genes examined (*GAL1, GAL10, PHO5, URA3, SUC2*). These nucleosomes are likely to block access to basal transcription factors.

NUCLEOSOMES REPRESS TRANSCRIPTION INITIATION

The above observations argue that nucleosomes are general repressors of gene activity not because they prevent the binding of activators to UAS elements, but because they block the basal transcription machinery from the downstream promoter or TATA element. In fact, a nucleosome at a transcription initiation site prevents transcription initiation in vitro (Lorch et al. 1987). As described below, this is also likely to occur in vivo.

Nucleosome removal from a particular site can be obtained in vivo by placing a histone H2B gene (Han et al. 1987) or H4 gene (Han et al. 1988) under control of the *GAL1* promoter in a yeast strain whose natural chromosomal copies of that gene are disrupted. In the presence of galactose, the promoter is active and such a strain grows relatively normally. However, in glucose-containing media, the promoter is inactive, and histone synthesis is blocked. Most of the cells continue to replicate once, thus depleting histone and nucleosome synthesis by approximately 50%. It is likely that the remaining nucleosomes are free to move and are no longer positioned at most sites since the depletion of nucleosomes results in DNA that resembles naked DNA in its sensitivity to nucleases. Thus, 50% nucleosome depletion resembles much greater nucleosome loss when measured by nuclease sensitivity.

The nucleosome-depleted cells arrest very synchronously as singly budded cells with a 2N or greater quantity of DNA that fails to segregate to the bud. The reason for the inability to segregate chromosomes is unclear but could reflect the effect of nucleosome depletion on centromeres in yeast. For example, H4 or H2B depletion greatly alters the sensitivity of centromeric DNA to nuclease (Saunders et al. 1990). Depletion of centromeric nucleosomes may very well block chromosomal segregation.

Fusion of regulatory DNA sequences to reporter constructs under conditions of nucleosome depletion demonstrates that nucleosome loss specifically activates TATA elements but not UAS elements (Han et al. 1988; Han and Grunstein 1988). Therefore, we may conclude that nucleosomes repress transcription initiation elements but not the activator-binding sites. This suggests that activators may normally bind to the nucleosome-free UAS elements and then stimulate the disruption of downstream chromatin, enabling the activation of the TATA elements.

DISRUPTING CHROMATIN MAY REPRESENT THE DOMINANT FORM OF ACTIVATION FOR SOME GENES

Nucleosome depletion activates most downstream promoter elements to a similar absolute level (when measured by reporter gene activity). Yet these levels can vary greatly as a fraction of total activated transcription. For example, different promoter elements (CUP1, HIS3, PHO5, CYC1, and GAL1) have been fused to the lacZ reporter gene. Their absolute levels of activity upon nucleosome depletion in each case are within approximately a factor of two. Yet the level of β-galactosidase generated by histone depletion as a percentage of that produced after full induction can differ greatly for different genes (Han and Grunstein 1988; Durrin et al. 1992). These results indicate that for some genes (e.g., HIS3 and CUP1), alleviating nucleosomal repression can approach high levels of activation. For others (PHO5, CYC1, and GAL1), additional nonnucleosomal mechanisms are more important to obtain their high activation levels.

HISTONE H4 AND H3 SITES OF ACETYLATION REGULATE GENE ACTIVITY

Posttranslational acetylation of lysine residues occurs at the histone amino termini and was first found to be correlated with increased gene activity by Allfrey (1977). That the histone amino termini regulate genes in a causal manner is evident from lesions constructed at the H4 amino terminus. Deletions and mutations affecting the H4 sites of acetylation decrease GAL1 activation from 10- to 20-fold (Durrin et al. 1991).

Insight into the reason for this decrease is provided by nucleosome-mapping experiments at the GAL1 promoter. We believe that a change in nucleosome positioning at or near the GAL1 TATA element may help explain the decreased activation of GAL1. The TATA element is found between two closely packed nucleosomes downstream from the GAL1 UAS. Just 15 bp upstream of this site is a Sau3AI site (GATC) which is accessible to ectopically produced Escherichia coli dam methylase even before activation of GAL1. Interestingly, this site becomes strongly inaccessible to dam methylase when the H4 amino terminus is deleted or its sites of acetylation are mutagenized to glycines. The absence of the charged lysine residues appears to cause a nucleosome to slide onto sites recognized by the basal transcription machinery.

This has been confirmed by high-resolution nucleosome mapping using ligation-mediated polymerase chain reaction (PCR) (Fisher-Adams and Grunstein 1995). These data support earlier work showing that the H4 amino terminus is responsible for nucleosome positioning at the α2 operator (Roth et al. 1992). They also support the role of the H4 amino terminus in affecting the downstream promoter but not the upstream promoter at GAL1 (Wan et al. 1995). These results suggest that natural nucleosome placement is important to allow its disruption during initiation.

The role of the H3 amino terminus is less clear. Deletion or mutagenesis of its sites of acetylation does not alter nucleosomal positioning at the TATA. In contrast to H4, these lesions in H3 increase GAL1 gene activation three- to fourfold (Mann and Grunstein 1992) via the UAS element (Wan et al. 1995). We do not know how this occurs mechanistically, but it may reflect a novel role for the amino terminus in repressing certain genes.

It is interesting to note that there is a redundancy in the function of the H3 and H4 amino termini with regard to nucleosome assembly but not gene regulation. For example, either the H3 or H4 amino termini can mediate nucleosome assembly. The absence of either tail does not appear to have much affect on assembly in vivo or in vitro, but a deletion of both tails in the same cell does decrease assembly (Ling et al. 1996). Surprisingly, even when the H3 and H4 amino termini are exchanged to the carboxyl termini of their respective proteins, nucleosome assembly is relatively normal. Thus, the H3 and H4 amino termini provide redundant signals that allow recognition by the chromatin assembly machinery. The actual positions of the H3 and H4 tails in relation to the nucleosomal structure are not important for assembly. However, when the H3 and H4 tails are transposed, nucleosome displacement at the TATA element and reduced transcription of GAL1 resemble the effects occurring in a strain lacking the H4 amino terminus altogether (Ling et al. 1996). Therefore, the natural position of the H4 amino terminus in the nucleosome is very important in enabling GAL1 gene regulation. This may allow the appropriate orientation of the nucleosome at the initiation site, which enables histone acetylation for subsequent transcription initiation.

UNACETYLATED HISTONE AMINO TERMINI ALSO REGULATE TRANSCRIPTION BY PREVENTING ACCESS TO THE BASAL TRANSCRIPTION MACHINERY

Regulation of transcription by histones is not mediated solely through the sites of acetylation of histones H3 and H4. Deletion of residues carboxy-terminal to the sites of acetylation at histones H2A, H2B, H3, and H4 causes derepression of basal transcription. These deletions, adjacent to the structured histone-fold domains (H2AΔ17-20, H2BΔ30-37, H3Δ36-39, and H4Δ25-28) activate a GAL1 promoter-URA3 construct under otherwise repressive conditions, allowing detection of URA3 synthesis using 5-fluoroorotic acid (5-FOA) sensitivity. These deletions also preferentially decrease plasmid superhelical density, suggesting the release of supercoils from nucleosomes in the episomes (Lenfant et al. 1996). Therefore, the histone-fold adjacent residues are likely to sequester downstream regulatory elements in nucleosomal DNA. Their release from the nucleosome then allows greater access to the basal transcription machinery and increased transcription. Interestingly, muta-genizing all four sites of acetylation (K5, K8, K12, K16) in H4 to glutamines does not cause much increase in basal transcription (10-fold increase in 5-FOA sensitivity vs. 10^6-fold increase in 5-FOA sensitivity when residues 25–28 are deleted). Therefore, we do not believe that the acetylated portion of the H4 amino terminus directly regulates access of the basal transcription machinery to the *GAL1* TATA. Acetylation may cause a conformational change in the nucleosome which spreads outside of the sites of acetylation enabling access of basal transcription factors to the TATA element.

HISTONE ACETYLATION AND DEACETYLATION

With the identification of histone acetyltransferase and deacetylase genes, it has become possible to use genetic and biochemical analyses to define more carefully how acetylation and deacetylation regulate individual promoters. The first histone acetyltransferase gene cloned was *HAT1*, whose enzyme has intrinsic acetyltransferase activity that acetylates histone H4 K12 in vitro. Since this enzyme acetylates free histones but does not appear to acetylate nucleosomal histones (in vitro), it may have a role in histone deposition (Kleff et al. 1995; Parthun et al. 1996). This was followed by the discovery that a *Tetrahymena* histone acetyltransferase has homology with GCN5, a yeast (*Saccharomyces cerevisiae*) protein required for the activation of a number of genes (Brownell et al. 1996). GCN5 is the catalytic subunit of a complex that includes ADA2 and ADA3, and its HAT activity is necessary for its function as a coactivator (Candau et al. 1997; Wang et al. 1998). The recombinant GCN5 enzyme also has intrinsic activity and acetylates yeast histone H4 K8 and histone H3 K14 in vitro (Kuo et al. 1996). In humans, a number of enzymes with histone acetyltransferase activity and coactivator function have been discovered, including p300/CBP (Bannister and Kouzarides 1996; Ogryzko et al. 1996); pCAF (Yang et al. 1996; Blanco et al. 1998); and the ligand-inducible steroid receptor coactivators Src-1 (Spencer et al. 1997) and ACTR (Chen et al. 1997). These findings demonstrate that the function of many activators depends at least in part on their recruitment of a histone acetyltransferase activity. Even the activation domain of the GAL4-herpesvirus fusion protein (GAL4-VP16), whose function is similar to the activation domain of GAL4, interacts genetically and physically with histone-acetyltransferase-containing complexes to activate transcription (Berger et al. 1992; Pina et al. 1993; Utley et al. 1998).

Conversely, histone deacetylases can prevent gene activity by their interaction with DNA-binding repressor proteins. The mammalian heterodimeric repressor Mad-Max is a repressor in large part because it recruits human homologs of the yeast SIN3 adaptor protein and its associated deacetylase RPD3 (Taunton et al. 1996) to the DNA regulator (Ayer et al. 1995; Hassig et al. 1997; Laherty et al. 1997). Other human repressors including the unliganded thyroid hormone receptor function in a similar manner (Alland et al. 1997; Heinzel et al. 1997; Nagy et al. 1997). In yeast, repression by histone deacetylase is typified by the repression of UME6-regulated genes. UME6 represses genes such as *INO1*, *IME2*, and *SPO13* by its tethering of SIN3, which in turn interacts with a complex that contains RPD3 (Kadosh and Struhl 1997). Interestingly, in both yeast and mammalian cells, this SIN3 complex has some residual repressor activity even in the absence of RPD3 (Kadosh and Struhl 1997; Laherty et al. 1997). In conclusion, the SIN3-RPD3 complex (like the histone acetyltransferases mentioned above) is recruited to DNA elements to repress transcription. Repression occurs in large part (although not exclusively) through the histone deacetylase activity of RPD3 (Kadosh and Struhl 1998).

SEQUENCE OF YEAST HISTONE DEACETYLASE HDA1 UNCOVERS FOUR OTHER STRUCTURALLY DISTINCT HISTONE DEACETYLASES (RPD3, HOS1, HOS2, HOS3)

Our own laboratory has pursued the function of histone deacetylases by the purification of the yeast histone deacetylase complex we termed HDA. Chromatographically, this activity partially overlaps a second activity (HDB) (Fig. 1). The HDA complex was shown to contain four separable protein species, p75, p73, p72, and p71 (Carmen et al. 1996). When peptides of these species were sequenced, it was found that p73 and p72 had identical peptide sequences and were therefore likely to be posttranslationally modified variants of the same protein. That only one such sequence is present in yeast was confirmed by examining the sequence of the entire yeast genome. Hence, the HDA complex appears to contain only three proteins which we named HDA1, HDA2, and HDA3 representing p75, p73/p72, and p71, respectively (Carmen et al. 1996).

Interestingly, the encoded *HDA1* sequence was shown to be similar to those of four other yeast proteins which included RPD3, HOS1, HOS2, and HOS3 and which varied in length from 432 to 696 amino acids (Fig. 2) (Rundlett et al. 1996). Western blots using antibodies to these proteins demonstrate that all non-HDA1 proteins are present in the broad HDB activity (Fig. 1) and that HDA1 and RPD3 are present in different histone deacetylase complexes (Rundlett et al. 1996). Each of the five histone deacetylases is likely to be the catalytic subunit of a different deacetylase complex. These complexes vary in size from approximately 160 kD (HOS3), 350 kD (HDA1), to greater than 600 kD for each of the complexes containing either RPD3, HOS1, or HOS2 (data not shown).

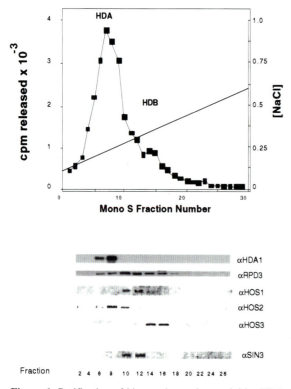

Figure 1. Purification of histone deacetylase activities HDA and HDB. The enzymatic activities were purified from nuclear extracts and assayed as described previously (Carmen et al. 1996; Rundlett et al. 1996) and chromatographed in the final step on Mono-S HR5/5. Antibodies raised against each of the five histone deacetylase were used in Western blots to illustrate the separation of each of the deacetylases. HOS1 and RPD3 cochromatograph under these conditions but are functionally distinct. Although SIN3 cochromatographs with RPD3 and HOS1 (fractions 10, 12), its function is required only for RPD3 function as described in the text.

RPD3 COCHROMATOGRAPHS AND IS FUNCTIONALLY ASSOCIATED WITH SIN3

RPD3 and SIN3 cochromatograph (Fig. 1), which is consistent with their function as corepressors. However, the HOS1 and RPD3 activities also cochromatograph on Mono S after fractionation through multiple chromatography steps (Carmen et al. 1996). Therefore, from these data, it cannot be excluded that SIN3 is associated with the HOS1 complex. However, SIN3 is required only for RPD3 function and not for HOS1 function as shown below.

A strain containing only chromosomal *RPD3* and disruptions in *HDA1, HOS1, HOS2, HOS3* is viable (data not shown). Disruption of *SIN3* in this strain is lethal since disruption of all five histone deacetylase complexes is lethal. Therefore, we introduced a plasmid into this strain containing the *HDA1* and *URA3* genes to generate strain YSY002 (Fig. 3). Plasmid (pD2416) loss may be assayed in YSY002 in the presence of the drug 5-FOA since the presence of URA3 expression results in 5-FOA sensitivity. Although plasmid pD2416 may be lost without loss of viability (in YSY002), this is not the case if *SIN3* is also disrupted (in NSY114). Therefore, SIN3 is required for RPD3 function in maintaining normal cell growth.

To test whether SIN3 is required for HOS1 activity, we used a similar approach. A strain containing *HOS1* and whose remaining four deacetylase genes are disrupted is inviable. Both *HOS1* and *HOS3* must be present in the same strain when the remaining three histone deacetylases are disrupted to prevent lethality. Using the *HDA1*-containing plasmid as described above, we demonstrated that whether *SIN3* was present (in NSY101) or disrupted (in NSY102), the strains were able to grow similarly on 5-FOA when plasmid pDS2416 was lost (Fig. 3). Therefore, SIN3 is not likely to be required for HOS1 function.

HISTONE DEACETYLASES IN YEAST ARE FUNCTIONALLY DISTINCT

Individual disruptions of each of the five histone deacetylase genes in yeast does not prevent growth. Moreover, either *HDA1, RPD3,* or *HOS2* can allow cell viability when the other deacetylases are disrupted. There is clearly overlap in the ability of different deacetylases to maintain cellular viability. However, their function is only partially redundant with respect to transcriptional repression. Micro-DNA arrays (DeRisi et al. 1997) done in collaboration with Vichy Iyer and Pat Brown (Stanford

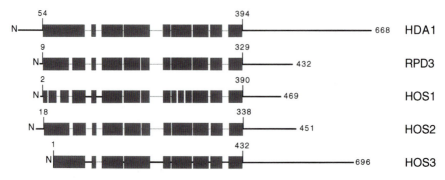

Figure 2. Core regions of greatest similarity between the five histone deacetylase proteins of yeast. Sequences were aligned using the PILEUP Program (Genetics Computer Group, Madison, WI) as described previously (Rundlett et al. 1996). The rectangles represent regions of similarity and gaps are represented by dashed lines. Regions of similarity outlined above were analyzed by the BESTFIT program. Compared to HDA1, RPD3 shows approximately 28.2% identity and 42.9% similarity, HOS1 shows approximately 31.1% identity and 40.9% similarity, HOS2 shows approximately 30% identity and 44.3% similarity, and HOS3 shows approximately 38.7% identity and 47.1% similarity. This figure also illustrates the longer less-conserved carboxy-terminal extensions on HDA1 and HOS3.

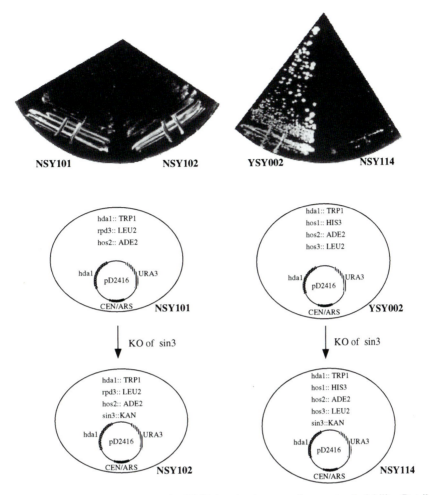

Figure 3. SIN3 is required for RPD3 function but not for HOS1 function in supporting yeast cell viability. Details are described in the text.

ROLE OF RPD3-SIN3 AT UME6-REGULATED GENES

University) have demonstrated that mutations in each of the five histone deacetylases affect the regulation of a distinct and only partially overlapping subset of the yeast genome (data not shown). These observations argue that the histone deacetylase complexes have different functions and may be targeted to different genes. The function of a deacetylase at a specific gene is best illustrated by the study of RPD3.

We first wished to know whether RPD3 affects histone acetylation at genes that it regulates. We also asked whether the effect is localized to the promoter only or throughout the entire gene. The procedure we used to examine histone acetylation level at specific genes is based on chromatin immunoprecipitation. This approach involves the cross-linking of yeast cells with formaldehyde, after which chromatin is sonicated extensively to yield DNA fragments from 300 to 800 bp in length. Antibodies to chromosomal proteins are then used to immunoprecipitate DNA fragments at which they interact. The DNA fragments that are purified may be identified by hy-

bridization (Braunstein et al. 1993; Orlando and Paro 1993) or by the use of PCR (Hecht et al. 1996; Strahl-Bolsinger et al. 1997). In this latter approach, primer sets may be used that include promoter sequences or sequences adjacent to the regulators. With the knowledge of the yeast genome sequence and the appropriate primer pairs, the amplification of many primer sets may be carried out in the same reaction. Utilizing antibodies that are specific to individual sites of acetylation in histone H4 (Turner et al. 1989), it is then possible to determine which RPD3-regulated genes or gene fragments are selectively immunoprecipitated by the antibody in wild-type cells or those mutant for *RPD3* (Rundlett et al. 1998). For comparison, we also examined the effect of the related histone deacetylase HDA1 at these genes. DNA fragments that are selectively immunoprecipitated by the antibody against a particular site of acetylation are assumed to be hyperacetylated at that site in chromatin.

Antibodies to acetylated H4 sites K5, K8, K12, and K16 were used to immunoprecipitate chromatin fragments from wild-type cells and those disrupted for *hda1* and *rpd3*. Primer sets were chosen that encompassed the regulatory sequences of each of the genes whose expression was analyzed. We also used primer sets to telomeric

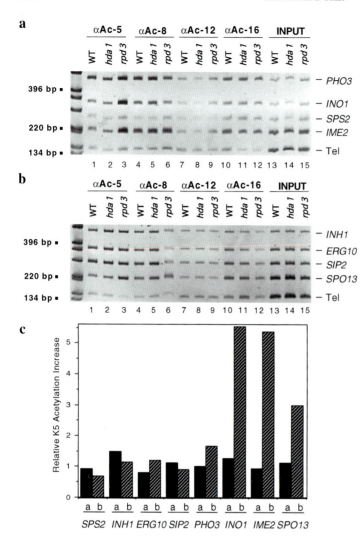

Figure 4. *RPD3* disruption results in K5 hyperacetylation of chromatin at the promoter regions of *INO1*, *IME2*, and *SPO13*. Shown are PCR analyses of DNA primer sets of various regulatory DNA sequences immunoprecipitated by antibodies against H4 lysines 5, 8, 12, or 16. DNAs from WT (lanes *1, 4, 7, 10*), *hda1* (lanes *2, 5, 8, 11*), and *rpd3* (lanes *3, 6, 9, 12*) strains, along with input DNA obtained prior to immunoprecipitation (lanes *13–15*) are shown. (*a*) *PHO3*, *INO1*, *SPS2*, and *IME2* genes were analyzed. (*b*) *INH1*, *ERG10*, *SIP2*, and *SPO13* genes were analyzed in a separate experiment. PCR amplification of a fragment 0.5 kb from the telomere of chromosome VI-R was used as reference to ensure equal loading of samples. (*c*) Quantitation of increases in H4 K5 acetylation in *hda1* (*a*) and *rpd3* (*b*) cells relative to wild type. (Reprinted, with permission, from Rundlett et al. 1998 [copyright Macmillan].)

sequences, whose acetylation we found to be relatively unaffected, to control for loading differences on the polyacrylamide gels. As shown in Figure 4, a and b, the antibodies to histone H4 K5-Ac immunoprecipitate the *INO1*, *IME2,* and *SPO13* promoters in *rpd3* mutant cells but not in *hda1* mutant cells. This effect was most pronounced with the K5-Ac antibody than with other antibodies used. The quantitation of K5 acetylation by densitometry is shown in Figure 4c. Thus, genes known to be regulated by RPD3 (*INO1*, *IME2*, *SPO13*) are all hyperacetylated when *RPD3* is disrupted (Rundlett et al. 1998).

Interestingly, both K5 and K12 are strongly hyperacetylated in a *sin3* mutant strain (Rundlett et al. 1998). Why *SIN3* deletion has a greater effect on hyperacetylation in these experiments and can mediate repression even in the absence of *RPD3* (Kadosh and Struhl 1997) is unclear. Although SIN3 does not appear to associate during chromatography with other histone deacetylases other than RPD3, it cannot be excluded that another deacetylase is attracted to SIN3 in chromatin when *RPD3* is disrupted.

To ask whether RPD3 deacetylates histone H4 K5 only at the promoter or throughout the gene, primer sets were used at sites from +0.14 to +1.9 in and downstream from the coding region and sites –0.09 to –2.0 upstream of the coding region (Fig. 5). These were used to amplify DNA immunoprecipitated by antibody to H4 K5. We found that only the region at the promoter (–0.09) was strongly hyperacetylated at histone H4 K5. From the boundaries of the primer sets used, we could conclude that RPD3 deacetylates histone H4 largely within 386 bp of the UME6-binding site, a region encompassing some two nucleosomes.

The mechanism of repression by RPD3 may involve the two nucleosomes. Luger et al. (1997) have demonstrated that the H4 amino-terminal region 16–25 interacts with a negatively charged face of the H2A-H2B dimer in the adjacent nucleosome of the crystal lattice. Since the acetylated H4 sites are adjacent to this region, they may also bind in the neighboring nucleosome in chromatin. This may explain why the H4 amino terminus can repress transcription. Perhaps it serves to link two adjacent nucleosomes at the *INO1* downstream promoter, thus preventing access of the basal transcription machinery to the TATA-containing region at the nucleosome edge (Fig. 6).

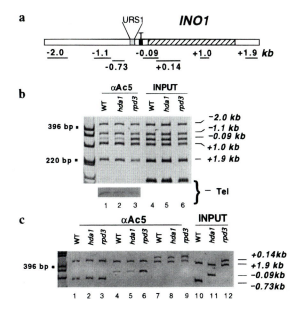

Figure 5. The *INO1* gene is hyperacetylated at histone H4 K5 mainly at two nucleosomes or less, adjacent to the UME6-binding site (URS1). (*a*) The distance from the start of translation (ATG) for the 5′ end of each PCR product is shown in kilobases (kb). T refers to the TATA element. The coding region of *INO1* is hatched. (*b*) Chromatin immunoprecipitation using anti-H4 K5-Ac antibody and PCR analysis of DNA from WT (lane *1*), *hda1* (lane *2*), and *rpd3* (lane *3*) cells using primer sets located at approximately 1-kb intervals across the *INO1* gene. WT, *hda1*, and *rpd3* input DNAs are shown in lanes *4*, *5*, and *6*. Tel is an overexposure of the hypoacetylated telomere region to demonstrate similar loading. (*c*) Higher-resolution chromatin immunoprecipitation analysis of the region adjacent to the URS1. PCR primers were used at distances of –0.73 kb upstream of the translation start (lanes *1–3*), at the promoter (–0.09) (lanes *4–6*), and +0.14 kb downstream from the translation start (lanes *7–9*). A primer set +1.9 kb upstream of the gene was used as an internal control for each PCR. It is evident that in wild-type cells, only the region at the promoter (–0.09) is hypoacetylated (*b*, lane *1*). This region becomes hyperacetylated upon *rpd3* disruption but not after *hda1* disruption (*b*, compare lanes *2* and *3*). When compared to the control region at +1.9 kb, PCR amplification at –0.73 kb and +0.14 kb shows much less effect of *rpd3* disruption than at –0.09 kb. From the downstream boundary of the URS1 to the upstream boundary of the primer set at +0.14 is 386 bp (which would represent DNA wound around approximately two nucleosomes). (Reprinted, with permission, from Rundlett et al. 1998 [copyright Macmillan].)

DISCUSSION AND CONCLUSIONS

We now have a much more comprehensive view of the role of chromatin structure in gene activity. TATA elements are often repressed by nucleosomes through the histone amino termini and sequences adjacent to the acetylation sites, but UAS elements are not. When activators bind the UAS elements, one of their roles is to alleviate the repression of basal transcription by histones. For some genes, antagonizing a repressive chromatin structure may be the main function of certain activators. For other genes, the removal of a repressive chromatin structure is likely to represent a necessary but transcriptionally modest step toward full gene activity. Even here

the position of nucleosomes near the downstream promoter element may be important in order to allow proper derepression and activation. Nucleosomes improperly positioned by mutations at the histone amino termini (Fisher-Adams and Grunstein 1995) may prevent efficient nucleosome displacement or interfere with the recruitment of basal transcription factors by activators (Chatterjee and Struhl 1995; Farrell et al. 1996)

Histone acetylation is in many cases required for nucleosome displacement. However, we have much to learn regarding its molecular details. Major problems to be addressed include the targeting of different histone acetyltransferase and deacetylase complexes to their loci of action; how this targeting is regulated; the specificity of the enzymes for various histone sites in chromatin; and the mechanism by which acetylation and deacetylation alters nucleosome or chromatin structure to regulate gene activity.

Despite instances of targeted acetylation (see above), it appears likely that most histone acetylation in yeast is untargeted. For example, we have demonstrated that except for a limited domain (less than 386 bp) at the *INO1* promoter, regions upstream and downstream from the promoter (4 kb) are hyperacetylated. In addition, many genes examined are hyperacetylated even prior to their induction (Rundlett et al. 1998). It is unlikely that all hyperacetylation occurs through acetyltransferases tethered at promoters. Perhaps hyperacetylation is the default state after assembly of acetylated histones or acetylation of most chromatin occurs through nontargeted acetylation in a nuclear subcompartment rich in histone acetyltransferases. When genes are regulated by tethered histone acetyltransferases and deacetylases, these enzymes often function at the same promoter. Even for *INO1*, the promoter region that is hypoacetylated through RPD3 action is subsequently strongly acetylated when *RPD3* is disrupted. Given the localized function of the histone acetyltransferase GCN5 at other promoters (Kuo et al. 1998), it is tempting to speculate that GCN5 helps maintain the balance between the acetylation and deacetylation required to regulate *INO1*. How the remaining yeast histone deacetylases (HDA1, HOS1, HOS2 and HOS3) are brought to their target genes remains to be determined. These mechanisms and their possible regulation will be especially interesting to decipher given the very different structures of the histone deacetylase complexes.

Antibodies to specific sites of acetylation have been used to demonstrate that the SIN3-RPD3 complex affects histone H4 sites K5 and K12 as well as a number of sites in histone H3 (Rundlett et al. 1998). GCN5 is believed to affect histone H3 at K14 and to a lesser extent, histone H4 at K8 in vitro (Kuo et al. 1996). It will be important to determine whether histone acetyltransferases and deacetylases in yeast have additional lysine targets in all four core histones in vivo. This can only be done using chromatin immunoprecipitation at this stage with the available set of antibodies to sites of acetylation. Such antibodies vary considerably in their titer and specificity for individual sites (Turner et al. 1989). It is possible that as different antibodies are generated, new targets of the acetyltransferases and deacetylases may emerge. At that

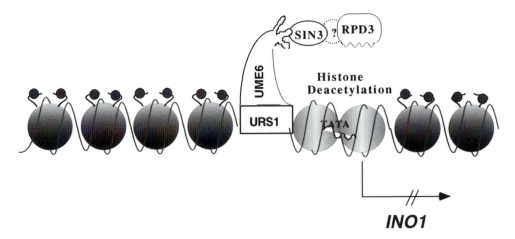

Figure 6. Repression of *INO1* by RPD3-SIN3 complex. UME6, SIN3 and RPD3 interact as described previously (Kadosh and Struhl 1997). The question mark illustrates the absence of data to suggest a direct interaction between SIN3 and RPD3. Deacetylation of histone H4 by the RPD3-SIN3 complex affects a limited region believed to contain two nucleosomes or less. It is speculated from the X-ray crystal structure (Luger et al. 1997) that the H4 amino terminus extends into and interacts with a negatively charged face of the H2A-H2B dimer of the adjacent nucleosome. The H4 amino terminus may in this manner link two nucleosomes. Since the TATA element is found at the inner nucleosomal boundaries of the two nucleosomes as shown, histone deacetylation may promote nucleosome-nucleosome contacts, thus precluding access of basal transcription factors to the TATA element. Our data do not address the possibility that the RPD3-SIN3 complex affects also a short region upstream of the URS1 to the boundary of the –0.73 kb primer set.

point, it may be easier to determine the biological reasons behind the use of these targets.

Finally, it appears quite possible that the function of histone H4 acetylation is not to disrupt histone amino-terminal–DNA interactions within the nucleosome but to prevent histone-histone interactions between nucleosomes. However, this must be tested experimentally in vitro and in chromatin. Only then will it be possible to determine mechanistically why acetylation of the histone tails activates and deacetylation represses transcription initiation.

ACKNOWLEDGMENTS

The authors thank the members of the Grunstein laboratory, in particular Amy Wang, Kunheng Luo, Stephanie Gussak, and Maria Vogelauer for their critical review of the manuscript. This work was supported by National Institutes of Health grant GM-23674.

REFERENCES

Alland L., Muhle R., Hou H., Jr., Potes J., Chin L., Schreiber-Agus N., and DePinho R.A. 1997. Role for N-CoR and histone deacetylase in Sin3-mediated repression. *Nature* **387:** 49.

Allfrey V.G. 1977. Post-synthetic modifications of histone structure: A mechanism for the control of chromosome structure by the modulation of histone-DNA interactions. In *Chromatin and chromosome structure* (ed. H.J. Li and R. Eckhardt), p. 167. Academic Press, New York.

Almer A. and Hörz W. 1986. Nuclease hypersensitive regions with adjacent positioned nucleosomes mark the gene boundaries of the *PHO5/PHO3* locus in yeast. *EMBO J.* **5:** 2681.

Ayer D.E., Lawrence Q.A., and Eisenman R.N. 1995. Mad-Max transcriptional repression is mediated by ternary complex formation with mammalian homologs of yeast repressor Sin3. *Cell* **80:** 767.

Bannister A.J. and Kouzarides T. 1996. The CBP co-activator is a histone acetyltransferase. *Nature* **384:** 641.

Benezra R., Cantor C.R., and Axel R. 1986. Nucleosomes are

phased along the mouse beta-major globin gene in erythroid and nonerythroid cells. *Cell* **14:** 697.

Berger S.L., Pina B., Silverman N., Marcus G.A., Agapite J., Regier J.L., Triezenberg S.J., and Guarente L. (1992). Genetic isolation of ADA2: A potential transcriptional adaptor required for function of certain acidic activation domains. *Cell* **70:** 251.

Blanco J.C.G., Minucci S., Lu J., Yang X.-J., Walker K.K., Chen H., Evans R.M., Nakatani Y., and Ozato K. 1998. The histone acetylase PCAF is a nuclear receptor coactivator. *Genes Dev.* **12:** 1638.

Braunstein M., Rose A.B., Holmes S.G., Allis C.D., and Broach J.R. 1993. Transcriptional silencing in yeast is associated with reduced nucleosome acetylation. *Genes Dev.* **7:** 592.

Brownell J.E., Zhou J., Ranalli T., Kobayashi R., Edmondson D.G., Roth S.Y., and Allis C.D. 1996. *Tetrahymena* histone acetyltransferase A: A homolog to yeast Gcn5p linking histone acetylation to gene activation. *Cell* **84:** 843.

Candau R., Zhou J-X., Allis C.D., and Berger S.L. 1997. Histone acetyltransferase activity and interaction with ADA2 are critical for GCN5 function in vivo. *EMBO J.* **16:** 555.

Carmen A.A., Rundlett S.E., and Grunstein M. 1996. HDA1 and HDA3 are components of a yeast histone deacetylase (HDA) complex. *J. Biol. Chem.* **271:** 15837.

Chasman D.I., Lue N.F., Buchman A.R., LaPointe J.W., Lorch Y., and Kornberg R.D. 1990. A yeast protein that influences the chromatin structure of UASg and functions as a powerful auxiliary gene activator. *Genes Dev.* **4:** 503.

Chatterjee S. and Struhl K. 1995. Connecting a promoter-bound protein to TBP bypasses the need for a transcriptional activation domain. *Nature* **374:** 820.

Chen H., Lin R.J., Schiltz R.L., Chakravarti D., Nash A., Nagy L., Privalsky M.L., Nakatani Y., and Evans R.M. 1997. Nuclear receptor coactivator ACTR is a novel histone acetyltransferase and forms a multimeric activation complex with P/CAF and CBP/p300. *Cell* **90:** 569.

DeRisi J.L., Iyer V.R., and Brown P.O. 1997. Exploring the metabolic and genetic control of gene expression on a genomic scale. *Science* **278:** 680.

Durrin L., Mann R., and Grunstein M. 1992. Nucleosome loss activates *CUP1* and *HIS3* promoters to fully induced levels in the yeast *Saccharomyces cerevisiae*. *Mol. Cell. Biol.* **12:** 1621.

Durrin L., Mann R., Kayne P., and Grunstein M. 1991. Yeast

histone H4 N-terminal sequence is required for promoter activation *in vivo*. *Cell* **65**: 1023.

Farrell S., Simkovich N., Wu Y., Barberis A., and Ptashne M. 1996. Gene activation by recruitment of the RNA polymerase II holoenzyme. *Genes Dev.* **10**: 2359.

Fedor M.J., Lue N.F., and Kornberg R.D. 1988. Statistical positioning of nucleosomes by specific protein-binding to an upstream activating sequence in yeast. *J. Mol. Biol.* **204**: 109.

Fisher-Adams G. and Grunstein M. 1995. Yeast histone H4 and H3 N-termini have different effects on the chromatin structure of the *GAL1* promoter. *EMBO J.* **14**: 1468.

Han M. and Grunstein M. 1988. Nucleosome loss activates yeast downstream promoters *in vivo*. *Cell* **55**: 1137.

Han M., Chang M., Kim U.-J., and Grunstein M. 1987. Histone H2B repression causes cell-cycle-specific arrest in yeast: Effects on chromosomal segregation, replication and transcription. *Cell* **48**: 589.

Han M., Kim U.-J., Kayne P., and Grunstein M. 1988. Depletion of histone H4 and nucleosomes activate the *PHO5* gene in *Saccharomyces cerevisiae*. *EMBO J.* **7**: 2221.

Hecht A., Strahl-Bolsinger S., and Grunstein M. 1996. Spreading of transcriptional repression by SIR3 from telomeric heterochromatin. *Nature* **383**: 92.

Hassig C.A., Fleischer T.C., Billin A.N., Schreiber S.L., and Ayer D.E. 1997. Histone deacetylase activity is required for full transcriptional repression by mSin3A. *Cell* **89**: 341.

Heinzel T., Lavinsky R.M., Mullen T.M., Soderstrom M., Laherty C.D., Torchia J., Yang W.M., Brard G., Ngo S.G., Davie J.R., Seto E., Eisenman R.N., Rose D.W., Glass C.K., and Rosenfeld M.G. 1997. N-CoR, mSin3 and histone deacetylase-containing complexes in nuclear receptor and mad repression. *Nature* **387**: 43.

Kadosh D. and Struhl K. 1997. Repression by Ume6 involves recruitment of a complex containing Sin3 corepressor and Rpd3 histone deacetylase to target promoters. *Cell* **89**: 365.

―――. 1998. Histone deacetylase activity of Rpd3 is important for transcriptional repression *in vivo*. *Genes Dev.* **12**: 797.

Kleff S., Andrulis E.D., Anderson C.W., and Sternglanz R. 1995. Identification of a gene encoding a yeast histone H4 acetyltransferase. *J. Biol. Chem.* **270**: 24674.

Kuo M., Zhou J., Churchill M.E.A., and Allis C.D. 1998. Histone acetyltransferase activity of yeast GCN5 is required for the activation of target genes *in vivo*. *Genes Dev.* **12**: 627.

Kuo M., Brownell J.E., Sobel R.E., Ranalli T.A., Cook R.G., Edmondson D.G., Roth S.Y., and Allis C.D. 1996. Transcription-linked acetylation by Gcn5p of histones H3 and H4 at specific lysines. *Nature* **383**: 269.

Laherty C.D., Yang W.M., Sun J.M., Davie J.R., Seto E., and Eisenman R.N. 1997. Histone deacetylases associated with the mSin3 corepressor mediate mad transcriptional repression. *Cell* **89**: 349.

Lenfant F., Thomsen B., Mann R., Ling X., and Grunstein M. 1996. N-terminal sequences of all four core histones are involved in the repression of basal transcription in yeast. *EMBO J.* **15**: 3974.

Ling X., Harkness T.A.A., Schultz M.C., Fisher-Adams G., and Grunstein M. 1996. Yeast histone H3 and H4 amino termini are important for nucleosome assembly *in vivo* and *in vitro*: Redundant and position-independent functions in assembly but not in gene regulation. *Genes Dev.* **10**: 686.

Lohr D. 1984. Organization of the *GAL1-GAL10* intergenic control region chromatin. *Nucleic Acids Res.* **12**: 8457.

Lorch Y., LaPointe J.W., and Kornberg R.D. 1987. Nucleosomes inhibit the initiation of transcription but allow chain elongation with the displacement of histones. *Cell* **49**: 203.

Luger K., Mader A.W., Richmond R.K., Sargent D.F., and Richmond T.J. 1997. Crystal structure of the nucleosome core particle at 2.8 Å resolution. *Nature* **389**: 251.

Mann R. and Grunstein M. 1992. Histone H3 N-terminal mutations allow hyperactivation of the yeast *GAL1* gene *in vivo*. *EMBO J.* **11**: 3297.

Matallana E., Franco L., and Pérez-Ortín J.E. 1992. Chromatin structure of the yeast *SUC2* promoter in regulatory mutants. *Mol. Gen. Genet.* **231**: 395.

Nagy L., Kao H.Y., Chakravarti D., Lin R.J., Hassig C.A., Ayer D.E., Schreiber S.L., and Evans R.M. 1997. Nuclear receptor repression mediated by a complex containing SMRT, mSIN3A and histone deacetylase. *Cell* **89**: 373.

Ogryzko V.V., Schiltz R.L., Russanova V., Howard B.H., and Nakatani Y. 1996. The transcriptional co-activators p300 and CBP are histone acetyltransferases. *Cell* **87**: 953.

Orlando V. and Paro R. 1993. Mapping polycomb-repressed domains in the bithorax complex using *in vivo* formaldehyde cross-linked chromatin. *Cell* **75**: 1187.

Parthun M.R., Widom J., and Gottschling D.E. 1996. The major cytoplasmic histone acetyltransferase in yeast: Links to chromatin replication and histone metabolism. *Cell* **87**: 85.

Pina B., Berger S., Marcus G.A., Silverman N., Agapite J., and Guarente L. 1993. ADA3: A gene, identified by resistance to GAL4-VP16, with properties similar to and different from those of ADA2. *Mol. Cell Biol.* **13**: 5981.

Roth S.Y., Shimizu M., Johnson L., Grunstein M., and Simpson R.T. 1992. Stable nucleosome positioning and complete repression by the yeast α2 repressor are disrupted by amino-terminal mutations in histone H4. *Genes Dev.* **6**: 411.

Rundlett S.R., Carmen A.A., Suka N., Turner B.T., and Grunstein M. 1998. RPD3 deacetylates histone H4 lysine 5 to repress UME6 regulated genes. *Nature* **393**: 831.

Rundlett S., Carmen A.A., Turner B., Kobayashi R., Bavykin S., and Grunstein M. 1996. HDA1 and RPD3 are members of functionally distinct yeast histone deacetylase complexes. *Proc. Natl. Acad. Sci.* **93**: 14503.

Saunders M.J., Yeh, E., Grunstein M., and Bloom K. 1990. Nucleosome depletion alters the chromatin structure of *Saccharomyces cerevisiae* centromeres. *Mol. Cell. Biol.* **10**: 5721.

Spencer T.E., Jenster G., Burcin M.M., Allis C.D., Zhou J., Mizzen C.A., McKenna N.J., Onate S.A., Tsai S.Y., Tsai M.-J., and O'Malley B.W. 1997. Steroid receptor coactivator-1 is a histone acetyltransferase. *Nature* **389**: 194.

Strahl-Bolsinger S., Hecht A., Luo K., and Grunstein M. 1997. SIR2 and SIR4 interactions differ in core and extended telomeric heterochromatin of yeast. *Genes Dev.* **11**: 83.

Tanaka S., Livingstone-Zatchej M., and Thoma F. 1996. Chromatin structure of the yeast *URA3* gene at high resolution provides insight into structure and positioning of nucleosomes in the chromosomal context. *J. Mol. Biol.* **257**: 919.

Taunton J., Hassig C.A., and Schreiber S.L. 1996. A mammalian histone deacetylase related to the yeast transcriptional regulator Rpd3p. *Science* **272**: 408.

Turner B.M., O'Neill L.P., and Allan I.M. 1989. Histone H4 acetylation in human cells. Frequency of acetylation at different sites defined by immunolabeling with site-specific antibodies. *FEBS Lett.* **253**: 141.

Utley R.T., Ikeda K., Grant P.A., Cote J., Steger D.J., Eberharter A., John S., and Workman J.L. 1998. Transcriptional activators direct histone acetyltransferase complexes to nucleosomes. *Nature* **394**: 498.

Wan J., Mann R., and Grunstein M. 1995. Yeast histone H3 and H4 N-termini function through different *GAL1* regulatory elements to repress and activate transcription. *Proc. Natl. Acad. Sci.* **92**: 5664.

Wang L., Liu L., and Berger S.L. 1998. Critical residues for histone acetylation by Gcn5, functioning in Ada and SAGA complexes, are also required for transcriptional function *in vivo*. *Genes Dev.* **12**: 640.

Yang X., Ogryzko V.V., Nishikawa J., Howard B., and Nakatani Y. 1996. A p300/CBP-associated factor that competes with the adenoviral oncoprotein E1A. *Nature* **382**: 319.

Targeting Sir Proteins to Sites of Action: A General Mechanism for Regulated Repression

M. Cockell, M. Gotta,* F. Palladino,† S.G. Martin, and S.M.Gasser

Swiss Institute for Experimental Cancer Research, CH-1066 Epalinges/Lausanne, Switzerland

In yeast, a position-dependent chromatin-mediated silencing affects several regions of the genome, including the silent mating-type loci, telomeric regions, and rDNA. At *HM* and telomeric loci, repression requires a complex of the silent information regulator proteins (Sir2-Sir4p) that binds nucleosomes via the amino-terminal tails of histones H3 and H4. Nucleation of this complex requires interaction with sequence-specific DNA-binding proteins, among which figure Rap1, Abf1, and ORC. We show here that both the dosage and balance of Sir proteins are critical for silencing. Different domains of these highly modular proteins are involved in regulating Sir protein distribution among potential sites of repression. In particular, two factors that bind the amino-terminal 271 amino acids of Sir4p (i.e., Sir1p, Sif2p) compete for limiting pools of Sir factors, antagonizing the formation of repressed chromatin at telomeres. In contrast, the interaction of Ku70/80 with the carboxyl terminus of Sir4p appears to help recruit Sir proteins to telomeres and perhaps under specific conditions to internal sites.

Position-mediated transcriptional silencing in yeast shares many characteristics with heterochromatin-mediated gene repression in higher eukaryotes. Like inactive chromatin regions from other species, the histones associated with these regions are underacetylated on specific amino-terminal lysine residues, which distinguishes them from bulk chromatin (Turner et al. 1992; Braunstein et al. 1993, 1996). In addition, repressed domains of the yeast genome are less accessible to a variety of chemical and enzymatic probes (for review, see Thompson et al. 1993). The silenced state at *HM* loci is stably inherited from mother to daughter cells, whereas telomere proximal repression is metastable, giving rise to clonal expression patterns, again analogous to the variegated phenotypes observed for genes translocated near centromeric heterochromatin in flies (Sandell and Zakian 1992; Henikoff 1996).

Sir3p and Sir4p are able to form homo- and heterodimers and both interact with Rap1, as well as with the histones H3 and H4 (Marshall et al. 1987; Chien et al. 1991; Moretti et al. 1994; Cockell et al. 1995; Hecht et al. 1995; Liu and Lustig 1996). Pull-down and coimmunoprecipitation experiments also indicate that Sir4p interacts directly with Sir2p, whereas data showing direct interaction between Sir3p and Sir2p are contradictory (Moazed and Johnson 1996; Moazed et al. 1997; Strahl-Bolsinger et al. 1997). Fluorescence in situ hybridization (FISH) with subtelomeric probes and immunolocalization of Rap1 and Sir proteins show that silencing proteins and sites of repression colocalize in six to ten perinuclear foci (Palladino et al. 1993; Gotta et al. 1996; Maillet et al. 1996). Based on this, two hybrid, and genetic evidence for Rap1-Sir-histone interactions, it was proposed that Sir complexes polymerize along the chromatin fiber, through interaction with histone tails (Hecht et al. 1995). Polymerization appears to initiate from telomeric clusters of Rap1 consenses (e.g., the TG_{1-3} repeat) and the E and I silencer elements flanking the *HM* loci (Hecht et al. 1995). Cross-linking studies confirm that Sir2/3/4p complexes are bound to repressed regions of the genome and extended along the chromatin fiber (Hecht et al. 1996; Strahl-Bolsinger et al. 1997).

Besides being bound at telomeres, a substantial proportion of Sir2p is associated with the rDNA, in a manner independent of Sir3p and Sir4p (Gotta et al. 1997; Smith et al. 1998). Thus, the enhancement of rDNA recombination and loss of silencing observed in *sir2* mutants of pol II genes inserted into rDNA (Gottlieb and Esposito 1989; Bryk et al. 1997; Smith and Boeke 1997) probably reflect direct interactions of Sir2p with the nucleolar chromatin (Fritze et al. 1997; Gotta et al. 1997). It is not known what targets Sir2p to the nucleolus, although under normal conditions, it appears to be limiting at both rDNA and telomeres. This suggests that there are mechanisms which regulate its distribution between different loci. Intriguingly, when yeast cells reach an advanced state of senescence, they become sterile due to derepression of the *HM* loci (Smeal et al. 1996). This correlates with the accumulation of Sir3p in the nucleolus (Kennedy et al. 1997).

The redistribution of Sir3p to the nucleolus depends on both *SIR2* and *UTH4*, a homolog of the *Drosophila pumillo* gene (Kennedy et al. 1995, 1997). A factor such as Uth4p may either target Sir3p to the nucleolus, mask default domains which tether Sir proteins to telomeres, or uncover new binding domains. These possibilities are not mutually exclusive, and the observations that both an amino-terminal fragment of Sir3p and a fragment of Sir4p lacking its carboxyl terminus accumulate in the nucleolus when overexpressed support either model.

Sir proteins are thought to function at other unidentified sites within the nucleus. It has been recently pro-

---

Present addresses: *Department of Genetics, University of Cambridge, Cambridge, United Kingdom; †Department of Zoology, Perolles, University of Fribourg, Fribourg, Switzerland.

Table 1. Yeast Strains

AJL275-2AV$_R$	(*MATα, ade2-1, can1-100, his3-11, leu2-3, 112, trp1-1, ura3-1, V$_R$::URA3-Tel*)
EG37	(*MAT**a**, leu2-3, 112, ura3-52, trp1-289, his3, gal2, HML::E>I*)
UCC518	*MAT**a**, ade2-101, his3-Δ200, leu2-Δ1, lys2-801, trp1-Δ1, ura3-52, ppr1::HIS3, V$_R$::URA3*)
UCC520	*MAT**a**, ade2-101, his3-Δ200, leu2-Δ1, lys2-801, trp1-Δ1, ura3-52, ppr1::HIS3, V$_R$::URA3*)
UCC522	(*MAT**a**, ade2-101, his3-Δ200, leu2-Δ1, lys2-801, trp1-Δ1, ura3-52, ppr1::HIS3, V$_R$::URA3*)
UCC3107	(*MATα ade2::hisG can1::hisG his3-11 leu2 trp-1Δ ura3-52 TEL V-R::ADE2*)
UCC3505	(*MATα ura3-52 lys2-801 ade2-101 trp1-63, his3-Δ200 leu2-1 ppr1::HIS3 adh4::URA3-TEL VII-L; TEL V-R::ADE2*)
JS341	(*MATα, ade2, trp1-63, leu2-1, his3-Δ200, ura3-167, sir2::kanMX; RDN::mURA3-HIS3*)
GA492	(*MATα his3 leu2 trp1 ade2-1 ura3-52 adh4::URA3-TEL VII-L lys2::LYS2-dam$^+$ TEL V-R::ADE2*)
MC118	(*MATα his3 leu2 trp1 ade2-1 ura3-52 adh4::URA3-TEL VII-L lys2::LYS2-dam$^+$ TEL V-R::ADE2 sif2::kanMX2*)

posed that a Sir2/3/4p complex may be recruited to sites of double-strand break repair through interaction with the heterodimeric DNA end-binding protein Ku70/80 (Tsukamoto et al. 1997). In this respect, it is interesting to note that Ku70/80 was shown to bind telomeres by immunolocalization and that loss of either gene disrupts telomeric silencing and the localization of telomeres at the nuclear periphery (Boulton and Jackson 1998; Gravel et al. 1998; Laroche et al. 1998; S.G. Martin, unpubl.). In addition, we have identified a novel Sir4p ligand, called Sif2p, that competes for telomeric silencing and the localization of Sir4p at telomeric sites (Cockell et al. 1998). Evidence suggesting that multiple mechanisms control the distribution of Sir proteins among telomeres, rDNA, *HM* loci, and other sites of action is summarized below.

EXPERIMENTAL PROCEDURES

For standard molecular biological techniques, published protocols were used (Sambrook et al. 1989).

Standard growth media were used for *Saccharomyces cerevisiae* (Gottschling et al. 1990; Rose et al. 1990), and all yeast strains are described in Table 1. Repression assays were performed as described previously (Singer and Gottschling 1994; Cockell et al. 1998; Gotta et al. 1998). Values for each repression assay were obtained by making serial dilutions of several independent transformants and calculating the mean value (figures show serial dilutions for a single representative colony). The cloning of *SIR3, SIR4*, their amino- and carboxy-terminal fragments, and the *SIR2* into overexpression vectors has been described previously (Cockell et al. 1995, 1998; Maillet et al. 1996; Gotta et al. 1998). A DNA fragment encoding the *SIR2* gene was obtained by polymerase chain reaction (PCR) from genomic DNA and inserted between the unique *Eco*RI and *Xho*I sites of the vector pJG45 (Golemis et al. 1996) to encode a protein with a weak transcriptional activation domain fused to the amino terminus of Sir2p. Immunofluorescence was performed as described previously (Gotta et al. 1996, 1998).

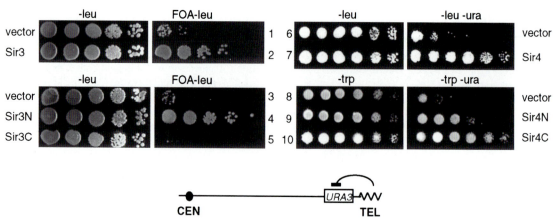

Figure 1. Overexpression of Sir3p, Sir4p, and subdomains affect TPE. Strain AJL2275-2AV$_R$ which carries *ADE2* adjacent to the VII-L telomere (rows *1–5*) was transformed with vector alone (pADH, rows *1* and *3*) or with plasmids carrying either full-length Sir3p (pADH-SIR3, row *2*), Sir3N (pADH-SIR3N, row *4*), or Sir3C (pADH-SIR3C, row *5*). Colonies were grown 3–5 days at 30ºC on medium lacking leucine to ensure maintenance of the plasmids. Large colonies were resuspended in H$_2$O, and the same tenfold serial dilutions were plated onto medium lacking leucine and onto medium lacking leucine and containing 5-FOA. Strain UCC3505 (rows *6–10*) was transformed with vectors alone (pADH, row *6* and pJG45, row *8*) or with plasmids carrying either full-length Sir4p (pADH-SIR4C, row *7*), Sir4N (pJG45-SIR4N, row *9*), or Sir4C (pJG45-SIR4C, row *10*). Colonies were grown 3–5 days at 30ºC on medium lacking leucine or tryptophan as indicated, to ensure maintenance of the plasmids and resuspended in H$_2$O, and the same tenfold serial dilutions were plated onto medium lacking tryptophan and medium lacking both tryptophan and uracil.

RESULTS

Effects of *SIR3* and *SIR4* Gene Dosage on Silencing

It is well established that the absence of any one of the Sir2-Sir4 proteins results in complete derepression at telomeres and *HM* loci (Aparicio et al. 1991; Laurenson and Rine 1992). However, because both the dosage and balance of Sir proteins are critical for silencing, the effects of overexpression of individual Sir proteins or subdomains are complex. As shown in Figure 1 (rows 1 and 2), the overexpression of Sir3p can improve telomeric silencing of a telomere-proximal *URA3* gene, enhancing the ability of cells to grow on 5-fluoro-orotic acid (5-FOA) (Boeke et al. 1987; Renauld et al. 1993). This improvement in telomeric position effect (TPE) reflects both a higher frequency and greater extent of silencing, allowing one to conclude that Sir3p seems to be limiting for TPE (Renauld et al. 1993).

Formaldehyde cross-linking techniques have shown that the Sir repression complex propagates inward from the telomeric repeats (Strahl-Bolsinger et al. 1997). Within a 4-kb core region, Sir2p, Sir3p, Sir4p, and Rap1p bind uniformly to nucleosomes. However, upon Sir3p overexpression, only Sir3p and Sir4p coimmunoprecipitate with the extended region of repression (up to 18 kb from the chromosome end). Since far more Sir3p was recovered than Sir4p, it was suggested that once core heterochromatin is formed, Sir3p can further propagate a repressed structure with a limited participation of the intact Sir complex.

In contrast to *SIR3*, overexpression of *SIR4* disrupts silencing and the silencing complex at both *HM* and telomeric loci (Marshall et al. 1987; Cockell et al. 1995). This effect is monitored by growth on media lacking uracil, so that enhanced growth reflects derepression of the telomere-proximal *URA3* gene (Fig. 1, rows 6 and 7). This derepression is presumably due to the titration of one or more essential silencing components. One ligand responsible for the dominant negative effect of full-length *SIR4* overexpression is clearly Sir3p, since the co-overexpression of Sir3p and Sir4p can partially restore silencing (see Fig. 5) (Maillet et al. 1996; Gotta et al. 1998). *SIR4* overexpression might also compete for an unidentified component required for Sir polymerization along the chromatin fiber or for essential silencing factors that can bind uniquely to Sir4p (e.g., Ku70 or Sir2p; see Moazed et al. 1997; Strahl-Bolsinger et al. 1997; Tsukamoto et al. 1997). Two-hybrid studies (Cockell et al. 1998; and see below) and results showing that Sir4p can enhance the mitotic stability of plasmids (Ansari and Gartenberg 1997) both indicate that Sir4p can mediate interactions and provide functions other than binding to Sir3p.

Functional studies of the amino- and carboxy-terminal domains of Sir3p and Sir4p demonstrate that the proteins have highly modular organization with specialized ligand-binding sites in the amino and carboxyl termini. A summary of these interactions is presented in Figure 2. Most striking is the observation that a small amino-terminal domain of Sir4p can complement a *sir4*-deficient strain when it is expressed in *trans* with the carboxy-terminal half of the protein (Marshall et al. 1987). Similarly, coexpression in *trans* of the amino and carboxyl domains of Sir3p also partially complements a complete *sir3* null allele for mating, although not for telomeric silencing (Gotta et al. 1998).

Amino-terminal and Carboxy-terminal Domains of Sir3p and Sir4p Affect Silencing through Interaction with Multiple Ligands

In an attempt to understand the different roles of the Sir3 amino and carboxyl termini, we have overexpressed these subdomains and monitored mating type and telomeric silencing in a *ura3* strain carrying a *URA3*-marked telomere. Overexpression of the carboxy-terminal half of Sir3p, which binds to itself, Sir4p, and the amino termini of histones H3 and H4, causes strong derepression of both telomeric and mating-type repression (Fig. 1, rows 3 and 5) (see Gotta et al. 1998). Intriguingly, overexpression of a 503-amino-acid Sir3p amino-terminal domain improves telomeric silencing with approximately the same efficiency as full-length Sir3p (Fig. 1, compare rows 1, 2, 3, and 4). This is not a bypass of the regular silencing mechanism, as the Sir3N-enhanced repression requires Sir2p, Sir4p, the histone amino termini, and, importantly, full-length Sir3p (Gotta et al. 1998). The effect is identical in the presence or absence of Sir1p. Expression of the same domain derepresses transcription of reporter genes inserted in the rDNA array, and immunolocalization of Sir3N-GFP and Sir2p suggests that Sir3N directly antagonizes nucleolar Sir2p, releasing an rDNA-bound population of Sir2p that can enhance repression at telomeres (Gotta et al. 1998).

Expression of either amino- or carboxy-terminal fragments of Sir4p also results in derepression at telomeres and *HM* loci (Fig. 1, rows 8, 9, and 10; summarized in Fig. 2). Most published work to date has focused on the coiled-coil motif found at the extreme carboxyl terminus, which is sufficient for both homodimerization of Sir4p and heterodimerization with Sir3p and Rap1 (Moretti et al. 1994). The carboxyl terminus also appears to bind Ku70 and Ubp3p (summarized in Fig. 2) (Moazed and Johnson 1996; Tsukamoto et al. 1997), whereas interaction with Sir2p involves an adjacent region between amino acids 743–1114 (Moazed et al. 1997). The ability of the carboxy-terminal domain to disrupt silencing could be explained as competition for any of these interactions (see Fig. 2).

In view of the lack of ligands known for the amino-terminal domain, it was surprising to observe that ectopic expression of amino acids 9–271 of Sir4p efficiently derepresses silencing both at telomeres and at the *HML* locus (shown for TPE in Fig. 1, compare rows 8 and 9). Through a series of two-hybrid assays, we find that this fragment interacts specifically with several components of the silencing machinery, notably, Sir1p and the amino-terminal tails of histones H3 and H4 (Fig. 2). A nonproductive binding of the amino-terminal fragment to Sir1p

Functional domains of Sir3p

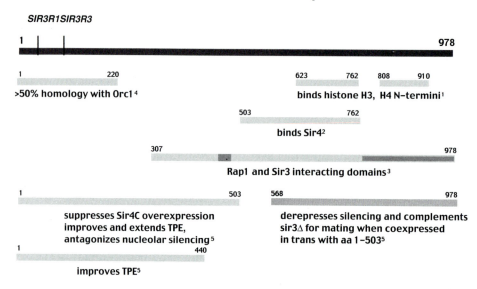

Functional domains of Sir4p

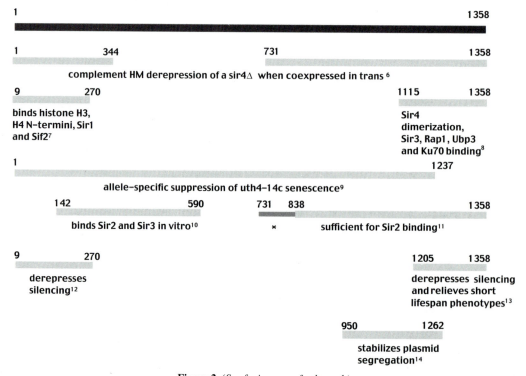

Figure 2. (*See facing page for legend.*)

may impair the nucleation of silent complexes at *HM* loci, and competition for histone amino termini may reduce repression generally. However, we do not exclude the presence of other unidentified Sir4N ligands necessary for the assembly of Sir complexes.

Sir2p Is Limiting at Telomeres, but Strong Overexpression Derepresses

In contrast to previously published findings (Renauld et al. 1993), we observe that the normal cellular level of Sir2p, like Sir3p, is limiting for TPE. One explanation for the apparent discrepancy may be the fact that low- and high-level overexpression of *SIR2* affects repression differently. Thus, when wild-type strains express low levels of Sir2p from plasmids, repression of a telomere-proximal *URA3* reporter is enhanced by 10–100-fold (Fig. 3, compare rows 1 and 2). High-level Sir2p expression (*GAL* upstream activating sequence [UAS] induction on a high-copy vector) gives the opposite result, reducing repression of the same reporter by approximately two orders of magnitude (Fig. 3, compare rows 7 and 8). Thus, although normal Sir2p levels are limiting for TPE, high-level expression has a dominant negative effect, like that of Sir4p. An intermediate level of Sir2p might be expected to give no apparent change in TPE.

The data shown in Figure 3 were obtained using a Sir2p fusion protein under the control of a galactose-inducible promoter on a high-copy vector (see Experimental Procedures). Glucose repression of this construct appears to be leaky since it is able to complement *sir2Δ* strains for their various silencing deficiencies on glucose (data not shown). Enhancement of TPE for a telomere-proximal *ADE2* reporter was also observed when *SIR2* per se (nonfused) was expressed from the *ADH* promoter on both high- and low-copy vectors on glucose (data not shown).

Silent Chromatin Does Not Spread from the Telomere upon Sir2p Overexpression

To determine whether enhanced repression due to Sir2p overexpression correlates with an increased spread of TPE, we examined the effects of *SIR2* overexpression in strains carrying the *URA3* reporter at different distances from the telomere (see diagram in Fig. 3). Weak *SIR2* overexpression enhances repression of the reporters placed at 2 kb (Fig. 3, compare rows 1 and 2) and at 3.5 kb (Fig. 3, compare rows 3 and 4), but we see no increase in repression when the reporter is positioned at 6.5 kb from the telomere (Fig. 3, compare rows 5 and 6). Thus, in contrast to the effects of *SIR3* overexpression, which can extend silencing up to 18 kb from the end of the chromosome (Renauld et al. 1993; Strahl-Bolsinger et al. 1997), *SIR2* overexpression appears to be unable to extend silencing beyond the region of core heterochromatin (Strahl-Bolsinger et al. 1997). We deduce that the derepression resulting from strong *SIR2* overexpression occurs when either Sir3p or Sir4p becomes limiting for the Sir2/3/4p complex, and we conclude that Sir2p, unlike Sir3p, is only able to propagate along subtelomeric nucleosomes when complexed to the other Sir proteins.

The rDNA Locus and Subtelomeric Regions Compete for a Limiting Pool of Sir2p

Recently, it has been shown that the variegated repression of pol II genes inserted into an rDNA repeat unit requires *SIR2*, but not *SIR3* and *SIR4* (Bryk et al. 1997; Fritze et al. 1997; Smith and Boeke 1997). Consistently, in logarithmically growing wild-type cells, Sir2p, but not Sir3p and Sir4p, is found bound to the rDNA within the nucleolus (Gotta et al. 1997). To see if *SIR2* overexpression has a dominant negative effect in the nucleolus, we monitored the effects of weak and strong Sir2p expression on a *URA3* reporter inserted at the rDNA locus. When serial dilutions of this strain are plated on 5-FOA in media that support low-level Sir2p overexpression, repression increases approximately tenfold (Fig. 4, rows 1 and 2). Consistently, approximately tenfold fewer Sir2p-expressing cells are able to grow on medium lacking uracil, when compared to cells carrying the vector alone. Similar results were observed when the levels of Sir2p or the Sir2p fusion protein are increased two- to threefold in strains that carry a wild-type *SIR2* locus (Smith et al. 1998; M. Cockell, data not shown), suggesting that Sir2p is normally limiting for repression at the rDNA.

Figure 2. Functional domains of Sir3p and Sir4p. Shown are schematic representations of full-length Sir3p and Sir4p, and the functional domains revealed by genetic, two-hybrid and biochemical studies (Moretti et al. 1994; Hecht et al. 1995; Strahl-Bolsinger et al. 1997; Gotta et al. 1998; P. Moretti and D. Shore; M. Grunstein; both pers. comm.). *SIR3R1* and *SIR3R3* are the two mutations isolated as suppressors of histone H4 mutants (Johnson et al. 1990). Footnotes indicate the following: (*1*) Hecht et al. 1995. (2) Two-hybrid data indicate that the Sir4p-binding domain is 3′ of amino acid 494 (M. Cockell and M. Gotta, unpubl.), and unpublished pull-down data indicate that there is only one site of interaction, not two as previously suggested (Strahl-Bolsinger et al. 1997; M. Grunstein, pers. comm.). (3) (Moretti et al. 1994) As indicated by the shaded boxes, the domain necessary and sufficient for Rap1p interaction has been narrowed down to amino acids 455–481 of Sir3p, and the Sir3p homodimerization domain has been defined from amino acid 762 to the end of the protein (D. Shore, pers. comm. (5) Gotta et al. (1998). The bottom half shows a schematic representation of Sir4p and the functional domains that have been assigned from genetic, two-hybrid, and biochemical studies. (6) Marshall et al. (1987). (7) Cockell et al. (1998); M. Cockell (unpubl.); n.b. the amino-terminal fragment 9–270 does not contain sufficient information for Sir2p or Sir3p binding, although it has significant overlap with a larger fragment (142–590) that has been shown to bind these two proteins in vitro. (8) Chien et al. (1991); Moretti et al. (1994); Moazed and Johnson (1996); Tsukamoto et al. (1997). (9) Kennedy et al. (1995). (*10*) Strahl-Bolsinger et al. (1997). (*11*) Strahl-Bolsinger et al. (1997); Moazed et al. (1997). (*) Deletion of amino acids 731–838 indicated by the shaded area renders the fragment unable to bind Sir2p; M. Cockell (unpubl.). (*12*) Cockell et al. (1998); M. Cockell (unpubl.). (*13*) Marshall et al. (1987); Cockell et al. (1995). (*14*) Ansari and Gartenberg (1997).

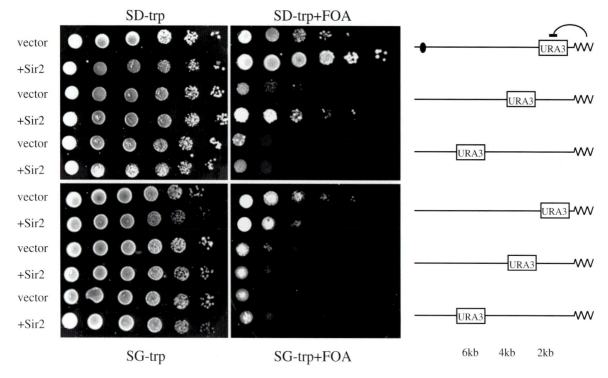

Figure 3. Sir2p is limiting for TPE but strong overexpression derepresses. Strains UCC518, UCC520, and UCC522 (Renauld et al. 1993) were transformed with a vector alone (pJG45) and with a plasmid (pJG45-Sir2) expressing a Sir2 fusion protein under a galactose-inducible promoter as described in Experimental Procedures. Colonies were grown for 3–5 days at 30ºC on medium lacking tryptophan to ensure maintenance of the plasmids. Colonies were resuspended in H$_2$O, and the same tenfold serial dilutions were plated onto SD and SG medium lacking tryptophan and SD and SG medium lacking tryptophan and containing 5-FOA. Next to the appropriate rows, schematic representations are drawn of the telomeres marked with *URA3* at 2, 3.5, and 6.5 kb in strains UCC518, UCC520, and UCC522, respectively.

Strong Sir2 Overexpression Improves rDNA Silencing

Immunoprecipitation of reversibly cross-linked protein-DNA complexes suggests that Sir2p may interact with chromatin throughout the length of the rDNA units, although it is not clear if nucleolar Sir2p is part of a larger complex or if its binding involves interaction with another limiting component. To investigate this, we examined the effect of inducing strong overexpression of our Sir2 fusion protein on the *URA3* reporter at the rDNA by plating the marked strain on media containing galactose (Fig. 4, lower panels). Under conditions of high Sir2p overexpression, repression in the rDNA increases nearly 10^5-fold over the control strain (see SG-trp-ura). When the same strain is plated on galactose-containing medium supplemented with 5-FOA, we detect only a tenfold increase in silencing, probably because a large proportion of this *sir2Δ* host strain is already able to survive on 5-FOA due to recombination and loss of *URA3*. Importantly, we observe no dominant negative effect of high Sir2p levels at the rDNA locus by either assay. These data are consistent with a model in which Sir2p coats the rDNA array, rather than binding at sequence-specific signals that are limiting within each rDNA repeat. Moreover, unlike the situation at telomeres, there does not ap-

pear to be a readily titrated ligand for Sir2p in the nucleolus.

Balanced Expression of Sir3p and Sir4p Domains Restores TPE and Telomeric Foci

As described above, the overexpression of full-length Sir4p or its carboxy-terminal domain (amino acids 743–1358, Sir4C) relieves silencing at both *HM* and telomeric loci. Coincident with the loss of silencing, the punctate staining patterns of Rap1p and Sir3p are disrupted upon Sir4C overexpression and Sir proteins are found diffuse throughout the nucleus (Cockell et al. 1995). To identify the limiting components potentially titrated by the overexpressed domains, a YEp13-based high-copy-number genomic library was introduced into the Sir4C overexpressing strain. Transformants were screened for red sectoring, indicative of restored repression of a telomere-proximal *ADE2* gene. Of 30 sectored transformants, each carrying a different plasmid, only 1 reproducibly restored the sectored phenotype following plasmid isolation and retransformation. Southern blot analysis and sequencing revealed that this clone contains the first 1.8 kb of *SIR3*, encoding the amino-terminal 503 amino acids of the protein (Sir3N, see Fig. 2), ex-

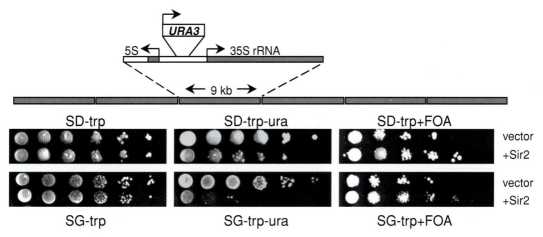

Figure 4. Sir2 overexpression increases rDNA silencing. Strain JS341 (*sir2Δ*) carrying the *URA3* gene inserted at the rDNA as indicated in the diagram was transformed with either the vector (pJG45) or the plasmid carrying a full-length *SIR2* gene fusion (pJG45-SIR2) under control of a galactose-inducible promoter. Colonies were grown for 3–5 days at 30°C on medium lacking histidine, uracil, and tryptophan and then resuspended in H$_2$O, and the same tenfold serial dilutions were plated onto glucose (SD) or galactose (SG) media that either lacked tryptophan, lacked tryptophan and uracil, or lacked tryptophan and contained 5-FOA.

pressed under the control of its own promoter (Gotta et al. 1998).

This fragment was cloned from an independent *SIR3* vector such that it could be expressed under the control of the *ADH* promoter. The resulting plasmid, pADH-SIR3N, was able to restore sectoring to levels higher than those of wild type in a Sir4C overexpressing strain (Fig. 5A). The efficiency of restoration of silencing correlated with the level of Sir3N expression. Immunofluorescence with anti-Sir4C antibodies indicates no destabilization or down-regulation of the Sir4C fragment when Sir3N is expressed (data not shown).

Sir3p and Sir4p Overexpression Restores Telomeric Foci

The relief of Sir4C derepression by Sir3N appears not to be mediated by the direct titration of one protein by the other, since the Sir3p amino-terminal fragment does not bind Sir4C by any of the standard assays, i.e., two-hybrid, coimmunoprecipitation, or GST pull-down (see Fig. 2). Since the Sir3N-provoked suppression requires intact Sir3p, one alternative model is that Sir3N "activates" or allows recruitment of an inactive Sir3p population. If true, then the overexpression of full-length Sir3p might be expected to restore TPE in a strain overexpressing Sir4C, as shown for Sir3N. This is indeed the case (Gotta et al. 1998). Moreover, the derepression provoked by overexpression of full-length Sir4p can be suppressed, and repression can even be improved beyond control levels, by balancing the higher levels of Sir4p with elevated levels of Sir3p. This is visualized as an enhancement in the frequency of dark red colonies when both pGPD-SIR3 and pADH-SIR4 are introduced into a strain carrying an *ADE2*-marked telomere (Fig. 5B). These results demonstrate that the disruptive effect of Sir4C or Sir4p overexpression can be compensated by increasing the amounts of Sir3p available for the assembly of silent chromatin.

Loss of silencing tightly correlates with the delocalization of Rap1p and Sir proteins from telomeric foci (Palladino et al. 1993; Gotta et al. 1996). In strains overexpressing either Sir4C or full-length Sir4p, telomeric silencing is relieved and Rap1p and Sir3p are delocalized from telomeric foci (Cockell et al. 1995). Here, we show that the simultaneous overexpression of both Sir3N and Sir4C (Fig. 5C, panel f), like the simultaneous and balanced overexpression of full-length Sir3p and Sir4p, restores the punctate staining of full-length Sir3p (Fig. 5C, panels c and h). This correlates with the restoration of repression at telomeres. Similarly, Rap1p and Sir4p foci are restored under conditions of balanced Sir protein overexpression (Fig. 5C, panels g and i). Western blot analysis confirms that the levels of Sir4p and Sir4C remain high when Sir3p or Sir3N are overexpressed, ruling out a trivial effect of Sir3N or Sir3p on *SIR4* expression or protein stability (Maillet et al. 1996). Thus, we extend the tight correlation between the restoration of TPE, balanced Sir3p and Sir4p levels, and a characteristic focal staining pattern, which correlates tightly with the formation of silencing-competent complexes.

Sif2p, a Novel Ligand for Sir4p, Competes for TPE

In a two-hybrid screen for novel ligands specific for the Sir4 amino terminus, we have identified Sif2p, a novel nuclear WD40 protein that is antagonistic to telomeric silencing when overexpressed (Cockell et al. 1998). The relevance of the Sif2p-Sir4N interaction for silencing was analyzed in isogenic *SIF2* and *sif2::kan*MX2 strains by scoring the expression of telomere-proximal *URA3* or *ADE2* reporters. When *SIF2* is disrupted, TPE improves between 10-fold and 100-fold. This was neither background- nor reporter-dependent and is shown for the Tel VII-L::*ADE2* reporter in Figure 6A. The enhanced repression reflects the Sir-mediated pathway, since the *sif2::kan*MX2 deletion does not improve repression in a *sir3* or *sir4* deletion background (data not shown).

To see whether *SIF2* deletion affects the *HM* loci, we deleted *SIF2* in a strain carrying a *LEU2´´lacZ* reporter at *HML* (Boscheron et al. 1996). When the reporter is flanked by only one silencer, the construct is partially repressed in *SIF2*[+] cells, allowing a maximum degree of sensitivity for either increase or loss of repression. Com-

plete deletion of *SIF2* does not derepress at *HML*, but rather slightly improves repression (~10%), showing that Sif2p, unlike Sir1p, is not necessary for *HM* silencing (Cockell et al. 1998). Together, these studies suggest that Sif2p competes for Sir4p recruitment or assembly at both telomeric and *HM* loci, with an effect more pronounced at telomeres.

Consistent with this hypothesis, we note that the overexpression of *SIF2* in a Tel VIIL::*ADE2* marked strain sig-

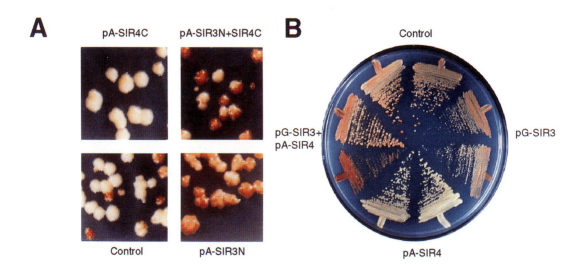

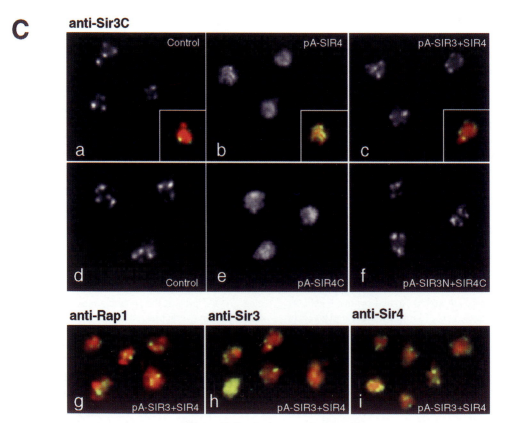

Figure 5. (*See facing page for legend.*)

nificantly weakens telomeric repression (Fig. 6B). The simplest model that could account for these phenotypes is that Sif2p regulates *SIR* gene expression or Sir protein turnover, yet this was ruled out by an ECL-Western, showing that the levels of Sir4p and Sir2p are not altered when *SIF2* is overexpressed or deleted (Cockell et al. 1998). Consistently, immunofluorescence signals of Rap1, Sir3p, and Sir4p remain constant upon *SIF2* overexpression. Alternatively, Sif2p could interfere with silencing by regulating the assembly of Sir4p into a Sir complex or disrupting silent chromatin. To test this, we have evaluated the subnuclear localization of Sif2p and Sir4p under conditions in which Sif2p interferes with telomeric repression; indeed, Sif2p overexpression provokes the mislocalization of Sir proteins from their usual concentrated localization at telomeres (Cockell et al. 1998).

In addition to the data presented here, other investigators have substantiated the idea that telomeres, *HM* loci, the rDNA, and other internal sites compete for silencing factors (Chien et al. 1993; Buck and Shore 1995; Kennedy et al. 1995; Boscheron et al. 1996; Lustig et al. 1996; Maillet et al. 1996; Marcand et al. 1996; Smith et al. 1998). We and others have recently shown that the yeast Ku proteins, which play an important part in Rad52-independent double-strand break repair, are directly involved in telomeric silencing, although not in repression at *HM* loci (Boulton and Jackson 1998; Laroche et al. 1998). Since the perinuclear clustering of telomeres and telomere length maintenance are also perturbed when either Ku subunit is mutated, Ku may bind at the chromosomal end to help nucleate silent chromatin (Laroche et al. 1998). Indeed, two-hybrid data suggest direct interaction between Ku70 and the Sir4 carboxyl terminus (Tsukamoto et al. 1997).

The apparent redundancy of interactions among Sir proteins, the growing number of Sir-binding factors that are implicated in targeting Sir complexes to different locations (see Fig. 7), and the complicated effects of Sir imbalance suggest that silencing complexes may assemble by several alternative pathways. We propose that Sir4p is a pivotal site of contact that helps integrate the alternative targeting pathways.

DISCUSSION

The dosage and balance of the proteins encoded by *SIR2*, *SIR3*, and *SIR4* are critical for the maintenance of position-dependent transcriptional silencing in yeast. Overexpression of some Sir proteins and Sir domains (Sir4p, Sir4N, Sir4C, Sir3C, and high levels of Sir2p) provoke derepression, whereas others (Sir3N, Sir3p, and low levels of Sir2p) improve repression at telomeres. The derepression results at least in part from competition for limited binding sites on other Sir proteins or for the amino-terminal tails of histones H3 and H4. Improved TPE may, in some cases, result from activation of a population of inactive Sir3p or from recruitment of Sir2p from the rDNA (Gotta et al. 1998). Importantly, improved repression is not achieved by bypassing the normal pathways of Sir-dependent repression.

Silencing can be restored in many cases by restoring the balance between overexpressed domains. As shown here, the depression provoked by Sir4C or Sir4p overexpression can be suppressed by overexpression of either Sir3p or its amino-terminal domain. The restoration of silencing correlates with a reformation of the typical focal staining pattern of Sir proteins at the nuclear periphery. The elevated balanced expression of Sir3p and Sir4p also allows the cell to repress silencer-flanked reporters at internal chromosomal sites, which are normally derepressed due to their distance from telomeric repeats (Maillet et al. 1996).

Competition between Domains of the Nucleus for Silencing Factors

In 1991, Aparicio et al. pointed out that the factors involved in telomeric position effect were a subset of those required to mediate repression of the silent mating-type cassettes and suggested the idea of a hierarchy of transcriptional silencing pathways (Aparicio et al. 1991). In support of this, it has been demonstrated that *HM* loci and telomeres compete for the recruitment of silencing factors, since both deletion of the Rap1 interacting factor, Rif1p, and tethering of Sir1p to telomeres increase TPE at

---

Figure 5. (*A*) Overexpression of Sir3N restores sectoring in Sir4C overexpressing cells. Strain AJL275-2A$_{VIIL}$, which carries *ADE2* adjacent to the VII-L telomere, was transformed with pAAH5 and p2GH (labeled Control), pAAH5 and pADH-SIR4C (labeled pA-SIR4C), pADH-SIR3N and p2GH (labeled pA-SIR3N), and pADH-SIR4C and pADH-SIR3N (labeled pA-SIR3N+SIR4C). In all the cases, two plasmids were present, and isolated colonies were streaked onto medium lacking histidine and leucine to ensure maintenance of the plasmids. Adenine concentrations are limiting. Colonies were allowed to grow for 5 days at 30°C. Following incubation, the plates were stored at 4°C to enhance the pigmentation of the cells. (*B*) Overexpression of full-length Sir3p restores silencing in a strain that overexpresses full-length Sir4p. Strain UCC3107, which carries *ADE2* adjacent to the V-R telomere, was transformed with plasmids as indicated as well as a second control plasmid. The control plasmids are the backbone vector without *SIR* gene inserts, namely, pAAH5 and p2HG. Plasmids with *SIR* gene inserts are pADH-SIR4 (labeled pA-SIR4) and pGPD-SIR3 (labeled pG-SIR3). Colonies (two independent transformants for each case) were streaked onto medium lacking histidine and leucine to ensure maintenance of the plasmids and allowed to grow 3–5 days at 30°C. Following incubation, the plates were stored at 4°C for 1 week to enhance the pigmentation of the cells. (*C*) Overexpression of Sir3N restores the focal staining pattern of Sir3p. (*a,b,c*) Strain EG37 (Maillet et al. 1996) transformed with the vectors pAAH5 and pRS316 (panel *a*), with pC-ASir4 (Maillet et al. 1996) and pAAH5 (panel *b*), and with pC-ASir4 and p2μ-ASir3 (panel *c*). (*d,e,f*) Strain AJL275-2AVII$_L$ transformed with the vectors pAAH5 and p2HG (panel *d*), with pADH-SIR4C (pFP340) and pAAH5 (panel *e*), and pADH-SIR4C (pFP340) and pADH-SIR3N (panel *f*). The cells were stained with anti-Sir3C antibodies detected by a DTAF-conjugated secondary antibody (green or white signal) that are superimposed on the red genomic DNA staining in the inserts (see Experimental Procedures). (*g,h,i*) Strain EG37 transformed with pC-ASir4 and p2μ-ASir3. Immunofluorescence was performed with anti-Rap1p (*g*), anti-Sir3p (*h*), and anti-Sir4p (*i*). (Data reprinted, with permission, from Gotta et al. 1998 [copyright ASM].)

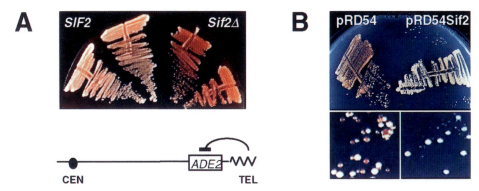

Figure 6. Deletion of *SIF2* enhances telomeric silencing and overexpression of Sif2p derepresses telomeric silencing. (*A*) *ADE2* repression of isogenic *SIF2* (strain GA492) and *sif2::kanMX2* deletion strains (MC118) that carry *ADE2* at the V-R telomere was compared by streaking independent single colonies onto synthetic media with limiting adenine concentrations. A higher frequency of red colonies indicates improved telomeric repression. On 5-FOA media, the improved repression of Tel VIIL::*URA3* was quantified as 25-fold (Cockell et al. 1998). (*B*) Effect of overexpression of full-length HA-tagged Sif2p (pRD54-SIF2) was compared with the effect of the vector alone (pRD54), in a Tel V-R::*ADE2* strain (UCC3107). Transformants were streaked onto synthetic selective medium lacking uracil, but with limiting adenine and 2% galactose/1% raffinose to induce transcription from the *GAL1* promoter. Upon Sif2p overexpression, colonies are white, indicating derepression of the telomeric *ADE2*; red and white sectors indicate metastable repression. (Data reprinted, with permission, from Cockell et al. 1998 [copyright ASM].)

the expense of repression at the *HM* loci (Chien et al. 1993; Buck and Shore 1995).

We have observed that a significant proportion of Sir2p is found in the nucleolus in addition to its telomeric localization (Gotta et al. 1997). In the nucleolus, Sir2p represses recombination, transcription, and accessibility to *dam* methylase (Bryk et al. 1997; Fritze et al. 1997; Smith and Boeke 1997). Disruption of *SIR4*, as well as overexpression of *SIR2*, increases the efficiency of rDNA silencing while disrupting TPE, suggesting that there is competition between the nucleolus and telomeres for limiting amounts of Sir2p (Figs. 3 and 4) (Fritze et al. 1997; Smith

and Boeke 1997). Finally, several groups have shown that Sir-dependent repression achieved either by targeted Sir proteins or by silencer elements themselves is influenced by the location of the reporter gene with respect to the ends of chromosomes (Thompson et al. 1994; Lustig et al. 1996; Maillet et al. 1996; Marcand et al. 1996). The spatial restrictions on repression can be overcome by elevating the concentration of Sir proteins in the nucleus, leading to the proposal that telomeres function as reservoirs for high concentrations of silencing factors, which can be targeted elsewhere under specific physiological conditions (Maillet et al. 1996; Marcand et al. 1996).

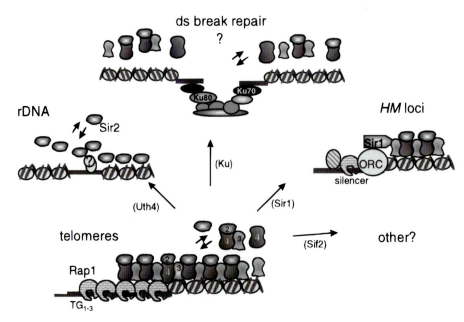

Figure 7. A schematic model for targeting of limiting pools of Sir proteins to its sites of action. The model proposes that telomeres act as a reservoir for the binding of Sir proteins which prevents them from indiscriminate repression at other sites in the genome. Sir proteins from this telomeric pool may be recruited to other sites by interaction with additional targeting factors such as Sir1p, Sif2p, Ku70p, or Uth4p.

Regulating Sir Protein Distribution

What regulates the distribution and equilibrium of silencing factors within the yeast nucleus? We have presented evidence that there are specific proteins involved in the recruitment of the silencing complex to different sites of action. The best characterized of these is Sir1p, although immunological and genetic arguments implicate other Sir4p ligands, Sif2p and Uth4p, in similar functions. Moreover, mutations in either subunit of the Ku70/80 heterodimer specifically relieve TPE but not *HM* silencing (Laroche et al. 1998), showing that it is not a general component of silent chromatin. Ku70 may, however, be instrumental in the recruitment of Sir proteins both to telomeres and possibly to sites of double-strand break repair. The levels of these putative recruitment proteins and their affinity for Sir ligands may be regulated by physiological signals, providing a means for chromatin to respond to environmental change.

There appears to be a tight control over the abundance of Sir proteins, although the mechanisms that achieve this remain obscure. Certain cell cycle controls may restrict assembly of the chromatin-bound Sir2/3/4p complex at particular loci or at specific points in the cell cycle. It can be expected that certain Sir domains are both positively and negatively regulated by phosphorylation. Like the Sir4 amino terminus, the Sir4 carboxyl terminus appears to be an important binding site for factors that regulate the targeting and turnover of silencing complexes, since it binds Ku, Rap1, and Ubp3p, a ubiquitin-binding protein involved in deubiquitination (Moazed and Johnson 1996; Ansari and Gartenberg 1997; Tsukamoto et al. 1997). Some of the physiologically relevant interactions mediated by Sir4p are illustrated in Figure 2.

In Figure 7, we depict our current model for Sir protein function within the nucleus. Subtelomeric sequences appear to serve as a sort of reservoir or pool of repressive Sir complexes, which, however, have only a minor role in stabilizing telomeric structure (Palladino et al. 1993). The equilibrium of Sir factors that we show exists between different sites of action is modulated by a series of proteins that may help recruit the silencing complexes to specific sites of action, such as Sir1p, Ku, and possibly Sif2p and Uth4p. Through mutually exclusive interactions, complexes containing subsets of Sir proteins may form that also shift the balance from one site of repression to another. Finally, at another level of control, physiological conditions may restrict assembly of the chromatin-bound Sir2/3/4p complex to particular loci. Examples of competition between different domains of heterochromatin have also been described for *Drosophila* (Lloyd et al. 1997), suggesting that competition between different assembly pathways or sites for limiting factors involved in chromatin structure is a conserved mechanism for regulating gene repression.

ACKNOWLEDGMENTS

We acknowledge the expert technical assistance of T. Laroche. We thank H. Renauld for strains and D. Shore and M. Grunstein for allowing us to cite unpublished results. We extend our acknowledgments to many laboratories and scientists with whom we have shared many fruitful collaborations and lively discussions, namely, Andreas Hecht, Sabine Strahl Bolsinger, Michael Grunstein, Brian Kennedy, Leonard Guarente, Laurent Maillet, Eric Gilson, and Edward Louis. M.G. thanks ISREC for a Ph.D. fellowship. Research in the Gasser laboratory is supported by the Swiss National Science Foundation, Human Frontiers Science Program, and the Swiss League against Cancer.

REFERENCES

Ansari A. and Gartenberg M.R. 1997. The yeast silent information regulator Sir4p anchors and partitions plasmids. *Mol. Cell. Biol.* **12:** 7061.

Aparicio O.M., Billington B.L., and Gottschling D.E. 1991. Modifiers of position effect are shared between telomeric and silent mating-type loci in *S. cerevisiae. Cell* **66:** 1279.

Boeke J.D., Trueheart J., Natsoulis G., and Fink G.R. 1987. 5-Fluoroorotic acid as a selective agent in yeast molecular genetics. *Methods Enzymol.* **154:** 164.

Boscheron C., Maillet L., Marcand S., Tsai-Pflugfelder M., Gasser S.M., and Gilson E. 1996. Cooperation at a distance between silencers and proto-silencers at the yeast *HML* locus. *EMBO J.* **15:** 2184.

Boulton S.J. and Jackson S.P. 1998. Components of the Ku-dependent non-homologous end-joining pathway are involved in telomeric length maintenance and telomeric silencing. *EMBO J.* **17:** 1819.

Braunstein M., Rose A.B., Holmes S.G., Allis C.D., and Broach J.R. 1993. Transcriptional silencing in yeast is associated with reduced nucleosome acetylation. *Genes Dev.* **7:** 592.

Braunstein M., Sobel R.E., Allis C.D., Turner B.M., and Broach J.R. 1996. Efficient transcriptional silencing in *Saccharomyces cerevisiae* requires a heterochromatin histone acetylation pattern. *Mol. Cell. Biol.* **16:** 4349.

Bryk M., Banerjee M., Murphy M., Knudsen K.E., Garfinkel D.J., and Curcio M.J. 1997. Transcriptional silencing of Ty elements in the RDN1 locus of yeast. *Genes Dev.* **11:** 255.

Buck S.W. and Shore D. 1995. Action of a RAP1 carboxy-terminal silencing domain reveals an underlying competition between HMR and telomeres in yeast. *Genes Dev.* **9:** 370.

Chien C.T., Bartel, P.L., Sternglanz R. and Fields S. 1991. The two-hybrid system: A method to identify and clone genes for proteins that interact with a protein of interest. *Proc. Natl. Acad. Sci.* **88:** 9578.

Chien C.T., Buck S., Sternglanz R., and Shore D. 1993. Targeting of SIR1 protein establishes transcriptional silencing at HM loci and telomeres in yeast. *Cell* **75:** 531.

Cockell M., Renauld H., Watt P., and Gasser S.M. 1998. Sif2p interacts with the Sir4p amino-terminal domain and antagonizes telomeric silencing in yeast. *Curr. Biol.* **8:** 787.

Cockell M., Palladino F., Laroche T., Kyrion G., Liu C., Lustig A.J., and Gasser S.M. 1995. The carboxy termini of Sir4 and Rap1 affect Sir3 localization: Evidence for a multi-component complex required for yeast telomeric silencing. *J. Cell Biol.* **129:** 909.

Fritze C.E., Verschueren K., Strich R., and Esposito R.E. 1997. Direct evidence for SIR2 modulation of chromatin structure in yeast rDNA. *EMBO J.* **16:** 6495.

Golemis E.A., Gyuris J., and Brent R. 1996. Interaction trap/two hybrid system to identify interacting proteins. In *Current protocols in molecular biology* (ed. F.M. Ausubel et al.), p. 20.1.1. Wiley, New York.

Gotta M., Palladino F., and Gasser S.M. 1998. A functional analysis of the Sir3 N-terminal domain. *Mol. Cell. Biol.* **18:** 6110.

Gotta M., Laroche T., Formenton A., Maillet L., Scherthan H., and Gasser S.M. 1996. The clustering of telomeres and colocalization with Rap1, Sir3 and Sir4 proteins in wild-type *Sac*-

charomyces cerevisiae. J. Cell Biol. **134:** 1349.

Gotta M., Strahl-Bolsinger S., Renauld H., Laroche T., Kennedy B.K., Grunstein M., and Gasser S.M. 1997. Localization of Sir2p: The nucleolus as a compartment for silent information regulators. *EMBO J.* **16:** 3243.

Gottlieb S. and Esposito R.E. 1989. A new role for a yeast transcriptional silencer gene, SIR2, in regulation of recombination in ribosomal DNA. *Cell* **56:** 771.

Gottschling D.E., Aparicio O.M., Billington B.L., and Zakian V.A. 1990. Position effect at *S. cerevisiae* telomeres: Reversible repression of Pol II transcription. *Cell* **63:** 751.

Gravel S., Larrivée M., Labrecque P., and Wellinger R.J. 1998. Yeast Ku as a regulator of chromosomal DNA end-structure. *Science* **280:** 741.

Hecht A., Strahl-Bolsinger S., and Grunstein M. 1996. Spreading of transcriptional repressor SIR3 from telomeric heterochromatin. *Nature* **383:** 92.

Hecht A., Laroche T., Strahl-Bolsinger S., Gasser S.M., and Grunstein M. 1995. Histone H3 and H4 N-termini interact with SIR3 and SIR4 proteins: A molecular model for the formation of heterochromatin in yeast. *Cell* **80:** 583.

Henikoff S. 1996. Dosage-dependent modification of position-effect variegation in *Drosophila. BioEssays* **18:** 401.

Johnson L.M., Kayne P.S., Kahn E.S., and Grunstein M. 1990. Genetic evidence for an interaction between SIR3 and histone H4 in the repression of the silent mating loci in *Saccharomyces cerevisiae. Proc. Natl. Acad. Sci.* **87:** 6286.

Kennedy B.K., Austriaco N.R., Jr., Zhang J., and Guarente L. 1995. Mutation in the silencing gene SIR4 can delay aging in *S. cerevisiae. Cell* **80:** 485.

Kennedy B.K., Gotta M., Sinclair D.A., Mills K., McNabb D.S., Murthy M., Pak S.M., Laroche T., Gasser S.M., and Guarente L. 1997. Redistribution of silencing proteins from telomeres to the nucleolus is associated with extension in life span in *S. cerevisiae. Cell* **89:** 381.

Laroche T., Martin S.G., Gotta M., Gorham H.C., Pryde F.E., Louis E.J., and Gasser S.M. 1998. Mutations of yeast Ku genes disrupt the subnuclear organization of telomeres. *Curr. Biol.* **8:** 653.

Laurenson P. and Rine J. 1992. Silencers, silencing and heritable transcriptional states. *Microbiol. Rev.* **56:** 543.

Liu C. and Lustig A.J. 1996. Genetic analysis of Rap1p/Sir3p interactions in telomeric and HML silencing in *Saccharomyces cerevisiae. Genetics* **143:** 81.

Lloyd V.K., Sinclair D.S., and Grigliatti T.A.. 1997. Competition between different variegating rearrangements for limited heterochromatic factors in *Drosophila melanogaster. Genetics* **145:** 945.

Lustig A.J., Liu C., Zhang C., and Hanish J.P. 1996. Tethered Sir3p nucleates silencing at telomeres and internal loci in *Saccharomyces cerevisiae. Mol. Cell. Biol.* **16:** 2483.

Maillet L., Boscheron C., Gotta M., Marcand S., Gilson E., and Gasser S.M. 1996. Evidence for silencing compartments within the yeast nucleus: A role for telomere proximity and Sir protein concentration in silencer-mediated repression. *Genes Dev.* **10:** 1796.

Marcand S., Buck S.W., Moretti P., Gilson E., and Shore D. 1996. Silencing of genes at nontelomeric sites in yeast is controlled by sequestration of silencing factors at telomeres by Rap 1 protein. *Genes Dev.* **10:** 1297.

Marshall M., Mahoney D., Rose A., Hicks J.B., and Broach J.R. 1987. Functional domains of SIR4, a gene required for position effect regulation in *S. cerevisiae. Mol. Cell. Biol.* **7:** 4441.

Moazed D. and Johnson A.D. 1996. A deubiquitinating enzyme interacts with SIR4 and regulates silencing in *S. cerevisiae. Cell* **86:** 667.

Moazed D., Kistler A., Axelrod A., Rine J., and Johnson A.D. 1997. Silent information regulator protein complexes in *Saccharomyces cerevisiae:* A SIR2/SIR4 complex and evidence for a regulatory domain in SIR4 that inhibits its interaction with SIR3. *Proc. Natl. Acad. Sci.* **94:** 2186.

Moretti P., Freeman K., Coodly L., and Shore D. 1994. Evidence that a complex of SIR proteins interacts with the silencer and telomere-binding protein RAP1. *Genes Dev.* **8:** 2257.

Palladino F., Laroche T., Gilson E., Axelrod A., Pillus L., and Gasser S.M. 1993. SIR3 and SIR4 proteins are required for the positioning and integrity of yeast telomeres. *Cell* **75:** 543.

Renauld H., Aparicio O.M., Zierath P.D., Billington B.L., Chhablani S.K., and Gottschling D.E. 1993. Silent domains are assembled continuously from the telomere and are defined by promoter distance and strength, and by SIR3 dosage. *Genes Dev.* **7:** 1133.

Rose M.D., Winston F., and Hieter P. 1990. *Methods in yeast genetics.* Cold Spring Harbor Laboratory Press, Cold Spring Harbor, New York.

Sambrook J., Fritsch E.F., and Maniatis T. 1989. *Molecular cloning: A laboratory manual,* 2nd edition. Cold Spring Harbor Laboratory Press, Cold Spring Harbor, New York.

Sandell L. and Zakian V.A. 1992. Telomeric position effect in yeast. *Trends Cell Biol.* **2:** 10.

Singer M.S. and Gottschling D.E. 1994. TLC1: Template RNA component of *Saccharomyces cerevisiae* telomerase. *Science* **266:** 404.

Smeal T., Claus J., Kennedy B., Cole F., and Guarente L. 1996. Loss of transcriptional silencing causes sterility in old mother cells of *S. cerevisiae. Cell* **84:** 633.

Smith J.S. and Boeke J.D. 1997. An unusual form of transcriptional silencing in yeast ribosomal DNA. *Genes Dev.* **11:** 241.

Smith J.S., Brachmann C.B., Pillus L., and Boeke J.D. 1998. Distribution of a limited Sir2 protein pool regulates the strength of yeast rDNA silencing and is modulated by Sir4p. *Genetics* **149:** 1205.

Strahl-Bolsinger S., Hecht A., Kunheng L., and Grunstein M. 1997. SIR2 and SIR4 interactions differ in core and extended telomeric heterochromatin in yeast. *Genes Dev.* **11:** 83.

Thompson J.S., Hecht A., and Grunstein M. 1993. Histones and the regulation of heterochromatin in yeast. *Cold Spring Harbor Symp. Quant. Biol.* **58:** 247.

Thompson J.S., Ling X., and Grunstein M. 1994. Histone H3 amino terminus is required for telomeric and silent mating locus repression in yeast. *Nature* **369:** 245.

Tsukamoto Y., Kato J., and Ikeda H. 1997. Silencing factors participate in DNA repair and recombination in *Saccharomyces cerevisiae. Nature* **388:** 900.

Turner B.M., Birley A.J., and Lavender J. 1992. Histone H4 isoforms acetylated at specific lysine residues define individual chromosomes and chromatin domains in *Drosophila* polytene nuclei. *Cell* **69:** 375.

Activation and Repression Mechanisms in Yeast

K. Struhl, D. Kadosh, M. Keaveney, L. Kuras, and Z. Moqtaderi

*Department Biological Chemistry and Molecular Pharmacology, Harvard Medical School,
Boston, Massachusetts 02115*

In eukaryotes, gene expression depends on activator proteins that bind enhancer elements and stimulate transcription by RNA polymerase II (pol II) (Struhl 1995; Zawel and Reinberg 1995). This general requirement for activators is inferred from numerous observations in vivo that intact promoters are much more efficiently transcribed than core promoter derivatives containing only the TATA and initiator elements. The pol II transcription machinery is complex and has a molecular weight comparable to that of a ribosome. The pol II machinery is composed of two basic components, TFIID and the pol II holoenzyme. The TFIID complex, which contains the TATA-binding protein (TBP) and TBP-associated factors (TAFs), specifically binds the core promoter region; TBP interacts with high affinity and specificity for TATA elements, whereas certain TAFs can interact with some specificity for initiator and downstream elements (Burley and Roeder 1996; Verrijzer and Tjian 1996; Burke and Kadonaga 1997). The pol II holoenzyme contains the core subunits of the enzyme, basic transcription factors (e.g., TFIIB), as well as Srb, Med, and a variety of other proteins (Koleske and Young 1995; Myers et al. 1998).

Activator proteins generally bind their cognate promoter elements with high specificity and affinity, and they can often bind their target sites in the context of nucleosomal templates, the physiologically relevant substrate (Kingston et al. 1996; Polach and Widom 1996). In contrast, the TBP moiety of the TFIID complex is virtually unable to bind TATA elements in nucleosomal templates, although weak binding is observed when chromatin is disrupted by histone acetylation or by nucleosome remodeling (Imbalzano et al. 1994). The pol II holoenzyme does not appear to recognize specific DNA sequences, and its association with promoters reflects protein-protein interactions with TFIID and/or activators.

Activators contain a DNA-binding domain that specifically recognizes enhancer elements and a physically separate activation domain that stimulates transcription (Struhl 1996; Ptashne and Gann 1997). Activation domains are functionally autonomous; they retain their functional activity when fused at different positions to a wide variety of heterologous DNA-binding domains and when tethered at different positions in the promoter region. Activation domains can interact directly with many components of the pol II machinery, and they can affect multiple steps in the assembly of an active transcription complex. However, the molecular mechanisms of transcriptional activation in vivo, particularly the physiological significance and relative importance of specific protein-protein interactions and mechanistic steps, remain to be clarified. For example, activators can interact with TBP or isolated TAFs, but there is no evidence for activator-TAF interactions in the context of TFIID or for activator-TBP interactions when TBP is bound to TATA elements.

This paper reviews our efforts to understand the molecular mechanism of transcriptional activation in yeast. These studies take advantage of the power of yeast genetics and molecular biology, and the experiments are typically performed under conditions where all proteins are present at physiological concentrations, and the DNA template is in the form of chromatin. In addition, we discuss activation and repression mechanisms in which changes in chromatin structure have a direct and active role in transcriptional regulation.

QUALITY OF ACTIVATION DOMAINS AND TATA ELEMENTS ARE LIMITING FOR TRANSCRIPTION IN VIVO

In considering the physiological mechanism of transcriptional activation, a critical issue is the nature of the limiting component or step in the process. By definition, a component of a chemical or biological process is limiting if small decreases in functional concentration or activity decrease the output of the process. The question of whether individual components of the pol II machinery are limiting for transcriptional activation in vivo is completely separate from the issue of whether such components are absolutely required for pol II transcription. Even if a component is essential (i.e., removal or complete inactivation eliminates transcription), it is not limiting if large decreases in its activity do not significantly affect the overall output.

Progressive deletion of the Gcn4 activation domain causes a series of small step-wise reductions of activity, rather than defining a position where there is a precipitous loss of activity (Hope et al. 1988). The strong correlation between the length of the Gcn4 activation region and the level of transcriptional activity is strongly suggestive of a repeating structure consisting of units that act additively. More specifically, the boundaries defining the step-wise reductions in transcription occur every seven residues, suggesting that α-helical character is important for activation domain. Along with many other results, these observations indicate that transcriptional activation regions do not have a defined tertiary structure such as found in active sites or domains in a protein. Indeed, X-ray structural analysis indicates that activation domains become structured only upon specific interaction with another protein (Uesugi et al. 1997). Most importantly, the observation that progressive and subtle changes in the activation domain result in a gradual decrease in transcriptional output strongly argues that the quality of the activation domain is limiting for transcription in vivo.

A similar line of evidence suggests that the quality of the TATA element is also limiting for transcription in vivo. Detailed mutational analyses of the canonical TATA element in the *his3* promoter indicates that single-base-pair substitutions in this element results in a wide range of transcriptional outputs in vivo (Chen and Struhl 1988; Harbury and Struhl 1989) and in vitro (Wobbe and Struhl 1990). TATA elements in natural yeast promoters vary considerably in sequence, indicating that TATA element quality is physiologically important in determining relative levels of gene expression. It should be noted, however, that TATA element quality is not necessarily equivalent to TBP-TATA-binding affinity and that the level of TBP (or TFIID) may not be limiting. In particular, TFIIA and TFIIB recognize the TBP-TATA complex (Nikolov et al. 1995; Geiger et al. 1996; Tan et al. 1996), suggesting that the quality of the TATA element (particularly its ability to be structurally deformed) may influence the formation of transcriptionally relevant protein-DNA complexes that involve proteins in addition to TBP.

ACTIVATOR-DEPENDENT RECRUITMENT OF THE POL II MACHINERY TO PROMOTERS

Kinetic Evidence

The rate at which TBP interacts with the TATA element and promotes transcription in vivo was determined by rapidly inducing an altered-specificity TBP derivative (Strubin and Struhl 1992) and measuring transcription from promoters with appropriately mutated TATA elements (Klein and Struhl 1994). In the absence of an activator, transcription dependent on the altered-specificity TBP occurs only after a lag of several hours. In contrast, Gcn4-activated transcription occurs rapidly upon induction of the TBP derivative. This strongly suggests that accessibility of TBP to the chromatin template in vivo is limiting and that the Gcn4 activation domain can increase recruitment of TBP to the promoter.

Note that this experiment only measures the initial access of TBP to promoters, i.e., the difference in lag times in the nonactivated versus the activated situation. At steady state, the relative increase in transcription dependent on the altered-specificity TBP is equivalent in both situations (although, of course, the actual transcription level is higher in the activated case). This suggests that there is a difference in the ability of TBP to access a previously inactive template as opposed to a template that has been recently utilized.

Artificial Recruitment of the pol II Machinery Bypasses the Need for an Activation Domain

We (Chatterjee and Struhl 1995) and others (Klages and Strubin 1995; Xiao et al. 1995) showed that efficient activation can occur simply by physically connecting TBP to a heterologous DNA-binding domain; i.e., artificial recruitment of TBP to the promoter can bypass the normal requirement for an activation domain (Fig. 1). This suggests that interaction of TBP with the TATA element can be a limiting step for transcription in vivo, that natural activation domains can increase recruitment of TBP to the promoter, and that interactions between activation domains and general factors that function after TBP recruitment (e.g., TFIIB, TFIIF, and pol II) are not absolutely required for transcriptional activation.

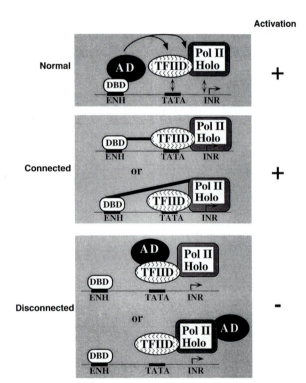

Figure 1. Transcriptional activation in yeast occurs predominantly by recruitment of the pol II machinery. (*Top*) In the physiologically relevant situation, activators bind enhancer elements via DNA-binding domains (DBD) and stimulate transcription via activation domains (AD). Arrows indicate the interactions between activation domains and the TFIID and/or the pol II holoenzyme complexes, although the direct targets within these complexes are not specified. Activator-dependent recruitment of the pol II machinery is depicted by arrows between TFIID and the TATA element and the pol II holoenzyme and the mRNA initiation site. (*Middle*) In the connected situation, activation is achieved in the absence of an activation domain by physically connecting (*thick bold line*) a component of either TFIID or the pol II holoenzyme to an enhancer-bound protein, thereby artificially recruiting the pol II machinery to promoters. (*Bottom*) In the disconnected situation, activation does not occur when the activation domain is transferred from its normal location on the enhancer-bound protein to a component of either TFIID or the pol II holoenzyme. (Reprinted, with permission, from Keaveney and Struhl 1998 [copyright Cell Press].)

Although initially demonstrated for TBP, activation by artificial recruitment occurs when an enhancer-bound protein is connected to virtually any individual component of the pol II machinery. Examples of such components include TFIIB (Gonzalez-Couto et al. 1997; Lee and Struhl 1997), TAFs (Apone et al. 1996; Gonzalez-Couto et al. 1997; Keaveney and Struhl 1998), and pol II holoenzyme subunits (Barberis et al. 1995; Farrell et al. 1996). However, the relationship of these artificial recruitment experiments to the physiological mechanism by which activation domains enhance transcription by the pol II machinery is unclear. Because the direct connection between the enhancer-bound protein and the pol II machinery is equivalent to exceptionally strong protein-protein interactions, artificial recruitment experiments might represent a bypass mechanism that is distinct from the physiological process that occurs with natural activators; i.e., the interaction of an activation domain with a single target in the pol II machinery might not be sufficient to mediate a significant degree of activation in vivo.

Activator-dependent Recruitment of the pol II Machinery Is the Predominant Mechanism for Activation in Yeast

If the physiological role of an activation domain is simply to recruit the pol II machinery, an activation domain within the machinery itself should not overcome the inherent inability of the pol II machinery to associate with promoters. To examine whether an activation domain in the preinitiation complex is sufficient for activation, we transferred activation domains from their normal location on the enhancer-bound protein to a variety of components of the pol II machinery (Keaveney and Struhl 1998). In this situation, the activation domain is physically disconnected from the enhancer-bound protein, and transcriptional stimulation does not occur (Fig. 1). However, complementation experiments indicate that the pol II machinery harboring a strong activation domain is transcriptionally competent and supports normal cell growth. This strongly suggests that the presence of an activation domain within the pol II machinery does not affect the transcriptional status of the majority of yeast genes.

In comparing the normal and "disconnected" situations, the components of the pol II machinery, the domains of the activator, the promoter, and cell physiology are identical, yet transcriptional output is dramatically different. As all of the ingredients for activation are available in the disconnected situation, the failure to activate almost certainly reflects an inability of the pol II machinery to interact with the promoter in vivo, not an inherent inactivity of the pol II machinery itself. Furthermore, unlike the activation domain, the requirement for the DNA-binding domain of the enhancer-bound protein cannot be bypassed, even though such DNA-binding domains are not usually involved in the transcriptional initiation process per se other than bringing activation domains to promoters.

These considerations indicate that (1) efficient activation requires firmly anchoring of the pol II machinery at the promoter, (2) the pol II machinery is inherently unable to associate stably with the promoter even if it carries an activation domain, (3) the DNA-binding domain provides the anchor for the pol II machinery to associate stably with the promoter, and (4) the predominant role of the activation domain is to provide the connection between the anchor and the enzymatically active entity. Thus, the location of the activation domain is important because most enhancer-binding proteins can directly associate with nucleosomal templates, whereas TFIID and the pol II holoenzyme cannot.

Activators Increase TBP Occupancy at Promoters

More recently, we have directly analyzed TBP occupancy at promoters in vivo by chromatin immunoprecipitation (L. Kuras and K. Struhl, in prep.). Specifically, cells containing an epitope-tagged TBP grown under appropriate conditions were treated with formaldehyde to cross-link proteins to DNA in situ. Following fragmentation of the DNA to an average length of 350 bp, protein-DNA complexes were immunoprecipitated with antibodies to the epitope, and the resulting DNA was quantitated by polymerase chain reaction (PCR). In all cases tested (>10 promoters, including those responsive to the Gal4, Ace1, Gcn4, and Hsf1 activators), TBP is not present at promoters in the absence of a functional activator. Moreover, the level of transcription correlates well with the degree of TBP occupancy. These results provide direct evidence that TBP association with promoters is a major limiting step in vivo, and they suggest that activators permit TBP to access chromatin templates. However, these experiments do not address the issue of whether TBP is a direct target of activators.

ROLE OF TBP, TFIIA, AND TFIIB IN RESPONSE TO ACTIVATORS

Mutations That Weaken the TBP-TATA Interaction Specifically Affect the Response to Strong Activators

Using a genetic strategy based on an altered-specificity TBP, we identified TBP derivatives that are impaired in the response to three acidic activators (Gcn4, Gal4, Ace1) but otherwise appear normal for pol II transcription (Lee and Struhl 1995). These activation-defective mutants affect residues that directly contact DNA and are defective for binding TATA elements. Similar activation-defective derivatives with mutations on the DNA-binding surface and a defect in TATA-element binding were identified in an independent genetic screen (Arndt et al. 1995). Thus, interactions at the TBP-TATA element interface can specifically affect the response to acidic activator proteins. However, activation deficiency does not simply reflect reduced affinity for the TATA element but rather involves more specific perturbations of the TBP-TATA interface (Lee and Struhl 1995).

The importance of the TBP-TATA interaction in the response to activators is also seen in a complementary set of experiments involving the two *his3* TATA elements (Struhl 1986; Iyer and Struhl 1995a). The downstream TATA element contains the canonical TATAAA sequence (Chen and Struhl 1988; Wobbe and Struhl 1990), whereas the upstream TATA element is an extended region that lacks a conventional TATA sequence (Mahadevan and Struhl 1990) and is functionally equivalent to a weak TATA element (Iyer and Struhl 1995a). Differential *his3* TATA element utilization does not depend on specific properties of activator proteins, but rather is determined by the overall level of *his3* transcription (Iyer and Struhl 1995a). At low levels of transcription, the upstream TATA element is preferentially utilized even though it is inherently weaker than the downstream TATA element; this reflects an intrinsic preference for using upstream TATA elements. The TATA elements are utilized equally at intermediate levels, whereas the canonical TATA sequence is strongly preferred at high levels of transcription. These and other observations indicate that differential TATA utilization results from the functional saturation of weak TATA elements at low levels of transcriptional stimulation.

The importance of the TBP-TATA interaction for responding to strong activators might be related to transcriptional reinitiation (Struhl 1996). For promoters that depend on strong activators, it is likely that some of the assembled pol II machinery remains upon disengagement of the core enzyme and subsequent transcriptional elongation. In other words, efficient transcriptional activation in vivo might require the ability of an assembled preinitiation complex to initiate multiple rounds of transcription. We suggest that promoters with a compromised TBP-TATA interaction will result in fewer rounds of initiation per complex and increased reliance on assembling the entire triad on an unoccupied promoter.

Role of TFIIA in the Response to Acidic Activators

Using a different genetic strategy, we identified a TBP mutant that is specifically defective in the interaction with TFIIA (Stargell and Struhl 1995). This mutant supports transcription of most genes, but it is significantly impaired for the response to three different acidic activators, Gal4, Gcn4, and Ace1. Fusion of a TFIIA subunit to this TBP derivative corrects the phenotypic defects, indicating that the transcriptional activation defect is caused by the inability of this TBP derivative to interact efficiently with TFIIA. Interestingly, this TFIIA interaction mutant of TBP supports normal cell growth, suggesting that strong acidic activators may not be required for transcription of many yeast genes and for viability of the organism.

The properties of this TBP derivative suggests that the TBP-TFIIA interaction, and presumably TFIIA itself, is important for the response to acidic activators in vivo. In vitro, TFIIA stabilizes the TBP-TATA interaction, alters the conformation of TBP, and extends the DNase I footprint upstream of the TATA element and increases activator-dependent assembly of a TFIID-TFIIA-TATA complex (Lee et al. 1992; Lieberman and Berk 1994; Chi et al. 1995). Taken together, these observations suggest that the role of the TBP-TFIIA interaction in transcriptional activation reflects the ability of TFIIA to stabilize the interaction of TBP to the TATA element and to increase recruitment of TFIID to the promoter. However, TFIIA may also counteract repressor proteins (e.g., Mot1) that interact with TBP and block its interaction with the TATA element (Auble et al. 1994).

TFIIB Does Not Appear to be Generally Limiting for Transcriptional Activation

We analyzed the transcriptional properties of TBP derivatives in which residues that directly interact with TFIIB are replaced by alanines (Lee and Struhl 1997). A derivative with a 50-fold defect in forming TBP-TFIIB-TATA complexes in vitro supports viability and efficiently responds to activators in vivo. Another derivative, which is even more defective in the TBP-TFIIB interaction, retains the ability to respond to activators even though it does not support cell viability. Thus, a severely defective TBP-TFIIB interaction does not preclude transcriptional activation of most yeast genes in vivo.

In a complementary set of experiments, we analyzed the transcriptional effects caused by mutations on the DNA-binding surface of TFIIB that severely affect both TBP-TFIIB-TATA complex formation and interaction with the VP16 activation domain (Chou and Struhl 1997). In accord with the properties of the TFIIB-defective mutants of TBP, these TFIIB derivatives support viability, and they efficiently respond to Gal4-VP16 and natural acidic activators in different promoter contexts. One TFIIB derivative shows reduced transcription of *GAL4*, indicative of a selective transcriptional effect.

Taken together, these results argue that TFIIB recruitment is not generally a limiting step for transcriptional activation in wild-type cells. The growth phenotypes of the TFIIB mutants and the TFIIB-defective TBP mutants indicate that recruitment of TFIIB is limiting at some promoters in the mutant strains. Thus, even under conditions where TFIIB is artificially made to be limiting at a subset of promoters by virtue of mutations, there is little effect on a range of activated promoters. Nevertheless, the mutant TFIIB derivatives must be sufficiently stabilized at promoters in vivo, because TFIIB is generally required for pol II transcription (Moqtaderi et al. 1996b). Such stabilization might reflect TFIIB interactions with TAFs, TFIIF, and pol II (Zawel and Reinberg 1995) and/or recruitment as part of the pol II holoenzyme (Koleske and Young 1995). Finally, the mutant TFIIB derivatives might be stabilized at promoters simply because the concentration of TFIIB is sufficiently high to saturate ternary complex formation.

TBP Mutants Define Two Distinct Steps in the Transcriptional Activation Process

Steps in a complex biological process are often defined by mutations or inhibitors that block the process at distinct stages. In this vein, we used activation-defective TBP mutants to define two steps in the process of activation in vivo (Stargell and Struhl 1996). Specifically, we asked whether artificial recruitment of these TBP mutants could correct their transcriptional activation defects. Consistent with the ability of acidic activators to increase recruitment of TBP to the promoter, the activation defect of some TBP derivatives can be corrected by artificial recruitment. In contrast, the activation defect of the other TBP derivatives is not bypassed by artificial recruitment, suggesting that they are blocked in a postrecruitment step. Thus, these TBP mutants define two steps in the process of transcriptional stimulation by acidic activators: efficient recruitment to the TATA element and a postrecruitment interaction with a component(s) of the initiation complex.

The existence of mutations that block at distinct stages of the process indicates that the steps occur under physiological conditions in wild-type yeast cells. However, because mutations perturb the natural process, they do not provide information about which steps are limiting in wild-type cells. The two steps we have defined in vivo might correspond to the ability of acidic activators in vitro to stimulate the formation of a TFIID-TFIIA-TATA complex (Lieberman and Berk 1994; Chi et al. 1995) and to increase recruitment of TFIIB or subsequent factors to TBP(or TFIID)-TATA complexes (Lin and Green 1991; Choy and Green 1993). Another possibility, which is not mutually exclusive, is that the two steps of activation defined in vivo might correspond to in vitro activation reactions that depend either on TAFs or on the pol II holoenzyme. In this view, both the TAF- and the holoenzyme-dependent activation mechanisms observed in vitro would be required for the full response to acidic activators observed under physiological conditions.

PHYSIOLOGICAL ROLE OF TAFs

TFIID Is Not Generally Required for Transcriptional Activation

With the exception of TAF110, yeast cells contain homologs of all TAFs found in TFIID complexes in flies and humans (Moqtaderi et al. 1996a). Although all of these TAFs are essential for yeast cell growth, individual depletion of a variety of TAFs does not significantly affect transcriptional activation of the vast majority of genes, including those responsive to activators such as Gcn4, Gal4, Ace1, and Hsf1 (Apone et al. 1996; Moqtaderi et al. 1996b; Walker et al. 1996). Furthermore, depletion of TAF130 and TAF60 results in the dissolution of the TFIID complex in vivo (Moqtaderi et al. 1998). This suggests that the transcription observed in such TAF-depleted cells is mediated by the isolated TBP subunit, presumably in a manner related to TAF-independent activation in vitro. As transcription of essentially all yeast genes requires activator proteins, this result indicates that TAFs are not generally required for activation. This conclusion does not exclude the possibility that TAFs are targets for a limited subset of activators or that activator-TAF contacts are redundant with other protein-protein interactions mediated by activators (Struhl 1996).

TFIID Is Required for Core Promoter Function, Particularly at Promoters with Weak TATA Elements

Depletion of four TAFs (TAF130, TAF19, TAF40, and TAF67) causes a distinct profile of promoter-selective effects (Moqtaderi et al. 1996b, 1998). In particular, depletion of any of these TAFs differentially affects *his3* TATA element utilization; transcription from the nonconsensus TATA element is significantly reduced, whereas transcription from the consensus TATA sequence is unaffected. In addition, transcription of *trp3*, which contains a nonconsensus TATA element, is strongly decreased in these TAF-depletion strains. This subset of four TAFs is important for core promoter function, particularly at certain promoters containing weak TATA elements. In general accord with the role of TAFs in core promoter function, depletion of TAF130 also affects transcription of certain cell cycle and ribosomal protein genes, and analysis of hybrid promoters indicates that TAF function is associated with the core promoter, not the enhancer (Shen and Green 1997).

Very recently, it has been discovered that a subset of TAFs is present in the yeast SAGA and human PCAF histone acetylase complexes (Grant et al. 1998; Ogryzko et al. 1998; Struhl and Moqtaderi 1998). As a consequence, for these TAFs, physiological functions inferred from mutations or depletions could result from their presence in TFIID, SAGA, or both. Strikingly, the four TAFs with a common function at core promoters are exclusively found in TFIID. Moreover, TFIID is virtually devoid of TAFs upon TAF130 depletion, suggesting that the core promoter defects reflect the properties of the isolated TBP subunit. In accord with this suggestion, the transcriptional profile in these TAF-depleted strains is remarkably similar to that observed in yeast cells containing human TBP, which presumably interacts poorly with yeast TAFs (Cormack et al. 1994). Taken together, these observations suggest that the primary essential function of TAFs in TFIID is to facilitate transcription from certain kinds of core promoters. In weak promoters lacking consensus TATA elements, TAF interactions with initiator and/or downstream promoter elements (Burley and Roeder 1996; Burke and Kadonaga 1997) are likely to compensate for the weakened TBP-TATA interaction.

The Histone H3-like TAF Is Broadly, but not Universally, Required for Transcription

Unlike the case for all other TAFs tested, depletion of TAF17, which structurally resembles histone H3, causes a decrease in transcription of most genes (Moqtaderi et al. 1998). Although depletion of TAF17 causes the disintegration of TFIID in vivo, the results discussed above for TAF130-depleted or human TBP-dependent cells suggest that such disintegration is insufficient to account for the broad transcriptional affects. Instead, we suggest that the

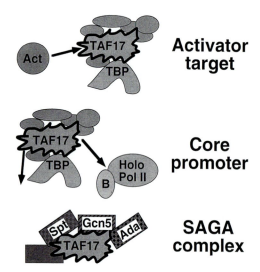

Figure 2. Models for TAF17 function. (*Top*) TAF17 in the context of TFIID interacting with the activation domain. (*Middle*) TAF17 in the context of TFIID interacting with promoter DNA and/or components of the pol II machinery such as TFIIB. (*Bottom*) TAF17 in the SAGA complex which could affect interactions with activators, TBP, or could affect histone acetylase activity. These models are not mutually exclusive.

TAF17-dependent effects on transcription are at least partly due to the presence of TAF17 in the SAGA histone acetylase complex (Fig. 2). Although mutational analyses suggest that the SAGA complex is nonessential for yeast cell growth (Roberts and Winston 1997), it is possible that SAGA has an essential role mediated by the TAFs. Alternatively, the broad transcriptional defects upon TAF17 depletion might reflect the simultaneous inactivation of the SAGA and TFIID histone acetylase complexes.

Although depletion of TAF17 broadly decreases transcription, copper-inducible (i.e., Ace1-dependent) transcription of *CUP1* is unaffected. More convincingly, TAF17-depleted cells efficiently activate heat shock genes after a brief temperature shift. Furthermore, a modified *his3* gene dependent on heat shock factor (Hsf) is inducible in TAF17-depleted cells, indicating that the immunity of the heat shock response to TAF17 depletion is due to Hsf itself, not some special property of the heat shock transcripts. Thus, TAF17-depleted cells are not fundamentally crippled for pol II transcription, and they can mediate de novo transcriptional activation by heat shock factor (Maqtaderi et al. 1998).

Our results with TAF17 are strikingly similar to those obtained previously for Kin28 (the CTD kinase subunit of TFIIH) and Srb4 (a pol II holoenzyme component) in that these proteins have broad transcriptional consequences but minimal affect on activation by Ace1 or Hsf (Lee and Lis 1998). We speculate that certain strong activators might efficiently use any of several targets and thus be less strictly dependent on any one. In contrast, a typical activator might entirely rely on a particular target or it might require multiple targets to generate a significant

transcriptional response. In this view, TAF17 (or a closely associated protein such as the other histone TAFs) might be a general target of activators; loss of TAF17 would therefore affect most genes. In this regard, recent evidence has suggested that Srb4 might be an activator target (Koh et al. 1998).

ACTIVATION AND REPRESSION MECHANISMS THAT DIRECTLY INVOLVE CHROMATIN

Poly(dA:dT), a Ubiquitous Promoter Element That Stimulates Transcription via Its Intrinsic Structure

Many yeast promoters contain homopolymeric dA:dT sequences that affect nucleosome formation in vitro and are required for wild-type levels of transcription in vivo (Struhl 1985). Although typical promoter elements function as recognition sites for activator proteins, several lines of evidence indicate that poly(dA:dT) is a novel promoter element whose function depends on its intrinsic structure, not its interaction with activators (Iyer and Struhl 1995b). First, poly(dA:dT) stimulates Gcn4-activated transcription in a manner that is length-dependent and inversely related to intracellular Gcn4 levels. Second, Datin, the only known poly(dA:dT)-binding protein, behaves as a repressor through poly(dA:dT) sequences. Third, poly(dG:dC), a structurally dissimilar homopolymer that also affects nucleosomes, has transcriptional properties virtually identical to those of poly(dA:dT). Fourth, poly(dA:dT) function improves continuously when its length is increased by small increments. Fifth, *Hin*fI endonuclease cleavage in vivo indicates that poly(dA:dT) increases accessibility of the Gcn4-binding site and adjacent sequences in physiological chromatin. Thus, the intrinsic structure of poly(dA:dT) locally affects nucleosomes and increases the accessibility of transcription factors bound to nearby sequences.

The observed effects on chromatin structure in vivo are directly due to the effects of poly(dA:dT) on nucleosomes in vitro. The similar micrococcal nuclease cleavage patterns in the presence or absence of poly(dA:dT) suggest that altered nucleosome phasing or nucleosome-free DNA is not involved. The local perturbation of chromatin structure extends over a region of approximately 200 bp, which is somewhat larger than a single nucleosome. From these observations, we have suggested that a nucleosome covering poly(dA:dT) will be destabilized relative to adjacent and otherwise normal nucleosomes such that it will be less effective in competing with transcription factors for DNA (Fig. 3). In this view, longer dA:dT tracts would be more destabilizing to the relevant nucleosome, and the repressive effects of Datin might be rationalized by its occupancy of nucleosome-perturbing sequences. Aside from TATA elements, poly(dA:dT) is the most common sequence in yeast promoter regions; thus, it is very likely that poly(dA:dT) sequences are relevant for the expression of a significant fraction of yeast genes and hence have a major role in cell physiology.

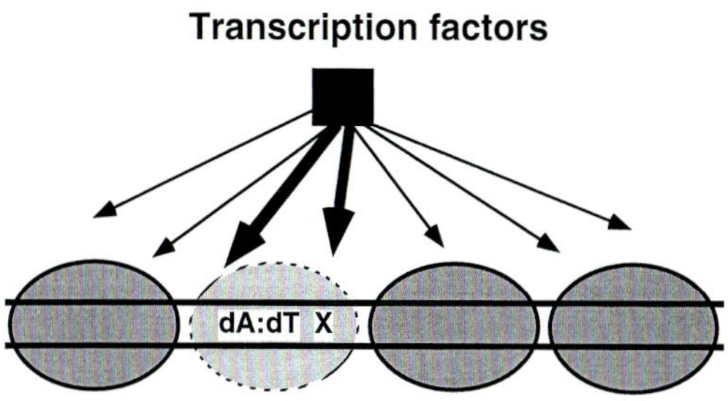

Transcription factors

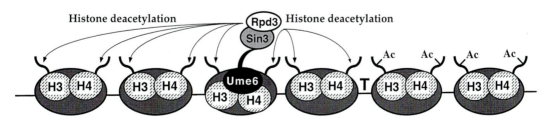

Figure 3. Model for poly(dA:dT) function. A stretch of DNA containing a poly(dA:dT) sequence and a binding site for a transcription factor (X) is coated by nucleosomes (*shaded ovals*); the position of the nucleosomes with respect to the dA:dT tract is arbitrarily drawn to reflect the apparent lack of nucleosome phasing. The nucleosome covering the dA:dT tract is shown as being perturbed (*lighter shading* and *dashed line*) in comparison to adjacent nucleosomes; this perturbation could reflect decreased stability and/or altered conformation of the nucleosome. DNA sequences covered by this nucleosome (e.g., X) on either side of poly(dA:dT) are preferentially accessible (*thicker arrows*) to transcription factors (*black box*). (Reprinted, with permission, from Iyer and Struhl 1995b.)

Transcriptional Repression by Targeted Recruitment of the Sin3-Rpd3 Histone Deacetylase Complex and Generation of a Locally Repressed Domain of Chromatin

The yeast Sin3-Rpd3 histone deacetylase complex is required for transcriptional repression by Ume6, a zinc finger protein that binds URS1 elements and regulates genes involved in meiosis and arginine catabolism (Kadosh and Struhl 1997). A short region of Ume6 interacts directly with Sin3 corepressor, and this region is necessary and sufficient for recruitment of the complex to promoters and for transcriptional repression. The Sin3-Rpd3 complex is not required for the function of the Tup1 and Acr1 transcriptional repressors under equivalent experimental conditions, indicating that repression by Sin3-Rpd3 requires recruitment to target promoters (Kadosh and Struhl 1997). Histone deacetylase activity is important for repression; Rpd3 mutants that are catalytically impaired in vitro, but competent for Sin3-Rpd3 complex

formation, are severely or completely defective for transcriptional repression in vivo (Kadosh and Struhl 1998a). These observations strongly suggest that transcriptional repression occurs by targeted histone deacetylation. This mechanism is highly conserved, and it accounts for repression in mammalian cells by Mad, Rb, YY1, and steroid hormone corepressors (Pazin and Kadonaga 1997; Struhl 1998).

Direct analysis of the chromatin structure of repressed promoters in yeast cells indicates that recruitment of the Sin3-Rpd3 histone deacetylase complex and transcriptional repression are associated with localized histone deacetylation (Kadosh and Struhl 1998b; Rundlett et al. 1998). Decreased acetylation of histones H3 and H4 (preferentially lysines 5 and 12) is observed in wild-type strains but not in strains lacking the DNA-binding repressor (Ume6), Sin3 corepressor, and Rpd3 histone deacetylase. Mapping experiments indicate that the domain of histone deacetylation is highly localized, occurring over a range of one to two nucleosomes. The limited spread of histone deacetylation from the site of recruitment suggests that localized chromatin modification is an inherent property of the Sin3-Rpd3 complex that is relatively insensitive to the presence or absence of other promoter elements. Furthermore, the tethered Sin3-Rpd3 complex has a limited degree of flexibility that permits it to modify the nucleosome at the recruitment site and perhaps the neighboring nucleosome. Thus, the Sin3-Rpd3 complex defines a novel mechanism of transcriptional repression that involves targeted recruitment of a histone-modifying activity and localized perturbation of chromatin structure (Fig. 4).

Although the magnitude of histone deacetylation of individual lysines is modest (two- to threefold), the overall effect on chromatin structure is likely to be more substantial because at least two histones (H3 and H4) and multiple lysine residues are affected. The simplest model for transcriptional repression is that localized histone deacetylation generates a repressive chromatin structure that inhibits the binding of activator proteins or TFIID to their cognate promoter elements. However, in the promoter we have examined (Kadosh and Struhl 1998b), we

Histone deacetylation **Histone deacetylation**

Figure 4. Transcriptional repression by targeted recruitment of the Sin3-Rpd3 histone deacetylase complex. The Ume6 repressor binds URS1 (shown as occurring in the context of a nucleosomal template) and recruits the Sin3-Rpd3 corepressor complex to the promoter. As a consequence, histones H3 and H4 (lysines 5, 12, and to a lesser extent 16) are deacetylated (lack of Ac) over a range of one to two nucleosomes from the site of recruitment. (*Arrows*) For the promoter tested, the region of local histone deacetylation includes the UAS element, but probably ends upstream of the TATA elements (T). Analogous regions of other Sin3-Rpd3-repressed promoters might vary in length and position. The figure is not intended to suggest any particular mechanism of repression (e.g., inhibiting access of activators, TFIID, or the pol II holoenzyme or inhibiting the communication between these components). (Reprinted, with permission, from Kadosh and Struhl 1998b.)

disfavor a direct effect on TBP binding because the domain of localized histone deacetylation is unlikely to extend as far as the TATA elements. Alternatively, locally deacetylated chromatin might not reduce the accessibility of activators or TBP per se, but rather interfere with the communication of these components with each other or with the pol II holoenzyme. More detailed information on the mechanism of transcriptional repression will require measurements of promoter occupancy of activators, TFIID, and pol II holoenzyme in vivo.

CONCLUDING COMMENTS

During the past few years, it has become increasingly clear that transcriptional activation and repression mechanisms are intimately connected with chromatin structure. For example, some histone acetylases are components of the pol II transcription machinery itself, whereas other histone acetylases are present in large multiprotein complexes that interact with activation domains, TBP, or the pol II holoenzyme (Struhl 1998). In addition, there is some evidence that the Swi/Snf nucleosome remodeling complex might interact with the carboxy-terminal tail of pol II (Wilson et al. 1996). Finally, as discussed here and elsewhere (Pazin and Kadonaga 1997; Struhl 1998), the Sin3-Rpd3 histone deacetylase complex mediates transcriptional repression in yeast and mammals by modifying chromatin upon being directly recruited to promoters by DNA-binding repressor proteins. With respect to the mechanism of transcriptional activation, it is clear that nucleosomal templates significantly block access of the pol II machinery (particularly TBP) to promoters in vivo. As such, a major function of activation domains is to increase recruitment of the pol II machinery to promoters in the context of chromatin. Recruitment is likely to involve both direct interactions to the pol II machinery itself and interactions with chromatin-modifying activities (which may or may not be directly associated with the pol II machinery) that alter the properties of the promoter template. However, it is still unclear which proteins are direct and physiological targets of natural activation domains.

ACKNOWLEDGMENTS

This work was supported by a predoctoral fellowship to D.K. from the National Science Foundation, a postdoctoral fellowship from the National Institutes of Health (GM-17930) to M.K., a postdoctoral fellowship from the Human Frontiers Program to L.K., and research grants from the National Institutes of Health (GM-30186 and GM-53720) to K.S.

REFERENCES

Apone L.M., Virbasius C.A., Reese J.C., and Green M.R. 1996. Yeast TAF$_{II}$90 is required for cell-cycle progression through G$_2$/M but not for general transcription activation. *Genes Dev.* **10:** 2368.

Arndt K.M., Ricupero-Hovasse S., and Winston F. 1995. TBP mutants defective in activated transcription *in vivo*. *EMBO J.* **14:** 1490.

Auble D.T., Hansen K.E., Mueller C.G.F., Lane W.S., Thorner

J., and Hahn S. 1994. Mot1, a global repressor of RNA polymerase II transcription, inhibits TBP binding to DNA by an ATP-dependent mechanism. *Genes Dev.* **8:** 1920.

Barberis A., Pearlberg J., Simkovich N., Farrell S., Reinagel P., Bamdad C., Sigal G., and Ptashne M. 1995. Contact with a component of the polymerase II holoenzyme suffices for gene activation. *Cell* **81:** 359.

Burke T.W. and Kadonaga J.T. 1997. The downstream core promoter element, DPE is conserved from *Drosophila* to humans and is recognized by TAF$_{II}$60 of *Drosophila*. *Genes Dev.* **11:** 3020.

Burley S.K. and Roeder R.G.. 1996. Biochemistry and structural biology of transcription factor IID (TFIID). *Annu. Rev. Biochem.* **65:** 769.

Chatterjee S. and Struhl K. 1995. Connecting a promoter-bound protein to the TATA-binding protein overrides the need for a transcriptional activation region. *Nature* **374:** 820.

Chen W. and Struhl K. 1988. Saturation mutagenesis of a yeast *his3* TATA element: Genetic evidence for a specific TATA-binding protein. *Proc. Natl. Acad. Sci.* **85:** 2691.

Chi T., Lieberman P., Ellwood K., and Carey M. 1995. A general mechanism for transcriptional synergy by eukaryotic activators. *Nature* **377:** 254.

Chou S. and Struhl K. 1997. Transcriptional activation by TFIIB mutants that severely impair the interaction with promoter DNA and acidic activation domains. *Mol. Cell. Biol.* **17:** 6794.

Choy B. and Green M.R. 1993. Eukaryotic activators function during multiple steps of preinitiation complex assembly. *Nature* **366:** 531.

Cormack B.P., Strubin M., Stargell L.A., and Struhl K. 1994. Conserved and nonconserved functions of yeast and human TATA-binding proteins. *Genes Dev.* **8:** 1335.

Farrell S., Simkovich N., Wu Y.B., Barberis A., and Ptashne M.. 1996. Gene activation by recruitment of the RNA polymerase II holoenzyme. *Genes Dev.* **10:** 2359.

Geiger J.H., Hahn S., Lee S., and Sigler P.B. 1996. Crystal structure of the yeast TFIIA/TBP/DNA complex. *Science* **272:** 830.

Gonzalez-Couto E., Klages N., and Strubin M. 1997. Synergistic and promoter-selective activation of transcription by recruitment of TFIID and TFIIB. *Proc. Natl. Acad. Sci* **94:** 8036.

Grant P.A., Schieltz D., Pray-Grant M.G., Steger D.J., Reese J.C., Yates J.R., and Workman J.L. 1998. A subset of TBP-associated factors, TAF$_{II}$s are integral components of the SAGA complex that are required for nucleosomal acetylation and transcription stimulation. *Cell* **94:** 45.

Harbury P.A.B. and Struhl K. 1989. Functional distinctions between yeast TATA elements. *Mol. Cell. Biol.* **9:** 5298.

Hope I.A., Mahadevan S., and Struhl K. 1988. Structural and functional characterization of the short acidic transcriptional activation region of yeast GCN4 protein. *Nature* **333:** 635.

Imbalzano A.N., Kwon H., Green M.R., and Kingston R.E. 1994. Facilitated binding of TATA-binding protein to nucleosomal DNA. *Nature* **370:** 481.

Iyer V. and Struhl K. 1995a. Mechanism of differential utilization of the *his3* T$_R$ and T$_C$ TATA elements. *Mol. Cell. Biol.* **15:** 7059.

———. 1995b. Poly(dA:dT), a ubiquitous promoter element that stimulates transcription via its intrinsic structure. *EMBO J.* **14:** 2570.

Kadosh D. and Struhl K. 1997. Repression by Ume6 involves recruitment of a complex containing Sin3 corepressor and Rpd3 histone deacetylase to target promoters. *Cell* **89:** 365.

———. 1998a. Histone deacetylase activity of Rpd3 is important for transcriptional repression *in vivo*. *Genes Dev.* **12:** 797.

———. 1998b. Targeted recruitment of the Sin3-Rpd3 histone deacetylase complex generates a highly localized domain of repressed chromatin *in vivo*. *Mol. Cell. Biol.* **18:** 5121.

Keaveney M. and Struhl K. 1998. Activator-mediated recruitment of the RNA polymerase II machinery is the predominant mechanism for transcriptional activation in yeast. *Mol. Cell* **1:** 917.

Kingston R.E., Bunker C.A., and Imbalzano A.N. 1996. Repression and activation by multiprotein complexes that alter chromatin structure. *Genes Dev.* **10:** 905.

Klages N. and Strubin M. 1995. Stimulation of RNA polymerase II transcription initiation by recruitment of TBP in vivo. *Nature* **374:** 822.

Klein C. and Struhl K. 1994. Increased recruitment of TATA-binding protein to the promoter by transcriptional activation domains *in vivo. Science* **266:** 280.

Koh S.S., Ansari A.Z., Ptashne M., and Young R.A. 1998. An activator target in the RNA polymerase II holoenzyme. *Mol. Cell* **1:** 895.

Koleske A.J. and Young R.A.. 1995. The RNA polymerase II holoenzyme and its implications for gene regulation. *Trends Biochem. Sci.* **20:** 113.

Lee D. and Lis J.T. 1998. Transcriptional activation independent of TFIIH kinase and the RNA polymerase II mediator *in vivo. Nature* **393:** 389.

Lee D.K., Dejong J., Hashimoto S., Horikoshi M., and Roeder R.G.. 1992. TFIIA induces conformational changes in TFIID via interactions with the basic repeat. *Mol. Cell. Biol.* **12:** 5189.

Lee M. and Struhl K. 1995. Mutations on the DNA-binding surface of TBP can specifically impair the response to acidic activators *in vivo. Mol. Cell. Biol.* **15:** 5461.

———. 1997. A severely defective TATA-binding protein-TFIIB interaction does not preclude transcriptional activation *in vivo. Mol. Cell. Biol.* **17:** 1336.

Lieberman P.M. and Berk A.J. 1994. A mechanism for TAFs in transcriptional activation: Activation domain enhancement of TFIID-TFIIA-promoter DNA complex formation. *Genes Dev.* **8:** 995.

Lin Y.-S. and Green M.R. 1991. Mechanism of action of an acidic transcriptional activator *in vitro. Cell* **64:** 971.

Mahadevan S. and Struhl K. 1990. T_C, an unusual promoter element required for constitutive transcription of the yeast *his3* gene. *Mol. Cell. Biol.* **10:** 4447.

Moqtaderi Z., Yale J.D., Struhl K., and Buratowski S. 1996a. Yeast homologues of higher eukaryotic TFIID subunits. *Proc. Natl. Acad. Sci.* **93:** 14654.

Moqtaderi Z., Bai Y., Poon D., Weil P.A., and Struhl K. 1996b. TBP-associated factors are not generally required for transcriptional activation in yeast. *Nature* **382:** 188.

Moqtaderi Z., Keaveney M., and Struhl K. 1998. The histone H3-like TAF17 is broadly required for transcription in yeast cells. *Mol. Cell* **2:** 675.

Myers L.C., Gustafsson C.M., Bushnell D.A., Lui M., Erdjument-Bromage H., Tempst P., and Kornberg R.D. 1998. The Med proteins of yeast and their function through the RNA polymerase II carboxy-terminal domain. *Genes Dev.* **12:** 45.

Nikolov D.B., Chen H., Halay E.D., Usheva A.A., Hisatake K., Lee D.K., Roeder R.G., and Burley S.K. 1995. Crystal structure of a TFIIB-TBP-TATA-element ternary complex. *Nature* **377:** 119.

Ogryzko V.V., Kotani T., Zhang X., Schlitz R.L., Howard T., Yang X.-J., Howard B.H., Qin J., and Nakatani Y. 1998. Histone-like TAFs within the PCAF histone acetylase complex. *Cell* **94:** 35.

Pazin M.J. and Kadonaga J.T.. 1997. What's up and down with histone deacetylation and transcription? *Cell* **89:** 325.

Polach K.J. and Widom J. 1996. A model for the cooperative binding of eukaryotic regulatory proteins to nucleosomal target sites. *J. Mol. Biol.* **258:** 800.

Ptashne M. and Gann A. 1997. Transcriptional activation by recruitment. *Nature* **386:** 569.

Roberts S.M. and Winston F. 1997. Essential functional interactions of SAGA, a *Saccharomyces cerevisiae* complex Of Spt, Ada, and Gcn5 proteins, with the Snf/Swi and Srb/mediator complexes. *Genetics* **147:** 451.

Rundlett S.E., Carmen A.A., Suka N., Turner B.M., and Grunstein M. 1998. Transcriptional repression by UME6 involves deacetylation of lysine 5 of histone H4 by RPD3. *Nature* **392:** 831.

Shen W.C. and Green M.R. 1997. Yeast TAF$_{II}$145 functions as a core promoter selectivity factor, not a general coactivator. *Cell* **90:** 615.

Stargell L.A. and Struhl K. 1995. The TBP-TFIIA interaction in the response to acidic activators *in vivo. Science* **269:** 75.

———. 1996. A new class of activation-defective TATA-binding protein mutants: Evidence for two steps of transcriptional activation *in vivo. Mol. Cell. Biol.* **16:** 4456.

Strubin M. and Struhl K. 1992. Yeast TFIID with altered DNA-binding specificity for TATA elements. *Cell* **68:** 721.

Struhl K. 1985. Naturally occurring poly(dA-dT) sequences are upstream promoter elements for constitutive transcription in yeast. *Proc. Natl. Acad. Sci.* **82:** 8419.

———. 1986. Constitutive and inducible *Saccharomyces cerevisiae* promoters: Evidence for two distinct molecular mechanisms. *Mol. Cell. Biol.* **6:** 3847.

———. 1995. Yeast transcriptional regulatory mechanisms. *Annu. Rev. Genet.* **29:** 651.

———. 1996. Chromatin structure and RNA polymerase II connection: Implications for transcription. *Cell* **84:** 179.

———. 1998. Histone acetylation and transcriptional regulatory mechanisms. *Genes Dev.* **12:** 599.

Struhl K. and Moqtaderi Z. 1998. The TAFs in the HAT. *Cell* **94:** 1.

Tan S., Hunziker Y., Sargent D.F., and Richmond T.J. 1996. Crystal structure of a yeast TFIIA/TBP/DNA complex. *Nature* **381:** 127.

Uesugi M., Nyanguile O., Lu H., Levine A.J., and Verdine G.L. 1997. Induced α helix in the VP16 activation domain upon binding to a human TAF. *Science* **277:** 1310.

Verrijzer C.P. and Tjian R. 1996. TAFs mediate transcriptional activation and promoter selectivity. *Trends Biochem. Sci.* **21:** 338.

Walker S.S., Reese J.C., Apone L.M., and Green M.R. 1996. Transcription activation in cells lacking TAF$_{II}$s. *Nature* **382:** 185.

Wilson C.J., Chao D.M., Imbalzano A.N., Schnitzler G.R., Kingston R.E., and Young R.A. 1996. RNA polymerase II holoenzyme contains SWI/SNF regulators involved in chromatin remodeling. *Cell* **84:** 235.

Wobbe C.R. and Struhl K. 1990. Yeast and human TATA-binding proteins have nearly identical DNA sequence requirements for transcription *in vitro. Mol. Cell. Biol.* **10:** 3859.

Xiao H., Friesen J.D., and Lis J.T. 1995. Recruiting TATA-binding protein to a promoter: Transcriptional activation without an upstream activator. *Mol. Cell. Biol.* **15:** 5757.

Zawel L. and Reinberg D. 1995. Common themes in assembly and function of eukaryotic transcription complexes. *Annu. Rev. Biochem.* **64:** 533.

The Mad Protein Family Links Transcriptional Repression to Cell Differentiation

G.A. McArthur, C.D. Laherty, C. Quéva,* P.J. Hurlin, L. Loo, L. James, C. Grandori,
P. Gallant, Y. Shiio, W.C. Hokanson,† A.C. Bush, P.F. Cheng, Q.A. Lawrence,
B. Pulverer,‡ P.J. Koskinen,§ K.P. Foley,¶ D.E. Ayer,† and R.N. Eisenman
Division of Basic Sciences, Fred Hutchinson Cancer Research Center, Seattle, Washington 98109-1042

The ability of proliferating cells to exit the cell cycle is an important aspect of the ordered growth and development of tissues and organisms (for review, see Raff 1996). Many normal cells cease proliferation in response to a number of external cues including withdrawal of mitogens, cell-cell contact, specific soluble ligands (e.g., transforming growth factor-β, TGF-β), and damage. Cessation of proliferation is also a critical element in the terminal differentiation of diverse cell types. Importantly, failure to exit the cell cycle in response to appropriate signals is likely to play a key role in oncogenesis. Until recently, there has been scant information on the mechanisms and pathways involved in cell cycle exit. However, work within the last several years delineating the major molecular components of the cell cycle has revealed a plethora of both positive and negative regulators of cell cycle progression. The negative regulators comprise a wide range of proteins including tumor suppressors, checkpoint control proteins, and cyclin-dependent kinase inhibitors (CKIs) (for reviews, see Hartwell and Kastan 1994; Sherr 1996). Although many of these proteins appear to regulate the orderly progression of the cell cycle, several have also been linked to cell cycle exit related to differentiation. These include proteins in the retinoblastoma (RB) family and the CKIs $p21^{Cip1/WAF1}$, $p27^{Kip1}$, and $p57^{Kip2}$ (Lee et al. 1994; Halevy et al. 1995; Parker et al. 1995; de Nooij et al. 1996; Missero et al. 1996; Casaccia-Bonnefil et al. 1997; Yan et al. 1997; Di Cunto et al. 1998).

In principle, it might be expected that differentiation-related arrest of cell proliferation would occur through inhibition of the cell cycle machinery. Indeed, this appears to be the case for the CKI proteins which directly attenuate the activity of cyclin-dependent kinases. In addition, another aspect of the arrest and differentiation program would seem likely to involve regulation of gene expression. Terminal differentiation must involve the repression of genes provoking proliferation and the activation of genes encoding differentiation-specific proteins. Of the group of cell cycle exit mediators mentioned above, only

the proteins belonging to the RB family are believed to function as transcriptional regulators (Hamel et al. 1992; Weintraub et al. 1995; Tevosian et al. 1997), although recent evidence suggests that $p21^{Cip1/WAF1}$ may, in an as yet undefined manner, negatively influence the transcription of differentiation markers (Di Cunto et al. 1998). In this paper, we summarize our recent studies on the Mad family of transcriptional repressors. These proteins are induced during the terminal differentiation of a wide range of cell types and appear to be involved in limiting cell proliferation during the differentiation process. Research on the Mad family has provided a possible link between differentiation and chromatin structure.

THE MAD PROTEIN FAMILY

The Mad family comprises four paralogs: Mad1, Mxi1, Mad3, and Mad4. All of the family members were first identified in expression screens for proteins interacting with the small basic helix-loop-helix zipper (bHLHZ) protein Max (Ayer et al. 1993; Zervos et al. 1993; Hurlin et al. 1995b). Max itself had been initially discovered as a dimerization partner for members of the Myc oncoprotein family (the bHLHZ proteins c-Myc, N-Myc, and L-Myc) (Blackwood and Eisenman 1991). Association between Myc and Max had been previously shown to be mediated by their HLHZ domains, resulting in formation of a heterodimer with high affinity for the E-box DNA sequence CACGTG (Blackwell et al. 1990; Blackwood and Eisenman 1991; Prendergast and Ziff 1991; Prendergast et al. 1991). Myc possesses an amino-terminal tripartite *trans*-activation domain (Kato et al. 1990), and its ability to activate transcription at promoters with proximal E-box sites is dependent on Max (Amati et al. 1992; Kretzner et al. 1992). Indeed, there is evidence that Max is an obligate partner for Myc's biological function as a promoter of proliferation, apoptosis, and transformation, as well as an inhibitor of differentiation (for recent review, see Henriksson and Lüscher 1996). Max is a stable protein whose biosynthesis appears to be independent of cell proliferation and differentiation. In contrast, Myc has a half-life of 20–30 minutes, and its expression is highly regulated. It is rapidly induced during mitogenesis and down-regulated during growth arrest and differentiation. The studies on Myc and Max led to a model in which newly synthesized Myc combines with preexisting Max protein and activates transcription at Myc:Max-binding

Present addresses: *Astra Transgenic Centre, S-431 83 Mölndal, Sweden; †Huntsman Cancer Institute, University of Utah, Salt Lake City, Utah 84112; ‡Institut fur Biochemie, Universitat Innsbruck, A-6020 Innsbruck, Austria; §Turku Centre for Biotechnology, University of Turku/Abo Akademi University, 20520 Turku, Finland; ¶Department of Genetics, ZymoGenetics, Inc., Seattle, Washington 98102.

sites in the promoters of specific target genes (for review, see Grandori and Eisenman 1997).

The idea that stable Max protein transiently heterodimerizes with Myc family proteins prompted the search for other Max partners and the discovery of the four Mad family members. In addition, two other Max-binding proteins were identified in a two-hybrid screen with Max: Mnt (Hurlin et al. 1997) also known as Rox (Meroni et al. 1997) and Mga (P. Hurlin, unpubl.). Although all of these proteins belong to the bHLHZ class, the four Mad family members are highly related to each other in regions outside of their bHLHZ domains (Hurlin et al. 1995a), with overall similarities among family members ranging from 56% to 72% (P. Hurlin, unpubl.). The four *Mad* genes have distinct chromosomal localizations (Edelhoff et al. 1994; Hurlin et al. 1995a). The biochemical properties of the four Mad proteins parallel those of the Myc family: They do not homodimerize or bind DNA alone, they interact specifically with Max through their HLHZ domains, and the heterodimers bind the E-box sequence CACGTG (Ayer et al. 1993; Zervos et al. 1993; Hurlin et al. 1995a). One significant difference between the Mad:Max and Myc:Max heterodimers became apparent in transcription assays. Whereas Myc:Max complexes activate transcription from minimal promoters with proximal E-box sites, the Mad:Max heterodimers repress transcription from the same promoters (Ayer et al. 1993; Hurlin et al. 1995a; Schreiber-Agus et al. 1995). A dominant interfering Max protein (i.e., lacking only the basic region and therefore capable of forming non-DNA-binding dimers with both Myc and Mad) was found to abrogate both activation by c-Myc and repression by Mad1, suggesting that interaction with both Max and DNA is required for the distinct transcriptional activities of both Myc and Mad (Ayer et al. 1993). Complexes of Myc and Max as well as Mad and Max could be detected in cells by coimmunoprecipitation (Blackwood et al. 1992b; Ayer and Eisenman 1993). These findings raised the possibility that Mad proteins function as antagonists of Myc and led to further studies focused on understanding both the biology of Mad and its mechanism of transcriptional repression.

A Myc:Max to Mad:Max Heterocomplex Switch during Differentiation

Mad family gene expression is predominantly associated with differentiating compartments in many organs and across a broad spectrum of cell types. In situ hybridization to mouse brain, eye, neural tube, skin, bone, and thymus (situations where a clear demarcation exists between proliferating and differentiating compartments) shows that the expression patterns of the different *Mad* family transcripts are overlapping yet distinct (Hurlin et al. 1995a; Quéva et al. 1998). In general, *Mad1* is expressed late in differentiating tissues, whereas *Mxi1* and *Mad4* mRNAs are present at low levels in proliferating cells and increase during differentiation. Most striking is the expression of *Mad3*; this gene appears to be induced only transiently in cells during the proliferative cycles

just preceding growth arrest. In contrast, expression of *myc* family genes is confined primarily to proliferating cells, consistent with previous studies (Downs et al. 1989; Schmid et al. 1989; Hirning et al. 1991).

The in situ hybridization results are supported by biochemical studies using cell lines or primary cells that can be induced to differentiate in culture. In hematopoietic cell lines, *Mad1* and *Mxi1* RNAs are often detected as an "immediate early" response to differentiation inducers (Ayer and Eisenman 1993; Zervos et al. 1993; Larsson et al. 1994). The Mad1 protein is detected as a short-lived ($t1/2 = 30$ min) nuclear phosphoprotein. Importantly, it was demonstrated in both a hematopoietic cell line and in primary keratinocytes that Mad:Max complexes are detected as early as 2 hours following induction and that Myc:Max complexes are replaced by Mad:Max complexes between 24 and 48 hours following induction of differentiation (Ayer and Eisenman 1993; Hurlin et al. 1995b). In keratinocytes whose differentiation is compromised by introduction of human papillomavirus (HPV) early region DNA, expression of *Myc* persists and *Mad* expression is delayed and reduced, suggesting that induction of *Mad* is tightly coupled to differentiation (Hurlin et al. 1995b). The pattern of *Mad1* RNA expression is similarly altered in a mouse model system in which tumor progression is induced using an HPV E6/E7 transgene directed to the basal layer of the epidermis (Hurlin et al. 1995b).

Taken together, these studies indicate that *Mad* expression occurs in association with differentiation and that the *Mad* family genes are induced in what can be loosely described as a sequential fashion (although the exact order can vary among tissues). These types of experiments led to the idea that the transcription repression function of Mad proteins is manifested during differentiation when competition or switching between Myc:Max and Mad:Max complexes would direct down-regulation of proliferation-related genes normally activated by Myc-Max. In this model, Mad proteins would be general mediators of growth arrest associated with differentiation.

Ectopic Mad Expression Blocks Proliferation

One approach to understanding Mad function has been to ectopically overexpress Mad proteins in tissue culture cells and gauge the effects on cell proliferation. The consensus finding is that ectopic expression of Mad1 abrogates cotransformation by Myc plus activated Ras and also inhibits nontransformed cell proliferation primarily during G_1 progression (Lahoz et al. 1994; Chen et al. 1995; Koskinen et al. 1995; Vastrik et al. 1995; Roussel et al. 1996; McArthur and Eisenman 1997). Mxi1, Mad3, and Mad 4 were also found to inhibit growth of Myc/Ras transformants, and Mxi1 blocked proliferation of glioblastoma cell lines (Lahoz et al. 1994; Hurlin et al. 1995b; Wechsler et al. 1997). Of importance is that the inhibitory function of Mad1 is closely linked to its transcriptional repression activity. Thus, Mad1 mutants that do not associate with Max, do not bind DNA, or do not contain a functional repression domain (see below) are incapable of blocking cell proliferation in fibroblasts (Hurlin et al. 1995a; Koskinen et al. 1995; Schreiber-Agus et al. 1995; Roussel et al. 1996).

Because Mad family proteins can block cell cycle progression in fibroblasts, which are nondifferentiating cells, it seems likely that Mad function is more closely tied to cessation of cell proliferation per se rather than to specifically driving cell differentiation. In several cases, Mad has been expressed in cells capable of differentiation, such as keratinocytes or 3T3L1 cells, and found either to fail to induce differentiation (keratinocytes; P. Gallant, unpubl.) or to block induction of differentiation (3T3L1 cells; B. Pulverer, unpubl.). In the C2C12 myoblast line, Mad1 increases differentiation by increasing survival of normally apoptotic cells (W.C. Hokanson, unpubl.; see below). Similarly, in a murine erythroleukemia cell line (MEL), approximately 25% of Mad1-transfected cells differentiate in the absence of inducer (Cultraro et al. 1997). Although Mad1 could possibly be directly inducing differentiation in these cells, the relatively low levels of differentiation after introduction of the gene may indicate that differentiation in the MEL system is secondary to inhibition of proliferation. In other words, such cells may be "poised" to differentiate in such a manner that nearly any block to the cell cycle might trigger their differentiation.

Aside from the fact that Mad repression activity is critical, little is known at present concerning the mechanism by which Mad proteins inhibit cell proliferation. We have recently found that ectopic expression of Mad1 significantly increases the rate of G_1 accumulation in serum-deprived BALB/c fibroblasts and also delays S phase reentry following serum restimulation. Interestingly, the G_1 accumulation and S phase delay are completely alleviated in BALB/c fibroblasts expressing the HPV-E7, but not the HPV-E6, oncogene (McArthur and Eisenman 1997). Because the HPV-E7 protein can bind to and inactivate the RB protein (Dyson et al. 1989), this result would suggest that the RB family is a major functional target of Mad1. This idea fits well with studies showing that Myc drives cell proliferation by inhibition of p27^{KIP1} function and by directly inducing cdc25A, both of which would be expected to stimulate cyclin:cdk activities leading in turn to phosphorylation and inactivation of RB (for recent review, see Amati et al. 1998). Since Mad is thought to antagonize Myc function, the sequestration of RB by HPV-E7 would be expected as we have shown to abrogate Mad's block of the cell cycle (McArthur and Eisenman

1997). However, it is important to bear in mind that both HPV-E7 and cyclin E:cdk2 may have targets other than RB which could also influence Mad function.

Mad1 Promotes Cell Cycle Exit during Differentiation

The results described above are consistent with the model that Mad1 plays a part in cell cycle arrest during differentiation of numerous cell types. However, the experiments are not a definitive test of the model: The expression studies linking Mad1 to differentiation are at best correlative and the data demonstrating inhibition of proliferation rely on ectopic overexpression. A more rigorous test of Mad function has been to generate a targeted deletion of the murine *Mad1* gene (Foley et al. 1998). Mice carrying homozygous *Mad1* null alleles are viable with no obvious differences in behavior, size, fertility, incidence of neoplasia, or life span compared to wild-type littermates. However, examination of the bone marrows (BM) in 8–11-week old mice using FACS analysis with lineage-specific antibodies reveals a significantly increased frequency of granulocytes and decreased frequencies and absolute numbers of B lymphoid, erythroid, and macrophage cells in mutant as compared to wild-type littermates.

To explore the basis for the altered frequency of BM hematopoietic cells, we used in vitro colony assays employing recombinant cytokines as a measure of the stage of differentiation of BM precursor cells (for review, see Metcalf 1989). The results were quite striking in showing that BM cells derived from *Mad1*$^{-/-}$ mice display a significant increase in mature, limited proliferative potential, cluster-forming cells (Cluster-FC) in response to cytokines that stimulate the granulocyte lineage (G-CSF and GM-CSF). Importantly, the more immature lineage-committed colony-forming cells (CFC) (considered precursors to the Cluster-FC) are present at the same frequency in mutant and wild-type mice (Fig. 1). In contrast, other cytokines such as M-CSF and erythropoietin produce only minimal differences in colony abundance between wild-type and mutant BM cells. The increased precursor cells observed in the *Mad1*$^{-/-}$ BM are primarily derived from the granulocytic lineage as judged from morphology and cytokine response.

Other assays, including determination of the number of cell divisions by dye dilution, measurement of colony size, and assessment of the recovery rate of cells follow-

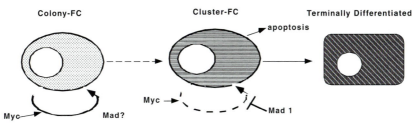

Figure 1. Balance between Myc and Mad is postulated to limit the proliferation of granulocytic cluster-forming cells. Cluster-forming cells in *Mad1*$^{-/-}$ mice undergo ectopic cell divisions and show increased sensitivity to apoptosis-inducing conditions when compared with cells from wild-type littermates (see Foley et al. 1998). Therefore, Mad1 may normally function to attenuate proliferation and apoptosis of cells at this stage.

ing BM ablation, indicated that granulopoiesis in *Mad1*[−/−] animals is marked by cell-autonomous ectopic cell divisions and delayed differentiation of limited proliferative potential cluster-forming precursor cells (Fig. 1) (Foley et al. 1998). The increase in Cluster-FC was puzzling in that it was not reflected in a correspondingly greater number of mature granulocytes in the BM or periphery. However, further work showed that the Cluster-FCs were more sensitive to a number of apoptosis-inducing conditions. This finding suggests, but does not prove, that programmed cell death of the excess granulocytic precursors acts as a compensatory mechanism to maintain the appropriate population size of mature granulocytes. This result is consistent with the finding that deregulated Myc can drive apoptosis in tissue culture cells. Figure 1 diagrams the major conclusions from these studies.

Although the most dramatic phenotype is confined to granulocyte precursors, it is possible that the effects of the *Mad1* knockout may be more widespread. Slightly increased frequencies are also reproducibly observed in low proliferative potential macrophage and erythroid precursors. These lineages also display reduced survival after cytokine withdrawal. Furthermore, we find that resting B lymphocytes from *Mad1*[−/−] mice enter the cell cycle more rapidly upon mitogenic stimulation. On the other hand, we are unable to detect significant changes in cell compartments in other embryonic tissues where *Mad1* is normally expressed. However, examination of spleen and ovaries in the *Mad1*[−/−] mice using in situ hybridization with specific *Mad* family probes reveals that although *Mad1* is absent, other Mad family members are now expressed (Foley et al. 1998). This finding raises the possibility that not only is there cross-talk among *Mad* family genes, but other *Mad* family members may functionally compensate for the loss of *Mad1*. Indeed, the results of the targeted deletion of *Mxi1*, indicating B-cell lymphomas as well as hyperplasia in a wide range of organs in aging mice (Schreiber-Agus et al. 1998), are at least consistent with related biological functions for Mad1 and Mxi1. Generation of other Mad family deletions is in progress.

Interestingly, the granulocyte phenotype in the *Mad1*[−/−] mice is reminiscent of that observed for B cells in transgenic mice overexpressing c-*myc* under control of the immunoglobulin heavy chain enhancer (Eμ). Eμ-*myc* mice possess an expanded compartment of apparently normal progenitor B cells that undergo increased apoptosis at the expense of differentiation (Langdon et al. 1986, 1988). With the caveat that the *Mad1* deficiency has been assessed primarily in granulocytes and c-*myc* overexpression was directed toward B cells, these findings are consonant with the model shown in Figure 1 that Mad1 functions to limit the proliferation-promoting, and possibly the apoptosis-promoting, activities of Myc.

MAD PROTEINS MEDIATE FORMATION OF A REPRESSION COMPLEX

Perhaps the most striking property of the Mad family proteins is their ability, as heterodimers with Max, to re-

press transcription at promoter-proximal Myc:Max DNA-binding sites (Ayer et al. 1993; Hurlin et al. 1995a; Schreiber-Agus et al. 1995). This repression activity appears to be required for Mad1 and Mxi1 to inhibit proliferation in normal and transformed cells and is therefore likely to be important for the biological roles played by Mad proteins (Hurlin et al. 1995a; Koskinen et al. 1995; Schreiber-Agus et al. 1995; Roussel et al. 1996). Furthermore, studies on the Mad proteins have led to the discovery of what appears to be a general mechanism of transcriptional repression.

Amino acid sequence comparison of the murine and human Mad family proteins (Mad1, 3, 4 and Mxi1) reveals two extensive regions of homology amid scattered overall homology (see Hurlin et al. 1995a). One region, the centrally located bHLHZ segment, is the domain responsible for interaction with Max and DNA binding. The other strikingly homologous region lies within the 35 amino-terminal amino acids (Fig. 2). A portion of this region was also found in another transcriptional repressor, the Max-binding bHLHZ protein Mnt (Hurlin et al. 1997). Outside of the amino-terminal and bHLHZ homology regions Mnt is unrelated to the Mad family, strengthening the idea that the amino-terminal segment may be functionally related to repression. Indeed, deletion of the amino-terminal homology region leads to loss of Mad1 repression, and a putative alternatively spliced form of Mxi1 lacking this region also fails to repress transcription, although binding to Max is not affected (Ayer et al. 1995; Hurlin et al. 1995a; Schreiber-Agus et al. 1995). Importantly, the 35-residue amino-terminal homology region of Mad1 alone is capable of acting as a repressor when fused to the Gal4 DNA-binding domain and assayed on several different promoters containing Gal4-binding sites. Furthermore, the Mad1 amino-terminal domain suppresses activators such as VP16 and Myc in *cis* (Ayer et al. 1996). The amino-terminal segment of Mad family proteins and Mnt is therefore a dominantly acting transcription repression domain.

The repression domain of the Mad family is now known to function through association with a complex comprising the corepressors mSin3A or mSin3B and histone deacetylases (HDAC), among other proteins that are still being characterized. The impetus for much of this work comes from the discovery that Mad1 and Mxi1 in-

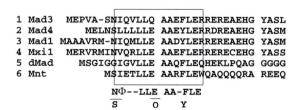

Figure 2. A conserved repression domain in Mad family proteins and Mnt. The amino-terminal sequences of murine Mad family proteins, *Drosophila* Mad (dMad; L. Loo, unpubl.), and murine Mnt are shown. The boxed region encompasses the putative α-helical region thought to comprise the minimal Sin3 interaction domain (SID). A consensus sequence is shown below the box.

teracted with a specific region of mSin3A and mSin3B in a two-hybrid screen and that the mSin3 proteins are mammalian orthologs of the *Saccharomyces cerevisiae* "global" corepressor Sin3p (Wang et al. 1990; Vidal et al. 1991; Ayer et al. 1995; Schreiber-Agus et al. 1995). The regions of highest homology between mSin3A/B and yeast Sin3p lie within four related *paired amphipathic helix* regions (PAH1-4) (these are imperfect repeats of approximately 75 residues that were each modeled as two helices separated by a nonhelical linker) as well as an approximately 400-amino-acid region separating PAH3 and PAH4. The association between Mad1/Mxi1 and mSin3A/B is mediated by the amino-terminal repression domain of Mad1/Mxi1 (dubbed the *Sin interaction domain* or SID) and the PAH2 region of mSin3 (Ayer et al. 1995; Schreiber-Agus et al. 1995). Because purified bacterial-expressed recombinant Mad1-SID and mSin3A-PAH2 can associate, it is likely that the interaction between these regions is direct and not mediated by other proteins (L. Loo, unpubl.). Interestingly, the Mad SID (comprising residues 9–21; see Fig. 2) can be modeled as an amphipathic α-helix, and putative disruption of this helix by substitution of proline for leucine residues 12 and 15 (Mad1Pro) inhibits interaction with mSin3. Importantly, these mutations also abrogate repression by both full-length Mad1 and the SID fused to Gal4 (Ayer et al. 1995, 1996). In addition, ternary complexes containing Mad:Max and mSin3 that can be formed in solution specifically recognize the E-box DNA-binding site. These data suggest, but do not formally prove, that Mad-Max dimers repress transcription by binding mSin3 through PAH2 and recruiting it to specific DNA sequences. As expected, mSin3 itself could substitute for the Mxi1-SID in repression assays and also repress transcription when fused to Gal4 (Ayer et al. 1996; Rao et al. 1996). Because *S. cerevisiae* does not contain *Myc*, *Max*, or *Mad* homologs, the finding that Mad:Max repression is mediated by a protein highly related to the yeast Sin3p suggests that mammalian cells have retained an evolutionarily conserved means of gene silencing.

Although these results point to an important role for mSin3 proteins in Mad-mediated repression, they do not explain the mechanism of repression. Nonetheless, it seemed likely that investigating mSin3's interactions with other proteins might provide some insight. Impetus also came from earlier studies in yeast which established an epistatic relationship between *SIN3* and another gene known as *RPD3* (Vidal and Gaber 1991; Stillman et al. 1994). Initially, this finding had scant impact because no function for Rpd3p was known and no homologs had been identified. However, in 1996, two highly related mammalian homologs of Rpd3p were identified and shown to be histone deacetylases (HDACs). HD1 (now called HDAC1) was isolated through its interaction with the HDAC inhibitor trapoxin and shown to possess histone deacetylase activity (Taunton et al. 1996). mRpd3 (now called HDAC2) was identified in a two-hybrid screen as a corepressor associated with the YY1 transcription factor (Yang et al. 1996). Subsequent experiments by several laboratories have established that both

HDAC1 and HDAC2 associate with mSin3 in vitro and in vivo (Alland et al. 1997; Hassig et al. 1997; Heinzel et al. 1997; Laherty et al. 1997; Nagy et al. 1997; Zhang et al. 1997). Furthermore, Sin3p and Rpd3 histone deacetylase in *S. cerevisiae* was shown to interact with the UME6 transcriptional repressor (Kadosh and Struhl 1997), providing strong evidence that the Sin3-HDAC corepressor complex is recruited by repressors in systems as diverse as yeast and mammals.

The interaction of HDACs with Sin3-related corepressors suggests a mechanism for gene silencing in which DNA-binding transcriptional repressors such as Mad:Max would recruit Sin3-HDAC to specific regions of DNA (Fig. 3). The HDAC activity in turn would remove acetyl groups from the lysine residues located within amino-terminal tails of histones H3 and H4. Acetylation neutralizes the positively charged lysines which are thought to interact with the phosphate backbone of DNA wound around the nucleosomes or to promote internucleosomal association. Such interactions are known to restrict accessibility of positively acting transcription factors, presumably resulting in gene silencing (for reviews, see Wolffe and Pruss 1996; Grunstein 1997; Wolffe 1997). Support for this model derives from the long-studied correlation between deacetylated nucleosomes and transcriptionally silenced regions such as heterochromatin in *Drosophila* and yeast and the X-inactivated chromosome in mammals. In contrast, transcriptionally active euchromatin is generally acetylated, and much recent work has demonstrated that a wide range of transcriptional activators function by recruiting complexes containing histone acetylases (HATs) (for reviews, see Wolffe and Pruss 1996; Grunstein 1997; Wolffe 1997). Of course, HDACs and HATs may target proteins other

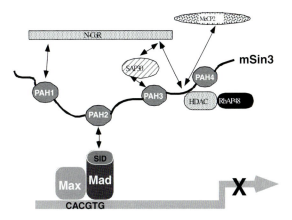

Figure 3. Summary of interactions in the mSin3-HDAC repression complex. Double-headed arrows indicate other proteins interacting with the complex. Mad:Max heterodimers recruit the mSin3 corepressor to Mad-Max CACGTG-binding sites on DNA. HDAC is histone deacetylases 1 and 2 which interact with mSin3 in the PAH3-4 intervening region. RbAp48 associates with HDAC (Hassig et al. 1997).The nuclear hormone receptor corepressor N-CoR interacts with PAH1 and the PAH3-PAH4 region; SAP30 may facilitate N-CoR interaction with PAH3; MeCP2 interacts with the PAH3-PAH4 region and recruits the corepressor complex to methylated DNA.

than histones. Indeed, some proteins contained in preinitiation complexes are known to be acetylated (Imhof et al. 1997). Their modification could also regulate gene expression. Nonetheless, there is increasing evidence, including work presented at this Symposium, linking histone acetylation-deacetylation with modulation of gene expression.

STABLE NUCLEAR mSin3 COREPRESSOR COMPLEXES ARE UTILIZED BY SEVERAL REPRESSION SYSTEMS

The relatively low levels of Max network proteins in cells appear to be maintained through synthesis balanced by protein turnover. Thus, although Max is highly stable and unregulated, all of the Max-binding proteins studied, including Myc family proteins Mad1 and Mnt, have half-lives on the order of 20–30 minutes (Lüscher and Eisenman 1987, 1988; Berberich et al. 1992; Blackwood et al. 1992a; Ayer and Eisenman 1993; Hurlin et al. 1997). This suggests that relatively stable pools of Max would be utilized for heterodimerzation and then recycled following the down-regulation of Myc, Mad, and Mnt (Blackwood et al. 1992b; Hurlin et al. 1994; Henriksson and Lüscher 1996). In this model, the newly synthesized and rapidly degraded Max-binding proteins are the rate-limiting components of the DNA-binding complex.

To better understand the dynamics of repression complex formation, we investigated the stability and localiza-

tion of mSin3A and HDAC2. Pulse-chase analyses of endogenous mSin3A and HDAC2 immunoprecipitated from HeLa cells demonstrated their half-lives to be greater than 4 hours, with substantial labeled protein still present 24 hours following a 30-minute pulse with [^{35}S]methionine (Fig. 4A). Pulse-chase analysis of the mSin3-HDAC complex also demonstrates a high degree of complex stability (D.E. Ayer, unpubl.). This contrasts sharply with the extremely labile nature of Mad1 whose half-life, when ectopically expressed in 293 cells, was approximately 20 minutes, in agreement with a previous study measuring the half-life of endogenous Mad1 (Fig. 4A) (Ayer and Eisenman 1993).

To establish the subcellular localization of endogenous mSin3A and HDAC2 proteins, indirect immunofluorescence was performed on HeLa cells using polyclonal antibodies specific for mSin3A and HDAC2. Both proteins were detected almost exclusively in the nucleus, the exception occurring in dividing cells when the nuclear membrane was disrupted (Fig. 4B). No fluorescence was observed when the antisera were blocked by preincubation with immunogen, indicating that the antibodies were specifically recognizing the cognate antigen (data not shown). The pattern of nuclear localization for mSin3A and HDAC2 is marked by exclusion from nucleoli and resembles that previously observed for c-Myc, Max, and Mad (Hann et al. 1983; Blackwood et al. 1992a; Koskinen et al. 1995).

These protein turnover and localization results suggest that stable pools of corepressor complexes exist in the nu-

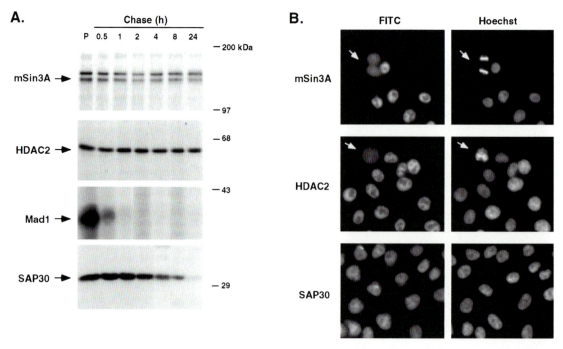

Figure 4. Half-lives and subcellular localization of mSin3A, HDAC2, and SAP30. (*A*) Immunoprecipitations were performed using anti-mSin3A, anti-HDAC2, and anti-SAP30 sera on HeLa cells harvested at the indicated time points following a 30-min pulse-labeling (P) with [^{35}S]methionine. Immunoprecipitations were also performed using anti-Mad1 serum with lysates from 293 cells transfected with a Mad1 expression plasmid at the same time points following pulse labeling. Bands corresponding to mSin3A, HDAC2, Mad1, and SAP30 are identified, and molecular weights are indicated. (*B*) Indirect immunofluorescence was performed on fixed HeLa cells using anti-mSin3A, anti-HDAC2, and anti-SinAP30 sera. FITC indicates fluorescence of the FITC-conjugated secondary antibody, and Hoescht indicates nuclear staining using Hoescht 33258. Arrows identify mitotic cells.

cleus, which could potentially be recruited to DNA by different rate-limiting transcription factors. This is consistent with recent experiments demonstrating that repression systems in addition to Mad:Max utilize the mSin3 corepressor complex. These include unliganded nuclear hormone receptors and MeCP2, the methylated DNA-binding repressor.

Nuclear Hormone Receptor Corepressors and SAP30

Nuclear hormone receptors, comprising heterodimers of RXR with thyroid hormone receptor (THR) or retinoic acid receptor (RAR), bind DNA and repress transcription in the absence of their specific ligands. Repression by unliganded receptors is mediated through binding of related corepressors known as N-CoR and SMRT (Chen and Evans 1995; Hörlein et al. 1995). Furthermore, the estrogen receptor (ER) in the presence of its antagonist transhydroxytamoxifen (TOT) binds DNA and represses transcription through association with N-CoR (Lavinsky et al. 1998). Interestingly N-CoR and SMRT both mediate repression by interacting with the mSin3-HDAC complex (Alland et al. 1997; Heinzel et al. 1997; Li et al. 1997; Nagy et al. 1997). N-CoR apparently associates with mSin3 at least in vitro through two of its repression domains. There is also some evidence that N-CoR may be present in Mad-mSin3-HDAC complexes (Heinzel et al. 1997).

SAP30, another protein detected in mSin3-HDAC complexes, associates with the PAH3 region of mSin3 and has been shown to contact N-CoR as well as HDAC (Laherty et al. 1998; Zhang et al. 1998). Figure 4B shows that, as expected for a protein associated with the mSin3-HDAC complex, SAP30 is localized to the nucleus. Antibody microinjection experiments show that N-CoR, HDAC, and mSin3 are involved in repression by Mad:Max as well as THR and RAR (Heinzel et al. 1997). However, similar experiments show SAP30 involvement to be restricted to ER-TOT repression as well as to repression by RPX and Pit1 (transcription factors also known to interact with N-CoR) (Laherty et al. 1998; Lavinsky et al. 1998; Xu et al. 1998). We interpret these experiments to mean that SAP30 functions as a specificity factor for a subset of repressors. In other words, specific transcriptional repressors (e.g., ER-TOT) may require SAP30 to facilitate or stabilize the association of N-CoR with mSin3. The relatively short half-life of SAP30 compared to mSin3 and HDAC (Fig. 4A) raises the possibility that SAP30 may be more highly regulated. However, it remains to be clarified how the putative interaction of SAP30 with HDAC (Zhang et al. 1998) relates to SAP30's association with N-CoR, PAH3, and a subset of transcriptional repressors.

Transcriptional Silencing of Methylated DNA

Methylation of cytosines at the 5-position in CpG base pairs is associated with silencing of specific genes (for review, see Bird 1992). The passive inheritance of methylated sequences is thought to underlie repression related to

genetic imprinting, X-chromosome inactivation, and the inactivation of endogenous retroviral genes (Groudine et al. 1981; Li et al. 1993; Beard et al. 1995). Methylation per se is unlikely to account for silencing and probably directs alterations in chromatin structure which in turn generate a repressive state (Buschhausen et al. 1987; Kass et al. 1997). During the last few years, several proteins with binding specificity for methyl-CpG base pairs have been identified (Meehan et al. 1989; Lewis et al. 1992; Nan et al. 1997). One of these methyl-CpG-binding proteins, termed MeCP2, possesses a transrepression domain (TRD) and is highly abundant in differentiated cells (e.g., 10^6 molecules/cell in rat brain) (Nan et al. 1997). Experiments have now shown that MeCP2 is another protein that associates with the mSin3-HDAC corepressor complex (Jones et al. 1988; Nan et al. 1998). The capacity of MeCP2 to interact with mSin3-HDAC has been demonstrated in vitro and in vivo. The interaction is mediated through the TRD of MeCP2 and the region between PAH3 and PAH4 of mSin3. The TRD of MeCP2 thus occupies the same general region on mSin3 as does HDAC (Laherty et al. 1997). As expected, anti-MeCP2 immunoprecipitates contain HDAC activity, and the deacetylase inhibitor TSA can partially block repression by the MeCP2 TRD (Jones et al. 1998; Nan et al. 1998). Furthermore, methylation-induced silencing in *Xenopus* oocytes is reversed following TSA treatment (Jones et al. 1998). It is still uncertain what fraction of genomic silencing through methylation is driven by deacetylation: MeCP2 is largely detected in differentiated cells, and other methyl-CpG-binding proteins such as MeCP1 have not yet been examined for their interactions with mSin3. Nonetheless, a plausible model is that HDAC, targeted to methylated DNA sequences through MeCP2, deacetylates local chromatin to induce repression.

REPRESSION AND DIFFERENTIATION

We have studied the proteins comprising the Max network in terms of their biological and molecular functions. Expression of the Myc or Mad family proteins produce profound but contrasting effects on cell behavior. Depending on biological "context," Myc can drive cell proliferation or apoptosis. Furthermore, deregulation of Myc expression is closely tied to oncogenesis. On the other hand, Mad family protein expression is normally linked to cell differentiation and the cessation of proliferation in many cell types. Indeed, ectopic overexpression of Mad leads primarily to inhibition of cell cycle progression and, in certain situations, decreased sensitivity to apoptosis-inducing conditions (C. Quéva et al., prep.). Mice lacking both *Mad1* alleles show normal development and differentiation but contain granulocytic precursor cells capable of undergoing ectopic cell divisions and showing increased sensitivity to apoptosis. Although more extensive studies on Mad family knockout mice are required, the results to date support the notion that Mad is involved in cell cycle exit related to differentiation.

The apparently antagonistic biological functions of Myc and Mad are reflected at the molecular level by the interactions and activities of these proteins. It has not been ruled out that there are functions of Myc and Mad that are Max-independent, but there is much evidence that heterodimerzation with Max has an important role in the activities of both proteins. Although Myc and Mad could potentially compete for available Max protein, and Myc:Max and Mad:Max heterodimers could compete for DNA-binding sites, the fact that Myc and Mad proteins are coexpressed for only a relatively short period during differentiation suggests that the major antagonistic effects result from the opposing transcriptional activities of the two heterodimers (this may be less true for Mnt and Mad3, repressors that are coexpressed with Myc). The fact that Mad1 transcriptional repression activity is required for inhibition of proliferation in Mad1-transfected cells suggests that repression is likely to be required for Mad's normal function during differentiation. However, further experiments, possibly using Mad SID mutants in mice, will be required to definitively establish this point. Furthermore, little is known concerning the mechanism by which Mad negatively influences cell progression. Clearly, one prediction, based on the Max network, is that Mad represses Myc target genes involved in cell cycle regulation and cell growth (for review, see Grandori and Eisenman 1997). However, this idea has not been subjected to a rigorous test. Preliminary studies using Myc-Mad chimeric proteins indicate that there may not be complete overlap between Myc and Mad targets (L. James and R.N. Eisenman, unpubl.).

The studies on Mad have led to a more general mechanism of transcription repression. Recruitment of HDAC to DNA through association of specific binding proteins with the mSin3 corepressor appears to underlie not only Mad:Max repression, but also repression by nuclear hormone receptors and MeCP2/methylation-induced silencing. Furthermore, HDAC is recruited by RB and the related p130 protein and may therefore account for repression mediated by these proteins (Brehm et al. 1998; Luo et al. 1998; Magnaghi-Jaulin et al. 1998). There is no evidence for mSin3 involvement, suggesting that RB and p130 are themselves acting as corepressors. Whether all repression mediated by these factors is dependent on the mSin3-HDAC complex is not known. Furthermore, there is a hint that not all silencing mediated by mSin3 is HDAC-dependent (Laherty et al. 1997; Nan et al. 1998). The corepressor complex is large and contains many proteins that have not been characterized (Grunstein 1997; Jones et al. 1998); some of these may regulate repression (e.g., SAP30) or possess novel repression functions.

The finding that Mad:Max targets the mSin3-HDAC corepressor complex to DNA suggests that Mad functions in differentiation by deacetylating histones and inducing a repressive chromatin state. This model could explain why Mad proteins are needed in addition to the down-regulation of Myc for normal cell cycle exit during differentiation. Loss of an activator may not be sufficient to effect a permanent shutdown of a critical gene. Induction of an active repressor to drive a change in chromatin structure could be required to reprogram gene expression during differentiation. It will be important to establish the extent and persistence of deacetylated chromatin at specific endogenous Myc and Mad target genes. It is of interest that many of the transcriptional regulators now thought to associate with histone deacetylases are involved in cell differentiation. Further study is likely to provide a more mechanistic link between transcriptional repression, chromatin structure, and terminal differentiation.

ACKNOWLEDGMENTS

We are grateful to Ms. Jenny Torgerson for assistance with the manuscript. We thank the following organizations for grant and fellowship support for the research discussed in this paper: the Academy of Finland, A. Kordelin Foundation, and Lady Tata Memorial Trust (P.J. K.); National Institutes of Health virology training grant T32CA-09229 (D.E.A.); Leukemia Society of America (C.D.L. and D.E.A.); Ladies Auxiliary to the Veterans of Foreign Wars Fellowship (to C.D.L.); NIH postdoctoral fellowships (P.J.H., C.D.L., and L.L.); Poncin Award (L.J.); Cancer Research Institute (C.G.); Swiss National Foundation (P.G.); Damon Runyon-Walter Winchell Foundation postdoctoral fellowship DRG076 (G.A.M.); INSERM and the Phillippe Foundation (C.Q.), Humboldt Foundation (B.P.); American Cancer Society (K.P.F.); an NCI Japanese Foundation for Cancer Research Training Program in the U.S.-Japan Cooperative Cancer Committee (Y.S.); and grants from the NIH/NCI RO1-57138, RO1-CA-20525, and P50HL54881 (R.N.E.). R.N.E. is an American Cancer Society Research Professor.

REFERENCES

Alland L., Muhle R., Hou H., Jr., Potes J., Chin L., Schreiber-Agus N., and DePinho R.A. 1997. Role for N-CoR and histone deacetylase in Sin3-mediated transcriptional repression. *Nature* **387:** 49.

Amati B., Alevizopoulos K., and Vlach J. 1998. Myc and the cell cycle. *Front. Biosci.* **3:** 250.

Amati B., Dalton S., Brooks M.W., Littlewood T.D., Evan G.I., and Land H. 1992. Transcriptional activation by the human c-Myc oncoprotein in yeast requires interaction with Max. *Nature* **359:** 423.

Ayer D.E. and Eisenman R.N. 1993. A switch from Myc:Max to Mad:Max heterocomplexes accompanies monocyte/macrophage differentiation. *Genes Dev.* **7:** 2110.

Ayer D.E., Kretzner L., and Eisenman R.N. 1993. Mad: A heterodimeric partner for Max that antagonizes Myc transcriptional activity. *Cell* **72:** 211.

Ayer D.E., Lawrence Q.A., and Eisenman R.N. 1995. Mad-Max transcriptional repression is mediated by ternary complex formation with mammalian homologs of yeast repressor Sin3. *Cell* **80:** 767.

Ayer D.E., Laherty C.D., Lawrence Q.A., Armstrong A.P., and Eisenman R.N. 1996. Mad proteins contain a dominant transcription repression domain. *Mol. Cell. Biol.* **16:** 5772.

Beard C., Li E., and Jaenisch R. 1995. Loss of methylation activates Xist in somatic but not in embryonic cells. *Genes Dev.* **9:** 2325.

Berberich S., Hyde-DeRuyscher N., Espenshade P., and Cole M. 1992. Max encodes a sequence-specific DNA binding protein and is not regulated by serum growth factors. *Oncogene* **7:** 775.

Bird A. 1992. The essentials of DNA methylation. *Cell* **70:** 5.

Blackwell T.K., Kretzner L., Blackwood E.M., Eisenman R.N.,

and Weintraub H. 1990. Sequence-specific DNA-binding by the c-Myc protein. *Science* **250:** 1149.

Blackwood E.M. and Eisenman R.N. 1991. Max: A helix-loop-helix zipper protein that forms a sequence-specific DNA binding complex with Myc. *Science* **251:** 1211.

Blackwood E.M., Kretzner L., and Eisenman R.N. 1992a. Myc and Max function as a nucleoprotein complex. *Curr. Opin. Genet. Dev.* **2:** 227.

Blackwood E.M., Lüscher B., and Eisenman R.N. 1992b. Myc and Max associate in vivo. *Genes Dev.* **6:** 71.

Brehm A., Miska E.A., McCance D.J., Reid J.L., Bannister A.J., and Kouzarides T. 1998. Retinoblastoma protein recruits histone deacetylase to repress transcription. *Nature* **391:** 597.

Buschhausen G., Witting B., Graessmann M., and Graessmann A. 1987. Chromatin structure is required to block transcription of the methylated herpes simplex virus thymidine kinase gene. *Proc. Natl. Acad. Sci.* **84:** 1177.

Casaccia-Bonnefil P., Tikoo R., Kiyokawa H., Friedrich V., Jr., Chao M., and Koff A. 1997. Oligodendrocyte precursor differentiation is perturbed in the absence of the cyclin-dependent kinase inhibitor p27^{Kip1}. *Genes Dev.* **11:** 2335.

Chen J., Willingham T., Margraf L.R., Schreiber-Agus N., DePinho R.A., and Nisen P.D. 1995. Effects of the MYC oncogene antagonist, MAD, on proliferation, cell cycling and the malignant phenotype of human brain tumor cells. *Nat. Med.* **1:** 638.

Chen J.D. and Evans R.M. 1995. A transcriptional co-repressor that interacts with nuclear hormone receptors. *Nature* **377:** 454.

Cultraro C.M., Bino T., and Segal S. 1997. Function of the c-Myc antagonist Mad1 during a molecular switch from proliferation to differentiation. *Mol. Cell. Biol.* **17:** 2353.

de Nooij J.C., Letender M.A., and Hariharan I.K. 1996. A cyclin-dependent kinase inhibitor, Dacapo, is necessary for timely exit from the cell cycle during *Drosophila* embryogenesis. *Cell* **87:** 1237.

Di Cunto F., Topley G., Calautti E., Hsiao J., Ong L., Seth P.K., and Dotto G.P. 1998. Inhibitory function of p21$^{Cip1/WAF1}$ in differentiation of primary mouse keratinocytes independent of cell cycle control. *Science* **280:** 1069.

Downs K.M., Martin G.R., and Bishop J.M. 1989. Contrasting patterns of myc and N-myc expression during gastrulation of the mouse embryo. *Genes Dev.* **3:** 860.

Dyson N., Howley P., Munger K., and Harlow E. 1989. The human papilloma virus-16 E7 oncoprotein is able to bind the retinoblastoma gene product. *Science* **243:** 934.

Edelhoff S., Ayer D.E., Zervos A.S., Steingrimsson E., Jenkins N.A., Copeland N.G., Eisenman R.N., Brent R., and Disteche C.M. 1994. Mapping of two genes encoding members of a distinct subfamily of MAX interacting proteins: MAD to human chromosome 2 and mouse chromosome 6, and MXI1 to human chromosome 10 and mouse chromosome 19. *Oncogene* **9:** 665.

Foley K.P., McArthur G.A., Quéva C., Hurlin P.J., Soriano P., and Eisenman, R.N. 1998. Targeted disruption of Mad1 inhibits cell cycle exit during granulocyte differentiation. *EMBO J.* **17:** 774.

Grandori C. and Eisenman R.N. 1997. Myc target genes. *Trends Biochem. Sci.* **22:** 177.

Groudine M., Eisenman R.N., and Weintraub H. 1981. Chromatin structure of endogenous retroviral genes and activation by an inhibitor of DNA methylation. *Nature* **292:** 311.

Grunstein M. 1997. Histone acetylation in chromatin structure and transcription. *Nature* **389:** 349.

Halevy O., Novitch B.G., Spicer D.B., Skapek S.X., Rhee J., Hannon G.J., Beach D., and Lassar A.B. 1995. Correlation of terminal cell cycle arrest of skeletal muscle with induction of p21 by MyoD. *Science* **267:** 1018.

Hamel P.A., Gill R.M., Phillips R.A., and Gallie B.L. 1992. Transcriptional repression of the E2-containing promoters EIIaE, c-myc, and RB1 by the product of the RB1 gene. *Mol. Cell. Biol.* **12:** 3431.

Hann S.R., Abrams H.D., Rohrschneider L.R., and Eisenman

R.N. 1983. Proteins encoded by v-*myc* and c-*myc* oncogenes: Identification and localization in acute leukemia virus transformants and bursal lymphoma cell lines. *Cell* **34:** 789.

Hartwell L.H. and Kastan M.B. 1994. Cell cycle control and cancer. *Science* **266:** 1821.

Hassig C.A., Fleischer T.C., Billin A.N., Schreiber S.L., and Ayer D.E. 1997. Histone deacetylase activity is required for full transcriptional repression by mSin3A. *Cell* **89:** 341.

Heinzel T., Lavinsky R.M., Mullen T.-M., Söderström M., Laherty C.D., Torchia J., Yang W.-M., Brard G., Ngo S.G., Davie J.R., Seto E., Eisenman R.N., Rose D.W., Glass C.K., and Rosenfeld M.G. 1997. N-CoR, mSin3, and histone deacetylase-containing complexes in nuclear receptor and Mad repression. *Nature* **387:** 43.

Henriksson M. and Lüscher B. 1996. Proteins of the Myc network: Essential regulators of cell growth and differentiation. *Adv. Cancer Res.* **68:** 109.

Hirning U., Schmid P., Schulz W.A., Rettenberger G., and Hameister H. 1991. A comparative analysis of N-myc and c-myc expression and cellular proliferation in mouse organogenesis. *Mech. Dev.* **33:** 119.

Hörlein A.J., Näär A.M., Heinzel T., Torchia J., Gloss B., Kurokawa R., Ryan A., Kamei Y., Söderstrom M., Glass C.K., and Rosenfeld M.G. 1995. Ligand-independent repression by the thyroid hormone receptor mediated by a nuclear receptor co-repressor. *Nature* **377:** 397.

Hurlin P., Quéva C., and Eisenman R.N. 1997. Mnt, a novel Max-interacting protein is coexpressed with Myc in proliferating cells and mediates repression at Myc binding sites. *Genes Dev.* **11:** 44.

Hurlin P., Ayer D.E., Grandori C., and Eisenman R.N. 1994. The Max transcription factor network: Involvement of Mad in differentiation and an approach to identification of target genes. *Cold Spring Harbor Symp. Quant. Biol.* **59:** 109.

Hurlin P.J., Foley K.P., Ayer D.E., Eisenman R.N., Hanahan D., and Arbeit J.M. 1995a. Regulation of Myc and Mad during epidermal differentiation and HPV-associated tumorigenesis. *Oncogene* **11:** 2487.

Hurlin P.J., Quéva C., Koskinen P.J., Steingrímsson E., Ayer D.E., Copeland N.G., Jenkins N.A., and Eisenman R.N. 1995b. Mad3 and Mad4: Novel Max-interacting transcriptional repressors that suppress c-Myc-dependent transformation and are expressed during neural and epidermal differentiation. *EMBO J.* **14:** 5646.

Imhof A., Yang X.-J., Ogryzko V.V., Nakatani Y., Wolffe A.P., and Ge H. 1997. Acetylation of general transcription factors by histone acetyltransferases. *Curr. Biol.* **7:** 689.

Jones P.L., Vennstra G.J.C., Wade P.A., Vermaak D., Kass S.U., Landsberger N., Strouboulis J., and Wolffe A.P. 1998. Methylated DNA and MeCP2 recruit histone deacetylase to repress transcription. *Nat. Genet.* **19:** 187.

Kadosh D. and Struhl K. 1997. Repression by UME6 involves recruitment of a complex containing Sin3 corepressor and Rpd3 histone deacetylase to target promoters. *Cell* **89:** 365.

Kass S.U., Landsberger N., and Wolffe A.P. 1997. DNA methylation directs a time-dependent repression of transcription initiation. *Curr. Biol.* **7:** 157.

Kato G.J., Barrett J., Villa-Garcia M., and Dang C.V. 1990. An amino-terminal c-Myc domain required for neoplastic transformation activates transcription. *Mol. Cell. Biol.* **10:** 5914.

Koskinen P.J., Ayer D.E., and Eisenman R.N. 1995. Repression of Myc-Ras co-transformation by Mad is mediated by multiple protein-protein interactions. *Cell Growth Differ.* **6:** 623.

Kretzner L., Blackwood E.M., and Eisenman R.N. 1992. The Myc and Max proteins possess distinct transcriptional activities. *Nature* **359:** 426.

Laherty C.D., Yang W.-M., Sun J.-M., Davie J.R., Seto E., and Eisenman R.N. 1997. Histone deacetylases associated with the mSin3 corepressor mediate Mad transcriptional repression. *Cell* **89:** 349.

Laherty C.D., Billin A.N., Lavinsky R.M., Yochum G.S., Bush A.C., Sun J.-M., Davie J.R., Rose D.W., Rosenfeld M.G., Ayer D.E., and Eisenman R.N. 1998. SAP30, a component of the mSin3 corepressor complex involved in N-CoR-mediated repression by specific nuclear hormone teceptor and homeodomain proteins. *Mol. Cell* **2:** 417.

Lahoz E.G., Xu L., Schreiber-Agus N., and DePinho R.A. 1994. Suppression of Myc, but not E1a, transformation activity by Max-associated proteins, Mad and Mxi1. *Proc. Natl. Acad. Sci.* **91:** 5503.

Langdon W.Y., Harris A.W., and Cory S. 1988. Growth of Em-*myc* B-lymphoid cells in vitro and their evolution towards autonomy. *Oncogene Res.* **3:** 271.

Langdon W.Y., Harris A.W., Cory S., and Adams J.M. 1986. The c-*myc* oncogene perturbs B lymphocyte development in E-myc transgenic mice. *Cell* **47:** 11.

Larsson L.-G., Pettersson M., Öberg F., Nilsson K., and Lüscher B. 1994. Expression of *mad*, *mxi1*, *max*, and c-*myc* during induced differentiation of hematopoietic cells: Opposite regulation of *mad* and c-*myc*. *Oncogene* **9:** 1247.

Lavinsky R.M., Jepsen K., Heinzel T., Torchia J., Mullen T.-M., Schiff R., Del-Rio A.L., Ricote M., Ngo S., Gemsch J., Hilsenbeck S.G., Osborne C.K., Glass C.K., Rosenfeld M.G., and Rose D.W. 1998. Diverse signaling pathways modulate nuclear receptor recruitment of N-CoR and SMRT complexes. *Proc. Natl. Acad. Sci.* **95:** 2920.

Lee E.Y., Hu N., Yuan S.S.-F., Cox L.A., Bradley A., Lee W.-H., and Herrup K. 1994. Dual roles of the retinoblastoma protein in cell cycle regulation and neuron differentiation. *Genes Dev.* **8:** 2008.

Lewis J.D., Meehan R.R., Henzel W.J., Maurer-Fogy I., Jeppesen P., Klein F., and Bird A. 1992. Purification, sequence, and cellular localization of a novel chromosomal protein that binds to methylated DNA. *Cell* **69:** 906.

Li E., Beard C., and Jaenisch R. 1993. Role for DNA methylation in genomic imprinting. *Nature* **366:** 362.

Li H., Leo C., Schroen D.J., and Chen J.D. 1997. Characterization of receptor interaction and transcriptional repression by the corepressor SMRT. *Mol. Endocrinol.* **11:** 2025.

Luo R.X., Postigo A.A., and Dean D.C. 1998. Rb interacts with histone deacetylase to repress transcription. *Cell* **92:** 463.

Lüscher B. and Eisenman R.N. 1987. Proteins encoded by the c-*myc* oncogene: Analysis of c-myc protein degradation. In *Oncogenes and cancer* (ed. S.A. Aaronson et al.), p. 291. Japan Society Press, Utrecht, The Netherlands.

———. 1988. c-myc and c-myb protein degradation: Effect of metabolic inhibitors and heat shock. *Mol. Cell. Biol.* **8:** 2504.

Magnaghi-Jaulin L., Groisman R., Naguibneva I., Robin P., Lorain S., Le Villain J.P., Troalen F., Trouche D., and Harel-Bellan A. 1998. Retinoblastima protein represses transcription by recruiting a histone deacetylase. *Nature* **391:** 601.

McArthur G.A. and Eisenman R.N. 1997. The Max transcription factor network: Mad1 inhibition of cell cycle progression is inhibited by the human papilloma virus E7 protein. In *American Society for Clinical Oncology educational book* (ed. M.C. Perry), p. 42. American Society for Clinical Oncology, San Diego, California.

Meehan R.R., Lewis J.D., Mckay S., Kleiner E.L., and Bird A.P. 1989. Identification of a mammalian protein that binds specifically to DNA containing methylated CpGs. *Cell* **58:** 499.

Meroni G., Reymond A., Alcalay M., Borsani G., Tanigami A., Tonlorenzi R., Lo Nigro C., Messali S., Zollo M., Ledbetter D.H., Brent R., Ballabio A., and Carrozzo R. 1997. Rox, a novel bHLHZip protein expressed in quiescent cells that heterodimerizes with Max, binds a non-canonical E box and acts as a transcriptional repressor. *EMBO J.* **16:** 2892.

Metcalf D. 1989. The molecular control of cell division, differentiation commitment and maturation in haemopoietic cells. *Nature* **339:** 27.

Missero C., Di Cunto F., Kiyokawa H., Koff A., and Dotto G.P. 1996. The absence of p21$^{Cip1/WAF1}$ alters keratinocyte growth and differentiation and promotes *ras*-tumor progression. *Genes Dev.* **10:** 3065.

Nagy L., Kao H.-Y., Chakravarti D., Lin R.J., Hassig C.A., Ayer D.E., Schreiber S.L., and Evans R.M. 1997. Nuclear receptor repression mediated by a complex containing SMRT, mSin3A, and histone deacetylase. *Cell* **89:** 373.

Nan X., Campoy F.J., and Bird A. 1997. MeCP2 is a transcriptional repressor with abundant binding sites in genomic chromatin. *Cell* **88:** 471.

Nan X., Ng H.-H., Johnson C.A., Laherty C.D., Turner B.M., Eisenman R.N., and Bird A. 1998. Transcriptional repression by the methyl-CpG binding protein MeCP2 involves a histone deacetylase complex. *Nature* **393:** 386.

Parker S.B., Eichele G., Zhang P., Rawls A., Sands A., Bradley A., Olson, E.N., Harper J.W., and Elledge S.J. 1995. p53-independent expression of p21^{Cip1} in muscle and other terminally differentiating cells. *Science* **267:** 1024.

Prendergast G.C. and Ziff E.B. 1991. Methylation-sensitive sequence-specific DNA binding by the c-Myc basic region. *Science* **251:** 186.

Prendergast G.C., Lawe D., and Ziff E.B. 1991. Association of Myn, the murine homolog of Max, with c-Myc stimulates methylation-sensitive DNA binding and Ras cotransformation. *Cell* **65:** 395.

Quéva C., Hurlin P.J., Foley K.P., and Eisenman R.N. 1998. Sequential expression of the MAD family of transcriptional repressors during differentiation. *Oncogene* **16:** 967.

Raff M.C. 1996. Size control: The regulation of cell numbers in animal development. *Cell* **86:** 173.

Rao G., Alland L., Guida G., Schreiber-Agus N., Chen K., Chin L., Rochelle J.M., Seldin M.F., Skoultchi A.I., and DePinho R.A. 1996. Mouse Sin3A interacts with and can functionally substitute for the amino-terminal repression domain of the Myc antagonist Mxi1. *Oncogene* **12:** 1165.

Roussel M.F., Ashmun R.A., Sherr C.J., Eisenman R.N., and Ayer D.E. 1996. Inhibition of cell proliferation by the Mad1 transcriptional repressor. *Mol. Cell. Biol.* **16:** 2796.

Schmid P., Schulz W.A., and Hameister H. 1989. Dynamic expression of the *myc* protooncogene in midgestation mouse embryos. *Science* **243:** 226.

Schreiber-Agus N., Chin L., Chen K., Torres R., Rao G., Guida P., Skoultchi A.I., and DePinho R.A. 1995. An amino-terminal domain of Mxi1 mediates anti-Myc oncogenic activity and interacts with a homolog of the yeast repressor SIN3. *Cell* **80:** 777.

Schreiber-Agus N., Meng Y., Hoang T., Hou H., Jr., Chen K., Greenberg R., Cordon-Cardo C., Lee H.-W., and DePinho R.A. 1998. Role of Mxi1 in ageing organ systems and the regulation of normal and neoplastic growth. *Nature* **393:** 483.

Sherr C.J. 1996. Cancer cell cycles. *Science* **274:** 1672.

Stillman D.J., Dorland S., and Yu Y. 1994. Epistatic analysis of suppressor mutations that allow HO expression in the absence of the yeast SWI5 transcriptional activator. *Genetics* **136:** 781.

Taunton J., Hassig C.A., and Schreiber S.L. 1996. A mammalian histone deacetylase related to the yeast transcriptional regulator Rpd3p. *Science* **272:** 408.

Tevosian S.G., Shih H.H., Mendelson K.G., Sheppard K.-A., Paulson K.E., and Yee A.S. 1997. HBP1: A HMG box transcriptional repressor that is targeted by the retinoblastoma family. *Genes Dev.* **11:** 383.

Vastrik I., Kaipainen A., Penttila T.-L., Lymboussakis A., Alitalo R., Parvinen M., and Alitalo K. 1995. Expression of the *mad* gene during cell differentiation in *vivo* and its inhibition of cell growth in *vitro*. *J. Cell Biol.* **128:** 1197.

Vidal M. and Gaber R.F. 1991. RPD3 encodes a second factor required to achieve maximum positive and negative transcriptional states in *Saccharomyces cerevisiae*. *Mol. Cell. Biol.* **11:** 6317.

Vidal M., Strich R., Esposito R.E., and Gaber R.F. 1991. RPD1 (SIN3/UME4) is required for maximal activation and repression of diverse yeast genes. *Mol. Cell. Biol.* **11:** 6306.

Wang H., Clark I., Nicholson P.R., Herskowitz I., and Stillman D.J. 1990. The *Saccharomyces cerevisiae* SIN3 gene, a negative regulator of HO, contains four paired amphipathic helix motifs. *Mol. Cell. Biol.* **10:** 5927.

Wechsler D.S., Shelly C.A., Petroff C.A., and Dang C.V. 1997. MXI1, a putative tumor suppressor gene, suppresses growth of human glioblastoma cells. *Cancer Res.* **57:** 4905.

Weintraub S.J., Chow K.N.B., Luo R.X., Zhang S.H., He S., and Dean D.C. 1995. Mechanism of active transcriptional repression by the retinoblastoma protein. *Nature* **375:** 812.

Wolffe A.P. 1997. Sinful repression. *Nature* **387:** 16.

Wolffe A.P. and Pruss D. 1996. Targeting chromatin disruption: Transcription regulators that acetylate histones. *Cell* **84:** 817.

Xu L., Lavinsky R.M., Dasen J., Flynn S.E., McInerney E., Mullen T.-M., Heinzel T., Szeto D., Korzus E., Kurokawa R., Aggarwal A.K., Rose D.W., Glass C.K., and Rosenfeld M.G. 1998. Regulated control of POU domain activators and homeodomain repressors by the N-CoR/corepressor and the CBP/coactivator integration complexes. *Nature* **395:** 301.

Yan Y., Frisen J., Lee M.-H., Massagué J., and Barbacid M. 1997. Ablation of the CDK inhibitor p57Kip2 results in increased apoptosis and delayed differentiation during mouse development. *Genes Dev.* **11:** 973.

Yang W.-M., Inouye C., Zeng Y., Bearss D., and Seto E. 1996. Transcriptional repression by YY1 is mediated by interaction with a mammalian homolog of the yeast global regulator RPD3. *Proc. Natl. Acad. Sci.* **93:** 12845.

Zervos A.S., Gyuris J., and Brent R. 1993. Mxi1, a protein that specifically interacts with Max to bind Myc-Max recognition sites. *Cell* **72:** 223.

Zhang Y., Iratni R., Erdjument-Bromage H., Tempst P., and Reinberg D. 1997. Histone deacetylases and SAP18, a novel polypeptide, are components of a human Sin3 complex. *Cell* **89:** 357.

Zhang Y., Sun Z.-W., Iratni R., Erdjument-Bromage H., Tempst P., Hampsey M., and Reinberg D. 1998. SAP30, a novel protein conserved between human and yeast, is a component of a histone deacetylase complex. *Mol. Cell* **1:** 1021.

Histone Deacetylase Directs the Dominant Silencing of Transcription in Chromatin: Association with MeCP2 and the Mi-2 Chromodomain SWI/SNF ATPase

P.A. WADE, P.L. JONES, D. VERMAAK, G.J.C. VEENSTRA, A. IMHOF, T. SERA, C. TSE,*
H. GE, Y.-B. SHI, J.C. HANSEN,* AND A.P. WOLFFE

*Laboratory of Molecular Embryology, National Institute of Child Health and Human Development,
National Institutes of Health, Bethesda, Maryland 20892-5431; *Department of Biochemistry,
The University of Texas Health Science Center, San Antonio, Texas 78284-7760*

Histone acetylation is intimately connected to transcriptional regulation (Brownell and Allis 1996; Wolffe 1996; Wolffe and Pruss 1996; Grunstein 1997). Acetylation of the core histone amino-terminal tails provides a means by which transcription factors can gain access to their recognition elements within nucleosomes (Lee et al. 1993; Wolffe et al. 1993; Vettese-Dadey et al. 1996). In addition, acetylation facilitates transcription from nucleosomal arrays (Ura et al. 1997; Nightingale et al. 1998; Tse et al. 1998). Coactivator complexes that function as histone acetyltransferases are required for transcription activation (Brownell et al. 1996; Ogryzko et al. 1996; Yang et al. 1996; Kuo et al. 1998), whereas corepressors containing histone deacetylases confer transcriptional repression (Taunton et al. 1996; Alland et al. 1997; Hassig et al. 1997, 1998; Nagy et al. 1997; Wong et al. 1998; D. Vermaak et al., in prep.). Histones are locally modified on target promoters (Kuo et al. 1998; Rundlett et al. 1998) and specific lysines in particular histones are functional targets for acetyltransferases and deacetylases (Rundlett et al. 1998; Zhang et al. 1998). Histone acetylation states are dynamic, with acetylated lysines in hyperacetylated histones turning over rapidly in transcriptionally active chromatin, but much less rapidly in the hypoacetylated histones of transcriptionally silent regions (Covault and Chalkley 1980; Zhang and Nelson 1988). The dynamics of histone acetylation might provide a mechanistic foundation for the reversible activation and repression of transcription (Wade et al. 1997; Wolffe 1997).

The modification of histones in vitro and in vivo does not prove that these abundant structural proteins are the only targets for regulatory activity in vivo. Other components of the transcriptional machinery such as p53, TFIIE, and TFIIF can be acetylated in vitro (Gu and Roeder 1997; Imhof et al. 1997). Modification of these more limiting factors in the eukaryotic nucleus might have the dominant control function for transcription. Likewise, coactivators or corepressors might influence the recruitment or function of the basal transcriptional machinery by mechanisms independent of the acetylation status of either the histones or any other proteins.

To understand the contribution that protein acetylation status makes to the transcription process, it is important (1) to define the functional consequences of modifying particular proteins to known extents, (2) to identify the determinants of specificity in the modification process, and (3) to understand the biological context in which particular acetyltransferases or deacetylases function. In this paper, we summarize progress in our laboratory toward achieving these goals.

STRUCTURAL AND FUNCTIONAL CONSEQUENCES OF HISTONE ACETYLATION

Acetylated histones wrap DNA less tightly in mononucleosomes (Norton et al. 1989; Bauer et al. 1994; Krajewski and Becker 1998; but see Lutter et al. 1992); the acetylated amino-terminal tails of the histones bind DNA less tightly (Hong et al. 1993; Puig et al. 1998) and are more mobile with respect to the DNA surface than unmodified tails (Cary et al. 1982). Nucleosomal arrays compact into higher-order chromatin structures less well when histones are acetylated (Fig. 1a) (but see Garcia-Ramirez et al. 1995; Tse et al. 1998; McGhee et al. 1983). Interactions between adjacent nucleosomal arrays are also reduced when histones are acetylated (Fig. 1b,d) (Tse et al. 1998) and chromatin solubility is increased (Fig. 1e) (Perry and Chalkley 1982). Within the 75 mg/ml nucleoprotein environment of the eukaryotic nucleus, all of these structural transitions might contribute to facilitating transcription.

Nucleosomes containing acetylated histones are more accessible to transcription factors (Lee et al. 1993; Wolffe et al. 1993; Vettese-Dadey et al. 1996; but see Howe and Ausio 1998). Proteolytic removal of the amino termini of the core histones leads to comparable increases in transcription factor access to nucleosomal DNA as histone hyperacetylation (Lee et al. 1993; Vettese-Dadey et al. 1994). Nucleosomal arrays assembled from acetylated histones are better templates for transcription than those containing unmodified histones (Ura et al. 1997; Nightingale et al. 1998). The level of histone modification required to facilitate the transcription process is relatively low, and a total of 12 acetylated lysines per histone octamer (out of 28 potential acetylated lysines) will promote in vitro transcription more than 15-fold (Fig. 1c). This same level of modification reduces chromatin com-

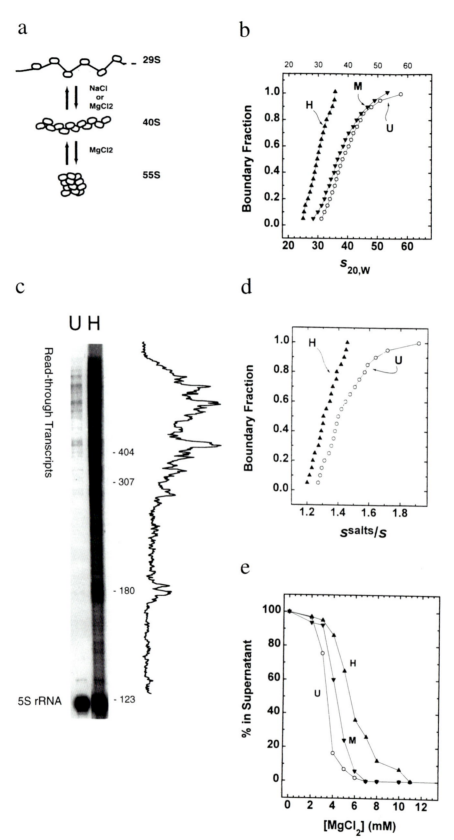

Figure 1. (*See facing page for legend.*)

paction to the same extent as proteolytic removal of the amino termini (Tse and Hansen 1997; Tse et al. 1998). The similarities in functional and structural consequences between histone hyperacetylation and proteolytic removal of the amino-terminal tails strongly suggest that one consequence of acetylation is to reduce the interaction of the tails with DNA (Hong et al. 1993) or potentially with adjacent nucleosomes (Luger et al. 1997). It is also probable that acetylation of the histones serves to illuminate particular nucleosomes and/or segments of chromatin for interaction with other chromatin remodeling factors or components of the transcriptional machinery (Lee et al. 1993; Turner 1993). Evidence exists for the interaction of specific repressive proteins with the amino-terminal tails (Hecht et al. 1995; Edmondson et al. 1996), but the exact requirements or influence of specific acetylation states on interactions with positively acting components of the transcriptional machinery have not been resolved. The potential combination of direct chromatin structural transitions and modulation of protein-protein interactions following acetylation or deacetylation of the histone tails provides a powerful means of regulating transcription.

FUNCTIONAL CONSEQUENCES OF MODIFYING TFIIE AND TFIIF IN VITRO

Several transcription factors are substrates for acetylation in vitro (Gu and Roeder 1997; Imhof et al. 1997). These include TFIIE and TFIIF which are modified by p300, PCAF, and TAF$_{II}$250 (Fig. 2) (Imhof et al. 1997). The TFIIEβ subunit is modified at a specific lysine residue (K52) in the amino terminus; however, mutagenesis of Lys-52 into an arginine does not interfere with TFIIEβ function (Imhof et al. 1997). In addition, although the coactivator/acetyltransferase PCAF (Yang et al. 1996) can modestly stimulate in vitro transcription in an acetyl CoA-dependent manner (Fig. 3), pretreatment of

TFIIE or TFIIF with PCAF, p300, and TAF$_{II}$250 to acetylate both transcription factors to 50% efficiency does not stimulate transcription (Imhof et al. 1997). Thus, the modification of TFIIE and TFIIF does not lead to major functional consequences under standard in vitro reaction conditions (Ge et al. 1996). It remains possible that TFIIE and TFIIF might require acetylation to function effectively in an in vivo context. For example, the topological constraints imposed by chromatin and preinitiation complex assembly might increase the dependence of the transcription reaction on protein-protein or protein-DNA interactions mediated by TFIIE and TFIIF (Tan et al. 1994; Dikstein et al. 1996).

RbAp48 IS A SPECIFICITY DETERMINANT FOR HISTONE DEACETYLATION

With the existence of diverse targets for acetylation by coactivators and for deacetylation by corepressors, it is important to identify specificity determinants. The greater the specificity of the protein-protein interactions that target a modification, the more likely their potential biological significance. Many proteins with functions in histone and chromatin biology contain a member of a family of WD40 repeat proteins. WD repeat proteins fold into a multibladed β propeller structure that provides a platform for interaction with other proteins (Neer and Smith 1996). Two mammalian proteins of this family (RbAp48 and RbAp46) were originally purified through their interaction with the retinoblastoma (Rb) protein (Qian et al. 1993; Qian and Lee 1995). Human chromatin assembly factor 1 (CAF1) contains RbAp48 (Verreault et al. 1996) and the cytoplasmic human histone acetyltransferase Hat1 contains RbAp46 (Verreault et al. 1997). RbAp46 and RbAp48 are also associated with mammalian histone deacetylases (Hassig et al. 1997; Zhang et al. 1997). In Drosophila, a WD40 repeat protein p55 is a component of chromatin assembly factor 1 (dCAF1;

Figure 1. Folding, solubility, and transcription of nucleosomal arrays dependent on histone acetylation. (*a*) Salt-dependent folding of an array of 12 nucleosomes as determined from sedimentation velocity experiments. Schematic representations of array conformations are shown, whose extent of compaction would yield the indicated sedimentation coefficients. (*b*) Sedimentation velocity analysis of underacetylated (U; *open circle*), moderately acetylated (M; *closed inverted tiangle*), and highly acetylated (H; *closed triangle*) nucleosomal arrays in 2 mM free Mg^{++} for 1 hr at room temperature. Samples were then sedimented at 18,000 rpm, and 20 boundary scans were collected. The temperature of the run was 21°C. Each boundary was divided into 20 equal fractions. The diffusion corrected sedimentation coefficient at each boundary division was determined, and the data were plotted as boundary fraction versus S$_{20}$,W to yield the integral distribution of sedimentation coefficients present in the sample. (*c*) Octamers containing underacetylated core histones (2 acetylated lysines) or highly acetylated core histones (12 acetylated lysines) were reconstituted into underacetylated (U) and highly acetylated (H) nucleosomal arrays which were than transcribed in *Xenopus* oocyte nuclear extracts. A histone-free plasmid containing one copy of the *Xenopus borealis* somatic 5S rRNA gene (which produces a 120-nucleotide transcript) was included in each reaction as an internal control. RNA products of the transcription reactions were electrophoresed in a denaturing 6% polyacrylamide gel. After electrophoresis, radioactive RNA was visualized using a Molecular Dynamics Phosphorimager. *Msp*I-digested pBR322 was utilized as a size marker. A densitometric trace of the RNA transcripts produced from the highly acetylated arrays is also shown. (*d*) Sedimentation velocity analysis of nucleosomal array folding in transcription buffer. Highly acetylated (*closed triangle*) and underacetylated (*open circle*) nucleosomal arrays were incubated in either transcription buffer or transcription buffer containing 50 mM KCl and 7 mM MgCl$_2$ for 1 hr at room temperature. For these experiments, the nucleoside triphosphates in transcription buffer were replaced with 2 mM Na$_5$PPPi to avoid interference with the absorbance optical system of the analytical ultracentrifuge. (*e*) Mg^{++}-dependent oligomerization of underacetylated, moderately acetylated, and highly acetylated nucleosomal arrays. Oligomerization was assayed by differential centrifugation. Shown is the percentage of underacetylated (U; *open circle*), moderately acetylated (M; *closed inverted triangle*), and highly acetylated (H; *closed triangle*) samples that remain in the supernatant after exposure to the indicated amounts of MgCl$_2$ for 10 min at room temperature and centrifugation at 16,000g for 10 min in an Eppendorf microcentrifuge. Each data point represents the mean plus or minus the standard deviation of three to four determinations.

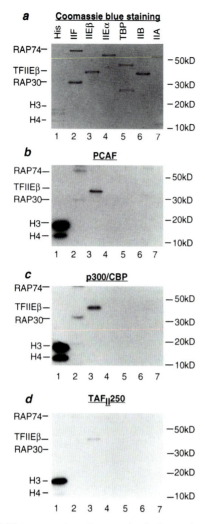

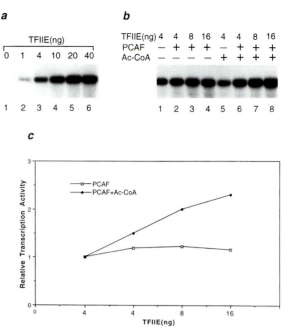

Figure 2. Histone acetyltransferases selectively acetylate general transcription factors. Core histones and purified recombinant general transcription factors for RNA polymerase II were normalized by SDS-PAGE and visualized by Coomassie blue staining (*a*), or acetylated in the presence of [³H]acetyl CoA by PCAF (*b*), by p300/CBP (*c*), or by TAF$_{II}$250 (*d*). (Lane *1*) Core histones; (lane *2*) RAP74 and RAP30 subunits of TFIIF; (lane *3*) TFIIEβ(p34); (lane *4*) TFIIEα(p56); (lane *5*) TBP; (lane *6*) TFIIB; (lane *7*) TFIIA(p55+p12). Protein molecular-mass standards in kilodaltons (kD) are indicated on the right. The smaller peptide (~28 kD) detected from the TBP fraction in an *Escherichia coli* protein copurified from Ni^{++} agarose column.

Figure 3. PCAF stimulates basal level of transcription in an acetyl CoA-dependent manner. (*a*) Titration of recombinant TFIIE and a TFIIE-dependent reconstituted transcription assay. Increased amounts of TFIIE are indicated on the top. (*b*) PCAF, acetyl CoA, and TFIIE affect basal transcription in vitro. Recombinant TFIIE (α+β) without preincubation (lane *1*) or preincubated with PCAF alone (lanes *2–4*), with acetyl CoA alone (lane *5*), or with both PCAF and acetyl CoA (lanes *6–8*) was added to the TFIIE-dependent transcription assay as indicated. (*c*) Quantitated relative transcription activities in which the relative activities from lanes *1* and *5* were normalized as *1* for lanes *2–4* (*open box*) and lanes *6–8* (*closed diamond*), respectively.

Tyler et al. 1996) and nucleosome remodeling factor (NURF; Martinez-Balbas et al. 1998). Human RbAp46 interacts with the first α-helix of the histone-fold domains of histone H4 (key contacts between amino acids 34 and 41) and H2A (Verreault et al. 1997). *Xenopus* RbAp48 interacts with histone H4 predominantly and only weakly with H2A (Fig. 4a). Deletion analysis indicates that amino acids at the amino-terminal boundary of the first α-helix of the histone-fold domain of H4 are essential for binding to RbAp48 (Fig. 4b). The two carboxy-terminal α helices of H4, which are essential for heterodimerization with histone H3 (Freeman et al. 1996), are not required for stable interaction with RbAp48 (Fig. 4c). We conclude that *Xenopus* RbAp48 is specifically targeted to

a small segment of histone H4, which is probably inaccessible in a nucleosomal context (Luger et al. 1997). Because the catalytic subunit of deacetylase RPD3 interacts with RbAp48 (D. Vermaak et al., in prep.) but not with histones directly (Fig. 4c) (D. Vermaak et al., in prep.), we suggest that RbAp48 has an essential role in targeting the histone deacetylase to a segment of the H4 directly adjacent to the acetylated lysines.

HISTONE DEACETYLASE DIRECTS THE DOMINANT SILENCING OF TRANSCRIPTION IN CHROMATIN

Expression of RPD3 in *Xenopus* oocytes leads to the transcriptional silencing of many promoters; however, transcriptional repression is not immediate (Wong et al. 1998; D. Vermaak et al., in prep.). Following the microinjection of double-stranded nonreplicating template DNA containing a promoter into the nucleus of a *Xenopus* oocyte, it takes more than 3 hours to assemble a physiological density of nucleosomes (Almouzni and Wolffe 1993). The assembly of a functional transcription complex takes less than 3 hours. *Xenopus* RPD3 only begins to repress transcription 4–5 hours after microinjection of template DNA, correlating with nucleosome assembly (Fig. 5b,c,d). Transcription is not repressed at earlier times (3

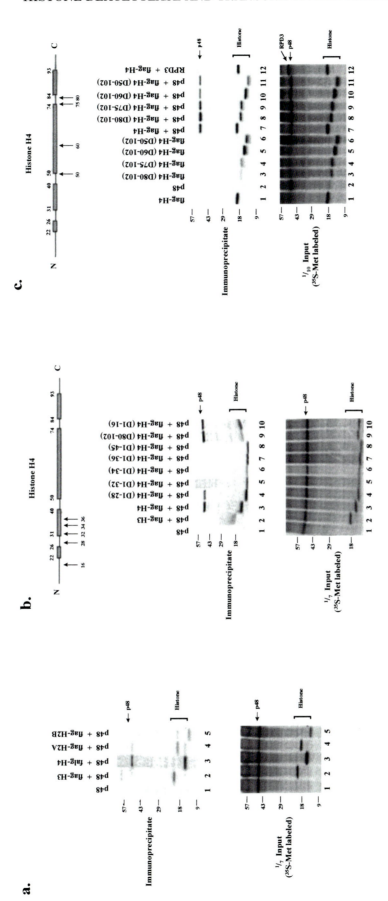

Figure 4. Rbp48 interacts with a specific sequences in the amino-terminal tail of histone H4. (*a*) Interaction of flag-tagged histones H3, H4, H2A, and H2B with Rbp48. (*Upper panel*) Immuno-precipitation with anti-flag antibodies; (*lower panel*) Rbp48 protein (p48) and histones (histones) synthesized after microinjection of mRNAs into *Xenopus* oocytes. (*b*) Interaction of flag-tagged amino-terminal deletion mutants of histone H4 with Rbp48. A scheme of the H4 protein is shown with amino acids at amino termini indicated. (*Upper panel*) Immunoprecipitation; (*lower panel*) synthesis of Rbp48 (p48) and the various histones (histones). (*c*) As in *b*, except carboxy-terminal deletions of H4 were used with Rbp48. (Lane *12*) Synthesis of RPD3 and H4 (*lower panel*) and the lack of coimmunoprecipitation (*upper panel*).

hr). Transcription run-on experiments indicate that the transcriptional machinery is erased from DNA concomitant with repression. This result suggests that a chromatin infrastructure is required for the dominant transcriptional silencing by histone deacetylase (Wong et al. 1998)

A *XENOPUS* HISTONE DEACETYLASE ASSOCIATES WITH THE METHYLATED CpG-TARGETED TRANSCRIPTIONAL REPRESSOR MeCP2

In *Xenopus laevis*, normal patterns of development and gene expression are perturbed by inhibiting histone deacetylase using Trichostatin A (Almouzni et al. 1994). This drug also prevents transcriptional repression mediated by nuclear receptors in the *Xenopus* oocyte nucleus (Wong et al. 1998). To understand the functional roles of acetylation in early vertebrate development, we initiated a biochemical survey of histone deacetylases in *Xenopus* oocytes and eggs. We developed an assay for deacetylases using histones modified by yeast Hat1p (Parthun et al. 1996) as a substrate. *Xenopus* egg and oocyte extracts hydrolyzed the lysine-acetate amide bond efficiently; this activity was inhibited using Trichostatin A (Wade et al. 1998a). The purification protocols for two deacetylase enzymes are described in detail elsewhere (Jones et al. 1998a; Wade et al. 1998a). The enzymes were initially fractionated through two ion-exchange steps (Fig. 6a). The initial cation-exchange chromatography fractionated the extract protein into two pools (0.1 M and 0.5 M NaCl), each of which contained abundant deacetylase activity (data not shown). The 0.5 M NaCl elution from the BioRex column, approximately 15% of the total extract protein, was further fractionated by gradient elution on Mono Q (Fig. 6b). This step resolves one major deacetylase activity peak from two minor peaks. The first minor deacetylase peak (fraction 27) contains the methyl CpG-binding protein MeCP2, Rbp48, and several associated polypeptides. The major peak of deacetylase contains the Mi-2 chromodomain SWI/SNF ATPase, RbAp48, and several distinct associated polypeptides (see below). The MeCP2-containing fractions were further fractionated on a heparin agarose column (Fig. 6c). MeCP2 and Sin3 cofractionate together with histone deacetylase activity. Eight additional polypeptides cofractionate with MeCP2 and Sin3 (Fig. 6d). Antibodies against MeCP2 immunoprecipitated Sin3 and vice versa, together with histone deacetylase activity (Jones et al. 1998b). These results are consistent with the existence of a soluble complex containing MeCP2, Sin3, and histone deacetylase. They also suggest that DNA methylation might rely upon chromatin assembly and histone modification to confer transcriptional silencing (Kass et al. 1997a,b).

Several experiments are consistent with a role for chromatin assembly and histone modification in transcriptional silencing on methylated DNA. Microinjection of methylated double-stranded templates into oocyte nuclei leads to their transcriptional repression only when a physiological density of nucleosomes in chromatin is completely assembled (Kass et al. 1997a). DNase-I-hypersen-

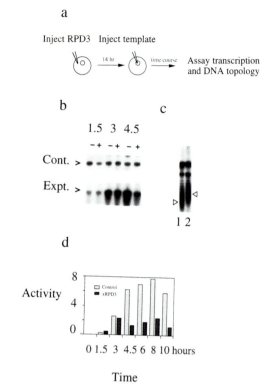

Figure 5. *Xenopus* histone deacetylase (xRPD3) requires a chromatin template to direct transcriptional repression. (*a*) Scheme of the experiment. Groups of oocytes were microinjected with (+) or without (–) xRPD3 mRNA (1 ng) and incubated 14 hr to allow expression of xRPD3p. After this time, double-stranded TRβA template was injected. At various times as indicated, half of the oocytes were extracted to assess mRNA synthesis and half to assess DNA topology. (*b*) Transcription repression on the TRβA promoter takes more than 3 hr to be established. Transcripts were analyzed by primer extension from oocytes injected with xRPD3 (+) and control oocytes (–). Time after injection of the TRβA template is indicated in hours. TRβA transcripts (Expt.) and the endogenous H4 mRNA (Internal Control) are indicated. (*c*) Chromatin assembly onto double-stranded TRβA. The DNA samples prepared from half of the oocytes were analyzed for chromatin assembly using a 1% agarose gel with chloroquine (90 μg/ml). After electrophoresis, DNAs were blotted to a nylon membrane and hybridized with a random primer-labeled TRβA promoter fragment. The arrowheads indicate the centers of distribution of the topoisomers assembled after 3 hr (lane *1*) or 4.5 hr (lane *2*). Increasing numbers of nucleosomes are indicated by the movement of the topoisomers toward the top of the gel (Wong et al. 1998). (*d*) Quantitation of relative transcription levels at different times after injection of TRβA promoter into oocytes injected with xRPD3 mRNA and controls.

sitive sites assembled on methylated templates are erased once chromatin matures, and the silent promoter is packaged into a nucleosomal array. The erasure of the transcriptional machinery from promoters is transmissible in *cis*, because it is possible to direct repression by methylating vector DNA, leaving the promoter itself unmethylated (Kass et al. 1997a).

We examined the role of histone deacetylase in transcriptional repression of methylated DNA using the inhibitor TSA. Concentrations of TSA between 30 and 300 nM inhibited histone deacetylase activity in the MeCP2

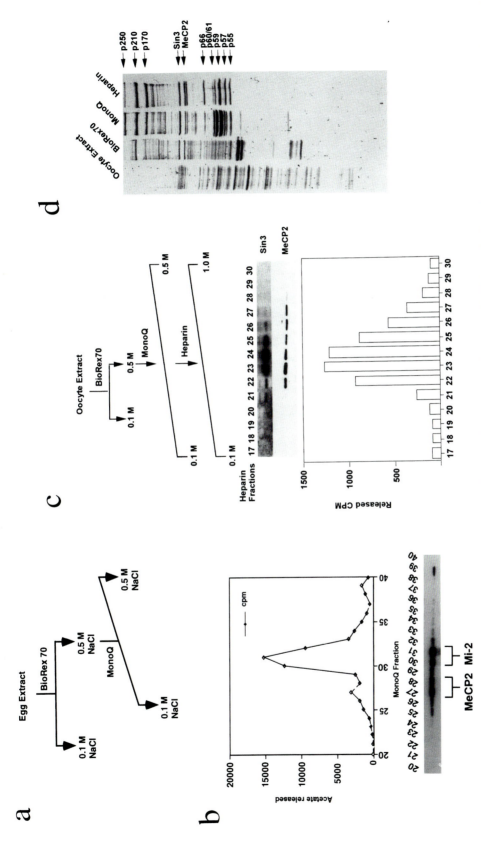

Figure 6. Fractionation of histone deacetylase activity associated with the methyl-CpG-specific transcriptional repressor MeCP2. (*a*) The flow chart summarizes the chromatographic procedures used (Wade et al. 1998a). (*b*) Fractionation of histone deacetylase activity and Rbp48 immunoreactive polypeptide on Mono Q. The fractions containing MeCP2 and Mi-2 are indicated. (*c*) Fractionation of MeCP2-containing deacetylase on Mono Q followed by Heparin. Immunoblots against Sin3 and MeCP2, together with histone deacetylase assays on the various fractions, are shown (Jones et al. 1998a). (*d*) Coomassie-blue-stained gel with each lane containing an equal amount of protein (7 μg) from each step of *Xenopus* extract purification. The apparent molecular weights for each of the polypeptides copurifying with MeCP2 and Sin3 are listed on the right.

complex by 80% and 100%, respectively. Microinjection of unmethylated *Xenopus* hsp70 promoter-containing templates results in strong transcription after an overnight incubation period (10 hr) during which chromatin is assembled (Fig. 7a). The addition of 300 nM TSA modestly enhances transcription by two- to threefold compared to a nonchromatinized human cytomegalovirus (CMV) promoter (Fig. 7a, lanes 1 and 2). Complete CpG methylation of the hsp70 promoter using *Sss*I methyltransferase leads to transcriptional silencing of more than 100-fold

after an overnight incubation compared with that of a CMV promoter control (Fig. 7a, lanes 3 and 4). The addition of increasing concentrations of TSA progressively relieves transcriptional repression on the methylated template relative to the CMV control (Fig. 7a, lanes 4-7). Transcription from the methylated hsp70 promoter was enhanced more than 100-fold, indicating that histone deacetylase has an active role in repressing transcription on methylated DNA. The *Xenopus* hsp70 promoter is assembled into a DNase-I-hypersensitive site within chro-

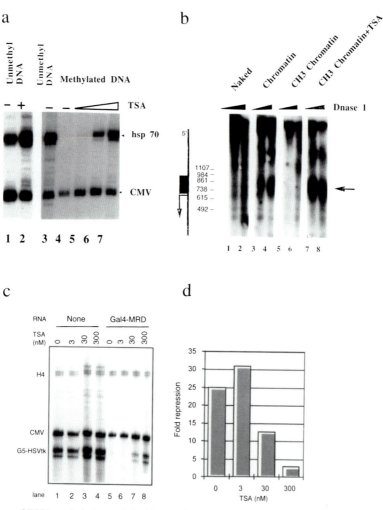

Figure 7. The influence of DNA methylation and the histone deacetylase inhibitor TSA on *Xenopus* hsp70 promoter activity. (*a*) Methylated or unmethylated plasmid pHSP-CAT (Landsberger and Wolffe 1995) was injected into oocyte nuclei, and transcription from the hsp70 promoter was assayed at 10 hr after injection as indicated. Injection of an unmethylated pCMVCAT (0.25 ng/oocyte), 3 hr before oocyte isolation and RNA extraction, serves as an internal standard. (Lanes *1,3,4*) No TSA was added; (lanes *2,5,6,7*) TSA was added to 300, 3, 30, and 300 nM, respectively. The positions of transcripts derived from the hsp70 and CMV promoters are indicated. (*b*) Methylated CH3 (lanes *5–8*) or mock-methylated (control) plasmid pHSP-CAT (lanes *3,4*) was injected into oocyte nuclei and assayed for the presence of DNase-I-hypersensitive sites 10 hr after injected, when oocytes were incubated in the presence (lanes *7,8*) or absence of 30 nM TSA (lanes *3–6*). The plasmid was linearized with *Nco*I and the Southern blot was probed with an *Nco*I-*Eco*RI fragment (+313 to +616 relative to the start site of transcription). The arrow indicates the major area of hypersensitivity. (*c*) TSA relieves transcriptional repression induced by Gal4-MRD. Oocytes were injected with 200 pg of Gal4-MRD RNA (lanes *5–8*), or not injected (lanes *1–5*), and subsequently incubated for 4 hr. pCMV-CAT and pG5HSVtk-CAT (the latter plasmid contains five Gal4 sites upstream of the promoter) were injected, and RNA was isolated and analyzed after 18 hr. Endogenous histone H4 RNA was used for normalized quantitation of RNA transcribed from the injected CMV and HSVtk promoters. Templates were incubated in the absence (lanes *1,5*) or presence of 3 nM (lanes *2,6*), 30 nM (lanes *3,7*), or 300 nM (lanes *4,8*) of TSA. (*d*) Graphic representation of result shown in *c*: 30 nM (or more) of Trichostatin A partially relieves repression of transcription by Gal4-MRD. The fold repression was calculated by comparing the normalized levels of RNA transcribed from the HSVtk promoter in lanes *1–4* and *5–8*.

matin following microinjection into *Xenopus* oocyte nuclei (Fig. 7b). Methylation of the promoter leads to the loss of hypersensitivity, but the addition of TSA is sufficient to restore hypersensitivity (Fig. 7b, lanes 3-8). Thus, inhibition of histone deacetylase activity facilitates the remodeling of chromatin structure concomitant with transcription competence on methylated DNA templates. MeCP2 consists of a methyl CpG recognition domain that binds DNA (Nan et al. 1993) and a repression domain that confers transcriptional silencing. The MeCP2 repression domain fused to the Gal4 DNA-binding domain is sufficient to silence transcription from a herpes simplex virus thymidine kinase gene promoter linked to five upstream Gal4 DNA-binding sites in a methylation-independent process (Fig. 7c). TSA reduces the capacity of the Gal4-tethered MeCP2-repression domain to confer transcriptional silencing by more than tenfold (Fig. 7d). These results suggest that the methylation-specific repressor MeCP2 recruits the Sin3 histone deacetylase complex to promoters by binding methylated DNA through its methyl-CpG binding domain (Jones et al. 1998b).

A *XENOPUS* HISTONE DEACETYLASE ASSOCIATES WITH THE MI-2 CHROMODOMAIN SWI/SNF ATPASE

The second peak of both histone deacetylase activity and of Rbp48 (fractions 30 and 31) was further purified by sedimentation on linear sucrose gradients (Fig. 8a), sedimenting at a rate (by comparison to proteins of known mass) consistent with a molecular mass of 1–1.5 MD (data not shown). The purified enzyme consisted of a series of cosedimenting polypeptides correlating precisely with histone deacetylase activity (Fig. 8a,b). We

observe major subunits of 250, 70, 64, 58, 55, and 35 kD, with the latter four presumptive subunits migrating as doublets (or a triplet in the case of p65) in SDS-PAGE.

The polypeptide of 250 kD was identified by protease digestion followed by peptide isolation, mass spectrometry, and amino acid sequencing (Fernandez et al. 1994). There is an amino acid sequence match to that of a previously described human protein, Mi-2 (Nilasena et al. 1995; Seelig et al. 1995). Mi-2, an autoantigen in the human disease dermatomyositis is a nuclear protein (Seelig et al. 1995) apparently encoded by two closely related human genes also known as HsCHD3 and HsCHD4 (Woodage et al. 1997). Sequence features of Mi-2 include PHD fingers, chromodomains, limited similarity to the telobox DNA-binding domain, and an SWI/SNF2 ATPase motif (Seelig et al. 1995; Woodage et al. 1997). Serum from patients afflicted with dermatomyositis coprecipitates from mammalian cells a set of polypeptides with a molecular weight remarkably similar to that of the deacetylase subunits found in the *Xenopus* protein (Seelig et al. 1995).

To experimentally test the possibility that the *Xenopus* Mi-2 complex functions as an ATPase, we performed ATPase assays with various cofactors. ATPase activity precisely cosedimented with deacetylase activity (Fig. 8c). Consistent with a role in modulation of chromatin structure, this activity was strongly stimulated by nucleosomes, but not by free DNA or histones. The physical coupling of histone deacetylase and nucleosome-stimulated ATPase activities in a single protein complex suggests participation in common functions. The abundance of this deacetylase activity suggests that it may have a general function in chromatin biology in the early *Xenopus* embryo. The presence of PHD fingers and the chromodomain provides the potential to allow protein-protein

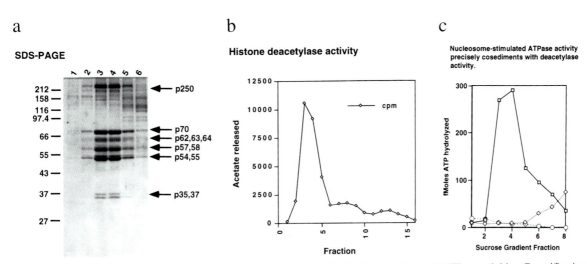

Figure 8. A histone deacetylase complex composed of six major subunits has both deacetylase and ATPase activities. Copurification of deacetylase activity with a multiple polypeptide complex. (*a*) Indicated fractions from a sucrose gradient (Wade et al. 1998a) were electrophoresed (10% SDS-PAGE) and stained with Coomassie blue. Migration of molecular-mass markers is given in kilodaltons on the left side of the gel. Arrows denote major components of the enzyme complex. (*b*) The same fractions were assayed for deacetylase activity. Hydrolyzed lysine-acetylate amide bounds are depicted as cpm. (*c*) Nucleosome-stimulated ATPase activity precisely cosediments with histones deacetylase activity. The sucrose gradient fractions were incubated with [γ-^{32}P]ATP and purified chicken erythrocyte mononucleosomes, salmon sperm DNA, or purified chicken erythrocyte core histones for 30 min at room temperature (Wade et al. 1998b).

interactions to target the deacetylase to particular sites for specific regulatory functions (Wade et al. 1998b).

CONCLUSION

Acetylation of the core histones has a potent influence on transcription. Chromatin structural transitions dependent on acetylation (see Fig. 1) facilitate both access of the transcriptional machinery to DNA and the processivity of RNA polymerase through a nucleosomal array. Although transcription factors can also be acetylated, the functional consequences currently described are small compared to modification of the histones (see Figs. 2 and 3). Histone deacetylases associate with a family of WD repeat proteins, RbAp48/p46, that have specificity for association with the histones. In particular, we can identify a segment of histone H4 that is exposed on the nucleosome surface as a target for *Xenopus* Rbp48 (Fig. 4). *Xenopus* RPD3, the catalytic subunit of histone deacetylase, associates with Rbp48, and we propose that RbAp48 targets the deacetylase activity to the histone tail domains. Overexpression of *Xenopus* RPD3 can direct the dominant silencing of transcription within a chromatin template. The deacetylase does not interfere with transcription when nucleosome assembly is incomplete (see Fig. 5). Taken together, we conclude that histone deacetylation mediated by the Rbp48/RPD3 complex can dominantly repress transcription.

We have characterized two deacetylase complexes from *Xenopus* oocytes and eggs. The methyl-CpG-binding transcriptional repressor MeCP2 and the Mi-2 chromodomain SWI/SNF ATPase are associated with distinct deacetylases. The presence of MeCP2 in a histone deacetylase complex links DNA methylation to the directed modification of chromatin structure and function. The presence of the Mi-2 chromodomain SWI/SNF ATPase links both long-range chromosomal organization and local ATP-dependent nucleosome disruption to histone modification. These biological connections should provide new insight into the molecular mechanisms by which histone acetylation contributes to chromosomal function and into the compartmentalization of the genome into active and inactive domains that can be stably maintained throughout the development of an organism.

ACKNOWLEDGMENTS

We thank Thuy Vo for manuscript preparation. P.L.J. is supported by a PRAT fellowship, G.J.C.V. by an NWO fellowship, and T. Sera by a JSPS fellowship.

REFERENCES

Alland L., Muhle R., Hou H., Jr., Potes J., Chin L., Schreiber-Agus N., and De Pinho R.A. 1997. Role of NCoR and histone deacetylase in Sin3-mediated transcriptional and oncogenic repression. *Nature* 387: 49.

Almouzni G. and Wolffe A.P. 1993. Replication coupled chromatin assembly is required for the repression of basal transcription *in vivo*. *Genes Dev.* 7: 2033.

Almouzni G., Khochbin S., Dimitrov S., and Wolffe A.P. 1994. Histone acetylation influences both gene expression and development of *Xenopus laevis*. *Dev. Biol.* 165: 654.

Bauer W.R., Hayes J.J., White J.H., and Wolffe A.P. 1994. Nucleosome structural changes due to acetylation. *J. Mol. Biol.* 236: 685.

Brownell J.E. and Allis C.D. 1996. Special HATs for special occasions: Linking histone acetylation to chromatin assembly and gene activation. *Curr. Opin. Genet. Dev.* 6: 176.

Brownell J.E., Zhou J., Ranalli T., Kobayashi R., Edmondson D.G., Roth S.Y., and Allis C.D. 1996. *Tetrahymena* histone acetyltransferase A: A homolog to yeast Gcn5p linking histone acetylation to gene activation. *Cell* 84: 843.

Cary P.D., Crane-Robinson C., Bradbury E.M., and Dixon G.H. 1982. Effect of acetylation on the binding of N-terminal peptides of histone H4 to DNA. *Eur. J. Biochem.* 127: 137.

Covault J. and Chalkley R. 1980. The identification of distinct populations of acetylated histone. *J. Biol. Chem.* 255: 9110.

Dikstein R., Ruppert S., and Tjian R. 1996. TAF$_{II}$250 is a bipartite protein kinase that phosphorylates the basal transcription factor RAP74. *Cell* 84: 781.

Edmonson D.G., Smith M.M., and Roth S.Y. 1996. Repression domain of the yeast global repressor Tup1 interacts directly with histones H3 and H4. *Genes Dev.* 10: 1247.

Fernandez J., Andrews L., and Mische S.M. 1994. An improved method for enzymatic digestion of polyvinylidene defluoride-bound proteins for internal sequence analysis. *Anal. Biochem.* 214: 112.

Freeman L., Kurumizaka H., and Wolffe A.P. 1996. Functional assays for assembly of histones H3 and H4 into the chromatin of *Xenopus* embryos. *Proc. Natl. Acad. Sci.* 93: 12780.

Garcia-Ramirez M., Rocchini C., and Ausio J. 1995. Modulation of chromatin folding by histone acetylation. *J. Biol. Chem.* 270: 17923.

Ge H., Martinez E., Chiang C.-M., and Roeder R.G. 1996. Activator-dependent transcription by mammalian RNA polymerase II: In vitro reconstitution with general transcription factors and cofactors. *Methods Enzymol.* 274: 57.

Grunstein M. 1997. Histone acetylation in chromatin structure and transcription. *Nature* 389: 349.

Gu W. and Roeder R.G. 1997. Activation of p53 sequence-specific DNA binding by acetylation of the p53 C-terminal domain. *Cell* 90: 595.

Hassig C.A., Fleischer T.C., Billin A.N., Schreiber S.L., and Ayer D.E. 1997. Histone deacetylase activity is required for full transcriptional repression by mSin 3A. *Cell* 89: 341.

Hassig C.A., Tong J.K., Fleischer T.C., Owa T., Grable P.G., Ayer D.E., and Schreiber S.L. 1998. A role for histone deacetylase activity in HDAC1-mediated transcriptional repression. *Proc. Natl. Acad. Sci.* 95: 3519.

Hecht A., Laroche T., Strahl-Bolsinger S., Gasser S.M., and Grunstein M. 1995. Histone H3 and H4 N-termini interact with SIR3 and SIR4 proteins: A molecular model for the formation of heterochromatin in yeast. *Cell* 80: 583.

Hong L., Schroth G.P., Matthews H.R., Yau P., and Bradbury E.M. 1993. Studies of the DNA binding properties of the histone H4 amino terminus. *J. Biol. Chem.* 268: 305.

Howe L. and Ausio J. 1998. Nucleosome translational position, not histone acetylation, determines TFIIIA binding to nucleosomal *Xenopus laevis* 5S rRNA genes. *Mol. Cell. Biol.* 18: 1156.

Imhof A., Yang, X.J., Ogryzko, V.V., Nakatani, Y., Wolffe, A.P., and Ge, H. 1997. Acetylation of general transcription factors by histone acetyltransferases: Identification of a major site of acetylation in TFIIEβ. *Curr. Biol.* 7: 689.

Jones P.L., Wade P.A., Vermaak D., and Wolffe A.P. 1998a. Purification of MeCP2 containing deacetylases. *Methods Mol. Biol.* (in press).

Jones P.L., Veenstra G.J.C., Wade P.A., Vermaak D., Kass S.U., Landsberger N., Strouboulis J., and Wolffe A.P. 1998b. Methylated DNA and MeCP2 recruit histone deacetylase to repress transcription. *Nat. Genet.* 19: 187.

Kass S.U., Landsberger N., and Wolffe A.P. 1997a. DNA methylation directs a time-dependent repression of transcription initiation. *Curr. Biol.* 7: 157.

Kass S.U., Pruss D., and Wolffe A.P. 1997b. How does DNA methylation repress transcription? *Trends Genet.* 13: 444.

Krajewski W.A. and Becker P.B. 1998. Reconstitution of hyperacetylated, DNase I-sensitive chromatin chracterized by high conformational flexibility of nucleosomal DNA. *Proc. Natl. Acad. Sci.* **95:** 1540.

Kuo M.-H., Zhou J., Jambeck P., Churchill M.E.A., and Allis C.D. 1998. Histone acetyltransferase activity of yeast Gcn5p is required for the activation of target genes *in vivo. Genes Dev.* **12:** 627.

Landsberger N. and Wolffe A.P. 1995. The role of chromatin and *Xenopus* heat shock transcription factor (XHSF1) in the regulation of the *Xenopus* hsp70 promoter *in vivo. Mol. Cell. Biol.* **15:** 6013.

Lee D.Y., Hayes J.J., Pruss D., and Wolffe A.P. 1993. A positive role for histone acetylation in transcription factor binding to nucleosomal DNA. *Cell* **72:** 73.

Luger K., Mader A.W., Richmond R.K., Sargent D.F., and Richmond T.J. 1997. X-ray structure of the nucleosome core particle at 2.8Å resolution. *Nature* **389:** 251.

Lutter L.C., Judis L., and Paretti R.F. 1992. The effects of histone acetylation on chromatin topology *in vivo. Mol. Cell. Biol.* **12:** 5004.

McGhee J.D., Nickol J.M., Felsenfeld G., and Rau D.C. 1983. Histone acetylation has little effect on the higher order following of chromatin. *Nucleic Acids Res.* **11:** 4065.

Martinez-Balbas M.A., Tsukiyama T., Gdula D., and Wu C. 1998. *Drosophila* NURF-55, a WD repeat protein involved in histone metabolism. *Proc. Natl. Acad. Sci.* **95:** 132.

Nagy L., Kao H.Y., Chakravarti D., Lin R.J., Hassig C.A., Ayer D.E., Schreiber S.L., and Evans R.M. 1997. Nuclear receptor repression mediated by a complex containing SMRT, mSin3A, and histone deacetylase. *Cell* **89:** 373.

Nan X., Meehan R.R., and Bird A.P. 1993. Dissection of the methyl-CpG binding domain from the chromosomal protein MeCP2. *Nucleic Acids Res.* **21:** 4886.

Neer E.J. and Smith T.F. 1996. G protein heterodimers: New structures propel new questions. *Cell* **84:** 175.

Nightingale K.P., Wellinger R.E., Sogo J.M., and Becker P.B. 1998. Histone acetylation facilitates RNA polymerase II transcription of the *Drosophila* hsp26 gene in chromatin. *EMBO J.* **17:** 2865.

Nilasena D.S., Trieu E.P., and Targoff I.N. 1995. Analysis of the Mi-2 autoantigen of dermatomyositis. *Arthritis Rheum.* **38:** 123.

Norton V.G., Imai B.S., Yau P., and Bradbury E.M. 1989. Histone acetylation reduces nucleosome core particle linking number change. *Cell* **57:** 449.

Ogryzko V.V., Schiltz R.L., Russanova V., Howard B.H., and Nakatani Y. 1996. The transcriptional coactivators p300 and CBP are histone acetyltransferases. *Cell* **87:** 953.

Parthun M.R., Widom J., and Gottschling D.E. 1996. The major cytoplasmic histone acetyltransferase in yeast: Links to chromatin assembly and histone metabolism. *Cell* **87:** 85.

Perry M. and Chalkley R. 1982. Histone acetylation increases the solubility of chromatin and occurs sequentially over most of the chromatin. *J. Biol. Chem.* **257:** 7336.

Puig O.M., Belles E., Lopez-Rodas G., Sendra R., and Tordera V. 1998. Interaction between N-terminal domain of H4 and DNA is regulated by the acetylation degree. *Biochim. Biophys. Acta* **1397:** 79.

Qian Y.W. and Lee E.Y.-H.P. 1995. Dual retinoblastoma-binding proteins with properties related to a negative regulator of Ras in yeast. *J. Biol. Chem.* **270:** 25507.

Qian Y.W., Wang Y.-C., Hollingsworth R.E.J., Jones D., Ling N., and Lee E.Y.-H.P. 1993. A retinoblastoma-binding protein related to a negative regulator of Ras in yeast. *Nature* **364:** 648.

Rundlett S.E., Carmen A.A., Suka N., Turner B.M., and Grunstein M. 1998. Transcriptional repression by UME6 involves deacetylation of lysine 5 of histone H4 by RPD3. *Nature* **392:** 831.

Seelig H.P., Moosbrugger I., Ehrfeld H., Fink T., Renz M., and Genth E. 1995. The major dermatomyositis-specific Mi-2 autoantigen is a presumed helicase involved in transcription activation. *Arthritis Rheum.* **38:** 1389.

Tan S., Aso T., Conaway R.C., and Conaway J.W. 1994. Roles for both the RAP30 and RAP74 subunits of transcription factor IIF in transcription initiation and elongation by RNA polymerase II. *J. Biol. Chem.* **269:** 25684.

Taunton J., Hassig C.A., and Schreiber S.L. 1996. A mammalian histone deacetylase related to a yeast transcriptional regulator Rpd3. *Science* **272:** 408.

Tse C. and Hansen J.C. 1997. Hybrid trypsinized nucleosomal arrays: Identification of multiple functional roles of the H2A/H2B and H3/H4 N-termini in chromatin fiber compaction. *Biochemistry* **36:** 11381.

Tse C., Sera T., Wolffe A.P, and Hansen J.C. 1998. Disruption of higher order folding by core histone acetylation dramatically enhances transcription of nucleosomal arrays by RNA polymerase III. *Mol. Cell. Biol.* **18:** 4629.

Turner B.M. 1993. Decoding the nucleosome. *Cell* **75:** 5.

Tyler J.K., Bulger M., Kamakaka R.T., Kobayashi R., and Kadonaga, J.T. 1996. The p55 subunit of *Drosophila* chromatin assembly factor 1 is homologous to a histone deacetylase-associated protein. *Mol. Cell. Biol.* **16:** 6149.

Ura K., Kurumizaka H., Dimitrov S., Almouzni G., and Wolffe A.P. 1997. Histone acetylation: Influence on transcription by RNA polymerase, nucleosome mobility and positioning, and linker histone dependent transcriptional repression. *EMBO J.* **16:** 2096.

Verreault A., Kaufman P.D., Kobayashi R., and Stillman, B. 1996. Nucleosome assembly by a complex of CAF-1 and acetylated histones H3/H4. *Cell* **87:** 95.

———. 1997. Nucleosomal DNA regulates the core-histone binding subunit of the human hat1 acetyltransferase. *Curr. Biol.* **8:** 96.

Vettese-Dadey M., Walter P., Chen H., Juan L-J., and Workman J.L. 1994. Role of the histone amino termini in facilitated binding of a transcription factor, GAL4-AH to nucleosome cores. *Mol. Cell. Biol.* **14:** 970.

Vettese-Dadey M., Grant P.A., Hebbes T.R., Crane-Robinson C., Allis C.D., and Workman J.L. 1996. Acetylation of histone H4 plays a primary role in enhancing transcription factor binding to nucleosomal DNA *in vitro. EMBO J.* **15:** 2508.

Wade P.A., Pruss D., and Wolffe A.P. 1997. Histone acetylation: Chromatin in action. *Trends Biochem. Sci.* **22:** 128.

Wade P.A., Jones P.L., Vermaak D., and Wolffe A.P. 1998a. Purification of a histone deacetylase complex from *Xenopus laevis:* Preparation of substrates and assay procedures. *Methods Enzymol.* (in press).

———. 1998b. A multiple subunit histone deacetylase from *Xenopus laevis* contains a Snf2 superfamily ATPase. *Curr. Biol.* **8:** 843.

Wolffe A.P. 1996. Histone deacetylase: A regulator of transcription. *Science* **272:** 371.

———. 1997. Sinful repression. *Nature* **387:** 16.

Wolffe A.P. and Pruss D. 1996. Targeting chromatin disruption: Transcription regulators that acetylate histones. *Cell* **84:** 817.

Wolffe A.P., Almouzni G., Ura K., Pruss D., and Hayes J.J. 1993. Transcription factor access to DNA in the nucleosome. *Cold Spring Harbor Symp. Quant. Biol.* **58:** 225.

Wong J., Patterton D., Imhof A., Shi Y.-B., and Wolffe A.P. 1998. Distinct requirements for chromatin assembly in transcriptional repression by thyroid hormone receptor and histone deacetylase. *EMBO J.* **17:** 520.

Woodage T., Basrai M.A., Baxevanis A.D., Hieter P., and Collins F.S. 1997. Characterization of the CHD family of proteins. *Proc. Natl. Acad. Sci.* **94:** 11472.

Yang X.-J., Ogryzko V.V., Nishikawa J.-I., Howard B., and Nakatani Y. 1996. A p300/CBP-associated factor that competes with the adenoviral E1A oncoprotein. *Nature* **382:** 319.

Zhang D.-E. and Nelson D.A. 1988. Histone acetylation in chicken erythrocytes. Rates of acetylation and evidence that histones in both active and potentially active chromatin are rapidly modified. *Biochem. J.* **250:** 233.

Zhang W., Bone J.R., Edmondson D.G., Turner B.M., and Roth, S.Y. 1998. Essential and redundant functions of histone acetylation revealed by mutation of target lysines and loss of the Gcn5p acetyltransferase. *EMBO J.* **17:** 3155.

Zhang Y., Iratni R., Erdjument-Bromage H., Tempst P., and Reinberg D. 1997. Histone deacetylases and SAP18, a novel polypeptide, are components of a human Sin3 complex. *Cell* **89:** 357.

Gene Regulation by the Yeast Ssn6-Tup1 Corepressor

M. WAHI,* K. KOMACHI,* AND A.D. JOHNSON*†

*Department of Biochemistry and Biophysics, †Department of Microbiology and Immunology,
University of California, San Francisco, California 94143

To respond rapidly to environmental changes, cells need to coordinate the expression of genes that function in specific biological pathways. For example, when some cell types are deprived of oxygen, they express a set of genes required for anaerobic metabolism. When a dividing cell receives a signal to enter the quiescent state, it ceases to express the set of genes needed for cell division. Such coordinated control of gene expression often occurs at the transcriptional level. This transcriptional coregulation is mediated by sequence-specific DNA-binding regulatory proteins that are brought to their target genes by conserved DNA sequences located in the upstream control region of each gene in the particular set. Once bound to the control region, they activate or repress transcription of the targeted gene. In contrast to transcriptional activation, which has been relatively well-studied, little is known about how specific DNA-binding proteins turn off transcription in a coordinated manner.

Studies in *Saccharomyces cerevisiae* are leading to a better understanding of this transcriptional repression process. One of the most pervasive repression systems in this yeast involves the concerted repression of genes belonging to several different biological pathways. At the heart of this repression system is the Ssn6-Tup1 corepressor. This corepressor, which is composed of three or four monomers of the Tup1 protein and one monomer of the Ssn6 protein (Varanasi et al. 1996; Redd et al. 1997), does not bind DNA itself but is brought to specific gene sets by a variety of sequence-specific DNA-binding proteins (see Fig. 1) (Keleher et al. 1992; Tzamarias and Struhl 1994; Treitel and Carlson 1995). Each of these DNA-binding proteins (or, in some cases, combinations of DNA-binding proteins) brings Ssn6-Tup1 to a specific set of genes (Mukai et al. 1991; Keleher et al. 1992; Trumbly 1992; Zitomer and Lowry 1992; Huang et al. 1998; Márquez et al. 1998). For example, α2-Mcm1 directs Ssn6-Tup1 to a group of cell-type-specific genes, the **a**-specific genes, and thereby represses them (Keleher et al. 1992). Likewise, the hypoxic genes are repressed when Rox1 brings Ssn6-Tup1 to them (Zitomer and Lowry 1992). When genes repressed by Ssn6-Tup1 are needed by the cell, the activities or synthesis of the appropriate DNA-binding proteins are altered, thus releasing the relevant genes from repression (Herskowitz et al. 1992; De Vit et al. 1997; Zitomer et al. 1997). In this way, each gene set controlled by Ssn6-Tup1 can be regulated independently. Results from DNA array monitoring suggest that Ssn6-Tup1 regulates up to approximately 180 genes, which is 3% of all yeast genes (DeRisi et al. 1997).

Although it is not known whether Ssn6-Tup1 directly controls all of these genes, the group of genes indicated in Figure 1 certainly does not represent the complete list of gene sets directly repressed by Ssn6-Tup1.

The best-understood DNA-binding proteins that bring Ssn6-Tup1 to target genes are the α2-Mcm1 and the **a**1-α2 transcriptional repressors of the cell-type-specific genes (Fig. 2). *S. cerevisiae* has three cell types, the **a** and α haploid and the **a**/α diploid (for review, see Herskowitz et al. 1992). α cells contain α2 and Mcm1, a protein found in all three cell types. α2 binds cooperatively with

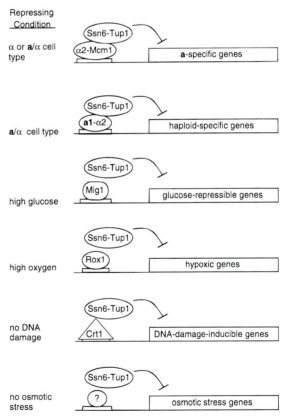

Figure 1. Ssn6-Tup1 mediates repression of diverse sets of genes. Under specific repressing conditions, Ssn6-Tup1 mediates repression of the six gene sets listed at the right. Ssn6-Tup1 is recruited to each gene set by sequence-specific DNA-binding proteins indicated to the left of each gene set. Ssn6-Tup1 interacts directly with α2-Mcm1 and **a**1-α2 and is presumed to interact directly with the other DNA-binding proteins indicated. The DNA-binding protein that directs repression of the osmotic stress genes is unknown.

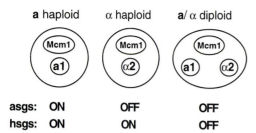

	a haploid	α haploid	**a**/ α diploid
asgs:	ON	OFF	OFF
hsgs:	ON	ON	OFF

Figure 2. Transcriptional repression of cell-type-specific genes in *S. cerevisiae*. Cell-type-specific gene repression is controlled by the *MAT* locus. **a** cells contain *MAT***a**, whereas α cells contain *MAT*α, and **a**/α diploids contain both. *MAT***a** encodes **a**1, which has no role in **a** cells. *MAT*α encodes two genes, one of which is α2. In α cells, α2 binds with the non-cell-type-specific protein, Mcm1, forming the α2-Mcm1 repressor. This repressor turns off the **a**-specific genes (asgs). In **a** cells, the **a**-specific genes are expressed because α2 is absent. In **a**/α cells, α2 binds with **a**1 to form the **a**1-α2 repressor. This repressor turns off the haploid-specific genes (hsgs). In **a**/α cells, α2 also forms the α2-Mcm1 repressor and turns off the **a**-specific genes. Ssn6-Tup1 is present in all three cell types.

Mcm1 to a conserved DNA sequence, called the α2/Mcm1 site (Johnson and Herskowitz 1985; Keleher et al. 1988), located in the upstream control region of the **a**-specific genes, i.e., those genes that are specifically required for **a** cells to mate. DNA-bound α2-Mcm1 then directs repression of the targeted **a**-specific gene. **a**/α cells contain **a**1 and α2, which form a heterodimer that binds tightly to a conserved **a**1/α2 site (Miller et al. 1985; Siliciano and Tatchell 1986; Goutte and Johnson 1988). This site is located in the upstream control region of the haploid-specific genes, i.e., those genes required for both **a** and α cells to mate, as well as those required for the negative control of sporulation and meiosis (for reviews, see Herskowitz et al. 1992; Johnson 1995). Both the α2-Mcm1 and the **a**1-α2 combination have been well characterized biochemically, genetically, and structurally (Goutte and Johnson 1988; Keleher et al. 1988; Herskowitz et al. 1992; Li et al. 1995; Tan and Richmond 1998).

α2-Mcm1 and **a**1-α2 bring Ssn6-Tup1 to the respective gene sets by direct protein-protein interactions. Both Ssn6 and Tup1 belong to well-characterized protein families. Ssn6 contains ten copies of the tetratricopeptide repeat (TPR) and Tup1 contains seven copies of the WD repeat (Schultz and Carlson 1987; Williams and Trumbly 1990; Komachi et al. 1994). The function of the Ssn6 TPRs and the Tup1 WD repeats is, at least in part, to mediate direct interactions with α2-Mcm1 and **a**1-α2 (Komachi et al. 1994; Smith et al. 1995). Although it is not known whether Ssn6-Tup1 interacts directly with other DNA-binding proteins, genetic evidence suggests that this is indeed the case (Tzamarias and Struhl 1995; Carrico and Zitomer 1998).

Once Ssn6-Tup1 is brought to its target genes by specific DNA-binding proteins, how does it exert repression? We know that Ssn6-Tup1-mediated repression is not specific for particular activators since inserting the α2/Mcm1 site in the control region of other yeast genes, namely, *CYC1*, *TRP1*, *URA3*, *GAL4*, and even an RNA

polymerase-I-transcribed gene, renders them repressed by α2-Mcm1 (Johnson and Herskowitz 1985; Roth et al. 1990; Herschbach and Johnson 1993; Redd et al. 1996; K. Komachi and A.D. Johnson, unpubl.). Moreover, Ssn6-Tup1 does not appear to act by displacing DNA-bound activators since α2-Mcm1 can fully repress test promoters containing both an α2/Mcm1 site and a Gal4-activator-binding site, even when Gal4 is bound to its site (Redd et al. 1996). These and other results have led to two models for Ssn6-Tup1-mediated repression: the "chromatin" and the "general transcription machinery" models. In the former, Ssn6-Tup1 represses by organizing chromatin at the promoter of its target genes. This organized chromatin occludes the promoter such that components of the general transcription machinery cannot assemble. In the second model, Ssn6-Tup1 exerts repression by interacting directly with component(s) of the general transcription machinery. This interaction interferes with the assembly of the general transcription machinery or with a later step during transcription initiation. These models are not mutually exclusive, as both mechanisms could contribute to full repression.

In this paper, we review evidence supporting each model and report new data pertaining to both. Our results reinforce the links between Ssn6-Tup1 and the general transcription machinery and quantitatively address the contribution of certain features of the chromatin model to steady-state levels of repression.

EXPERIMENTAL PROCEDURES

Media, growth conditions, and genetic methods. These have been described previously (Wahi and Johnson 1995 and references therein).

Yeast strains. *S. cerevisiae* strains used in this study are listed in Table 1. The strains used in the histone experiment were constructed as described in Tables 1 and 3, with the following specifications. Replacement of *MAT* with *mat*Δ*::URA3* was performed by transforming the parent strain with *Hin*dIII-cut pKK146 (Wahi and Johnson 1995). Replacement of *mat*Δ*::URA3* with *MAT*α was performed by transforming the parent strain with *Hin*dIII-cut pKK49 (Table 2) and selecting for 5-fluoroorotic acid (5-FOA)-resistant transformants.

All strains used in the *SRB8* studies were derived from 246-1-1 and EG123, which are isogenic except at *MAT* (Siliciano and Tatchell 1984). Strains SM1196 and SM1179 have been described previously (Hall and Johnson 1987). The **a** and α *srb8*Δ*::LEU2* strains MWY44 and MWY45 were created as follows. The *srb8*Δ*::LEU2* allele (see plasmids), which encodes only the amino-terminal 406 amino acids of the 1226-amino-acid Srb8 protein (Hengartner et al. 1995), was first introduced into diploid MWY33 (Wahi and Johnson 1995), generating MWY43, by transforming with a 3.9-kb *Bam*HI-*Pst*I fragment isolated from pMW26 cut with *Bam*HI and *Pst*I and then selecting for Leu⁺ transformants. This *leu2* disruption also deletes the first 62 nucleotides of a 384-nucleotide open reading frame, YCR082W, which is di-

Table 1. Yeast Strains

Strain	Genotype/Description	References
RMY200	*MAT***a** *ade2-101 (och) his3Δ200 lys2-801 (amb) trp1Δ901 ura3-52 hht1, hhf1Δ::LEU2 hht2, hhf2Δ::HIS3/pRM200*	Mann and Grunstein (1992)
KKY192	RMY200 in which *MAT***a** has been replaced with *matΔ::URA3*	this paper
KKY193	KKY192 in which *matΔ::URA3* has been replaced with *MATα*	this paper
KKY194	RMY200 transformed with pKK792	this paper
KKY195	RMY200 transformed with pKK793	this paper
KKY196	KKY193 transformed with pKK792	this paper
KKY197	KKY193 transformed with pKK793	this paper
KKY198	KKY194 transformed with pKK795 and cured of pRM200	this paper
KKY199	KKY195 transformed with pKK795 and cured of pRM200	this paper
KKY202	KKY198 transformed with pRM200 and cured of pKK795	this paper
KKY205	KKY199 transformed with pRM200 and cured of pKK795	this paper
KKY214	KKY202 transformed with pKK797 and cured of pRM200	this paper
KKY215	KKY205 transformed with pKK797 and cured of pRM200	this paper
KKY216	KKY214 × KKY196	this paper
KKY217	KKY215 × KKY197	this paper
KKY218	KKY216 cured of pRM200	this paper
KKY219	KKY217 cured of pRM200	this paper
KKY222	KKY218 transformed with pRM200 and cured of pKK797	this paper
KKY223	KKY218 transformed with pRM430 and cured of pKK797	this paper
KKY224	KKY218 transformed with pKK826 and cured of pKK797	this paper
KKY225	KKY218 transformed with pKK830 and cured of pKK797	this paper
KKY226	KKY219 transformed with pRM200 and cured of pKK797	this paper
KKY227	KKY219 transformed with pRM430 and cured of pKK797	this paper
KKY228	KKY219 transformed with pKK826 and cured of pKK797	this paper
KKY229	KKY219 transformed with pKK830 and cured of pKK797	this paper
246-1-1	*MATα can1 gal2 his4 leu2 suc2Δ trp1 ura3*	Siliciano and Tatchell (1984)
EG123	*MAT***a** *can1 gal2 his4 leu2 suc2Δ trp1 ura3*	Siliciano and Tatchell (1984)
SM1196	*MATα mfa2::lacZ can1 gal2 his4 leu2 suc2Δ trp1 ura3*	Hall and Johnson (1987)
SM1179	*MAT***a** *mfa2::lacZ can1 gal2 his4 leu2 suc2Δ trp1 ura3*	Hall and Johnson (1987)
MWY16	*MATα are2-13 mfa2::lacZ can1 gal2 his4 leu2 suc2Δ trp1 ura3*	Wahi and Johnson (1995)
MWY17	*MAT***a** *are2-13 mfa2::lacZ can1 gal2 his4 leu2 suc2Δ trp1 ura3*	Wahi and Johnson (1995)
MWY44	*MAT***a** *srb8Δ::LEU2 mfa2::lacZ can1 gal2 his4 leu2 suc2Δ trp1 ura3 YCR082WΔ*[a]	this paper
MWY45	*MATα srb8Δ::LEU2 mfa2::lacZ can1 gal2 his4 leu2 suc2Δ trp1 ura3 YCR082WΔ*[a]	this paper
MWY33	SM1179 × SM1196	Wahi and Johnson (1995)

[a]YCR082WΔ is explained in Experimental Procedures.

rectly downstream from *SRB8* (Oliver et al. 1992). YCR082W, the function of which is unknown, does not encode *ARE2* since pMW21, which contains YCR082W plus 3904 nucleotides of upstream sequence and 258 nucleotides of downstream sequence, fails to complement the *are2* defect (data not shown) and because complementation results which do not rely on the *srb8* deletion allele indicate that *ARE2* is identical to *SRB8* (see Results and Discussion). Correct integration was confirmed by

polymerase chain reaction (PCR) (see below). MWY43 was sporulated, and an **a** (MWY44) and α (MWY45) Leu[+] segregant were recovered from dissected tetrads. Mating type was confirmed by PCR (see below). MWY16 and MWY17 have been described previously (Wahi and Johnson 1995).

Plasmids. Plasmids used in the histone H3 and H4 study are listed in Table 2. pKK826 and pKK830 were

Table 2. Plasmids Used in the Histone H3 and H4 Study

Plasmid	Description	Reference
pRM200	*HHF2 HHT2/TRP1 ARS CEN*	Mann and Grunstein (1992)
pRM430	*HHF2 hht2Δ4-30/TRP1 ARS CEN*	Mann and Grunstein (1992)
pMH310	*HHF2/TRP1 ARS CEN*	Han and Grunstein (1988)
pPK613	*hhf2Δ(4-28)/URA3 ARS CEN*	Kayne et al. (1988)
pPK618	*hhf2Δ(4-19)/URA3 ARS CEN*	Kayne et al. (1988)
pR490	*ADE2/pBR322*	Beth Rockmill (pers. comm.)
pKK49	*Hind*III fragment of *MATα* cloned into *Hind*III site of pGEM3	this paper
pKK792	*cyc1::lacZ* (no α2/Mcm1 site)/*ADE2* integrating vector	this paper
PKK793	cyc1::*lacZ* + α2/Mcm1 site between UAS and TATA/*ADE2* integrating vector	this paper
pKK795	*hhf2Δ4-19 HHT2/URA3 ARS CEN*	this paper
pKK797	*HHF2 HHT2/URA3 ARS CEN*	this paper
pKK826	*hhf2 (K12Q, K16Q) HHT2/TRP1 ARS CEN*	this paper
pKK830	*hhf2 (K12Q, K16Q) hht2Δ4-30/TRP1 ARS CEN*	this paper

constructed by replacing the *Bam*HI-*Eco*RI fragment of pRM200 and pRM430 (Mann and Grunstein 1992), respectively, with the *Bam*HI-*Eco*RI fragment of pKK824. pKK824 is the *Eco*RI-*Hin*dIII fragment of pKK822 inserted into the *Eco*RI and *Hin*dIII sites of pPK613 (Kayne et al. 1988). pKK822 was constructed by inserting the double-stranded oligo 5′-GAT CTA AAG GTG GTA AAG GTC TAG GTC AAG GTG GTG CCC AGC GTC ACA-3′/ 5′-GAT CTG TGA CGC TGG GCA CCA CCT TGA CCT AGA CCT TTA CCA CCT TTA-3′ into the *Bgl*II site of pKK549. pKK549 is the *Eco*RI fragment of pKK548 ligated into pRS304 (Sikorski and Hieter 1989). pKK548 was constructed by ligating *Bgl*II-*Hin*dIII-cut PCR fragment 1 and *Bgl*II-*Eco*RI-cut PCR fragment 2 into *Hin*dIII-*Eco*RI-cut pUC18 (Yanisch-Perron et al. 1985). PCR fragment 1 was generated using the oligos 5′-AGA TAA TGG GGC TCT TTA CAT TTC-3′ and 5′-TTT ACC ACC TTT AGA TCT ACC GGA CAT TAT TTT ATT GTA-3′ as primers and pKK541 as template. PCR fragment 2 was generated using the oligos 5′-AGC ACG CTT ATC GCT CCA ATT TCC-3′ and 5′-AAG CGT CAC AGA TCT ATT CTA GAG GAT AAC ATC CAA GCT-3′ as primers and pKK541 as template. pKK541 is the *Hin*dIII fragment from pMH310 (Han and Grunstein 1988) inserted into *Hin*dIII-digested pΔSJ. pΔSJ was constructed by Andrew Vershon by deleting the *Xho*I-*Sal*I fragment containing the translational start of pSJ1 (Johnson 1991; Komachi et al. 1994). pSJ1 contains the *GAL10* promoter upsteam of a polylinker (Johnson et al. 1991).

pKK792 and pKK793 were constructed by inserting the *Bam*HI fragment from pR490 (B. Rockmill) into the *Bgl*II site of pKK833 and pKK834, respectively. pKK833 and pKK834 are pLG-Δ312S (Guarente and Hoar 1984) and pS1-19 (Keleher et al. 1988), respectively, in which a *Bgl*II linker has been inserted between the *Hin*dIII and *Sma*I sites. pKK795 was constructed by inserting the *Bam*HI-*Sal*I fragment from pRM200 into pPK618 (Kayne et al. 1988). pKK797 was constructed by replacing the *Bam*HI-*Eco*RI fragment of pKK795 with the *Bam*HI-*Eco*RI fragment of pKK561. pKK561 was constructed by inserting the *Bam*HI-*Eco*RI fragment from pMH310 into *Bam*HI-*Eco*RI-digested pRS313 (Sikorski and Hieter 1989). pGEM3 was obtained from Promega; pBR322 was obtained from New England Biolabs.

The following plasmids were used in the *SRB8* studies. pMW18, containing *ARE2*, was isolated from a YCp50-based yeast genomic library (Rose et al. 1987). pMW19 was constructed by deleting an approximately 5-kb *Hin*dIII fragment containing *TUP1* from pMW18. pMW22 was derived from pMW19 by deleting a 1.9-kb *Pvu*II fragment from within the *SRB8* locus. pMW26, used to disrupt *SRB8*, was constructed in two steps. First, a 4.5-kb *Eco*NI-*Nhe*I fragment (the ends filled in using Klenow fragment) from pMW19 was subcloned into the *Sma*I site of pRS316 (Sikorski and Hieter 1989), generating pMW21. (The fragment is oriented with the *Eco*NI end toward the *Bam*HI site in the pRS316 polylinker.) pMW21 was then cut with *Pac*I and *Bgl*II, deleting a 2.8-

kb fragment encoding the carboxy-terminal two thirds of Srb8, and then ligated to a 2.2-kb *LEU2* PCR fragment with PCR-introduced *Pac*I and *Bgl*II ends. The high-copy *TUP1*-bearing plasmid pFW28 has been described previously by Williams and Trumbly (1990). YEp24 has been described by Botstein et al. (1979)

***Cloning of* ARE2.** To clone *ARE2*, we used an α *are2* strain containing a chromosomal *mfa2::lacZ* reporter. Because *MFA2* is an **a**-specific gene, the *mfa2::lacZ* reporter is subject to α2-Mcm1 repression in α cells. Since this reporter is strongly repressed in wild-type α cells, yeast colonies derived from these cells remain white in an X-gal filter assay (see below). Because *are2* mutants are defective in α2-Mcm1 repression, β-galactosidase (β-gal) is produced and the mutant colonies appear blue. We transformed the α *are2-13 mfa2::lacZ* strain MWY16 with a YCp50-based yeast genomic library (Rose et al. 1987), plated transformants at approximately 65 colonies per SD⁻Ura plate, and then screened for white transformants on X-gal, i.e., for complementation of the *mfa2::lacZ* derepression phenotype. As a secondary screen, transformants that appeared white were tested for the flocculent phenotype in liquid SD⁻Ura medium. Of approximately 4000 transformants screened by X-gal filter assay, two transformants, 13-30 and 13-38, were both white on X-gal and nonflocculent in liquid medium. Both mutant phenotypes were restored upon curing the plasmids. The library plasmid (pMW18) was isolated from transformant 13-38 and analyzed further.

Linkage analysis. An **a** *srb8Δ::LEU2 mfa2::lacZ* strain (MWY44) was crossed to an α *are2-13 mfa2::lacZ* strain MWY16. The resulting diploid sporulated inefficiently unless transformed with a plasmid bearing *SRB8* (pMW19). Following tetrad dissection of the sporulated diploid, the segregants were cured of the plasmid, and the mutant phenotypes were then analyzed. Tetrad analysis (a total of 18 asci tested from two independent diploids) indicated that the *are2* and *srb8* mutations are tightly linked. All segregants of either mating type were flocculent and all α segregants were derepressed for *mfa2::lacZ* (data not shown). The *srb8* deletion mutant also contained a deletion in an open reading frame of unknown function directly downstream from *SRB8*; this open reading frame does not complement the *are2* mutation (see yeast strains).

β-galactosidase assays. The X-gal filter assay has been described previously(Schena et al. 1989; Wahi and Johnson 1995). Quantitative liquid β-gal assays were performed as described previously by Miller (1972). However, in the *SRB8* studies, modifications were carried out as described by Wahi and Johnson (1995), and in the histone studies, modifications were carried out as described by Komachi et al. (1994). Flocculent cells were dispersed by adding EDTA to 25 mM before measuring optical density. Numbers are averages; standard deviations are shown in Tables 3–4 and Figure 3.

Table 3. Repression of *cyc1::lacZ* Reporters by α2-Mcm1 in Wild-type Strains and in Strains Having Mutations in *HHF2* and/or *HHT2*

Strain	*HHF2* allele	*HHT2* allele	Reporter	β-gal activity
KKY222	wild-type	wild-type	no α2/Mcm1 site	131.8 ± 6
KKY223	wild-type	Δ4-30	no α2/Mcm1 site	53.4 ± 3
KKY224	K12Q, K16Q	wild-type	no α2/Mcm1 site	34.8 ± 0.3
KKY225	K12Q, K16Q	Δ4-30	no α2/Mcm1 site	12.4 ± 0.1
KKY226	wild-type	wild-type	+ α2/Mcm1 site	0.12 ± 0.03
KKY227	wild-type	Δ4-30	+ α2/Mcm1 site	0.47 ± 0.1
KKY228	K12Q, K16Q	wild-type	+ α2/Mcm1 site	0.03 ± 0.005
KKY229	K12Q, K16Q	Δ4-30	+ α2/Mcm1 site	0.18 ± 0.1

*MAT*a/*MAT*α *hhf1, hht1Δ::LEU2/hhf1, hht1Δ::LEU2 hhf2, hht2Δ::HIS3/hhf2, hht2Δ:: HIS3* strains carrying an integrated *cyc1::lacZ* reporter (pKK792 or pKK793) and wild-type or mutant *HHF2* and wild-type or mutant *HHT2* on a *CEN ARS* plasmid were constructed and assayed for β-gal activity.

PCR assays. The PCR amplification protocol and three primers used to determine the *MAT* allele have been described previously (Huxley et al. 1990), with modifications described by Wahi and Johnson (1995). To confirm integration of *srb8Δ::LEU2*, a PCR amplification protocol described in Wahi and Johnson (1995) was used.

RESULTS AND DISCUSSION

The Chromatin Model

It had been previously proposed that Ssn6-Tup1 exerts repression by positioning a nucleosome at the promoter of its target genes; this nucleosome would presumably interfere with the assembly of the general transcription machinery. In vivo nuclease protection studies indicated that DNA-bound α2-Mcm1 stably positions a nucleosome adjacent to the α2/Mcm1 site at endogenous **a**-specific genes in an Ssn6-Tup1-dependent manner (Shimizu et al. 1991; Cooper et al. 1994). More recent work from three different laboratories, however, has demonstrated that full (or at least nearly full) repression can be seen in the absence of positioned nucleosomes (Redd et al. 1996; Gavin and Simpson 1997; Huang et al. 1997).

Evidence nevertheless suggests that histones are involved in Ssn6-Tup1-mediated repression. Tup1 interacts directly with purified histones H3 and H4, but not histones H2A or H2B (Edmondson et al. 1996). Of particular importance to this interaction are certain lysine residues that lie in the amino-terminal "tails" of histones H3 and H4 (Edmondson et al. 1996). These residues are known to be reversibly acetylated, and there is a general correlation between deacetylation of these lysines and transcriptional repression (Turner 1991; Braunstein et al. 1996; Kuo et al. 1996). Importantly, Tup1 was found to bind only to the underacetylated forms of histones H3 and H4; in fact, Tup1 did not bind at all to the tri- and tetra-acetylated forms of either histone (Edmondson et al. 1996). These results indicate that deacetylated lysine residues in the histone H3 and H4 tails are critical for the direct interaction with Tup1. Consistent with this interpretation, an H4 mutation in which two of these lysines (12 and 16) were changed to glutamines (thus mimicking the acetylated state) severely weakened the interaction with Tup1, as did a deletion of the first six lysines from histone H3 (Edmondson et al. 1996). These findings have led to a revised chromatin model in which Ssn6-Tup1 mediates repression through organizing chromatin in a way that involves a direct interaction between Tup1 and the underacetylated histone H3 and H4 tails, but that does not necessarily involve nucleosome positioning.

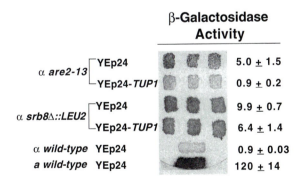

β-Galactosidase Activity

α *are2-13* YEp24 — 5.0 ± 1.5 / YEp24-*TUP1* — 0.9 ± 0.2

α *srb8Δ::LEU2* YEp24 — 9.9 ± 0.7 / YEp24-*TUP1* — 6.4 ± 1.4

α *wild-type* YEp24 — 0.9 ± 0.03

a *wild-type* YEp24 — 120 ± 14

Figure 3. Allele-specific suppression of *srb8/are2* by *TUP1* overexpression. α *mfa2::lacZ* strains (MWY16 and MWY45) containing the indicated *srb8/are2* allele were transformed with either the high-copy *URA3* plasmid, YEp24, or a YEp24 vector bearing *TUP1* (pFW28). Transformants were patched on an SD plate lacking uracil, and then an X-gal filter assay was performed (see Experimental Procedures). Patches of strains expressing *mfa2::lacZ* turn blue (gray in photo) on X-gal, whereas those repressing *mfa2::lacZ* remain white. For comparison, isogenic wild-type **a** and α *mfa2::lacZ* strains (SM1179 and SM1196, respectively) carrying YEp24 were also assayed. In the quantitative β-gal assays, two to four independent colonies for each strain were assayed in triplicate on different days.

Table 4. Expression of *mfa2::lacZ* in Wild-type and *srb8/are2* Strains

Genotype	Units of β-gal activity *Mat*a	*Mat*α	Repression
ARE+	130 ± 20	1.1 ± 0.3	120
are2-13	97 ± 25	9.4 ± 2.2	10
srb8Δ::LEU2	120 ± 20	7.7 ± 0.7	16

Numbers represent units of β-gal activity in *mfa2::lacZ* strains of the indicated genotype. We define repression as the ratio of β-gal activity in an **a** *mfa2::lacZ* cell to that in the isogenic α cell. Results are averages of assays performed in triplicate. At least three independent colonies for each strain were assayed. Strains (SM1196, SM1179, MWY16, MWY17, MWY44, and MWY45) are described in Table 1.

A key observation in support of this chromatin model is the reported derepression of **a**-specific genes in certain histone H3 and H4 mutants. Amino-terminal mutations in histones H3 and/or H4 which reduce or abolish the in vitro interaction with Tup1 also cause derepression of **a**-specific reporter genes in cells containing α2 (Roth et al. 1992; Edmondson et al. 1996). For instance, a histone H4 mutation (H4 K12Q,K16Q) was reported to cause 3-fold derepression of a *cyc1::lacZ* reporter possessing an α2/Mcm1 site, and a histone H3/H4 double mutation (H3 Δ1-28, H4 K12Q, K16Q) caused 13-fold derepression of the same reporter (Edmondson et al. 1996). However, interpretation of the repression data is complicated by the fact that these histone mutations cause pleiotropic effects. For example, histone H4 tail mutations derepress the silent mating-type loci and thereby cause expression of both *MAT***a** and *MAT*α (Kayne et al. 1988; Johnson et al. 1990; Park and Szostak 1990). (*S. cerevisiae* contains one silent copy of *MAT***a** and *MAT*α at the *HMR***a** and *HML*α locus, respectively; see Herskowitz et al. 1992.) Therefore, the strains containing the histone H4 mutations express both *MAT***a** and *MAT*α and behave as **a**/α cells with respect to cell-type-specific gene repression (see Fig. 2). α2-Mcm1 repression is usually severalfold lower in **a**/α cells than in α cells (Goutte and Johnson 1988), in part because in **a**/α cells, *MAT*α2 mRNA is present at only 10–20% of its level in α cells (Siliciano and Tatchell 1986). Moreover, in **a**/α cells, α2 combines both with Mcm1 and with **a**1, perhaps further reducing the effective levels of α2-Mcm1 (Goutte and Johnson 1988). In addition, most of the previous studies utilized reporter genes carried on plasmids either in all tested strains (Edmondson et al. 1996; Huang et al. 1997) or in some strains (Roth et al. 1992); the effects of the histone H3 and H4 mutations on plasmid copy number have not been well established. Because of these complications, we quantitated the effects of histone H3 and H4 mutations on α2-Mcm1 repression using congenic sets of yeast strains that allow us to control for these variables.

Effect of Mutations in Histones H3 and H4 on α2-Mcm1 Repression

To determine the effect of mutations in histones H3 and H4 on α2-Mcm1 repression, we constructed a set of **a**/α strains that are deleted for the genes encoding histones H3 and H4 and that carry wild-type or mutant *HHF2* (histone H4) and wild-type or mutant *HHT2* (histone H3) on a *CEN* plasmid. We use the same *hhf2* allele (H4 K12Q,K16Q) that was used in the previous repression studies (Edmondson et al. 1996; Huang et al. 1997). This *hhf2* allele causes derepression of the silent mating loci; however, in our experiment, this derepression is controlled for by the fact that all strains—mutant and wild-type—are **a**/α. The *hht2* allele (H3 Δ4-30) used in our study encodes a histone H3 that has an amino-terminal deletion of amino acids 4 through 30 (Mann and Grunstein 1992), compared to an amino-terminal deletion of amino acids 1 through 28 used in the previous work (Edmondson et al. 1996; Huang et al. 1997). Although this

histone H3 mutant contains four additional amino-terminal amino acids, it lacks all of the amino-terminal lysine residues known to undergo reversible acetylation. Since these deacetylated lysine residues are critical for Tup1 interaction with histone H3, this mutant allows for an assessment of the role of these lysines in repression in vivo. These strains also carry a *cyc1::lacZ* reporter that has either no α2/Mcm1 site or one site located between the UAS and TATA. This reporter is similar to that used in one of the previous studies (Edmondson et al. 1996); however, the reporter we use has been integrated.

With respect to repression of the reporter possessing the α2/Mcm1 site, Table 3 shows that the histone H4 mutation causes no significant derepression, the single H3 mutation has a fourfold effect, and the double H3/H4 mutation has little if any effect. Regarding this last point, expression of the *cyc1::lacZ* reporter containing an α2/Mcm1 site was approximately equal in the **a**/α *hhf2 hht2* and **a**/α wild-type strains, indicating no significant derepression of the reporter in the histone H3/H4 double mutant. We also note that the single *hhf2* and the double *hhf2 hht2* mutations cause decreased levels of expression of the nonrepressed (i.e., no α2/Mcm1 site) reporter in the **a**/α strains (KKY224 and KKY225). However, as discussed above, the repressed levels change little if at all (see KKY228 and KKY229).

Our results differ from those reported earlier in several respects (Roth et al. 1992; Edmondson et al. 1996). First, our results were obtained using congenic **a**/α strains and integrated reporters in all cases where the level of repression was being compared. Second, we observed that the histone mutations had the most dramatic effect on the expression levels of the nonrepressed (i.e., no α2/Mcm1 site) reporter. The effects of histone mutations on activated transcription have been noted previously (Durrin et al. 1991; Mann and Grunstein 1992). Third, as discussed above, the double H3/H4 mutant strain showed little or no derepression.

These experiments challenge the importance of the amino-terminal lysines of histones H3 and H4 for steady-state repression by Ssn6-Tup1. It is possible that the histone tail lysines affect the kinetics of repression and/or derepression, but the contribution of the histone tails (and by inference the contribution of the Tup1 interaction with the underacetylated tails of histones H3 and H4) to steady-state repression is modest at best. The available data suggest that Ssn6-Tup1 represses transcription primarily through a mechanism independent of an interaction between Ssn6-Tup1 and the amino-terminal tail lysines of histones H3 and H4.

The General Transcription Machinery Model

Although there is currently no proof that Ssn6-Tup1 interacts directly with the transcription machinery, several lines of evidence suggest that it may be a target for Ssn6-Tup1. First, α2-Mcm1 represses a test promoter containing α2/Mcm1 sites two- to fivefold in vitro (Herschbach et al. 1994; Redd et al. 1997). This in vitro repression depends on Ssn6-Tup1 and takes place in a crude system to which naked DNA template is added. Since the template promoter lacks known activator-binding sites, the repression observed in vitro is likely to affect basal transcription, i.e., transcription in the absence of activator proteins bound upstream. These results suggest that Ssn6-Tup1 directs α2-Mcm1 repression, at least in part, by inhibiting the assembly or activities of the general transcription machinery. Consistent with this idea, Ssn6-Tup1 acts at an early step in transcription, before the first phosphodiester bond is formed (M. Arnaud and A.D. Johnson, unpubl.).

The second line of evidence supporting the transcription machinery model comes from genetic screens for components required for Ssn6-Tup1-mediated repression. These screens, which were carried out in several laboratories, have identified a specific subset of genes that encode proteins tightly associated with RNA polymerase II (pol II) (for review, see Carlson 1997). We previously identified mutations in four genes (designated ARE for alpha2 repression) which cause partial loss of α2-Mcm1 repression and pleiotropic phenotypes (similar to those caused by ssn6 or tup1 mutations), such as flocculence and sporulation defects (Wahi and Johnson 1995). ARE1 was found to be identical to SRB10, which is associated genetically and physically with pol II (Nonet and Young 1989; Thompson et al. 1993; Liao et al. 1995; Wahi and Johnson 1995). Coincidentally, it was discovered that many of the genes required for glucose repression, another Ssn6-Tup1 repression pathway, also encode pol-II-associated proteins (see below). In this section, we report results that strengthen the connection between α2-Mcm1 repression and the general transcription machinery. We have found that ARE2 is identical to SRB8 and that ARE3 is likely identical to ROX3, both of which encode proteins associated with pol II. In addition, we report a genetic interaction between SRB8 and TUP1 that reinforces the links between Ssn6-Tup1 and the general transcription machinery.

ARE2 Is Identical to SRB8

A low-copy yeast genomic library plasmid (pMW18) that complements both the mfa2::lacZ derepression and the flocculent phenotype of an α are2 mutant was isolated (see Experimental Procedures). Given that ARE1 is identical to SRB10 and that the SRB genes encode proteins that are physically associated in a large complex containing pol II (Thompson et al. 1993; Kim et al. 1994; Koleske and Young 1994; Hengartner et al. 1995; Liao et al. 1995; Myers et al. 1998), it seemed possible that ARE2 might be another SRB gene. Previous linkage analysis had indicated that ARE2 resides near TUP1 (data not shown). We learned that SRB8 is approximately 2-kb proximal to TUP1 (S.-M. Liao and R. Young, pers. comm.); subsequent restriction map analysis indicated that both SRB8 and TUP1 are located on the genomic insert of pMW18. Deleting TUP1 from the genomic insert of pMW18 had no effect on complementation, indicating that TUP1, on this low-copy plasmid, was not simply suppressing the are2 mutation. On the other hand, deleting an internal 1.9-kb PvuII fragment from SRB8 carried

on pMW19 (TUP1-deleted pMW18) abolished complementation of both the mfa2::lacZ derepression and the flocculent phenotype, indicating that SRB8 is ARE2. Diploids resulting from mating an a srb8Δ::LEU2 mfa2::lacZ strain to an α are2-13 mfa2::lacZ strain also showed complete lack of complementation, and tetrad analysis (a total of 18 asci tested from two independent diploids) demonstrated tight linkage (see Experimental Procedures). Taken together, these results indicate that ARE2 is identical to SRB8, thus supporting a connection between Ssn6-Tup1 and the pol II machinery.

TUP1 and SRB8 Interact Genetically

The connection between Ssn6-Tup1 and the general transcription machinery is further supported by an allele-specific genetic interaction we have observed between TUP1 and SRB8. Two srb8 alleles were used in this analysis: One is are2-13, derived from our genetic screen, and the other is srb8Δ::LEU2, an allele that deletes DNA sequence encoding the carboxy-terminal two thirds of Srb8 and also 62 nucleotides of a small (384-nucleotide) downstream open reading frame (Oliver et al. 1992) of unknown function (see Experimental Procedures). It is unlikely that deletion of this second open reading frame affects our results, as the open reading frame has no known role in α2-Mcm1 repression. As indicated in Table 4, the srb8Δ and are2-13 strains have nearly equivalent repression defects. In the srb8Δ mutant, repression is reduced to 13% of wild-type repression and in the are2-13 mutant, repression is reduced to 8%.

Although these mutations cause similar decreases in α2-Mcm1 repression of mfa2::lacZ, they respond differently to TUP1 overexpression. As shown in Figure 3, overexpression of TUP1 suppresses the mfa2::lacZ derepression in the α are2-13 strain but not in the α srb8Δ::LEU2 strain. TUP1 overexpression in the α are2-13 strain reduces the level of mfa2::lacZ expression from 5 units of β-gal activity to 0.9 units, the level seen in the wild-type α strain. In contrast, TUP1 overexpression does not significantly restore repression of mfa2::lacZ in the α srb8Δ::LEU2 strain (9.9 units of β-gal activity in the absence of TUP1 overexpression and 6.4 units in its presence). This allele-specific genetic interaction suggests, but in no way proves, that Tup1 and Srb8 physically interact. One possibility is that the protein encoded by are2-13 is a point mutant that binds to Tup1 with lower affinity than does the wild-type protein. TUP1 overexpression may then drive this weakened interaction forward. Of course, other models are also consistent with these results, and proof of a direct interaction between Tup1 and Srb8 awaits biochemical analysis. Nevertheless, these genetic results support the idea that Ssn6-Tup1 repression is mediated, in part, through components that are associated with pol II.

The ARE screen was one of several that identified known pol-II-associated components required for Ssn6-Tup1-mediated repression. The results of these screens are summarized in Table 5. SRB8, SRB9, SRB10, and SRB11 were originally identified in a genetic screen for suppressors of a growth defect caused by a carboxy-terminal domain (CTD) deletion of the largest subunit of pol II (Nonet and Young 1989). The *SRB* genes, of which there are nine, were subsequently found to encode components of a multiprotein complex physically associated with pol II through an interaction with the CTD (Thompson et al. 1993; Kim et al. 1994; Koleske and Young 1994; Hengartner et al. 1995; Liao et al. 1995). *SRB8-SRB11* are identical to *SSN* genes (Kuchin et al. 1995; Liao et al. 1995; Song et al. 1996). The *SSN* genes were identified in a search for genes required for glucose repression, another Ssn6-Tup1-mediated process (Carlson et al. 1984). Two of these four *SRB*s are also identical to *ARE* genes (Wahi and Johnson 1995; this paper). *SIN4* is required for full levels of α2-Mcm1 repression (Chen et al. 1993a; Jiang and Stillman 1995; Wahi and Johnson 1995) and is identical to *SSN4* (Song et al. 1996). *RGR1* was not identified in the *SSN* screen but was identified in a similar screen for genes required for glucose repression (Sakai et al. 1988, 1990). Whether *RGR1* is required for α2-Mcm1 repression has not been formerly tested, but it does not complement the remaining *ARE* gene, *ARE4* (M. Wahi and A.D. Johnson, unpubl.). *ROX3* is identical to *SSN7* (Song et al. 1996) and is likely identical to *ARE3* as well. We have isolated a low-copy yeast genomic library plasmid containing *ROX3* that complements both the α2-Mcm1 repression defect and the flocculence of an *are3* mutant (A. Szidon et al., unpubl.).

The seven pol-II-associated components listed in Table 5 have several properties in common. Mutations in each of the genes encoding these components cause pleiotropic phenotypes similar to those caused by *ssn6* and *tup1* mutations (see above). Furthermore, mutations in each of these genes cause only partial loss of Ssn6-Tup1-mediated repression. For example, in the case of α2-Mcm1 repression, an *srb10* deletion reduces repression to 15% and a *sin4* deletion to 3–11% of wild-type levels (Wahi and Johnson 1995). In addition, most of these components are known to be general regulators, i.e., both positive and negative regulators depending on the specific promoter (for review, see Carlson 1997).

Despite these similarities, these components can be divided into two functionally distinct groups. One consists of Srb8, Srb9, Srb10, and Srb11 and the other consists of Sin4, Rgr1, and Rox3. Several lines of evidence support this division. First, unlike mutations in *SRB8-SRB11*, mutations in *SIN4, RGR1,* and *ROX3* strongly activate basal transcription, i.e, transcription in the absence of known UAS elements, in vivo (Jiang and Stillman 1992; Chen et al. 1993b; Covitz et al. 1994; Jiang et al. 1995; Wahi and Johnson 1995; Song et al. 1996). Second, no mutation in *RGR1, SIN4,* or *ROX3* was identified in the *SRB* screen, suggesting that this group has a different transcriptional regulatory role, at least with respect to the CTD. Third, these groups belong to distinct biochemical complexes. Rgr1, Sin4, and Rox3 are components of a complex, termed the mediator, which can be purified away from a larger complex that contains pol II and other pol-II-associated proteins (Kim et al. 1994; Li et al. 1995; Gustafsson et al. 1997; Myers et al. 1998). Srb8-Srb11, although associated with pol II by other purification protocols, are the only SRB proteins (out of the nine different ones) that are absent from the mediator (Myers et al. 1998). Thus, both biochemical and genetic evidence indicates that there are two functionally distinct groups of pol-II-associated proteins required for Ssn6-Tup1-mediated repression.

Although the exact role of these two groups in Ssn6-Tup1-mediated repression is unknown, there are intriguing clues. The Srb8-Srb11 group may phosphorylate one or more as yet unknown factors in response to regulation by Ssn6-Tup1. Whereas the protein sequences of Srb8 and Srb9 reveal no obvious function (Hengartner et al. 1995), *SRB10* encodes a protein kinase and *SRB11* encodes a cyclin (Kuchin et al. 1995; Liao et al. 1995). The connection between Ssn6-Tup1-mediated repression and the kinase activity of this subcomplex was recently supported by the finding that repression of a promoter containing LexA-binding sites by LexA-Tup1 is virtually abolished in the absence of Srb10 kinase (Kuchin and Carlson 1998). LexA-Tup1 represses the test promoter 14-fold, but in the absence of Srb10, it represses the promoter 1.7-fold, the level of repression produced by LexA alone. Importantly, the catalytic activity of Srb10 is required for this LexA-Tup1-mediated repression (Kuchin and Carlson 1998). Possible targets of the Srb10 kinase when regulated by Ssn6-Tup1 include the CTD, which is known to be a target of Srb10 (Kuchin et al. 1995; Liao et al. 1995), general transcription factors, and/or Ssn6-Tup1 itself.

Sin4, Rgr1, and Rox3, on the other hand, are components of the mediator, which is a complex composed of approximately 20 proteins and which is associated with pol II through an interaction with the CTD (Kim et al. 1994; Li et al. 1995; Gustafsson et al. 1997; Myers et al. 1998). The mediator is required for response to specific DNA-binding transcriptional activators in an in vitro transcription system in which pol II, various basal factors, and naked DNA template are added; the mediator also stimulates basal transcription of naked templates in vitro (Kim et al. 1994; Li et al. 1995; Myers et al. 1998). Sin4 and Rgr1 exist in a distinct subcomplex within the mediator (Li et al. 1995; Myers et al. 1998). This subcomplex contains two additional proteins, Gal11 and Hrs1 (Li et al. 1995; Myers et al. 1998), that have genetic and biochem-

Table 5. At Least Seven Genes Encoding pol-II-associated Components Are Required for Ssn6-Tup1-mediated Repression

Gene	Aliases
SRB8	*SSN5/ARE2*
SRB9	*SSN2/SCA1*[a]
SRB10	*SSN3/ARE1*
SRB11	*SSN8*
SIN4	*SSN4*
RGR1	
ROX3	*SSN7/ARE3*[b]

[a]*SCA1* was identified in a screen for suppressors of lethality caused by mutations in the pol II CTD (Yuryev and Corden 1996).

[b]*ARE3* is likely to be identical to *ROX3* on the basis of complementation data (see text).

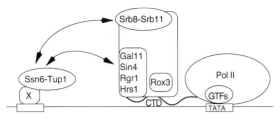

Figure 4. Model for Ssn6-Tup1-mediated repression. Sequence-specific DNA-binding proteins bring Ssn6-Tup1 to diverse gene sets. Ssn6-Tup1 then mediates repression by interacting with multiple downstream targets. These targets include components of a large multiprotein complex (termed mediator; indicated by large box associated with CTD) that is tightly associated with RNA polymerase II (pol II). The pol-II-associated repression targets can be divided into two functionally distinct groups of general regulators: Srb8-Srb9-Srb10-Srb11 and Sin4-Rgr1-Rox3. (Rgr1 and Sin4 are known to exist in a distinct subcomplex with two other proteins Gal11 and Hrs1, see text.) Ssn6-Tup1 regulates the activities of these groups such that they in turn inhibit the assembly of the general transcription machinery or inhibit transcription initiation. Activities of both groups are required for full levels of Ssn6-Tup1-directed repression. Srb8-Srb11 is not a mediator component but is tightly associated with pol II by another purification protocol. (CTD) Carboxy-terminal domain of the largest subunit of pol II; (GTF) general transcription factor; (X) sequence-specific DNA-binding protein.

ical characteristics similar to those of Sin4 and Rgr1 (Santos-Rosa et al. 1996; Piruat et al. 1997; Myers et al. 1998). This subcomplex is not required for the activation function of the mediator, indicating that mediator components have redundant functions with respect to this in vitro transcriptional activation, or more interestingly, that the subcomplex has a distinct function (Li et al. 1995). The fact that the mediator has a role in transcriptional activation in vitro does not exclude the possibility that some mediator components, including Sin4, Rgr1, and Rox3, also respond to regulation by transcriptional repressors in vivo.

Assuming that both Srb8-Srb11 and Sin4-Rgr1-Rox3 have direct roles in Ssn6-Tup1-mediated repression, it appears that Ssn6-Tup1 has multiple repression targets. *ssn6* or *tup1* null mutations completely abolish repression of diverse gene sets. In contrast, despite numerous screens, no mutations in any other gene have been identified that completely eliminate this global repression. Various genetic screens and mutant studies have identified components that each partially contributes to Ssn6-Tup1-mediated repression, indicating that these components constitute a group (possibly not complete) of repression targets. These groups are Srb8-Srb11, Sin4-Rgr1-Rox3, and the histone H3 and H4 amino-terminal "tails" (although this third group appears to have a more modest role, at least with respect to steady-state levels of α2-Mcm1 repression). Converging work from several laboratories has lead to the idea that these individual target interactions add up to complete repression by Ssn6-Tup1 (see Fig. 4). Tests of this working model will include multiple mutant combinations, in vitro reconstitution of repression, and detailed in vitro interaction studies of Tup1 and Ssn6 with these repression components.

ACKNOWLEDGMENTS

We thank Yvette Castro for help in cloning *ARE2*; Alex Szidon for work on *ARE3*; Burk Braun for help in analyzing DNA sequences; Danesh Moazed for useful discussions; Ira Herskowitz, Andrew Murray, and Nancy Hollingsworth for advice; Sha-Mei Liao and Richard Young, and Marian Carlson for communicating unpublished results; the M. Grunstein lab and Beth Rockmill for providing plasmids; and lab members, especially Burk Braun, for critical comments on the manuscript. This work was supported by a National Institutes of Health (NIH) grant to A.D.J. K.K. was supported by a National Science Foundation predoctoral fellowship and by a training grant from the NIH to the Department of Biochemistry and Biophysics at the University of California, San Francisco. M.W. was supported by an NIH-NIGMS MSTP grant.

REFERENCES

Botstein D., Falco S.C., Stewart S.E., Brennan M., Scherer S., Stinchcomb D.T., Struhl K., and Davis R.W. 1979. Sterile host yeasts (SHY): A eukaryotic system of biological containment for recombinant DNA experiments. *Gene* **8:** 17.

Braunstein M., Sobel R.E., Allis C.D., Turner B.M., and Broach J.R. 1996. Efficient transcriptional silencing in *Saccharomyces cerevisiae* requires a heterochromatin histone acetylation pattern. *Mol. Cell. Biol.* **16:** 4349.

Carlson M. 1997. Genetics of transcriptional regulation in yeast: Connections to the RNA polymerase II CTD. *Annu. Rev. Cell Dev. Biol.* **13:** 1.

Carlson M., Osmond B.C., Neigeborn L., and Botstein D. 1984. A suppressor of SNF1 mutations causes constitutive high-level invertase synthesis in yeast. *Genetics* **107:** 19.

Carrico P.M. and Zitomer R.S. 1998. Mutational analysis of the Tup1 general repressor of yeast. *Genetics* **148:** 637.

Chen S., West R., Jr., Johnson S.L., Gans H., Kruger B., and Ma J. 1993a. TSF3, a global regulatory protein that silences transcription of yeast GAL genes, also mediates repression by α2 repressor and is identical to SIN4. *Mol. Cell. Biol.* **13:** 831.

Chen S., West R.W., Jr., Ma J., Johnson S.L., Gans H., and Woldehawariat G. 1993b. TSF1 to TSF6, required for silencing the *Saccharomyces cerevisiae* GAL genes, are global regulatory genes. *Genetics* **134:** 701.

Cooper J.P., Roth S.Y., and Simpson R.T. 1994. The global transcriptional regulators, SSN6 and TUP1, play distinct roles in the establishment of a repressive chromatin structure. *Genes Dev.* **8:** 1400.

Covitz P.A., Song W., and Mitchell A.P. 1994. Requirement for RGR1 and SIN4 in RME1-dependent repression in *Saccharomyces cerevisiae*. *Genetics* **138:** 577.

De Vit M.J., Waddle J.A., and Johnston M. 1997. Regulated nuclear translocation of the Mig1 glucose repressor. *Mol. Biol. Cell* **8:** 1603.

DeRisi J.L., Iyer V.R., and Brown P.O. 1997. Exploring the metabolic and genetic control of gene expression on a genomic scale. *Science* **278:** 680.

Durrin L.K., Mann R.K., Kayne P.S., and Grunstein M. 1991. Yeast histone H4 N-terminal sequence is required for promoter activation in vivo. *Cell* **65:** 1023.

Edmondson D.G., Smith M.M., and Roth S.Y. 1996. Repression domain of the yeast global repressor TUP1 interacts directly with histones H3 and H4. *Genes Dev.* **10:** 1247.

Gavin I.M. and Simpson R.T. 1997. Interplay of yeast global transcriptional regulators Ssn6p-Tup1p and Swi-Snf and their effect on chromatin structure. *EMBO J.* **16:** 6263.

Goutte C. and Johnson A.D. 1988. a1 protein alters the DNA binding specificity of alpha 2 repressor. *Cell* **52:** 875.

Guarente L. and Hoar E. 1984. Upstream activation sites of the

CYC1 gene of *Saccharomyces cerevisiae* are active when inverted but not when placed downstream of the "TATA box." *Proc. Natl. Acad. Sci.* **81:** 7860.

Gustafsson C.M., Myers L.C., Li Y., Redd M.J., Lui M., Erdjument-Bromage H., Tempst P., and Kornberg R.D. 1997. Identification of Rox3 as a component of mediator and RNA polymerase II holoenzyme. *J. Biol. Chem.* **272:** 48.

Hall M.N. and Johnson A.D. 1987. Homeo domain of the yeast repressor α2 is a sequence-specific DNA-binding domain but is not sufficient for repression. *Science* **237:** 1007.

Han M. and Grunstein M. 1988. Nucleosome loss activates yeast downstream promoters in vivo. *Cell* **55:** 1137.

Hengartner C.J., Thompson C.M., Zhang J., Chao D.M., Liao S.M., Koleske A.J., Okamura S., and Young R.A. 1995. Association of an activator with an RNA polymerase II holoenzyme. *Genes Dev.* **9:** 897.

Herschbach B.M. and Johnson A.D. 1993. The yeast α2 protein can repress transcription by RNA polymerases I and II but not III. *Mol. Cell. Biol.* **13:** 4029.

Herschbach B.M., Arnaud M.B., and Johnson A.D. 1994. Transcriptional repression directed by the yeast α2 protein in vitro. *Nature* **370:** 309.

Herskowitz I., Rine J., and Strathern J.N. 1992. Mating-type determination and mating-type interconversion in *Saccharomyces cerevisiae*. In *The molecular and cellular biology of the yeast* Saccharomyces: *Gene expression* (ed. E.W. Jones et al.), vol. 2, p. 583. Cold Spring Harbor Laboratory Press, Cold Spring Harbor, New York.

Huang L., Zhang W., and Roth S.Y. 1997. Amino termini of histones H3 and H4 are required for **a**1-α2 repression in yeast. *Mol. Cell. Biol.* **17:** 6555.

Huang M., Zhou Z., and Elledge S.J. 1998. The DNA replication and damage checkpoint pathways induce transcription by inhibition of the Crt1 repressor. *Cell* **94:** 595.

Huxley C., Green E.D., and Dunham I. 1990. Rapid assessment of *S. cerevisiae* mating type by PCR. *Trends Genet.* **6:** 236.

Jiang Y.W. and Stillman D.J. 1992. Involvement of the SIN4 global transcriptional regulator in the chromatin structure of *Saccharomyces cerevisiae*. *Mol. Cell. Biol.* **12:** 4503.

———. 1995. Regulation of HIS4 expression by the *Saccharomyces cerevisiae* SIN4 transcriptional regulator. *Genetics* **140:** 103.

Jiang Y.W., Dohrmann P.R., and Stillman D.J. 1995. Genetic and physical interactions between yeast RGR1 and SIN4 in chromatin organization and transcriptional regulation. *Genetics* **140:** 47.

Johnson A.D. 1995. Molecular mechanisms of cell-type determination in budding yeast. *Curr. Opin. Genet. Dev.* **5:** 552.

Johnson A.D. and Herskowitz I. 1985. A repressor (MATα2 product) and its operator control expression of a set of cell type specific genes in yeast. *Cell* **42:** 237.

Johnson L.M., Kayne P.S., Kahn E.S., and Grunstein M. 1990. Genetic evidence for an interaction between SIR3 and histone H4 in the repression of the silent mating loci in *Saccharomyces cerevisiae*. *Proc. Natl. Acad. Sci.* **87:** 6286.

Johnson S.L. 1991. "Structure and function analysis of the yeast CDC4 gene product." Ph.D. thesis, University of Washington, Seattle.

Kayne P.S., Kim U.J., Han M., Mullen J.R., Yoshizaki F., and Grunstein M. 1988. Extremely conserved histone H4 N terminus is dispensable for growth but essential for repressing the silent mating loci in yeast. *Cell* **55:** 27.

Keleher C.A., Goutte C., and Johnson A.D. 1988. The yeast cell-type-specific repressor α2 acts cooperatively with a non-cell-type-specific protein. *Cell* **53:** 927.

Keleher C.A., Redd M.J., Schultz J., Carlson M., and Johnson A.D. 1992. Ssn6-Tup1 is a general repressor of transcription in yeast. *Cell* **68:** 709.

Kim Y.J., Bjorklund S., Li Y., Sayre M.H., and Kornberg R.D. 1994. A multiprotein mediator of transcriptional activation and its interaction with the C-terminal repeat domain of RNA polymerase II. *Cell* **77:** 599.

Koleske A.J. and Young R.A. 1994. An RNA polymerase II holoenzyme responsive to activators. *Nature* **368:** 466.

Komachi K., Redd M.J., and Johnson A.D. 1994. The WD repeats of Tup1 interact with the homeo domain protein α2. *Genes Dev.* **8:** 2857.

Kuchin S. and Carlson M. 1998. Functional relationships of Srb10-Srb11 kinase, carboxy-terminal domain kinase CTDK-I, and transcriptional corepressor Ssn6-Tup1. *Mol. Cell. Biol.* **18:** 1163.

Kuchin S., Yeghiayan P., and Carlson M. 1995. Cyclin-dependent protein kinase and cyclin homologs SSN3 and SSN8 contribute to transcriptional control in yeast. *Proc. Natl. Acad. Sci.* **92:** 4006.

Kuo M.H., Brownell J.E., Sobel R.E., Ranalli T.A., Cook R.G., Edmondson D.G., Roth S.Y., and Allis C.D. 1996. Transcription-linked acetylation by Gcn5p of histones H3 and H4 at specific lysines. *Nature* **383:** 269.

Li T., Stark M.R., Johnson A.D., and Wolberger C. 1995. Crystal structure of the MAT**a**1/MATα2 homeodomain heterodimer bound to DNA. *Science* **270:** 262.

Li Y., Bjorklund, S. Jiang, Y.W., Kim Y.J., Lane W.S., Stillman D.J., and Kornberg R.D. 1995. Yeast global transcriptional regulators Sin4 and Rgr1 are components of mediator complex/RNA polymerase II holoenzyme. *Proc. Natl. Acad. Sci.* **92:** 10864.

Liao S.M., Zhang J., Jeffery D.A., Koleske A.J., Thompson C.M., Chao D.M., Viljoen M., van Vuuren H.J., and Young R.A. 1995. A kinase-cyclin pair in the RNA polymerase II holoenzyme. *Nature* **374:** 193.

Mann R.K. and Grunstein M. 1992. Histone H3 N-terminal mutations allow hyperactivation of the yeast GAL1 gene in vivo. *EMBO J.* **11:** 3297.

Márquez J.A., Pascual-Ahuir A., Proft M., and Serrano R. 1998. The Ssn6-Tup1 repressor complex of *Saccharomyces cerevisiae* is involved in the osmotic induction of HOG-dependent and -independent genes. *EMBO J.* **17:** 2543.

Miller A.M., MacKay V.L., and Nasmyth K.A. 1985. Identification and comparison of two sequence elements that confer cell-type specific transcription in yeast. *Nature* **314:** 598.

Miller J.H. 1972. *Experiments in molecular genetics*. Cold Spring Harbor Laboratory, Cold Spring Harbor, New York.

Mukai Y., Harashima S., and Oshima Y. 1991. AAR1/TUP1 protein, with a structure similar to that of the beta subunit of G proteins, is required for **a**1-α2 and α2 repression in cell type control of *Saccharomyces cerevisiae*. *Mol. Cell. Biol.* **11:** 3773.

Myers L.C., Gustafsson C.M., Bushnell D.A., Lui M., Erdjument-Bromage H., Tempst P., and Kornberg R.D. 1998. The Med proteins of yeast and their function through the RNA polymerase II carboxy-terminal domain. *Genes Dev.* **12:** 45.

Nonet M.L. and Young R.A. 1989. Intragenic and extragenic suppressors of mutations in the heptapeptide repeat domain of *Saccharomyces cerevisiae* RNA polymerase II. *Genetics* **123:** 715.

Oliver S.G., van der Aart Q.J., Agostoni-Carbone M.L., Aigle M., Alberghina L., Alexandraki D., Antoine G., Anwar R., Ballesta J.P., Benit P., et al. 1992. The complete DNA sequence of yeast chromosome III. *Nature* **357:** 38.

Park E. C. and Szostak J.W. 1990. Point mutations in the yeast histone H4 gene prevent silencing of the silent mating type locus HML. *Mol. Cell. Biol.* **10:** 4932.

Piruat J.I., Chavez S., and Aguilera A. 1997. The yeast HRS1 gene is involved in positive and negative regulation of transcription and shows genetic characteristics similar to SIN4 and GAL11. *Genetics* **147:** 1585.

Redd M.J., Arnaud M.B., and Johnson A.D. 1997. A complex composed of Tup1 and Ssn6 represses transcription in vitro. *J. Biol. Chem.* **272:** 11193.

Redd M.J., Stark M.R., and Johnson A.D. 1996. Accessibility of α2-repressed promoters to the activator Gal4. *Mol. Cell. Biol.* **16:** 2865.

Rose M.D., Novick P., Thomas J.H., Botstein D., and Fink G.R. 1987. A *Saccharomyces cerevisiae* genomic plasmid bank based on a centromere-containing shuttle vector. *Gene* **60:** 237.

Roth S.Y., Dean A., and Simpson R.T. 1990. Yeast α2 repressor positions nucleosomes in TRP1/ARS1 chromatin. *Mol. Cell. Biol.* **10:** 2247.

Roth S.Y., Shimizu M., Johnson L., Grunstein M., and Simpson R.T. 1992. Stable nucleosome positioning and complete repression by the yeast α 2 repressor are disrupted by amino-terminal mutations in histone H4. *Genes Dev.* **6:** 411.

Sakai A., Shimizu Y., and Hishinuma F. 1988. Isolation and characterization of mutants which show an oversecretion phenotype in *Saccharomyces cerevisiae. Genetics* **119:** 499.

Sakai A., Shimizu Y., Kondou S., Chibazakura T., and Hishinuma F. 1990. Structure and molecular analysis of RGR1, a gene required for glucose repression of *Saccharomyces cerevisiae. Mol. Cell. Biol.* **10:** 4130.

Santos-Rosa H., Clever B., Heyer W.D., and Aguilera A. 1996. The yeast HRS1 gene encodes a polyglutamine-rich nuclear protein required for spontaneous and hpr1-induced deletions between direct repeats. *Genetics* **142:** 705.

Schena M., Freedman L.P., and Yamamoto K.R. 1989. Mutations in the glucocorticoid receptor zinc finger region that distinguish interdigitated DNA binding and transcriptional enhancement activities. *Genes Dev.* **3:** 1590.

Schultz J. and Carlson M. 1987. Molecular analysis of SSN6, a gene functionally related to the SNF1 protein kinase of *Saccharomyces cerevisiae. Mol. Cell. Biol.* **7:** 3637.

Shimizu M., Roth S.Y., Szent-Gyorgyi C., and Simpson R.T. 1991. Nucleosomes are positioned with base pair precision adjacent to the α2 operator in *Saccharomyces cerevisiae. EMBO J.* **10:** 3033.

Sikorski R.S. and Hieter P. 1989. A system of shuttle vectors and yeast host strains designed for efficient manipulation of DNA in *Saccharomyces cerevisiae. Genetics* **122:** 19.

Siliciano P.G. and Tatchell K. 1984. Transcription and regulatory signals at the mating type locus in yeast. *Cell* **37:** 969.

———. 1986. Identification of the DNA sequences controlling the expression of the MATα locus of yeast. *Proc. Natl. Acad. Sci.* **83:** 2320.

Smith R.L., Redd M.J., and Johnson A.D. 1995. The TPRs of SSN6 interact with the homeodomain of α2. *Genes Dev.* **9:** 2903.

Song W., Treich I., Qian N., Kuchin S., and Carlson M. 1996. SSN genes that affect transcriptional repression in *Saccharomyces cerevisiae* encode SIN4, ROX3, and SRB proteins associated with RNA polymerase II. *Mol. Cell. Biol.* **16:** 115.

Tan S. and Richmond T.J. 1998. Crystal structure of the yeast MATα2/MCM1/DNA ternary complex. *Nature* **391:** 660.

Thompson C.M., Koleske A.J., Chao D.M., and Young R.A. 1993. A multisubunit complex associated with the RNA polymerase II CTD and TATA-binding protein in yeast. *Cell* **73:** 1361.

Treitel M.A. and Carlson M. 1995. Repression by SSN6-TUP1 is directed by MIG1, a repressor/activator protein. *Proc. Natl. Acad. Sci.* **92:** 3132.

Trumbly R.J. 1992. Glucose repression in the yeast *Saccharomyces cerevisiae. Mol. Microbiol.* **6:** 15.

Turner B.M. 1991. Histone acetylation and control of gene expression. *J. Cell Sci.* **99:** 13.

Tzamarias D. and Struhl K. 1994. Functional dissection of the yeast Cyc8-Tup1 transcriptional co-repressor complex. *Nature* **369:** 758.

———. 1995. Distinct TPR motifs of Cyc8 are involved in recruiting the Cyc8-Tup1 corepressor complex to differentially regulated promoters. *Genes Dev.* **9:** 821.

Varanasi U.S., Klis M., Mikesell P.B., and Trumbly R.J. 1996. The Cyc8 (Ssn6)-Tup1 corepressor complex is composed of one Cyc8 and four Tup1 subunits. *Mol. Cell. Biol.* **16:** 6707.

Wahi M. and Johnson A.D. 1995. Identification of genes required for α2 repression in *Saccharomyces cerevisiae. Genetics* **140:** 79.

Williams F.E. and Trumbly R.J. 1990. Characterization of TUP1, a mediator of glucose repression in *Saccharomyces cerevisiae. Mol. Cell. Biol.* **10:** 6500.

Yanisch-Perron C., Vieira J., and Messing J. 1985. Improved M13 phage cloning vectors and host strains: Nucleotide sequences of the M13mp18 and pUC19 vectors. *Gene* **33:** 103.

Yuryev A. and Corden J.L. 1996. Suppression analysis reveals a functional difference between the serines in positions two and five in the consensus sequence of the C-terminal domain of yeast RNA polymerase II. *Genetics* **143:** 661.

Zitomer R.S. and Lowry C.V. 1992. Regulation of gene expression by oxygen in *Saccharomyces cerevisiae. Microbiol. Rev.* **56:** 1.

Zitomer R.S., Carrico P., and Deckert J. 1997. Regulation of hypoxic gene expression in yeast. *Kidney Int.* **51:** 507.

In Vivo Functions of Histone Acetylation/Deacetylation in Tup1p Repression and Gcn5p Activation

D.G. Edmondson,* W. Zhang,* A. Watson,* W. Xu,* J.R. Bone,* Y. Yu,†
D. Stillman,† and S.Y. Roth*

*Department of Biochemistry and Molecular Biology, University of Texas M.D. Anderson Cancer Center,
Houston, Texas 77030; †Division of Molecular Biology and Genetics, Department of Oncological Sciences,
University of Utah Health Science Center, Salt Lake City, Utah 84132

Histone acetylation is a dynamic process that affects chromatin structure and transcriptional regulation at multiple levels. Acetylation occurs exclusively on lysine residues in the amino-terminal "tails" of the core histones (Turner 1991; Wade et al. 1997). Since these tail domains are external to the core particle, they are in a unique position to affect DNA-histone interactions, nucleosome-nucleosome interactions, and interactions between nonhistone regulatory factors and the histones (Hansen 1997; Luger et al. 1997). Acetylation could easily influence any of these interactions by changing the charge and the structure of the histone tails.

The identification of Gcn5p and its *Tetrahymena* homolog, p55, as the catalytic subunits of nuclear histone acetyltransferase type-A (HAT A) activities provided a molecular basis for the long-standing correlation between histone acetylation and transcriptional activation (Brownell et al. 1996). Numerous other transcriptional proteins, including $TAF_{II}250$ (Mizzen et al. 1996), P/CAF (X.-J. Yang et al. 1996), p300/CBP (Bannister and Kouzarides 1996; Ogryzko et al. 1996), SRF (Spencer et al. 1997), and ACTR (Chen et al. 1997) among others, have been subsequently determined to possess HAT activity. Moreover, several histone deacetylase activities have been identified and linked to specific transcriptional repressors (Taunton et al. 1996; Rundlett et al. 1996; W.-M. Yang et al. 1996; Alland et al. 1997; Nagy et al. 1997; Brehm et al. 1998; Luo et al. 1998; Magnaghi-Jaulin et al. 1998). These observations suggest a simple paradigm wherein HATs are needed to activate transcription, and deacetylases are needed for repression. Many questions remain, however. Among these are whether histones are really the pertinent substrate of the HATs for transcriptional activation in vivo. If so, how are particular chromatin domains enriched in acetylated or deacetylated histones established or maintained, given that the half-life of an acetyl moiety on the histones can be on the order of seconds (Davie 1997)?

In this paper, we review data from our lab that support an important role for histone acetylation/deacetylation in transcriptional regulation in vivo and which suggest that particular regulatory proteins may have an active role in maintaining the acetylation state of a given chromatin domain. We begin with our studies of the Tup1p repressor in yeast, which directly interacts with histones to create a repressive chromatin environment. We then review ge-

netic experiments which suggest that specific patterns of histone acetylation may act as circuit boards for coordination of cellular processes such as transcription, cell cycle progression, and chromatin compaction.

METHODS

Far-Western dot blots. Serial dilutions of synthetic peptides were dot-blotted on PVDF or nitrocellulose membranes. The blots were then stained with India ink to verify that equivalent amounts of peptides had been loaded or processed for far-Western as described previously (Edmondson et al. 1996). Briefly, blots were blocked for 2 hours in 0.05% Tween-20 in phosphate-buffered saline (PBS), followed by 1% bovine serum albumin (BSA) in PBS for 2 hours. The blots were briefly rinsed with PBS and then incubated in 1% goat serum, 0.3% BSA in PBS containing the Tup1p probe. The probe was generated by in vitro translation of Tup1p in the presence of $[^{35}S]$methionine. After a 2-hour incubation, the blots were washed four times for 5 minutes each in PBS, allowed to dry, and then exposed to X-ray film or phosphoimager screens. Quantitation was performed with ImageQuant™ software (Molecular Dynamics).

Immunoprecipitation and Western analysis. Yeast were grown to a density of approximately 1×10^7 to 2×10^7, pelleted by centrifugation, and resuspended in extraction buffer containing 25 mM Tris (pH 7.5), 5 mM $MgCl_2$, 15 mM EGTA, 10% glycerol, 0.3% Triton X-100, 2% NaN_3, 250 mM NaCl, and 1 mM DTT. A half-volume of glass beads were added to the resuspended cells in Eppendorf tubes, and the yeast were vortexed at high speed for 45 minutes at 4°C. The resulting extract was centrifuged for 5 minutes and then 15 minutes at 4°C to remove insoluble debris. Antibody or preimmune sera were added to the resulting extract, and the mixture was rotated at 4°C overnight. The immune complexes were captured by incubation with protein A-Sepharose (Pierce) for 1 hour, followed by centrifugation at $1300g$. The immune complexes were washed three times for 5 minutes each with extraction buffer, and the proteins were eluted with SDS-PAGE loading buffer.

Samples were separated by SDS-PAGE and blotted to PVDF membrane. The membranes were blocked for 1 hour in 1% milk dissolved in Tris-buffered saline containing 0.05% Tween-20 (TBST), and then incubated overnight at 4°C with a polyclonal antibody to Tup1p (amino acids 253–653), diluted to 1:5000 in blocking solution. Following extensive washing in TBST, a 1-hour incubation in anti-rabbit-HRP (diluted 1:20,000 in TBST) was performed. The blots were washed again and developed using Super Signal™ (Pierce) according to manufacturers protocols. Quantitation was performed using AlphaEase™ software (Alpha Innotech).

Yeast strains and β-galactosidase assays. *Saccharomyces cerevisiae* strains used in experiments reported here are listed in Table 1 and were propagated according to standard procedures (Rose et al. 1990). β-galactosidase activities of reporter genes were quantitated by standard assays following preparation of whole-cell extracts by a glass bead procedure (Rose et al. 1990).

RESULTS

Organization of Chromatin by Tup1p

Tup1p is required for repression of several different gene families in yeast, including cell-type-specific genes, glucose-repressible genes, genes responsive to oxygen, DNA-damage-inducible genes, and genes sensitive to osmotic changes. In vivo, Tup1p interacts with Ssn6p (Varanasi et al. 1996; Redd et al. 1997), and this corepressor complex is recruited to promoters by specific DNA-binding proteins such as the α2 repressor (Komachi et al. 1994). Ssn6p contains ten TPR motifs (Schultz et al. 1990) that apparently facilitate interactions with multiple DNA-binding factors (Smith et al. 1995; Tzarmarias and Struhl 1995). Tup1p contains six WD40 motifs, also thought to provide interaction surfaces for a variety of factors. Both Ssn6p and Tup1p can confer repression independently of the DNA-binding proteins if they are fused to a heterologous DNA-binding domain (such as the LexA DNA-binding domain), and provided an appropriate reporter gene (Keleher et al. 1992; Tzarmarias and Struhl 1994). Interestingly, *SSN6*-LexA fusions require *TUP1* for repression (Keleher et al. 1992), but *TUP1*-LexA fusions do not require *SSN6* (Tzarmarias and Struhl 1994), leading to the idea that Tup1p provides the repressor activity to endogenous Ssn6p-Tup1p complexes.

How Tup1p mediates repression is unclear, but at least two modes of action have been suggested, and both may be operational. Tup1p may directly inhibit some component of the basal transcription machinery, as suggested by the finding that Srb10p and Srb11p which form a cyclin-dependent kinase complex associated with the pol II mediator are required for full repression by Tup1p (Kuchin et al. 1995; Wahi and Johnson 1995; Carlson 1997). Mutations in either *SRB10* (*SSN3, UME5, ARE1*) or *SRB11* (*SSN8*) cause a five- to sevenfold derepression of Tup1p-regulated genes such as *SUC2* (Kuchin et al. 1995) or an **a**-cell-specific reporter (*MFA2::lacZ*) (Wahi and Johnson 1995). However, it is unclear whether *SRB10* and *SRB11* are required directly or indirectly for repression. Tup1p can establish a modest degree of repression in vitro (Herschbach et al. 1994; Redd et al. 1997). This repression occurs in the absence of any activator protein, again supporting the notion that Ssn6p and Tup1p may act directly on the basal transcription machinery.

Tup1p might also inhibit transcription through effects on chromatin. Most promoters that are repressed by Tup1p exhibit a high degree of nucleosomal organization, and repression of all Tup1p-regulated promoters tested to date is sensitive to combined mutations in histones H3 and H4. Tup1p interacts directly with these histones in vitro, and the region of Tup1p required for interaction with H3 and H4 is the same as that determined by others (Tzamarias and Struhl 1994) to be capable of independently conferring repression when fused to the LexA DNA-binding domain. Together, these results suggest that Tup1p-histone interactions may trigger organization of a chromatin domain that is refractory to transcription.

The Amino-terminal Tails of H3 and H4 Are Necessary and Sufficient for Tup1p Binding In Vitro

Our previous experiments indicated that the amino-terminal tails of H3 and H4 are required for interaction with Tup1p. Deletion mutations that remove amino acids 1–28 (del 1–28) of H3 or 4–19 (del 4–19) of H4 abolish Tup1p binding in a far-Western assay (Edmondson et al. 1996). To determine whether the amino-terminal domains of these histones are sufficient for binding, we examined the ability of Tup1p to interact with synthetic peptides corresponding to the first 20 amino acids of H3, H4, or H2B (Fig. 1). Peptides were spotted onto PVDF membranes and probed with [^{35}S]methionine-labeled in-vitro-translated Tup1p. The Tup1p probe bound both the H3 and H4 peptides readily, but did not bind to the amino-terminal

Table 1. Yeast Strains

Strain	Genotype[a]
DY4549	*MATa, ade2, can1, his3, leu2, lys2, trp1, ura3, rpd7:TRP1*
DY4551	*MATa, ade2, can1, his3, leu2, lys2, trp1, ura3, rpd16:HIS3*
DY4554	*MATa, ade2, can1, his3, leu2, lys2, trp1, ura3, rpd3:LEU2, rpd7:TRP1*
DY4556	*MATa, ade2, can1, his3, leu2, lys2, trp1, ura3, rpd3:LEU2, rpd16:HIS3*
DY4560	*MATa, ade2, can1, his3, leu2, lys2, trp1, ura3, rpd16:HIS3, rpd7:TRP1*
DY4563	*MATa, ade2, can1, his3, leu2, lys2, trp1, ura3, rpd3:LEU2, rpd16:HIS3, rpd7:TRP1*
FY250	*MAT α, ura3-52, his3-Δ200, leu2-Δ1, trp1-Δ63*
MCY3651	*MATα, ura3-52, his3-Δ200, leu2-Δ1, trp1-Δ63, ssn3Δ1::HIS3*
MCY3655	*MATα, ura3-52, his3-Δ200, leu2-Δ1, trp1-Δ63, ssn8Δ1::LEU2*

[a]All DY strains are isogenic and based on a W303 background. FY250, MCY3651, and MCY3655 are isogenic and are in an S288C background.

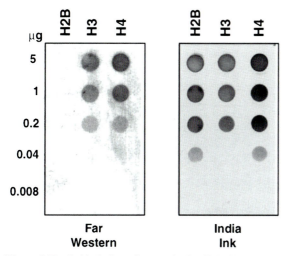

Figure 1. Tup1p binds the amino-terminal "tails" of histones H3 and H4. Synthetic peptides corresponding to the first 20 amino acids of histone H3, H4, and H2B were blotted onto PVDF or nitrocellulose membranes. The blot in the left panel was processed for far-Western analysis, and the blot in the right panel was stained with India ink to assess whether peptides were loaded equivalently.

domain of H2B (Fig. 1), consistent with our previous results indicating that Tup1p does not interact with either H2B or H2A. These data indicate that the first 20 amino acids of H3 or H4 are sufficient as well as necessary for Tup1p binding.

Amino-terminal Mutations in H3 and H4 Synergistically Compromise Repression In Vitro

To test the importance of the in vitro interactions observed above to transcriptional repression in vivo, we tested the effects of several mutations in the H3 and/or H4 amino termini on repression of reporter genes in vivo.

Deletion of either the H3 or the H4 tail alone causes only a modest (three- to fivefold) reduction in Tup1p-mediated repression of various reporter genes in vivo (Roth et al. 1992; Edmondson et al. 1996; Huang et al. 1997). These results raise the possibility that the amino-terminal domains of H3 and H4 serve redundant functions for Tup1p binding and repression. If so, combinations of mutations in these two histone tails might synergistically relieve repression. Ideally, we would like to test the effects of a combination of the H3 and H4 amino-terminal deletion mutations described above, since we would predict that Tup1p binding to both histones would be abolished in the absence of both tails, leading to a dramatic decrease in repression. Unfortunately, this combination of mutations is lethal (Morgan et al. 1991), so we cannot test this hypothesis. However, combination of the del (1–28) mutation in H3 with a less severe mutation in H4 (K12QK16Q; a double point mutation substituting glutamines for lysines at positions 12 and 16) is not lethal. This H4 mutation reduces Tup1p binding approximately 50% in vitro and at most, compromises repression three-fold on its own (Fig. 2) (Edmondson et al. 1996; Huang et

al. 1997) . However, in combination with the H3 amino-terminal deletion, a synergistic loss of repression (from 9- to 35-fold) is observed for three different reporter genes, representing three separate classes of genes (**a**-cell-specific, DNA-damage-inducible, and haploid-specific) regulated by Tup1p (Fig. 2) (Edmondson et al. 1996; Huang et al. 1997). These data support an important but redundant role for the H3 and H4 amino-terminal domains in Tup1p-mediated repression in vivo.

Histone Hyperacetylation Interferes with Tup1p Binding to Histones In Vitro

Both the H3 and the H4 amino-terminal tails contain multiple lysines subject to acetylation. This modification might directly affect Tup1p binding to these histones. Indeed, our previous experiments indicated that Tup1p bound only weakly to highly acetylated forms of H3 and H4 in a far-Western assay (Edmondson et al. 1996). To further examine the effects of acetylation on Tup1p binding,

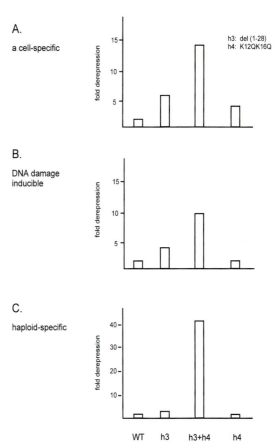

Figure 2. Mutations in the H3 and H4 amino-terminal tails synergistically compromise Tup1p-mediated repression. Three different β-galactosidase reporter genes subject to repression by Tup1p were assayed in the presence of single and double mutations in the H3 and H4 amino termini, as indicated. In all three cases, a synergistic loss of repression was observed in the presence of double mutations in the two histone tails. (*A*) α2-op-CYC1-LacZ reporter; (*B*) RNR2-LacZ reporter; (*C*) **a**1/α2-CYC1-LacZ reporter. (Data are summarized from Edmondson et al. [1996] and Huang et al. [1997].)

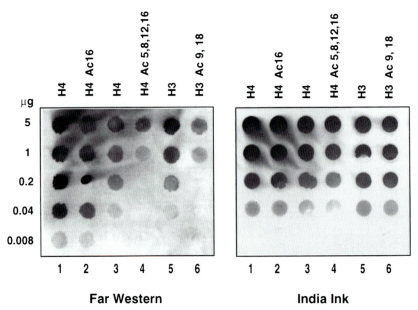

Figure 3. Acetylation of histone peptides decreases Tup1p binding. Synthetic peptides corresponding to amino acids 11–21 of *Drosophila* H4 (lanes *1* and *2*) or the first 20 amino acids of yeast H3 and H4 (lanes *3–6*) were dot-blotted onto PVDF membranes. Some peptides were synthesized with acetyl-lysine groups at the indicated positions. The blot in the left panel was processed for far-Western analysis, and the blot in the right panel was stained with India ink to assess whether peptides were loaded equivalently.

we compared the ability of Tup1p to bind to synthetic peptides bearing acetyl moieties at defined positions (Fig. 3).

Tup1p bound a peptide corresponding to amino acids 11–21 in H4 quite well, and this binding is not changed by the introduction of acetyl-lysine at position 16 (Fig. 3, lanes 1 and 2). A longer peptide corresponding to amino acids 1–20 in H4 also bound Tup1p quite well, as expected (Fig. 3, lane 3). However, the presence of acetyl-lysine at positions 5, 8, 12, and 16 decreased binding at least 25-fold (Fig. 3, lane 4). Tup1p binding could be detected with as little as 0.04 μg of the unacetylated peptide, whereas 1 μg of the acetylated peptide was required to achieve a similar degree of binding.

Inhibition of Tup1p binding was also observed with synthetic peptides corresponding to a multiply acetylated form of H3 (Fig. 3, lanes 5 and 6). Again, Tup1p binding was observed with as little as 0.04 μg of an unacetylated peptide corresponding to the first 20 amino acids of H3 (Fig. 3, lane 5), but binding to an analogous peptide containing acetyl-lysine at positions 9 and 18 could not be detected at less than 1 μg (Fig. 3, lane 6). Together, these data indicate that hyperacetylation of H3 or H4 inhibits Tup1p binding, consistent with our previous observations (Edmondson et al. 1996).

Loss of Histone Deacetylase Activities Compromises Repression by Tup1p In Vivo

The above results predict that alteration of histone acetylation levels would alter the level of repression achieved by Tup1p. To test this hypothesis, we examined Tup1p-mediated repression in cells lacking specific histone deacetylase activities. Five histone deacetylase genes have been identified in yeast: *RPD3, HDA1, HOS1 (RPD16), HOS2 (RPD7)*, and *HOS3* (Rundlett et al.

1996). We found that deletion of any single deacetylase gene had little effect on the overall (bulk) level of histone acetylation in the cell (data not shown). Correspondingly, these mutations had little effect on Tup1p-mediated repression of an a-cell-specific reporter gene (Fig. 4). Deletion of pairs of deacetylase genes also had little effect on repression. Deletion of three genes (*HDA1, HOS1* [*RPD16*], and *HOS2* [*RPD7*]) simultaneously, however, resulted in a significant change in bulk histone acetylation (data not shown) and in a more than 30-fold loss in repression of the reporter gene (Fig. 4). Importantly, levels of Tup1p, Ssn6p, or of RNA encoding the α2 repressor were not altered by this combination of mutations. Although all components for repression are present in these cells, repression is apparently compromised (directly or indirectly) by the increased acetylation state of the histones.

Altered Binding of Tup1p to Histones Isolated from Cells Lacking *SRB10* or *SRB11*

H3 is subject to posttranslation phosphorylation at serine residues 10 and 28. Changes in H3 phosphorylation might well affect Tup1p binding, as does acetylation. In addition, Tup1p itself is a phosphoprotein (Redd et al. 1997), and changes in this modification might alter Tup1p-histone interactions. The kinases responsible for H3 or Tup1p phosphorylation in vivo are not well defined. Given the requirement of the Srb10/11 kinase complex for complete repression by Tup1p (Kuchin et al. 1995; Wahi and Johnson 1995; Carlson 1997), we examined whether H3 phosphorylation, Tup1p phosphorylation, or Tup1p-H3 interactions were altered in cells containing null alleles of *SRB10* or *SRB11*.

Tup1p runs as two closely spaced bands on SDS gels

A.

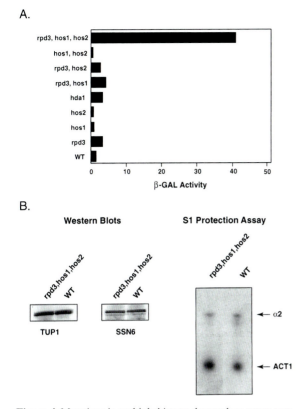

B.

Western Blots   **S1 Protection Assay**

TUP1   SSN6

← α2

← ACT1

Figure 4. Mutations in multiple histone deacetylase genes compromise Tup1p repression in vivo. (*A*) An **a**-cell-specific reporter gene (Edmondson et al. 1996) subject to Tup1p-mediated repression was assayed for activity in the presence of single, double, and triple mutations in histone deacetylase genes as indicated. (*B*) Levels of Tup1p and Ssn6p in wild-type cells or cells containing the triple deacetylase mutations were compared by Western blot. Since the α2 protein is very short lived and difficult to detect, levels of α2 RNA were compared in the wild-type and mutant cells by nuclease protection assay.

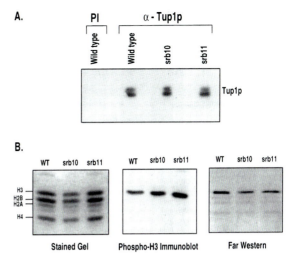

Figure 5. Effects of mutations in *SRB10* or *SRB11* on Tup1p and H3. (*A*) Phosphorylation of Tup1p is not changed in *srb10* or *srb11* mutant cells. Tup1p was immunoprecipitated from wild-type or mutant cell protein extracts. (PI) Preimmune serum control. (*B*) Phosphorylation of histone H3 is increased in *srb10* and *srb11* cells, whereas binding of Tup1p to histone H3 is decreased. (*Left panel*) Coomassie blue stain of bulk histones separated on a SDS-PAGE gel; (*center panel*) Western blot of an equivalent gel stained for phosphorylated histone H3 showing increased levels of phosphorylation in *srb10* and *srb11* cells; (*right panel*) far-Western blot indicating that the binding of Tup1p to histone H3 isolated from *srb10* or *srb11* cells is reduced.

(Fig. 5A), and previous experiments indicate that the upper band corresponds to a phosphorylated form of Tup1p (Redd et al. 1997; D. Edmondson, unpubl.). Whole-cell extracts were prepared from isogenic wild-type, *srb10*, or *srb11* cells, immunoprecipitated with a Tup1p-specific antisera, and examined by Western blot. Although slight variations were observed from experiment to experiment, the ratio of the two Tup1p bands was not significantly altered by the loss of either *SRB10* or *SRB11* (Fig. 5A). The kinase complex is not required, then, for Tup1p phosphorylation in bulk, although it might have localized affects on Tup1p molecules associated with particular genes.

To examine whether the kinase complex affects H3 phosphorylation, histones were isolated from the above cells, and the phosphorylation state of H3 was probed using an antibody specific for phospho-H3. No decrease in H3 phosphorylation was observed in the histones from the *srb10* or *srb11* cells (Fig. 5B). Interestingly, a reproducible twofold increase in H3 phosphorylation was observed in the presence of these mutations. *SRB10* and *SRB11*, then, are not required for bulk H3 phosphorylation but may regulate levels of this histone modification.

The above histones were next examined by far-Western assay for their ability to bind to Tup1p. Tup1p binding was reproducibly decreased by 50% to histones isolated from *srb10* or *srb11* cells (Fig. 5B). Although we do not yet know the basis for this decreased binding, these alterations in H3 interactions might well contribute to the five- to sevenfold loss of repression by Tup1p observed in *srb10* or *srb11* cells.

Model for Tup1-mediated Repression

Together, the above results suggest a model wherein Ssn6p/Tup1p complexes are recruited to promoters by specific DNA-binding factors and once in place, interact with histones H3 and H4 to organize a repressive chromatin environment (Fig. 6). Since Tup1p-histone interactions are sensitive to the acetylation state of the histones, this modification might provide a major point of regulation. When Tup1p-regulated genes are active, they are predicted to be associated with more highly acetylated histones, as are most other active genes. This acetylation likely contributes to an "open" chromatin structure conducive to transcription. Upon repression, histone acetylation is predicted to be decreased, facilitating interactions with Tup1p and the folding of chromatin into higher-order structures. Tup1p might initiate and/or stabilize this folded state, thereby inhibiting interactions of promoter sequences with general transcription factors. The association of the histone tails with Tup1p may also shield the histones from further acetylation to stabilize the repressed state. Alternatively, Tup1p might not prevent

EDMONDSON ET AL.

464

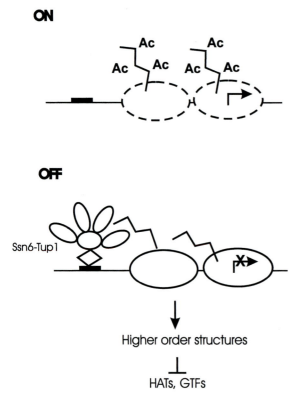

ON

OFF

Ssn6-Tup1

Higher order structures

⊥

HATs, GTFs

Figure 6. Model for Tup1p-mediated repression. In the active state, genes repressed by Tup1p are predicted to be associated with highly acetylated histones. Upon repression, histone acetylation is reduced, facilitating interactions with Tup1p and formation of higher-order chromatin structures that limit access to general transcription factors (GTFs) and, potentially, histone acetyltransferases (HATs).

acetylation of the tails, but rather, this modification may be used to reverse repression through release of Tup1p from the histones.

Several major questions remain: How is the modification state of the histones "set" for interaction with Tup1p? Does Tup1p also interact with and directly inhibit components of the basal transcription machinery? One intriguing possibility is that Tup1p interacts with Srb10p/Srb11p (or other components of the pol II mediator) and that these interactions not only interfere with the activity or recruitment of the polymerase, but also lead to a change in the modification state of the histones. In this way, Tup1p might initiate repression through interactions with the transcription machinery and then maintain the repressed state through the organization of chromatin.

Histone Acetylation, Cell Growth, and Gene Activation

The yin/yang between histone acetylation and association of repressors such as Tup1p with chromatin illustrates an important way the modulation of chromatin structure can influence gene expression. Although the identification of multiple HAT proteins in the last 2 years has suggested much about how histone acetylation might

be accomplished and regulated, the functions of these enzymes in vivo have not been established clearly. For example, even though recombinant Gcn5p (rGcn5p) exhibits robust HAT activity in vitro (Brownell et al. 1996) with a discrete substrate specificity (Kuo et al. 1996), the specificity of Gcn5p-containing complexes in vivo is not defined. rGcn5p preferentially acetylates H3 at a single site, lysine (K) 14. Two lysines (K8 and K16) in H4 also serve as substrate if this histone is provided in isolation to the enzyme (Kuo et al. 1996). Interestingly, the recombinant enzyme cannot recognize histones incorporated into nucleosomes, the predicted physiological substrate. This finding indicates that other subunits might be required for nucleosomal acetylation by Gcn5p. It also raises the possibility that additional substrates (histone and nonhistone) not recognized by the recombinant enzyme might be utilized in vivo. Consistent with this idea, multiple complexes containing Gcn5p have been isolated from yeast extracts, and these complexes acetylate nucleosomal histones, with a preference for H3 and H2B (Grant et al. 1997). Still, it is not clear whether histones are the pertinent substrates utilized by Gcn5p for transcriptional activation in vivo or whether other nonhistone substrates might contribute to this function.

To approach these questions, we examined the acetylation states of histones H3 and H4 isolated from cells containing (*GCN5* cells) or lacking (*gcn5* cells) *GCN5* to determine which acetylation sites in these histones are affected in vivo by loss of the enzyme (Zhang et al. 1998). Western blots probed with antibodies specific for acetylation at individual sites in H4 (Turner and Fellows 1989) or pairs of sites in H3 revealed that acetylation of multiple sites in both histones is decreased in the absence of *GCN5* (summarized in Fig. 7A) (Zhang et al. 1998). Specifically, we found that acetylation at all four sites (K5, K8, K12, and K16) in H4 and at least three sites in H3 (K9, K14, and K18) were reduced by 50–90% in *gcn5* cells. These findings raise two interesting possibilities. The first is that Gcn5p may directly acetylate these sites in vivo. This again would imply that association of Gcn5p with other proteins may alter its specificity. An intriguing extension of this idea is that different subunits may associate with Gcn5p at different times or at different promoters in vivo to create unique patterns of acetylation. The second possibility is that Gcn5p affects the acetylation of these sites indirectly in vivo, perhaps by influencing the expression or activity or other HATs or histone deacetylases. This leads to the attractive idea that multiple histone-modifying activities might be coordinately regulated.

To further probe the importance of acetylatable lysines to Gcn5p functions, we mutated a series of lysines in H3 and H4 and examined the effects of these mutations on cell growth and transcriptional activation in cells containing or lacking *GCN5* (Zhang et al. 1998). These genetic studies indicate that acetylation of certain lysines may provide redundant functions. However, specific patterns of histone acetylation are important for transcriptional activation and cell growth. Mutation of K14 in H3, for example, has minimal affects on cell growth in the

A.

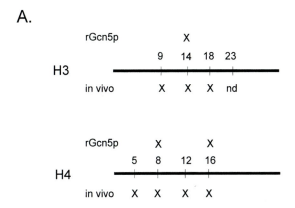

B.

H3	H4	β-GAL	
		GCN5	gcn5
WT	WT	663	71
K14Q	WT	1270	126
K14R	WT	514	217
K9Q	WT	1205	457
K9R	WT	1463	395
WT	K8,16Q	2175	262
WT	K8,16R	2010	292
WT	K5,12Q	1719	188
Wt	K5,12R	1582	312
K14Q	K8,16Q	1259	857
K14R	K8,16R	1041	lethal
K9Q	K5,12Q	1082	187
K9R	K5,12R	1394	282

Figure 7. In vivo functions of *GCN5*. (*A*) In vitro, recombinant Gcn5p (rGcn5p) acetylates K14 in H3 and K8, K16 in H4. In vivo, however, loss of *GCN5* reduces acetylation at K9, K14, and K18 in H3 and at all four acetylatable lysines in H4. (*B*) Mutations in various lysines in the H3 and H4 amino termini were tested for their effects on cell growth and activation of a reporter gene by GAL4-VP16, which requires Gcn5p. Note the high level of reporter activity in *gcn5* cells containing the K14Q mutation in H3 together with the K8,16Q mutations in H4. Note also that conversion of these same sites to R is lethal in *gcn5* cells. No other combination of mutations in H3 and H4 exhibited these effects, indicating unique patterns of acetylation are needed for Gcn5p-mediated transcriptional activation and normal cell growth. (All data are summarized from Zhang et al. [1998].)

presence of *GCN5*, but in its absence, K14 H3 mutations (to arginine, glutamine, or glycine) result in very slow cell-doubling times (more than 250 min; Zhang et al. 1998). These results indicate that acetylation of some other substrate by Gcn5p can suppress the effects of the K14 mutations, but loss of K14 combined with loss of this redundant acetylation event (upon loss of Gcn5p) results in severe growth defects. Not all acetylation events are redundant, though, as illustrated by our finding that a combination of K14Q mutation in H3 and K8,16Q mutations

in H4 largely bypasses the requirement of Gcn5p for activation of a Gal4-VP16-sensitive reporter gene, whereas other combinations of H3 and H4 mutations, such as K9Q in H3 together with K5,12Q in H4, do not (Fig. 7B) (Zhang et al. 1998). These results not only emphasize the independent functions of particular combinations of acetylation sites, they also establish the importance of K14 in H3 and K8,16 in H4 to transcriptional activation by Gcn5p.

The idea that specific combinations of acetylation events are important to cell growth is strikingly illustrated by our finding that a combination of the K14R mutation in H3 with the K8,16R mutations in H4 is lethal in *gcn5* cells (Fig. 7B) (Zhang et al. 1998). No other combination of mutations tested exhibited such synthetic lethality. We are currently investigating whether the lethal state of the above combination reflects an inability to properly fold chromatin or an inability to activate one or more essential genes (or both). Importantly, our genetic studies indicate that particular patterns of acetylation, not just the overall level of acetylation, are important modulators of histone functions.

Mammalian Gcn5p Homologs

Two human homologs of Gcn5p have been described: hsGCN5 and hsP/CAF (p300/CBP-associated factor) (X.-J. Yang et al. 1996; Wang et al. 1997). The human GCN5 cDNA was predicted to encode a protein similar in size to that of yeast Gcn5p. P/CAF, however, is a much larger protein. Although the carboxy-terminal half of this protein is homologous to both yeast and human GCN5, the amino-terminal domain of P/CAF is unique. We have cloned murine homologs of both Gcn5p and P/CAF, and surprisingly, we found that mouse GCN5 also contains a unique amino-terminal domain very homologous to that of P/CAF (Xu et al. 1998). Upon further characterization, we found that this longer form of GCN5 is present in both human and mouse cells and that the shorter form reported previously may arise from an alternative or incomplete splicing event. Interestingly, recombinant proteins corresponding to the longer form of GCN5 or to P/CAF can acetylate nucleosomal histones, whereas the shorter human or yeast proteins cannot. Moreover, murine GCN5 interacts with p300/CBP in a manner similar to that of P/CAF. Thus, the unique amino-terminal domains of these mammalian enzymes apparently provide additional functions that are important to the recognition of chromatin substrates as well as to the regulation of gene expression.

DISCUSSION

The long-standing correlation between increased histone acetylation and transcriptional activation was molecularly validated by the discovery that the Gcn5p transcriptional adaptor possessed HAT activity (Brownell et al. 1996). That finding, together with the development of acetylation site-specific antibodies (Turner and Fellows 1989; Lin et al. 1989) and the ability to mutate specific acetylation sites in yeast (Kayne et al. 1988; Megee et al. 1990), has provided a new realm of experimentation to probe the functional significance of histone acetylation. There is now little doubt that histone acetylation contributes to transcriptional regulation, but fundamental questions remain as to the importance of individual acetylation sites and to the structural consequences of changes in acetylation states in vivo.

Acetylation of lysines neutralizes the positive charge of the ε-amino group, and this change in charge is thought to alter interactions between the histones and the negatively charged phosphate backbone of the DNA (Turner 1991; Hansen 1997; Wade et al. 1997). In addition, acetylation may alter the structure of the histone tail, and thus its path between nucleosomes or between the nucleosome and the DNA. Thus, acetylation likely affects not only the structure of individual nucleosomes, but also nucleosome-nucleosome interactions and thus higher-order chromosome structures. Higher overall levels of acetylation are predicted to cause more drastic alterations in structure in vivo, as is observed in vitro (Hansen 1997). What is not clear is whether the overall level of acetylation is the important parameter or whether acetylation at different individual sites or different combinations of sites has specialized functions.

Our genetic studies in yeast indicate that only certain lysine mutations, and by inference, certain patterns of histone acetylation, are able to bypass Gcn5p functions in transcriptional activation. Different patterns of acetylation also elicit different effects on cell growth. These data support the idea that different combinations of acetylation events have different functions. These differences might reflect different structural consequences or different effects on the association of nonhistone proteins (such as Tup1p) with chromatin. It is striking that we were able to observe such differences upon examination of only a few different combinations of sites in H3 and H4. Given that there are 16,384 possible combinations of acetylation events among the four core histones, acetylation provides a wealth of regulatory potential. Indeed, differentially acetylated histone tails may serve as circuit boards to effect very specific changes in chromatin structure and/or protein associations needed for transcription of particular genes, as well as other processes such as replication, recombination, or chromatin condensation/decondensation.

Many of the HATs identified so far exhibit overlapping or redundant activities in vitro (Bannister and Kouzarides 1996; Kuo et al. 1996; Mizzen et al. 1996; Ogryzko et al. 1996; X.-J. Yang et al. 1996; Grant et al. 1997). Another outstanding question is whether each of these enzymes has a unique role in transcription or other processes in vivo. For example, our discovery that mammalian GCN5 possesses an extended amino-terminal domain homologous to that of P/CAF and that it interacts with CBP suggests these two proteins may provide redundant functions (Xu et al. 1998). Indeed, in every way tested so far, the specificities of the murine P/CAF and GCN5 proteins are identical in vitro. However, these proteins exhibit differential, and often inverse, levels of expression in primary mouse tissues, suggesting they might have different roles in dif-

ferent cells or at different times during development. In some ways, the similarities between GCN5 and P/CAF are reminiscent of those between CBP and p300. Interestingly, although CBP/p300 are interchangeable in many functional assays, mutations in these two genes are associated with distinct disease states (Petrij et al. 1995; Muraoka et al. 1996), suggesting that CBP and p300 proteins have separable functions in vivo. Future studies of GCN5- and P/CAF-deficient mice should determine whether these two proteins also exhibit specialized functions.

Histone acetylation is a dynamic process (Davie 1997), and histone deacetylation is just as important to the regulation of gene expression and other processes as is acetylation. Another fundamental question, then, concerns the maintenance of acetylated states. Perhaps acetyl groups are constantly added and removed in a given domain, so that there is no constant structure but rather an ongoing flux in acetylation states. This constant flux might be true for active genes or even potentially active genes, but it seems unlikely for genes that are stably repressed, and even less likely for genes that are constitutively silenced. Such genes appear to be stably associated with hypoacetylated histones. Recruitment of deacetylase activities by specific repressor proteins may establish this hypoacetylated state, but how is this state maintained? Are these histones somehow shielded from acetylation?

Perhaps our studies with Tup1p provide some clues. Our in vitro studies indicate that Tup1p preferentially interacts with underacetylated forms of H3 and H4, and our genetic findings indicate that these interactions are required for repression in vivo (Edmondson et al. 1996). Tup1p-histone interactions may induce or stabilize chromatin folding, and these structures may potentially shield the tails from further acetylation. Although no homologs to *TUP1* have yet been found in multicellular organisms, proteins in *Drosophila* and human cells with structural and functional features similar to those of Tup1p have been described previously (Stifani et al. 1992). The *Drosophila* repressor groucho, for example, is a WD repeat protein important to neural development. Both groucho and the related mammalian TLE (*t*ransducin-*l*ike *e*nhancer of split) proteins associate with chromatin in vivo and interact with the amino-terminal terminal tail of H3 (Palaparti et al. 1997). Tup1p may then serve as a paradigm for repressor activities such as these that function through stable organization of chromatin domains.

ACKNOWLEDGMENTS

We thank Karen Hensley for help with many graphics and Aurora Diaz for secretarial assistance. We thank C. David Allis (University of Virginia) for the gift of antiphosphorylated H3 antibodies and for some of the histone H3 and H4 peptides. We also thank Mitzi Kuroda (Baylor College of Medicine) for additional H4 peptides. We thank Marian Carlson for the gift of *ssn3* (*srb10*) and *ssn8* (*srb11*) cells. This work was supported by grants from the National Institutes of Health (GM-51189), the Robert A. Welch Foundation (G1371), and the USAM-RMC (BC960173) to S.Y.R. D.E. was supported by the

Theodore Law UCF Scientific Achievement fellowship. J.R.B. is supported by a fellowship from the American Cancer Society, and W.X. is supported by a Rosalie B. Hite fellowship.

REFERENCES

Alland L., Muhle R., Hou H., Jr., Potes J., Chin L., Schreiber-Agus N., and DePinho R.A. 1997. Role for N-CoR and histone deacetylase in Sin3-mediated transcriptional repression. *Nature* **387:** 49.

Bannister A.J. and Kouzarides T. 1996. The CBP co-activator is a histone acetyltransferase. *Nature* **384:** 641.

Brehm A., Miska E.A., McCance D.J., Reid J.L., Bannister A.J., and Kouzarides T. 1998. Retinoblastoma protein recruits histone deacetylase to repress transcription. *Nature* **391:** 597.

Brownell J.E., Zhou J., Ranalli T., Kobayashi R., Edmondson D.G., Roth S.Y., and Allis C.D. 1996. *Tetrahymena* histone acetyltransferase A: A homolog to yeast Gcn5p linking histone acetylation to gene activation. *Cell* **84:** 843.

Carlson M. 1997. Genetics of transcriptional regulation in yeast: Connections to the RNA polymerase II CTD. *Annu. Rev. Cell Biol.* **13:** 1.

Chen H., Lin R.J., Schiltz R.L., Chakravarti D., Nash A., Nagy L., Privalsky M.L., Nakataini Y., and Evans R.M. 1997. Nuclear receptor coactivator ACTR is a novel histone acetyltransferase and forms a multimeric activation complex with P/CAF and CBP/p300. *Cell* **90:** 569.

Davie J.R. 1997. Nuclear matrix, dynamic histone acetylation and transcriptionally active chromatin. *Mol. Biol. Rep.* **24:** 197.

Edmondson D.G., Smith M.M., and Roth S.Y. 1996. Repression domain of the yeast global repressor Tup1 interacts directly with histones H3 and H4. *Genes Dev.* **10:** 1247.

Grant P.A., Duggan L., Cote J., Roberts S., Brownell J.E., Candau R., Ohba R., Owen-Hughes T., Allis C.D., Winston F., Berger S.L., and Workman J.L. 1997. Yeast GCN5 functions in two multisubunit complexes to acetylate nucleosomal histones: Characterization of an ADA complex and the SAGA (SPT/ADA) complex. *Genes Dev.* **11:** 1640.

Hansen J.C. 1997. The core histone amino-termini: Combinatorial interaction domains that link chromatin structure with function. *ChemTracts Biochem. Mol. Biol.* **10:** 56.

Herschbach B.M., Arnaud M.B., and Johnson A.D. 1994. Transcriptional repression directed by the yeast alpha 2 protein in vitro. *Nature* **370:** 309.

Huang L., Zhang W., and Roth S.Y. 1997. Amino termini of histones H3 an dH4 are required for **a**1/α2 repression in yeast. *Mol. Cell. Biol.* **17:** 6555.

Kayne P.S., Kim U., Han M., Mullen J.R., Yoshizaki F., and Grunstein M. 1988. Extremely conserved histone H4 amino terminus is dispensable for growth but essential for repressing the silent mating loci in yeast. *Cell* **55:** 27.

Keleher C.A., Redd M.J., Schultz J., Carlson M., and Johnson A.D. 1992. Ssn6-Tup1 is a general repressor of transcription in yeast. *Cell* **68:** 709.

Komachi K., Redd M.J., and Johnson A.D. 1994. The WD repeats of Tup1 interact with the homeo domain protein α2. *Genes Dev.* **8:** 2857.

Kuchin S., Yeghiayan P., and Carlson M. 1995. Cyclin-dependent protein kinase and cyclin homologs *SSN3* and *SSN8* contribute to transcriptional control in yeast. *Proc. Natl. Acad. Sci.* **92:** 4006.

Kuo M.H., Brownell J.E., Sobel R.E., Ranalli T.A., Cook R.G., Edmondson D.G., Roth S.Y., and Allis C.D. 1996. Transcription-linked acetylation by Gcn5p of histones H3 and H4 at specific lysines. *Nature* **383:** 269.

Lin R., Leone J.W., Cook R.G., and Allis C.D. 1989. Antibodies specific to acetylated histones document the existence of deposition- and transcription-related histone acetylation in *Tetrahymena. J. Cell Biol.* **108:** 1577.

Luger K., Mader A.W., Richmond R.K., Sargent D.F., and Richmond T.J. 1997. Crystal structure of the nucleosome core particle at 2.8 Å resolution. *Nature* **389:** 251.

Luo R.X., Postigo A.A., and Dean D.C. 1998. Rb interacts with histone deacetylase to repress transcription. *Cell* **92:** 463.

Magnaghi-Jaulin L., Groisman R., Naguibneva I., Robin P., Lorain S., Le Villain J.P., Troalen F., Trouche D., and Harel-Bellan A. 1998. Retinoblastoma protein represses transcription by recruiting a histone deacetylase. *Nature* **391:** 601.

Megee P.C., Morgan B.A., Mittman B.A., and Smith M.M. 1990. Genetic analysis of histone H4: Essential role of lysines subject to reversible acetylation. *Science* **247:** 841.

Mizzen C.A., Yang X.J., Kokubo T., Brownell J.E., Bannister A.J., Owen-Hughes T., Workman J., Wang L., Berger S.L., Kouzarides T., Nakatani Y., and Allis C.D. 1996. The TAF(II)250 subunit of TFIID has histone acetyltransferase activity. *Cell* **87:** 1261.

Morgan B.A., Mittman B.A., and Smith M.M. 1991. The highly conserved N-terminal domains of histones H3 and H4 are required for normal cell cycle progression. *Mol. Cell. Biol.* **11:** 4111.

Muraoka M., Konishi M., Kikuchi-Yanoshita R., Tanaka K., Shitara N., Chong J., Iwama T., and Miyaki M. 1996. p300 gene alterations in colorectal and gastric carcinomas. *Oncogene* **12:** 1565.

Nagy L., Kao H.Y., Chakravarti D., Lin R.J., Hassig C.A., Ayer D.E., Schreiber S.L., and Evans R.M. 1997. Nuclear receptor repression mediated by a complex containing SMRT, mSin3A, and histone deacetylase. *Cell* **89:** 373.

Ogryzko V.V., Schlitz R.L., Russanova V., Howard B.H., and Nakatani Y. 1996. The transcriptional coactivators p300 and CBP are histone acetyltransferases. *Cell* **87:** 953.

Palaparti A., Baratz A., and Stifani S. 1997. The Groucho/transducin-like enhancer of split transcriptional repressors interact with the genetically defined amino terminal silencing domain of histone H3. *J. Biol. Chem.* **272:** 26604.

Petrij F., Giles R.H., Dauwerse H.G., Saris J.J., Hennekam R.C.M., Masuno M., Tommerup N., van Ommen G.-J.B., Goodman R.H., Peters D.J.M., and Breuning M.H. 1995. Rubinstein-Taybi syndrome caused by mutations in the transcriptional co-activator CBP. *Nature* **376:** 348.

Redd M.J., Arnaud M.B., and Johnson A.D. 1997. A complex composed of Tup1 and Ssn6 represses transcription in vitro. *J. Biol. Chem.* **272:** 11193.

Rose M.D., Winston F., and Hieter P. 1990. *Methods in yeast genetics: A laboratory manual.* Cold Spring Harbor Laboratory Press, Cold Spring Harbor, New York.

Roth S.Y., Shimizu M., Johnson L., Grunstein M., and Simpson R.T. 1992. Stable nucleosome positioning and complete repression by the yeast α 2 repressor are disrupted by amino-terminal mutations in histone H4. *Genes Dev.* **6:** 411.

Rundlett S.E., Carmen A.A., Kobayashi R., Bavykin S., Turner B.M., and Grunstein M. 1996. HDA1 and RPD3 are members of distinct yeast histone deacetylase complexes that regulate silencing and transcription. *Proc. Natl. Acad. Sci.* **93:** 14503.

Schultz J., Marshall-Carlson L., and Carlson M. 1990. The N-terminal TPR region is the functional domain of SSN6, a nuclear phosphoprotein of *Saccharomyces cerevisiae. Mol. Cell. Biol.* **10:** 4744.

Smith R.L., Redd M.J., and Johnson A.D. 1995. The tetratricopeptide repeats of Ssn6 interact with the homeo domain of α2. *Genes Dev.* **9:** 2903.

Spencer T.E., Jenster G., Burcin M.M., Allis C.D., Zhou J., Mizzen C.A., McKenna N.J., Onate S.A., Tsai M.J., and O'Malley B.W. 1997. Steroid receptor coactivator-1 is a histone acetyltransferase. *Nature* **389:** 194.

Stifani S., Blaumueller C.M., Redhead N.J., Hill R.E., and Artavani-Tsakonas S. 1992. Human homologs of a *Drosophila* Enhancer of split gene product define a novel family of nuclear proteins. *Nat. Genet.* **2:** 119.

Taunton J., Hassig C.A., and Schreiber S.L. 1996. A mammalian histone deacetylase related to the yeast transcriptional regulator Rpd3p. *Science* **272:** 408.

Turner B.M. 1991. Histone acetylation and control of gene expression. *J. Cell Sci.* **99:** 13.

Turner B.M. and Fellows G. 1989. Specific antibodies reveal ordered and cell-cycle-related use of histone-H4 acetylation sites in mammalian cells. *Eur. J. Biochem.* **179:** 131.

Tzamarias D. and Struhl K. 1994. Functional dissection of the yeast Cyc8-Tup1 corepressor complex. *Nature* **369:** 758.

———. 1995. Distinct TPR motifs of CYC8 are involved in recruiting the CYC8-Tup1 corepressor complex to differentially regulated promoters. *Gene Dev.* **9:** 821.

Varanasi U.S., Klis M., Mikesell P.B., and Trumbly R.J. 1996. The Cyc8 (Ssn6) Tup1 corepressor complex is composed of one Cyc8 and four Tup1 subunits. *Mol. Cell. Biol.* **16:** 6707.

Wade P.A., Pruss D., and Wolffe A.P. 1997. Histone acetylation: Chromatin in action. *Trends Biochem. Sci.* **22:** 128.

Wahi M. and Johnson A.D. 1995. Identification of genes required for α2 repression in *Saccharomyces cerevisiae*. *Genetics* **140:** 79.

Wang L., Mizzen C., Ying C., Candau R., Barlev N., Brownell J., Allis C.D., and Berger S.L. 1997. Histone acetyltransferase activity is conserved between yeast and human GCN5 and is required for complementation of growth and transcriptional activation. *Mol. Cell. Biol.* **17:** 519.

Xu W., Edmondson D.G., and Roth S.Y. 1998. Mammalian GCN5 and P/CAF acetyltransferases share homologous amino-terminal domains important for the recognition of nucleosomal substrates. *Mol. Cell. Biol.* **18:** 5659.

Yang W.-M., Inouye C., Zeng Y., Bearss D., and Seto E. 1996. Transcriptional repression by YY1 is mediated by interaction with a mammalian homolog of the yeast global regulator RPD3. *Proc. Natl. Acad. Sci.* **93:** 12845.

Yang X.-J., Ogryzko V.V., Nishikawa J.-I., Howard B.H., and Nakatani Y. 1996. A p300/CBP associated factor that competes with the adenoviral oncoprotein E1A. *Nature* **382:** 319.

Zhang W., Bone J.R., Edmondson D.G., Turner B.M., and Roth S.Y. 1998. Essential and redundant functions of histone acetylation revealed by mutation of target lysines and loss of the Gcn5p acetyltransferase. *EMBO J.* **17:** 3155.

Signaling to Chromatin through Histone Modifications: How Clear Is the Signal?

C. Mizzen,* M.-H. Kuo,* E. Smith,* J. Brownell,* J. Zhou,* R. Ohba,* Y. Wei,*
L. Monaco,† P. Sassone-Corsi,† and C.D. Allis*
†Institut de Génétique et de Biologie Moléculaire et Cellulaire, CNRS, INSERM, ULP, B.P. 163, France;
*Department of Biochemistry and Molecular Genetics, University of Virginia Health Sciences Center,
Charlottesville, Virginia 22902

In eukaryotes, DNA is assembled with histones to form nucleosomes, the basic subunit of chromatin structure. The wrapping of DNA around histone octamers to form nucleosomal filaments and the further folding of these filaments into higher-order structures are necessary to package eukaryotic genomes within nuclei. However, the intimate association of DNA with histones in nucleosomes and the dense packing of nucleosomal filaments in supranucleosomal structures can restrict access to DNA by proteins involved in DNA-templated processes. Transcription, replication, repair, recombination, and segregation are all influenced by the remarkable topological complexity of DNA in chromatin and require precise organization of DNA within chromatin fibers and modulation of higher-order and nucleosomal structure in order to occur. Although important aspects of chromatin structure such as the organization of nucleosomal filaments in heterochromatin and chromosomes remain obscure, a rapidly expanding body of knowledge provides clear mechanistic links between dynamic modulation of chromatin architecture and the regulation of transcription. Thus, as discussed below, advances in our understanding of histones and the regulation of their function through posttranslational modifications are likely to impinge on a wide variety of basic questions in cell biology and genetics.

OUR OVERALL WORKING HYPOTHESIS

The fundamental problem to which my laboratory has been committed, like many in the chromatin field, ultimately resolves into molecular analyses of protein:protein and protein:DNA interactions. All core histones possess a globular core domain, required for histone:histone interactions central to nucleosome formation, and highly positively charged, unstructured tail domains that protrude from the nucleosome. These "tails" are important for histone:DNA interactions and for interactions with other proteins. Decreasing the density of positive charge in these tails, either through the direct neutralization of charge by acetylation of lysines or, similarly, by the addition of negative charge to serine/threonine residues by phosphorylation, represents a mechanism to regulate these interactions reversibly.

The broad canvas upon which many of our experiments are best framed is described in an essay that Sharon Roth and I wrote several years ago (Roth and Allis 1992). In it,

we proposed an alternative viewpoint to the "textbook" idea that linker histone (H1-type) phosphorylation is causally related to chromosome condensation. Instead, we suggested that H1 phosphorylation actually promotes decondensation of higher-order chromatin fibers, allowing *trans*-acting nonhistone proteins to gain access to repressed chromatin templates. A version of this model, revised to include possible roles of acetylation and phosphorylation of core histones, is shown in Figure 1. The model proposes that histone acetylation and phosphorylation can act, singly or synergistically, to increase the accessibility and subsequent binding of *trans*-acting factors to target DNA sequences by weakening histone:DNA interactions and potentially also through localized destabilization of supranucleosomal structure. Moreover, certain factors may recognize specifically modified histone tails themselves. We stipulate that the final outcome of this process depends on the nature of the *trans*-acting factors involved. Processes as disparate as transcriptional activation, replication, and possibly chromatin condensation (see below) may occur in response to histone acetylation and phosphorylation depending on the activities of the factors recruited in response to these modifications. In general, though, our model proposes that the addition of these modifications initially serves to "loosen" or derepress chromatin structure and that histone dephosphorylation and deacetylation are involved in the generation and/or stabilization of condensed or repressed chromatin. Since nearly all published reports examining aspects of one of these modifications have done so to the exclusion of the other, we discuss them separately below. However, we urge readers to consider that the combinatorial use of these modifications potentially represents a regulatory mechanism capable of greater specificity and amplification than either alone (see below).

CHIPPING AWAY AT THE ENZYMOLOGY OF HISTONE ACETYLATION

Exciting recent developments, particularly the realization that a surprisingly large number of histone acetyltransferases (HATs) and deacetylases (HDACs) were actually previously identified as proteins with roles in transcriptional regulation, have provided critical mechanistic links between chromatin modification and gene expression (for review, see Hassig and Schreiber 1997;

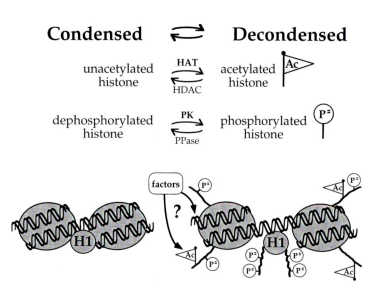

Condensed ⇌ Decondensed

unacetylated histone — HAT / HDAC → acetylated histone

dephosphorylated histone — PK / PPase → phosphorylated histone

Figure 1. Proposed roles of histone acetylation and phosphorylation in modulating chromatin structure. Condensed chromatin, inactive with respect to DNA transcription and replication, is characterized by low levels of posttranslationally modified histones as shown in the left portion of the figure. Acetylation of the amino-terminal tails of core histones by histone acetyltransferases (HAT) and phosphorylation of core and linker histone tails by protein kinases (PK) attenuate their interactions with DNA or other chromatin components, leading to decondensation of chromatin fibers and increased accessibility of DNA recognized by factors that regulate chromatin activity as shown in the right portion of the figure. Modified histone tails themselves may also be recognized by regulatory factors. Histone deacetylases (HDAC) and protein phosphatases (PPase) remove acetyl and phosphoryl groups, respectively, from the histone tails, enhancing their binding to nucleosomal DNA or other chromatin components and promoting a return to the condensed, quiescent, basal state. The positions shown for the globular domain of H1 and modified core histone tails are arbitrary and are not intended to reflect experimental data.

Mizzen and Allis 1998; Struhl 1998). Our initial report (Brownell et al. 1996) that the *Tetrahymena* macronuclear protein p55 and the related yeast transcriptional coactivator Gcn5p possessed HAT activity in vitro was rapidly followed by reports demonstrating HAT activity for the transcriptional coactivators PCAF (Yang et al. 1996), p300 (Ogryzko et al. 1996), and CBP (Bannister and Kouzarides 1996), the TAF$_{II}$250 subunit of TFIID (Mizzen et al. 1996), and the nuclear receptor coactivators ACTR (Chen et al. 1997) and SRC-1 (Spencer et al. 1997). This rapid expansion of knowledge was matched by groups investigating deacetylase activities. The initial identification and cloning of the human deacetylase HDAC1 (Taunton et al. 1996), a protein similar to the yeast transcriptional regulator Rpd3, were contemporaneous with the isolation of yeast deacetylase complexes containing either Rpd3 or the related protein HDAC1 (Carmen et al. 1996; Rundlett et al. 1996). Subsequent efforts have identified other HDAC1/Rpd3 homologs in yeast and mammals which are presumed to possess deacetylase activity (for review, see Pazin and Kadonaga 1997; Davie 1998).

This initial "discovery phase" has been followed by studies aimed at determining the functional requirements for acetyltransferase and deacetylase activities. Although these studies are in their infancy, it is already apparent that the benefits from further investigation will be substantial. A recent series of experiments investigating the role of deacetylase activity in *trans*-repression by Rb protein (Brehm et al. 1998; Magnaghi-Jaulin et al. 1998) have dramatically demonstrated that the architecture of chromatin is likely to impinge on a wide variety of fundamental processes, including cellular transformation (for review, see DePinho 1998). Below, we discuss recent progress in the characterization of histone acetyltransferases and elucidation of their roles in transcriptional regulation.

The best characterized acetyltransferase activities are the HATs represented by the GCN5 family, a group of enzymes highly conserved between yeast and humans (see Fig. 2A). Approximately 2 years ago, we cloned and characterized the first nuclear histone acetyltransferase from *Tetrahymena* and found that it shares striking similarity to the transcriptional coactivator GCN5, which was, itself, shown to possess intrinsic HAT activity (Brownell et al. 1996). This HAT activity is likely to be relevant in vivo because *GCN5*-responsive genes are dependent on catalytically active Gcn5p for activation of transcription (Kuo et al. 1998; see below). PCAF, a human homolog of GCN5, was found to have a novel amino-terminal domain that binds to the developmental regulator CBP/p300 in a manner that can be disrupted by the adenoviral protein E1a (Yang et al. 1996). A similar amino-terminal domain was found in a second human homolog of GCN5 (hGCN5) and in a *Drosophila* homolog (dGCN5), but not in plant, fungal, or protozoan GCN5s (Fig. 2A) (Smith et al. 1998a). Recently, mouse homologs of PCAF and hGCN5 were found to possess this amino-terminal domain (Fig. 2A) (S. Roth, pers. comm.,). Interestingly, several cDNAs of mouse and human GCN5 retain a small portion of an intron that interrupts the hGCN5/mGCN5 open reading frame between the amino-terminal (putative CBP binding) and the carboxy-terminal HAT domain (Smith et al. 1998a; S. Roth, pers. comm.). Several investigators have found smaller polypeptides that cross-react with PCAF and hGCN5 antibodies (Forsberg et al. 1997; Barlev et al. 1998; Currie 1998), suggesting that multiple protein products (i.e., hGCN5-S and hGCN5-L in Fig. 2A) can be produced. It should be noted that hGCN5-S does not bind CBP and may be targeted to genes in a manner more similar to that of yGCN5 (Yang et al. 1996).

Another protein family with members recently shown to possess histone acetyltransferase activity is the MYST

A. GCN5 family members

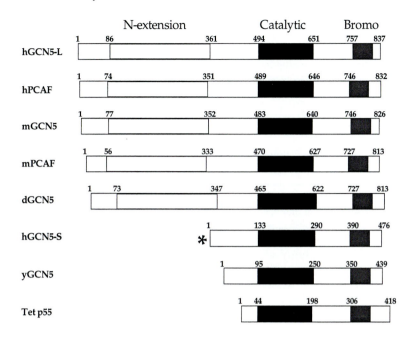

B. MYST family members

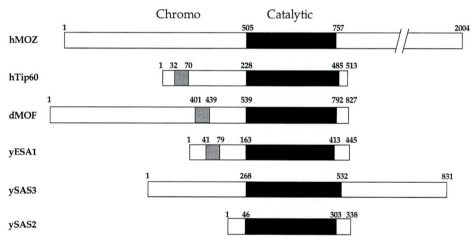

Figure 2. Features of recently described members of the GCN5 and MYST families of histone acetyltransferases. The proteins are aligned according to homology of a domain that contains a previously described putative acetyl-CoA-binding motif (Neuwald and Landsman 1997). The asterisk beside hGCN5-S denotes the fact that evidence of multiple forms of hGCN5 has been described. See text for further details.

family (see Fig. 2B) (Borrow et al. 1996). MYST proteins include the Sas2/3 proteins implicated in transcriptional silencing in yeast (Reifsnyder et al. 1996); MOF, which is required for the hypertranscription of X-linked genes in *Drosophila* males (Hilfiker et al. 1997); MOZ, which when fused in frame to CBP is responsible for a form of acute myeloid leukemia (Borrow et al. 1996); Tip60, which has been shown to interact with HIV-tat in the two-hybrid system (Kamine et al. 1996); and Esa1p, which is a yeast gene required for growth (Smith et al. 1998b). All of these proteins share extensive homology, including

zinc-finger-like sequences and an acetyl-CoA-binding motif in the presumed catalytic domain depicted in Figure 2B. A subset of MYST family members (Esa1p, Tip60, MOF) have a chromodomain (Fig. 2B) (Hilfiker et al. 1997), possibly indicating a common role in transcriptional activation.

Of the known MYST family members, Esa1p (Smith et al. 1998b) and Tip60 (Yamamoto and Horikoshi 1997) have been shown to acetylate histones with a preference for histone H4 in vitro. Furthermore, mutations in MOF's acetyltransferase domain lead to loss of H4 Lys-16 acetylation at MOF responsive genes (Hilfiker et al. 1997). Thus, for both GCN5 and MOF family members, biological data complement in vitro biochemical data, suggesting that at least some of the GCN5 and MYST family members actually acetylate histones in vivo (see below). However, some HATs can acetylate nonhistone proteins in vitro, and there are several reports of lysine ε-N acetylation of nonhistone proteins in vivo (for review, see Mizzen and Allis 1998).

Two important and related issues have been addressed recently. The first considers whether histones are physiological targets of these activities in vivo. This question is particularly relevant given that the rationale for designating these enzymes to be histone acetyltransferases stems largely from their activity on histone and chromatin substrates in vitro. Other investigators have shown that several of these acetyltransferases are capable of acetylating nonhistone proteins in vitro. The identities of these alternate substrates (e.g., TFIIE, TFIIF, p53, and Sin1p) suggest potentially important functional consequences (Gu and Roeder 1997; Imhof et al. 1997; Pollard and Peterson 1997). Two lines of evidence suggest that for the activities examined, histones are likely to be bona fide substrates in vivo. The most compelling data are that overexpression of functional Gcn5p leads to core histone hyperacetylation in vivo and that expression of a low-copy functional *GCN5* allele in *gcn5Δ* yeast restored the specific association of hyperacetylated histones with a Gcn5p-regulated promoter (Kuo et al. 1998). Significant support also comes from analyses of the acetylation site preferences displayed by these enzymes in vitro. In every report using histone or histone peptide substrates published to date, these enzymes acetylated only sites known to be acetylated in vivo, leaving other potential sites in these lysine-rich substrates unmodified (Kuo et al. 1996; Mizzen et al. 1996; Ogryzko et al. 1996; Smith et al. 1998b). A possible explanation for these observations is that specific substrate features, present both in vivo and in vitro, are recognized by these enzymes, but this has not been investigated systematically. In this regard, it may be significant that the purported sites of acetylation by p300 in p53 show similarities to acetylation sites in H4 (Gu and Roeder 1997). Together, these data strongly suggest that histones are physiological (although perhaps not exclusive) substrates for the respective enzymes.

The second issue concerns whether the HAT activity of these proteins is required for transcriptional regulation. Experimental evidence suggests that many or all of these activities exist in multisubunit complexes in vivo (e.g., the SAGA complex; see Grant et al. 1997; Roberts and Winston 1997; Pollard and Peterson 1997), raising the possibility that other subunits may contribute to transcriptional regulation in ways that remain to be determined. Recently, we addressed this question by analyzing the expression of two genes known to be targeted by Gcn5p in strains bearing Gcn5p mutants within which conserved amino acid residues in the putative catalytic domain were individually replaced with alanine (Kuo et al. 1998). In all cases where the HAT activity of a given Gcn5p mutant was diminished in vivo (as demonstrated by Gcn5p overproduction, see above), transcriptional activation of *HIS3* and a UASgal-CYC1-*lacZ* reporter was correspondingly down-regulated. Furthermore, all mutants showing normal or nearly normal levels of expression of these two genes exhibited normal levels of histone acetylation. Thus, it is highly likely that one of the central functions of yGcn5p-containing complexes is to acetylate nucleosomes which, in turn, facilitates transcriptional activation. However, genetic studies (Roberts and Winston 1997) have demonstrated Gcn5p-independent functions of the SAGA complex, indicating that, depending on the genetic context (*cis*- and/or *trans*-acting elements, chromatin structure, etc.), the importance of HAT activity of Gcn5p-containing complexes may differ.

Although the detailed mechanism underlying how histone acetylation/deacetylation affects transcription remains to be defined, recent work provides an important clue: Site-specific acetylation/deacetylation of nucleosomes within or around promoters at an early stage in transcription is likely to have a key role in regulating gene activity. This is exemplified by studies of the yeast enzymes Gcn5p and Rpd3p. Using antibodies specific for acetylated and nonacetylated histones pair-wise in a powerful chromatin immunoprecipitation technique (Fig. 3), it has recently been shown that histones are targets for acetylation and deacetylation by Gcn5p and Rpd3p, respectively, in vivo. Notably, both these activities appear to be recruited specifically to gene promoter chromatin (Kadosh and Struhl 1998; Kuo et al. 1998; Rundlett et al. 1998). As shown in Figure 3, this assay procedure reveals that *HIS3* promoter chromatin is enriched for acetylated H3 in yeast strains overexpressing either wild-type Gcn5p or the HAT-active L192A Gcn5p mutant compared to cells transfected with vector only or cells expressing a HAT-inactive F221A Gcn5p mutant. We predict that the next generation of chromatin immunoprecipitation experiments is likely to address to what extent histone acetylation and deacetylation is "targeted" to specific promoters through interaction with gene-specific regulatory proteins (for discussion, see Struhl 1998). If these enzymes are indeed targeted to select promoters through these interactions, then an important question remains: What activities and mechanisms bring about general acetylation, marked by acetylation of histone H4, throughout widespread domains of "poised" chromatin (Hebbes et al. 1988, 1994)?

The recent discoveries of acetyltransferase complexes in yeast and *Tetrahymena* containing HATs distinct from Gcn5p that catalyze robust acetylation of nucleosomal H4 at all of the recognized in vivo sites (Grant et al. 1997; Utley et al. 1998; R. Ohba et al., unpubl.) ensure that many more surprises and twists remain in this relatively young field. The finding that Esa1p, the first essential acetyltransferase to be described in budding yeast (Smith et al. 1998b), catalyzes a pattern of H4 acetylation remarkably similar to that of the complexes described above suggests the intriguing possibility that Esa1p is the unidentified catalytic activity in the above complexes. The conserved and essential nature of MYST family proteins (see above, Fig. 2B) provides further support for an acetyltransferase system distinct from that of the GCN5

A

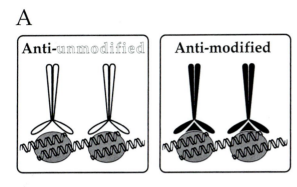

Anti-unmodified | Anti-modified

B

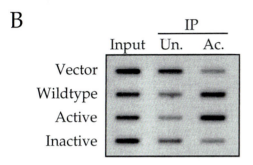

	Input	IP Un.	IP Ac.
Vector			
Wildtype			
Active			
Inactive			

Figure 3. Assessment of the association of modified histones with specific DNA sequences by a chromatin immunoprecipitation assay. (*A*) Chromatin fragments generated by sonication of formaldehyde cross-linked nuclei are immunoprecipitated in parallel by antibodies specific for unmodified (*left*) and modified (e.g., acetylated or phosphorylated) histones (*right*). Following deproteinization, the distribution of specific DNA sequences can be assessed by quantitative PCR or standard hybridization analyses. (*B*) Hybridization analyses reveals that the HAT activity of Gcn5p is required for histone acetylation at the *HIS3* promoter under inducing conditions (for further details, see Kuo et al. 1998; see text).

family, but putative roles for these activities in gene activation and dosage compensation remain to be defined.

WHAT ABOUT OTHER HISTONE MODIFICATIONS?

The Function of Histone Phosphorylation Remains Enigmatic

Long-standing observations in the histone/chromatin field suggest that acetylation is not the only covalent modification of histones that matters. Histones are known to undergo an impressive array of posttranslational modifications including phosphorylation, methylation, ubiquitination, poly-ADP-ribosylation, and glycosylation (van Holde 1989; Wolffe 1995; Davie 1998). However, the functional significance of these modifications is poorly understood. Gracing the pages of most cell biology texts, for example, is the fact that linker or H1-type histones are extensively phosphorylated just prior to a cell's entry into mitosis, a finding central to the formerly widely accepted hypothesis that linker histone hyperphosphorylation is causally linked to chromosome con-

densation (for review, see Bradbury 1992; Koshland and Strunnikov 1996). However, recent observations have made this view controversial. Below, we discuss recent progress in understanding the significance of phosphorylation of linker histones and mitogen-stimulated phosphorylation of core histone H3.

A significant obstacle to understanding the role of linker histone phosphorylation is the lack of consensus regarding the function of H1 in chromatin. In contrast to core histones, linker histones display considerable structural diversity throughout the eukaryotic kingdom (for review, see Zlatanova and van Holde 1996; Wolffe et al. 1997), and it is possible that these different forms fulfill somewhat different functions. For the purpose of this discussion, we define H1/linker histones as abundant basic proteins that interact with linker DNA. In general, linker histones of metazoans possess a tripartite structure in which a central globular domain is flanked by flexible amino- and carboxy-terminal domains that contain a preponderance of basic amino acids. Long-standing observations from nuclease digestion and chemical cross-linking experiments suggest that the globular domain of metazoan linker histones binds DNA on the exterior of the nucleosome at or near the dyad axis so as to "seal" the nucleosome at the point where the DNA strands enter and exit the particle (for review, see van Holde 1989; Zlatanova and van Holde 1996). However, recent analyses of mononucleosomes reconstituted on a defined DNA show that the globular domain binds at an asymmetric position approximately 68 bp away from the dyad, possibly at a site in the major groove facing the histone octamer where it can also contact histones H3 and H2A (Hayes and Wolffe 1993; Hayes et al. 1994; Hayes 1996; Pruss et al. 1996; for review, see Crane-Robinson 1997).

Less is known regarding the disposition of the amino- and carboxy-terminal domains. The data available are consistent with binding of linker DNA by both tails (for review, see Zlatanova and van Holde 1996). Linker histones in macronuclei of ciliated protozoans such as *Tetrahymena* lack a globular domain and presumably represent simply a "tail" domain (Wu et al. 1986). The protein encoded by the *HHO1* gene of *Saccharomyces cerevisiae*, predicted to represent the long-sought yeast H1 based on homology with the globular domain of metazoan linker histones (Landsman 1996), localizes to yeast nuclei in vivo (Ushinsky et al. 1997) and interacts with linker DNA of chromatin in vitro in a manner consistent with linker histone function, suggesting that Hho1p represents a bona fide linker histone (Patterton et al. 1998). Note that Hho1p is predicted to contain two globular domains joined by a short lysine, alanine, proline-rich "tail" domain (Wolffe et al. 1997; Patterton et al. 1998). Although it seems likely that the presence or absence of one or two globular domains could impart distinct functionality to linker histones, insufficient information is available at present to know if this is the case.

In general, linker histones repress transcription in vitro, supporting the notion that they are general repressors (see, e.g., Laybourn and Kadonaga 1991). Linker histones are required for the salt-dependent conformational transition of extended "10-nm" oligonucleosomal filaments to a regular "30-nm" fiber in vitro (Thoma et al. 1979), supporting the hypothesis that they have a role in the formation and maintenance of higher-order chromatin structure. Although mechanisms by which H1 may mediate these transitions and the nature of higher-order chromatin structures themselves have been much debated (for review, see van Holde and Zlatanova 1995, 1996; Zlatanova and van Holde 1996), clear evidence that linker histones are involved in chromatin condensation in vivo has been obtained in *Tetrahymena*. Macronuclear H1 knockout strains are viable but have macronuclei that are enlarged twofold compared to wild-type cells. Conversely, micronuclear linker histone knockout strains have enlarged micronuclei (Shen et al. 1995). Subsequent analyses revealed that although transcription by pol I and pol III were unaffected, both positive and negative changes in the transcription of specific genes by pol II occurred in response to the loss of macronuclear H1 (Shen and Gorovsky 1996). These results suggest that linker histones can act as positive or negative gene-specific transcriptional regulators, and further evidence of selective effects of H1 on transcription has been obtained in analyses of embryogenesis in *Xenopus* (Steinbach et al. 1997; Wolffe et al. 1997). Similarly, preferential repression of pol-I-transcribed rRNA genes and the pol-II-transcribed *ACT1* and *URA3* genes, but not the pol-II-transcribed Ty gene, occurred when a sea urchin H1 was overexpressed in yeast (Linder and Thoma 1994). However, disruption of the endogenous *HHO1* gene did not lead to any overt changes in transcription nor did it lead to detectable changes in chromatin organization (Patterton et al. 1998). This discrepancy may relate to the unique structure of Hho1p itself or to specific features of chromatin structure in yeast which possess an unusually short repeat length (~160 bp) and hence minimal linker DNA.

The nearly universal correlation of linker histone hyperphosphorylation with mitosis has suggested to many that H1 hyperphosphorylation directly promotes higher degrees of chromatin compaction and, by extension, repression of transcription. However, recent experiments demonstrating that H1 is not required for chromosome condensation in vitro (Ohsumi et al. 1993) and that chromosome condensation induced by phosphatase treatments does not require H1 hyperphosphorylation (Guo et al. 1995) suggest that mitotic H1 phosphorylation is distinct from mechanisms with more direct roles in causing chromosome condensation. Moreover, other studies have shown that condensed chromatin structures in interphase nuclei are frequently enriched in dephosphorylated isoforms of histone H1 (for review, see Roth and Allis 1992). Using antibodies raised against highly phosphorylated macronuclear H1, which are highly selective for phosphorylated H1 in a variety of organisms (Lu et al. 1994), and a complementary antibody made against dephosphorylated macronuclear H1, we have shown that in *Tetrahymena*, dephosphorylated H1 localizes to condensed chromatin within macronuclear chromatin bodies, whereas phosphorylated H1 localizes to euchromatin peripheral to the chromatin bodies. Parallel staining with other antisera revealed that the distributions of TBP and hv1 (an H2A variant thought to be enriched in transcriptionally active chromatin) resembled that of phosphorylated H1, whereas the major H2A protein was distributed equivalently throughout euchromatin and chromatin bodies (Lu et al. 1995). These data suggest that phosphorylated H1 is enriched in transcriptionally competent chromatin, and dephosphorylated H1 is enriched in condensed, presumably transcriptionally inactive chromatin. Our phosphorylated H1-specific antisera have also been used to show that transformation of cells by a variety of means is accompanied by increases in the levels of phosphorylated forms of H1 and relaxation or decondensation of bulk chromatin structure compared to parental cells (Chadee et al. 1995; Taylor et al. 1995; Herrera et al. 1996). Similarly, phosphorylation of H1 is required for glucocorticoid-receptor-mediated disruption of mouse mammary tumor virus (MMTV) promoter chromatin in vivo (Lee and Archer 1998), arguing collectively that linker histone phosphorylation has a poorly appreciated role in decondensing chromatin which may be important for altering gene expression linked to cell cycle progression and potentially, oncogenesis (see DePinho 1998).

H3 Phosphorylation: A Powerful New Antibody Leads the Way

Like H1 phosphorylation, H3 phosphorylation correlates with mitotic chromosome condensation (for review, see Bradbury 1992). However, in response to mitogenic stimulation, a rapid and transient phosphorylation of a fraction of the H3 in mammalian cells occurs that correlates well with an immediate-early response (Mahadevan et al. 1991). These observations suggest the intriguing possibility that H3 phosphorylation, like histone acetylation and linker histone phosphorylation, also has a causal role in decondensing the chromatin fiber for transcription-related functions. To test this hypothesis, we adapted the strategy discussed above for acetylation and generated an antiserum directed against a short synthetic peptide mimic of the H3 amino-terminal tail singly phosphorylated at Ser-10. This antiserum recognizes H3 as it becomes phosphorylated during mitotic chromosome condensation in organisms ranging from yeast to humans and, as such, is an exceptional marker for mitosis (Hendzel et al. 1997; Wei and Allis 1998).

In excellent agreement with the pioneering studies of Mahadevan and colleagues (1991), a dramatic induction of H3 phosphorylation, indicated by strong reactivity with our phosphorylated H3 antibodies, occurs after treatment of cultured cells with mitogens (e.g., TPA or EGF), and this induction is both rapid and transient. Immunoreactivity was not detected in histone isolated from nonstimulated control cells. We next examined the immunolocalization pattern of phosphorylated H3 in cells treated with mitogens. Unlike mitosis-associated H3 phosphorylation (Hendzel et al. 1997), mitogen-stimulated phosphorylation of H3 localizes to numerous small dots and aggregates of dots. These dots show very little overlap with DAPI-stained chromatin, but rather localize to DAPI-depleted regions of the interchromatin space and are not observed in serum-starved control cells or in mitogen-treated cells stained after the immediate-early response has diminished (M. Hendzel, unpubl.). These data suggest that Ser-10 in H3 is phosphorylated in a mitogen-dependent fashion with kinetics that resemble the transcriptional induction of immediate-early response genes. Thus, our Ser-10-phosphorylated H3 antibody reveals a dynamic response at the chromatin level to the activation of mitogenic signaling pathways that is clearly distinct from mitotic H3 phosphorylation.

A Histone Modification Activity Gel Detects an H3 Kinase Induced by Mitogens

An in-gel HAT assay proved to play a critical part in the identification of *Tetrahymena* p55/yeast Gcn5p as the first catalytic subunit of a nuclear HAT activity (Brownell and Allis 1995; Brownell et al. 1996), and this assay has subsequently been used to detect other polypeptides that possess intrinsic HAT activity (see, e.g., Mizzen et al. 1996). As shown in Figure 4, potentially any histone-modifying activity can be detected in complex samples following resolution on SDS-PAGE gels containing histones and subsequent incubation of gels with radioactively labeled cofactor appropriate for the transferase activity under study. Thus, the molecular weight of potentially catalytically active polypeptides can be determined. We adapted this basic procedure to detect H3 ki-

nases by using purified H3 as the substrate and providing $[\gamma\text{-}^{32}\text{P}]$ATP as the cofactor. Using nuclear extracts from mitogen-stimulated and quiescent mouse fibroblasts, we consistently identified a polypeptide with an apparent molecular mass of approximately 90 kD that possessed strong kinase activity for H3. Importantly, this species was only detected in extracts prepared from mitogen-stimulated cells; the 90-kD polypeptide was not observed when extracts were prepared and assayed identically from quiescent, serum-starved cells.

Due to its molecular weight and induction by mitogens, we speculated that the catalytically active species being detected in our in-gel histone kinase assay might be a member of the p90^{rsk} family, a group of structurally related, mitogen-activated kinases implicated in cell proliferation. To begin to test this hypothesis, parallel in-gel H3 kinase assays were performed, except that the active 90-kD band was excised, re-electrophoresed into a second SDS one-dimensional gel and probed with p90^{rsk} antibodies. Consistent with the well-documented nuclear localization of p90^{rsk} upon mitogenic stimulation (Chen et al. 1992; Zhao et al. 1995), immunoblotting analysis detected an immunoreactive band only in the mitogen-stimulated cells, which exhibited an electrophoretic mobility similar to that of the detected kinase subunit in the activity assay, further supporting the hypothesis that a p90^{rsk} family member has kinase activity for H3 in this assay.

We next sought to evaluate the substrate specificity of p90^{rsk} toward various histone or chromatin (nucleosomal)

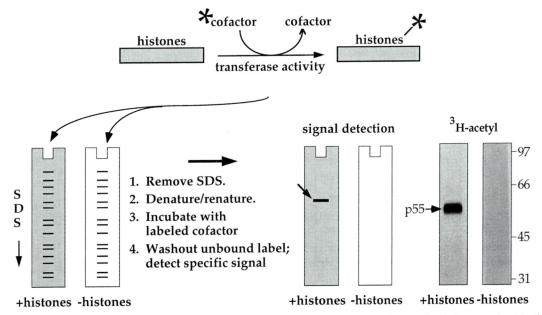

Figure 4. Principle of the activity gel assay employed to detect histone-modifying enzymes. The method is based on the detection of modification through the use of radioisotopic cofactor (e.g., $[\gamma\text{-}^{32}\text{P}]$ATP or [^{3}H]acetyl-CoA) as outlined in the upper portion. To determine the molecular weight of polypeptides with intrinsic modifying activity, samples are electrophoresed on SDS gels containing histones as shown in the lower portion of the figure. Following treatments to denature and renature the resolved proteins, gels are incubated with radioisotopic cofactor and processed to detect modified histones surrounding active species in the gels by autoradiography. An example showing the detection of the *Tetrahymena* p55 HAT in crude macronuclear extract is shown in the lower right portion of the figure. As described in the text, we used this method to demonstrate that Rsk-2 possesses intrinsic H3 kinase activity. We note that, at least in theory, this technique could also be used to identify other histone-modifying enzymes such as histone methyltransferases, using *S*-adenosylmethionine as the radioactively labeled cofactor.

substrates. Intact H3, whether assayed as a purified histone or in combination with other histones or within nucleosomal particles, was efficiently phosphorylated by
rabbit skeletal muscle Rsk-2 in vitro (C. Mizzen, unpubl.). Importantly, reactivity with our phosphorylated
H3 antibodies occurs in a Rsk-2-dependent fashion, suggesting that Ser-10 is a strongly preferred, if not exclusive, site of phosphorylation by Rsk-2. We confirmed this
observation by microsequence analysis of H3 phosphorylated by Rsk-2 in vitro.

Mitogenic H3 Phosphorylation Does Not Occur in RSK-2 (CLS)-deficient Cells

It could be argued that the above results do little more
than demonstrate that the amino terminus of histone H3
contains a consensus motif preferred by p90^{rsk}. Are
these in vitro results physiologically relevant? We
sought to obtain evidence that p90^{rsk} is responsible for
mitogen-stimulated H3 phosphorylation in vivo. In humans, the p90^{rsk} kinase family comprises three closely
related members, Rsk-1, Rsk-2, and Rsk-3 (Moller et al.
1994; Zhao et al. 1995), and tissue-specific differences
in gene expression suggest distinct physiological roles
for the various members of the p90^{rsk} family (Moller et
al. 1994; Zhao et al. 1995, 1996; Trivier et al. 1996). Recently, X-linked mutations in the Rsk-2 kinase, but not
Rsk-1 or Rsk-3, have been linked directly to Coffin-
Lowry syndrome (CLS) (Trivier et al. 1996). Therefore,
we analyzed fibroblasts derived from CLS patients to
determine whether a specific relationship existed between Rsk-2 activity and mitogen-stimulated H3 phosphorylation.

To begin to address this issue, epidermal growth factor
(EGF)-stimulated human fibroblasts were stained with
our phosphorylated H3 antibodies. A large proportion of
nuclei are positively stained, exhibiting multiple bright
dots or aggregates of dots. In striking contrast, essentially
no positive nuclei are detected in CLS cells carried
through the same experimental procedures (P. Sassone-
Corsi, unpubl.). Importantly, we observed normal immunostaining of mitotic figures in CLS cells, indicating
that H3 phosphorylation in mitotic cells is seemingly normal and Rsk-2-independent. These data suggest that H3
is a downstream target of at least two distinct signaling
pathways: One is Rsk-2-dependent and linked to activation by growth factors, and the second is Rsk-2-indepen-
dent and associated with mitosis.

Immunoblotting analyses, again using the phosphorylated H3 antibodies, lend further support to the notion that
Rsk-2 is the kinase responsible for mitogen-stimulated
H3 phosphorylation. Efficient phosphorylation of H3 was
observed in normal fibroblasts following a diverse collection of inductive treatments such as UV irradiation,
serum stimulation, or EGF addition. In contrast, none of
these treatments induced H3 phosphorylation in CLS
cells. Since cells from CLS patients express Rsk-1 and
Rsk-3 at levels comparable to normal cells (Trivier et al.
1996), these results indicate that Rsk-2 is the kinase responsible for mitogen-activated H3 phosphorylation.

H3 Kinase Activity Is Missing in CLS Cells

To further test the hypothesis that Rsk-2 is the kinase
involved in H3 phosphorylation following mitogenic
stimulation, antibodies selective for Rsk-2 were used to
immunoprecipitate Rsk-2 from extracts prepared from
normal and CLS cells after mitogenic stimulation. Each
immunoprecipitate was then assayed directly for kinase
activity using H3 as substrate. Although there is efficient
phosphorylation of H3 using the Rsk-2 immunoprecipitated from normal cells, there is little, if any, activity in
immunoprecipitates from CLS cells. To rigorously identify Rsk-2 as the responsible kinase, immunoprecipitates
were also analyzed in H3 kinase activity gels. As expected, a 90-kD band was readily detected in immunoprecipitates from normal cell extracts that was not apparent in immunoprecipitates from CLS cells. These results
indicate that Rsk-2 is the major mitogen-induced kinase
responsible for H3 phosphorylation. Interestingly, recent
experiments have shown the EGF-induced expression of
c-*fos* is severely impaired in CLS cells despite normal
phosphorylation of several other transcription factors involved in this gene's activation (D. De Cesare and P. Sassone-Corsi, unpubl.). We look forward to experiments using the chromatin immunoprecipitation technique (Fig. 3)
with phosphorylation-specific H3 antibodies to test to
what extent mitogen-stimulated genes like c-*fos* are associating with this histone modification.

CONCLUSIONS, IMPACT, AND PERSPECTIVE

Signaling to Chromatin through Histone Modifications

A central question in transcriptional regulation asks:
What are the mechanisms by which the transcriptional
machinery gains access to DNA in chromatin? In turn, a
central question in signal transduction asks: How is transcription regulated in response to upstream activity in relevant signaling pathways? A growing body of evidence
indicates that a wide variety of transcription factors are
substrates for MAP kinases, suggesting a mechanism
wherein MAPK cascades regulate transcription through
direct effects on transcription factor activity (for review,
see Treisman 1996). Most often, regulation of candidate
factors is thought to occur by reversible phosphorylation
impacting at multiple levels: translocation to the nucleus,
DNA binding, and transcriptional activation. However,
until recently, little information was available to suggest
how transcription factors accessed cognate DNAs in
chromatin.

The discoveries that enzymes possessing intrinsic HAT
or HDAC activity were already known components of the
pol II transcription machinery have led to a major
"paradigm shift" in the transcription field (for review, see
Struhl 1998). Clearly, the existence of highly conserved,
promoter-directed, acetyltransferase and deacetylase activities is evidence of a primary role for chromatin structure in transcriptional regulation which is consistent with
the evolutionary conservation and longevity of the histones themselves. Further support for this view comes

from the structural similarities noted between histones and proteins involved in transcriptional regulation (for review, see Burley et al. 1997) and the similarities to nucleosome structure that have been noted for TFIID-promoter complexes (Hoffmann et al. 1997).

The data described here provide the first in vivo evidence that histone H3 is a physiologically relevant target of the MAP kinase effector Rsk-2 and underscore the importance of a chromatin remodeling step, and H3 phosphorylation in particular, as part of a critical downstream event in MAPK signaling. Our findings reinforce the view that chromatin structure and histone proteins in particular are targets of signaling pathways as initially proposed by Mahadevan and colleagues (1991). The fact that H3 phosphorylation and immediate-early gene induction following mitogen stimulation can occur in the absence of transcription per se (Barratt et al. 1994) further suggests that chromatin remodeling is an essential prerequisite for gene activation.

Is Histone Acetylation Synergistically Linked to Histone Phosphorylation?

As discussed above, a rapidly increasing body of evidence suggests that eukaryotic cells have evolved conserved gene regulatory pathways wherein chromatin-modifying activities, notably histone acetylases and deacetylases, are recruited to specific promoters through selective interactions with regulatory proteins (for review, see Wolffe and Pruss 1996; Mizzen and Allis 1998; Struhl 1998). We suggest that similar mechanisms may

also determine the targeted recruitment of histone kinases. p90^{rsk} has been shown to interact with the transcriptional coactivator CBP (CREB-binding protein) in a growth-factor-dependent fashion (Nakajima et al. 1996) using a CBP-interacting domain (CID) that also interacts with the human histone acetyltransferase PCAF (Yang et al. 1996). Overexpression of this CID domain in PC12 cells blocks H3 phosphorylation induced by nerve growth factor (M. Montminy and C.D. Allis, unpubl.), suggesting that recruitment of Rsk-2 to CBP is required to bring about this nuclear response. These observations suggest models for the regulation of H3 phosphorylation similar to that shown in Figure 5. It is not known at present whether p90^{rsk} kinases and PCAF compete for the same binding site or can bind simultaneously to p300/CBP. However, the specificities of these interactions represent one mechanism capable of providing mutually exclusive or synergistic patterns of H3 phosphorylation and histone acetylation. The notion that specific acetylases (e.g., PCAF; see Puri et al. 1998) and/or histone kinases associate with CBP to bring about localized remodeling of the chromatin template and facilitate gene expression is consistent with evidence that CBP ordinarily acts to integrate and coordinate diverse physiological stimuli (Kamei et al. 1996; for review, see Shikama and La Thangue 1997).

Whether histone acetylation is synergistically linked to histone phosphorylation in mitogen-responsive promoters such as c-*fos* and c-*jun*, as first suggested by Mahadevan and colleagues (Mahadevan et al. 1991; Barratt et al. 1994) remains an intriguing and relatively unexplored possibility (also see Alberts et al. 1998). Of immediate in-

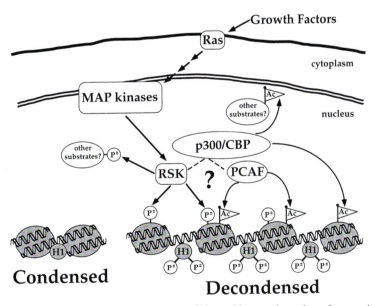

Figure 5. Possible interactions involved in coordinated recruitment of histone kinase and acetyltransferase activities to chromatin. Upon activation by MAP kinases, p90^{rsk} proteins are translocated to the nucleus where Rsk-2, in particular, phosphorylates H3 in response to mitogenic stimuli. This phosphorylation may be targeted to nucleosomes acetylated by p300/CBP and/or PCAF, since p90^{rsk} proteins are known to bind to a site on p300/CBP that is near an interaction site for PCAF. As discussed in the text, both of these events are thought to promote chromatin decondensation and facilitate transcriptional activation (see Fig. 1). Note that each of these modifying activities may target other substrates in addition to histones. See text for further details. If some of these interactions are correct, we note that chromatin immunoprecipitation, using our modification-specific antibodies (see Fig. 3) provides a powerful means to recover DNA sequences (i.e., mitogen-stimulated target genes) that are closely associated with modified histones (see Alberts et al. 1998).

terest is whether chromatin immunoprecipitation assays (see Fig. 3) can be used with phosphorylated H3 antibodies to test this possibility directly. Of equal interest is whether histone acetylation is disturbed or altered when H3 phosphorylaton does not occur, a question open to direct experimental test using Rsk-2-deficient CLS cells. It is worth comment that the preferred site of acetylation by GCN5 family members is Lys-14 in H3, a site only four amino acids removed from the Ser-10 site of phosphorylation catalyzed by Rsk-2. Since both the acetyltransferase PCAF (Yang et al. 1996) and p90^{rsk} (Nakajima et al. 1996) are believed to interact with CPB in a signaling-dependent fashion, it is of interest to determine if multiple histone-modifying activities exist together in multisubunit coactivator complexes.

Have We Seen Only the Tip of the Iceberg?

1997 has been a "titanic year" for the chromatin field with far-reaching progress made in areas ranging from the crystal structure of the nucleosome core particle to the identification of defined complexes that compact and pair the chromatin fiber in its most condensed form (for review, see Allis and Gasser 1998). With respect to covalent histone modifications, we may have only seen the tip of the iceberg. It seems likely that other histone modifications will be found to have important roles in mediating DNA-templated processes. For example, when mammalian cells are exposed to ionizing radiation that introduces double-strand breaks into DNA, a specific variant of H2A is uniquely phosphorylated at Ser-139 (Rogakou et al. 1998). Interestingly, this conserved serine is located in the carboxyl terminus of H2A on the same side of the nucleosome core particle as the H3 amino-terminal tail (Luger et al. 1997; Luger and Richmond 1998). Most chromatin researchers know that the carboxyl terminus of H2A is also the site of ubiquitin attachment at a highly conserved lysine (see Davie 1998). Thus, the poorly appreciated carboxy-terminal tail in H2A and this side of the nucleosome particle may be exposed to a wealth of covalent modifications whose functional importance remains to be determined.

What Are Histone Acetylation and Phosphorylation Doing Anyway?

Although most current reviews on the subject (our own included) are quick to mention that changes in net charge on the histone tails upon acetylation or phosphorylation are likely to influence histone:DNA interactions and alter higher-order chromatin structures, clear evidence that this is indeed the case is not easy to come by in the literature (for review, see Hansen 1997). The paucity of assays for evaluating higher-order chromatin folding represents a real obstacle to progress in this general area, although several promising developments suggest that this situation is changing (see, e.g., Li et al. 1998; Tse et al. 1998). Although the current crystal structure of the core particle localizes only portions of the histone amino-

terminal tails (Luger et al. 1997; Luger and Richmond 1998), several important clues into tail function are suggested. For example, the crystal structure suggests that Lys-16 in H4, a well-documented, functionally important acetylation site, may contact a conserved "acidic patch" in H2A of an adjacent nucleosome. This suggests the intriguing possibility that acetylation of H4 at this position could disrupt this contact, possibly destabilizing higher-order chromatin packing.

Three lines of evidence suggest that the effects of acetylation, at least in some cases, depend on the specific sites modified. The most compelling evidence comes from analyses of the transcriptional effects of systematic mutagenesis of acetylation sites within histone tails. These studies provide strong evidence that the effects of acetylation of H3 and H4 are both acetylation-site-specific and gene-specific (for review, see Grunstein 1997; Mizzen and Allis 1998). This notion is further supported by the fact that specific HATs which function in different pathways display distinct site specificities in vitro (Kuo et al. 1996; Mizzen et al. 1996; Ogryzko et al. 1996; Smith et al. 1998b). Although it is often difficult to judge the validity of in vitro studies, as discussed above, we note that aberrant acetylation at nonphysiological sites has not been detected; moreover, the results are consistent with site usage in vivo in different biological settings (Chicoine et al. 1986; Sobel et al. 1994; 1995). Finally, nonrandom localization of specifically acetylated forms of histones using acetylation site-specific antisera provides strong support for site-specific effects of acetylation (for review, see Turner 1998).

Recent evidence suggests that histone tails, projecting away from the histone octamer with unique, combinatorial markings of acetylation, phosphorylation, etc., may serve as "flags" or "receptor" surfaces for interaction with other nonhistone proteins (Hecht et al. 1995; Edmondson et al. 1996). We anticipate that specifically modified histone peptides will be useful in the isolation of histone interacting factors by affinity chromatography. Along this line, it seems prudent to examine chromatin remodeling complexes that appear to require histone tails for the ability to function, such as NURF (Georgel et al. 1997), Swi/Snf (Logie and Peterson 1997), and possibly chromosome condensation factors like condensins (Hirano et al. 1997), to determine if specific interactions occur between uniquely modified histone tails and specific components of these complexes.

In sum, interest in chromatin has been rejuvenated, in part, by the discovery that certain transcriptional regulators are enzymes that modify highly conserved, often invariant, residues in histones. The conserved nature of these modifications underscores the general utility in continuing to develop immunological reagents directed at other poorly understood histone modifications. The remarkable selectivity of these antibodies has proven to be extremely valuable in dissecting the potential part that these modifications play in a variety of biological processes. As mechanisms of transcriptional activation and repression are elucidated in greater detail, it has become apparent that modulation of chromatin structure through reversible modification has a critical role in transcriptional regulation. Similarly, we anticipate that more discoveries of upstream signal transduction pathways acting on histones or other chromatin components are on the horizon. Although faint, the signals to chromatin are becoming increasingly loud and clear.

ACKNOWLEDGMENTS

C.D.A. and P.S.-C. thank all former and present members of our labs whose contributions have made this story possible. Their enthusiasm, dedication, and hard work have been inspirational. We are grateful to Drs. Richard Cook, Jim Davie, Michael Hendzel, Louis Mahadevan, Martin Gorovsky, and Sharon Roth for helpful and stimulating discussions and for permitting us to cite unpublished results. Throughout this paper, reference is made to histone modification-specific antibodies. All of these antisera are commercially available through Upstate Biotechnology Inc. (Lake Placid, New York); their continued interest and support of this line of reagents is also gratefully acknowledged.

REFERENCES

Alberts A.S., Geneste O., and Treisman R. 1998. Activation of SRF-regulated chromosomal templates by Rho-family GTPases requires a signal that also induces H4 hyperacetylation. *Cell* **92:** 475.

Allis C.D. and Gasser S.M. 1998. New excitement over an old word: "Chromatin." *Curr. Opin. Genet. Dev.* **8:** 137.

Bannister A.J. and Kouzarides T. 1996. The CBP co-activator is a histone acetyltransferase. *Nature* **384:** 641.

Barlev N.A., Poltoratsky V., Owen-Hughes T., Ying C., Liu L., Workman J.L., and Berger S.L. 1998. Repression of GCN5 histone acetyltransferase activity via bromodomain- mediated binding and phosphorylation by the Ku-DNA-dependent protein kinase complex. *Mol. Cell. Biol.* **18:** 1349.

Barratt M.J., Hazzalin C.A., Cano E., and Mahadevan L.C. 1994. Mitogen-stimulated phosphorylation of histone H3 is targeted to a small hyperacetylation-sensitive fraction. *Proc. Natl. Acad. Sci.* **91:** 4781.

Borrow J., Stanton V.P., Jr., Andresen J.M., Becher R., Behm F.G., Chaganti R.S., Civin C.I., Disteche C., Dube I., Frischauf A.M., Horsman D., Mitelman F., Volinia S., Watmore A.E., and Housman D.E. 1996. The translocation t(8;16)(p11;p13) of acute myeloid leukaemia fuses a putative acetyltransferase to the CREB-binding protein. *Nat. Genet.* **14:** 33.

Bradbury E.M. 1992. Reversible histone modifications and the chromosome cell cycle. *BioEssays* **14:** 9.

Brehm A., Miska E.A., McCance D.J., Reid J.L., Bannister A.J., and Kouzarides T. 1998. Retinoblastoma protein recruits histone deacetylase to repress transcription. *Nature* **391:** 597.

Brownell J.E. and Allis C.D. 1995. An activity gel assay detects a single, catalytically active histone acetyltransferase subunit in *Tetrahymena* macronuclei. *Proc. Natl. Acad. Sci.* **92:** 6364.

Brownell J.E., Zhou J., Ranalli T., Kobayashi R., Edmondson D.G., Roth S.Y., and Allis C.D. 1996. *Tetrahymena* histone acetyltransferase A: A homolog to yeast Gcn5p linking histone acetylation to gene activation. *Cell* **84:** 843.

Burley S.K., Xie X., Clark K.L., and Shu F. 1997. Histone-like transcription factors in eukaryotes. *Curr. Opin. Struct. Biol.* **7:** 94.

Carmen A.A., Rundlett S.E., and Grunstein M. 1996. HDA1 and HDA3 are components of a yeast histone deacetylase (HDA) complex. *J. Biol. Chem.* **271:** 15837.

Chadee D.N., Taylor W.R., Hurta R.A.R., Allis C.D., Wright J.A., and Davie J. 1995. Increased phosphorylation of histone H1 in mouse fibroblasts transformed with oncogenes of constitutively active mitogen-activated protein kinase kinase. *J. Biol. Chem.* **270:** 20098.

Chen H., Lin R.J., Schiltz R.L., Chakravarti D., Nash A., Nagy L., Privalsky M.L., Nakatani Y., and Evans R.M. 1997. Nuclear receptor coactivator ACTR is a novel histone acetyltransferase and forms a multimeric activation complex with P/CAF and CBP/p300. *Cell* **90:** 569.

Chen R.H., Sarnecki C., and Blenis J. 1992. Nuclear localization and regulation of erk- and rsk-encoded protein kinases. *Mol. Cell. Biol.* **12:** 915.

Chicoine L.G., Schulman I.G., Richman R., Cook R.G., and Allis C.D. 1986. Nonrandom utilization of acetylation sites in histones isolated from *Tetrahymena:* Evidence for functionally distinct H4 acetylation sites. *J. Biol. Chem.* **261:** 1071.

Crane-Robinson C. 1997. Where is the globular domain of linker histone located on the nucleosome? *Trends Biochem. Sci.* **22:** 75.

Currie R.A. 1998. NF-Y is associated with the histone acetyltransferases GCN5 and P/CAF. *J. Biol. Chem.* **273:** 1430.

Davie J. 1998. Covalent modifications of histones: Expression from chromatin templates. *Curr. Opin. Genet. Dev.* **8:** 173.

DePinho R.A. 1998. Transcriptional repression: The cancer-chromatin connection. *Nature* **391:** 533.

Edmondson D.G., Smith M.M., and Roth S.Y. 1996. Repression domain of the yeast global repressor Tup1 interacts directly with histones H3 and H4. *Genes Dev.* **10:** 1247.

Forsberg E.C., Lam L.T., Yang X.J., Nakatani Y., and Bresnick E.H. 1997. Human histone acetyltransferase GCN5 exists in a stable macromolecular complex lacking the adapter ADA2. *Biochemistry* **36:** 15918.

Georgel P.T., Tsukiyama T., and Wu C. 1997. Role of histone tails in nucleosome remodeling by *Drosophila* NURF. *EMBO J.* **16:** 4717.

Grant P.A., Duggan L., Cote J., Roberts S.M., Brownell J.E., Candau R., Ohba R., Owen-Hughes T., Allis C.D., Winston F., Berger S.L., and Workman J.L. 1997. Yeast Gcn5 functions in two multisubunit complexes to acetylate nucleosomal histones: Characterization of an Ada complex and the SAGA (Spt/Ada) complex. *Genes Dev.* **11:** 1640.

Grunstein M. 1997. Histone acetylation in chromatin structure and transcription. *Nature* **389:** 349.

Gu W. and Roeder R.G. 1997. Activation of p53 sequence-specific DNA binding by acetylation of the p53 C-terminal domain. *Cell* **90:** 595.

Guo X.W., Th'ng J.P., Swank R.A., Anderson H.J., Tudan C., Bradbury E.M., and Roberge M. 1995. Chromosome condensation induced by fostriecin does not require p34cdc2 kinase activity and histone H1 hyperphosphorylation, but is associated with enhanced histone H2A and H3 phosphorylation. *EMBO J.* **14:** 976.

Hansen J.C. 1997. The core histone amino-termini: Combinatorial interaction domains that link chromatin structure with function. *Chemtracts-Biochem. Mol. Biol.* **10:** 737.

Hassig C.A. and Schreiber S.L. 1997. Nuclear histone acetylases and deacetylases and transcriptional regulation: HATs off to HDACs. *Curr. Opin. Chem. Biol.* **1:** 300.

Hayes J.J. 1996. Site-directed cleavage of DNA by a linker histone-Fe(II) EDTA conjugate: Localization of a globular domain binding site within a nucleosome. *Biochemistry* **35:** 11931.

Hayes J.J. and Wolffe A.P. 1993. Preferential and asymmetric interaction of linker histones with 5S DNA in the nucleosome. *Proc. Natl. Acad. Sci.* **90:** 6415.

Hayes J.J., Pruss D., and Wolffe A.P. 1994. Contacts of the globular domain of histone H5 and core histones with DNA in a "chromatosome." *Proc. Natl. Acad. Sci.* **91:** 7817.

Hebbes T.R., Thorne A.W., and Crane-Robinson C. 1988. A direct link between core histone acetylation and transcriptionally active chromatin. *EMBO J.* **7:** 1395.

Hebbes T.R., Clayton A.L., Thorne A.W., and Crane-Robinson

C. 1994. Core histone hyperacetylation co-maps with generalized DNase I sensitivity in the chicken beta-globin chromosomal domain. *EMBO J.* **13:** 1823.

Hecht A., Laroche T., Strahl-Bolsinger S., Gasser S.M., and Grunstein M. 1995. Histone H3 and H4 N-termini interact with SIR3 and SIR4 proteins: A molecular model for the formation of heterochromatin in yeast. *Cell* **80:** 583.

Hendzel M.J., Wei Y., Mancini M.A., Van Hooser A., Ranalli T., Brinkley B.R., Bazett-Jones D.P., and Allis C.D. 1997. Mitosis-specific phosphorylation of histone H3 initiates primarily within pericentromeric heterochromatin during G2 and spreads in an ordered fashion coincident with mitotic chromosome condensation. *Chromosoma* **106:** 348.

Herrera R.E., Chen F., and Weinberg R.A. 1996. Increased histone H1 phosphorylation and relaxed chromatin structure in Rb-deficient fibroblasts. *Proc. Natl. Acad. Sci.* **93::** 11510.

Hilfiker A., Hilfiker-Kleiner D., Pannuti A., and Lucchesi J.C. 1997. mof, a putative acetyl transferase gene related to the Tip60 and MOZ human genes and to the SAS genes of yeast, is required for dosage compensation in *Drosophila. EMBO J.* **16:** 2054.

Hirano T., Kobayashi R., and Hirano M. 1997. Condensins, chromosome condensation protein complexes containing XCAP- C, XCAP-E and a *Xenopus* homolog of the *Drosophila* Barren protein. *Cell* **89:** 511.

Hoffmann A., Oelgeschlager T., and Roeder R.G. 1997. Considerations of transcriptional control mechanisms: Do TFIID-core promoter complexes recapitulate nucleosome-like functions? *Proc. Natl. Acad. Sci.* **94:** 8928.

Imhof A., Yang X.J., Ogryzko V.V., Nakatani Y., Wolffe A.P., and Ge H. 1997. Acetylation of general transcription factors by histone acetyltransferases. *Curr. Biol.* **7:** 689.

Kadosh D. and Struhl K. 1998. Targeted recruitment of the Sin3-Rpd3 histone deacetylase complex generates a highly localized domain of repressed chromatin in vivo. *Mol. Cell. Biol.* **18:** 5121.

Kamei Y., Xu L., Heinzel T., Torchia J., Kurokawa R., Gloss B., Lin S.C., Heyman R.A., Rose D.W., Glass C.K., and Rosenfeld M.G. 1996. A CBP integrator complex mediates transcriptional activation and AP-1 inhibition by nuclear receptors. *Cell* **85:** 403.

Kamine J., Elangovan B., Subramanian T., Coleman D., and Chinnadurai G. 1996. Identification of a cellular protein that specifically interacts with the essential cysteine region of the HIV-1 Tat transactivator. *Virology* **216:** 357.

Koshland D. and Strunnikov A. 1996. Mitotic chromosome condensation. *Annu. Rev. Cell Dev. Biol.* **12:** 305.

Kuo M.H., Zhou J., Jambeck P., Churchill M.E., and Allis C.D. 1998. Histone acetyltransferase activity of yeast Gcn5p is required for the activation of target genes *in vivo. Genes Dev.* **12:** 627.

Kuo M.H., Brownell J.E., Sobel R.E., Ranalli T.A., Cook R.G., Edmondson D.G., Roth S.Y., and Allis C.D. 1996. Transcription-linked acetylation by Gcn5p of histones H3 and H4 at specific lysines. *Nature* **383:** 269.

Landsman D. 1996. Histone H1 in *Saccharomyces cerevisiae:* A double mystery solved? *Trends Biochem. Sci.* **21:** 287.

Laybourn P.J. and Kadonaga J.T. 1991. Role of nucleosomal cores and histone H1 in regulation of transcription by RNA polymerase II. *Science* **254:** 238.

Lee H.L. and Archer T.K. 1998. Prolonged glucocorticoid exposure dephosphorylates histone H1 and inactivates the MMTV promoter. *EMBO J.* **17:** 1454.

Li G., Sudlow G., and Belmont A.S. 1998. Interphase cell cycle dynamics of a late-replicating, heterochromatic homogeneously staining region: Precise choreography of condensation/decondensation and nuclear positioning. *J. Cell Biol.* **140:** 975.

Linder C. and Thoma F. 1994. Histone H1 expressed in *Saccharomyces cerevisiae* binds to chromatin and affects survival, growth, transcription, and plasmid stability but does not change nucleosomal spacing. *Mol. Cell. Biol* **14:** 2822.

Logie C. and Peterson C.L. 1997. Catalytic activity of the yeast

SWI/SNF complex on reconstituted nucleosome arrays. *EMBO J.* **16:** 6772.

Lu M.J., Mpoke S.S., Dadd C.A., and Allis C.D. 1995. Phosphorylated and dephosphorylated linker histone H1 reside in distinct chromatin domains in *Tetrahymena* macronuclei. *Mol. Biol. Cell* **6:** 1077.

Lu M.J., Dadd C.A., Mizzen C.A., Perry C.A., McLachlan D.R., Annunziato A.T., and Allis C.D. 1994. Generation and characterization of novel antibodies highly selective for phosphorylated linker histone H1 in *Tetrahymena* and HeLa cells. *Chromosoma* **103:** 111.

Luger K. and Richmond T.J. 1998. The histone tails of the nucleosome. *Curr. Opin. Genet. Dev.* **8:** 140.

Luger K., Mader A.W., Richmond R.K., Sargent D.F., and Richmond T.J. 1997. Crystal structure of the nucleosome core particle at 2.8 Å resolution. *Nature* **389:** 251.

Magnaghi-Jaulin L., Groisman R., Naguibneva I., Robin P., Lorain S., Le Villain J.P., Troalen F., Trouche D., and Harel-Bellan A. 1998. Retinoblastoma protein represses transcription by recruiting a histone deacetylase. *Nature* **391:** 601.

Mahadevan L.C., Willis A.C., and Barratt M.J. 1991. Rapid histone H3 phosphorylation in response to growth factors, phorbol esters, okadaic acid, and protein synthesis inhibitors. *Cell* **65:** 775.

Mizzen C.A. and Allis C.D. 1998. Linking histone acetylation to transcriptional regulation. *Cell. Mol. Life Sci.* **54:** 6.

Mizzen C.A., Yang X.J., Kokubo T., Brownell J.E., Bannister A.J., Owen-Hughes T., Workman J., Wang L., Berger S.L., Kouzarides T., Nakatani Y., and Allis C.D. 1996. The TAF(II)250 subunit of TFIID has histone acetyltransferase activity. *Cell* **87:** 1261.

Moller D.E., Xia C.H., Tang W., Zhu A.X., and Jakubowski M. 1994. Human rsk isoforms: Cloning and characterization of tissue-specific expression. *Am. J. Physiol.* **266:** C351.

Nakajima T., Fukamizu A., Takahashi J., Gage F.H., Fisher T., Blenis J., and Montminy M.R. 1996. The signal-dependent coactivator CBP is a nuclear target for pp90RSK. *Cell* **86:** 465.

Neuwald A.F. and Landsman D. 1997. GCN5-related histone N-acetyltransferases belong to a diverse superfamily that includes the yeast SPT10 protein. *Trends Biochem. Sci.* **22:** 154.

Ogryzko V.V., Schiltz R.L., Russanova V., Howard B.H., and Nakatani Y. 1996. The transcriptional coactivators p300 and CBP are histone acetyltransferases. *Cell* **87:** 953.

Ohsumi K., Katagiri C., and Kishimoto T. 1993. Chromosome condensation in *Xenopus* mitotic extracts without histone H1. *Science* **262:** 2033.

Patterton H.G., Landel C.C., Landsman D., Peterson C.L., and Simpson R.T. 1998. The biochemical and phenotypic characterization of Hho1p, the putative linker histone H1 of *Saccharomyces cerevisiae. J. Biol. Chem.* **273:** 7268.

Pazin M.J. and Kadonaga J.T. 1997. What's up and down with histone deacetylation and transcription? *Cell* **89:** 325.

Pollard K.J. and Peterson C.L. 1997. Role for ADA/GCN5 products in antagonizing chromatin-mediated transcriptional repression. *Mol. Cell. Biol.* **17:** 6212.

Pruss D., Bartholomew B., Persinger J., Hayes J., Arents G., Moudrianakis E.N., and Wolffe A.P. 1996. An asymmetric model for the nucleosome: A binding site for linker histones inside the DNA gyres. *Science* **274:** 614.

Puri P.L., Sartorelli V., Yang X.-J., Hamamori Y., Ogryzko V.V., Howard B.H., Kedes L., Wang J.Y.J., Graessmann A., Nakatani Y., and Levrero M. 1998. Differential roles of p300 and PCAF acetyltransferases in muscle differentiation. *Mol. Cell* **1:** 35.

Reifsnyder C., Lowell J., Clarke A., and Pillus L. 1996. Yeast SAS silencing genes and human genes associated with AML and HIV-1 Tat interactions are homologous with acetyltransferases. *Nat. Genet.* **14:** 42.

Roberts S.M. and Winston F. 1997. Essential functional interactions of SAGA, a *Saccharomyces cerevisiae* complex of Spt, Ada and Gcn5 proteins, with the Snf/Swi and Srb/Mediator complexes. *Genetics* **147**: 451.

Rogakou E.P., Pilch D.R., Orr A.H., Ivanova V.S., and Bonner W.M. 1998. DNA double-stranded breaks induce histone H2AX phosphorylation on serine 139. *J. Biol. Chem.* **273**: 5858.

Roth S.Y. and Allis C.D. 1992. Chromatin condensation: Does histone H1 dephosphorylation play a role? *Trends Biochem. Sci.* **17**: 93.

Rundlett S.E., Carmen A.A., Suka N., Turner B.M., and Grunstein M. 1998. Transcriptional repression by UME6 involves deacetylation of lysine 5 of histone H4 by RPD3. *Nature* **392**: 831.

Rundlett S.E., Carmen A.A., Kobayashi R., Bavykin S., Turner B.M., and Grunstein M. 1996. HDA1 and RPD3 are members of distinct yeast histone deacetylase complexes that regulate silencing and transcription. *Proc. Natl. Acad. Sci.* **93**: 14503.

Shen X. and Gorovsky M.A. 1996. Linker histone H1 regulates specific gene expression but not global transcription in vivo. *Cell* **86**: 475.

Shen X., Yu L., Weir J.W., and Gorovsky M.A. 1995. Linker histones are not essential and affect chromatin condensation in vivo. *Cell* **82**: 47.

Shikama N., Lyon J., and La Thangue N.B. 1997. The p300/CBP family: Integrating signals with transcription factors and chromatin. *Trends Cell Biol.* **7**: 230.

Smith E.R., Belote J.M., Schiltz R.L., Yang X.-J., Moore P.A., Berger S.L., Nakatani Y., and Allis C.D. 1998a. Cloning of *Drosophila* GCN5: Conserved features among metazoan GCN5 family members. *Nucleic Acids Res.* **26**: 2948.

Smith E.R., Eisen A., Gu W., Sattah M., Pannuti A., Zhou J., Cook R.G., Lucchesi J.C., and Allis C.D. 1998b. ESA1 is a histone acetyltransferase that is essential for growth in yeast. *Proc. Natl. Acad. Sci.* **95**: 3561.

Sobel R.E., Cook R.G., and Allis C.D. 1994. Non-random acetylation of histone H4 by a cytoplasmic histone acetyltransferase as determined by novel methodology. *J. Biol. Chem.* **269**: 18576.

Sobel R.E., Cook R.G., Perry C.A., Annunziato A.T., and Allis C.D. 1995. Conservation of deposition-related acetylation sites in newly synthesized histones H3 and H4. *Proc. Natl. Acad. Sci.* **92**: 1237.

Spencer T.E., Jenster G., Burcin M.M., Allis C.D., Zhou J., Mizzen C.A., McKenna N.J., Onate S.A., Tsai S.Y., Tsai M.J., and O'Malley B.W. 1997. Steroid receptor coactivator-1 is a histone acetyltransferase. *Nature* **389**: 194.

Steinbach O.C., Wolffe A.P., and Rupp R.A. 1997. Somatic linker histones cause loss of mesodermal competence in *Xenopus*. *Nature* **389**: 395.

Struhl K. 1998. Histone acetylation and transcriptional regulatory mechanisms. *Genes Dev.* **12**: 599.

Taunton J., Hassig C.A., and Schreiber S.L. 1996. A mammalian histone deacetylase related to the yeast transcriptional regulator Rpd3p. *Science* **272**: 408.

Taylor W.R., Chadee D.N., Allis C.D., Wright J.A., and Davie J.R. 1995. Fibroblasts transformed by combinations of *ras*, *myc* and mutant p53 exhibit increased phosphorylation of histone H1 that is independent of metastatic potential. *FEBS Lett.* **377**: 51.

Thoma F., Koller T., and Klug A. 1979. Involvement of histone H1 in the organization of the nucleosome and of the salt-dependent superstructures of chromatin. *J. Cell Biol.* **83**: 403.

Treisman R. 1996. Regulation of transcription by MAP kinase cascades. *Curr. Opin. Cell Biol.* **8**: 205.

Trivier E., De Cesare D., Jacquot S., Pannetier S., Zackai E., Young I., Mandel J.L., Sassone-Corsi P., and Hanauer A. 1996. Mutations in the kinase Rsk-2 associated with Coffin-Lowry syndrome. *Nature* **384**: 567.

Tse C., Sera T., Wolffe A.P., and Hansen J.C. 1998. Disruption of higher order folding by core histone acetylation dramatically enhances transcription of nucleosomal arrays by RNA polymerase III. *Mol. Cell. Biol.* (in press).

Turner B.M. 1998. Histone acetylation as an epigenetic determinant of long-term transcriptional competence. *Cell. Mol. Life Sci.* **54::** 21.

Ushinsky S.C., Bussey H., Ahmed A.A., Wang Y., Friesen J., Williams B.A., and Storms R.K. 1997. Histone H1 in *Saccharomyces cerevisiae*. *Yeast* **13**: 151.

Utley R.T., Ikeda K., Grant P.A., Cote J., Steger D.J., Eberharter A., John S., and Workman J.L. 1998. Transcriptional activators direct histone acetyltransferase complexes to nucleosome. *Nature* **394**: 498.

van Holde K.E. 1989. *Chromatin.* Springer-Verlag, New York.

van Holde K. and Zlatanova J. 1995. Chromatin higher order structure: Chasing a mirage? *J. Biol. Chem.* **270**: 8373.

———. 1996. What determines the folding of the chromatin fiber? *Proc. Natl. Acad. Sci.* **93**: 10548.

Wei Y. and Allis C.D. 1998. A new marker for mitosis. *Trends Cell Biol.* **8**: 266.

Wolffe A. 1995. *Chromatin structure and function.* Academic Press, London.

Wolffe A.P. and Pruss D. 1996. Targeting chromatin disruption: Transcription regulators that acetylate histones. *Cell* **84**: 817.

Wolffe A.P., Khochbin S., and Dimitrov S. 1997. What do linker histones do in chromatin? *BioEssays* **19**: 249.

Wu M., Allis C.D., Richman R., Cook R.G., and Gorovsky M.A. 1986. An intervening sequence in an unusual histone H1 gene of *Tetrahymena thermophila*. *Proc. Natl. Acad. Sci.* **83**: 8674.

Yamamoto T. and Horikoshi M. 1997. Novel substrate specificity of the histone acetyltransferase activity of HIV-1-Tat interactive protein Tip60. *J. Biol. Chem.* **272**: 30595.

Yang X.J., Ogryzko V.V., Nishikawa J., Howard B.H., and Nakatani Y. 1996. A p300/CBP-associated factor that competes with the adenoviral oncoprotein E1A. *Nature* **382**: 319.

Zhao Y., Bjorbaek C., and Moller D.E. 1996. Regulation and interaction of pp90(rsk) isoforms with mitogen-activated protein kinases. *J. Biol. Chem.* **271**: 29773.

Zhao Y., Bjorbaek C., Weremowicz S., Morton C.C., and Moller D.E. 1995. RSK3 encodes a novel pp90rsk isoform with a unique N-terminal sequence: Growth factor-stimulated kinase function and nuclear translocation. *Mol. Cell. Biol.* **15**: 4353.

Zlatanova J. and van Holde K. 1996. The linker histones and chromatin structure: New twists. *Prog. Nucleic Acid Res. Mol. Biol.* **52**: 217.

Regulation of Transcription by Multisubunit Complexes That Alter Nucleosome Structure

D.J. STEGER,* R.T. UTLEY,* P.A. GRANT,* S. JOHN,* A. EBERHARTER,* J. CÔTÉ,†
T. OWEN-HUGHES,‡ K. IKEDA,§ AND J.L. WORKMAN*

*Howard Hughes Medical Institute and Department of Biochemistry and Molecular Biology, Pennsylvania
State University, University Park, Pennsylvania 16802; †Laval University Cancer Research Center,
Quebec City, Quebec, Canada; ‡Department of Biochemistry, University of Dundee, Dundee, Scotland;
§Department of Biology, Jichi Medical School, Japan

An emerging picture within the field of eukaryotic gene expression is beginning to highlight critical interactions between transcriptional regulatory proteins and chromatin structures in the activation and repression of transcription. The integral part played by chromatin in transcription is exemplified by studies demonstrating that genes normally turned off in yeast under standard growth conditions become activated by reducing nucleosomal density throughout the genome by depleting histone levels in the cell (Han and Grunstein 1988; Han et al. 1988). In agreement with these and other in vivo observations, a multitude of in vitro analyses using nucleosome-assembled templates have revealed that nucleosomes repress transcription and strengthen the requirement for sequence-specific regulators in the induction of transcription (for review, see Felsenfeld 1992; Kornberg and Lorch 1992; Owen-Hughes and Workman 1994; Paranjape et al. 1994).

How upstream activator proteins interact with nucleosomal DNA to acquire access to binding sites is an important step in transcriptional regulatory pathways (for review, see Wolffe 1994; Varga-Weisz and Becker 1995; Steger and Workman 1996; Svaren and Horz 1996). In vitro, the binding of sequence-specific activators and histones to DNA is not mutually exclusive. However, consistent with the idea that histones repress transcription, the affinities transcription factors have for sites positioned in nucleosomes are often dramatically reduced compared to those for naked DNA. An important determinant for factor binding to nucleosomal DNA involves the location of recognition elements relative to the underlying histone octamer. All of the DNA comprising a nucleosome is not complexed with equal affinity to the octamer, such that sequences positioned near the nucleosome edge are more accessible to nucleases than those located close to the dyad axis (Simpson 1979). In agreement with this early observation, DNA-binding proteins have greater affinities for sites positioned near the edge of the nucleosome versus its center (Li and Wrange 1993; Vettese-Dadey et al. 1994). Factor binding to sites deep within the nucleosome can be achieved in vitro, however, through cooperative binding among factors targeting multiple sites contained within the same nucleosome (Taylor et al. 1991; Adams and Workman 1995). It

has been proposed that cooperative interactions derive from the initial binding of a protein at the edge of a nucleosome, which stabilizes the transient release of DNA sequences from the surface of the histone octamer and increases the probability of exposure of adjacent sites for other proteins (Polach and Widom 1996). Given the presence of a histone sink such as the histone chaperon nucleoplasmin, the binding of several activators to a single nucleosome may sufficiently disrupt histone-DNA contacts so that the underlying octamer is removed from the DNA (Owen-Hughes and Workman 1996). Thus, at least in some circumstances in vitro, the binding of sequence-specific activators to a nucleosome initiates histone displacement and results in high-affinity binding of the activators to naked DNA.

A conclusion drawn from biochemical analyses is that the nucleosomal structure encountered by a sequence-specific activator determines whether effective binding occurs. Importantly, nucleosomal structure is not static. Distinct subpopulations of nucleosomes isolated from cells have nuclease sensitivities and chromatographic properties different from those of bulk nucleosomes (van Holde 1989). As a result, the interactions DNA-binding proteins have with chromatin are necessarily dynamic, which suggests that these interactions could serve as regulatory steps in gene activation.

Genetic screens that have implicated histones in transcriptional regulation have provided a major advance in understanding how transcription factors productively interact with chromatin. These studies, along with subsequent biochemical analyses, have identified multiprotein complexes whose primary function is to help activate gene expression by altering chromatin so that its DNA sequences become more accessible to *trans*-acting factors (for review, see Kadonaga 1998; Workman and Kingston 1998). For example, five protein complexes that are able to modify nucleosomal structure in an ATP-dependent fashion have been described. These are SWI/SNF, NURF, RSC, CHRAC and ACF, and each complex can increase either transcription factor binding or restriction enzyme access to nucleosomal DNA (Côté et al. 1994; Kwon et al. 1994; Tsukiyama and Wu 1995; Cairns et al. 1996; Ito et al. 1997; Varga-Weisz et al. 1997). Recent discoveries demonstrating that many transcriptional coactivators possess histone acetyltransferase (HAT) activity and that several transcriptional corepressors contain histone deacetylase function have further emphasized the important part played by histones as targets for transcriptional regulators (for review, see Grunstein 1997; Pazin and Kadonaga 1997; Wade et al. 1997; Wolffe et al. 1997). A correlation between the acetylation of lysine residues within the amino-terminal tails of the core histones and the transcriptional activity of cellular chromatin has been recognized for many years (for review, see Turner and O'Neill 1995). These findings now directly link histone acetylation with transcription.

Studies involving chromatin remodeling complexes provide solid support to the idea that the repressive effects of chromatin on transcription can be counteracted by cellular activities which directly modify nucleosomal structure. We address this idea here by presenting recent data with SWI/SNF and native HAT complexes purified from the yeast *Saccharomyces cerevisiae*. Evidence is described indicating that SWI/SNF facilitates transcription factor binding by disrupting histone-DNA contacts. SWI/SNF is demonstrated to further remodel unstable transcription factor–nucleosome complexes by participating in removing histones from the DNA. Our studies focusing on histone acetylation involve the identification and characterization of four distinct nucleosomal HAT activities from yeast. One of these multiprotein complexes, termed SAGA, is revealed to be composed of subunits belonging to the TAF, SPT, and Ada protein families of transcriptional regulators. Consistent with the idea that histone acetylation regulates gene expression, SAGA and the remaining three complexes termed Ada, NuA4, and NuA3 are revealed to stimulate transcription in vitro from nucleosomal templates in an acetyl-CoA-dependent manner. To address targeting of HAT complexes to particular genes via interactions with sequence-specific DNA-binding proteins, data from a gel-shift assay are presented illustrating the preferential acetylation of factor-bound nucleosomes. We summarize the data described here by discussing potential mechanisms of SWI/SNF and HAT function in the regulation of transcription.

EXPERIMENTAL PROCEDURES

Detailed methods for all of the experiments described below can be found elsewhere. Protocols involving yeast SWI/SNF are presented by Côté et al. (1998) and Utley et al. (1997). Experimental procedures describing the purification and characterization of yeast HAT complexes can be found in Grant et al. (1997, 1998a,c) and Eberharter et al. (1998). A procedure for analyzing transcription in vitro from nucleosome-assembled templates in the presence and absence of purified HATs is provided by Steger et al. (1998). Experimental methods designed to address the targeting of HAT complexes to particular genes via sequence-specific activators are described by Utley et al. (1998).

RESULTS

SWI/SNF Enhances the Binding of Transcription Factors to Nucleosomal DNA by Disrupting Nucleosomal Structure

SWI/SNF is a 2-MD multiprotein complex required for the function of several gene-specific transcriptional activators and for the transcriptional induction of a subset of yeast genes (for review, see Carlson and Laurent 1994; Peterson and Tamkun 1995). Multiple studies in vivo have led to the proposal that SWI/SNF facilitates activator function by alleviating chromatin-mediated repression. Mutations in genes that encode histones (Hirschhorn et al. 1992, 1995; Kruger et al. 1995) or nonhistone chromosomal proteins (Kruger and Herskowitz 1991) partially suppress transcription defects associated with *swi/snf* mutations. Moreover, SWI/SNF is required for changes in chromatin structure at the SUC2 promoter that coincide with the transcriptional activation of this gene in vivo (Hirschhorn et al. 1992, 1995). This remodeling may result in part from SWI/SNF antagonizing the activity of the Ssn6-Tup1 repressor complex (Gavin and Simpson 1997).

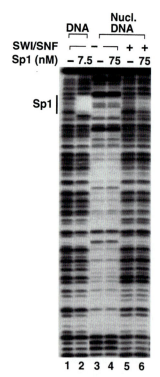

Figure 1. SWI/SNF disrupts histone-DNA contacts from nucleosome core particles and stimulates Sp1 binding. A 167-bp fragment containing an Sp1 recognition element 35 bp from one end was assembled into nucleosome core particles (Nucl. DNA) or mock-reconstituted (DNA). Core particles and histone-free DNA were incubated with and without Sp1 and purified SWI/SNF as indicated. All reactions contained 1 mM ATP. Samples were subsequently digested with DNase I and electrophoresed through an acrylamide denaturing gel. Protection by Sp1 from DNase I cleavage is indicated.

To investigate potential mechanisms of SWI/SNF function, we and others have biochemically examined how the complex alters nucleosomal structure, and what effect this may have on the interactions between sequence-specific activators and chromatin. These in vitro studies have revealed that SWI/SNF from either yeast or mammals is able to disrupt DNase I digestion patterns of nucleosome core particles and stimulate the binding of a diverse set of activators including GAL4 derivatives, TBP, Sp1, NF-κB, and USF (Côté et al. 1994; Imbalzano et al. 1994; Kwon et al. 1994; Wang et al. 1996; Utley et al. 1997). An example of this result is illustrated in Figure 1. Addition of SWI/SNF to nucleosome cores (Fig. 1, lane 5), which in this case harbor a single Sp1 recognition element, produces a pattern of DNase I cutting dramatically different from that generated with core particles in the absence of SWI/SNF (Fig. 1, lane 3), which more closely resembles the digestion profile from histone-free DNA (Fig. 1, lane 1). Furthermore, the appearance of an Sp1 footprint on nucleosomal DNA in the presence of both Sp1 and SWI/SNF (Fig. 1, lane 6), but not Sp1 alone (Fig. 1, lane 4), indicates that SWI/SNF stimulates the binding of this protein to nucleosome cores.

In all cases tested, both the disruption of nucleosomal structure and stimulation of factor binding by SWI/SNF require ATP. This is consistent with the fact that the SWI2/SNF2 subunit contains a DNA-dependent ATPase domain which is required for SWI/SNF function in vivo (Laurent et al. 1993). However, ATP hydrolysis is not continuously required for SWI/SNF to enhance factor binding to nucleosome core particles. Following SWI/SNF disruption of nucleosome cores, ATP can be depleted without loss of the disrupted state, and the remaining SWI/SNF-bound nucleosomes retain an increased affinity for GAL4-AH (Imbalzano et al. 1996). Our recent results demonstrate that SWI/SNF-mediated changes in nucleosome core structure can persist for an extended period after detachment of the complex from core particles (Côté et al. 1998). Removing SWI/SNF from disrupted nucleosomes does not reduce the affinity of GAL4-AH for nucleosomal sites, indicating that the stimulation of transcription factor binding does not require SWI/SNF-transcription factor interactions. Instead, the stimulation of factor binding is caused solely by SWI/SNF reconfiguring nucleosome core structure. Given time, altered nucleosomes revert back to their original configurations in the absence of further SWI/SNF action, arguing against a loss of histone components during SWI/SNF remodeling. Together, these studies suggest that SWI/SNF uses the energy of ATP hydrolysis to loosen histone-DNA contacts without removing histones from nucleosomes, which in turn increases access to the DNA by trans-acting factors.

The ability of SWI/SNF to remodel nucleosome core particles also extends to arrays of nucleosomes. Addition of SWI/SNF to a chromatin template composed of several positioned nucleosomes causes the defined nucleosomal boundaries to break down as determined by MNase digestion, suggesting that the complex is able to disrupt multiple nucleosomes within an array (Owen-Hughes et al. 1996). This disruption stimulates transcription factor binding (Owen-Hughes et al. 1996) and restriction endonuclease cutting (Logie and Peterson 1997).

We have also described a permanent effect of SWI/SNF action. The complex is able to mediate the formation of a transcription-factor-dependent DNase-I-hypersensitive site within an array of nucleosomes in vitro, which remains upon removal of SWI/SNF (Owen-Hughes et al. 1996). The combined actions of SWI/SNF and five GAL4 dimers binding to either a nucleosome core particle or a nucleosome within an array of nucleosomes were demonstrated to sufficiently destabilize the underlying histone octamer to displace histones from the DNA. Other transcription factor–nucleosome complexes, however, resist histone displacement from SWI/SNF (Steger and Workman 1997). In this study, factors targeting a human immunodeficiency virus type-1 (HIV-1) nucleosome were found to co-occupy the DNA along with histones. An example of SWI/SNF-mediated displacement of histones from GAL4-AH-bound nucleosome core particles, but not from HIV-1 nucleosome cores bound by Sp1, NF-κB p50, LEF-1, ETS-1, and USF, is presented in Figure 2. In this experiment, 5-GAL4 site nucleosome cores and HIV-1 core particles were incubated with amounts of activators sufficient to achieve complete binding in the absence or presence of SWI/SNF. Following binding, activators were removed from complexes by oligonucleotide competition, and the resulting products were analyzed by gel electrophoresis.

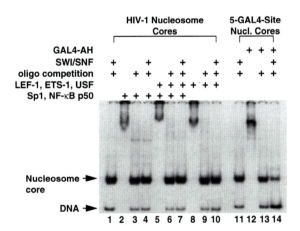

Figure 2. Factor-bound HIV-1 nucleosome core particles resist histone displacement by SWI/SNF, whereas histones are removed by SWI/SNF from GAL4-AH-bound nucleosome cores. Sp1 (23 nM), NF-κB p50 (150 nM), LEF-1 (50 nM), ETS-1 (100 nM), and USF (150 nM) were added to HIV-1 nucleosome cores (composed of HIV-1 5′ LTR sequence from –199 to –49) with and without purified SWI/SNF as indicated. Following the binding-reaction-incubation period, the designated samples were passed through an oligonucleotide competition step to remove bound factors from the mononucleosomes. Binding of GAL4-AH to the 5-GAL4-site core particles was performed similarly to the HIV-1 reactions. All samples were subjected to EMSA. (Reprinted, with permission, from Steger and Workman 1997 [copyright Oxford University Press].)

The appearance of free DNA, rather than nucleosome cores, after oligonucleotide challenge indicates that histone displacement occurred. As illustrated in Figure 2, only the combination of GAL4-AH and SWI/SNF leads to the disappearance of nucleosome cores and the corresponding appearance of free DNA following removal of bound factors (Fig. 2, lane 14). The data suggest that transcription factor–nucleosome complexes have distinct stabilities that are differentially effected by SWI/SNF. Interestingly, the formation of stable transcription factor–nucleosome complexes in vitro may indicate continued functions for histones at active promoters in vivo.

Identification and Characterization of Native HAT Activities from *S. cerevisiae*

The cloning of a nuclear HAT from *Tetrahymena* led to its identification as a homolog of the yeast transcriptional coactivator protein, Gcn5 (Brownell and Allis 1995; Brownell et al. 1996). Gcn5 (Ada4), along with Ada1, Ada2, Ada3, and Ada5/Spt20, is a member of the Ada family of proteins (Berger et al. 1992; Piña et al. 1993; Grant et al. 1998b). Genetic and biochemical data indicate that Ada proteins function together as a complex that serves as an intermediary between gene-specific and general transcriptional activators (see, e.g., Georgakopoilos and Thireos 1992; Melcher and Johnston 1995; Candau and Berger 1996; Martens et al. 1996).

Although recombinant yeast Gcn5 acetylates free histones, it cannot modify histones in the context of nucleosomes. This suggests that additional factors, potentially including other Ada proteins, are required by Gcn5 for the acetylation of chromosomal histones. We have examined this issue and investigated the possibility that non-Gcn5-containing HAT complexes exist in yeast by fractionating whole-cell extracts and assaying fractions for the ability to acetylate nucleosomal histones (Grant et al. 1997). It was found that Ni-agarose binds substantial HAT activity from yeast extracts, and further fractionation of the Ni-agarose-bound material using ion exchange chromatography separates this activity into four distinct native HATs. An example of this result is presented Figure 3. Activities 1 and 4, which primarily acetylate nucleosomal histone H3 but also modify H2B to a lesser extent, have been named Ada and SAGA, respectively. Complexes 2 and 3, which predominantly acetylate his-

Table 1. Partial Subunit Compositions for Native Yeast HAT Complexes

Ada (0.8 MD)	SAGA (1.8 MD)	NuA4 (1.3 MD)	NuA3 (0.4 MD)
Ada2, Ada, 3 **Gcn5**	Ada1, Ada2, Ada3, **Gcn5**, Ada5 (spt20) Spt3, Spt7, Spt8, Spt20 (Ada5) TAF$_{II}$90, TAF$_{II}$68, TAF$_{II}$60, TAF$_{II}$25, TAF$_{II}$20	**Esa1**	?

Catalytic subunits are indicated in bold type.

tones H4 and H3, respectively, have been named NuA4 (*nu*cleosome *a*cetyltransferase of histone H*4*) and NuA3 according to their acetylation preferences.

Size-exclusion chromatography revealed that the four HAT activities exist as large multiprotein complexes with molecular masses ranging from 0.4 MD for NuA3 to 1.8 MD for SAGA (Grant et al. 1997; S. John, unpubl.). Through a combination of biochemical and genetic approaches, the subunit compositions for the SAGA and Ada complexes have been partially determined (Grant et al. 1997, 1998c; Roberts and Winston 1997), and the data are summarized in Table 1. Both contain subunits belonging to the Ada protein family, and Gcn5 is the primary if not the only subunit possessing catalytic HAT activity. Although not part of the Ada complex, several Spt proteins are present in SAGA. *SPT* genes were identified by mutations that suppress Ty and δ insertion mutations in the promoter regions of particular yeast genes. A subset of *SPT* genes composed of *SPT3*, *SPT7*, *SPT8*, and *SPT20* are believed to help TBP function at specific promoters (Winston 1992). As its name implies, SAGA is an *S*pt, *A*da, *G*cn5 *a*cetyltransferase.

It is quite striking that SAGA also contains components that are TBP-associated factors, TAF$_{II}$s. TBP and TAF$_{II}$s make up TFIID, which upon binding to the TATA box of core promoters is thought to initiate the sequential loading of other general transcription factors and RNA polymerase II itself (for review, see Burley and Roeder 1996; Tansey and Herr 1997). Interestingly, depletion of TAF$_{II}$68 from SAGA disrupts its in vitro abilities to acetylate nucleosomal histones and stimulate transcription (Grant et al. 1998c). This suggests a role independent of TFIID for the subset of TAF$_{II}$ proteins present in SAGA that is involved in the regulation of transcription by histone acetylation. Members from the Spt and Ada families of proteins have also been demonstrated to be essential for SAGA function (Grant et al. 1997; Wang et al. 1998), indicating that three classes of transcriptional regulators are united within a single multiprotein complex. The recent purification of two native HAT activities isolated from human cells that contain Gcn5 homologs (i.e., PCAF or hGcn5) has revealed dramatic parallels between the subunits present in these complexes and those in SAGA. The PCAF and hGcn5 complexes contain human versions of Ada2, Ada3, and Spt3, as well as a subset of TAF proteins nearly identical to that present in SAGA (Ogryzko et al. 1998).

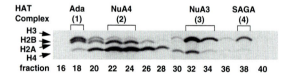

Figure 3. Biochemical separation of native yeast HAT complexes with distinct acetylation specificities for nucleosomal histones. Yeast whole-cell extract was loaded onto a Ni^{++}-NTA agarose column, and the eluate applied to a Mono Q column. Mono Q fractions were assayed for nucleosomal HAT activity by incubating with HeLa nucleosomes and ^{3}H-labeled acetyl CoA, followed by SDS-PAGE. Shown is a fluorograph of the gel, with arrows indicating the positions of the core histones.

NuA4 and NuA3 are Gcn5-independent HAT complexes (Grant et al. 1997). Although the subunit compositions for these complexes remain largely unknown, recent evidence indicates that the catalytic component for NuA4 is a protein termed *essential SAS2-related acetyl-transferase* or Esa1 (J. Côté et al., in prep.). Recombinant Esa1 acetylates H4, H2A, and H3 when presented with free histones, but like recombinant Gcn5, it does not modify nucleosomal histones (Smith et al. 1998). Esa1 is believed to have a role in transcription based on its strong homology with the *Drosophila* Mof protein, which was discovered in a genetic screen for gene products necessary for the increased transcription and acetylation of H4 associated with dosage compensation (Hilfiker et al. 1997).

Functional Analysis of HAT Complexes

As mentioned above, the state of histone acetylation at genomic loci is correlated with transcription. Hyperacetylated histones appear to accumulate in actively transcribed chromatin (Hebbes et al. 1994), whereas hypoacetylated histones are frequently enriched in silent domains (Braunstein et al. 1993; Jeppesen and Turner 1993).

In addition to acetylating nucleosomal histones, at least two of the HAT complexes discussed above are composed of proteins that function as transcriptional regulators. This strongly suggests a role for these complexes in transcription. We have investigated this idea by analyzing transcription in vitro from both histone-free and nucleosome-assembled DNA templates in the presence of purified SAGA, Ada, NuA4, and NuA3 (Steger et al. 1998). A promoter chosen for these experiments is from the HIV-1 5′ long terminal repeat (LTR), since histone acetylation has been implicated in the control of HIV-1 gene transcription in cultured cell lines (Van Lint et al. 1996). Our results demonstrate that all four of the native yeast HAT complexes are capable of enhancing HIV-1 transcription from nucleosomal DNA. Furthermore, transcriptional stimulation by these HATs is dependent on the presence of acetyl CoA, indicating that acetylation is required for the effect. Figure 4 illustrates this result for SAGA. In this example, SAGA stimulates HIV-1 transcription 16-fold in the presence of acetyl CoA (Fig. 4, compare lanes 3 and 4) and has little or no effect on transcription in the absence of acetyl CoA (Fig. 4, lanes 1 and 2). Although not shown, SAGA, in addition to the other HAT activities, does not regulate HIV-1 transcription from histone-free DNA (Steger et al. 1998). For Ada and NuA4, we have demonstrated that acetylation of only the histones associated with the HIV-1 template mediates enhanced transcription, suggesting that these complexes facilitate transcription, at least in part, by modifying histone proteins. In addition, assembly of the HIV-1 promoter into chromatin greatly represses transcription. Compared to naked DNA, the level of transcription directed from the nucleosomal DNA is reduced approximately 200-fold. Collectively, the data demonstrate that SAGA, Ada, NuA4, and NuA3 can function in vitro to relieve nucleosome-mediated transcriptional repression.

That some genes contain hyperacetylated histones, whereas others contain hypoacetylated histones implies that HATs and histone deacetylases are specifically recruited to particular regions of the genome. Indeed, a recent investigation of GCN5-dependent genes has revealed that histone H3 acetylation is specifically enriched at the promoter regions of these genes (Kuo et al. 1998). It is possible that the targeting of HATs and deacetylases is mediated by physical interactions with sequence-specific DNA-binding proteins. Previous studies have identified interactions between Ada2 and the VP16 activation domain (Silverman et al. 1994; Barlev et al. 1995), and Ada2 and Ada3 have been shown to potentiate transcription from VP16- and Gcn4-dependent reporter genes in yeast (Berger et al. 1992; Piña et al. 1993). In light of these results, we have examined whether NuA4, NuA3, and the two Ada-containing complexes, SAGA and Ada, directly interact with Gcn4 and the VP16 activation domain (Utley et al. 1998). GST pull-down assays revealed that SAGA and NuA4 interact with both Gcn4 and the VP16 activation domain, whereas Ada and NuA3 do not. The lack of binding between the Ada HAT and the VP16

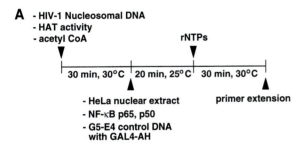

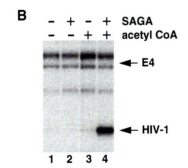

Figure 4. SAGA stimulates HIV-1 transcription from chromatin templates. (*A*) Schematic representation of the in vitro transcription protocol. The HAT complex acetylates nucleosomal histones during the first incubation period when provided with acetyl CoA. Preinitiation complexes are subsequently formed at the promoter during a 20-min incubation with HeLa nuclear extract at room temperature. The p65 and p50 subunits of NF-κB were added to enhance HIV-1 transcription levels. G5-E4 DNA (i.e., five GAL4 sites upstream of a minimal E4 promoter) and GAL4-AH are included to serve as a control for RNA recovery through subsequent steps. Transcription is initiated upon the addition of rNTPs, and RNA transcripts are detected by primer extension. (*B*) HIV-1 nucleosomal DNA (10 ng) was incubated in the absence and presence of acetyl CoA (1 μM) and SAGA, followed by the addition of HeLa nuclear extract (75 μg) for transcription. Primer extension products from HIV-1 and E4 RNA transcripts are indicated.

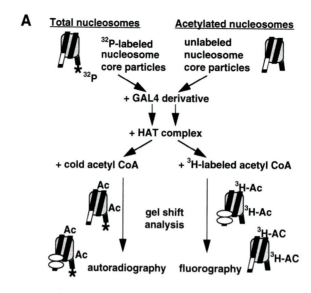

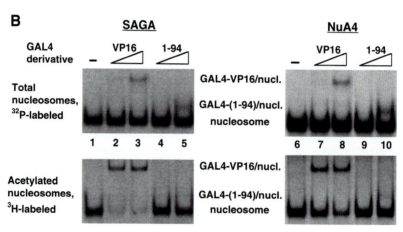

Figure 5. HAT complex-GAL4-VP16 interactions target histone acetylation to activator-bound nucleosome core particles. (*A*) Schematic diagram of the protocol designed to examine targeting of histone acetylation. Nucleosome cores containing a single GAL4 recognition element (*white rectangle*) positioned at the nucleosome edge were incubated with GAL4 derivatives, HATs, and acetyl CoA in the order indicated, followed by gel mobility shift analysis. Dimers of GAL4 derivatives (*paired ovals*) are shown bound to nucleosome core particles in the final step. Ac denotes hyperacetylated nucleosome cores. (*B*) Comparison of bulk ^{32}P-labeled nucleosomes (*top panels*) with ^{3}H-labeled, acetylated nucleosomes (*bottom panels*) in the presence of 5 and 15 nM GAL4-VP16 or 24 and 80 nM GAL4-(1-94). SAGA was present in lanes 1–5 and NuA4 in lanes 1–10. The positions in the gels of free nucleosome cores and factor-bound nucleosomes are indicated.

activation domain suggests that Ada2 may be masked by other subunits within the complex. Alternatively, components other than Ada proteins in SAGA may be mediating interactions with the VP16 activation domain, which is reasonable given that NuA4-VP16 interactions cannot be mediated by Ada proteins since these are not part of the NuA4 complex.

Interactions between DNA-binding activators and HATs suggest that histone acetylation may be targeted to regions of chromatin bound by activators. To test this hypothesis, we developed the targeting assay outlined in Figure 5A and have found that acetylation by SAGA and NuA4, but not Ada and NuA3, is preferentially directed to nucleosome cores bound by GAL4-VP16 relative to those not bound by the activator (Utley et al. 1998). An example of this result presented in Figure 5B demon-

strates that the fraction of GAL4-VP16-bound nucleosomes acetylated by SAGA (bottom panel, lanes 2 and 3) or NuA4 (bottom panel, lanes 7 and 8) is greater than the fraction of total nucleosomes bound by GAL4-VP16 in the presence of SAGA (top panel, lanes 2 and 3) or NuA4 (top panel, lanes 7 and 8). In other words, although SAGA and NuA4 are capable of acetylating nucleosomal histones in the absence of bound activator (Fig. 5, bottom panels, lanes 1 and 6), the enriched acetylation of the factor-bound nucleosomes indicates that they are a preferred target for this modification. Directed acetylation by SAGA (Fig. 5, top and bottom panels, lanes 4 and 5) or NuA4 (Fig. 5, top and bottom panels, lanes 9 and 10) is not observed with only the GAL4-DNA-binding domain. This suggests that interactions between the HATs and the VP16 activation domain, rather than those between the

GAL4-DNA-binding domain and the nucleosome core, are responsible for targeted acetylation.

Similar to the findings discussed above for the HIV-1 promoter, SAGA and NuA4 stimulate GAL4-VP16-dependent transcription from chromatin templates in an acetyl-CoA-dependent fashion (Utley et al. 1998). As a whole, the functional data suggest that the recruitment of complexes possessing HAT activity via direct interactions with transcription factors bound to nucleosomal DNA targets histone acetylation to specific genes, which in turn facilitates transcription from those genes.

DISCUSSION

This paper highlights our recent data with two distinct types of nucleosome-modifying activities. For yeast SWI/SNF, we have attempted to provide evidence indicating that the complex relieves chromatin-mediated repression by altering nucleosomal structure, which subsequently facilitates the binding of sequence-specific transcription factors. Through a mechanism not fully understood, the complex harnesses the energy from ATP hydrolysis to disrupt histone-DNA contacts in nucleosomes without removing histones. Furthermore, SWI/SNF remodeling renders nucleosomal arrays sensitive to interactions with sequence-specific activators, such that their binding can lead to displacement of the underlying histones in some cases.

To summarize the data involving native yeast HAT complexes, we present a schematic model illustrating potential mechanisms for the functions of these activities in transcriptional regulation (Fig. 6). Through direct interactions with sequence-specific activators bound at promoter regions, HATs are recruited to particular genes, leading to the acetylation of histones at these sites. Proteins other than histones may also be targeted by HATs, since it has been demonstrated that the tumor suppressor protein, p53 (Gu and Roeder 1997), the general transcription factors, TFIIEβ and TFIIF (Imhof et al. 1997), and the HMG1-like protein, Sin1 (Pollard and Peterson 1997) are acetylated in vitro by activities also modifying nucleosomal histones. The in vitro transcription experiments discussed here strongly support the notion that histone acetylation facilitates gene transcription. However, as no direct evidence exists describing how this works, we provide a few speculations, none of which are mutually exclusive. Histone acetylation may stabilize low-affinity and/or transient binding of certain activators in chromatin. Indeed, the acetylation of histone H4 has been reported to enhance the nucleosome binding of USF (Vettese-Dadey et al. 1996). Histone acetylation may stimulate transcription by functioning as a signal to recruit chromatin-specific regulators such as ATP-dependent nucleosome remodeling complexes or activities yet to be defined. It is also possible that general transcription factors function more efficiently in acetylated chromatin, leading to facilitated initiation and/or elongation. Alternatively, acetylation of sequence-specific activators and/or general transcription factors rather than, or in addition to, histones may account for stimulated activities of these proteins in nucleosomal

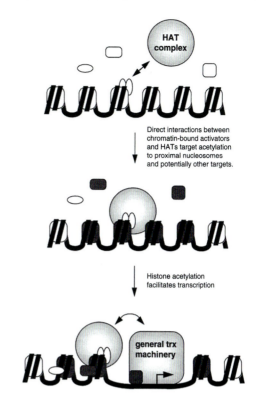

Figure 6. Hypothetical model for the regulation of transcription by HATs. See text for details. Protein-protein interactions are indicated by double-ended arrows. Acetylation of transcription factors and nucleosomal histones is indicated by dark shading.

DNA. Functions not involving acetylation may also be performed by some HAT complexes. SAGA and Ada, which are composed of coactivator subunits, may participate in the regulation of transcription by bridging upstream activators to the basal transcription machinery through direct protein-protein interactions.

An interesting direction for future study involves a biochemical analysis of potential relationships between different chromatin remodeling activities such as SWI/SNF and HATs. Mutations in ADA2, ADA3, and GCN5 cause phenotypes similar to that of *swi/snf* mutants, and these Ada genes are required for full expression of SWI/SNF-dependent genes in vivo (Pollard and Peterson 1997). In addition, a genetic screen for mutations that cause lethality in an *ada5/spt20* mutant identified a member of the SWI/SNF complex (Roberts and Winston 1997). From the first study, these authors suggested that ADA/GCN5 and SWI/SNF complexes enhance activator function by acting together to alter chromatin structure. However, from the second genetic analysis, the authors interpreted their findings to indicate that SWI/SNF and ADA/GCN5 complexes act in parallel pathways and, as a result, provide a mutual backup system for RNA polymerase II transcription. Whether these complexes function in a synergistic or redundant fashion can be directly addressed in vitro by examining transcription from nucleosomal templates in the presence of purified SWI/SNF and HATs.

ACKNOWLEDGMENTS

Support for this work was provided by a grant from NIGMS to J.L.W. D.J.S. and S.J. are postdoctoral associates of the Howard Hughes Medical Institute. P.A.G. is funded by postdoctoral fellowship PF-98-017-01-GMC from the American Cancer Society. A.E. is a recipient of a postdoctoral fellowship from the Austrian Science Foundation (FWF). J.L.W. is an associate investigator of the Howard Hughes Medical Institute.

REFERENCES

Adams C.C. and Workman J.L. 1995. The binding of disparate transcription factors to nucleosomal DNA is inherently cooperative. *Mol. Cell. Biol.* **15:** 1405.

Barlev N.A., Candau R., Wang L., Darpino P., Silverman N., and Berger S.L. 1995. Characterization of physical interactions of the putative transcriptional adaptor, ADA2, with acidic activation domains and TATA-binding protein. *J. Biol. Chem.* **270:** 19337.

Berger S.L., Pina B., Silverman N., Marcus G.A., Agapite J., Reigier J.L., Triezenberg S.J., and Guarente L. 1992. Genetic isolation of ADA2: A potential transcriptional adaptor required for function of certain acidic activation domains. *Cell* **70:** 251.

Braunstein M., Rose A.B., Holmes S.G., Allis C.D., and Broach J.R. 1993. Transcriptional silencing in yeast is associated with reduced nucleosome acetylation. *Genes Dev.* **7:** 592.

Brownell J.E. and Allis C.D. 1995. An activity gel assay detects a single, catalytically active histone acetyltransferase subunit in *Tetrahymena* macronuclei. *Proc. Natl. Acad. Sci.* **92:** 6364.

Brownell J.E., Zhou J., Ranalli T., Kobayashi R., Edmondson D.G., Roth S.Y., and Allis C.D. 1996. *Tetrahymena* histone acetyltransferase A: A homolog to yeast Gcn5p linking histone acetylation to gene activation. *Cell* **84:** 843.

Burley S.K. and Roeder R.G. 1996. Biochemistry and structural biology of transcription factor IID (TFIID). *Annu. Rev. Biochem.* **65:** 769.

Cairns B.R., Lorch Y., Li Y., Zhang M., Lacomis L., Erdjument-Bromage H., Tempst P., Du J., Laurent B., and Kornberg R.D. 1996. RSC, an essential, abundant chromatin-remodeling complex. *Cell* **87:** 1249.

Candau R. and Berger S.L. 1996. Structural and functional analysis of yeast putative adaptors: Evidence for an adaptor complex *in vivo. J. Biol. Chem.* **271:** 5237.

Carlson M. and Laurent B.C. 1994. The SNF/SWI family of global transcriptional activators. *Curr. Opin. Cell Biol.* **6:** 396.

Côté J., Peterson C.L., and Workman J.L. 1998. Perturbation of nucleosome core structure by the SWI/SNF complex persists after its detachment, enhancing subsequent transcription factor binding. *Proc. Natl. Acad. Sci.* **95:** 4947.

Côté J., Quinn J., Workman J.L., and Peterson C.L. 1994. Stimulation of GAL4 derivative binding to nucleosomal DNA by the yeast SWI/SNF complex. *Science* **265:** 53.

Eberharter A., John S., Grant P.A., Utley R.T., and Workman J.L. 1998. Identification and analysis of yeast nucleosomal histone acetyltransferase complexes. *Methods* **15:** 315.

Felsenfeld G. 1992. Chromatin as an essential part of the transcriptional mechanism. *Nature* **355:** 219.

Gavin I.M. and Simpson R.T. 1997. Interplay of yeast global transcriptional regulators Ssn6p-Tup1p and Swi-Snf and their effect on chromatin structure. *EMBO J.* **16:** 6263.

Georgakopoilos T. and Thireos G. 1992. Two distinct yeast transcriptional activators require the function of the GCN5 protein to promote normal levels of transcription. *EMBO J.* **11:** 4145.

Grant P.A., Berger S.L., and Workman J.L. 1998a. Identification and analysis of native nucleosomal histone acetyltransferase complexes. In *Chromatin protocols* (ed. P.B. Becker). Humana Press, Totowa, New Jersey. (In press.)

Grant P.A., Sterner D.E., Duggan L.J., Workman J.L., and Berger S.L. 1998b. The SAGA unfolds: Convergence of transcription regulators in chromatin-modifying complexes. *Trends Cell Biol.* **8:** 193.

Grant P.A., Schieltz D., Pray-Grant M.G., Steger D.J., Reese J.C., Yates J.R., and Workman J.L. 1998c. A subset of TAF(II)s are integral components of the SAGA complex required for nucleosome acetylation and transcriptional stimulation. *Cell* **94:** 45.

Grant P.A., Duggan L., Côté J., Roberts S.M., Brownell J.E., Candau R., Ohba R., Owen-Hughes T., Allis C.D., Winston F., Berger S.L., and Workman J.L. 1997. Yeast Gcn5 functions in two multisubunit complexes to acetylate nucleosomal histones: Characterization of an Ada complex and the SAGA (Spt/Ada) complex. *Genes Dev.* **11:** 1640.

Grunstein M. 1997. Histone acetylation in chromatin structure and transcription. *Nature* **389:** 349.

Gu W. and Roeder R.G. 1997. Activation of p53 sequence-specific DNA binding by acetylation of the p53 C-terminal domain. *Cell* **90:** 595.

Han M. and Grunstein M. 1988. Nucleosome loss activates downstream promoters in vivo. *Cell* **55:** 1137.

Han M., Kim U.J., Kayne P., and Grunstein M. 1988. Depletion of histone H4 and nucleosomes activates the PHO5 gene in *Saccharomyces cerevisiae* . *EMBO J.* **7:** 2221.

Hebbes T.R., Clayton A.L., Thorne A.W., and Crane-Robinson C. 1994. Core histone hyperacetylation co-maps with generalized DNase I sensitivity in the chicken β-globin chromosomal domain. *EMBO J.* **13:** 1823.

Hilfiker A., Hilfiker-Kleiner D., Pannuti A., and Lucchesi J.C. 1997. *mof* , a putative acetyltransferase gene related to the Tip60 and MOZ human genes and to the SAS genes of yeast, is required for dosage compensation in *Drosophila. EMBO J.* **16:** 2054.

Hirschhorn J.N., Bortvin A.L., Ricupero-Hovasse S.L., and Winston F. 1995. A new class of histone H2A mutations in *Saccharomyces cerevisiae* causes specific transcriptional defects in vivo. *Mol. Cell. Biol.* **15:** 1999.

Hirschhorn J.N., Brown S.A., Clark C.D., and Winston F. 1992. Evidence that SNF2/SWI2 and SNF5 activate transcription in yeast by altering chromatin structure. *Genes Dev.* **6:** 2288.

Imbalzano A.N., Schnitzler G.R., and Kingston R.E. 1996. Nucleosome disruption by human SWI/SNF is maintained in the absence of continued ATP hydrolysis. *J. Biol. Chem.* **271:** 20726.

Imbalzano A.N., Kwon H., Green M.R., and Kingston R.E. 1994. Facilitated binding of TATA-binding protein to nucleosomal DNA. *Nature* **370:** 481.

Imhof A., Yang X.-J., Ogryzko V.V., Nakatani Y., Wolffe A.P., and Ge H. 1997. Acetylation of general transcription factors by histone acetyltransferases. *Curr. Biol.* **7:** 689.

Ito T., Bulger M., Pazzin M.J., Kobayashi R., and Kadonaga J.T. 1997. ACF, an ISWI-containing and ATP-utilizing chromatin assembly and remodeling factor. *Cell* **90:** 145.

Jeppesen P. and Turner B.M. 1993. The inactive X chromosome in female mammals is distinguished by a lack of histone H4 acetylation, a cytogenic marker for gene expression. *Cell* **74:** 281.

Kadonaga J.T. 1998. Eukaryotic transcription: An interlaced network of transcription factors and chromatin-modifying machines. *Cell* **92:** 307.

Kornberg R.D. and Lorch Y. 1992. Chromatin structure and transcription. *Annu. Rev. Cell Biol.* **8:** 563.

Kruger W., and Herskowitz I. 1991. A negative regulator of *HO* transcription, SIN1 (SPT2), is a nonspecific DNA-binding protein related to HMG1. *Mol. Cell. Biol.* **11:** 4135.

Kruger W., Peterson C.L., Sil A., Coburn G., Arents G., Moudrianakis E.N., and Herskowitz I. 1995. Amino acid substitutions in the structured domains of histones H3 and H4 partially relieve the requirement of the yeast SWI/SNF complex for transcription. *Genes Dev.* **9:** 2770.

Kuo M.-H., Zhou J., Jambeck P., Churchill M.E.A., and Allis C.D. 1998. Histone acetyltransferase activity of yeast Gcn5 is required for the activation of target genes *in vivo*. *Genes Dev.* **12:** 627.

Kwon H., Imbalzano A.N., Khavarl P.A., Kingston R.E., and Green M.R. 1994. Nucleosome disruption and enhancement of activator binding by a human SWi/SNF complex. *Nature* **370:** 477.

Laurent B.C., Treich I., and Carlson M. 1993. The yeast SNF2/SWI2 protein has DNA-stimulated ATPase activity required for transcriptional activation. *Genes Dev.* **7:** 583.

Li Q. and Wrange O. 1993. Translational positioning of a nucleosomal glucocorticoid response element modulates glucocorticoid receptor affinity. *Genes Dev.* **7:** 2471.

Logie C. and Peterson C.L. 1997. Catalytic activity of the yeast SWI/SNF complex on reconstituted nucleosome arrays. *EMBO J.* **16:** 6772.

Martens J.A., Genereaux J., Saleh A., and Brandl C.J. 1996. Transcriptional activation by yeast PDR1p is inhibited by its association with NGG1p/ADA3p. *J. Biol. Chem.* **271:** 15884.

Melcher K. and Johnston S.A. 1995. GAL4 interacts with TATA-binding protein and coactivators. *Mol. Cell. Biol.* **15:** 2839.

Ogryzko V.V., Kotani T., Zhang X., Schiltz R.L., Howard T., Yang X.-J., Howard B.H., Qin J., and Nakatani Y. 1998. Histone-like TAFs within the PCAF histone acetylase complex. *Cell* **94:** 35.

Owen-Hughes T.A. and Workman J.L. 1994. Experimental analysis of chromatin function in transcription control. *Crit. Rev. Eukaryot. Gene Expr.* **4:** 403.

———. 1996. Remodeling the chromatin structure of a nucleosome array by transcription factor-targeted *trans*-displacement of histones. *EMBO J.* **15:** 4702.

Owen-Hughes T., Utley R.T., Côté J., Peterson C.L., and Workman J.L. 1996. Persistent site-specific remodeling of a nucleosome array by transient action of the SWI/SNF complex. *Science* **273:** 513.

Paranjape S.M., Kamakaka R.T., and Kadonaga J.T. 1994. Role of chromatin structure in the regulation of transcription by RNA polymerase II. *Annu. Rev. Biochem.* **63:** 265.

Pazin M.J. and Kadonaga J.T. 1997. What's up and down with histone deacetylation and transcription? *Cell* **89:** 325.

Peterson C.L. and Tamkun J.W. 1995. The SWI/SNF complex: A chromatin remodeling machine? *Trends Biochem. Sci.* **20:** 143.

Piña B., Berger S., Marcus G.A., Silverman N., Agapite J., and Guarente L. 1993. ADA3: A gene, identified by resistance to GAL4-VP16, with properties similar to and different from those of ADA2. *Mol. Cell. Biol.* **13:** 5981.

Polach K.J. and Widom J. 1996. A model for the cooperative binding of eukaryotic regulatory proteins to nucleosomal target sites. *J. Mol. Biol.* **258:** 800.

Pollard K.J., and Peterson C.L. 1997. Role for ADA/GCN5 products in antagonizing chromatin-mediated transcriptional repression. *Mol. Cell. Biol.* **17:** 6212.

Roberts S.M. and Winston F. 1997. Essential functional interactions of SAGA, a *Saccharomyces cerevisiae* complex of Spt, Ada, and Gcn5 proteins, with the Snf/Swi and Srb/Mediator complexes. *Genetics* **147:** 451.

Silverman N., Agapite J., and Guarente L. 1994. Yeast Ada2 protein binds to the VP16 protein activation domain and activates transcription. *Proc. Natl. Acad. Sci.* **91:** 11665.

Simpson R.T. 1979. Mechanism of a reversible, thermally induced conformational change in chromatin core particles. *J. Biol. Chem.* **254:** 10123.

Smith E.R., Eisen A., Gu W., Sattah M., Pannuti A., Zhou J., Cook R.G., Lucchesi J.C., and Allis C.D. 1998. ESA1 is a histone acetyltransferase that is essential for growth in yeast. *Proc. Natl. Acad. Sci.* **95:** 3561.

Steger D.J. and Workman J.L. 1996. Remodeling chromatin structures for transcription: what happens to the histones? *BioEssays* **18:** 875.

———. 1997. Stable co-occupancy of transcription factors and histones at the HIV-1 enhancer. *EMBO J.* **16:** 2463.

Steger D.J., Eberharter A., John S., Grant P.A., and Workman J.L. 1998. Purified histone acetyltransferase complexes stimulate HIV-1 transcription from preassembled nucleosomal arrays. *Proc. Natl. Acad. Sci.* **95:** 12924.

Svaren J. and Horz W. 1996. Regulation of gene expression by nucleosomes. *Curr. Opin. Genet. Dev.* **6:** 164.

Tansey W.P. and Herr W. 1997. TAFs: Guilt by association. *Cell* **88:** 729.

Taylor C.A., Workman J.L., Schuetz T.J., and Kingston R.E. 1991. Facilitated binding of GAL4 and heat shock factor to nucleosomal templates: Differential function of DNA-binding domains. *Genes Dev.* **5:** 1285.

Tsukiyama T. and Wu C. 1995. Purification and properties of an ATP dependent nucleosome remodeling factor. *Cell* **83:** 1011.

Turner B.M. and O'Neill L.P. 1995. Histone acetylation in chromatin and chromosomes. *Semin. Cell Biol.* **6:** 229.

Utley R.T., Côté J., Owen-Hughes T., and Workman J.L. 1997. SWI/SNF stimulates the formation of disparate activator-nucleosome complexes but is partially redundant with cooperative binding. *J. Biol. Chem.* **272:** 12642.

Utley R.T., Ikeda K., Grant P.A., Côté J., Steger D.J., Eberharter A., John S., and Workman J.L. 1998. Transcriptional activators direct histone acetyltransferase complexes to nucleosomes. *Nature* **394:** 498.

van Holde K.E. 1989. *Chromatin*. Springer-Verlag, New York.

Van Lint C., Emiliani S., Ott M., and Verdin E. 1996. Transcriptional activation and chromatin remodeling of the HIV-1 promoter in response to histone acetylation. *EMBO J.* **15:** 1112.

Varga-Weisz P.D. and Becker P.B. 1995. Transcription factor-mediated chromatin remodeling: Mechanisms and models. *FEBS Lett.* **369:** 118.

Varga-Weisz P.D., Wilm M., Bonte E., Dumas K., Mann M., and Becker P.B. 1997. Chromatin-remodeling factor CHRAC contains the ATPases ISWI and topoisomerase II. *Nature* **388:** 598.

Vettese-Dadey M., Walter P., Chen H., Juan L.-J., and Workman J.L. 1994. Role of the histone amino termini in facilitated binding of a transcription factor, GAL4-AH, to nucleosome cores. *Mol. Cell. Biol.* **14:** 970.

Vettese-Dadey M., Grant P.A., Hebbes T.R., Crane-Robinson C., Allis C.D., and Workman J.L. 1996. Acetylation of histone H4 plays a primary role in enhancing transcription factor binding to nucleosomal DNA *in vitro*. *EMBO J.* **15:** 2508.

Wade D.A., Pruss D., and Wolffe A.P. 1997. Histone acetylation: Chromatin in action. *Trends Biochem. Sci.* **22:** 128.

Wang L., Liu L., and Berger S.L. 1998. Critical residues for histone acetylation by Gcn5, functioning in Ada and SAGA complexes, are also required for transcriptional function in vivo. *Genes. Dev.* **12:** 640.

Wang W., Côté J., Xue Y., Zhou S., Khavari P.A., Biggar S.R., Muchardt C., Kalpana G.V., Goff S.P., Yaniv M., Workman J.L., and Crabtree G.R. 1996. Purification and biochemical heterogeneity of the mammalian SWI-SNF complex. *EMBO J.* **15:** 5370.

Winston F. 1992. Analysis of SPT genes: A genetic approach towards analysis of TFIID, histones and other transcription factors of yeast. In *Transcriptional regulation* (eds. S.L. McKnight and K.R. Yamamoto), p. 1271. Cold Spring Harbor Laboratory Press, Cold Spring Harbor, New York.

Wolffe A.P. 1994. Nucleosome positioning and modification: Chromatin structures that potentiate transcription. *Trends Biochem. Sci.* **19:** 240.

Wolffe A.P., Wong J., and Pruss D. 1997. Activators and repressors: Making use of chromatin to regulate transcription. *Genes Cells* **2:** 291.

Workman J.L. and Kingston R.E. 1998. Alteration of nucleosome structure as a mechanism of transcriptional regulation. *Annu. Rev. Biochem.* **67:** 545.

TBP-associated Factors in the PCAF Histone Acetylase Complex

T. Kotani,*† X. Zhang,‡ R.L. Schiltz,* V.V. Ogryzko,* T. Howard,* M.J. Swanson,*
A. Vassilev,* H. Zhang,* J. Yamauchi,* B.H. Howard,* J. Qin,‡ and Y. Nakatani*

*Laboratory of Molecular Growth Regulation, National Institute of Child Health and Human Development and
‡Laboratory of Biophysical Chemistry, National Heart, Lung and Blood Institute, National Institutes of Health,
Bethesda, Maryland 20892

PCAF (p300/CBP-associated factor) histone acetylase has an important role in regulation of transcription, cell cycle progression, and differentiation in conjunction with p300/CBP. To investigate PCAF function at the molecular level in greater detail, we purified PCAF in its native state. PCAF is found in a complex with more than 20 associated polypeptides. Strikingly, some polypeptides associated with PCAF are identical to the TATA-box-binding polypeptide (TBP)-associated factors (TAFs) which are subunits of TFIID. Furthermore, some polypeptides show significant sequence similarity to other TAFs. Taken together, these results lead to the conclusion that a histone octamer-like domain may be present within the PCAF complex, as previously demonstrated in the TFIID complex. Although the function of the histone octamer-like structure in the PCAF complex is unclear, it may replace the histone octamer after relaxation of nucleosomal structure by acetylation of the histone tails. Importantly, the histone-like domains in the PCAF complex lack regions corresponding to histone amino-terminal tails. In this regard, if it replaces the histone octamer, the histone-like structure in the PCAF complex may have an architectural role in the maintenance of a transcriptionally active chromatin state regardless of histone deacetylase activity. The fact that PCAF is found in a complex with more that 20 associated polypeptides suggests that E1A, by competing with PCAF for p300/CBP interaction, perturbs access of the PCAF complex to promoters. Thus, subunits in the PCAF complex may be involved in cellular events mediated by PCAF, i.e., regulation of transcription, cell cycle progression, and differentiation.

PCAF

The transforming proteins encoded by adenovirus and several other small DNA tumor viruses disturb host-cell growth control by interacting with cellular factors that normally function to repress cell proliferation. One of the most intensively studied of these viral proteins, the product of the adenovirus E1A gene, is itself sufficient for transformation under certain conditions (for review, see Moran 1993). E1A transforming activity resides in two distinct domains that bind to the p300/CBP and retinoblastoma susceptibility gene product (RB) families (for reviews, see Dyson and Harlow 1992; Moran 1993). Interactions of E1A with p300/CBP and RB are thought to influence functionally distinct growth regulatory pathways, allowing the two domains to contribute additively to transformation (for review, see Moran 1993).

The paradigm for how E1A and functionally related viral proteins perturb cell growth regulation derives in large part from studies on their interactions with RB (for reviews, see Dyson and Harlow 1992; Moran 1993; Nevins 1994). RB down-regulates, in a cell-cycle-dependent fashion, a subset of cellular transcription activators involved in cell proliferation. The underphosphorylated form of RB, which is predominant in resting (G_0/G_1) cells, binds to cellular targets to inhibit their activity. RB is dissociated from these activators upon its hyperphosphorylation by cyclin-dependent protein kinases at the end of the G_1 stage (for review, see Livingston et al. 1993; Weinberg 1995). Recent findings regarding the recruitment of histone deacetylases by RB indicate that the latter inhibits transcription, at least in part, through modification of chromatin (Brehm et al. 1998; Luo et al. 1998; Magnaghi-Jaulin et al. 1998). Importantly, E1A perturbs interactions between RB and its cellular targets by competing for binding to the RB site called the pocket domain, leading to constitutive activation of RB-binding factors (for review, see Dyson and Harlow 1992; Livingston et al. 1993).

The second class of cellular factors implicated in E1A-dependent transformation, p300 and CBP, are transcriptional coactivators that are recruited to a subset of promoters through interaction with particular gene-specific activators, including MyoD, Jun, Fos, Myb, p53, and nuclear hormone receptors (for reviews, see Janknecht and Hunter 1996; Goldman et al. 1997). E1A inhibits p300/CBP-mediated transcriptional activation of many promoters (for review, see Rochette-Egly et al. 1990). In a case that has been thoroughly examined, that of p300 and YY1, E1A inhibits transcription without disrupting the p300-YY1 complex (Lee et al. 1995). From these lines of evidence, we considered that the function of a cellular factor, which is important for p300/CBP-dependent activation, is perturbed by E1A. We have cloned such a factor in a unique way based on an analogy between species. In humans, CBP binds to Jun in a phosphoryla-

†*Present address:* Nara Institute of Science and Technology, 8916-5 Takayama, Ikoma, Nara 630-01, Japan.

tion-dependent manner in association with stimulation of transcription (Arias et al. 1994). On the other hand, in yeast, GCN4 is believed to be a Jun counterpart on the basis of similarities in DNA recognition (Struhl 1987) as well as the participation of both proteins in UV response pathways (Engelberg et al. 1994). Yeast genetic screening has allowed isolation of various coactivators for GCN4, including GCN5, ADA2, and ADA3 (Berger et al. 1992; Georgakopoulos and Thireos 1992; Horiuchi et al. 1995). On the basis of the analogy between GCN4 and Jun, we considered that similar coactivators might be involved in the human p300/CBP pathway. In support of this view, p300/CBP and ADA2 exhibit a significant sequence similarity within a 100-amino-acid region including a Zn^{++} finger motif (Chrivia et al. 1993).

We found two distinct GCN5-related sequences from the human EST database. The sequence of the full-length clones revealed that one polypeptide contains, in addition to the yGCN5-related carboxy-terminal region, an amino-terminal region with no obvious sequence similarity to any proteins in the databases, whereas the second encodes only the yGCN5-related region (Fig. 1). Given that the p300/CBP-binding activity resides in the amino terminus of the former polypeptide, we named it PCAF and the latter human GCN5 (hGCN5) (Yang et al. 1996). However, Smith et al. (1998) recently isolated a longer form of hGCN5 (hGCN5-L) which possesses the region related to the amino terminus of PCAF. Importantly, the protein level of hGCN5-L is decreased in PCAF-overexpressing cells (R.L. Schiltz et al., unpubl.), whereas mouse GCN5-L protein is increased in PCAF knock-out cells (J. Yamauchi and Y. Nakatani, unpubl.). Thus, it is likely that PCAF and GCN5-L, at least in part, functionally complement each other.

Interaction experiments with various p300/CBP mutants indicate that E1A and PCAF bind to the same or very closely spaced sites in p300/CBP. Due to the difference of affinities, E1A binding to p300/CBP inhibits access of PCAF in vitro. Moreover, overexpression of E1A disrupts the endogenous PCAF-p300 interaction in vivo, as revealed by coimmunoprecipitation of these proteins from cell extracts. Importantly, exogenous expression of PCAF slows the rate of G_1/S transit in HeLa cells. In support of competition between PCAF and E1A, exogenous expression of PCAF significantly counteracts the mitogenic activity of E1A (Yang et al. 1996). This conclusion is supported by recent observations in a muscle differentiation system. Exogenous expression of PCAF in myoblasts promotes MyoD-mediated p21 expression and terminal cell cycle arrest. Conversely, E1A inhibits activation of the myogenic program through competition with PCAF for access to p300/CBP (Puri et al. 1997). Consistently, biochemical analysis demonstrated that MyoD, p300/CBP, and PCAF form a protein complex on promoters and that expression of E1A disrupts the complex without affecting the interaction between MyoD and DNA.

PCAF IS A HISTONE ACETYLASE

The assembly of DNA into nucleosomes and higher-order chromatin represents a repressive block to transcription that must be overcome during transcriptional activation. One mechanism of alleviating chromatin repression involves the acetylation of histones. The core histones consist of a histone-fold domain, which has a role in nucleosome formation, and an amino-terminal basic tail. The sequences of these amino-terminal tails are well conserved from yeast to humans, and specific lysine residues within these tails are targets of acetylation. Neutralization of the positively charged lysine residues of the histone tail by acetylation was reported to decrease the affinity of the amino-terminal tail of H4 for DNA (Hong et al. 1993) and alter the conformation of nucleosomes in vitro (Norton et al. 1989). A consequence of histone acetylation is proposed to be a more relaxed chromatin structure that is less transcriptionally repressive. Crystallographic data suggest that histone tails may also be involved in contacts between adjacent nucleosomes and thus have a role in the formation of higher-order chromatin structures (Luger et al. 1997). Finally, the histone tails may contribute to repression of transcription by recruiting such factors as Tup1 and Sir3/Sir4 (Hecht et al. 1995; Edmondson et al. 1996).

Several lines of evidence discussed above imply that E1A perturbs normal cellular transcription by preventing the access of PCAF to p300/CBP. Our finding that PCAF has intrinsic histone acetylase activity (Yang et al. 1996) allowed us to propose the simple model that E1A inhibits recruitment of the histone acetylase to promoters. In support of this view, the PCAF-binding site in p300 is crucial for coactivation of p53-dependent transcription by p300 (Lill et al. 1997). However, later findings complicate this simple view. Unexpectedly, we identified p300 and CBP as histone acetylases (Ogryzko et al. 1996). Furthermore, in collaboration with the Allis and Evans labs (Mizzen et al. 1996; Chen et al. 1997), we have demonstrated that a subunit of TFIID ($TAF_{II}250$) and a coactivator of nuclear hormone receptors (ACTR), respectively, are histone

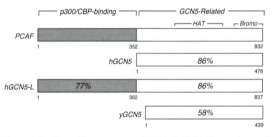

Figure 1. Overall structure of PCAF. PCAF consists of a yGCN5-related carboxy-terminal region, which contains the histone acetylase and bromo domains, and the amino-terminal region. Whereas the amino-terminal region is responsible for interaction with p300/CBP, the carboxy-terminal region is responsible for multimeric complex formation (see Fig. 2). hGCN5-L (the long form of GCN5) and hGCN5 are made from the same gene presumably by alternative splicing. Although hGCN5-L is homologous to PCAF, hGCN5 contains only the yGCN5-related region. Sequence similarities to PCAF are indicated.

acetylases. It is especially intriguing that three histone acetylases, PCAF, p300/CBP, and ACTR, interact with one another (Chen et al. 1997). Although it is still not clear why multiple histone acetylases are required, some data indicate that the recruitment of p300/CBP is not sufficient for gene activation and that PCAF is a crucial coactivator for p300/CBP-dependent activation. In several cases, E1A does not disturb the interaction between gene-specific activators and p300/CBP, yet it still represses transcription (Lee et al. 1995). Moreover, a differential requirement for the acetylase activity of PCAF and p300/CBP has been observed, in that myogenic transcription and differentiation are dependent on the histone acetylase domain of PCAF, but not on that of p300 (Puri et al. 1997).

Although the histone tails contain many lysine residues, only a subset of these can be acetylated in vivo (for review, see Mizzen and Allis 1998). The distribution of the acetylation sites, shown in the Figure 2, is obtained by analyzing total histone preparations, and therefore does not reflect the acetylation patterns of individual histone molecules. It is considered that positions of acetylated lysines could be related to specific functions. For instance, Kuo et al. (1996) have demonstrated that cytoplasmic histone acetylases, which are involved in chromatin deposition, and yGCN5 acetylate distinct lysines (K5, K12 and K8, K16, respectively) in histone H4. In this regard, we considered that the differential requirement of the acetylase activities of p300 and PCAF (Puri et al. 1997; Korzus et al. 1998) might be due to differences in their substrate specificities. To test this possibility, we acetylated HeLa nucleosomes with recombinant PCAF or p300, purified the individual core histones and analyzed their amino-terminal protein sequences in order to see which lysines were acetylated (Schiltz et al. 1999). PCAF displayed a strong preference for acetylation of K14 of H3 but was also able to acetylate K8 of H4. In contrast, p300 was able to acetylate all known in vivo acetylation sites of each of the core histones (Fig. 2). Furthermore, several lysines within the histone tails are not

acetylated in vivo, and these were not targeted by p300 in vitro. These data suggest that histones are a bona fide target of p300/CBP's histone acetylase activity in vivo. Unlike PCAF, p300 also targets K5 and K12 on H4; therefore, the acetylation of these lysines may be involved in both histone deposition and transcriptional activation in higher eukaryotes. The broader substrate specificity of p300 than that of PCAF is particularly intriguing due to the requirement of the PCAF acetylase activity but not that of p300 in myogenic transcription and differentiation (Puri et al. 1997). These results indicate that the overall level of histone acetylation may not directly correlate with transcriptional activation and further suggest that additional mechanisms may be involved.

THE PCAF COMPLEX

As discussed above, PCAF has an important role in transcriptional activation and differentiation. To understand PCAF function at the molecular level in greater detail, we purified PCAF in its native state (Ogryzko et al. 1998). PCAF was purified from FLAG-epitope-tagged PCAF-expressing HeLa cells by anti-FLAG antibody immunoprecipitation, followed by anti-PCAF antibody immunoprecipitation. As a control, we performed a mock purification from the parental HeLa cell line. As shown in Figure 3, PCAF is present in an unexpectedly complicated multisubunit complex consisting of more than 20 distinct polypeptides ranging from 10 to 400 kD. EST clones encoding these polypeptides were identified from peptide sequences, and molecular masses of tryptic fragments were determined by mass spectrometry. When EST clones were incomplete, clones encoding the full open reading frame were isolated by screening a cDNA library. Similar to the (y)GCN5-containing SAGA complex of yeast (Grant et al. 1997, 1998), the PCAF complex possesses ADA-like and SPT-like proteins as subunits. Two distinct polypeptides that are related to yADA2 and yADA3 migrate as a 53 kD band. Moreover, the 37-kD band is related to ySPT3.

TAFS IN THE PCAF COMPLEX

Strikingly, the 34-, 30-, 22-, and 16-kD bands in the complex were identified as human (h)TAF$_{II}$31, hTAF$_{II}$30, hTAF$_{II}$20, and hTAF$_{II}$15, respectively (Fig. 4) (see Ogryzko et al. 1998). These findings are remarkable in that they provide the first evidence of a non-TBP-containing TAF complex. In support of mass sequencing data, immunoblotting analysis shows that these TAFs are present in both the PCAF and TFIID complexes. In contrast, hTAF$_{II}$250, hTAF$_{II}$130, hTAF$_{II}$100, hTAF$_{II}$80, and TBP are specific to TFIID, whereas PCAF is specific to the PCAF complex.

Previously, in collaboration with the Burley and Roeder laboratories, we demonstrated that TFIID contains a histone octamer-like structure (Hoffmann et al. 1996; Xie et al. 1996; for review, see Burley and Roeder 1996). The amino-terminal regions of hTAF$_{II}$80 and hTAF$_{II}$31 have sequence similarity with those of histones

```
H2A   S G R G K Q G G K A R A K A K T R S S R
              5       9       13 15

H2B   P E P A K S A P A P K K G S K K A V T K A Q K K D G K E R K
                      1112    15      20      24  27      30

H3    A R T K Q T A R K S T G G K A P R K Q L A T K A A R K S A P
          4       9       14      18      23      27

H4    S G R G K G G K G L G K G G A K R H R K
              5     8       12    16      20
```

	H2A	H2B	H3	H4
PCAF	None	None	14	8
p300	5	5,12,15,20	14,18,23	5,8,12

Figure 2. Summary of p300 and PCAF nucleosomal acetylation sites. (*Top*) Peptide sequence of the histone basic tails targeted by histone acetylases. (*Bottom*) The acetylation sites of nucleosomal core histones targeted by p300 and PCAF, as well as the known in vivo acetylation sites. Among the many basic residues in the histone tails, only specific lysine residues are acetylated in vivo. In support of the view that histones are a bona fide substrate of p300 and PCAF, residues acetylated by p300 and PCAF in vitro overlap with known in vivo acetylation sites.

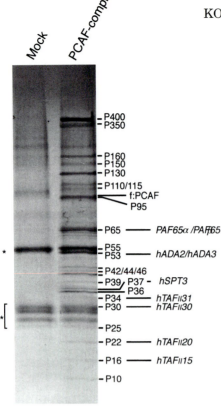

Figure 3. Purification of the PCAF complex. The PCAF complex was purified from FLAG epitope-tagged PCAF-expressing HeLa cells by anti-FLAG immunoaffinity chromatography followed by anti-PCAF antibody immunoprecipitation. Mock purification was performed from nontransduced HeLa cells. Purified proteins were separated by SDS-PAGE and stained with Coomassie brilliant blue. Whereas some polypeptides are human counterparts of yeast coactivators (ADA/SPT proteins), some are identified as TAFs and novel TAF-related factors (PAFs).

Human		Yeast		
TFIID	PCAF	SAGA	TFIID	Feature
hTAF250	PCAF	GCN5	TAF145(130)	Histone acetylase
hTAF150			TAF150	
hTAF130			—	
hTAF100	PAF65β	yTAF90	TAF90	WD40
hTAF80	PAF65α	yTAF60	TAF60	H4
hTAF55			TAF67	
			TAF47	
hTAF31(32)	hTAF31	yTAF20	TAF17(20)	H3
hTAF30	hTAF30	yTAF25	TAF25(23)	
hTAF20/15	hTAF20/15	yTAF61	TAF61(68)	H2B
hTAF28			TAF19	
hTAF18	hSPT3	SPT3	TAF18	
TBP	—	—	TBP	
	?	ADA1		
	hADA2	ADA2		
	hADA3	ADA3		
	?	SPT7		
	?	SPT8		
	?	SPT20		

Figure 4. Comparison between TFIID and the PCAF/GCN5-containing complex in human and yeast. Whereas the PCAF complex (Ogryzko et al. 1998) is compared with human TFIID, the yeast SAGA complex (Grant et al. 1998) is compared with yeast TFIID. The PCAF/GCN5-containing complexes and TFIID have common features, i.e., intrinsic histone acetylase activity, histone octamer-like TAFs (or related polypeptide), WD40-repeat containing TAF (or related polypeptide) hTAF$_{II}$30/yTAF$_{II}$25, and SPT3-related polypeptides. Whereas human and yeast PCAF/GCN5-containing complexes are strikingly similar, an important difference is that the PCAF complex contains TAF-related factors (PAF65α and β) and the SAGA complex contains corresponding bona fide TAFs. In light of direct interactions of hTAF$_{II}$80 and hTAF$_{II}$100 with TBP, PAF65α and PAF65β may have a role in formation of a complex lacking TBP. In support of this view, ADA and TBP are found in the same complex. Although the SAGA complex does not have TBP, the latter is likely to be dissociated from the complex during chromatographic steps.

H4 and H3, respectively. Impressively, the crystal structure of the histone-fold domains of *Drosophila* (d)TAF$_{II}$62-dTAF$_{II}$42 (homologous to hTAF$_{II}$80-hTAF$_{II}$31) is almost identical to that of the histone H3-H4 heterotetramer. Although TFIID has no histone H2A counterpart, the H2B-like hTAF$_{II}$20 (or hTAF$_{II}$15, the product of alternatively spliced mRNA) forms a homodimer (Hoffmann et al. 1996) that is analogous to the H2A-H2B heterodimer in the histone octamer. These observations, together with the stoichiometry of this set of TAFs in TFIID, suggest the presence of a histone octamer-like structure within TFIID that is composed of two dimers of hTAF$_{II}$20/hTAF$_{II}$15 attached to a tetramer of hTAF$_{II}$80 and hTAF$_{II}$31 (for review, see Burley and Roeder 1996; Hoffmann et al. 1996). It is particularly intriguing that the PCAF complex possesses a subset of histone-like TAFs, namely, the H3-like hTAF$_{II}$31 and the H2B-like hTAF$_{II}$20/15. Although the PCAF complex lacks the H4-like hTAF$_{II}$80, the 65-kD band in the complex contains a novel polypeptide with similarity to hTAF$_{II}$80 (24% identity and 42% similarity), referred to as *PCAF-associated factor* (PAF)65α. As expected from sequence conservation between the histone-fold domains of PAF65α and hTAF$_{II}$80, the H4-like region of PAF65α

stoichiometrically interacts with the H3-like region of hTAF$_{II}$31 in vitro.

Yeast suppressor screening revealed that the H4-like TAF genetically interacts with the WD40-repeat containing TAF, which is homologous to hTAF$_{II}$100 (M.J. Swanson and Y. Nakatani unpubl.). Although the PCAF complex lacks hTAF$_{II}$100, a second polypeptide in the 65-kD band, PAF65β, encodes a novel polypeptide with similarity to hTAF$_{II}$100 (Ogryzko et al. 1998). PAF65β and hTAF$_{II}$100 share homology throughout their entire sequence with significant gaps (46% identity and 58% similarity). In particular, the WD40 repeats are well conserved between these factors.

THE PCAF COMPLEX AND TFIID

The subunit compositions of the PCAF complex and TFIID are compared in Figure 4. An important common feature is that both contain histone acetylases: PCAF and hTAF$_{II}$250 subunits have intrinsic histone acetylase activity (Mizzen et al. 1996; Yang et al. 1996). Moreover, both complexes have a histone octamer-like structure and a WD40-repeat-containing subunit. Whereas the H3-like hTAF$_{II}$31 and H2B-like hTAF$_{II}$20/15 are common in both complexes, the PCAF complex has specialized H4-like and WD40-repeat-containing subunits (i.e., PAF65α and β, respectively). Furthermore, both complexes have hTAF$_{II}$30, originally identified as a TAF that is responsible for estrogen-receptor-mediated activation (Jacq et al. 1994). Finally, hTAF$_{II}$18 has sequence similarity to the amino terminus of SPT3 (Mengus et al. 1995).

It is particularly interesting that the PCAF complex has novel TAF-related factors, PAF65α and β. Why must the PCAF complex have these specialized factors instead of conventional TAFs? A plausible reason is that the H4-like TAF and WD40-repeat containing TAF interact directly with TBP (for review, see Burley and Roeder 1996). Therefore, PAF65α and β might be required for formation of complexes lacking TBP (further discussed below). Although hTAF$_{II}$20/15 also weakly interacts with TBP in vitro (Hoffmann et al. 1996), it is apparent that this interaction per se is not sufficient for formation of a complex containing TBP (i.e., in light of the lack of TBP in the PCAF complex).

THE PCAF COMPLEX AND THE YEAST SAGA COMPLEX

The SAGA complex, a yeast GCN5-containing complex, resembles the PCAF complex to a remarkable degree (Fig. 4). As noted earlier, the PCAF complex contains polypeptides related to yADA2, yADA3, and ySPT3 found in the SAGA complex (Grant et al. 1997, 1998). Counterparts for yADA1, ySPT7, ySPT8, and ySPT20 have not yet been identified to date in the PCAF complex, but we suspect that corresponding polypeptides may be found among the unidentified subunits. Moreover, SAGA contains the histone H4-like yTAF$_{II}$60, the H3-like yTAF$_{II}$20/17, the H2B-like yTAF$_{II}$61, the WD40 repeat-containing yTAF$_{II}$90, and yTAF$_{II}$25 (which is homologous to hTAF$_{II}$30). An important distinction between the PCAF and SAGA complexes is that the latter contains the bona fide H4-like TAF and the WD40-repeat TAF, rather than TAF-related subunits. As discussed above, PAF65α and β might have a role in formation of a complex lacking TBP because the corresponding TAFs interact directly with TBP. Seemingly inconsistent with this hypothesis is evidence that the SAGA complex has no TBP. However, TBP still might be associated with the SAGA complex in a native context, but be lost upon the purification of the SAGA complex. In support of this view, Saleh et al. (1997) showed that TBP is coimmunoprecipitated with ADA3. Thus, it is possible that the SAGA complex associates with TBP in a native context, although further investigation will be required to clarify this issue. Drysdale et al. (1998) showed that the GCN4 activation domain recruits a SAGA-like complex from yeast extract in vitro. However, the GCN4 activation domain does not interact with purified TFIID (see Fig. 5). In this regard, it will be important to test the necessity of TFIID for SAGA-dependent transcription.

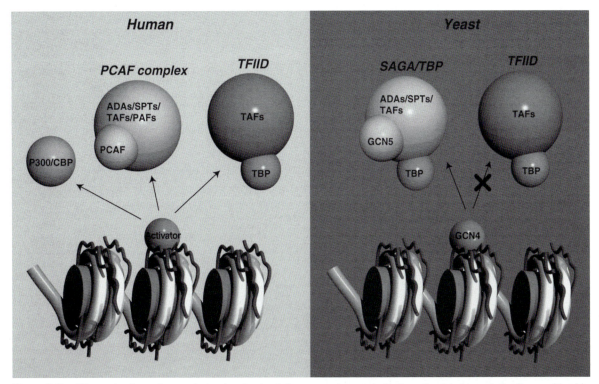

Figure 5. Possible mechanisms of transcriptional activation in humans and yeast. Activators stimulate transcription by recruiting histone acetylases and RNA polymerase II transcriptional machinery. In humans, various activators bind to p300/CBP, the PCAF complex, and TFIID. In yeast, GCN4 recruits the SAGA complex and TBP, which presumably associates with the same complex in a native context. Importantly, GCN4 does not recruit TFIID. In this respect, the SAGA complex and TBP might be able to direct transcription without TFIID.

HISTONE ACETYLASE ACTIVITY OF THE PCAF COMPLEX

We have compared the acetylation activity of the PCAF complex to that of recombinant PCAF and have found that although the complex acetylates nucleosomal substrates more efficiently, its specificity is apparently not changed. Acid-urea gel analysis reveals that as in the case of recombinant PCAF, predominantly a single residue in histone H3, presumably K14, is acetylated by the PCAF complex. It is noteworthy that this residue is very efficiently acetylated by the complex when compared to either recombinant PCAF or p300. Conceivably, acetylation of K14 of H3 has a critical role in PCAF-mediated transcriptional activation.

POSSIBLE ROLES OF PCAF-ASSOCIATED FACTORS

A number of activators recruit TFIID through direct interaction with various TAFs (Burley and Roeder 1996). hTAF$_{II}$80 and/or the homologous dTAF$_{II}$62 (also known as dTAF$_{II}$60) interacts directly with the activation domains of p53 (Thut et al. 1995) and NF-κB/p65 (Burley and Roeder 1996). hTAF$_{II}$31 and/or the homologous dTAF$_{II}$42 (dTAF$_{II}$40) interacts with the activation domains of p53 (Lu and Levine 1995; Thut et al. 1995) and VP16 (Goodrich et al. 1993). Similarly, hTAF$_{II}$30 interacts with the estrogen receptor (Jacq et al. 1994). At present, it is believed that these activators recruit TFIID through interaction with TAFs. However, theoretically, these activators also can recruit the PCAF complex. In this regard, it would be intriguing to determine the mechanism by which activators distinguish TAFs within TFIID or the PCAF complex (see Fig. 5).

The remarkable similarities between the TFIID and PCAF complex, including acetylase activity, a histone-like substructure, and a WD40 repeat subunit, suggest that the function(s) of the histone-like factors in the PCAF complex could be similar to that of TFIID. It was suggested that TFIID bound to certain class II core promoters might serve as a specialized chromatin component that fulfills the topological requirements necessary to mediate and maintain the inducibility of genes (Kokubo et al. 1994; Hoffmann et al. 1997). Similarly, if an octamer-like structure is present in the PCAF complex, it can play an analogous part in the maintenance of an active chromatin state by replacing the transcriptionally repressive histone octamer.

In this respect, the absence of regions corresponding to histone amino-terminal tails in the histone-like subunits might be an essential feature of both PCAF and TFIID complexes. Unacetylated histone tails contribute to the repressed chromatin state by interacting with nucleosomal DNA and adjacent nucleosomes and/or by recruiting repressor factors (for review, see Grunstein 1997; Luger et al. 1997; Wade and Wolffe 1997; Mizzen and Allis 1998; Struhl 1998). Although histone acetylases convert chromatin to an active state, the acetylated histone tails are subject to very rapid deacetylation by histone deacetylases (Jackson et al. 1975). Whereas this rapid turnover may be of advantage for dynamic gene regulation, in the case of constitutively active genes such as housekeeping or lineage-specific genes, it might be more useful to abolish the turnover altogether. The lack of regions corresponding to histone amino-terminal tails in the histone-like factors might be important in this regard, as it would yield a nucleosome-like structure resistant to deacetylation and would thus facilitate maintenance of the active chromatin state.

CONCLUSIONS

A central question in eukaryotic transcription is how the transcriptional machinery retrieves specific genetic information from tightly packed chromatin. The recent identification of a number of transcriptional regulators as histone acetylases or deacetylases provides convincing evidence that acetylation of histones is an important aspect of eukaryotic transcription. The PCAF histone acetylase, cloned in our laboratory, is a human homolog of yeast GCN5. It has a role in regulation of transcription, cell cycle progression, and differentiation by targeting select promoters. PCAF forms a multimeric protein complex consisting of more than 20 distinct polypeptides, ranging from 400 kD to 10 kD. Identification of the histone-like subunits within the PCAF complex provides an unusual twist to the PCAF story, as it suggests, in parallel with the general transcription factor TFIID, the presence of a histone octamer-like structure within the PCAF complex. Further work will clarify the role of PCAF-associated factors and the octamer-like structure in PCAF function.

REFERENCES

Arias J., Alberts A.S., Brindle P., Claret F.X., Smeal T., Karin M., Feramisco J., and Montminy M. 1994. Activation of cAMP and mitogen responsive genes relies on a common nuclear factor. *Nature* **370:** 226.

Berger S.L., Pina B., Silverman N., Marcus G.A., Agapite J., Regier J.L., Triezenberg S.J., and Guarente L. 1992. Genetic isolation of ADA2: A potential transcriptional adaptor required for function of certain acidic activation domains. *Cell* **70:** 251.

Brehm A., Miska E., McCance D., Reid J., Bannister A., and Kouzarides T. 1998. Retinoblastoma protein recruits histone deacetylase to repress transcription. *Nature* **391:** 597.

Burley S. and Roeder R. 1996. Biochemistry and structural biology of transcription factor IID (TFIID). *Annu. Rev. Biochem.* **65:** 769.

Chen H., Lin R., Schiltz R., Chakravarti D., Nash A., Nagy L., Privalsky M., Nakatani Y., and Evans R. 1997. Nuclear receptor coactivator ACTR is a novel histone acetyltransferase and forms a multimeric activation complex with P/CAF and CBP/p300. *Cell* **90:** 569.

Chrivia J.C., Kwok R.P., Lamb N., Hagiwara M., Montminy M.R., and Goodman R.H. 1993. Phosphorylated CREB binds specifically to the nuclear protein CBP. *Nature* **365:** 855.

Drysdale C., Jackson B., McVeigh R., Klebanow E., Bai Y., Kokubo T., Swanson M., Nakatani Y., Weil P., and Hinnebusch A. 1998. The Gcn4p activation domain interacts specifically in vitro with RNA polymerase II holoenzyme, TFIID, and the Adap-Gcn5p coactivator complex. *Mol. Cell. Biol.* **18:** 1711.

Dyson N. and Harlow E. 1992. Adenovirus E1A targets key regulators of cell proliferation. *Cancer Surv.* **12:** 161.

Edmondson D.G., Smith M.M., and Roth S.Y. 1996. Repression domain of the yeast global repressor Tup1 interacts directly with histones H3 and H4. *Genes Dev.* **10:** 1247.

Engelberg D., Klein C., Martinetto H., Struhl K., and Karin M. 1994. The UV response involving the Ras signaling pathway and AP-1 transcription factors is conserved between yeast and mammals. *Cell* **77:** 381.

Georgakopoulos T. and Thireos G. 1992. Two distinct yeast transcriptional activators require the function of the GCN5 protein to promote normal levels of transcription. *EMBO J.* **11:** 4145.

Goldman P.S., Tran V.K., and Goodman R.H. 1997. The multifunctional role of the co-activator CBP in transcriptional regulation. *Recent Prog. Horm. Res.* **52:** 103.

Goodrich J., Hoey T., Thut C., Admon A., and Tjian R. 1993. *Drosophila* TAFII40 interacts with both a VP16 activation domain and the basal transcription factor TFIIB. *Cell* **75:** 519.

Grant P.A., Schieltz, D., Pray-Grant, M.G., Steger, D.J., Reese, J.C., Yates J.R., III, and Workman, J.L. 1998. A subset of TAFIIs are integral components of the SAGA complex required for nucleosome acetylation and transcriptional stimulation. *Cell* **94:** 45.

Grant P.A., Duggan L., Cote J., Roberts S., Brownell J., Candau R., Ohba R., Owen-Hughes T., Allis C., Winston F., Berger S., and Workman J. 1997. Yeast Gcn5 functions in two multisubunit complexes to acetylate nucleosomal histones: Characterization of an Ada complex and the SAGA (Spt/Ada) complex. *Genes Dev.* **11:** 1640.

Grunstein M. 1997. Histone acetylation in chromatin structure and transcription. *Nature* **389:** 349.

Hecht A, Laroche T., Strahl-Bolsinger S., Gasser S.M., and Grunstein M. 1995. Histone H3 and H4 N-termini interact with SIR3 and SIR4 proteins: A molecular model for the formation of heterochromatin in yeast. *Cell* **80:** 583.

Hoffmann A., Oelgeschlager T., and Roeder R. 1997. Considerations of transcriptional control mechanisms: Do TFIID-core promoter complexes recapitulate nucleosome-like functions? *Proc. Natl. Acad. Sci.* **94:** 8928.

Hoffmann A., Chiang C., Oelgeschlager T., Xie X., Burley S., Nakatani Y., and Roeder R. 1996. A histone octamer-like structure within TFIID. *Nature* **380:** 356.

Hong L., Schroth, G., Matthews H., Yau P., and Bradbury E. 1993. Studies of the DNA binding properties of histone H4 amino terminus. Thermal denaturation studies reveal that acetylation markedly reduces the binding constant of the H4 "tail" to DNA. *J. Biol. Chem.* **268:** 305.

Horiuchi J., Silverman N., Marcus G.A., and Guarente L. 1995. ADA3, a putative transcriptional adaptor, consists of two separable domains and interacts with ADA2 and GCN5 in a trimeric complex. *Mol. Cell. Biol.* **15:** 1203.

Jackson V., Shires A., Chalkley R., and Granner D. 1975. Studies on highly metabolically active acetylation and phosphorylation of histones. *J. Biol. Chem.* **250:** 4856.

Jacq X., Brou C., Lutz Y., Davidson I., Chambon P., and Tora L. 1994. Human TAFII30 is present in a distinct TFIID complex and is required for transcriptional activation by the estrogen receptor. *Cell* **79:** 107.

Janknecht R. and Hunter T. 1996. Versatile molecular glue. Transcriptional control. *Curr. Biol.* **6:** 951.

Kokubo T., Gong D.W., Wootton J.C., Horikoshi M., Roeder R.G., and Nakatani Y. 1994. Molecular cloning of *Drosophila* TFIID subunits. *Nature* **367:** 484.

Korzus E., Torchia, J., Rose, D., Xu, L., Kurokawa, R., McInerney, E., Mullen, T., Glass, C., and Rosenfeld, M. 1998. Transcription factor-specific requirements for coactivators and their acetyltransferase functions. *Science* **279:** 703.

Kuo M.-H., Brownell J.E., Sobel R.E., Ranalli T.A., Cook R.G., Edmondson, D.G., Roth, S.Y., and Allis C.D. 1996. Transcription-linked acetylation by Gcn5p of histones H3 and H4 at specific lysines. *Nature* **382:** 269.

Lee J.S., Galvin K.M., See R.H., Eckner R., Livingston D., Moran E., and Shi Y. 1995. Relief of YY1 transcriptional repression by adenovirus E1A is mediated by E1A-associated protein p300. *Genes Dev.* **9:** 1188.

Lill N., Grossman S., Ginsberg D., DeCaprio J., and Livingston D. 1997. Binding and modulation of p53 by p300/CBP coac-

tivators. *Nature* **387:** 823.

Livingston D., Kaelin W., Chittenden T., and Qin X. 1993. Structural and functional contributions to the G1 blocking action of the retinoblastoma protein. *Br. J. Cancer* **68:** 264.

Lu H. and Levine A. 1995. Human TAFII31 protein is a transcriptional coactivator of the p53 protein. *Proc. Natl. Acad. Sci.* **92:** 5154.

Luger K., Mader A., Richmond R., Sargent D., and Richmond T. 1997. Crystal structure of the nucleosome core particle at 2.8 Å resolution. *Nature* **389:** 251.

Luo R., Postigo A., and Dean D. 1998. Rb interacts with histone deacetylase to repress transcription. *Cell* **92:** 463.

Magnaghi-Jaulin L., Groisman R., Naguibneva I., Robin P., Lorain S., Le Villain J., Troalen F., Trouche D., and Harel-Bellan A. 1998. Retinoblastoma protein represses transcription by recruiting a histone deacetylase. *Nature* **391:** 601.

Mengus G., May M., Jacq X., Staub A., Tora L., Chambon P., and Davidson I. 1995. Cloning and characterization of hTAFII18, hTAFII20 and hTAFII28: Three subunits of the human transcription factor TFIID. *EMBO J.* **14:** 1520.

Mizzen C. and Allis C. 1998. Linking histone acetylation to transcriptional regulation. *Cell. Mol. Life Sci.* **54:** 6.

Mizzen C., Yang X., Kokubo T., Brownell J., Bannister A., Owen-Hughes T., Workman J., Wang L., Berger S., Kouzarides T., Nakatani Y., and Allis C. 1996. The TAF(II)250 subunit of TFIID has histone acetyltransferase activity. *Cell* **87:** 1261.

Moran E. 1993. DNA tumor virus transforming proteins and the cell cycle. *Curr. Opin. Genet. Dev.* **3:** 63.

Nevins J. 1994. Cell cycle targets of the DNA tumor viruses. *Curr. Opin. Genet. Dev.* **4:** 130.

Norton V. Imai B., Yau P., and Bradbury E. 1989. Histone acetylation reduces nucleosome core particle linking number change. *Cell* **57:** 449.

Ogryzko V.V., Schiltz R., Russanova V., Howard B., and Nakatani Y. 1996. The transcriptional coactivators p300 and CBP are histone acetyltransferases. *Cell* **87:** 953.

Ogryzko V.V., Kotani T., Zhang X., Schiltz R.L., Howard T., Yang X.J., Howard B.H., Qin J., and Nakatani Y. 1998. Histone-like TAFs within the PCAF histone acetylase complex. *Cell* **94:** 35.

Puri P.L., Sartorelli V., Yang X.-J., Hamamori Y., Ogryzko V.V., Howard B.H., Kedes L., Wang J.Y., Graessmann A., Nakatani Y., and Levrero M. 1997. The PCAF histone acetylase activity is essential for muscle differentiation. *Mol. Cell* **1:** 35.

Rochette-Egly C., Fromental C., and Chambon P. 1990. General repression of enhancer activity by the adenovirus-2 E1A proteins. *Genes Dev.* **5:** 1200.

Schiltz R.L., Mizzen C.A., Vassilev A., Cook R.G., Allis C.D., and Nakatani Y. 1999. Overlapping, but distinct, patterns of histone acetylation by the human coactivators p300 and PCAF within nucleosomal substrates. *J. Biol. Chem.* (in press).

Smith E., Belote J., Schiltz R., Yang X., Moore P., Berger S., Nakatani Y., and Allis C. 1998. Cloning of *Drosophila* GCN5: Conserved features among metazoan GCN5 family members. *Nucleic Acids Res.* **26:** 2948.

Struhl K. 1987. The DNA-binding domains of the jun oncoprotein and the yeast GCN4 transcriptional activator protein are functionally homologous. *Cell* **50:** 841.

———. 1998. Histone acetylation and transcriptional regulatory mechanisms. *Genes Dev.* **12:** 599.

Thut C., Chen J., Klemm R., and Tjian R. 1995. p53 transcriptional activation mediated by coactivators TAFII40 and TAFII60. *Science* **267:** 100.

Wade P. and Wolffe A. 1997. Histone acetyltransferases in control. *Curr. Biol.* **7:** R82.

Weinberg R. 1995. The retinoblastoma protein and cell cycle control. *Cell* **81:** 323.

Xie X., Kokubo T., Cohen S., Mirza U., Hoffmann A., Chait B., Roeder R., Nakatani Y., and Burley S. 1996. Structural similarity between TAFs and the heterotetrameric core of the histone octamer. *Nature* **380:** 316.

Yang X., Ogryzko V., Nishikawa J., Howard B., and Nakatani Y. 1996. A p300/CBP-associated factor that competes with the adenoviral oncoprotein E1A. *Nature* **382:** 319.

Structure of the Yeast Histone Acetyltransferase Hat1: Insights into Substrate Specificity and Implications for the Gcn5-related *N*-acetyltransferase Superfamily

R.N. Dutnall, S.T. Tafrov,* R. Sternglanz,* and V. Ramakrishnan

Department of Biochemistry, University of Utah School of Medicine, Salt Lake City, Utah 84132;
Department of Biochemistry and Cell Biology, State University of New York, Stony Brook, New York 11794

It was discovered well over three decades ago that the amino-terminal tails of the four core histones (H2A, H2B, H3, and H4) are posttranslationally modified by the addition of an acetyl group to the ε-amino group of specific lysine side chains (Phillips 1963). Soon afterward, Allfrey et al. (1964) suggested that such a seemingly minor modification to histones could be involved in the control of gene expression via its "affect on the capacity of the histones to inhibit ribonucleic acid synthesis." In the subsequent years, many pieces of evidence have accumulated to support this hypothesis. Histone acetylation has indeed been found to be correlated with levels of gene activity (for review, see Turner 1998). For example, potentially active euchromatin regions are associated with hyperacetylated histones, whereas inactive heterochromatin is associated with hypoacetylated histones (Hebbes et al. 1988; O'Neill and Turner 1995). In both *Saccharomyces cerevisiae* and *Drosophila*, heterochromatin contains histone H4 uniquely acetylated at Lys-12 (Turner et al. 1992; Braunstein et al. 1993, 1996). The inactive X chromosome in humans is also associated with hypoacetylated histones (Jeppesen and Turner 1993). Histone acetylation has also been shown to be involved in the propagation of imprinted chromosomal structure at centromeres during cell division (Ekwall et al. 1997). The finding that in some forms of cancer, chromosomal translocations create novel fusion proteins with known histone acetyltransferase (HAT) enzymes (Borrow et al. 1996; Sobulo et al. 1997; Carapeti et al. 1998), along with the observation that the p300 HAT enzyme can regulate the DNA-binding acitivity of the tumor suppressor gene p53 via acetylation (Gu and Roeder 1997), indicates that histone acetylation could have an important role in the control of cell proliferation and tumorigenesis.

Several genes that encode HATs and histone deacetylases (HDACs) have now been characterized. Many of the HAT genes turned out to be previously known transcriptional coactivators such as Gcn5 (Brownell et al. 1996), p300/CBP (Bannister and Kouzarides 1996; Ogryzko et al. 1996), the TAFII250 subunit of TFIID (Mizzen et al. 1996) and the SRC-1 family (Chen et al. 1997; Spencer et al. 1997). Likewise, the first HDAC gene to be cloned turned out to be highly related to the yeast repressor protein Rpd3 (Taunton et al. 1996). These observations not only provided a direct link to the corre-

lation between histone acetylation and gene activity, but also sparked a renewed interest in the roles of histone acetylation and chromatin structure in gene activity.

Three general mechanisms have been proposed to explain the effect of histone acetylation on chromatin structure and its ability to act as a transcription template. The first is that it neutralizes the positively charged amino-terminal tails of the histones that lie toward the outside of the structure of the nucleosome, reducing their affinity for DNA, with the result that the accessibility of regulatory sites within nucleosomes to DNA-binding proteins is increased (Wolffe and Pruss 1996). The second is that acetylation of the histone tails could modulate the ability of nucleosomes to form higher-order chromatin structures, thus rendering chromatin more generally accessible to components of the DNA replication, repair, or transcription machinery (Allan et al. 1982; Garcia-Ramirez et al. 1995). Third, acetylation could act as a specific flag for regulatory or enzymatic factors that bind to chromatin, in a fashion similar to protein phosphorylation in signal transduction pathways (Loidl 1988; Dutnall and Ramakrishnan 1997). For example, it has been proposed that association of the Sir3 protein with nucleosomes is favored by the heterochromatin-specific acetylation state of histone H4 (only Lys-12 acetylated) and prevented by acetylation of H4 Lys-16 (Grunstein 1998).

The core histone tails contain several possible sites for acetylation, as well as sites of modification by phosphorylation, methylation, ADP-ribosylation and ubiquitination (Fig. 1) (van Holde 1989). Some sites of histone acetylation are correlated with particular cellular activities (for review, see Brownell and Allis 1996; Roth and Allis 1996). For example, acetylation of histone H4 at Lys-5 and Lys-12 and histone H3 at Lys-9 is the pattern found soon after these histones are newly synthesized and is thought to have an important role in replication-dependent chromatin assembly. Conversely, histone H4 Lys-8 and Lys-16 and histone H3 Lys-14 are the sites associated with transcription-related histone acetylation. In yeast, Lys-12 is found to be the unique site of histone H4 acetylation in regions of heterochromatin and may have an important role in gene silencing (Braunstein et al. 1996; Grunstein 1998).

Just as patterns of histone acetylation vary, the HAT enzymes vary with respect to their specificity in a number

```
              10           20          30-
H2A  SGRGKQGGKA RAKAKTRSSR AGLQFPVGRV-

H2B  PEPAKSAPAP KKGSKKAVTK AQKKDGKKRK-

H3   ARTKQTARKS TGGKAPRKQL ATKAARKSAP-

H4   SGRGKGGKGL GKGGAKRHRK VLRDNIQGIT-
```

Figure 1. Sites of lysine acetylation within the amino-terminal tails of the four core histones. The first 30 residues of each of the four core histones H2A, H2B, H3, and H4 are shown. Lysine residues that have been found to be acetylated are shown in bold (van Holde 1989).

of ways. For example, the enzyme p300/CBP is capable of acetylating all four core histones (Bannister and Kouzarides 1996; Ogryzko et al. 1996). Others, like Hat1, Gcn5, or Esa1, can only acetylate one or two histones. These three enzymes also vary with respect to the specific lysine residues they modify. For example, Hat1 is capable of modifying histone H4 at Lys-12 and Lys-5 and histone H2A at Lys-5 (Kleff et al. 1995; Parthun et al. 1996; Kolle et al. 1998; Verreault et al. 1998), whereas Gcn5 modifies histone H3 at Lys-14 and also histone H4 at Lys-8 and Lys-16 (Kuo et al. 1996). They also have varying preferences within the sites they modify. Hat1 prefers histone H4 Lys-12, whereas Gcn5 prefers histone H3 Lys-14. Esa1 prefers to modify histone H4 Lys-5 but will also modify other sites in histone H4 as well as histones H3 and H2A (Smith et al. 1998).

Another aspect that varies among HAT enzymes is their ability to modify histones when bound to DNA within the context of the nucleosome. Hat1 and Esa1 cannot modify nucleosomal substrates (Parthun et al. 1996; Smith et al. 1998), whereas p300/CBP, P/CAF, ACTR, and SRC-1 can modify both free and nucleosomal histones (Bannister and Kouzarides 1996; Ogryzko et al. 1996; Yang et al. 1996; Chen et al. 1997; Spencer et al. 1997). Gcn5 can also acetylate nucleosomal histones but only when it is part of the SAGA complex (Grant et al. 1997). The substrate specificity can also be influenced by a nucleosomal context, as illustrated by the enzyme P/CAF which can acetylate both free histones H3 and H4 while it apparently only modifies histone H3 within a nucleosome (Yang et al. 1996). Some HAT enzymes are even capable of acetylating nonhistone substrates. p300 is able to stimulate the DNA-binding activity of the p53 tumor suppressor protein by acetylating its carboxy-terminal domain (Gu and Roeder 1997). In vitro, P/CAF can acetylate some of the RNA polymerase II general transcription factors (Imhof et al. 1997), and Gcn5 can modify the Sin1 protein (Pollard and Peterson 1997).

The diversity of potential HAT activities present within a cell make it clear that in order to understand more fully the biological role of histone acetylation, we need to have a clear picture of the true substrate specificity of these enzymes in vivo. It is important to note that thus far, only one enzyme (Gcn5) has been shown to be a *bone fide* histone acetyltransferase in vivo (Kuo et al.

1998; Wang et al. 1998; Zhang et al. 1998). We have determined the structure of the yeast HAT enzyme Hat1, in complex with its cofactor acetyl coenzyme A (AcCoA) (Dutnall et al. 1998). The structure of Hat1 not only serves as a template for understanding the function of many related HAT enzymes, but also provides a plausible model for the substrate recognition specificity of Hat1.

THE STRUCTURE OF HAT1

S. cerevisiae Hat1 was the first HAT gene to be identified (Kleff et al. 1995; Parthun et al. 1996). Homologous Hat1 proteins exist in humans (Verreault et al. 1998) and maize (Eberharter et al. 1996; Kolle et al. 1998), and possible relatives have also been found in *Schizosaccharomyces pombe* and *Caenorhabditis elegans*. The protein that we crystallized contained the first 320 residues of the 374-residue *S. cerevisiae* protein. We found that deleting the carboxy-terminal 54 residues did not affect catalytic activity but was needed to obtain crystals that diffracted X-rays to high resolution. Presumably, the carboxy-terminal residues are flexible in solution and interfere with the formation of a highly ordered crystalline lattice.

Hat1 is composed of a mixture of α helices and β sheets arranged in two domains referred to as the amino- and carboxy-terminal domains (Fig. 2). Overall, the structure is quite elongated with the protein adopting a curved shape. The AcCoA molecule is bound in a cleft in the carboxy-terminal domain, situated approximately in the middle of the concave surface of the protein. Many direct and indirect (water-mediated) contacts are made to the AcCoA from residues that are highly conserved with other HAT enzymes (see below). The acetyl group and pantetheine arm, which resemble in many respects part of a polypeptide chain, make many interactions with the edge of the central six-stranded β-sheet of the carboxy-terminal domain. Some of these interactions mimic the main-chain hydrogen-bonding pattern found in a typical protein β-sheet structure.

THE ACTIVE SITE AND POSSIBLE MODES OF CATALYSIS

The acetyl group itself, which marks the active site of the protein, is located in a small hydrophobic pocket but is nevertheless accessible to the surface. It seems likely that the protein itself does not provide any single catalytic residue in the reaction to transfer the acetyl group from AcCoA onto the substrate lysine, but instead it relies upon the high acyl transfer potential of AcCoA (Dutnall et al. 1998). In this scheme, the lysine to be modified enters the active site and the reaction proceeds via a direct nucleophilic attack from its ε-amino group against the carbonyl carbon of the acetate group. Hat1 could facilitate this reaction not only by ordering the reacting molecules (the entropic cost of which would presumably be paid for by binding energy), but also by providing an environment in which the lysine to be modified assumes an uncharged state. The structure of the Hat1-AcCoA

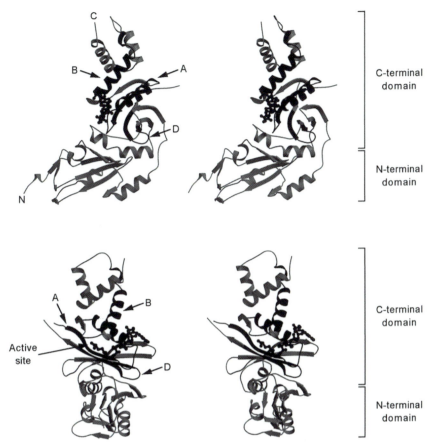

Figure 2. Structure of Hat1. Two schematic views of the protein are shown as stereo pairs. The views in the upper and lower panels are related by a 90° rotation about the vertical axis. The amino- and carboxy-terminal domains of the protein and the active site of the protein are indicated. The darker sections of the protein indicate the position of conserved sequence motifs (A, B, and D) found in a family of acetyltransferase enzymes (see text). The AcCoA molecule is shown as a ball and stick representation.

complex shows how the protein orders the AcCoA molecule. The surface of the molecule to which the histone substrate most likely binds contains many acidic side chains that could serve as acceptors for a proton from the lysine during the binding reaction to create a better nucleophile. Finally, in the vicinity of the acetate group are several main-chain carbonyl groups for which no hydrogen bond donors exist in the current structure. Some of these are suitably oriented to help position the ε-mino group of the lysine via hydrogen bond interactions.

AN INSIGHT INTO THE MECHANISM OF SUBSTRATE SPECIFICITY OF HAT1

With the structure of Hat1 in hand, we were interested in investigating possible modes of histone binding by this enzyme and in determining whether they revealed any clues to account for the particular pattern of histone modification produced by this enzyme. Purified yeast Hat1 will specifically acetylate histone H4 at Lys-12 (Parthun et al. 1996). The recombinant version of this enzyme, produced in *Escherichia coli*, acetylates histone H4 Lys-12 and Lys-5, albeit to a lesser extent, and weakly acetylates histone H2A (probably at Lys-5) (Kleff et al. 1995;

Parthun et al. 1996). By comparing the ability of yeast Hat1 to acetylate histones from a variety of sources (yeast, chicken, and *Tetrahymena*), Parthun et al. identified a possible recognition motif, G*x*GK*x*G (where *x* is any amino acid), in the vicinity of the modified lysine (Fig. 3). Subsequent studies of the Hat1 enzymes from humans (Verreault et al. 1998) and maize (Kolle et al. 1998) revealed that they had very similar substrate specificities which also conform to this recognition motif.

From the structure of Hat1, we have been able to develop a plausible model to account for its observed specificity. This model was produced by attempting to fit a short peptide, based on the sequence of histone H4 around Lys-12, into the Hat1 structure, focusing primarily on positioning this lysine in a situation where its ε-amino group would be close to the carbonyl group of the AcCoA. During the process, it was immediately noticeable that a channel of varying width and depth runs across the surface of Hat1 intersecting with the vicinity of the carbonyl group of AcCoA. The characteristics of this channel not only are ideal for binding a peptide of about the same length as the proposed Hat1 recognition motif, but could also explain why Hat1 prefers to acetylate histone H4 Lys-12.

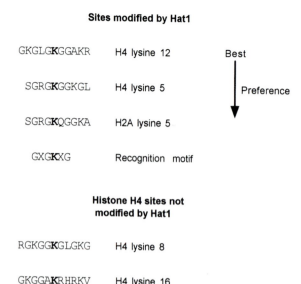

Sites modified by Hat1

GKGLG**K**GGAKR	H4 lysine 12	Best
SGRG**K**GGKGL	H4 lysine 5	Preference
SGRG**K**QGGKA	H2A lysine 5	
GXG**K**XG	Recognition motif	

Histone H4 sites not modified by Hat1

RGKGG**K**GLGKG	H4 lysine 8
GKGGA**K**RHRKV	H4 lysine 16

Figure 3. Hat1 substrate recognition motif. The sequences adjacent to lysine residues that can be modified by Hat1 are shown for histone H4 Lys-12 and Lys-5 and H2A Lys-5 (the modified lysine is shown in bold). The order of preference of Hat1 and the recognition motif derived by Parthun et al. (1996) are indicated. For comparison, the sequences around histone H4 lysine residues that are not modified by Hat1 are shown in a similar way.

The principal characteristics of this channel and the way they could account for the specificity of Hat1 are summarized schematically in Figure 4. In general, the channel is sufficiently long to accommodate a 6–7-residue peptide with an extended main-chain conformation, the same length as the proposed recognition motif. An extended conformation seems reasonable as it would allow Hat1 to maximize the number of contacts it could make to the peptide, particularly contacts to the peptide main-chain groups. In such a conformation, and with histone H4 Lys-12 in the active site, the characteristics of the side chains on either side of Lys-12 present a pattern of characteristics complementary to those of the proposed binding channel. In particular, Leu-10 of histone H4 would be placed next to a small hydrophobic pocket in the deepest part of the binding channel. Meanwhile, Gly-13 and Gly-14 would be positioned in the shallowest and narrowest part of the binding channel where it would be difficult to accommodate larger side chains. Finally, Lys-8 and Lys-16, which lie outside of the originally proposed recognition motif, would take up positions opposite two acidic patches at the ends of the channel where they could take part in electrostatic interactions.

Further credence for this model comes from considering other lysine residues in the tail of H4 which are either modified to a lesser degree by Hat1 (Lys-5) or not at all (Lys-8 and Lys-16). If Lys-5 is placed in the active site (Fig. 4), it can be seen that the H4 side chains surrounding it display some complementary features to the binding channel, but the overall fit is not quite as good as described for Lys-12. Although Gly-6 and Gly-7 would be in the narrowest part of the channel, Arg-3 is placed next to the hydrophobic pocket. It is possible that Gly-2 and Gly-4 provide enough flexibility in the main chain to allow the side chain of Arg-3 to reach over to the acidic patch that would be contacted by Lys-8 if Lys-12 is placed in the active site. If either Lys-8 or Lys-16 are placed in the active site, their surrounding residues display an even poorer fit to the binding channel. In particular, they each have neighboring large side chains that could not be accommodated in the narrowest part of the channel (Leu-10 in the case of Lys-8, and Arg-17 and His-18 in the case of Lys-16). In addition, they lack a residue suitable to interact in the hydrophobic pocket (in both cases, a glycine is positioned in this part of the channel).

The general conclusion is that Lys-12 is the preferred substrate of Hat1, because when it is in the active site, the tail of H4 can form a complementary fit to the enzyme. Lys-5 of H4 and Lys-5 of H2A are less suitable modification sites as they provide only a partial fit to the enzyme. Conversely, Hat1 cannot modify other lysine residues as the surrounding sequences either do not display side chains that could make any kind of binding interaction or, even worse, display side chains that clash with the enzyme in some way. This kind of stereochemical discrimination is found in any intermolecular interaction that displays some kind of specificity, a good example being the discrimination of DNA-binding sites by the DNA-binding domains of transcription factors (Steitz 1990).

Another more general aspect of this model is that it implies that modifications to side chains within or near the recognition sequence of a HAT enzyme could influence its ability to modify its target residue (in either a positive or negative fashion). Using acetylation of histone H4 Lys-12 by Hat1 as an example, if Lys-8 or Lys-16 of histone H4 is acetylated, it would lack a positive charge and therefore might not be able to interact productively with the acidic patches at the ends of its putative peptide-binding channel. The close primary sequence proximity of the acetylation sites in the histone tails (Fig. 1) means that it is therefore possible that the activities of various HAT and HDAC enzymes could be coupled. Furthermore, as suggested by C.D. Allis at the Symposium (see Mizzen et al., this volume), this same effect could couple histone acetylation to other modifications such as methylation or phosphorylation. This could have an important role in building up specific patterns of histone modifications.

IMPLICATIONS FOR OTHER HAT ENZYMES

Hat1 and many of the known HAT enzymes such as Gcn5, P/CAF, p300/CBP, and the MYST/SAS group of HATs (which includes Esa1, Mof, and Tip60; Reifsnyder et al. 1996) belong to a superfamily of enzymes that includes other acetyltransferases (Neuwald and Landsman 1997). This family, the Gcn5-related *N*-acetyltransferase (GNAT) superfamily, encompasses a diverse set of enzymes whose substrates include not only proteins as in the case of HATs and amino-terminal acetyltransferases, but also molecules such as diamines, aminoglycosides, and various antibiotic compounds. Most of the family

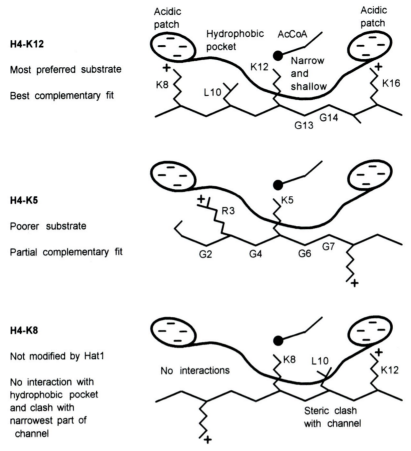

Figure 4. How the degree of complementary fit may explain the substrate specificity of Hat1. A highly schematic representation of the peptide-binding channel of Hat1 is shown, along with a stick diagram of the histone H4 tail for three cases: Lys-12, the most preferred substrate, Lys-5 (acetylated by Hat1 but to a lesser degree), and Lys-8, which is not modified by Hat1. The features of the channel that may have a role in substrate discrimination are indicated.

members share four conserved sequence motifs (A–D) spanning more than 100 residues. Motif A is the longest and most conserved, and the HAT members of the family apparently lack motif C.

In Hat1, the GNAT motifs are all found within the carboxy-terminal domain of the protein and make up much of the architecture involved in AcCoA binding (Fig. 2). This agrees with the previous suggestion that this would be the case based on the fact that the common factor among members of the GNAT superfamily is the ability to bind AcCoA (Neuwald and Landsman 1997). Many of the most highly conserved residues of the GNAT motifs are involved in contacts to AcCoA, including the motif Q/RxxGxG (within motif A) which makes up part of a loop involved in binding the pyrophosphate moiety of AcCoA.

It is highly likely that the catalytic domains of many members of the GNAT superfamily will resemble the carboxy-terminal domain of Hat1. This is based not only on the observed sequence similarity among these enzymes, but also on biochemical studies of GNAT superfamily members. Particularly relevant are the results of recent mutagenesis experiments on the HAT enzymes Gcn5

(Kuo et al. 1998; Wang et al. 1998) and CBP (Martinez-Balbas et al. 1998). These studies identified a number of residues important for the HAT activity of these proteins, as revealed by the reduction of HAT activity when they were mutated. Strikingly, many of these residues correspond to residues in Hat1 that are involved in AcCoA binding, whereas most of the others can be explained by their effect on structural stability (Dutnall et al. 1998). Some of this correspondence is to be expected, particularly for residues within the conserved GNAT family motifs. However, some of the mutations correspond to Hat1 regions that contact AcCoA but lie outside of the previously identified GNAT family motifs. In addition, when we aligned the sequences of Hat1 and Gcn5 (using the Hat1 structure as a guide to improve the basis for sequence comparisons), we found that the borders of the carboxy-terminal domain of Hat1 (resides 120–320) lie very close to the borders of the minimal domain of Gcn5 needed for full HAT activity in vitro (Candau et al. 1997).

It is not clear whether other HAT enzymes such as TAF$_{II}$250 or the p160 family (which includes SRC-1 and ACTR) will share structural similarity to Hat1. Although they do not share obvious sequence similarity with the

GNAT superfamily, it is possible that the level of sequence similarity falls below the cut-off level used in many similarity searches. This is certainly true for the CBP protein which was only identified as a GNAT superfamily member recently (Martinez-Balbas et al. 1998). Alternatively, they may have followed a separate evolutionary path to develop a distinct fold capable of binding both AcCoA and histones. This possibility is supported by the fact that there are a variety of protein folds capable of binding AcCoA (Engel and Wierenga 1996).

CONCLUSIONS AND PERSPECTIVES

The structure of Hat1 has provided us with our first look at the structure of a histone acetyltransferase. It shows how the enzyme binds its cofactor AcCoA and has given us a glimpse of a possible mode of histone substrate recognition for this enzyme. The structure also serves as a template for rationalizing biochemical studies of other HAT enzymes and other members of the GNAT superfamily and for designing future experiments. When the structures of other HAT enzymes become available, it will be interesting to examine structural similarities and differences between them in terms of their varied substrate specificities and biological functions.

REFERENCES

Allan J., Harborne N., Rau D.C., and Gould H. 1982. Participation of core histone "tails" in the stabilization of the chromatin solenoid. *J. Cell Biol.* **93:** 285.

Allfrey V., Faulkner R.M., and Mirsky A.E. 1964. Acetylation and methylation of histone and their possible role in the regulation of RNA synthesis. *Proc. Natl. Acad. Sci.* **51:** 786.

Bannister A.J. and Kouzarides T. 1996. The CBP co-activator is a histone acetyltransferase. *Nature* **384:** 641.

Borrow J., Stanton V.P., Jr., Andresen J.M., Becher R., Behm F.G., Chaganti R.S., Civin C.I., Disteche C., Dube I., Frischauf A.M., Horsman D., Mitelman F., Volinia S., Watmore A.E., and Housman D.E. 1996. The translocation t(8;16)(p11;p13) of acute myeloid leukaemia fuses a putative acetyltransferase to the CREB-binding protein. *Nat. Genet.* **14:** 33.

Braunstein M., Rose A.B., Holmes S.G., Allis C.D., and Broach J.R. 1993. Transcriptional silencing in yeast is associated with reduced nucleosome acetylation. *Genes Dev.* **7:** 592.

Braunstein M., Sobel R.E., Allis C.D., Turner B.M., and Broach J.R. 1996. Efficient transcriptional silencing in *Saccharomyces cerevisiae* requires a heterochromatin histone acetylation pattern. *Mol. Cell. Biol.* **16:** 4349.

Brownell J.E. and Allis C.D. 1996. Special HATs for special occasions: Linking histone acetylation to chromatin assembly and gene activation. *Curr. Opin. Genet. Dev.* **6:** 176.

Brownell J.E., Zhou J., Ranalli T., Kobayashi R., Edmondson D.G., Roth S.Y., and Allis C.D. 1996. *Tetrahymena* histone acetyltransferase A: A homolog to yeast Gcn5p linking histone acetylation to gene activation. *Cell* **84:** 843.

Candau R., Zhou J.X., Allis C.D., and Berger S.L. 1997. Histone acetyltransferase activity and interaction with ADA2 are critical for GCN5 function in vivo. *EMBO J.* **16:** 555.

Carapeti M., Aguiar R.C.T., Goldman J.M., and Cross N.C.P. 1998. A novel fusion between MOZ and the nuclear receptor coactivator TIF2 in acute myeloid leukemia. *Blood* **91:** 3127.

Chen H., Lin R.J., Schiltz R.L., Chakravarti D., Nash A., Nagy L., Privalsky M.L., Nakatani Y., and Evans R.M. 1997. Nuclear receptor coactivator ACTR is a novel histone acetyltransferase and forms a multimeric activation complex with

P/CAF and CBP/p300. *Cell* **90:** 569.

Dutnall R.N. and Ramakrishnan V. 1997. Twists and turns of the nucleosome: Tails without ends. *Structure* **5:** 1255.

Dutnall R.N., Tafrov S.T., Sternglanz R., and Ramakrishnan V. 1998. Structure of the histone acetyltransferase Hat1: A paradigm for the GCN5-related N-acetyltransferase superfamily. *Cell* **94:** 427.

Eberharter A., Lechner T., Goralik-Schramel M., and Loidl P. 1996. Purification and characterization of the cytoplasmic histone acetyltransferase B of maize embryos. *FEBS Lett.* **386:** 75.

Ekwall K., Olsson T., Turner B.M., Cranston G., and Allshire R.C. 1997. Transient inhibition of histone deacetylation alters the structural and functional imprint at fission yeast centromeres. *Cell* **91:** 1021.

Engel C. and Wierenga R. 1996. The diverse world of coenzyme A binding proteins. *Curr. Opin. Struct. Biol.* **6:** 790.

Garcia-Ramirez M., Rocchini C., and Ausio J. 1995. Modulation of chromatin folding by histone acetylation. *J. Biol. Chem.* **270:** 17923.

Grant P.A., Duggan L., Cote J., Roberts S.M., Brownell J.E., Candau R., Ohba R., Owen-Hughes T., Allis C.D., Winston F., Berger S.L., and Workman J.L. 1997. Yeast Gcn5 functions in two multisubunit complexes to acetylate nucleosomal histones: Characterization of an Ada complex and the SAGA (Spt/Ada) complex. *Genes Dev.* **11:** 1640.

Grunstein M. 1998. Yeast heterochromatin: Regulation of its assembly and inheritance by histones. *Cell* **93:** 325.

Gu W. and Roeder R.G. 1997. Activation of p53 sequence-specific DNA binding by acetylation of the p53 C-terminal domain. *Cell* **90:** 595.

Hebbes T.R., Thorne A.W., and Crane-Robinson C. 1988. A direct link between core histone acetylation and transcriptionally active chromatin. *EMBO J.* **7:** 1395.

Imhof A., Yang X.J., Ogryzko V.V., Nakatani Y., Wolffe A.P., and Ge H. 1997. Acetylation of general transcription factors by histone acetyltransferases. *Curr. Biol.* **7:** 689.

Jeppesen P. and Turner B.M. 1993. The inactive X chromosome in female mammals is distinguished by a lack of histone H4 acetylation, a cytogenetic marker for gene expression. *Cell* **74:** 281.

Kleff S., Andrulis E.D., Anderson C.W., and Sternglanz R. 1995. Identification of a gene encoding a yeast histone H4 acetyltransferase. *J. Biol. Chem.* **270:** 24674.

Kolle D., Sarg B., Lindner H., and Loidl P. 1998. Substrate and sequential site specificity of cytoplasmic histone acetyltransferases of maize and rat liver. *FEBS Lett.* **421:** 109.

Kuo M.H., Zhou J., Jambeck P., Churchill M.E., and Allis C.D. 1998. Histone acetyltransferase activity of yeast Gcn5p is required for the activation of target genes in vivo. *Genes Dev.* **12:** 627.

Kuo M.H., Brownell J.E., Sobel R.E., Ranalli T.A., Cook R.G., Edmondson D.G., Roth S.Y., and Allis C.D. 1996. Transcription-linked acetylation by Gcn5p of histones H3 and H4 at specific lysines. *Nature* **383:** 269.

Loidl P. 1988. Towards an understanding of the biological function of histone acetylation. *FEBS Lett.* **227:** 91.

Martinez-Balbas M.A., Bannister A.J., Martin K., Haus-Seuffert P., Meisterernst M., and Kouzarides T. 1998. The acetyltransferase activity of CBP stimulates transcription. *EMBO J.* **17:** 2886.

Mizzen C.A., Yang X.J., Kokubo T., Brownell J.E., Bannister A.J., Owen-Hughes T., Workman J., Wang L., Berger S.L., Kouzarides T., Nakatani Y., and Allis C.D. 1996. The TAF(II)250 subunit of TFIID has histone acetyltransferase activity. *Cell* **87:** 1261.

Neuwald A.F. and Landsman D. 1997. GCN5-related histone N-acetyltransferases belong to a diverse superfamily that includes the yeast SPT10 protein. *Trends Biochem. Sci.* **22:** 154.

Ogryzko V.V., Schiltz R.L., Russanova V., Howard B.H., and Nakatani Y. 1996. The transcriptional coactivators p300 and CBP are histone acetyltransferases. *Cell* **87:** 953.

O'Neill L.P. and Turner B.M. 1995. Histone H4 acetylation distinguishes coding regions of the human genome from heterochromatin in a differentiation-dependent but transcription- independent manner. *EMBO J.* **14:** 3946.

Parthun M.R., Widom J., and Gottschling D.E. 1996. The major cytoplasmic histone acetyltransferase in yeast: Links to chromatin replication and histone metabolism. *Cell* **87:** 85.

Phillips D.M.P. 1963. The presence of acetyl groups in histones. *Biochem. J.* **87:** 258.

Pollard K.J. and Peterson C.L. 1997. Role for ADA/GCN5 products in antagonizing chromatin-mediated transcriptional repression. *Mol. Cell. Biol.* **17:** 6212.

Reifsnyder C., Lowell J., Clarke A., and Pillus L. 1996. Yeast SAS silencing genes and human genes associated with AML and HIV-1 Tat interactions are homologous with acetyltransferases. *Nat. Genet.* **14:** 42.

Roth S.Y. and Allis C.D. 1996. Histone acetylation and chromatin assembly: A single escort, multiple dances? *Cell* **87:** 5.

Smith E.R., Eisen A., Gu W., Sattah M., Pannuti A., Zhou J., Cook R.G., Lucchesi J.C., and Allis C.D. 1998. ESA1 is a histone acetyltransferase that is essential for growth in yeast. *Proc. Natl. Acad. Sci.* **95:** 3561.

Sobulo O.M., Borrow J., Tomek R., Reshmi S., Harden A., Schlegelberger B., Housman D., Doggett N.A., Rowley J.D., and Zeleznik-Le N.J. 1997. MLL is fused to CBP, a histone acetyltransferase, in therapy-related acute myeloid leukemia with a t(11;16)(q23;p13.3). *Proc. Natl. Acad. Sci.* **94:** 8732,

Spencer T.E., Jenster G., Burcin M.M., Allis C.D., Zhou J., Mizzen C.A., McKenna N.J., Onate S.A., Tsai S.Y., Tsai M.J., and O'Malley B.W. 1997. Steroid receptor coactivator-1 is a histone acetyltransferase. *Nature* **389:** 194.

Steitz T.A. 1990. Structural studies of protein-nucleic acid interaction: The sources of sequence-specific binding. *Q. Rev. Biophys.* **23:** 205.

Taunton J., Hassig C.A., and Schreiber S.L. 1996. A mammalian histone deacetylase related to the yeast transcriptional regulator Rpd3. *Science* **272:** 408.

Turner B.M. 1998. Histone acetylation as an epigenetic determinant of long-term transcriptional competence. *Cell. Mol. Life Sci.* **54:** 21.

Turner B.M., Birley A.J., and Lavender J. 1992. Histone H4 isoforms acetylated at specific lysine residues define individual chromosomes and chromatin domains in *Drosophila* polytene nuclei. *Cell* **69:** 375.

van Holde K.E. 1989. *Chromatin*. Springer-Verlag, New York.

Verreault A., Kaufman P.D., Kobayashi R., and Stillman B. 1998. Nucleosomal DNA regulates the core-histone-binding subunit of the human Hat1 acetyltransferase. *Curr. Biol.* **8:** 96.

Wang L., Liu L., and Berger S.L. 1998. Critical residues for histone acetylation by Gcn5, functioning in Ada and SAGA complexes, are also required for transcriptional function in vivo. *Genes Dev.* **12:** 640.

Wolffe A.P. and Pruss D. 1996. Targeting chromatin disruption: Transcription regulators that acetylate histones. *Cell* **84:** 817.

Yang X.J., Ogryzko V.V., Nishikawa J., Howard B.H., and Nakatani Y. 1996. A p300/CBP-associated factor that competes with the adenoviral oncoprotein E1A. *Nature* **382:** 319.

Zhang W., Bone J.R., Edmondson D.G., Turner B.M., and Roth S.Y. 1998. Essential and redundant functions of histone acetylation revealed by mutation of target lysines and loss of the Gcn5p acetyltransferase. *EMBO J.* **17:** 3155.

The Establishment of Active Chromatin Domains

A. Bell, J. Boyes,* J. Chung,† M. Pikaart, M.-N. Prioleau, F. Recillas,
N. Saitoh, and G. Felsenfeld

*Laboratory of Molecular Biology, National Institute of Diabetes, Digestive and Kidney Diseases,
National Institutes of Health, Bethesda, Maryland 20892-0540*

It has become obvious during the past several years that the true template for transcription in eukaryotes is chromatin, not bare DNA, and that the processes that modify chromatin structure are integral parts of the transcription mechanism. Modification of histones by acetylation, or the ATP-dependent disruption of nucleosomes by a variety of protein complexes, can alter transcription rates at specific genes. Enzymes responsible for histone acetylation or deacetylation can be part of a complex which also contains DNA sequence-specific transcription factors, and activation or repression is coupled to binding of the appropriate acetylase or deacetylase. The involvement of the transcription factor assures that the histone modification is targeted to the proper gene and may also mean that the effects of histone modification are confined to a relatively small region in the neighborhood of the promoter.

This is a comfortable idea because it involves only a small extension of the classic model of transcriptional activation at promoters on histone-free DNA. One can imagine that the nucleosomes bound at promoters and enhancers are disrupted or displaced and that the problem is then quite similar to what one encounters on prokaryotic templates. However, the picture is complicated by observations suggesting that the reorganization of chromatin from an inactive to an active state can extend over many kilobases of DNA on either side of the genes to be activated and that this transformation can be independent of the formation of nuclease-hypersensitive sites at promoters and enhancers near individual genes within the region. At the very least, this suggests that there may be regulatory mechanisms involving long-range chromatin structure that involve components and principles different from those required for regulation of specific genes.

Evidence for such mechanisms comes from studies of silencing near yeast telomeres and mating-type loci. Specialized proteins that recognize these sites interact with histones to form a modified and repressive chromatin structure which appears to propagate along the DNA from a region of initiation. Recent studies of the *Polycomb* group genes in *Drosophila* also point to a repressive mechanism (in this case affecting certain homeotic genes) arising from association of a complex of Poly-

comb and probably trithorax group proteins with DNA Polycomb response elements (PREs). Here, it seems likely that the Polycomb proteins do not coat large continuous stretches of chromatin but rather bind at intervals along the suppressed region, creating a suppressed chromatin domain by interactions among the binding sites (Pirrotta 1998).

The existence of such domains raises a set of questions separate from the issue of individual gene activation. How are such active domains established and maintained during development? How are boundaries determined so that the active region does not encroach on adjacent inactive regions, or vice versa? For a number of years, we have been addressing this question in the chicken β-globin locus, which is one of the best characterized in terms of chromatin structure as well as control of expression of the individual genes of the cluster.

The β-globin locus contains coding regions for four genes, two of which, ρ- and ε-globin, are expressed in embryonic erythroid cells, and the other two, β^H- and β^A-globin, are expressed in adult cells (Fig. 1). The notable regulatory regions of the locus are a strong β^A/ϵ enhancer located between these two genes and operating on them both. This enhancer is in itself capable of conferring position-independent expression in erythroid cells of transgenic mice, so that it qualifies as a locus control element (Reitman et al. 1990). It does not, however, comprise the entire locus control region (LCR) (Reitman et al. 1995). Upstream of the ρ gene a series of three DNase-I-hypersensitive sites mark the remaining LCR elements. All three of these hypersensitive sites are erythroid-specific; they are not present in nonerythroid cells. Together with the downstream β/ε enhancer (also a hypersensitive site), they are capable of directing position-independent, developmentally regulated expression of all four genes when introduced with the genes and their local regulatory elements into transgenic mice.

INVASION OF NUCLEOSOMES BY AN ERYTHROID TRANSCRIPTION FACTOR

We first wished to explore possible steps in establishment of an active chromatin structure over individual regulatory elements in the globin locus. We chose a simple system involving activation by the erythroid transcription factor GATA-1 (Boyes et al. 1998). Binding sites for GATA-1 are found near all of the globin genes (as well as other erythroid-specific genes), often in multiple copies,

Present addresses: *Section of Gene Function and Regulation, Chester Beatty Laboratories, Institute of Cancer Research, 237 Fulham Road, London, SW3 6JB; †National Heart, Lung and Blood Institute, National Institutes of Health, Building 10 Room 7D13, Bethesda, Maryland 20892.

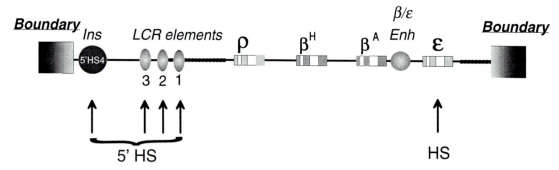

Figure 1. Map of the chicken β-globin locus. The arrows mark the DNase-I-hypersensitive sites. The sites marked 1,2,3 and Enh comprise the LCR. The site marked 5'HS4 marks the insulator element. Work of Hebbes et al. (1994) has shown that the active chromatin domain terminates rather abruptly just on the 5' side of the insulator (see text).

as well as in components of the LCR. Chicken GATA-1 is a 31-kD protein with two zinc fingers which binds to its site on DNA as a monomer. The carboxy-terminal finger is both sufficient and necessary for binding to simple sites (Omichinski et al. 1993a). We have shown that six tandem copies of a GATA-1 site constitute a strong enhancer (Reitman and Felsenfeld 1988), comparable to the naturally occurring β/ε enhancer (which itself contains two GATA sites). We reconstituted a nucleosome on a 167-bp fragment containing six GATA sites and explored the effects of binding the carboxy-terminal GATA finger on the stability of this nucleosome. We found to our surprise that it was possible to form a complex in which all of the GATA sites were occupied but the histones were not displaced (Boyes et al. 1998). The resulting complex was unusually sensitive to digestion by micrococcal nuclease (Fig. 2). Removal of GATA-1 by addition of DNA competitor carrying GATA sites resulted in regeneration of an intact nucleosome. Similar results were obtained with full-length GATA-1 and with nucleosomes constructed with the naturally occurring chicken α-globin gene enhancer, which contains three GATA sites spread over 120 bp.

The micrococcal nuclease digestion results as well as DNase I footprinting (Boyes et al. 1998) confirm that within the nucleosome-GATA-1 complex, the DNA at the ends of the nucleosome has been lifted from the nucleosome surface. A model in which binding sites become accessible when DNA spontaneously lifts from the ends of the nucleosome has been proposed by Widom and his collaborators (Polach and Widom 1995,1996), who have also presented supporting data. It is remarkable that the GATA-nucleosome complex remains stable under such conditions, presumably held together by strong electrostatic contacts between DNA and histones near the nucleosome's dyad axis. We suggest that GATA-1 mediates this perturbation of nucleosome structure because GATA-1 binding to DNA involves simultaneous interactions in the major and minor grooves (Omichinski et al. 1993b). Other transcription factors (e.g., the trithorax-like protein [GAGA-binding protein] of *Drosophila* (Pedone et al. 1996) with similar modes of binding could also behave in this way.

The DNase-cutting pattern (Boyes et al. 1998) makes it clear that were this complex integrated into a chromatin

structure, it would be detected as a hypersensitive site. Is this the final state of an activated enhancer carrying GATA-1? We believe that the perturbed nucleosome structure we observe may be a first step in the process to displace or completely disrupt the nucleosome, but it is not the final state. Experiments with the same six-GATA sequence carried on a replicating plasmid indicate that in vivo a further step displaces or totally disrupts the nucleosome (J. Boyes et al., unpubl.). Additional factors, perhaps related to the ATP-dependent disruptors of chromatin structure, are probably responsible for this last step.

ACTIVATION OF A CHROMATIN DOMAIN

The activation of individual genes in the β-globin locus is almost certainly preceded by potentiation of the chromatin structure of the entire domain. The activity of the chicken β/ε enhancer derives from a pair of binding sites for GATA-1 and a single site for the erythroid factor NFE-2 (Emerson et al. 1985,1987; Andrews et al. 1993). What controls the establishment of an active chromatin structure (i.e., a hypersensitive site) over the enhancer? Studies in our laboratory (Reitman et al. 1993) have shown that when constructions containing the enhancer are introduced into transgenic mice, the establishment of a hypersensitive site over the enhancer requires the cooperation of a promoter. This enhancer is not in itself sufficiently "strong" to establish an active domain autonomously. This suggests that the establishment of an active domain might require either cooperation among elements in the domain (either promoters or other LCR elements), certainly consistent with what is known about the cooperative mode of action of LCR elements (Wijgerde et al. 1995).

We have also asked what role each of the transcription-factor-binding sites within the β/ε enhancer has in the establishment of hypersensitivity. We stably transformed 6C2 cells (a chicken erythroid cell line arrested at the CFU-E stage) with constructions carrying the β-globin promoter and gene but with an enhancer carrying various mutations in the GATA-1 and NF-E2 sites (Boyes and Felsenfeld 1996). We found that all erythroid-factor-binding sites contributed to the full hypersensitive site. We expected that loss of the individual factor-binding

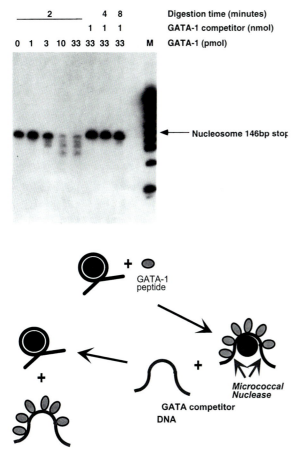

| 2 | 4 | 8 | Digestion time (minutes) |
| 1 | 1 | 1 | GATA-1 competitor (nmol) |

0 1 3 10 33 33 33 33   M   GATA-1 (pmol)

← Nucleosome 146bp stop

Figure 2. (*Top*) Micrococcal nuclease digestion of a nucleosome complex with the carboxy-terminal DNA-binding finger of GATA-1. The DNA contains six GATA-binding sites. Complexes carrying increasing amounts of GATA-1 were digested for 2 min (lanes *1–5*). The DNA was electrophoresed, blotted, and probed with same sequence. In the three right-hand lanes, GATA-1 was removed from the complex by adding a cold DNA competitor of a different size before digestion. The resulting nucleosome is seen to be resistant to digestion beyond the canonical size characteristic of the nucleosome core, showing that the effects of GATA-1 binding are reversible. (*Top* reprinted from Boyes et al. 1998.) (*Bottom*) Diagram showing the reactions in the *top* panel.

sites would diminish the hypersensitivity to nuclease digestion, probably because each hypersensitive site became physically less accessible. When we studied the kinetics of digestion with the enzyme *Pvu*II, however, we found that the rate of digestion for the mutants was the same as that for the wild type; what had altered was the number of sites that were accessible (Fig. 3). The establishment of this hypersensitive site is therefore an all-or-none event: Either the histone octamer is displaced or it is not. Presumably, the probes we are using do not detect intermediates of the kind described above for the six GATA site nucleosomes, or in this case, such intermediates do not exist. We have suggested (Boyes and Felsenfeld 1996; Felsenfeld 1996) that this kind of behavior may serve to establish a sharper response of gene activation to changes in transcription factor concentration within the nucleus.

ESTABLISHMENT OF DOMAIN BOUNDARIES

The β-globin locus is distinguished by the presence of distant upstream elements that constitute the LCR. (In the case of the chicken, as noted above, one of these elements has been translocated downstream.) There is ample evidence that a major role of the LCR is to help establish an active chromatin conformation over the entire β-globin locus (Forrester et al. 1990). This may be distinct from any direct enhancer-like action that LCR elements may also possess.

Active chromatin domains can exist independent of the transcriptional state of the individual genes within them. The classical method (Stalder et al. 1980) for detecting active domains takes advantage of the increased general sensitivity of such a region to DNase I. Over a region of about 30 kb, the region containing the chicken β-globin genes is detectably more sensitive than are surrounding sequences (Hebbes et al. 1994). Over the identical region, higher levels of histone acetylation are also observed, consistent with the presence of an active chromatin domain.

The abruptness of the boundaries at the ends of the domain raises the question: How does the chromatin know where to switch over from an active to an inactive conformation? At the 5′ end, we observed some years ago a novel hypersensitive site (5′HS4) that, unlike those of the LCR, is constitutive, i.e., it is present in all cells and tissues (Reitman and Felsenfeld 1990). Although at that time the work of Hebbes et al. had not yet appeared, we speculated on the basis of less refined data that this site might mark the boundary of the active globin locus and have a role in its establishment or maintenance. The first insulator elements (*scs* and *scs′* of *Drosophila*) had been described previously (Kellum and Schedl 1992), and we decided to investigate whether the region containing 5′HS4 had insulating properties.

For this purpose, we employed an assay to assess enhancer blocking activity (Chung et al. 1993). A reporter gene was constructed by coupling the human γ-globin promoter to a neomycin resistance gene; an enhancer derived from a mouse globin locus control element was added downstream (Fig. 4). After stable transformation with this DNA, K562 cells were plated on G418 and resistant colonies were counted and compared to the number of hygromycin-resistant colonies due to the activity of an adjacent hygromycin-resistant gene. To test for enhancer blocking effects, we subcloned a 1.2-kb fragment containing 5′HS4 and placed it on both sides of the promoter and *neo* gene so that it lay between the promoter and enhancer in tandemly integrated inserts; the hygromycin resistance gene remained unprotected. The effect of the 1.2-kb fragment was to reduce by an order of magnitude (Fig. 4) the ratio of G418-resistant colonies to hygromycin-resistant colonies. Double copies of the 1.2-kb fragment were even more effective. Constructs in which the element was placed outside the space between enhancer and promoter had only a very small repressive effect on transcription (Chung et al. 1997).

To identify more precisely the sequences within the 1.2-kb region that are responsible for this activity, we

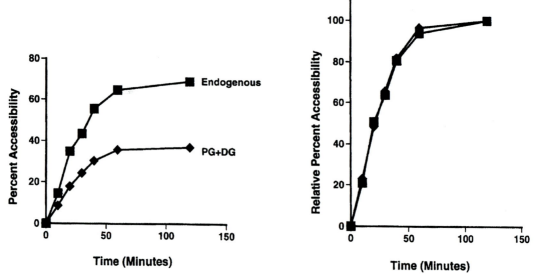

Figure 3. Kinetics of digestion of the *Pvu*II site within the chicken β/ε enhancer. (*Left*) Rate of cutting of the endogenous hypersensitive site compared with a stably integrated construct (in 6C2 cells) carrying a double mutation (PG +DG) at the two GATA-1 sites within the enhancer. In the mutant, only about half the sites are accessible. (*Right*) Same data normalized to the plateaus, showing that the digestible sites in the mutant are cut at the same rate as the endogenous site. (Reprinted from Boyes and Felsenfeld 1996.)

have repeated the enhancer blocking assay (without the internal hygromycin control) with subfragments derived from the region (Chung et al. 1997). We find that a considerable amount of activity is present in a 250-bp "core" sequence at the 5′ end of the 1.2-kb region. Two copies of the core are as effective as one copy of the 1.2-kb fragment, and six copies form a much more effective insulator (Fig. 5). More recently, we have reduced the size of the necessary DNA still further and are attempting to identify protein components that bind to this region and whose binding can be correlated with the enhancer-blocking activity (A. Bell et al., unpubl.). We have also developed new transient expression assays that should be more rapid and help us to unravel the mechanism of action (F. Recillas et al., unpubl.).

The β-globin insulator also is capable of conferring position-independent expression on stably transformed reporter genes that it surrounds. We have found that this insulator protects against position effects in *Drosophila*. When a mini-*white* reporter gene carrying a minimal promoter was surrounded by two copies of the 1.2-kb fragment and stably transformed into *Drosophila*, all of the transformed lines showed pale yellow eye color, indicating that the insulator elements had protected the reporter against stimulatory influences of nearby endogenous enhancers (Chung et al. 1993). This experiment was carried out in exact parallel to the earlier study of Kellum and Schedl (1992), who showed a similar effect with the *Drosophila scs* element. Other investigators have used the β-globin insulator to generate transgenic mouse lines with a reproducibly high level of expression of a gene coding for a *trans*-activator protein (Wang et al. 1997). Quite recently, we have used the β-globin insulator to surround a reporter gene prior to insertion into 6C2 cells.

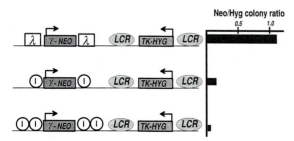

Figure 4. Enhancer-blocking experiments in which the 1.2-kb β-globin insulator is placed on both sides of a reporter carrying a gene for neomycin resistance, protecting it from a strong enhancer (LCR). A second gene, for hygromycin resistance, is driven by a separate promoter, and this reporter is not surrounded by insulator sequences. After stable transformation into a human erythroleukemia line (K562), the ratio of G418 to hygromycin-resistant colonies is measured. (Reprinted from Chung et al. 1993.)

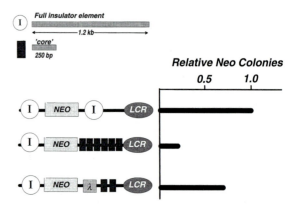

Figure 5. Effects in enhancer blocking assays of a 250-bp "core" sequence derived from the 5′ end of the 1.2-kb insulator element. (*Rectangles*) Core; (*circles*) full 1.2-kb sequence (Chung et al. 1997).

We find that this protects against both silencing and unusual activation of the reporter (M. Pikaart et al., unpubl.).

These properties of the β-globin insulator seem to be at least consistent with a role in the establishment and maintenance of a chromatin boundary at the 5′ end of the globin locus (see Fig. 1). Recent work in our laboratory has shown that the transcriptionally inactive region extends about 16 kb upstream from the insulator before another gene is encountered (M.-N. Prioleau et al., unpubl.). One possibility that follows directly from the enhancer-blocking activity of the insulator is that it serves to prevent interaction between the β-globin locus and this upstream gene. Less is known about the 3′ end of the locus, but we have found that it is marked by a constitutive hypersensitive site with properties that could reflect a role in boundary formation (N. Saitoh, unpubl.). In any case, the heterochromatic regions that lie on either side of the β-globin locus are likely to be quite different in structure from those seen at yeast telomeres or *Drosophila* centromeres and require quite different constraints to limit their expansion. For this reason, there is little reason to expect that the proteins bound at the β-globin locus will resemble the *Drosophila* proteins shown to be associated with insulating activity: the suppressor of hairy wing protein that binds to the *gypsy* element (Geyer and Corces 1992) or the BEAF-32 protein that binds to the *scs′* element (Zhao et al. 1995).

MODELS OF INSULATOR ACTION

A variety of models have been proposed for the mechanism of insulator action, and especially for directional enhancer-blocking activity (see, e.g., Chung et al. 1993). The two major classes of models involve either some form of steric hindrance or some interference with a processive mechanism presumed to be required for activation. Implicit in each of these models is an assumed mechanism of enhancer action. In steric hindrance models, the enhancer-bound transcription factors are blocked from recruiting the transcriptional machinery to the promoter by restricting physical access (Fig. 6). This could occur if the proteins bound to the insulator were bulky enough to prevent enhancer-promoter contact (Fig 6A), but might not explain why blocking is observed (in the case of the *gypsy* element) when the enhancer is distant from the insulator. A modified steric hindrance model that takes this into account creates a separate chromatin domain with insulators at the boundaries: A pair of insulator elements at opposite ends of a domain could, for example, help to establish a closed loop containing the promoter that might prevent the enhancer from reaching it (Fig. 6C). Alternatively, a single insulator element might be immobilized, for example, in the nuclear envelope and prevent contact between elements on opposite sides of the insulator (Fig. 6B). The other class of models presupposes that the activating signal tracks along the DNA from the enhancer to the promoter and that the insulator derails this advancing signal. Processive mechanisms that are consistent with this model could include transcription

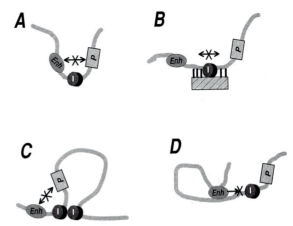

Figure 6. Models that have been proposed to explain the enhancer-blocking activity of insulators (see text).

(transcripts have been detected in the noncoding regions of the human β-globin domain; Ashe et al. 1997). In a variant of this model, the enhancer-bound complex itself forms a loop to DNA, and the point of contact moves toward the promoter (Fig. 6D). Our own data suggest that a simple steric hindrance model might be sufficient to explain results obtained with small linearized plasmids, but other experiments that reflect protection against position effect are more consistent with protection against a processive or cooperative deacetylase.

At this point, there is no reason to require that a single model explain both enhancer blocking and protection against position effects (particularly protection against silencing). It may well be that more than one kind of blocking mechanism is involved. We are aided by the discovery of new examples of insulator elements for which clear biological functions can be established (Hagstrom et al. 1996; Zhong and Krangel 1997). It seems likely that more of these elements will be found, particularly in gene clusters where it is necessary to prevent inappropriate crosstalk of enhancers with the wrong promoters during development. Elucidation of their modes of action may reveal a great deal not only about the establishment of chromatin domains, but also about the detailed mechanisms by which enhancers and promoters find each other.

REFERENCES

Andrews N.C., Erdjument-Bromage H., Davidson M.B., Tempst P., and Orkin S.H. 1993. Erythroid transcription factor NF-E2 is a haematopoietic-specific basic-leucine zipper protein. *Nature* **362:** 722.

Ashe H.L., Monks J., Wijgerde M., Fraser P., and Proudfoot N.J. 1997. Intergenic transcription and transinduction of the human β-globin locus. *Genes Dev.* **11:** 2494.

Boyes J. and Felsenfeld G. 1996. Tissue-specific factors additively increase the probability of the all-or-none formation of a hypersensitive site. *EMBO J.* **15:** 2496.

Boyes J., Omichinski J., Clark D., Pikaart M., and Felsenfeld G. 1998. Perturbation of nucleosome structure by the erythroid transcription factor GATA-1. *J. Mol. Biol.* **279:** 529.

Chung J.H., Bell A.C., and Felsenfeld G. 1997. Characterization of the chicken β-globin insulator. *Proc. Natl. Acad. Sci.* **94:** 575.

Chung J. H., Whiteley M., and Felsenfeld G. 1993. A 5′ element of the chicken β-globin domain serves as an insulator in human erythroid cells and protects against position effect in *Drosophila*. *Cell* **74:** 505.

Emerson B.M., Lewis C.D., and Felsenfeld G. 1985. Interaction of specific nuclear factors with the nuclease-hypersensitive region of the chicken adult beta-globin gene: Nature of the binding domain. *Cell* **41:** 21.

Emerson B.M., Nickol J.M., Jackson P.D., and Felsenfeld G. 1987. Analysis of the tissue-specific enhancer at the 3′ end of the chicken adult beta-globin gene *Proc. Natl. Acad. Sci.* **84:** 4786.

Felsenfeld G. 1996. Chromatin unfolds. *Cell* **86:** 13.

Forrester W.C., Epner E., Driscoll M.C., Enver T., Brice M., Papayannopoulou T., and Groudine M. 1990. A deletion of the human β-globin locus activation region causes a major alteration in chromatin structure and replication across the entire β-globin locus. *Genes Dev.* **4:** 1637.

Geyer P.K. and Corces V.G. 1992. DNA position-specific repression of transcription by a *Drosophila* zinc finger protein. *Genes Dev.* **6:** 1865.

Hagstrom K., Muller M., and Schedl P. 1996. Fab-7 functions as a chromatin domain boundary to ensure proper segment specification by the *Drosophila* bithorax complex. *Genes Dev.* **10:** 3202.

Hebbes T.R., Clayton A.L., Thorne A.W., and Crane-Robinson C. 1994. Core histone hyperacetylation co-maps with generalized DNase I sensitivity in the chicken β-globin chromosomal domain. *EMBO J.* **13:** 1823.

Kellum R. and Schedl P. 1992. A group of scs elements function as domain boundaries in an enhancer-blocking assay. *Mol. Cell. Biol.* **12:** 2424.

Omichinski J.G., Trainor C., Evans T., Gronenborn A.M., Clore G.M., and Felsenfeld G. 1993a. A small single-"finger" peptide from the erythroid transcription factor GATA-1 binds specifically to DNA as a zinc or iron complex. *Proc. Natl. Acad. Sci.* **90:** 1676.

Omichinski, J.G., Clore G.M., Schaad O., Felsenfeld G., Trainor C., Appella E., Stahl S.J., and Gronenborn A.M. 1993b. Solution structure of the specific DNA complex of the zinc containing DNA binding domain of the erythroid transcription factor GATA-1 by multidimensional NMR. *Science* **261:** 438.

Pedone P.V., Ghirlando R., Clore G.M., Gronenborn A.M., Felsenfeld G., and Omichinski J.G. 1996. The single Cys₂-His₂ zinc finger domain of the GAGA protein flanked by basic residues is sufficient for high-affinity specific DNA binding. *Proc. Natl. Acad. Sci.* **93:** 2822.

Pirrotta V. 1998. Polycombing the genome: PcG, trxG, and chromatin silencing. *Cell* **93:** 333.

Polach K.J. and Widom J. 1995. Mechanism of protein access to specific DNA sequences in chromatin: A dynamic equilibrium model for gene regulation. *J. Mol. Biol.* **254:** 130.

———. 1996. A model for the cooperative binding of eukaryotic regulatory proteins to nucleosomal target sites. *J. Mol. Biol.* **258:** 800.

Reitman M. and Felsenfeld G. 1988. Mutational analysis of the chicken β-globin enhancer reveals two positive-acting domains. *Proc. Natl. Acad. Sci.* **85:** 6267.

———. 1990. Developmental regulation of topoisomerase II sites and DNase I-hypersensitive sites in the chicken β-globin locus. *Mol. Cell. Biol.* **10:** 2774.

Reitman M., Lee E., and Westphal H. 1995. Function of the upstream hypersensitive sites of the chicken β-globin gene cluster in mice. *Nucleic Acids Res.* **23:** 1790.

Reitman M., Lee E., Westphal H., and Felsenfeld G. 1990. Site-independent expression of the chicken βᴬ-globin gene in transgenic mice. *Nature* **348:** 749.

———. 1993. An enhancer/locus control region is not sufficient to open chromatin. *Mol. Cell. Biol.* **13:** 3990.

Stalder W.E., Larsen A., Engel J.D., Dolan M., Groudine M., and Weintraub H. 1980. Tissue-specific DNA cleavage in the globin chromatin domain introduced by DNase I. *Cell* **20:** 451.

Wang Y., DeMayo F.J., Tsai S.Y., and O'Malley B.W. 1997. Ligand-inducible and liver-specific target gene expression in transgenic mice. *Nat. Biotechnol.* **15:** 239.

Wijgerde M., Grosveld F., and Fraser P. 1995. Transcription complex stability and chromatin dynamics *in vivo*. *Nature* **377:** 209.

Zhao K., Hart C.M., and Laemmli U.K. 1995. Visualization of chromosomal domains with boundary element-associated factor BEAF-32. *Cell* **81:** 879.

Zhong X.P. and Krangel M.S. 1997. An enhancer-blocking element between alpha and delta gene segments within the human T cell receptor alpha/delta locus. *Proc. Natl. Acad. Sci.* **94:** 5219.

Nuclear Matrix Attachment Regions Confer Long-range Function upon the Immunoglobulin μ Enhancer

L.A. Fernández, M. Winkler, W. Forrester,* T. Jenuwein,† and R. Grosschedl

Howard Hughes Medical Institute and Department of Microbiology and Immunology, University of California, San Francisco, California 94143

Transcriptional enhancers and locus control regions (LCR) can augment the activity of linked promoters over large distances in nuclear chromatin (for review, see Blackwood and Kadonaga 1998). Enhancers, which have been typically identified and characterized in tissue culture transfection assays, are often functionally compromised in transgenic mice (Palmiter and Brinster 1986). In particular, the activity of enhancers can vary with the chromosomal position of the integrated transgene (Wilson et al. 1990). Using mouse germ-line transformation assays, LCRs have been defined as regulatory elements that confer proper developmental regulation on linked genes, including independence of their site of chromosomal position (for review, see Dillon and Grosveld 1994; Kioussis and Festenstein 1997). LCRs have been found in many genetic loci, and they are composite elements that include transcriptional enhancers and other poorly defined regulatory sequences.

Multiple mechanisms may account for the regulation of promoters by enhancers and LCRs. The potential of enhancer- and promoter-binding proteins to interact with each other by DNA looping can result in reciprocal stabilization of the protein complexes, recruitment of regulatory proteins or RNA polymerases, and stimulation of transcription (Stargell and Struhl 1996; Ptashne and Gann 1997). Direct evidence of interactions between enhancer- and promoter-bound proteins has been obtained by electron microscopy (Su et al. 1991) (Mastrangelo et al. 1991). Moreover, enhancers have been shown to augment the activity of topologically unlinked promoters in transfection assays, supporting a looping versus a tracking mechanism of action (Dunaway and Dröge 1989; Mueller-Storm et al. 1989). However, in nuclear chromatin, in which the binding of proteins to DNA is impaired, additional mechanisms may be required to mediate enhancer function (Felsenfeld 1992; Armstrong and Emerson 1998). A tracking-based model of enhancer function is consistent with studies of insulators, which interfere with enhancer-promoter activation and establish chromatin domain boundaries (Kellum and Schedl 1991; Chung et al. 1993; Felsenfeld 1996). In particular, the placement of insulator sequences between two divergently transcribed promoters prevents the enhancer from

activating the distal promoter, but not the proximal promoter (Cai and Levine 1995).

The immunoglobulin μ heavy chain locus contains an intragenic enhancer region, located about 1.5 kb downstream from the variable region (V_H) promoter in the rearranged μ gene, which can function as an LCR in transgenic mice. The intragenic μ enhancer region includes three functional components. First, the enhancer region contains a well-characterized 100-bp transcriptional enhancer core that interacts with multiple DNA-binding proteins (Ernst and Smale 1995). Second, sequences 3′ of the enhancer core display promoter activity and direct heterogeneous initiation of noncoding Iμ transcripts (Lennon and Perry 1985; Su and Kadesch 1990). Finally, both the enhancer core and the Iμ promoter are flanked on either side by A-T-rich sequences that have been shown to associate with the nuclear matrix (Cockerill et al. 1987). These sequences are referred to as nuclear matrix attachments regions (MARs) and are operationally defined through their association with a proteinaceous scaffold or "nuclear matrix" in histone-depleted nuclei (Laemmli et al. 1992; Davie 1995; Pederson 1998). MARs have been found to colocalize with boundaries of nuclease-sensitive chromatin domains and with transcriptional control sequences and may thus subserve multiple functions (Cockerill and Garrard 1986; Gasser and Laemmli 1986; Bode and Maass 1988; Phi-Van et al. 1990; Klehr et al. 1991).

Here, we study the contribution of MARs to the enhancer-mediated control of the immunoglobulin μ gene expression. We analyze transgenic mice carrying gene constructs in which the enhancer region alone or together with flanking MARs was linked to the μ gene or to heterologous sequences. In addition, we examine the effects of premethylation of DNA, prior to transfection of tissue culture cells, on the activity of μ genes that contain or lack the MARs. These experiments indicate that the MARs are required for the accessibility and activation of distal promoters for bacteriophage RNA polymerases or RNA polymerase II. Finally, we show that the MARs, in combination with the μ enhancer, can target heterologous gene sequences to the nuclear matrix of transgenic B cells.

EXPERIMENTAL PROCEDURES

Cell lines, cell culture, and transfections. Lymphoid cells were grown in RPMI medium containing 5% fetal bovine serum (complement-inactivated by incubation at

---

Present addresses: *Harvard Medical School, Detartment of Pathology, 200 Longwood Ave., Boston, Massachusetts 02115; †Institute of Molecular Pathology, Dr. Bohrgasse 7, A-1030 Vienna, Austria.

56°C) and 50 μM β-mercaptoethanol, at 37°C and 5% CO_2. The pre-B-cell lines derived from transgenic animals used in these studies have been described elsewhere (Jenuwein et al. 1993, 1997; Forrester et al. 1994). The B-cell lines used for transfections were M12 (Kim et al. 1979) and S194 (Blasquez et al. 1989). Cells were transfected by electroporation (0.5 ml at a concentration of 2×10^7 cells/ml in growth medium, using 0.4 mm-gapped cuvette [Bio Rad] and a 0.25 kV, 960 μF pulse at room temperature). Approximately 10–20 μg of vector-free immunoglobulin gene and 1 μg of EcoRI-linearized SV2neo were cotransfected. Selection was obtained by adding G418 (GIBCO-BRL) 1 μg/ml (active) 24 hours after electroporation. Clones were selected by diluting to density of 10^4 to 10^5 cells/ml and seeding in 1-ml aliquots into 24-well plates.

DNA techniques. Plasmid DNA preparation and general DNA manipulation were performed following standard protocols (Sambrook et al. 1989). Genomic DNA was isolated as described earlier (Forrester et al. 1990). Methylation status of genomic DNA was tested by HpaII and MspI restriction enzyme digestion (New England Biolabs). In vitro methylation at CpG dinucleotides of vector-free DNA fragments, containing μ and ΔMAR genes, was carried out using SssI methylase (New England Biolabs). SssI-methylation was tested after the incubation period by blocking of HpaII digestion of these DNA fragments. For Southern blot, the genomic DNA samples were digested with restriction enzymes (as indicated), fractionated in 1% TAE agarose gels, transferred to Hybond N+ (Amersham), and hybridized with 2×10^6 cpm/ml of specific ^{32}P-labeled DNA probes, which were generated by random priming with degenerated hexanucleotides and $[\alpha$-^{32}P]dCTP (Amersham). Unincorporated dNTPs were removed by mini-spin column chromatography in Sephadex G-50 (Pharmacia).

RNA analysis. Total RNA was isolated as described previously (Grosschedl et al. 1984). For the in vitro synthesis of T7- and T3-specific transcripts from chromatin, we incubated 2×10^7 nuclei with 200 units of bacteriophage RNA polymerase (Jenuwein et al. 1993). Probes for S1 analysis of specific transcripts were produced by unidirectional polymerase chain reaction (PCR) extension from gene-specific oligonucleotides using Taq DNA polymerase (Boehringer) and linearized DNA plasmid templates. All S1 probes were subsequently purified by elution from polyacrylamide 6%/7 M urea gels. The preparation of S1 probes for transcripts initiated at the transgene V_H cap site, Iμ, mb-1 gene, and bacteriophage promoters T7 and T3 have been described previously (Jenuwein et al. 1993, 1997; Forrester et al. 1994). The mouse β-actin transcripts were detected by ^{32}P 5′-end-labeling of a mouse β-actin-specific oligonucleotide (5′GGC CAT CTC CTG CTC GAA GTC), which was subsequently used as a primer in a unidirectional PCR using as a template a PvuII-digested mouse actin cDNA-containing plasmid (pmβactin).

Nuclei preparation and DNase I digestion. Nuclei were isolated as described earlier (Jenuwein et al. 1993,

1997). Isolated nuclei were digested (1 μg/ml at OD_{260}) with DNase I (Worthington) at a concentration range of 1.25–15 μg/ml and incubated at 37°C for approximately 10 minutes (Forrester et al. 1990, 1994).

In vivo footprinting. In vivo footprinting was performed essentially as described by P. Mueller, P. Garrity, and B. Wold (Ausubel et al. 1993) by dimethylsulfate (DMS; Aldrich) treatment of pre-B cells growing in cell culture at 37°C. Genomic DNA was isolated from DMS-treated and untreated control cells, cleaved at methylated G residues in 1 M piperidine (Aldrich), and subjected to LM-PCR using Vent DNA polymerase (New England Biolabs) (Garrity and Wold 1992) and specific sets of nested primers to amplify μ enhancer sequences. A complete description of the protocol and the oligonucleotides used will be described elsewhere.

Fluorescence in situ hybridization of nuclear halos. The nuclear halos were prepared as described by Gerdes et al. (1994) and de Belle et al. (1998). For detection of transgene sequences, a plasmid containing the T3 large T and T7-VP1 sequences from SV40 was linearized and labeled with biotin-14 dCTP using the Bioprime-Kit (GIBCO). Hybridization was performed according to standard protocols. After hybridization, the probe signal was amplified by catalyzed reporter deposition (CARD) using the TSA-indirect system (New England Nuclear Life Science) and finally detected with FITC-conjugated ExtrAvidin (Sigma). The DNA was stained with DAPI (Sigma).

RESULTS

MARs Are Required for μ Enhancer Function at a Distance

To examine the role of MARs for the expression of the immunoglobulin μ gene in vivo, we have previously generated transgenic mice carrying a wild-type μ gene or a mutant ΔMAR μ gene in which the MARs flanking the enhancer on either side have been deleted (Fig. 1A). Analysis of the expression of the transgenes in splenic B cells by an S1 nuclease protection assay indicated that the activity of the V_H promoter in the ΔMAR μ gene is reduced 30–1000-fold relative to the wild-type μ gene (Forrester et al. 1994). Moreover, the ΔMAR gene, but not the wild-type μ gene, showed a significant variability in the levels of expression in different mouse lines, suggesting that the MARs contribute to the position independence of transgene expression. Notably, the levels of expression of the wild-type and ΔMAR μ genes were similar in stably transfected B cells, indicating that MARs are required for μ gene expression in transgenic mice but are dispensable in tissue culture transfection assays. Analysis of the chromatin structure of the wild-type and ΔMAR transgenes revealed a decrease in the general DNase I sensitivity of the mutant transgene, although the enhancer of both transgenes formed a DNase-I-hypersensitive site (Forrester et al. 1994). Moreover, transcripts from the enhancer-proximal Iμ promoter were detected in transgenic

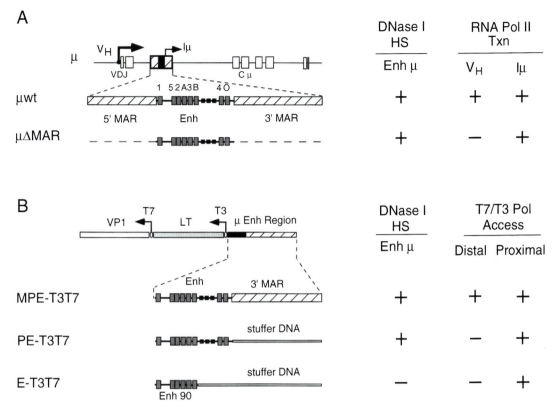

Figure 1. Structure of transgenes containing μ LCR sequences. (*A*) Schematic diagram of the rearranged μ transgenes. The structure of the enhancer region is shown for the wild-type and ΔMAR constructs. The positions of the V_H and Iμ promoters and the directions of transcription are indicated by arrows. The intragenic μ LCR region, located between VDJ and Cμ exons, is enlarged to show flanking MARs (*hatched box*) and the μ enhancer (Enh) with multiple binding sites for transcriptional activators (Ernst and Smale 1995). (*B*) Structure of T3T7 transgenes containing various fragments of the μ enhancer region linked to T3 and T7 bacteriophage promoters at enhancer-proximal and -distal (1 kb) positions, respectively. The composition of the enhancer region is indicated: MPE contains a single MAR, the Iμ promoter, and the enhancer core; PE contains Iμ promoter and the enhancer core; and E contains the 90-bp enhancer core. Arrows indicate the position and direction of the bacteriophage promoters. DNA fragments form the large T (LT) and VP1 genes of SV40 were used as reporters for transcription from the bacteriophage promoters. "Stuffer" sequences, replacing enhancer components, were used to maintain the spatial relationship of the transgene components. The lengths of the VP1 and LT reporters were 1 kb each to provide a similar distance of the bacteriophage promoters in multicopy transgene inserts. On the right hand of each transgene (*A* and *B*), a summary of data is presented from DNase I digestion and S1 nuclease protection assays of transcripts generated by endogenous RNA polymerase II or by exogenous T7 and T3 RNA polymerases. The formation of DNase-I-hypersensitive sites (HS) at the μ enhancer is indicated by a plus sign. Detection of transcripts from the enhancer-distal V_H or T7 promoter and the enhancer-proximal Iμ or T3 promoters are indicated. (Data are summarized from Forrester et al. [1994] and Jenuwein et al. [1997].)

ΔMAR B cells, raising the possibility that the MARs are required for long-range function of the μ enhancer.

To further show the specific requirement of the MARs for the activation of the distal V_H promoter in transgenic mice, we compared the generation of V_H and Iμ transcripts from the ΔMAR μ gene in transgenic and stably transfected B cells (Fig 2). By using labeled DNA probes that visualize V_H- or Iμ-specific transcripts in S1 nuclease protection assays, we found that no V_H transcripts were detected in five transgenic lines examined, whereas Iμ transcripts were present in four out of the five lines. In contrast, equivalent levels of transcription from the V_H and Iμ promoters were detected in all stably transfected B-cell lines analyzed. In these experiments, we also probed for mb-1 transcripts as an internal control for the integrity and quantity of the RNA preparations. These data indicate that the μ enhancer is dependent on flanking MARs to activate the distal V_H promoter in germ-line transformation assays.

MARs Extend Enhancer-induced Factor Accessibility in Nuclear Chromatin

In principle, multiple mechanisms may account for the MAR dependence of μ enhancer function at a distance. MARs could augment the function of the μ enhancer by propagating an enhancer-induced alteration in chromatin structure or, alternatively, by facilitating long-range interactions between enhancer- and promoter-bound factors. To examine a possible role of MARs in regulating chromatin accessibility, we introduced into the mouse germ-line gene constructs in which the μ enhancer, alone or together with a MAR, was linked to binding sites for bacteriophage RNA polymerases (see Fig. 1B). We also generated transgenic mice with a T3T7 construct containing the 90-bp μ enhancer core, alone or in combination with a MAR. In these gene constructs, a T3 promoter is inserted immediately adjacent to the μ enhancer, and a T7 promoter resides 1 kb downstream from the enhancer,

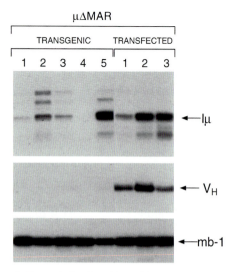

Figure 2. Transcription from V_H and $I\mu$ promoters in ΔMAR μ genes in transgenic B cells and stably transfected B cells. S1 nuclease protection assays of total cytoplasmic RNA, using specific 5′-labeled DNA probes that detect $I\mu$ (*top panel*), V_H (*middle panel*), and mb-1 (*bottom panel*) transcripts. Five independent transgenic ΔMAR lines and three stable ΔMAR cell clones are shown. (Data adapted, in part, from Forrester et al. 1994.)

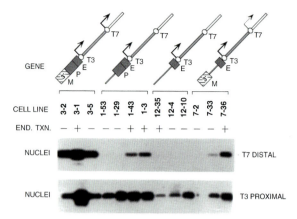

Figure 3. MARs extend enhancer-induced chromatin accessibility. S1 nuclease protection assays of total RNA from transgenic pre-B cell nuclei, incubated with exogenous T7 or T3 RNA polymerase and rNTPs. Correctly initiated transcripts were detected with DNA probes for SV40 large T (T3; *lower panel*) or VP1 (T7; *upper panel*) sequences. The structure of the transgenes is shown above. DNA fragments containing MARs (M), $I\mu$ promoter region (P), and Eμ-90 core element (E) are indicated. (*Open circles*) Positions of the enhancer-proximal T3 and the enhancer-distal T7 promoters. The numbers of the transgenic lines are indicated. Plus and minus symbolize an active or inactive transcriptional state of the transgene prior to addition of the bacteriophage RNA polymerases. Transcription of the transgene by endogenous RNA polymerases (end.txn) was determined by S1 nuclease protection assays. (Data adapted from Jenuwein et al. 1997.)

mimicking the natural positions of the $I\mu$ and V_H promoter in the rearranged μ gene (Jenuwein et al. 1997). The rationale for replacing RNA polymerase II promoters with binding sites for bacteriophage RNA polymerases was to assess chromatin accessibility in the absence of interactions between enhancer- and promoter-binding proteins. Moreover, this experimental strategy allows for the analysis of chromatin accessibility independent of transcription by endogenous RNA polymerases (Jenuwein et al. 1993, 1997).

For each gene construct, we generated multiple transgenic mice and obtained pre-B-cell lines by transformation with Abelson murine leukemia virus. To analyze factor access at the T3 and T7 promoters in native chromatin, we isolated nuclei from transgenic pre-B cells and incubated them with the respective RNA polymerases. We detected T7- and T3-specific transcripts by S1 nuclease protection assays with SV40 large T (LT)- and VP1-specific probes, respectively (Fig. 3, two bottom panels). Transcripts initiating at the enhancer-proximal T3 promoter were found in all lines, consistent with previous work showing that the μ enhancer core is necessary and sufficient to establish local accessibility in chromatin (Jenuwein et al. 1993). Moreover, the relative numbers of T3-specific transcripts were proportional both to the numbers of transgene copies and to T7-specific transcripts generated from naked genomic DNA (Fig. 3; data not shown). We also detected abundant T7-specific transcripts in all lines containing transgenes that include the μ enhancer core, the $I\mu$ promoter region, and a single MAR at either 5′ or 3′ position. In contrast, T7 RNA polymerase access in transgenes that include the enhancer core and $I\mu$ promoter, but lack both MARs, was reduced five- to tenfold in two out of the four lines examined, and no T7-specific transcripts were detected in

the other two lines. Transgenes carrying the μ enhancer core, but lacking the MARs and $I\mu$ promoter, failed to generate T7-specific transcripts in all three lines examined, whereas T3-specific transcripts were detected at modestly reduced levels. In combination with a single MAR, the μ enhancer core lacking the $I\mu$ promoter region mediated T7 accessibility in two out of three lines, albeit at five- to tenfold reduced levels relative to transgenes also containing the $I\mu$ promoter region. Since changes in chromatin structure typically reflect an active transcriptional state, we also determined the transcriptional state of the transgene (Fig. 3). In transgenes containing the MAR, transcription by endogenous polymerases was not required for factor access at the distal T7 promoter. However, endogenous transcription augmented the accessibility at a distance, in particular in transgenes containing the enhancer and $I\mu$ promoter. (Fig. 3). Taken together, the μ enhancer with a MAR generates an extended domain of chromatin accessibility independent of an active transcriptional state.

Assembly of a Nucleoprotein Complex at the μ Enhancer Is Independent of MARs

The impaired activation of the distal V_H promoter and the reduced accessibility of the T7 promoter in transgenes containing the enhancer alone could reflect the assembly of an incomplete protein complex at the enhancer lacking the flanking MARs. To examine the in vivo factor occupancy of the enhancer in the wild-type and ΔMAR μ

transgenes, we performed genomic footprinting assays on intact pre-B cells from transgenic mice. Comparison of the in vitro DMS methylation pattern of naked DNA with the in vivo methylation protection pattern of a wild-type μ transgene shows a specific pattern of protected and enhanced G modifications which is similar to that previously reported for endogenous μ locus (data not shown) (Ephrussi et al. 1985). Interestingly, the ΔMAR transgene displayed the same in vivo DMS methylation pattern as the wild-type transgene, indicating that the assembly of a stable complex at the μ enhancer is independent of the flanking MARs. However, the entire μ enhancer, including the Iμ promoter, is required for full factor occupancy, since a 90-bp enhancer core fragment generates only a partial footprint in vivo (data not shown).

MARs Target the μ Transgene to the Nuclear Matrix

The nuclear matrix has been shown to represent a proteinaceous scaffold that is resistant to high salt and detergent extraction of nuclei (Pederson 1998). As shown in the schematic diagram in Figure 4, this treatment of nuclei, which removes all histones, leads to the formation of a nuclear halo in which DNA loops extend from a scaffold. In histone-depleted nuclei, some nucleotide sequences such as the ribosomal RNA repeats localize in the nuclear matrix, whereas other gene sequences can be identified in the extending DNA loops (Marilley and Gassend-Bonnet 1989; Gerdes et al. 1994).

Although the MAR sequences in the μ enhancer region have been shown to associate with the nuclear matrix in vitro (Cockerill et al. 1987), the question arises as to whether these sequences can target linked sequences to the nuclear matrix in a heterologous chromosomal context. Toward this end, we performed fluorescence in situ hybridization (FISH) analysis of nuclear halo preparations of transgenic pre-B cells carrying T3T7 constructs in which the μ enhancer alone or in combination with a single MAR is inserted. The transgene containing the μ enhancer together with a MAR is predominantly localized to the nuclear matrix (Fig. 4, left). In contrast, the T3T7 gene construct containing only the enhancer region is predominantly localized to the DNA loops extending from the nuclear matrix (Fig. 4, right). In these experiments, the DNA is visualized by DAPI, which stains the nuclear matrix containing a high concentration of DNA. These data suggest that the MARs of the immunoglobu-

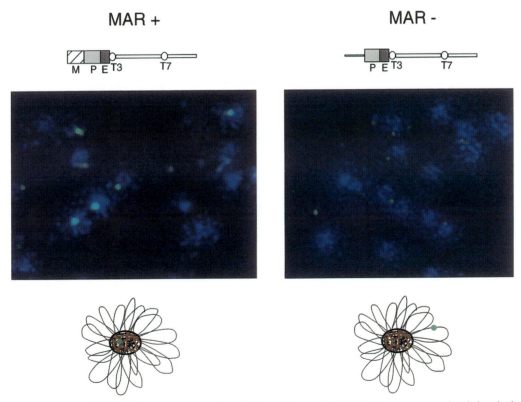

Figure 4. Transgenes containing MAR sequences associate with the nuclear matrix. FISH microscopy on nuclear halos obtained from transgenic B-cell lines carrying T3T7 constructs with or without MAR sequences. The structure of the transgenes is shown. (*Hatched box*) MAR (M); (*dark gray box*) μ enhancer core (E); Iμ promoter region (P). Nuclear halos from transgenic B-cell lines harboring MPE-T3T7 (*left*) or PE-T3T7 (*right*) constructs were prepared and hybridized with a biotin-labeled DNA probe specific for the transgenic T3-LT and T7-VP1 sequences. The probe was detected after hybridization by using Avidin-FITC (*green*). The DNA was stained with DAPI (*blue*). Transgene sequences (*green*) from the MAR-containing construct (*left panel*) colocalize with the bulk DNA signal from the nuclear matrix (*blue*), whereas the signal from the transgene lacking MAR sequences (*right panel*) is detected in a diffuse DNA halo outside the DAPI staining (*blue*). A simplified drawing of the hybridization signals is shown at the bottom of each photograph.

lin locus can localize linked heterologous sequences to the nuclear matrix in vivo.

DNA Methylation In Vitro Imparts MAR Dependence on μ Gene Expression in Transfected B Cells

One intriguing observation concerning the function of MARs in the immunoglobulin gene is their selective requirement for gene expression in transgenic mice. In contrast, the MARs are virtually dispensable in transfected B-cell lines. A major difference between germline transformation and transfection assays is the general methylation that occurs in early embryogenesis (Monk et al. 1987; Li et al. 1993; Razin and Cedar 1994; Bestor and Tycko 1996). DNA methylation has been shown to provide a repressive state of gene expression (Kass et al. 1997a,b). Recently, it has been shown that binding of the methyl-CpG-binding protein MeCP-2 to methylated DNA recruits a Sin3/histone deacetylase that silences gene expression (Jones et al. 1998; Nan et al. 1998).

With the aim of examining the possibility that methylation of DNA may provide conditions that require the function of MARs for the expression of the μ gene, we used the approach of methylating DNA in vitro prior to transfection of tissue culture cells (Fig. 5) (Lichtenstein et al. 1994). We isolated wild-type and ΔMAR μ genes from

plasmid DNA and treated the DNA fragments with SssI methyltransferase, which methylates CpG dinucleotides. The methylated DNA fragments were mixed with unmethylated DNA encoding the neomycin resistance marker and stably transfected into B cells. This strategy allows for complete premethylation of the DNA and should mimic the methylated state of the μ gene in mouse development. Analysis of the expression of the premethylated wild-type and ΔMAR μ gene indicated a 5–20-fold reduction of V_H promoter activity in clones carrying the ΔMAR gene relative to clones carrying the wild-type gene. This observation indicates that premethylation of the μ gene imparts a requirement for MARs in tissue culture transfection assays similar to that observed in transgenic mice. In addition, we examined the methylation state of the transfected genes and found that the wild-type gene was demethylated at HpaII sites in most but not all cell clones, whereas no demethylation was detected in clones carrying the ΔMAR gene. Consistent with previous data on immunoglobulin κ (Lichtenstein et al. 1994; Kirillov et al. 1996), these results suggest that the MARs contribute to demethylation. However, demethylation does not appear to be required for expression of the transfected μ gene. Taken together, these data indicate that the MARs are required for enhancer function at a distance in methylated DNA templates and suggest that premethylation of DNA may allow for the study of LCRs in standard transfection assays.

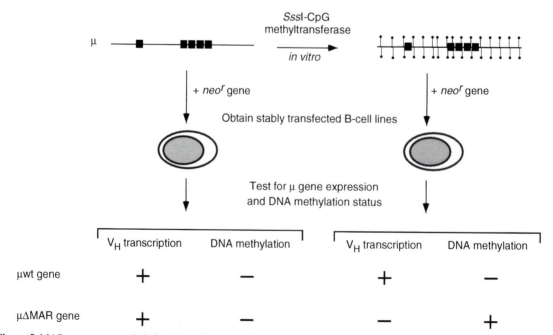

Figure 5. MARs overcome methylation-dependent repression of long-range enhancer promoter activation. Schematic diagram of the experimental strategy to examine the effect of premethylation of DNA on μ gene expression. Wild-type and ΔMAR μ gene fragments, unmethylated or methylated in vitro with SssI methyltransferase, were cotransfected with an unmethylated neomycin resistance marker (neo[r]) into a B-cell line. Stable cell clones were selected and analyzed for μ gene expression by S1 nuclease protection assays and for methylation state by digestion of genomic DNA with the methylation-sensitive enzyme HpaII or the methylation-insensitive isoschizomere MspI. Transgene sequences were detected by DNA blot analysis. As described in the text, the MAR sequences were required for activation of the V_H promoter in cell clones carrying a premethylated μ gene. In contrast, cell clones containing unmethylated wild-type and ΔMAR genes displayed similar levels of V_H transcription. The premethylated wild-type μ gene, but not the premethylated ΔMAR gene, is demethylated. These data implicate the MAR sequences in demethylation and in transcriptional activation of the V_H promoter on premethylated DNA templates.

DISCUSSION

In this study, we examine the mechanism by which enhancers stimulate promoters at a distance. We find that the μ enhancer alone can activate a distal V_H promoter in stably transfected tissue culture cells, but not in transgenic mice. In combination with flanking MARs, however, the μ enhancer mediates efficient activation of the V_H promoter in both transfected cells and transgenic mice. Second, we show that the enhancer alone is sufficient to induce an alteration in nuclear chromatin that can be detected by the formation of the DNase-I-hypersensitive site. Moreover, the enhancer is fully occupied by proteins in vivo, as determined by genomic footprinting analysis. Third, we find that the enhancer, in combination with a flanking MAR, induces an extended domain of chromatin accessibility independent of DNA looping. Fourth, the MARs of the immunoglobulin locus can target linked heterologous transgene sequences to the nuclear matrix. Finally, we show that in vitro methylation of the μ gene prior to transfection of tissue culture cells imparts a repression of long-range μ enhancer action that can be overcome by the flanking MARs, suggesting that the distance-dependent enhancer effects can be regulated.

Role of Enhancer and MAR Sequences in Mediating Chromatin Accessibility In Vivo

We have previously shown that the 90-bp enhancer core is necessary and sufficient to confer access upon a proximal bacteriophage promoter in nuclear chromatin (Jenuwein et al. 1993). This alteration in chromatin structure is not accompanied by the formation of a DNase-I-hypersensitive site and appears to be independent of transcription by endogenous RNA polymerases. Moreover, genomic footprinting analysis of the transgenic 90-bp μ enhancer revealed weak protection, suggesting that this enhancer fragment induces only a transient alteration in chromatin or an alteration that occurs only in a fraction of gene copies and/or cells. In contrast, the 220-bp μ enhancer fragment generated strong footprints in vivo, indicating that the μ enhancer is fully occupied by DNA-binding proteins in vivo. Moreover, this enhancer fragment induces the formation of a DNase-I-hypersensitive site. The 220-bp enhancer includes the cryptic Iμ promoter, which may be required for the formation of a DNase-I-hypersensitive site similar to the requirement of a promoter for the formation of a DNase-I-hypersensitive site at the β-globin enhancer (Reitman et al. 1993). Notably, the 220-bp enhancer fragment does not mediate accessibility at a distance. In contrast, the 220-bp enhancer, in combination with flanking MARs, can induce an extended domain of chromatin accessibility as determined by the binding of a bacteriophage RNA polymerase to a site 1 kb away from the enhancer/MAR fragment. In these experiments, no known RNA polymerase II promoter sequences are present in the vicinity of the T7 RNA polymerase-binding site, suggesting that the alterations in chromatin structure, mediated by MARs and the μ enhancer, are independent of interactions between enhancer-bound proteins and RNA polymerase II by DNA

looping. Together, these data are consistent with a multistep model of enhancer function in regulating chromatin accessibility and promoter activation in vivo (Fig. 6).

Potential Mechanisms for the Regulation of Enhancer Function by MARs

Multiple mechanisms might account for the collaboration between the μ enhancer and MARs in mediating chromatin accessibility at a distance. In principle, proteins that bind to the MARs might alter chromatin structure. MARs have been found to be recognized by multiple DNA-binding proteins in vitro. First, members of the

NO ACCESSIBILITY

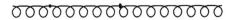

UNSTABLE FACTOR BINDING

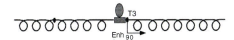

STABLE ENHANCER COMPLEX ASSEMBLY AND LOCAL ACCESSIBILITY

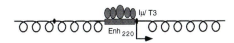

EXTENDED ACCESSIBILITY AND TRANSCRIPTION

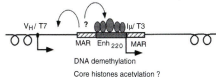

DNA demethylation
Core histones acetylation ?
Chromatin remodeling complex recruitment ?

Figure 6. Multistep model for the generation of a domain of accessible chromatin. Schematic summary is shown of experiments examining the accessibility of various transgenes by binding of bacteriophage RNA polymerases, digestion with DNase I, and analysis of in vivo occupancy by genomic footprinting. An individual factor-binding site is not accessible in nuclear chromatin. The 90-bp μ enhancer core mediates accessibility of an enhancer-proximal T3 promoter but does not generate complete factor occupancy at the enhancer as determined by genomic footprint analysis. The enhancer complex is presumably unstable and formed only in a subset of transgene copies and/or cells. The 220-bp μ enhancer, including the enhancer core and cryptic Iμ promoter, is sufficient to mediate stable enhancer complex formation that can be detected by full occupancy of the binding sites in vivo. The 220-bp enhancer also forms a DNase-I-hypersensitive site and mediates proximal, but not distal, accessibility of bacteriophage promoters. The enhancer flanked by one or two MARs mediates extended chromatin accessibility and stimulation of the enhancer-distal V_H promoter. Moreover, this combination of regulatory sequences mediates accessibility at a T7 promoter located 1 kb away from the enhancer. Finally, the μ enhancer and MARs together induce demethylation at distal CpG residues. The function of MARs in extending enhancer-induced chromatin accessibility could involve acetylation of core histones and/or recruitment of chromatin-remodeling complexes to the enhancer.

CAAT displacement proteins (CDP), which are expressed predominantly in pre-B and non-B cells (Scheuermann and Chen 1989), special A-T-binding protein (SATB1), which is expressed in T cells (Dickinson et al. 1992; Liu et al. 1997), and the transcription factor Bright (Herrscher et al. 1995), which is found in late-stage B cells, recognize the same sequences in the MARs of the μ locus. These proteins repress or activate the function of the μ enhancer and may augment the difference of enhancer activity in different cell types. However, the cell-type distribution of these proteins and their functional properties make it unlikely that they participate in the regulation of the enhancer by MARs in transgenic pre-B cells. Second, MARs contain AT-rich sequences that are recognized by HMG-I/Y in nuclear extracts of *Xenopus* oocytes (Zhao et al. 1993). Binding of HMG-I/Y has been found to displace histone H1 in vitro, raising the possibility that HMG-I/Y might antagonize the association of H1 with chromatin flanking the MARs. Finally, MAR sequences have been shown to be recognized by MeCP-2 (Weitzel et al. 1997). MeCP-2 typically recognizes methylated CpG dinucleotides that are associated with repressed chromatin domains (Nan et al. 1996, 1997). MeCP-2 has been recently shown to recruit histone deacetylase (Jones et al. 1998; Nan et al. 1998), which mediates deacetylation of core histones and silencing of chromatin (Struhl 1998). Therefore, it is unlikely that the binding of MeCP-2 is involved in augmenting the function of the μ enhancer at a distance. However, MeCP-2 has also been shown to bind to MARs, independent of methylated CpG dinucleotides, and we cannot rule out the possibility that MeCP-2 can exert different functions.

Another mechanism by which MARs may effect long-range enhancer function involves changes in DNA topology. MARs have been shown to contain sequences that have a high propensity for unwinding (Kohwi-Shigematsu and Kohwi 1990; Bode et al. 1992; Dickinson and Kohwi-Shigematsu 1995), and changes in DNA topology may effect chromatin structure at a distance and activation of promoters. In particular, the V_H promoter has been shown to be sensitive to changes in DNA topology (Parvin and Sharp 1993).

A third mechanism may involve the recruitment of protein complexes that modify histones or remodel nucleosomes. MARs may facilitate the recruitment histone acetylases (Klehr et al. 1992; Davie 1998). Hyperacetylation of core histones has been correlated with an accessible chromatin structure (Lee et al. 1993; Hebbes et al. 1994; Davie 1996). Histone acetylation has been detected predominantly in the immediate vicinity of active promoters and enhancers (D. Allis, pers. comm.), and MARs may be involved in extending a domain of histone acetylation to distal positions. Alternatively, the chromatin structure may be altered by chromatin remodeling enzymes that are recruited to the enhancer and MAR sequences. A role of MARs in extending the domain of histone modification or chromatin remodeling is consistent with our observations that MARs can confer long-range chromatin accessibility independent of interactions between enhancer- and promoter-binding proteins.

Finally, the immunoglobulin MARs may target DNA to specialized compartments in the nucleus (Razin and Gromova 1995) enriched in various enzymes involved in transcription and RNA processing (Mortillaro et al. 1996; Pederson 1998).

The Combination of Enhancer and MAR Sequences Constitutes a Functional LCR

LCRs are typically composed of multiple sequence elements that include transcriptional enhancers and other regulatory sequences (Dillon and Grosveld 1994). The common functional feature of these LCRs is their ability to mediate correct developmental expression patterns independent of chromosomal position. The requirement of "nonenhancer" elements for the function of LCRs has been well demonstrated in the β-globin locus, the ADA locus, and the CD2 locus (Aronow et al. 1995; Festenstein et al. 1996). The LCR of the ADA locus contains an enhancer that mediates efficient gene expression in transfected tissue culture cells and flanking "facilitator" sequences that are required for gene expression in transgenic mice. This functional organization is similar to that of the intragenic μ enhancer region, which can act as an LCR in transgenic mice. In the LCR of the ADA locus, the position and orientation of the facilitator sequences relative to the enhancer are important and suggest a functional collaboration between proteins found at the facilitators and the enhancer (Aronow et al. 1995; Kioussis and Festenstein 1997). In the μ enhancer region, we have not examined the spatial relationship between MARs and enhancer. However, the MARs by themselves are unable to mediate chromatin accessibility and are dependent on the juxtaposition with the μ enhancer to participate in the alteration of chromatin structure and activation of linked promoters (Jenuwein et al 1993, 1997; Forrester et al. 1994).

The collaboration between MARs and enhancers may require specific elements, but the selectivity appears to be rather broad. Experiments in which the MAR element from the intronic enhancer in the Igκ locus was replaced with MARs from different origins indicated that these MARs could collaborate with a κ enhancer to mediate an extended demethylation (Kirillov et al. 1996). However, the effects of the MAR substitution on activation of distal promoters have not been examined. Conversely, the MARs of the immunoglobulin locus are unable to confer activity upon a linked SV40 enhancer in transgenic mice (W. Forrester et al., unpubl.). Thus, the collaboration between enhancer and MARs has some selectivity. The "nonenhancer" sequences of other LCRs have not yet been well defined, and it is unclear whether MAR-like sequences may also act in the context of other LCRs or whether diverse regulatory elements collaborate with enhancers to mediate position-independent gene expression in vivo.

ACKNOWLEDGMENTS

We thank Gary Felsenfeld, Mark Groudine, and Alan Wolffe for valuable discussions and Nancy Biles for preparing the manuscript. L.A.F. is supported by a postdoctoral fellowship of the Ministerio de Educacion y Ciencia of Spain, and M.W. is supported by a postdoctoral fellowship from the DFG. This work was supported by a grant from the National Institutes of Health to R.G.

REFERENCES

Armstrong J.A. and Emerson B. 1998. Transcription of chromatin: These are complex times. *Curr. Opin. Genet. Dev.* **8:** 165.

Aronow B.J., Ebert C.A., Valerius M.T., and Potter S.S. 1995. Dissecting a locus control region: Facilitation of enhancer function by extended enhancer-flanking sequences. *Mol. Cell. Biol.* **15:** 1123.

Ausubel F.M., Brent R., Kingston R.E., Moore D.D., Seidman J.G., Smith J.A., and Struhl K. 1993. *Current protocols in molecular biology.* Greene and Wiley-Interscience, New York.

Bestor T.H. and Tycko B. 1996. Creation of genomic methylation patterns. *Nat. Genet.* **12:** 363.

Blackwood E.M. and Kadonaga J.T. 1998. Going the distance: A current view of enhancer action. *Science* **281:** 60.

Blasquez V.C., Xu M., Moses S.C., and Garrard W.T. 1989. Immunoglobulin κ gene expression after stable integration. I. Role of the intronic MAR and enhancer in plasmacytoma cells. *J. Biol. Chem.* **264:** 21183.

Bode J. and Maass K. 1988. Chromatin domain surrounding the human interferon-β gene as defined by scaffold-attached regions. *Biochemistry* **27:** 4706.

Bode J., Kohwi Y., Dickinson L., Joh T., Klehr D., Mielke C., and Kohwi-Shigematsu T. 1992. Biological significance of unwinding capability of nuclear matrix-associating DNAs. *Science* **255:** 195.

Cai H. and Levine M. 1995. Modulation of enhancer-promoter interactions by insulators in the *Drosophila* embryo. *Nature* **376:** 533.

Chung J.H., Whiteley M., and Felsenfeld G. 1993. A 5′ element of the chicken β-globin domain serves as an insulator in human erythroid cells and protects against position effect in *Drosophila.* *Cell* **74:** 505.

Cockerill P. N. and Garrard W.T. 1986. Chromosomal loop anchorage of the κ immunoglobulin gene occurs next to the enhancer in a region containing topoisomerase II sites. *Cell* **44:** 273.

Cockerill P. N., Yuen M.-H., and Garrard W.T. 1987. The enhancer of the immunoglobulin heavy chain locus is flanked by presumptive chromosomal loop anchorage elements. *J. Biol. Chem.* **262:** 5394.

Davie J.R. 1995. The nuclear matrix and the regulation of chromatin organization and function. *Int. Rev. Cytol.* **162A:** 191.

———. 1996. Histone modifications, chromatin structure, and the nuclear matrix. *J. Cell. Biochem.* **62:** 149.

———. 1998. Covalent modification of histones: Expression from chromatin templates. *Curr. Opin. Genet. Dev.* **8:** 173.

de Belle I., Cai S., and Kohwi-Shigematsu T. 1998. The genomic sequences bound to special AT-rich sequence-binding protein 1 (SATB1) in vivo in Jurkat T cells are tightly associated with the nuclear matrix at the bases of the chromatin loops. *J. Cell Biol.* **141:** 335.

Dickinson L.A. and Kohwi-Shigematsu T. 1995. Nucleolin is a matrix attachment region DNA-binding protein that specifically recognizes a region with high base-unpairing potential. *Mol. Cell. Biol.* **15:** 456.

Dickinson L.A., Joh T., Kohwi Y., and Kohwi-Shigematsu T. 1992. A tissue-specific MAR/SAR DNA-binding protein with unusual binding site recognition. *Cell* **70:** 631.

Dillon N. and Grosveld F. 1994. Chromatin domains as potential units of eukaryotic gene function. *Curr. Opin. Genet. Dev.* **4:** 260.

Dunaway M. and Dröge P. 1989. Trans activation of the *Xenopus* rRNA promoter by its enhancer. *Nature* **341:** 657.

Ephrussi A., Church G., Tonegawa S., and Gilbert W. 1985. B

lineage-specific interations of an immunoglobulin enhancer with cellular factors *in vivo.* *Science* **227:** 134.

Ernst P. and Smale S.T. 1995. Combinatorial regulation of transcription II: The immunoglobulin μ heavy chain gene. *Immunity* **2:** 427.

Felsenfeld G. 1992. Chromatin as an essential part of the transcriptional mechanism. *Nature* **355:** 219.

———. 1996. Chromatin unfolds. *Cell* **86:** 13.

Festenstein R., Tolaini M., Corbella P., and Mamalaki C. 1996. Locus control region function and heterochromatin-induced position effect variegation. *Science* **271:** 1123.

Forrester W.C., Genderen C.V., Jenuwein T., and Grosschedl R. 1994. Dependence of enhancer-mediated transcription of the immunoglobulin μ gene on nuclear matrix attachment regions. *Science* **265:** 1221.

Forrester W., Epner E., Driscoll M., Enver T., Brice M., Papayannopoulou T., and Groudine M. 1990. A deletion of the human β-globin locus activation region causes a major alteration in chromatin structure and replication across the entire β-globin locus. *Genes Dev.* **4:** 1637.

Garrity P.A. and Wold B. 1992. Effects of different DNA polymerases in ligation-mediated PCR: Enhanced genomic sequencing and *in vivo* footprinting. *Proc. Natl. Acad. Sci.* **89:** 1021.

Gasser S.M. and Laemmli U.K. 1986. Cohabitation of scaffold binding regions with upstream/enhancer elements of three developmental regulated genes of *D. melanogaster.* *Cell* **46:** 521.

Gerdes M.G., Carter K.C., Moen P.T., and Lawrence J.B. 1994. Dynamic changes in the higher-level chromatin organization of specific sequences revealed by *in situ* hybridization to nuclear halos. *J. Cell Biol.* **126:** 289.

Grosschedl R., Weaver D., Baltimore D., and Costantini F. 1984. Introduction of a μ immunoglobulin gene into the mouse germ line: Specific expression in lymphoid cells and synthesis of functional antibody. *Cell* **38:** 647.

Hebbes T., Clayton A., Thorne A., and Crane-Robinson C. 1994. Core histone hyperacetylation co-maps with generalized DNase I sensitivity in the chicken beta-globin chromosomal domain. *EMBO J.* **13:** 1823.

Herrscher R.F., Kaplan M.H., Lelsz D.L., Das C., Scheuermann R., and Tucker P.W. 1995. The immunoglobulin heavy-chain matrix-associating regions are bound by Bright: A B cell-specific *trans*-activator that describes a new DNA-binding protein family. *Genes Dev.* **9:** 3067.

Jenuwein T., Forrester W.C., Qiu R.G., and Grosschedl R. 1993. The immunoglobulin μ enhancer core establishes local factor access in nuclear chromatin independent of transcriptional stimulation. *Genes Dev.* **7:** 2016.

Jenuwein T., Forrester W.C., Fernandez-Herrero L.A., Laible G., Dull M., and Grosschedl R. 1997. Extension of chromatin accessibility by nuclear matrix attachment regions. *Nature* **385:** 269.

Jones P.L., Veenstra G.J., Wade P.A., Vermaak D., Kass S.U., Landsberger N., Strouboulis J., and Wolffe A.P. 1998. Methylated DNA and MeCP2 recruit histone deacetylase to repress transcription. *Nat. Genet.* **19:** 187.

Kass S.U., Landsberger N., and Wolffe A.P. 1997a. DNA methylation directs a time-dependent repression of transcription initiation. *Curr. Biol.* **7:** 157.

Kass S.U., Pruss D., and Wolffe A.P. 1997b. How does DNA methylation repress transcription? *Trends Genet.* **13:** 444.

Kellum R. and Schedl P. 1991. A position-effect assay for boundaries of higher order chromosomal domains. *Cell* **64:** 941.

Kim K.., Kanellopoulos-Langevin C., Merwin R.M., Sachs D.H., and Asofsky R. 1979. Establishment and characterization of BALB/c lymphoma lines with B cell properties. *J. Immunol.* **122:** 549.

Kioussis D. and Festenstein R. 1997. Locus control regions: Overcoming heterochromatin-induced gene inactivation in mammals. *Curr. Opin. Genet. Dev.* **7:** 614.

Kirillov A., Kistler B., Mostoslavsky R., Cedar H., Wirth T., and

Bergman Y. 1996. A role for nuclear NF-κB in B-cell-specific demethylation of the Igκ locus. *Nat. Genet.* **13:** 435.

Klehr D., Maas K., and Bode J. 1991. Scaffold-attached regions from the human interferon β domain can be used to enhance the stable expression of genes under the control of various promoters. *Biochemistry* **30:** 1264.

Klehr D., Schlake T., Maass K., and Bode J. 1992. Scaffold-attached regions (SAR elements) mediate transcriptional effects due to butyrate. *Biochemistry* **31:** 322.

Kohwi-Shigematsu T. and Kohwi Y. 1990. Torsional stress stabilizes extended base unpairing in suppressor sites flanking immunoglobulin heavy chain enhancer. *Biochemistry* **29:** 9551.

Laemmli U.K., Käs E., Poljak L., and Adachi Y. 1992. Scaffold-associated regions: *cis*-acting determinants of chromatin structural loops and functional domains. *Curr. Opin. Genet. Dev.* **2:** 275.

Lee D.Y., Hayes J.J., Pruss D., and Wolffe A.P. 1993. A positive role for histone acetylation in transcription factor access to nucleosomal DNA. *Cell* **72:** 73.

Lennon G.G. and Perry R.P. 1985. Cμ-containing transcripts initiate heterogenously within the *IgH* enhancer region and contain a novel 5′-nontranslatable exon. *Nature* **318:** 475.

Li E., Beard C., and Jaenisch R. 1993. Role for DNA methylation in genomic imprinting. *Nature* **366:** 362.

Lichtenstein M., Keini G., Cedar H., and Bergman Y. 1994. B cell-specific demethylation: A novel role for the intronic κ chain enhancer sequence. *Cell* **76:** 913.

Liu J., Bramblett D., Zhu Q., Lozano M., Kobayashi R., Ross S., and Dudley J. 1997. The matrix attachment region-binding protein SATB1 participates in negative regulation of tissue-specific gene expression. *Mol. Cell. Biol.* **17:** 5275.

Marilley M. and Gassend-Bonnet G. 1989. Supercoiled loop organization of genomic DNA: A close relationship between loop domains, expression units, and replicon organization in rDNA from *Xenopus laevis*. *Exp. Cell Res.* **180:** 475.

Mastrangelo I.A., Courey A.J., Wall J.S., Jackson S.P., and Hough P.V.C. 1991. DNA looping and Sp1 multimer links: A mechanism for transcriptional synergism and enhancement. *Proc. Natl. Acad. Sci.* **88:** 5670.

Monk P., Boubelik M., and Lehnert S. 1987. Temporal and regional changes in DNA methylation in the embryonic, extraembryonic and germ cell lineages during mouse embryo development. *Development* **99:** 371.

Mortillaro M.J., Blencowe B.J., Wei X., Nakayasu H., Du, L., Warren S.L., Sharp P.A., and Berezney R. 1996. A hyperphosphorylated form of the large subunit of RNA polymerase II is associated with splicing complexes and the nuclear matrix. *Proc. Natl. Acad. Sci.* **93:** 8253.

Mueller-Storm H.P., Sogo J.M., and Schaffner W. 1989. An enhancer stimulates transcription in *trans* when attached to the promoter via a promoter bridge. *Cell* **58:** 767.

Nan X., Campoy F.J., and Bird A. 1997. MeCP2 is a transcriptional repressor with abundant binding sites in genomic chromatin. *Cell* **58:** 471.

Nan X., Tate P., Li E., and Bird A. 1996. DNA methylation specifies chromosomal localization of MeCP2. *Mol. Cell. Biol.* **16:** 414.

Nan X., Ng H.-H., Johnson C.A., Laherty C.D., Turner B.M., Eisenman R.N., and Bird A. 1998. Transcriptional repression by the methyl-CpG-binding protein MeCP2 involves a histone deacetylase complex. *Nature* **393:** 386.

Palmiter R.D. and Brinster R.L. 1986. Germline transformation of mice. *Annu. Rev. Genet.* **20:** 465.

Parvin J. and Sharp P. 1993. DNA topology and a minimal set of basal factors for transcription by RNA polymerase II. *Cell* **73:** 533.

Pederson T. 1998. Thinking about a nuclear matrix. *J. Mol. Biol.* **277:** 147.

Phi-Van L., von Kries J.P., Ostertag W., and Strätling W.H. 1990. The chicken lysozyme 5′ matrix attachment region increases transcription from a heterologous promoter in heterologous cells and dampens position effects on the expression of transfected genes. *Mol. Cell. Biol.* **10:** 2302.

Ptashne M. and Gann A. 1997. Transcriptional activation by recruitment. *Nature* **386:** 569.

Razin A. and Cedar H. 1994. DNA methylation and genomic imprinting. *Cell* **77:** 473.

Razin S. and Gromova I.I. 1995. The channels model of nuclear matrix structure. *BioEssays* **17:** 443.

Reitman M., Lee E., Westphal H., and Felsenfeld G. 1993. An enhancer/locus control region is not sufficient to open chromatin. *Mol. Cell. Biol.* **13:** 3990.

Sambrook J., Fritsch E.F., and Maniatis T. 1989. *Molecular cloning: A laboratory manual*, 2nd edition. Cold Spring Harbor Laboratory Press, Cold Spring Harbor, New York.

Scheuermann R.H. and Chen U. 1989. A developmental-specific factor binds to suppressor sites flanking the immunoglobulin heavy chain enhancer. *Genes Dev.* **3:** 1255.

Stargell L.A. and Struhl K. 1996. Mechanisms of transcriptional activation *in vivo*: Two steps forward. *Trends Genet.* **12:** 311.

Struhl K. 1998. Histone acetylation and transcription regulatory mechanisms. *Genes Dev.* **12:** 599.

Su L. and Kadesch T. 1990. The immunoglobulin heavy-chain enhancer functions as the promoter for Iμ sterile transcription. *Mol. Cell. Biol.* **10:** 2619.

Su W., Jackson S., Tjian R., and Echols H. 1991. DNA looping between sites for transcriptional activation: Self association of DNA-bound Sp1. *Genes Dev.* **5:** 820.

Weitzel J.M., Buhrmester H., and Stratling W.H. 1997. Chicken MAR-binding protein ARBP is homologous to rat methyl-CpG- binding protein MeCP2. *Mol. Cell. Biol.* **17:** 5656.

Wilson C., Belen H.., and Gehring W.J. 1990. Position effects on eukaryotic gene expression. *Annu. Rev. Cell Biol.* **6:** 679.

Zhao K., Kas E., Gonzalez E., and Laemmli U.K. 1993. SAR-dependent mobilization of histone H1 by HMG-I/Y *in vitro*: HMG-I/Y is enriched in H1-depleted chromatin. *EMBO J.* **12:** 3237.

ATP-dependent Remodeling of Chromatin

C. Wu, T. Tsukiyama,* D. Gdula, P. Georgel,† M. Martínez-Balbás,‡ G. Mizuguchi,
V. Ossipow, R. Sandaltzopoulos, and H.-M. Wang
*Laboratory of Molecular Cell Biology, National Cancer Institute, National Institutes of Health,
Bethesda, Maryland 20892-4255*

The organization of DNA in chromatin constrains the eukaryotic genome in the nucleus and provides global mechanisms for the suppression of gene activity. This genetic repression extends from the nucleosome, the primary unit of chromatin organization, to the higher-order arrangement of nucleosome arrays (for reviews, see Grunstein 1990; Felsenfeld 1992; Kornberg and Lorch 1992; Weintraub 1993). We are interested in the mechanisms by which interactions between DNA and chromosomal proteins, particularly the nucleosomal histones, are disrupted to allow access by the transcriptional machinery. Our studies began almost 20 years ago with the discovery of DNase-I-hypersensitive sites in cellular chromatin at the uninduced heat shock gene promoters of *Drosophila* (Wu 1980). This finding, revealed by the indirect end-labeling technique, indicated extensive alterations in the accessibility of promoter DNA in chromatin and suggested a local perturbation of histone-DNA or histone-histone interactions.

The local disruption of chromatin structure could, in principle, be attributed to the binding of sequence-specific transcription factors or to structural pecularities of the underlying DNA sequence. Microinjection of plasmid DNA carrying the *Drosophila hsp70* promoter in *Xenopus* oocytes and the assembly of these plasmids in chromatin failed to generate a hypersensitive site (S.L. McKnight and C. Wu, unpubl.), indicating that the intrinsic structure of *hsp70* promoter DNA had no substantial effect on nucleosome organization when assembled in chromatin. Consequently, our attention shifted to the role of sequence-specific DNA-binding proteins. Since very little was known about transcription factors in eukaryotes in the early 1980s, we set out to identify protein factors specific for *Drosophila* heat shock promoters by developing exonuclease footprinting techniques and by biochemical fractionation of cell extracts (Wu 1984a,b, 1985).

A decade of studies from our laboratory and the laboratories of Sarah Elgin, John Lis, Carl Parker, and others elaborated the small number of general and specific factors interacting with the promoters of *Drosophila* heat shock genes. Under normal (nonshock) conditions, the uninduced *hsp70* promoter was bound by the GAGA

transcription factor at multiple upstream sites, the TATA-binding protein (TBP) with its associated factors (TFIID), and a paused RNA polymerase (for review, see Lis and Wu 1993, 1995; Wilkins and Lis 1997). One or more of these three components, particularly the GAGA factor, were likely to be responsible for generating an accessible chromatin structure so as to facilitate binding of the heat shock transcription factor (HSF) at sites located between the GAGA elements. Activation of HSF in response to the heat stress signal involves a monomer to trimer transition and derepression of an activator domain, leading to release of the paused polymerase and the recruitment of additional polymerase molecules (for review, see Lis and Wu 1993; Wu 1995). Knowledge of these *hsp70* promoter-specific factors, combined with the timely development of a robust chromatin assembly system derived from a high-speed supernatant of *Drosophila* embryo extracts (Becker and Wu 1992), set the stage for our attempts to reconstruct nucleosome disruption at the *hsp70* promoter in vitro.

NUCLEOSOME DISRUPTION MEDIATED BY BINDING OF GAGA FACTOR

The *Drosophila* GAGA factor is constitutively expressed and binds to multiple GAGA elements present in *hsp70* and many other *Drosophila* genes (Biggin and Tjian 1988; Soeller et al. 1988, 1993). We introduced recombinant GAGA factor at different times during reconstitution of an *hsp70* plasmid with the *Drosophila* chromatin assembly extract. The reconstituted plasmid chromatin was analyzed for nucleosome organization by digestion with micrococcal nuclease (MNase), followed by gel electrophoresis and blot hybridization (Tsukiyama et al. 1994). When DNA was reconstituted in the chromatin assembly reaction without GAGA factor, the MNase digestion pattern showed the assembly of an intact nucleosome at sequences corresponding to the oligonucleotide probe (positions –115 to –132; partially overlapping two of four GAGA elements on the *hsp70* promoter) (Fig. 1). In addition to the 146-bp fragment derived from the nucleosome core particle, a ladder of discrete fragments corresponding to nucleosome oligomers was observed at intermediate stages of MNase digestion. This indicated that the DNA surrounding the *hsp70* promoter was organized in a regularly spaced array of nucleosomes with a characteristic repeat length of approximately 180 bp for *Drosophila*.

Present addresses: *Division of Basic Sciences, Fred Hutchinson Cancer Research Center, 1100 Fairview Avenue North, P.O. Box 19024, Seattle, Washington 98109; †University of Rochester, Center for Oral Biology, Room 55711, 575 Elmwood Ave., Rochester, New York 14642; ‡Wellcome/CRC Institute, Tennis Court Road, Cambridge CB2 1QR, United Kingdom.

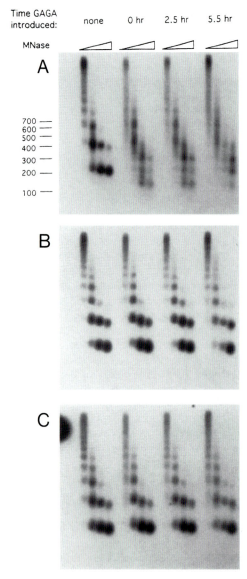

Time GAGA
introduced: none   0 hr   2.5 hr   5.5 hr

MNase

A

700 —
600 —
500 —
400 —
300 —
200 —

100 —

B

C

Figure 1. GAGA factor-dependent chromatin disruption in vitro (*A–C*). Micrococcal nuclease (MNase) digestion patterns of *hsp70* plasmid chromatin reconstituted with GAGA factor. DNA blots were hybridized sequentially with (*A*) oligonucleotide (−115 to −132 of *hsp70*), (*B*) (+1803 to +1832 of *hsp70*), and (*C*) (2499 to 2528 of pBluescript SK−).

A dramatically different cleavage pattern at the *hsp70* promoter was observed when GAGA factor was added at the onset of nucleosome assembly (0 hr). Upon extensive digestion with MNase, the abundance of the 146-bp fragment was decreased up to fivefold, and subnucleosomal fragments shorter than 146 bp were observed. These changes in the digestion pattern suggested an invasion and cleavage by MNase of the DNA within the nucleosome core particle, providing evidence for disruption in nucleosome organization. Furthermore, as a consequence of this cleavage, the pattern of MNase digestion of nucleosome oligomers degenerated into a smear at intermediate stages of digestion. A similar alteration of chromatin structure was observed when GAGA factor was

introduced when the assembly of regularly spaced nucleosomes was nearly complete (2.5 hr), and significant disruption was even observed when GAGA factor was introduced after completion of nucleosome assembly over the entire plasmid (5.5 hr). These results showed that GAGA factor could not only compete with the deposition of nucleosomes by a preemptive mechanism, but more importantly also alter the structure of existing nucleosomes. No effects of GAGA factor on regions distant from the *hsp70* promoter were observed (Fig. 1B,C).

RECONSTITUTION OF DNase I HYPERSENSITIVITY

In concurrence with the results of MNase digestion, a broad DNase-I-hypersensitive site with a major peak at approximately −100 was reconstituted on the *hsp70* promoter when GAGA factor was introduced in the assembly reaction (Fig. 2). DNase I hypersensitivity was observed when GAGA factor was introduced at 0, 2.5, and 5.5 hours of chromatin assembly. Although the major hypersensitive peak at approximately −100 was close to or coincident with the natural peak of hypersensitivity at about −93 previously mapped on the endogenous *hsp70* genes (Fig. 2) (Wu 1984a), differences in the fine structure and span of the hypersensitive region could be observed. This is likely to be due to a deficiency of TFIID and RNA polymerase II (the two other protein complexes constitutively bound in vivo) that may be necessary for faithful reconstitution of the entire hypersensitive structure. The DNase I cleavages mapped to single-nucleotide resolution show modest protection over the GAGA repeats and strong hypersensitivity of the sequences between and flanking the GAGA repeats (Tsukiyama et al. 1994). The TATA box and two heat shock control elements are included within the DNase-I-hypersensitive sequences.

MULTIPLE GAGA ELEMENTS REQUIRED FOR DISRUPTION

The sequence requirements for nucleosome disruption were evaluated by reconstituting plasmids carrying deletions of the *hsp70* upstream region (Fig. 3, top). The effect of GAGA factor on plasmid pdhspΔ186, which includes the four GAGA elements within positions −186 to +296, was essentially the same as the effect on the original *hsp70* plasmid containing the entire coding region and 1.5 kb of upstream DNA (Fig. 3, bottom). However, when the template carried a 5′ deletion to −90, which removes the two distal GAGA elements (pdhspΔ90), disruption was observed when GAGA factor was added at the start of, but not at the completion of, assembly (Fig. 3). When the *hsp70* promoter was deleted to −50, which removes the third GAGA element (pdhspΔ50), no discernible disruption by GAGA factor was observed (Fig. 3). Therefore, promoter sequences including at least two GAGA elements are necessary for nucleosome displacement when GAGA factor is competing directly with nucleosome deposition, and sequences including three to four GAGA elements are required for the disruption of an existing nucleosome.

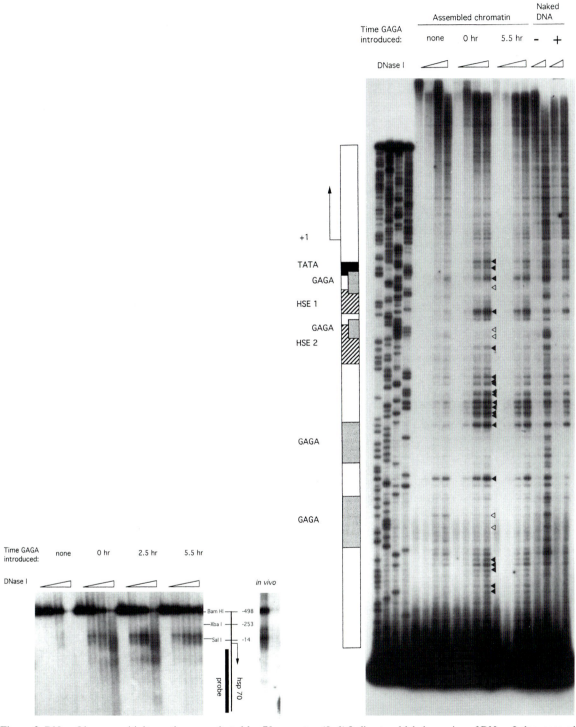

Figure 2. DNase I hypersensitivity on the reconstituted *hsp70* promoter. (*Left*) Indirect end-label mapping of DNase I cleavages relative to a *Bam*HI site at +1258. The location of the probe on *hsp70* is indicated by the solid bar. The in vivo sample shows the cleavage pattern of the endogenous *hsp70* genes in 0–24-hr *Drosophila* embryo nuclei. (*Right*) Primer extension–linear amplification analysis of DNase I cleavages on a 6% sequencing gel. (*Closed triangles*) Hypersensitive nucleotide; (*open triangles*) protected nucleotides; (*stippled bars*) GAGA elements;(*hatched bars*) heat shock elements; (*solid bars*) TATA box. The first four lanes on the left are sequencing reactions: T,C,G,A.

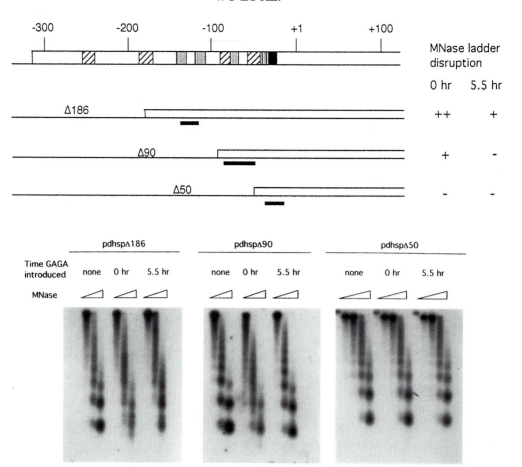

Figure 3. Promoter elements for chromatin disruption. *(Top)* Restriction map showing *hsp70* promoter deletions. Vector DNA: *(thin line; open bars) hsp70*, oligonucleotide probes: *(solid bar)*. All constructs contain up to +296 bp of *hsp70*. The relative extent of chromatin disruption is given: (++) arbitrary designation of disruption observed on plasmid dhspXX3.2, which carries 1.5 kb of upstream DNA. *(Bottom)* Effects of GAGA factor on *hsp70* promoter constructs carrying upstream sequences to –186 , –90, and –50, respectively. Plasmid DNAs were reconstituted with GAGA factor as indicated, digested with MNase, blotted, and hybridized with oligonucleotides corresponding to positions –132 to –115 (pdhspΔ186), –89 to –50 (pdhspΔ90), and –36 to –17 (pdhspΔ50).

DISCOVERY OF ATP REQUIREMENT FOR NUCLEOSOME DISRUPTION

Previous studies indicated that the process of nucleosome assembly in crude extracts requires ATP and an energy regeneration system (Almouzni and Méchali 1988; Shimamura et al. 1988; Becker and Wu 1992). We questioned whether nucleosome disruption mediated by GAGA factor was also dependent on ATP by adding the ATP-hydrolyzing enzyme apyrase after the completion of nucleosome assembly but before the introduction of GAGA factor. Treatment with apyrase suppressed nucleosome disruption by GAGA factor on the *hsp70* promoter (Fig. 4, top); similar results were obtained when ATP was removed by gel filtration chromatography (Fig. 4, bottom). Moreover, nucleosome disruption could be restored upon the introduction of fresh ATP, which could not be substituted with GTP, ADP, or the nonhydrolyzable analogs ATP-γ-S and AMP-PCP (Fig. 4, bottom). Hence, the specific disruption of nucleosome structure by GAGA factor requires ATP hydrolysis. Since there was no evidence that GAGA factor itself utilized ATP, our results indicated the presence of an unknown cofactor in the *Drosophila* chromatin assembly extract (Tsukiyama et al. 1994).

A SARKOSYL-SENSITIVE, ATP-DEPENDENT COFACTOR FOR NUCLEOSOME DISRUPTION

To detect the ATP-dependent activity acting together with GAGA factor, we sought conditions that would inactivate or eliminate it in reconstituted chromatin. A brief treatment of reconstituted chromatin with Sarkosyl abolished nucleosome disruption at the *hsp70* promoter, even when GAGA factor and ATP were present for remodeling (Fig. 5, top) (Tsukiyama and Wu 1995). Nucleosome disruption was restored by the addition of fresh chromatin assembly extract to the Sarkosyl-treated chromatin, and this restoration was dependent on the presence of GAGA factor and ATP. Hence, three critical components are required for nucleosome disruption: (1) a DNA-binding factor, (2) multiple cognate elements in close proximity, and (3) a Sarkosyl-sensitive factor that utilizes ATP. We have named this Sarkosyl-sensitive, ATP-dependent component *nu*cleosome *r*emodeling *f*actor (NURF).

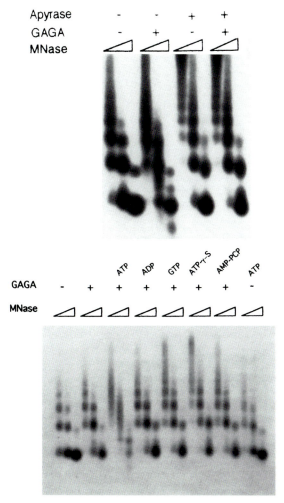

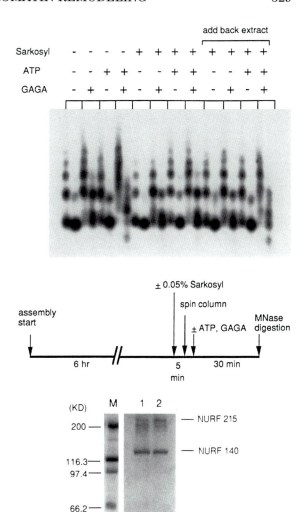

Figure 4. ATP requirement for disruption of chromatin structure. (*Top*) Effect of apyrase. *hsp70* plasmid chromatin reconstituted with GAGA factor and treated with apyrase was analyzed by MNase digestion as in Fig. 1. (*Bottom*) Restoration of nucleosome disruption with fresh ATP. *hsp70* chromatin was purified by gel filtration and incubated with GAGA factor, nucleotides, and analogs as indicated, digested with MNase, and processed as above.

PURIFICATION AND CHARACTERIZATION OF NURF

To purify NURF, we employed Sarkosyl-inactivated *hsp70* plasmid chromatin as a substrate and high-salt nuclear extracts prepared from *Drosophila* embryos as starting material (Tsukiyama and Wu 1995). These nuclear extracts (Wampler et al. 1990) contained abundant NURF activity and were fractionated by chromatography on DE-52, BioRex, Q Sepharose, hydroxyapatite, and phosphocellulose P-11 (Tsukiyama and Wu 1995). At the seventh and final step of purification on a glycerol gradient, SDS-PAGE revealed four major polypeptides of 215, 140, 55, and 38 kD (Fig. 5, bottom). The overall purification from the nuclear extract to the glycerol gradient fraction was estimated to be approximately 1200-fold, with about 12% yield.

Introduction of increasing amounts of purified NURF in the presence of a fixed amount of GAGA factor

Figure 5. (*Top*) Sarkosyl-sensitive cofactor for GAGA factor-mediated nucleosome disruption. MNase digestion of *hsp70* plasmid chromatin was treated with 0.05% Sarkosyl as indicated and assayed for GAGA factor and ATP-dependent nucleosome disruption by digestion with MNase and DNA blot hybridization with an oligonucleotide (–113 to –142 of *hsp70*). Two samples corresponding to intermediate and high levels of MNase digestion are given. (*Bottom*) SDS-PAGE and Coomassie blue staining of purified NURF. Two independently purified preparations of NURF are shown in lanes *1* and *2*. The 215-kD polypeptide is susceptible to degradation.

showed increasing disruption of nucleosome structure at the *hsp70* promoter, with no disruption at the coding sequences (Fig. 6A). A 20–40-fold higher level of NURF alone produced a general smearing of the oligonucleosomal DNA ladder for both promoter and coding regions and an apparent reduction in the nucleosome repeat length, showing that global effects on chromatin can be

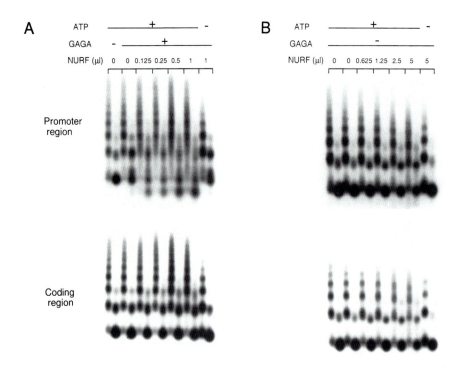

Figure 6. Activity of NURF on nucleosome arrays. (*A*) MNase digestion of reconstituted *hsp70* plasmid chromatin pretreated with Sarkosyl to inactivate NURF. An increasing amount of NURF (single-stranded DNA cellulose fraction) was used in the presence or absence of GAGA factor and ATP; samples were processed as in Fig. 1. The probes correspond to *hsp70* –113 to –142 (promoter), and +1803 to +1832 (coding). (*B*) MNase digestion of plasmid chromatin with increasing amounts of NURF (single-stranded DNA cellulose fraction) but without GAGA factor, analyzed as above.

obtained in the absence of sequence-specific DNA-binding proteins (Fig. 6B). Both local and global effects of NURF were dependent on the presence of ATP. In concurrence with these results, the addition of NURF to a single nucleosome reconstituted on a 161-bp *hsp70* promoter fragment resulted in ATP-dependent changes in the pattern of DNase I digestion of nucleosomal (but not free) DNA. DNase I hypersensitivity was induced at several distinct sites on the mononucleosome, whereas protection from DNase I cleavage was found at other sites (Fig. 7A, lanes 10–13).

On the 161-bp nucleosome, GAGA factor alone showed weak affinity for its cognate sites (Fig. 7B, lanes 11–13). The inclusion of NURF in the reaction resulted in enhanced footprinting over the GAGA elements and the induction of strong DNase I hypersensitivity in the intervening region (Fig. 7B, lanes 15–17). Interestingly, despite NURF-assisted GAGA factor binding and nucleosome reconfiguration, the GAGA factor footprint retained

internal DNase I cleavages with approximately 10-bp periodicity, suggesting that the integrity of the 161-bp nucleosome was somewhat maintained (Fig. 7B, lanes 15–17; see positions –122 to –146). Similar conclusions were drawn from restriction enzyme analyses (Tsukiyama and Wu 1995). The decreased severity of disruption on the 161-bp nucleosome, perhaps reflecting an intermediate state in the remodeling process, may be due to an abnormal nucleosome conformation caused by the short linker DNA (Usachenko et al. 1994) to insufficient linker length for nucleosome sliding, or to the absence of adjacent nucleosomes and torsional constraints.

NURF HAS A NUCLEOSOME-STIMULATED ATPase ACTIVITY

To investigate the ATP requirement for NURF activity, we examined the ability of the NURF complex to hydrolyze ATP to inorganic phosphate in the presence of

Figure 7. Action of NURF on mononucleosomes. (*A*) DNase I digestion of the 161-bp nucleosome and naked DNA in the presence of an increasing amount of NURF (P11 fraction: 100 ng/μl). The 161-bp DNA fragment is from –185 to –30 of the *Drosophila hsp70* promoter with 6 bp from the plasmid vector. (*Open triangles*) DNase I protection; (*closed triangles*) enhancements; (*large triangles*) sites where significant changes were observed at 0.11 μl or less of NURF; (*small triangles*) sites where the maximum effects were observed at the highest amount (1.0 μl) of NURF. (*B*) DNase I digestion of the 161-bp nucleosome and naked DNA in the presence of an increasing amount of GAGA factor, with or without NURF (1.0 μl of P11 fraction). All reactions contained 1 mM ATP. (*Open circles*) DNase I protection; (*closed circles*) hypersensitivity. (*C*) ATPase activity of NURF. The percentage of ATP hydrolysis to free phosphate was analyzed by thin-layer chromatography. The NURF P-11 fraction was tested for ATPase activity in the presence of equivalent concentrations of free *E. coli* DNA, purified *Drosophila* core histones, nucleosomes reconstituted from the same DNA, and core histones by dialysis from high salt. Conditions for the ATPase assay were the same as those for DNase I footprinting.

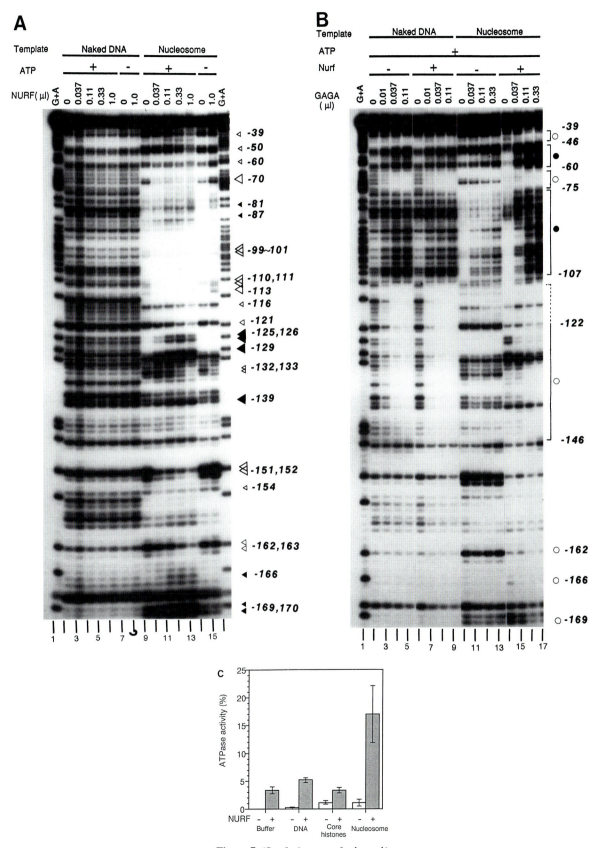

Figure 7. (*See facing page for legend.*)

free DNA or reconstituted nucleosomes. Purified NURF has a constitutive ATPase activity (~1.7 mole Pi/mole NURF/min) which was not enhanced significantly by the inclusion of free DNA or core histones in the reaction. However, the ATPase activity was stimulated about five-fold in the presence of reconstituted nucleosomes (Fig. 7C). Moreover, when the same core histones and *Escherichia coli* DNA were directly mixed without undergoing proper reconstitution, the stimulation of ATPase activity was not observed, indicating that NURF recognizes nucleosomes. An element of nucleosome recognition by NURF includes the flexible histone tails, as shown by partial proteolysis and inhibition of ATPase activity by GST-histone tail peptides (Georgel et al. 1997).

TRANSCRIPTIONAL ACTIVATION OF CHROMATIN IS DEPENDENT ON NURF

Although the combined actions of NURF and a sequence-specific DNA-binding factor are required for promoter-specific disruption of nucleosomes in vitro, it was unclear whether disruption was necessary for transcriptional activation of chromatin. To address this issue, we employed a model plasmid (pGIE-0) for in vitro transcription that carries five GAL4-binding sites upstream of the TATA box and the adenovirus E4 minimal core promoter (Pazin et al. 1994). GAL4-HSF, containing essentially the GAL4 DNA-binding domain (GAL4 1-147) fused to the constitutive activator of HSF, was used as the transcriptional regulator (Wisniewski et al. 1996). A *Drosophila* soluble nuclear extract prepared in physiological salt was the source of the RNA polymerase II transcriptional apparatus (Kamakaka et al. 1991; Kamakaka and Kadonaga 1994). This nuclear extract (NE) is deficient in histones and has weak ATP-dependent chromatin remodeling activity (Pazin et al. 1994; Mizuguchi et al. 1997), thereby allowing evaluation of purified NURF (Fig. 8A, top).

As might be expected, no significant transcription was observed when chromatin inactivated for endogenous NURF was analyzed (Fig. 8A, lanes 4–6), and little perturbation of nucleosome organization by GAL4-HSF or GAL4 1-147 was detected (Fig. 8B, sections 1–3). Control experiments using chromatin associated with endogenous, active NURF showed strong transcriptional activation by GAL4-HSF (Fig. 8A, lane 3), and this activation was dependent on the presence of the activator domain of HSF (Fig. 8A, lanes 1 and 2). When purified NURF was included at an approximately physiological level for remodeling of the Sarkosyl-inactivated template, a significant restoration of transcription by GAL4-HSF (but not GAL4 1-147) was observed (Fig. 8A, lanes 9 and 10). This was accompanied by a restoration of nucleosome remodeling at the promoter region (Fig. 8B, section 6). Interestingly, nucleosome remodeling by GAL4 1-147 was also observed (Fig. 9B, section 5), although this was insufficient to allow transcription. Hence, the repression of basal transcription imposed by nucleosomes could be significantly relieved by the essential actions of NURF, the GAL4 DNA-binding domain,

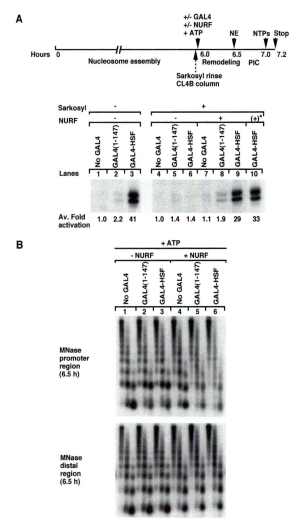

Figure 8. NURF is able to activate chromatin. (*A*) Experimental scheme is at the top. Primer extension analysis shows transcriptional activation of chromatin templates by GAL4 derivatives. Chromatin assembled for 6.0 hr was either not treated (*left panel*) or treated (*right panel*) with 0.05% Sarkosyl, purified by gel filtration, and incubated with GAL4 derivatives, +/– 0.5 µl NURF (100 ng/µl P11 fraction or 50 ng/µl glycerol gradient fraction), + 0.5 mM ATP for 30 min to allow nucleosome remodeling. Chromatin templates were then incubated for 30 min with nuclear extract (NE) to form preinitiation complexes, followed by addition of NTPs for 10 min of transcription. (*B*) MNase digestion assay. Chromatin samples processed as for transcription (6.5 hr) were analyzed by MNase digestion and sequential Southern blot hybridization using oligonucleotide probes specific for the adenovirus E4 TATA box (promoter region), and –900/–874 bp (distal region).

and, importantly, the HSF activator domain. However, once nucleosome remodeling had occurred, a high level of NURF appeared not to be continuously required for recruitment of the general transcriptional machinery and initiation (Mizuguchi et al. 1997). These results provide direct evidence that NURF is able to remodel nucleosomes for transcriptional initiation in vitro and underscore an emerging role for activator domains in alleviating nucleosomal repression by a mechanism independent of NURF.

CLONING THE GENES CODING FOR NURF

To elucidate the composition and mechanism of NURF action, we have sought to clone the genes for NURF proteins. We purified a sufficient quantity of NURF for peptide microsequencing and reverse genetic analysis. Three of the four major NURF components have been cloned and their assignments as integral subunits of NURF were verified by coimmunoprecipitation (Tsukiyama et al. 1995; Martínez-Balbás et al. 1998; Gdula et al. 1998). Cloning of the remaining largest subunit of NURF is nearly complete.

The 140-kD NURF component was identified biochemically and immunologically as ISWI, related in the ATPase domain to Brahma, a transcriptional regulator of homeotic genes in *Drosophila*, and its counterpart SWI2/SNF2 in *Saccharomyces cerevisiae* (Tsukiyama et al. 1995). This result converged with genetic and biochemical findings indicating that the SWI2/SNF2 protein could antagonize histone-mediated repression in vivo and remodel nucleosomes in vitro (for review, see Peterson and Tamkun 1995; Wu 1997; Cairns 1998). Interestingly, ISWI is contained in two other *Drosophila* chromatin remodeling complexes ACF (Ito et al. 1997) and CHRAC (Varga-Weisz et al. 1997), indicating repeated usage of this protein as the energy-transducing component of distinct nucleosome remodeling machines.

The 55-kD subunit of NURF contains seven WD repeat motifs (Martínez-Balbás et al. 1998) and is also a component of the CAF-1 complex; related proteins are found in complexes containing histone acetyltransferase and histone deacetylase (Parthun et al. 1996; Taunton et al. 1996; Tyler et al. 1996; Verrault et al. 1996, 1998; Hassig et al. 1997; Zhang et al. 1997). Its mammalian counterpart, RbAp48, was originally isolated by virtue of binding to the retinoblastoma protein (Qian et al. 1993), although the significance of this interaction is unclear. The conserved WD repeat motifs in NURF-55 suggests that it may fold as a seven-bladed propeller as described for the β-subunit of the G-protein heterotrimer (Neer and Smith 1996). Such a structure may provide a platform for the assembly of multiple complexes involved in histone metabolism (Martínez-Balbás et al. 1998).

The 38-kD subunit of NURF is strikingly homologous to known inorganic pyrophosphatases (PPases) over the entire protein sequence, including 14 invariant residues in the active site cavity (Gdula et al. 1998). PPases are required for a broad spectrum of biosynthetic pathways, including nucleic acid synthesis; the enzyme hydrolyzes the pyrophosphate by-product of nucleotide incorporation to phosphate, thereby driving the polymerization reaction in the forward direction (Kornberg 1962). Both recombinant NURF-38 and the native NURF complex were shown to have PPase activity. Interestingly, this activity could be inhibited by sodium fluoride without affecting chromatin remodeling in vitro. This surprising connection between PPase and NURF raises the possibility that *Drosophila* PPase may be utilized for structural or regulatory purposes within the NURF complex or be adapted as a vehicle to target PPase to chromatin to faciliate transcription or replication.

SUMMARY AND PERSPECTIVES

Transcriptional regulation is inextricably linked to chromatin structure. At the first level of chromatin organization, the structure of the nucleosome presents a defined barrier that must be lifted for transcription. DNase-I-hypersensitive sites in cellular chromatin reflect a disruption of the nucleosomal barrier and are informative markers of the location of gene control sequences. We now understand that multiple mechanisms can be brought into play to perturb nucleosome structure with consequences for transcription. In addition to the binding by sequence-specific proteins and the effects of histone modifications discussed elsewhere in this volume, our biochemical studies have elucidated an energy-dependent pathway that perturbs nucleosome structure. Purification of the NURF complex and identification of its component genes have provided a means to further explore the nature and mechansim of the dynamic changes in DNA architecture and histone arrangement during nucleosome remodeling.

These advances in turn pose a long list of additional questions to be addressed. What is the precise nucleosome remodeling function of NURF in vivo? Does it have a role for transcription and for other processes requiring nucleosome disorder, such as DNA replication? How many chromatin remodeling machines like NURF or the SWI/SNF complex exist, how are their activities regulated in growth and development, and what are the rules by which each complex is brought into play for any particular gene? Is NURF recruited to specific chromosomal sites or does it diffuse along the chromatin fiber, creating transient windows of opportunity for sequence-specific factors? How is the activity of NURF integrated with the activities of the histone-modifying enzymes? What other mechanisms are utilized by the RNA polymerase apparatus to cross the threshold of activation, when recruited by an activator domain to a site of grossly perturbed but still repressed chromatin? These and other questions offer a continuing challenge for our efforts to elucidate the role of chromatin structure in transcriptional regulation.

ACKNOWLEDGMENTS

This work has been supported by the Intramural Research Program of the National Cancer Institute, National Institutes of Health. Project Z01CB05263: Eukaryotic Chromatin Structure and Gene Regulation.

REFERENCES

Almouzni G. and Méchali M. 1988. Assembly of spaced chromatin: Involvement of ATP and DNA topoisomerase activity. *EMBO J.* **7:** 4355.

Becker P.B. and Wu C. 1992. Cell-free system for assembly of transcriptionally repressed chromatin from *Drosophila* embryos. *Mol. Cell. Biol.* **12:** 2241.

Biggin M.D. and Tjian R. 1988. Transcription factors that activate the Ultrabithorax promoter in developmentally staged extracts. *Cell* **53:** 699.

Cairns B.R. 1998. Chromatin remodeling machines: Similar motors, ulterior motives. *Trends Biochem. Sci.* **23:** 20.

Felsenfeld G. 1992. Chromatin as an essential part of the transcriptional mechanism. *Nature* **335:** 219.

Gdula D., Sandalizopoulos R., Tsukiyama T., Ossipow V., and Wu C. 1998. Inorganic pyrophosphatase is a component of *Drosophila* nucleosome remodeling factor complex. *Genes Dev.* **12:** 3206.

Georgel P.T., Tsukiyama T., and Wu C. 1997. Role of histone tails in nucleosome remodeling by *Drosophila* NURF. *EMBO J.* **16:** 4717.

Grunstein M. 1990. Histone function in transcription. *Annu. Rev. Cell Biol.* **6:** 643.

Hassig C.A., Fleischer T.C., Billin A.N., Schreiber S.L., and Ayer D.E. 1997. Histone deacetylase activity is required for full transcriptional repression by mSin3A. *Cell* **89:** 341.

Ito T., Bulger M., Pazin M.J., Kobayashi R., and Kadonaga J.T. 1997. ACF, an ISWI-containing and ATP-utilizing chromatin assembly and remodeling factor. *Cell* **90:** 145.

Kamakaka R.T. and Kadonaga J.T. 1994. The soluble nuclear fraction, a highly efficient transcription extract from *Drosophila* embryos. *Methods Cell Biol.* **44:** 225.

Kamakaka R.T., Tyree C.M., and Kadonaga J.T. 1991. Accurate and efficient RNA polymerase II transcription with a soluble nuclear fraction derived from *Drosophila* embryos. *Proc. Natl. Acad. Sci.* **88:** 1024.

Kornberg A. 1962. On the metabolic significance of phosphorylytic and pyrophosphorylytic reactions. In *Horizons in biochemistry* (ed. H. Kasha and P. Pullman), p. 251. Academic Press, New York.

Kornberg R.D. and Lorch Y. 1992. Chromatin structure and transcription. *Annu. Rev. Cell Biol.* **8:** 563.

Lis J.T. and Wu C. 1993. Protein traffic on the heat shock promoter: Parking, stalling and trucking along. *Cell* **74:** 1.

———. 1995. Promoter potentiation and activation: Chromatin structure and transcriptional induction of heat shock genes. In *Chromatin structure and gene expression* (ed. S.C.R. Elgin), p. 71. IRL Press, Oxford, United Kingdom.

Martínez-Balbás M.A., Tsukiyama T., Gdula D., and Wu C. 1998. *Drosophila* NURF-55, a WD repeat protein involved in histone metabolism. *Proc. Natl. Acad. Sci.* **95:** 132.

Mizuguchi G., Tsukiyama T., Wisniewski J., and Wu C. 1997. Role of nucleosome remodeling factor NURF in transcriptional activation of chromatin. *Mol. Cell* **1:** 141.

Neer E.J. and Smith T.F. 1996. G protein heterodimers: New structures propel new questions. *Cell* **84:** 175.

Parthun M.R., Widom J., and Gottschling D.E. 1996. The major cytoplasmic histone acetyltransferase in yeast, links to chromatin replication and histone metabolism. *Cell* **87:** 85.

Pazin M.J., Kamakaka R.T., and Kadonaga J.T. 1994. ATP-dependent nucleosome reconfiguration and transcriptional activation from preassembled chromatin templates. *Science* **266:** 2007.

Peterson C.L. and Tamkun J.W. 1995. The SWI-SNF complex, a chromatin remodeling machine? *Trends Biochem. Sci.* **20:** 143.

Qian Y.-W., Wang Y.-C., Hollingsworth R.E., Jones D., Jr., Ling N., and Lee E.Y.-H. 1993. A retinoblastoma-binding protein related to a negative regulator of Ras in yeast. *Nature* **364:** 648.

Shimamura A., Tremethick D., and Worcel A. 1988. Characterization of the repressed 5S DNA minichromosomes assembled *in vitro* with a high speed supernatant of *Xenopus laevis* oocytes. *Mol. Cell. Biol.* **8:** 4257.

Soeller W.C., Oh C.E., and Kornberg, T. 1993. Isolation of cDNAs encoding the *Drosophila* GAGA transcription factor. *Mol. Cell. Biol.* **13:** 7961.

Soeller W.C., Poole S.J., and Kornberg T. 1988. In vitro tran-

scription of the *Drosophila engrailed* gene. *Genes Dev.* **2:** 68.

Taunton J., Hassig C.A., and Schreiber S.L. 1996. A mammalian histone deacetylase related to the yeast transcriptional regulator Rpd3p. *Science* **272:** 408.

Tsukiyama T. and Wu C. 1995. Purification and properties of an ATP-dependent nucleosome remodeling factor. *Cell* **83:** 1011.

Tsukiyama T., Becker P.B., and Wu C. 1994. ATP-dependent nucleosome disruption at a heat-shock promoter mediated by binding of GAGA transcription factor. *Nature* **367:** 525.

Tsukiyama T., Daniel C., Tamkun J., and Wu C. 1995. ISWI, a member of the SWI2/SNF2 ATPase family, encodes the 140 kDa subunit of the nucleosome remodeling factor. *Cell* **83:** 1021.

Tyler J.K., Bulger M., Kamakaka R.T., Yashi R., and Kadonaga J.T. 1996. The p55 subunit of *Drosophila* chromatin assembly factor 1 is homologous to a histone deacetylase-associated protein. *Mol. Cell. Biol.* **16:** 6149.

Usachenko S.I., Bavykin S.G., Gavin I.M., and Bradbury E.M. 1994. Rearrangement of the histone H2A C-terminal domain in the nucleosome. *Proc. Natl. Acad. Sci.* **91:** 6845.

Varga-Weisz P.D., Wilm M., Bonte E., Dumas K., Mann M., and Becker P.B. 1997. Chromatin-remodelling factor CHRAC contains the ATPases ISWI and topoisomerase II. *Nature* **388:** 598.

Verrault A., Kaufman P.D., Kobayashi R., and Stillman B. 1996. Nucleosome assembly by a complex of CAF-1 and acetylated histones H3/H4. *Cell* **87:** 95.

———. 1998. Nucleosomal DNA regulates the core-histone-binding subunit of the human Hat1 acetyltransferase. *Curr. Biol.* **8:** 96.

Wampler S.L., Tyree C.M., and Kadonaga J.T. 1990. Fractionation of the general RNA polymerase II transcription factors from *Drosophila* embryos. *J. Biol. Chem.* **265:** 21223.

Weintraub H. 1993. Summary: Genetic tinkering—Local problems, local solutions. *Cold Spring Harbor Symp. Quant. Biol.* **58:** 819.

Wilkins R.C. and Lis J.T. 1997. Dynamics of potentiation and activation: GAGA factor and its role in heat shock gene regulation. *Nucleic Acids Res.* **25:** 3963.

Wisniewski J., Orosz A., Allada R., and Wu C. 1996. The C-terminal region of *Drosophila* heat shock factor (HSF) contains a constitutively functional transactivation domain. *Nucleic Acids Res.* **24:** 367.

Wu C. 1980. The 5′ ends of *Drosophila* heat shock genes in chromatin are hypersensitive to DNase I. *Nature* **286:** 854.

———. 1984a. Two protein-binding sites in chromatin are implicated in the activation of heat shock genes. *Nature* **309:** 229.

———. 1984b. Activating protein factor binds in vitro to upstream control sequences in heat shock gene chromatin. *Nature* **311:** 81.

———. 1985. An exonuclease protection assay reveals heat shock element and TATA box DNA-binding proteins in crude nuclear extracts. *Nature* **317:** 84.

———. 1995. Heat shock transcription factors: Structure and function. *Annu. Rev. Cell Dev. Biol.* **11:** 441.

———. 1997. Chromatin remodeling and the control of gene expression. *J. Biol. Chem.* **272:** 28171.

Zhang Y., Iratni R., Erdjument-Bromage H., Tempst P., and Reinberg D. 1997. Histone deacetylases and SAP18, a novel polypeptide, are components of a human Sin3 complex. *Cell* **89:** 357.

A Model for Chromatin Remodeling by the SWI/SNF Family

G.R. SCHNITZLER, S. SIF, AND R.E. KINGSTON

Department of Molecular Biology, Massachusetts General Hospital, Boston, Massachusetts 02114
and Department of Genetics, Harvard Medical School, Boston, Massachusetts

During the past few years, our knowledge of how chromatin structure can be made more accessible to transcription and other reactions on the DNA has grown rapidly. One prominent mechanism for increasing access to DNA in chromatin is via remodeling of chromatin structure by ATP-dependent processes. A wide variety of ATP-dependent nucleosome remodeling complexes have been identified. Each of these contains a homolog to the yeast SWI2/SNF2 ATPase and has been shown to alter the structure of chromatin in vitro.

The first remodeling complex to be discovered, yeast SWI/SNF, was initially identified as a set of mutations that caused SUC2, HO, and a variety of other promoters to become active under repressing conditions (for review, see Winston and Carlson 1992). Suppressor mutations were found in histone genes and other genes involved in chromatin-dependent repression, such as SPT4 and SPT6, Sin1 (see Winston and Carlson 1992), and SSN6 (for discussion, see Gavin and Simpson 1997), indicating a role for SWI/SNF proteins in relieving repression by chromatin. Indeed, the disruption of positioned nucleosomes over the SUC2 promoter in inducing conditions fails to occur in SWI and SNF mutant cells (Hirschhorn et al. 1992; Matallana et al. 1992), and SWI/SNF is critical for activation of a synthetic promoter with a nucleosome over the TATA element (Ryan et al. 1998).

The yeast SWI and SNF gene products were found to exist in a large protein complex that could disrupt mononucleosome cores in vitro (Côté et al. 1994). At the same time, mammalian homologs of the SWI/SNF proteins were shown to exist in a set of complexes that could also disrupt mononucleosome cores (Imbalzano et al. 1994; Kwon et al. 1994). The human SWI/SNF complex(es) has been implicated in gene regulation through hormone receptors (Khavari et al. 1993; Chiba et al. 1994; Ichinose et al. 1997; Fryer and Archer 1998) and growth control through interaction with pRb (Dunaief et al. 1994; Singh et al. 1995; Strober et al. 1996; Sumi et al. 1997; Trouche et al. 1997).

More recently, a second yeast chromatin-disrupting complex, RSC, has been identified that is highly related to both the yeast SWI/SNF complex and the human SWI/SNF complexes (Cairns et al. 1996). Mutation of subunits of the RSC complex results in different phenotypes than does mutation of the SWI/SNF genes. Notably, yeast strains with deletions in RSC subunits are generally not viable, whereas strains with deletions in SWI/SNF

subunits are viable under many growth conditions. It is not clear whether the human SWI/SNF complex(es) are functional homologs of the yeast SWI/SNF complex and/or the RSC complex.

Each of these three complexes (RSC, and yeast and human SWI/SNF) contains an ATPase subunit(s) that is homologous to SWI2/SNF2 throughout its length. In addition, each complex contains subunits that are homologous to three other yeast SWI/SNF-subunits: SNF5, Swp73, and SWI3 (Cairns et al. 1996; Wang et al. 1996a,b). Because of these similarities, we refer to these three complexes as the SWI/SNF family of remodeling complexes. The existence of subunits that are not conserved between the complexes, and the different phenotypes resulting from their impairment, imply distinct biological roles for each; however, the extreme conservation of four of the subunits suggests that they may function by similar mechanisms.

Where tested, the SWI/SNF complexes have been shown to have the following abilities: DNA and nucleosome-stimulated ATP hydrolysis, alteration of the DNase footprint of mononucleosome cores, disruption of nucleosomes on 5S DNA, enhancement of transcription factor binding to cores and arrays, transcriptional stimulation of the HSP70 promoter, and reduction of negative supercoils on polynucleosome plasmid minichromosomes (Côté et al. 1994; Imbalzano et al. 1994, 1996; Kwon et al. 1994; Brown et al. 1996; Cairns et al. 1996; Owen-Hughes et al. 1996; Logie and Peterson 1997; Utley et al. 1997).

The ATPase of the helicase-related SWI2/SNF2 gene product is critical for function of the yeast SWI/SNF complex (Côté et al. 1994) and has been shown to have DNA-stimulated ATPase activity on its own (Laurent et al. 1993). This protein is the prototypical member of a large family of ATPases involved in gene regulation, DNA repair, cell cycle control, and many other functions (Eisen et al. 1995). The *Drosophila* protein ISWI, which is homologous to SWI2/SNF2 only over its ATPase domain, is present in a second family of chromatin-remodeling complexes: ACF, CHRAC, and NURF. Each of the three complexes has a distinct subset of chromatin-remodeling abilities, including disruption on arrays and/or mononucleosome cores, restriction enzyme accessibility, and/or spacing of nucleosomes during or after chromatin assembly (for review, see Cairns 1998). The ATPase of each of the ISWI-containing complexes is stimulated by nucleosomes, but not bare DNA, which contrasts with the SWI2/SNF2 homologs, whose ATPase activity is stimulated by both substrates.

The various in vivo and in vitro effects of the chromatin-remodeling complexes suggest that each complex might be specialized for different purposes, perhaps by differential recognition of transcription factors, chromatin-associated proteins or modified chromatin components, and/or by variations on their mechanisms of action. The roles attributable to known complexes are just beginning to be discovered, and there are a large number of SWI2/SNF2 homologs from yeast to humans for which chromatin disruption has not yet been tested. Thus, it appears likely that cells make use of a battery of specialized chromatin-remodeling complexes to carry out specific functions.

Nonetheless, the conservation of the central region of the critical ATPase suggests that these complexes might all perform a single, central enzymatic function that is targeted, modified, or elaborated on by amino- or carboxy-terminal regions of the ATPase or other complex constituents. Despite the homology of the ATPase domain to bacterial helicases, no DNA-unwinding activity has been detected for SNF2/SWI2 or yeast SWI/SNF (Laurent et al. 1993; Côté et al. 1994). Models for how these complexes might function have included weakening the interactions of H2A/H2B dimers with DNA and the H3/H4 tetramer (Côté et al. 1994) or tracking along DNA to disrupt or reposition nucleosomes (Pazin and Kadonaga 1997).

We have used purified human SWI/SNF preparations to address the mechanism by which these complexes disrupt chromatin. The observation that disruption is stable following depletion of ATP from the reaction (Imbalzano et al. 1996) suggested either the existence of a stable disrupted product or a disrupted-nucleosome–SWI/SNF complex. Here we review our data which show that SWI/SNF acts enzymatically to create a stably disrupted mononucleosome core and to reconvert this product back to the nucleosomal base state. We also present data to suggest that the stably disrupted state is a natural alternative form of the nucleosome. We compare these experiments to recent results on other SWI/SNF family members and discuss their implications for the mechanism of action and potential biological functions of SWI/SNF.

METHODS

Proteins and chromatin templates. Human SWI/SNF was affinity-purified from cell lines containing a FLAG-tagged version of the Ini1 subunit (Sif et al. 1998), or purified by standard chromatography to yield A and B fractions of about 10% purity (Kwon et al. 1994). Bulk mononucleosome cores were purified from HeLa cells by sequential micrococcal nuclease digestions and glycerol gradient centrifugations (Schnitzler et al. 1998). Mononucleosome cores were assembled by salt dialysis onto a labeled TPT (or TPTGal41XX) template and purified by glycerol gradient centrifugation (Imbalzano et al. 1996; Schnitzler et al. 1998). The "dimer" species ana-

lyzed in Figure 5 is a common minor product of this assembly that runs faster than cores on the gradient. Plasmid chromatin was assembled onto the internally labeled closed, circular plasmid pG_5HC_2AT in a reaction containing heat-treated *Xenopus* extract assembly factors and topoisomerase I and was purified by glycerol gradient centrifugation (Imbalzano et al. 1996).

Remodeling assays. SWI/SNF remodeling assays on mononucleosomes were performed at 30°C for the indicated times and either loaded directly onto nucleoprotein gels (EMSA) or treated with DNase I and prepared for denaturing electrophoresis as described previously (Imbalzano et al. 1996). Reactions were scaled up or down in size while maintaining the concentrations of all buffer components and were all adjusted to 60 mM KCl, except where noted. Micrococcal nuclease digestion was performed as described previously (Schnitzler et al. 1998). For remodeling of plasmid chromatin, similar assays that also contained wheat germ topoisomerase I were stopped with SDS/EDTA and deproteinated with proteinase K (Imbalzano et al. 1996) before separation on a 1.5% agarose TBE gel.

Isolation and analysis of novel species. The novel species was separated from undisrupted cores and SWI/SNF by one of two methods. (1) A 100-µl disruption reaction containing 4.4 µg of tagged hSWI/SNF and 0.9 µg of bulk cores and 9 ng of labeled TPT mononucleosome cores was treated for 2 minutes at room temperature with 0.6 units of DNase and separated by EMSA. Bands corresponding to novel core positions were cut out, and proteins were eluted as described previously (Schnitzler et al. 1998). Isolated bands were brought to equal volumes with elution buffer, phenol-extracted, ethanol-precipitated, and prepared for denaturing electrophoresis. (2) A similar disruption reaction was adjusted to 240 mM KCl and separated by glycerol gradient centrifugation in buffers containing 180 mM KCl (Schnitzler et al. 1998). For two-dimensional analysis, a large-scale disruption reaction was separated by EMSA, and the whole lane was cut out and placed above the stacking gel of a 14% SDS-PAGE gel (Schnitzler et al. 1998).

RESULTS

Remodeling of Nucleosomes Is Stable in the Absence of Association with SWI/SNF

ATP is required for hSWI/SNF to disrupt chromatin; however, disruption is stable for many hours after ATP is depleted from the reaction (Imbalzano et al. 1996). To determine whether this is due to an independent stably disrupted nucleosomal product or a disrupted-nucleosome–SWI/SNF complex, we developed conditions that prevented SWI/SNF from interacting with nucleosomes. We found that the coimmunoprecipitation of nucleosomes and SWI/SNF by antibodies to BRG1 was blocked by either competitor DNA or a KCl concentration greater than approximately 180 mM. Furthermore, these treatments cause release of coimmuno-

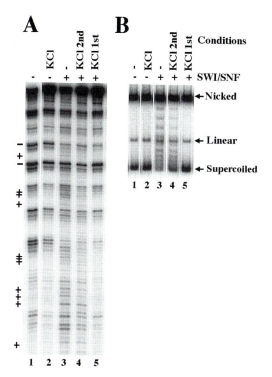

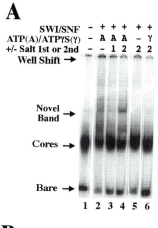

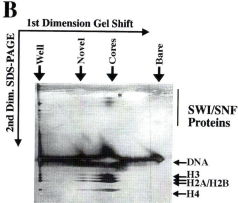

Figure 1. Remodeling remains after SWI/SNF binding to nucleosomes is disrupted. (*A*) 25-μl disruption reactions contained 180 ng of SWI/SNF B fraction, 360 ng of bulk mononucleosome cores, and 0.3 ng of labeled TPT cores. KCl was added to 150 mM either before SWI/SNF (1st) or after 40 min of incubation (2nd). At 60 min, reactions were DNase-treated and prepared for electrophoresis. (*B*) 12.5-μl disruption reactions contained 2.5 ng of labeled plasmid chromatin, 100 ng of immunopurified tagged SWI/SNF fraction, and 1.5 units of Topo I. KCl was adjusted to 220 mM either before SWI/SNF (1st) or after 30 min (2nd). Reactions were stopped at 60 min.

Figure 2. A novel nucleosome species is created by SWI/SNF. (*A*) Mononucleosome cores assembled with a TPT fragment were incubated with or without 350 ng of SWI/SNF A fraction in 12.5-μl remodeling reactions in the presence of ATP (A), ATPγS (γ), or water (−). KCl concentration was adjusted from 60 mM to 180 mM either before (1) or 20 min after (2) the addition of SWI/SNF to the reactions. At 40 min, reactions were separated by EMSA. (*B*) A 125-μl remodeling reaction with 3 mM MgCl₂, no BSA, and containing 1.8 μg of bulk mononucleosome cores and 2.47 μg of affinity-purified hSWI/SNF, was challenged with 240 mM KCl after 55 min of incubation. At 70 min, the reaction was separated by EMSA (1st dimension), and the excised whole lane was run into a 14% SDS-PAGE gel (2nd dimension), which was then silver stained. Asterisk indicates bands that comigrate with cores, but appear in reactions without SWI/SNF. (Adapted, with permission, from Schnitzler et al. 1998 [copyright Cell Press].)

precipitated cores from pellets (Schnitzler et al. 1998; G. Schnitzler and R. Kingston, unpubl.).

Incubation of mononucleosomes with hSWI/SNF and ATP disrupts the regular 10-bp periodicity on a rotationally phased mononucleosome core, resulting in characteristic enhancements and protections (e.g., Fig. 1A, lanes 1 and 3). Incubation of hSWI/SNF with a plasmid that has been assembled into nucleosomes results in a striking loss of the negative supercoils that result from the wrapping of DNA around a standard nucleosome (e.g., Fig. 1B, lanes 1 and 3). As would be expected for conditions that break SWI/SNF-nucleosome interactions, when salt is added to 150 or 220 mM before SWI/SNF, no disruption is seen in either assay (Fig. 1, lanes 5). However, if salt is added after disruption has occurred (lanes 4) the disrupted state is maintained. Similar results are seen when competitor DNA is used (G. Schnitzler and R. Kingston, unpubl.). In the supercoiling assay, the topoisomerase I used to relax supercoils not constrained by the chromatin is not inhibited by salt or DNA competitor at the concentrations used (G. Schnitzler and R. Kingston, unpubl.).

These experiments are consistent with the existence of an altered nucleosomal product that is stable in the absence of association with SWI/SNF. Similar experiments with

the yeast SWI/SNF and RSC complexes have reached the same conclusion (Côté et al. 1998; Lorch et al. 1998).

SWI/SNF Catalyzes the Formation of an Altered Nucleosomal Product

When reactions are analyzed by native gel electrophoresis, SWI/SNF often causes the nucleosome probe to shift to the wells in a smear, consistent with binding of the complex to the cores. When SWI/SNF disruption reactions are adjusted to high salt before electrophoresis, the reaction products can be resolved into two bands, one running at the position of untreated nucleosomes and a second "novel" band (Fig. 2A, lane 4). The novel band is seen only under conditions that generate remodeling of nucleosome structure, not if salt or competitor is added first, and in the presence of both SWI/SNF and ATP, but not ATPγS (Fig. 2A) (G. Schnitzler and R. Kingston, unpubl.), consistent with its being a product of the ATP-dependent remodeling reaction catalyzed by SWI/SNF.

To examine the composition of the novel species, a scaled-up disruption reaction was separated by EMSA and run in a second dimension by SDS-PAGE. Control nucleosomes and the band running the size of nucleosomes in disruption reactions contain DNA and the four core histones as expected. The novel species also contains only DNA and the four core histones in apparently correct stoichiometry (Fig. 2B) (Schnitzler et al. 1998). The only detectable SWI/SNF proteins are seen in the well. Thus, the novel species is not a factor-bound DNA, but an altered nucleosome with a full complement of histones. Further analysis of this species demonstrated that it has the mass, peptide composition, and DNA composition of a dinucleosome, suggesting that it results from the noncovalent association of two remodeled nucleosomes (Schnitzler et al. 1998). The ability of the yeast RSC complex to create an altered nucleosome with similar properties has recently been shown (Lorch et al. 1998).

To detect the remodeled nucleosomes in this protocol, a ratio of approximately 4 nucleosomes to 1 SWI/SNF molecule was used, and this generated both near-maximal percentage of the novel band and near-maximal disruption (G. Schnitzler and R. Kingston, unpubl.). We also see equivalent disruption at more than a 30:1 ratio of nucleosomes to SWI/SNF (Fig. 1A) (Schnitzler et al. 1998; G. Schnitzler and R. Kingston, unpubl.), demonstrating that human SWI/SNF complexes can act catalytically to form this species. The ability of SWI/SNF to function catalytically has also been shown for the yeast SWI/SNF complex on polynucleosome arrays (Logie and Peterson 1997) and for the ISWI-containing complexes ACF and NURF (for review, see Cairns 1998).

SWI/SNF Creates a Dynamic Equilibrium between Nucleosomal Forms

We wanted to understand why we were never able to generate more than approximately 25% novel species in SWI/SNF disruption reactions. The isolated novel species was stable for weeks at 4°C and hours at 30°C (G. Schnitzler and R. Kingston, unpubl.), suggesting that it was not a metastable product that automatically converted back to cores. One possibility was that SWI/SNF could also catalyze a reverse reaction to produce standard cores when starting with the remodeled species. We found that in the presence of SWI/SNF and ATP, the isolated novel species was converted back to cores to generate about the same ratio of cores:novel as in reactions which started with cores (Fig. 3A). These data demonstrate that SWI/SNF can interconvert nucleosomes between a standard state and a remodeled state in an ATP-dependent reaction. Similar experiments with the RSC complex demonstrated that it could also perform this reverse reaction (Lorch et al. 1998).

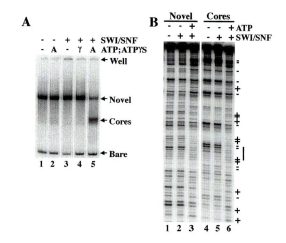

Figure 3. SWI/SNF interconverts novel species and normal cores. (*A*) 2.5 μl of gradient-isolated novel species (see Schnitzler et al. 1998) was incubated in 12.5 μl of remodeling reactions containing 3 mM MgCl₂, 25 ng of cold nucleosome cores, and, where indicated, ATP or ATPγS, and 34 ng of tagged SWI/SNF fraction, for 60 min before EMSA. (*B*) DNase: The same novel-species fraction or equal amounts of labeled cores were analyzed in 125-μl remodeling reactions with 248 ng of bulk mononucleosome cores and, where indicated, 544 ng of tagged SWI/SNF fraction and/or ATP. After 30 min at 30°C, reactions were DNase-treated and analyzed by standard methods. Bar indicates approximate position of the nucleosome dyad axis. (Reprinted, with permission, from Schnitzler et al. 1998 [copyright Cell Press].)

Changes to the DNA Path on the Surface of Altered Nucleosomes

Remodeling of nucleosome structure by the SWI/SNF family of complexes does not appear to require removal of any of the histones from the DNA either during or following the reaction (Côté et al. 1994; Imbalzano et al. 1996; Owen-Hughes et al. 1996; Logie and Peterson 1997; Utley et al. 1997; Schnitzler et al. 1998; and see above). However, many experiments suggest dramatic alterations to the DNA path on the surface of the histone core as a result of SWI/SNF function. In reactions containing any of the SWI/SNF-family of complexes and ATP, DNase accessibility is altered and the affinity of many transcription factors is increased (Côté et al. 1994; Imbalzano et al. 1994; Kwon et al. 1994; Cairns et al. 1996; Utley et al. 1997). Yeast SWI/SNF has also been shown to increase the accessibility of nucleosomal arrays to restriction enzyme digestion (Logie and Peterson 1997).

Analysis of the novel species may explain the origin of some of these SWI/SNF-dependent changes in DNA accessibility. When an isolated novel species is digested with DNase, it generates a pattern quite different from that of untreated cores (Fig. 3B, lane 1). When either cores or novel species are treated with SWI/SNF and ATP, an identical "disrupted" pattern is generated (Fig. 3B, lanes 3 and 6). Most interestingly, this disrupted pattern appears to be a mixture of the pattern seen for novel species and untreated cores. Thus, the remodeled state, as judged from DNase access in reactions with SWI/SNF, appears to reflect the SWI/SNF-catalyzed equilibrium between cores and remodeled species.

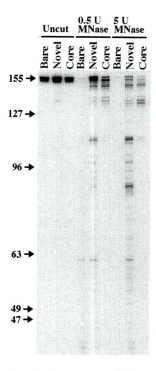

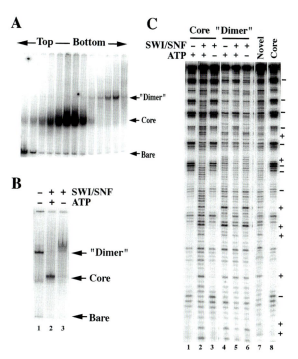

Figure 4. Novel species is more accessible to micrococcal nuclease. Gradient-isolated novel species (TPTGal41-XX), untreated control cores, or bare DNA was digested with the indicated amount of micrococcal nuclease, then treated as per footprint reactions. Positions (in bp) of the uncut and cut species are indicated from the *Mlu*I (labeled) end. (Reprinted, with permission, from Schnitzler et al. 1998 [copyright Cell Press].)

Figure 5. "Dimer" from nucleosome assembly is similar to novel species. (*A*) Glycerol gradient fractions from the salt-dilution assembly of TPT fragment into cores were assayed on a nucleoprotein gel. (*B*) 5 µl of 25-µl reactions containing about 0.2 ng of TPT dimer species, 174 ng of bulk mononucleosomes, 2.5 mM MgCl₂, and, where indicated, 180 ng of SWI/SNF B were separated on a nucleoprotein gel after about 30 min. All reactions contained ATP initially, but this was hydrolyzed before SWI/SNF addition by 0.5 units of Apyrase in "–ATP" lanes. (*C*) The remaining 20 µl of the three reactions above (lanes *4–6*) and otherwise identical control reactions containing 0.3 ng of TPT cores (lanes *1–3*) were treated with DNase and analyzed by denaturing PAGE. Lanes *7* and *8* are gel-isolated TPT novel species and control cores, respectively. To generate these, scaled-up disruption reactions were treated with DNase and separated by EMSA. "Cores" and "novel" were eluted from the appropriately shifted bands in reactions lacking or containing SWI/SNF, respectively. The isolated novel species is expected to have some contamination of cores due to the conversion of DNase-cut cores to novel during the 2-min DNase digestion.

The novel species also has an increased affinity for GAL4, which may explain the increased affinity of GAL4 for nucleosomes in reactions containing active SWI/SNF (Kwon et al. 1994; Imbalzano et al. 1996; Schnitzler et al. 1998). It also has altered accessibility to restriction enzymes (Schnitzler et al. 1998). In addition, the accessibility of micrococcal nuclease (Fig. 4) is altered on the novel species. Micrococcal nuclease is a double-stranded endonuclease that cannot normally cut DNA in a nucleosome, and only trims in from the ends, yielding an approximately 140-bp product. Strikingly, the novel species is digested at several sites within the span of DNA protected by the normal core. These cuts suggest the presence of areas where DNA-histone contacts are very weak or even where the DNA is looped away from the surface of the octamer.

A Nucleosomal Species Similar to "Novel" Is Generated in In Vitro Assemblies

If the disrupted form of the octamer is stable, why is it not observed more frequently? We find that in a normal salt-dilution nucleosome assembly, there is a small amount of a product, which we refer to as a "dimer," that runs at the size of the remodeled species on a glycerol gradient (Fig. 5A). Like the novel species generated by

SWI/SNF, this dimer species can be converted to nucleosome cores by SWI/SNF and ATP (Fig. 5B). It has a DNase digestion pattern similar to that of gel-isolated novel species (Fig. 5C, compare lanes 4 and 7). When incubated with SWI/SNF and ATP, the DNase pattern is very similar to that of normal cores incubated with SWI/SNF, and seemingly intermediate between the pattern of cores and dimer (compare lanes 2 and 5 to lanes 1 and 4). These results suggest that the novel species created by SWI/SNF is a stable alternative conformation that can arise during the assembly of a nucleosome in a salt-dilution protocol. The fact that it is far less prevalent than normal cores after assembly, however, suggests that it is a somewhat less favorable conformation. Noncovalent dimers of mononucleosomes, similar in mobility to that of the novel and/or dimer species observed here, have been observed as a result of freeze-drying or long storage (Chiu et al. 1980).

DISCUSSION

A Model for SWI/SNF Function

The results from recent experiments on human SWI/SNF, yeast SWI/SNF, and RSC are in general agreement on the following aspects of SWI/SNF-complex function: (1) Remodeling of nucleosome structure does not require removal of any of the histone core proteins; (2) SWI/SNF acts catalytically; (3) remodeling significantly alters the path of the DNA; (4) SWI/SNF catalyzes the formation of an altered species with the same components as the standard nucleosome; and (5) SWI/SNF can also convert this product back to a standard core, facilitating a dynamic equilibrium between the two forms (see experiments and references above).

These observations lead to the following hypothetical model for the function of the SWI/SNF family of complexes. We propose that the nucleosome can adopt two different conformations that have different contact points between the histone octamer core and the DNA. The standard nucleosome core is somewhat more energetically favorable of these two alternative conformations (see Fig. 5) (Lorch et al. 1998) and may be formed preferentially by chromatin assembly factors. The SWI/SNF family of complexes can use the energy of ATP hydrolysis to effectively lower the energy barrier between these two forms. SWI/SNF may do nothing to affect the intrinsic equilibrium between these two forms, but may simply act to greatly enhance the rate of exchange between them. Whether this hypothetical mechanism is unique to the SWI/SNF family or can be generalized to include the ISWI-containing remodeling complexes is a critical unresolved issue.

Biological Implications of the Model

An important direct prediction from this model indicates how SWI/SNF complexes might function in vivo. If SWI/SNF is active in the presence of factors, such as transcription factors, that bind more readily to the remodeled structure than to the standard structure of the nucleosome, this binding will drive the reaction toward the remodeled structure based on equilibrium considerations (Fig. 6). Similarly, if SWI/SNF is active in the presence of factors that bind more readily to the standard structure, this binding will drive the reaction toward the standard nucleosome structure.

The key point is that SWI/SNF action can drive an array of standard nucleosomes toward a remodeled state, but it can also drive an array of remodeled nucleosomes toward a standard state in the presence of factors that preferentially bind to the standard state. Thus, the direction that SWI/SNF will function in will be determined by the starting condition of a nucleosomal array and by the local environment of regulatory molecules (Fig. 6). In essence, SWI/SNF would create a rapid equilibrium between two different nucleosomal states that will be "fixed" by other factors that serve to drive the reaction either to the standard state or to the remodeled state.

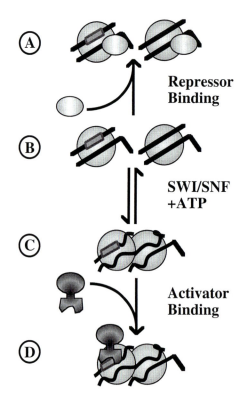

Figure 6. Model for SWI/SNF function in vivo. SWI/SNF and ATP interconvert the pairs of nucleosomes between two stable states: normal cores (*B*) and altered novel species (*C*). Factors that bind preferentially to one form or the other would tip the reaction to one side or the other. Thus, the preferential binding of activators to the novel species would act to stabilize the disrupted state (*D*), whereas factors that preferentially bind to the normal octamer (candidates for which might include H1 or SSN6/TUP1 complexes) would act to stabilize the normal state (*A*).

Knowing whether this model truly describes the situation within cells will require considerable additional work. A few studies, however, have shown that under certain conditions, nucleosomes can be disrupted with regard to nuclease digestion, yet the histones appear to remain associated (Lee and Garrard 1991; Mymryk and Archer 1995). These disrupted nucleosomes are in active transcription units or associated with activated transcription factors, conditions which might effectively stabilize the disrupted state. In the other direction, repression by putative chromatin-coating proteins such as SSN6/TUP1 or SPT4/6 in yeast results in highly ordered nucleosomes with strong micrococcal nuclease boundaries, which appear to be inconsistent with the presence of altered nucleosomes (for review, see Hartzog and Winston 1997).

Nature of the Disrupted Species: Dimerization

What is the disrupted species? Our results indicate that it is two separate nucleosome cores that have been altered and dimerized. It is not a normal dinucleosome, since the DNA has not been ligated and it can be converted back to mononucleosomes by SWI/SNF. To form this species,

DNA from one nucleosome might partially wrap around a second to form bridges. Alternatively, the DNA could remain associated and a dimer could form through histone-histone contacts. SWI/SNF might be a dimeric complex that can bind two nucleosomes at once and generate the double-nucleosomal disrupted product in a concerted action (Schnitzler et al. 1998). An alternative possibility is that SWI/SNF generates the disrupted mononucleosomes that subsequently dimerize to give the stable disrupted form. We have been unable to detect a disrupted mononucleosome (after disruption, the species running as a mononucleosome on gels, gradients, and gel filtration has the DNase digestion pattern or normal cores). This does not rule out the potential existence of such a species, however. The disrupted mononucleosome might either have a high affinity for itself and dimerize rapidly and/or it could be metastable and revert to cores rapidly unless it is dimerized.

Potential Role of ATP Hydrolysis

The model proposed above for SWI/SNF action leads to a suggestion for how the energy of ATP hydrolysis might be involved. Generally, the coupling of ATP hydrolysis makes an enzyme work unidirectionally, since going the other direction requires the highly unfavorable reconversion of ADP and Pi to ATP. For SWI/SNF, the immediate product after ATP hydrolysis may be a metastable state that then decays stochastically into either the normal nucleosome core or novel species. In this manner, SWI/SNF would act to reduce the activation barrier between two relatively stable forms of the octamer. This is similar to the proposed mechanism for some ATP-dependent molecular chaperones (for review, see Hartl 1996), although without the added preference for action on one product versus the other.

Effect of Novel Species in Polynucleosomal Arrays

Our study of mononucleosomes has allowed an analysis of the basic disrupted species, but it has left open the question of whether the novel species exists on, and is the cause of disruption effects seen on, polynucleosome templates. The following data, however, suggest that this may be the case. First, disruption of plasmid chromatin is stable after removal of ATP from the reaction or breaking of SWI/SNF-nucleosome interactions with competitors or high salt (Fig. 1) (Imbalzano et al. 1996). As for mononucleosomes, these results demonstrate the presence of SWI/SNF-independent stable disrupted state, which is likely to be due to an accumulation of altered dinucleosomes similar to those seen with cores. Second, if the novel species does exist on plasmid chromatin, it helps to explain the curious observation that even at very high concentrations of SWI/SNF, a wide range of topoisomers between relaxed and fully negatively supercoiled (normal chromatin) is seen, with most species being around the middle (Fig. 1) (Kwon et al. 1994; Imbalzano

et al. 1996). A dynamic equilibrium between novel and normal nucleosome species would result in each plasmid harboring a mixture of normal and disrupted nucleosomes, with the relative amount of each (and hence amount of supercoiling) being a random distribution around the mean.

Reduced Supercoiling and the Novel Species

The most intriguing effect of SWI/SNF disruption of plasmid chromatin is the reduction of negative supercoil itself. In the normal nucleosome core, the histone octamer is arranged as a left-handed gyre, and the wrapping of DNA around it results in the generation of a single, constrained, negative supercoil. Human SWI/SNF does not strip histones from mononucleosomes, and yeast SWI/SNF does not normally do this to either mononucleosomes or arrays of nucleosomes (see references above). Thus, the loss of negative supercoils is probably not due to loss of octamers. Rather, it appears to be due to accumulation of disrupted products that do not create the normal negative supercoil. Despite altered sensitivity, DNA in the novel species is still much more protected from nuclease cutting than bare DNA. Thus, extensive DNA unwrapping also cannot account for the loss of supercoiling. One possible explanation stems from studies of nucleosome assembly on small, positively or negatively supercoiled minicircles. Octamers and H3/H4 tetramers assemble readily on minicircles with −1 superhelicity, since the negatively supercoiled final state makes this energetically favorable. Tetramers, however, were also shown to assemble onto minicircles with +1 superhelicity, indicating that the tetramer could form a right-handed gyre (Hamiche et al. 1996). Although octamers would not form on +1 circles by this salt-dilution technique, it is interesting to speculate that the action of SWI/SNF and ATP might effect such a change to generate an octamer with positive or neutral superhelicity.

Potential for Nucleosome Repositioning

If SWI/SNF does catalyze an interconversion between the standard nucleosome structure and a remodeled conformation on polynucleosomes, one possible mechanistic "by-product" of this reaction is increased mobility of the histone octamer relative to DNA. Since the remodeled state has an altered DNA path relative to the normal state, the intermediate between these two states presumably has loosened DNA contacts. These loosened contacts might promote effective "sliding" of the octamer along the DNA. SWI/SNF function is catalytic and appears to have a reasonably rapid turnover number, as judged from the rapid equilibrium that can be established between standard and remodeled states. Thus, one important result of SWI/SNF function on arrays might be dramatically increased movement of nucleosomes along the array. Such a function is consistent with observations that two of the ISWI-containing ATP-dependent remodeling complexes, ACF and CHRAC, can facilitate movement of nucleosomes surrounding DNA-binding factors and nucleosome spacing (for review, see Cairns 1998).

An ability of SWI/SNF to effectively slide octamers would help to explain an apparent inconsistency in the mechanism of SWI/SNF complexes. Whereas in all assays on mononucleosomes and some on polynucleosome templates disruption is stable after removal of ATP and/or SWI/SNF, one assay clearly shows an opposite result. The increased accessibility of restriction enzymes to an array of 5S nucleosomes by yeast SWI/SNF requires continuous ATP hydrolysis (Logie and Peterson 1997). If SWI/SNF promoted sliding of octamers at the same time it facilitated the interchange between novel and normal cores, it would result in the effective movement of the bare linker DNA between nucleosomes, uncovering restriction sites that were normally inaccessible a fraction of the time. If ATP was depleted, repositioning would stop and any site now covered by either a normal or novel nucleosome (since the novel species still impairs restriction enzyme access to DNA; Schnitzler et al. 1998) would be much less accessible.

CONCLUSIONS

Recently, a great deal of mechanistic information about chromatin-remodeling complexes has been discovered. These findings have led us to propose that SWI/SNF functions catalytically to interconvert two different conformations of the nucleosome core. This indicates that histones have evolved to allow the formation of more than one nucleosomal state and that the transition between these states is an important aspect of chromatin regulation. Information on whether the altered conformation of the nucleosome that has been detected in vitro is also found in vivo is limited, however. An increased understanding of the structure of remodeled nucleosomes in vitro and in vivo will allow more precise models for the mechanisms that regulate chromatin structure in a dynamic manner in vivo.

ACKNOWLEDGMENTS

We thank A. Imbalzano for construction of the pTPT series of templates and for comments, and Geeta Narlikar and members of the laboratory for discussion and comments. This work was supported by grants from the National Institutes of Health (R.E.K. and S.S.). G.R.S. is a Helen Hay Whitney Fellow.

REFERENCES

Brown S.A., Imbalzano A.N. ,and Kingston R.E. 1996. Activator-dependent regulation of transcriptional pausing on nucleosomal templates. *Genes Dev.* **10:** 1479.

Cairns B.R. 1998. Chromatin remodeling machines: Similar motors, ulterior motives. *Trends Biochem. Sci.* **23:** 20.

Cairns B.R., Lorch Y., Li Y., Zhang M., Lacomis L., Erdjument B.H., Tempst P., Du J., Laurent B., and Kornberg R.D. 1996. RSC, an essential, abundant chromatin-remodeling complex. *Cell* **87:** 1249.

Chiba H., Muramatsu M., Nomoto A., and Kato H. 1994. Two human homologues of *Saccharomyces cerevisiae* SWI2/SNF2

and *Drosophila* Brahma are transcriptional coactivators cooperating with the estrogen receptor and the retinoic acid receptor. *Nucleic Acids Res.* **22:** 1815.

Chiu S.S., Lee K.P., and Lewis P.N. 1980. Rearrangements and reconstitution of mononucleosomes as monitored by thermal denaturation measurements. *Can. J. Biochem.* **58:** 73.

Côté J., Peterson C.L., and Workman J.L. 1998. Perturbation of nucleosome core structure by the SWI/SNF complex persists after its detachment, enhancing subsequent transcription factor binding. *Proc. Natl. Acad. Sci.* **95:** 4947.

Côté J., Quinn J., Workman J.L., and Peterson C.L. 1994. Stimulation of GAL4 derivative binding to nucleosomal DNA by the yeast SWI/SNF complex. *Science* **265:** 53.

Dunaief J.L., Strober B.E., Guha S., Khavari P.A., Ålin K., Luban J., Begemann M., Crabtree G.R., and Goff S.P. 1994. The retinoblastoma protein and BRG1 form a complex and cooperate to induce cell cycle arrest. *Cell* **79:** 119.

Eisen J.A., Sweder K.S., and Hanawalt P.C. 1995. Evolution of the SNF2 family of proteins: Subfamilies with distinct sequences and functions. *Nucleic Acids Res.* **23:** 2715.

Fryer C.J. and Archer T.K. 1998. Chromatin remodeling by the glucocorticoid receptor requires the BRG1 complex. *Nature* **393:** 88.

Gavin I.M. and Simpson R.T. 1997. Interplay of yeast global transcriptional regulators Ssn6-Tup1 and Swi-Snf and their effect on chromatin structure. *EMBO J.* **16:** 6263.

Hamiche A., Carot V., Alilat M., De Lucia F., O'Donohue M.-F., Revet B., and Prunell A. 1996. Interaction of the histone (H3-H4)$_2$ tetramer of the nucleosome with positively supercoiled DNA minicircles: Potential flipping of the protein from a left- to a right-handed superhelical form. *Proc. Natl. Acad. Sci.* **93:** 7588.

Hartl F.U. 1996. Molecular chaparones in cellular protein folding. *Nature* **381:** 571.

Hartzog G.A. and Winston F. 1997. Nucleosomes and transcription: Recent lessons from genetics. *Curr. Opin. Genet. Dev.* **7:** 192.

Hirschhorn J.N., Brown S.A., Clark C.D., and Winston F. 1992. Evidence that SNF2/SWI2 and SNF5 activate transcription in yeast by altering chromatin structure. *Genes Dev.* **6:** 2288.

Ichinose H., Garnier J.-M., Chambon P., and Losson R. 1997. Ligand-dependent interaction between the estrogen receptor and the human homologues of SWI2/SNF2. *Gene* **188:** 95.

Imbalzano A.N., Schnitzler G.R., and Kingston R.E. 1996. Nucleosome disruption by human SWI/SNF is maintained in the absence of continued ATP hydrolysis. *J. Biol. Chem.* **271:** 20726.

Imbalzano A.N., Kwon H., Green M.R., and Kingston R.E. 1994. Facilitated binding of TATA-binding protein to nucleosomal DNA. *Nature* **370:** 481.

Khavari P.A., Peterson C.L., Tamkun J.W., and Crabtree G.R. 1993. BRG1 contains a conserved domain of the SWI2/SNF2 family necessary for normal mitotic growth and transcription. *Nature* **366:** 170.

Kwon H., Imbalzano A.N., Khavari P.A., Kingston R.E., and Green M.R. 1994. Nucleosome disruption and enhancement of activator binding by a human SWI/SNF complex. *Nature* **370:** 477.

Laurent B.C., Treich I., and Carlson M. 1993. The yeast SNF2/SWI2 protein has DNA-stimulated ATPase activity required for transcriptional activation. *Genes Dev.* **7:** 583.

Lee M.-S. and Garrard W.T. 1991. Transcription-induced nucleosome 'splitting': An underlying structure for DNase I sensitive chromatin. *EMBO J.* **10:** 607.

Logie C. and Peterson C.L. 1997. Catalytic activity of the yeast SWI/SNF complex on reconstituted nucleosome arrays. *EMBO J.* **16:** 6772.

Lorch Y., Cairns B.R., Zhang M., and Kornberg R.D. 1998. Activated RSC-nucleosome complex and persistently altered form of the nucleosome. *Cell* **94:** 29.

Matallana E., Franco L., and Perez-Ortin J.E. 1992. Chromatin structure of the yeast *SUC2* promoter in regulatory mutants. *Mol. Gen. Genet.* **231:** 395.

Mymryk J.S. and Archer T.K. 1995. Dissection of progesterone receptor-mediated chromatin remodeling and transcriptional activation in vivo. *Genes Dev.* **9:** 1366.

Owen-Hughes T., Utley R.T., Côté J., Peterson C.L., and Workman J.L. 1996. Persistent site-specific remodeling of a nucleosome array by transient action of the SWI/SNF complex. *Science* **273:** 513.

Pazin M.J. and Kadonaga J.T. 1997. SWI2/SNF2 and related proteins: ATP-driven motors that disrupt protein- DNA interactions? *Cell* **88:** 737.

Ryan M.P., Jones R., and Morse R.H. 1998. SWI-SNF complex participation in transcriptional activation at a step subsequent to activator binding. *Mol. Cell. Biol.* **18:** 1774.

Schnitzler G., Sif S., and Kingston R.E. 1998. Human SWI/SNF interconverts a nucleosome between its base state and a stable remodeled state. *Cell* **94:** 17.

Sif S., Stukenberg P.T., Kirschner M.W., and Kingston R.E. 1998. Mitotic inactivation of human SWI/SNF chromatin remodeling complex. *Genes Dev.* **12:** 2842.

Singh P., Coe J., and Hong W. 1995. A role for retinoblastoma protein in potentiating transcriptional activation by the glucocorticoid receptor. *Nature* **374:** 562.

Strober B.E., Dunaief J.L., Guha S., and Goff S.P. 1996. Functional interactions between the hBRM/hBRG1 transcriptional activators and the pRB family of proteins. *Mol. Cell. Biol.* **16:** 1576.

Sumi I.C., Ichinose H., Metzger D., and Chambon P. 1997. SNF2β-BRG1 is essential for the viability of F9 murine embryonal carcinoma cells. *Mol. Cell. Biol.* **17:** 5976.

Trouche D., Le Chalony C., Muchardt C., Yaniv M., and Kouzarides T. 1997. RB and hbrm cooperate to repress the activation functions of E2F1. *Proc. Natl. Acad. Sci.* **94:** 11268.

Utley R.T., Côté J., Owen-Hughes T., and Workman J.L. 1997. SWI/SNF stimulates the formation of disparate activator-nucleosome complexes but is partially redundant with cooperative binding. *J. Biol. Chem.* **272:** 12642.

Wang W., Xue Y., Zhou S., Kuo A., Cairns B.R., and Crabtree G.R. 1996a. Diversity and specialization of mammalian SWI/SNF complexes. *Genes Dev.* **10:** 2117.

Wang W., Côté J., Xue Y., Zhou S., Khavari P.A., Biggar S.R., Muchardt C., Kalpana G.V., Goff S.P., Yaniv M., Workman J.L., and Crabtree G.R. 1996b. Purification and biochemical heterogeneity of the mammalian SWI-SNF complex. *EMBO J.* **15:** 5370.

Winston F. and Carlson M. 1992. Yeast SNF/SWI transcriptional activators and the SPT/SIN chromatin connection. *Trends Genet.* **8:** 387.

SWI/SNF Complex: Dissection of a Chromatin Remodeling Cycle

C.L. Peterson

Program in Molecular Medicine and Department of Biochemistry and Molecular Biology,
University of Massachusetts Medical Center, Worcester, Massachusetts 01605

The organization of eukaryotic DNA into chromatin necessitates mechanisms to rapidly, and reversibly, unfold or decompact specific loci so that DNA sequences are accessible to enzymes that must "read" the cell's genetic material. Two types of "chromatin-remodeling" enzymes may have evolved to contend with this difficult task. The ATP-dependent remodelers use the energy of ATP hydrolysis to disrupt the structure of nucleosome core particles or to influence the spacing or mobility of nucleosomes within long arrays. The prototype of this class is the yeast SWI/SNF complex, which was initially identified in genetic screens as a positive regulator of transcription (for review, see Peterson and Tamkun 1995). SWI/SNF (and its *Drosophila* and mammalian counterparts) remains the only ATP-dependent remodeling enyzme that is clearly involved in the transcription process in vivo. The second group of chromatin remodelers includes the histone acetyltransferases which acetylate lysine residues in the flexible amino-terminal domains of the histone proteins. The covalent modification of histones by acetyltransferases may cause subtle changes in nucleosome core structure, disrupt or promote the interaction of nonhistone proteins with the chromatin fiber, or control the folding of nucleosome arrays.

A hallmark of chromatin remodelers is that they are complex multisubunit enyzmes. In the case of SWI/SNF, it is composed of 11 different polypeptide subunits and has an apparent native molecular mass of 2 MD. Likewise, the yeast histone acetyltransferase, GCN5, is a subunit of at least two large complexes of 0.8 MD (the ADA complex) and 1.8 MD (the SAGA complex) (Grant et al. 1997; Pollard and Peterson 1997). The enormous sizes of these enzymes have led to their being labeled as chromatin-remodeling "machines." Why do these enzymes have such complex subunit organization? The implication is that the chromatin-remodeling reaction is itself complex, composed of multiple steps or subreactions that must be completed to achieve the "remodeled" state. Each of these steps in the reaction must be defined and dissected in order to understand how a chromatin disruption event is initiated, what exactly the structural features of the remodeling state are, and of course how the remodeling cycle can be regulated by extrinsic or intrinsic signals. In this paper, I use the SWI/SNF complex as a conceptual framework to illustrate the steps in a simple chromatin-remodeling cycle (Fig. 1). In the case of SWI/SNF, each of the four steps that describe the cycle has been identified and each has been studied to some de-

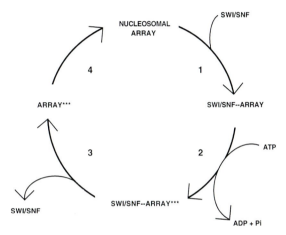

Figure 1. The SWI/SNF remodeling cycle. Steps in the cycle (1–4) are described in detail in the text. Asterisks are used to denote the remodeled state of the array.

gree in isolation from the overall remodeling reaction. In the following sections, I describe in detail what is known about each of these steps and how they might be regulated both in vitro and in vivo. Although much is known about the SWI/SNF remodeling cycle, it is clear that much of the reaction still remains a mystery.

STEP 1: INTERACTION WITH THE NUCLEOSOMAL SUBSTRATE

The SWI/SNF complex is a rare enzyme in yeast, present at only 100–500 copies per cell (Côté et al. 1994; B. Cairns, pers. comm.). The transcriptional defects due to inactivation of SWI/SNF only result in a decrease in transcription of a small subset of yeast genes, indicating that SWI/SNF activity must be targeted to particular genes and to specific nucleosomes in vivo. In the case of GAL4-dependent reporter genes, SWI/SNF dependence does not correlate with promoter strength but requires that the GAL4-binding sites be of low inherent affinity *and* be encompassed by nucleosomes (Fig. 2). Likewise, the SWI/SNF-dependent genes, *HO* and *SUC2*, are characterized by a highly ordered nucleosome arrangement that covers their upstream regulatory regions (Gavin and Simpson 1997; Wu and Winston 1997; J. Krebs and C.L. Peterson, unpubl.). Although SWI/SNF has been suggested to be associated with a RNA polymerase II holoenzyme (Wilson et al. 1996), SWI/SNF can be biochemi-

	Promoter Strength	SWI/SNF Dependence
4 x Low	++++	NO
2 x High	+++	NO
2 x Low	++	YES
2 x Low	++	NO

Figure 2. SWI/SNF-dependent genes. GAL4-dependent reporter genes were assayed for expression in isogenic *SWI+* and *swi−* yeast strains. (*Open boxes*) GAL4-binding sites located either in nucleosome-free regions or assembled into nucleosomes. The low-affinity sites refer to the endogenous GAL4 sites from the *GAL1* locus; the high-affinity sites refer to a synthetic, 17-mer consensus GAL4 site. Data are summarized from Burns and Peterson (1997).

cally separated from the holoenzyme in two different purification schemes (Côté et al. 1994; Cairns et al. 1996). This targeting model remains controversial. Alternatively, targeting of SWI/SNF activity to specific chromosomal loci may involve interactions with upstream activator proteins. Such an interaction has been demonstrated for mammalian SWI/SNF and the glucocorticoid receptor (GR) in vivo (Fryer and Archer 1998) and yeast SWI/SNF and the rat GR in vitro in yeast extracts (Yoshinaga et al. 1992).

SWI/SNF might also be targeted to specific chromosomal loci by a special chromatin structure of the locus. For instance, SWI/SNF may recognize nucleosome arrays that harbor a distinct type of histone posttranslational modification, such as site-specific acetylation or phosphorylation (for discussion, see Pollard and Peterson 1998). The histone acetyltransferase, GCN5, is required for expression of many of the same genes that require SWI/SNF activity, and GCN5 and SWI/SNF show similar genetic interactions with chromatin components (Pollard and Peterson 1997). One possibility is that SWI/SNF is targeted to nucleosome arrays that have been acetylated by GCN5 HAT complexes, or alternatively, the remodeling activity of SWI/SNF might be modulated by histone acetylation (see below).

In vitro studies of this first step in the remodeling cycle have focused on the untargeted interaction of SWI/SNF with DNA (Quinn et al. 1996) or with nucleosomes (Côté et al. 1994, 1998). In all cases examined, this step in the cycle is not influenced by addition of ATP. SWI/SNF binds with nanomolar affinity to DNA fragments larger than 140 bp in a sequence-nonspecific fashion (Quinn et al. 1996). Furthermore, SWI/SNF binds preferentially to synthetic four-way junction DNA, which led to the hypothesis that SWI/SNF might recognize the entry-exit face of a nucleosome (Quinn et al. 1996). SWI/SNF binding to DNA fragments or to four-way junctions is sensitive to low concentrations of distamycin, indicating that DNA binding involves crucial minor groove interactions (Quinn et al. 1996).

Synthetic four-way junction DNAs (see Fig. 3) present several different types of structural features to a DNA-binding protein. First, they contain two short stretches of DNA helix arranged in a crossover "X" structure (Pohler et al. 1994). Second, each pair of junction "arms" mimics a DNA bend with a different bending angle. In the latter case, proteins that bind to four-way junction DNAs via recognition of bent DNA also bind with high affinity to short DNA duplexes that contain one or more mismatched adenosine residues (Fig. 3, bulged substrates) (Bhattacharyya et al. 1991). In the case of SWI/SNF, however, only the four-way junction substrate is bound with high affinity; a variety of short DNA duplexes that contain one to four mismatched adenosine residues (A1–A4 bulged templates) are bound with at least an order of magnitude lower affinity (Fig. 3). Even a Y junction, which is more similar to a four-way junction, is bound with at least fivefold lower affinity (Fig. 3). These results indicate that SWI/SNF prefers to bind to substrates that present two independent DNA duplexes or that contain a DNA crossover. Consistent with the former possibility, electron microscopy visualization of SWI/SNF binding to DNA demonstrates a propensity for SWI/SNF to loop DNA at high frequency (>60% of the imaged molecules; Bazett-Jones et al. 1998). How relevant is the binding of multiple DNA duplexes to the interaction of SWI/SNF with nucleosomes? As DNA wraps

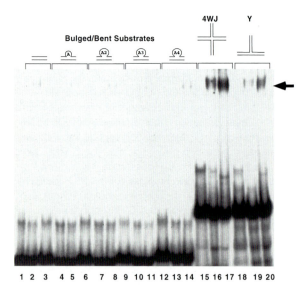

Figure 3. SWI/SNF binds preferentially to four-way-junction DNA. Shown are native gel-retardation assays using purified yeast SWI/SNF and various ^{32}P-labeled DNA probes. Arrow to right of figure denotes SWI/SNF-DNA complexes. Probes used in lanes 1–14 are 40-mer oligonucleotide duplexes that contain the indicated number of mismatched A residues (Bhattacharyya et al. 1991). Four-way junction and Y junction probes were prepared from three or four oligonucleotides, as described previously (Bhattacharyya et al. 1991; Quinn et al. 1996). Gel-retardation assays were performed as describe by Quinn et al. (1996). Reactions in lanes *3, 6, 9, 12, 15,* and *18* contained no SWI/SNF; reactions in lanes *1, 4, 7, 10, 13, 16,* and *19* contained 10 nM SWI/SNF; reactions in lanes *2, 5, 8, 11, 14, 17,* and *20* contained 20 nM SWI/SNF.

twice around the histone octamer, two gyres of DNA are in close apposition (Luger et al. 1997). The close proximity of the two DNA gyres may be one reason why SWI/SNF has a three- to fourfold higher affinity for a DNA fragment when it is incorporated into a nucleosome (Côté et al. 1998).

To determine exactly how SWI/SNF interacts with nucleosomes, we have measured the efficiency with which SWI/SNF subunits can be cross-linked to nucleosomal DNA that has been modified at different positions with photoaffinity DNA probes (for details, see Pruss et al. 1996). Whereas SWI/SNF binds nonspecifically to "free" DNA, SWI/SNF can be efficiently cross-linked to nucleosomal DNA only when the cross-linking nucleotides are positioned about 3.5 helical turns of DNA from the nucleosome edge (Fig. 4) (S.M. D'Cruz et al., unpubl.). This preferred SWI/SNF-binding site overlaps the position at which the amino-terminal domain of histone H2B protrudes between the two DNA gyres (Fig. 4, white dot within outlined box) (Luger et al. 1997), suggesting that SWI/SNF might directly interact with this histone tail. Thus, the first step in the remodeling cycle appears to involve a precise positioning of the nucleosomal substrate. Positioning is likely to involve interactions with the two DNA gyres as well as with the local histone environment (e.g., the H2B amino terminus).

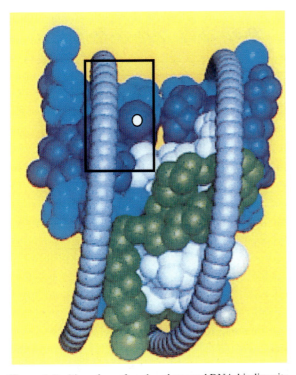

Figure 4. Position of a preferred nucleosomal DNA-binding site for SWI/SNF. The histone octamer is shown wrapped with a 10 Å tube which represents a projection of the DNA helix axis. The plane of the paper contains the dyad axis and entry/exit points of the DNA. (*Open box*) DNA position where SWI/SNF is efficiently cross-linked; (*white dot*) origin of the H2B amino-terminal tail. Histone octamer structure is adapted from Arents and Moudrianakis (1993). (*Green*) Histone H3; (*white*) histone H4; (*light blue*) histone H2A; (*dark blue*) histone H2B.

STEP 2: ATP-DEPENDENT REMODELING

Once SWI/SNF binds to its nucleosomal substrate, it then uses the binding and hydrolysis of ATP to promote nucleosome remodeling. What is meant by the term "remodeling"? In the generic definition, remodeling refers to a change in nucleosome structure or positioning. Operationally, what is meant by remodeling depends on the assay that is used to detect the event. For instance, in the case of SWI/SNF, nucleosome remodeling usually refers to the ATP-dependent increase in accessibility of nucleosomal DNA to DNase I (Côté et al. 1994), restriction enzymes (Logie and Peterson 1997), or DNA-binding transcription factors (Côté et al. 1994; Utley et al. 1997). For example, the affinity of GAL4 for a nucleosomal binding site is increased by at least 30-fold in the presence of SWI/SNF and ATP (Côté et al. 1994). In contrast, a remodeling assay for the *Drosophila* NURF (Tsukiyama and Wu 1995) or ACF (Ito et al. 1997) complexes involves MNase digestion of plasmid minichromosomes and Southern blotting in order to detect a transcription-factor-dependent change in positioning of one nucleosome in the array. In most instances, the same assays have not been used to analyze "remodeling" by different ATP-dependent enzymes.

How SWI/SNF action leads to such dramatic changes in DNA accessibility throughout both gyres of nucleosomal DNA remains largely a mystery. Three models have been proposed in the past 6 years to describe SWI/SNF remodeling. (1) The processive helicase model (Travers 1992) was based on the sequence of the SWI2/SNF2 subunit that contains seven sequence motifs diagnostic of DNA-stimulated ATPases and DNA helicases. This model posits that SWI/SNF moves processively along the chromosomal fiber disrupting chromatin loop topology and nucleosomes due to a strand unwinding (helicase) activity. (2) A similar model has been proposed by Pazin and Kadonaga (1997) in which the complex also moves or tracks along DNA but transfers the histone octamer behind the advancing enzyme, much like the "spooling" model proposed for passage of some RNA polymerases through nucleosomes (Studitsky et al. 1994, 1997). The SWI2/SNF2 subunit (Laurent et al. 1993) and the intact SWI/SNF complex (Côté et al. 1994) do have potent DNA-stimulated ATPase activity; for the intact complex, it can hydrolyze about 1000 ATP molecules per minute. However, SWI/SNF does not exhibit DNA helicase activity on classical helicase substrates (Côté et al. 1994), nor can it unwind four-way junction molecules (C.L. Peterson, unpubl.). Furthermore, incubation of nucleosome core particles with SWI/SNF and ATP does not lead to increased reactivity of nucleosomal DNA to potassium permaganate, indicating that there is no unwinding of DNA during a remodeling reaction (Côté et al. 1998). Several observations indicate that SWI/SNF also does not function as a processive "spooling" or tracking enzyme. First, SWI/SNF action does not appear to remove DNA from the surface of the histone octamer (Côté et al. 1998). Second, SWI/SNF does not remove or destabilize the histone octamer unless five dimers of GAL4 are also bound to the remodeled nucleosome (Owen-Hughes et al. 1996). Finally, the position-dependent binding of nucleosome core particles by SWI/SNF (Fig. 4) predicts that SWI/SNF functions by a more classical enyzme-substrate interaction rather than by tracking along DNA. (3) A third model proposed that SWI/SNF action leads to displacement of one or more histone H2A/H2B dimers (Peterson and Tamkun 1995). However, SWI/SNF can remodel a mononucleosome in which the histones are chemically cross-linked together (Bazett-Jones et al. 1998), and thus this model also does not accurately describe the reaction. It remains a possibility that SWI/SNF action may cause a more subtle disruption of the H2A/H2B dimer–H3/H4 tetramer interface that does not involve histone eviction.

How do we currently view the the ATP-dependent step in the SWI/SNF-remodeling reaction? The only dramatic affect on nucleosome structure appears to be the loosening of histone-DNA interactions (Table 1). When nucleosomes contain photoaffinity DNA probes at different positions on the two DNA gyres, SWI/SNF and ATP lead to a large (>10-fold) decrease in histone-DNA cross-linking efficiency at almost every tested nucleosome position (S.M. D'Cruz et al., unpubl.). Surprisingly, these changes in histone-DNA interactions do not require ATP hydrolysis; decreases in histone-DNA cross-linking efficiencies are induced by either ATPγS (a very slowly hydrolyzed nucleotide derivative) or ADP. In contrast, the enhanced accessibility of nucleosomal DNA to nucleases and DNA-binding proteins does require ATP hydrolysis. Thus, the histone-DNA cross-linking method appears to have identified an intermediate in the remodeling reaction that only requires nucleotide binding. The production of this putative intermediate is not consistent with processive action models, as these models require multiple rounds of ATP hydrolysis for the enzyme to move along DNA. Our current, speculative view is that SWI/SNF first binds and positions the nucleosome core (Fig. 4), and subsequent binding of ATP causes a conformational change in SWI/SNF that leads to a change in the path of nucleosomal DNA on the octamer surface (a path of lowered affinity). Either ATP hydrolysis releases the remodeled substrate for interaction with nucleases or DNA-binding proteins or it leads to further changes in nucleosome structure that are required to create the fully remodeled state.

STEP 3: TERMINATION AND DISSOCIATION

The low abundance of SWI/SNF in vivo suggests that the enzyme must undergo a catalytic remodeling cycle (Fig. 1); i.e., the complex must be able to terminate one

Table 1. Properties of a SWI/SNF-remodeled Nucleosome

- Nucleosomal DNA is more accessible to DNase I, restriction enzymes, and DNA-binding proteins
- Histone-DNA interactions are weakened throughout both DNA gyres
- DNA remains associated with histone octamer surface
- DNA helix is not unwound during the reaction
- Histones are not evicted

remodeling cycle and dissociate from the first target array so that it can relocate to a new target array to begin a new remodeling cycle. We have recently described a novel nucleosome array remodeling assay where we were able to quantify this dissociation reaction step (Logie and Peterson 1997). This assay measures the SWI/SNF-dependent stimulation of the rate of cleavage of a unique restriction site within a well-defined nucleosome array. The DNA template that we have used is composed of 11 head-to-tail repeats of a *L. variegatus* 5S rRNA gene (Fig. 5, top). Each repeat can rotationally and translationally position a nucleosome after in vitro reconstitution with purified histone octamers, yielding a homogeneous array of positioned nucleosomes. The sixth repeat of our nucleosome array template contains an *L. variegatus* 5S se-

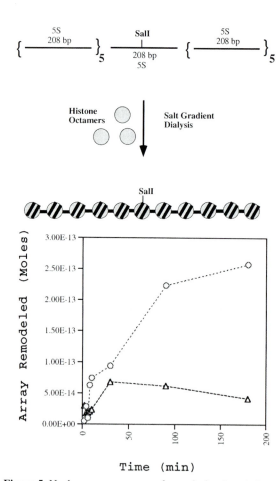

Figure 5. Nucleosome array assay for analysis of catalytic nucleosome remodeling. (*Top*) Schematic of array reconstitution. DNA template is composed of 11 head-to-tail repeats of a 5S RNA gene. Histone octamers were purified to homogeneity from chicken erythrocytes. (*Bottom*) Catalytic remodeling by SWI/SNF requires the histone amino-terminal domains. Remodeling assays contained 0.1 pmole of SWI/SNF, 1.2 pmole of either intact or trypsinized nucleosome array substrate, 1 mM ATP, and 2000 units/ml *Hinc*II. The moles of remodeled array are calculated from the moles of array that are cleaved by *Hinc*II per unit time in the presence of SWI/SNF, substracting the moles of array that are cleaved in the absence of SWI/SNF. (*Open circles*) + Tails SWI/SNF; (*open triangles*) – Tails SWI/SNF.

quence engineered by Polach and Widom (1995) that bears a unique *Sal*I/*Hin*cII restriction site close to the dyad axis of symmetry of this central, positioned nucleosome. In the absence of SWI/SNF, reconstituted arrays are cleaved by *Sal*I with slow first-order kinetics that are about 300-fold slower than the cleavage rate of the unassembled DNA template. However, in the presence of SWI/SNF and ATP, cleavage of the arrays is stimulated 100-fold. In this two-enzyme-coupled reaction, we were able to determine an approximate turnover number for SWI/SNF remodeling and to illustrate that SWI/SNF can perform a catalytic remodeling cycle (Logie and Peterson 1997). This catalytic nature of SWI/SNF becomes apparent when the reconstituted arrays are present in large molar excess over SWI/SNF in the assay (Fig. 5, bottom) (Logie and Peterson 1997). In this case, a single SWI/SNF complex is able to remodel multiple arrays at a rate of about one array per 50 minutes (Fig. 5, bottom) (see also Logie and Peterson 1997).

A recent study indicates that the dissociation step of the SWI/SNF-remodeling cycle may be regulated by the histone amino-terminal domains (C. Logie et al., unpubl.). These domains are not required for assembly of nucleosomes, but extend from the nucleosome core and are the sites of histone posttranslational modifications, such as acetylation (for review, see Hansen 1998). To determine their role in SWI/SNF remodeling, nucleosome arrays were reconstituted from two types of histone octamers: intact histone octamers and trypsinized histone octamers. The trypsinized octamers lack most of their amino-terminal and carboxy-terminal tail domains, but they contain the histone-fold domains that are sufficient to assemble nucleosome core particles. These two array substrates were used in two types of restriction-enzyme-coupled assays. In the first set of reactions, equal molar concentrations of SWI/SNF and substrate array were used; this represents a "single-round" remodeling assay that tests whether the amino-terminal tails affect the kinetics of the first two steps in the remodeling cycle (Fig. 1; Array binding and ATP-dependent remodeling). Surprisingly, both substrate arrays were remodeled with nearly identical rates by SWI/SNF. Thus, the histone tail domains are not required for SWI/SNF remodeling per se. However, strikingly different results were obtained in the second type of assay. In these reactions, a 12-fold molar excess of substrate array was incubated with SWI/SNF, and multiple rounds of remodeling were analyzed (Fig. 5, bottom). Whereas 0.1 pmole of SWI/SNF was able to remodel 0.26 pmole of the intact array after 180 minutes (2.5 rounds), less than 0.1 pmole of the trypsinized arrays were remodeled after 180 minutes (Fig. 5, bottom). Thus, it appears that SWI/SNF requires the histone tail domains to carry out multiple rounds of remodeling.

To determine why catalytic remodeling is blocked on the trypsinized arrays, template commitment assays were performed. In these assays, SWI/SNF was preincubated with a molar excess of either an intact array or a trypsinized array. These preincubation substrates were not radioactively labeled and thus their remodeling was not scored. After five minutes of preincubation, a second

aliquot of radioactively labeled, intact array was added to each reaction, and remodeling of this second array was monitored. When SWI/SNF was preincubated with the intact array, it was competent to dissociate from this array and remodel the second aliquot of array. However, when SWI/SNF was preincubated with the trypsinized array, the second aliquot of intact array was not efficiently remodeled. Thus, SWI/SNF was unable to effectively dissociate from an array that lacked histone amino- and carboxy-terminal tails.

The observed requirement for the histone tail domains in catalytic remodeling might be due to a direct effect of the tails (e.g., an allosteric effector) or to an indirect affect on the structure of the arrays. To investigate these possibilities, we tested whether GST-histone tail fusion proteins (Hecht et al. 1995) might inhibit SWI/SNF-remodeling activity, presumably by "trapping" the complex in the dissociation state. Addition of GST, GST-H4^{tail}, or GST-H2A^{tail} had no affect on the remodeling kinetics of SWI/SNF on an intact substrate array. In contrast, addition of GST-H2B^{tail} reduced remodeling by 40%, whereas the GST-H3^{tail} fusion protein reduced remodeling by 60% (C. Logie et al., unpubl,). These results are consistent with a model in which the H2B and H3 tails have important roles in the termination of SWI/SNF activity or in the dissociation of SWI/SNF at the end of a remodeling cycle. It is unlikely to be a coincidence that the H2B tail appears to be in close proximity to the nucleosome-bound SWI/SNF complex (Fig. 4).

Surprisingly, the GST-H3 fusion also inhibited the DNA-stimulated ATPase activity of SWI/SNF, whereas the GST-H2B fusion had no effect in this assay. Thus, the H2B and H3 tails may have distinct roles in the SWI/SNF-remodeling cycle. The GST-H3 tail is not a nonspecific inhibitor of all reactions, since the GST-H3 tail stimulates the ATPase activity of the related chromatin remodeling complex, RSC (Cairns et al. 1996; C. Logie et al., unpubl.). The functional interations between SWI/SNF and the histone amino-terminal domains suggests many opportunities for regulation of SWI/SNF activity. Acetylation or phosphorylation of histone H3 might mimic trypsinization, slowing the rate of SWI/SNF dissociation. In essence, this would lead to an increased dwell time for SWI/SNF at a specific chromosomal locus. Consistent with this view, the rate of multiple-round remodeling by SWI/SNF is reproducibly slower on arrays that contain hyperacetylated histones (C. Logie et al., unpubl.). If acetylation modulates the SWI/SNF remodeling cycle, perhaps this provides an explanation for the functional interactions between the GCN5 histone acetyltransferase and SWI/SNF (Pollard and Peterson 1997).

STEP 4: REVERSAL FROM THE REMODELED STATE

Once SWI/SNF has released the remodeled array, the array has two possible fates. One, the remodeled state could remain indefinitely, creating a persistent change in chromatin structure. Alternatively, the remodeled nucleosomes might revert back to their normal, inaccessible state at a measurable rate. On mononucleosome substrates, the remodeled state is indeed quite stable; reversion back to the original state requires incubations in excess of 4 hours (Côté et al. 1998). In these studies, SWI/SNF was removed from the remodeled nucleosome core particle by competition with the excess nucleosome array. Thus, the stable nature of the remodeled nucleosome core was not due to stabilization by bound SWI/SNF. In contrast to the nucleosome core particles, remodeled nucleosomes that are in the context of a nucleosome array revert to the inaccessible state with rates that are too rapid to measure accurately (Logie and Peterson 1997). On these types of substrates, a persistent change in nucleosome structure is only achieved when multiple copies of a transcription factor are bound to the remodeled nucleosome (Owen-Hughes et al. 1996). These results suggest that SWI/SNF remodeling of a nucleosome array initiates a "window of opportunity" for the enhanced binding of transcription factors. If the factor is present when SWI/SNF releases the remodeled substrate, site occupancy is favored. However, if the factor is not present, or present in limiting concentrations, then the remodeled state may be lost before the site is occupied. It seems likely that this stage in the reaction cycle might also be a target of regulation. The histone amino-terminal domains do not seem to contribute to this reaction, as remodeled, trypsinized nucleosome arrays still revert back to the inaccessible state quite rapidly (C. Logie et al., unpubl.). Other factors might exist, however, that either increase or decrease the reversion rate; such effectors are predicted to have a profound influence on the efficiency of transcription factor binding and the potency of SWI/SNF action.

REGULATION OF THE CYCLE: A ROLE FOR MULTIPLE ATPase SUBUNITS?

In many complex biochemical reactions, proofreading steps are incorporated into the cycle to ensure that steps in the cycle are initiated only when the prior step has been successfully completed. In several cases, ATP hydrolysis is used to drive conformational changes in either proteins or RNAs which irreversibly move reaction cycles forward or into abortive reactions. As discussed in more detail in the previous sections, the histone amino-terminal domains appear to control the dissociation of SWI/SNF from the remodeled array. However, this function of the amino-terminal tails must only be exerted after the ATP-dependent remodeling step or else the complex would dissociate too early in the cycle. How does SWI/SNF control the functional interaction with the tail domains? One clue may be forthcoming from the recent identification of genes that encode the last two uncharacterized SWI/SNF subunits, SWP59 and SWP61.

We used MALDI-TOF mass spectrometric peptide mapping coupled with database searches to identify the genes that encode the SWP59 and SWP61 subunits of SWI/SNF (Peterson et al. 1998). Twenty tryptic peptides from the SWP59 subunit and 22 tryptic peptides from the SWP61 subunit matched the calculated molecular masses

of peptides from the *ARP9* and *ARP7* genes, respectively. Furthermore, strains harboring deletion alleles of *ARP7* or *ARP9* show typical growth and transcriptional defects characteristic of mutations in SWI/SNF subunit genes. Thus, it appears that *ARP7* and *ARP9* encode SWI/SNF subunits.

What are ARPs? The *ARP7* and *ARP9* genes are members of the *a*ctin-*r*elated *p*rotein family which is a branch of a larger actin superfamily of proteins that includes conventional actins, heat shock protein 70 (Hsp70), heat shock cognate 70 (Hsc70), sugar kinases, glycerol kinase, and other ATP-binding proteins from prokaryotic and eukaryotic sources (for reviews, see Frankel and Mooseker 1996; Poch and Winsor 1997). ARP7 and ARP9 show similarities over the entire length of actin, spanning 13 blocks of homology that encompass sequences that are known to be important for actin structure or function, including the ATPase domain. Thus, ARP7 and ARP9 are likely to have maintained the overall actin-fold and probably bind and hydrolyze ATP. The identification of ARP7 and ARP9 as crucial subunits of SWI/SNF suggests that the complex may contain a total of three different ATPase subunits (ARP7, ARP9, and SWI2/SNF2). The presence of ARP subunits is not unique to the yeast SWI/SNF complex, as the *Drosophila* counterpart of SWI/SNF, the brm complex, also contains an ARP subunit and a related, heat shock cognate (hsc) subunit (O. Papoulas and J. Tamkun, pers. comm.).

What might be the role of ARP7 and ARP9 in a SWI/SNF remodeling cycle? A hallmark of the actin/Hsp70 family of ATPases is that binding of ATP, and its subsequent slow hydrolysis (e.g., 0.02 min^{-1} for DnaK), is linked to large protein conformational changes (Holmes et al. 1993). One attractive model is that the ARP subunits may control changes in SWI/SNF conformations that are required for the remodeling step (step 2) or for coordinating the remodeling and dissociation steps (steps 2 and 3). For instance, the ATP-bound form of ARP7/9 may represent a SWI/SNF complex that is unable to bind the histone amino-terminal tails; subsequent ATP hydrolysis by ARP7/9 might be functionally coupled to binding of the histone tails to SWI/SNF (Fig. 6). In this model, a slow rate of ATP hydrolysis by the ARPs may act as an internal "clock" that governs when the dissociation step is activated. Although this model is purely speculative at this point, it can be directly tested in the near future. A prediction of this "ARP clock" model is that mutations in the ATPase domain of either ARP7 or ARP9 should be equivalent to a trypsinized nucleosome array: Multiple rounds of remodeling should be eliminated. If, on the other hand, the ARPs are involved in

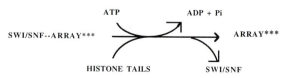

Figure 6. Speculative termination/dissociation model involving ATP hydrolysis by the ARP7/9 subunits and the histone tails.

ATP-dependent remodeling per se, then these ATPase mutants should eliminate remodeling even under "single round" conditions.

I have tried in this chapter to provide an overall view of not only what we know about the SWI/SNF remodeling cycle, but also what types of models will be tested in the near future. Although many of the biochemical properties of SWI/SNF are known, questions still abound: What is the structure of a remodeled nucleosome? How do the histone tails lead to dissociation of SWI/SNF? What are the roles of the ARP subunits? It is anticipated that SWI/SNF will continue to provide a paradigm for ATP-dependent chromatin remodeling. Certainly its subunit complexity, the presence of multiple ATPase subunits, and the regulated interactions with the histone amino-terminal tails earn SWI/SNF the title of Chromatin Remodeling Machine.

ACKNOWLEDGMENTS

I thank David Bazett-Jones, Jerry Workman, Blaine Bartholomew, and Colin Logie for sharing unpublished work, and other members of my laboratory for stimulating discussions that have led to the models proposed in this chapter. Research on SWI/SNF has been funded by a grant from the National Institutes of Health (GM-49650). I am currently a Scholar of the Leukemia Society of America.

REFERENCES

Arents G. and Moudrianakis E.N. 1993 Topography of the histone octamer surface: Repeating structural motifs utilized in the docking of nucleosomal DNA. *Proc. Natl. Acad. Sci.* **90:** 10489.

Bazett-Jones D.P., Côté J., Landel C.C., Peterson C.L., and Workman J.L. 1998. SWI/SNF complex creates loop domains in DNA and polynucleosome arrays and can disrupt DNA:histone contacts within these domains. *Mol. Cell. Biol.* (in press).

Bhattacharyya A., Murchie A.I.H., von Kitzing E., Diekmann S., Kemper B., and Lilley D.M.J. 1991. A model for the interaction of DNA junctions and resolving enzymes. *J. Mol. Biol.* **221:** 1191.

Burns L. and Peterson C.L. 1997. Yeast SWI/SNF complex facilitates the binding of a transcriptional activator to nucleosomal sites in vivo. *Mol. Cell. Biol.* **17:** 4811.

Cairns B.R., Lorch Y., Li Y., Zhang M., Lacomis L., Erdjument-Bromage H., Tempst P., Du J., Laurent B., and Kornberg R.D. 1996. RSC, an essential, abundant chromatin-remodeling complex. *Cell* **87:** 1249.

Côté J., Peterson C.L., and Workman J.L. 1998. Perturbation of nucleosome core structure by the SWI/SNF complex persists following its detachment, enhancing subsequent transcription factor binding. *Proc. Natl. Acad. Sci.* **95:** 4947.

Côté J., Quinn J., Workman J., and Peterson C.L. 1994. Stimulation of GAL4 derivative binding to nucleosomal DNA by the yeast SWI/SNF protein complex. *Science* **265:** 53.

Frankel S. and Mooseker M.S. 1996 The actin-related proteins. *Curr. Opin. Cell Biol.* **8:** 30.

Fryer C.J. and Archer T.K. 1998. Chromatin remodelling by the glucocorticoid receptor requires the BRG1 complex. *Nature* **393:** 88.

Gavin I.M. and Simpson R.T. 1997. Interplay of yeast global transcriptional regulators Ssn6p-Tup1p and Swi-Snf and their effect on chromatin structure. *EMBO J.* **16:** 6263.

Grant P.A., Duggan L., Côté J., Roberts S.M., Brownell J.E., Candau R., Ohba R., Owen- Hughes T., Allis C.D., Winston F., Berger S.L., and Workman J.L. 1997. Yeast Gcn5 functions in two multisubunit complexes to acetylate nucleosomal histones: Characterization of an Ada complex and the SAGA (Spt/Ada) complex. *Genes Dev.* **11:** 1640.

Hansen J.C. 1998. The core histone amino termini: Combinatorial interaction domains link chromatin structure with function. *Chemtracts-Biochem. Mol. Biol..* **10:** 56.

Hecht A., Laroche T., Strahl-Bolsinger S., Gasser S.M., and Grunstein M. 1995. Histone H3 and H4 N-termini interact with SIR3 and SIR4 proteins: A molecular model for the formation of heterochromatin in yeast. *Cell* **80:** 583.

Holmes K.C., Sander C., and Valencia A. 1993. A new ATP-binding fold in actin, hexokinase and Hsc70. *Trends Cell Biol.* **3:** 53.

Ito T., Bulger M., Pazin M.J., Kobayashi R., and Kadonaga J.T. 1997. ACF, an ISWI-containing and ATP-utilizing chromatin assembly and remodeling factor. *Cell* **90:** 145.

Laurent B.C., Treich I., and Carlson M. 1993. The yeast SNF2/SWI2 protein has DNA-stimulated ATPase activity required for transcriptional activation. *Genes Dev.* **7:** 583.

Logie C. and Peterson C.L. 1997. Catalytic nucleosome remodeling by the yeast SWI/SNF complex on nucleosome arrays. *EMBO J.* **16:** 6772.

Luger K., Mader A.W., Richmond R.K., Sargent D.F., and Richmond T.J. 1997. Crystal structure of the nucleosome core particle at 2.8 Å resolution. *Nature* **389:** 251.

Owen-Hughes T.A., Utley R.T., Côté J., Peterson C.L., and Workman J.L. 1996. Persistent site-specific remodeling of a nucleosome array by transient action of the SWI/SNF complex. *Science* **273:** 513.

Pazin M.J. and Kadonaga J.T. 1997. SWI2/SNF2 and related proteins: ATP-driven motors that disrupt protein-DNA interactions? *Cell* **88:** 737.

Peterson C.L. and Tamkun J.W. 1995. The SWI/SNF complex: A chromatin remodeling machine? *Trends Biochem. Sci.* **20:** 143.

Peterson C.L., Zhao Y., and Chait B. 1998. Subunits of the yeast SWI/SNF complex are members of the actin related protein (ARP) family. *J. Biol. Chem.* **273:** 23641.

Poch O. and Winsor B. 1997. Who's who among the *Saccharomyces cerevisiae* actin-related proteins? A classification and nomenclature proposal for a large family. *Yeast* **13:** 1053.

Pohler J.R.G., Duckett D.R., and Lilley D.M.J. 1994. Structure of four-way DNA junctions containing a nick in one strand. *J. Mol. Biol.* **238:** 62.

Polach K.J. and Widom J. 1995. Mechanism of protein access to specific DNA sequences in chromatin: A dynamic model for gene regulation. *J. Mol. Biol.* **254:** 130.

Pollard K.J. and Peterson C.L. 1997. Role for ADA/GCN5 products in antagonizing chromatin-mediated transcriptional repression. *Mol. Cell. Biol.* **17:** 6212.

———. 1998. Chromatin remodeling: A marriage between two families? *BioEssays* **20:** 771.

Pruss D., Bartholomew B., Persinger J., Hayes J., Arents G., Moudrianakis E.N., and Wolffe A.P. 1996. An asymmetric model for the nucleosome: A binding site for linker histones inside the DNA gyres. *Science* **274:** 614.

Quinn J., Fyrberg A., Ganster R.W., Schmidt M.C., and Peterson C.L. 1996. DNA-binding properties of the yeast SWI/SNF complex. *Nature* **379:** 844.

Studitsky V.M., Clark D.J., and Felsenfeld G. 1994 A histone octamer can step around a transcribing polymerase without leaving the template. *Cell* **76:** 371.

Studitsky V.M., Kassavetis G.A., Geiduschek E.P., and Felsenfeld G. 1997. Mechanism of transcription through the nucleosome by eukaryotic RNA polymerase. *Science* **278:** 1960.

Travers A.A. 1992. The reprogramming of transcriptional competence. *Cell* **69:** 573.

Tsukiyama T. and Wu C. 1995. Purification and properties of an ATP-dependent nucleosome remodeling factor. *Cell* **83:**

1011.

Utley R.T., Côté J., Owen-Hughes T., and Workman J.L. 1997. SWI/SNF stimulates the formation of disparate activator-nucleosome complexes but is partially redundant with cooperative binding. *J. Biol. Chem.* **272:** 12642.

Wilson C.J., Chao D.M., Imbalzano A.N., Schnitzler G.R., Kingston R.E., and Young R.A. 1996. RNA polymerase II holoenzyme contains SWI/SNF regulators involved in chromatin remodeling. *Cell* **84:** 235.

Wu L. and Winston F. 1997. Evidence that Snf-Swi controls chromatin structure over both the TATA and UAS regions of the SUC2 promoter in *Saccharomyces cerevisiae. Nucleic Acids Res.* **25:** 4230.

Yoshinaga S.K., Peterson C.L., Herskowitz I., and Yamamoto K. 1992. Roles of SWI1, SWI2, and SWI3 proteins for transcriptional enhancement by steroid receptors. *Science* **258:** 1598.

The SAGA of Spt Proteins and Transcriptional Analysis in Yeast: Past, Present, and Future

F. WINSTON AND P. SUDARSANAM

Department of Genetics, Harvard Medical School, Boston, Massachusetts 02115

The isolation and analysis of mutations that alter the regulation of gene expression has taught us about many of the fundamental aspects of transcriptional control. In *Escherichia coli*, this type of genetic analysis led to the concepts of regulatory proteins and the promoter (Beckwith and Silhavy 1992). Following in the footsteps of the analysis of *E. coli*, many genetic studies in the yeast *Saccharomyces cerevisiae* have been performed to identify mutants altered in transcriptional regulation. However, in contrast to *E. coli*, where most factors identified were operon-specific, the factors identified in yeast have often been more global, controlling the transcription of large, apparently unrelated, sets of genes (for review, see Hampsey 1998). In addition, mutant selections or screens in yeast often identified large numbers of regulatory factors, virtually indistinguishable from each other by mutant phenotypes. The large number and global nature of the yeast factors have made the clarification of their functions elusive. Results from the past several years have revealed that many of these factors are present in large multiprotein complexes that are involved in controlling different aspects of transcription, ranging from modulating repression by nucleosomes to direct interactions with transcriptional activators. However, we are only now beginning to gain a clearer understanding of how these complexes function and interact in vivo to provide normally regulated transcription.

In this paper. we review one set of regulatory factors discovered in yeast by genetic analysis. The genes that encode these factors were identified by suppressors of promoter mutations caused by the insertion of yeast transposable (Ty) elements. The suppressor mutations identified more than 20 genes named *SPT* (Spt = *Su*ppressor of *T*y). Some *spt* mutations occurred in genes encoding histones (Clark-Adams et al. 1988) and TATA-binding protein (TBP) (Eisenmann et al. 1989), providing the first evidence for the requirements for these factors for normal transcription in vivo. We review the isolation and analysis of *spt* mutants and focus on a subset of Spt proteins. These Spt proteins have recently been demonstrated to be in a multiprotein complex, conserved between yeast and humans, that is critical for normal transcription in vivo. Finally, we discuss a number of unresolved issues regarding the roles of large transcription complexes.

SUPPRESSORS OF INSERTION MUTATIONS IN *S. CEREVISIAE*

A large number of transcription factors have been identified by selecting for suppressors of promoter mutations.

The promoter mutations used in the selections described in this paper were insertion mutations caused by the yeast transposable elements Ty1 or Ty2 (hereafter referred to as Ty) or by Ty long terminal repeat sequences, named δ sequences. Ty elements are retrotransposons, similar to mammalian retroviruses in their gene organization and mode of transposition (Boeke and Sandmeyer 1991). The δ sequences flanking Ty1 elements are approximately 335 bp in length and contain signals for both transcription initiation and termination. Ty1 elements are highly transcribed, with transcription initiating in the 5′ δ and terminating in the 3′ δ. An insertion of a Ty element or a Ty δ sequence into the promoter region of a gene can exert a strong effect on transcription of the adjacent gene.

Ty insertion mutations were initially identified by two opposite effects on adjacent gene expression—causing constitutive transcription or causing inhibition of transcription. Whether a Ty insertion causes one phenotype or the other depends on several factors, including the position and orientation of the Ty insertion and the DNA sequence of the particular Ty that causes the insertion mutation (Boeke and Sandmeyer 1991). Ty insertion mutations that cause constitutive transcription were shown to be controlled by the mating-type genes. In these cases, the transcription of a gene under Ty control was constitutive in haploids of either mating type; however, transcription was greatly reduced in *MATa/MATα* diploids. This observation suggested that Ty elements might have a role in the control of mating-specific genes (Errede et al. 1980). Little support for this hypothesis has been discovered.

The strong similarities between yeast Ty elements and mammalian retroviral proviruses made it of great interest to learn more about how Ty elements control gene expression. This effort was aided by the class of Ty insertion mutations that inhibited adjacent gene transcription, as it allowed for the selection of suppressor mutations that could overcome this inhibition. The study of genes identified by suppressor analysis constituted a genetic approach, much more feasible in yeast than in larger eukaryotes, to identify and analyze the genes that modify the control of transcription by this class of eukaryotic transposable elements. In addition, studies of suppressors of insertion mutations in other organisms, including *E. coli* (Das et al. 1976), maize (Federoff 1989), and *Drosophila* (Modolell et al. 1983), had resulted in many interesting discoveries related to transcription and transposition. Thus, there was strong motivation to pursue similar analysis in yeast to further our understanding of Ty biology.

Two practical aspects of the Ty insertion mutations that inhibited adjacent gene transcription made the isolation of suppressors attractive. First, the insertion mutations were in the promoter region of *HIS4*, an extremely well-studied gene that is required for histidine biosynthesis (Roeder et al. 1980). These insertions resulted in histidine auxotrophy-(His⁻ phenotype). Using the available detailed knowledge of *HIS4*, different types of selections for *HIS4* expression could be performed, allowing the isolation of different classes of revertants (Chaleff and Fink 1980). Second, since yeast can grow with *HIS4* expressed at only 10% of its wild-type level, the selections were quite sensitive, enabling the identification of partial suppressors.

Starting with Ty insertions that inhibited *HIS4* transcription, selection for His⁺ revertants identified two general classes of events: genomic rearrangements and unlinked suppressor mutations. The rearrangements encompassed a wide variety of changes, including deletions, inversions, and translocations (Chaleff and Fink 1980). Study of these rearrangements suggested that Ty elements might be important for genome organization and/or stability. One of the most common rearrangements observed was a homologous recombination event between the Ty δ sequences, resulting in a solo δ sequence in the promoter region of *HIS4*. This particular solo δ insertion mutation, *his4-912δ*, is a conditional mutant, His⁺ at 37°C and His⁻ at 23°C (Chaleff and Fink 1980; Roeder and Fink 1980). The second class of His⁺ revertants, unlinked suppressors, was identified by selecting for suppressors of both Ty and solo δ insertion mutations. The genes identified by these unlinked suppressors were subsequently shown to be involved in transcription. These genes, *SPT* genes, are the focus of this article.

THE IDENTIFICATION AND ANALYSIS OF *SPT* GENES: MANY GENES, SEVERAL CLASSES

Several hunts for *spt* mutants have been performed, and more than 20 *SPT* genes have been identified (Winston et al. 1984b, 1987; Fassler and Winston 1988; Natsoulis et al. 1991; Roberts and Winston 1996; Zhang et al. 1997). Early analysis of unlinked suppressors of Ty insertion mutations identified three genes, named *SPT1*, *SPT2*, and *SPT3* (Chaleff 1981; Winston et al. 1984b). The mechanism of suppression was shown to be at the level of transcription by demonstrating that *spt* mutations allow the production of wild-type *HIS4* mRNA (Fig. 1) (Silverman and Fink 1984). The realization that *SPT* genes were involved in transcription and that more *SPT* genes were likely to be discovered motivated large-scale mutant hunts for new *spt* mutants. Unexpectedly, these genes were found to be involved in more than *HIS4* or Ty transcription. All *spt* mutants show pleiotropic phenotypes, suggesting a more global control of transcription. Molecular analyses of *SPT* gene function has confirmed this conclusion. Thus, the *spt* mutant hunts, although targeted at Ty and δ insertion mutations, have provided insight into general transcriptional regulatory mechanisms, as described below.

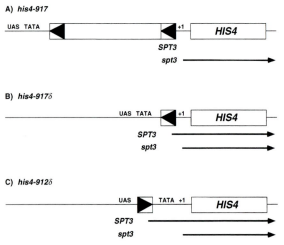

Figure 1. Diagram of Ty and δ insertion mutations used to select and characterize *spt* mutations. Shown are three different insertion mutations in the promoter of the *S. cerevisiae HIS4* gene. The δ sequences are represented by squares with filled triangles. The direction of the triangles shows the direction of the δ promoter. The UAS, TATA, and +1 sites of the *HIS4* promoter are indicated. The δ promoter also has UAS, TATA, and +1 sites (Turkel et al. 1997; A.M. Dudley et al., in prep.) but these are not shown. (*A*) The insertion mutation *his4-917* contains a Ty element at position –7 relative to the *HIS4* +1. In an otherwise wild-type background, no *HIS4* mRNA is detected. In an *spt3* mutant, a wild-type length *HIS4* mRNA is produced. (*B*) *his4-917δ* was derived from *his4-917*, and it still causes a nonconditional His⁻ phenotype (Roeder and Fink 1982). In an *SPT3* wild-type background, transcription initiates in the δ sequence, producing an aberrant *HIS4* mRNA. Since this strain is His⁻, the longer *HIS4* mRNA is believed to be nonfunctional for translation of *HIS4*, due to the presence of translation initiation and termination codons 5′ of the normal *HIS4* initiation codon. In an *spt3* mutant, a wild-type length mRNA is produced (Silverman and Fink 1984). (*C*) *his4-912δ* was derived from a different Ty insertion, *his4-912* (Farabaugh and Fink 1980). This insertion is at position –98 with respect to the *HIS4* +1. Studies of *his4-912δ* have suggested that the pattern of transcription is caused by a competition between the *HIS4* and δ TATA sequences (Hirschman et al. 1988).

The identification of many *SPT* genes raised two obvious possibilities. Either each Spt function contributes to a common activity or Spt proteins are involved in distinct activities. The latter possibility was strongly supported by the observation that *SPT* genes could be grouped into classes based on distinct mutant phenotypes, including different suppression patterns with respect to Ty and δ insertion mutations, defects in mating and sporulation, and double mutant phenotypes (Winston et al. 1984b). The discoveries that one class contained the gene encoding TBP (Eisenmann et al. 1989) and a second class contained genes encoding histones (Clark-Adams et al. 1988) strongly reinforced this view, since TBP and histones are believed to be involved in different aspects of transcription. One group of *SPT* genes contains five members that are functionally related to TBP function. Four of these *SPT* genes (*SPT3*, *SPT7*, *SPT8*, and *SPT20*) have recently been shown to encode members of the large protein complex SAGA, described below. The fifth gene, *SPT15*, encodes TBP itself.

THE TBP CLASS OF *SPT* GENES

Several mutant hunts identified these five *SPT* genes and established their relationship. The first large-scale selection for *SPT* mutants (Winston et al. 1984b) identified seven different genes, *SPT1–SPT7*. It was the analysis of this set of mutants that demonstrated the existence of different phenotypic classes of *spt* mutations. From this analysis, only *spt3* mutations suppressed a particular insertion mutation, *his4-917δ*, a solo δ insertion mutation in the *HIS4* promoter. In addition, *spt3* mutations reduced Ty mRNA levels (Winston et al. 1984b). These results made *SPT3* a gene of great interest, as it was the first gene to be implicated in the transcription of a eukaryotic transposable element. The other *SPT* genes identified, *SPT1, SPT2, SPT4, SPT5,* and *SPT6,* have also been subjects of further study (see, e.g., Swanson and Winston 1992; Bortvin and Winston 1996; Hartzog et al. 1998), but they are not discussed in this paper.

To learn more about the role of Spt3 in vivo, the subsequent mutant hunts were refined to identify other genes in the *SPT3* class (Winston et al. 1987). In a second mutant selection, suppressors of *his4-917δ* were selected directly, and four genes were identified: *SPT3, SPT7, SPT8,* and *SPT15*. Of these four genes, *SPT3* and *SPT7* had been identified in the earlier selections. *SPT3* was an expected class; however, *SPT7* mutations were unexpected, since the earlier *spt7* allele tested did not suppress *his4-917δ* and therefore had not been placed in the *SPT3* class. This misclassification illustrates the danger of basing any strong conclusions on the analysis of a single mutant allele. The *spt7* mutant tested from the earlier selection was a weak allele and did not display *spt3*-like phenotypes.

In this new mutant hunt, representative mutations were also screened for another *spt3* mutant phenotype, a reduced level of Ty mRNA. *spt3, spt7,* and *spt8* mutants all had greatly reduced levels of Ty mRNA. However, the single *spt15* mutation tested did not display a significantly reduced level of Ty mRNA. On this basis, *spt15* mutations were mistakenly placed in a separate class (cited in the Discussion of Winston et al. 1987). This error again illustrates the hazards of basing a conclusion from the analysis of a single mutant allele. Subsequent analysis of another *spt15* mutation revealed a severe defect in Ty mRNA levels, and the close relationship between *SPT15* and *SPT3, SPT7,* and *SPT8* was fully realized (Eisenmann et al. 1989, 1992; Arndt et al. 1992).

The discovery that *SPT15* encodes TBP was an extremely exciting finding because it provided the initial demonstration that TBP is essential for growth and that it is required for normal transcription in vivo. It also provided the basis for a model for the function of Spt3, Spt7, and Spt8. The analysis of *SPT3, SPT7,* and *SPT8* demonstrated that they are not essential for growth. However, null mutations in these genes do cause poor growth and the same mutant phenotypes as particular missense mutations in *SPT15*, including defective expression from certain RNA polymerase II-dependent promoters (Winston et al. 1984a; Hirschhorn and Winston 1988; Eisenmann et al. 1994; Gansheroff et al. 1995). These *SPT* genes do not regulate each other's transcription or protein levels. The

extreme phenotypic similarity between the *spt15* missense mutations and the *spt3, spt7,* and *spt8* null mutants suggested that Spt3, Spt7, and Spt8 facilitate normal TBP function at certain promoters in vivo.

This model for the role of these four Spt proteins was strongly supported by both genetic and biochemical analyses that indicated a direct interaction between Spt3 and TBP. The genetic evidence came from the identification of allele-specific mutations in both *SPT15* and *SPT3* that suppress each other (Eisenmann et al. 1992). The *spt15* mutations that suppress particular *spt3* mutations cause amino acid changes in a small region of TBP predicted to interact with Spt3 (Fig. 2) (Nikolov et al. 1992). This type of allele specificity suggests a direct interaction between these two proteins. In support of these genetic results, Spt3 and TBP were shown to coimmunoprecipitate (Eisenmann et al. 1992; Lee and Young 1998). Recent work has demonstrated that Spt3 is conserved between yeast and human cells and is also similar to two human TBP-associated factors (Tafs; Birck et al. 1998). Tafs are proteins that interact with TBP in a complex named TFIID, and they are believed to be required for transcriptional activation (Struhl and Moqtaderi 1998). Taken together, these results suggest that Spt3 may directly interact with TBP to modulate its function at certain promoters. The other Spt proteins in this group may also act, directly or indirectly, with respect to TBP function. Genetic results suggest that the Spt3-TBP interaction is modulated by Spt8 (Eisenmann et al. 1994), and recent in vitro analysis, discussed below, suggests a direct interaction between TBP and Spt8.

The putative role for Spt proteins in assisting TBP function encouraged us to study the role of Spt proteins in controlling transcription in vitro. The predicted sequences of the proteins did not provide any clues regarding their functions. Experiments were performed to determine if we could observe a transcriptional defect in vitro using extracts from either *spt3Δ* or *spt7Δ* mutants. These experiments were also motivated by the success of similar experiments with *srb2Δ* mutants (Koleske et al. 1992). However, despite extensive attempts, no defect in basal or activated levels of transcription could be observed in vitro, even when using a δ promoter that is

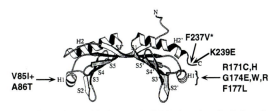

Figure 2. Amino acid changes in TBP that alter Spt3 function. Mutations in *SPT15*, encoding TBP, were isolated that suppress particular mutations in *SPT3* (Eisenmann et al. 1992). The clustering of the amino acid changes in TBP suggests that this region of TBP, including H1′ and the carboxy-terminal region, interacts with Spt3. The change F237V, marked by an asterisk, suppresses the change G174E, further supporting an interaction between H1′ and the carboxy-terminal region of TBP. (Structure of TBP, adapted, with permission, from Nikolov et al. 1992 [copyright Macmillan].)

strongly Spt-dependent in vivo (Gansheroff 1995; J. Madison et al., unpubl.). Thus, the in vitro transcription systems available for yeast were not sensitive to the requirement for Spt proteins. One obvious difference between in vitro and in vivo conditions is the lack of normal chromatin structure of the in vitro template. In addition, the lack of any evidence for the ability of Spt proteins to bind DNA or for the presence of Spt proteins in other identified transcription complexes implied that little biochemical information was available on the roles of these proteins.

In a continuing attempt to learn more about the possible functions of these Spt proteins, we performed another mutant hunt designed to identify other *SPT* genes in this class (Roberts and Winston 1996). We repeated the selection for suppressors of the insertion mutation *his4-917δ*. However, in the yeast strain used for the selection, we first duplicated the *SPT3* and *SPT15* genes to increase our chances of finding new genes. These gene duplications hid any new recessive mutations in *SPT3* and *SPT15*, thus sensitizing the screen toward finding mutations in previously unstudied genes. This strategy was successful and resulted in the identification of a single mutation in a previously unstudied gene that we named *SPT20* (Roberts and Winston 1996). Analysis of an *spt20Δ* mutant demonstrated that it was extremely similar to the other *spt* mutants described. Since only one *spt20* allele was identified, it remained possible that other genes of this class existed. This was shown to be true by the analysis of *ADA1* (Horiuchi et al. 1997), described in the next section. As described below, the discovery of *SPT20* was a critical turning point in our understanding of Spt proteins.

In summary, three mutant hunts identified five *SPT* genes that appeared to be functionally related. One of them, *SPT15*, encodes TBP, one of the most widely studied transcription factors in eukaryotes. The other four, *SPT3*, *SPT7*, *SPT8*, and *SPT20,* encode proteins believed to be functionally related to TBP. The Spt3 protein was shown to associate with TBP and to be homologous to two human Tafs, although Spt3 is not itself a Taf. Based on mutant phenotypes, all four Spt proteins are believed to be required for transcription from particular promoters.

SPT, ADA, GCN5 PROTEINS AND THE SAGA COMPLEX

The identification of *SPT20* was a very important step in the study of this set of *SPT* genes, as it established a connection between Spt proteins and another group of transcription factors. In addition to its discovery by an Spt⁻ selection, *SPT20* was simultaneously discovered in an independent selection, designed to identify transcriptional adapters or coactivators (Marcus et al. 1996). This selection identified five genes, designated *ADA1–ADA5* (Berger et al. 1992; Marcus et al. 1994, 1996; Horiuchi et al. 1997). The *ADA4* gene was shown to be the same as *GCN5*, previously identified in a screen for transcriptional activators (Georgakopoulos and Thireos 1992) and more recently shown to encode a histone acetyltrans-

ferase (HAT) (Brownell et al. 1996). Another *ADA* gene, *ADA5*, was shown to be the same as *SPT20* (Marcus et al. 1996; Roberts and Winston 1996). *SPT20/ADA5* was the only common gene between the *SPT* and *ADA* genes. Thus, this surprising discovery linked two sets of factors previously believed to be unrelated and implied that they function together. This connection was strengthened by the subsequent analysis of *ADA1*, as *ada1Δ* mutants were shown to have phenotypes extremely similar to those of *spt7Δ* and *spt20Δ* mutants (Horiuchi et al. 1997; Sterner et al. 1998).

A physical connection between Ada and Spt proteins was demonstrated in the process of characterizing Gcn5. The biochemical characterization of Gcn5 had shown that it could acetylate free histones but not its presumed in vivo substrate, histones in nucleosomes (Brownell et al. 1996; Kuo et al. 1996). The purification of HAT complexes that could acetylate nucleosomal histones revealed several Gcn5-containing complexes (Grant et al. 1997; Horiuchi et al. 1997; Pollard and Peterson 1997; Ruiz-Garcia et al. 1997; Saleh et al. 1997). One of these complexes was shown to contain Spt3, Spt7, Spt8, and Spt20 as well as Ada proteins (Grant et al. 1997). This complex was named SAGA (*Spt-Ada-Gcn5 a*cetyltransferase). Very recent work has demonstrated that SAGA also contains five Tafs (Grant et al. 1998b). SAGA is conserved between yeast and humans (Ogryzko et al. 1998; Yu et al. 1998).

Both genetic and biochemical evidence suggests that SAGA contains multiple classes of proteins and activities important for transcription both in vivo and in vitro. Genetic analysis of null mutations in some of the known SAGA genes has divided them into three groups (Table 1) (Gansheroff 1995; Roberts and Winston 1996, 1997; Horiuchi et al. 1997). Mutations in group 1 (*SPT7*, *SPT20*, and *ADA1*) cause the broadest and most severe set of mutant phenotypes, suggesting that they are the most crucial for SAGA functions. Mutations in group 2 (*SPT3* and *SPT8*) and group 3 (*GCN5*, *ADA2*, and *ADA3*) each cause a smaller set of phenotypes, distinct from each other. Most of the phenotypes of groups 2 and 3 are subsets of the group 1 phenotypes. These results suggest that group 1 is required for all SAGA function, whereas groups 2 and 3 are required for distinct subsets of functions. This is supported by phenotypic analysis of double mutants

Table 1. Distinct SAGA Mutant Phenotypes

Group	Representative mutant	Spt	Gal	Ino	Min	HU	Ada
	wild type	+	+	+	+	+	+
1	*spt20Δ*	–	–	–	+	–	–
2	*spt3Δ*	–	+/–	+	+	+	+
3	*gcn5Δ*	+	+	+	–	+	–

The phenotypes are as follows: Spt, suppression of insertion mutations; Gal, ability to grow on galactose as a carbon source; Ino, ability to grow in the absence of inositol; Min, growth on minimal media; HU, sensitivity to hydroxyurea; Ada, sensitivity to high levels of Gal4-VP16. Other members of the *spt20* class include *spt7* and *ada1*; other members of the *spt3* class includes *spt8*; and other members of the *gcn5* class include *ada2* and *ada3*.

between groups 2 and 3: An *spt3Δ gcn5Δ* double mutant strongly resembles an *spt20Δ* mutant (Sterner et al. 1998). It is not yet clear how the Tafs in SAGA fit into these classes. Analysis of *taf* mutants with respect to SAGA function has not yet been reported; however, this promises to be complicated, given the essential nature of Tafs and their presence in both SAGA and TFIID.

The biochemical analysis of SAGA fits well with the genetic analysis and supports the idea of distinct SAGA functions. The group 1 SAGA components (Spt7, Spt20, and Ada1) are required for the integrity of SAGA, as no form of SAGA can be detected in *spt7Δ*, *spt20Δ*, or *ada1Δ* mutants (Grant et al. 1997; Sterner et al. 1998). Other biochemical roles for these three proteins have not yet been elucidated.

Similar biochemical analysis demonstrated that the group 2 SAGA components (Spt3 and Spt8) are not required for SAGA integrity. However, SAGA is smaller in both *spt3Δ* and *spt8Δ* mutants (Sterner et al. 1998). SAGA has been shown to bind to TBP in vitro (Sterner et al. 1998). SAGA prepared from an *spt8Δ* mutant fails to bind to TBP in vitro, suggesting a direct role for Spt8 in SAGA-TBP interactions (Sterner et al. 1998). On the other hand, SAGA prepared from an *spt3Δ* mutant retains SAGA-TBP interactions. This result is in contrast to the genetic and biochemical analyses described earlier that suggested an Spt3-TBP interaction in vivo. This discrepancy between the in vitro and in vivo data remains to be resolved. The in vitro experiments may not assay all the functional interactions between SAGA and TBP that occur in vivo. The interactions in vivo likely occur in the presence of nucleosomes and other transcriptional activators that may impose a more stringent requirement for Spt3 in modulating SAGA-TBP interactions. Although the exact biochemical functions of Spt3 and Spt8 and their respective functions in the SAGA-TBP interaction remain to be elucidated, they appear to have key roles in an important aspect of TBP function at SAGA-dependent promoters.

The group 3 SAGA components (Gcn5, Ada2, and Ada3) are required for SAGA HAT activity, but not for SAGA integrity. Gcn5, as mentioned earlier, is a HAT and there is strong evidence that Ada2 and Ada3 are both required for Gcn5 HAT activity on nucleosomal histones (Grant et al. 1998a). The presence of biochemically distinct SAGA activities is demonstrated by the fact that SAGA prepared from both *spt3Δ* and *spt8Δ* mutants still possesses HAT activity (Sterner et al. 1998). The roles of the Tafs in SAGA remain to be determined, although it has been demonstrated that Taf61 is required for a number of SAGA activities in vitro (Grant et al. 1998b). Tafs are the only known members of SAGA that are essential for growth. However, since the group 1 SAGA components are required for SAGA integrity, yet are not essential for growth, the essential nature of the Tafs is probably due to their role in TFIID rather than in SAGA. Finally, thirteen SAGA proteins have been identified, but there are more than 20 proteins in purified SAGA (Grant et al. 1997, 1998b). It seems likely that other SAGA proteins with distinct biochemical activities and genetic phenotypes remain to be discovered.

THE FUNCTIONAL INTERACTION OF SAGA WITH OTHER TRANSCRIPTION COMPLEXES

Studies in yeast have led to the identification of several large complexes important for transcription in vivo (Hampsey 1998). However, the role of each complex in terms of target genes and possible redundancy with other complexes is not understood. Genetic studies of SAGA have provided some insight into the relationship of SAGA to two other large transcription complexes in yeast, Snf/Swi, an ATP-dependent nucleosome remodeling complex required for normal transcription of many genes (Kingston et al. 1996), and Srb/mediator, required for several different aspects of transcriptional activation (Carlson 1997; Hampsey 1998). As a genetic approach to understand the roles of SAGA in vivo, a screen for mutations that cause lethality in an *spt20Δ* genetic background was performed. There is no detectable SAGA in *spt20Δ* mutants, yet the cells are still alive. In theory, a "synthetic lethal" screen would identify functions that are essential for growth specifically in the absence of SAGA. Thus, the screen might identify factors or complexes that are at least partially redundant with SAGA.

We obtained surprising results in this screen. Mutations were identified in genes that encode members of Snf/Swi and Srb/mediator (Roberts and Winston 1997). The synthetic lethality occurred in a pattern that supported the idea that SAGA contains multiple functions that are redundant with those of Snf/Swi and with at least some functions of Srb/mediator. Mutations that abolish SAGA, for example *spt20Δ*, cause synthetic lethality in combination with any *snf/swi* or *srb*/mediator mutations tested, whereas mutations that affect only a subset of SAGA functions, such as *spt3Δ* or *gcn5Δ*, do not cause this double mutant lethality (Table 2) (Roberts and Winston 1997; Sterner et al. 1998).

What is the reason for the synthetic lethality between SAGA, Snf/Swi, and Srb/mediator members? Conceivably, these complexes each contributes to the transcription of one essential gene, and that gene is expressed at a level too low to support viability in the double mutants. However, we favor a different model in which these three complexes contribute in broad and partially overlapping ways toward normal transcription of a large number of genes by RNA polymerase II (Fig. 3) (Roberts and Winston 1997). In the absence of any one of these complexes, transcription is generally impaired, but not enough to cause inviability. However, in the absence of two complexes, transcription is severely defective, causing death.

Table 2. SAGA Double Mutant Phenotypes

Mutant	Phenotype in combination with	
	snf/swi	*srb*/mediator
Wild type	alive	alive
spt20Δ	dead	dead
spt3Δ	alive	alive
gcn5Δ	alive, sick	alive

Examples of double mutant phenotypes that distinguish SAGA members (Roberts and Winston 1997).

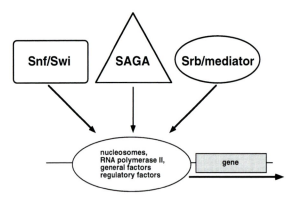

Figure 3. Large transcription complexes have overlapping targets in vivo. Genetic and molecular analyses have suggested that three complexes, SAGA, Snf/Swi, and Srb/mediator, contribute in a partially redundant fashion to enable proper transcription by RNA polymerase II in *S. cerevisiae*. (Adapted from Roberts and Winston 1997.)

One recent result supports the model of partial redundancy between SAGA and Snf/Swi. In these experiments, we took advantage of the viability of *snf2Δ gcn5Δ* double mutants. Although this double mutant is viable, it is extremely sick, forming only very small colonies after several days of growth (Roberts and Winston 1997). This poor growth suggests that the activities of Snf/Swi and Gcn5 (part of SAGA) may be partially redundant. Indeed, both the activities of Snf/Swi and Gcn5 are believed to be dedicated to helping transcription factors bind to their sites in nucleosomes, Snf/Swi as an ATP-dependent nucleosome remodeling activity and Gcn5 as a HAT. Thus, although the two activities are biochemically distinct, their biological consequences are similar, and the synthetic phenotype of a *snf2Δ gcn5Δ* mutant could be due to a general inability of the cell to overcome nucleosomal repression. This idea has been tested at one gene that is strongly Snf/Swi-dependent, the *SUC2* gene. In single mutants, a *snf/swi* mutation decreases *SUC2* mRNA levels approximately tenfold, whereas a *gcn5Δ* mutation causes no detectable effect on *SUC2* mRNA levels. However, in the *snf2Δ gcn5Δ* double mutant, there is a severe decrease in *SUC2* mRNA levels, indicating that the remaining *SUC2* transcription in a *snf2Δ* mutant is strongly dependent on Gcn5 (P. Sudarsanam et al., in prep.). These data strongly support the idea that different complexes are partially redundant for transcription. Further genetic and biochemical studies dissecting the relationship between these complexes at different promoters should enhance our understanding of the contribution of each complex toward transcription.

ISSUES FOR FUTURE STUDIES

The discovery and analysis of SAGA and other large transcription complexes in yeast have raised many interesting and perplexing issues concerning their functions. These issues are relevant to all eukaryotes due to the strong conservation of these complexes across species. In this section, we discuss six closely related issues currently facing those who study these complexes.

1. *Why are there so many complexes?* Several different complexes have been identified in yeast (Chang and Jaehning 1997; Hampsey 1998; Lee and Young 1998). Most of these complexes must have different roles in transcription, since a mutation that impairs any one complex causes distinct mutant phenotypes. Our evidence suggests that some transcription complexes with distinct biochemical activities are partially redundant (Roberts and Winston 1997). Cells may have evolved a system of partial redundancy among these complexes in order to have flexibility in exerting transcriptional control. In addition, possessing multiple mechanisms to achieve the same biological goal could help protect cells against certain classes of inhibitors or mutations. Some complexes, such as histone acetylases, possess similar biochemical activities yet are still likely to carry out distinct roles in vivo (Grant et al. 1997, 1998a). In these cases, each complex may possess distinct substrate specificities, may interact with different transcription factors, or may respond in different fashion to environmental stimuli.

2. *Why do the complexes contain so many different proteins?* Possessing many proteins allows a complex to have a greater number of activities. For some complexes, such as Srb/mediator and SAGA, different activities within the complex have been shown to be dependent on distinct sets of subunits. However, other complexes, such as Snf/Swi, have not been shown to have multiple activities. Possessing many subunits also creates the potential for interactions with proteins that might control the activity of the complex either by modification or direct interaction, with direct targets of the complex and with other transcription factors. Some subunits may also serve a structural role within a complex. Determining the individual roles of proteins within a complex, particularly when the proteins apparently contribute to the same activity, represents a formidable challenge that will require both direct biochemical analysis and a more complete understanding of the roles and regulation of each complex in vivo.

3. *For complexes that contain multiple activities, are all of the different activities required at a dependent promoter?* For example, are all SAGA activities required at a SAGA-dependent promoter or are only a subset of SAGA activities required? Current evidence, based on distinct mutant phenotypes and extensive genetic and biochemical studies, strongly suggests that the Spt3/Spt8-dependent activity of SAGA acts independently of the Gcn5-dependent activity at many promoters (Gansheroff et al. 1995; Roberts and Winston 1996, 1997; Horiuchi et al. 1997; Sterner et al. 1998; A.M. Dudley et al., in prep.). However, there are other promoters at which both the Spt- and Gcn5-dependent SAGA activities contribute to activation (Sterner et al. 1998). Thus, SAGA may represent a multifunc-

tional complex, able to provide one or more activities, as needed, to help transcription.

4. *What determines the degree to which a promoter is dependent on any of these complexes?* Many possible determinants could confer dependence of a promoter on a particular complex or on a specific activity of a complex. These include specific promoter sequences that bind a component of the complex, gene-specific regulatory proteins that interact with the complex, components of the general transcription machinery, or a particular chromatin structure. There will likely be multiple determinants that will vary depending on the complex and the promoter. Identification of all the genes that are strongly dependent on a particular complex will help to address whether particular promoter sequences or factors might confer complex dependence. It seems likely that elucidating the targeting and coordination of large transcription complexes at a promoter will uncover new mechanisms of transcriptional control.

5. *What is the functional relationship between different complexes that act at the same promoter?* In one case, described in this paper, Snf/Swi and Gcn5 both have roles in transcription of *SUC2*. What confers dependence of the *SUC2* promoter on both activities? One possibility is that each activity acts independently at the promoter. However, the two activities may also function in a dependent fashion. For example, Snf/Swi remodeling activity might depend on the acetylation state of histones as controlled by Gcn5, or, conversely, Gcn5 HAT activity might depend on the nucleosome structure as determined by Snf/Swi. There is also evidence that Gcn5 and other HATs can act to acetylate proteins other than histones (Pollard and Peterson 1997). Therefore, it is possible that Gcn5 directly controls the activity of a complex such as Snf/Swi by modification of a subunit. Biochemical assays incorporating activities from both complexes on chromatin templates, as well as in vivo analysis of promoters, should help to delineate these relationships.

6. *Why are some proteins present in more than one complex?* A surprising finding was that a subset of Tafs is present in SAGA as well as in TFIID. Therefore, these proteins, previously thought to interact only in a complex with TBP, have a more general role in transcription. The most economical model is that Tafs perform a common function that is required for both SAGA and TFIID. For example, they may help TBP bind to DNA or interact with a common set of transcription factors. However, the Tafs may also have distinct roles in each complex. Although the interactions among the Tafs are probably the same in each complex, they must interact with different sets of proteins within SAGA and TFIID.

The presence of some Tafs in more than one complex complicates in vivo analysis of their function. For example, it becomes more difficult to discover why these particular *TAF* genes are essential. Current evidence suggests that TFIID is essential for growth and that SAGA is

not, but it could be the combined loss of both complexes that causes inviability in *taf* null mutants. The identification of *taf* mutations that affect only one complex could help sort out their different roles. However, the identification of such mutations will require a combination of careful genetics, rigorous biochemistry, and some luck. The presence of Tafs in more than one complex raises the general issue of how many other proteins will be present in just one or in more than one complex. In most cases, biochemistry has not yet clearly distinguished between these possibilities. Therefore, sorting out the roles in vivo will be a significant challenge.

FINAL CONSIDERATIONS

The isolation and analysis of *spt* mutations, along with many other productive mutant hunts, have contributed significantly toward a greater understanding of eukaryotic transcription in vivo. The convergence of these genetic approaches with biochemical analysis has been gratifying, illuminating, and confusing, all at the same time. Our current image of transcriptional control in vivo is a fascinating but crowded picture, filled with a number of large protein complexes acting at promoters in different ways. Both genetic and biochemical analyses should help bring this picture into sharper focus and reveal the dynamic interactions of these complexes with each other and with other regulatory factors in the regulation of transcription.

ACKNOWLEDGMENT

Work in our laboratory is supported by grants from the National Institutes of Health.

REFERENCES

Arndt K.M., Ricupero S.L., Eisenmann D.M., and Winston F. 1992. Biochemical and genetic characterization of a yeast TFIID mutant that alters transcription in vivo and DNA binding in vitro. *Mol. Cell. Biol.* **12:** 2372.

Beckwith J. and Silhavy T. 1992. *The power of bacterial genetics: A literature-based course.* Cold Spring Harbor Laboratory Press, Cold Spring Harbor, New York.

Berger S.L., Pina B., Silverman N., Marcus G.A., Agapite J., Regier J.L., Triezenberg S.J., and Guarente L. 1992. Genetic isolation of ADA2: A potential transcriptional adaptor required for function of certain acidic activation domains. *Cell* **70:** 251.

Birck C., Poch O., Romier C., Ruff M., Mengus G., Lavigne A.C., Davidson I., and Moras D. 1998. Human TAF(II)28 and TAF(II)18 interact through a histone fold encoded by atypical evolutionary conserved motifs also found in the SPT3 family. *Cell* **94:** 239.

Boeke J.D. and Sandmeyer S.B. 1991. Yeast transposable elements. In *The molecular and cellular biology of the yeast* Saccharomyces: *Genome dynamics, protein synthesis, and energetics* (ed. J.R. Broach et al.), p. 193. Cold Spring Harbor Laboratory Press, Cold Spring Harbor, New York.

Bortvin A. and Winston F. 1996. Evidence that Spt6 controls chromatin structure by a direct interaction with histones. *Science* **272:** 1473.

Brownell J.E., Zhou J., Ranalli T., Kobayashi R., Edmondson D.G., Roth S.Y., and Allis C.D. 1996. *Tetrahymena* histone acetyltransferase A: A homolog to yeast Gcn5p linking histone acetylation to gene activation. *Cell* **84:** 843.

Carlson M. 1997. Genetics of transcriptional regulation in yeast: Connections to the RNA polymerase II CTD. *Annu. Rev. Cell Dev. Biol.* **13:** 1.

Chaleff D.T. 1981. "The genetic analysis of an insertion mutation in yeast." Ph.D. thesis, Cornell University, Ithaca, New York.

Chaleff D.T. and Fink G.R. 1980. Genetic events associated with an insertion mutation in yeast. *Cell* **21:** 227.

Chang M. and Jaehning J.A. 1997. A multiplicity of mediators: Alternative forms of transcription complexes communicate with transcriptional regulators. *Nucleic Acids Res.* **25:** 4861.

Clark-Adams C.D., Norris D., Osley M.A., Fassler J.S., and Winston F. 1988. Changes in histone gene dosage alter transcription in yeast. *Genes Dev.* **2:** 150.

Das A., Court D., and Adhya S. 1976. Isolation and characterization of conditional lethal mutants of *Escherichia coli* defective in transcription termination factor rho. *Proc. Natl. Acad. Sci.* **73:** 1959.

Eisenmann D.M., Dollard C., and Winston F. 1989. SPT15, the gene encoding the yeast TATA binding factor TFIID, is required for normal transcription initiation in vivo. *Cell.* **58:** 1183.

Eisenmann D.M., Arndt K.M., Ricupero S.L., Rooney J.W., and Winston F. 1992. SPT3 interacts with TFIID to allow normal transcription in *Saccharomyces cerevisiae. Genes Dev.* **6:** 1319.

Eisenmann D.M., Chapon C., Roberts S.M., Dollard C., and Winston F. 1994. The *Saccharomyces cerevisiae* SPT8 gene encodes a very acidic protein that is functionally related to SPT3 and TATA-binding protein. *Genetics* **137:** 647.

Errede B., Cardillo T.S., Sherman F., Dubois E., Deschamps J., and Wiame J.M. 1980. Mating signals control expression of mutations resulting from insertion of a transposable repetitive element adjacent to diverse yeast genes. *Cell.* **22:** 427.

Farabaugh P.J. and Fink G.R. 1980. Insertion of the eukaryotic transposable element Ty1 creates a 5-base pair duplication. *Nature* **286:** 352.

Fassler J. S. and Winston F. 1988. Isolation and analysis of a novel class of suppressor of Ty insertion mutations in *Saccharomyces cerevisiae. Genetics* **118:** 203.

Federoff N. 1989. Maize transposable elements. In *Mobile DNA* (ed. D.E. Berg and M.M. Howe), p. 375. American Society for Microbiology, Washington, D.C.

Gansheroff L. J. 1995. "Characterization of *SPT7*, a gene important for transcription in *S. cerevisiae.*" Ph.D. thesis, Harvard University, Cambridge, Massachusetts.

Gansheroff L.J., Dollard C., Tan P., and Winston F. 1995. The *Saccharomyces cerevisiae* SPT7 gene encodes a very acidic protein important for transcription in vivo. *Genetics* **139:** 523.

Georgakopoulos T. and Thireos G. 1992. Two distinct yeast transcriptional activators require the function of the GCN5 protein to promote normal levels of transcription. *EMBO J.* **11:** 4145.

Grant P.A., Sterner D.E., Duggan L.J., Workman J.L., and Berger S.L. 1998a. The SAGA unfolds: Convergence of transcription regulators in chromatin-modifying complexes. *Trends Cell Biol.* **8:** 193.

Grant P.A., Schieltz D., Pray-Grant M.G., Steger D.J., Reese J.C., Yates J.R., and Workman J.L. 1998b. A subset of TAFIIs are integral components of the SAGA complex required for nucleosome acetylation and transcriptional stimulation. *Cell* **94:** 45.

Grant P.A., Duggan L., Cote J., Roberts S.M., Brownell J.E., Candau R., Ohba R., Owen-Hughes T., Allis C.D., Winston F., Berger S.L., and Workman J.L. 1997. Yeast Gcn5 functions in two multisubunit complexes to acetylate nucleosomal histones: Characterization of an Ada complex and the SAGA (Spt/Ada) complex. *Genes Dev.* **11:** 1640.

Hampsey M. 1998. Molecular genetics of the RNA polymerase II general transcriptional machinery. *Microbiol. Mol. Biol. Rev.* **62:** 465.

Hartzog G.A., Wada T., Handa H., and Winston F. 1998. Evidence that Spt4, Spt5, and Spt6 control transcription elongation by RNA polymerase II in *Saccharomyces cerevisiae. Genes Dev.* **12:** 357.

Hirschhorn J.N. and Winston F. 1988. SPT3 is required for normal levels of **a**-factor and α-factor expression in *Saccharomyces cerevisiae. Mol. Cell. Biol.* **8:** 822.

Hirschman J.E., Durbin K.J., and Winston F. 1988. Genetic evidence for promoter competition in *Saccharomyces cerevisiae. Mol. Cell. Biol.* **8:** 4608.

Horiuchi J., Silverman N., Pina B., Marcus G.A., and Guarente L. 1997. ADA1, a novel component of the ADA/GCN5 complex, has broader effects than GCN5, ADA2, or ADA3. *Mol. Cell. Biol.* **17:** 3220.

Kingston R.E., Bunker C.A., and Imbalzano A.N. 1996. Repression and activation by multiprotein complexes that alter chromatin structure. *Genes Dev.* **10:** 905.

Koleske A.J., Buratowski S., Nonet M., and Young R.A. 1992. A novel transcription factor reveals a functional link between the RNA polymerase II CTD and TFIID. *Cell* **69:** 883.

Kuo M.H., Brownell J.E., Sobel R.E., Ranalli T.A., Cook R.G., Edmondson D.G., Roth S.Y., and Allis C.D. 1996. Transcription-linked acetylation by Gcn5p of histones H3 and H4 at specific lysines. *Nature* **383:** 269.

Lee T.I. and Young R.A. 1998. Regulation of gene expression by TBP-associated proteins. *Genes Dev.* **12:** 1398.

Marcus G.A., Horiuchi J., Silverman N., and Guarente L. 1996. ADA5/SPT20 links the ADA and SPT genes, which are involved in yeast transcription. *Mol. Cell. Biol.* **16:** 3197.

Marcus G.A., Silverman N., Berger S.L., Horiuchi J., and Guarente L. 1994. Functional similarity and physical association between GCN5 and ADA2: Putative transcriptional adaptors. *EMBO J.* **13:** 4807.

Modolell J., Bender W., and Meselson M. 1983. *Drosophila melanogaster* mutations suppressible by the suppressor of Hairy-wing are insertions of a 7.3-kilobase mobile element. *Proc. Natl. Acad. Sci.* **80:** 1678.

Natsoulis G., Dollard C., Winston F., and Boeke J.D. 1991. The products of the SPT10 and SPT21 genes of *Saccharomyces cerevisiae* increase the amplitude of transcriptional regulation at a large number of unlinked loci. *New Biol.* **3:** 1249.

Nikolov D.B., Hu S.H., Lin J., Gasch A., Hoffmann A., Horikoshi, M., Chua N.H., Roeder R.G., and Burley S.K. 1992. Crystal structure of TFIID TATA-box binding protein. *Nature* **360:** 40.

Ogryzko V.V., Kotani T., Zhang X., Schiltz R.L., Howard T., Yang X.-J., Howard B.H., Quin J., and Nakatani Y. 1998. Histone-like TAFs within the PCAF histone acetylase complex. *Cell* **94:** 35.

Pollard K.J. and Peterson C.L. 1997. Role for ADA/GCN5 products in antagonizing chromatin-mediated transcriptional repression. *Mol. Cell. Biol.* **17:** 6212.

Roberts S.M. and Winston F. 1996. SPT20/ADA5 encodes a novel protein functionally related to the TATA- binding protein and important for transcription in *Saccharomyces cerevisiae. Mol. Cell. Biol.* **16:** 3206.

———. 1997. Essential functional interactions of SAGA, a *Saccharomyces cerevisiae* complex of Spt, Ada, and Gcn5 proteins, with the Snf/Swi and Srb/mediator complexes. *Genetics* **147:** 451.

Roeder G.S. and Fink G.R. 1980. DNA rearrangements associated with a transposable element in yeast. *Cell* **21:** 239.

———. 1982. Movement of yeast transposable elements by gene conversion. *Proc. Natl. Acad. Sci.* **79:** 5621.

Roeder G.S., Farabaugh P.J., Chaleff D.T., and Fink G.R. 1980. The origins of gene instability in yeast. *Science* **209:** 1375.

Ruiz-Garcia A.B., Sendra R., Pamblanco M., and Tordera V. 1997. Gcn5p is involved in the acetylation of histone H3 in nucleosomes. *FEBS Lett.* **403:** 186.

Saleh A., Lang V., Cook R., and Brandl C.J. 1997. Identification of native complexes containing the yeast coactivator/repressor proteins NGG1/ADA3 and ADA2. *J. Biol. Chem.* **272:** 5571.

Silverman S.J. and Fink G.R. 1984. Effects of Ty insertions on HIS4 transcription in *Saccharomyces cerevisiae. Mol. Cell. Biol.* **4:** 1246.

Sterner D.E., Grant P.A., Roberts S.M., Duggan L.J., Belotserkovskaya R., Pacella L.A., Winston F., Workman J.L., and

Berger S.L. 1998. Functional organization of the yeast SAGA complex: Distinct components involved in structural integrity, nucleosome acetylation, and TBP binding. *Mol. Cell. Biol.* (in press).

Struhl K. and Moqtaderi Z. 1998. The TAFs in the HAT. *Cell* **94:** 1.

Swanson M.S. and Winston F. 1992. SPT4, SPT5 and SPT6 interactions: Effects on transcription and viability in *Saccharomyces cerevisiae. Genetics* **132:** 325.

Turkel S., Liao X.B., and Farabaugh P.J. 1997. GCR1-dependent transcriptional activation of yeast retrotransposon Ty2-917. *Yeast* **13:** 917.

Winston F., Durbin K.J., and Fink G.R. 1984a. The *SPT3* gene is required for normal transcription of Ty elements in *S. cerevisiae. Cell* **39:** 675.

Winston F., Chaleff D. T., Valent B., and Fink G.R. 1984b. Mutations affecting Ty-mediated expression of the HIS4 gene of *Saccharomyces cerevisiae. Genetics* **107:** 179.

Winston F., Dollard C., Malone E.A., Clare J., Kapakos J.G., Farabaugh P., and Minehart P.L. 1987. Three genes are required for trans-activation of Ty transcription in yeast. *Genetics* **115:** 649.

Yu J., Madison J.M., Mundlos S., Winston F., and Olsen B.R. 1998. Characterization of a human homologue of the *Saccharomyces cerevisiae* transcription factor Spt3 (SUPT3H). *Genomics* **53:** (in press).

Zhang S., Burkett T.J., Yamashita I., and Garfinkel D.J. 1997. Genetic redundancy between SPT23 and MGA2: Regulators of Ty-induced mutations and Ty1 transcription in *Saccharomyces cerevisiae. Mol. Cell. Biol.* **17:** 4718.

Specificity of ATP-dependent Chromatin Remodeling at the Yeast *PHO5* Promoter

E.S. HASWELL AND E.K. O'SHEA

Department of Biochemistry and Biophysics, University of California, San Francisco,
San Francisco, California 94143-0448

A basic problem in biology is how the protein machinery for transcription gains access to specific DNA sequences that are packaged into chromatin. The repeating structural subunit of chromatin is the nucleosome, a globular histone octamer around which 146 bp of DNA are wrapped (McGhee and Felsenfeld 1980; Luger and Richmond 1998). To occupy promoter sequences that are packaged into a nucleosome, many regulators and components of the transcription machinery must compete with histones for DNA binding. Both biochemical and genetic analyses demonstrate that the packaging of DNA into nucleosomes can inhibit transcription initiation and that this inhibition can be overcome at the level of factor binding (Paranjape et al. 1994; Elgin 1995).

Recent studies suggest that two classes of enzymatic activities may help transcriptional regulators bind nucleosomal DNA: histone acetyltransferases (Mizzen and Allis 1998) and ATP-dependent chromatin-remodeling machines (Armstrong and Emerson 1998; Cairns 1998). Covalent modification of the flexible tails of histones through acetylation may affect protein/DNA interactions within a nucleosome or protein/protein interactions between nucleosomes. ATP-dependent chromatin-remodeling machines are thought to change chromatin structure by reversibly altering DNA-histone contacts in an ATP-dependent manner.

A number of ATP-dependent chromatin-remodeling activities have been described in yeast (Cairns et al. 1994, 1996; Peterson et al. 1994), flies (Tsukiyama and Wu 1995; Ito et al. 1997; Varga-Weiss et al. 1997), and mammals (Orphanides et al. 1988), and it is likely that there are many more yet to be discovered. It is not known how many of these complexes are involved in the regulation of transcription, how specificity for particular chromosomal loci might be created, or how ATP hydrolysis is employed to alter chromatin structure.

One approach to the investigation of chromatin remodeling during transcription is to study in vitro chromatin rearrangements that are well characterized in vivo. Biochemical analysis of such a system can be used to identify components of the chromatin-remodeling activity and to elucidate the mechanism of the reaction. If a genetically malleable organism is used for in vitro studies, results from biochemical studies can be readily tested for in vivo relevance. Taking such an approach, we have developed an in vitro system to study chromatin remodeling of the *PHO5* promoter in *Saccharomyces cerevisiae*.

The yeast *PHO5* promoter serves as a model system for

the study of the effects of chromatin on gene regulation (Svaren and Horz 1995, 1997). The *PHO5* gene, which encodes a secreted acid phosphatase, is repressed when yeast are grown under high phosphate conditions and is induced upon phosphate starvation. Under repressing conditions, two pairs of positioned nucleosomes flank an 80-bp hypersensitive site in the *PHO5* promoter. Upon activation of the gene, these four nucleosomes no longer protect the DNA from digestion with nucleases (Fig. 1). Two transcription factors are required for *PHO5* expression: Pho4, a helix-loop-helix protein with an acidic activation domain, and Pho2, a homeodomain protein. Pho4 and Pho2 bind to two elements in the promoter, UASp1 and UASp2, that are required for *PHO5* expression (Barbaric et al. 1998).

The mechanism underlying the alteration in *PHO5* chromatin structure upon phosphate starvation is not known. The chromatin transition requires Pho2 and Pho4 (Fascher et al. 1990), as well as the Pho4-binding sites in UASp1 and UASp2 (Fascher et al. 1993). Additionally, the DNA-binding domain of Pho4 does not suffice to remodel chromatin in low phosphate (Svaren et al. 1994). Previous studies have demonstrated that the generation of active chromatin at the *PHO5* promoter is independent of active transcription (Fascher et al. 1993) or replication (Schmid et al. 1992). *PHO5* chromatin remodeling in spheroplasts requires the presence of glucose (Schmid et al. 1992), which suggests that remodeling requires energy such as ATP or acetyl CoA. A genetic approach has shown that neither SWI/SNF, an ATP-dependent chromatin-remodeling complex (Schneider 1995; Gaudreau et al. 1997), nor the histone acetylase GCN5 (Pollard and Peterson

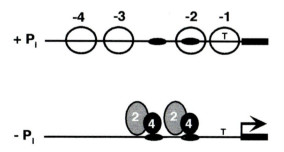

Figure 1. Chromatin structure of the *PHO5* promoter under high- and low-phosphate conditions. (*Open circles*) Positioned nucleosomes detected in high phosphate; (*closed ellipses*) location of the Pho4- and Pho2-binding sites. (T) TATA box; (*thick line*) *PHO5* open reading frame.

1997) is required for *PHO5* chromatin rearrangement in low phosphate (however, see Gregory et al. 1998).

We have established an in vitro system capable of *PHO5* chromatin remodeling that we will use to identify and characterize factors required for this process. We use minichromosomes carrying the *PHO5* promoter and gene, purified from yeast with intact chromatin, as a source of chromatin template. Remodeling of the *PHO5* promoter in vitro occurs only in the presence of ATP, Pho4, and Pho2, and a fraction derived from yeast nuclear extract termed S(0.3). Remodeling in vitro recapitulates several hallmarks of the physiological *PHO5* chromatin rearrangements, including restriction to the promoter region and a requirement for Pho4-binding sites (E.S. Haswell and E.K. O'Shea, in prep.).

As mentioned above, the SWI/SNF complex is not required in vivo for the generation of active chromatin at *PHO5*. Therefore, we tested the requirement for SWI/SNF in our in vitro assay by using S(0.3) extract and *PHO5* minichromosomes prepared from an snf6Δ strain. Normal levels of remodeling were obtained with these reagents (E.S. Haswell and E.K. O'Shea, in prep.), indicating that the SWI/SNF complex is not required for *PHO5* chromatin rearrangements in vitro.

That the S(0.3)-dependent in vitro chromatin-remodeling reaction so closely resembles the *PHO5* chromatin transition in vivo suggests that the factor responsible may be a physiologically relevant activity. If this is true, SWI/SNF, which is dispensable for *PHO5* activation in vivo, should not remodel chromatin in vitro. We tested this prediction by examining the activity of purified SWI/SNF in our in vitro *PHO5* chromatin-remodeling system.

METHODS

PHO5 minichromosome and transcription factor preparations are described at length in another report (E.S. Haswell and E.K. O'Shea, in prep.).

In vitro chromatin remodeling. Minichromosomes (10 μl) (~10 ng DNA in 250 mM NaCl, 5 mM MgCl$_2$, 10 mM PIPES at pH 7.3, 0.5 mM EGTA) were incubated in 50-μl reactions containing 2 μg/ml dCdG, 0.1 mg/ml bovine serum albumin, 12 mM HEPES at pH 7.5, 6 mM MgCl$_2$, 1 mM dithiothreitol, 5% glycerol, and 0.5 mM CaCl$_2$. After addition of Pho4 or Pho4 DBD peptide to 90 nM, Pho2 to approximately 20 nM, and SWI/SNF complex to approximately 1 nM, reactions were incubated at room temperature for 15 minutes. An ATP regeneration mix (final concentrations: 0.2 μg/ml of creatine kinase in 10 mM glycine at pH 8, 30 mM creatine phosphate, and 0.5 mM ATP) was added, and reactions were incubated at 30°C for 30 minutes. Reactions were split in two and digested with micrococcal nuclease (Worthington) at 37°C for 5 minutes. Digestion was stopped with 0.5% SDS and 25 mM EDTA, and DNA was purified by overnight treatment with proteinase K at 37°C, followed by phenol extractions, chloroform extraction, and ethanol precipitation.

Southern blotting. Samples were loaded on 1.2% agarose gels and run in 0.5× TBE at 4 V/cm for 4 hours. Gels were prepared for blotting as described previously (Sambrook et al. 1989) and blotted to nylon membrane (Amersham) overnight in 20× SSC. Prehybridization and hybridization with random prime labeled probes were performed in Rapid Hyb Buffer (Amersham). Typically, fragments of 100–300 bp embedded in low-melting-point agarose were random prime-labeled overnight at room temperature. Probes for analysis of nucleosome –2 and the open reading frame corresponded to a *Cla*I/*Bst*EII fragment and an *Xmn*I to *Mfe*I fragment of *PHO5*, respectively.

RESULTS AND DISCUSSION

Although SWI/SNF is not required for remodeling in vitro, we wished to determine whether SWI/SNF is sufficient to remodel chromatin in our in vitro system. Purified SWI/SNF was incubated with *PHO5* minichromosomes and an ATP-regeneration system in the presence or absence of the transcription factors Pho2 and Pho4. After the remodeling reaction, chromatin was digested with micrococcal nuclease and the DNA was purified. Samples were separated by agarose gel electrophoresis, transferred to nylon membrane, and hybridized to a ^{32}P-labeled probe derived from *PHO5* promoter sequence.

Much to our surprise, Pho4-dependent remodeling occurred in the presence of purified SWI/SNF and ATP (Fig. 2A). This SWI/SNF-dependent chromatin-remodeling activity resembled the in vivo chromatin transition in that little remodeling occurred when the Pho4 DNA-binding domain alone was used in place of the full-length protein, and the remodeling activity was predominantly restricted to the *PHO5* promoter region (Fig. 2B). We conclude that SWI/SNF and an activity distinct from SWI/SNF contained in the S(0.3) fraction are functionally indistinguishable in our in vitro chromatin-remodeling system.

Specificity of In Vitro Remodeling Reactions

One model to explain our observations that SWI/SNF is not required for the *PHO5* chromatin transition in vivo, yet is sufficient to remodel *PHO5* minichromosomes in our in vitro system, is that the in vitro assay is lacking a specificity factor. For example, the repressed state of the *PHO5* promoter might be maintained by a specific repressor, and chromatin remodeling in the presence of this repressor might require a specific remodeling complex. If such a protein were removed during template purification, chromatin remodeling would no longer be specific. Although we have taken considerable care to duplicate the physiological reaction in vitro, there are many ways in which such important components might be omitted from the reaction.

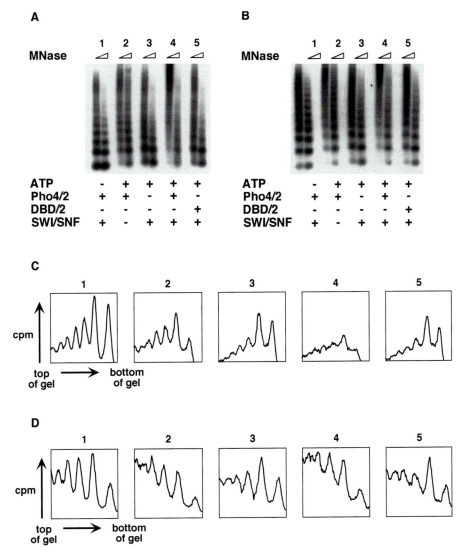

Figure 2. SWI/SNF-dependent chromatin remodeling of *PHO5* minichromosomes in vitro. In vitro remodeling reactions were performed, and samples were processed as described in Methods. In *A*, a Southern blot was probed with a radioactively labeled sequence derived from the *PHO5* promoter sequence packaged into nucleosome –2. Chromatin remodeling of *PHO5* minichromosomes occurs in the presence of SWI/SNF, Pho2, full-length Pho4, and ATP. Little remodeling is seen when the Pho4 DNA-binding domain is added in place of full-length Pho4. (*B*) Blot shown in *A* was stripped and reprobed with a sequence corresponding to the *PHO5* sequence approximately 1 kb from its ATG. Much less remodeling is seen in lane *4* when this probe is used. This suggests that most SWI-SNF-dependent chromatin remodeling is localized to the *PHO5* promoter region of the minichromosomes. *C* and *D* are quantitative line analyses of *A* and *B*, respectively.

Redundancy in *PHO5* Chromatin-remodeling Activities

A more interesting possibility is that several ATP-dependent factors are capable of remodeling chromatin at the *PHO5* promoter. In this scenario, neither SWI/SNF nor the S(0.3) fraction are required for in vitro remodeling of *PHO5* minichromosomes, as each is sufficient without the other. Functional redundancy between two or more ATP-dependent chromatin-remodeling machines would explain the *PHO5* inducibility of *swi⁻* or *snf⁻* mutants, because *PHO5* chromatin could still be remodeled by the factor contributed by the S(0.3) extract fraction.

Our biochemical analyses and previous genetic experiments are consistent with a nonspecific mode of action for ATP-dependent remodeling machines at the *PHO5* promoter. If the Pho4 DNA-binding domain alone is expressed in yeast in place of full-length Pho4, no chromatin remodeling is seen (Svaren et al. 1994). Thus, an activation domain appears to be required for remodeling in vivo and in our in vitro remodeling reactions. However, this function is not specific to Pho4, as the VP16 activation domain can substitute when fused to the Pho4 DNA-binding domain (Svaren et al. 1994), and Gal4 can completely replace Pho4 if UASp1 is replaced with a Gal4 site (Gregory et al. 1998).

Models for *PHO5* Chromatin Remodeling

The current paradigm for activation of transcription is that many activators promote transcription through recruitment of the preinitiation complex. Perhaps the same mechanism is used for targeting chromatin remodeling to specific promoters. Activation domains might recruit remodeling factors to promoters through direct interactions or indirectly through the holoenzyme (Gaudreau et al. 1997). This model is supported by evidence that the glucocorticoid receptor interacts with SWI/SNF in yeast (Yoshinaga et al. 1992), and with BRG in mammalian cells (Fryer and Archer 1998).

Figure 3A illustrates how this type of mechanism might play out at the *PHO5* promoter. According to this model, the Pho4 activation domain contacts remodeling complexes directly and increases the stability of their interaction with the *PHO5* promoter. The ability to interact with remodeling complexes may be a common property of acidic activation domains, as VP16 and Gal4 can substitute for Pho4 in this role. Perhaps acidic activators interact with ATP-dependent chromatin-remodeling complexes through the Snf2 family member found in each complex. However, no yeast activator required at a SWI/SNF-dependent promoter has been shown to interact physically with SWI/SNF. In addition, none of the Snf2-like proteins in RSC (Cairns et al. 1996), NURF (Tsukiyama and Wu 1995), or ACF (Ito et al. 1997) have been reported to interact with transcription factors. It is also notable that SWI/SNF from budding yeast (Côté et al. 1994) and humans (Kwon et al. 1994), as well as RSC (Cairns et al. 1996) and NURF (Tsukiyama and Wu 1995), is capable of remodeling a mononucleosome without a sequence-specific DNA-binding protein. This implies that ATP-dependent chromatin-remodeling factors do not require the presence of a transcriptional activator to function.

Given the paucity of evidence for a direct interaction between sequence-specific DNA-binding proteins and SWI/SNF, other mechanisms for *PHO5* remodeling must be considered. In a second model (Fig. 3B), ATP-dependent remodeling machines are not recruited to the *PHO5* pro-

moter but constitutively modify all histone/DNA contacts (Mizuguchi et al. 1997). The resulting generalized loss of nucleosome stability facilitates DNA binding by Pho4, and as a result, an open chromatin conformation is created. The insufficiency of the DNA-binding domain of Pho4 to remodel chromatin might be explained if the ability of Pho4 to bind chromatin is enhanced through interaction with Pho2, as this requires a different domain of Pho4 (Barbaric et al. 1998). Improved binding of Pho4 to chromatinized UASp2 might facilitate stabilization of an open conformation created by the ATP-dependent activities.

SWI/SNF-dependent Promoters

At least two distinct ATP-dependent remodeling complexes can act at the *PHO5* promoter in vitro. However, expression of a number of genes in yeast, including the *HO*, *SUC2*, and *INO1* genes, requires the SWI/SNF complex (for review, see Winston and Carlson 1992). These promoters appear to be able to distinguish between ATP-dependent remodeling activities. Perhaps the particular characteristics of the interaction between these promoters and their transcriptional activators create a requirement for the SWI/SNF complex. The ATP-dependent remodeling machines so far characterized differ from one another in a number of biochemical assays (see, e.g., Georgel and Tsukiyama 1997; Mizuguchi et al. 1997). In addition, dependence on SWI/SNF is not a fixed state. A SWI/SNF-independent promoter can be made SWI/SNF-dependent by destabilizing the interaction between an activator and its DNA-binding site, either by employing low-affinity DNA-binding sites or by placing them in a nucleosome (Burns and Peterson 1997). Thus, a complex interplay between chromatin structure, DNA binding by transcriptional activators, and the remodeling activities themselves may determine which ATP-dependent chromatin-remodeling factor acts at a particular promoter.

Studies of chromatin-remodeling systems have raised a number of intriguing questions: How many ATP-dependent chromatin-remodeling activities are there in budding yeast? Which of these activities are involved in transcriptional regulation? How do these activities find the pro-

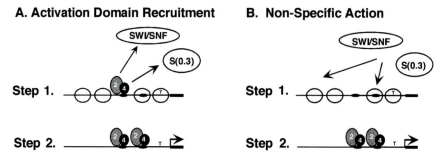

Figure 3. Two models to explain the sufficiency of SWI/SNF to remodel *PHO5* chromatin in vitro. (*A*) The Pho4 and Pho2 transcriptional activators are able to bind the nucleosome-free region of the *PHO5* promoter in the absence of remodeling activities. The activation domain of Pho4 then recruits either the SWI/SNF complex or the complex contained in the S(0.3) fraction. These complexes remodel chromatin and allow occupation of the nucleosomal Pho4 site. Stable occupation of these sites allows recruitment of the RNA polymerase holoenzyme and activates transcription initiation. (*B*) Constitutive activity of either the SWI/SNF complex or the activity in the S(0.3) fraction at the *PHO5* promoter allows Pho4 and Pho2 to bind the promoter. The binding of Pho4 and Pho2 stabilizes a conformation of the promoter that is accessible to RNA polymerase, and transcriptional activation ensues.

moters on which they act? We hope to address these questions using a combination of biochemistry and genetics. The in vitro *PHO5* chromatin-remodeling assay developed in our laboratory and available tools for the genetic manipulation of budding yeast should support complementary approaches to the investigation of *PHO5* promoter chromatin remodeling.

ACKNOWLEDGMENTS

We thank Craig Peterson for providing purified SWI/SNF complex, Mary Maxon for purified Pho2, and David King and Robert Tjian for the Pho4 DNA-binding domain peptide. We are also grateful to Nicole Miller Rank and Keith Yamamoto for comments on the manuscript. This work was supported by National Institutes of Health grant GM-51377 to E.K.O. and a National Science Foundation graduate fellowship to E.S.H.

REFERENCES

Armstrong J.A. and Emerson B.M. 1998. Transcription of chromatin: These are complex times. *Curr. Opin. Genet. Dev.* **8:** 165.

Barbaric S., Munsterkotter M., Goding C., and Horz W. 1998. Cooperative Pho2-Pho4 interactions at the *PHO5* promoter are critical for binding of Pho4 to UASp1 and for efficient transactivation by Pho4 at UASp2. *Mol. Cell. Biol.* **18:** 2629.

Burns L.G. and Peterson C.L. 1997. The yeast SWI/SNF complex facilitates binding of a transcriptional activator to nucleosomal sites *in vivo. Mol. Cell. Biol.* **17:** 4811.

Cairns B.R. 1998. Chromatin remodeling machines: Similar motors, ulterior motives. *Trends Biochem. Sci.* **23:** 20.

Cairns B.R., Kim Y.J., Sayre M.H., and Laurent B.C. 1994. A multisubunit complex containing the Swi1/Adr6, Swi2/Snf2, Swi3, Snf5, and Snf6 gene products isolated from yeast. *Proc. Natl. Acad. Sci.* **91:** 1950.

Cairns B.R., Lorch Y., Li Y., Zhang M., Lacomis L., Erdjument-Bromage H., Tempst P., Du J., Laurent B., and Kornberg R. 1996. RSC, an essential, abundant chromatin-remodeling complex. *Cell* **87:** 1249.

Côté J., Quinn J., Workman J.L., and Peterson C.L. 1994. Stimulation of GAL4 derivative binding to nucleosomal DNA by the yeast SWI/SNF complex. *Science* **265:** 53.

Elgin S.C.R., Ed. 1995. *Chromatin structure and gene expression.* IRL Press, Oxford, United Kingdom.

Fascher K.-D., Schmitz J., and Horz W. 1990. Role of *trans*-activating proteins in the generation of active chromatin at the *PHO5* promoter in *S. cerevisiae. EMBO J.* **9:** 2523.

———. 1993. Structural and functional requirements for the chromatin transition at the *PHO5* promoter in *Saccharomyces cerevisiae* upon *PHO5* activation. *J. Mol. Biol.* **231:** 658.

Fryer C.J. and Archer T.K. 1998. Chromatin remodeling by the glucocorticoid receptor requires the BRG1 complex. *Nature* **393:** 88.

Gaudreau L., Schmid A., Blaschke D., Ptashne M., and Horz W. 1997. RNA polymerase holoenzyme recruitment is sufficient to remodel chromatin at the yeast *PHO5* promoter. *Cell* **89:** 55.

Georgel P.T. and Tsukiyama T. 1997. Role of histone tails in nucleosome remodeling by *Drosophila* NURF. *EMBO J.* **16:** 4717.

Gregory P.D., Schmid A., Zavari M., Lui L., Berger S.L., and Horz W. 1998. Absence of Gcn5 HAT activity defines a novel state in the opening of chromatin at the *PHO5* promoter in yeast. *Mol. Cell* **1:** 495.

Ito T., Bulger M., Pazin M.J., Kobayashi R., and Kadonaga J.T. 1997. ACF, an ISWI-containing and ATP-utilizing chromatin assembly and remodeling factor. *Cell* **90:** 145.

Kwon H., Imbalzano A.N., Khavari P.A., Kingston R.E., and Green M.R. 1994. Nucleosome disruption and enhancement of activator binding by a human SWI/SNF complex. *Science* **370:** 477.

Luger K. and Richmond T.J. 1998. DNA binding within the nucleosome core. *Curr. Opin. Struct. Biol.* **8:** 33.

McGhee J.D. and Felsenfeld G. 1980. Nucleosome structure. *Annu. Rev. Biochem.* **49:** 1115.

Mizuguchi G., Tsukiyama T., Wisniewski J., and Wu C. 1997. Role of nucleosome remodeling factor NURF in transcriptional activation of chromatin. *Mol. Cell* **1:** 141.

Mizzen C.A. and Allis C.D. 1998. Linking histone acetylation to transcriptional regulation. *Cell. Mol. Life Sci.* **54:** 6.

Orphanides G., LeRoy G., Chang C.H., Luse D.S., and Reinberg D. 1988. FACT, a factor that facilitates transcript elongation through nucleosomes. *Cell* **92:** 105.

Paranjape S.M., Kamakaka R.T., and Kadonaga J.T. 1994. Role of chromatin structure in the regulation of transcription by RNA polymerase II. *Annu. Rev. Biochem.* **63:** 265.

Peterson C.L., Dingwall A., and Scott M.P. 1994. Five SWI/SNF gene products are components of a large multisubunit complex required for transcriptional enhancement. *Proc. Natl. Acad. Sci.* **91:** 2905.

Pollard K.J. and Peterson C.L. 1997. Role of ADA/GCN5 products in antagonizing chromatin-mediated transcriptional repression. *Mol. Cell. Biol.* **17:** 6212.

Sambrook J., Fritsch E.F., and Maniatis T. 1989. *Molecular cloning: A laboratory manual,* 2nd edition. Cold Spring Harbor Laboratory Press, Cold Spring Harbor, New York.

Schmid A., Fascher K.-D., and Horz W. 1992. Nucleosome disruption at the yeast *PHO5* promoter upon *PHO5* induction occurs in the absence of DNA replication. *Cell* **71:** 853.

Schneider K. 1995. "The regulation of *PHO5* expression." Ph.D. thesis, University of California, San Francisco.

Svaren J. and Horz W. 1995. Interplay between nucleosomes and transcription factors at the yeast *PHO5* promoter. *Semin. Cell Biol.* **6:** 177.

———. 1997. Transcription factors vs nucleosomes: Regulation of the *PHO5* promoter in yeast. *Trends Biochem. Sci.* **22:** 93.

Svaren J., Schmitz J., and Horz W. 1994. The transactivation domain of Pho4 is required for nucleosome disruption at the *PHO5* promoter. *EMBO J.* **13:** 4856.

Tsukiyama T. and Wu C. 1995. Purification and properties of an ATP-dependent nucleosome remodeling factor. *Cell* **83:** 1011.

Varga-Weiss P.D., Wilm M., Bonte E., Dumas K., Mann M., and Becker P.B. 1997. Chromatin-remodeling factor CHRAC contains the ATPases ISWI and topoisomerase II. *Nature* **388:** 598.

Winston F. and Carlson M. 1992. Yeast SNF/SWI transcriptional activators and the SPT/SIN chromatin connection. *Trends Biochem. Sci.* **8:** 387.

Yoshinaga S.K., Peterson C.L., Herskowitz I., and Yamamoto K.R. 1992. Roles of SWI1, SWI2, and SWI3 proteins for transcriptional enhancement by steroid receptors. *Science* **258:** 1598.

Role of Chromatin Structure and Distal Enhancers in Tissue-specific Transcriptional Regulation In Vitro

R. Bagga, J.A. Armstrong, and B.M. Emerson

Regulatory Biology Laboratory, The Salk Institute for Biological Studies, La Jolla, California 92037

Appropriate differentiation and development of higher organisms require precisely regulated expression of multiple genes. The primary control for most genes is exerted at the level of transcription. This involves the combinatorial action of tissue-specific and ubiquitous transcription factors acting at regulatory sequences that are proximal (promoters) or distal (enhancers, insulators, silencers, and locus control regions [LCRs]) to a gene. The existence of functionally distinct *cis*-acting elements indicates that the high degree of regulation involved in coordinated gene expression within a complex organism requires more intricate circuitry than a simple promoter can provide to turn genes on and off. A critical aspect of this circuitry and coordination is the regulation imposed upon genes within a complex nuclear environment.

The packaging of DNA into chromatin within the eukaryotic nucleus is highly organized and plays a critical part in regulating gene expression and other nuclear processes, such as DNA replication, recombination, repair, and cell cycle progression. The basic structural unit of chromatin is the nucleosome which consists of approximately 145 bp of DNA wrapped in 1.75 superhelical turns around a histone octamer containing two molecules each of histones H2A, H2B, H3, and H4. This unit is repeated once every 200 +/–40 bp as a nucleosomal array in chromosomal DNA. The array is further compacted into a higher-order structure by the association of histone H1 with nucleosomes within the array. The functional consequence of chromatin packaging, in general, is to restrict access of DNA to a variety of DNA-binding proteins that regulate gene activity. Biochemical and genetic evidence amply demonstrates that nucleosomes are normally repressive for transcription. However, several elegant mechanisms have evolved that modulate chromatin structure to increase the accessibility of DNA for protein interaction (for review, see Armstrong and Emerson 1998; Cairns 1998; Kadonaga 1998). In this way, genes are effectively programmed to be active or inactive in a particular cell type or poised for expression at a specific stage of development or in response to environmental signals.

Chromatin structural changes can occur at several levels: either globally by the decondensation (active genes) or condensation (inactive genes) of a large chromosomal domain or locally by the disruption (active) or formation (inactive) of one or more nucleosomes on a promoter or enhancer region. Global chromatin structural changes have been shown to occur in the human β-globin gene locus by the action of the distal LCR (Forrester et al. 1990).

In addition, active genes are characterized by containing hyperacetylated histones and undermethylated DNA. Interestingly, both global and local levels of chromatin structural perturbation often require the interaction of regulatory proteins with histone amino-terminal tails within the nucleosome, which are also the main targets of post-translational modification (for review, see Davie 1998). Two critical pathways that facilitate this interaction involve distinct protein complexes that function either as motors to disrupt nucleosomes (ATP-driven chromatin remodeling complexes) or as enzymatic machinery to modify histones chemically and alter their affinity for DNA (histone acetylases, HATs/deacetylases, HDACs/kinases).

It is clear that multiple levels of control are involved in regulated gene expression: from the activation of the chromosomal domain in which a gene resides to the formation of a basal initiation complex on a given promoter within the domain. Questions remain as to how tissue- or developmental-stage-specific expression is established and how coordinate expression of multiple genes is achieved. In addition, the mechanism by which critical DNA control elements, often acting at long-range, such as enhancers, insulators, silencers, and LCRs, regulate transcription is still poorly understood. This paper focuses upon our recent findings concerning the role of tissue-specific DNA-binding proteins and chromatin remodeling complexes in generating regulated gene expression within a complex chromosomal locus and the mechanism by which long-range enhancers and DNA architectural proteins control promoter activity.

ERYTHROID-SPECIFIC REGULATION OF THE HUMAN β-GLOBIN GENE LOCUS

The human β-globin family provides an excellent system in which to decipher the mechanisms involved in tissue-specific and developmental gene regulation. This family resides within a chromosomal locus spanning 100 kb and is composed of five genes (5′-LCR-ε-Gγ-Aγ-δ-β-3′) whose individual members are expressed at different times during erythroid maturation (Fig. 1). The epsilon gene, ε, is expressed in the embryonic yolk sac, then the fetal gamma genes, Gγ-Aγ are activated in the fetal liver, and finally the adult δ- and β-globin genes are expressed in the bone marrow from the late fetal period onward (for review, see Stamatoyannopoulos and Nienhuis 1994). Transcriptional control of this gene family occurs at sev-

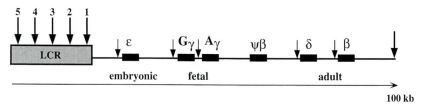

Figure 1. Human β-globin gene locus. Arrows designate nuclease-hypersensitive sites in chromatin that are constitutive (HS 5), erythroid-specific (HSs 1–4), or erythroid-stage-specific (small arrows).

eral levels and involves well-characterized changes in chromatin structure (for review, see Orkin 1995; Martin et al. 1996). For example, the 100-kb chromosomal locus is maintained in an active (DNase-sensitive) configuration in erythroid cells by the action of the LCR, and localized structural changes (DNase-hypersensitive sites) are associated with individual promoters as they are activated during development (Groudine et al. 1983; Forrester et al. 1990). The LCR resides 5–20 kb upstream of the embryonic ε gene (Grosveld et al. 1987) and adopts an activated chromatin structure consisting of five Dnase-hypersensitive sites (HSs 1–5) only in erythroid cells (Tuan et al. 1985). Linkage of the LCR to isolated β-globin genes confers high levels of integration site-independent expression in transgenic mice (Grosveld et al. 1987). However, gene-proximal regulatory regions are sufficient to regulate correct tissue-specific and developmental expression (Starck et al. 1994). The LCR can be dissected into subregions (HSs 1–5) that have distinct functions. For example, when examined individually, HS 2 contains the strongest enhancer activity, whereas HS 3 has the most pronounced insulator activity, i.e., protection against position-effect variation of transgene expression (for review, see Orkin 1995; Martin et al. 1996). Taken together, most studies are in agreement that the LCR functions as a holocomplex to establish and maintain an active gene locus. How stage-specific gene expression is achieved once the β-globin locus is activated by the LCR remains to be elucidated.

ACTIVATION OF β-GLOBIN CHROMATIN STRUCTURE BY ERYTHROID PROTEINS AND THE ROLE OF CHROMATIN REMODELING COMPLEXES

We have previously established in vitro transcription and chromatin assembly systems that accurately reproduce many important aspects of chick β^A-globin gene regulation in order to define the mechanisms by which they occur. Our past studies have shown that two erythroid-specific proteins, GATA-1 and NF-E4, are sufficient to activate transcription of a chromatin-assembled chick β^A-globin gene by generating an accessible nucleosomal structure in the promoter (Barton et al. 1993). We have also reproduced erythroid-specific, developmentally regulated β^A-globin transcription using cosmids containing the entire 38-kb chick β-globin chromosomal locus when assembled into synthetic nuclei in the presence of stage-specific erythroid proteins (Barton and

Emerson 1994). We have recently applied a similar chromatin reconstitution approach to analyze the mechanism of transcriptional regulation in the human β-globin gene family.

The LCR and individual globin genes within the human β-globin locus contain binding sites for all of the well-characterized erythroid-restricted DNA-binding proteins, such as GATA-1, AP-1/NF-E2, EKLF, and other CACC-binding factors. NF-E2 is responsible for HS 2 enhancer activity (Ney et al. 1990; Talbot and Grosveld 1991), and EKLF is a critical component of HS 3 insulator function within the LCR (Tewari et al. 1998). Interestingly, EKLF binding to a CACC element at –90 in the human β-globin promoter is essential for stage-specific expression of that gene (Bieker and Southwood 1995; Nuez et al. 1995; Perkins et al. 1995). We have examined the ability of NF-E2 and EKLF to generate an active chromatin structure within the LCR HS 2 enhancer and the β-globin promoter, respectively, to gain insight into how these critical erythroid factors establish regulated transcription in a complex gene locus. In these studies, we employ a biochemical approach to reconstitute cloned β-globin genes or loci into chromatin, using assembly extracts from *Drosophila* embryos (Bulger and Kadonaga 1994) and analyze the ability of specific erythroid proteins to generate an altered nucleosomal structure that is correlated with transcriptional activation.

We initially analyzed whether the interaction of NF-E2 with the LCR-HS 2 enhancer could generate the altered chromatin structure observed in erythroid cells when the LCR is tissue-specifically activated. NF-E2 is a basic leucine zipper heterodimer consisting of the hematopoietic-specific subunit, p45, and its ubiquitous DNA-binding partner, p18 (Andrews et al. 1993a,b). NF-E2 binds to two tandem AP1-like sites which form the core of the HS 2 enhancer. Our experiments revealed that recombinant NF-E2 forms a DNase-I-hypersensitive site in the HS 2 enhancer similar to the one observed in erythroid cells. Moreover, NF-E2 binding in vitro results in a disruption of nucleosome structure of up to 200 bp (Fig. 2). Surprisingly, the ubiquitous DNA-binding subunit p18 requires the hematopoietic-specific protein p45 to stabilize its interaction with chromatin, but not DNA. This implies that tissue-specific NF-E2 activation of the HS 2 enhancer through chromatin is conferred by the p45 non-DNA-binding subunit. Nucleosome disruption by NF-E2 in this reconstituted system requires ATP, suggesting the involvement of energy-dependent nucleosome remodeling factors within the *Drosophila* embryonic assembly extract (Armstrong and Emerson 1996).

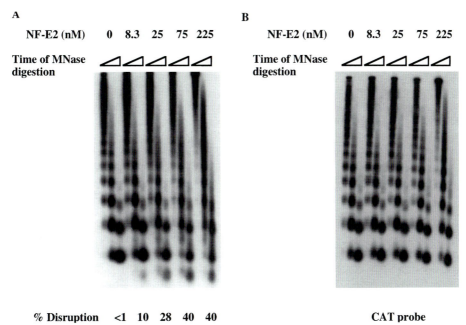

A

NF-E2 (nM)   0   8.3   25   75   225

Time of MNase digestion

B

NF-E2 (nM)   0   8.3   25   75   225

Time of MNase digestion

% Disruption   <1   10   28   40   40

CAT probe

Figure 2. Chromatin assembly in the presence of NF-E2 results in nucleosome disruption in the LCR/HS2 enhancer. LCR/HS2βCAT plasmid DNA was incubated with recombinant NF-E2 prior to chromatin assembly. Nucleosome structural analysis was performed by digestion with micrococcal nuclease (MNase) and Southern blot transfer. (*A*) The blotted membrane was hybridized with an oligonucleotide probe corresponding to the NF-E2 site; (*B*) membrane was stripped and rehybridized with an oligonucleotide probe within the CAT reporter gene, 1 kb downstream from the NF-E2 sites.

We next examined the ability of the erythroid-restricted DNA-binding protein, EKLF, to activate transcription of chromatin-assembled β-globin genes in vitro. Several important studies have demonstrated that EKLF is essential for β-globin activation through its interaction with a –90 CACC box in the proximal promoter. Naturally occurring mutations in this region are found in human thalessemias and result in a significant decrease in β-globin expression (Feng et al. 1994). EKLF (erythroid Krüppel-like factor) is a zinc-finger-containing transcriptional activator that is present throughout erythroid development (Miller and Bieker 1993). Interestingly, EKLF knock-out experiments have shown that expression of the embryonic ε- and fetal γ-globin genes occurs normally in mice during the embryonic stage of hematopoiesis in the yolk sac. However, these mice later develop a fatal anemia and die in utero during definitive erythropoiesis in the fetal liver, indicating that EKLF is critical for the final stages of red cell maturation (Nuez et al. 1995; Perkins et al. 1995). Since this is the stage in which fetal γ- and adult β-globin gene competition occurs, an EKLF deletion results in a globin chain imbalance due to a decrease in β-globin synthesis and concomitant increase in γ-globin transcription (Wijgerde et al. 1996). Thus, EKLF has been clearly shown to have a major role in the developmentally regulated switch from fetal to adult β-globin activation. Two models have been proposed to explain how EKLF mediates this switch: (1) by altering the affinity of the distal LCR so that it preferentially interacts with the β-globin promoter (Wijgerde et al. 1996) or (2) by acting in a gene-proximal manner to establish a chromatin-accessible, transcriptionally active β-globin promoter (for discussion, see Martin et al. 1996). Indeed, in the absence of EKLF, a DNase-hypersensitive

chromatin structure fails to appear either in the β-globin promoter or in HS 3 of the LCR (Wijgerde et al. 1996; Tewari et al. 1998).

Our results demonstrate that unlike NF-E2, EKLF cannot bind to its recognition site within the β-globin promoter when assembled into chromatin. Instead, EKLF requires a separate coactivator to facilitate its interaction. This indicates that different mechanisms exist for NF-E2 and EKLF to gain accessibility to their sites within the β-globin gene locus. We have purified the EKLF-dependent coactivator and found that it is an ATP-dependent chromatin remodeling complex. In combination with its coactivator, EKLF binds to the –90 CACC site in chromatin and generates a nuclease-hypersensitive region throughout the proximal promoter. This accessible chromatin structure can then efficiently form an initiation complex and be actively transcribed. Neither EKLF nor the coactivator alone can remodel chromatin or transcriptionally activate the β-globin gene. Thus, stage-specific expression of β-globin depends not only on EKLF, but also on a mammalian chromatin remodeling complex that facilitates its interaction with a nucleosome-repressed promoter. Since the ε- and γ-globin promoters also contain CACC motifs, yet the absence of EKLF in vivo does not directly affect their expression, other CACC-binding proteins including Sp1 and members of the Krüppel-like zinc finger protein family, such as TFE-2/BKLF (Crossley et al. 1996), may regulate distinct globin genes. An intriguing question is whether the EKLF coactivator can function with these CACC-binding proteins to facilitate their interaction with other globin promoters in a developmental-stage-specific manner. Interestingly, EKLF binding to chromatin does not result in staph ladder disruption as observed with NF-E2, indicating that the two erythroid factors differ mechanistically in their ability to interact with chromatin and to structurally remodel it.

The existence of complexes that remodel or enzymatically modify nucleosomal structure illustrates another level of control which is exerted upon DNA-binding proteins to regulate their accessibility in chromatin. Questions remain as to whether specific chromatin remodeling complexes are functionally specialized. For example, do distinct complexes regulate the chromatin activation of the LCR and individual promoters within the globin gene locus? How is the activity of these complexes developmentally controlled? Do the complexes act preferentially with particular genes or classes of transcription factors? Are the complexes specific with regard to tissue type or nuclear function? Indeed, much diversity in subunit composition has been detected within the family of mammalian chromatin remodeling complexes (Wang et al. 1996a,b) and a future challenge is to demonstrate functional specificity or selectivity among its members. Another challenge is to analyze the relative contributions of chromatin enzymatic activities such as HATs, HDACs, kinases, and phosphatases in the regulation of a complex gene locus such as β-globin.

LONG-RANGE TRANSCRIPTIONAL REGULATION BY TISSUE-SPECIFIC ENHANCERS

A critical aspect of transcription regulation in eukaryotes is the ability of *cis*-acting elements that function at long-range to control the activity of target promoters. The most widely studied *cis*-acting elements are enhancers, but LCRs, silencers, and insulators also regulate gene expression at a distance. Historically, it has been difficult to decipher the mechanism of action of natural enhancers that function at long-range since this effect was not easy to reproduce in vitro, either on DNA or on chromatin templates. Because of this, the vast majority of available information pertaining to long-range enhancers is derived from in vivo analyses. These studies have yielded a wealth of information about enhancer-dependent transcription, but they have not provided mechanistic data to explain how these control elements actually work. For example, it is currently unknown how a gene must be packaged in the nucleus for long-range enhancers to be functional or whether interaction with specific nuclear structures such as the nuclear matrix is required. It also remains to be determined how promoters communicate with distal enhancers. In this regard, several models have been proposed: DNA looping by the association of proteins bound at distal sites, DNA tracking by protein translocation, and long-range interactions affected by DNA topology (for review, see Ptashne 1986; Rippe et al. 1995). Evidence supporting each model has been demonstrated in different systems. Another compelling question is whether enhancers regulate promoters by activation or derepression. The prevailing view is that enhancers function by increasing the local concentration of transcriptional activators in the vicinity of the promoter through DNA looping, rather than by operating in an entirely distinct manner from that of upstream promoter elements. However, evidence exists showing that the primary role of enhancers injected into mouse preimplantation embryos is to relieve promoter repression (Majumder et al. 1993). Thus, open questions remain concerning the nuclear structure required to generate an enhancer-responsive gene, the relationship of enhancer-bound proteins to the transcription initiation complex, how promoter-enhancer communication is achieved (looping, tracking, topology, or a combination thereof), and what the functional consequences are (activation or derepression or both).

We have been very interested in understanding how distal regulation by LCRs and enhancers occurs since it is a critical determinant of tissue- and developmental-stage-specific expression of the genes we study. Thus far, we have been able to recreate erythroid-specific, developmentally regulated β^A-globin transcription that is completely enhancer-dependent in vitro using either plasmids containing the individual chick β^A-globin gene or cosmids containing the 38-kb chick β-globin chromosomal locus. In each DNA template, the β-ε-enhancer resides in its natural location at a distance of 2 kb 3′ of the β^A-globin promoter. Enhancer-dependent transcription was achieved by reconstitution of these templates into chromatin or incorporation into synthetic nuclei in the presence of stage-specific erythroid proteins (Barton and Emerson 1994). In an effort to delineate the mechanism of long-range enhancer function, we simplified our in vitro system and demonstrated that enhancer regulation of cloned chick β^A-globin genes could be generated in the absence of chromatin assembly using erythroid transcription extracts (Barton et al. 1997).

We have also reproduced long-range enhancer-dependent regulation using the well-characterized tissue-specific T-cell receptor α-chain (TCRα) gene. This gene undergoes V(D)J recombinase-mediated rearrangement at a discrete stage of T-cell development in a process that requires the TCRα enhancer (Capone et al. 1993). Once productively rearranged, the TCRα gene is tissue-specifically regulated by the 3′ α enhancer from distances up to 50 kb. The minimal enhancer consists of 116 bp that interacts with four proteins, CREB/ATF, LEF-1, PEBP2α, and Ets-1, and functions in a context-dependent manner (for review, see Leiden 1993). The lymphoid-restricted factor, LEF-1, contains an HMG-binding domain (Waterman et al. 1991) and has been shown to act as an "architectural" factor by its ability to induce a sharp bend in the enhancer which brings the other bound proteins into close juxtaposition to produce a functional complex (Giese et al. 1995). How the protein-enhancer complex then regulates its cognate Vα promoter at a distance is unknown. To examine the mechanistic processes involved in these long-range interactions, we have developed an in vitro system that accurately recreates distal enhancer-dependent transcriptional regulation of cloned TCRα genes using protein extracts from α/β-expressing Jurkat T cells without a separate chromatin assembly step (Bagga and Emerson 1997). In our in vitro experiments, the natural TCRα enhancer is located 2.5 kb 3′ of its cognate Vα

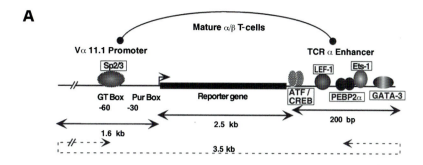

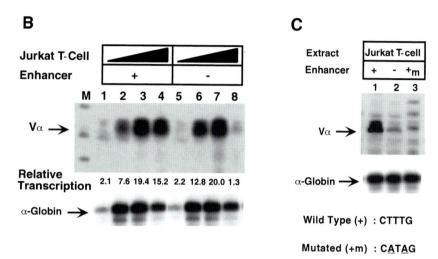

Wild Type (+) : CTTTG

Mutated (+m) : CATAG

Figure 3. Distal enhancer-dependent transcription of the T-cell receptor α-chain gene in vitro. (*A*) Diagram of the TCRα cat reporter construct used in the in vitro transcription studies with protein-DNA complexes designated on the Vα 11.1 promoter and TCRα enhancer. (*Solid arrow*) Distance from the enhancer to the promoter through the gene is 2.5 kb; (*dashed arrow*) distance through the vector sequences on a closed circular plasmid is 3.5 kb. (*B*) In vitro transcription of TCRα genes using a titration of Jurkat T-cell extract. Supercoiled TCRα gene plasmids were transcribed with increasing amounts of a Jurkat T-cell extract: (Lanes *1–4*) +enhancer; (lanes *5–8*) –enhancer. (*Lower panel*) A human α-globin plasmid was included in each reaction as an internal transcription and recovery control. Relative transcription (in arbitrary units) was determined by phosphorimage analysis of the gel. (*C*) Site-directed mutagenesis of the LEF-1-binding site in the TCRα enhancer abrogates enhancer-dependent transcription in vitro. Supercoiled TCRα gene plasmids were transcribed with Jurkat T-cell extracts. (Lane *1*) TCRα + enhancer; (lane *2*) TCRα – enhancer; (lane *3*) TCRα +_m mutant enhancer. (*Lower panel*) A human α-globin plasmid was included in each reaction as an internal transcription and recovery control. Sequences of both wild-type and mutated LEF-1-binding sites with mutated residues underlined are shown below.

promoter (Fig. 3A), and mutations that abolish LEF-1 binding to the enhancer completely eliminate promoter derepression at a distance (Fig. 3C).

Our studies with both the β^A-globin and TCRα genes revealed that each enhancer functions in vitro primarily by promoter derepression instead of activation (Fig. 3B). This occurs by increasing the number of active promoters rather than the rate of transcription, in agreement with recent in vivo analyses (Walters et al. 1995). Both β^A-globin- and TCRα enhancer-mediated gene regulation requires the action of repressor complexes that bind to topologically sensitive DNA structures. In the absence of repressors, transcription is independent of enhancers and DNA topology. We find that the architectural protein HMG I/Y is the critical component (1) of promoter repression in the absence of an enhancer and (2) in generating a DNA topology that enables promoters to respond to

distal enhancers (Fig. 4). Moreover, we find that HMG I/Y is a far more effective structural determinant of enhancer activity than is nucleosome formation, even though enhancers must ultimately function within a chromatin context in vivo. The importance of HMG I/Y and other DNA architectural proteins in gene expression has been reviewed recently (Bustin and Reeves 1996).

One well-studied example of transcriptional regulation by HMG I/Y is the interferon-β (IFN-β) gene (Thanos and Maniatis 1995). HMG I/Y interacts with a virus-inducible enhancer located 104 bp upstream of the transcription start site and facilitates the binding of transcription factors ATF2, c-*jun*, IRF, and NF-κB to form an active "enhanceosome." Synergistic activation of the IFN-β promoter, both in vitro and in vivo, requires all of these factors and HMG I/Y. The appropriately assembled IFN-β enhanceosome stimulates the recruitment of the

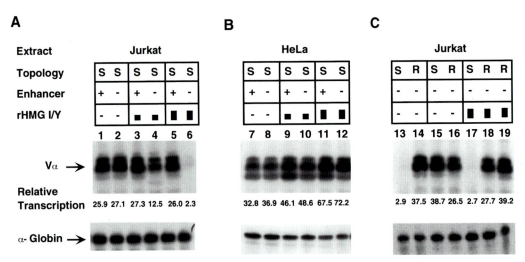

Figure 4. Architectural protein HMG I/Y can restore TCRα enhancer-dependent transcription by mediating Vα promoter repression. (*A*) In vitro transcription of supercoiled TCRα genes using Jurkat T-cell extracts complemented with recombinant HMG I/Y. Supercoiled TCRα plasmids +/– enhancer were each incubated with increasing concentrations of FPLC-purified, recombinant HMG I/Y (lanes *3–6*). These reactions were transcribed with an amount of Jurkat T-cell extract containing limiting concentrations of repressor activity (see Fig. 3) which efficiently expressed TCRα genes in the presence or absence of the enhancer (lanes *1* and *2*). This enabled the reactions to be complemented by HMG I/Y to test its effect as a selective repressor of enhancerless templates. Relative transcription in arbitrary units is listed below each lane. (*Lower panel*) A human α-globin plasmid was included in each reaction as an internal control for transcription and recovery. (*B*) In vitro transcription of supercoiled TCRα genes using HeLa extracts complemented with recombinant HMG I/Y. Supercoiled TCRα plasmids +/– enhancer were each incubated with increasing concentrations of FPLC-purified HMG I/Y (lanes *9–12*). Transcriptions were performed with an amount of HeLa cell extract that shows equivalent expression of TCRα genes +/– enhancer in the absence of HMG I/Y (lanes *7* and *8*). Relative transcription in arbitrary units is listed below each lane. (*Lower panel*) A human α-globin plasmid was included in each reaction as an internal control. (*C*) In vitro transcription of relaxed TCRα genes using Jurkat T-cell extracts complemented with recombinant HMG I/Y. Topoisomerase-I-relaxed (R) and supercoiled (S) TCRα plasmid DNA (minus enhancer) were each incubated with increasing concentrations of FPLC-purified rHMG I/Y (lanes *17–19*). Transcriptions were performed in Jurkat T-cell extracts: high conc. (repressor excess, lanes *13–14, 19*); or low conc. (repressor limiting, lanes *15–18*). Relative transcription in arbitrary units is listed below each lane. (*Lower panel*) A human α-globin plasmid was included in each reaction as an internal control.

basal transcription machinery to the promoter (Kim and Maniatis 1997). Recent experiments suggest that transcriptional synergism of the IFN-β enhanceosome is mediated by a defined surface composed of *trans*-acting factors which are stabilized by HMG I/Y in a precise stereospecific configuration. This novel surface specifically juxtaposes with CBP, a component of the holo RNA polymerase II, thus providing a lock-and-key fit (Merika et al. 1998). These studies provide insight into how specificity of enhancer function is achieved, particularly at short-range. Another compelling question that we hope to address in our in vitro system is how enhancers communicate with their target promoters at considerable distances, especially if several promoters are present, and the role of architectural proteins in this process.

In the nucleus, DNA is believed to be organized into independent loops composed of topologically closed domains. These closed domains, like supercoiled DNA, are structurally dynamic and can adopt many different configurations depending on the total bending and twisting deformations (for review, see Ner et al. 1994; Yang et al. 1995). These configurations can dramatically alter the structure of the entire DNA molecule or may exist at discrete sequences. Localized non-B-form conformations include cruciforms, hairpin structures, four-way junctions, untwisted regions, and crossovers that can be generated by supercoiling. Specific proteins, such as HU, HMG-1, -2, and HMG I/Y may either induce or stabilize these irregular DNA structures and act as "chaperones" in the formation of particular DNA topologies that are active for specific nuclear processes (Ner et al. 1994). In addition, other regulatory proteins may only recognize irregular DNA structures and be unable to bind to B-form duplex DNA (Duncan et al. 1994). Thus, another level of gene regulation is imposed by the existence of a class of proteins that binds unusual DNA structures which are generated by architectural proteins that stabilize a preferred template configuration. If the DNA is unable to exist in various topologies, then this hierarchy of control will presumably be lost.

On the basis of these observations, it is reasonable to predict that promoters and distal enhancers within topologically closed DNA domains can communicate through a particular structural configuration that occurs over kilobases, which can be stabilized by proteins like HMG I/Y or other architectural factors. We know from the example of the IFN-β enhanceosome that synergistic activation requires the close proximity of the enhancer to the promoter to recruit the basal transcription machinery. In this regard, HMG I/Y could stabilize a particular DNA topology which brings the TCRα enhancer and Vα promoter into contact, perhaps by looping the intervening DNA. In the absence of a functional enhancer, HMG I/Y may interact with the promoter alone and potentially create a DNA topology that is refractory to transcription. An example of such a refractory topology conferred by another DNA architectural protein, HU, has been documented in the case of the *gal* operon (Aki and Adhya 1997). Thus, HMG I/Y may have a dual role in generating DNA topologies that favor either activation or repression depending on its target sites of interaction and how they are modulated.

PERSPECTIVES

This paper has focused on two major aspects of transcriptional regulation: the modulation of chromatin structure by tissue-specific DNA-binding proteins and nucleosome remodeling complexes and the mechanism by which long-range enhancers control promoter activity. These examples illustrate the complexity and multiple levels of control that are exerted upon a gene in specific cellular environments. They also illustrate the power of biochemical systems to reconstruct critical processes in transcriptional regulation and to superimpose several levels of control upon each other. These systems will become increasingly sophisticated so that a thorough mechanistic understanding of many complex events involved in gene expression can eventually be deciphered. In the future, it will be important to examine gene regulation in a very broad and comprehensive manner, albeit one step at a time. This means designing experiments to understand not only how a gene's promoter is activated, but also how the chromosomal locus containing the gene is initially designated to be active upon differentiation of the tissue in which the gene is expressed. This presumably requires LCR and insulator activity. How do these elements work? Once the chromosomal locus is activated, how is promoter function managed by chromatin accessibility and long-range *cis*-acting elements such as enhancers and silencers? For example, it remains to be determined what the relative contributions of long-range LCR/enhancer interactions, gene linkage, short-range gene proximal effects, specific transcription factors, and chromatin modifying activities are to developmentally regulated gene activation within the β-globin locus. Finally, what is the interrelationship of multiple genes in a specific tissue or complex gene locus and how are patterns of gene expression established and maintained upon differentiation? Even if one cannot analyze all regulatory aspects of a particular gene, it is important to have an appreciation of the complexity involved so that the full range of transcriptional control can be considered. Viewing one's own work with that perspective makes the field of chromatin and transcription a very exciting place to be.

ACKNOWLEDGMENTS

This work was supported by grants from the Mathers Foundation and the National Institutes of Health to B.M.E. (RO1 GM-38760 and PO1 CA-54418). R.B. is a recipient of an Arthritis Foundation investigator award.

REFERENCES

Aki T. and Adhya S. 1997. Repressor induced site-specific binding of HU for transcriptional regulation. *EMBO J.* **16:** 3666.

Andrews N.C., Erdjument-Bromage H., Davidson M.B., Tempst P., and Orkin S.H. 1993a. Erythroid transcription factor NF-E2 is a haematopoietic-specific basic-leucine zipper protein. *Nature* **362:** 722.

Andrews N.C., Kotkow K.J., Ney P.A., Erdjument-Bromage H., Tempst P., and Orkin S.H. 1993b. The ubiquitous subunit of erythroid transcription factor NF-E2 is a small basic-leucine zipper protein related to the v-*maf* oncogene. *Proc. Natl. Acad. Sci.* **90:** 11488.

Armstrong J.A. and Emerson B.M. 1996. NF-E2 disrupts chromatin structure at human β-globin locus control region hypersensitive site 2 in vitro. *Mol. Cell. Biol.* **16:** 5634.

———. 1998. Transcription of chromatin: These are complex times. *Curr. Opin. Genet. Dev.* **8:** 165.

Bagga R. and Emerson B.M. 1997. An HMG I/Y-containing repressor complex and supercoiled DNA topology are critical for long-range enhancer-dependent transcription in vitro. *Genes Dev.* **11:** 629.

Barton M.C. and Emerson B.M. 1994. Regulated expression of the β-globin gene locus in synthetic nuclei. *Genes Dev.* **8:** 2453.

Barton M.C., Madani N., and Emerson B.M. 1993. The erythroid protein cGATA-1 functions with a stage-specific factor to activate transcription of chromatin-assembled β-globin genes. *Genes Dev.* **7:** 1796.

———. 1997. Distal enhancer regulation by promoter derepression in topologically constrained DNA in vitro. *Proc. Natl. Acad. Sci.* **94:** 7257.

Bieker J.J. and Southwood C.M. 1995. The erythroid Krüppel-like factor transactivation domain is a critical component for cell-specific inducibility of the β-globin promoter. *Mol. Cell. Biol.* **15:** 852.

Bulger M. and Kadonaga J.T. 1994. Biochemical reconstitution of chromatin with physiological nucleosome spacing. *Methods Mol. Genet.* **5:** 241.

Bustin M. and Reeves R. 1996. High-mobility-group chromosomal proteins: Architectural components that facilitate chromatin function. *Prog. Nucleic Acid Res. Mol. Biol.* **54:** 35.

Cairns B.R. 1998. Chromatin-remodeling machines: Similar motors, ulterior motives. *Trends Biochem. Sci.* **23:** 20.

Capone M., Watrin F., Fernex C., Horvat B., Krippl B., Wu L., Scollay R., and Ferrier P. 1993. TCR beta and TCR alpha gene enhancers confer tissue- and stage-specificity on V(D)J recombination events. *EMBO J.* **12:** 4335.

Crossley M., Whitelaw E., Perkins A., Williams G., Fujiwara Y., and Orkin S.H. 1996. Isolation and characterization of the cDNA encoding BKLF/TEF-2, a major CACCC-box-binding protein in erythroid cells and selected other cells. *Mol. Cell. Biol.* **16:** 1695.

Davie J.R. 1998. Covalent modifications of histones: Expression from chromatin templates. *Curr. Opin. Genet. Dev.* **8:** 173.

Duncan R., Bazar L., Michelotti G., Tomonaga T., Krutzsch H., Avigan M., and Levens D. 1994. A sequence-specific, single-strand binding protein activates the far upstream element of c-*myc* and defines a new DNA-binding motif. *Genes Dev.* **8:** 465.

Feng W.C., Southwood C.M., and Bieker J.J. 1994. Analyses of beta-thalassemia mutant DNA interactions with erythroid Krüppel-like factor (EKLF), an erythroid cell-specific transcription factor. *J. Biol. Chem.* **269:** 1493.

Forrester W.C., Epner E., Driscoll M.C., Enver T., Brice M., Papayannopoulou T., and Groudine M. 1990. A deletion of the human β-globin locus control region causes a major alteration in chromatin structure and replication across the entire β-globin gene locus. *Genes Dev.* **4:** 1637.

Giese K., Kingsley C., Kirschner J.R., and Grosschedl R. 1995. Assembly and function of a TCRα enhancer complex is dependent on LEF-1 induced DNA bending and multiple protein-protein interactions. *Genes Dev.* **9:** 995.

Grosveld F., Assendelft G.B.V., Greaves D., and Kollias G. 1987. Position-independent, high-level expression of the human β-globin gene in transgenic mice. *Cell* **51:** 975.

Groudine M., Kohwi-Shigematsu T., Gelinas R., Stamatoyannopoulos G., and Papayannopoulou T. 1983. Human fetal to adult hemoglobin switching: Changes in chromatin structure of the β-globin gene locus. *Proc. Natl. Acad. Sci.* **80:** 7551.

Kadonaga J.T. 1998. Eukaryotic transcription: An interlaced network of transcription factors and chromatin-modifying machines. *Cell* **92:** 307.

Kim T.K. and Maniatis T. 1997. The mechanism of transcriptional synergy of an in vitro assembled interferon-β enhanceosome. *Mol. Cell* **1:** 119.

Leiden J.M. 1993. Transcriptional regulation of T-cell receptor genes. *Annu. Rev. Immunol.* **11:** 539.

Majumder S., Miranda M., and DePamphilis M.L. 1993. Analysis of gene expression in mouse preimplantation embryos demonstrates that the primary role of enhancers is to relieve repression of promoters. *EMBO J.* **12:** 1131.

Martin D.I.K., Fiering S., and Groudine M. 1996. Regulation of β-globin gene expression: Straightening out the locus. *Curr. Opin. Genet. Dev.* **6:** 488.

Merika M., Williams A.J., Chen G., Collins T., and Thanos D. 1998. Recruitment of CBP/p300 by the IFNβ enhanceosome is required for synergistic activation of transcription. *Mol. Cell* **1:** 277.

Miller I.J. and Bieker J.J. 1993. A novel, erythroid cell-specific murine transcription factor that binds to the CACCC element and is related to the Krüppel family of nuclear proteins. *Mol. Cell. Biol.* **3:** 2776.

Ner S.S., Travers A.A., and Churchill M.E. 1994. Harnessing the writhe: A role for DNA chaperones in nucleoprotein-complex formation. *Trends Biochem. Sci.* **19:** 185.

Ney P., Sorrentino B.P., McDonagh K.T., and Nienhuis A.W. 1990. Tandem AP-1-binding sites within the human β-globin dominant control region function as an inducible enhancer in erythroid cells. *Genes Dev.* **4:** 993.

Nuez B., Michalovich D., Bygrave A., Ploemacher R., and Grosveld F. 1995. Defective haematopoiesis in fetal liver resulting from inactivation of the EKLF gene. *Nature* **375:** 316.

Orkin S.H. 1995. Regulation of globin gene expression in erythroid cells. *Eur. J. Biochem.* **231:** 271.

Perkins A.C., Sharpe A.H., and Orkin S.H. 1995. Lethal beta-thalassaemia in mice lacking the erythroid CACCC-transcription factor EKLF. *Nature* **375:** 318.

Ptashne M. 1986. Gene regulation by proteins acting nearby and at a distance. *Nature* **322:** 697.

Rippe K., von Hippel P.H., and Langowski J. 1995. Action at a distance: DNA-looping and initiation of transcription. *Trends Biochem. Sci.* **20:** 500.

Stamatoyannopoulos G. and Nienhuis A. 1994. Hemoglobin switching. In *The molecular basis of blood diseases* (ed. G. Stamatoyannopoulos), p. 107. W.B. Saunders, Philadelphia.

Starck J., Sarkar R., Romana M., Bhargava A., Scarpa A.L., Tanaka M., Chamberlain J.W., Weissman S.M., and Forget B.G. 1994. Developmental regulation of human γ- and β-globin genes in the absence of the locus control region. *Blood* **84:** 1656.

Talbot D. and Grosveld F. 1991. The 5´ HS2 of the globin locus control region enhances transcription through the interaction of a multimeric complex binding at two functionally distinct NF-E2 binding sites. *EMBO J.* **10:** 1391.

Tewari R., Gillemans N., Wijgerde M., Nuez B., von Lindern M., Grosveld F., and Philipsen S. 1998. Erythroid Krüppel-like factor (EKLF) is active in primitive and definitive erythroid cells and is required for the function of 5´HS3 of the β-globin locus control region. *EMBO J.* **17:** 2334.

Thanos D. and Maniatis T. 1995. Virus induction of human IFN beta gene expression requires the assembly of an enhanceosome. *Cell* **83:** 1091.

Tuan D., Solomon W., Li Q., and London I.M. 1985. The β-like-globin gene domain in human erythroid cells. *Proc. Natl. Acad. Sci.* **82:** 6384.

Walters M.C., Fiering S. Eidemiller J., Magis W., Groudine M., and Martin D.I. 1995. Enhancers increase the probability but not the level of gene expression. *Proc. Natl. Acad. Sci.* **92:** 7125.

Wang W., Xue Y., Zhou S., Kuo A., Cairns B.R., and Crabtree G.R. 1996a. Diversity and specialization of mammalian SWI/SNF complexes. *Genes Dev.* **10:** 2117.

Wang W., Côté J., Xue Y., Zhou S., Khavari P.A., Biggar S.R., Muchardt C., Kalpana G.V., Goff S.P., Yaniv M., and Workman J.L. 1996b. Purification and biochemical heterogeneity of the mammalian SWI-SNF complex. *EMBO J.* **15:** 5370.

Waterman M.L., Fischer W.H., and Jones K.A. 1991. A thymus-specific member of the HMG protein family regulates the human T-cell receptor Cα enhancer. *Genes Dev.* **5:** 656.

Wijgerde M., Gribnau J., Trimborn T., Nuez B., Philipsen S., Grosveld F., and Fraser P. 1996. The role of EKLF in human β-globin gene competition. *Genes Dev.* **10:** 2894.

Yang Y., Westcott T.P., Pedersen S.C., Tobias I., and Olson W.K. 1995. Effects of localized bending on DNA supercoiling. *Trends Biochem. Sci.* **20:** 313.

The Transcriptional Basis of Steroid Physiology

R.J. LIN,* H.-Y. KAO,* P. ORDENTLICH,* AND R.M. EVANS*†
†Howard Hughes Medical Institute, The Salk Institute for Biological Studies, La Jolla, California 92037

The actions of steroids, the active metabolites of vitamins A and D and thyroid hormones, are mediated by intracellular nuclear receptors whose coordinate activity defines the physiological response. During the past 80 years, eight distinct classes of bioactive hormonal lipids have been identified, whose comprise is generally viewed as a key part of the endocrine system. In the early part of this century, the endocrine glands were looked upon as an isolated group of structures, secreting substances which, in some mysterious fashion, were able to influence embryonic and postembryonic development, sexual maturation, biochemical pathways involving protein and sugar metabolism, behavior, and reproductive physiology.

A consideration of the action of hormones relates directly to the nature of the physiologic process itself. As a recurrent event, every positive effect, including induction of gene expression must be countered by a process that reverses the activation to reestablish the ground state. Thus, the counter of transcriptional activation by hormones is transcriptional repression by the nonliganded receptor. As hormone levels drop, the generation of an increasing number of nonoccupied receptors leads to a reversal of the activated state. Indeed, we argue that transcriptional repression is an intrinsic part of endocrine physiology and contributes an essential feature to feedback regulation associated with the inhibition of the physiologic response. It is this subject that is the focus of this paper.

The inhibitory capacity of receptors was first recognized as part of an analysis of the thyroid hormone receptor that is converted to an oncogene, the v-erbA, by mutations that block hormone binding (Damm et al. 1989; Sap et al. 1989). These mutations reveal that the nonliganded receptor possesses a bona fide activity that functions as a potent transcriptional repressor. Multiple studies on transcriptional silencing by v-erbA and the nonliganded thyroid hormone receptor suggest that repression is required for oncogenesis (Zenke et al. 1990; Sharif and Privalsky 1991) and that this process is mediated by a diffusible cofactor that associates with the ligand-binding domain (LBD) of the receptors (Casanova et al. 1994).

A second example of the importance of transcriptional repression relates to the vitamin A signaling pathway. As with the thyroid hormone receptor, mutant forms of the retinoic acid receptor α (RARα) are created by a variety of chromosomal translocations that produce oncogenic derivatives linked to human acute promyelocytic leukemia (APL) (de Thé et al 1991; Kakizuka et al. 1991; Chen et al. 1993). Multiple studies reveal that the mutant retinoic acid receptors interfere with vitamin-A-mediated granulocytic differentiation (Grignani et al. 1993; Chen et al. 1994; Rousselot et al. 1994; Jansen et al. 1995; Ruthardt et al. 1997) and are oncogenic when expressed in transgenic mice (Brown et al. 1997; Grisolano et al. 1997; He et al. 1997, 1998).

We and other investigators have identified transcriptional corepressors (SMRT and N-CoR) that associate with nonliganded receptors and result in suppression of basal transcriptional activity (Chen and Evans 1995; Horlein et al. 1995). The mechanism of this repression is not fully understood, but recent advances suggest that an important feature of both transcriptional activation and transcriptional repression relates to controlled modulation of chromatin structure.

Chromatin remodeling has long been suggested to be a critical component of transcriptional regulation, and it has been suggested that promoter-specific transcriptional activation may involve local changes in chromatin structure (Felsenfeld 1992). We and others have demonstrated that nuclear hormone receptors may utilize the CREB-binding protein (CBP) or its homolog, p300, to function as a nuclear receptor cofactor (Chakravarti et al. 1996; Kamei et al. 1996). In addition to CBP/p300, multiple hormone-dependent and -independent associated cofactors have been characterized (Glass et al. 1997). Of particular interest is the recent demonstration that CBP/p300 associates with the histone acetylase PCAF, which displays significant sequence homology with the yeast transcription activator Gcn5p, also known to be a histone acetylase (Yang et al. 1996). Furthermore, CBP/p300 harbors intrinsic histone acetyltransferase (HAT) activity, resulting in alternative or perhaps simultaneous histone acetylation (Bannister and Kouzarides 1996; Ogryzko et al. 1996). The notion that multiple transcriptional coactivators possess acetylase activity suggests that their recruitment to a DNA template would destabilize nucleosomes locally, creating a permissive state for promoter activation (Grunstein 1997; Struhl 1998).

With this view in mind, we have speculated that the reversal of transcriptional activation by the reestablishment of a repressive chromatin state might be a critical feature in physiologic homeostatis. The cloning of the first mammalian histone deacetylase suggested a possible role for this protein in transcriptional repression (Taunton et al. 1996). Recently, a considerable advance has been made toward understanding transcriptional repression by the demonstration that nuclear receptor corepressors SMRT and N-CoR associate with both mSin3A and human histone deacetylases (HDACs; Alland et al. 1997; Heinzel et al. 1997; Nagy et al. 1997). mSin3A is a homolog of the yeast HO gene repressor, Sin3p, first cloned by Sternberg et al. (1987) and subsequently shown to act through Rpd3p. The discovery that HDAC-1 is homologous to Rpd3p provided a conceptual link between histone deacetylation and transcriptional repression. Our demonstration that HDAC-1 and Sin3A are present in these same protein complexes and associate with nuclear receptor corepressors suggests that a repressor complex is targeted to chromatin by association with specific DNA-binding proteins.

The balance of this review is divided into two related sections. The first describes recent studies providing direct evidence that corepressors are physiologically relevant and directly contribute to human disease and may be bona fide therapeutic targets. The second section shows that the corepressor complex is not unique to nuclear receptors, but rather represents a general regulatory mechanism that integrates diverse signaling pathways.

COREPRESSORS, MUTANT RECEPTORS, AND APL

Human APLs of the M3 subtype of acute myeloid leukemias are characterized by chromosomal translocations involving the RARα gene on chromosome 17 (Grignani et al. 1994). Two major translocation partners, PML and PLZF, have been described previously (de Thé et al. 1991; Kakizuka et al. 1991; Chen et al. 1993). The resulting fusion proteins, PML-RARα and PLZF-RARα, interfere with normal retinoid-mediated *trans*-activation and granulocytic differentiation (Grignani et al. 1993; Chen et al. 1994; Rousselot et al. 1994; Jansen et al. 1995; Ruthardt et al. 1997) and are oncogenic in transgenic

mice (Brown et al. 1997; Grisolano et al. 1997; He et al. 1997, 1998). One unique feature of APL is that retinoic acid, the natural ligand for RARα, can trigger complete remission in patients with PML-RARα fusion proteins, thus providing the first example of retinoid differentiation therapy (Chomienne et al. 1996). Interestingly, although PLZF-RARα leukemias display clinical phenotypes similar to those of PML-RARα leukemias, they respond poorly to retinoid treatment (Licht et al. 1995). Despite intense investigation, the molecular basis of the differentiation block caused by these fusion proteins and the marked difference in clinical response to retinoic acid among APLs with different fusion proteins is unknown.

As both PML-RARα and PLZF-RARα retain most of the sequence from the wild-type RARα, including an intact DNA-binding domain and ligand-binding domain, they are likely able to associate with RAR cofactors. Using a series of biochemical and genetic assays, we demonstrated that both PML-RARα and PLZF-RARα interact with the nuclear receptor corepressor SMRT (Chen and Evans 1995) and its associated histone deacetylase complex that contains mSin3A and HDAC1 (Alland et al. 1997; Heinzel et al. 1997; Nagy et al. 1997). Interestingly, in contrast to PML-RARα, which can only interact with SMRT via its RAR moiety, PLZF-RARα has acquired a novel corepressor binding site at its amino-terminal PLZF moiety. This region, the BTB/POZ domain, can independently associate with multiple components of the corepressor complex and function as an autonomous repression domain which is essential to the oncogenic activity of PLZF-RARα (Dong et al. 1996; Ruthardt et al. 1997). These results are summarized in Figure 1. As mentioned above, PML-RARα and PLZF-RARα leukemias differ in their response to retinoic acid.

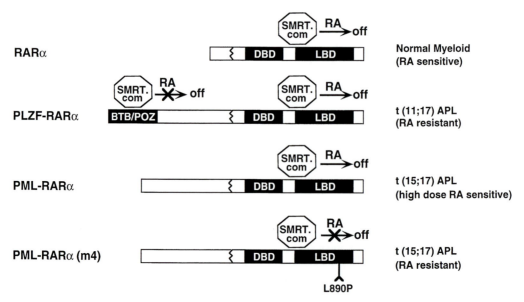

Figure 1. Schematic representation of mutant RARs in APL and their abnormal association with nuclear receptor corepressors. The structure of PLZF-RARα and PML-RARα is shown and the translocation fusion point is indicated. The corepressor-histone deacetylase complex is denoted SMRT.com, whose dissociation is triggered by retinoic acid. Although PML-RARα/SMRT.com interaction is sensitive to retinoic acid; PLZF-RARα and PML-RARα(m4) lose the ability to dissociate this complex even under high doses of retinoic acid.

Because SMRT and the histone deacetylase complex dissociate from wild-type RARs upon ligand binding (Chen and Evans 1995; Nagy et al. 1997), we investigated whether abnormal association of the corepressor complex with chimeric receptors could explain this difference in retinoid sensitivity. Using several in vitro binding assays, we demonstrated that the interaction between corepressors and the BTB/POZ domain of PLZF-RARα is resistant to even high doses of retinoic acid. In contrast, PML-RARα is able to dissociate corepressors, although significantly higher concentrations of retinoic acid are required (Fig. 1). This striking correlation of corepressor dissociation from the chimeric receptors in vitro with the distinct retinoid sensitivity in APL patients prompted us to examine the interaction between SMRT and PML-RARα in retinoid-resistant t(15;17) APLs. A mutation in PML-RARα has been identified in a retinoid-resistant subclone of the NB4 cell line that carries the t(15;17) translocation (NB4-R4; Shao et al. 1997). We found that this PML-RARα mutant is unable to release SMRT even under pharmacological concentrations of retinoic acid (Fig. 1), presumably due to impaired ligand binding (Shao et al. 1997).

THE ONCOGENIC FUNCTION OF MUTANT RARS REQUIRES HISTONE DEACETYLASE ACTIVITY

On the basis of the above results, we hypothesize that the common differentiation block caused by RARα fusion proteins is mediated by the association of the corepressor complex and the inability to dissociate this complex under physiological concentrations of retinoic acid. We further speculate that the potential of APL patients to respond to retinoic acid is determined by the retinoid sensitivity of the interaction between the chimeric receptors and the corepressor complex. The discovery of specific inhibitors of mammalian histone deacetylases (Yoshida et al. 1995) and the subsequent demonstration that these inhibitors block transcriptional repression mediated by HDACs (Heinzel et al. 1997; Nagy et al. 1997) allowed us to directly test the contribution of these proteins in gene regulation. Whereas PML-RARα and PLZF-RARα expression blocks responsiveness of target genes to ATRA, coaddition of histone deacetylase inhibitors trichostatin A (TSA) or Na-butyrate (NaB) restored the retinoid response. This suggested that the inhibition of HDAC activity removes the block to retinoid responsiveness. These results further led us to explore whether TSA has any effect on retinoid-induced differentiation of NB4 cells. As expected, we found that low concentrations of TSA dramatically enhanced the differentiation effect of ATRA, which had little effect alone. This differentiation-enhancing effect is likely mediated through the retinoid signaling pathway since we observed strong synergistic effect of TSA on retinoid induction of the endogenous retinoid target gene CD18. Taken together, our results strongly suggest that HDAC-mediated transcriptional repression plays a central role in PML-RARα- and PLZF-RARα-induced leukemias and that relief of this repres-

sion may potentiate retinoid responses in APL cells. Interestingly, although t(15;17) APL patients initially respond to pharmacological doses of retinoic acid and achieve complete remission, many of them relapse and acquire permanent retinoid resistance (Grignani et al. 1994). We wish to determine whether TSA could overcome such resistance using NB4-R4 cells as a model. Strikingly, we found that high concentrations of TSA partially restored RA-induced differentiation. Although this is a surprising result considering the reduced ligand binding ability of the PML-RARα mutant in these cells, it is possible that endogenous wild-type RARs could mediate the differentiation effects. This speculation is supported by our finding that TSA restored retinoid response of βRARE reporters in NB4-R4 cells. Collectively, these results indicate that inhibition of the dominant negative activity of the PML-RARα mutant by histone deacetylase inhibitors could increase net retinoid transcriptional responses in these resistant cells.

ABERRANT CHROMATIN ACETYLATION MAY CONTRIBUTE TO THE DEVELOPMENT OF APL

The importance of retinoic acid in hematopoiesis and the invariable association of RARα fusion proteins with APL suggest that the disruption of the retinoid signaling pathway plays a major role in the initiation of this disease. Although the structure and function of RARα fusion proteins have been studied extensively, the molecular mechanism underlying their ability to block retinoid-mediated transcription and granulocytic differentiation remains elusive. Here, we summarize that the oncogenic potential of PML-RARα and PLZF-RARα correlates not with their trans-activating function, but rather with their trans-repressing function. Normally, wild-type RARα undergoes a ligand-induced conformational change that facilitates the release of corepressors and subsequent association of coactivators. In contrast, PML-RARα and PLZF-RARα are defective in this switch under physiological concentrations of ligands and function as constitutive repressors. This in turn blocks terminal differentiation of myeloid precursors and causes leukemias. Moreover, additional mutations may occur in PML-RARα that further impair its ligand responsiveness in either relapsed APL patients (Ding et al. 1998; Imaizumi et al. 1998) or NB4 cells (Shao et al. 1997). It is also apparent from this study that relieving the repressor activity of mutant RARs abolishes their oncogenicity. Although pharmacological doses of retinoic acid trigger the release of corepressors from PML-RARα and induce remission in APL patients, they fail to do so in PLZF-RARα leukemias and in retinoid-resistant PML-RARα leukemias. However, this resistance may be overcome by histone deacetylase inhibitors in combination with retinoic acid and thus may provide therapeutic benefit to APL patients. In conclusion, our findings support a unifying model for APL pathogenesis mediated by oncogenic RARs which repress transcription through constitutive association with the corepressor-histone deacetylase complex (Fig. 2). It should also be noted, however, that these fusion receptors may contribute to leukemogenesis through additional mechanisms. For instance, since the same deacetylase complex is also required for Mad/Max function associated with cellular differentiation (Alland et al. 1997; Hassig et al. 1997; Heinzel et al. 1997; Laherty et al. 1997), RARα fusion proteins may inhibit Mad/Max activity through titration of these factors and in turn promote Myc/Max function in cellular proliferation.

It has been thought for some time that deregulation of histone acetylation and alteration of chromatin structure during development may lead to cellular transformation and neoplasia. Several leukemogenic translocations that fuse histone acetyltransferase CBP and p300 to *MLL* and *MOZ* genes, respectively, have been reported (Borrow et al. 1996; Ida et al. 1997; Taki et al. 1997). Moreover, biallelic inactivating somatic mutations of the p300 gene have been observed in gastric and colon cancers (Muraoka et al. 1996), and amplification of another HAT, AIB1 (ACTR, pCIP), is frequently seen in breast and ovarian cancers (Anzick et al. 1997). Our findings further strengthen the importance of chromatin in transcriptional regulation and establish for the first time a direct link between the nuclear receptor corepressor-histone deacetylase complex and human cancer.

A NOTCH CONNECTION

The Notch signal transduction pathway is a key regulator of multiple differentiation programs. Originally identified in *Drosophila*, the *Notch* gene encodes a transmembrane protein that contains multiple epidermal growth factor (EGF)-like repeats, a cysteine-rich region in the extracellular domain, and cdc10/ankyrin repeats in the intracellular domain. The Notch receptor, upon ligand binding, transduces a signal to the nucleus that results in the expression of several downstream genes, including those of the Enhancer of Split [E(spl)] complex. The mechanism by which the Notch signal is transmitted has been a point of controversy. Recent work from several groups, however, demonstrates that the most likely scenario is that a cleavage product of the Notch receptor, which contains the intracellular domain, translocates to the nucleus where it forms a transcriptionally active complex with the CSL (CBF1/Suppressor of Hairless/Lag-1) family of proteins (for review, see Bray 1998).

In the absence of Notch signaling, the vertebrate CSL protein, CBF1/RBP-Jκ, functions as a transcriptional repressor by binding to target sequences in the interleukin-6 (IL-6) and NF-κB promoters (Kannabiran et al. 1997; Plaisance et al. 1997; Miyazawa et al. 1998). The ade-

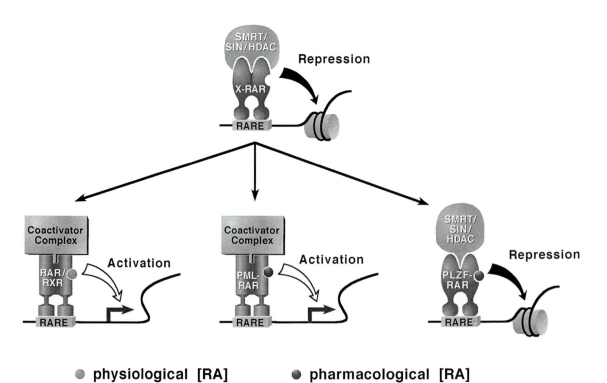

Figure 2. Model for the role of transcription silencing in the pathogenesis and hormonal response of APL. In the absence or presence of physiological concentrations of retinoic acid, the transcription of retinoid target genes is repressed by X-RAR (PML-RARα and PLZF-RARα) proteins through the action of the corepressor complex. This results in the block of granulocytic differentiation. Under therapeutic doses of retinoic acid, PML-RARα, but not PLZF-RARα, dissociates the corepressor complex and activates retinoid target genes and induces differentiation of leukemic blasts in APL patients.

novirus pIX gene has also been shown to be repressed through a CBF1/RBP-Jk-binding site (Dou et al. 1994). Consistent with a role as a transcriptional repressor, a GAL4-CBF1 fusion protein can repress the activity of a TK (thymidine kinase) promoter when linked to GAL4-binding sites . A repression domain in CBF1 maps to a region involved in DNA binding and interaction with Notch (Hsieh and Hayward 1995; Hsieh et al. 1996). It is thought that Notch binding to CBF1 masks the repression domain and leads to a transcriptionally active complex. Indeed, a Notch/CBF1 complex has been shown to bind and activate the promoter of the *Hairy enhancer of split* (*HES-1*) gene, a downstream component of the Notch signaling pathway (Jarriault et al. 1995). The switching of CBF1 from a repressor to an activator therefore represents a critical molecular step in Notch signaling.

The mechanism by which extracellular signals regulate transcription has long been an object of intense study. As multiprotein corepressor complexes containing histone deacetylase activity have been implicated in transcriptional repression by nuclear hormone receptors, we wished to know whether such a complex may regulate Notch signaling.

To determine whether the SMRT/N-CoR corepressors are involved in CBF1-mediated repression, yeast two hybrid assays and GST-pulldown experiments were used to assess the interaction of SMRT with CBF1 fusion proteins. As positive results were achieved in both assays, it was possible to map the domain in SMRT to amino acids

649–811 as shown in the schematic in Figure 3. A mutant of CBF1 that does not repress EEF233AAA (Hsieh and Hayward 1995) fails to interact with SMRT (see Fig. 3). In addition to the yeast two-hybrid assays, GST-pulldown experiments confirm that a GST-SMRT (649–811) fusion protein interacts with wild-type CBF1 but not the mutant CBF1 (EEF233AAA). The interaction of SMRT with CBF1 in mammalian cells was confirmed by coimmunoprecipitation experiments, as well as mammalian two-hybrid assays (not shown).

To determine the functional significance of an interaction between SMRT and CBF1, we asked if SMRT and Notch are antagonists. The ability of SMRT to inhibit Notch signaling through a GAL4-CBF fusion protein was examined by cotransfecting increasing amounts of SMRT in the presence of TAN-1 and GAL4-CBF1. As shown in Figure 4A, SMRT strongly inhibited the activity of TAN-1, a truncated form of Notch1 that contains only the cytoplasmic domain. The related isoforms of N-CoR, RIP13a, and RIP13d1 also inhibited TAN-1 activity through CBF1 (not shown). The significance of these findings to Notch signaling was furthered by asking whether SMRT could affect the activity of the HES-1 promoter. The enhancer of split genes have been shown in *Drosophila* and *Xenopus laevis* to be regulated by Notch through activation of the CSL proteins (Wettstein et al. 1997). As expected, coexpression of SMRT inhibited TAN-1 activation of this promoter (Fig. 4B).

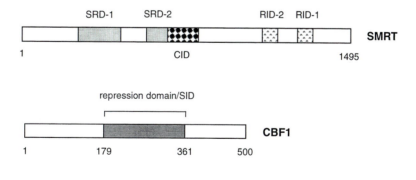

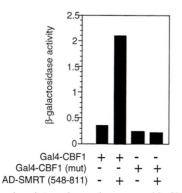

Figure 3. (*Top*) Schematic representing the various interaction regions mapped in SMRT. The CID corresponds to the CBF1 interacting domain. The region of CBF1 that interacts with SMRT is also delineated. (*Bottom*) Results of a yeast two-hybrid assay demonstrating that SMRT interacts with wild-type CBF1 but not a repression-defective mutant of CBF1.

The activity of TAN-1 has been demonstrated to be through the formation of a multiprotein complex with CBF1 on DNA. To address whether SMRT could antagonize the formation of this complex, 293T cells were transfected with FLAG-CBF1 and TAN-1 in the presence or absence of increasing amounts of SMRT. In the absence of SMRT, TAN-1 forms a stable complex with FLAG-CBF. With increasing amounts of SMRT, the complex is disrupted, suggesting that SMRT and TAN-1 compete for binding to CBF1. These results therefore demonstrate that SMRT, and presumably the SMRT-related factor N-CoR, can interact with CBF1 in vitro and in vivo and thus perturb the formation of a TAN-1/CBF1 complex.

SMRT has been demonstrated to form a multiprotein complex containing mSin3A and the histone deacetylase HDAC1 (Nagy et al. 1997). It has been suggested that the transcription repression mediated by this complex is due to histone deacetylation and the consequential remodeling of the chromatin into a transcriptionally nonpermissive state. To determine if the interaction of SMRT with CBF1 implicates a histone-deacetylase-containing complex in CBF1-mediated repression, we tested a known inhibitor of histone deacetylases, Trichostatin A (TSA) on the expression of ESR-1, an Enhancer-of-split-related gene from *Xenopus*. This gene was demonstrated to be induced by activated forms of *Xenopus* Notch or by the Notch ligand, X-Delta-1, presumably through the *Xenopus* Suppressor of Hairless (X-Su[H]). If histone deacetylation was involved in regulation of ESR-1, we would expect that treatment with TSA would enhance the induction of ESR-1 expression. Indeed, TSA treatment results in a two- to threefold increase in the levels of ESR-1 at each dose of X-Delta-1. This finding is consistent with our prediction that a histone-deacetylase-containing complex recruited by X-Su(H) to the ESR-1 promoter

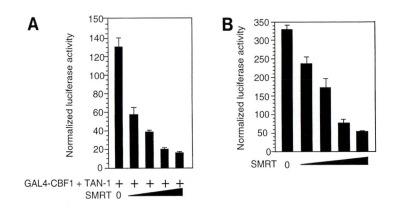

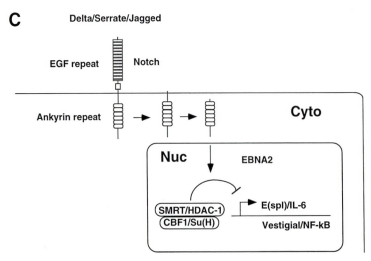

Figure 4. (*A*) SMRT inhibits TAN-1 activation through a GAL4-CBF1 construct. Increasing amounts of SMRT plasmid were co-transfected with TAN-1, resulting in inhibition of TAN-1 activity. (*B*) The ability of TAN-1 to activate through endogenous CBF1 is inhibited by increasing amounts of SMRT. (*C*) Model summarizing the role a SMRT/HDAC-1 complex may have in regulating Notch signaling. In the absence of signaling, the SMRT/HDAC-1 complex is bound to CBF1 and functions as a corepressor of transcription. Upon Notch signaling, this complex dissociates and Notch interacts with CBF1 to form a transcriptionally active complex.

may function to repress the expression of this gene until binding by an activated form of the Notch receptor. We have extended these results to demonstrate that the histone deacetylase, HDAC-1, could interact with wild-type CBF1 but not a repression-defective mutant CBF1 (not shown).

We propose a model (Fig. 4C) that serves to illustrate our findings with respect to the role a SMRT/HDAC-1 corepressor complex may have in the regulation of Notch signaling. In the absence of active Notch signaling, this complex is bound to CBF1 and through histone deacetylation maintains the local chromatin structure in a transcriptionally nonpermissive state, thus repressing transcription. Upon activation of Notch by its ligand Delta, an active form of the Notch receptor containing the intracellular domain translocates to the nucleus, where it competes for binding to CBF1. Binding of Notch to CBF1 displaces the corepressor complex and leads to transcriptional activation. It will be interesting to see if Notch binding to CBF1 serves to recruit a multiprotein coactivating complex.

We have demonstrated here that a corepressor complex containing SMRT/N-CoR and HDAC-1 is, in part, responsible for the transcriptional repression by CBF1. We further show that this complex antagonizes the ability of TAN-1, an active form of the Notch receptor, to interact with and stimulate CBF1-mediated transcription, suggesting a critical role for this complex in Notch signaling. Finally, experiments performed in *Xenopus* animal caps treated with the histone deacetylase inhibitor, TSA, provide evidence that the Notch target gene ESR-1 may be regulated by a histone deacetylase containing complex.

DISCUSSION

A major question concerning nuclear receptor function is to understand how nuclear receptors mediate activation and repression while remaining bound to DNA. A considerable advance was made toward understanding this when it was shown that ligand binding causes the dissociation of corepressor proteins and promotes association of coactivators. This notion was substantially advanced by the subsequent demonstration that corepressors, such as SMRT and N-CoR, associate with both mSin3A and the histone deacetylase HDAC-1. These studies provided the first direct evidence that nuclear receptor cofactors may function to remodel chromatin in a hormone-dependent fashion. They also suggested that the substrate for hormone signaling may be the addition or removal of acetate groups from the histone tail.

Transcriptional repression in metazoans was first described in the thyroid hormone receptor and its oncogenic counterpart, v-*erbA*, and subsequently in the RARs. This work led to the speculation that repression was linked directly to absence of ligand and was a critical feature underlying the oncogenic phenotype. In pursuing these observations, we wish to establish a direct link between transcriptional repression and human disease. This came about during the characterization of the regulatory properties of the PML-RARα translocation which is the genetic product resulting in APL. Similarly, a second translocation fusion, the PLZF gene with RAR also gives rise to APL, suggesting a direct link between the RAR and cancer. The characterization of these two fusion proteins revealed that they both recruit the histone deacetylase repression complex, which establishes a direct molecular link between corepressors and cancer. Second, the identification of deacetylases as molecular targets of oncogenes and the effect of growth suppression of RA-sensitive as well as RA-resistant leukemic cells by inhibitors of histone deacetylases suggest that pharmacologic manipulation of nuclear-receptor cofactors is a viable approach to the treatment of human leukemias. Thus, both chromatin and chromatin remodeling can be critical features generating human disease, and the molecules involved in these processes are bona fide therapeutic targets.

Subsequent studies have established a direct molecular link between nuclear receptor corepressors and other transcriptional pathways. This work led us to explore the possibility that such factors may also underlie the transcriptional switch that mediates neuronal differentiation. This pathway, controlled by the Notch protein, is mediated via a transcriptional regulatory protein termed CBF-1 as described above. Indeed, in the absence of the Notch switch, CBF-1 binds the corepressor complex and functions as a potent transcriptional repressor. Thus, we have been able to establish a direct molecular link from an extracellular signaling pathway through a cell surface receptor to a nuclear factor that is gaited by the switch from transcriptional repression to transcriptional activation. Not only does this reveal the basis for a novel genetic switch, but it also extends the proposal that the transcriptional regulatory pathway is widely conserved and that corepressors may function as integrators of multiple cell growth and signal transduction pathways.

ACKNOWLEDGMENTS

R.M.E. is an investigator of the Howard Hughes Medical Institute at the Salk Institute for Biological Studies. R.J.L. is a Lucille P. Markey predoctoral fellow at the University of California San Diego. H.-Y.K. is a fellow of the Leukemia Society of America. P.O. is a fellow of the Hewlitt Foundation for Medical Research

REFERENCES

Alland L., Muhle R., Hou H., Jr., Potes J., Chin L., Schreiber-Agus N., and DePinho R.A. 1997. Role for N-CoR and histone deacetylase in Sin3-mediated transcriptional repression. *Nature* **387:** 49.

Anzick S.L., Kononen J., Walker R.L, Azorsa D.O., Tanner M.M., Guan X.Y., Sauter G., Kallioniemi O.P., Trent J.M., and Meltzer P.S. 1997. AIB1, a steroid receptor coactivator amplified in breast and ovarian cancers. *Science* **277:** 965.

Bannister A.J. and Kouzarides T. 1996. The CBP co-activator is a histone acetyltransferase. *Nature* **384:** 641.

Borrow J., Stanton V.P., Jr., Andresen J.M., Becher R., Behm F.G., Chaganti R.S., Civin C.I., Disteche C., Dube I., Frischauf A.M., Horsman D., Mitelman F., Volinia S., Watmore A.E., and Housman D.E. 1996. The translocation t(8;16)(p11;p13) of acute myeloid leukemia fuses a putative acetyltransferase to the CREB-binding protein. *Nat. Genet.* **14:** 33.

Bray S. 1998. A Notch affair. *Cell* **93**: 499.

Brown D., Kogan S., Lagasse E., Weissman I., Alcalay M., Pelicci P.G., Atwater S., and Bishop J.M. 1997. A PMLRARα transgene initiates murine acute promyelocytic leukemia. *Proc. Natl. Acad. Sci.* **94**: 2551.

Casanova J., Helmer E., Selmi-Ruby S., Qi J.S., Au-Fliegner M., Desai-Yajnik V., Koudinova N., Yarm F., Raaka B.M., and Samuels H.H. 1994. Functional evidence for ligand-dependent dissociation of thyroid hormone and retinoic acid receptors from an inhibitory cellular factor. *Mol. Cell. Biol.* **14**: 5756.

Chakravarti D., LaMorte V.J., Nelson M.C., Nakajima T., Juguilon H., Montminy M., and Evans R.M 1996. Role of CBP/p300 in nuclear receptor signalling. *Nature* **383**: 99.

Chen J.D. and Evans R.M. 1995. A transcriptional co-repressor that interacts with nuclear hormone receptors. *Nature* **377**: 454.

Chen Z., Brand N.J., Chen A., Chen S.J., Tong J.H., Wang Z.Y., Waxman S., and Zelent A. 1993. Fusion between a novel Krüppel-like zinc finger gene and the retinoid acid receptor-α locus due to a variant t(11;17) translocation associated with acute promyelocytic leukemia. *EMBO* . **12**: 1161.

Chen Z., Guidez F., Rousselot P., Agadir A., Chen S.J., Wang Z.Y., Degos L., Zelent A., Waxman S., and Chomienne C. 1994. PLZF-RARα fusion proteins generated from the variant t(11;17) (q23;q21) translocation in acute promyelocytic leukemia inhibit ligand-dependent transactivation of wild type retinoid acid receptors. *Proc. Natl Acad. Sci.* **91**: 1178.

Chomienne C., Fenaux P., and Degos L. 1996. Retinoid differentiation therapy in promyelocytic leukemia. *FASEB J.* **10**: 1025.

Damm K., Thompson C.C., and Evans R.M. 1989. Protein encoded by v-erbA functions as a thyroid-hormone receptor antagonist. *Nature* **339**: 593.

de Thé H., Lavau C., Marchi A., Chomienne C., Degos L., and Dejean A. 1991. The PML-RARα fusion mRNA generated by the t(15;17) translocation in acute promyelocytic leukemia encodes a functionally altered RAR. *Cell* **66**: 675.

Ding W., Li W.-P., Nobile L.M., Grills G., Carrera I., Paiett E., Tallman M.S., Wiernik P.H., and Gallagher R.E. 1998. Retinoic acid receptor alpha (RARα)-region mutations in the PML-RARα fusion gene of acute promyelocytic leukemia (APL) patients after relapse from *all-trans* retinoic acid (ATRA) therapy. *Blood* **90**: 415a.

Dong S., Zhu J., Reid A., Strutt P., Guidez F., Zhong H.J., Wang Z.Y., Licht J., Waxman S., Chomienne C., Chen A., Zelent A., and Chen. S.-J. 1996. Amino-terminal protein-protein interaction motif (POZ-domain) is responsible for activities of the promyelocytic leukemia zinc finger-retinoic acid receptor-α fusion protein. *Proc. Natl. Acad. Sci.* **93**: 3624.

Dou S., Zeng X., and Cortes P. 1994. The recombination signal sequence-binding protein RBP-2N functions as a transcriptional repressor. *Mol. Cell. Biol.* **14**: 3310.

Felsenfeld G. 1992. Chromatin as an essential part of the transcriptional mechanism. *Nature* **355**: 219.

Glass C.K., Rose D.W., and Rosenfeld M.G. 1997. Nuclear receptor coactivators. *Curr. Opin. Cell. Biol.* **9**: 222.

Grignani F., Fagioli M., Alcalay M., Longo L., Pandolfi P.P., Donti E., Biondi A., Lo Coco F., Grignani F., and Pelicci P.G. 1994. Acute promyelocytic leukemia: From genetics to treatment. *Blood* **83**: 10.

Grignani F., Ferrucci P.F., Testa U., Talamo G., Fagioli M., Alcalay M., Mencarelli A., Grignani F., Peschle C., Nicoletti I., and Pelicci P.G. 1993. The acute promyelocytic leukemia-specific PML-RARα fusion protein inhibits differentiation and promotes survival of myeloid precursor cells. *Cell* **74**: 423.

Grisolano J.L., Wesselschmidt R.L., Pelicci P.G., and Ley T.J. 1997. Altered myeloid development and acute leukemias in transgenic mice expressing PML-RARα under control of cathepsin G regulatory sequences. *Blood* **89**: 376.

Grunstein M. 1997. Histone acetylation in chromatin structure and transcription. *Nature* **389**: 349.

Hassig C.A., Fleischer T.C., Billin A.N., Schreiber S.L., and Ayer D.E. 1997. Histone deacetylase activity is required for full transcriptional repression by mSin3A. *Cell* **89**: 341.

He L.-Z., Guidez F., Triboli C., Peruzzi D., Ruthardt M., Zelent A., and Pandolfi P.P. 1998. Distinct interactions of PML-RARα and PLZF-RARα with co-repressors determine differential responses to RA in APL. *Nat. Genet.* **18**: 126.

He L.-Z., Tribioli C., Rivi R., Peruzzi D., Pelicci P.G., Soareas V., Cattoretti G., and Pandolfi P.P. 1997. Acute leukemia with promyelocytic features in PML/RARα transgenic mice. *Proc. Natl. Acad. Sci.* **94**: 5302.

Heinzel T., Lavinsky R.M., Mullen T.M., Soderstrom M., Laherty C.D., Torchia J., Yang W.M., Brard G., Ngo S.D., Davie J.R., Seto E., Eisenman R.N., Rose D.W., Glass C.K., and Rosenfeld M.G. 1997. A complex containing N-CoR, mSin3 and histone deacetylase mediates transcriptional repression. *Nature* **387**: 43.

Horlein A.J., Naar A.M., Heinzel T., Torchia J., Gloss B., Kurokawa R., Ryan A., Kamei Y., Soderstrom M., Glass C.K., and Rosenfeld M.G. 1995. Ligand-independent repression by the thyroid hormone receptor mediated by a nuclear receptor co-repressor. *Nature* **377**: 397.

Hsieh J.J.-D. and Hayward S.D. 1995. Masking of the CBF1/RBPjk transcriptional repression domain by Epstein-Barr virus EBNA2. *Science* **268**: 560.

Hsieh J.J.-D., Henkel T., Salmon P., Robey E., Peterson M.G., and Hayward D. 1996. Truncated mammalian Notch 1 activated CBF1/RBPJk-repressed genes by a mechanism resembling that of Epstein-Barr Virus EBNA2. *Mol. Cell. Biol.* **16**: 952.

Ida K,. Kitabayashi I., Taki T., Taniwaki M., Noro K., Yamamoto M., Ohki M., and Hayashi Y. 1997. Adenoviral E1A-associated protein p300 is involved in acute myeloid leukemia with t(11;22)(q23;q13). *Blood* **90**: 4699.

Imaizumi M., Suzuki H., Yoshinari M., Sato A., Saito T., Sugawara A., Tsuchiya S., Hatae Y., Fujimoto T., Kakizuka A., Konno T., and Iinuma K. 1998. Mutations in the E-domain of RARα portion of the PML/RARα chimeric gene may confer clinical resistance to all-*trans* retinoic acid in acute promyelocytic leukemia. *Blood* **92**: 374.

Jansen J.H., Mahfoudi A., Rambaud S., Lavau C., Wahli W., and Dejean A. 1995. Multimeric complexes of the PML-retinoic acid receptor α fusion protein in acute promyelocytic leukemia cells and interference with retinoid and peroxisome-proliferator signaling pathways. *Proc. Natl. Acad. Sci.* **92**: 7401.

Jarriault S., Brou C., Logeat F., Schroeter E.H., Kopan R., and Israel A. 1995. Signaling downstream of activated mammalian Notch. *Nature* **377**: 355.

Kakizuka A., Miller W.H., Jr., Umesono K.,Warrell R.P., Frankel S.R., Murty V.V.V.S., Dmitrovsky E., and Evans R.M. 1991. Chromosomal translocation t(15;17) in human acute promyelocytic leukemia (APL) fuses the RARα receptor with a novel putative transcription factor PML. *Cell* **66**: 663.

Kamei Y., Xu L., Heinzel T., Torchia J., Kurokawa R., Gloss B., Lin S.C., Heyman R.A., Rose D.W., Glass C.K., and Rosenfeld M.G. 1996. A CBP integrator complex mediates transcriptional activation and AP-1 inhibition by nuclear receptors. *Cell* **85**: 403.

Kannabiran C., Zeng X. Y., and Vales L.D. 1997. The mammalian transcriptional repressor RBP (CBF1) regulates interleukin-6 gene expression. *Mol. Cell. Biol.* **17**: 1.

Laherty C.D., Yang W.M., Sun J.M., Davie J.R., Seto E., and Eisenman R.N. 1997. Histone deacetylases associated with the mSin3 corepressor mediate mad transcriptional repression. *Cell* **89**: 349.

Licht J.D., Chomienne C., Goy A., Chen A., Scott A.A., Head D.R., Michaux J.L., Wu Y., DeBlasio A., Miller W.H., Jr., Zelentz A.D., Willman C.L., Chen A., Chen S.-J., Zelent A., Macintyre E., Veil A., Cortes J., Kantarjian H., and Waxman S.. 1995. Clinical and molecular characterization of a rare syndrome of acute promyelocytic leukemia associated with translocation (11;17). *Blood* **85**: 1083.

Miyazawa K., Mori A., Yamamoto K., and Okudaira H. 1998. Transcriptional roles of CCAAT/enhancer binding protein-β, nuclear factor-κB, and C-promoter binding factor 1 in interleukin (IL)-1β-induced IL-6 synthesis by human rheumatoid fibroblast-like synoviocytes. *J. Biol. Chem.* **273:** 7620.

Muraoka M., Konishi M., Kikuchi-Yanoshita R., Tanaka K., Shitara N., Chong J.M., Iwama T., and Miyaki M. 1996. p300 gene alterations in colorectal and gastric carcinomas. *Oncogene* **12:** 1565.

Nagy L., Kao H.-Y., Chakravarti D., Lin R., Hassig C.A., Ayer D.E., Schreiber S.L., and Evans R.M. 1997. Nuclear receptor repression mediated by a complex containing SMRT, Sin3, and histone deacetylase. *Cell* **89:** 373.

Ogryzko V.V., Schiltz R.L., Russanova V., Howard B.H., and Nakatani Y. 1996. The transcriptional coactivators p300 and CBP are histone acetyltransferases. *Cell* **87:** 953.

Plaisance S., Vanden Berghe, W., Boone, E., Fiers, W., and Haegeman, G. 1997. Recombination signal sequence binding protein Jκ is constitutively bound to the NF-κB site of the Interleukin-6 promoter and acts as a negative regulatory factor. *Mol. Cell. Biol.* **17:** 3733.

Rousselot P., Hardas B., Patel A., Guidez F., Gaken J., Castaigne S., Dejean A., de Thé H., Degos L., Farzaneh F., and Chomienne C. 1994. The PML-RARα gene product of t(15;17) translocation inhibits retinoic acid-induced granulocytic differentiation and mediated transactivation in human myeloid cells. *Oncogene* **9:** 545.

Ruthardt M.,Testa U., Nervi C., Ferrucci P.F., Grignani F., Puccetti E., Grignani F., Peschle C., and Pelicci P.G. 1997. Opposite effects of the acute promyelocytic leukemia PML-retinoic acid receptor α (RARα) and PLZF-RARα fusion proteins on retinoic acid signaling. *Mol. Cell. Biol.* **17:** 4859.

Sap J., Munoz A., Schmitt J., Stunnenberg H., and Vennstrom B. 1989. Repression of transcription mediated at a thyroid hormone response element by the v-*erb-A* oncogene product. *Nature* **340:** 242.

Shao W., Benedetti L., Lamph W.W., Nervi C., and Miller W.H. 1997. A retinoid-resistant acute promyelocytic leukemia subclone expresses a dominant negative PML-RARα mutation. *Blood* **89:** 4282.

Sharif M. and Privalsky M.L. 1991. v-*erbA* oncogene function in neoplasia correlates with its ability to repress retinoic acid receptor action. *Cell* **66:** 885.

Sternberg P.W., Stern M.J., Clark I., and Herskowitz I. 1987. Activation of the yeast *HO* gene by release from multiple negative controls. *Cell* **48:** 567.

Struhl K. 1998. Histone acetylation and transcriptional regulatory mechanisms. *Genes Dev.* **12:** 599.

Taki T., Sako M., Tsuchida M., and Hayashi Y. 1997. The t(11;16)(q23;p13) translocation in myelodysplastic syndrome fuses the MLL gene to the CBP gene. *Blood* **89:** 3945.

Taunton J., Hassig C.A., and Schreiber S.L. 1996. A mammalian histone deacetylase related to the yeast transcriptional regulator Rpd3p. *Science* **272:** 408.

Wettstein D.A., Turner D.L., and Kintner C.R. 1997. The *Xenopus* homolog of *Drosophila Suppressor of Hairless* mediates Notch signaling during primary neurogenesis. *Development* **124:** 693.

Yang X-J., Ogryzko V.V., Nishikawa J., Howard B.H., and Nakatani Y. 1996. A p300/CBP-associated factor that competes with the adenoviral oncoprotein E1A. *Nature* **382:** 319.

Yoshida M., Horinouchi S., and Beppu T. 1995. Trichostatin A and trapoxin: Novel chemical probes for the role of histone acetylation in chromatin structure and function. *BioEssays* **17:** 423.

Zenke M., Munoz A., Sap J., Vennstrom B., and Beug H. 1990. v-*erbA* oncogene activation entails the loss of hormone-dependent regulator activity of c-*erbA*. *Cell* **61:** 1035.

Building Transcriptional Regulatory Complexes: Signals and Surfaces

K.R. Yamamoto, B.D. Darimont, R.L. Wagner,* and J.A. Iñiguez-Lluhí
*Department of Cellular and Molecular Pharmacology, *Graduate Group in Biophysics, University of California, San Francisco, San Francisco, California 94143-0450*

Unlike DNA replication, in which every base pair of the genome is copied precisely once per cell per generation, transcription of DNA to RNA is differential and highly selective. Many transcripts are rare, accumulating to one copy per cell or less, whereas others are massively expressed. Perhaps only 7% of the mammalian genome is *ever* transcribed. As first established in elegant studies in phage and bacterial systems (Beckwith and Zipser 1970; Ptashne 1986), the selectivity and extent of eukaryotic mRNA synthesis are specified in part by two classes of genomic sites: *promoter* sequences at which the transcription machinery (including RNA polymerase II) assembles and initiates RNA polymerization (Losick and Chamberlin 1976) and *response elements* at which regulatory factors bind and alter the efficiency of promoter function (McKnight and Yamamoto 1992).

The functions of regulatory factors for RNA polymerase II are not "hard-wired"; rather, they are highly sensitive to context, responding differentially to specific developmental, environmental, or physiologic cues (Yamamoto 1997a; Lefstin and Yamamoto 1998). Indeed, transcriptional regulators reside at the endpoints of many, probably most, signal transduction pathways (Yamamoto 1989). At the molecular level, signaling reflects at least one of three classes of modulated events: noncovalent interactions with small molecules, as in hormone binding to intracellular receptors; covalent interactions with small molecules, as in protein phosphorylation; and noncovalent interactions with macromolecules, as in protein:protein association. These signaling events alter the activities of the regulators, likely by affecting the stability or accessibility of "functional surfaces." Such surfaces may correspond to enzymatic activities or perhaps more typically to interfaces for protein:protein interactions that give rise to multiprotein regulatory complexes.

The context dependence of transcriptional regulation implies that functional regulatory complexes are structurally dynamic, their composition governed by specific protein-binding sites arrayed within a given response element, by the combination of regulatory factors in a given cell type, and by the physiologic status of that cell as represented by the activities of signaling networks that influence regulatory factor activities (Yamamoto et al. 1992). Such a "mixed assembly" model for combinatorial regulation implies that DNA-binding regulatory factors, and their various coactivators and corepressors, must interact rather flexibly, to enable assembly into multiple final complexes, yet also quite specifically, to ensure precise assembly into appropriate complexes. As a result, a given regulator might activate transcription in one context, repress in another, and bind but exert no regulatory effect in a third.

The intracellular receptor (IR) superfamily comprises the largest group of metazoan transcriptional regulators and includes receptors for steroids, thyroid hormone, retinoids, vitamin D3, and other small lipophilic ligands (Mangelsdorf et al. 1996). The glucocorticoid receptor (GR), for example, resides in the cytoplasm as an inactive aporeceptor in the absence of its steroidal ligands (Picard and Yamamoto 1987; Bohen and Yamamoto 1994); upon hormone binding, the GR-hormone complex dissociates from a molecular chaperone complex bound to the aporeceptor, traverses to the nucleus, and binds to glucocorticoid response elements (GREs), from which it potentially regulates nearby promoters. In addition, GR is also phosphorylated (Krstic et al. 1997) and can associate with various nonreceptor regulators (Diamond et al. 1990; Imai et al. 1993) with chromatin remodeling complexes (Cairns et al. 1996) and with various coactivators and general transcription factors. Hence, GR receives multiple signals and can assemble into an array of distinct multifactor regulatory complexes. The integration of this spectrum of inputs specifies the net activities of at least three functional transcriptional activation surfaces, denoted Enh2/AF1, τ2, and AF2 (Fig. 1); repression domains have not yet been well delineated.

An essential feature of IR action is that each is highly context-dependent in its activities. GR, for example, mediates glucocorticoid effects in virtually all cell types, but different genes are regulated in each cell type. Thyroid hormone receptor (TR), in contrast, controls a virtually distinct set of genes in a subset of tissues. The implication is that specific and distinct regulatory complexes must form in different contexts and, in turn, that fundamental and likely general principles must govern the assembly of these complexes.

In this paper, we illustrate some of these principles. To do so, we divide IR regulatory complexes into functional subcomplexes, one sensitive to *physiological context* imposed by the hormonal ligand, and the other responsive to *gene-specific context* conferred by the response element. After summarizing recent studies with those subcomplexes, we consider the findings in relation to two distinct and broader frames of reference: We propose a rationale for the complexity of regulation by small molecules, and we define the contextual parameters required to specify uniquely the activity of a transcriptional regulator.

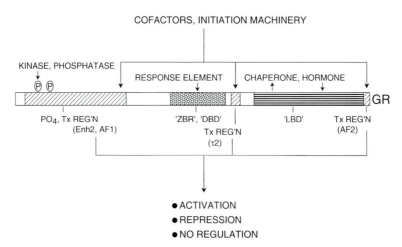

Figure 1. Functional domains of the glucocorticoid receptor and signaling components that affect its transcriptional regulatory functions. Diagram depicts the functional domains arrayed along the glucocorticoid receptor, including the conserved zinc-binding region (ZBR) which includes the DNA-binding domain (DBD), the ligand binding domain (LBD), and three separate transcriptional regulatory regions, Enh2/AF1, τ2, and AF2. Above the diagram are shown various interacting components (chaperone complex binds reciprocally with hormone), which can be viewed as signaling inputs that provide the receptor with physiological, cellular, and gene-specific context information. Integration of the signals leads to structural configuration of the regulatory domains, and their next activities lead to transcriptional activation, repression, or binding with no regulation.

RESULTS

Hormone-driven Formation and Selectivity of an Activator:Coactivator Interface

IRs differ dramatically in their functions as aporeceptors, agonist-bound or antagonist-bound species; indeed, the ligand-binding domains (LBDs) themselves (Fig. 1) function as compact "molecular switches" (Rusconi and Yamamoto 1987; Picard et al. 1988). Interestingly, the LBDs from various IRs (Bourguet et al. 1995; Renaud et al. 1995; Wagner et al. 1995; Brzozowski et al. 1997; Williams and Sigler 1998) share a common overall fold, despite substantial sequence divergence. Comparisons of structures of agonist- and antagonist-bound ER LBDs (Brzozowski et al. 1997) and of apo- and agonist-bound PPARγ LBDs (Nolte et al. 1998) demonstrate that ligand binding induces conformational changes, affecting especially the position of an α-helix denoted helix 12, at or near the LBD carboxyl terminus. Helix 12 is an essential component of the agonist-dependent AF2 transcriptional activation domain (Fig. 1) (Danielian et al. 1992; Barettino et al. 1994; Durand et al. 1994).

Functional AF2 domains can bind coactivators from the p160 family, which includes at least three distinct members: SRC-1 (also NcoA-1), p/CIP (also AIB1, TRAM-1, RAC3, ACTR), and TIF2 (also GRIP1, NcoA-2) (for review, see Moras and Gronemeyer 1998 and references therein). These approximately 160-kD proteins include amino-terminal βHLH and PAS domains, domains for the interaction with IRs (denoted NID) and CBP, and carboxy-terminal activation domains (Fig. 2) (Ding et al. 1998; Kalkhoven et al. 1998; Voegel et al. 1998). The NID contains three LxxLL motifs, each residing within distinct patches of conserved sequence termed NR boxes 1, 2, and 3 (Fig. 2) (Le Douarin et al. 1996; Heery et al. 1997; Ding et al 1998; Voegel et al. 1998). IR

LBDs interact with the NID, and individual receptors display apparent preferences for particular NR boxes (Torchia et al. 1997; Ding et al. 1998; Voegel et al. 1998). To investigate the role of signals as determinants of regulatory surfaces, Darimont et al. (1998) analyzed the interactions of the TR and GR LBDs with a coactivator from the p160 family, GRIP1 (Hong et al. 1996, 1997).

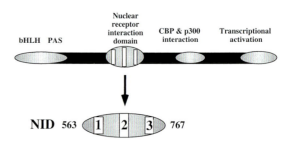

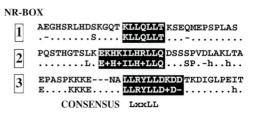

Figure 2. Functional domains of p160 family coactivators, focusing on nuclear receptor interaction domain (NID) of GRIP1. The NID of GRIP1 (563–767) contains predicted α helices that include the conserved LxxLL motifs of NR boxes 1, 2, and 3. Sequence alignment of the LxxLL motifs in members of the p160 coactivator family (GRIP1 [shown], Hong et al. 1996, 1997; pCIP, Torchia et al. 1997; and SRC-1, Oñate et al. 1995) revealed that each NR box has a box-specific consensus sequence that extends slightly beyond the overall LxxLL consensus.

Biochemical and structural analyses of IR LBD:p160 interaction. We measured the affinities of IR:GRIP1 interactions in vitro using a "quantitative GST pull-down" assay (Darimont et al. 1998). In these experiments, fusions bearing glutathione *S*-transferase (GST) and NID or NID derivatives bearing mutant NR boxes were constructed, expressed, purified, and used to monitor interactions with ^{35}S-labeled TRβ or TRβ LBD by retention on glutathione agarose; full-length TRβ or TRβ LBD bound the NID of GRIP1 with equal affinity and specificity (data not shown). Purification of the GST-NID fusions to near homogeneity enabled quantitative comparisons of interactions between different GRIP1 derivatives and receptors; each assay was performed with known concentrations of the GST-NID and receptor derivatives.

The TRβ LBD:NID interaction was strongly hormone-dependent (Fig. 3a), and consistent with in vivo studies (Ding et al. 1998), a preference for interaction with NR-box2 was evident; i.e., the extent of interaction declined when NR-box2 was mutated (NID2⁻), whereas mutation of NR-box3 (NID3⁻) was without apparent effect, and a double mutant of NR boxes 2 and 3 failed to bind. In titrations with increasing concentrations of wild-type or mutant NIDs, we found that the binding of the TRβ LBD to NR-box2 ($EC_{50} = 0.9 \pm 0.2$ μM) was three- to fourfold stronger than to NR-box3 ($EC_{50} = 3.2 \pm 0.9$ μM) and that binding to NR boxes 2 and 3 appeared to be independent (Fig. 3b).

We then tested the ability of peptides containing individual NR boxes and different lengths of adjacent sequences to compete the interaction of the NR-box2-containing NID fragment NID3⁻ with the TRβ LBD. The central hexapeptide of NR-box3, LLRYLL, competed the interaction of NID3⁻ with the TRβ LBD with very low efficiency, but extension of this hexapeptide to the 14-mer KENALLRYLLDKDD produced a strong inhibitor (Fig. 3c). Similarly, the two NR-box2 peptides EKHKILHRLLQDS or TSLKEKHKILHRLLQDSS were potent com-

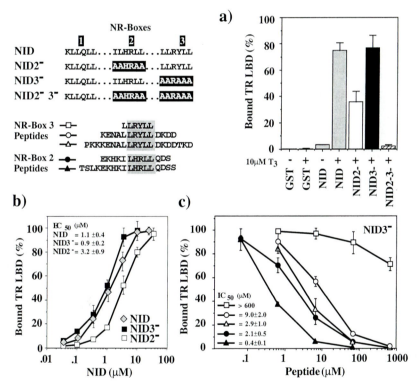

Figure 3. (*a*) GRIP1 NR boxes 1, 2, and 3 interact differentially with the TRβ LBD. GST denotes an isolated GST domain, and NID is the GST fusion of the GRIP1 fragment (563–767)His₆ that contains NR boxes 1, 2, and 3 (*gray*). In NID2⁻ (*white*) or NID3⁻ (*black*), the bulky hydrophobic residues of NR-box2 (ILHRLL) or NR-box3 (LLRYLL) were replaced by alanine yielding (AAHRAA) and (AARAAA), respectively. NID2⁻3⁻ (*hatched*) contains replacement of both NR boxes 2 and 3. Assays were carried out using 10 nM labeled TRβ LBD and 1.6 μM purified, glutathione-agarose bound GST-NID proteins in either the absence (–) or the presence (+) of 10 μM T₃. The yield of bound receptor is given as the percentage relative to the input. The data show the average of ≥3 independent experiments together with the standard deviation. (*b*) NR-box2 interacts with TRβ LBD with a fourfold higher affinity than NR-box3. Labeled TR_ LBD (10 nM) was incubated in the presence of 10 μM T₃ and various concentrations of purified and glutathione-agarose-bound GST-NID (*gray diamond*) and GST-NID2⁻ (*open square*) or GST-NID3⁻ (*closed square*) lacking either a functional NR box 2 or 3, respectively. As in *a*, the amount of bound receptor is relative to the receptor input. The data represent the average and range of at least two independent experiments. (*c*) Peptides containing NR boxes compete the interaction of TRβ LBD with the NID. Labeled TR_ LBD (10 nM) was incubated with 1.6 μM glutathione-agarose-bound GST-NID3⁻ in the presence of 10 μM T₃ and increasing concentrations of NR-box2 peptides EKHKILHRLLQDS (*closed circle*) or TSLKEKHKILHRLLQDSS (*closed triangle*), or NR-box3 peptides LLRYLL (*open square*), KENALLRYLLDKDD (*open circle*) or PKKKENALLRYLLDKDDTKD (*open triangle*). Bound receptor is shown relative to the amount retained in the absence of peptide. The data and calculated IC_{50} values represent the average and standard deviation of three independent experiments. (Reprinted, with permission, from Darimont et al. 1998.)

petitors of the TRβ LBD:NID3⁻ interaction, demonstrating that the 13- and 14-amino-acid peptides encompassing NR boxes 2 and 3, respectively, are sufficient to interact with the TRβ LBD. In contrast, NR-box1-containing peptides competed weakly and only at very high concentrations (not shown). Strikingly, the apparent affinities of the TRβ LBD for the NR-box2 peptide EKHKILHRLLQDS ($K_D$app = 0.8 ± 0.3 μM) and for NR-box3 peptide KENALLRYLLDKDD ($K_D$app = 3.2 ± 1.2 μM) were indistinguishable from the affinities for the full domain derivatives NID3⁻ (functional NR-box2) or NID2⁻ (functional NR-box3), respectively. Thus, rather than requiring the structural context of the full NID, the LxxLL motif and sequences immediately adjacent appear to be sufficient to account fully for the interaction of these NR boxes with the TRβ LBD.

Structural determination of cocrystals of the 13-amino-acid NR-box2 peptide KHKILHRLLQDSS with the TRβ

LBD revealed that, as with the TRα LBD (Wagner et al. 1995), the LBD consisted of 12 α helices and four β strands organized in three layers. The bound peptide was an amphipathic α helix of nearly three turns with its hydrophobic face, bearing the conserved leucine residues of the LxxLL motif, packed into complementary hydrophobic pockets in a groove of the LBD defined by residues on H3, H4, H5, and H12 (Fig. 4). The peptide helix appears to be stabilized by helix-capping side chain interactions with two conserved LBD residues, K288 and E457. Mutation of these and other LBD residues on the interface resulted in a loss of GRIP1 binding in vitro and activation activity in vivo (Feng et al. 1998). Similarly, substitution mutations of the motif leucine residues, even to other bulky hydrophobic side chains, severely compromised binding affinity, whereas mutation of the "xx" residues in NR-box2, His Arg, was without effect (Darimont et al. 1998). These findings are consistent with the strong com-

a)

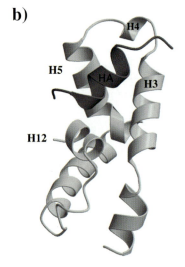

b)

c)

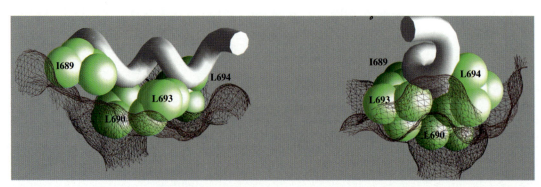

Figure 4. Interface between the NR-box2 peptide and the TRβ LBD. (*a*) Surface of the TRβ LBD. The side chains of the leucine residues from the NR-box2 peptide fit within a hydrophobic groove on the surface of the TRβ LBD, whereas the side chain of the box2-specific isoleucine residue packs against the outside edge of the groove. The remainder of the peptide is shown as main chain. (*b*) Cα trace of the interface. The α-helical NR-box2 peptide contacts a surface of the TRβ LBD formed from helices H3, H4, H5, and H12. (*c*) Surface complementarity of the TRβ LBD:LxxLL interface. The side chains of the NR-box2 ILxxLL motif are shown in a CPK representation, with the main chain of the peptide drawn as a Cα worm. The three leucine residues fit into pockets on the molecular surface of the TRβ LBD, depicted as mesh, whereas the box2-specific isoleucine residue rests on the edge of the surface cleft.

plementarity of the interface and with the α-helicity of the peptide.

To determine if the residues adjacent to the LxxLL motif participate in the LBD interaction, we constructed chimeric peptides with various adjacent sequences linked to the NR-box2 or NR-box3 motif. We found that the affinity of the TRβ LBD interaction with these peptides was indeed affected, positively by some adjacent sequences and negatively by others (Darimont et al. 1998).

Parallel studies with the GR LBD supported the primacy of the GRIP1 LxxLL motif as the specificity determinant, revealing that GR interacts preferentially with NR-box3, rather than with NR-box2, and that sequences within the motif itself, rather than the adjacent residues, have a crucial role in determining affinity (Darimont et al. 1998). Additionally, unlike TRβ, in which the interaction affinity appears to be determined solely by the motif and adjacent sequences, Hong et al. (1999) identified a GRIP1 domain well outside of the NID that also contributes to the affinity of the GR:GRIP1 interaction.

Common protein-protein interfaces. Our studies demonstrate that interactions between IR AF2 domains and p160 family members are specified by a simple sequence motif in amphipathic α helices within the p160 NIDs, LxxLL, and that the relative affinities of the interactions can be influenced by a combination of three determinants: the motif sequence itself, the residues immediately adjacent to the motif, or a segment of the coactivator outside of the NID. Interestingly, the relative contributions of these three determinants differ among different receptor LBDs.

The AF2 hydrophobic groove is present in all known structures of receptor:agonist complexes, and many residues within and surrounding the groove are functionally conserved across the family (Tone et al. 1994; Renaud et al. 1995; Collingwood et al. 1997; Henttu et al. 1997; Jurutka et al. 1997; Masuyama et al. 1997; Saatcioglu et al. 1997; Feng et al. 1998). Although we are only just beginning to analyze the mechanisms by which the affinity determinants exert their effects (Darimont et al. 1998), it is apparent that the use of different combinations of multiple determinants provides a way to generate selectivity in the formation of IR:p160 complexes in different contexts.

More generally, we noticed that the TRβ LBD:GRIP1 structural interface is strikingly similar to the three other protein-protein interfaces that have been defined in transcriptional regulatory complexes: p53:MDM2 (Kussie et al. 1996), VP16:TAF_{II}31 (Uesugi et al. 1997), and CREB:CBP (Radhakrishnan et al. 1997). All of these complexes share a common theme: An amphipathic α-helix containing a conserved hydrophobic motif interacts with a complementary, hydrophobic surface (Fig. 5). As with GRIP1, a functional interaction appears to require specific hydrophobic residues; e.g., Fxxhh has been proposed as a motif that interacts selectively with TAF_{II}31 (Uesugi et al. 1997). We suggest that a common "structural language" is used to define protein-protein interactions in transcriptional regulatory complexes and that the

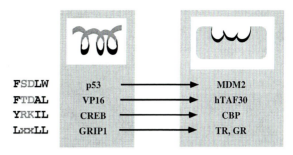

Figure 5. A common structural interface for protein-protein interactions in transcriptional regulatory complexes? All structurally defined complexes of protein pairs (see text for references) that interact during transcriptional regulation are characterized by an amphipathic α-helix contacting, through its hydrophobic surface, a hydrophobic groove on the surface of its partner protein. Note that the structural pairs do not assort by function. For example, p53 and CREB, which are DNA-binding factors, contain amphipathic helices that contact their cofactors, whereas TR and GR bind to DNA, but contain hydrophobic grooves that contact the GRIP1 coactivator.

use of multiple determinants to define common interactions provides mechanisms both for flexible assembly and for specificity.

Ligand effects on LBD structure and function. With respect to the LBD, agonists are ligands that support its interaction with coactivators, presumably by promoting formation of the AF2 coactivator interaction surface. Antagonists, such as the GR ligand RU486 or the estrogen receptor ligand raloxifene, fail to support LBD interactions with p160 coactivators (Hong et al. 1997; Norris et al. 1998). Comparison of agonist- and antagonist-bound structures of the ERα LBD reveals strikingly different positions of helix 12 (Brzozowski et al. 1997; Shiau et al. 1998). In the ER:antagonist complexes, helix 12 packs against the hydrophobic residues of helices 3 and 5 in a configuration remarkably similar to that of the NR-box2 peptide of GRIP1 with the agonist-bound TR LBD. Thus, antagonists such as raloxifene preclude the interaction of ER with coactivators in two ways: (1) eliminating the part of the coactivator interaction surface that involves H12 and (2) occluding with H12 the remaining part of the coactivator interaction surface (Fig. 6) (Darimont et al. 1998).

Many IR antagonists, such as raloxifene and RU486, appear to be structurally related to agonists that have been modified by a bulky adduct. We suggest that such antagonists may commonly function by interfering with positioning of H12 in the coactivator interface. It will be intriguing to assess the generality of the position of H12 that was seen in both the ER:raloxifene (Brzozowski et al. 1997) and the ER:tamoxifen (Shiau et al. 1998) complexes. In that structure, H12 contacts H3 and H5 through a hydrophobic surface (LLxxML) similar to the p160 LxxLL motif. The corresponding face of H12 among other IRs is less similar to LxxLL, although it typically contains bulky hydrophobic residues. Conceivably, the H12:H3/H5 interface might resemble a "default structure" adopted when the "agonist conformation" is precluded.

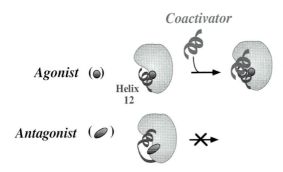

Figure 6. Ligand-specific effects on helix 12 positioning in the LBD determine formation of the AF2 surface. Classical agonists induce the helix 12 disposition desribed in Fig. 4b and therefore promote coactivator binding. Classical antagonists preclude the "agonist" position for helix 12, thereby eliminating its potential contact with coactivators, and in addition, induce helix 12 to contact residues in helices 3 and 5, occluding the remaining potential interface for coactivators. Importantly, the *biological* functions of small molecules as agonists and antagonists are context-dependent (see text).

For GR, various ligands have been characterized as agonist, antagonist, or partial agonist. Importantly, however, these distinctions are not receptor-intrinsic; whether a given ligand is an agonist or an antagonist depends on the cell type in which the receptor is expressed (Garabedian and Yamamoto 1992; Webb et al. 1995). In cellular contexts in which RU486 is an agonist rather than an antagonist, it is clear that the LBD does not configure an AF2 surface that GRIP1 can recognize (Hong et al. 1997; Norris et al. 1998; B.D. Darimont, unpubl.); similarly, ER:tamoxifen complexes fail to interact with GRIP1, although tamoxifen is a potent agonist in some settings (Webb et al. 1995).

The implications of these findings are that closely related molecules, each of which can occupy the same specific ligand-binding site on a receptor, must evoke clearly distinct conformational effects that are interpreted by cell-specific nonreceptor factors. Within a regulatory complex, this "interpretive" function somehow determines whether a receptor will associate with coactivators, corepressors, or neither. Putative cellular factors that interpret ligand-induced structural effects on regulators, and the nature of the surfaces that are generated to produce functionally alternative regulatory complexes, are unknown. In any case, our results support the notion that small molecule ligands define distinct physiologic contexts for IRs.

Response Elements as Determinants of Direction and Magnitude of Regulation

Beginning with elegant studies of the *Escherichia coli* AraC and bacteriophage λ cI proteins, it has been apparent that many transcriptional regulators can activate transcription in some contexts and repress in others. How do IRs "decide" whether to activate or to repress? The first hint came from studies of a prolactin gene response element that conferred glucocorticoid-mediated repression when fused to a test promoter (Sakai et al. 1988). GR bound to that element, but to a sequence unrelated to the "simple GREs" first identified as elements that confer activation by glucocorticoids (Chandler et al. 1983). In addition, the prolactin element, in the absence of hormone, and even in the absence of GR, *activated* the reporter and was bound by a nonreceptor factor. The important conclusions were that GR can bind specifically to different DNA sequences and that the regulatory consequences of those interactions could themselves be distinct.

Subsequently, Lefstin et al. (1994) discovered that GRE binding relieves an intramolecular inhibition of the GR transcriptional regulatory domains; two point mutations in the zinc-binding region were isolated that prevent this functional inhibition. Recent structural determinations have established that each mutation evokes conformational changes in solution that normally accompany GRE binding (M. van Tilborg et al., in prep.). These studies implied that GR transcriptional regulatory domains do not adopt their final structures until the receptor binds to a GRE and that, in principle, there could exist distinct functional classes of GREs, each triggering a characteristic mode of regulation by GR (Lefstin and Yamamoto 1998).

Multiple classes of GREs. Following our analysis of the prolactin GRE, we studied another nonconsensus GR-binding site (differing both from the palindromic simple GRE and from the prolactin element), associated with the proliferin gene (Mordacq and Linzer 1989). Like the prolactin site, the 26-bp proliferin element, plfG, displayed enhancer activity in the absence of GR, and hormone-bound GR then "repressed," reversing the constitutive activation. Mordacq and Linzer (1989) showed that plfG was also bound by AP1, thus accounting for the enhancement in the absence of GR. We denoted plfG as a "composite" GRE (Diamond et al. 1990; Yamamoto et al. 1992), a class of GREs at which hormonal regulation is specified not by GR alone, as at simple GREs, but rather by GR together with a nonreceptor factor, both of which bind to the element and together define the expression pattern. Indeed, in the absence of AP1, GR bound but was inactive at plfG; cells expressing AP1 predominantly as c-Jun homodimers activated from plfG in response to dexamethasone; and cells in which AP1 was composed mostly of c-Jun–c-Fos heterodimers repressed from plfG upon hormone addition (Diamond et al. 1990; Starr et al. 1996). Immunoprecipitation experiments demonstrated an interaction in vitro between GR and c-Jun, and genetic studies indicated that GR assumed distinct conformations in its positive and negative regulatory modes at plfG (Yamamoto et al. 1992).

Subsequently, numerous additional composite GREs have been described, with factors other than AP1 serving as the nonreceptor components in the different elements (see, e.g., Mittal et al. 1994; Scott et al. 1998). It seems likely that composite elements will be the rule rather than the exception at natural genes, as they provide a DNA-based mechanism for at least the first step in formation of multiprotein regulatory complexes.

In parallel with our studies of the plfG composite GRE, other investigators characterized repression by GR at the collagenase gene AP1 site (Jonat et al. 1990; Schüle et al. 1990; Heck et al. 1994). Although contentious for a time, it is now generally agreed that GR represses collagenase by associating with the bound AP1 protein (either the c-Jun homodimer or c-Jun–c-Fos heterodimer), but not with DNA. This behavior is transferable to heterologous promoters merely by insertion of the AP1-binding site, which we denote colA. Therefore, colA defines a third class of element, a "tethering GRE" (Miner and Yamamoto 1991; Starr et al. 1996), at which GR associates through protein-protein contacts alone. Thus, specific DNA contacts were dispensable for repression from colA but not from plfG, but like plfG, repression from colA was insensitive to mutations that compromise activation at simple or composite elements (Yamamoto et al. 1992).

Taken together, these studies revealed three classes of GREs (Fig. 7): *simple*, where GR alone binds to DNA and activates transcription; *composite*, where GR binds to DNA and also interacts with a bound coregulator, specifying either activation or repression depending on the subunit composition of the coregulator; and *tethering*, where GR, rather than binding to DNA, makes protein-protein contacts with a DNA-bound activator and re-presses transcription. These three types of GREs together generate two ways to activate and two ways to repress transcription.

Starr et al. (1996) isolated two point mutations in plfG that displayed a gain-of-function phenotype in which the mutant elements acquired simple GRE activity (i.e., GR-mediated activation in the absence of AP1). These same mutants also displayed increased composite enhancement in the presence of c-Jun, and in the presence of c-Jun–c-Fos, enhanced rather than repressed transcription. These findings demonstrate that the plfG sequence specifies at least in part the mode of regulation by GR.

Receptors decode response-element-based signals. The first evidence that GR could "read" and interpret GRE-associated signals came from studies of a site-directed mutant, K461A. K461 is one of three GR residues shown by crystallographic analysis to contact specific DNA bases in a simple GRE (Luisi et al. 1991). The K461A mutant displayed remarkable phenotypes: At simple GREs, activity declines only modestly despite a substantially reduced affinity; at the plfG composite GRE, increased synergy is seen under enhancing conditions (Jun-Jun) and activation is observed under repressing conditions (Jun-Fos); finally, the colA-tethering GRE mediates activation rather than repression (Starr et al. 1996). In contrast to its reduced binding to the simple GRE, the K461A mutant protein displayed near wild-type interactions with plfG DNA and AP1 protein. The lysine at this position is conserved throughout virtually the entire IR superfamily, and equivalent mutations on MR, ER, AR, TR, RXR, and RAR similarly produced derivatives that activated under all conditions (Starr et al. 1996).

These results suggest that K461 interacts with GRE components (DNA, protein, or both) and triggers allosteric transitions in GR structure that generate alternative receptor surfaces involved in transcription activation or repression (Starr et al. 1996; Yamamoto 1997a). It will likely prove general that response elements serve as allosteric ligands that provide gene-specific context information to bound regulators (Lefstin and Yamamoto 1998).

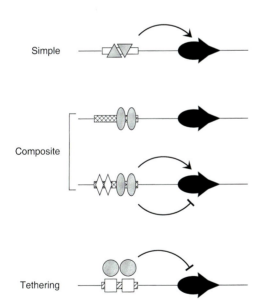

Figure 7. Three classes of GREs. At *simple* GREs, GR (*shaded triangle*) binds at an imperfect palindromic DNA sequence (*open rectangle*) and enhances transcription from nearby promoters (*solid arrow*). At a *composite* GRE, GR (*shaded oval*) binds to a sequence (*crosshatched rectangle*) that can be unrelated to that of simple GREs; in the case of the plfG composite GRE, for example (see text), GR binds, but in the absence of AP1 bound to the same element, GR fails to regulate transcription. In the presence of AP1, GR either enhances or represses, depending on the subunit composition of AP1. At *tethering* GREs, GR (*shaded circle*) fails to bind DNA, and instead binds to a specifically bound nonreceptor activator and represses transcription. Shape differences for GR at each class of GRE are meant to represent its distinct activities at each, presumably reflecting distinct conformations.

A Putative Cross-talk Regulator at Compound Composite Elements

The response-element-mediated switch in IR function between enhancement and repression is experimentally facile because the switch point is unambiguous. However, it should be apparent that these findings establish a broader principle: Response elements could specify any mode of intermolecular communication by IRs, or indeed any regulators, e.g., homotypic IR:IR interactions, functional contacts between IRs and coactivators or corepressors, heterotypic ("cross-talk") interactions between IRs and nonreceptor regulators in regulatory complexes, and the like. We illustrate this principle with a recent, still preliminary, study, which is intriguing because it implies a functional relationship between signaling by response elements and cell-specific factors.

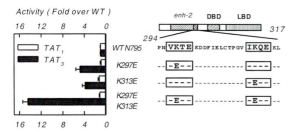

Figure 8. Synergy control motif sequences in GR. Mutants shown, and others not shown (J. Iñiguez-Lluhí, unpubl.), define a repeated motif in the GR amino-terminal region, but distinct from the Enh2 transcriptional regulatory region, that reduces the magnitude of GR-GR synergy in appropriate cell contexts. In the results shown, CV-1 cells were transfected by lipofection with 200 ng of luciferase reporter plasmid linked to one or two copies of a GR-binding element from the tyrosine aminotransferase (TAT) gene. Cultures were treated, or not, with 10 nM dexamethasone. Data represent averages ± S.E.M. of three to four independent transfections, performed in triplicate.

Receptors limit synergy at compound response elements. In the course of screening a library bearing the amino-terminal domain of GR containing multiple random substitutions, we isolated a new class of GR mutants with more than tenfold increased activity relative to wild-type GR (J. Iñiguez-Lluhí, unpubl.). Strikingly, this phenotype was observed at "compound GREs," i.e., GREs bearing multiple GR-binding sites, whereas the mutations had no effect at "solo" GREs (Fig. 8). The effect on compound GREs was dependent neither on the arrangement of the binding sites within the compound element nor on the separation between the response element and the promoter (J. Iñiguez-Lluhí, unpubl.). Thus, rather than affecting intrinsic activation, these novel mutations relieve an inhibitory effect on GR-GR synergy.

The mutations responsible for the phenotype were found to reside within a short region (amino acids 296–315) of rat GR. This region lies outside of the Enh2/AF1 transcriptional activation domain (see Fig. 1), which is defined by mutations in a 16-amino-acid region, amino acids 219–234 (Iñiguez-Lluhí et al. 1997). Further analysis (J. Iñiguez-Lluhí, unpubl.) revealed that the mutations map to two closely spaced copies of a conserved sequence motif, (I/V)KxE (Fig. 8). These sequences are conserved in other IRs as well and yield similar up-synergy phenotypes when mutated. Similar motifs have also been seen (but have not yet been tested) in a subset of nonreceptor transcription regulators (J. Iñiguez-Lluhí, unpubl.).

Synergy control is not receptor-intrinsic. In yeast expressing rat GR, strong synergy is observed at a compound GRE carrying a pair of GR-binding sites (Table 1). Interestingly, however, synergy in yeast was not increased by mutation of the (I/V)KxE motifs, implying that mammalian cells may contain a factor, perhaps lacking in yeast, that interacts with wild-type (I/V)KxE motifs and restrains synergy. Whatever the mechanism, these results demonstrate that the effect of the (I/V)KxE motifs is not receptor-intrinsic.

A different sort of context effect is observed with the mouse mammary tumor virus (MMTV) GRE, a com-

pound composite element containing four GR sites plus NF1 and Oct1 sites. All of the GR sites participate in hormonal activation, and important functional interactions occur between GR and the nonreceptor factors, as NF1 and Oct1 bind to their sites only in the presence of bound GR (Archer et al. 1991); indeed, direct GR-Oct1 contacts have been described (Prefontaine et al. 1998). In this case, our (I/V)KxE mutations produce a net *decrease* in overall enhancement (data not shown). One interpretation of this result is that the *increased* homotypic (GR-GR) synergy produced by the mutations is accompanied by *decreased* heterotypic synergy and that the latter interactions are more important for net activation from this compound element.

Our working model is that the I/VKxE mutants may identify an IR surface that is recognized by a putative "synergy control factor" (SCF; Fig. 9), which limits homotypic synergy and thereby alters the assembly or actions of regulatory complexes. If the expression or the activity of this factor were itself regulated, receptor activity at compound response elements could be selectively and dramatically altered, perhaps reminiscent of a sharp transition in IR activity at a critical stage of fibrosarcoma tumorigenesis (Vivanco et al. 1995).

DISCUSSION AND PERSPECTIVES

Rationalizing the Complexity of Transcriptional Regulation

Viewed from close range, transcriptional regulation seems to be remarkably baroque, adorned by a burgeoning array of factors operating in various combinations as

Table 1. Activities of Wild-type GR and K297E/K313E Mutant in *Saccharomyces cerevisiae* from Solo or Compound GREs

Receptor	β-galactosidase activity (units ± S.E.M.)	
	GRE$_1$	GRE$_2$
Wild type	203 ± 55	1699 ± 363
K297E/K313E	244 ± 45	1838 ± 207

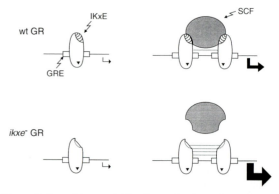

Figure 9. A working model for the mechanism of synergy control motifs. GRs bound at compound GREs undergo homotypic interactions, resulting in synergistic activity. In cells containing functional synergy control factor (SCF), however, SCF interacts with the IkxE motifs, limiting the magnitude of the synergy. Mutation of the IkxE motifs abrogates SCF action, permitting strong synergy.

multiprotein complexes that bind, bend, remodel, coactivate, or corepress, with each factor serving architectural, interactive, or enzymatic roles. The expression, intracellular locations, or activities of regulatory factors are commonly modulated by physiologic or environmental cues.

The requirements for regulation, as well as its consequences, are seen most dramatically in higher eukaryotes: metazoans and higher plants. These organisms are unique among living creatures in containing terminally differentiated cells, in which each cell type excludes permanently the option of expressing genes specific to other cell types. In turn, this remarkable commitment demands efficient and effective intercellular communication if the cells are to coordinate their functions as a single organism; i.e., higher eukaryotes require a level of intercellular signaling that is uniquely stringent. Notably, the intracellular receptor gene superfamily is a relatively recent evolutionary elaboration: It is found in all metazoans but is absent from all other organisms. It is reasonable to assume that IRs evolved to satisfy, at least in part, the special signaling requirements of metazoans (Yamamoto 1997b).

Lipophilic small molecules, which serve as IR ligands, seem ideal as intercellular signals: They are simple to synthesize or acquire from the environment, relatively stable, and may readily enter target cells. Their structural simplicity, however, generates an "information capacity paradox": How can molecules that lack chemical complexity symbolize complex physiological states? How can a steroid hormone, with a molecular weight less than one third of that of a single base pair of DNA, control the expression of cell-specific batteries of genes? Receptors are the first step toward resolving this paradox: Their association with a ligand can be viewed as a large-scale "chemical modification" of the ligand, greatly increasing its complexity and thereby allowing it, for example, to discriminate and bind to response elements (Yamamoto 1985, 1997b).

The complexity of combinatorial regulation can then be derived as a direct extension of this simple logical framework. For example, glucocorticoid-GR complexes, although much more complex than the hormone alone, cannot account for the differential actions of glucocorticoids in different cell types. A solution is for the hormone-receptor complex to associate with additional protein factors, in particular with combinations of factors that are cell-type-specific or that couple the receptor and hormone to other signaling pathways. In effect, these additional factors increase further the complexity of the ligand, enabling a single hormone-receptor complex (thus a single hormone) to operate on different genes in different cell types.

Similarly, hormonal association with a component within multiprotein complexes cannot account for gene-specific actions, such as activation at some loci and repression at others. However, we know that IRs extract signaling information from response elements (Lefstin and Yamamoto 1998) and that IR regulatory domains are held dormant prior to response element binding (Lefstin et al. 1994), implying that functional associations with cofactors may occur on response elements and not free in solution. Thus, response elements, like the cofactors, add further complexity to hormone-IR complexes (and ultimately to the hormone itself) in a manner that enables the

hormone to confer distinct effects even within a single nucleus. Thus, the receptor together with the various components that interact directly or indirectly with it (see Fig. 1) comprise the full "spectrum of complexity" of the hormonal signal, and the scope of its potential actions is limited only by number of distinct complexes that can be specifically formed.

Achieving Flexibility and Precision in Regulatory Complex Assembly

Implicit in the concept of combinatorial regulation is the dynamic nature of regulatory complexes. The requirements for complex formation seem at first glance to be incompatible: The component parts must be sufficiently *flexible* in their interactions to allow mixed assembly, yet must be sufficiently *precise* to ensure decisiveness in their regulatory effects. In the interactions of IR LBDs with p160 family coactivators, and in their structural similarity to interactions observed in other complexes, we may begin to perceive molecular solutions to this issue. All reported interactions involve an amphipathic α-helix forming complementary contacts through its hydrophobic face with a hydrophobic groove. We suggest that these simple components, bearing similar but distinguishable signature sequence motifs, serve as "interaction addresses" that mark potential interacting partners by permitting specific, albeit low-affinity, interactions. Embedded within or close to these interaction addresses are additional "affinity elements": interacting surfaces or conformational components that increase or decrease the intrinsic affinity of the base interaction.

Even among interactions between two families of factors, the interaction addresses can differ, and the roles of the various affinity elements can differ in different complexes. Thus, TR interacts preferentially with GRIP1 NR-box2 using affinity elements that are immediately adjacent to its LHRLL motif, whereas GR binds to NR-box3 using affinity elements within its LRYLL motif, together with a domain downstream from the NR boxes. For both receptors, the hydrophobic groove that comprises the functional AF2 domain forms only upon agonist binding. In contrast, antagonist binding precludes formation of that interaction address. It seems likely that other ligands may confer more subtle differences in formation of the AF2 hydrophobic groove, such that the various affinity elements are recognized or utilized differently. Likewise, a p160 factor, through phosphorylation or interactions with other proteins, could alter the position, activity, or accessibility of both its interaction addresses and its affinity elements.

We thus suggest that regulators such as IRs receive and integrate multiple signaling inputs—hormonal ligands, phosphorylation, response elements, interacting proteins—to define and refine the precision of otherwise quite flexible cofactor interactions.

Three "Context Coordinates" That Uniquely Define Regulator Activity

On the basis of even early investigations of *E. coli* AraC and bacteriophage λ cI, the imprudence of ascribing intrinsic activities to transcriptional regulators should have been apparent. Rather, their activities are context-dependent, and the complexities of context increase with the complexity of organisms. This is not to say, of course, that regulator activities are ambiguous but instead that there are many ways to affect them. The mechanisms and modes of action of these factors are highly context-dependent by design: IRs and other regulators are immersed in signal transduction pathways, integrating incoming information.

As a heuristic device, it may be useful to consider three broad contextual classes that affect regulator actions. *Physiological* context refers to developmental and homeostatic status, as governed, for example, by neural and endocrine signals. *Cellular* context refers to differentiation state, and cell and tissue type, as governed by the expressed array of intracellular transcription factors and regulators, by determinants of chromatin structural domains, and by paracrine signals from neighboring cells and extracellular matrix. Both physiological and cellular contexts can be profoundly altered by pathological effects and by pharmacologic intervention. Finally, *gene* context refers to the array and structure of operative response elements and the organization of the promoter. Given this array of contextual elements, it seems unsurprising that regulatory complexes contain multiple components.

We have shown here that elements of physiological and gene context can be examined by probing functional subcomplexes of IR regulatory complexes. Importantly, however, a full description of receptor activity requires information from all three context classes. This is analogous to a spatial representation, in which all three coordinates, x, y, and z, must be defined to uniquely specify a position. As we begin to understand better the parameters affecting each context class, and their structural consequences on regulatory factors and complexes, we shall better understand the nature and mechanisms of metazoan transcriptional regulation.

ACKNOWLEDGMENTS

We thank previous and current colleagues in the Yamamoto lab, notably, Barry Starr and Jeff Lefstin, some of whose published studies are summarized here, who contributed ideas, discussions, and experiments that motivated the present work. We also thank Brian Freeman for help with some of the figures. Our research is supported by grants from the National Institutes of Health and the National Science Foundation; postdoctoral fellowship support was from the Leukemia Society of America (J.A.I.-L.) and the Helen Hay Whitney Foundation (B.D.D.).

REFERENCES

Archer T.K., Cordingley M.G., Wolford R.G., and Hager G.L. 1991. Transcription factor access is mediated by accurately positioned nucleosomes on the mouse mammary tumor virus promoter. *Mol. Cell. Biol.* **11:** 688.

Barettino D., Vivanco Ruiz M.M., and Stunnenberg H.G. 1994. Characterization of the ligand-dependent transactivation do-

main of thyroid hormone receptor. *EMBO J.* **13:** 3039.

Beckwith J.R. and Zipser D., Eds. 1970. *The lactose operon.* Cold Spring Harbor Laboratory, Cold Spring Harbor, New York.

Bohen S.P. and Yamamoto K.R. 1994. Modulation of steroid receptor signal transduction by heat shock proteins. In *The biology of heat shock proteins and molecular chaperones* (ed. R.I. Morimoto et al.), p. 313. Cold Spring Harbor Laboratory Press, Cold Spring Harbor, New York.

Bourguet W., Ruff M., Chambon P., Gronemeyer H., and Moras D. 1995. Crystal structure of the ligand-binding domain of the human nuclear receptor RXR-α. *Nature* **375:** 377.

Brzozowski A.M., Pike A.C.W., Dauter Z., Hubbard R.E., Bonn T., Engström O., Öhman L., Greene G.L., Gustafsson J.Å., and Carlquist M. 1997. Molecular basis of agonism and antagonism in the oestrogen receptor. *Nature* **389:** 753.

Cairns B.R., Levinson R.S., Yamamoto K.R., and Kornberg R.D. 1996. Essential role of Swp73p in the function of yeast Swi/Snf complex. *Genes Dev.* **10:** 2131.

Chandler V.L., Maler B.A., and Yamamoto K.R. 1983. DNA sequences bound specifically by glucocorticoid receptor *in vitro* render a heterologous promoter hormone responsive *in vivo*. *Cell* **33:** 489.

Collingwood T.N., Rajanayagam O., Adams M., Wagner R., Cavailles V., Kalkhoven E., Matthews C., Nystrom E., Stenlof K., Lindstedt G., Tisell L., Fletterick R.J., Parker M.G., and Chatterjee V.K.K. 1997. A natural transactivation mutation in the thyroid hormone beta receptor: Impaired interaction with putative transcriptional mediators. *Proc. Natl. Acad. Sci.* **94:** 248.

Danielian P.S., White R., Lees J.A., and Parker M.G. 1992. Identification of a conserved region required for hormone dependent transcriptional activation by steroid hormone receptors. *EMBO J.* **11:** 1025.

Darimont B.D., Wagner R.L., Apriletti J.W., Stallcup M.R., Kushner P.J., Baxter J.D., Fletterick R.L., and Yamamoto K.R. 1998. Structure and specificity of nuclear receptor-coactivator interactions. *Genes Dev.* **12:** 3343.

Diamond M., Miner J.N., Yoshinaga S.K., and Yamamoto K.R. 1990. Transcription factor interactions: Selectors of positive or negative regulation from a single DNA element. *Science* **249:** 1266.

Ding X.F., Anderson C.M., Ma H., Hong H., Uht R.M., Kushner P.J., and Stallcup M.R. 1998. Nuclear receptor-binding sites of coactivators glucocorticoid receptor interacting protein 1 (GRIP1) and steroid receptor coactivator 1 (SRC-1): Multiple motifs with different binding specificities. *Mol. Endocrinol.* **12:** 302.

Durand B., Saunders M., Gaudon C., Roy B., Losson R., and Chambon P. 1994. Activation function 2 (AF-2) of retinoic acid receptor and 9-*cis* retinoic acid receptor: Presence of a conserved autonomous constitutive activation domain and influence of the nature of the response element on AF-2 activity. *EMBO J.* **13:** 5370.

Feng W., Ribeiro R.C.J., Wagner R.L., Nguyen H., Apriletti W., Fletterick R.J., Baxter J.D., Kushner P.J., and West B.L. 1998. Hormone-dependent coactivator binding to a hydrophobic cleft on nuclear receptors. *Science* **280:** 1747.

Garabedian M.J. and Yamamoto K.R. 1992. Genetic dissection of the signaling domain of a mammalian steroid receptor in yeast. *Mol. Biol. Cell* **3:** 1245.

Heck S., Kullmann M., Gast A., Ponta H., Rahmsdorf H.J., Herrlich P., and Cato A.C.B. 1994. A distinct modulating domain in glucocorticoid receptor monomers in the repression of activity of the transcription factor AP-1. *EMBO J.* **13:** 4087.

Heery D.M., Kalkhoven E., Hoare S., and Parker M.G. 1997. A signature motif in the transcriptional co-activators mediate binding to nuclear receptors. *Nature* **387:** 733.

Henttu P.M.A., Kalkhoven E., and Parker M.G. 1997. AF-2 activity and recruitment of steroid receptor coactivator 1 to the estrogen receptor depend on a lysine residue conserved in nuclear receptors. *Mol. Cell. Biol.* **17:** 1832.

Hong H., Kohli K., Garabedian M.J., and Stallcup M.R. 1997. GRIP1, a transcriptional coactivator for the AF-2 transactivation domain of steroid, thyroid, retinoid, and vitamin D receptors. *Mol. Cell. Biol.* **17**: 2735.

Hong H., Kohli K., Trivedi A., Johnson D.L., and Stallcup M.R. 1996. GRIP1, novel mouse protein that serves as a transcriptional coactivator in yeast for the hormone binding domains of steroid receptors. *Proc. Natl. Acad. Sci.* **93**: 4948.

Hong H., Darimont B.D., Ma H., Yang L., Yamamoto K.R., and Stallcup M.R. 1999. An additional region of coactivator GRIP1 required for interaction with the hormone binding domains of a subset of nuclear receptors. *J. Biol. Chem.* **274**: 3496.

Imai E., Miner J.N., Mitchell J.A., Yamamoto K.R., and Granner D.K. 1993. Glucocorticoid receptor-cAMP response element-binding protein interaction and the response of the phosphoenolpyruvate carboxykinase gene to glucocorticoids. *J. Biol. Chem.* **268**: 5353.

Iñiguez-Lluhí J.A., Lou D.Y., and Yamamoto K.R. 1997. Three amino acid substitutions selectively disrupt the activation but not the repression function of the glucocorticoid receptor N terminus. *J. Biol. Chem.* **272**: 4149.

Jonat C., Rahmsdorf H.J., Herrlich P., Park K.-K., Cato A.C.B., Gebel S., and Ponta H. 1990. Antitumor promotion and anti-inflammation: Down-modulation of AP-1 (*fos/jun*) activity by glucocorticoid hormone. *Cell* **62**: 1189.

Jurutka P.W., Hsieh J.C., Remus L.S., Whitfield G.K., Thompson P.D., Haussler C.A., Blanco J.C., Ozato K., and Haussler M.R. 1997. Mutations in the 1,25-dihydroxyvitamin D3 receptor identifying C-terminal amino acids required for the transcriptional activation that are functionally dissociated from hormone binding, heterodimeric DNA binding, and interaction with basal transcription factor IIb, in vitro. *J. Biol Chem.* **272**: 14592.

Kalkhoven E., Valentine J.E., Heery D.M., and Parker M.G. 1998. Isoforms of steroid receptor coactivator 1 differ in their ability to potentiate transcription by the oestrogen receptor. *EMBO J.* **17**: 232.

Krstic M.D., Rogatsky I., Yamamoto K.R., and Garabedian M.J. 1997. Mitogen-activated and cyclin-dependent protein kinases selectively and differentially modulate transcriptional enhancement by the glucocorticoid receptor. *Mol. Cell. Biol.* **17**: 3947.

Kussie P.H., Gorina S., Marechal V., Elenbaas B., Moreau J., Levine A.J., and Pavletich N.P. 1996. Structure of the MDM2 oncoprotein bound to the p53 tumor suppressor transactivation domain. *Science* **274**: 948.

Le Douarin B., Nielsen A.L., Garnier J.M., Ichinose H., Jeanmougin F., Losson R., and Chambon P. 1996. A possible involvement of TIF1α and TIFβ in the epigenetic control of transcription by nuclear receptors. *EMBO J.* **15**: 6701.

Lefstin J.A. and Yamamoto K.R. 1998. Allosteric effects of DNA on transcriptional regulators. *Nature* **392**: 885.

Lefstin J.A., Thomas J.R., and Yamamoto K.R. 1994. Influence of a steroid receptor DNA-binding domain on transcriptional regulatory functions. *Genes Dev.* **8**: 2842.

Losick R. and Chamberlin M., Eds. 1976. *RNA polymerase.* Cold Spring Harbor Laboratory, Cold Spring Harbor, New York.

Luisi B.F., Xu W.X., Otwinowski Z., Freedman L.P., Yamamoto K.R., and Sigler P.B. 1991. Crystallographic analysis of the interaction of the glucocorticoid receptor with DNA. *Nature* **352**: 497.

Mangelsdorf D.J., Thummel C., Beato M., Herrlich P., Schütz G., Umesono K., Blumberg B., Kastner P., Mark M., Chambon P., and Evans R.M. 1996. The nuclear receptor family: The second decade. *Cell* **86**: 835.

Masuyama H., Brownfield C.M., St-Arnaud R., and MacDonald P.N. 1997. Evidence for ligand-dependent intramolecular folding of the AF-2 domain in vitamin D receptor-activated transcription and coactivator interaction. *Mol. Endocrinol.* **11**: 1507.

McKnight S.L. and Yamamoto K.R., Eds. 1992. *Transcriptional regulation.* Cold Spring Harbor Laboratory Press, Cold Spring Harbor, New York.

Miner J.N. and Yamamoto K.R. 1991. Regulatory crosstalk at composite response elements. *Trends Biochem. Sci.* **16**: 423.

Mittal R., Kumar K.U., Pater A., and Pater M.M. 1994. Differential regulation by c-*jun* and c-*fos* protooncogenes of hormone response from composite glucocorticoid response element in human papilloma virus type 16 regulatory region. *Mol. Endocrinol.* **8**: 1701.

Moras D. and Gronemeyer H. 1998. The nuclear receptor ligand-binding domain: Structure and function. *Curr. Opin. Cell Biol.* **10**: 384.

Mordacq J.C. and Linzer D.I.H. 1989. Co-localization of elements required for phorbol ester stimulation and glucocorticoid repression of proliferin gene expression. *Genes Dev.* **3**: 760.

Nolte R.T., Wisely G.B., Westin S., Cobb J.E., Lambert M.H., Kurokawa R., Rosenfeld M.G., Willson T.M., Glass C.K., and Milburn M.V. 1998. Ligand binding and co-activator assembly of the peroxisome proliferator-activated receptor-gamma. *Nature* **395**: 137.

Norris J.D., Fan D., Stallcup M.R., and McDonnell D.P. 1998. Enhancement of the estrogen receptor transcriptional activity by the coactivator GRIP1 highlights the role of activation function 2 in determining estrogen receptor pharmacology. *J. Biol. Chem.* **273**: 6679.

Oñate S.A., Tsai S.Y., Tsai M.J., and O'Malley B.W. 1995. Sequence and characterization of a coactivator for the steroid hormone receptor superfamily. *Science* **270**: 1354.

Picard D. and Yamamoto K.R. 1987. Two signals mediate hormone-dependent nuclear localization of the glucocorticoid receptor. *EMBO J.* **6**: 3333.

Picard D., Salser S. J., and Yamamoto, K.R. 1988. A movable and regulable inactivation function within the steroid binding domain of the glucocorticoid receptor. *Cell* **54**: 1073.

Prefontaine G.G., Lemieux M.E., Gifin W., Schild-Poulter C., Pope L., LaCasse E., Walker P., and Hache R.J.G. 1998. Recruitment of octamer transcription factors to DNA by glucocorticoid receptor. *Mol. Cell. Biol.* **18**: 3416.

Ptashne M. 1986. *A genetic switch: Gene control and phage lambda.* Cell Press & Blackwell Scientific, Boston.

Radhakrishnan I., Perez-Alvarado G.C., Parker D., Dyson H.J., Montminy M.R., and Wright P.E. 1997. Solution structure of the KIX domain of CBP bound to the transactivation domain of CREB: A model for activator:coactivator interactions. *Cell* **91**: 741.

Renaud J.P., Rochel N., Ruff M., Vivat V., Chambon P., Gronemeyer H., and Moras D. 1995. Crystal structure of the RAR-γ ligand-binding domain bound to all-trans retinoic acid. *Nature* **378**: 681.

Rusconi S. and Yamamoto K.R. 1987. Functional dissection of the hormone and DNA binding activities of the glucocorticoid receptor. *EMBO J.* **6**: 1309.

Saatcioglu F., Lopez G., West B.L., Zandi E., Feng W., Lu H., Esmaili A., Apriletti J.W., Kushner P.J., Baxter J.D., and Karin M. 1997. Mutations in the conserved C-terminal sequence in thyroid hormone receptor dissociate hormone-dependent activation from interference with AP-1 activity. *Mol. Cell. Biol.* **17**: 4687.

Sakai D.D., Helms S., Carlstedt-Duke J., Gustafsson J.-Å., Rottman F.M., and Yamamoto K.R. 1988. Hormone-mediated repression of transcription: A negative glucocorticoid response element from the bovine prolactin gene. *Genes Dev.* **2**: 1144.

Schüle R., Rangarajan P., Kliewer S., Ransone L.J., Bolado J., Yang N., Verma I.M., and Evans R.M. 1990. Functional antagonism between oncoprotein c-Jun and the glucocorticoid receptor. *Cell* **62**: 1217.

Scott D.K., Strömstedt P.E., Wang J.C., and Granner D.K. 1998. Further characterization of the glucocorticoid response unit in the phosphoenolpyruvate carboxykinase gene. The role of the glucocorticoid receptor-binding sites. *Mol. Endocrinol.* **12**: 482.

Shiau A.K., Barstad D., Loria P.M., Cheng L., Kushner P.J., Agard D.A., and Greene G.L. 1998. The structural basis of estrogen receptor/coactivator recognition and antagonism of this interaction by tamoxifen. *Cell* **95**: 927.

Starr D.B., Matsu W., Thomas J.R., and Yamamoto K.R. 1996.

Intracellular receptors use a common mechanism to interpret signaling information at response elements. *Genes Dev.* **10:** 1271.

Tone Y., Collingwood T.N., Adams M., and Chatterjee V.K. 1994. Functional analysis of a transactivation domain in the thyroid beta receptor. *J. Biol. Chem.* **269:** 31157.

Torchia J., Rose D.W., Inostroza J., Kamei Y., Westin S., Glass C.K., and Rosenfeld M.G. 1997. The transcriptional co-activator p/CIP binds CBP and mediates nuclear-receptor function. *Nature* **387:** 677.

Uesugi M., Nyanguile O., Lu H., Levine A.J., and Verdine G.L. 1997. Induced a helix in the VP16 activation domain upon binding to a human TAF. *Science* **277:** 1310.

Vivanco M.D., Johnson R., Galante P.E., Hanahan D., and Yamamoto K.R. 1995. A transition in transcriptional activation by the glucocorticoid and retinoic acid receptors at the tumor stage of dermal fibrosarcoma development. *EMBO J.* **14:** 2217.

Voegel J.J., Heine M.J.S., Tini M., Vivat V., Chambon P., and Gronemeyer H. 1998. The coactivator TIF2 contains three nuclear receptor-binding motifs and mediates transactivation through CBP binding-dependent and -independent pathways. *EMBO J.* **17:** 507.

Wagner R.L., Apriletti J.W., McGrath M.E., West B.L., Baxter J.D., and Fletterick R.J. 1995. A structural role for hormone in the thyroid hormone receptor. *Nature* **378:** 690.

Webb P., Lopez G.N., Uht R.M., and Kushner P.J. 1995. Tamoxifen activation of the estrogen receptor/AP-1 pathway: Potential origin for the cell-specific estrogen-like effects of antiestrogens. *Mol. Endocrinol.* **9:** 443.

Williams S.P. and Sigler P.B. 1998. Atomic structure of progesterone complexed with its receptor. *Nature* **393:** 392.

Yamamoto K.R. 1985. Steroid receptor regulated transcription of specific genes and gene networks. *Annu. Rev. Genetics* **19:** 209.

———. 1989. A conceptual view of transcriptional regulation. *Am. Zool.* **29:** 537.

———. 1997a. Multilayered control of intracellular receptor function. *Harvey Lect.* **91:** 1.

———. 1997b. Intracellular receptors: New instruments for a symphony of signals. In *The molecular biology of steroid and nuclear hormone receptors* (ed. L.P. Freedman), p. vii. Birkhauser Press, Boston.

Yamamoto K.R., Pearce D., Thomas J., and Miner J.N. 1992. Combinatorial regulation at a mammalian composite response element. In *Transcriptional regulation* (ed. S.L. McKnight and K.R. Yamamoto), p. 1169. Cold Spring Harbor Laboratory Press, Cold Spring Harbor, New York.

The Herpes Simplex Virus VP16-induced Complex: Mechanisms of Combinatorial Transcriptional Regulation

W. HERR

Cold Spring Harbor Laboratory, Cold Spring Harbor, New York 11724

At any one time, a cell transcribes only a small subset of all the genes in its genome. To provide specificity to the regulation of transcription, each cell relies on the assembly of particular sets or combinations of transcription factors on different promoters. Through such combinatorial mechanisms, individual transcription factors can acquire different activities depending on the presence or absence of other coregulators in the same cell, thus reducing the total number of regulators needed to control cellular transcription.

In this paper, I describe mechanisms of combinatorial transcriptional regulation that have been revealed from studies of a transcriptional regulatory complex that regulates human herpes simplex virus (HSV) gene transcription: the VP16-induced complex. These studies have disclosed properties of site-specific transcriptional regulators and ways in which transcriptional regulators can modify each other's activities through combinatorial interactions.

THE VP16-INDUCED COMPLEX AND ITS COMPONENTS

When HSV enters a cell, it deposits a virion protein called VP16 (also known as Vmw65 and αTIF) into the infected cell. Upon entry into the infected cell, VP16 assembles a transcriptional regulatory complex—the VP16-induced complex—with two cellular proteins, Oct-1 and HCF, on HSV immediate-early (IE) promoters. VP16 first binds to HCF, a protein involved in cell proliferation. This heterodimeric VP16-HCF complex then associates with Oct-1, a POU-domain transcription factor, on VP16 *cis*-regulatory targets, which are called "TAATGARAT" elements because often within the element is the sequence TAATGARAT (where R = purine).

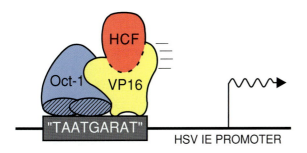

Figure 1. Illustration of the VP16-induced complex.

Once assembled on HSV IE promoters, the VP16-induced complex is able to activate IE gene transcription owing to a very potent transcriptional activation domain.

Figure 1 shows a cartoon of the VP16-induced complex on an HSV IE gene promoter. The VP16-induced complex contains four components: the viral protein VP16, the two cellular proteins Oct-1 and HCF, and the *cis*-regulatory TAATGARAT element. Figure 2 shows a schematic representation of each of the three protein components of the complex, with the minimal regions involved in complex formation colored.

The viral component of the VP16-induced complex, VP16, is a 490-amino-acid protein. It is synthesized late during HSV lytic infection and is incorporated into the tegument of the virion, between the capsid and the outer membrane of the virus (see Roizman and Sears 1990). Like the cellular components, VP16 is a modular protein composed of separable regions involved in transcriptional activation and VP16-induced complex formation (Sadowski et al. 1988; Cousens et al. 1989; Greaves and O'Hare 1990). The transcriptional activating region or "domain" (AD) lies within the carboxy-terminal 80 amino acids and is enriched in acidic residues (Triezenberg et al. 1988), whereas the VP16-induced complex-forming region (VIC) lies within residues 49 to 385 (Greaves and O'Hare 1990).

Of the two cellular components, Oct-1 is the only one known to provide DNA-binding specificity in the VP16-induced complex. Oct-1 is present in many different cell types where it is implicated in activating transcription of genes encoding small nuclear (sn)RNAs, which are involved in pre-mRNA splicing, and the histone H2B (Singh et al. 1986; Sive and Roeder 1986; Bohmann et al. 1987). Oct-1 is also implicated in B-cell-specific activation of immunoglobulin-gene transcription with the cell-specific transcriptional coregulator OCA-B (also referred to as OBF-1 and Bob-1; Luo et al. 1992; Gstaiger et al. 1995; Luo and Roeder 1995; Strubin et al. 1995). Oct-1 contains a POU DNA-binding domain and amino- and carboxy-terminal transcriptional activation domains, which are enriched in glutamines (Q) or serines and threonines (S/T), respectively (Tanaka and Herr 1990; Tanaka et al. 1992). Only its POU DNA-binding domain, however, is required to stabilize the VP16-induced complex (Kristie et al. 1989; Stern et al. 1989).

The POU domain is a bipartite DNA-binding domain (Sturm and Herr 1988) which was discovered as a conserved region among four transcription factors: the pitu-

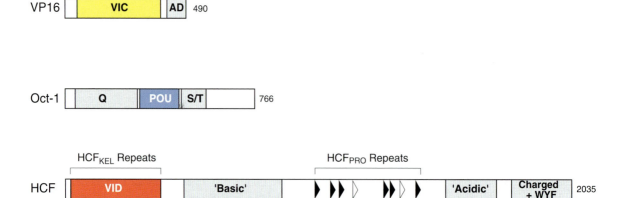

Figure 2. Illustration of the three protein components of the VP16-induced complex. See text for details.

itary transcription factor Pit-1, the octamer-motif-binding proteins Oct-1 and Oct-2, and the *Caenorhabditis elegans* protein UNC-86 (Herr et al. 1988). The POU domain consists of two DNA-binding structures, an amino-terminal POU-specific (POU$_S$) domain and a carboxy-terminal POU-homeo (POU$_H$) domain, joined by a flexible linker (for a review, see Herr and Cleary 1995). Figure 3 shows an illustration of the Oct-1 POU domain bound to its high-affinity binding site, the octamer motif ATGC-AAAT, in the histone H2B gene promoter (Klemm et al. 1994). The POU$_H$ domain contains three α helices and is very similar in structure to other homeodomains such as

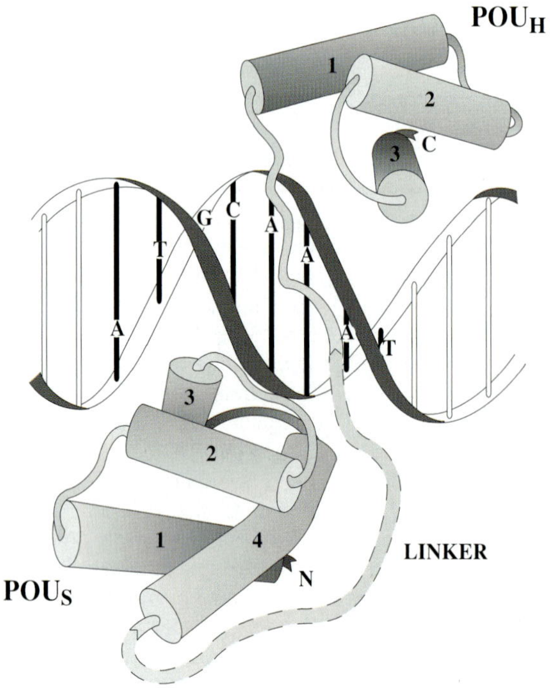

those of the yeast protein Matα2 and the *Drosophila* protein engrailed (Klemm et al. 1994; Sivaraja et al. 1994). In contrast, the POU$_S$ domain, which contains four α helices, is more similar to the structures of the bacteriophage λ and 434 repressor, and 434 Cro DNA-binding domains (Assa-Munt et al. 1993; Dekker et al. 1993).

The second cellular component of the VP16-induced complex, HCF (also referred to as C1, CFF, and VCAF), is also present in many cell types, particularly in proliferating cells (Wilson et al. 1995b; Kristie 1997), but its cellular function is less well understood. HCF was discovered by virtue of its association with VP16 (Gerster and Roeder 1988; Kristie et al. 1989; Katan et al. 1990; Kristie and Sharp 1990; Xiao and Capone 1990; Stern and Herr 1991) and was subsequently shown to promote cell proliferation in uninfected cells (Goto et al. 1997). In contrast to Oct-1, HCF is not known to provide any specificity to VP16-induced complex formation. Its only known role is to stabilize the association of VP16 and Oct-1 on TAATGARAT elements.

HCF is an unusual protein. As illustrated in Figure 4, it is synthesized as a large precursor protein of approximately 300 kD (Wilson et al. 1993; Kristie et al. 1995). After synthesis, it moves to the nucleus, where it is cleaved at a series of six near-perfect 26-amino-acid-long repeats, called HCF$_{PRO}$ repeats, located near the middle of the protein (Wilson et al. 1993, 1995a; Kristie et al. 1995). Curiously, the resulting amino- and carboxy-terminal fragments, which range in size from approximately 110 kD to 150 kD, are stable and remain noncovalently associated with one another (Wilson et al. 1993, 1995a). The role of HCF processing in the function of HCF is unknown, and HCF processing is apparently not highly conserved because a homolog of HCF in *C. elegans* does not contain any evident HCF$_{PRO}$ repeats (Kristie 1997; LaBoissière et al. 1997; Liu et al. 1999).

The primary sequence of HCF suggests that, like VP16 and Oct-1, it is modular (see Fig. 2) (Wilson et al. 1993). The amino- and carboxy-terminal regions contain relatively high concentrations of charged residues as well as the large hydrophobic residues phenylalanine, tryptophan, and tyrosine (WYF), whereas two centrally located

Figure 3. Illustration of the Oct-1 POU domain bound to the histone H2B octamer site. The octamer sequence ATG-CAAAT in the Oct-1-binding site is indicated. (Adapted from Klemm et al. [1994] and Cleary et al. [1997].)

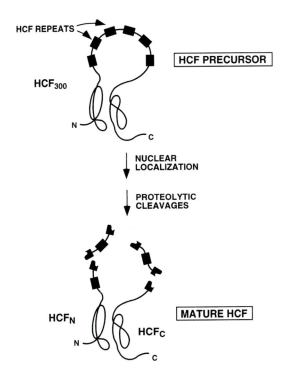

Figure 4. HCF processing. (Reprinted, with permission, from Wilson et al. 1995a.)

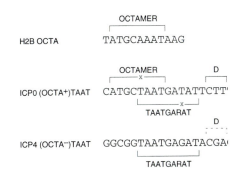

Figure 5. Comparison of the nucleotide sequence of VP16-responsive (OCTA⁺)TAATGARAT and (OCTA⁻)TAATGARAT sites, with the high-affinity octamer Oct-1-binding site in the histone H2B promoter.

regions contain a preponderance of either basic or acidic residues as shown in Figure 2. The amino-terminal "Charged/WYF" region (red) contains six approximately 50-amino-acid-long repeats related to sequence repeats found in the *Drosophila* egg chamber protein Kelch, and therefore called HCF$_{KEL}$ repeats (LaBoissière et al. 1997; Wilson et al. 1997). Sequence comparison with proteins of known structure suggests that each HCF$_{KEL}$ repeat forms a four-stranded β-sheet and that the six β sheets come together like blades of a propeller to form a so-called β-propeller structure (Bork and Doolittle 1994). The region encompassing the six HCF$_{KEL}$ repeats can associate with VP16 and is necessary and sufficient to stabilize a VP16-induced complex (Wilson et al. 1997). This "VP16-interaction domain" (VID) is also involved in promoting cell proliferation (Goto et al. 1997). Thus, VP16 targets a region of HCF that is involved in cell proliferation.

The fourth component of the VP16-induced complex is its binding site on the DNA—the TAATGARAT element. TAATGARAT elements can be divided into two classes, depending on the presence, (OCTA⁺)TAATGARAT, or absence, (OCTA⁻)TAATGARAT, of an overlapping octamer (OCTA) sequence (see apRhys et al. 1989). Figure 5 shows a comparison of the consensus octamer sequence found in the human histone H2B promoter (H2B OCTA), and (OCTA⁺)TAATGARAT and (OCTA⁻)TAATGARAT sites found in the promoters of the genes encoding the ICP0 and ICP4 IE proteins, respectively. As shown, the (OCTA⁺)TAATGARAT element contains a near-perfect octamer consensus sequence overlapping the TAAT-

GARAT sequence. In contrast to (OCTA⁺)TAATGARAT elements, (OCTA⁻)TAATGARAT elements share little similarity with an octamer sequence. Nevertheless, they are surprisingly still recognized by Oct-1 as a binding site, albeit with less affinity than a consensus octamer site (Baumruker et al. 1988; apRhys et al. 1989).

Octamer- and TAATGARAT-related sequences are not the only important part of TAATGARAT elements, however. VP16-responsive TAATGARAT elements also contain sequences 3′ of the TAATGARAT core sequence that are important for the VP16 response, particularly a "D element" which is involved in the selectivity of VP16 interaction with TAATGARAT elements (Huang and Herr 1996; Misra et al. 1996). The D element has been mapped to a 3-bp CTT sequence (bracket) in the ICP0 (OCTA⁺)TAATGARAT site (Huang and Herr 1996). A D element has not been mapped in the ICP4 (OCTA⁻)TAATGARAT site, but this site is also likely to contain one as suggested by the dashed bracket in Figure 5.

FLEXIBLE DNA SEQUENCE RECOGNITION BY OCT-1

The ability of Oct-1 to bind to (OCTA⁻)TAATGARAT elements is one example of how some sequence-specific transcription factors can bind to very dissimilar sequences (Baumruker et al. 1988; apRhys et al. 1989). Oct-1 is able to achieve flexible DNA sequence recognition by two mechanisms. In the first mechanism, an important and conserved arginine side chain in the Oct-1 POU$_S$ domain can adapt to changes in its contact site in the DNA (Cleary and Herr 1995). Such a mechanism has been observed also with the related DNA-binding domain of the phage 434 repressor (Rodgers and Harrison 1993). In the second mechanism, which is a direct result of the flexibility of the POU domain linker, the relative positions of the Oct-1 POU$_S$ and POU$_H$ domains can vary depending on the sequence of the Oct-1-binding site (Cleary and Herr 1995; Cleary et al. 1997).

Figure 6 illustrates this point. It shows the positions of the Oct-1 POU$_S$ and POU$_H$ domains bound to octamer, (OCTA⁺)TAATGARAT, and (OCTA⁻)TAATGARAT sites as revealed by protein-DNA photocross-linking (Cleary et al. 1997). On the octamer and (OCTA⁻)TAAT-GARAT sites, the Oct-1 POU domain adopts very different conformations, with the POU_S domain lying on one or the other side of the POU_H domain (Fig. 6, top and bottom). Surprisingly, on an (OCTA⁺)TAATGARAT site the Oct-1 POU domain can adopt both octamer-like or (OCTA⁻)TAATGARAT-like conformations. Thus, the Oct-1 POU domain can adopt multiple conformations on a single *cis*-regulatory site. In this instance, VP16 can associate with Oct-1 in either of its conformations on the (OCTA⁺)TAATGARAT site, suggesting that the structure of the POU domain within a VP16-induced complex can be dynamic (Cleary et al. 1997). This second mode of flexible DNA sequence recognition by Oct-1 illustrates how a DNA-binding domain can display considerable structural versatility and adaptability: As the DNA sequence changes, the DNA-binding domain can adapt by changing its structure.

VP16 RECOGNIZES THE OCT-1 POU_H DOMAIN

In addition to binding DNA flexibly, the Oct-1 POU domain is responsible for the ability of Oct-1 to stabilize the VP16-induced complex (Kristie et al. 1989; Stern et al. 1989). VP16 recognizes the Oct-1 POU_H domain and can distinguish the Oct-1 POU_H domain from the POU_H domain of a close relative of Oct-1 called Oct-2 (Stern et al. 1989; Stern and Herr 1991). Oct-2 is a B-cell POU do-

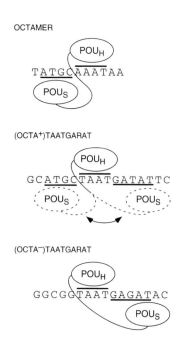

Figure 6. Flexibility of the Oct-1 POU domain on Oct-1 binding sites. The position of the Oct-1 POU_S and POU_H domains is shown on the octamer (*top*), (OCTA⁺)TAATGARAT (*middle*), and (OCTA⁻)TAATGARAT (*bottom*) sites. The dashed POU_S domain on the (OCTA⁺)TAATGARAT site indicates that the POU_S domain can adopt either position on this site. (Adapted from Cleary et al. 1997.)

main protein that displays the same DNA-binding specificity as Oct-1 (Staudt et al. 1986), but it is unable to associate with VP16 effectively (Gerster and Roeder 1988). VP16 fails to associate with the Oct-2 POU domain as a result of one of the seven amino acid differences—a glutamic acid residue in Oct-1 for an alanine residue in Oct-2—on the solvent-exposed surface of the DNA-bound Oct-1 and Oct-2 POU_H domains (Lai et al. 1992; Pomerantz et al. 1992). This single-amino-acid difference is sufficient to confer different combinatorial transcriptional specificities to Oct-1 and Oct-2 by virtue of their selective response to VP16. Indeed, if the glutamic acid residue in Oct-1 is replaced by the alanine residue of Oct-2, Oct-1 no longer activates transcription with VP16, and if the alanine residue in Oct-2 is replaced by the glutamic acid residue of Oct-1, Oct-2 can activate transcription with VP16 (Lai et al. 1992). Thus, the differential transcriptional response of homeodomain proteins to a coregulator can hinge on the identity of a single-amino-acid residue.

VP16 ASSOCIATION WITH THE OCT-1 POU_H DOMAIN RESEMBLES THE INTERACTION OF MATα2 WITH THE HOMEODOMAIN OF ITS COREGULATOR MATa1

The structure of VP16 that binds the Oct-1 POU_H domain and other components of the VP16-induced complex is unknown. A region of VP16 that is critical for VP16-induced complex formation lies at the carboxyl terminus of the 340-amino-acid segment of VP16 that is sufficient to form the VP16-induced complex (see Fig. 2) (for references, see Lai and Herr 1997). Individual residues within this region of VP16 are selectively involved in interaction with all three other components of the VP16-induced complex: HCF, Oct-1, and the DNA (Lai and Herr 1997). Consistent with this region contacting other components of the VP16-induced complex, the region is accessible to proteases in free VP16, but not within the VP16-induced complex (Hayes and O'Hare 1993).

The structure of VP16 within the VP16-induced complex is unknown, but Li et al. (1995) have presented a persuasive model that the interaction of VP16 with the Oct-1 POU_H domain resembles the interaction of the yeast transcriptional repressor Matα2 with the homeodomain of one of its coregulators, Mata1. The *Saccharomyces cerevisiae* Matα2 and Mata1 proteins are both homeodomain-containing proteins that join together to repress expression of haploid-specific genes. Cooperative binding to DNA by these two proteins is mediated by a carboxy-terminal "tail" of Matα2 which binds to a solvent-exposed surface of the DNA-bound Mata1 homeodomain (Li et al. 1995). The surface of the Mata1 homeodomain that is contacted by Matα2 is similar to the surface of the Oct-1 homeodomain that is contacted by VP16 (Pomerantz et al. 1992). Furthermore, Li et al. (1995) noted that a region of VP16 that is involved in contacting the Oct-1 POU_H domain (Stern and Herr 1991; Walker et al. 1994; Lai and Herr 1997) bears sequence similarity to the carboxy-terminal tail of Matα2. These observations led to the suggestion that VP16 contacts Oct-1 in a manner similar to the way Matα2 contacts Mata1 (Li et al. 1995). Thus, in associating with a cellular homeodomain protein, VP16 may use a strategy for homeodomain recognition that is also used by cellular proteins.

VP16 CAN MODIFY THE ACTIVITY OF OCT-1 BY ALTERING ITS SPECIFICITY OF PROMOTER ACTIVATION AND DNA-BINDING SITE SELECTION

When VP16 associates with Oct-1, it can modify the activity of Oct-1 in two ways, as illustrated in Figure 7. First, VP16 provides a transcriptional activation domain that, unlike those of Oct-1, can activate transcription from mRNA promoters effectively, and second, VP16 can stabilize Oct-1 binding to DNA. Compared to Oct-2, the transcriptional activation domains in Oct-1 are more effective activators of snRNA gene promoters than of mRNA-type promoters, whereas the transcriptional activation domains in Oct-2 display the opposite preference (see Fig. 7A) (Tanaka et al. 1992). This differing promoter-selective function of transcriptional activation domains results from differences in the core elements of snRNA- and mRNA-type promoters (Das et al. 1995). In contrast to mRNA-type promoters, which often possess a TATA box (labeled T/A in Fig. 7) which binds the basal transcription complex TFIID, RNA polymerase II snRNA gene promoters lack a TATA box and instead possess a PSE element, which binds the snRNA-basal complex SNAP$_c$ (also known as PTF; see Henry et al., this volume). Probably, Oct-1 activates snRNA gene transcription more effectively than Oct-2 because its transcriptional activation domains can associate better with the basal transcriptional machinery assembled on snRNA gene promoters (e.g., SNAP$_c$), and Oct-2 activates mRNA-type promoters more effectively than Oct-1 because its transcriptional activation domains associate better with the basal transcriptional machinery assembled on mRNA-type promoters (e.g., TFIID) (see Tanaka et al. 1988). In contrast to Oct-1 transcriptional activation domains, however, the VP16 transcriptional activation domain activates mRNA-type promoters very effectively, although it fails to activate snRNA gene promoters effectively (Das et al. 1995). Thus, by associating with VP16 (Fig. 7C, bottom), Oct-1 can activate transcription from mRNA-type promoters such as those of HSV IE genes.

VP16 also alters the *cis*-regulatory specificity of Oct-1 function by recruiting Oct-1 to (OCTA⁻)TAATGARAT sites. Although Oct-1 (and Oct-2) can recognize with low affinity an (OCTA⁻)TAATGARAT element in vitro, neither Oct-1 nor Oct-2 activates transcription effectively from snRNA- or mRNA-type promoters containing the (OCTA⁻)TAATGARAT site (Fig. 7B) (Cleary et al. 1993), probably because neither transcription factor is able to bind to the (OCTA⁻)TAATGARAT element effectively in vivo. VP16, however, can stabilize Oct-1 on the (OCTA⁻)TAATGARAT site, and the resulting VP16-induced complex can activate transcription from either an

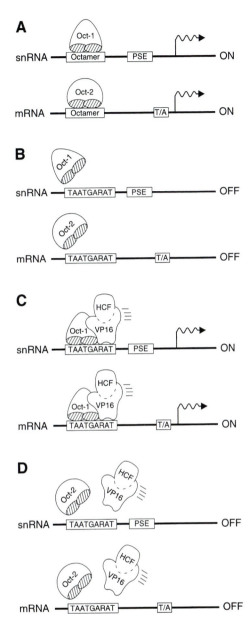

Figure 7. Two modes by which VP16 alters the transcriptional activity of Oct-1: Promoter-selective transcriptional activation domain function and corecruitment to a novel *cis*-regulatory element. (*A*) Oct-1 and Oct-2 activate transcription from snRNA- and mRNA-type promoters, respectively, from high-affinity octamer sites through the use of differing promoter-selective activation domains. (*B*) Neither Oct-1 nor Oct-2 activates transcription from snRNA- or mRNA-type promoters containing low-affinity TAATGARAT sites. (*C*) In the presence of VP16, Oct-1 can activate transcription from both snRNA- and mRNA-type promoters containing low-affinity TAATGARAT sites because VP16 stabilizes Oct-1 on this binding site and in the case of the mRNA-type promoter provides an mRNA-type promoter transcriptionl activation domain. (*D*) Even in the presence of VP16, Oct-2 does not activate transcription from either snRNA- or mRNA-type promoters containing low-affinity TAAT-GARAT sites because Oct-2 fails to associate with VP16.

snRNA-type promoter (Fig. 7C, top) or an mRNA-type promoter (Fig. 7C, bottom) through the action of the Oct-1 or VP16 transcriptional activation domains, respectively (Cleary et al. 1993). In contrast to Oct-1, however, even though Oct-2 possesses mRNA-type transcriptional activation domains, Oct-2 fails to activate transcription from the TAATGARAT-site-containing snRNA- or mRNA-type promoters even in the presence of VP16 because Oct-2 cannot be stabilized on the site by VP16 (Fig. 7D) (Cleary et al. 1993). Thus, through selective association with Oct-1 but not Oct-2, VP16 is able to selectively alter the transcriptional activity of one but not the other of two transcription factors that on their own possess the same DNA-binding specificity.

THE B-CELL OCT-1 COREGULATOR OCA-B IS A FUNCTIONAL ANALOG OF VP16 BUT TARGETS A SEPARATE SURFACE OF THE OCT-1 POU DOMAIN

The B-cell-specific coregulator of Oct-1 OCA-B also associates with the Oct-1 POU domain (Luo et al. 1992; Gstaiger et at. 1995; Luo and Roeder 1995; Strubin et al. 1995) and, like VP16, alters the activity of Oct-1 in two ways: It provides a transcriptional activation domain that displays preferential activity on mRNA-type promoters, and it stabilizes Oct-1 on octamer elements (Babb et al. 1997). Thus, both a viral and a cellular coregulator can alter the transcriptional activation and DNA-binding properties of Oct-1. Unlike VP16, however, OCA-B associates with a different surface of the Oct-1 POU domain, a surface that includes both the POU_H domain and the POU_S domain (Babb et al. 1997). Indeed, VP16 and OCA-B can associate with the Oct-1 POU domain simultaneously (Babb et al. 1997). Thus, as with its association with DNA, the Oct-1 POU domain displays considerable versatility in how it can associate with coregulatory proteins.

THE OCT-1 POU DOMAIN ALSO INTERACTS DIRECTLY WITH THE BASAL TRANSCRIPTIONAL MACHINERY

In addition to interacting with viral and cellular coregulators, the Oct-1 POU domain also displays versatility by interacting directly with a basal transcription complex. On snRNA promoters, the POU_S domain contacts the largest subunit of the PSE-binding basal complex $SNAP_c$, SNAP190, and through this interaction is able to activate snRNA gene transcription (see Ford et al. 1998 and references therein). Interestingly, the SNAP190 component of $SNAP_c$ contacts a region of the Oct-1 POU domain that is also contacted by OCA-B (Mittal et al. 1996; Babb et al. 1997) and the region of SNAP190 that contacts the Oct-1 POU domain shares sequence similarity to OCA-B (Ford et al. 1998). These results suggest that the broadly expressed transcription factor Oct-1 can associate with either a cell-specific coregulator (OCA-B) or the basal transcriptional machinery ($SNAP_c$) through similar protein-protein interactions (Ford et al. 1998).

DIFFERENTIAL TRANSCRIPTIONAL REGULATION BY OCT-1 THROUGH ASSOCIATION WITH HOMOLOGOUS VP16 PROTEINS POSSESSING DIFFERENT DNA-BINDING SPECIFICITIES

The association of VP16 with Oct-1 but not Oct-2 illustrates how selective association with a coregulator can result in differential transcriptional regulation by two transcription factors that, on their own, display the same DNA-binding specificity. The association of Oct-1 with VP16 proteins from different herpesviruses shows how association of a single transcription factor with homologous coregulators can result in differential transcriptional regulation by a single transcription factor as illustrated in Figure 8.

HSV is not the only herpesvirus with a VP16-like protein. For example, VP16 homologs from bovine herpesvirus (BHV), the chicken pox varicella zoster virus, and equine herpesvirus activate IE gene transcription from their cognate virus and form VP16-induced complexes (for references, see Huang and Herr 1996; Misra et al. 1996). Comparison of the HSV and BHV VP16 proteins showed that they recognize Oct-1 and HCF similarly, but display different DNA-binding specificities (Huang and Herr 1996; Misra et al. 1996). For example, as illustrated in Figure 8A, the HSV VP16 protein (H-VP16) forms a VP16-induced complex and activates

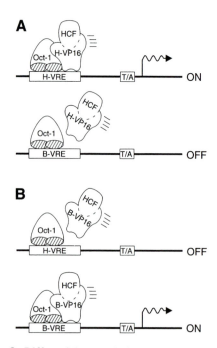

Figure 8. Differential transcriptional regulation by Oct-1 through selective corecruitment to *cis*-regulatory sites by homologous VP16 coregulators that possess different DNA-binding specificities. (*A*) HSV VP16 (H-VP16) activates transcription from an HSV-related VP16-response element (H-VRE) but not a BHV-related VP16-response element (B-VRE). (*B*) BHV VP16 (B-VP16) activates transcription from a BHV-related VP16-response element (B-VRE) but not an HSV-related VP16-response element (H-VRE). (Adapted from Huang and Herr 1996.)

transcription effectively from an HSV VP16-response element (H-VRE), but not from a BHV VP16-response element (B-VRE). In contrast, as illustrated in Figure 8B, the BHV VP16 protein (B-VP16) forms a VP16-induced complex and activates transcription effectively from a BHV VP16-response element, but not from an HSV VP16-response element. The region of the VP16-response elements responsible for this difference in VP16-response element recognition is the short 3-bp D-element sequence that lies 3′ of the TAATGARAT core sequence (see Fig. 5) (Huang and Herr 1996). Thus, VP16 proteins can selectively alter the transcriptional activity of Oct-1, demonstrating how the activity of a broadly expressed homeodomain protein can be selectively directed to different promoters by different but related coregulators.

HCF: AN ON-OFF SWITCH FOR VP16-INDUCED COMPLEX ASSEMBLY

Unlike Oct-1 and VP16, HCF is not known to impart any selectivity in protein-protein and protein-DNA interactions within the VP16-induced complex. It is not known how HCF stabilizes the complex; perhaps, in binding VP16, HCF stabilizes the conformation of VP16 that associates with the Oct-1 POU_H domain. Through whichever mechanism HCF uses, however, by stabilizing the VP16-induced complex, HCF plays a unique part in combinatorial transcriptional regulation by serving as an "on-off" switch as opposed to a selectivity determinant. The stabilizing role of HCF may provide important temporal and spacial specificity to transcriptional activation by the VP16-induced complex. Such a mechanism may allow HSV to regulate whether the virus undergoes lytic infection by activating IE gene transcription or remains latent in the infected cell. For example, VP16 may sense the proliferative or cell cycle status of the infected cell by its ability to bind to HCF (Wilson et al. 1993). Consistent with this hypothesis, a temperature-sensitive mutation in HCF can cause cells to arrest in a G_0/G_1 state at the nonpermissive temperature (Goto et al. 1997), and the same mutation prevents VP16 interaction with HCF (Goto et al. 1997; Wilson et al. 1997).

Just as, in addition to the viral protein VP16, Oct-1 interacts with cellular coregulators (e.g., OCA-B), so apparently does HCF. HCF binds a cellular basic leucine zipper transcription factor called LZIP (Freiman and Herr 1997) or Luman (Lu et al. 1997), and enhances the ability of LZIP to activate transcription (Lu et al. 1998). The mechanism of HCF interaction with VP16 and LZIP is very similar: The same surface of HCF binds VP16 and LZIP, and a shared four-amino-acid motif in VP16 and LZIP binds HCF (Freiman and Herr 1997; Lu et al. 1998). Thus, unlike VP16 and OCA-B, which target different surfaces of the Oct-1 POU domain, VP16 and LZIP contact the same surface of HCF. Perhaps, VP16 can serve as a better gauge of the status of the infected cell if it mimics precisely a cellular interaction involved in cell proliferation. It will be interesting to determine the nature of the cellular function that VP16 is mimicking.

CONCLUSIONS

In conclusion, the VP16-induced complex and its related cellular complexes reveal the many intricate ways in which transcription factors can be combined to result in new transcriptional regulatory patterns. The VP16-induced complex further shows how subtle differences in structure—a difference as subtle as a single-amino-acid difference in a homeodomain or a few base pairs in a *cis*-regulatory site—can result in very different transcriptional outcomes. Thus, although part of a human viral pathogen, VP16 has revealed much about how the cells in our bodies regulate transcription. Further studies of VP16 will undoubtedly teach us much more.

ACKNOWLEDGMENTS

I am indebted to the past and present members of my laboratory and to my collaborators for many of the results described here. I thank Robert Babb and Nouria Hernandez for critical readings of the manuscript. The research of my laboratory was funded by U.S. Public Health Service grants CA-13106 and GM-54598.

REFERENCES

apRhys C.M.J., Ciufo D.M., O'Neill E.A., Kelly T.J., and Hayward G.S. 1989. Overlapping octamer and TAATGARAT motifs in the VF65-response elements in herpes simplex virus immediate-early promoters represent independent binding sites for cellular nuclear factor III. *J. Virol.* **63:** 2798.

Assa-Munt N., Mortishire-Smith R.J., Aurora R., Herr W., and Wright P.E. 1993. The solution structure of the Oct-1 POU-specific domain reveals a striking similarity to the bacteriophage λ repressor DNA-binding domain. *Cell* **73:** 193.

Babb, R., Cleary M.A., and Herr W. 1997. OCA-B is a functional analog of VP16 but targets a separate surface of the Oct-1 POU domain. *Mol. Cell. Biol.* **17:** 7295.

Baumruker T., Sturm R., and Herr W. 1988. OBP100 binds remarkably degenerate octamer motifs through specific interactions with flanking sequences. *Genes Dev.* **2:** 1400.

Bohmann D., Keller W., Dale T., Scholer H.R., Tebb G., and Mattaj I.W. 1987. A transcription factor which binds to the enhancers of SV40, immunoglobulin heavy chain and U2 snRNA genes. *Nature* **325:** 268.

Bork P. and Doolittle R.F. 1994. *Drosophila* kelch motif is derived from a common enzyme fold. *J. Mol. Biol.* **236:** 1277.

Cleary M.A. and Herr W. 1995. Mechanisms for flexibility in DNA sequence recognition and VP16-induced complex formation by the Oct-1 POU domain. *Mol. Cell. Biol.* **15:** 2090.

Cleary M.A., Pendergrast P.S., and Herr W. 1997. Structural flexibility in transcription complex formation revealed by protein-DNA photocrosslinking. *Proc. Natl. Acad. Sci.* **94:** 8450.

Cleary M.A., Stern S., Tanaka M., and Herr W. 1993. Differential positive control by Oct-1 and Oct-2: Activation of a transcriptionally silent motif through Oct-1 and VP16 corecruitment. *Genes Dev.* **7:** 72.

Cousens D.J., Greaves R., Goding C.R., and O'Hare P. 1989. The C-terminal 79 amino acids of the herpes simplex virus regulatory protein, Vmw65, efficiently activate transcription in yeast and mammalian cells in chimeric DNA-binding proteins. *EMBO J.* **8:** 2337.

Das G., Hinkley C.S., and Herr W. 1995. Basal promoter elements as a selective determinant of transcriptional activator function. *Nature* **374:** 657.

Dekker N., Cox M., Boelens R., Verrijzer C., van der Vliet P.C., and Kaptein R. 1993. Solution structure of the POU-specific DNA-binding domain of Oct-1. *Nature* **362:** 852.

Ford E., Strubin M., and Hernandez N. 1998. The Oct-1 POU domain activates snRNA gene transcription by contacting a region in the SNAP$_C$ largest subunit that bears sequence similarities to the Oct-1 coactivator OBF-1. *Genes Dev.* **12:** 3528.

Freiman R.N. and Herr W. 1997. Viral mimicry: Common mode of association with HCF by VP16 and the cellular protein LZAIP. *Genes Dev.* **11:** 3122.

Gerster T. and Roeder R.G. 1988. A herpesvirus *trans*-activating protein interacts with transcription factor OTF-1 and other cellular proteins. *Proc. Natl. Acad. Sci.* **85:** 6347.

Goto H., Motomura S., Wilson A.C., Freiman R.N., Nakabeppu Y., Fukushima K., Fujishima M., Herr W., and Nishimoto T. 1997. A single-point mutation in HCF causes temperature-sensitive cell-cycle arrest and disrupts VP16 function. *Genes Dev.* **11:** 726.

Greaves R.F. and O'Hare P. 1990. Structural requirements in the herpes simplex virus type 1 transactivator Vmw65 for interaction with the cellular octamer-binding proteins and target TAATGARAT sequences. *J. Virol.* **64:** 2716.

Gstaiger M., Knoepfel L., Georgiev O., Schaffner W., and Hovens C.M. 1995. A B-cell coactivator of octamer-binding transcription factors. *Nature* **373:** 360.

Hayes S. and O'Hare P. 1993. Mapping of a major surface-exposed site in herpes simplex virus protein Vmw65 to a region of direct interaction in a transcription complex assembly. *J. Virol.* **67:** 852.

Herr W. and Cleary M.A. 1995. The POU domain: Versatility in transcriptional regulation by a flexible two-in-one DNA binding domain. *Genes Dev.* **9:** 1679.

Herr W., Sturm R.A., Clerc R.G., Corcoran L.M., Baltimore D., Sharp P.A., Ingraham H.A., Rosenfeld M.G., Finney M., Ruvkun G., and Horvitz H.R. 1988. The POU domain: A large conserved region in the mammalian *pit-1*, *oct-1*, *oct-2*, and *Caenorhabditis elegans unc-86* gene products. *Genes Dev.* **2:** 1513.

Huang C.C. and Herr W. 1996. Differential control of transcription by homologous homeodomain coregulators. *Mol. Cell. Biol.* **16:** 2967.

Katan M., Haigh A., Verrijzer C.P., van der Vliet P.C., and O'Hare P. 1990. Characterization of a cellular factor which interacts functionally with Oct-1 in the assembly of a multicomponent transcription complex. *Nucleic Acids Res.* **18:** 6871.

Klemm J.D., Rould M.A., Aurora R., Herr W., and Pabo C.O. 1994. Crystal structure of the Oct-1 POU domain bound to an octamer site: DNA recognition with tethered DNA-binding modules. *Cell* **77:** 21.

Kristie T.M. 1997. The mouse homologue of the human transcription factor C1 (host cell factor). *J. Biol. Chem.* **272:** 26749.

Kristie T.M. and Sharp P.A. 1990. Interactions of the Oct-1 POU subdomains with specific DNA sequences and the HSV α-*trans*-activator protein. *Genes Dev.* **4:** 2383.

Kristie T.M., LeBowitz J.H., and Sharp P.A. 1989. The octamer-binding proteins form multi-protein-DNA complexes with the HSV α-TIF regulatory protein. *EMBO J.* **8:** 4229.

Kristie T.M., Pomerantz J.L., Twomey T.C., Parent S.A., and Sharp P.A. 1995. The cellular C1 factor of the herpes simplex virus enhancer complex is a family of polypeptides. *J. Biol. Chem.* **270:** 4387.

LaBoissière S., Walker S., and O'Hare P. 1997. Concerted activity of host cell factor subregions in promoting stable VP16 complex assembly and preventing interference by the acidic activation domain. *Mol. Cell. Biol.* **17:** 7108.

Lai J.-S. and Herr W. 1997. Interdigitated residues within a small region of VP16 interact with Oct-1, HCF, and DNA. *Mol. Cell. Biol.* **17:** 3937.

Lai J.-S., Cleary M.A., and Herr W. 1992. A single amino acid exchange transfers VP16-induced positive control from the Oct-1 to the Oct-2 homeo domain. *Genes Dev.* **6:** 2058.

Li T., Stark M.R., Johnson A.D., and Wolberger C. 1995. Crystal structure of the MAT**a**1/α2 homeodomain heterodimer

bound to DNA. *Science* **270:** 262.

Liu Y., Hengartner M.O., and Herr W. 1999. Selected elements of herpes simplex virus accessory factor HCF are highly conserved in *Caenorhabditis elegans*. *Mol. Cell. Biol.* **19:** 909.

Lu R., Yang P., O'Hare P., and Misra V. 1997. Luman, a new member of the CREB/ATF family, binds to herpes simplex virus VP16-associated host cellular factor. *Mol. Cell. Biol.* **17:** 5117.

Lu R., Yang P., Padmakumar S., and Misra V. 1998. The herpesvirus transactivator VP16 mimics a human basic domain leucine zipper protein, Luman, in its interaction with HCF. *Mol. Cell. Biol.* **72:** 6291.

Luo Y. and Roeder R.G. 1995. Cloning, functional characterization and mechanism of action of the B cell-specific transcriptional coactivator OCA-B. *Mol. Cell. Biol.* **15:** 4115.

Luo Y., Fujii H., Gerster T., and Roeder R.G. 1992. A novel B cell-derived coactivator potentiates the activation of immunoglobulin promoters by octamer-binding transcription factors. *Cell* **71:** 231.

Misra V., Walker S., Yang P., Hayes S., and O'Hare P. 1996. Conformational alteration of Oct-1 upon DNA binding dictates selectivity in differential interactions with related transcriptional coactivator. *Mol. Cell. Biol.* **16:** 4404.

Mittal V., Cleary M.A., Herr W., and Hernandez N. 1996. The Oct-1 POU-specific domain can stimulate small nuclear RNA gene transcription by stabilizing the basal transcription complex SNAP$_C$. *Mol. Cell. Biol.* **16:** 1955.

Pomerantz J.L., Kristie T.M., and Sharp P.A. 1992. Recognition of the surface of a homeo domain protein. *Genes Dev.* **6:** 2047.

Rodgers D.W. and Harrison S.C. 1993. The complex between phage 434 repressor DNA-binding domain and operator sites O_R3: Structural differences between consensus and non-consensus half-sites. *Structure* **1:** 227.

Roizman B. and Sears A.E. 1990. Herpes simplex viruses and their replication. In *Virology*, 2nd edition (ed. B.N. Fields et al.), p. 1795. Raven Press, New York.

Sadowski I., Ma J., Triezenberg S.J., and Ptashne M. 1988. GAL4-VP16: An unusual potent transcriptional activator. *Nature* **335:** 551.

Singh H., Sen R., Baltimore D., and Sharp P.A. 1986. A nuclear factor that binds to a conserved sequence motif in transcriptional control elements of immunoglobulin genes. *Nature* **319:** 154.

Sivaraja M., Botfield M.C., Mueller M., Jancso A., and Weiss M.A. 1994. Solution structure of a POU-specific homeodomain: 3D-NMR studies of human B-cell transcription factor Oct-2. *Biochemistry* **33:** 9845.

Sive H.L. and Roeder R.G. 1986. Interaction of a common factor with conserved promoter and enhancer rsequences in histone H2B, immunoglobulin, and U2 small nuclear RNA (snRNA) genes. *Proc. Natl. Acad. Sci.* **83:** 6382.

Staudt L.M., Singh H., Sen R., Wirth T., Sharp P.A., and Baltimore D. 1986. A lymphoid-specific protein binding to the octamer motif of immunoglobulin genes. *Nature* **323:** 640.

Stern S. and Herr W. 1991. The herpes simplex virus *trans*-activator VP16 recognizes the Oct-1 homeo domain: Evidence for a homeo domain recognition subdomain. *Genes Dev.* **5:** 2555.

Stern S., Tanaka M., and Herr W. 1989. The Oct-1 homeodomain directs formation of a multiprotein–DNA complex with the HSV transactivator VP16. *Nature* **341:** 624.

Strubin M., Newell J.W., and Matthias P. 1995. OBF-1, a novel B cell-specific coactivator that stimulates immunoglobulin promoter activity through association with octamer-binding proteins. *Cell* **80:** 497.

Sturm R.A. and Herr W. 1988. The POU domain is a bipartite DNA-binding structure. *Nature* **336:** 601.

Tanaka M. and Herr W. 1990. Differential transcriptional activation by Oct-1 andf Oct-2: Interdependent activation domains induce Oct-2 phosphorylation. *Cell* **60:** 375.

Tanaka M., Grossniklaus U., Herr W., and Hernandez N. 1988. Actcivation of the U2 snRNA promoter by the octamer motif defines a new class of RNA polymerase II enhancer elements.

Genes Dev. **2:** 1764.

Tanaka M., Lai J.-S., and Herr W. 1992. Promoter-selective activation domains in Oct-1 and Oct-2 direct differential activation of an snRNA and mRNA promoter. *Cell* **68:** 755.

Triezenberg S.J., Kingsbury R.G., and McKnight S.L. 1988. Functional dissection of VP16, the trans-activator of herpes simplex virus immediate early gene expression. *Genes Dev.* **2:** 718.

Walker S., Hayes S., and O'Hare P. 1994. Site-specific conformational alteration of the Oct-1 POU domain-DNA complex as the basis for differential recognition by Vmw65 (VP16). *Cell* **79:** 841.

Wilson A.C., Peterson M.G., and Herr W. 1995a. The HCF repeat is an unusual proteolytic cleavage signal. *Genes Dev.* **9:** 2445.

Wilson A.C., LaMarco K., Peterson M.G., and Herr W. 1993. The VP16 accessory protein HCF is a family of polypeptides processed from a large precursor protein. *Cell* **74:** 115.

Wilson A.C., Freiman R.N., Goto H., Nishimoto T., and Herr W. 1997. VP16 targets an amino-terminal domain of HCF involved in cell cycle progression. *Mol. Cell. Biol.* **17:** 6139.

Wilson A.C., Parrish J.E., Massa H.F., Nelson D.L., Trask B.J., and Herr W. 1995b. The gene encoding the VP16-accessory protein HCF (HCFC1) resides in human Xq28 and is highly expressed in fetal tissues and the adult kidney. *Genomics* **25:** 462.

Xiao P. and Capone J.P. 1990. A cellular factor binds to the herpes simplex virus type 1 *trans*-activator Vmw65 and is required for Vmw65-dependent protein-DNA complex assembly with Oct-1. *Mol. Cell. Biol.* **10:** 4974.

Structure and Function of the Interferon-β Enhanceosome

T. Maniatis, J.V. Falvo, T.H. Kim, T.K. Kim, C.H. Lin,*
B.S. Parekh, and M.G. Wathelet

*Department of Molecular and Cellular Biology, *Department of Chemistry and Chemical Biology,
Harvard University, Cambridge, Massachusetts 02138*

A central problem in eukaryotic gene regulation is understanding how the transcription machinery is targeted to specific sets of genes in response to extracellular signals. A number of studies have shown that multicomponent transcription enhancer complexes are key regulatory components in this process (for recent review, see Carey 1998). These complexes, consisting of both transcriptional activator proteins and architectural proteins (Grosschedl 1995; Werner and Burley 1997), are assembled on specific DNA sequences when the appropriate sets of transcriptional activator proteins are coordinately activated in response to a signal. The assembly of these enhancer complexes is highly cooperative and promotes high levels of transcriptional synergy (Carey 1998). This transcriptional synergy is the consequence of multiple interactions within the enhancer complex, and between this complex, general transcription factors and the multicomponent RNA polymerase II complex (Ptashne and Gann 1997; Carey 1998).

A remarkable feature of this process is that many transcription factors can be activated in response to more than one signal, and they are capable of activating many different genes. For example, the transcriptional activator NF-κB is typically sequestered in the cytoplasm of unstimulated cells, bound to a member of the IκB family of transcriptional inhibitory proteins (Verma et al. 1995). An extraordinarily large number of different inducers, including cytokines, phorbol esters, UV irradiation, bacterial products, and viruses, can activate distinct signaling pathways, all of which result in the degradation of IκB proteins and the nuclear translocation of NF-κB. Because many different genes are regulated by NF-κB, once activated, this transcription factor could, in principle, activate all genes containing an NF-κB-binding site. However, only a specific subset of such genes is activated in response to a given signal. This specificity is achieved through the coordinate activation of NF-κB and other transcription factors resulting in the assembly of unique enhancer complexes; i.e., each gene has a specific key encoded by the unique array of transcription-factor-binding sites in its regulatory region. Thus, only those genes containing enhancers composed of the appropriate combination of transcription-factor-binding sites are activated in response to a given signal.

The human interferon-β (IFN-β) gene provides a particularly well-characterized example of this combinatorial mechanism. This gene is normally silent, but it is activated when cells are infected with virus. The virus-inducible IFN-β enhancer contains binding sites for three distinct transcriptional activator complexes and for the architectural protein HMG I(Y) (high mobility group protein I[Y]). Studies have shown that the relative rotational positions of these binding sites with respect to each other on the DNA helix are essential for enhancer activity. This and other observations led to the view that not only is the specific array of transcription-factor-binding sites important for enhancer function, but also the enhancer complex forms a higher-order structure that is required for the efficient recruitment of the transcription machinery to the IFN-β promoter. On the basis of this requirement, the IFN-β enhancer complex has been termed an "enhanceosome," a higher-order nucleoprotein complex. In this paper, we summarize our current understanding of the structure and function of the IFN-β enhanceosome.

Another intriguing problem in the regulation of gene expression is understanding how individual genes are capable of responding to different signals. Interferon-inducible genes provide an excellent example of this phenomenon. The binding of IFNs to their cell surface receptors results in the activation of a large set of genes whose products interfere with virus replication. Interestingly, there are two classes of IFN-inducible genes: those that are activated by IFN treatment but not by virus infection, and those that are activated both by IFN treatment and by virus infection (for review, see Wathelet et al. 1992). Remarkably, a single regulatory element, the ISRE (interferon-stimulated response element), is required to activate the second class of genes in response to either virus infection or IFN treatment. As discussed below, distinct transcriptional activator complexes that bind to the ISRE are activated by IFN or by virus infection.

THE IFN-β GENE ENHANCER COMPLEX

Virus infection of cells leads to the transient, high-level expression of the IFN-β gene. Typically, maximal levels of expression are achieved at approximately 6–12 hours after infection, and expression ceases by 24 hours postinfection. The regulatory element required for this sensitive on/off switch was one of the first inducible transcriptional enhancer sequences to be characterized (Maniatis et al. 1992). A DNA fragment of less than 60 bp located immediately upstream of the TATA box is a virus-inducible enhancer: It functions regardless of its orientation or position relative to the start site of transcription. Virtually every nucleotide of the IFN-β enhancer is required for maximal levels of virus induction and is in contact with a DNA-binding protein. Four positive regulatory domains,

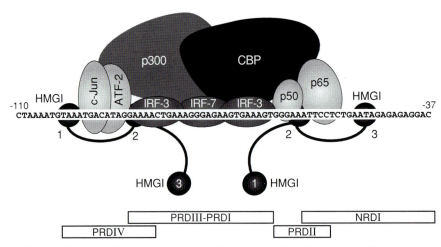

Figure 1. Diagram showing the composition and organization of the human IFN-β enhanceosome. Shown is the nucleotide sequence of the IFN-β enhancer which is located between –47 and –105 nucleotides from the start site of transcription. The boxes beneath the sequence show the location of the PRDs. The transcriptional activator proteins that bind to each of the PRDs are shown, as well as the architectural protein HMG I(Y) and the transcriptional coactivator proteins p300 and CBP.

designated PRDI through PRDIV, interact with distinct transcriptional activator proteins. In vivo and in vitro characterization of proteins that bind to the enhancer led to the model of the IFN-β enhancer complex illustrated in Figure 1. The p300/CBP coactivators have an architectural role in the formation of the nucleoprotein complex by interacting with each transcriptional activator through distinct domains. In addition, HMG I(Y) binds to the minor groove of DNA at four sites within the enhancer complex and is required for maximal levels of IFN-β gene expression. The following is our current understanding of each positive regulatory domain.

PRDII

NF-κB binds to PRDII upon virus infection, and mutations in PRDII that interfere with the binding of NF-κB block virus induction of the IFN-β gene. Induction of the IFN-β gene in vivo is blocked by cotransfection of antisense RNA of the p65 subunit of NF-κB (Thanos and Maniatis 1992, 1995a). Induction of the IFN-β gene is also blocked by inhibition of the nuclear translocation of NF-κB, for example, by transfection of a dominant-negative mutant of IκB-α (S32A, S36A) (F.S. Lee and T. Maniatis, unpubl.) or by treatment with the highly specific proteasome inhibitor lactacystin (M.G. Wathelet and T. Maniatis, unpubl.). Furthermore, virus induction of the IFN-β gene is abolished by targeted disruption of both the p50 and p65 genes in mice (D. Baltimore, pers. comm.). These observations together with in vivo cross-linking experiments (see below) definitively establish that NF-κB is an essential component of the IFN-β enhanceosome.

NF-κB binds to PRDII through contacts in the major groove in a fixed orientation, such that the p50 and p65 subunits bind to the 5′ and 3′ half-sites, respectively, as shown in Figure 1 (Urban et al. 1991; Thanos and Maniatis 1995a). HMG I(Y), on the other hand, binds via the minor groove to the A-T-rich region in PRDII, as well as to a second site 10 bp downstream, NRDI (negative reg-

ulatory domain I). Mutations in either binding site that disrupt HMG I(Y) binding decrease the level of IFN-β gene expression after virus infection (Thanos and Maniatis 1992; Yie et al. 1997). Moreover, expression of antisense HMG I(Y) RNA significantly reduces the level of virus induction of the IFN-β gene (Thanos and Maniatis 1992). In-vitro-binding studies show that the binding of NF-κB and HMG I(Y) to PRDII is cooperative, and HMG I(Y) increases the affinity of NF-κB for PRDII by a factor of 10 (Thanos and Maniatis 1992).

HMG I(Y) consists of three positively charged DNA-binding domains, each of which fits into the minor groove of A-T-rich DNA (Reeves and Nissen 1990). Remarkably, the binding of HMG I(Y) to PRDII involves intramolecular cooperativity between two of the three DNA-binding domains. Specifically, domains 2 and 3 bind to PRDII and NRDI, binding sites that are separated from each other by exactly one turn of the DNA helix (Yie et al. 1997). This intramolecular cooperativity is required for specific binding to PRDII/NRDI (Maher and Nathans 1996; Yie et al. 1997).

In summary, these results indicate the formation of a cooperative, specifically oriented complex of the p50/p65 heterodimer of NF-κB and HMG I(Y) at PRDII/NRDI.

PRDIV

PRDIV is recognized by a heterodimer of ATF-2/c-Jun and a second molecule of HMG I(Y). The ATF-2/c-Jun heterodimer is virus-inducible, at least in some cell lines. Antisense ATF-2 and c-Jun RNA were shown to decrease the level of virus induction of the IFN-β gene in cotransfection experiments, thus implicating this heterodimer in IFN-β gene regulation (Du and Maniatis 1992, 1994; Du et al. 1993). Virus induction of the IFN-β gene is not affected in ATF-2 knock-out mice; this is likely due to the compensatory binding of another ATF family member (Reimold et al. 1996). Similar to the situation with NF-κB at PRDII, HMG I(Y) promotes the binding of ATF-2/c-

Jun to PRDIV (Du et al. 1993). The HMG I(Y)-binding sites flanking PRDIV are separated by exactly one helical turn of DNA, and in this case, the two sites are recognized by the first and second DNA-binding domains of HMG I(Y) (Yie et al. 1997).

The cooperative binding of ATF-2 and HMG I(Y) to PRDIV is due, at least in part, to direct interactions between the two proteins (Du et al. 1993). Evidence that this interaction is required for maximal enhancer activity was provided by an analysis of an ATF-2 isoform (ATF-2_{192}), which binds to PRDIV, but does not interact with HMG I(Y). This isoform of ATF-2 does not bind cooperatively to DNA with HMG I(Y). In fact, HMG I(Y) decreases the binding of ATF-2_{192}/c-Jun to PRDIV (Du and Maniatis 1994). In addition, this isoform does not functionally synergize with other enhanceosome components in an in vitro transcription system (Kim and Maniatis 1997). Thus, both protein-DNA and protein-protein interactions play a part in the formation of a stable ATF-2/c-Jun/HMG I(Y) complex at PRDIV.

Interestingly, the ATF-2/c-Jun heterodimer has a preferred binding orientation at PRDIV, but unlike the situation with NF-κB, the orientation of ATF-2/c-Jun is fixed by other proteins in the enhanceosome. In vitro cross-linking studies revealed that the binding orientation of the heterodimer on naked PRDIV DNA is nearly random. However, in the presence of an IRF protein bound to the adjacent PRDIII-I site, the heterodimer adopts a unique orientation as illustrated in Figure 1 (J.V. Falvo et al., unpubl.).

In summary, the ATF-2/c-Jun heterodimer and one molecule of HMG I(Y) bind cooperatively to PRDIV, and the resulting complex exhibits a specific conformation on the DNA.

PRDIII-I

This region of the IFN-β promoter is recognized by proteins of the interferon regulatory factor (IRF) family. IRF-1, the first transcriptional activator identified in this family, recognizes both PRDIII and PRDI (Fujita et al. 1988; Miyamoto et al. 1988). A number of observations suggested but did not prove that IRF-1 is required for virus induction of the IFN-β gene. For example, transient transfection of IRF-1 can activate the endogenous IFN-β gene (Fujita et al. 1989). In addition, IRF-1 can synergize with the other IFN-β enhancer components in cotransfection experiments (Thanos and Maniatis 1995a,b) and can participate in the cooperative assembly of a functional enhanceosome in vitro (Kim and Maniatis 1997). Although IRF-1 has been valuable for studies of the IFN-β enhanceosome, it does not appear to function in the virus-induced expression of the IFN-β gene (Wathelet et al. 1998 and references therein). For example, IRF-1 is synthesized de novo upon virus induction with approximately the same kinetics as IFN-β, whereas IFN-β can be induced by virus in the absence of protein synthesis (Harada et al. 1989; Pine et al. 1990; Maniatis et al. 1992).

As described in more detail below, recent studies have shown that PRDIII and PRDI function as a single virus-inducible enhanson, P31 (Wathelet et al. 1998), and that two other members of the IRF family are actually involved in virus induction of the IFN-β gene in vivo, namely, IRF-3 (Lin et al. 1998; Sato et al. 1998; Schafer et al. 1998; Wathelet et al. 1998; Weaver et al. 1998; Yoneyama et al. 1998) and IRF-7 (Wathelet et al. 1998). At present, the stoichiometry and configuration of IRF proteins on P31, and the conformation of DNA before and after IRF binding, are unknown.

EVIDENCE FOR A HIGHER-ORDER STRUCTURE: THE IFN-β ENHANCEOSOME

A striking feature of the PRDs is that their relative rotational positions with respect to each other on the DNA helix are critical for virus inducibility of the IFN-β enhancer. As illustrated in Figure 2, insertion of a half-helical turn of DNA (6 bp) between PRDI and PRDII results in a dramatic decrease in the level of virus induction. However, insertion of a full helical turn of DNA (10 bp) at the same position, which restores the helical phasing of the two sites, reconstitutes wild-type levels of virus induction. Similar effects were observed when 4- and 10-bp insertions were made between PRDIV and PRDIII (Thanos and Maniatis 1995b). The orientation of PRDII and the core ATF-2/c-Jun-binding site in PRDIV in the context of the IFN-β enhancer are also critical for virus inducibility (Falvo et al. 1995; J.V. Falvo and B.S. Parekh, unpubl.). In addition, PRDII and PRDIV contain intrinsic, in-phase DNA bends which are reversed by the binding of NF-κB, ATF-2/c-Jun, and HMGI(Y), and this effect can be correlated with inducibility of the IFN-β enhancer (Falvo et al. 1995). The requirement for precise alignment of the PRDs along the DNA helix for enhancer function in vivo, along with protein-induced remodeling of DNA conformation and the multiple protein-protein interactions that can occur among the various transcriptional activators and HMG I(Y) in vitro (see below), is consistent with the formation of a higher-order nucleoprotein complex.

Recently, cotransfection experiments with chimeric versions of the IFN-β transcriptional activators demonstrated that the arrangement of activation domains within the enhanceosome is critical for its function. In the same study, the transcriptional coactivator CBP was shown to enhance the level of IFN-β expression induced by virus or cotransfected activators. Moreover, this effect was dependent on the orientation and helical phasing of the PRDs (Merika et al. 1998). As described below, these results are consistent with a model of a unique activating surface presented by the enhanceosome for interactions with coactivator proteins and the general transcription machinery.

MECHANISMS OF TRANSCRIPTIONAL SYNERGY

Cotransfection of plasmids expressing ATF-2/c-Jun, p50/p65 of NF-κB, and IRF-1 along with an IFN-β promoter-driven reporter gene results in a high level of tran-

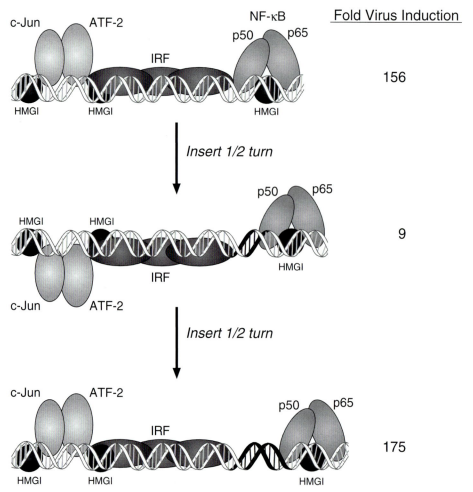

Figure 2. Effect of helical phasing mutations on virus induction of the IFN-β promoter. (*Left*) Model of the core IFN-β enhanceosome; (*right*) effect of inserting 1/2 helical turn (6 bp) of DNA between PRDI and PRDII on virus induction.

scriptional synergy (Thanos and Maniatis 1995b). In vitro assembly of the corresponding enhanceosome revealed that in the presence of limiting amounts of the enhanceosome components, the assembly of the functional complex is highly cooperative (Thanos and Maniatis 1995b; Kim and Maniatis 1997). In addition, DNA competition experiments indicated that the assembled enhanceosome is extraordinarily stable. These conclusions were confirmed and extended using a well-characterized in vitro transcription system and purified recombinant enhanceosome components: Only low levels of transcription were observed when recombinant activators were added individually. However, when all of the components were added together to the reaction, a high level of transcriptional synergy was observed. Significantly, HMG I(Y) and the correct helical phasing were required for this synergy (Kim and Maniatis 1997).

Interestingly, synergy was observed only when the appropriate heterodimer components were present. For example, even though the ATF-2 homodimer can bind to PRDIV, synergy was observed only when the ATF-2/c-Jun heterodimer was used. Similarly, the p65 homodimer can bind to PRDII, but synergy was observed only when

the p50/p65 heterodimer was used (Kim and Maniatis 1997). Thus, a first component of transcriptional synergy is the cooperative assembly of DNA-binding proteins on the enhancer, presenting a specific transcriptional activation surface.

Studies of the general transcription factor requirements for enhanceosome-dependent transcription initiation indicated that a minimal complex consisting of TFIIB, TFIIA, and TFIID and the transcriptional coactivator complex USA (upstream stimulatory activity) is required for synergistic preinitiation complex assembly. In addition, DNA competition experiments revealed a mutual cooperativity between the enhanceosome and the TFIIBAD/USA complex (Kim and Maniatis 1997). Thus, a second component of enhanceosome-dependent transcriptional synergy is multiple interactions between the enhanceosome and the general transcription apparatus.

We conclude that the targeting of the transcriptional machinery to the IFN-β promoter requires a complex array of specific protein-protein and protein-DNA interactions, within the enhanceosome, between the enhanceosome and the general transcription factors TFIIBAD/USA, and the RNA polymerase II complex (Fig. 3).

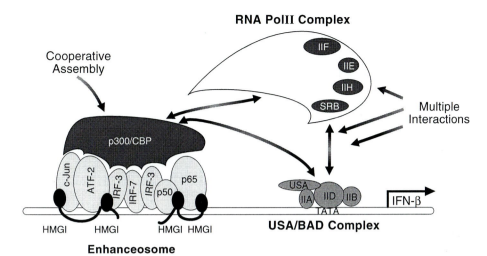

Figure 3. Diagram showing the multiple protein-protein interactions required for transcriptional synergy from the IFN-β promoter. This model is based on in vitro transcription studies in which IRF-1 rather than IRF-3 and IRF-7 was used (Kim and Maniatis 1997; T.K. Kim et al., unpubl.). In contrast, in virus-infected cells, p300/CBP appears to be stably associated with IRF-3/IRF-7 (Wathelet et al. 1998). The figure is drawn to reflect the latter situation. The synergy derives from the cooperative assembly of the enhanceosome, cooperativity between the enhanceosome and the USA/BAD complex, interactions with the RNA polymerase II complex and the overall stability of the fully assembled preinitiation complex.

IDENTIFICATION AND CHARACTERIZATION OF A VIRUS-ACTIVATED FACTOR THAT BINDS TO VIRUS-INDUCIBLE ENHANCER ELEMENTS

As mentioned above, although IRF-1 can function as a component of the IFN-β enhanceosome in transfection experiments and in vitro, the evidence that it is actually involved in the activation of IFN-β gene expression was not definitive. Thus, it was possible that another activator that binds to P31 is virus-inducible but could not be detected in vitro because of its weak affinity for the site. For example, this putative factor may bind only to the intact enhancer in the presence of the other enhanceosome components.

The first evidence for this possibility was provided by studies of genes that are inducible by IFN. The promoters of these genes contain an element, the ISRE, that binds to the IFN-inducible complex called ISGF-3. This complex consists of the IRF family protein ISGF-3γ and two transcriptional activator proteins STAT-1 and STAT-2. Each of these proteins exists in the cytoplasm prior to treatment with IFN. When IFN binds to the IFN receptor complex, the Jak-1 and Tyk-2 kinases are activated and phosphorylate STAT-1 and STAT-2, leading to the assembly of the ISGF-3 complex and its translocation to the nucleus, where it binds to and activates a large number of IFN-inducible genes (Darnell et al. 1994; Darnell 1997).

As mentioned earlier, IFN-inducible genes can be divided into two classes: those that are inducible by IFN but not virus (e.g., the 9-27 gene), and those that are inducible by both IFN and virus (e.g., the IFI-15K gene). When the cells were infected by virus and the extracts used in electrophoretic mobility shift experiments, a virus-inducible protein complex was identified that binds to the ISRE of the IFI-15K but not that of the 9-27 gene. This complex,

termed VAF, was shown to contain two IRF family members, IRF-3 and IRF-7, and at least one copy each of the transcriptional coactivators p300 and CBP (Wathelet et al. 1998). Remarkably, in virus-infected cells, all IRF-3/IRF-7 detected by electrophoretic mobility shift assay is associated with p300/CBP, indicating that the interaction between these proteins is unusually strong.

A striking similarity exists between the ISRE of IFN-inducible genes which are also virus-inducible and the P31 sequence of the IFN-β gene enhancer (Fig. 4A) (Wathelet et al. 1987). In fact, partially purified VAF was shown to bind to the P31 element, albeit with an affinity that is 100 times less than its affinity for the IFI-15K ISRE (Wathelet et al. 1998). Thus, the binding of ISGF-3 and VAF to different promoters could be correlated with IFN or virus inducibility, respectively. As illustrated in Figure 4B, the 9-27 gene, which is inducible by IFN but not by virus, binds to ISGF-3 but not VAF. Similarly, the IFN-β gene, which is inducible by virus but not IFN, binds to VAF but not to ISGF-3. In contrast, the IFI-15K and IFI-56K gene promoters, which are inducible by both IFN and virus, bind to both VAF and ISGF-3. Interestingly, the sequence GAAANN corresponds to the core IRF-binding motif (Escalante et al. 1998), and the number of GAAANN repeats present in an ISRE-like sequence correlates with the ability to respond to different inducers. Thus, ISREs that contain only two such repeats are only inducible by IFNs, whereas ISREs that contain three or more repeats are inducible by both IFNs and viruses (Fig. 4B). It should be noted that other virus-inducible complexes recognizing the ISRE have been described, but their exact relationship to VAF remains to be determined (Fujita et al. 1988; Whiteside et al. 1992; Daly and Reich 1993; Bovolenta et al. 1995; Genin et al. 1995; Navarro et al. 1998).

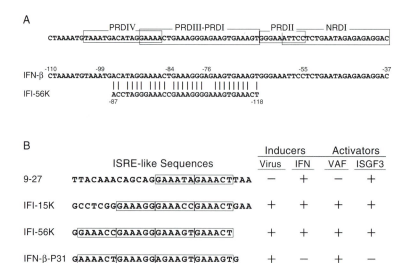

Figure 4. (*A*) Organization of the IFN-β gene promoter and alignment with the virus-inducible IFI-56K gene promoter are shown. (*B*) Comparison of ISRE-like sequences is shown along with inducibility by virus and/or IFN and capacity to bind the indicated activators. The GAAANN repeat corresponds to the core IRF-binding motif (Escalante et al. 1998) and is boxed in the ISRE-like sequences. Note that ISGF-3 can bind to P31, although with a much reduced affinity compared to an ISRE.

A number of additional studies implicated VAF components in the process of virus induction of the IFN-β gene. First, ectopic expression of IRF-3 (Sato et al. 1998; Wathelet et al. 1998; Yoneyama et al. 1998) and IRF-7 (Wathelet et al. 1998) dramatically enhanced virus induction of the IFN-β promoter. Second, Gal4/IRF-3 or Gal4/IRF-7 fusion proteins, but not Gal4/IRF-1 proteins, can activate the virus-dependent expression of a reporter gene driven by Gal4-binding sites (Wathelet et al. 1998). Third, virus induction of the IFN-β promoter is suppressed by dominant negative mutants of IRF-3 (Wathelet et al. 1998; Yoneyama et al. 1998) or IRF-7 (Wathelet et al. 1998). Fourth, virus infection results in phosphorylation of IRF-3 (Lin et al. 1998; Sato et al. 1998; Wathelet et al. 1998; Weaver et al. 1998; Yoneyama et al. 1998) and of IRF-7 (Wathelet et al. 1998), their association with p300/CBP, and their accumulation in the nucleus (Lin et al. 1998; Sato et al. 1998; Wathelet et al. 1998; Weaver et al. 1998; Yoneyama et al. 1998).

These observations are consistent with the following model (Fig. 5). Prior to virus infection, IRF-3 and IRF-7 are associated as homo- or hetero-oligomers, in an inactive conformation (Wathelet et al. 1998). Virus infection results in the activation of an unknown kinase(s) that phosphorylates specific serine/threonine residues at the carboxyl termini of both IRF-3 and IRF-7. This phosphorylation induces a conformational change allowing specific DNA binding and association with p300/CBP to form VAF. Cytoplasmic VAF would then translocate into the nucleus and bind to the ISRE-like sequences of virus-inducible genes. In the case of the ISRE, VAF binds strongly in the absence of any additional proteins. In contrast, the binding of VAF to P31 in the IFN-β promoter requires cooperative interactions with other components of the enhancer complex (Fig. 5). Indeed, components of the enhanceosome are known to interact with each other: HMG I(Y) interacts with both ATF-2/c-Jun and NF-κB,

ATF-2 interacts with IRF-3 and IRF-7, and p300 and CBP interact with each other and with ATF-2, c-Jun, IRF-3, IRF-7, and p65 (Thanos and Maniatis 1992; Du et al. 1993; Wathelet et al. 1998; C.H. Lin et al., unpubl.).

ASSEMBLY OF THE IFN-β ENHANCEOSOME IN VIVO IN RESPONSE TO VIRUS INFECTION

To directly test the model described above, and to confirm the association of NF-κB and ATF-2/c-Jun with the IFN-β enhancer, we carried out in vivo protein-DNA cross-linking studies. Mock- and virus-infected cells were treated with formaldehyde, and the cross-linked protein-DNA complexes were purified and immunoprecipitated with antibodies directed against individual enhanceosome components. The DNA present in the immunoprecipitates was released by reversal of the formaldehyde cross-links and used as a template in PCRs with primers complementary to the promoters of the virus- and IFN-inducible gene IFI-56K or the virus-inducible IFN-β promoter (Wathelet et al. 1998). A DNA fragment corresponding to the IFI-56K promoter could be amplified from the DNA immunoprecipitated from infected cells with either the α-IRF-3 or α-IRF-7 antibodies. This DNA fragment was not observed in the immunoprecipitates from uninfected cells. The IFI-56K promoter could not be amplified from α-IRF-1, α-p50, α-p65, α-ATF-2, and α-c-Jun immunoprecipitates. Thus, in virus-infected cells, IRF-3 and IRF-7 associate with the IFI-56K promoter.

The IFN-β promoter could also be amplified from α-IRF-3 or α-IRF-7 immunoprecipitates obtained from virus-infected cells. However, in contrast to the IFI-56K promoter, the IFN-β promoter could also be amplified from α-p50, α-p65, α-ATF-2, and α-c-Jun immunoprecipitates derived from virus-infected cells. In addition, the IFN-β promoter could not be amplified from α-IRF-1

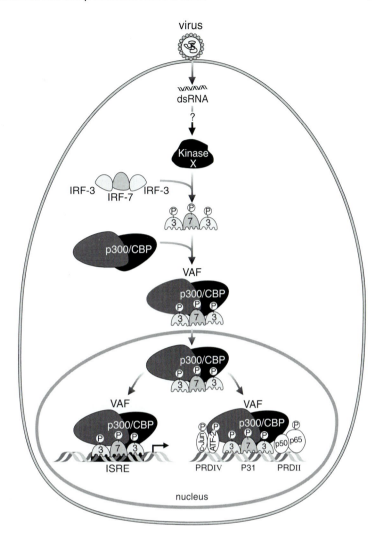

Figure 5. Model for the activation of VAF. When a cell is infected by virus, double-stranded RNA is produced, which activates a putative kinase X. Kinase X phosphorylates IRF-3 and IRF-7, which are present in a complex. The phosphorylated IRF-3 and IRF-7 associate with the transcriptional coactivators p300/CBP to form a stable VAF complex, which then translocates to the nucleus. The VAF complex then associates with ISRE-like sequences and activates transcription.

immunoprecipitates. However, a DNA fragment corresponding to the IRF-2 promoter could be amplified from α-IRF-1 immunoprecipitates (B.S. Parekh and T. Maniatis, unpubl.). The levels of IRF-2 DNA amplified from control and virus-infected cells paralleled the levels of IRF-2 mRNA extracted from these cells: The constitutive expression was further stimulated upon virus infection.

With the exception of p50, none of the antibodies tested could immunoprecipitate the IFN-β promoter DNA from uninfected cells. The binding of p50 homodimers to the IFN-β promoter prior to virus infection is consistent with the possibility that p50 is involved in the preinduction repression of the IFN-β promoter (Maniatis et al. 1992).

Taken together, these experiments indicate that the IFN-β enhanceosome is assembled on the endogenous promoter in vivo upon virus induction and that this enhanceosome contains ATF-2/c-Jun, VAF, and NF-κB (Wathelet et al. 1998).

DISCUSSION

Studies of the IFN-β gene promoter have provided insights into the mechanisms by which the transcription machinery is targeted to specific genes in a signal-dependent manner. As shown in the model of Figure 6A, the multiprotein complex required to activate the IFN-β gene consists of three subcomplexes: the enhanceosome, the TFIIBAD/USA complex, and the RNA polymerase II complex. Association of ATF-2/c-Jun, VAF, and NF-κB with the enhancer is dependent on their intrinsic affinities for their cognate sequences and on protein-protein interactions between these DNA-binding proteins, the architectural protein HMG I(Y), and p300/CBP. Consistent with this view, specific regions of p300/CBP have been shown to interact with NF-κB p65 (Zhong et al. 1998), c-Jun (Bannister et al. 1995), ATF-2 (Kawasaki et al. 1998), and IRF-3 and IRF-7 (Lin et al. 1998; Wathelet et al. 1998), as illustrated in Figure 6B. Interestingly, at least in the case of p65 (Zhong et al. 1998) and IRF-3/IRF-7 (Lin et al. 1998; Wathelet et al. 1998; M.G. Wathelet et al., unpubl.), these interactions are further stimulated by signal-dependent phosphorylation. These multiple interactions lead to the cooperative assembly and high stability of the enhanceosome complex. Moreover, interactions between the enhanceosome, TFIIBAD/USA proteins, and the RNA polymerase II complex provide additional contacts to further stabilize the preinitiation complex.

The transcriptional coactivators p300 and CBP have multiple roles in the enhanceosome and preinitiation complex assembly: (1) They function as architectural factors, as they allow the recruitment of IRF-3/IRF-7 to the low-affinity P31 site through interactions with ATF-2, c-Jun, and p65 bound on the flanking PRDIV and PRDII sites, (2) they function as coactivators, i.e., they mediate interactions between the transcriptional activators bound to the promoter and the basic transcriptional machinery, and (3) they function in chromatin remodeling at the IFN-β promoter through the localized acetylation of histones H3 and H4 in the immediate vicinity of the transcription initiation start

site (B.S. Parekh and T. Maniatis, unpubl.). Taken together, these mechanisms ensure the highly specific activation of the IFN-β gene in response to virus infection.

Our current understanding of the signaling pathways leading to the activation of the IFN-β gene is summarized in Figure 7. When a virus enters a cell, double-stranded RNA is usually generated during its replication. The presence of double-stranded RNA leads to the activation of multiple signaling pathways by an unknown mechanism. At least three distinct signaling pathways are activated, each leading to the activation of distinct transcription factors. The coordinate activation of ATF-2/c-Jun, VAF, and NF-κB in turn leads to the activation of virus-inducible promoters.

A

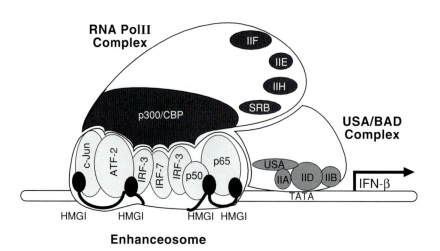

B

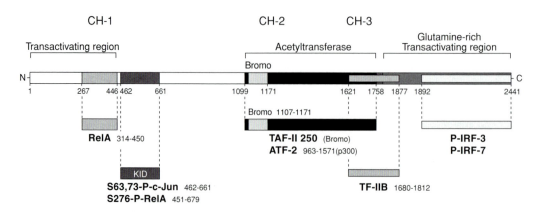

Figure 6. Role of p300/CBP in the formation of a stable IFN-β preinitiation complex. (*A*) Interactions between the enhanceosome and other components of the preinitiation complex. p300 and CBP function as architectural factors in the assembly of the enhanceosome and provide an interface for the USA/BAD complex as well as the RNA polymerase II complex. (*B*) Regions of p300/CBP known to interact specifically with the transcriptional activators present in the IFN-β enhanceosome and other components of the preinitiation complex.

Models for Specificity in Signal Transduction

Our studies have also provided new insights into the problem of how a single gene can respond to distinct extracellular signals. Specifically, we have shown that certain ISRE or ISRE-like sequences are capable of interacting with two entirely different transcription factor complexes, each of which is activated by a different signaling pathway. The ISREs of genes that respond to both virus and IFN induction are recognized by the ISGF-3 complex in response to IFN treatment and by the VAF complex in response to virus infection.

Comparison of the virus induction of the IFI-15K or IFI-56K genes with that of the IFN-β gene provides a striking contrast in the mechanisms of signal-dependent gene activation. In the former case, the ISRE has a high affinity for VAF and is sufficient for virus induction. ISREs with high affinity for VAF also have high affinity for ISGF-3 and are thus IFN-inducible. In contrast, the ISRE-like sequence of the IFN-β promoter (P31) interacts weakly with VAF; thus, additional contacts between the p300/CBP component of VAF and the NF-κB and ATF-2/c-Jun transcriptional activators bound to the adjacent PRDII and PRDIV sequences are required in order for VAF to bind to the IFN-β promoter in virus-infected cells. This requirement for p300/CBP as an architectural factor provides the specificity mechanism that ensures the IFN-β promoter is activated only upon virus infection. Indeed, both IRF-1 and ISGF-3 recognize the P31 element with higher affinity than VAF. Yet signals that induce IRF-1 (IFNs, UV, TNF, etc.) and ISGF-3 (IFNs) do not induce the IFN-β gene. The case of TNF is particularly intriguing since TNF induces not only IRF-1, but ATF-2/c-Jun and NF-κB as well. Together, these transcription factors could in principle form an enhanceosome in cells exposed to TNF. However, unlike IRF-3/IRF-7, both IRF-1 (Merika et al. 1998) and ISGF-3 (Bhattacharya et al. 1996; Zhang et al. 1996) interact with the same region of p300/CBP as c-Jun and p65. Thus, in the case of IRF-1 and ISGF-3, p300/CBP may not be able to play its architectural part: These activators, ATF-2/c-Jun and NF-κB, cannot simultaneously interact with p300/CBP on the IFN-β promoter under physiological conditions. These differences ensure that the IFN-β gene is expressed only in response to virus infection but not to TNF or IFN, preventing the establishment of a positive feedback loop that would be toxic to the organism.

The Enhanceosome Concept

How general is the enhanceosome concept? Several complex tissue-specific and inducible transcriptional enhancers, such as the immunoglobulin μ heavy chain enhancer (for review, see Ernst and Smale 1995) and the interleukin-2 gene enhancer (for review, see Rothenberg and Ward 1996), are thought to be recognized by coordinately regulated sets of transcription factors. In some cases, the correct rotational position of transcription-factor-binding sites is known to be required for enhancer activity, including the tryptophan oxygenase gene (Schule et al. 1988), the human proenkephalin gene (Comb et al.

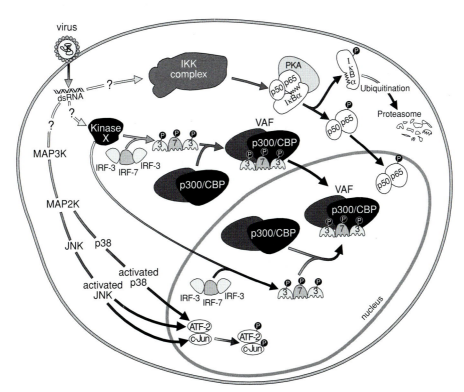

Figure 7. Three signal transduction pathways triggered by virus replication are shown. The identity of the MAP3K, MAP2K, and IKK activated by virus infection remains to be determined. Once IκB is phosphorylated and ubiquitinated, it is degraded by the proteasome, allowing protein kinase A to phosphorylate NF-κB, which then translocates to the nucleus.

1988), the major histocompatibility class II HLA-DRA gene (Reith et al. 1994), the E-selectin gene (Meacock et al. 1994), and the T-cell receptor-α (TCR-α) enhancer (Giese et al. 1995). The E-selectin promoter is most similar to the IFN-β gene; its activity is regulated by ATF-2/c-Jun, NF-κB, HMG I(Y), and p300/CBP (Gerritsen et al. 1997; Read et al. 1997; for review, see Collins et al. 1995). The TCR-α enhancer, on the other hand, contains transcriptional activator binding sites separated by a binding site for the HMG domain protein lymphoid enhancer-binding factor-1 (LEF-1) (Giese et al. 1995; for review, see Grosschedl 1995; Carey 1998). LEF-1 introduces a radical bend in DNA (130°) (Love et al. 1995), which is thought to bring the transcription factors bound to flanking DNA sequences into proximity. A third component, ALY, is thought to mediate interactions between LEF-1 and adjacent activators (Bruhn et al. 1997). Thus, unlike the IFN-β enhanceosome where HMG I(Y) promotes subtle changes in DNA conformation and facilitates the binding of individual activators, the TCR-α enhanceosome involves the induction of a radical DNA bend by LEF-1 and interaction of transcription factors bound to well-separated binding sites.

In summary, multiple levels of specificity are required for the assembly of signal-dependent enhanceosomes: (1) the type and organization of the transcription-factor-binding sites in the enhancer, (2) the presence of the appropriate sets of coordinately activated transcriptional activator proteins that bind to the enhancer, (3) the presence of architectural proteins that facilitate protein-DNA interactions and influence DNA conformation, and (4) the presence of transcriptional coactivator proteins that recognize the activation domain surface presented by the protein-DNA complex. It is possible that the assembly of an enhanceosome is required only for those genes whose expression must be stringently regulated, and/or for those genes whose expression requires transcription factors that are used by more than one signaling pathway. Thus, enhanceosomes not only provide a mechanism for highly specific gene activation, but also have a critical role in the integration of multiple signaling pathways. The organization of enhanceosomes also provides the flexibility to evolve new regulatory circuits through novel combinations of regulatory sequences, transcription factors, and the corresponding signaling pathways. Because of the combinatorial nature of this process, a virtually unlimited number of unique enhanceosomes can be assembled from a relatively small set of transcriptional activator proteins. The weak linkage of the regulatory components and the high degree of flexibility inherent in this system promote the "evolvability" of signal transduction/transcriptional control circuits (Kirschner and Gerhart 1998). Thus, the evolution of new regulatory ciruits could be achieved by changes and rearrangements within enhanceosomes. In addition, it is possible that complex regulatory elements, such as those of the *Drosophila evenskipped* gene and various homeotic genes, consist of multiple, distinct enhanceosomes, and the evolution of new developmental pathways occurs through the generation of novel combinations of enhanceosomes.

ACKNOWLEDGMENTS

This work was supported by a grant from the National Institutes of Health (AI-20642) to T.M. J.V.F. was supported by a National Defense Science and Engineering Graduate Fellowship. T.K.K. was supported by a fellowship from the Cancer Research Fund of the Damon Runyon–Walter Winchell Foundation, DRG-1329.

REFERENCES

Bannister A.J., Oehler T., Wilhelm D., Angel P., and Kouzarides T. 1995. Stimulation of c-Jun activity by CBP: c-Jun residues Ser63/73 are required for CBP induced stimulation in vivo and CBP binding in vitro. *Oncogene* **11:** 2509.

Bhattacharya S., Eckner R., Grossman S., Oldread E., Arany Z., D'Andrea A., and Livingston D.M. 1996. Cooperation of Stat2 and p300/CBP in signalling induced by interferon-alpha. *Nature* **383:** 344.

Bovolenta C., Lou J., Kanno Y., Park B.K., Thornton A.M., Coligan J.E., Schubert M., and Ozato K. 1995. Vesicular stomatitis virus infection induces a nuclear DNA-binding factor specific for the interferon-stimulated response element. *J. Virol.* **69:** 4173.

Bruhn L., Munnerlyn A., and Grosschedl R. 1997. ALY, a context-dependent coactivator of LEF-1 and AML-1, is required for TCRα enhancer function. *Genes Dev.* **11:** 640.

Carey M. 1998. The enhanceosome and transcriptional synergy. *Cell* **92:** 5.

Collins T., Read M.A., Neish A.S., Whitley M.Z., Thanos D., and Maniatis T. 1995. Transcriptional regulation of endothelial cell adhesion molecules: NF-κB and cytokine-inducible enhancers. *FASEB J.* **9:** 899.

Comb M., Mermod N., Hyman S.E., Pearlberg J., Ross M.E., and Goodman H.M. 1988. Proteins bound at adjacent DNA elements act synergistically to regulate human proenkephalin cAMP inducible transcription. *EMBO J.* **7:** 3793.

Daly C. and Reich N.C. 1993. Double-stranded RNA activates novel factors that bind to the interferon-stimulated response element. *Mol. Cell. Biol.* **13:** 3756.

Darnell J.E., Jr. 1997. STATs and gene regulation. *Science* **277:** 1630.

Darnell J.E., Jr., Kerr I.M., and Stark G.R. 1994. Jak-STAT pathways and transcriptional activation in response to IFNs and other extracellular signaling proteins. *Science* **264:** 1415.

Du W. and Maniatis T. 1992. An ATF/CREB binding site is required for virus induction of the human interferon beta gene (corrected) (erratum appears in *Proc. Natl. Acad. Sci.* [1992], **89:** 5700). *Proc. Natl. Acad. Sci.* **89:** 2150.

———. 1994. The high mobility group protein HMG I(Y) can stimulate or inhibit DNA binding of distinct transcription factor ATF-2 isoforms. *Proc. Natl. Acad. Sci.* **91:** 11318.

Du W., Thanos D., and Maniatis T. 1993. Mechanisms of transcriptional synergism between distinct virus-inducible enhancer elements. *Cell* **74:** 887.

Ernst P. and Smale S.T. 1995. Combinatorial regulation of transcription II: The immunoglobulin mu heavy chain gene. *Immunity* **2:** 427.

Escalante C.R., Yie J., Thanos D., and Aggarwal A.K. 1998. Structure of IRF-1 with bound DNA reveals determinants of interferon regulation. *Nature* **391:** 103.

Falvo J.V., Thanos D., and Maniatis T. 1995. Reversal of intrinsic DNA bends in the IFNβ gene enhancer by transcription factors and the architectural protein HMG I(Y). *Cell* **83:** 1101.

Fujita T., Reis L.F., Watanabe N., Kimura Y., Taniguchi T., and Vilcek J. 1989. Induction of the transcription factor IRF-1 and interferon-β mRNAs by cytokines and activators of second-messenger pathways. *Proc. Natl. Acad. Sci.* **86:** 9936.

Fujita T., Sakakibara J., Sudo Y., Miyamoto M., Kimura Y., and Taniguchi T. 1988. Evidence for a nuclear factor(s), IRF-1, mediating induction and silencing properties to human IFN-β gene regulatory elements. *EMBO J.* **7:** 3397.

Genin P., Braganca J., Darracq N., Doly J., and Civas A. 1995. A novel PRD I and TG binding activity involved in virus-induced transcription of IFN-A genes. *Nucleic Acids Res.* **23:** 5055.

Gerritsen M.E., Williams A.J., Neish A.S., Moore S., Shi Y., and Collins T. 1997. CREB-binding protein/p300 are transcriptional coactivators of p65. *Proc. Natl. Acad. Sci.* **94:** 2927.

Giese K., Kingsley C., Kirshner J.R., and Grosschedl R. 1995. Assembly and function of a TCR alpha enhancer complex is dependent on LEF-1-induced DNA bending and multiple protein-protein interactions. *Genes Dev.* **9:** 995.

Grosschedl R. 1995. Higher-order nucleoprotein complexes in transcription: Analogies with site-specific recombination. *Curr. Opin. Cell Biol.* **7:** 362.

Harada H., Fujita T., Miyamoto M., Kimura Y., Maruyama M., Furia A., Miyata T., and Taniguchi T. 1989. Structurally similar but functionally distinct factors, IRF-1 and IRF- 2, bind to the same regulatory elements of IFN and IFN-inducible genes. *Cell* **58:** 729.

Kawasaki H., Song J., Eckner R., Ugai H., Chiu R., Taira K., Shi Y., Jones N., and Yokoyama K.K. 1998. p300 and ATF-2 are components of the DRF complex, which regulates retinoic acid- and E1A-mediated transcription of the c-*jun* gene in F9 cells. *Genes Dev.* **12:** 233.

Kim T.K. and Maniatis T. 1997. The mechanism of transcriptional synergy of an in vitro assembled interferon-β enhanceosome. *Mol. Cell* **1:** 119.

Kirschner M. and Gerhart J. 1998. Evolvability. *Proc. Natl. Acad. Sci.* **95:** 8420.

Lin R., Heylbroeck C., Pitha P.M., and Hiscott J. 1998. Virus-dependent phosphorylation of the IRF-3 transcription factor regulates nuclear translocation, transactivation potential, and proteasome-mediated degradation. *Mol. Cell. Biol.* **18:** 2986.

Love J.J., Li X., Case D.A., Giese K., Grosschedl R., and Wright P.E. 1995. Structural basis for DNA bending by the architectural transcription factor LEF-1. *Nature* **376:** 791.

Maher J.F. and Nathans D. 1996. Multivalent DNA-binding properties of the HMG-I proteins. *Proc. Natl. Acad. Sci.* **93:** 6716.

Maniatis T., Whittemore L.-A., Du W., Fan C.-M., Keller A.D., Palombella V.J., and Thanos D. 1992. Positive and negative control of human interferon-β gene expression. In *Transcriptional regulation* (ed. S.L. McKnight and K.R. Yamamoto), p. 1193. Cold Spring Harbor Laboratory Press, Cold Spring Harbor, New York.

Meacock S., Pescini-Gobert R., DeLamarter J.F., and Hooft van Huijsduijnen R. 1994. Transcription factor-induced, phased bending of the E-selectin promoter. *J. Biol. Chem.* **269:** 31756.

Merika M., Williams A.J., Chen G., Collins T., and Thanos D. 1998. Recruitment of CBP/p300 by the IFNβ enhanceosome is required for synergistic activation of transcription. *Mol. Cell* **1:** 277.

Miyamoto M., Fujita T., Kimura Y., Maruyama M., Harada H., Sudo Y., Miyata T., and Taniguchi T. 1988. Regulated expression of a gene encoding a nuclear factor, IRF-1, that specifically binds to IFN-β gene regulatory elements. *Cell* **54:** 903.

Navarro L., Mowen K., Rodems S., Weaver B., Reich N., Spector D., and David M. 1998. Cytomegalovirus activates interferon immediate-early response gene expression and an interferon regulatory factor 3-containing interferon-stimulated response element-binding complex. *Mol. Cell. Biol.* **18:** 3796.

Pine R., Decker T., Kessler D.S., Levy D.E., and Darnell J.E., Jr. 1990. Purification and cloning of interferon-stimulated gene factor 2 (ISGF2): ISGF2 (IRF-1) can bind to the promoters of both beta interferon- and interferon-stimulated genes but is not a primary transcriptional activator of either. *Mol. Cell. Biol.* **10:** 2448.

Ptashne M. and Gann A. 1997. Transcriptional activation by recruitment. *Nature* **386:** 569.

Read M.A., Whitley M.Z., Gupta S., Pierce J.W., Best J., Davis R.J., and Collins T. 1997. Tumor necrosis factor alpha-induced E-selectin expression is activated by the nuclear factor-kappa B and c-JUN N-terminal kinase/p38 mitogen-activated protein kinase pathways. *J. Biol. Chem.* **272:** 2753.

Reeves R. and Nissen M.S. 1990. The A·T-DNA-binding domain of mammalian high mobility group I chromosomal proteins. A novel peptide motif for recognizing DNA structure. *J. Biol. Chem.* **265:** 8573.

Reimold A.M., Grusby M.J., Kosaras B., Fries J.W., Mori R., Maniwa S., Clauss I.M., Collins T., Sidman R.L., Glimcher M.J., and Glimcher L.H. 1996. Chondrodysplasia and neurological abnormalities in ATF-2-deficient mice. *Nature* **379:** 262.

Reith W., Siegrist C.A., Durand B., Barras E., and Mach B. 1994. Function of major histocompatibility complex class II promoters requires cooperative binding between factors RFX and NF-Y. *Proc. Natl. Acad. Sci.* **91:** 554.

Rothenberg E.V. and Ward S.B. 1996. A dynamic assembly of diverse transcription factors integrates activation and cell-type information for interleukin 2 gene regulation. *Proc. Natl. Acad. Sci.* **93:** 9358.

Sato M., Tanaka N., Hata N., Oda E., and Taniguchi T. 1998. Involvement of the IRF family transcription factor IRF-3 in virus-induced activation of the IFN-β gene. *FEBS Lett.* **425:** 112.

Schafer S.L., Lin R., Moore P.A., Hiscott J., and Pitha P.M. 1998. Regulation of type I interferon gene expression by interferon regulatory factor-3. *J. Biol. Chem.* **273:** 2714.

Schule R., Muller M., Otsuka-Murakami H., and Renkawitz R. 1988. Cooperativity of the glucocorticoid receptor and the CACCC-box binding factor. *Nature* **332:** 87.

Thanos D. and Maniatis T. 1992. The high mobility group protein HMG I(Y) is required for NF-κB-dependent virus induction of the human IFN-β gene. *Cell* **71:** 777.

———. 1995a. Identification of the rel family members required for virus induction of the human beta interferon gene. *Mol. Cell. Biol.* **15:** 152.

———. 1995b. Virus induction of human IFNβ gene expression requires the assembly of an enhanceosome. *Cell* **83:** 1091.

Urban M.B., Schreck R., and Baeuerle P.A. 1991. NF-κB contacts DNA by a heterodimer of the p50 and p65 subunit. *EMBO J.* **10:** 1817.

Verma I.M., Stevenson J.K., Schwarz E.M., Van Antwerp D., and Miyamoto S. 1995. Rel/NF-kappa B/I kappa B family: Intimate tales of association and dissociation. *Genes Dev.* **9:** 2723.

Wathelet M.G., Berr P.M., and Huez G.A. 1992. Regulation of gene expression by cytokines and virus in human cells lacking the type-I interferon locus. *Eur. J. Biochem.* **206:** 901.

Wathelet M.G., Clauss I.M., Nols C.B., Content J., and Huez G.A. 1987. New inducers revealed by the promoter sequence analysis of two interferon-activated human genes. *Eur. J. Biochem.* **169:** 313.

Wathelet M.G., Lin C.H., Parekh B.S., Ronco L.V., Howley P.M., and Maniatis T. 1998. Virus infection induces the assembly of coordinately activated transcription factors on the IFN-β enhancer in vivo. *Mol. Cell* **1:** 507.

Weaver B.K., Kumar K.P., and Reich N.C. 1998. Interferon regulatory factor 3 and CREB-binding protein/p300 are subunits of double-stranded RNA-activated transcription factor DRAF1. *Mol. Cell. Biol.* **18:** 1359.

Werner M.H. and Burley S.K. 1997. Architectural transcription factors: Proteins that remodel DNA. *Cell* **88:** 733.

Whiteside S.T., Visvanathan K.V., and Goodbourn S. 1992. Identification of novel factors that bind to the PRD I region of the human beta-interferon promoter. *Nucleic Acids Res.* **20:** 1531.

Yie J., Liang S., Merika M., and Thanos D. 1997. Intra- and intermolecular cooperative binding of high-mobility-group protein I(Y) to the beta-interferon promoter. *Mol. Cell. Biol.* **17:** 3649.

Yoneyama M., Suhara W., Fukuhara Y., Fukuda M., Nishida E.,

and Fujita T. 1998. Direct triggering of the type I interferon system by virus infection: Activation of a transcription factor complex containing IRF-3 and CBP/p300. *EMBO J.* **17:** 1087.

Zhang J.J., Vinkemeier U., Gu W., Chakravarti D., Horvath C.M., and Darnell J.E., Jr. 1996. Two contact regions between

Stat1 and CBP/p300 in interferon gamma signaling. *Proc. Natl. Acad. Sci.* **93:** 15092.

Zhong H., Voll R.E., and Ghosh S. 1998. Phosphorylation of NF-κB p65 by PKA stimulates transcriptional activity by promoting a novel bivalent interaction with the coactivator CBP/p300. *Mol. Cell* **1:** 661.

Autoinhibition as a Transcriptional Regulatory Mechanism

B.J. Graves, D.O. Cowley, T.L. Goetz, J.M. Petersen,* M.D. Jonsen,† and M.E. Gillespie‡

*Huntsman Cancer Institute, University of Utah School of Medicine, Salt Lake City, Utah 84132; *Department of Biomolecular Chemistry, University of Wisconsin, Madison, Wisconsin 53706; †Eleanor Roosevelt Institute, Denver, Colorado 80206; ‡Sloan-Kettering Memorial Cancer Center, New York, New York 10021*

The modularity of transcription factors provides unique opportunities for biological regulation. Functional domains for DNA binding, activation, repression, and subunit association are often delineated as discrete regions in both primary and tertiary protein structures. In many cases, these domains can function autonomously in a heterologous context. Furthermore, positive regulatory pathways, involving posttranslational modifications and protein associations, often target a specific functional domain. Intramolecular interactions that negatively regulate a specific domain provide an additional level of control. This phenomenon, termed autoinhibition, is characterized by the observation that the deletion or mutation of sequences outside of a domain can enhance the activity of that domain. Indeed, regions that negatively control the activity of functional domains are frequently found in transcription factors within both prokaryotic and eukaryotic systems. Figure 1 schematically illustrates the role of inhibitory sequences within the context of a generic transcription factor and lists some of the general and regulatory transcription factors in which autoinhibition regulates DNA binding, activation, or subunit configuration. This growing number of reports indicates that the autoinhibition phenomenon is widespread and represents an important regulatory strategy to modulate the activity of transcription factors.

We study the *ets* family of eukaryotic regulatory transcription factors to investigate transcriptional control mechanisms (Sharrocks et al. 1997; Graves and Petersen 1998). A current focus is the problem of specificity faced by such multigene families. The *ets* genes clearly illustrate this problem. The ETS domain, an 85-amino-acid region that is conserved in all *ets*-encoded proteins, directs DNA binding. There is at least approximately 40% sequence identity within this domain among all family members. A core recognition sequence 5′-GGA-3′ is used by all *ets*-encoded proteins. The binding site extends on two flanks to include at least 9 bp. The consensus sequence for this extended region also is remarkably similar for most family members. Due to this common mode of DNA binding, multiple regulatory pathways are required to dictate specificity within this system. Most *ets*-encoded proteins function in conjunction with other regulatory transcription factors, often in partnerships that display cooperative DNA binding. In addition, many family members are modulated by MAPK-dependent signaling. Autoinhibition also figures strongly in the regulation of the *et*s family (see Fig. 1, *ets*-encoded proteins listed include ERM, Elk-1, Sap-1, NET, Ets-1, and Ets-2). We propose that this added level of control has an important role in programming specificity into this gene family.

Our studies focus on the autoinhibition of Ets-1 DNA binding. Ets-1, a regulatory transcription factor that controls gene expression in vertebrate T cells (Bories et al. 1995; Muthusamy et al. 1995), is modular with at least three functional domains (Fig. 2). The ETS domain mediates DNA binding (Karim et al. 1990), an activation domain provides activator activity (Schneikert et al. 1992; Yang et al. 1998), and the Pointed domain is implicated

Figure 1. Autoinhibition as a regulatory mechanism. Examples of autoinhibition of *activation:* C/EBPβ (Williams et al. 1995), HSF (Newton et al. 1996; Kline and Morimoto 1997), ATF-2 (Li and Green 1996), CBFα2/AML-1 (Kanno et al. 1998), B-myb (Ansieau et al. 1997), glucocorticoid receptor (Lefstin et al. 1994; Starr et al. 1996; Lefstin and Yamamoto 1998), ERM (Laget et al. 1996); *DNA binding:* p53 (Hupp et al. 1992; Hansen et al. 1998), TATA-binding protein (TBP) (Kuddus and Schmidt 1993; Mittal and Hernandez 1997), *E. coli* RNA polymerase σ (Dombroski et al. 1993), PU.1 interaction protein (Pip) (Brass et al. 1996), *Drosophila* Hoxb-1 (Chan and Mann 1996; Chan et al. 1996); ERM (Laget et al. 1996), Elk-1 (Price et al. 1995), SAP-1b (Dalton and Treisman 1992), Net (Giovane et al. 1994), Ets-1 (Hagman and Grosschedl 1992; Lim et al. 1992; Fisher et al. 1994; Petersen et al. 1995; Jonsen et al. 1996); and *subunit association:* HSF (Orosz et al. 1996), Sin1 (Perez-Martin and Johnson 1998).

Figure 2. Functional modules of transcription factor Ets-1. The ETS domain and Pointed domain are defined by both sequence conservation and structural studies. Extensive functional data are available for DNA binding by the ETS domain, whereas the function of the Pointed domain in protein-binding remains controversial. The activation domain is mapped by transcription assays using transient expression in mammalian cells. Phosphorylation by MAPK near the amino terminus stimulates the transcription activity of Ets-1 by an unknown mechanism. Ca-dependent phosphorylation of Ets-1 near the inhibitory sequences represses DNA-binding activity, possibly by reinforcing autoinhibition.

in a variety of protein-protein interactions. The sequences that are necessary and sufficient for autoinhibition of DNA binding lie in two regions that flank the ETS domain (Hagman and Grosschedl 1992; Lim et al. 1992; Nye et al. 1992; Fisher et al. 1994; Petersen et al. 1995; Jonsen et al. 1996).

This paper describes the development of a structural and mechanistic model for the autoinhibition of Ets-1 DNA binding. In brief, a structural element, termed the inhibitory module, allosterically interferes with DNA binding. A conformational change, which includes disruption of the inhibitory module, must accompany DNA binding. Coupling DNA binding with a conformational change accounts for the observed negative effects of the inhibitory module. The data that support this model, which were obtained from a combination of structural, biochemical, and genetic approaches, are summarized. We also propose how Ets-1 autoinhibition is utilized within several biological regulatory pathways. Finally, the broader biological and structural implications of autoinhibition as a regulatory mechanism are discussed.

METHODS

The discovery and characterization of the autoinhibitory phenomenon in Ets-1 required an accurate and precise measurement of the DNA-binding activity of both native Ets-1 and deletion mutants. We used quantitative electrophoretic mobility shift assays (EMSA) to obtain these types of data. This methodology, described briefly here, can be applied to other problems that require accurate measurement of the relative affinities of DNA-protein interactions.

Expression and purification. The purification for all murine *ets-1*-encoded proteins discussed here was described previously (Nye et al. 1992; Petersen et al. 1995; Jonsen et al. 1996). In brief, proteins encoded by *ets-1* were synthesized in a T7-polymerase-dependent bacterial expression system. The expression vectors were engineered such that full-length Ets-1 and deletion polypeptides were expressed without tags or leader sequences. Polypeptides were purified to greater than 90% purity by conventional methods including ion exchange and gel filtration chromatography. The total protein concentration was measured by spectroscopy using empirically determined extinction coefficients. The concentration of active protein was determined by performing DNA titration experiments and EMSA. In this strategy, the concentration of protein-DNA complex formed in the presence of a large excess of DNA is assumed to represent the concentration of protein active in DNA binding. In all cases, greater than 90% of the protein was judged to be active.

Quantitative EMSA. Standard buffer and electrophoresis conditions were used for EMSA (Nye et al. 1992). To obtain equilibrium binding constants, a set of equilibrium binding reactions were set up that contained equal amounts of DNA and varying amounts of protein (Petersen et al. 1995; Jonsen et al. 1996). The fraction of DNA bound was plotted versus the concentration of free protein to generate a binding curve. The equilibrium dissociation constant is equal to the concentration of free protein that provides half-maximal binding of the DNA, K_D equals $[P][D]/[PD]$; however, under conditions in which $[D] = [PD]$, then $K_D = [P]$. Nonlinear least-squares analysis on the binding curves also provided a K_D determination using the formulation $[DP]/Dt = 1/1 + (K_D/[P])$. The curve fit corroborated the values obtained from the point of half-maximal binding.

To obtain values with accuracy and reproducibility, we minimized the parameters that needed to be measured. Binding curves required only two variables for each equilibrium reaction: first, the concentration of active protein (total input) and, second, the fraction of DNA unshifted relative to the total DNA. The DNA concentration was at least tenfold lower than the K_D value, usually in the picomolar range. This strategy allowed us to assume that the total protein concentration is approximately equal to the free protein concentration. In practical terms, this required that only the maximal possible concentration of DNA be set, with any minor losses being inconsequential to the experimental design. To avoid problems of dissociation during electrophoresis, the fraction bound was deduced from the amount of DNA unshifted relative to total DNA. The use of such low concentrations of DNA required high specific activity to facilitate detection, even by phosphorimaging. This was achieved by radioactively labeling at both 5′ ends using $[\gamma\text{-}^{32}P]$ATP (7000 Ci/mmole) and T4 polynucleotide kinase.

Partial proteolysis. Partial proteolysis of Ets-1 was performed as described previously (Jonsen et al. 1996). In brief, reaction mixtures included 10 μg of Ets-1 and trypsin at varying concentrations. After a 2-minute incubation at 20°C, trypsin was inhibited either by the addition of SDS-PAGE sample buffer or by the addition of 1 mM phenylmethylsulfonyl fluoride (PMSF) and storage at 4°C. DNA-binding activity of the pool of PMSF-treated tryptic fragments was assayed by EMSA under standard conditions.

RESULTS

Mapping the Inhibitory Sequences

The inhibitory sequences of Ets-1 initially were discovered by assaying the DNA-binding activity of the products of Ets-1 partial proteolysis. Trypsin cleavage of Ets-1 generated five major peptides, three of which contained the ETS domain (Fig. 3a,c). The DNA-binding activity of the tryptic fragments was assayed by EMSA (Fig. 3b). As evidenced by the relative depletion of free DNA, the pool of tryptic fragments showed higher binding activity than undigested Ets-1. These results demonstrated that the ETS domain displays a higher DNA-binding activity when released from full-length Ets-1. These findings fit the empirical definition of autoinhibition, demonstrating the existence of regions within native Ets-1 that repress the full potential of the DNA-binding domain.

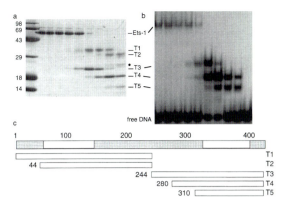

Figure 3. Demonstration of autoinhibition by partial proteolysis. (*a*) Protease digestions of equal amounts of Ets-1 and increasing amounts of trypsin were analyzed by SDS-PAGE. (*Closed circle*) Trypsin band. (*b*) Pools of tryptic fragments from each digestion in *a* were tested for DNA-binding activity in an electrophoretic mobility shift assay. (*c*) Structure of tryptic fragments (Jonsen et al. 1996).

Sequences necessary for autoinhibition were mapped more accurately by deletion mutagenesis in combination with quantitative EMSA. The affinities of four mutants illustrate conclusions from the mapping studies (Fig. 4). First, two regions participate in inhibition because deletion from either the amino terminus (ΔN331) or carboxyl terminus (ΔC428) led to activation. Second, the two regions appeared to function cooperatively since deletion of both regions (Ets-1$^{331-428}$) did not cause a greater activation than either single deletion. Finally, the deletion mutant ΔN280 retained all of the inhibitory sequences necessary for full repression. This mutant, whose struc-

ture is similar to the tryptic fragment T4, was expressed as a soluble polypeptide in bacteria and formed a single structural domain as determined by thermal unfolding experiments (Petersen 1996). The inhibited mutant ΔN280 and the activated mutant ΔN331, which is also expressed as a soluble protein, were used extensively for further structural and mechanistic studies of autoinhibition.

Structural Model of the Inhibitory Module

High-resolution structural information for the ETS domain and the inhibitory sequences greatly facilitated the development of the mechanistic model of Ets-1 autoinhibition. As determined by both nuclear magnetic resonance (NMR) and crystallographic studies of several *ets* family members (Liang et al. 1994; Donaldson et al. 1996; Kodandapani et al. 1996; Werner et al. 1997; Batchelor et al. 1998), the ETS domain is composed of three α helices and four β strands (Fig. 5a). These secondary structural elements fold into a HTH motif and a β-sheet (Fig. 5b). This combined β-sheet and α-helix motif, known as winged helix or winged HTH, is found in a large number of DNA-binding proteins (Brennan 1993; Burley 1994). As described by structural analysis of three DNA-ETS domain complexes, both the HTH and β-sheet interact directly with DNA (Kodandapani et al. 1996; Werner et al. 1997; Batchelor et al. 1998). Additional details of the ETS domain DNA interface are discussed below.

The secondary structure of the inhibitory sequences, as determined by NMR analyses, includes two α helices (HI-1 and HI-2) in the amino-terminal region and one α-helix (H4) in the carboxy-terminal region (Fig. 5a) (Skalicky et al. 1996). A proposed model of the three-dimensional structure of ΔN280 packs these three inhibitory helices with helix H1 of the ETS domain to form a four-helix bundle, which we term the inhibitory module

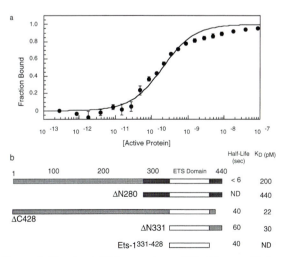

Figure 4. Mapping inhibitory regions of Ets-1 by quantitative DNA-binding assays. (*a*) Binding curve was generated from a set of equilibrium binding reactions that contained a constant amount of duplex DNA and increasing amounts of Ets-1. (*b*) Native Ets-1 and deletion polypeptides were characterized in both equilibrium studies (K_D) and kinetic experiments (complex half-life). All data are summarized from Jonsen et al. (1996), except K_D values of ΔN331 and ΔN280 which are from Petersen et al. (1995). Dark shading represents active inhibitory sequences.

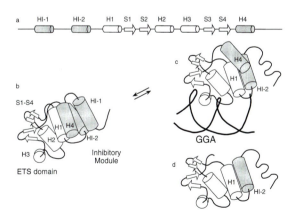

Figure 5. Structural and mechanistic model of Ets-1 autoinhibition. (*a*) Secondary structure of the repressed polypeptide, ΔN280, as determined by NMR spectroscopy (Donaldson et al. 1994; Skalicky et al. 1996). (*b*) Structure of ETS domain and helix H4 as determined by NMR (Donaldson et al. 1996). The structure of the inhibitory module is modeled as a four-helix bundle (Skalicky et al. 1996). (*c*) Disruption of the inhibitory module upon DNA binding (Petersen et al. 1995). (*d*) Disruption of the inhibitory module by deletion helix H4 (Jonsen et al. 1996).

(Fig. 5b). Because the solution structure of ΔN280 is not available, this model is based on four indirect criteria. (1) The NMR-based solution structure of ΔN331 demonstrated that helix H4 packs against H1 of the ETS domain (Donaldson et al. 1996). We have assumed that this same configuration exists in ΔN280. (2) The hydrophobic helical surfaces of HI-1 and HI-2 have the expected limited solvent exposure in this model. (3) Comparative NMR studies of ΔN331 and ΔN280 demonstrated that helices H1 and H4 were impacted by the presence of HI-1 and HI-2. Specifically, chemical shift changes were detected in H4 and H1 but not other regions of the ETS domain (Skalicky et al. 1996). (4) Partial proteolysis experiments implicated a structural coupling between H4 and the amino-terminal inhibitory region. The susceptibility of the amino-terminal inhibitory region to trypsin was enhanced by a deletion of helix H4 (Jonsen et al. 1996). This altered protease sensitivity is consistent with the model of the inhibitory module that connects the carboxy-terminal and the amino-terminal inhibitory regions.

This structural picture of the inhibitory module explains the functional analysis of the inhibitory elements. Both the carboxy-terminal and amino-terminal regions are required for inhibition, suggesting that the two regions function cooperatively. In the structural corollary, the four-helix bundle must be present for the proper packing of any individual inhibitory element. Deletion of any helix would disrupt the packing of the other helices in the inhibitory module and thus activate DNA binding.

Role of Conformational Change in Autoinhibition

Conformational change has a critical role in the mechanism of autoinhibition of Ets-1. Ets-1 polypeptides that display autoinhibition undergo a structural transition upon DNA binding that appears to disrupt the inhibitory module. This conformational change was discovered by two independent approaches (Petersen et al. 1995). Circular dichroism detected a loss of helicity of the repressed ΔN280 fragment upon addition of DNA. It was hypothesized that this structural change occurred in the amino-terminal inhibitory region because ΔN331 showed no comparable change. In the second approach, protease sensitivity at four sites within helix HI-1 was enhanced by DNA binding. These findings indicated that HI-1 was unfolded in the presence of DNA. Thus, autoinhibited Ets-1 polypeptides sample at least two conformational states, one in which a four-helix bundle mediates intramolecular interactions between different parts of the protein (Fig. 5a) and one in which helical ordering is disrupted (Fig. 5b).

This model of a structural switch is consistent with the functional and structural probing of inhibited and activated Ets-1 polypeptides. Specifically, the phenotype of ΔC428 in which helix H4 is deleted supports this model (Fig. 4b) (Jonsen et al. 1996). The high affinity of ΔC428 is accompanied by enhanced protease sensitivity of helix HI-1. The pattern of protease cleavage within HI-1 of ΔC428 was identical to the cleavage observed with inhibited forms of Ets-1 in the presence of DNA. Inhibited forms of Ets-1 (e.g., full-length or ΔN280) transiently sample an alternate conformation during DNA binding (Fig. 5c), whereas the activated species ΔC428 constitutively displays this conformation (Fig. 5d). These structural data and experimental observations suggest that the energetic basis for the lower affinity of Ets-1 is the coupling of this conformational change with DNA binding. Thus, autoinhibition is strongly dependent on the conformational flexibility of Ets-1.

Coupling the ETS Domain to the Inhibitory Module: A Role for Helix H1

We propose that helix H1 provides the link between DNA binding and the function of the inhibitory module. Helix H1 is uniquely positioned, being a part of the proposed four-helix bundle as well as being an integral element of the ETS domain. This hypothesis provides a mechanistic model that explains the role of the conformational flexibility. A complete description of the model requires a more detailed presentation of the mode of DNA binding by the ETS domain.

Helix H1 has a direct role in DNA binding that complements other DNA contacts made by the ETS domain (Fig. 6). Like other HTH proteins, the ETS domain contacts DNA directly in the major groove by hydrogen bonding to the functional groups of the edges of the base pairs. In the case of the ETS domain, helix H3 of the HTH makes these contacts to the conserved sequences of the binding site, 5′-GGA-3′. Phosphate contacts are another conserved feature of the ETS domain-DNA interface. Four specific phosphates were identified by ethylation interference as positions of close contact for four different ets-encoded proteins and are expected to be critical contacts for all ets-encoded proteins (Fig. 6b,c) (Nye et al. 1992; Gunther and Graves 1994; Graves et al. 1996). Indeed, each of these phosphates is in contact with protein in the two high-resolution crystallographic studies of ETS domain-DNA interactions (Kodandapani et al. 1996; Batchelor et al. 1998). One set of phosphate contacts lies 5′ to the GGA motif and involves ionic interactions with basic residues within the β-sheet. On the lower strand and 3′ to the GGA motif, helix H1 makes one of the two conserved phosphate contacts. The contact is mediated by a hydrogen bond between the amide of the peptide backbone and a phosphate oxygen. This constellation of base pair and phosphate contacts generates a high-affinity DNA-protein interaction.

We propose that the inhibitory module allosterically inhibits the function of helix H1 in DNA binding. In this model, interaction between helix H1 and DNA is not possible when the helix is packed in the inhibitory module; thus, conformational change is necessary to facilitate helix H1-phosphate contact. Two lines of evidence suggest that the helix H1 contact is highly positional, which would make it susceptible to such allosteric effects of the inhibitory module. First, the structural data suggest that the dipole moment of the helix H1 drives this interaction (Fig. 7). The dipole of an α-helix generates a positive charge distribution at the amino terminus that facilitates

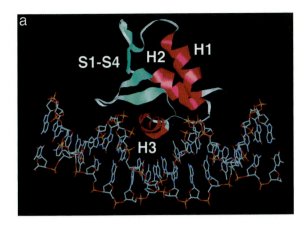

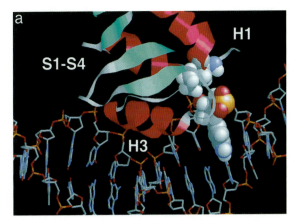

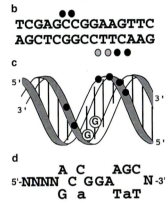

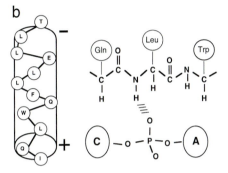

Figure 6. DNA binding by the Ets-1 ETS domain. (*a*) Structure of the ETS domain of Ets-1 in complex with DNA as determined by NMR spectroscopy (Werner et al. 1997). Model was derived from Brookhaven coordinates, model 8 for entry 2STT, and displayed using RasMol v. 2.6. (*b*) DNA sequence from structure in *a*. (*Circles*) Positions of ethylation interference for Ets-1. Only four of the positions (*closed circles*) appear to be invariant among all *ets*-encoded proteins. (*c*) Summary of alkylation interference data for Ets-1 (Nye et al. 1992). The DNA helix is rotated 90° around the helical axis with respect to DNA in *a*. (G) Major interfering N7 methyl guanines; (*circles*) interfering ethylated phosphates. (*d*) Selected consensus (*top strand*) for ΔN331 (Nye et al. 1992). Lowercase letters show less strongly preferred sequences. N denotes positions that showed no sequence preference; however, DNA duplex in these regions was necessary for maximal binding activity (Gillespie 1998).

Figure 7. DNA contact by helix H1 of the Ets-1 ETS domain. (*a*) Proposed hydrogen bond between backbone amide of leucine of helix H1 and phosphate oxygen is highlighted by space-filling modeling of relevant residues. Molecular model was generated as described in Fig. 6a. (*b*) Detail of helix H1 with dipole moment and the proposed hydrogen bond.

the formation of a hydrogen bond by the usually inert, backbone amide. In addition, the amino terminus of an α-helix, with its positive charge, can neutralize the highly negative charge of the phosphate backbone. Thus, regions that are distant from the site of DNA interaction could affect the strength of this contact by affecting the position of the helix or the strength of the dipole. Second, there is a preferred DNA sequence at the position of helix H1 contact as determined by selected binding site consensus experiments (Fig. 6d). The data suggest that a purine-pyrimidine dinucleotide is highly preferred at this position in high-affinity sites. This sequence preference is observed despite the lack of any direct protein contact to the base pairs at this position. These data suggest that the preferred sequence dictates a configuration of the DNA backbone, and then protein functional groups recognize

the DNA conformation. Thus, the protein and DNA are precisely positioned to facilitate the dipole-driven DNA contact.

This mechanistic model of Ets-1 autoinhibition is based on the mode of DNA binding observed with ETS domain-DNA interactions involving fragments for which autoinhibition is not reported: (1) a high-affinity fragment of Ets-1 (Werner et al. 1997), (2) a fragment of PU.1 which shows no autoinhibition (Kodandapani et al. 1996), and (3) GABPα whose possible autoinhibition has not been investigated (Batchelor et al. 1998). However, our model requires that a conformational change allows helix H1 to make these same DNA contacts in low-affinity Ets-1 fragments. Indeed, multiple approaches demonstrate that the high- and low-affinity forms of Ets-1 make the same DNA contacts. The DNA interface for Ets-1 has been characterized by DNase I footprinting, alkylation interference, and consensus sequence selection (Fisher et al. 1991; Nye et al. 1992; Woods et al. 1992). All assays were performed for Ets-1 polypeptides of both high and low affinities. In every case, these qualitative approaches mapped identical contacts on the DNA. Most important to our discussion, the helix H1-DNA contact was directly implicated by ethylation interference experiments performed on both a high-affinity fragment ΔN322 and the low-affinity native Ets-1 (Nye et al. 1992). These biochemical data complement the structural and genetic approaches and support our proposal that the helix H1-DNA contact is the key sensor of the inhibitory module.

DISCUSSION

Structural and Mechanistic Model of Ets-1 Autoinhibition

Structural and biochemical data led to a detailed model of Ets-1 autoinhibition. Helix H1 contact with DNA is proposed to be highly dependent on the conformation of both the DNA and protein. The helix H1-DNA interaction is incompatible with the role of helix H1 in the helical packing of the inhibitory module. Thus, DNA binding is accompanied by a dramatic, yet reversible, conformational change. This coupling of conformational change and DNA binding leads to low affinity. The energetic expense could be in the unfolding reaction, the refolding reaction, or both. The conformational flexibility of the inhibitory module constitutes a structural switch.

Regulation of the Structural Switch in Ets-1: Phosphorylation and Partners

The structural switch of Ets-1 could modulate DNA binding in either a negative or a positive direction. Experiments in progress in our laboratory are designed to investigate both of these possibilities. The DNA-binding activity of Ets-1 is repressed by Ca^{++}-signaling-dependent phosphorylation (Rabault and Ghysdael 1994; D. Cowley, unpubl.). The sites of phosphorylation map close to the inhibitory module, and this regulation could function through the autoinhibition pathway. From our detailed structural model, we predict that Ets-1 phosphorylated at these sites would have a more stable inhibitory module. We are pursuing both genetic and biochemical analyses to test this hypothesis.

We predict that the inhibitory module of Ets-1 is counteracted by interactions with other proteins; the Ets-1 structural switch provides a mechanism by which partner proteins can enhance DNA-binding activity. The conformational change induced by DNA binding could be stabilized by an interacting protein. Thus, the high-affinity binding potential of the ETS domain could be realized within the ternary complex. We are investigating the synergy between Ets-1 and the transcription factor CBFα2/AML-1 that functions in both viral and cellular transcriptional regulation in T cells (Wotton et al. 1994; Giese et al. 1995; Sun et al. 1995). In quantitative experiments, the presence of CBFα2 causes a tenfold enhancement of Ets-1 DNA-binding activity (T. Gu et al., unpubl.). Preliminary mapping studies indicate that the sequences necessary for this enhancement are included within the low-affinity fragment, ΔN280. Furthermore, the high-affinity species, ΔN331, shows no cooperative DNA binding with CBFα2. These findings are consistent with the proposal that the inhibitory module is involved in the modulation of Ets-1 DNA binding by CBFα2. It is possible that helix HI-1 is retained in an unfolded state in the presence of CBFα2. Alternatively, direct protein interactions with CBFα2 could refold helix HI-1 yet prevent the helical packing of a stable inhibitory module. Genetic and biochemical approaches are being used to test the state of the structural switch in this cooperative partnership.

Autoinhibition in Other *ets*-encoded Proteins: Helix H1 as the Keystone

Other members of the *ets* family of transcription factors also display autoinhibition. The phenomenon is expected to be almost identical in Ets-2 because there is sequence similarity between Ets-2 and Ets-1 in all three inhibitory helices. Furthermore, the approach shown in Figure 3 has been used to detect autoinhibition of DNA binding in Ets-2 (D. Cowley, unpubl.). ERM is reported to control both *trans*-activation and DNA binding by an autoregulatory mechanism (Laget et al. 1996). Autoinhibition of DNA binding is reported for the ternary complex factors (TCFs: Elk-1 and SAP-1a as well as the related *ets*-encoded protein NET) which interact with SRF. The inhibitory region overlaps sequences necessary for interactions with SRF, and this interaction stabilizes the DNA-binding activity of Elk-1 and Sap-1a (Dalton and Treisman 1992; Ling et al. 1997). None of these cases have been investigated to the same structural level that we have described for Ets-1. Furthermore, none of the inhibitory sequences of ERM or the TCFs show similarity to the Ets-1 inhibitory elements. Nevertheless, we speculate that the helix H1-DNA contact, which is predicted to be a conserved feature of the family, may be a common target for each *ets*-encoded protein that autoregulates DNA binding.

The role of helix H1 as a rheostat of ETS domain function is supported by biochemical and structural analyses of the *ets*-encoded protein GABPα. GABP binds DNA as a heterotetramer with two α subunits that contain ETS domains (Thompson et al. 1991; Virbasius et al. 1993). The two β subunits enhance the DNA-binding activity of the ETS domain by two mechanisms. On one hand, the β-subunit provides a homotypic dimerization interface, thus contributing to the stability of the heterotetramer. More relevant to our discussion, the β-subunit interacts directly with the α-subunit and enhances the DNA-binding activity of the ETS domain.

Crystallographic studies of a GABP ternary complex shows that GABPβ, which itself does not bind DNA, interacts directly with the helix H1 of the ETS domain (Batchelor et al. 1998). Of particular interest is the interaction at the amino terminus of helix H1 where a lysine residue from the β-subunit hydrogen bonds with the glutamine just amino-terminal to helix H1 in the α-subunit (see Fig. 7b for homologous glutamine in Ets-1). These interactions could stabilize a dipole-driven DNA contact of helix H1 by buttressing the amino terminus. More experiments are necessary to test whether this stimulation of GABPα DNA binding is counteracting inhibitory intramolecular interactions. Nevertheless, this powerful picture of a ternary complex strongly suggests that helix H1 can be a target of regulation (Graves 1998).

Biological Implications of Autoinhibition

The layering of positive signaling over repression is an efficient way to tightly control activity of a transcription factor. For example, there are many positive-regulatory mechanisms that impinge on transcription factor function (e.g., nuclear transport, protein interactions, and post-translational modifications). In each case, a set of intramolecular interactions could inhibit the activity until the positive signal is available. Indeed, a variety of control pathways that regulate transcription factors interface with autoinhibition. We have discussed our investigations of Ets-1 regulation and the possible case of GABP. Another case is the regulation of heat shock factor, HSF, by the heat shock response. Both activation (Newton et al. 1996; Kline and Morimoto 1997) and the subunit association (Orosz et al. 1996) of HSF are negatively regulated by intramolecular mechanisms. In each case, heat treatment, which mimics heat shock, releases the autoinhibition. There are many examples of protein partners counteracting autoinhibition. Specifically, DNA binding of Pip requires PU.1 (Brass et al. 1996). Likewise, Hoxb is activated by Pbx (Chan and Mann 1996; Chan et al. 1996), CBFα2 by CBFβ (Kanno et al. 1998; T. Gu et al., unpubl.), Elk-1 by SRF (Ling et al. 1997), and SAP-1a by SRF (Dalton and Treisman 1992). In another interesting case, the σ-subunit of *Escherichia coli* RNA polymerase only binds DNA in association with core polymerase. However, truncation of the amino-terminal region derepresses the DNA-binding activity of the carboxy-terminal region of σ (Dombroski et al. 1993). Proteolytic cleavage of inhibitory sequences is a simple route to relieve autoinhibition. For example, the DNA-binding activity of the NF-κB protein p105 is inhibited by amino-terminal ankyrin repeats. The inhibitory amino terminus is proteolytically processed to generate the p50 subunit of NF-κB which is active in DNA binding (Liou et al. 1992). Post-translational modifications also can counteract autoinhibitory sequences. For example, phosphorylation and, possibly, acetylation reverse the autoinhibition of p53 DNA binding (Ko and Prives 1996; Gu and Roeder 1997). Thus, autoinhibition is an integral part of many biological regulatory mechanisms.

Loss of autoinhibition may have a role in the dysregulation of transcription factors. Many chromosome translocations that cause human cancers create fusion proteins in which one or both components are transcription factors (Cleary 1991; Rabbitts 1994). Truncated or chimeric proteins also are carried by acutely transforming retroviruses. The modularity of transcription factors is critical to the function of these altered genes as oncogenes. Because individual modules can function as autonomous domains, the activities of most modules are retained in the altered proteins. The presence of autoinhibitory sequences adds an interesting twist to these phenomena. If regions that encode inhibitory elements are disrupted, the autonomous domain may acquire enhanced or dysregulated activity. In fact, the *ets-1* gene was originally discovered as part of the oncogene of the avian retrovirus, E26. This oncogenic version of Ets-1 has a carboxy-terminal alteration that disrupts the inhibitory module (Lim et al. 1992).

Structural Implications of Autoinhibition

Our model of autoinhibition implicates basic biophysical features of proteins, including intramolecular interactions, conformational change, and allosteric effects. These general events are expected to be involved in all other cases of autoinhibition, although the structural detail of each regulated protein is expected to differ. For example, the conformational change could involve only subtle rearrangements of a few residues or a single structural element. In other cases, as occurs in Ets-1, unfolding of a structural element could be involved. The resulting flexible region could transduce additional changes to other parts of the protein. Another interesting case would include a transition from an unfolded state to a defined structure. This type of transition has been reported for several activation domains and is required for functionality (Uesugi et al. 1997). We propose that the maintenance of an unstructured state is a form of autoinhibition. In this case, an unstructured activation domain autoregulates its own activity. Despite this diversity of possible conformational changes, the common theme among these structural switches clearly defines a single phenomenon.

We predict that structural switches related to autoinhibition help orchestrate the assembly of DNA-protein and multiprotein complexes that have vital roles in transcriptional regulation. Many transcription factors are proposed to undergo conformational change and display allosteric effects that can mediate autoinhibition, as discussed here and reviewed in other reports (Lefstin and Yamamoto 1998). Not surprisingly, other classes of proteins also utilize these regulatory strategies. Specifically, there are many well-studied descriptions of autoinhibitory elements in a variety of kinases and phosphatases (Soderling 1990; Kemp and Pearson 1991). Crystal structures of Src (Xu et al. 1997), Ca++-calmodulin-dependent protein kinase I (Goldberg et al. 1996), and twitchin kinase (Hu et al. 1994) precisely illustrate the inhibitory intramolecular interactions that autoregulate the enzymatic activity of these proteins. As more components of the transcriptional machinery become available for structural studies, the detailed architecture of inhibitory intramolecular interactions should emerge to illustrate the full potential of autoinhibition as a transcriptional regulatory strategy.

ACKNOWLEDGMENTS

Our investigation of autoinhibition of Ets-1 was done as a collaboration with the laboratory of Dr. Lawrence McIntosh at the University of British Columbia. NMR spectroscopy experiments performed on Ets-1 by the McIntosh laboratory provided both secondary and tertiary structures that were essential to the modeling presented in this report. Drs. Logan Donaldson and Jack Skalicky of the McIntosh laboratory participated in this collaboration. We acknowledge the critical discussions and sharing of experimental findings that contributed to the development of the model presented here. Work in the Graves laboratory was funded by the National Institutes of Health (NIH): GM-38663 to B.J.G., NIH training grant HD-07491 to J.M.P. and D.O.C., as well as NIH training grant CA-09602 to T.L.G.

REFERENCES

Ansieau S., Kowenz-Leutz E., Dechend R., and Leutz A. 1997. B-Myb, a repressed *trans*-activating protein. *J. Mol. Med.* **75:** 815.

Batchelor A., Piper D., Charles de la Brousse F., Mcknight S., and Wolberger C. 1998. The structure of GABPα/β: An ETS domain-ankryin repeat heterodimer bound to DNA. *Science* **279:** 1037.

Bories J.-C., Willerford D.M., Grevin D., Davidson L., Camus A., Martin P., Stehelin D., and Alt F.W. 1995. Increased T-cell apoptosis and terminal B-cell differentiation induced by inactivation of the *Ets-1* proto-oncogene. *Nature* **377:** 635.

Brass A.L., Kehrli E., Eisenbeis C.F., Storb U., and Singh H. 1996. Pip, a lymphoid-restricted IRF, contains a regulatory domain that is important for autoinhibition and ternary complex formation with the *ets* factor PU.1. *Genes Dev.* **10:** 2335.

Brennan R.G. 1993. The winged-helix DNA-binding motif: Another helix-turn-helix takeoff. *Cell* **74:** 773.

Burley S.K. 1994. DNA-binding motifs from eukaryotic transcription factors. *Curr. Opin. Struct. Biol.* **4:** 3.

Chan S.K. and Mann R.S. 1996. A structural model for a homeotic protein-extradenticle-DNA complex accounts for the choice of HOX protein in the heterodimer. *Proc. Natl. Acad. Sci.* **93:** 5223.

Chan S.K., Popperl H., Krumlauf R., and Mann R.S. 1996. An extradenticle-induced conformational change in a HOX protein overcomes an inhibitory function of the conserved hexapeptide motif. *EMBO J.* **15:** 2476.

Cleary M. 1991. Oncogenic conversion of transcription factors by chromosomal translocations. *Cell* **66:** 619.

Dalton S. and Treisman R. 1992. Characterization of SAP-1, a protein recruited by serum response factor to the c-*fos* serum response element. *Cell* **68:** 597.

Dombroski A.J., Walter W.A., and Gross C.A. 1993. Amino-terminal amino acids modulate sigma-factor DNA-binding activity. *Genes Dev.* **7:** 2446.

Donaldson L.W., Petersen J.M., Graves B.J., and McIntosh L.P. 1994. Secondary structure of the ETS domain places murine Ets-1 in the superfamily of winged helix-turn-helix DNA-binding proteins. *Biochemistry* **33:** 13509.

———. 1996. Solution structure of the ETS domain from murine Ets-1: A winged helix-turn-helix DNA binding motif. *EMBO J.* **15:** 125.

Fisher R.J., Mavrothalassitis G., Kondoh A., and Papas T.S. 1991. High-affinity DNA-protein interactions of the cellular ETS1 protein: The determination of the ETS binding motif. *Oncogene* **6:** 2249.

Fisher R.J., Favash M., Casas-Finet J., Erickson J.W., Kondoh A., Bladen S.V., Fisher C., Watson D.K., and Papas T.S. 1994. Real-time DNA binding measurements of the ETS1 recombinant oncoproteins reveal significant kinetic differences between the p42 and p51 isoforms. *Protein Sci.* **3:** 257.

Giese K., Kingsley C., Kirshner J.R., and Grosschedl R. 1995. Assembly and function of a TCRα enhancer complex is dependent on LEF-1 induced DNA bending and multiple protein-protein interactions. *Genes Dev.* **9:** 995.

Gillespie M.E. 1998. "A structural and genetic investigation of DNA binding by the murine Ets-1 ETS domain, a winged HTH transcription factor." Ph.D. thesis, University of Utah, Salt Lake City.

Giovane A., Pintzas A., Maira S., Sobieszczuk P., and Wasylyk B. 1994. Net, a new *ets* transcription factor that is activated by Ras. *Genes Dev.* **8:** 1502.

Goldberg J., Nairn A.C., and Kuriyan J. 1996. Structural basis for the autoinhibition of calcium/calmodulin-dependent protein kinase I. *Cell* **84:** 875.

Graves B.J. 1998. Inner workings of a transcription factor partnership. *Science* **279:** 1000.

Graves B.J. and Petersen J.M. 1998. Specificity within the *ets* family of transcription factors. *Adv. Cancer Res.* **75:** 1.

Graves B.J., Gillespie M.E., and McIntosh L.P. 1996. DNA binding by the ETS domain (letter). *Nature* **384:** 322.

Gu W. and Roeder R.G. 1997. Activation of p53 sequence-specific DNA binding by acetylation of the p53 C-terminal domain. *Cell* **90:** 595.

Gunther C.V. and Graves B.J. 1994. Identification of ETS domain proteins in murine T lymphocytes that interact with the Moloney murine leukemia virus enhancer. *Mol. Cell. Biol.* **14:** 7569.

Hagman J. and Grosschedl R. 1992. An inhibitory carboxyl terminal domain in Ets-1 and Ets-2 mediates differential binding of ETS family factors to promoter sequences of the *mb-1* gene. *Proc. Natl. Acad. Sci.* **89:** 8889.

Hansen S., Lane D.P., and Midgley C.A. 1998. The N terminus of the murine p53 tumour suppressor is an independent regulatory domain affecting activation and thermostability. *J. Mol. Biol.* **275:** 575.

Hu S.H., Parker M.W., Lei J.Y., Wilce M.C., Benian G.M., and Kemp B.E. 1994. Insights into autoregulation from the crystal structure of twitchin kinase. *Nature* **369:** 581.

Hupp T.R., Meek D.W., Midgley C.A., and Lane D.P. 1992. Regulation of the specific DNA binding function of p53. *Cell* **71:** 875.

Jonsen M.D., Petersen J.M., Xu Q., and Graves B.J. 1996. Characterization of the cooperative function of inhibitory sequences of Ets-1. *Mol. Cell. Biol.* **16:** 2065.

Kanno T., Kanno Y., Chen L.F., Ogawa E., Kim W.Y., and Ito Y. 1998. Intrinsic transcriptional activation-inhibition domains of the polyomavirus enhancer binding protein 2/core binding factor alpha subunit revealed in the presence of the beta subunit. *Mol. Cell. Biol.* **18:** 2444.

Karim F.D., Urness L.D., Thummel C.S., Klemsz M.J., McKercher S.R., Celeda A., Van Beveren C., Maki R.A., Gunther C.V., Nye J.A., and Graves B.J. 1990. The ETS-domain: A new DNA binding motif that recognizes a purine-rich core DNA sequence. *Genes Dev.* **4:** 1451.

Kemp B.E. and Pearson R.B. 1991. Intrasteric regulation of protein kinases and phosphatases. *Biochim. Biophys. Acta* **1094:** 67.

Kline M.P. and Morimoto R.I. 1997. Repression of the heat shock factor 1 transcriptional activation domain is modulated by constitutive phosphorylation. *Mol. Cell. Biol.* **17:** 2107.

Ko L.J. and Prives C. 1996. p53: Puzzle and paradigm. *Genes Dev.* **10:** 1054.

Kodandapani R., Pio F., Ni C.-Z., Piccialli G., Klemsz M., McKercher S., Maki R.A., and Ely K.R. 1996. A new pattern for helix-turn-helix recognition revealed by the PU.1 ETS-domain-DNA complex. *Nature* **380:** 456.

Kuddus R. and Schmidt M.C. 1993. Effect of the non-conserved N terminus on the DNA binding activity of the yeast TATA binding protein. *Nucleic Acids Res.* **21:** 1789.

Laget M.-P., Defossez P.-A., Albagli O., Baert J.-L., Dewitte F., Stehelin D., and de Launoit Y. 1996. Two functionally distinct domains responsible for transactivation by the Ets family member ERM. *Oncogene* **12:** 1325.

Lefstin J.A. and Yamamoto K.R. 1998. Allosteric effects of DNA on transcriptional regulators. *Nature* **392:** 885.

Lefstin J.A., Thomas J.R., and Yamamoto K.R. 1994. Influence of a steroid receptor DNA-binding domain on transcriptional regulatory functions. *Genes Dev.* **8:** 2842.

Li X.Y. and Green M.R. 1996. Intramolecular inhibition of activating transcription factor-2 function by its DNA-binding domain. *Genes Dev.* **10:** 517.

Liang H., Mao X., Olejniczak E., Nettesheim D.G., Yu L., Meadows R.P., Thompson C.B., and Fesik S.W. 1994. Solution structure of the ETS domain of Fli-1 when bound to DNA. *Nat. Struct. Biol.* **1:** 871.

Lim F., Kraut N., Frampton J., and Graf T. 1992. DNA binding by c-Ets-1, but not v-Ets, is repressed by an intramolecular mechanism. *EMBO J.* **11:** 643.

Ling Y., Lakey J.H., Roberts C.E., and Sharrocks A.D. 1997. Molecular characterization of the B-box protein-protein interaction motif of the ETS-domain transcription factor Elk-1. *EMBO J.* **16:** 2431.

Liou H.C., Nolan G.P., Ghosh S., Fujita T., and Baltimore D. 1992. The NF-κB p50 precursor, p105, contains an internal IκB-like inhibitor that preferentially inhibits p50. *EMBO J.* **11:** 3003.

Mittal V. and Hernandez N. 1997. Role for the amino-terminal region of human TBP in U6 snRNA transcription. *Science* **275:** 1136.

Muthusamy N., Barton K., and Leiden J.M. 1995. Defective activation and survival of T-cells lacking the Ets-1 transcription factor. *Nature* **377:** 639.

Newton E.M., Knauf U., Green M., and Kingston R.E. 1996. The regulatory domain of human heat shock factor 1 is sufficient to sense heat stress. *Mol. Cell. Biol.* **16:** 839.

Nye J.A., Petersen J.M., Gunther C.V., Jonsen M.D., and Graves B.J. 1992. Interaction of murine Ets-1 with GGA-binding sites establishes the ETS domain as a new DNA-binding motif. *Genes Dev.* **6:** 975.

Orosz A., Wisniewski J., and Wu C. 1996. Regulation of *Drosophila* heat shock factor trimerization: Global sequence requirements and independence of nuclear localization. *Mol. Cell. Biol.* **16:** 7018.

Perez-Martin J. and Johnson A.D. 1998. The C-terminal domain of Sin1 interacts with the SWI-SNF complex in yeast. *Mol. Cell. Biol.* **18:** 4157.

Petersen J. 1996. "Ets-1 DNA binding and autoinhibition." Ph.D. thesis, University of Utah, Salt Lake City.

Petersen J.M., Skalicky J.J., Donaldson L.W., McIntosh L.P., Alber T., and Graves B.J. 1995. Modulation of transcription factor Ets-1 DNA binding: DNA-induced unfolding of an alpha helix. *Science* **269:** 1866.

Price M.A., Rogers A.E., and Treisman R. 1995. Comparative analysis of the ternary complex factors Elk-1, SAP-1a and SAP-2 (ERP/NET). *EMBO J.* **14:** 2589.

Rabault B. and Ghysdael J. 1994. Calcium-induced phosphorylation of ETS1 inhibits its specific DNA binding activity. *J. Biol. Chem.* **269:** 28143.

Rabbitts T.H. 1994. Chromosomal translocations in human cancer. *Nature* **372:** 143.

Schneikert J., Lutz Y., and Wasylyk B. 1992. Two independent activation domains in c-Ets-1 and c-Ets-2 located in non-conserved sequences of the *ets* gene family. *Oncogene* **7:** 249.

Sharrocks A., Brown A., Ling Y., and Yates P. 1997. The ETS domain transcription factor family. *Int. J. Biochem. Cell Biol.* **29:** 1371.

Skalicky J.J., Donaldson L.W., Petersen J.M., Graves B.J., and McIntosh L.P. 1996. Structural coupling of the inhibitory regions flanking the ETS domain of murine Ets-1. *Protein Sci.* **5:** 296.

Soderling T.R. 1990. Protein kinases: Regulation by auto-inhibitory domains. *J. Biol. Chem.* **265:** 1823.

Starr D.B., Matsui W., Thomas J.R., and Yamamoto K.R. 1996. Intracellular receptors use a common mechanism to interpret signaling information at response elements. *Genes Dev.* **10:** 1271.

Sun W., Graves B.J., and Speck N.A. 1995. Transactivation of the Moloney murine leukemia virus and T-cell receptor β-chain enhancers by *cbf* and *ets* requires intact binding sites for both proteins. *J. Virol.* **69:** 4941.

Thompson C.C., Brown T.A., and McKnight S.L. 1991. Convergence of *Ets*- and *Notch*-related structural motifs in a heteromeric DNA binding complex. *Science* **253:** 762.

Uesugi M., Nyanguile O., Lu H., Levine A.J., and Verdine G.L. 1997. Induced alpha helix in the VP16 activation domain upon binding to a human TAF. *Science* **277:** 1310.

Virbasius J.V., Virbasius C.A., and Scarpulla R.C. 1993. Identity of GABP with NRF-2, a multisubunit activator of cytochrome oxidase expression, reveals a cellular role for an ETS domain activator of viral promoters. *Genes Dev.* **7:** 380.

Werner M.H., Clore G.M., Fisher C.L., Fisher R.J., Trinh L., Shiloach J., and Gronenborn A.M. 1997. Correction of the NMR structure of the ETS1/DNA complex. *J. Biomol. NMR* **10:** 317.

Williams S.C., Baer M., Dillner A.J., and Johnson P.F. 1995. CRP(C/EBPβ) contains a bipartite regulatory domain that controls transcriptional activation, DNA binding and cell specificity. *EMBO J.* **14:** 3170.

Woods D.B., Ghysdael J., and Owen M.J. 1992. Identification of nucleotide preferences in DNA sequence recognized specifically by c-Ets-1 protein. *Nucleic Acids Res.* **20:** 699.

Wotton D., Ghysdael J., Wang S., Speck N.A., and Owen M.J. 1994. Cooperative binding of Ets-1 and core binding factor to DNA. *Mol. Cell. Biol.* **14:** 840.

Xu W., Harrison S.C., and Eck M.J. 1997. Three-dimensional structure of the tyrosine kinase c-Src. *Nature* **385:** 595.

Yang C., Shapiro L.H., Rivera M., Kumar A., and Brindle P.K. 1998. A role for CREB binding protein and p300 transcriptional coactivators in Ets-1 transactivation functions. *Mol. Cell. Biol.* **18:** 2218.

Mechanisms of Activation by CREB and CREM: Phosphorylation, CBP, and a Novel Coactivator, ACT

G.M. Fimia, D. De Cesare, and P. Sassone-Corsi

Institut de Génétique et de Biologie Moléculaire et Cellulaire, CNRS, INSERM,
Université Louis Pasteur, Illkirch, Strasbourg, 67404 France

The regulation of gene expression by specific signal transduction pathways is tightly connected to the cell phenotype, and conversely, the response elicited by a given transduction pathway varies depending on the cell type. Several molecules implicated in intracellular signaling are encoded by oncogenes, directly linking their possible aberrant expression to cellular transformation or altered proliferation. A complete analysis of these processes will help to unravel the profound changes that cause cancer and by the same token understand the physiology of normal growth. A fundamental stride has been the discovery that many transcription factors constitute final targets of specific transduction pathways (Hunter and Karin 1992). It is evident that because of the large array of kinases and transcription factors involved, and the multitude of their interactions, the complexity and versatility of the nuclear response results enormously expanded.

COUPLING SIGNALING TO TRANSCRIPTION

It is reasonable to think that a panoply of nuclear proteins involved in transcription, RNA processing, and chromatin remodeling is under direct control of intracellular responses to external signals. One of the best examples of transcription factor controlled by transduction pathways is AP-1, which is composed by the Fos and Jun oncoproteins (Vogt and Bos 1989). AP-1 activity may be increased by inducing c-*fos* gene transcription, a process mediated by the ERK-1 and ERK-2 mitogen-activated protein (MAP) kinases, which directly phosphorylate the transcription factor Elk-1/TCF responsible for stimulation of c-*fos* expression through the serum response element in the promoter (Treisman 1996). Alternatively, AP-1 activity may be enhanced by direct phosphorylation of Jun by a different type of MAPK, the stress-activated protein kinases (JNK/SAPK) (Davis 1994). Transcription factor ATF-2, a dimerization partner of Jun, is also a target of the JNK kinase (Hazzalin et al. 1996). Interestingly, ATF-2 was first cloned as member of the ATF/CREB family, a group of transcription factors binding to cAMP-responsive elements (CREs) (Hai et al. 1989).

The ATF/CREB family includes several members: The products of the CREB (CRE-binding protein), CREM (CRE modulator), and ATF-1 (activating transcription factor 1) genes have been shown to also be direct targets of intracellular signaling as they are phosphorylated by the cAMP-dependent protein kinase A (PKA) (Sassone-Corsi 1995). In addition, cross-talks between cAMP and mitogenic signaling pathways have been established (Ginty et al. 1994), which reinforces the notion of converging signaling within the PKA and PKC pathways in the cytoplasm (Nishizuka 1986; Cambier et al. 1987; Yoshimasa et al. 1987; Frodin et al. 1994) and in the nucleus (Hai and Curran 1991; Masquilier and Sassone-Corsi 1992).

Intracellular levels of cAMP are regulated primarily by adenylyl cyclase. This enzyme is in turn modulated by various extracellular stimuli mediated by receptors and their interaction with G proteins (McKnight et al. 1988). The binding of a specific ligand to a receptor results in the activation or inhibition of the cAMP-dependent pathway, ultimately affecting the transcriptional regulation of various genes through distinct promoter responsive sites. Increased cAMP levels directly affect the function of the tetrameric PKA complex. Binding of cAMP to two PKA regulatory subunits releases the catalytic subunits. These are translocated from cytoplasmic and Golgi complex anchoring sites and phosphorylate a number of cytoplasmic and nuclear proteins on serines in the context X-Arg-Arg-X-*Ser*-X (McKnight et al. 1988; Roesler et al. 1988). A number of isoforms for both the regulatory and catalytic subunits have been identified, suggesting a further level of complexity in this response (McKnight et al. 1988).

The analysis of regulatory sequences of several genes allowed the identification of promoter elements that mediate the transcriptional response to increased levels of intracellular cAMP (Lalli and Sassone-Corsi 1994). A number of sequences have been identified of which the best characterized is the CRE. The consensus CRE site is constituted by the 8-bp palindromic sequence TGACGTCA (Sassone-Corsi 1988; Ziff 1990). Several genes that are regulated by a variety of endocrinological stimuli contain similar sequences in their promoter regions, although at different positions.

CREB, CREM, and ATF-1 proteins belong to the bZip (basic domain-leucine zipper) transcription factor class (Busch and Sassone-Corsi 1990) and act as dimers. They are also able to heterodimerize with each other but only in certain combinations. Indeed, a "dimerization code" exists that seems to be a property of the leucine zipper structure of each factor. CRE-binding proteins may act as both activators and repressors of transcription. The activators mediate transcriptional induction upon their phosphorylation by PKA (see below; Gonzalez and Montminy

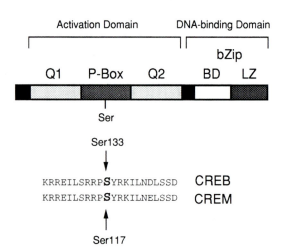

Figure 1. Structure of activators CREB and CREM. The two glutamine-rich domains (Q1 and Q2) and the bZip region (BD and LZ) are indicated in addition to the P box. The amino acid sequence of the area of the CREM and CREB P box containing Ser-133 and Ser-117 is shown. Phosphorylation at this serine turns CREB and CREM into activators through interaction with the coactivator CBP. Ser-133 and Ser-117 have been demonstrated to be phosphorylated by various kinases, including PKA and RSK-2.

1989; Rehfuss et al. 1991; de Groot et al. 1993b; Sassone-Corsi 1995) and the presence of an activation domain (Fig. 1). Their expression is constitutive and widely distributed in various tissues in a housekeeping fashion. Among the repressors, the cAMP-inducible ICER (inducible cAMP early repressor) product deserves special mention. It is generated from a cAMP-inducible alternative promoter of the CREM gene (Molina et al. 1993; Stehle et al. 1993). Thus, ICER is an early response CRE-binding factor and is involved in the dynamics of cAMP-responsive transcription (Lamas and Sassone-Corsi 1997). Here, we focus primarily on the molecular mechanims by which the activators CREB and CREM elicit their regulatory function.

THE DOGMA

Phosphorylation: A Prerequisite for Activation

It is classically thought that transcriptional activation by CREB and CREM requires a phosphorylation event at a serine residue located in the activation domain of the protein (Fig. 1). This serine (Ser-133 in CREB; Ser-117 in CREM) is the phosphoacceptor site for PKA, and, as such, it constitutes the direct link between signaling and activation of gene expression. Importantly, the same serine can also be phosphorylated by other kinases, and thus it represents a site where various signaling pathways converge and cross-talk (Sheng et al. 1990; Dash et al. 1991; Ginty et al. 1994; Sassone-Corsi 1995; De Cesare et al. 1998).

An important example of signaling cross-talk in the nucleus involves the pathway coupled to the nerve growth factor (NGF) receptor, Trk, which results in the activation of several kinases. Trk is a receptor tyrosine kinase which, once activated, stimulates the activity of the small GTP-binding protein Ras (Gomez and Cohen 1991). Activation of Ras triggers the MAPK pathway, which includes the MAP kinase kinase (MEK) and the ribosomal S6 kinase $pp90^{rsk}$ (Cobb and Goldsmith 1995). Interestingly, constitutively activated expression of MAPK and MEK is sufficient to induce neurite outgrowth in PC12 cells (Cowley et al. 1994; Fukuda et al. 1995), indicating a direct role of this pathway in eliciting the changes in gene expression required for the neuronal differentiation program. Although MAPK and MEK have not been shown to directly phosphorylate CREB, the use of cells expressing a dominant-interfering Ras mutant has revealed the involvement of this pathway for CREB phosphorylation upon NGF induction (Ginty et al. 1994). Indeed, the involvement of a CREB-kinase with characteristics similar to those of $pp90^{rsk}$ has been proposed. $pp90^{rsk}$ is likely to be responsible for CREB phosphorylation in human melanocytes (Böhm et al. 1995), whereas a distinct member of the RSK family, $p70^{s6k}$, also possesses CREB phosphorylation activity (de Groot et al. 1994). Recently, we have used cells originated from patients with the Coffin-Lowry syndrome which carry mutations in the gene encoding the RSK-2 protein, one of the three isoforms of $pp90^{rsk}$ (Trivier et al. 1996). We have demonstrated that RSK-2 is responsible for CREB phosphorylation in response to epidermal growth factor (EGF) and for the consequent transcriptional induction of c-*fos* (De Cesare et al. 1998). Finally, CREB has been shown to be phosphorylated upon activation of the stress pathway involving the p38/MAPKAP-2 kinases (Tan et al. 1996). Thus, various signaling pathways may converge to modulate gene expression via the same transcriptional regulator, CREB. The complexity of the signaling pathways controlling transcription factors is a demonstration of the pleiotropic functions that these molecules have in the regulation of physiology and metabolism.

The Activation Domain

The structure of the activation domain in CREB and CREM is basically identical (Fig. 1). Ser-133 and Ser-117 are located in a domain identified as the P box (phosphorylation box), which contains other phosphoacceptor sites for various kinases (de Groot et al. 1993a,b). The role of phosphorylation at these additional sites is not fully understood and appears to be secondary to Ser-133 and Ser-117. Two domains identified as Q1 and Q2 flank the P box (Fig. 1); they contain about three times more glutamine residues than the remainder of the protein. Glutamine-rich domains have been characterized in other factors, such as AP-2 and Sp1 (Williams et al. 1988; Courey and Tjian 1989), where they function as transcriptional activation domains. The current notion is that they constitute surfaces of the protein that can interact with other components of the transcriptional machinery, such as RNA polymerase II cofactors. The Q2 domain appears to make a more significant contribution to the *trans*-activation function than Q1. This is demonstrated by the properties of the two naturally occurring CREM isoforms CREMτ1 and CREMτ2 (Laoide et al. 1993) and artificially generated deletion mutants of CREB (Brindle et al. 1993). CREMτ1 and τ2 singly incorporate the Q1 and Q2 domains, respectively, CREMτ2 being a stronger transcriptional activator upon phosphorylation at Ser-117 (Laoide et al. 1993). In agreement with these results on CREM, deletion of the Q2 region in CREB dramatically abolishes activation function (Brindle et al. 1993). Furthermore, ATF-1 lacks a counterpart of the Q1 domain and still functions as an efficient transcription activator (Hai et al. 1989). Thus, the P box and Q2 are sufficient to mediate cAMP-induced transcription.

It is apparent that the activation domain is inherently a modular structure. Indeed, each component is encoded by an individual exon, so that differential splicing as it occurs in CREM results in the generation of factors with different activating properties (Laoide et al. 1993). Interestingly, the Q2 domain fused to the heterologous GAL4 DNA-binding domain still retains a noninducible activation function (Brindle et al. 1993). In addition, the P box is able to confer PKA inducibility on a heterologous acidic activation domain (e.g., GAL4) both in *trans* and in *cis* (Brindle et al. 1993). Thus, theoretically, the P box could be involved not only in the regulation of the adjacent Q domains, but also in controlling the activation function of other factors bound to separate promoter elements.

Role of CBP in Transcriptional Activation

An important step toward the understanding of the CREB transcriptional mechanism of activation has come with the identification of a 265K, 2441-amino-acid protein, CBP (CREB-binding protein) which is able to interact specifically with the phosphorylated P-box domain (Chrivia et al. 1993). The CBP sequence reveals two zinc finger domains, a glutamine-rich domain at its carboxyl terminus and a single consensus PKA recognition site. Phosphorylation of Ser-133 promotes binding to CBP and consequently the interaction with TFIIB, a general transcription factor involved in RNA polymerase II activity (Kwok et al. 1994). Thus, CBP appears to function as a link between CREB and the transcription preinitiation complex. This interaction is likely to implicate some additional RNA polymerase II cofactors. Indeed, the identification of different proteins binding to the CREB activation domain has partially elucidated the physical interaction with the basal transcriptional apparatus. The Q2 domain constitutively interacts with the TBP-associated factor hTAF130, a subunit of the TFIID complex (Ferreri et al. 1994). Interestingly, CREM isoforms containing only the Q2 or the P-box domain behave as transcriptional repressors (Foulkes et al. 1991; Delmas et al. 1992), suggesting that interaction with both CBP and TAF130 is required for efficient transcriptional activation. To date, no proteins have been identified that interact with the Q1 domain.

Another protein, p300 (Eckner et al. 1994), is closely related to CBP and appears to have similar regulatory functions and, as CBP, appears to be involved in a variety of cellular processes, such as cell growth, differentiation, DNA repair, and apoptosis. At least two different mechanisms are required for CBP and p300 coactivator functions. First, they interact with the basal transcription factors such as TFIIB, TBP, and RNA helicase A (Kwok et al. 1994; Swope et al. 1996; Nakajima et al. 1997). Second, they have either an intrinsic or associated histone acetyltransferase (HAT) activity that is thought to induce chromatin remodeling at the level of regulatory promoter elements (Bannister and Kouzarides 1996; Ogryzko et al. 1996; Yang et al. 1996). Notably, CBP and p300 regulate the acetylation levels not only of histones, but also of activators and basal transcription factors, possibly modulating their function (Gu and Roeder 1997; Imhof et al. 1997).

CBP and p300 appear to constitute the convergence point of various transduction pathways, by integrating multiple signals into modulation of gene expression. They exert this function through a multitude of possible interactions with various nuclear proteins. For example, p300 associates with the adenoviral E1A oncoprotein, possibly playing a part in preventing the cell cycle G_0/G_1 transition. Both CBP and p300 appear to have intrinsic activating properties that are inhibited by E1A (Arany et al. 1995).

CBP and p300 also interact with several different transcription factors such as Jun, Fos, Myb, MyoD, p53, E2F, NF-κB, and nuclear receptors (for review, see Jankneckt and Hunter 1996), demonstrating their general importance in many nuclear regulatory functions other than in CREB-mediated transcription regulation. Moreover, CBP/p300 interact with another class of coactivators whose members, SRC-1/NCoA-1, TIF-2/NCoA-2, p/CIP, and ACTR (for review, see Glass et al. 1997), were initially identified for their ability to bind and stimulate the activity of different nuclear receptors. Interestingly, it has been demonstrated that the interaction of one of these coactivators, p/CIP, with CBP is also required for CBP-mediated transcriptional pathways different from nuclear receptors (Torchia et al. 1997). This indicates that formation of multicoactivator complexes is necessary for an efficient transcriptional initiation. Thus, studies of transcriptional activation by CREB and CREM continue to provide important insights into the function of transcription factors in general.

A BIOLOGICAL SCENARIO

High Levels of CREM in Germ Cells

An important feature of CREB and CREM is their ubiquitous and low-level expression in all tissues (for review, see Sassone-Corsi 1995). There is, however, a notable exception: CREM expression in male germ cells. CREM is the subject of a developmental switch in expression as it is highly abundant in adult testis, whereas it is expressed at very low levels in prepubertal animals (Foulkes et al. 1992; Sassone-Corsi 1997). By a process of alternative splicing of the exons encoding the activation domain, different CREM isoforms are expressed at different times during the differentiation program of the germ cells. The abundant CREM transcript in the adult encodes the activator form exclusively, whereas in prepubertal testis, only the repressor forms are detected at low levels. Thus, the CREM developmental switch also constitutes a reversal of function (Foulkes et al. 1992).

Spermatogenesis is a process occurring in a precise and coordinated manner within the seminiferous tubules. During this entire developmental process, the germ cells are maintained in intimate contact with the somatic Sertoli cells. As the spermatogonia mature, they move from the periphery toward the lumen of the tubule until the mature spermatozoa are conducted from the lumen to the collecting ducts. Complex events of gene regulation occur during this developmental program (Sassone-Corsi 1997).

A remarkable aspect of the CREM developmental switch in germ cells is constituted by its exquisite hormonal regulation. The spermatogenic differentiation program is under the tight control of the hypothalamic-pituitary axis (Jégou et al. 1992). The regulation of CREM function in testis seems to be intricately linked to FSH (follicle-stimulating hormone) signaling both at the level of the control of transcript processing and at the level of protein activity (Foulkes et al. 1993). Injection of FSH leads to a rapid and significant induction of the CREM transcript. The hormonal induction of CREM transcript levels by FSH is not transcriptional. Instead, by a mechanism of alternative polyadenylation, AUUUA destabilizer elements present in the 3´-untranslated region of the gene are excluded, dramatically increasing the stability of the CREM message. CREM is the first example of a gene whose expression is modulated by a pituitary hormone

during spermatogenesis (Foulkes et al. 1993). The implication of these findings is that hormones can regulate gene expression at the level of RNA processing and stability. Importantly, the effect of FSH cannot be direct because germ cells do not have FSH receptors. We have postulated that another hormonal message originating from the Sertoli cells upon FSH stimulation may be mediating CREM activation in germ cells.

CREM: Regulator of Haploid Gene Expression

The first hint as to the role of CREM during spermatogenesis was indicated by its protein expression pattern. In the seminiferous epithelium, CREM transcripts accumulate in spermatocytes and spermatids, but CREM protein is detected only in haploid spermatids (Delmas et al. 1993). The absence of CREM protein in spermatocytes reflects a strict translational control and the multiple levels of gene regulation in testis.

The expression of CREM activator protein in spermatids coincides with the transcriptional activation of several genes containing a CRE motif in their promoter region. These genes mainly encode structural proteins required for spermatozoon differentiation, suggesting a role for CREM in the activation of genes required for the late phase of spermatid differentiation. This observation implies that the transcription of some key structural genes is directly linked to hormonal control and consequently to

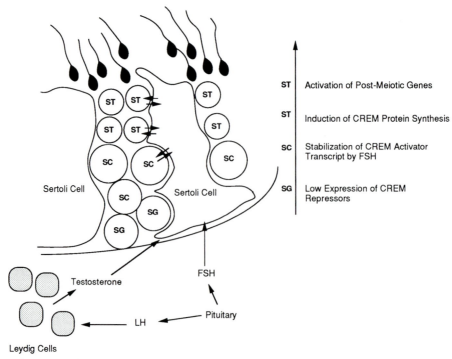

Figure 2. Schematic representation of a section of a seminiferous tubule where the CREM expression pattern is indicated. CREM expression is regulated at multiple levels during spermatogenesis. Premeiotic germ cells (spermatogonia, SG) express a low level of CREM repressor isoforms. During meiotic prophase, the pituitary follicle stimulating hormone (FSH) is responsible for the stabilization of CREM activator transcripts in spermatocytes (SC); CREM protein, on the other hand, is detected only after meiosis in haploid spermatids (ST). Note the strict relationships between the Sertoli and germ cells (arrows). In the haploid spermatids, CREM proteins activate a number of cellular genes expressed specifically during spermatid maturation (Delmas et al. 1993).

the level of cAMP present in seminiferous epithelium. Various genes, such as RT7 (Delmas et al. 1993), transition protein-1 (Kistler et al. 1994), angiotensin converting enzyme (Zhou et al. 1996), and calspermin (Sun et al. 1995), have been shown to be CREM targets by various experimental approaches, including in vitro transcription experiments with germ cell nuclear extracts. These experiments indicated that CREM participates in testis- and developmental-specific regulation of postmeiotic genes during spermiogenesis (Fig. 2).

Genetic evidence has demonstrated that CREM is absolutely required for postmeiotic gene expression. We have generated mutant mice with targeted disruption of the CREM gene by homologous recombination (Nantel et al. 1996). Comparison of the homozygous CREM-deficient mice with their normal littermates revealed a reduction of 20–25% in testis weight and a complete absence of spermatozoa. The homozygous males are sterile. Spermatogenesis is interrupted at the stage of very early spermatids. Neither elongating spermatids, nor spermatozoa, are observed, whereas somatic Sertoli cells appear to be normal. This demonstrates the necessity of a functional CREM transcription factor for male fertility.

In CREM-deficient animals, we have also observed significant numbers of multinucleated giant cells, normally present at very low frequency in wild-type animals. As demonstrated by in situ terminal transferase 3′-end labeling, we have shown that these cellular bodies correspond to apoptotic cells (Nantel et al. 1996), which in the CREM-deficient animals are 10–20-fold more abundant than in normal mice. Thus, these analyses indicate that the lack of CREM causes germ cells to cease differentiation and to undergo apoptosis.

The analysis of the expression of various putative CREM target genes confirms the key part played by this transcription factor in the activation of genes such as protamine 1 and 2, transition proteins 1 and 2, and calspermin (Fig. 2). The lack of expression of these genes may explain the impairment in the structuring of a mature spermatozoon in the CREM-deficient mice (Nantel et al. 1996).

A NOVEL MECHANISM OF ACTIVATION

The Conundrum

The high abundance of the CREM activator in testis and its role in regulating the expression of postmeiotic genes beg the question on the mechanisms by which it exerts its function. Analysis of the phosphorylation state of CREM at various stages of the spermatogenic differentiation cycle revealed a surprising pattern: At the time CREM transcriptionally activates postmeiotic genes, it is unphosphorylated (L. Monaco and N.S. Foulkes, pers. comm.). This notion strongly suggested that the molecular mechanism by which CREM activates transcription in male germ cells must be different from the classical scenario, which involves phosphorylation at Ser-117 and interaction with CBP and TAF130.

ACT, a Tissue-specific Partner of CREM

What could make CREM work as transcriptional activator even when unphosphorylated? We reasoned that a putative partner could modulate CREM function by turning it into an activator. We used a yeast two-hybrid approach to identify factors that are able to interact with and modulate CREM transcriptional activity. To search for a germ-cell-specific partner of CREM, we decided to screen a murine adult testis cDNA expression library. As bait, we utilized the first 229 amino acids of CREM, which includes the two glutamine-rich domains, Q1 and Q2, and the P box. When expressed in yeast, the CREM activation domain, fused to a GAL4 DNA-binding domain (1–147 amino acids), is completely inactive, most likely because of the lack of yeast homologs of CBP and TAF130. A screening of 3×10^6 primary transformants yielded eight different positive clones. One of these, in-

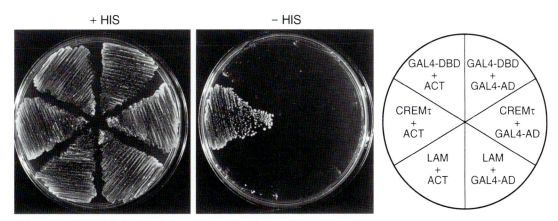

Figure 3. Isolation of a CREM partner in testis by two-hybrid assay. Shown is growth on selective medium of yeast transformants co-expressing CREM (the activator isoform CREMτ) and ACT. Yeast cells (CG1945 strain) were transformed with the indicated plasmids (*right*). CREMτ here indicates the plasmid encoding the CREMτ activation domain fused to the GAL4 DNA-binding domain (DBD); ACT indicates the clone obtained from the screening of the testis library. A lamin expression vector (LAM) and GAL4 DNA-binding domain (GAL4-DBD) and GAL4-AD (activation domain) expression vectors were included as negative controls (see scheme on the right panel). Individual Leu+ Trp+ trasformants were streaked on synthetic medium plates lacking tryptophan, leucine, and histidine (–HIS) and on synthetic medium plates lacking leucine and tryptophan (+HIS).

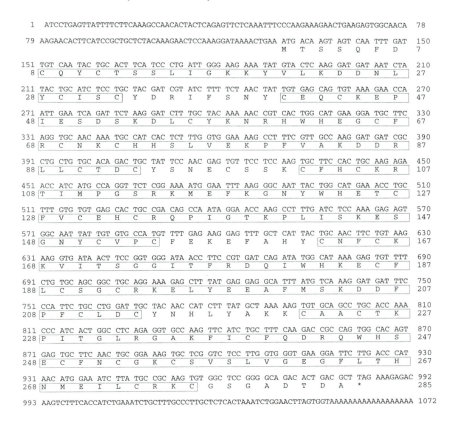

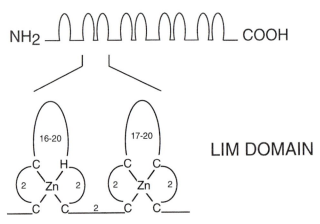

Figure 4. The ACT sequence. (*Top*) Nucleotide and amino acid sequence of ACT. LIM domains are boxed within the amino acid sequence, which is indicated in the one-letter code. (*Bottom*) Schematic representation of the structure of ACT. The double zinc finger motif corresponding to the LIM domain is represented in the lower part.

dependently isolated 10 times out of the 52 clones obtained from the screening, encoded a product that was found to interact with high affinity with CREM, as tested both for nutritional selection and for β-galactosidase activity (Fig. 3).

The sequence of this clone revealed a cDNA whose open reading frame (ORF) encodes a novel protein of 284 amino acids (Fig. 4). We named this protein ACT (activator of CREM in testis). The distinctive feature of this protein is the presence of four complete LIM domains and one amino-terminal half LIM motif (Fig. 4). The LIM domain is a con-

served cysteine- and histidine-rich structure of two repeated zinc fingers, first identified in homeodomain transcription factors and subsequently found in a variety of proteins with different functions (Dawid et al. 1998). This structural motif has been shown to function as a protein-protein interaction domain (Schmeichel and Beckerle 1994). Because of the lack of other structural domains, ACT belongs to the class of the LIM-only proteins (LMO).

Databank searches for sequence comparison revealed that ACT has a high degree of homology with a family of proteins, whose members (DRAL/FHL-2/SLIM3, SLIM2,

SLIM1/KyoT1) are expressed in heart and skeletal muscles (Morgan and Madgwick 1996; Genini et al. 1997; Chan et al. 1998; Taniguchi et al. 1998). Namely, ACT shows, with respect to the amino acid sequence, 60% identity and 80% similarity to DRAL, a protein of unknown function expressed in heart (Genini et al. 1997).

Coordinated Expression of CREM and ACT

Using a specific antisense riboprobe and RNase protection analysis, we determined the expression pattern of ACT in a variety of mouse tissues. ACT is abundantly and exclusively expressed in testis (Fig. 5). Remarkably, we did not detect any signal in heart and muscle, where the DRAL, SLIM2, and SLIM1 genes are expressed. We performed a collection of analyses to identify the population of testicular cells where ACT is expressed. In situ hybridization studies indicate that ACT is expressed specifically in the inner rim of the seminiferous tubule; expression varies depending on the stage of the tubule, indicating regulation during differentiation (Fig. 5). This result was confirmed by the use of an anti-ACT-specific antibody which revealed the presence of a protein of the expected size (33 kD) that accumulates at high levels in spermatid cells. Immunohistochemical analysis of testis

sections confirmed that ACT is present in round and elongated spermatids and showed nuclear localization.

CREM and ACT are colocalized in spermatids and follow the same expression pattern during testis development (Fig. 5). During the first wave of spermatogenesis, germ cells are synchronized in their development, and the temporal appearance of different cell types is well characterized. The levels of ACT transcript dramatically increased between the third and the fourth week after birth in the mouse. This period corresponds, during testis development, to the end of meiosis and the accumulation of spermatid cells. Analysis of CREM expression shows that ACT and CREM have overlapping expression patterns.

Association of CREM and ACT

The two-hybrid approach and the colocalized expression of CREM and ACT strongly suggested that the two proteins physically interact. Additional experiments demonstrate that CREM and ACT associate both in vitro and in vivo, in a 1:1 ratio. After generation of a purified ACT–glutathione-S-transferase (GST) fusion protein, and by using various CREM deletion proteins, we have found that the P-box domain of CREM is necessary and sufficient for the association. Similarly, we have gener-

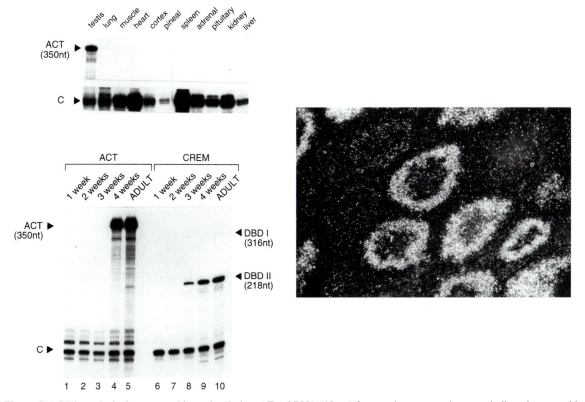

Figure 5. ACT is exclusively expressed in testis. (*Left top*) Total RNA (10 µg) from various mouse tissues as indicated was used in each lane; the size of the protected RNA fragment is shown in nucleotides. C indicates the protected fragment from a mouse β-actin riboprobe used as an internal control. (*Left bottom*) Coexpression of ACT and CREMτ during testis development. RNA was extracted from testes of mice at different ages and analyzed by the RNase protection assay, using ACT- and CREM-specific ribprobes; 10 µg of total RNA from mouse testis was used in each lane. C indicates the β-actin protected fragment used as an internal control. DBD I and DBD II refer to the two alternative DNA-binding domains of CREMτ (Foulkes et al. 1992) detected by the CREM riboprobe p6N/1, which has been described previously (Molina et al. 1993). (*Right*) Histological analysis of ACT transcript distribution in adult mouse testis by in situ hybridization, using an ACT-specific probe.

ated a collection of ACT mutants where the LIM domains have been progressively deleted from either the amino or carboxyl terminus. Results obtained in GST pull-down assays reveal that the third LIM domain of ACT is necessary for efficient interaction with CREM. Importantly, the CREM-ACT association occurs also in vivo, as demonstrated upon coexpression of the two proteins in mammalian cells and coimmunoprecipitation. Thus, at least three lines of evidence demonstrate that the CREM and ACT proteins interact: the two-hybrid in yeast, the in vitro GST pull-down experiments, and the in vivo coimmunoprecipitation in mammalian cells.

ACT Is a Potent Coactivator

Sequence analysis of ACT cDNA obtained from the two-hybrid screening revealed, surprisingly, that the ORF of ACT was not in frame with the sequence encoding the GAL4 activation domain within the yeast expression vector. This observation suggested that the ACT protein could

be translated from its own AUG and thus have intrinsic properties of a coactivator. To investigate this possibility, we tested a series of constructs where ACT was placed in combination or not with the GAL4 activation domain (Fig. 6A). We established that ACT, per se, has a potent activation capacity as it turns the inactive CREM into a powerful transcriptional activator. Analogous results have been obtained with CREB (Fig. 6B), although the activity of Sp1, another activator containing a glutamine-rich domain (Courey and Tjian 1989), is not enhanced by the presence of ACT. Our results also demonstrate the presence of an autonomous activation domain within the protein. Indeed, by fusing ACT with the GAL4 DNA-binding domain, we have shown that its recruitment to DNA is sufficient per se to elicit transcriptional activation (Fig. 6A).

Is the potent activation function of ACT capable of turning a repressor into an activator? We have previously described the CREM gene encoding various isoforms, which behave as both activators (CREMτ, CREMτ1, CREMτ2) and repressors (CREMα, CREMβ, CREMγ) (Laoide et al.

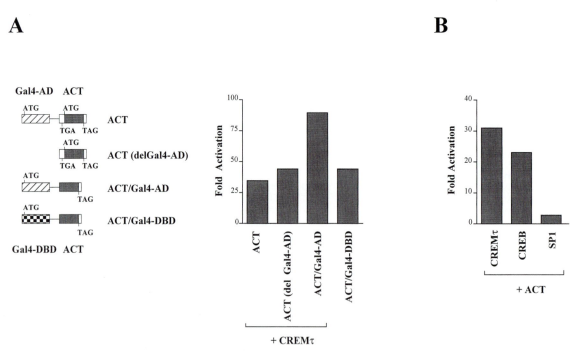

Figure 6. ACT behaves as a coactivator. (A) Analysis of the *trans*-activating properties of ACT. (*Left*) ACT indicates the clone obtained from the screening of the testis library; the ACT/GAL4-AD construct was obtained by cloning the ACT ORF in frame with the GAL4 activation domain, in a yeast expression vector; ACT/GAL4-DBD was generated by cloning ACT ORF in frame with the GAL4-DNA-binding domain. The ACT(delGAL4-AD) construct was generated by inserting the ACT ORF in a yeast expression vector lacking the GAL4-AD-encoding sequence. Initiation (ATG) and stop codons (TGA and TAG) within ACT, GAL4-AD, and GAL4-DBD ORFs are indicated. (*Closed boxes*) ACT ORF; (*hatched boxes*) GAL4-AD ORF; (*checkered boxes*) GAL4-DBD ORF. (*Right*) Constructs shown at the left were used to cotransform yeast (Y190 strain) with a plasmid encoding the GAL4-DBD-CREMτ fusion protein, where indicated. The GAL4-AD-ACT fusion protein was coexpressed with the GAL4-AD alone. β-galactosidase activity from three independent cotransformants for each combination of constructs was calculated in Miller units and reported as a ratio between the values obtained in presence of GAL4-DBD-CREMτ and the values obtained in absence of the bait protein (Fold Activation). For the ACT/GAL4-DBD construct, the Fold Activation was calculated with respect to the β-galactosidase activity obtained when yeast were cotransformed with the GAL4-DBD and GAL4-AD vectors. The results represent the mean from two independent transformations. Variation from mean values was less than 10%. (B) ACT binds to the CREB but not the SP1 activation domain. CREMτ, CREB, and SP1 activation domains were fused to the GAL4-DBD and coexpressed with GAL4-AD-ACT in yeast strain Y190. β-galactosidase activity was monitored, and the values were reported as Fold Activation, which was calculated as a ratio of the value obtained in presence of both GAL4-AD-ACT and GAL4-DBD fusion proteins versus the activity detected in presence of bait proteins alone. Results represent the mean of triplicate determinations from two independent transformations. Variation from mean values was less than 10%.

1993; Sassone-Corsi 1995). Therefore, we investigated whether ACT could elicit *trans*-activation even when coupled to one of the repressor isoforms. We have found that a CREM isoform lacking the glutamine-rich domains, but retaining the P box (i.e., CREMα; Foulkes et al. 1991), is not able to drive transcription in the presence of ACT, despite its ability to interact with ACT in vitro. Additional results show that the Q2 domain, but not the Q1 domain, contains some activity when coexpressed with ACT. Thus, ACT-mediated induction of transcription takes place only in the presence of CREM activator isoforms. The presence of the P box in combination with at least one glutamine-rich domain seems to be required.

These results supported the view that in male germ cells, ACT would provide the activation function that is lacking by the absence of CREM phosphorylation in Ser-117. Indeed, in yeast where both CREB and CREM are inactive because of the lack of CBP, the addition of ACT to these proteins results in a powerful transcriptional activation. Additional experiments in F9 and COS cells indicate that ACT is also able to coactivate both CREB and CREM upon transfection. Thus, ACT functions as a powerful coactivator in mammalian cells.

ACT Bypasses the Requirement of Phosphorylation

The absence of CREM phosphorylation at the time of transcriptional activation of postmeiotic genes suggested that ACT may be able to exert its function on a CREM protein with the Ser-117 mutated. As mentioned above, phosphorylation at Ser-117 in CREM and at Ser-133 in CREB is necessary for association with CBP and transcriptional activation (Chrivia et al. 1993). In the two-hybrid assay, we tested a CREM mutant protein bearing a serine to alanine substitution at position 117, a mutation that prevents phosphorylation of the protein at this site (de Groot et al. 1994). As shown in Figure 7, a similar degree of activation by ACT is observed with the wild-type CREM protein and the Ser>Ala-117 mutant, demonstrating that CREM phosphorylation is not required for ACT to exert its function. Thus, as demonstrated by the experiments in yeast, ACT is a coactivator whose action is CBP-, TAF130- and Ser-133/117-phosphorylation-independent.

Our results reveal that transcriptional activation by CREB and CREM may be obtained through at least two different molecular pathways. The first classical scenario

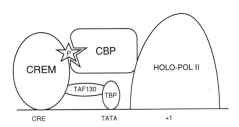

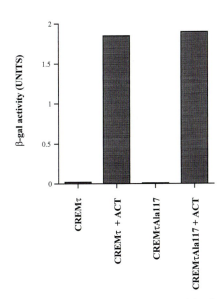

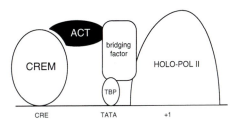

Figure 8. CREM-mediated transcription is promoted by interaction with different coactivators. (*Top*) Schematic representation of the classical view by which, through interaction with CBP, activators such as CREB and CREM elicit their function. A key event in this scenario is phosphorylation at Ser-117 (P) since it is required for binding to CBP and subsequent transcriptional activation. Interaction with TAF130 is constitutive and occurs via the Q2 domain of CREB/CREM. (*Bottom*) Representation of how ACT may elicit its coactivator function via interaction with CREM. In yeast, CREB and CREM are inactive because of the lack of CBP and TAF130. ACT elicits its function and interacts with CREM also in the absence of Ser-117 phosphorylation. Thus, ACT provides an alternative activation pathway which appears to work in a signaling-independent manner. A hypothetical bridging factor linking ACT to the basal transcription machinery is represented.

Figure 7. ACT stimulates CREMτ activity independently of phosphorylation at Ser-117. A CREM activation domain carrying a Ser-117 substitution to alanine (CREMτAla-117; Laoide et al. 1993) was fused to GAL4-DBD and expressed in the presence or absence of ACT in yeast strain Y190, in comparison with the wild-type CREM activation domain.

involves phosphorylation at Ser-133/Ser-117, subsequent association with CBP, and interaction of TAF130 with the Q2 region. The second, which occurs in a cell-specific manner, involves ACT and appears to bypass the above-mentioned requirements (Fig. 8).

CONCLUSION

The manner in which ACT works identifies a novel pathway of transcriptional activation by CREM and CREB. Indeed, these two transcription factors, whose function has been directly coupled to specific transduction systems (see above), appear to act in a signaling-independent manner in some specific circumstances (Fig. 8). This is the case of male germ cells where at a distinct time in the differentiation program, the coordinated expression of ACT provides a novel activation signal.

The finding of ACT raises various questions: (1) How does it work? Is it interacting with other factors of the basal transcriptional machinery, and which ones? (2) Is ACT able to stimulate the activation potential of other transcription factors, and if so, how? (3) Are there other ACT-like molecules and would they function in a similar manner? (4) What is the influence of ACT on the differentiation process of the germ cells?

These and other questions are awaiting answers in the near future. At any rate, the presence of ACT-like proteins of the LMO class, such as DRAL, suggests that ACT defines a novel class of coactivators whose function is signaling-independent and specific of distinct steps of a cellular differentiation program.

ACKNOWLEDGMENTS

We thank all of the members of the Sassone-Corsi laboratory for discussions and help. G.M.F and D.D.C. were supported by postdoctoral fellowships from the European Community. This work was supported by grants from Centre National de la Recherche Scientifique, Institut National de la Santé et de la Recherche Médicale, Centre Hospitalier Universitaire Régional, Fondation de la Recherche Médicale, Université Louis Pasteur, and Association pour la Recherche sur le Cancer and Rhône-Poulenc Rorer.

REFERENCES

Arany Z., Newsome D., Oldread E., Livingston D.M., and Eckner R. 1995. A family of transcriptional adaptor proteins targeted by the E1A oncoprotein. *Nature* **374:** 81.

Bannister A.J. and Kouzarides T. 1996. The CBP co-activator is a histone acetyltransferase. *Nature* **384:** 641.

Böhm M., Moellmann G., Cheng E., Alvarez-Franco M., Wagner S., Sassone-Corsi P., and Halaban R. 1995. Identification of p90rsk as the probable CREB-Ser[133] kinase in human melanocytes. *Cell Growth Differ.* **6:** 291.

Brindle P., Linke S., and Montminy M. 1993. Protein-kinase A-dependent activator in transcription factor CREB reveals new role for CREM repressors. *Nature* **364:** 821.

Busch S.J. and Sassone-Corsi P. 1990. Dimers, leucine-zippers and DNA binding domains. *Trends Genet.* **6:** 36.

Cambier J.C., Newell N.K., Justement L.B., McGuire J.C., Leach K.L., and Chen Z.Z. 1987. Iα binding ligands and cAMP stimulate nuclear translocation of PKC in B lymphocytes. *Nature* **327:** 629.

Chan K.K., Tsui S.K., Lee S.M., Luk S.C., Liew C.C., Fung K.P., Waye M.M., and Lee C.Y. 1998. Molecular cloning and characterization of FHL2, a novel LIM domain protein preferentially expressed in human heart. *Gene* **210:** 345.

Chrivia J.C., Kwok R.P.S., Lamb N., Haniwawa M., Montminy M.R., and Goodman R.H. 1993. Phosphorylated CREB binds specifically to the nuclear protein CBP. *Nature* **365:** 855.

Cobb M.H. and Goldsmith E.J. 1995. How MAP kinases are regulated. *J. Biol. Chem.* **270:** 14843.

Courey A.J. and Tjian R. 1989. Analysis of Sp1 *in vivo* reveals multiple transcriptional domains, including a novel glutamine activation motif. *Cell* **55:** 887.

Cowley S., Paterson H. , Kemp P., and Marshall C.J. 1994. Activation of MAP kinase kinase is necessary and sufficient for PC12 differentiation and for transformation of NIH 3T3 cells. *Cell* **77:** 841.

Dash P.K., Karl K.A., Colicos M.A., Prywes R., and Kandel E.R. 1991. cAMP response element-binding protein is activated by Ca^{2+}/calmodulin, as well as cAMP-dependent protein kinase. *Proc. Natl. Acad. Sci.* **88:** 5061.

Davis R.J. 1994. MAPKs: New JNK expands the group. *Trends Biochem. Sci.* **19:** 470.

Dawid I.B., Breen J.J., and Toyama R. 1998. LIM domains: Multiple roles as adapters and functional modifiers in protein interactions. *Trends Genet.* **14:** 156.

De Cesare D., Jacquot S., Hanauer A., and Sassone-Corsi P. 1998. RSK-2 activity is necessary for EGF-induced CREB phosphorylation and c-*fos* transcription. *Proc. Natl. Acad. Sci.* **95:** 12202.

de Groot R.P., Ballou L.M., and Sassone-Corsi P. 1994. Positive regulation of the cAMP-responsive activator CREM by the p70 S6 kinase: An alternative route to mitogen-induced gene expression. *Cell* **79:** 81.

de Groot R. P., Derua R., Goris J., and Sassone-Corsi P. 1993a. Phosphorylation and negative regulation of the transcriptional activator CREM by p34cdc2. *Mol. Endocrinol.* **7:** 1495.

de Groot R.P., den Hertog J., Vandenheede J.R., Goris J., and Sassone-Corsi P. 1993b. Multiple and cooperative phosphorylation events regulate the CREM activator function. *EMBO J.* **12:** 3903.

Delmas V., van der Hoorn F., Mellström B. , Jégou B., and Sassone-Corsi P. 1993. Induction of CREM activator proteins in spermatids: Downstream targets and implications for haploid germ cell differentiation. *Mol. Endocrinol.* **7:** 1502.

Delmas V., Laoide B.M., Masquilier D., de Groot R.P., Foulkes N.S., and Sassone-Corsi P. 1992. Alternative usage of initiation codons in mRNA encoding CREM generates regulators with opposite functions. *Proc. Natl. Acad. Sci.* **89:** 4226.

Eckner R., Ewen M.E., Newsome D., Gerdes M., DeCaprio J.A., Lawrence J.B., and Livingston D.M. 1994. Molecular cloning and functional analysis of the adenovirus E1A-associated 300-kD protein (p300) reveals a protein with properties of a transcriptional adaptor. *Genes Dev.* **8:** 869.

Ferreri K., Gill G., and Montminy M. 1994. The cAMP-regulated transcription factor CREB interacts with a component of the TFIID complex. *Proc. Natl. Acad. Sci.* **91:** 1210.

Foulkes N.S., Borrelli E., and Sassone-Corsi P. 1991. CREM gene: Use of alternative DNA binding domains generates multiple antagonists of cAMP-induced transcription. *Cell* **64:** 739.

Foulkes N.S., Mellström B., Benusiglio E., and Sassone-Corsi P. 1992. Developmental switch of CREM function during spermatogenesis: From antagonist to transcriptional activator. *Nature* **355:** 80.

Foulkes N. S., Schlotter F., Pévet P., and Sassone-Corsi P. 1993. Pituitary hormone FSH directs the CREM functional switch during spermatogenesis. *Nature* **362:** 264.

Frodin M., Peraldi P., and Van Obberghen E. 1994. Cyclic AMP activates the mitogen-activated protein cascade in PC12 cells. *J. Biol. Chem.* **269:** 6207.

Fukuda M., Gotoh Y., Tachibana T., Dell K., Hattori S., Yoneda

Y., and Nishida E. 1995. Induction of neurite outgrowth by MAP kinase in PC12 cells. *Oncogene* **11:** 239.

Genini M., Schwalbe P., Scholl F.A., Remppis A., Mattei M.G., and Schafer B.W. 1997. Subtractive cloning and characterization of DRAL, a novel LIM-domain protein down-regulated in rhabdomyosarcoma. *DNA Cell Biol.* **16:** 433.

Ginty D.D., Bonni A., and Greenberg M.E. 1994. Nerve growth factor activates a Ras-dependent protein kinase that stimulates c-*fos* transcription via phosphorylation of CREB. *Cell* **77:** 713.

Glass C.K., Rose D.W., and Rosenfeld M.G. 1997. Nuclear receptor coactivators. *Curr. Opin. Cell Biol.* **9:** 222.

Gomez N. and Cohen P. 1991. Dissection of the protein kinase cascade by which nerve growth factor activates MAP kinases. *Nature* **353:** 170.

Gonzalez G.A. and Montminy M.R. 1989. Cyclic AMP stimulates somatostatin gene transcription by phosphorylation of CREB at Ser 133. *Cell* **59:** 675.

Gu W. and Roeder R.G. 1997. Activation of p53 sequence-specific DNA binding by acetylation of the p53 C-terminal domain. *Cell* **90:** 595.

Hai T. Y. and Curran T. 1991. Cross-family dimerization of transcription factors Fos/Jun and ATF/CREB alters DNA binding specificity. *Proc. Natl. Acad. Sci.* **88:** 3720.

Hai T.Y., Liu F., Coukos W.J., and Green M.R. 1989. Transcription factor ATF cDNA clones: An extensive family of leucine zipper proteins able to selectively form DNA binding heterodimers. *Genes Dev.* **3:** 2083.

Hazzalin C.A., Cano E., Cuenda A., Barratt M.J., Cohen P., and Mahadevan L.C. 1996. p38/RK is essential for stress-induced nuclear responses: JNK/SAPKs and c-Jun/ATF-2 phosphorylation are insufficient. *Curr. Biol.* **6:** 1028.

Hunter T. and Karin M. 1992. The regulation of transcription by phosphorylation. *Cell* **70:** 375.

Imhof A., Yang X.J., Ogryzko V.V., Nakatani Y., Wolffe A.P., and Ge H. 1997. Acetylation of general transcription factors by histone acetyltransferases. *Curr. Biol.* **7:** 689.

Janknecht R. and Hunter T. 1996. Transcription. A growing coactivator network. *Nature* **383:** 22.

Jégou B., Syed V., Sourdaine P., Byers S., Gérard N., Velez de la Calle J., Pineau C., Garnier D.H., and Bauché F. 1992. The dialogue between late spermatids and Sertoli cells in vertebrates: A century of research. In *Spermatogenesis, fertilization, contraception: Molecular, cellular and endocrine events in male reproduction* (Schering Foundation) (ed. E. Nieschlag and U.-F. Habenicht), p 56. Springer Verlag, New York.

Kistler M.K., Sassone-Corsi P., and Kistler S.W. 1994. Identification of a functional cAMP-responsive element in the 5´-flanking region of the gene for transition protein 1 (TP1) a basic chromosomal protein of mammalian spermatids. *Biol. Reprod.* **51:** 1322.

Kwok R.P., Lundblad J.R., Chrivia J.C., Richards J.P., Bachinger H.P., Brennan R.G., Roberts S.G.E., Green M.R., and Goodman R.H. 1994. Nuclear protein CBP is a coactivator for the transcription factor CREB. *Nature* **370:** 223.

Lalli E. and Sassone-Corsi P. 1994. Signal transduction and gene regulation: The nuclear response to cAMP. *J. Biol. Chem.* **269:** 17359.

Lamas M. and Sassone-Corsi P. 1997. The dynamics of the transcriptional response to cAMP: Recurrent inducibility and refractory phase. *Mol. Endocrinol.* **11:** 1415.

Laoide B.M., Foulkes N.S., Schlotter F., and Sassone-Corsi P. 1993. The functional versatility of CREM is determined by its modular structure. *EMBO J.* **12:** 1179.

Masquilier D. and Sassone-Corsi P. 1992. Transcriptional crosstalk: Nuclear factors CREM and CREB bind to AP-1 sites and inhibit activation by Jun. *J. Biol. Chem.* **267:** 22460.

McKnight S.G., Clegg C.H., Uhler M.D., Chrivia J.C., Cadd G.G., Correll L.L., and Otten A.D. 1988. Analysis of the cAMP-dependent protein kinase system using molecular genetic approaches. *Recent Prog. Horm. Res.* **44:** 307.

Molina C.A., Foulkes N.S., Lalli E., and Sassone-Corsi P. 1993. Inducibility and negative autoregulation of CREM: An alternative promoter directs the expression of ICER, an early response repressor. *Cell* **75:** 875.

Morgan M.J. and Madgwick A.J. 1996. Slim defines a novel family of LIM- proteins expressed in skeletal muscle. *Biochem. Biophys. Res. Commun.* **225:** 632.

Nakajima T., Uchida C., Anderson S.F., Lee C.G., Hurwitz J., Parvin J.D., and Montminy M. 1997. RNA helicase A mediates association of CBP with RNA polymerase II. *Cell* **90:** 1107.

Nantel F., Monaco L., Foulkes N.S., Masquilier D., LeMeur M., Henriksén K., Dierich A., Parvinen M., and Sassone-Corsi P. 1996. Spermiogenesis deficiency and germ-cell apoptosis in CREM-mutant mice. *Nature* **380:** 159.

Nishizuka Y. 1986. Studies and perspectives of protein kinase C. *Science* **233:** 305.

Ogryzko V.V., Schiltz R.L., Russanova V., Howard B.H., and Nakatani Y. 1996. The transcriptional coactivators p300 and CBP are histone acetyltransferases. *Cell* **87:** 953.

Rehfuss R.P., Walton K.M., Loriaux M.M., and Goodman R.H. 1991. The cAMP-regulated enhancer binding protein ATF-1 activates transcription in response to cAMP-dependent protein kinase A. *J. Biol. Chem.* **266:** 18431.

Roesler W.J., Vanderbark G.R., and R.W. Hanson. 1988. Cyclic AMP and the induction of eukaryotic gene expression. *J. Biol. Chem.* **263:** 9063.

Sassone-Corsi P. 1988. Cyclic AMP induction of early adenovirus promoters involves sequences required for E1A transactivation. *Proc. Natl. Acad. Sci.* **85:** 7192.

———. 1995. Transcription factors responsive to cAMP. *Annu. Rev. Cell Dev. Biol.* **11:** 355.

———. 1997. Transcriptional checkpoints determining the fate of male germ cells. *Cell* **88:** 163.

Schmeichel K.L. and Beckerle M.C. 1994. The LIM domain is a modular protein-binding interface. *Cell* **79:** 211.

Sheng M., McFadden G., and Greenberg M.E. 1990. Membrane depolarization and calcium induce c-*fos* transcription via phosphorylation of transcription factor CREB. *Neuron* **4:** 571.

Stehle J.H., Foulkes N.S., Molina C.A., Simonneaux V., Pévet P., and Sassone-Corsi P. 1993. Adrenergic signals direct rhythmic expression of transcriptional repressor CREM in the pineal gland. *Nature* **365:** 314.

Sun Z., Sassone-Corsi P., and Means A.R. 1995. Calspermin gene transcription is regulated by two cyclic AMP response elements contained in an alternative promoter in the calmodulin kinase IV gene. *Mol. Cell. Biol.* **15:** 561.

Swope D.L., Mueller C.L., and Chrivia J.C. 1996. CREB-binding protein activates transcription through multiple domains. *J. Biol. Chem.* **271:** 28138.

Tan Y., Rouse J., Zhang A., Cariati S., Choen P., and Comb M. J. 1996. FGF and stress regulate CREB and ATF-1 via a pathway involving p38 MAP kinase and MAPKAP kinase-2. *EMBO J.* **15:** 4629.

Taniguchi Y., Furukawa T., Tun T., Han H., and Honjo T. 1998. LIM protein KyoT2 negatively regulates transcription by association with the RBP- J DNA-binding protein. *Mol. Cell. Biol.* **18:** 644.

Torchia J., Rose D.W., Inostroza J., Kamei Y., Westin S., Glass C.K., and Rosenfeld M.G. 1997. The transcriptional co-activator p/CIP binds CBP and mediates nuclear-receptor function. *Nature* **387:** 677.

Treisman R. 1996. Regulation of transcription by MAP kinase cascades. *Curr. Opin. Cell Biol.* **8:** 205.

Trivier E., De Cesare D., Jacquot S., Pannetier S., Zackai E., Young I., Mandel J.L., Sassone-Corsi P., and Hanauer A. 1996. Mutations in the kinase Rsk-2 associated with Coffin-Lowry syndrome. *Nature* **284:** 567.

Vogt P.K. and Bos T.J. 1989. The oncogene *jun* and nuclear signaling. *Trends Biochem. Sci.* **14:** 172.

Williams T., Admon A., Luscher B., and Tjian R. 1988. Cloning and expression of AP-2, a cell-type-specific transcription factor that activates inducible enhancer elements. *Genes Dev.* **2:** 1557.

Yang X.J., Ogryzko V.V., Nishikawa J., Howard B.H., and Nakatani Y. 1996. A p300/CBP-associated factor that competes with the adenoviral oncoprotein E1A. *Nature* **382:** 319.

Yoshimasa T., Sibley D.R., Bouvier M., Lefkowitz R.J., and Caron M.G. 1987. Cross-talk between cellular signalling pathways suggested by phorbol ester adenylate cyclase phosphorylation. *Nature* **327:** 67.

Ziff E. B. 1990. Transcription factors: A new family gathers at the cAMP response site. *Trends Genet.* **6:** 69.

Zhou Y., Sun Z., Means A.R., Sassone-Corsi P., and Bernstein K.E. 1996. CREMτ is a positive regulator of testis ACE transcription. *Proc. Natl. Acad. Sci.* **93:** 12262.

Regulation of SRF Activity by Rho Family GTPases

R. Treisman, A.S. Alberts,* and E. Sahai†

Transcription Laboratory, Imperial Cancer Research Fund, London WC2A 3PX, United Kingdom

The serum response element (SRE) is a conserved regulatory element found in the promoters of many cellular "immediate-early" genes, which are inducible following stimulation of susceptible cells with mitogens such as serum, polypeptide growth factors, and phorbol esters or stress stimuli such as UV irradiation (for review, see Treisman 1990). The core of the c-*fos* SRE is a binding site for the transcription factor SRF, a founding member of the MADS box family (Fig. 1A). SRF binds to the c-*fos* SRE as a complex with a member of the ternary complex factor (TCF) family of Ets domain proteins, and the full regulatory properties of the SRE require binding of both proteins (Treisman 1990; Graham and Gilman 1991; Kortenjann et al. 1994; Hill and Treisman 1995). In addition, SRF also acts constitutively at a number of muscle-specific promoters, apparently independently of the TCFs (Sartorelli et al. 1990; Chen et al. 1996; Sepulveda et al. 1998). Work in our laboratory is oriented toward understanding both the nature of the signal pathways involved in SRE regulation and the means by which they alter the activity of transcription factors at the SRE.

THE SRF-TCF TERNARY COMPLEX

Three TCF proteins, Elk-1, SAP-1, and SAP-2/ERP/NET, have been identified; each contains two conserved regions in addition to the Ets domain (Fig. 1A). In the complex, the Ets domain contacts a CAGGAA motif adjoining the SRF-binding site, whereas the conserved 20-residue B-box region directly contacts SRF (for references, see Treisman 1994; Ling et al. 1997). The third conserved C-box region, which contains copies of the S/T-P core consensus sequence for MAP kinase phosphorylation, acts as a regulatory domain. Following growth factor stimulation, this region is phosphorylated at each of these sites, potentiating the ability of the TCFs to activate transcription and, under some conditions, DNA binding (Janknecht et al. 1993; Marais et al. 1993; Gille et al. 1995). The TCFs regulate SRE activity in response to activation of the ras-raf-ERK pathway (Hill et al. 1994; Kortenjann et al. 1994) and cooperate with the SRF carboxy-terminal activation domain to activate transcription (Hill et al. 1993). In addition, the TCFs are targets for other MAP kinase signaling pathways such as the SAPK/JNK and p38 pathways, although these kinases vary in the efficacy with which they

phosphorylate the different S/T-P motifs (for references, see Price et al. 1996).

SRF ACTIVITY IS CONTROLLED BY RHO FAMILY GTPases

The biochemistry of TCF activation and the signaling pathways involved are quite well understood, but much about SRF regulation remains unclear. Although mutants of the c-*fos* promoter that cannot bind TCF do not respond to the ras-raf-ERK pathway, they remain respon-

A.

Serum Response Factor (SRF)

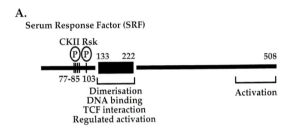

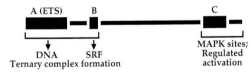

Ternary Complex Factors: Elk-1, SAP-1, SAP-2

B.

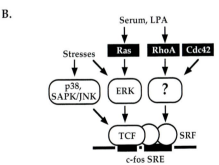

Figure 1. (*A*) SRF and the TCFs. (*Thick lines*) SRF and TCF polypeptide sequences; (*shaded boxes*) SRF DNA-binding domain; (*vertical lines*) phosphorylation sites for CKII and RSK. Regions of assigned function are indicated. (*Shaded boxes*) Three TCF homology region boxes A (Ets domain), B, and C. Functional domains are indicated. (*B*) Signaling pathways to the c-*fos* SRE. (*Paired circles*) SRF; (*rounded rectangle*) TCF; (*closed boxes*) their DNA contact sites. (*Reverse-shaded rectangles*) Ras superfamily GTPases; (*open rounded boxes*) putative MAPK cascades and the putative signal cascade downstream from Rho-family GTPases. Mutation of the TCF-binding site in the SRE selectively disrupts signaling through the TCFs.

Present addresses: *University of California Cancer Center, 2340 Sutter St. N326, San Francisco, California 94115; †CRC Centre for Cell and Molecular Biology, Chester Beatty Laboratories, Institute of Cancer Research, 237 Fulham Road, London SW3 6JB United Kingdom.

sive to signals elicited by whole serum, indicating that SRF is itself responsive to cellular signaling pathways (Graham and Gilman 1991; Kortenjann et al. 1994; Hill and Treisman 1995). Subsequent studies demonstrated that lysophosphatidic acid (LPA), a major serum mitogen, can also activate the c-*fos* promoter independently of TCF, as can intracellular activation of heterotrimeric G proteins by aluminum tetrafluoride ion (AlF$_4^-$; Hill and Treisman 1995) and defined serpentine receptors such as the m1 muscarinic receptor and the *mas* oncogene (Fromm et al. 1997; Zohn et al. 1998; R. Treisman, unpubl.).

One element in common among these stimuli is their ability to activate the small GTPase RhoA, and in transfection experiments, activated forms of RhoA strongly potentiate activity of SRF- but not TCF-controlled reporter genes (Fig. 1B) (Hill et al. 1995). Inactivation of cellular RhoA with C3 transferase abolishes the serpentine receptor-induced activation of SRF; activated forms of two other Rho-related GTPases, Rac1 and Cdc42, also activate SRF reporters, although the relevance of this to signal-regulated SRF activity remains unclear (Hill et al. 1995). No obvious correlation exists between the activity of known MAP kinase pathways and signaling to SRF, although activated Cdc42 and Rac1, but not RhoA, can activate stress-activated MAP kinase pathways in transfection assays (for review, see Minden and Karin 1997). Two major unresolved issues concerning regulation of SRF activity by Rho GTPases are (1) the nature of the signaling pathway utilized by these proteins to activate SRF and (2) the connection if any of this pathway to other processes involving RhoA, such as cytoskeletal rearrangements, cytokinesis, and oncogenic transformation.

ROLE OF RHOA EFFECTOR PROTEINS IN SRF FUNCTION

Generation of RhoA Effector Mutants

Upon exchange of GDP for GTP, the effector loop region of RhoA adopts a conformation that allows its interaction with effector molecules, thereby altering their subcellular localization or enzymatic activity and transmitting down-

stream signals (for review, see Van Aelst and D'Souza-Schorey 1997; Ihara et al. 1998). Many different RhoA effectors have been identified, and in some cases, clear connections between effector binding and downstream function have been made (see Table 1) (for reviews, see Lim et al. 1996; Narumiya et al. 1997). To examine the relationship between SRF activation, effector binding, and downstream functions of RhoA, we used the two-hybrid system to identify RhoA effector loop mutants impaired in the interaction with effectors.

The screen measured the interaction of the Gal4-RhoA fusion protein Gal4-RhoA.V14/S190, which carries an activated RhoA derivative lacking the carboxy-terminal CAAX motif, with fusion proteins comprising derivatives of the PKN and ROCK-I kinases, each tagged with the Gal4 activation domain. Mutations were generated either by cassette mutagenesis of positions previously implicated in effector binding or by introduction of mutations implicated in the interaction of other RhoA GTPases with their effectors. A sequential screening strategy was used to identify RhoA effector mutants that selectively impair interaction of RhoA with its protein kinase effectors PKN and ROCK-I. Three mutants, F39L, E40L, and E40W, were identified as able to interact with PKN but not ROCK-I, and three further mutants, F39V, E40N, E40T and Y42C, were identified as able to interact with ROCK-I but not PKN. Mutant F39A was identified as defective in interaction with both kinase families (Table 1). Analysis of the binding of these mutants to other known RhoA effectors revealed that each exhibited distinct effects on effector binding: In particular, the citron protein, recently also identified as a protein kinase (Madaule et al. 1998), revealed patterns of interaction distinct from those of the other kinases (Table 1).

SRF Regulation Does Not Correlate with Binding to Known RhoA Effectors

Serum- or LPA-stimulated activity of SRF-controlled reporter genes in NIH-3T3 cells is dependent on RhoA, and SRF activity is potentiated by activated RhoA (Hill et al. 1995). To gain insight into which downstream effec-

Table 1. RhoA Effectors, Functions, and Blocking Mutations

Effector protein [a]	Process	Blocking mutant [b]
Kinases		
PRK family (PKN, PRK2, PRK3)	endocytosis	F39A, F39V, Y42C
ROCK family (ROCK-I, ROCK-II)	cytoskeletal reorganization	E40L, E40W, F39A, F39L
citron	cytokinesis?	F39A, F39V, F39L
Adaptor proteins		
rhophilin	unknown	F39A, F39V, F39L
rhotekin	unknown	not tested
Formin domain proteins		
P140mDia, mDia2	cytokinesis, actin polymerization	not affected by mutation studies
Others		
kinectin	vesicle traffic	E40L, E40W, E40N, F39A, F39V, F39L
MLC phosphatase regulatory subunit	cytoskeletal reorganization	not tested

[a]Effectors are grouped by common functional domains. Rhophilin and rhotekin are distinguished by their interaction with RhoA through a common sequence motif shared with PKN, but they do not contain other known functional motifs. Effectors such as PI-4P 5-kinase are not included (for review, see Van Aelst and D'Souza-Schorey 1997).
[b]Mutants were identified using the two-hybrid assay, scoring sequentially for interaction with PKN and ROCK-I. Mutants were then tested for interaction with other known RhoA effectors. None of the mutants studied affected interactions with mDia2. Similar results were obtained in biochemical assays (Sahai et al. 1998).

tors are involved in these phenomena, we examined the ability of effector loop mutants to activate SRF. To facilitate comparison with other functional assays for RhoA.V14, we tested the mutants using both transfection and microinjection assays. For the microinjection assay, we used the SRF-controlled reporter SRE-FosHA, in which a minimal SRF-controlled promoter regulates expression of a hemagglutinin epitope-tagged Fos protein that can be monitored by indirect immunofluorescence (Hill et al. 1995; Alberts et al. 1998). Microinjection of expression plasmids encoding RhoA.V14 and effector loop mutants F39A, F39V, F39L, E40N, E40T, and Y42C all efficiently activate the reporter in this assay, whereas E40L did not lead to significant activation above background (Fig. 2A). All of the RhoA mutants were expressed at similar high levels and showed similar subcellular distributions (data not shown). We also tested the mutants using a transfection assay with the reporter 3D.ACAT (Hill et al. 1995), which utilizes the same promoter as SRE-FosHA. As in the microinjection experiments, mutants F39V and Y42C activated the reporter as effectively as RhoA.V14 itself, whereas activation by mutants E40L and E40W was at background levels (data

not shown). Mutants F39A, F39L, E40N, and E40T exhibited intermediate abilities to activate the reporter.

Taken together with the microinjection results, these data indicate that the identity of residue 40 within the effector loop is of particular importance for SRF activation. However, the data do not allow association of any one downstream effector protein with the ability to activate SRF. For example, the E40L mutant exhibits unimpaired interaction with PKN yet fails to activate SRF, whereas the Y42C and F39V mutants reduce PKN interaction yet activate SRF strongly. Similarly, although the inactive E40L mutant fails to bind ROCK normally, mutants F39A and F39L also fail to bind ROCK yet show at least some ability to activate SRF.

SRF Activation Does Not Correlate with Transformation or Cytoskeletal Rearrangement

Having identified effector mutations that affect RhoA's ability to activate SRF, we exploited these mutations to test whether the ability of RhoA to induce cytoskeletal rearrrangements correlates with its ability to potentiate activity of SRF. In the microinjection assay, the effector mutants differed in their ability to induce actin stress fibers. Binding to ROCK-I appeared to be necessary but not sufficient for stress fiber formation (data not shown; see Sahai et al. 1998), in agreement with previous studies with the ROCKs, which have implicated these kinases in cytoskeletal reorganization events (for review, see Lim et al. 1996; Narumiya et al. 1997). However, mutants F39L, F39A, and F39V all activated SRF reporters in the microinjection assay but failed to induce significant cytoskeletal rearragements in the microinjected cells, indicating that cytoskeletal rearrangements are not a prerequisite for SRF activation (summarized in Fig. 2A; Table 2).

Constitutively GTP-bound RhoA mutants such as RhoA.V14 can cooperate with activated Raf derivatives to induce transformation (for discussion, see Van Aelst and D'Souza-Schorey 1997). We therefore tested whether the ability of different RhoA effector mutants to potentiate SRF activity correlates with their ability to induce formation of transformed foci of NIH-3T3 cells in cooperation with activated Raf. In this assay, RhoA.V14 and its mutant derivatives F39V, E40N, E40T, and Y42C (each of which can interact with ROCK-I) promoted focus formation in cooperation with ΔNRaf (Table 2). Transformation by the other effector mutants was not significantly greater than that observed with ΔNRaf alone; in particular, mutant F39A, which activates SRF with an efficiency comparable to that of mutants E40N and E40T, was inactive in the transformation assay. These results suggest that it is interaction with ROCK kinases rather than SRF activation that provides the cooperating signal with ΔNRaf. Consistent with this idea, transformation by RhoA or its exchange factors, but not activation of SRF reporters or c-fos transcription, is inhibited by the small-molecule ROCK kinase inhibitor Y27632 (Uehata et al. 1997; E. Sahai and R. Treisman, in prep.).

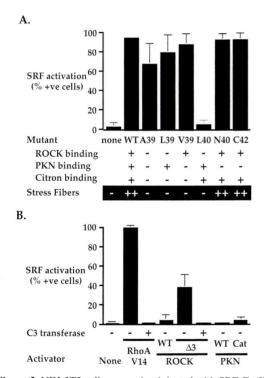

Figure 2. NIH-3T3 cells were microinjected with SRE-FosHA reporter plasmid together with plasmids expressing the indicated proteins; reporter expression was visualized by immunofluorescence 5 hr later. The proportion of reporter-positive injected cells is indicated (error bars indicate s.e.m.; n = 3). (A) Regulation of SRF by RhoA mutants. Summarized below the figure are the interaction properties of each mutant, together with their ability to induce actin stress fibers: (++) 50–100%; (−) ≤20% wild-type activity. (B) SRE activation by ROCK-I and PKN derivatives. (ROCK-WT) Codons 1–1354; (ROCKΔ3) codons 1–727; (PKN-WT) codons 1–942; (PKN-Cat) codons 539–942.

Table 2. Summary of RhoA Effector Mutant Experiments

| RhoA protein | Interaction[a] | | Activity | | |
	PKN (1–551)	ROCK-I (348–1018)	actin[b]	SRF[c]	focus[d] formation
Rho.AV14(WT)	++	++	++	++	++
F39A	–	–	–	+	–
F39V	–	++	–	++	+
F39L/C20R	++	–/+	–	+	–
E40L[e]	++	–	–	–	–
E40N[e]	++	++	++	+	+
E40T[e]	++	++	+	+	++
Y42C	–	++	++	++	++

[a]Interactions in the two-hybrid assay, scored by *lacZ* activity and His3 auxotrophy. (++) Wild-type interaction; (–/+) weak interaction detected by His3 auxotrophy but only background level by *lacZ* assay; (–) no significant interaction or activity.

[b]Filamentous actin was visualized with TRITC-conjugated phalloidin, and the proportion of injected cells exhibiting long stress fibers was counted. (++) 50–100% wild-type activity; (+) 20–50% wild-type activity; (–) ≤20% wild-type activity.

[c]SRF activity was measured either by microinjection or by transfection assay using reporters 3DA.CAT and SRE-FosHA, respectively. (++) Wild-type activity in both transfection and microinjection assay; (+) 50–100% wild-type activity in microinjection assay, 10–30% wild type in transfection assay; (–) ≤10% wild type in transfection assay.

[d]NIH-3T3 cells were transfected with expression plasmids encoding RhoA derivatives together with a plasmid encoding activated Raf. Posttransfection (15 days), the number of foci of size greater than >1.5 mm in diameter was counted. The combination of RhoA.V14 and ΔNRaf produced on average 38 foci/plate transfected. (++) 50–100% wild-type activity; (+) 20–50% wild-type activity; (–) ≤20% wild-type activity.

[e]These mutants did not interact with intact PKN in the two-hybrid assay.

The ROCK-I and PKN Kinase Domains Cannot Activate SRF Independently of RhoA Function

The data presented above suggest that either a novel RhoA effector is involved in signaling to SRF or multiple effectors are involved. In light of the latter possibility, we noted that all of the mutants that bound to ROCK-I were able to signal to the SRE, suggesting the possibility that although interaction with ROCK-I might not be required for RhoA to signal to SRF, it might nevertheless be able to affect SRF activity transcription at the SRE. Moreover, it has been reported that PKN family kinase PRK2 domain can potentiate SRF activity (Quilliam et al. 1996). To address these questions and to provide an alternative approach to the study of the role of the kinases in SRF activation, we tested the effect of expression of their isolated kinase domains on SRF activity.

In the microinjection assay, expression of intact ROCK-I had no effect on SRF activity (Fig. 2B). In contrast, ROCKΔ3 (ROCK-I codons 1–727), a truncated ROCK-I derivative that exhibits elevated kinase activity compared to intact ROCK-I, induced SRF activation in approximately one third of the injected cells (Fig. 2B). Surprisingly, in this case, SRF activation was dependent on endogenous Rho function, since it did not occur upon inactivation of RhoA by C3 transferase (see Fig. 2A). Both intact ROCK-I and ROCKΔ3 induced formation of thick actin fibers distinct from those induced by activated RhoA itself; as previously reported, this activity was not dependent on functional RhoA (data not shown; for re-

view, see Lim et al. 1996; Narumiya et al. 1997). Both reporter gene activity and cytoskeletal reorganization were not observed in cells injected with a kinase-inactive ROCKΔ3 derivative (data not shown). We also investigated whether expression of a constitutively active form of PKN is sufficient for SRF activation or stress fiber formation. Immune complex kinase assays using extracts from NIH-3T3 cells transfected with an expression plasmid encoding the PKN kinase domain (PKN.Cat; amino acids 539–942) showed that this protein possesses substantial constitutive histone kinase activity (data not shown). However, neither intact PKN nor PKN (539–942) could activate the SRF reporter gene or induce actin stress fibers (Fig. 2B).

Although the results with the ROCKΔ3 mutant suggest that ROCK may be capable of potentiating SRF activity, three lines of evidence argue against this kinase being the sole RhoA effector involved in SRF regulation. First, the effector mutant data show no simple correlation between ROCK binding and SRF activation; second, activation by ROCKΔ3 remains dependent on RhoA; and third, the specific ROCK inhibitor Y27632 (Uehata et al. 1997) does not block activation of SRF by serum-induced signals (E. Sahai and R. Treisman, in prep.). Taken together with the effector mutant data, the results are consistent with the view that neither ROCK-I nor PKN constitutes single RhoA effectors that mediate SRF activation and supports the hypothesis that RhoA.V14-induced SRF activation is mediated by a novel effector or that multiple RhoA effectors are involved in the response. Mutant E40L, which is inactive in SRF regulation, should be useful for the identification of novel RhoA effectors involved in activation of the pathway.

CDC42 BUT NOT RHOA CAN ACTIVATE A CHROMOSOMAL TEMPLATE

A classic microinjection study by Stacey et al. (1987) was one of the first to establish that signaling by oncogenic *ras* is sufficient for activation of the chromosomal c-*fos* gene. We therefore used a similar approach to investigate whether activation of Rho-controlled signal pathways can suffice for c-*fos* activation. The short-term nature of the microinjection assay minimizes autocrine effects that recent studies have demonstrated can complicate the interpretation of transfection assays (see, e.g., Minden et al. 1994). Mutationally activated forms of RhoA (RhoA.V14) and Cdc42 (Cdc42.V12) were introduced into NIH-3T3 cells by microinjection of either recombinant proteins or expression plasmids, and c-*fos* expression was assayed by indirect immunofluorescence 2 hours later. Surprisingly, Cdc42.V12 induced c-*fos* expression, but RhoA.V14 did not (Fig. 3A). In contrast, both GTPases could induce expression of an epitope-tagged *fos* gene (Fos-9E10) coinjected into the nuclei of NIH-3T3 cells, even when the plasmid concentration was reduced to limiting levels (Fig. 3B; data not shown). Thus, the ability of RhoA.V14 to activate c-*fos* gene expression differs according to the context of the DNA template.

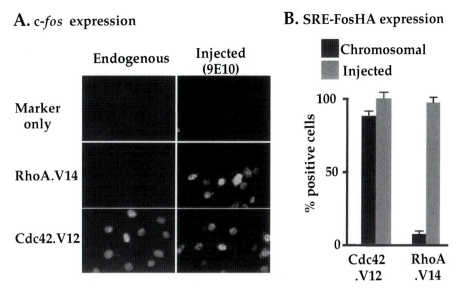

Figure 3. Differential activation of chromosomal and injected c-*fos* genes by Rho family proteins. (*A*) c-*fos* gene expression. (*Left panels*) RhoA.V14 and Cdc42.V12 proteins were injected into the cytoplasm of serum-deprived NIH-3T3 cells; 2 hr later, endogenous Fos was detected by indirect immunofluorescence. (*Right panels*) RhoA.V14 and Cdc42.V12 proteins were injected into the nuclei of serum-deprived NIH-3T3 cells together with a plasmid containing a 9E10-epitope-tagged human c-*fos* gene (pF711-9E10); 2 hr later, expression of the injected c-*fos* gene was detected by immunofluorescence with 9E10 antibody. (*B*) Differential activation of chromosomal and injected SRF-controlled reporter genes by Rho family proteins. Recombinant Cdc42.V12 and RhoA.V14 GTPases were injected into the nuclei and cytoplasm of SRE-FosHA cells together with SRE-Fos9E10, the same reporter but tagged with the 9E10 epitope. Two hours later, activity of each reporter gene was scored by immunofluorescence. Error bars indicate s.e.m. (*n* = 3).

RhoA-dependent Signals Are Necessary but not Sufficient for Activation of a Chromosomal SRE

Differential activation of chromosomal and injected c-*fos* templates by Rho family proteins might reflect a property of either the SRE itself or its interaction with other c-*fos* promoter elements. To address this issue, we examined two cell lines carrying chromosomal SRF-controlled reporter genes. SRE-LacZ cells contain an integrated SRE-LacZ reporter gene comprising three SRF-binding sites, Rous sarcoma virus (RSV) TATA box and the bacterial *lacZ* gene, whereas SRE-FosHA cells contain an integrated copy of the SRE-FosHA reporter described above. Both reporter genes are efficiently activated following serum or LPA stimulation (data not shown). This induction requires functional RhoA: It was abolished upon inactivation of RhoA by C3 transferase but could be restored upon coexpression of the C3-resistant RhoA mutant RhoA.I41 (data not shown). As with the chromosomal c-*fos* gene, neither chromosomal reporter gene was activated upon microinjection of RhoA.V14, although both were activated by Cdc42.V12; in contrast, both RhoA.V14 and Cdc42.V12 could activate coinjected SRE-LacZ and SRE-FosHA templates (Fig. 3B). Thus, in contrast to transfected templates, RhoA is necessary, but not sufficient, for activation of the chromosomal SRE in response to serum and LPA stimulation.

The SAPK/JNK Pathway Cooperates with RhoA.V14 and Is Necessary for Induction of the Chromosomal SRE by Cdc42.V12

The above results can be explained by the hypothesis that although Cdc42.V12 and RhoA.V14 can activate a signaling pathway required for activation of injected SRF reporters, only Cdc42.V12 can supply a signal specifically required for the activation of chromosomal templates. Since the activated forms of Cdc42 and Rac1, but not RhoA, can induce stress-activated MAP kinases (for review, see Minden and Karin 1997), we tested whether independent activation of these pathways by stress stimuli would allow activated RhoA to induce chromosomal reporter gene expression. SRE-*lacZ* cells were injected with recombinant RhoA.V14 and stimulated with anisomycin or UV irradiation before analysis of reporter gene activity. In agreement with our transfection studies (Hill et al. 1995), neither anisomycin nor UV irradiation alone could activate the chromosomal SRE-*lacZ* reporter gene; however, both could activate the reporter in the presence of RhoA.V14 (Fig. 4A; data not shown) Coexpression of RhoA.V14 with either MEKK1 or v-*src*, both of which can activate the SAPK/JNK pathway, also activated the chromosomal reporters (data not shown).

To confirm the involvement of the SAPK/JNK pathway in activation of the chromosomal SRF-controlled reporters, we performed inhibitor studies. To inhibit the SAPK/JNK pathway, we used either recombinant

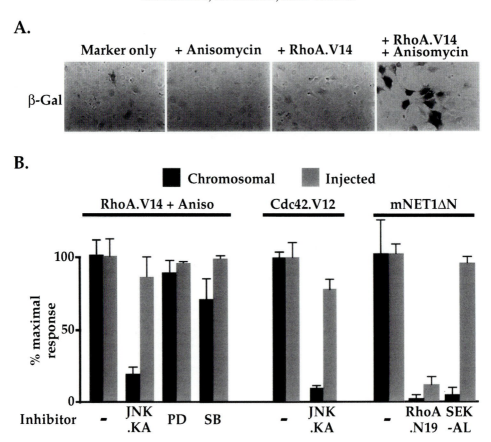

Figure 4. (*A*) RhoA.V14 can cooperate with stress stimuli to activate a chromosomal SRE. RhoA.V14 protein was injected into the cytoplasm of serum-deprived SRE-*lacZ* cells, which were stimulated 20 min later as indicated and processed for injection marker and LacZ activity after a further 2 hr. Representative fields are shown. (*B*) Chromosomal SRE activation by anisomycin/RhoA.V14 Cdc42.V12, and the RhoA GEF mNET1 requires the SAPK/JNK pathway. The effects of the indicated inhibitors on chromosomal or injected SRF-controlled reporter genes were compared for each stimulus. In each case, bars represent the percentage of positive injected cells, normalized to the number of positive cells obtained without coexpressed inhibitor (error bars indicate S.E.M. or half range). (*Gray bars*) Chromosomal reporter gene; (*black bars*) injected reporter gene. For analysis of RhoA.V14 and Cdc42.V12, serum-deprived SRE-*lacZ* cells were injected with recombinant GTPases either alone, together with kinase-inactive p54SAPKβ (JNK-KA), or treated with 50 μM PD98059 or 10 μM SB203580 as indicated; RhoA.V14-injected cells were immediately treated with anisomycin. For analysis of injected template, the SRF-controlled reporter SRE-Fos9E10 was included. For analysis of mNET1, cells were injected with the mNET1ΔN expression plasmid together with RhoA.N19 protein or SEK-AL expression plasmid as indicated. For chromosomal gene expression studies, SRE-FosHA cells were used; for analysis of injected gene expression, NIH-3T3 cells were injected with the SRE-FosHA reporter.

JNK.KA, a kinase-inactive mutant of p54SAPKβ, or an expression plasmid encoding SEK-AL, a nonactivatable SEK mutant. To inhibit the ERK and p38 MAP kinase pathways, we used the small-molecule inhibitors PD98059 and SB203580, respectively (for discussion, see Cohen 1997). Both JNK.KA and SEK-AL proteins inhibited activation of chromosomal SRF-controlled reporter genes by either Cdc42.V12 or the combination of RhoA.V14 and anisomycin, but they did not affect induction of SRE-Fos9E10, a reporter gene identical to SRE-FosHA but containing the 9E10 epitope (Fig. 4B). In contrast, inhibition of the ERK or p38 MAP kinase pathways had no effect on either template. Thus, in addition to RhoA-controlled signals, activation of chromosomal SRF-controlled templates requires activation of the SAPK/JNK pathway, and it is its ability to additionally activate this pathway that allows Cdc42.V12 to activate chromosomal targets.

Since these experiments were performed with activated mutant GTPases, we also tested whether activation of endogenous RhoA would require simultaneous activation of the SAPK/JNK pathway to induce expression of chromosomal reporter genes. To do this, we used an activated form of the mNET1 guanine nucleotide exchange factor (GEF), mNET1ΔN, which although specific for RhoA, also has the ability to activate the SAPK/JNK pathway (Alberts and Treisman 1998). Expression of mNET1 activated both chromosomal and injected SRF reporter genes. In both cases, this required functional RhoA, since it was blocked by coexpression of RhoA.N19, a dominant interfering RhoA mutant. However, as with the activated GTPases, inhibition of SAPK/JNK signaling by coexpression of SEK-AL blocked mNETΔ1-induced activation of the chromosomal but not the injected reporter gene (Fig. 4B).

What Is the Target for the Cooperating Signal?

The fact that chromosomally located templates require additional signals in addition to those mediating activation of injected templates suggests that these signals are required in order to allow transcriptional activation by SRF in a chromatin environment. Although transfected (and presumably injected) plasmid DNAs do become packaged into repeating nucleoprotein structures (Cereghini and Yaniv 1984), several studies have shown that such structures can be distinguished biochemically from cellular chromatin (for further discussion, see Jeong and Stein 1994; Alberts et al. 1998). In principle, cooperating signals might target either transcription factors or the chromatin template itself. A precedent for the latter is provided by the mitogen-induced phosphorylation of histone H3 and HMG14 (for references, see Barratt et al. 1994). We have found that deacetylase inhibitors such as sodium butyrate or trichostatin A also cooperate with RhoA to activate chromosomal templates and that signals that cooperate with RhoA can induce hyperacetylation of histone H4 (Alberts et al. 1998). However, H4 hyperacetylation is also induced by stimuli that do not cooperate with RhoA, and so although H4 acetylation may be necessary for activation of chromosomal templates, it cannot be sufficient. Taken together, these results point to additional roles for signaling in transcription of the chromosomal c-*fos* gene.

OUTSTANDING QUESTIONS

How Do RhoA-dependent Signals Regulate SRF Activity?

Despite our increased understanding of the signaling pathways that control SRF activity, the means by which these signals reach SRF remains obscure. Any model must be consistent with the following observations. Overexpression of SRF does not potentiate activation of SRE-controlled reporters in our assay systems (Hill et al. 1993), and fusion proteins containing SRF linked to heterologous DNA-binding domains are not serum-inducible (Johansen and Prywes 1993; Hill et al. 1994). These observations suggest that SRF levels are not limiting and that SRF must be in direct contact with DNA for regulation to occur. Consistent with this, certain inactivating SRE mutations that strongly reduce binding of endogenous SRF can be rescued by overexpression of the wild-type protein (Hill et al. 1994; Johansen and Prywes 1994). In this assay, serum-regulated SRF activity can be abolished by mutations within the DNA-binding domain that do not affect dimerization or DNA-binding affinity, but remains unaffected by mutation of a growth factor-regulated phosphorylation site at serine 103 (Hill et al. 1994; Johansen and Prywes 1994). Finally, although TCF is not required for regulation of SRF via the Rho pathway, inactive TCF bound to the c-*fos* SRE can interfere with the operation of this pathway, and some SRF mutants defective in TCF interaction are also impaired in their response to TCF-independent signals (Hill et al. 1994, 1995).

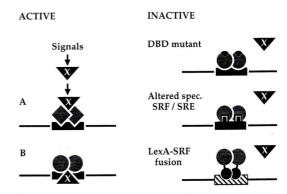

Figure 5. Models for SRF activation. SRF interacts with a non-TCF accessory factor X (*closed triangle*) which is a target for Rho-induced signals. Recruitment of this factor to SRF requires either a conformation of SRF dependent on its binding to DNA (*A*), interaction with both SRF and the SRE DNA (*B*), or both. On the right, three situations in which access of the accessory factor is blocked are illustrated. (*Closed rectangle*) SRE; (*open box*) LexA operator.

To explain these observations, we propose that activation of SRF via the Rho pathway involves recruitment of a novel non-TCF accessory factor (Fig. 5). According to this model, SRF is susceptible to signaling either because the DNA-binding domain adopts an altered conformation upon DNA binding that allows recruitment of such a factor (Fig. 5, model A) or because specific sequences in the DNA-binding domain are required to recruit it to the SRE in partnership with SRF (Fig. 5, model B). Conformation models have been proposed for activation by MCM1, the likely *Saccharomyces cerevisiae* homolog of SRF (Tan and Richmond 1990). Competition between TCF and the putative accessory factor for the same surface on SRF provides a potential explanation for the ability of defective TCFs to interfere with SRF-linked signaling (Hill et al. 1994, 1995). Several proteins have been found to interact with SRF and/or the SRE, including Phox1/mhox (Grueneberg et al. 1992), TFII-I/BAF135 (Grueneberg et al. 1997; Kim et al. 1998), ATF6 (Zhu et al. 1997), Nkx2.5 (Chen et al. 1996), YY1 (Natesan and Gilman 1995), and C/EBP (Sealy et al. 1997). However, there is as yet no direct evidence that any of these factors is actually a target of Rho-controlled signaling pathways. We are currently seeking to identify novel SRF interaction partners, by use of both modified yeast interaction screens and direct expression cloning of SRF activators in mammalian cells.

How Does SRF Participate in Both Inducible and Tissue-specific Transcription?

In addition to its role in expression of immediate-early genes such as c-*fos*, SRF is also involved in the regulation of a number of genes expressed in smooth, cardiac, and skeletal muscle types. Many of these sites possess no obvious TCF-binding site in their vicinity, and it is likely that their activity does not involve TCF proteins. Several interrelated questions arise from this. First, how is muscle specificity achieved? Second, is the activity of SRF at such promoters truly constitutive or signal-dependent? Third, why are immediate-early genes still signal-regulated in muscle cells? One demonstrated mechanism of muscle specificity is that SRF acts in concert with other muscle-specific and ubiquitous cofactors, including the myogenic bHLH factors, GATA4, and members of the tinman group of homeodomain proteins (for references, see Sartorelli et al. 1990; Chen et al. 1996; Sepulveda et al. 1998). It has generally been assumed that these factors render SRF constitutive, substituting for and presumably inhibiting the activity of the SRF-linked signal pathway. A radically different but equally plausible model which has not been considered previously, however, is that such interactions might somehow sensitize SRF to the basal activity of SRF-linked signaling pathways and that the "constitutive" activity of SRF at muscle-specific promoters is in fact illusory. Finally, it remains possible that in certain muscle (or nonmuscle) cell types, constitutive activation of the SRF-linked signal pathway maintains SRF activity at a high level, the immediate-early genes remaining signal-dependent because additional signals are required for their activation. As our understanding of the role of signal transduction pathways in the regulation of SRF becomes clearer, it will be interesting to elucidate their involvement in SRF-mediated muscle-specific transcription.

ACKNOWLEDGMENTS

We thank numerous colleagues for gifts of plasmids essential to this work and apologize to those whose papers we have not cited owing to space constraints. We thank members of the laboratory, and Julian Downward, Alan Hall, Chris Marshall, and Shuh Narumiya for helpful discussions. A.A. was partially funded by a postdoctoral fellowship from the Howard Hughes Medical Institute. The research described here was funded by the Imperial Cancer Research Fund.

REFERENCES

Alberts A.S. and Treisman R. 1998. Activation of RhoA and SAPK/JNK signalling pathways by the RhoA-specific exchange factor mNET1. *EMBO J.* **17:** 4075.

Alberts A., Geneste O., and Treisman R. 1998. Activation of SRF-regulated chromosomal templates by Rho-family GTPases requires a signal that also induces H4 hyperacetylation. *Cell* **92:** 475.

Barratt M.J., Hazzalin C.A., Cano E., and Mahadevan L.C. 1994. Mitogen-stimulated phosphorylation of histone H3 is targeted to a small hyperacetylation-sensitive fraction. *Proc. Natl. Acad. Sci.* **91:** 4781.

Cereghini S. and Yaniv M. 1984. Assembly of transfected DNA into chromatin: Structural changes in the origin-promoter-enhancer region upon replication. *EMBO J.* **3:** 1243.

Chen C.Y., Croissant J., Majesky M., Topouzis S., McQuinn T., Frankovsky M.J., and Schwartz R.J. 1996. Activation of the cardiac α-actin promoter depends upon serum response factor, Tinman homologue, Nkx-2.5, and intact serum response elements. *Dev. Genet.* **19:** 119.

Cohen P. 1997. The search for physiological substrates of MAP and SAP kinases in mammalian cells. *Trends Cell Biol.* **7:** 353.

Fromm C., Coso O.A., Montaner S., Xu N., and Gutkind J.S. 1997. The small GTP-binding protein Rho links G protein-coupled receptors and Gα12 to the serum response element and to cellular transformation. *Proc. Natl. Acad. Sci.* **94:** 10098.

Gille H., Kortenjann M., Thomae O., Moomaw C., Slaughter C., Cobb M.H., and Shaw P.E. 1995. ERK phosphorylation potentiates Elk-1 mediated ternary complex formation and transactivation. *EMBO J.* **14:** 951.

Graham R. and Gilman M. 1991. Distinct protein targets for signals acting at the c-*fos* serum response element. *Science* **251:** 189.

Grueneberg D., Natesan S., Alexandre C., and Gilman M.Z. 1992. Human and *Drosophila* homeodomain proteins that enhance the binding activity of serum response factor. *Science* **257:** 1089.

Grueneberg D.A., Henry R.W., Brauer A., Novina C.D., Cheriyath V., Roy A.L., and Gilman M. 1997. A multifunctional DNA-binding protein that promotes the formation of serum response factor/homeodomain complexes: Identity to TFII-I. *Genes Dev.* **11:** 2482.

Hill C.S. and Treisman R. 1995. Differential activation of c-*fos* promoter elements by serum, lysophosphatidic acid, G proteins and polypeptide growth factors. *EMBO J.* **14:** 5037.

Hill C.S., Wynne J., and Treisman R. 1994. Serum-regulated transcription by serum response factor (SRF): A novel role for the DNA-binding domain. *EMBO J.* **13:** 5421.

———. 1995. The Rho family GTPases RhoA, Rac1 and CDC42hs regulate transcriptional activation by SRF. *Cell* **81:** 1159.

Hill C.S., Marais R., John S., Wynne J., Dalton S., and Treisman R. 1993. Functional analysis of a growth factor-responsive transcription factor complex. *Cell* **73:** 395.

Ihara K., Muraguchi S., Kato M., Shimizu T., Shirakawa M., Kuroda S., Kaibuchi K., and Hakoshima T. 1998. Crystal structure of human RhoA in a dominantly active form complexed with a GTP analogue. *J. Biol. Chem.* **273:** 9656.

Janknecht R., Ernst W.H., Pingoud V., and Nordheim A. 1993. Activation of ternary complex factor Elk-1 by MAP kinases. *EMBO J.* **12:** 5097.

Jeong S. and Stein A. 1994. Micrococcal nuclease digestion of nuclei reveals extended nucleosome ladders having anomalous DNA lengths for chromatin assembled on non-replicating plasmids in transfected cells. *Nucleic Acids Res.* **22:** 370.

Johansen F.-E. and Prywes R. 1993. Identification of transcriptional activation and inhibitory domains in serum response factor (SRF) by using GAL4-SRF constructs. *Mol. Cell. Biol.* **13:** 4640.

———. 1994. Two pathways for serum regulation of the c-*fos* serum response element require specific sequence elements and a minimal domain of serum response factor. *Mol. Cell. Biol.* **14:** 5920.

Kim D.W., Cheriyath V., Roy A.L., and Cochran B.H. 1998. TFII-I enhances activation of the c-*fos* promoter through interactions with upstream elements. *Mol. Cell. Biol.* **18:** 3310.

Kortenjann M., Thomae O., and Shaw P.E. 1994. Inhibition of v-*raf*-dependent c-*fos* expression and transformation by a kinase-defective mutant of the mitogen-activated protein kinase Erk 2. *Mol. Cell. Biol.* **14:** 4815.

Lim L. Manser, E., Leung, T., and Hall, C. 1996. Regulation of phosphorylation pathways by p21 GTPases. The p21 Ras-related Rho subfamily and its role in phosphorylation signalling pathways. *Eur. J. Biochem.* **242:** 171.

Ling Y., Lakey J.H., Roberts C.E., and Sharrocks A.D. 1997. Molecular characterization of the B-box protein-protein interaction motif of the ETS-domain transcription factor Elk-1. *EMBO J.* **16:** 2431.

Madaule P., Eda M., Watanabe N., Fujisawa K., Matsuoka T., Bito H., Ishizaki T., and Narumiya S. 1998. Role of citron kinase as a target of the small GTPase Rho in cytokinesis. *Nature* **394:** 491.

Marais R., Wynne J., and Treisman R. 1993. The SRF accessory protein Elk-1 contains a growth factor-regulated transcriptional activation domain. *Cell* **73:** 381.

Minden A. and Karin M. 1997. Regulation and function of the JNK subgroup of MAP kinases. *Biochim. Biophys. Acta* **1333:** F85.

Minden A., Lin A., McMahon M., Lange C.C., Derijard B., Davis R.J., Johnson G.L., and Karin M. 1994. Differential activation of ERK and JNK mitogen-activated protein kinases by Raf-1 and MEKK. *Science* **266:** 1719.

Narumiya S., Ishizaki T., and Watanabe N. 1997. Rho effectors and reorganization of actin cytoskeleton. *FEBS Lett.* **410:** 68.

Natesan S. and Gilman M. 1995. YY1 facilitates the association of serum response factor with the c-*fos* serum response element. *Mol. Cell. Biol.* **15:** 5975.

Price M.A., Cruzalegui F.H., and Treisman R.H. 1996. The p38 and ERK MAP kinase pathways cooperate to activate ternary complex factors and c-*fos* transcription in response to UV light. *EMBO J.* **15:** 6552.

Quilliam L.A. Lambert Q.T., Mickelson-Young L.A., Westwick J.K., Sparks A.B., Kay B.K., Jenkins N.A., Gilbert D.J., Copeland N.G., and Der C.J. 1996. Isolation of a NCK-associated kinase, PRK2, an SH3-binding protein and potential effector of Rho protein signaling. *J. Biol. Chem.* **271:** 28772.

Sahai E., Alberts A.S., and Treisman R. 1998. RhoA effector mutants reveal distinct effector pathways for cytoskeletal reorganization, SRF activation and transformation. *EMBO J.* **17:** 1350.

Sartorelli V., Webster K.A., and Kedes L. 1990. Muscle-specific expression of the cardiac α-actin gene requires MyoD1, CArG-box binding factor, and Sp1. *Genes Dev.* **4:** 1811.

Sealy L., Malone D., and Pawlak M. 1997. Regulation of the c-*fos* serum response element by C/EBPβ. *Mol. Cell. Biol.* **17:** 1744.

Sepulveda J.L., Belaguli N., Nigam V., Chen C.Y., Nemer M., and Schwartz R.J. 1998. GATA-4 and Nkx-2.5 coactivate Nkx-2 DNA binding targets: Role for regulating early cardiac gene expression. *Mol. Cell. Biol.* **18:** 3405.

Stacey D.W., Watson T., Kung H.F., and Curran T. 1987. Microinjection of transforming ras protein induces c-*fos* expression. *Mol. Cell. Biol.* **7:** 523.

Tan S. and Richmond T.J. 1990. DNA binding-induced conformational change of the yeast transcriptional activator PRTF. *Cell* **62:** 367.

Treisman R. 1990. The SRE: A growth factor responsive transcriptional regulator. *Semin. Cancer Biol.* **1:** 47.

———. 1994. Ternary complex factors: Growth factor regulated transcriptional activators. *Curr. Opin. Genet. Dev.* **4:** 96.

Uehata M., Ishizaki T., Satoh H., Ono T., Kawahara T., Morishita T., Tamakawa H., Yamagami K., Inui J., Maekawa M., and Narumiya S. 1997. A key role for p160ROCK-mediated Ca^{++} sensitization of smooth muscle in hypertension. *Nature* **389:** 990.

Van Aelst L. and D'Souza-Schorey C. 1997. Rho GTPases and signaling networks. *Genes Dev.* **11:** 2295.

Zhu C., Johansen F.E., and Prywes R. 1997. Interaction of ATF6 and serum response factor. *Mol. Cell. Biol.* **17:** 4957.

Zohn I.E., Symons M., Chrzanowska-Wodnicka M., Westwick J.K., and Der C.J. 1998. Mas oncogene signaling and transformation require the small GTP-binding protein Rac. *Mol. Cell. Biol.* **18:** 1225.

Summary: Three Decades after Sigma

R. LOSICK

Department of Molecular and Cellular Biology, The Biological Laboratories,
Harvard University, Cambridge, Massachusetts 02138

The molecular structure of proteins is determined by specific elements, the structural genes. These act by forming a cytoplasmic "transcript" of themselves, the structural messenger, which in turn synthesizes the protein. The synthesis of the messenger by the structural gene is a sequential replicative process, which can be initiated only at certain points on the DNA strand...

F. Jacob and J. Monod
Genetic regulatory mechanisms in the synthesis of proteins. *Journal of Molecular Biology, 3* (1961) p. 354.

In formulating the messenger RNA hypothesis for the intermediate that carries information from the gene to the ribosome, Jacob and Monod galvanized attention on the problem of the "sequential replicative process" by which the DNA template is transcribed. Less than a decade later, an entire Cold Spring Harbor Symposium was devoted to this subject. Coming on the heels of several seminal discoveries, particularly in prokaryotic systems (see below), the historic 1970 Symposium on Transcription of Genetic Material was held at a time of much excitement. The Symposium's summarizer, Michael Chamberlin, noted that "....we have made a great deal of progress in understanding the process of bacterial and phage transcription and the manner in which it is controlled. While many questions remain to be answered, the questions themselves seem clear and it seems likely that many answers will be found in a reasonably short time. In contrast the study of transcription in eucaryotic systems has just begun to uncover the outlines of that subject. It seems likely that future symposia will find eucaryotic transcription a profitable subject for inquiry for some time."

Looking back from the vantage point of the Mechanisms of Transcription Symposium, we see that it is prokaryotic transcription which has continued to be a "profitable subject for inquiry" but that it is eukaryotic transcription which has moved under the spotlights. Studies with prokaryotic systems have opened the door to the inner workings of the transcription machinery and the mechanisms by which it is modulated. Meanwhile, investigations with eukaryotic systems have revealed extraordinary complexity and diversity in the transcription machinery of higher cells but have also forced us to view transcription in the context of the chromatin template upon which RNA synthesis takes place, the enzymatic machinery that converts nascent transcripts into mature messengers, and the organization of the nucleus itself.

HISTORY

Studies on the mechanism of transcription begin with the discovery of RNA polymerase in nuclei from rat liver by Weiss and Gladstone in 1959. RNA polymerase was discovered in *Escherichia coli* shortly thereafter by Stevens and independently by Hurwitz and coworkers, who showed, importantly, that the synthetic capacity of the enzyme is dependent on a DNA template. Still, it was the discovery of σ factor by Burgess and Travers at Harvard University and Dunn and Bautz at Rutgers University in late 1968 (almost precisely three decades prior to the publication of this volume) that finally made it possible to investigate the mechanism by which the synthesis of the messenger is "initiated at only certain points on the DNA strand." The concept that the capacity of RNA polymerase to recognize and bind to the promoter could be assigned to a particular subunit that could be reversibly associated with the enzyme had a profound impact on the transcription field because it rendered the problem of promoter recognition and its regulation accessible to investigation. Additionally, the discovery of σ heralded the existence of an important new class of regulatory proteins (alternative σ factors) that would bind to RNA polymerase in place of σ and direct it to particular classes of promoters.

The 2-year period from the discovery of σ up through the publication of the 1970 Symposium volume was a time of extraordinary progress for studies on the mechanism of transcription. Shortly after the announcement of σ, Roberts reported the discovery of a companion protein, Rho factor, which causes RNA polymerase to terminate transcription at particular sites on the template. Just as the discovery of σ opened the way for studies of initiation, the discovery of rho factor made it possible to investigate

All authors cited here without dates refer to papers in this volume.

elongation and its regulation. The next year, Chamberlin made the electrifying discovery (not disclosed at the time of the Symposium but included in the publication of the volume) that rather than relying entirely on the transcription machinery of its host, coliphage T7 specified its own RNA polymerase, a single polypeptide enzyme dedicated to the transcription of phage genes. It was also during this 2-year period that Roeder and Rutter and, independently, Chambon discovered that eukaryotes have three distinct forms of RNA polymerase (designated pol I, II, and III by Roeder), each devoted to the transcription of a different category of genes (ribosomal genes, protein-encoding genes, and genes for small RNAs, respectively). This seminal discovery paved the way for studies on the mechanisms of transcription in eukaryotes. No less significant, and indeed later recognized with the award of a Nobel Prize, was the exciting announcement in 1970 by Baltimore and by Temin and Mizutani of reverse transcriptases, RNA-dependent DNA polymerases that, opposite to RNA polymerase, copy the RNA genomes of certain tumor viruses into DNA.

Many important discoveries have, of course, been made since the period of the 1970 Symposium, but several landmarks stand out as having had the greatest impact on contemporary studies of transcription mechanisms in eukaryotes (leaving aside in this context the related subject of the control of gene transcription). One such landmark was the reconstruction in cell-free systems of correct initiation of transcription by pol III (in 1977), pol II (in 1979), and pol I (in 1982) and the use of these systems beginning in 1980 to discover and purify the panoply of "general transcription factors" that facilitate correct transcription from promoters. Seemingly mundane names for chromatographic column fractions, such as the TFIID complex of Roeder, are to this day deeply imbedded in the vernacular of molecular biology. Another landmark was the discovery of eukaryotic transcription factors in the early 1980s, including the pol III transcription factor TFIIIA (Roeder), the pol II factor Sp1 (Tjian), and hormone receptors (Yamamoto, Evans, and Chambon). Once again, just as the discovery of σ factor paved the way for studies of transcription initiation in bacteria, the isolation of eukaryotic transcription factors, including the demonstration that TFIIIA and Sp1 could activate RNA synthesis in vitro in a promoter-selective fashion, opened the floodgates to extraordinary progress in our understanding of the mechanisms of transcription in higher cells. Also contributing importantly to our understanding of transcription mechanisms was the elucidation of the modular nature of the eukaryotic promoters (as revealed in yeast by several groups and in higher cells by McKnight and others) and the characterization of the yeast transcription factors Gal4 and GCN4, including the discovery of their modular nature (Ptashne and Struhl).

1998

How has the world of transcription changed in the three decades since the discovery of σ? Perhaps the biggest surprise is the remarkable, and sometimes bewildering, complexity of the transcription machinery of eukaryotic cells. Whereas promoter recognition in bacteria often requires nothing more than core RNA polymerase and σ (the holoenzyme), transcription initiation in higher cells is mediated by three different RNA polymerases, each one of which depends on a series of general transcription factors, which are themselves often composed of multiple subunits. Nonetheless, the underlying principles of transcription are similar between prokaryotes and eukaryotes, so much so that it no longer makes sense to discuss mechanisms of transcription in the two systems separately. Whereas in 1970, topics were organized phylogenetically, with bacteria, phage, mammals, and animal viruses grouped in different sections of the Symposium, in 1998 the emphasis was on an integrated view of underlying mechanisms.

One way in which prokaryotic and eukaryotic transcription differ sharply is in the nature of the template. Whereas in 1970 little consideration was given to the chromatin template upon which eukaryotic transcription takes place, a major theme in 1998 was the issue of how RNA polymerase and its regulators gain access to the DNA through the covalent modification of histones and chromatin remodeling. Thus, in 1998, we see that the field of transcription mechanisms has begun to merge in part with the chromatin field. We may also be witnessing an intersection of transcription with cytology as some presentations pointed to the importance of the spatial distribution of transcriptional control proteins within the nucleus.

With these considerations in mind, I have organized this summary around three principal themes of the Symposium: RNA polymerase, activation and repression, and nucleosomes and chromatin.

RNA POLYMERASE

Phage and "Grown-up" Polymerases

The simplest and arguably best understood RNA polymerase is the transcribing enzyme of phage T7, which is able to carry out promoter recognition, initiation, elongation, and termination in a package of only a single, 110-kD polypeptide chain. Like the "grown-up" polymerases (to use the memorable phrase of Steitz) of bacteria and eukaryotic cells (but unlike their DNA polymerase cousins), the phage T7 enzyme has the capacity to initiate de novo, starting RNA synthesis from a single nucleotide. Steitz (see Cheetham et al.) reported on the crystal structure of the phage polymerase, performing a "thumbectomy" on its hand-like structure to allow us to peer into its catalytic center where initiation and elongation take place. Steitz reported that the catalytic center of the phage enzyme is rigidly fixed in space within the structure and would seem to be incapable of the spring-like flexing hypothesized in the inchworming model for *E. coli* RNA polymerase. Steitz was also able to put an upper limit on the length (6 bp) of the DNA-RNA hybrid in the catalytic center, with the probable length being only 2–3 bp.

Phage T7 RNA polymerase exhibits little evidence of evolutionary relatedness to cellular RNA polymerases.

Thus, for mechanistic studies of multisubunit transcribing machines, we must turn to the *E. coli* enzyme. Darst (see Darst et al.) reminded us that the *E. coli* enzyme is simpler than eukaryotic polymerases (4 versus 12 subunits in the case of yeast pol II). Yet, the α, β, and β′ subunits of the prokaryotic core RNA polymerase are homologous to two thirds of the mass of the eukaryotic enzymes, making *E. coli* RNA polymerase an excellent model for multisubunit RNA polymerases in general (for a valuable compendium of the pol II subunits and a summary of relatedness to the components of the *E. coli* enzyme, see Woychik in this volume). On the basis of electron microscopy of negatively stained crystals and the recent improvement of crystal preservation by flash freezing (cryo-electron microscopy), Darst (Darst et al.) has obtained low-resolution structures for *E. coli* holoenzyme (core plus σ) and for the core RNA polymerase. Both structures reveal an approximately 25 Å channel wide enough to accommodate double-stranded DNA and a thumb-like projection. Interestingly, in the holoenzyme, the channel is exposed with the thumb protruding away from the body of the enzyme, which could correspond to the configuration involved in promoter binding. In the absence of σ, however, the thumb is folded over the channel to form a ring of protein density that completely surrounds the channel. This could correspond to the elongation configuration, with the ring corresponding to a sliding clamp that holds the elongating polymerase onto the template and accounts for the processivity of the enzyme.

In what was perhaps the most inspiring and inspired moment of the Symposium, Darst superimposed his recently solved crystal structure for the amino-terminal domain of the α-subunit (the portion of the protein that interacts with the β and β′ subunits) on the low-resolution image of the core enzyme. It fit in a satisfying and convincing manner. To this, Darst added Steitz's structure for the CAP activator protein. Magically, one of the activation regions (activation region 2) on the surface of CAP juxtaposed precisely against the site on α that it contacts during transcription activation!

Kornberg (see Kornberg) brought us up to date on the structure of pol II from yeast, where a possible explanation for the approximately 30-bp distance between the start site and the TATA box emerges from the relative position of TFIIB (which lies adjacent to the TATA-binding protein, TBP) and the catalytic center. Satisfyingly, the yeast and bacterial enzymes share certain structural features in common, such as the putative clamp (see above) created by a thumb-like projection that wraps around the DNA channel. Finally, Kornberg presented intriguing images of the RNA polymerase in a complex with the mediator (see below). In these images, similar structures can be seen protruding from the polymerase in complexes from both yeast and the mouse.

Using as a starting point the low-resolution structure of yeast pol II and the mapping on it of the CTD and other landmarks, Ebright presented a daringly detailed model for the trajectory of DNA around and through the polymerase, including a proposed location for the transcription bubble and the path of the template around the enzyme. Next, he presented his progress in developing biophysical tools for testing the model, such as introducing fluorescent probes at defined positions and taking advantage of fluorescence resonance energy transfer (FRET) for measuring distances between landmark sites on the polymerase. These methods, Ebright argued, should be adaptable for real-time measurements at the millisecond time scale from single RNA polymerase molecules during the transcription process.

A Word on Nomenclature

For bacterial RNA polymerases, the term "holoenzyme" refers to core enzyme plus σ, the form of the polymerase that is capable of promoter recognition and transcription initiation. For pol II of eukaryotic cells, in contrast, holoenzyme is used to refer to the approximately 12-subunit enzyme complex (more or less corresponding to the core enzyme of bacteria) in association with other proteins (i.e., the mediator or Srb/Med complex; see below) that help confer responsiveness to activators. However, the pol II holoenzyme is not itself capable of initiating transcription from promoter sites. Rather, specific transcription additionally requires TFIID or other TBP-containing complexes, which direct pol II to the promoter but are not generally considered to be part of the holoenzyme.

Eukaryotes do not have a true equivalent of σ factors (ignoring in this context, the polymerase-binding patch in TFIIF that closely resembles region 2 of σ). Instead, the closest analogs to σ are the protein complexes, such as TFIID, that directly interact with core promoter sequences. For this and other reasons, the pol II holoenzyme is more closely equivalent to the bacterial core RNA polymerase than to the bacterial holoenzyme. In light of these considerations and to avoid confusion, I simply refer to the complex of pol II with the mediator as the pol II–mediator complex rather than the holoenzyme.

σ Factors, TFIID, and Other σ-like Complexes

An elegant presentation by Gross (see Gross et al.) revealed how *E. coli* σ factor "orchestrates" the initiation of transcription. Using genetic and biochemical approaches to dissect the 70-kD protein, Gross found that σ is subject to autoinhibition such that free σ (i.e., free of core RNA polymerase) adopts a conformation in which its DNA-binding surfaces are masked. This autoinhibition is mediated by the amino-terminal region (region 1) of σ, which probably forms a loop by contacting a site in the carboxy-terminal portion of the protein. Autoinhibition is relieved by the binding of σ to core RNA polymerase, which triggers a conformational change that exposes both region 2.4, which is responsible for contacting the promoter "−10" sequence, and region 4.2, which mediates recognition of the "−35" sequence. Amusingly, as reported by Graves (see Graves et al.), ETS-1, a vertebrate transcription factor that governs gene expression in T cells, is similarly subject to autoinhibition. A structural element called the inhibitory module allosterically interferes with DNA binding and must be disrupted in order for ETS-1 to bind to DNA.

Gross (see Gross et al.) also reported that the binding surface between σ, both $σ^{70}$ and the alternative factor $σ^{32}$, and core RNA polymerase encompasses an extended portion of the proteins that includes regions 2, 3, and 4. Meanwhile, Burgess (see Burgess et al.) described the application of a "fragment ladder" procedure for mapping complementary sites on core RNA polymerase that contact σ. Using this procedure, he has identified a strong binding site near the amino terminus of β´. Although a high-resolution structure for the RNA polymerase holoenzyme is still in the future, prospects for a detailed description of the contacts between σ and core polymerase seem close at hand.

At the heart of promoter recognition in eukaryotes is the TATA-binding protein (TBP), which is involved in the recruitment of pol II both to promoters containing a TATA box and to promoters lacking the signature element of eukaryotic promoters (for a review of the structure of the TBP-TATA complex and associated proteins, see Burley in this volume). Insights into TBP and the transcription machinery of higher cells are beginning to emerge from studies with members of the third kingdom of life, the Archaea. Like bacteria, Archaea have only a single form of RNA polymerase. However, in all other respects, the archaeal transcription machinery more closely resembles that of eukaryotes than bacteria, in that its promoters frequently have TATA boxes, which are recognized by a homolog of TBP. The archaeal transcription machinery also has a TFIIB ortholog (TFB) but none of the other TBP-binding proteins, general transcription factors, or the pol-II-associated Srb/Med proteins found in higher cells. Presentations by Jackson (see Bell and Jackson) and by Sigler (see Tsai et al.) focused on the problem of how the direction of transcription is fixed: Archaeal TBP is highly symmetric around its twofold axis of symmetry, even more so than its eukaryotic homologs. These workers reported that it is not recognition of the TATA box by TBP that fixes the orientation of transcription. Rather, transcriptional polarity is determined by the interaction of the TFIIB ortholog TFB with a sequence just upstream of the TATA box. Because the TFIIB factor of eukaryotes is known to exhibit DNA sequence specificity, perhaps the direction of transcription in higher cells is similarly determined by TFIIB. Because of its simplicity, the "stripped down" RNA synthesizing machinery of Archaea is a robust system for investigating the mechanisms of the basal transcription complex.

In eukaryotes, TBP is found in association with an assemblage of proteins known as TAFs (TBP-associated factors), the ensemble of the TAFs and TBP for pol II corresponding to the famous TFIID complex. In a beautiful analysis that combined structural and genetic approaches, Davidson (see Davidson et al.) described the region of contact between TBP and one of the TAFs in the TFIID complex, $TAF_{II}28$, which like $TAF_{II}18$, is known to exhibit a histone-like fold. We now know that TBP is a critical component of the transcription machinery for pol I and pol III as well, where it is found in association with other distinct sets of TAFs ($TAF_{I}s$ and $TAF_{III}s$). Thus, the basal transcription factor SL1 for pol I consists of TBP and three $TAF_{I}s$ and that the TFIIIB factor for pol III consists of TBP and two other proteins.

These TBP–TAF complexes are like σ factors in that they serve as a bridge between polymerase and the promoter. In addition, like σ, the TFIID (TBP-TAF_{II}) complex for pol II makes extensive contacts with the promoter. Whereas σ mediates contact with the "–10" and "–35" regions of prokaryotic promoters, TFIID is capable of contacting not only the TATA element through its TBP component, but also (depending on the promoter) the "initiator" (Inr) element at the start site and the so-called "downstream promoter element" (DPE) in a TAF_{II}-dependent manner, as discussed in presentations by Smale (see Smale et al.) and by Kadonaga (see Burke et al.). Roeder described additional proteins known as TICs that are involved in initiator function in a TAF_{II}-dependent manner.

Another way in which TBP-TAF complexes resemble σ comes from studies with TFIIIB. Kassavetis (see Kumar et al.) described elegant genetic and chemical cross-linking experiments that implicate the TBP-containing complex not only in the recruitment of pol III to the promoter, but also in the subsequent process of initiation. Because of its relative simplicity, the pol III system promises to be of increasing importance in studies of the mechanisms of eukaryotic transcription and its regulation.

Not only are there distinct TBP-TAF complexes for each polymerase, but the canonical TFIID (TBP-TAF_{II}) complex for pol II may be only one of several parallel TAF-containing complexes for pol II. Thus, Workman (see Steger et al.) drew an exciting parallel between TFIID and the newly identified SAGA complex of yeast. SAGA contains a subset of the $TAF_{II}s$ found in TFIID, Ada proteins, which have been implicated as targets of certain transcriptional activators, and GCN5. GCN5 is a member of a family of enzymes (histone acetyltransferases or "HATs") that facilitate transcription by targeted acetylation of nucleosomes at the promoter (about which more later). Although it lacks TBP, SAGA contains a protein (Spt3) that interacts with TBP and that contains two histone-like folds similar to those of $TAF_{II}28$ and $TAF_{II}18$. Nakatani (see Kotani et al.) described a similar SAGA-like complex from mammalian cells that contains $TAF_{II}s$, a homolog to GCN5 or the related protein PCAF, and homologs to SPT3 and Ada proteins. Meanwhile, one of the TFIID-specific $TAF_{II}s$, $TAF_{II}250$ (and its yeast equivalent), is now known to have HAT activity. Thus, the important concept that emerges is of two (or more?) "coactivator" complexes: TFIID and SAGA, which are recruited by transcriptional activators, thereby delivering both TBP and nucleosome-modifying enzymes to the promoter.

Taking the parallel with σ factors one step further, Tjian (Näär et al.) has characterized an alternative, TBP-like protein, TRF, which is present in *Drosophila* brain cells and is itself associated with a constellation of additional (TAF-like?) proteins. It will be interesting to learn whether the

TRF complex has promoter specificity and whether it heralds the existence of additional σ-like complexes.

Finally, a poster presentation by Tora challenged accepted notions about TBP and the TFIID complex. Tora described a TAF$_{II}$-containing complex called TFTC that lacks TBP but yet is capable of substituting for TFIID in promoting transcription from TATA-containing as well as TATA-less promoters. Given the large body of evidence implicating TBP in the formation of preinitiation complexes of (almost) all kinds, it will be interesting to learn whether the TFTC complex possesses a TRF-like replacement for TBP or some other means to bypass TBP function.

Elongation and Termination

Understanding mechanisms of elongation is important if for no other reason than that most of the transcription process is devoted to RNA chain extension. In addition, however, elongation is being recognized increasingly as a target of regulation in systems ranging from bacteriophage λ to the human immunodeficiency virus (HIV) (see below).

Goldfarb (see Korzheva et al.) described clever chemical cross-linking experiments that pinpoint the locations of three principal landmarks on the elongating transcription complex of *E. coli* RNA polymerase: the double-stranded DNA-binding site, which lies just downstream from the transcription bubble, the RNA-DNA heteroduplex-binding site at the transcription bubble, and the upstream RNA-binding site from which the newly synthesized transcript exits the elongation complex. The DNA-binding site is a sliding clamp (see above) that holds the elongating polymerase to the template. Evidence indicates that it covers approximately 9 bp of duplex DNA and that it lies in a region of polymerase corresponding to the extreme amino terminus of β′ and the extreme carboxyl terminus of β. Unexpectedly, this same junction region of β′ and β forms part of the RNA-binding site. Thus, the DNA- and RNA-binding sites could constitute a single structure. The significance of this finding is that it suggests how hairpins in RNA can act as termination signals: If the two binding sites are a unit, then dislodging the nascent transcript from the RNA-binding site by the formation of a hairpin could simultaneously disrupt the DNA-binding site, causing the polymerase to release both the transcript and the template.

Further evidence for the existence of the RNA-binding site and its role in termination came from the presentation of Roberts (Roberts et al.), who showed that a synthetic oligonucleotide corresponding to the sequence upstream of the transcription bubble would trigger termination in the absence of a hairpin. Just as a hairpin is believed to pry transcript from the RNA-binding site, so too the oligonucleotide releases nascent transcript from the RNA-binding site by hybridizing to it.

Roberts (see Roberts et al.) also described an unexpected connection between σ, which is responsible for promoter binding and initiation, and an elongation pause site at position +16 relative to the start site for the bacteriophage λ promoter for *late* gene expression. The pause site consists of an iteration of the promoter −10 sequence

with which σ interacts on the nontemplate strand of the transcription bubble, thereby trapping the elongating polymerase. The significance of the pause site is that the bacteriophage λ regulatory protein Q loads on the stalled RNA polymerase, enabling it to escape from the pause site and continue elongation. Regulatory proteins that act on engaged polymerases to facilitate elongation were a recurring theme at the Symposium and were echoed in three mechanistically unrelated systems that are considered further below: λ N (Greenblatt et al.), *Drosophila* heat shock factor (Lis), and HIV Tat (Jones [Garber et al.] and Price [Peng et al.]).

Transcription elongation is also an important focus of mechanistic studies with eukaryotic RNA polymerases. Luse (see Luse and Samakurashvili) presented experiments showing that ternary complexes of pol II that have been artificially stalled at various points downstream from the start site have a propensity to slide backward along the template. This backsliding is principally observed in complexes stalled close to the start site (up to position +40 or so) but not with complexes stalled further downstream. This indicates that pol II undergoes a transition in clearing the promoter to a more productive elongating form of the enzyme. Complementary work along the same lines presented by Conaway et al. showed that in clearing the promoter, RNA polymerase is converted to a stable elongating form at about 40–50 bp downstream from the start site and that this transition requires ATP and TFIIE and TFIIH. Interestingly, these general transcription factors can enter the reaction and promote the transition to productive complexes after the initiation step of transcription.

Price (see Peng et al.) described termination factors that cause newly initiated pol II molecules to abort elongation and produce short transcripts. The effect of these "*n*egative *t*ranscription *e*longation *f*actors" (N-TEF) can be overcome by "*p*ositive *t*ranscription *e*longation *f*actors" (P-TEF), which, similar to the antitermination proteins of bacteriophage λ, allow elongation to proceed past termination sites. One such factor, P-TEFb, is of considerable medical interest because of its involvement in antitermination by HIV, a topic to which I return below. P-TEFb is composed of the cyclin-dependent kinase CDK9 and cyclin T. Price (Peng et al.) showed that, like the general transcription factor TFIIH, P-TEFb phosphorylates the carboxy-terminal domain (CTD) of the large subunit of pol II (indeed much more extensively than does TFIIH) and that this is the basis for its ability to overcome the effect of N-TEF.

Finally, after RNA polymerase has undergone the transition into a productive complex, additional factors, such as SII, ELL, and elongin, increase the efficiency of transcription elongation further. SII is part of a proofreading mechanism that allows misincorporated nucleotides to be excised by cleavage of the nascent transcript, after which the polymerase backs up and resumes elongation from the 3′ end of the cleavage site. Interestingly, pol III has a similar proofreading mechanism built into the enzyme itself. Sentenac (Chédin et al.) reported that pol III can excise nucleotides incorporated by mispairing by an intrinsic ability to cleave nascent transcripts. He described a mutant form of pol III lacking one of its small units (C11) that initiates normally but has lost the ability to cleave nascent transcripts. The mutant enzyme also fails to pause and reads through certain terminators.

Why Polymerase Has a Tail

Perhaps the oddest feature of pol II is its tail, the repeated seven-amino-acid sequence found at the carboxyl terminus of the largest subunit of polymerase. Gratifyingly, an integrated view of the function of the CTD is beginning to emerge from the work of Symposium participants Bentley, Greenleaf, Dahmus, Kornberg, Price, and Young. Prior to the start of transcription, the CTD serves to anchor the mediator (the Srb/Med complex) to the polymerase. As discussed below, the mediator is a coactivator that helps certain activators recruit polymerase to the promoter. Next, as we have just seen, the CTD is phosphorylated by TFIIH, P-TEFb, and other CDK kinases. This allows initiated polymerases to clear the promoter and elongate. Here, too, the CTD is involved in regulation because, as shown below, the HIV Tat protein exerts its effect on transcription elongation at the step of phosphorylation of the tail.

Finally, as proposed by Bentley (McCracken et al.), the mediator exits elongating polymerase and is replaced on the CTD by mRNA maturation complexes for polyadenylation and capping, which act on the nascent transcript. This idea is based on previous evidence from Bentley and others indicating that the CTD is responsible for recruiting factors involved in 3′ cleavage and polyadenylation as well as enzymes involved in forming the CAP at the 5′ end of the transcript. (Mutant pol II harboring a truncated CTD also exhibits defects in splicing, but, according to Bentley, this may be a secondary consequence of the 3′ processing defect.)

Linkage between transcription and processing is an important concept because it indicates that the machinery for RNA synthesis should not be considered in isolation of other processes in the cell nucleus. Yet another such connection is illustrated by TFIIH, the kinase complex that plays a part in phosphorylation of the CTD. Egly (see Coin and Egly) explained that the 9-subunit complex is intimately involved not only in transcription, but also (astonishingly) in nucleotide excision repair. Some of the TFIIH subunits are components of the nucleotide excision repair machinery, and the genes for the two largest subunits are the sites of mutations causing genetic diseases, such as xeroderma pigmentosum, which have been attributed to a failure to repair DNA damage. Conceivably, RNA polymerase is capable of delivering the nucleotide excision repair machinery to sites of damage on the DNA that it encounters during the course of transcription elongation.

ACTIVATION AND REPRESSION

The intense interest in the subject of this Symposium derives in no small measure from the fact that much of biological regulation is exerted by switching transcription on and off. The vast and fundamental field of developmental biology, for example, which was the subject of last year's Symposium, is in the main devoted to understanding how genes are turned on and turned off at the right time and in the right place. In addition, mechanisms of homeostasis, which govern such processes as energy input and consumption and the cell cycle, operate in part through the control of gene transcription. For this reason, gene control mechanisms are also pertinent to pharmaceutical research, which is increasingly concerned with the discovery of small molecules that directly or indirectly influence normal and abnormal transcription. Just how gene transcription is activated and repressed requires an understanding of the interplay between regulatory proteins and the transcription machinery, the topic of this section of the summary.

Activation without an Activator

Some activation mechanisms are so simple they do not involve an auxiliary regulatory protein. A classic problem in molecular biology dating back to the earliest days of the field is the question of how the transcription of rRNA operons in *E. coli* is coupled to growth rate. It turns out that the answer is exquisitely simple and elegant. Solving this historic problem, Gourse (see Gourse et al.) has discovered that the rate of transcription from the major rRNA operon promoters, which form unusually unstable initiation complexes, is strongly dependent on the cellular concentration of the initiating nucleotide, ATP or GTP. Moreover, the cellular concentration of ATP and GTP, in turn, varies with the growth rate. Thus, cells growing rapidly on rich medium have high ATP and GTP levels and as a consequence transcribe from rRNA operon promoters at high rates. This in conjunction with other mechanisms enables the cells to produce ribosomes in abundance. Conversely, cells growing slowly on poor medium have low ATP and GTP levels, transcribe their rRNA operons at low rates, and produce relatively few ribosomes. Thus, growth rate control is in large part mediated by characteristics of the rRNA operon promoters without the intervention of an auxiliary transcription factor. Growth rate control by NTP sensing is also an exquisite homeostatic mechanism in that high amounts of ATP and GTP are consumed during protein biosynthesis by ribosomes!

Activation by Single-stranded Binding Protein

Coliphage N4 is unusual because it carries within its virion the RNA polymerase that directs transcription of phage *early* genes. Thus, in infecting a host cell, the phage introduces not only a copy of its genome, but also the machinery for transcribing phage genes (a situation not unlike that of certain RNA tumor viruses, which package reverse transcriptase within their virions). Elegant experiments by Rothman-Denes et al. show that the N4 virion polymerase recognizes not duplex DNA, but rather a hairpin on the promoter template strand. Extrusion of the hairpin is facilitated by negative supercoiling, but the unusual stability of the hairpin derives from the well-ordered, stacked structure of its loop. Subsequent to extrusion, the single-stranded DNA-binding protein (SSB) of the *E. coli* host melts the nontemplate hairpin to create the "activated promoter." Evidence suggests that SSB is not only an architectural protein that helps to create the activated promoter but that SSB may also interact directly with the virion RNA polymerase. As discussed by Rothman-Denes (see Rothman-Denes et al.), DNA structural transitions and SSBs may be more widely involved in transcriptional regulation than has been generally appreciated.

Enhanceosomes and Repressosome

The control of gene transcription requires mechanisms to achieve precision, such that genes are switched on or off at the correct time and place. One of the ways in which this precision is achieved is through the assembly on DNA sites, such as enhancers, of multiprotein complexes that can activate or repress transcription. As only the fully assembled complex is active, its assembly on the DNA ensures that transcription is not activated or repressed adventitiously. Viewed from this perspective, enhancers serve two functions: They tether activators to the vicinity of promoters and they nucleate the formation of the protein complexes that activate transcription.

In no system are these principles more beautifully illustrated than in the prokaryotic enhanceosome that governs the activation of RNA polymerase containing the alternative σ factor σ^{54}. The σ^{54}-holoenzyme forms closed complexes at promoters under its control but cannot initiate transcription unless triggered to do so by the activator protein NtrC, which, in turn, is regulated by its state of phosphorylation. Phosphorylated NtrC assembles into a multisubunit structure at the enhancer, which is distantly located from the promoter. The structure contains as many as six or eight NtrC molecules, not all of which are in contact with the DNA. Kustu (see Rombel et al.) presented scanning force micrographs that reveal the topography of the structure sitting on the DNA in impressive fashion. Kustu (Rombel et al.) also presented evidence showing that the NtrC structure is a kind of machine that uses energy from ATP hydrolysis to drive the isomerization of the closed promoter complexes of σ^{54}-holoenzyme into transcriptionally active, open complexes. This happens through the formation of a DNA loop in which enhancer-bound NtrC touches the promoter-bound holoenzyme. Thus, physiological signals (i.e., nitrogen limitation) governing σ^{54}-directed transcription are channeled into the assembly of a multisubunit molecular machine of precise architecture.

The formation of multicomponent complexes can serve to repress transcription as well as to activate it. One such "repressosome" was described by Adhya (see Adhya et al.) for the galactose operon of *E. coli*. Transcription of the *gal* operon is controlled by two operators, which are separated by 113 bp. The operators nucleate the formation of a heteromeric complex of GalR repressor molecules bound at the operators and in contact with each other through the formation of a DNA loop and a molecule of the histone-like protein HU sandwiched in between. Interestingly, HU ordinarily exhibits little sequence specificity in its binding to DNA, but protein-protein interactions in the repressosome hold it at a specific site between the operators.

The principle illustrated above by the NtrC enhanceosome of channeling a signal into the assembly of an activating structure on the DNA is elaborated upon by the enhanceosome for the human interferon-β (IFN-β) gene. The culmination of intensive investigation, Maniatis's discovery of the enhanceosome explains how the IFN-β gene is stringently controlled (see Maniatis et al.). The IFN-β enhanceosome is a multicomponent edifice of precise architecture that assembles on the 60-bp enhancer for the IFN-β gene. The components of the enhanceosome (nine protein species and a dozen proteins in total) assemble on the enhancer in three subdomains: the p50 and p65 subunits of NF-κB and the "architectural" protein HMG I(Y) at one end of the enhancer, the ATF-2/c-Jun heterodimer and an additional molecule of HMG I(Y) at the other end, and the virus-activated factor or VAF (which consists of two IRF family members IRF-3 and IRF-7 in association with the coactivator p300/CBP) in the middle. These three subdomains assemble into a higher-order nucleoprotein complex through numerous protein-protein interactions. Stringent control of IFN-β is explained by the fact that three signal transduction pathways, which emanate from virus infection, are separately responsible for recruiting ATF-2/c-Jun, NF-κB, and VAF to the enhancer. Recruitment culminates in the assembly of the enhanceosome, and only the enhanceosome is competent to activate transcription.

Work with viral activators and the serum response factor (SRF) provided two other telling examples of the enhanceosome concept. Herr (see Herr) summarized his work on the association of the herpes simplex activator protein VP16, which associates with two cellular proteins: cell proliferation factor HCF and the POU homeodomain protein Oct-1. This heterotrimeric complex assembles on so-called "TAATGARAT" DNA elements. Meanwhile, Carey (see Ellwood et al.) described the enhanceosome of Epstein-Barr virus, which consists of an enhancer-bound complex of the viral activator protein ZEBRA and three cellular proteins: Sp1 and the architectural proteins HMG-1 and HMG-2. Finally, SRF binds with TCF, a member of the Ets domain family of proteins, to form a complex at the serum response element (SRE) that is activated by several signal transduction pathways (see Treisman et al.).

Tethering Activation Complexes with RNA Rather Than DNA

Whereas most activators bind to DNA enhancers and contact the transcription machinery through the formation of a DNA loop, some activators bind to RNA enhancers in nascent transcripts and interact with the transcription machinery through the formation of an RNA loop. In a stunning example of convergent evolution, the N protein of bacteriophage λ and the Tat protein of the HIV both bind to enhancer sites in RNA known as *nut* and TAR, respectively. In both cases, the enhancer-bound activator protein works by enabling RNA polymerase to read through a termination or pause site downstream from the promoter. As described by Greenblatt (see Greenblatt et al.), λ N does so by recruiting to the surface of RNA polymerase four proteins (NusA, NusB, NusG, and S10) of the *E. coli* host. This antitermination complex is stabilized by multiple protein-protein and protein-RNA interactions, including the binding of NusB to a *nut* element known as box A and N to a *nut* hairpin known as box B. Bringing us up to date on the web of interactions in the complex, Greenblatt described detailed genetic and structural studies elucidating how N recognizes the box B hairpin and how this interaction facilitates the interaction of N with NusA on the RNA polymerase.

Jones (see Garber et al.) described the parallel case of RNA loop formation by HIV regulatory protein Tat. Tat binds to an RNA enhancer called TAR, which like the λ box B site, is a hairpin in the nascent transcript. Tat, in turn, interacts with, and recruits to the polymerase, the transcription elongation factor P-TEFb. P-TEFb, it will be recalled (above), is a CDK9/cyclin T protein kinase that acts by extensively phosphorylating the CTD of pol II. Thus, as in the case of λ N, Tat uses an RNA tether to assemble a protein complex on the RNA polymerase that facilitates transcription elongation.

Activator Targets and Recruitment

At the heart of the problem of gene activation are two questions: How do activators stimulate transcription and what are the functional contacts between activators and the transcription machinery. One answer is that many sites on the surface of the pol II–mediator complex can, and sometimes do, serve as targets (the "love handle" model) and that activators work by recruiting the transcription machinery to the promoter ("location, location, location" in the felicitous words of Ptashne). Compelling presentations by Hochschild and Ptashne demonstrated that activation can work in this fashion. The α, β′, and σ subunits of *E. coli* holoenzyme are known to be natural targets. Hochschild (see Dove and Hochschild) showed that arbitrary fusions to one or another subunit of RNA polymerase, including the cryptic ω subunit, can mediate activation by creating a contact site for a DNA-bound protein. Indeed, direct fusions of DNA-binding proteins to polymerase subunits can activate transcription from promoters containing an appropriate recognition sequence. Ptashne (see Zaman et al.) described similar "activator bypass" experiments in which DNA tethering of TFIID or components of the pol II–mediator complex efficiently activates transcription in yeast.

Of course, not all activation works by recruitment of polymerase. The *Drosophila* heat shock factor (see Lis) and the HIV TAT protein (Garber et al.; Peng et al.) are known to, or seem to, exert their effects at steps subsequent to promoter binding, and certain eukaryotic activa-

tors (see below) work in part by modifying or removing histones from the promoter (see also the discussion by Hahn, this volume). Nonetheless, the concept of achieving a high effective molar concentration at the promoter of one or another of the components of the transcription machinery is a powerful idea. Even in cases in which the polymerase is not the direct target of an activator, it is likely that activation works by recruitment of a component of the transcription machinery (Zamen et al.). The clearest exception to this generalization is the bacterial activator NtrC (Rombel et al.), which, as discussed above, works by causing RNA polymerase that is already bound at the promoter to undergo isomerization.

One of the frustrations of the eukaryotic transcription field has been the difficulty in identifying definitively (i.e., genetically as well as biochemically) functional contact sites between activators and the transcription machinery. Part of the reason may be that activators often have multiple targets. Such redundancy would render any one target dispensable. Conversely, as pointed out by Struhl (see Struhl et al.), a widely used target might be needed for viability and hence presumed to be an essential component of the transcription machinery. Certainly, no pol II target has been pinned down with the clarity of that between the *E. coli* activators CAP and λcI, for example, and the α and σ subunits of polymerase, respectively. Whatever the nature of the targets for activators of pol II, experiments by Struhl (Struhl et al.) and by Green (pers. comm.) in which the occupancy of promoters by TBP was assessed in vivo by use of the chromatin immunoprecipitation (CHIP) procedure demonstrate a strong correlation between recruitment of TBP and activation for many different promoters in yeast.

Because of their relative simplicity, pol I and pol III are attractive systems in which to understand mechanisms of transcription activation in eukaryotes. In the case of pol I, a single activator, UBF1, acts in conjunction with the TBP-containing complex SL1 to activate transcription of rRNA genes. In the case of pol III, work presented by Hernandez (see Henry et al.) shows that the activator protein Oct-1 for the U6 snRNA gene acts by contacting the 190-kD subunit of SNAP$_c$. SNAP$_c$ is a basal transcription factor that binds to a promoter element known as PSE. SNAP$_c$ is not a general transcription factor. Instead, it is dedicated to the U6 snRNA gene and certain other genes, including certain pol-II-transcribed genes, such as the U2 snRNA gene (!). SNAP$_c$ is composed of five subunits and interacts with TBP. (Interestingly, TBP is not present as part of the TFIIIB complex at the U6 promoter, unlike the situation for most other pol-III-transcribed genes.) Thus, just as the λcI activator works by contacting σ, so too Oct-1 activates transcription of the U6 gene by contacting a site within the SNAP$_c$-TBP assembly.

pol II, in contrast, is more complicated because its activity is modulated by many different and many kinds of activators and because evidence points to several components of the transcription machinery as potential targets. These include TFIIB (Green [Shen et al.]), the mediator (Srb/Med) complex of R. Young (pers. comm.) and Kornberg, the TBP-associated TAF$_{II}$s of Tjian (Näär et al.) and

Roeder, and TBP (Greenblatt [see Greenblatt et al.]) itself. Despite the difficulties in the definitive identification of target sites within these complexes, a safe inference seems to be that many activators work, at least in part, by recruiting TFIID, TFIIB, or the pol II–mediator complex or perhaps all three in a concerted process. The best evidence for this comes from experiments in which a correlation has been established between the level of activation observed in vivo and the interaction of an activator and its activation-defective variants with a general transcription factor, such as TBP and TFIIB, as measured in vitro.

Prominently featured in several presentations at the Symposium was the mediator (Srb/Med) complex (R. Young [pers. comm.] and Kornberg). The mediator complex can be purified in association with pol II as a pol II–mediator complex, which is often referred to as the holoenzyme (see comment on nomenclature above). In his presentation, Kornberg (see Kornberg) showed that the yeast mediator consists of 18 subunits, including proteins known as Srb and Med as well as Gal11, Sin4, and others, and that homologs of at least some of these proteins are present in an equivalent complex from mouse. Kornberg presented evidence showing that a Med2 mutant is impaired in VP16- but not Gcn4-mediated activation. Meanwhile, Ptashne (Zaman et al.) summarized genetic and biochemical evidence pointing to Srb4 (also a component of the mediator) as a target for the yeast activator Gal4. Briefly put, the capacity of certain suppressor mutants of SRB4 to partially restore activated transcription by an activator-defective mutant of Gal4 correlates with enhanced affinity of the variant SRB4 proteins for the activation domain of the mutant transcription factor. Finally, in a provocative new finding, Berk (see also Berk et al.) reported the discovery of a 153-kD component (called CR3BP) of a human mediator-like complex (i.e., containing the human homologs [Cdk8 and cyclin C] to Srb10 and Srb11 as well as several other proteins) to which the CR3 activation domain of the adenovirus E1A activator protein binds but to which activation defective mutants of E1A-CR3 are unable to bind. Intriguingly, CR3BP is homologous to the product of a *C. elegans* gene (*Sur2*), mutations in which suppress activated Ras. Roeder (see Roeder) described a similar complex that contains homologs of SRBs and other mediator components and that exhibits mediator function.

Also featured at the Symposium was the subject of the role of TAF$_{II}$s in transcriptional activation. Biochemical experiments by Tjian, Roeder, Berk, and others have demonstrated a requirement for TAF$_{II}$s in the activation of transcription in vitro and had implicated particular TAF$_{II}$s as targets of certain activators. On the basis of these experiments, it was thought that TAF$_{II}$s play a special part in linking activators to the transcription machinery, giving rise to the concept of "coactivators" (Tjian). Meanwhile, a variety of additional possible functions and activities have been attributed to TAF$_{II}$s, including interaction with the core promoter, stabilization of the TBP/promoter complex, phosphorylation, and histone acetylation. This has complicated the challenge of distinguishing a role for TAF$_{II}$s in activation from an essential function in the formation of the preinitiation complex. Moreover, at the Symposium, Roeder (see Roeder) reported that the requirement for TAF$_{II}$s for certain activators can be bypassed by the use of nuclear extracts rather than highly purified protein fractions. One interpretation of this finding is that TAF$_{II}$s are redundant to other targets in the transcription machinery that are lost upon extensive purification.

For these reasons, much effort has been devoted to the use of yeast cells to investigate the function of TAF$_{II}$s genetically. Because TAF$_{II}$s are essential, Green (see Shen et al.) and Struhl (see Struhl et al.) have made use of temperature-sensitive mutants and strategies for the conditional depletion of specific TAF$_{II}$s to assess the function of these proteins in gene activation. Their experiments indicate that depletion of certain TAF$_{II}$s, such as the yeast homolog (TAF145) of mammalian TAF250, does not generally affect activator-dependent transcription but does impair the transcription of certain genes strongly. On the other hand, recent work by Green (see Shen et al.), R. Young (pers. comm.), Struhl (see Struhl et al.), and Buratowski (pers. comm.) reveals widespread but not universal inhibitory effects of inactivation or depletion of other TAF$_{II}$s (namely, the histone-like TAF$_{II}$s, TAF17, TAF60, and TAF61) on general transcription as well as activator-dependent transcription. Another example of this was provided in a poster presentation by Natarajan, Jackson, and Hinnebusch, who described a truncation mutant of yeast TAF61 that impairs GCN4-directed transcription in vivo.

A further twist on the TAF story has come from the analysis of genes whose transcription exhibits a selective dependence on certain TAF$_{II}$s in yeast. Using a chimeric promoter, Green (Shen et al.) was able to attribute the dependence on TAF145 for one gene to an interaction of TAF145 with the core promoter rather than with an activator (in contrast to earlier work in mammalian cells where dependence on the TAF145 homolog, TAF250, had been found to map to the enhancer as well as to the promoter). On the other hand, similar experiments by Green (Shen et al.) and Struhl (Struhl et al.) with a gene whose activation depends on TAF17 show that in this case inducibility maps to the upstream activating region (UAS), a finding consistent with the possibility that TAF17 is an activator target. Complicating matters yet further, the role of TAFs in activation has recently taken a surprising turn with the discovery that the histone-like TAFs are components of one or more alternative complexes to TFIID, such as SAGA. Thus, some of the effects of depleting cells of TAFs could be due in whole or in part to effects on the SAGA complex, rather than the TFIID complex.

What are we to make of this state of affairs? TAF$_{II}$s, like other components of the general transcription machinery, may have multiple, promoter-specific functions. In certain promoters, certain TAF$_{II}$s may function as essential components of the preinitiation complex, independent of activators. In other promoters, TAF$_{II}$s may function as targets of activators. Indeed, it would be surprising if many available surfaces on TAF$_{\text{II}}$s as on TBP, TFIIB, and the pol II–mediator complex (the holoenzyme) are not exploited as targets for activators in one context or another. The challenge now is to use biochemistry and genetics in a variety of promoter contexts to pin down and evaluate the functional significance of contact sites between activators and numerous, potential target sites in the transcriptional machinery. My advice is to stay tuned!

Coactivators That Are Not Part of the Basal Transcription Machinery

As described above, activation often requires "coactivators" or "cofactors" in addition to the minimal machinery for basal transcription. Sometimes coactivators are found in association with the basal machinery, where they may serve in part as contact sites (targets) for the activators, but are dispensable for basal transcription. In other cases, however, coactivators are not associated with TBP or polymerase. Rather, some coactivators are associated with DNA-bound activators, such as CBP, the cell proliferation factor HCF, TRAPs, and the B-cell-specific coactivator Oca-B. In some of these cases, the coactivator may act as a bridge between DNA-bound proteins and the transcription machinery. In an amusing twist to this category of coactivators, Sassone-Corsi (see Fimia et al.) reported the discovery of another tissue-specific coactivator called ACT. ACT is only found in male germ cells, where it binds to, and is needed for transcriptional activation by, the cAMP response element binding protein CREM.

Yet other coactivators seem to be associated neither with DNA-bound activators nor with the basal transcription machinery. Two examples are the positive control (PC) factors PC2 and PC4, which appear to act as general coactivators. As reported by Roeder, PC4 is a 15-kD polypeptide that seems to work by acting as a bridge between certain activators and the transcription machinery. Meanwhile, PC2 is a large (~500 kD) complex of ill-defined composition and unknown mechanism. Tjian (Näär et al.) reported the discovery of a new cofactor called CRSP that is needed for activation by Sp1 but not p53. Like PC2, CRSP is a large complex, and Tjian has shown that it consists of at least five previously unknown polypeptides. Among the bewildering array of coactivators and cofactors, it would be a satisfying simplification were it to emerge that CRSP and PC2 are the same complex.

Repression

In comparison to activation, repression in eukaryotic cells is poorly understood. Just as coactivators facilitate transcription activation, so too repression is mediated by corepressor proteins. One such corepressor is the Ssn6-Tup1 complex of yeast. SSn6-Tup1 does not bind to DNA itself. Rather, it binds to DNA-binding proteins that tether the repression complex to the vicinity of genes subject to repression. Two models for Ssn6-Tup1-mediated repression are that it acts on chromatin and that it acts on the general transcription machinery. In the chro-

matin model, which was presented by Roth (see Edmondson et al.), Tup1 interacts directly with histones H3 and H4. In support of this model, H3 and H4 mutants defective in Tup1 binding exhibit derepression of Tup1-controlled genes. The interpretation of this evidence is, however, complicated by the fact that H3 and H4 mutants have pleiotropic effects. The general transcription machinery model, on the other hand, is favored by genetic evidence (see Wahi et al.; Zaman et al.) indicating that Tup1 interacts with components of the pol II–mediator complex.

A repressor that clearly does seem to act at the level of chromatin is Ume6, which, according to evidence presented by Struhl (see Struhl et al.), recruits the corepressor Sin3 and, significantly, a histone deacetylase Rpd3. More is said about histone modification below, but, briefly put, Struhl's evidence shows that Rpd3 causes highly localized deacetylation over a range of several nucleosomes in the vicinity of the site of recruitment. Paralleling the Ume6 story in yeast, the Mad:Max heterodimer of mammalian cells acts as a transcriptional repressor by recruiting a repression complex consisting of the mammalian homologs to the Sin3 corepressor, a protein called SMRT (or NCor) and, once again, histone deacetylases (see McArthur et al.). According to Eisenman (McArthur et al.), a *Drosophila* homolog of Mad interacts with a fly Sin3 homolog, a finding which suggests that the Mad:Max paradigm for repression may be conserved among metazoans.

From Repression to Activation

Particularly instructive is the case of the hormone nuclear receptors because they seem to embody in a single regulator mechanisms that involve corepressors and coactivators as well as mechanisms of chromatin modification and interaction with the transcription machinery. As explained by Evans (see Lin et al.), in the absence of ligand, the nuclear receptor associates with a corepressor/histone deacetylase complex similar to that described above for Mad:Max. Next, in response to ligand, the receptor undergoes a switch to a conformation that associates with a coactivator complex, which includes PCAF and CBP/p300 and which is thought to act in part at the level of histone acetylation. Thus, a ligand-induced conformational change switches nuclear receptor from associating with a corepressor complex to associating with a coactivator complex.

Insight into the molecular nature of this switch has come from the work of Yamamoto (see Yamamoto et al.), who reported that ligand induces a conformational change that creates a hydrophobic groove into which an amphipathic helix from the coactivator fits. Strikingly, a simple peptide corresponding to the helix binds to nuclear receptor in a ligand-dependent fashion.

Adding further to the story, Roeder reported a coactivator complex that mediates activation by nuclear hormone receptors called TRAP (for *t*hryroid *r*eceptor *a*ssociated *p*roteins) and that seems to work by interacting with the transcription machinery. How do the seemingly disparate findings of Evans and Roeder fit together? Roeder proposed an appealing model that reconciles the findings from California and New York. According to the Roeder model, nuclear hormone receptors activate transcription by switching sequentially from a repression complex that deacetylates nucleosome-associated histones in the vicinity of the promoter, to a ligand-induced activation complex that causes histone acetylation, and finally to a further activation complex that interacts with and recruits the transcription machinery to the promoter. Whether or not the sequential model proves to be correct, the findings with nuclear receptors underscores the importance of understanding transcriptional activation in the context of the chromatin template, which brings us to the third and last theme of the Symposium.

NUCLEOSOMES AND CHROMATIN

The compaction of DNA by nucleosomes presents an important challenge to regulatory proteins and the transcription machinery, which must gain entry to the DNA. Increasingly, investigations into the mechanisms of eukaryotic transcription are taking a holistic approach in which the transcription machinery is studied in the context of the chromatin template upon which RNA synthesis takes place. Indeed, as we have seen, the topics of activation and repression cannot be fully considered without reference to the chromatin template. Several seminal discoveries have contributed to this convergence of the transcription and chromatin fields, but in recent years, two developments seem to have been particularly influential. One is the *ch*romatin *i*mmuno*p*recipitation (CHIP) procedure of Grunstein and Paro, which, when used in conjunction with PCR, makes it possible to visualize the acetylation state of histones at the resolution of individual nucleosomes. The use of CHIP is responsible for the discovery that certain transcriptional regulatory proteins influence acetylation and deacetylation of nucleo-somes in a highly localized fashion. The second is the discovery of Allis (see Mizzen et al.) that a well-known "transcriptional" regulatory protein of yeast (GCN5) is an enzyme that modifies histones, i.e., a HAT or *h*istone *a*cetyl*t*ransferase. This finding ignited the search for, and identification of, additional HATs as components of the transcriptional machinery of yeast and higher cells.

Here I summarize four aspects of the nucleosome and chromatin problem that received particular attention at the Symposium: covalent modifications, including novel modifications; noncovalent modification (remodeling); transcription on chromatin templates; and, finally, silencing and other mysterious long-range effects of chromatin structure on gene transcription.

Covalent Modifications

Histones H3 and H4 are acetylated at specific lysines in their amino-terminal tails. Grunstein (Suka et al.) pointed out that in the recent nucleosome X-ray structure of Rich-

mond, the amino-terminal tail of H4 invades, and thereby cross-links with, H2A and H2B of the adjacent nucleosome. Thus, hypoacetylation of lysine residues in the H4 tail could repress transcription by linking adjacent nucleosomes as well as by (or instead of) facilitating interactions with the DNA. Grunstein reminded us that yeast has five histone deacetylases. Using CHIP in conjunction with antibodies against specific sites of lysine acetylation, Grunstein (Suka et al.) showed that one of the deacetylases (RPD3) causes hypoacetylation at Lys-5 of histone H4. Like Struhl (above), Grunstein finds that recruitment of RPD3 by the UME6 repressor and the Sin3 corepressor causes hypoacetylation in a highly localized fashion extending only two nucleosomes from the DNA-binding site for UME6.

Opposite to deacetylases, which have a role in repression, HATs promote transcription, presumably by disrupting the higher-order structure of nucleosomes or loosening their interaction with DNA or both. As we have seen, interest in the role of HATs in transcriptional control mechanisms has been heightened by the discovery that yeast GCN5 is a HAT and that it is part of the SAGA coactivator complex. As we have also seen, HATs are emerging as a feature of a wide variety of coactivators in yeast and higher cells. Prompted by these developments, Dutnall (see Dutnall et al.) reported the first X-ray structure of a HAT, yeast HAT1, in a complex with its cofactor acetyl coenzyme A. HAT1 is specific for Lys-12 of his-tone H4. An interesting feature of the structure is the identification of a possible substrate-binding site that helps to account for the specificity of the enzyme.

Importantly, acetylation is not the only covalent modification of histones. Winning my award (along with Darst) for one of the top two talks of the Symposium, Allis (see Mizzen et al.) demonstrated that histone H3 undergoes phosphorylation at Ser-10 and that this phosphorylation is a physiologically significant target of the Map kinase-signaling pathway. Using an antibody against an H3 peptide phosphorylated at Ser-10, Allis (Mizzen et al.) confirmed and extended earlier findings of a dramatic mitogen-induced increase in H3 phosphorylation. He then demonstrated that the H3 kinase is Rsk-2, a p90[rsk] family member, which is activated by the MAP kinase signal transduction pathway. Next, by carrying out immunostaining of human cells with his antipeptide antibody, he discovered that mitogen treatment induces a stunning, punctate pattern of H3 phosphorylation in the nucleus. Importantly, the *Rsk-2* gene is known to be linked to the genetic defect causing the Coffin-Lowry syndrome, an X-linked disorder. The punch line to all this is that cells derived from an individual with the Coffin-Lowry syndrome fail to exhibit punctate immunostaining in response to mitogen. This shows that Rsk-2 is responsible (or at least needed) for H3 phosphorylation in human cells and hence that H3 phosphorylation is a physiologically significant target of the MAP kinase signaling pathway! This raises the provocative possibility that histone phosphorylation contributes to mitogen-induced gene activation.

Finally, Johnson (Wahi et al.) presented intriguing evidence for a third covalent modification of chromatin. It is known that genes located near the telomere of yeast are maintained in a transcriptionally silent state due to the action of SIR gene products. Johnson has discovered that a de-ubiquitinating enzyme (UBP3) binds to SIR proteins and opposes silencing. Meanwhile, a ubiquitinating enzyme (UBC2) stimulates silencing. Does ubiquitination have a role in silencing and, if so, which (chromatin?) protein(s) is being modified?

Noncovalent Modifications

Also pertinent to the issue of how regulatory proteins and the transcription machinery gain access to the DNA are so-called "remodeling" enzymes. Remodeling enzymes catalyze the movement of nucleosomes and hence their displacement from DNA sequences important for gene transcription. One such remodeling enzyme is the SWI/SNF complex, the subject of presentations by Peterson, Kingston, and Winston. As explained by Peterson, SWI/SNF is an 11-subunit complex that catalyzes remodeling in an ATP-dependent manner. Using a clever assay based on accessibility to an endonuclease restriction site, Peterson demonstrated a requirement for the H3 tail in multiple rounds of the remodeling reaction. Kingston (Schnitzler et al.) reported evidence indicating that the remodeling reaction occurs by a mechanism involving the reversible formation of a novel species consisting of two nucleosomes in an altered conformation, which is stable in the absence of SWI/SNF.

A complication in evaluating the importance of the SWI/SNF complex as well as that of the HAT enzymes considered above, such as GCN5, is that mutations in the genes for these proteins have relatively modest effects. In what may be a telling finding, Winston (Winston and Sudarsanam) reported that *swi/snf* mutations have weak effects on their own but strong phenotypic effects when combined with a *gcn5* mutation. This finding points to the importance of remodeling in gene activation but suggests partial redundancy between covalent and noncovalent enzymes.

A second kind of remodeling enzyme is the NURF complex, which was discovered by Wu (see Wu et al.). NURF, which also acts in an ATP-dependent manner, is a 4-subunit complex. Wu reported that one subunit is related to a component of the SWI/SNF complex, thereby providing a link between the two kinds of remodeling complexes. Wu also reported the surprising discovery that one of the NURF subunits is inorganic pyrophosphatase. Pyrophosphatase is normally involved in thermodynamically driving the synthesis of polynucleotides, and its presence in the NURF complex is mysterious.

A further twist on remodeling enzymes is the discovery reported by Wolffe (see Wade et al.) of an additional SWI/SNF-like remodeling complex that also has histone deacetylase activity. The largest subunit (250 kD) of the complex corresponds to Mi-2, an autoantigen in the human disease dermatomyositis. This intriguing discovery raises the possibility that covalent and noncovalent remodeling mechanisms can act in a concerted fashion.

Emerging as an attractive system for studies of remodeling is the yeast *PHO5* gene. As explained by O'Shea (see Haswell and O'Shea), in its repressed state, the *PHO5* promoter is covered by four characteristically positioned nucleosomes, which are displaced upon the induction of transcription. O'Shea has found that the chromatin structure of *PHO5* contained in a yeast minichromosome faithfully mimics that of the chromosomal gene. Using this system, O'Shea has discovered a nuclear activity that remodels nucleosomes at *PHO5* in a manner that is dependent on ATP and *PHO4*, an activator that binds at the *PHO5* promoter.

Transcription on Chromatin Templates

Increasingly, mechanistic studies on eukaryotic transcription are emphasizing the use of chromatin as a physiologically relevant template. Providing a taste of things to come, Kadonaga, Reinberg, Tjian, and Wu described the use of chromatin templates for studies of activated transcription. Kadonaga, for example, described the use of chromatin templates prepared in vitro with nucleosome assembly factors. Kadonaga reported that the estrogen receptor acts cooperatively with the coactivator p300, which has HAT activity, to stimulate transcription from chromatin but not naked DNA. Likewise, Tjian (Näär et al.) was able to obtain transcription activation by the sterol response element-binding protein (SREBP) in a manner that depended on Sp1, the coactivator CRSP (above), $TAF_{II}s$, CBP, and, importantly in the present context, an in vitro assembled chromatin template containing the LDL receptor promoter. Reinberg (see Reinberg et al.) presented experiments describing the role of two remodeling factors designated RSF and FACT in transcription from a chromatin template. RSF enables pol II to initiate on the template, whereas FACT facilitates elongation past a block at about position +50, which is presumably due to the presence of a nucleosome. (Symposium participants will not quickly forget Roeder's affectionate complaint about his former postdoc, "Reinberg left with all my factors!")

In a lovely experiment, Wu demonstrated that the NURF remodeling factor is much more effective in promoting transcriptional activation by the Gal4-VP16 activator when the chromatin template is acetylated in vitro with p300 and acetyl CoA than when the template has not been covalently modified. This biochemical demonstration of cooperativity between covalent and noncovalent mechanisms of remodeling is reminiscent of the finding of Winston (Winston and Sudarsanam; see above) that yeast cells mutant for both SWI/SNF and GCN5 are much more severely impaired than an SWI/SNF or HAT mutant alone.

Silencing and Other Mysteries

The emphasis so far in this discussion of chromatin has been on the issue of how RNA polymerase and regulatory proteins gain access to the DNA in a highly localized fashion. But some observations indicate that chromatin organization can exert effects on transcription over distances of many kilobases. One of the premier systems in which these long-range chromatin effects have been studied is the chicken β-globin locus, which contains two embryonically expressed globin genes and two globin genes expressed in the adult. Felsenfeld (see Bell et al.) explained that the locus contains an enhancer and locus control elements (LCR), which together help to establish a so-called active chromatin structure over the entire locus so that individual globin genes within it can be transcribed. A striking feature of the β-globin locus is the sharpness of its boundaries, and Felsenfeld (Bell et al.) has discovered an "insulator" element at the 5′ end of the locus that is responsible for the transition from active to inactive chromatin. One category of models for how the insulator works is that it creates a barrier to chromatin inactivating proteins, such as histone deacetylases, that roam along the chromosome.

Emerson (see Bagga et al.) reported her latest findings on the human β-globin locus, which is 100 kb in length and contains five globin genes. The switch from fetal to adult globin gene expression is governed by the EKLF protein, and Emerson has discovered that EKLF requires a cell-specific SWI/SNF remodeling complex to gain access to it recognition sequence in the globin promoter.

In yeast, specific regions of the chromosome, such as the silent-mating-type loci and telomeres, are held in a transcriptional silent state by chromatin. This is a position-mediated effect, because otherwise active genes inserted into these regions succumb to the silencing mechanism. As explained by Gasser (see Cockell et al.), silencing is governed in part by silence information regulator (Sir) proteins that bind to nucleosomes via their tails in regions of chromatin-mediated repression (also see above and Wahi et al.). Amazingly, chromosomal sites of repression are sequestered together with Sir proteins at 6–10 foci located near the nuclear membrane. Gasser (Cockell et al.) documented this finding with a series of stunning fluorescence micrographs, which also showed that the loading of Sir proteins on telomeres and sequestration in foci depends on the Ku protein. Gasser's work drives home the point that mechanisms of transcription may need to be considered not only in the context of chromatin, but also in the cytological context of the organization of the nucleus.

Sir-mediated silencing may have a novel biological function in *Candida albicans*. Johnson (Wahi et al.) described a phenomenon of "phenotypic switching" in which the fungus reversibly switches between a variety of colony morphology types at a frequency too high to be due to mutation. Revealingly, the frequency of switching is enhanced by a *sir2* mutation. This suggests that Sir-mediated silencing may trap the expression of genes that determine the morphological types in a metastable state in which they are locked on or off. Switching may be a device to cope with environmental changes.

Perhaps the most mysterious of the elements involved in long-range effects on transcription is the so-called nuclear *m*atrix *a*ttachment *r*egion or MAR. As explained by Grosschedl (Fernández et al.), transcription of the immunoglobulin μ heavy-chain gene is governed by both an enhancer and a flanking MAR. The MAR is required for the enhancer to work over an extended distance. The nuclear matrix is the protein scaffold that remains after nuclei are extracted with high salt and detergent, which removes histones. When nuclei are treated in this fashion, a halo is formed consisting of the matrix and DNA loops that extend from the scaffold. Certain chromosomal regions are associated with the matrix, whereas others are found in the loops. Grosschedl (Fernández et al.) presented fluorescence images showing that the heavy-chain gene associates with the matrix in a manner that depends on the MAR. This leaves us with provocative questions: Does the matrix target DNA to sites in the nucleus enriched in components of the transcription machinery? Are we again faced with the prospect of considering the function of the transcription machinery in the context of the spatial organization of the nucleus?

PERSPECTIVES AND THE FUTURE

In his foreword to the 1970 Symposium, Watson explained that "....in November 1968 further molecular characterization of the enzyme RNA polymerase revealed the σ factor and suddenly everyone knew that a vast new field was opening up. An excitement, equal to that which accompanied the discovery of messenger RNA, immediately became apparent. Not only were important new control mechanisms in microbial cells about to be understood but the molecular basis of the embryological development of higher organisms might at last be open to real experimentation." Now, almost precisely 30 years after the discovery of σ, we see that Watson was correct. The discovery of σ heralded the beginning of an era of extraordinary progress in understanding transcription and its control in microbes to man that continues to this day. Looking ahead along the three lines of investigation considered in this Summary—RNA polymerase, activation and repression, and nucleosomes and chromatin—it seems that progress is unlikely to soon abate.

RNA Polymerase

The challenge now is to understand the inner workings of the multisubunit, RNA polymerase machine. In the past, mechanistic studies of transcription have largely not been grounded in the structure of the enzyme. But this is now changing. Progress in obtaining low-resolution images of the intact enzyme and high-resolution structures for its parts, the application of genetics and chemical cross-linking, and the use of biophysical tools, such as FRET, promise to quicken the pace of progress in understanding the mechanism of transcription.

The principal long-range goal is to understand how pol II works, because it is responsible for the transcription of protein-encoding genes and because it is subject to the most elaborate regulation. In the near term, however, most progress is likely to come chiefly from *E. coli* RNA polymerase, which closely resembles its eukaryotic counterparts. Increasingly, however, the stripped-down transcription machinery of Archaea will prove to be a fertile system for mechanistic studies. Among the eukaryotic enzymes, pol I and pol III offer the attraction of their relative simplicity (compared to pol II) and the important groundwork of mechanistic work that has already been carried out with these enzymes. Among pol II systems, the yeast enzyme is the most amenable to mechanistic investigations because of its accessibility to genetic approaches and the progress that has already been made in elucidating its structure.

Activation and Repression

One of the crowning achievements of investigations into transcriptional control mechanisms in bacteria has been the identification, at the resolution of individual amino acid side chains, of contact sites between regulatory proteins and the RNA polymerase and the ensuing progress in understanding the basis by which transcription is stimulated or repressed. Progress in understanding mechanisms of activation and repression in eukaryotes has proceeded more slowly for three reasons. First, the machinery for carrying out transcription and its control is vastly more complicated in eukaryotes than in bacteria. Not only is the basal transcription machinery more elaborate in eukaryotes (compare σ to the array of general transcription factors that facilitate promoter recognition, initiation, and elongation by pol II), but the action of activators and repressors involves a bewildering array of auxiliary proteins. Not only are the numbers and kinds of DNA-binding proteins in eukaryotic cells daunting, but the machinery for activation and repression involves numerous additional multisubunit complexes, such as those involved in the assembly of enhanceosomes (e.g., architectural proteins), coactivators (e.g., p300/CBP, P2, P4, and CRSP), TBP-containing complexes (TFIID and SAGA), the 18-subunit mediator, which is associated with polymerase, and the CTD, which undergoes cycles of phosphorylation. Second, both activation and repression involve modifications to the chromatin template as well as interactions with the transcription machinery, magnifying the problem of understanding mechanisms of gene control. The third reason for slow progress has been the frustrating problem of pinning down contact sites with certainty. Here, the possibility that regulatory proteins may have multiple targets (redundancy) and act at multiple steps has been a formidable obstacle.

Looking ahead, I believe that the era of factorology is drawing to a close. I venture to guess that a high proportion of all the characters in our passion play have at long last come on stage or will do so in the near future. If traditional biochemical approaches do not bring this outcome quickly, then surely the availability in yeast and *C. elegans* of entire genome sequences, in combination with the application of informatics and the systematic inactivation of open reading frames, will give us a comprehensive view of the gene control machinery before too much longer. Instead, the emphasis will be shifting to a detailed dissection of the interactions among these proteins: who touches who and what are the consequences. But protein chemistry will not be enough. We will continue to rely on traditional and molecular genetics in yeast, worms, and flies as well as DNA chip technologies (as featured in several Symposium presentations) to map out the web of interactions among regulatory proteins and their targets.

Nucleosomes and Chromatin

The 1998 Symposium has witnessed a merging of the fields of eukaryotic transcription and chromatin. A full understanding of the mechanisms of transcription and its regulation requires approaches that take into account the nature of the template. Covalent and noncovalent modifications of nucleosomes are inextricably linked to transcriptional control mechanisms, and the use of chromatin as a template is proving to be a fruitful approach in biochemical studies of activated transcription. In future conferences on transcription, consideration of chromatin may no longer be confined to separate sessions at the end of the meeting but may instead be interwoven throughout the conference. Looking yet further ahead, and taking a cue from the talks on silencing, long-range effects, and the nuclear matrix, the subject of mechanisms of transcription may intersect with cytological studies on the organization of the organelle, the nucleus in which the "sequential replicative process" of Jacob and Monod takes place.

ACKNOWLEDGMENTS

On behalf of all of the participants, many thanks to Bruce Stillman for his good taste in planning the Symposium and his gracious hospitality during the course of the meeting. Thanks also to the many Symposium participants who patiently tutored me on the subjects of eukaryotic transcription and chromatin, and to B. Dynlacht, M. Green, R. Gourse, T. Maniatis, M. Ptashne, R. Roeder, K. Struhl, and R. Tjian for comments on this summary. Research on transcription in my laboratory is supported by a grant from the National Institutes of Health.

Author Index

Subject Index

A

ACT
- coactivation capacity, 638–639
- CREM interaction
  - coordinated expression during testis differentiation, 637
  - CREM phosphorylation, bypassing requirement for activation, 639–640
  - evidence, 637–638
- discovery, 635–636
- LIM domains, 636, 638
- sequence homology with other proteins, 636–637, 640

Activator of CREM in testis. *See* ACT

Acute promyelocytic leukemia (APL)
- aberrant chromatin acetylation, 570–580
- retinoic acid receptor α mutations
  - histone deacetylase requirement for oncogenesis, 579, 583
  - SMRT interactions, 578–579
  - translocations, 577–578, 583
- retinoic acid therapy, 578

Ada proteins
- activation of transcription, 150
- functional analysis, 487–488
- histone acetyltransferase activity, 486

Adenovirus. *See* E1A

α2-Mcm1, direction of Ssn6-Tup1 repression
- histone mutation effects on repression, 452
- overview, 447–448

Antitermination. *See also* N; Q
- biological rationale for factors, 319
- discovery, 319

AP-1, signal transduction and transcriptional regulation, 631

APL. *See* Acute promyelocytic leukemia

ARE2
- cloning, 450
- identification as *SRB8*, 453
- *SRB8* genetic interactions with *TUP1*, 453–455

ARP7. *See* SWI/SNF

ARP9. *See* SWI/SNF

AsiA, inhibition of transcription, 150

ATF-1
- dimerization, 631
- phosphorylation, 631

ATF-2/c-Jun
- binding to PRDIV, 610–611
- signal transduction and transcriptional regulation, 631

ATP
- chromosome remodeling dependence. *See specific factors*
- promoter escape role, 358–359, 362

Autoinhibition
- biological implications, 627
- Ets-1 DNA binding
  - CBFα2 in switch regulation, 626
  - conformational change role in autoinhibition, 624

domains, 621–622
- electrophoretic mobility shift assay, 622–632
- expression and purification of protein, 622
- helix H1 coupling of DNA-binding domain and inhibitory module, 624–626
- inhibitory sequence mapping, 622–623
- partial proteolysis, 622
- phosphorylation in switch regulation, 626
- sequence recognition, 621
- structure of inhibitory module, 623–624
- ETS-2, 626
- overview and examples, 621, 626
- sigma, 655
- structural implications, 627

B

B´´. *See* TFIIIB

β-Globin locus
- erythroid-specific regulation
  - chromatin remodeling role
    - EKLF, 571–572, 665
    - NF-E2, 570
  - DNA-binding proteins in regulation, 570
  - locus control region, 569–570, 665
  - insulator elements, 511–513
  - regulatory regions, 509–511, 569–570
  - structure, 569–570

BRE. *See* TFIIB

Brf. *See* TFIIIB

Brn-3a, transcriptional activation, 63

BTF2. *See* TFIIH

C

cAMP, signal transduction and transcription regulation, 631

cAMP receptor protein (CAP; CRP), activation of transcription, 150, 173, 175

CAP. *See* cAMP receptor protein; Catabolite activator protein

Capping, mRNA and RNA polymerase II carboxy-terminal domain of large subunit
- interactions with capping enzymes, 305–306
- role, 304–307

Catabolite activator protein (CAP), α-subunit interactions in RNA polymerase, 273–274

CBF1
- histone deacetylase complex, 582–583
- N-CoR binding, 581–582
- Notch binding, 580–581, 583
- SMRT binding, 581–582
- TAN-1 binding, 581–582
- transcriptional repression, 580–583

CBFα2, autoinhibition switch regulation in Ets-1, 626

CBP/p300
- associated factor. *See* PCAF
- coactivator complexes, 192, 212–213
- E1A interactions, 493–494
- histone acetyltransferase activity, 213, 459, 662
- interferon-β enhanceosome role, 610, 615–617
- mutant studies, 466
- nuclear hormone receptor cofactor, 577
- p53 coactivation, 213–214
- structures, 633
- transcriptional activation
  - cofactor of Sp1, 194–195, 197
  - mechanism, 633

Cdc42, serum response element chromosomal template activation, 646–648

CDK7, phosphorylation and regulation of TFIIH transcriptional activity, 90–93

CDK9. *See* P-TEFb

Cell cycle. *See also specific regulators*
- overview of control, 423
- terminal differentiation, 423
- yeast TAF regulation, 222, 235–236

Chromatin. *See also* Histone; Nucleosome
- Archaea topology, 48–49
- ATP dependence of remodeling. *See specific factors*
- dependence of Sp1 cofactors, 193–194, 197–198
- disruption in gene activation, 392, 464–466, 483
- domains
  - activation, 510–511
  - boundaries, 511–513
  - insulator elements, 511–513
- immunoprecipitation, 663
- levels of structural change, 569
- nuclear matrix attachment regions with μ enhancer in structure alterations, 517–518, 521
- opening in transcription initiation, 347, 525
- *PHO5* promoter and remodeling
  - characteristics under different phosphate growth conditions, 563
  - energy requirements, 563
  - minichromosomes for in vitro remodeling, 564
  - models, 566
  - redundancy in remodeling activities, 565
  - Southern blot analysis, 564
  - specificity factors, 564
  - SWI/SNF activity in vitro, 564, 566–567
- Ssn6-Tup1 corepressor model, 448, 451–452, 459–460, 463–464

U